# SOLAR THERMAL ENGINEERING

**Space Heating and Hot Water Systems**

# SOLAR THERMAL ENGINEERING

## Space Heating and Hot Water Systems

**Peter J. Lunde**
Hartford Graduate Center
Hartford, Connecticut

**John Wiley & Sons**
New York   Chichester   Brisbane   Toronto

**Library of Congress Cataloging in Publication Data:**
　Lunde, Peter J
　　Solar thermal engineering.
　　Includes index.
　　1. Solar space heating.　2. Solar water heaters.
　I. Title.
　TH7413.L85　　697.78　　79-15389
　ISBN 0-471-03085–6

Printed in the United States of America
10 98 76 54321

# Preface

This book sets forth a unified theoretical base for the emerging field of solar thermal engineering. New theory and correlations are developed that provide an analysis for virtually every situation arising in the design of spa:e heating and hot water systems.

Parametric flexibility and sound theory have been held paramount so that the great majority of solar thermal applications can be treated directly, from beginning to end, without need for computer simulation.

The chapters are organized in a natural sequence for the analysis of a solar thermal system. Heat transfer and fluid flow relationships are developed as they are needed. The systems approach is evident throughout. Starting with basic relationships, the scope of the design problem is enlarged chapter-by-chapter, finally including all the components of the solar thermal system operating over longer and longer periods of time. Generalization at the theoretical level is used to reduce the number of concepts that need treatment, permitting thorough coverage of the subject material. Examples demonstrate every important point, and their progression throughout the book forms a complete solar system design.

The text is well suited for self-study or for classroom use. Although every chapter builds on those that precede, each can also stand alone. Readers already familiar with solar technology will find their background useful in understanding the new concepts and relationships that occur in almost every chapter.

There is adequate material here for a two-semester upper division or graduate level engineering course. The usual one-semester course can be taught by omitting specialized material in Chapters 3, 7, 8, and 10. For continuing education, a one-week format (30 hours) will work well as an introductory course. It can be reduced to three or even two days by concentrating on the descriptive material and the printed example solutions. A one-week follow-up course will permit most of the remainder of the book to be covered at an advanced level.

There are no prerequisites beyond high school maths and science necessary to use this textbook. Those without a science or engineering background will find the material difficult for self-study, however. Every chapter has enough depth that instructors can emphasize the material that they find interesting. Further theory will often be found in the problem assignments at the end of each chapter.

Out-of-class assignments for students in solar engineering can be formidable because the analysis must usually be repeated for every month. To reduce such tedium it is recommended that a programmable calculator be used as a part of the course. Instructors can then enlarge the scope of the problem assignments and make them more interesting. Solutions to the many examples have been designed to facilitate easy calculator programming and checkout. Preprogrammed magnetic cards for the TI-59 calculator (as well as FORTRAN computer programs) are available from the author. These speed data handling and the entire solar systems design, and permit rapid understanding of the text material for those already working actively in the field.

Every chapter of this book presents material germane to solar space heating or hot water systems. In Chapter 2 heat transfer fundamentals and load calculation techniques are detailed. A new correlation there permits calculation of degree-days using any temperature base. Chapter 3 covers solar radiation including a new method for elevation of horizontally measured data to a tilted surface, plus details of the ASHRAE clear sky technique and its new application to predict cloudy-sky radiation on the collector. The chapter has been written to apply to the southern as well as the northern hemispheres.

Chapters 4 and 5 treat flat plate collectors, introducing the *climatic* concept of performance with maximum storage. A new analytical approach is given that permits easy estimation of new collector characteristics as fluids and flow rates are changed. Chapter 6 describes concentrating collectors, using reflectors as an example to show how to predict long-term performance quantitatively when proper data are available. Chapter 7 describes application of the effectiveness approach to heat exchange within the solar thermal system. Chapter 8 is devoted to the daily analysis of systems having well-mixed or stratified finite storage. Chapter 9 extends that analysis to monthly and annual performances, while Chapter 10 explores the effect on performance caused by changing the system variables.

Life cycle cost analysis is thoroughly presented in Chapter 11, where the system previously examined is economically sized. In Chapter 12 the layout and sizing of the remaining portions of the system are discussed. Through application of previously developed techniques, passive systems are analyzed in the final chapter.

The Appendixes contain much new material dealing with solar radiation. Extra-terrestrial and clear sky radiation are given at close latitude increments to eliminate the need for interpolation. New monthly weather and radiation data from the National Weather Service are included for over 200 U.S. locations. Finally, detailed tabular monthly weather data at a number of radiation thresholds are given for Hartford, Atlanta, Denver, and Los Angeles. The data base for these tables was developed by Dr. Marshall A. Atwater at the Center for the Environment and Man, Inc., under a grant from the National Science Foundation. Additional weather tables of this type are available from the author, in either SI or Engineering units.

Most of the radiation data assembled for the text has been compiled only in SI units due to space limitations. Except for Chapter 9, 10, and 11—which use radiation data extensively—dual units have otherwise been used in the text and for most example solutions. This will be appreciated by the practicing engineer, for SI units have yet to be accepted at the practical level in the United States.

I am especially grateful to Donald B. Florek and Dr. Bronis R. Onuf of the Hartford Graduate Center for their early support of my solar courses, which gave intellectual and financial encouragement to the theoretical developments necessary for this book. I would also like to acknowledge the important contribution made by Dr. Arthur Rose, professor emeritus at the Pennsylvania State University, through his guidance of my research at Penn State.

*West Simsbury, Connecticut, 1979*                                **Peter J. Lunde**

# Contents

## 6    Optically Concentrating Collectors and Reflectors

## 7    Transfer of the Collected Heat

## 8    Storage of the Collected Heat

## 12    Solar Systems Design

## 13    Passive Heating Systems

## A    Solar Radiation Tables

# Nomenclature

| | |
|---|---|
| $A$ | apparent extraterrestrial solar intensity, W/m² (Btu/hr ft²) |
| $A_c$ | collector area, m² (ft²) |
| $A_x$ | heat exchanger area, m² (ft²) |
| $B$ | extinction coefficient for solar radiation, (air mass)$^{-1}$ |
| $c$ | specific heat at constant pressure, kJ/kg·°C (Btu/lb °F) |
| $C_c, C_x$ | cost of collector or heat exchanger, \$/m² (\$/ft²) |
| $C_{U_xA_x}$ | cost of a heat exchanger per unit $U_xA_x$, \$/(W/°C) [\$/(Btu/hr °F)] |
| $D$ | diameter of tube or pipe, m (ft) |
| $D_a, D_m$ | annual or monthly degree-day total, °C-day (°F-day) |
| $f_c$ | collector operating point or efficiency function, °C·m²/W (hr °F ft²/Btu) |
| $h$ | individual heat transfer coefficient, W/°C·m² (Btu/hr °F ft²) |
| $h, \Delta h$ | pressure or pressure drop relative to a liquid column, m (ft) |
| $H$ | solar flux or irradiance on a horizontal surface, global if not otherwise subscripted, W/m² (Btu/hr ft²) |
| $H_0$ | extraterrestrial irradiance on a horizontal surface, W/m² (Btu/hr ft²) |
| $H_T$ | monthly total global radiation on a horizontal surface, MJ/m² (Btu/ft²) |
| $\bar{H}_T$ | monthly daily-average total global radiation on a horizontal surface, MJ/m² (Btu/ft²) |
| $\bar{H}_{Td}, \bar{H}_{Tb}$ | monthly daily-average diffuse or beam (direct) radiation on a horizontal surface, MJ/m² (Btu/ft²) |
| $\bar{H}_{0T}$ | monthly daily-average extraterrestrial radiation on a horizontal surface, MJ/m² (Btu/ft²) |
| $I$ | global solar flux or irradiance on an inclined (tilted) surface, W/m² (Btu/hr ft²) |
| $I_d, I_b$ | diffuse or beam (direct) solar flux or irradiance on an inclined (tilted) surface, W/m² (Btu/hr ft²) |
| $I_0$ | extraterrestrial solar flux or irradiance normal to the incoming beam, W/m² (Btu/hr ft²) |
| $I_{DN}$ | direct solar flux or irradiance normal (perpendicular) to the incoming beam, W/m² (Btu/hr ft²) |
| $I_{th}$ | threshold flux or irradiance for operation of a solar collector—the radiation level at which losses are just equal to gains, W/m² (Btu/hr ft²) |

| | |
|---|---|
| $I_T$ | total global radiation falling on an inclined surface summed only when the irradiance is above a given threshold $I_{th}$, MJ/m² (Btu/ft²) |
| $\bar{I}_T$ | monthly daily-average total global radiation on an inclined surface, MJ/m² (Btu/ft²) |
| $\bar{I}_{Td}, \bar{I}_{Tb}$ | monthly daily-average diffuse or beam (direct) radiation total on an inclined surface, MJ/m² (Btu/ft²) |
| $\bar{I}_{0T}$ | monthly daily-average extraterrestrial radiation total on an inclined surface fixed in orientation at a particular terrestrial location, MJ/m² (Btu/ft²) |
| $\bar{I}_{T+}$ | utilizable radiation—total global radiation on an inclined surface falling beyond and in excess of the threshold irradiance $I_{th}$, MJ/m² (Btu/ft²) |
| $k$ | utilizability constant, m²/W (hr ft²/Btu) |
| $k$ | thermal conductivity, W/°C·m (Btu/hr °F ft or Btu in./hr °F ft²) |
| $l$ | thermal load or heat loss per unit collector area, W/m² (Btu/hr ft²) |
| $l_t$ | total thermal load per unit collector area that occurs during the time interval $t_T$, MJ/m² (Btu/ft²) |
| $l_{m,a}$ | total thermal loss or load per unit collector area during a monthly or annual time period, MJ/m² (Btu/ft²) |
| $L_{aW}, L_{aH}$ | total annual load or heat loss due to hot water or space heating, respectively, MJ (M Btu) |
| $L_m, L_a$ | total monthly or annual load or heat loss due to hot water or space heating, MJ (M Btu) |
| $L$ | length or thickness, m (ft) |
| $m, m_s$ | mass or weight of storage media per unit collector area, kg/m² (lb/ft²) |
| $mc, m_s c_s$ | storage capacity per unit collector area, MJ/°C·m² (Btu/°F ft²) |
| $M, M_s$ | mass or weight of storage media, kg (lb) |
| $Mc$ | storage capacity, MJ/°C (Btu/°F) |
| $p, \Delta p$ | pressure or pressure drop, kPa (lb/in.²) |
| $P$ | perimeter, m (ft) |
| $P$ | power, W (hp) |
| $q$ | rate of heat collection per unit collector area, W/m² (Btu/hr ft²) |
| $q_T$ | total heat collected per unit collector area, MJ/m² (Btu/ft²) |
| $q_N$ | net heat delivered to storage per unit collector area, MJ/m² (Btu/ft²) |
| $Q$ | rate of heat transfer or heat collection, W (Btu/hr) |
| $Q_T$ | total heat transferred or collected, MJ (Btu) |
| $Q_N$ | net heat transferred, or collected for storage, MJ (Btu) |
| $r$ | thermal resistivity, °C·m/W [hr °F ft²/(Btu in.)] |
| $R$ | thermal resistance, °C·m²/W (hr °F ft²/Btu) |
| $s$ | slope of collector efficiency curve, W/°C·m² (Btu/hr °F ft²) |
| $s$ | slope of annual performance curve, the *thermal return*, GJ/m² (M Btu/ft²) |
| $S$ | collector area per unit annual load, m²/GJ (ft²/M Btu) |
| $t_T$ | total time period for collector operation, Ms (hr) |
| $t_{m,a}$ | total elapsed time in a month or year, Ms (hr) |
| $t_s$ | time for turnover of the contents of a storage tank, Ms (hr) |
| $\bar{t}$ | radiation averaged operating time, Ms (hr) |

| | |
|---|---|
| $T_a$ | ambient temperature, °C (°F) |
| $\bar{T}_a$ | time-averaged ambient temperature, °C (°F) |
| $T_{CW}, T_{HW}$ | temperature of cold water supply or hot water usage, °C (°F) |
| $T_c$ | collector temperature, °C (°F) |
| $\bar{T}_c$ | time-averaged collector temperature, °C (°F) |
| $T_i, T_o, T_m$ | Inlet, outlet, or mean collector fluid temperature, °C (°F) |
| $\bar{T}_i, \bar{T}_o, \bar{T}_m$ | Time-averaged inlet, outlet, or mean collector fluid temperature, °C (°F) |
| $T_s$ | storage temperature, °C (°F) |
| $T_{s0}, T_{sf}$ | initial or final storage temperature, °C (°F) |
| $T_{sB}$ | base temperature for storage, °C (°F) |
| $u$ | velocity, mi/hr (m/s) |
| $U$ | overall heat transfer coefficient, W/°C·m² (Btu/hr °F ft²) |
| $U_L$ | overall heat transfer coefficient for heat loss from collector, W/°C·m² (Btu/hr °F ft²) |
| $UA$ | structure heat loss characteristic, MJ/°C-day or W/°C (Btu/°F-day or Btu/hr °F) |
| $U_f A_f$ | heat transfer characteristic for heat transfer through the fluid film inside collector tubes, W/°C (Btu/hr °F) |
| $U_s A_s$ | heat loss characteristic from storage, W/°C (Btu/hr °F) |
| $U_x A_x$ | heat transfer characteristic within a heat exchanger, W/°C (Btu/hr °F) |
| $U_x a_x$ | heat loss characteristic within a heat exchanger per unit collector area, W/°C·m (Btu/hr °F ft²) |
| $v$ | volumetric flow rate per unit collector area, L/s·m² (cfm/ft²) |
| $V$ | volumetric flow rate, L/s or m³/s (gal/min or ft³/min) |
| $V_T$ | total volumetric flow, L or m³ (gal or ft³) |
| $w$ | mass or weight flow per unit collector area, g/s·m² (lb/hr ft²) |
| $wc$ or $w_f c_f$ | flow capacity rate per unit collector area, W/°C·m² (Btu/hr °F ft²) |
| $W$ | mass or weight flow rate, g/s (lb/hr) |
| $Wc$ | flow capacity rate, W/°C (Btu/hr °F) |
| | |
| $\delta$ | thickness (if $L$ is used for length), m (ft) |
| $\mu$ | viscosity, mPa·s (centipoise) |
| $\nu$ | specific volume, m³/kg (ft³/lb) |
| $\rho$ | density, kg/m³ (lb/ft³) |

## UNITS and Conversion Factors

| | | Common Units | Multiply by →<br>Divide by ← | Preferred Units |
|---|---|---|---|---|
| $A$ | area | ft² | 0.0929 | m² |
| $c$ | specific heat | Btu/lb °F | 4.1868 | kJ/kg °C |
| $C$ | cost, usually per unit of the subscripted area | \$/ft² | 10.764 | \$/m² |
| $C_{U_x A_x}$ | cost per unit "$U_x A_x$" | \$/(Btu/hr °F) | 1.8956 | \$/(W/°C) |
| $D$ | diameter | ft | 0.3048* | m |
| | | in. | 0.0254* | m |
| $D$ | degree-days | °F-days | 5/9* | °C-days |
| $f_c$ | collector operating point or efficiency function | °F hr ft²/Btu | 0.17611 | °C·m²/W |
| $h$ | individual heat transfer coefficient | Btu/hr °F ft² | 5.6783 | W/°C·m² |
| $h, \Delta h$ | pressure or pressure drop relative to a liquid column | ft | 0.3048* | m |
| | | in. | 0.0254* | m |
| $H$ | solar flux or irradiance on a horizontal surface | Btu/hr ft² | 3.1546ᵃ | W/m² |
| $H_T$ | total (accumulated sum) of the radiation falling over a time period on a horizontal surface | Btu/ft²<br>Wh/m²<br>langley | 0.011357<br>0.0036*<br>0.04184* | MJ/m²<br>MJ/m²<br>MJ/m² |
| $I$ | solar flux or irradiance on an inclined surface | Btu/hr ft² | 3.1546ᵃ | W/m² |
| $I_T$ | total (accumulated sum) of the radiation falling over a time period on an inclined surface | Btu/ft²<br>Wh/m²<br>langley | 0.011357<br>0.0036*<br>0.04184* | MJ/m²<br>MJ/m²<br>MJ/m² |
| $k$ | utilizability constant | hr ft²/Btu | 0.31700 | m²/W |
| $k$ | thermal conductivity | Btu/hr °F ft | 1.7307 | W/°C·m |
| | | Btu in./hr °F ft² | 0.14423 | W/°C·m |
| $l$ | thermal load or heat loss per unit collector area | Btu/hr ft² | 3.1546 | W/m² |
| $l_T$ | total thermal load or heat loss per unit collector area area over a subscripted time period | Btu/ft² | 0.011357 | MJ/m² |
| $L$ | length or thickness | ft | 0.3048* | m |
| | | in. | 0.0254* | m |
| $L$ | thermal load or heat loss rate | Btu/hr | 0.29307 | W |
| $L_T$ | total thermal load or heat loss over a subscripted time period | Btu<br>M Btu | 0.0010551<br>1.0551 | MJ<br>MJ |

*(continued)*

## UNITS and Conversion Factors *(Continued)*

| | | Common Units | Multiply by →<br>Divide by ← | Preferred Units |
|---|---|---|---|---|
| $m$ | mass or weight of storage media per unit collector area | lb/ft² | 4.8824 | kg/ m² |
| $mc$ | storage capacity per unit collector area $(Mc/A_c)$ | Btu/°F ft² | 0.020442 | MJ/°C·m² |
| $M$ | mass or weight of storage media | lb | 0.45359 | kg |
| $Mc$ | storage capacity | Btu/°F | 0.0018991 | MJ/°C |
| $p, \Delta p$ | pressure or pressure drop | lb/in.² | 6.8948 | kPa |
| | | atm | 101.325* | kPa |
| | | mmHg | 0.13332 | kPa |
| | | mmH₂O | 9.8067 | Pa |
| | | in. Hg | 3.3769 | kPa |
| | | in. H₂O | 248.84 | Pa |
| $P$ | perimeter | ft | 0.3048* | m |
| | | in. | 0.0254 | m |
| $P$ | power | hp | 745.7 | W |
| $q$ | collected heat flux (per unit collector area) | Btu/hr ft² | 3.1546 | W/m² |
| $q_T$ | total heat collected (per unit collector area) over a subscripted time period | Btu/ft²<br>M Btu/ft² | 0.011357<br>11.357 | MJ/m²<br>GJ/m² |
| $Q$ | rate of heat transfer | Btu/hr | 0.29307 | W |
| $Q_T$ | total heat transferred over subscripted time period | Btu<br>M Btu<br>M Btu | 0.0010551<br>1055.1<br>1.0551 | MJ<br>MJ<br>GJ |
| $r$ | thermal resistivity | hr °F ft²/(Btu in.)<br>hr °F ft/Btu | 6.9335<br>0.57779 | °C·m/W<br>°C·m/W |
| $R$ | thermal resistance | hr °F ft²/Btu | 0.17611 | °C·m²/W |
| $s$ | slope of collector efficiency curve | Btu/hr °F ft² | 5.6783 | W/°C·m² |
| $s$ | slope of annual performance curve (thermal return) | M Btu/ft² | 11.357 | GJ/m² |
| $S$ | collector area per unit annual load | ft²/M Btu | 0.088055 | m²/GJ |
| $t$ | time | hr | 0.0036* | Ms |
| $T$ | temperature | °F | $°C = \frac{5}{9}(°F - 32)$*<br>$°F = \frac{9}{5}°C + 32$* | °C |
| $\Delta T$ | temperature difference | °F | 5/9* | °C |
| $u$ | velocity | mi/hr | 0.44704* | m/s |
| $U$ | overall heat transfer coefficient | Btu/hr °F ft² | 5.6783 | W/°C·m² |

*(continued)*

| | | Common Units | Multiply by →<br>Divide by<br>← | Preferred Units |
|---|---|---|---|---|
| $UA$ | heat exchanger size or | Btu/hr °F | 0.52753 | W/°C |
| | structure heat loss | Btu/hr °F | 24* | Btu/°F-day |
| | characteristic | Btu/°F day | 0.0018991 | MJ/°C-day |
| | | W/°C | 0.0864* | MJ/°C-day |
| $Ua$ | "$UA$" per unit collector area | Btu/hr °F ft² | 5.6783 | W/°C·m² |
| $U_L/mc$ | storage function | hr⁻¹ | 1/0.0036* | $(Ms)^{-1}$ |
| $v$ | volumetric flow rate per unit | cfm/ft² | 0.00508 | (m³/s)/m² |
| | collector area | cfm/ft³ | 5.08 | L/s·m² |
| $V$ or $Wv$ | volumetric flow rate | ft³/min | 0.47195 | L/s |
| | | ft³/hr | 0.0078658 | L/s |
| | | gal/min | 0.063090 | L/s |
| | | gal/hr | 0.0010515 | L/s |
| | | L/s | 0.001* | m³/s |
| $V_T$ | total volumetric flow over | ft³ | 28.317 | L |
| | subscripted time period | ft³ | 0.028317 | m³ |
| | | gal | 3.7854 | L |
| | | gal | 0.0037854 | m³ |
| | | ft³ | 7.481 | gal |
| $w$ | mass or weight flow per unit | lb/hr ft² | 1.3562 | g/s·m² |
| | collector area | | | |
| $wc$ | flow capacity rate per unit | Btu/hr °F ft² | 5.6783 | W/°C·m² |
| | collector area $(Wc/A_c)$ | | | |
| $W$ | mass or weight flow rate | lb/hr | 0.12600 | g/s |
| $Wc$ | flow capacity rate | Btu/hr °F | 0.52753 | W/°C |
| $Wv$ | volumetric flow rate (see $V$) | | | |
| $\delta$ | thickness (if $L$ is used for | ft | 0.3048* | m |
| | length) | in. | 0.0254* | m |
| $\mu$ | viscosity | lb/min ft | 24.803 | mPa·s (centipoise) |
| | | lb/hr ft | 0.41338 | mPa·s |
| | | lb/sec ft | 1488.2 | mPa·s |
| $\nu$ | specific volume $(1/\rho)$ | ft³/lb | 0.062428 | m³/kg |
| $\rho$ | density | lb/ft³ | 16.0185 | kg/m³ |

[a] 3.1525 Btu/hr ft² when based on the thermochemical calorie.
The asterisk (*) indicates conversion shown is exact.

## Important Constants

| | |
|---|---|
| $e$ | base for natural logarithms, equal to 2.71828+ |
| $g$ | acceleration due to gravity, 9.8 m/s² or 32.17 ft/sec² |
| $g_c$ | constant in mechanics, 1 kg·m/N·s² or 32.17 ft/sec² |
| $g/g_c$ | proportionality constant in mechanics equal to unity in the Engineering system and 9.80 W/(kg·m/s) in SI units |
| $I_{SC}$ | solar constant, 1353 W/m² (428.9 Btu/hr ft²)[a] |
| $\pi$ | ratio of the circumference of a circle to its radius, equal to 3.14159+ |

[a] 429.2 Btu/hr ft² using a conversion factor based on the thermochemical calorie.

## Dimensionless Ratios

$a_x$      heat exchanger area per unit collector area

$a_s$      area for heat loss from storage per unit collector area

$b_0$      slope of curve related to incident angle modifier

$B$      intermediate or dummy variable

$C$      concentration ratio for concentrating collector, the ratio of aperture area to absorber area

$C$      monthly average ratio of diffuse to direct normal radiation

$f$      Fanning friction factor

$f$      proportion of monthly demand met that month by solar energy. Also called the monthly solar contribution, solar participation, solar fraction.

$f_a$      proportion of the annual demand met during a particular month by solar energy

$\sum f_a$      the proportion of the annual demand met annually by solar energy. This is the 12-month sum of $f_a$ values. Also called the annual solar contribution, solar participation, or solar fraction.

$f'_B$      proportion of the monthly load that *could* be met by solar energy with maximum storage; the *potential* solar participation

$f_B$      proportion of monthly demand that is met by solar energy presuming maximum storage

$f'_{aB}$      proportion of the annual load that *could* be met in a given month by solar energy

$F_{i,m,o}$      heat transfer factor based on inlet, mean, or outlet temperature

$F_R$      heat transfer factor based on inlet temperature (eq. 5-28)

$F_x$      heat exchanger factor

$F'_i$      combined heat exchanger–heat transfer factor, equal to $F_x F_i$. Based on inlet temperature to the collector.

$j$      $j$-factor defined by eq. 5-5

$K_{\tau\alpha}$      the ratio of the optical efficiency at one incident angle to that when the radiation is normal to the collector

$\bar{K}_T$      total global radiation falling on a horizontal surface relative to that potentially available extraterrestrially over a monthly time period—the clearness index

$L_{aw}/L_a$      proportion of the annual demand due to hot water load

$L_m/L_a$      proportion of the annual load occuring in a given month—the demand ratio

$n$      refractive index

$N$      day number within a year

$N_m$      number of days in a given month

$N_{Pr}$      Prandtl number ($c\mu/k$)

$N_{Re}$      Reynolds number

$N_{TU}$      number of (heat) transfer units

$r$      proportion of absorber surface actually covered with tubes

$\bar{R}$      monthly elevation factor to transform total global radiation on a horizontal surface to total global radiation on a tilted surface

$\bar{R}_b$      monthly elevation factor to transform total direct radiation on a horizontal surface to that on a tilted surface

$X, X'$    ratios used for $f$-chart analysis
$Y, Y'$
$y$       intercept of collector efficiency curve on the ordinate axis

## Greek:

| | |
|---|---|
| $\alpha$ | absorptivity in the 0.2 to 2 micron wavelength range |
| $\gamma$ | proportion of reflected beam intercepted by the receiver |
| $\varepsilon$ | emissivity in the 3 to 100 micron wavelength range |
| $\varepsilon$ | heat exchanger effectiveness |
| $\lambda$ | constant defined by eq. 9-35 |
| $\eta$ | efficiency |
| $\rho$ | reflectivity |
| $\tau$ | transmissivity |
| $\tau\alpha$ | optical efficiency |
| $\overline{\tau\alpha}$ | optical efficiency averaged over the useful incident angles |
| $\Phi$ | enhancement factor for reflection of direct radiation |

## Angular Measure

| | |
|---|---|
| $\alpha$ | sun's altitude |
| $\beta$ | collector elevation or tilt angle, measured upward from horizontal |
| $\gamma$ | collector azimuth angle, measured " $-$ " to the east of south, " $+$ " to the west of south |
| $\delta$ | declination angle |
| $\theta, \theta_i$ | incident angle between the sun's radiation and a line normal to the collector |
| $\theta$ | acceptance half-angle for concentrating collectors |
| $\theta_z$ | zenith angle, between the line through the center of the earth and the sun's radiation |
| $\phi$ | local latitude |
| $\phi$ | rim angle for parabolic concentrators |
| $\psi$ | sun's azimuth, measured from due south, " $-$ " to east of south, " $+$ " to the west of south |
| $\omega$ | hour angle of the sun |
| $\omega_s$ | hour angle of sunrise/sunset |
| $\omega'_s$ | hour angle of sunrise/sunset on the collector |
| $\tilde{\omega}_s$ | hour angle of sunrise/sunset on the horizon or collector, whichever is smaller and therefore limits radiation |

## Symbols for Financial Analysis, Chapter 11

| | |
|---|---|
| $A$ | annual return, \$ |
| $C$ | capital investment, \$ |

| | |
|---|---|
| $F$ | future value of a given present sum $P$, \$ |
| $i, i_i$ | interest rate as a proportion to be paid annually |
| $i_c$ | interest rate on capital as a proportion |
| $i_d$ | discount rate as a proportion |
| $n$ | number of years of the investment |
| NPV | net present value of the investment, also called net worth |
| $P$ | present value of a sum to be worth $F$ in the future at a particular time, \$ |
| $R$ | payback ratio, or the investment justified in dollars per \$1 returned annually for the life of the investment |
| ROI | return on investment, generally expressed as a percentage |
| $t$ | fractional annual tax rate |

## Subscripts

| | |
|---|---|
| $a$ | ambient, annual, absorber, absorbed |
| $b$ | base for degree-days, beam (direct) radiation |
| $c$ | collector, coil |
| $CW$ | cold water |
| $d$ | diffuse, daily |
| $DN$ | direct normal |
| $e$ | equivalent |
| $f$ | final, fluid |
| $F$ | fin |
| $g$ | ground |
| $H$ | heating, higher temperature |
| $HW$ | hot water |
| $i$ | inlet, inside, incident |
| $l$ | larger stream, loss |
| $L$ | lower temperature |
| $m$ | mean, monthly, measured |
| $n$ | normal, nighttime |
| $N$ | net |
| $o$ | outlet, overall |
| $p$ | piping |
| $r$ | root, reflector, rated |
| $s$ | storage, smaller stream, sunrise or sunset, shell-side |
| $sr$ | sunrise |
| $ss$ | sunset |
| $SC$ | solar constant |
| $t$ | tube-side |
| $th$ | occurring while radiation is above the threshold level |
| $th+$ | occurrence in excess of the threshold radiation level |
| $T$ | total |
| $u$ | as used |

| | |
|---|---|
| $w$ | wind, water gauge |
| $W$ | water |
| $x$ | cross-sectional, heat exchange |
| * | a tabular value corresponding to another * value |
| 0 | reference, initial, or extraterrestrial |

**Superscripts:**

| | |
|---|---|
| $\bar{X}$ | average value, or daily value that gives monthly total when multiplied by the days in the month |
| $\tilde{x}$ | actual or noontime value |

# Introduction

Mankind has enjoyed the heat from the sun during all his existence. It is, however, a periodic and in most places a fickle source of energy; so much so that our most significant single technical accomplishment has been the development of fire, the first auxiliary energy source.

The more convenient fossil fuels—oil and gas—are now in short supply, and within a few decades even the remaining oil-rich areas will be exhausted. The present rate of consumption is so enormous that no discoveries conceivable can change this conclusion—they could only push the date a few more decades into the future.

There are still ample reserves of coal in the United States, but this fuel is becoming expensive and it is dangerous to mine or ruinous to the landscape if stripped from the surface. Pollution caused by coal combustion is unhealthy and unpleasant, and is difficult and expensive to control. The carbon dioxide from the combustion of coal (and other fossil fuels) is accumulating in the atmosphere and is likely to cause significant climatic changes within a few more decades.

Even nuclear power—which promised so much 20 years ago—has become controversial because of safety and waste disposal uncertainties as well as the proliferation problem. Failure to develop and use the breeder reactor on a mass scale to produce more nuclear fuel puts that energy source in the same category as the fossil fuels, with a fuel supply that is going, going, and will soon be gone.

Over the long run the continued consumption of either nuclear or fossil fuels at ever-increasing rates presents clear environmental hazards and escalating economic and social costs, which make the development of alternate sources of energy almost a certainty.

If we are clever enough, the sun will be the energy source to heat and cool our structures, provide us with hot water, and supply us with energy to continue industrialization.

1

## SOLAR THERMAL PROCESSES

This book deals with the engineering of solar thermal processes, a technology that has been developing and slowly improving since the turn of the century. Solar thermal processes treat the collection of the sun's energy for use as heat, generally at modest temperatures. They could be conceived only when other energy sources permitted the production of inexpensive, clear glass. As soon as this happened—in the late nineteenth century—solar energy began to fascinate inventors and scientists (as it still does) with its offer of inexhaustible energy to those who could tap the relatively feeble rays.

By 1915 several practical engines had been run from solar-generated steam, and most of the solar collector concepts used today had been developed. They all relied on glass, either for mirrors or for transparent covers to trap heat as in a greenhouse. But the timing was wrong, for the first decades of the twentieth century were the era of the major oil discoveries in the United States.

Some years later Florida and California provided the first commercial U.S. market for solar energy, in the form of flat plate collectors used for hot water heating. In the 1920 to 1940 period these states were sparsely populated, and they were far from the normal fuel distribution networks until natural gas was available. Electric power was expensive, and during the late 1920s in California and the 1930s in Florida tens of thousands of solar water heaters were installed. Either location was ideal: The high ambient temperatures restricted energy losses during collection, and sunshine was plentiful. There was little demand beyond hot water heating, of course; the investment was modest because space heating was never considered. These water heaters are in widespread use even today in Japan and Israel, which have mild climates and high energy costs.

The Indians in the southwestern United States used good solar-engineering principles to space heat their massive adobe structures even without using glass. Centuries later, in the early 1930s, the "solar house" became popular in U.S. architectural circles, but it was a far cry from the solar heating of today—just a lavish use of glass to let light and sun into the dwelling. A more legitimate forerunner to the "passive" solar structures now becoming popular was the Crystal House, built at the 1933 Chicago World's Fair by architects George and William Keck. This house had massive masonry floors that could absorb the radiant energy during the day and reradiate it at night when heat was needed (and even when it was not needed).

Since the use of solar heat in this type of structure is not controlled, engineers sought ways to better manage the collection and use of the sun's energy by combining them with central heating systems. From 1939 to 1956 such "active" solar houses were pioneered by researchers at the Massachusetts Institute of Technology under an endowment from Dr. Godfrey L. Cabot. Much of the basic flat plate collector theory used today was derived and experimentally confirmed

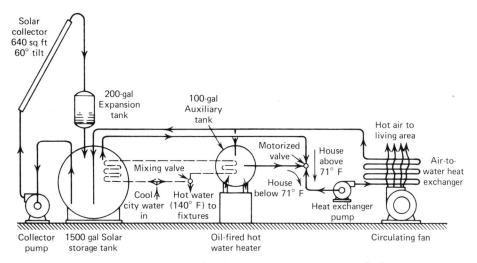

**FIGURE 1-1.** Schematic of MIT Solar House IV. (Ref. 1)

there by H. C. Hottel, B. B. Woertz, and A. Whillier, and several actual solar-heated houses were built. The final house, built in 1956, is the archetype for all solar-heated houses employing liquid-cooled collectors and a large storage tank for storing heated water. Figure 1-1 is a schematic diagram of that system. Flat plate collectors were tilted upward toward the sun, with glass covers like a gardener's "cold frame" to retain the sun's heat. Water was circulated from the storage tank through tubes attached to a black heat-absorbing surface to collect the heat for storage. The water tank heated slowly during the day to provide a reservoir of hot water for space or potable water heating at night or when the sun was covered with clouds. Space heat was delivered to the house by a fan blowing house air through a heating coil through which hot water from the storage tank was circulated.

Solar research was out of its time again by the 1950s, since oil discoveries in Venezuela and the Persian Gulf area had by then restored the traditional American faith that oil would be forever available at low prices. Nevertheless, U.S. drilled oil was clearly running out, and the local resource situation was getting worse because consumption was increasing exponentially while prices continued to *decrease*, in response to the plentiful supplies of foreign oil. In this economic climate George Löf, a chemical engineer from Denver, Colorado, constructed the first solar house to use active air collectors in 1957. Air instead of water was circulated through a solar collector (again much like a cold frame), and the hot air was used to heat the house directly or to heat a bed of pebbles that served as a thermal storage medium. If heat was needed when there was no sunlight, the air flow through the bed was reversed and room air was heated for distribution to the living spaces. Figure 1-2 is a schematic diagram of the air system in the Löf house, which is still in operation today. A system of dampers routes air in the proper direction to provide the desired mode of operation.

Two years after the Löf house was built, Harry Tomason, a patent examiner in Washington, D.C., also became interested in solar heating. He constructed several solar houses using his unique water "trickle" collector, within which the water ran downward along the collector surface and underneath the covering glass. The water storage tank was buried in a large rockbed to which it transferred heat slowly by conduction. Room air circulated to the rockbed was used to provide winter space heating.

Meanwhile in Odeillo, France, Dr. Felix Trombe and architect Jaques Michel had constructed a special kind of *passive* solar house in 1956. This solar house utilized the *Trombe wall*, which is a thick, concrete wall placed directly behind a large expanse of double vertical glazing, often with only a few inches air space between wall and window (Fig. 1-3). The wall is painted black and it absorbs the sun's radiant energy during the day. By nightfall, the concrete has heated considerably and radiation from the reverse side of the wall is available to heat the

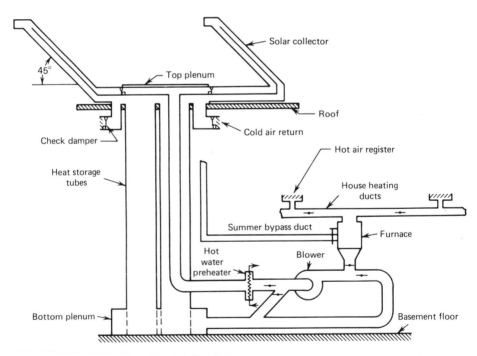

**FIGURE 1-2.** Sketch of George Löf's solar hot air heating system. (Ref. 2)

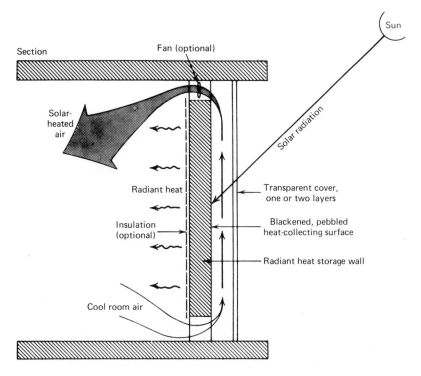

**FIGURE 1-3.** A south-facing, vertical, solar collector and heat storage wall. (Ref. 3)

living space. This concept has received much attention in architectural circles in the United States since the late 1960s and has become particularly popular in the Southwest where the sun is reliable.

Financial support for solar energy research or implementation was virtually nonexistent in the United States in the 1960s.* By the late 1960s, however, almost all the scientific community was aware of the magnitude of the coming energy shortfall. Government and public awareness of a potential problem in energy supply had to await the Arab oil embargo of 1973 to 1974. From that point onward there have been a plethora of government-sponsored programs aimed at encouraging the development of solar energy, particularly for space heating and hot water. By 1975 there were at least 140 solar-heated houses in the United States and, by the time this book is published, there should be several thousand.

---

* A *Resources for the Future* grant to the Colorado State Laboratories of George Löf to make cost estimations for the future use of solar energy is a notable exception.

A solar-heating industry has been spawned—complete with budding industrialists, serious researchers, and charlatans—partly as a result of government action and partly as a natural response to the inevitability of the advent of solar hot water and (eventually) solar space heating. Almost everyone is agreed on this: It is only a matter of time before solar energy will become commonplace for domestic and commercial hot water and space heating.

## Obstacles

Homeowners, developers, and industrialists have generally been slow to buy solar-heating equipment. This is because solar heating is capital intensive, that is, a large initial investment is required to save only part of the annual fuel costs. Design uncertainties also play a part in delaying solar implementation. No one knows exactly what a given system will save, since neither the weather nor the future cost of fuel can be predicted. Precious few can even predict the performance of solar heating or hot water systems under known conditions. Fewer yet can evaluate the performance of a given system even with full experimental data before them; there are an unusually large number of important variables that must be taken into account.

At the root of all the solar design problems is an inherent peculiarity of solar thermal systems—there is no design point, as there is in most other systems. An electrical-generating plant is built to give so many kilowatts, an aircraft engine for so much thrust cruise and so much at takeoff, and an ordinary home-heating system for so much heat on the coldest day of the year. Because provision is always made to store solar energy, the performance of a given solar system is dependent on all the previous weather conditions and energy demands. With alternate heat sources always available—at a price—it is seldom sensible to make the additional investment in solar equipment to carry the coldest days, for it will only sit idle for the rest of the year. Most analyses show that it is economical to design for one-half to three-quarters of the annual load to be carried by solar energy, and therefore the solar design problem expands beyond simple prediction of the system performance, requiring assurance that the predicted performance is the performance to be economically desired.

The number of variables to be optimized in a solar system is unusually large; however, they are ultimately tractable using techniques characterized as *systems engineering*. Most of this book is devoted to developing and demonstrating the appropriate systems engineering techniques to determine the overall performance of solar thermal systems.

## UNITS

In order to express quantities consistently in any engineering field a consistent system of units must be employed. In the United States the *English* system of units has been customary for many years. This system—better described as the *Engineering* system, since the English have abandoned it—is being replaced by the SI system (*Système International d'Unités*), a new system of units based on (but by no means interchangeable with) the original metric system familiar to many of us from chemistry and physics. The American Society of Heating, Refrigerating and Air Conditioning Engineers (ASHRAE) has established a policy that all publications shall be prepared using SI units given first followed by Engineering units until July 1, 1979. After that all publications will be prepared using SI units alone. Here we follow the former policy, using SI units first; but to help in the transition, where round numbers have been chosen, it has often been within the Engineering system. Because the general progression of examples throughout this text applies to one particular solar heat collection system, with many examples building on those previous, it has been necessary in most cases to present each example fully worked out in both systems of units.

The fundamental units for mass, length, time, and temperature for both SI and the Engineering system of units are listed in Table 1-1. Table 1-2 lists the derived units that are useful in solar engineering. Other than the size and consistency of the unit systems, there are three distinct differences between those that are of interest in solar engineering. First, the SI units encourage special names for important groupings of units. This practice is confusing to engineers, who depend on cancelling units as a check on accuracy in their work. The name conversions for these *derived* units must therefore be memorized, just as were the old numerical conversion factors in the Engineering system. Second, both the metric system and the Engineering system of units employed heat units (the calorie and the British thermal unit) based on the heat capacity of water. There is no such relationship between the basic SI unit for heat, the joule (itself a specially named derived unit), and the properties of water; therefore, the specific heat of water is not unity using SI units. Third, and most important, the basic SI derived unit for power (the watt) is exclusively defined as heat units per *second*, while heating rates expressed per *hour* are conventional for heating, ventilating, and solar work. The old and convenient metric unit of watt-hours or kilowatt-hours is not an SI

**TABLE 1-1.** Fundamental Units

| Dimensions | Units | |
| --- | --- | --- |
| | English Engineering System | International System |
| Mass | pound-mass (lb) | kilogram (kg) |
| Length | foot (ft) | meter (m) |
| Time | second (sec) | second (s) |
| Temperature | degree Fahrenheit (°F) | degree Kelvin (K) |

**TABLE 1-2.** Derived Units Frequently Used in Solar Engineering

| Unit Name | SI Units | Special Name | Engineering System Units |
|---|---|---|---|
| Area | m² | | ft² |
| Density | kg/m³ | | lb/ft³ |
| Energy | N·m | joule (J) | Btu |
| Force | kg·m/s² | newton (N) | lb |
| Power | J/s | watt (W) | Btu/hr |
| Pressure | N/m² | pascal (Pa) | lb/in² |
| Heat capacity | J/kg·K | | Btu/lb °F |
| Thermal conductivity | W/m·K | | Btu/hr °F ft |
| Thermal flux density | W/m² | | Btu/hr ft² |
| Viscosity | Pa·s | | lb/min ft |
| | (kg/m·s) | | lb/hr ft |
| Volume | m³ | | ft³, gal |

unit and has been specifically prohibited in publications of the International Solar Energy Society. Therefore, since power is to be expressed *per second*, time must always be expressed as seconds if the product of time and power is to equal energy. Consequently, hours must often be converted to seconds in the course of solving a problem using SI units.

The basic unit of temperature in the SI system is degrees Kelvin, an absolute system based on the Celsius or Centigrade scale just as absolute Rankine scale is based on the Fahrenheit scale. In most of our equations we deal with the tem-

**TABLE 1-3.** Multiple and Submultiple SI Prefixes

| Multiplying Factor | Prefix | Symbol |
|---|---|---|
| 1 000 000 000 000 = $10^{12}$ | tera | T |
| 1 000 000 000 = $10^{9}$ | giga | G |
| 1 000 000 = $10^{6}$ | mega | M |
| 1 000 = $10^{3}$ | kilo | k |
| 100 = $10^{2}$ | hecto[a] | h |
| 10 = $10^{1}$ | deka[a] | da |
| 0.1 = $10^{-1}$ | deci | d |
| 0.01 = $10^{-2}$ | centi[a] | c |
| 0.001 = $10^{-3}$ | milli | m |
| 0.000 001 = $10^{-6}$ | micro | $\mu$ |
| 0.000 000 001 = $10^{-9}$ | nano | n |
| 0.000 000 000 001 = $10^{-12}$ | pico | p |

[a] These prefixes are to be avoided if possible.

perature *difference*, which is the same whether expressed in degrees Kelvin (K) or in degrees Celsius or Centigrade (°C), and we will use the latter almost exclusively.

Since the SI system often produces inconveniently sized units, an elaborate system of multiple and submultiple prefixes has been developed for use with them. Table 1-3 lists these prefixes and their symbols.

The symbol M will be used exclusively for *million* in this text, in both SI and Engineering units. Other convenitional abbreviations for million are MM and $\overline{\text{M}}$, but these will not be used. In practice the requirement that only multiples of $10^3$ be used causes problems with certain derived units; kilograms per cubic meter ($kg/m^3$), for instance, is inconveniently large as a unit of density, while kilograms per cubic millimeter ($kg/mm^3$) is impossibly small. Unfortunately the convenient metric unit for density, grams per cubic centimeter ($g/cm^3$), is not permitted in the SI system.

Conversion of ordinary units such as feet or inches to SI units presents no problem for the student. Conversion of the units in this text, however, is much more difficult because we often use multiple units such as Btu/hr °F ft$^2$ or Btu/lb °F which in SI are W/°C·m$^2$ and kJ/kg·°C, respectively. To remove this impediment—which in practice can be surprisingly formidable—all ordinary combinations of units have been listed at the front of the book in both systems, and the proper conversion factor given. The reader should be aware that the "International Table" British thermal unit was used in preparing the conversion factors, in harmony with most solar work. Since the "thermochemical" British thermal unit differs by less than one part per thousand, there is no practical difference, but inconsistent conversions can be troublesome when numbers are being converted back and forth between the systems.

Another SI formality we follow in this textbook is the use of the "multidot" (·) within multiple SI units. Thus we write W/°C·m$^2$ to mean watts per degree Celsius per meter squared. Using this convention all units after the slash are in the denominator. The multidot separates them while correctly implying that they are multiplied together. This is done to avoid the ambiguity between, for instance, milliseconds (ms) and meter seconds (m·s).

In the Engineering system of units such ambiguities do not exist and a multidot has never been used. Instead we will follow the simple and common practice of using a space to separate these multiple units.

Conversion to SI can be a painful process for the practicing engineer. If one is hesitant, one should follow the example solutions to the more familiar work in Engineering units, but adjust to the SI system when radiation is involved; for space has not permitted dual tables and dual solution of problems involving actual radiation data. Since this information will probably be unfamiliar, it presents an opportunity to become familiar with the SI system without confusion.

On balance, most readers will come to prefer the SI units when they are used with appropriate prefixes so that their magnitudes are convenient. Their use will make recent scientific papers accessible and bring U.S. engineering practice into conformity with the rest of the world.

## Significant Figures

In past engineering practice the accuracy of data or calculations has often been reflected in the number of significant figures presented. This served the profession well until the computer age. A digital computer, however, always uses all its significant figures (usually six, or for a calculator up to twelve), and one must make one's own estimate of their accuracy. More important, results of computer simulation—an important new tool—quickly become meaningless unless at least four and usually more digits are retained.

Conversion to SI units creates yet another problem. Unless the conversion factors have one more significant figure than the other numbers, solutions using converted input data will not give the same results as the original data. True, either result generally will be within the true accuracy of the calculations, but when dual solutions are required, comparison becomes confusing unless the many dimensionless numbers are seen to be identical at each stage of calculation. Furthermore, if numbers are converted from one system to another and back, the original accuracy can be lost unless the conversion factors have the extra digit.

For these reasons, four digits—occasionally five—have been presented in the Examples in this book. The student should ordinarily be able to duplicate them on his own calculator to within three and one-half digits. Exact duplication is difficult—experience shows this would require both example and student to have carried six or even seven digits. Five-digit conversion factors are used and listed in the conversion tables.

The true accuracy of the calculations of course depends on the accuracy of the input data. This subject is not examined here. However, it is well to note that two-digit precision can be as poor as $\pm 10\%$, which is worse than the accuracy of most measurements. Therefore, three digits should always be recorded if the numbers are small. Consistent three-digit calculations require four significant figures to be carried, and that is what we use in this text.

## REFERENCES

### General

*Solar Dwelling Design Concepts.* U.S. Department of Housing and Urban Development and Office of Policy Development and Research. Prepared by AIA Research Corporation, GPO 023-000-00334-1, 1976.

J. D. Balcomb and J. E. Perry, Jr., *Assessment of Solar Heating and Cooling Technology.* Los Alamos Scientific Laboratory Informal Report LA-6379-MS, National Technical Information Service, 1977.

"Units and Symbols in Solar Energy." Technical Note produced by the Ad-Hoc Committee on Education and Standardization, International Solar Energy Society. *Solar Energy 21*, 65–68 (1978).

*ASTM/IEEE Standard Metric Practice*. Institute of Electrical and Electronic Engineers, Inc., New York, 1976.

*SI Units and Recommendations for the Use of Their Multiples and of Certain Other Units*. American National Standards Institute, 1430 Broadway, New York, 1973.

*ASHRAE/SI Metric Guide for Heating, Refrigeration, Ventilating and Air Conditioning*. American Society of Heating, Refrigerating and Air Conditioning Engineers, New York, 1976.

*An Economic Analysis of Solar Water & Space Heating*. Energy Research and Development Administration Report DSE-2322-1, U.S. Government Printing Office, Washington, D.C., 1976.

## Cited

1. C. D. Engebretson, "The Use of Solar Energy for Space Heating—M.I.T. Solar House IV." *Proceedings of the U.N. Conference on New Sources of Energy 5*, 159–169 (Rome, August 1961).

2. George O. G. Löf, M. M. El Wakil, and J. P. Chiou, "Design and Performance of Domestic Heating System Employing Solar Heated Air—The Colorado Solar House." *Proceedings of the U.N. Conference on New Sources of Energy 5*, 185–196 (Rome, August 1961).

3. B. N. Anderson, "Heat Transfer Mechanisms in the 1967 Odeillo House Integrated Collection and Storage Systems," *Passive Solar Heating and Cooling Conference and Workshop Proceedings*, May 18–19, 1976, Albuquerque, New Mexico, pp. 23–28.)

# Heat Transfer Fundamentals

O

## HEAT TRANSFER

Heat is transferred from one place to another by conduction, convection, and radiation. To this conventional list is added heat transfer by movement of the bulk of the fluid—a form of convection called *forced convection* that is important enough to warrant study by itself.

## Conduction

Substances that touch each other transmit heat by conduction. Heat is transferred through a substance by conduction. In pure conduction, there is no movement of the heat-transmitting substance beyond the normal molecular vibrations. The basic heat transfer relationship for conduction is the Fourier equation:

$$Q = \frac{kA\,\Delta T}{L} \tag{2-1}$$

where $Q$ is the rate of heat transfer, $k$ the thermal conductivity of the material, $L$ its thickness, $A$ the area for heat transfer, and $\Delta T$ the temperature difference for heat transfer. Figure 2-1 is a sketch showing the physical relationship of these variables.

### EXAMPLE 2-1

What is the heat transfer by conduction through a 2.44 m × 3.05 m (8 ft × 10 ft) window with glass 1.27 cm ($\frac{1}{2}$ in.) thick? Assume that the inside of the glass is held at 23.9 °C (75 °F) and the outside at −3.9 °C (25 °F). $k$ is about 0.865 W/°C·m [6 Btu/hr ft² (°F/in.)]

$$Q = \frac{kA\,\Delta T}{L} = \frac{0.865\text{ W}}{\text{°C·m}} \times \frac{2.44\text{ m} \times 3.05\text{ m}}{0.0127\text{ m}} \times [23.9 - (-3.9)]\,\text{°C}$$

$$Q = 14\,090\text{ W}$$

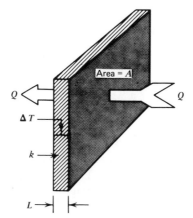

**FIGURE 2-1.** Heat transfer by conduction (sketch for Example 2-1).

In Engineering units,

$$Q = \frac{kA\,\Delta T}{L} = \frac{6\text{ Btu in.}}{\text{hr }^\circ\text{F ft}^2} \times \frac{8\text{ ft} \times 10\text{ ft}}{0.5\text{ in.}} \times (75 - 25)\,^\circ\text{F}$$

$$Q = 48\,000 \text{ Btu/hr}$$

Notice that in the Engineering system of units the thermal conductivity here is given in the "mixed" units of Btu/hr ft² (°F/in.). Since it is also commonly given as Btu/hr ft² (°F/ft) (often abbreviated to Btu/hr °F ft), users should be very wary of thermal conductivity numbers until it has been determined whether they are to be used with $L$ measured in feet or inches.

It is generally convenient to express heat transfer in terms of an overall heat transfer coefficient $U$:

$$Q = UA\,\Delta T \tag{2-2}$$

In Example 2-1, $U = k/L$, which is expressed as W/°C·m² and Btu/hr °F ft². If the $k/L$ term is only one of several heat transfer coefficients involved, it can be thought of as an individual coefficient:

$$h_{\text{conduction}} = k/L$$

which can be manipulated like the convection film coefficients described below.

## Convection

Heat transfer by convection necessarily involves a moving liquid or gas. Natural convection occurs when a warm surface is contacted by cold gas (or vice versa).

The gas is heated and it becomes lighter in density and therefore rises, its place taken by more cold gas, which then repeats the process. The same action occurs in reverse when a cold surface is contacted by a warm gas. Natural convection is an important part of the process by which heat is lost through a window to the outside air, or by which heat is lost from a solar collector. The circulating *air films* that form on both sides of a window in natural convection are much more effective in *reducing* heat transfer through the glass than is the glass itself. The conduction through glass is therefore often ignored and the heat transfer is said to be "controlled" by the convection process.

The convective heat transfer rate is also proportional to the difference in temperatures, and it therefore follows eq. 2-2. When a single convection film is considered, its individual film coefficient is assigned the letter $h$, and the heat transferred through that film under the $\Delta T$ driving force across that film can be expressed as

$$Q = hA\,\Delta T \qquad (2\text{-}3)$$

Figure 2-2 shows the physical relationship of these variables.

Most often, several $h$'s are combined into a single $U$, using techniques to be explained later in this chapter that depend on the overall heat transfer situation.

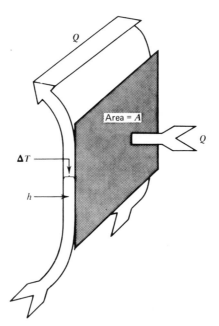

**FIGURE 2-2.** Heat transfer by convection (sketch for Example 2-2).

Then the overall coefficient $U$ is used with the overall $\Delta T$ driving force in eq. 2-2 to find the total heat flow through the element analyzed.

**EXAMPLE 2-2**

The 2.44 m × 3.05 m (8 ft × 10 ft) glass of Example 2-1 has a convection heat transfer coefficient on its outside surface of $h = 6.81$ W/°C·m² (1.2 Btu/hr °F ft²). If the surface is held at 23.9 °C (75 °F) and the outside ambient temperature is $-3.9$ °C (25 °F), what is the heat loss rate?

$$Q = UA\ \Delta T = hA\ \Delta T$$

$$Q = \frac{6.81\ \text{W}}{°\text{C}\cdot\text{m}^2} \times 2.44\ \text{m} \times 3.05\ \text{m} \times [23.9 - (-3.9)]\ °\text{C} = 1409\ \text{W}$$

In Engineering units,

$$Q = \frac{1.2\ \text{Btu}}{\text{hr}\ °\text{F ft}^2} \times 8\ \text{ft} \times 10\ \text{ft} \times (75 - 25)\ °\text{F} = 4800\ \text{Btu/hr}$$

Notice that the convection loss is much less than the rate at which even a $\frac{1}{2}$ in. thick glass can deliver heat to the convective film (see Example 2-1). Therefore, convection will control the heat loss from the 8 ft × 10 ft glass of Example 2-1.

## Radiation

When two solid surfaces touch, they assume the same temperature at the common surface. If they do not touch, the temperatures need not be the same. When they are not the same, there will be heat transfer from the hot surface to the cold surface by radiation (and by convection as well if there is a gas between them). The heat transfer by radiation with an infinitely large perfect absorber can be accurately predicted by a dimensional equation:

$$Q = 5.673\varepsilon A\left[\left(\frac{T_H}{100}\right)^4 - \left(\frac{T_L}{100}\right)^4\right] \tag{2-4}$$

where $Q$ is the heat transfer rate in watts, $A$ the area in m², and $T_H$ and $T_L$ the absolute temperatures in degrees Kelvin:

$$K = °C + 273.15 \tag{2-5}$$

In Engineering units the equation is

$$Q = 0.1713\varepsilon A\left[\left(\frac{T_H}{100}\right)^4 - \left(\frac{T_L}{100}\right)^4\right] \tag{2-6}$$

where $Q$ is the heat transfer rate in Btu/hr, $A$ the area in ft², and $T_H$ and $T_L$ the higher and lower temperatures, in °R. Rankine is the name of the Fahrenheit absolute temperature scale:

$$°R = °F + 459.67 \tag{2-7}$$

Expressions are available (4) for calculation of radiation exchange between surfaces with varying emissivities, areas, and orientations.

In each equation $\varepsilon$ is the emissivity of the surface. This is its propensity to lose heat by radiation to the surroundings, which are assumed to be perfectly absorbing. This is a good assumption for the sky, earth, and vegetation to which radiation loss usually occurs.

Emissivity varies from almost 1.0 for very black surfaces to as little as 0.02 for highly polished metals. Most common materials have an emissivity of about 0.8.

**EXAMPLE 2-3**

A 2.44 m × 3.05 m (8 ft × 10 ft) glass outside surface at 23.9 °C (75 °F) is radiating to the sky, which is at −3.9 °C (25 °F). If the emissivity of the glass is 0.8, what is the heat transfer rate by radiation? (The sky is a good absorber of radiation.)

From Eq. 2-6,

$$Q = 5.673 \times 2.44 \times 3.05 \times 0.8 \times \left[ \left( \frac{273 + 23.9}{100} \right)^4 - \left( \frac{273 + (-3.9)}{100} \right)^4 \right]$$

$$Q = 853.3 \text{ W}$$

In Engineering units, from eq. 2-4,

$$Q = 0.1713 \times 8 \times 10 \times 0.8 \times \left[ \left( \frac{460 + 75}{100} \right)^4 - \left( \frac{460 + 25}{100} \right)^4 \right]$$

$$Q = 2915 \text{ Btu/hr}$$

Notice that no units were used in these dimensional equations until the final answer. In a dimensional equation the units do *not* cancel and more care is necessary in their use. Nondimensional equations are always preferred if varying units are involved, but dimensional equations are often more convenient for routine use.

## Combined Radiation and Convection Coefficients

In comparing the heat transfer rates by conduction, convection, and radiation from Examples 2-1 to 2-3, notice that the losses from convection and radiation are of about the same magnitude and might well occur simultaneously. In most heat transfer work it is desirable to consider them together using an equation of the form of eq. 2-2. Consequently, combined radiation and convection coefficients are in common use and eqs. 2-4 to 2-7 are seldom used at all in building heat loss calculations, although they provide the basis for the calculation of the combined coefficient.

**EXAMPLE 2-4**

Assuming that the convection and radiation processes of Examples 2-2 and 2-3 take place simultaneously, what is the combined heat transfer coefficient?

Solving eq. 2-2 for $U$,

$$U = \frac{Q}{A\,\Delta T} = \frac{(1409 + 853)\ \text{W}}{(2.44\ \text{m} \times 3.05\ \text{m})[23.9 - (-3.9)]\ ^\circ\text{C}} = \frac{10.93\ \text{W}}{^\circ\text{C}\cdot\text{m}^2}$$

In Engineering units,

$$U = \frac{Q}{A\,\Delta T} = \frac{(4800 + 2915)\ \text{Btu/hr}}{80\ \text{ft}^2 \times (75 - 25)\ ^\circ\text{F}} = 1.929\ \text{Btu/hr}\ ^\circ\text{F}\ \text{ft}^2$$

Since the value of $U$ will change slowly as the temperatures vary, the base temperature of all combined coefficients should always be kept in mind. The use of combined radiation and convection heat transfer coefficients is conventional and should be well understood.

## Convection Coefficient due to Wind

A common situation in space heating and solar work involves heat loss from a large, flat surface to outside air at a rate varying with the wind velocity. The convection coefficient for this loss is about $U = 4.5\ \text{W}/^\circ\text{C}\cdot\text{m}^2$ (0.8 Btu/hr $^\circ$F ft$^2$) at zero velocity.* A dimensional equation can be written to correlate the convection coefficient $h_w$ with velocity at ordinary temperatures for smooth surfaces as

$$h_w = 4.5 + 2.9u \tag{2-8}$$

where $u$ is in m/s and $h_w$ is in W/$^\circ$C·m$^2$. In Engineering units,

$$h_w = 0.8 + 0.23u \tag{2-9}$$

where $u$ is the wind velocity in miles per hour and $h_w$ is in Btu/hr $^\circ$F ft$^2$.

## Heat Transfer by Fluid Flow

If water cools a solar collector or carries heat to a radiator, or if hot air heats a house, the heat transfer is by fluid flow or *forced convection*. The basic equation describing such heat transfer is

$$Q = Wc\,\Delta T \tag{2-10}$$

where $W$ is the mass flow rate of the fluid, $c$ the specific heat of the fluid at constant pressure,† and $\Delta T$ the temperature rise. The properties of air, water, several oils,

---

* For glass or a smooth painted surface. Coefficients for rough surfaces such as plaster or brick are 50% higher (1, 2). Note carefully that a radiation term must be added to the convection coefficient to predict the total loss from such a surface.

† This is often designated $c_p$ to distinguish it from $c_v$, the specific heat at constant volume. Since there is no application of $c_v$ in this book, the term $c$ will be used exclusively for $c_p$.

and two glycol-based heat transfer fluids are given in Table 5-1. They will be used extensively in this book.

**EXAMPLE 2-5**

0.0252 kg/s (200 lb/hr) of water is heated from 15.6 °C (60 °F) to 32.2 °C (90 °F) in a solar collector. What is the heat removal rate? The physical properties of water are given in Table 5-1.

$$Q = Wc \, \Delta T = \frac{0.0252 \text{ kg}}{\text{s}} \times \frac{4.19 \text{ kJ}}{\text{kg} \cdot {}^{\circ}\text{C}} \times (32.2 - 15.6) \,{}^{\circ}\text{C} \times \frac{\text{W} \cdot \text{s}}{\text{J}}$$

$$= 1.75 \text{ kW}$$

In Engineering units,

$$Q = Wc \, \Delta T = \frac{200 \text{ lb}}{\text{hr}} \times \frac{1 \text{ Btu}}{\text{lb} \,{}^{\circ}\text{F}} \times (90 - 60) \,{}^{\circ}\text{F} = 6000 \text{ Btu/hr}$$

**EXAMPLE 2-6**

Air infiltrates a 566 m³ (20 000 ft³) house with one change of inside air per hour. The outside temperature is 4.4 °C (40 °F), and the inside temperature is 21.1 °C (70 °F). What is the heat loss rate due to infiltration? The physical properties of air are given in Table 5-1.

$$Q = Wc \, \Delta T$$

$$Q = \frac{566 \text{ m}^3}{\text{hr}} \times \frac{\text{hr}}{3600 \text{ s}} \times \frac{1.187 \text{ kg}}{\text{m}^3} \times \frac{1.01 \text{ kJ}}{\text{kg} \cdot {}^{\circ}\text{C}} \times (21.1 - 4.4) \,{}^{\circ}\text{C} \times \frac{\text{W} \cdot \text{s}}{\text{J}}$$

$$= 3.15 \text{ kW}$$

In Engineering units,

$$Q = \frac{20 \, 000 \text{ ft}^3}{\text{hr}} \times \frac{\text{lb}}{13.5 \text{ ft}^3} \times \frac{0.242 \text{ Btu}}{\text{lb} \,{}^{\circ}\text{F}} \times (70 - 40) \,{}^{\circ}\text{F} = 10 \, 755 \text{ Btu/hr}$$

## THERMAL CHARACTERISTICS OF BUILDINGS

### Heat Losses

Residences require supplementary heat, on the average, only when the outside temperature drops perhaps 5.5 °C (10 °F) below that inside because heating provided by occupancy, electrical usage, and by solar heat warms exterior surfaces and enters the structure through windows. Office buildings can generate more heat relative to their outside area and, when they are well designed, they may require heat only as temperatures approach freezing.

It is, therefore, convenient to regard the inside of a building as having lower than its actual temperature in order to account for solar heat gain and internal heat generation. Heat losses from a building are then expressed by

$$Q = UA(T_b - T_a) \tag{2-11}$$

where $UA$ is a heat loss coefficient, $T_a$ is the outside ambient temperature, and $T_b$ is referred to as the *temperature base* for heat loss calculations. For a given structure the proper temperature base is the outside ambient temperature at which the heat losses are exactly balanced by the average solar gain plus the internal heat generated. By eq. 2-2, this is $Q_{gains}/UA$ degrees below the desired inside temperature. Its calculation is seldom attempted because of the difficulty in proper estimation of the solar component, which not only heats through windows but also reduces losses by heating exterior surfaces. Instead, the base has been taken as 65 °F (18.3 °C),* which was empirically determined nearly 50 years ago by U.S. utilities to correlate fuel demand with the weather. It is generally understood to be 10 °F below an assumed 75 °F (23.9 °C) inside ambient temperature.

Heat loss calculations continue to be done using this basis, despite better insulation and lower thermostat settings, presumably because one tenant or another may desire a 75 °F indoor temperature. Corrections can be made to account for a lower temperature preference, but this is usually done only after all calculations have been made, since all the available data use the 65 °F temperature base.

## Degree-Days

To avoid the need to analyze each day's heat losses individually, the basic heat transfer equation (2-2) is integrated over a day so that the total daily heat loss from a building is given by

$$Q_T = UA \left[ \int^{1 \, day} (T_b - T_a) \, dt \right]^+ = UA[T_b - \bar{T}_a]^+ \tag{2-12}$$

where $\bar{T}_a$ is the daily average temperature.

The term in the brackets is evaluated over a day-long period only when the result is positive, that is, when the average outside ambient temperature is less than $T_b$. This gives proper credit for natural heating on days when the hourly ambient temperature rises above the temperature base.

Examination of eq. 2-12 makes it clear that the weather-related temperature pattern and the structure-related, heat-loss characteristic $UA$ can be considered separately. When eq. 2-12 is applied over a longer period of time, a quantity called degree-days can be defined by

$$D = \sum_{\substack{time \\ period}} (T_b - \bar{T}_a)^+_{daily} \tag{2-13}$$

---

* The proposed temperature base for reporting using SI units is an even 18 °C.

where $\bar{T}_a$ is the average temperature during each day. Although the daily average ambient temperature is properly an average of the hourly temperatures throughout the day, the average of the day's high and low temperatures gives the same overall result.

If eqs. 2-12 and 2-13 are combined,

$$Q_T = UAD \tag{2-14}$$

which is the very simple relationship between heat loss characteristic $UA$ and the climate-related degree-days. Units must be carefully used in eq. 2-14, however, since $U$ will be per *second* (SI) and per hour (Eng.), while $D$ is expressed in degree-*days*.

**EXAMPLE 2-7**

The low and high Fahrenheit temperatures on five successive days were (5°, 15°), (15°, 35°), (30°, 60°), (65°, 75°), and (60°, 85°). How many degree-days were there?

| | Temperature | | | | Degree-Days = |
|---|---|---|---|---|---|
| Day | High | Low | Average | $(65° - \bar{T}_a)$ | $(65° - \bar{T}_a)^+$ |
| 1 | 15 | 5 | 10 | $(65 - 10)$ | 55 |
| 2 | 35 | 15 | 25 | $(65 - 25)$ | 40 |
| 3 | 60 | 30 | 45 | $(65 - 45)$ | 20 |
| 4 | 75 | 65 | 70 | $(65 - 70)$ | 0 |
| 5 | 85 | 60 | 72.5 | $(65 - 72.5)$ | 0 |
| | | | | | 115 Total °F-days |

**Changing the Degree-Day Base.** Table 2-1 gives 65 °F based annual and monthly degree-days for many U.S. locations. In Fig. 2-3 their geographical distribution is shown. Since the use of 65 °F as a degree-day base implies a 75 °F inside temperature, corrections may be desirable if lower inside temperatures are planned.

When the temperature does not go above the temperature base during a month, 1 degree-day is accumulated for each day per degree of temperature difference between the base and average temperatures. In milder weather, there will be more degree-days accumulated per degree of temperature difference because daily average temperatures above the temperature base have no effect on the heating load, yet they raise the average temperature. Figure 2-4 shows the degree-days accumulated per day as a function of $\Delta T_b$, the base to average ambient temperature difference. This has been derived from study of degree-days at a number of locations. A difference in location makes a small difference in the relationship,

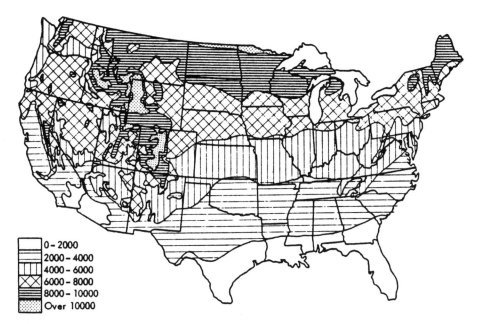

0 - 2000
2000 - 4000
4000 - 6000
6000 - 8000
8000 - 10000
Over 10000

**FIGURE 2-3.** Regional distribution of annual degree-days of heating. *Source: Climatic Atlas of U.S.*, Environmental Data Service, U.S. Department of Commerce, June 1968.

with milder climates departing less from the basic straight line correlation of 1 degree-day daily per degree temperature difference. Expressed analytically, the relationship in Fig. 2-4 is given in SI units by

$$D_m = \left\{ N_m \Delta T_b + \left[ 0.744 + 0.00387 D_a - \frac{0.5 D_a^2}{10^6} \right] N_m \exp - \left[ \frac{(\Delta T_b + 11.11)}{9.02} \right]^2 \right\}^+$$

(2-15a)

or in Engineering units by

$$D_m = \left\{ N_m \Delta T_b + \left[ 1.339 + 0.00387 D_a - \frac{0.277}{10^6} D_a^2 \right] N_m \exp - \left[ \frac{(\Delta T_b + 20)}{16.23} \right]^2 \right\}^+$$

(2-15b)

where $D_m$ are monthly C or F degree-days, $\Delta T_b$ the degrees C or F difference between degree-day base and monthly average temperature, $N_m$ the number of days in the month, and $D_a$ the total annual 18.33 °C (65 °F) base C or F degree-days at the location. Equations 2-15a and 2-15b are valid from 1000 to 4000 °C-days (1800 to 7200 °F-days). Use 4000 °C-days (7200 °F-days) in eq. 2-15 for $D_a$ at colder locations.

**TABLE 2-1.** Normal Total Heating Degree-Days (Base 65°F) (Ref. 6)

| State and Station | Jan | Feb | Mar | Apr | May | June | July | Aug | Sept | Oct | Nov | Dec | Annual |
|---|---|---|---|---|---|---|---|---|---|---|---|---|---|
| Ala., Birmingham | 592 | 462 | 363 | 108 | 9 | 0 | 0 | 0 | 6 | 93 | 363 | 555 | 2 551 |
| Huntsville | 694 | 557 | 434 | 138 | 19 | 0 | 0 | 0 | 12 | 127 | 426 | 663 | 3 070 |
| Mobile | 415 | 300 | 211 | 42 | 0 | 0 | 0 | 0 | 0 | 22 | 213 | 357 | 1 560 |
| Montgomery | 543 | 417 | 316 | 90 | 0 | 0 | 0 | 0 | 0 | 68 | 330 | 527 | 2 291 |
| Alaska, Anchorage | 1631 | 1316 | 1293 | 879 | 592 | 315 | 245 | 291 | 516 | 930 | 1284 | 1572 | 10 864 |
| Annette | 949 | 837 | 843 | 648 | 490 | 321 | 242 | 208 | 327 | 567 | 738 | 899 | 7 069 |
| Barrow | 2517 | 2332 | 2468 | 1944 | 1445 | 967 | 803 | 840 | 1035 | 1500 | 1971 | 2362 | 20 174 |
| Barter Is. | 2536 | 2369 | 2477 | 1923 | 1373 | 924 | 735 | 775 | 987 | 1482 | 1944 | 2337 | 19 862 |
| Bethel | 1903 | 1590 | 1655 | 1173 | 806 | 402 | 319 | 391 | 612 | 1042 | 1434 | 1866 | 13 196 |
| Cold Bay | 1153 | 1036 | 1122 | 951 | 791 | 591 | 474 | 425 | 525 | 772 | 918 | 1122 | 9 880 |
| Cordova | 1299 | 1086 | 1113 | 864 | 660 | 444 | 377 | 391 | 522 | 781 | 1017 | 1221 | 9 764 |
| Fairbanks | 2359 | 1901 | 1739 | 1063 | 555 | 222 | 171 | 332 | 642 | 1203 | 1833 | 2254 | 14 279 |
| Juneau | 1237 | 1070 | 1073 | 810 | 601 | 381 | 301 | 338 | 483 | 725 | 921 | 1135 | 9 075 |
| King Salmon | 1600 | 1333 | 1411 | 966 | 673 | 408 | 313 | 322 | 513 | 908 | 1290 | 1606 | 11 343 |
| Kotzebue | 2192 | 1932 | 2080 | 1554 | 1057 | 636 | 381 | 446 | 723 | 1249 | 1728 | 2127 | 16 105 |
| McGrath | 2294 | 1817 | 1758 | 1122 | 648 | 258 | 208 | 338 | 633 | 1184 | 1791 | 2232 | 14 283 |
| Nome | 1879 | 1666 | 1770 | 1314 | 930 | 573 | 481 | 496 | 693 | 1094 | 1455 | 1820 | 14 171 |
| Saint Paul | 1228 | 1168 | 1265 | 1098 | 936 | 726 | 605 | 539 | 612 | 862 | 963 | 1197 | 11 199 |
| Shemya | 1045 | 958 | 1011 | 885 | 837 | 696 | 577 | 475 | 501 | 784 | 876 | 1042 | 9 687 |
| Yakutat | 1169 | 1019 | 1042 | 840 | 632 | 435 | 338 | 347 | 474 | 716 | 936 | 1144 | 9 092 |
| Ariz., Flagstaff | 1169 | 991 | 911 | 651 | 437 | 180 | 46 | 68 | 201 | 558 | 867 | 1073 | 7 152 |
| Phoenix | 474 | 328 | 217 | 75 | 0 | 0 | 0 | 0 | 0 | 22 | 234 | 415 | 1 765 |
| Prescott | 865 | 711 | 605 | 360 | 158 | 15 | 0 | 0 | 27 | 245 | 579 | 797 | 4 362 |
| Tucson | 471 | 344 | 242 | 75 | 6 | 0 | 0 | 0 | 0 | 25 | 231 | 406 | 1 800 |
| Winslow | 1054 | 770 | 601 | 291 | 96 | 0 | 0 | 0 | 6 | 245 | 711 | 1008 | 4 782 |
| Yuma | 363 | 228 | 130 | 29 | 0 | 0 | 0 | 0 | 0 | 0 | 148 | 319 | 1 217 |
| Ark., Fort Smith | 781 | 596 | 456 | 144 | 22 | 0 | 0 | 0 | 12 | 127 | 450 | 704 | 3 292 |
| Little Rock | 756 | 577 | 434 | 126 | 9 | 0 | 0 | 0 | 9 | 127 | 565 | 716 | 3 219 |
| Texarkana | 626 | 468 | 350 | 105 | 0 | 0 | 0 | 0 | 0 | 78 | 345 | 561 | 2 533 |
| Calif., Bakersfield | 546 | 364 | 267 | 105 | 19 | 0 | 0 | 0 | 0 | 37 | 282 | 502 | 2 122 |
| Bishop | 874 | 666 | 539 | 306 | 143 | 36 | 0 | 0 | 42 | 248 | 576 | 797 | 4 227 |
| Blue Canyon | 865 | 781 | 791 | 582 | 397 | 195 | 34 | 50 | 120 | 347 | 579 | 766 | 5 507 |
| Burbank | 366 | 277 | 239 | 138 | 81 | 18 | 0 | 0 | 6 | 43 | 177 | 301 | 1 646 |
| Eureka | 546 | 470 | 505 | 438 | 372 | 285 | 270 | 257 | 258 | 329 | 414 | 499 | 4 643 |
| Fresno | 586 | 406 | 319 | 150 | 56 | 0 | 0 | 0 | 0 | 78 | 339 | 558 | 2 492 |
| Long Beach | 375 | 297 | 267 | 168 | 90 | 18 | 0 | 0 | 12 | 40 | 156 | 288 | 1 711 |
| Los Angeles | 372 | 302 | 288 | 219 | 158 | 81 | 28 | 22 | 42 | 78 | 180 | 291 | 2 061 |
| Mt. Shasta | 983 | 784 | 738 | 525 | 347 | 159 | 25 | 34 | 123 | 406 | 696 | 902 | 5 722 |
| Oakland | 527 | 400 | 353 | 255 | 180 | 90 | 53 | 50 | 45 | 127 | 309 | 481 | 2 870 |
| Point Arguello | 474 | 392 | 403 | 339 | 298 | 243 | 202 | 186 | 162 | 205 | 291 | 400 | 3 595 |
| Red Bluff | 605 | 428 | 341 | 168 | 47 | 0 | 0 | 0 | 0 | 53 | 318 | 555 | 2 515 |
| Sacramento | 614 | 442 | 360 | 216 | 102 | 6 | 0 | 0 | 12 | 81 | 363 | 577 | 2 773 |
| Sandberg | 778 | 661 | 620 | 426 | 264 | 57 | 0 | 0 | 30 | 202 | 480 | 691 | 4 209 |
| San Diego | 313 | 249 | 202 | 123 | 84 | 36 | 6 | 0 | 15 | 37 | 123 | 251 | 1 439 |
| San Francisco | 508 | 395 | 363 | 279 | 214 | 126 | 81 | 78 | 60 | 143 | 306 | 462 | 3 015 |
| Santa Catalina | 353 | 308 | 326 | 249 | 192 | 105 | 16 | 0 | 9 | 50 | 165 | 279 | 2 052 |
| Santa Maria | 459 | 370 | 363 | 282 | 233 | 165 | 99 | 93 | 96 | 146 | 270 | 391 | 2 967 |
| Colo., Alamosa | 1476 | 1162 | 1020 | 696 | 440 | 168 | 65 | 99 | 279 | 639 | 1065 | 1420 | 8 529 |
| Colorado Springs | 1128 | 938 | 893 | 582 | 319 | 84 | 9 | 25 | 132 | 456 | 825 | 1032 | 6 423 |
| Denver | 1132 | 938 | 887 | 558 | 283 | 66 | 6 | 9 | 117 | 428 | 819 | 1035 | 6 283 |
| Grand Junction | 1209 | 907 | 729 | 387 | 146 | 21 | 0 | 0 | 30 | 313 | 786 | 1113 | 5 641 |
| Pueblo | 1085 | 871 | 772 | 429 | 174 | 15 | 0 | 0 | 54 | 326 | 750 | 986 | 5 462 |
| Conn., Bridgeport | 1079 | 966 | 853 | 510 | 208 | 27 | 0 | 0 | 66 | 307 | 615 | 986 | 5 617 |
| Hartford | 1209 | 1061 | 899 | 495 | 177 | 24 | 0 | 6 | 99 | 372 | 711 | 1119 | 6 172 |
| New Haven | 1097 | 991 | 871 | 543 | 245 | 45 | 0 | 12 | 87 | 347 | 648 | 1011 | 5 897 |
| Del., Wilmington | 980 | 874 | 725 | 387 | 112 | 6 | 0 | 0 | 51 | 270 | 588 | 927 | 1 930 |
| Fla., Apalachicola | 347 | 260 | 180 | 33 | 0 | 0 | 0 | 0 | 0 | 16 | 153 | 319 | 1 308 |
| Daytona Beach | 248 | 190 | 140 | 15 | 0 | 0 | 0 | 0 | 0 | 0 | 75 | 211 | 879 |
| Fort Myers | 146 | 101 | 62 | 0 | 0 | 0 | 0 | 0 | 0 | 0 | 24 | 109 | 442 |
| Jacksonville | 332 | 246 | 174 | 21 | 0 | 0 | 0 | 0 | 0 | 12 | 144 | 310 | 1 239 |
| Key West | 40 | 31 | 9 | 0 | 0 | 0 | 0 | 0 | 0 | 0 | 0 | 28 | 108 |
| Lakeland | 195 | 146 | 99 | 0 | 0 | 0 | 0 | 0 | 0 | 0 | 57 | 164 | 661 |

| State and Station | Jan | Feb | Mar | Apr | May | June | July | Aug | Sep | Oct | Nov | Dec | Annual |
|---|---|---|---|---|---|---|---|---|---|---|---|---|---|
| Miami Beach | 56 | 36 | 9 | 0 | 0 | 0 | 0 | 0 | 0 | 0 | 0 | 50 | 141 |
| Orlando | 220 | 165 | 105 | 6 | 0 | 0 | 0 | 0 | 0 | 0 | 72 | 198 | 766 |
| Pensacola | 400 | 277 | 183 | 36 | 0 | 0 | 0 | 0 | 0 | 19 | 195 | 353 | 1 463 |
| Tallahassee | 375 | 286 | 202 | 36 | 0 | 0 | 0 | 0 | 0 | 28 | 198 | 360 | 1 485 |
| Tampa | 202 | 148 | 102 | 0 | 0 | 0 | 0 | 0 | 0 | 0 | 60 | 171 | 683 |
| West Palm Beach | 87 | 64 | 31 | 0 | 0 | 0 | 0 | 0 | 0 | 0 | 6 | 65 | 253 |
| Ga., Athens | 642 | 529 | 431 | 141 | 22 | 0 | 0 | 0 | 12 | 115 | 405 | 632 | 2 929 |
| Atlanta | 639 | 529 | 437 | 168 | 25 | 0 | 0 | 0 | 18 | 127 | 414 | 626 | 2 983 |
| Augusta | 549 | 445 | 350 | 90 | 0 | 0 | 0 | 0 | 0 | 78 | 333 | 552 | 2 397 |
| Columbus | 552 | 434 | 338 | 96 | 0 | 0 | 0 | 0 | 0 | 87 | 333 | 543 | 2 383 |
| Macon | 505 | 403 | 295 | 63 | 0 | 0 | 0 | 0 | 0 | 71 | 297 | 502 | 2 136 |
| Rome | 710 | 577 | 468 | 177 | 34 | 0 | 0 | 0 | 24 | 161 | 474 | 701 | 3 326 |
| Savannah | 437 | 353 | 254 | 45 | 0 | 0 | 0 | 0 | 0 | 47 | 246 | 437 | 1 819 |
| Thomasville | 394 | 305 | 208 | 33 | 0 | 0 | 0 | 0 | 0 | 25 | 198 | 366 | 1 529 |
| Idaho, Boise | 1113 | 854 | 722 | 438 | 245 | 81 | 0 | 0 | 132 | 415 | 792 | 1017 | 5 809 |
| Idaho Falls 46W | 1538 | 1249 | 1085 | 651 | 391 | 192 | 16 | 34 | 270 | 623 | 1056 | 1370 | 8 475 |
| Idaho Falls 42NW | 1600 | 1291 | 1107 | 657 | 388 | 192 | 16 | 40 | 282 | 648 | 1107 | 1432 | 8 760 |
| Lewiston | 1063 | 815 | 694 | 426 | 239 | 90 | 0 | 0 | 123 | 403 | 756 | 933 | 5 542 |
| Pocatello | 1324 | 1058 | 905 | 555 | 319 | 141 | 0 | 0 | 172 | 493 | 900 | 1166 | 7 033 |
| Ill., Cairo | 856 | 680 | 539 | 195 | 47 | 0 | 0 | 0 | 36 | 164 | 513 | 791 | 3 821 |
| Chicago | 1209 | 1044 | 890 | 480 | 211 | 48 | 0 | 0 | 81 | 326 | 753 | 1113 | 6 155 |
| Moline | 1314 | 1100 | 918 | 450 | 189 | 39 | 0 | 9 | 99 | 335 | 774 | 1181 | 6 408 |
| Peoria | 1218 | 1025 | 849 | 426 | 183 | 33 | 0 | 6 | 87 | 326 | 759 | 1113 | 6 025 |
| Rockford | 1333 | 1137 | 961 | 516 | 236 | 60 | 6 | 9 | 114 | 400 | 837 | 1221 | 6 830 |
| Springfield | 1135 | 935 | 769 | 354 | 136 | 18 | 0 | 0 | 72 | 291 | 696 | 1023 | 5 429 |
| Ind., Evansville | 955 | 767 | 620 | 237 | 68 | 0 | 0 | 0 | 66 | 220 | 606 | 896 | 4 435 |
| Fort Wayne | 1178 | 1028 | 890 | 471 | 189 | 39 | 0 | 9 | 105 | 378 | 783 | 1135 | 6 205 |
| Indianapolis | 1113 | 949 | 809 | 432 | 177 | 39 | 0 | 0 | 90 | 316 | 723 | 1051 | 5 699 |
| South Bend | 1221 | 1070 | 933 | 525 | 239 | 60 | 0 | 6 | 111 | 372 | 777 | 1125 | 6 439 |
| Iowa, Burlington | 1259 | 1042 | 859 | 426 | 177 | 33 | 0 | 0 | 93 | 322 | 768 | 1135 | 6 114 |
| Des Moines | 1398 | 1165 | 967 | 489 | 211 | 39 | 0 | 9 | 99 | 363 | 837 | 1231 | 6 808 |
| Dubuque | 1420 | 1204 | 1026 | 546 | 270 | 78 | 12 | 31 | 156 | 450 | 906 | 1287 | 7 376 |
| Sioux City | 1435 | 1198 | 989 | 483 | 214 | 39 | 0 | 9 | 108 | 369 | 876 | 1240 | 6 951 |
| Waterloo | 1460 | 1221 | 1023 | 531 | 229 | 54 | 12 | 19 | 138 | 428 | 909 | 1296 | 7 320 |
| Kans., Concordia | 1163 | 935 | 781 | 372 | 149 | 18 | 0 | 0 | 57 | 276 | 705 | 1023 | 5 479 |
| Dodge City | 1051 | 840 | 719 | 354 | 124 | 9 | 0 | 0 | 33 | 251 | 666 | 939 | 4 986 |
| Goodland | 1166 | 955 | 884 | 507 | 236 | 42 | 0 | 6 | 81 | 381 | 810 | 1073 | 6 141 |
| Topeka | 1122 | 893 | 722 | 330 | 124 | 12 | 0 | 0 | 57 | 270 | 672 | 980 | 5 182 |
| Wichita | 1023 | 804 | 645 | 270 | 87 | 6 | 0 | 0 | 33 | 229 | 618 | 905 | 4 620 |
| Ky., Covington | 1035 | 893 | 756 | 390 | 149 | 24 | 0 | 0 | 75 | 291 | 669 | 983 | 5 265 |
| Lexington | 946 | 818 | 685 | 325 | 105 | 0 | 0 | 0 | 54 | 239 | 609 | 902 | 4 683 |
| Louisville | 930 | 818 | 682 | 315 | 105 | 9 | 0 | 0 | 54 | 248 | 609 | 890 | 4 660 |
| La., Alexandria | 471 | 361 | 260 | 69 | 0 | 0 | 0 | 0 | 0 | 56 | 273 | 431 | 1 921 |
| Baton Rouge | 409 | 294 | 208 | 33 | 0 | 0 | 0 | 0 | 0 | 31 | 216 | 369 | 1 560 |
| Burrwood | 298 | 218 | 171 | 27 | 0 | 0 | 0 | 0 | 0 | 0 | 96 | 214 | 1 024 |
| Lake Charles | 381 | 274 | 195 | 39 | 0 | 0 | 0 | 0 | 0 | 19 | 210 | 341 | 1 459 |
| New Orleans | 363 | 258 | 192 | 39 | 0 | 0 | 0 | 0 | 0 | 19 | 192 | 322 | 1 385 |
| Shreveport | 552 | 426 | 304 | 81 | 0 | 0 | 0 | 0 | 0 | 47 | 297 | 477 | 2 184 |
| Maine, Caribou | 1690 | 1470 | 1308 | 858 | 468 | 183 | 78 | 115 | 336 | 682 | 1044 | 1535 | 9 767 |
| Portland | 1339 | 1182 | 1042 | 675 | 372 | 111 | 12 | 53 | 195 | 508 | 807 | 1215 | 7 511 |
| Md., Baltimore | 936 | 820 | 679 | 327 | 90 | 0 | 0 | 0 | 48 | 264 | 585 | 905 | 4 654 |
| Frederick | 995 | 876 | 741 | 384 | 127 | 12 | 0 | 0 | 66 | 307 | 624 | 955 | 5 087 |
| Mass., Blue Hill Obsy | 1178 | 1053 | 936 | 579 | 267 | 69 | 0 | 22 | 108 | 381 | 690 | 1085 | 6 368 |
| Boston | 1088 | 972 | 846 | 513 | 208 | 36 | 0 | 9 | 60 | 316 | 603 | 983 | 5 634 |
| Nantucket | 992 | 941 | 896 | 621 | 384 | 129 | 12 | 22 | 93 | 332 | 573 | 896 | 5 891 |
| Pittsfield | 1339 | 1196 | 1063 | 660 | 326 | 105 | 25 | 59 | 219 | 524 | 831 | 1231 | 7 578 |
| Worcester | 1271 | 1123 | 998 | 612 | 304 | 78 | 6 | 34 | 147 | 450 | 774 | 1172 | 6 969 |
| Mich., Alpena | 1404 | 1299 | 1218 | 777 | 446 | 156 | 68 | 105 | 273 | 580 | 912 | 1268 | 8 506 |
| Detroit (City) | 1181 | 1058 | 936 | 522 | 220 | 42 | 0 | 0 | 87 | 360 | 738 | 1088 | 6 232 |
| Escanaba | 1445 | 1296 | 1203 | 777 | 456 | 159 | 59 | 87 | 243 | 539 | 924 | 1293 | 8 481 |
| Flint | 1330 | 1198 | 1066 | 639 | 319 | 90 | 16 | 40 | 159 | 465 | 843 | 1212 | 7 377 |
| Grand Rapids | 1259 | 1134 | 1011 | 579 | 279 | 75 | 9 | 28 | 135 | 434 | 804 | 1147 | 6 894 |
| Lansing | 1262 | 1142 | 1011 | 579 | 273 | 69 | 6 | 22 | 138 | 431 | 813 | 1163 | 6 909 |

(*Continued*)

23

**TABLE 2-1.** (*Continued*)

| State and Station | Jan | Feb | Mar | Apr | May | June | July | Aug | Sep | Oct | Nov | Dec | Annual |
|---|---|---|---|---|---|---|---|---|---|---|---|---|---|
| Marquette | 1411 | 1268 | 1187 | 771 | 468 | 177 | 59 | 81 | 240 | 527 | 936 | 1268 | 8 393 |
| Muskegon | 1209 | 1100 | 995 | 594 | 310 | 78 | 12 | 28 | 120 | 400 | 762 | 1088 | 6 696 |
| Sault Ste. Marie | 1525 | 1380 | 1277 | 810 | 477 | 201 | 96 | 105 | 279 | 580 | 951 | 1368 | 9 048 |
| Minn., Duluth | 1745 | 1518 | 1355 | 840 | 490 | 98 | 71 | 109 | 330 | 632 | 1131 | 1581 | 10 000 |
| International Falls | 1919 | 1621 | 1414 | 828 | 443 | 174 | 71 | 112 | 368 | 701 | 1236 | 1724 | 10 606 |
| Minneapolis | 1631 | 1380 | 1166 | 621 | 288 | 81 | 22 | 31 | 189 | 505 | 1014 | 1454 | 8 382 |
| Rochester | 1593 | 1366 | 1150 | 630 | 301 | 93 | 25 | 34 | 186 | 474 | 1005 | 1438 | 8 295 |
| Saint Cloud | 1702 | 1445 | 1221 | 666 | 326 | 105 | 28 | 47 | 225 | 549 | 1065 | 1500 | 8 879 |
| Miss., Jackson | 546 | 414 | 310 | 87 | 0 | 0 | 0 | 0 | 0 | 65 | 315 | 502 | 2 239 |
| Meridian | 543 | 417 | 310 | 81 | 0 | 0 | 0 | 0 | 0 | 81 | 339 | 518 | 2 289 |
| Vicksburg | 512 | 384 | 282 | 69 | 0 | 0 | 0 | 0 | 0 | 53 | 279 | 462 | 2 041 |
| Mo., Columbia | 1076 | 874 | 716 | 324 | 121 | 12 | 0 | 0 | 54 | 251 | 651 | 967 | 5 046 |
| Kansas | 1032 | 818 | 682 | 294 | 109 | 0 | 0 | 0 | 39 | 220 | 612 | 905 | 4 711 |
| St. Joseph | 1172 | 949 | 769 | 348 | 133 | 15 | 0 | 6 | 60 | 285 | 708 | 1039 | 5 484 |
| St. Louis | 1026 | 848 | 704 | 312 | 121 | 15 | 0 | 0 | 60 | 251 | 627 | 936 | 4 900 |
| Springfield | 973 | 781 | 660 | 291 | 105 | 6 | 0 | 0 | 45 | 223 | 600 | 877 | 4 561 |
| Mont., Billings | 1296 | 1100 | 970 | 570 | 285 | 102 | 6 | 15 | 186 | 487 | 897 | 1135 | 7 049 |
| Glasgow | 1711 | 1439 | 1187 | 648 | 335 | 150 | 31 | 47 | 270 | 608 | 1104 | 1466 | 8 996 |
| Great Falls | 1349 | 1154 | 1063 | 642 | 384 | 186 | 28 | 53 | 258 | 543 | 921 | 1169 | 7 750 |
| Havre | 1584 | 1364 | 1181 | 657 | 338 | 162 | 28 | 53 | 306 | 595 | 1065 | 1367 | 8 700 |
| Helena | 1438 | 1170 | 1042 | 651 | 381 | 195 | 31 | 59 | 294 | 601 | 1002 | 1265 | 8 129 |
| Kalispell | 1401 | 1134 | 1029 | 639 | 397 | 207 | 50 | 99 | 321 | 654 | 1020 | 1240 | 8 191 |
| Miles City | 1504 | 1252 | 1057 | 579 | 276 | 99 | 6 | 6 | 174 | 502 | 972 | 1296 | 7 723 |
| Missoula | 1420 | 1120 | 970 | 621 | 391 | 219 | 34 | 74 | 303 | 651 | 1035 | 1287 | 8 125 |
| Nebr., Grand Island | 1314 | 1089 | 908 | 462 | 211 | 45 | 0 | 6 | 108 | 381 | 834 | 1172 | 6 530 |
| Lincoln | 1237 | 1016 | 834 | 402 | 171 | 30 | 0 | 6 | 75 | 301 | 726 | 1066 | 5 864 |
| Norfolk | 1414 | 1179 | 983 | 498 | 233 | 48 | 9 | 0 | 111 | 397 | 873 | 1234 | 6 979 |
| North Platte | 1271 | 1039 | 930 | 519 | 248 | 57 | 0 | 6 | 123 | 440 | 885 | 1166 | 6 684 |
| Omaha | 1355 | 1126 | 939 | 465 | 208 | 42 | 0 | 12 | 105 | 357 | 828 | 1175 | 6 612 |
| Scottsbluff | 1231 | 1008 | 921 | 552 | 285 | 75 | 0 | 0 | 138 | 459 | 876 | 1128 | 6 673 |
| Valentine | 1395 | 1176 | 1045 | 579 | 288 | 84 | 9 | 12 | 165 | 493 | 942 | 1237 | 7 425 |
| Nev., Elko | 1314 | 1036 | 911 | 621 | 409 | 192 | 9 | 34 | 225 | 561 | 924 | 1197 | 7 433 |
| Ely | 1308 | 1075 | 977 | 672 | 456 | 225 | 28 | 43 | 234 | 592 | 939 | 1184 | 7 733 |
| Las Vegas | 688 | 487 | 335 | 111 | 6 | 0 | 0 | 0 | 0 | 78 | 387 | 617 | 2 709 |
| Reno | 1073 | 823 | 729 | 510 | 357 | 189 | 43 | 87 | 204 | 490 | 801 | 1026 | 6 332 |
| Winnemucca | 1172 | 916 | 837 | 573 | 363 | 153 | 0 | 34 | 210 | 536 | 876 | 1091 | 6 761 |
| N.H., Concord | 1358 | 1184 | 1032 | 636 | 298 | 75 | 6 | 50 | 177 | 505 | 822 | 1240 | 7 383 |
| Mt. Wash. Obsy | 1820 | 1663 | 1652 | 1260 | 930 | 603 | 493 | 536 | 720 | 1057 | 1341 | 1742 | 13 817 |
| N.J., Atlantic City | 936 | 848 | 741 | 420 | 133 | 15 | 0 | 0 | 39 | 251 | 549 | 880 | 4 812 |
| Newark | 983 | 876 | 729 | 381 | 118 | 0 | 0 | 0 | 30 | 248 | 573 | 921 | 4 859 |
| Trenton | 989 | 885 | 753 | 399 | 121 | 12 | 0 | 0 | 57 | 264 | 576 | 924 | 4 980 |
| N. Mex., Albuquerque | 930 | 703 | 595 | 288 | 81 | 0 | 0 | 0 | 12 | 229 | 642 | 868 | 4 348 |
| Clayton | 986 | 812 | 747 | 429 | 183 | 21 | 0 | 6 | 66 | 310 | 699 | 899 | 5 158 |
| Raton | 1116 | 904 | 834 | 543 | 301 | 63 | 9 | 28 | 126 | 431 | 825 | 1048 | 6 228 |
| Roswell | 840 | 641 | 481 | 201 | 31 | 0 | 0 | 0 | 18 | 202 | 573 | 806 | 3 793 |
| Silver City | 791 | 605 | 518 | 261 | 87 | 0 | 0 | 0 | 6 | 183 | 525 | 729 | 3 705 |
| N.Y., Albany | 1311 | 1156 | 992 | 564 | 239 | 45 | 0 | 19 | 138 | 440 | 777 | 1194 | 6 875 |
| Binghamton (AP) | 1277 | 1154 | 1045 | 645 | 313 | 99 | 22 | 65 | 201 | 471 | 810 | 1184 | 7 286 |
| Binghampton (PO) | 1190 | 1081 | 949 | 543 | 229 | 45 | 0 | 28 | 141 | 406 | 732 | 1107 | 6 451 |
| Buffalo | 1256 | 1145 | 1039 | 645 | 329 | 78 | 19 | 37 | 141 | 440 | 777 | 1156 | 7 062 |
| Central Park | 986 | 885 | 760 | 408 | 118 | 9 | 0 | 0 | 30 | 233 | 540 | 902 | 4 871 |
| J. F. Kennedy Intl | 1029 | 935 | 815 | 480 | 167 | 12 | 0 | 0 | 36 | 248 | 564 | 933 | 5 219 |
| Laguardia | 973 | 879 | 750 | 414 | 124 | 6 | 0 | 0 | 27 | 223 | 528 | 887 | 4 811 |
| Rochester | 1234 | 1123 | 1014 | 597 | 279 | 48 | 9 | 31 | 126 | 415 | 747 | 1125 | 6 748 |
| Schenectady | 1283 | 1131 | 970 | 543 | 211 | 30 | 0 | 22 | 123 | 422 | 756 | 1159 | 6 650 |
| Syracuse | 1271 | 1140 | 1004 | 570 | 248 | 45 | 6 | 28 | 132 | 435 | 744 | 1153 | 6 756 |
| N.C., Asheville | 784 | 683 | 592 | 273 | 87 | 0 | 0 | 0 | 48 | 245 | 555 | 775 | 4 042 |
| Cape Hatteras | 580 | 518 | 440 | 177 | 25 | 0 | 0 | 0 | 0 | 78 | 273 | 521 | 2 612 |
| Charlotte | 691 | 582 | 481 | 156 | 22 | 0 | 0 | 0 | 6 | 124 | 438 | 691 | 3 191 |
| Greensboro | 784 | 672 | 552 | 234 | 47 | 0 | 0 | 0 | 33 | 192 | 513 | 778 | 3 805 |
| Raleigh | 725 | 616 | 487 | 180 | 34 | 0 | 0 | 0 | 21 | 164 | 450 | 716 | 3 393 |
| Wilmington | 546 | 462 | 357 | 96 | 0 | 0 | 0 | 0 | 0 | 74 | 291 | 521 | 2 347 |
| Winston Salem | 753 | 652 | 524 | 207 | 37 | 0 | 0 | 0 | 21 | 171 | 483 | 747 | 3 595 |

24

| State and Station | Jan | Feb | Mar | Apr | May | June | July | Aug | Sept | Oct | Nov | Dec | Annual |
|---|---|---|---|---|---|---|---|---|---|---|---|---|---|
| N. Dak., Bismarck | 1708 | 1442 | 1203 | 645 | 329 | 117 | 34 | 28 | 222 | 577 | 1083 | 1463 | 8 851 |
| Devil's Lake | 1872 | 1579 | 1345 | 753 | 381 | 138 | 40 | 53 | 273 | 642 | 1191 | 1634 | 9 901 |
| Fargo | 1789 | 1520 | 1262 | 690 | 332 | 99 | 28 | 37 | 219 | 574 | 1107 | 1569 | 9 226 |
| Williston | 1758 | 1473 | 1262 | 681 | 357 | 141 | 31 | 43 | 261 | 601 | 1122 | 1513 | 9 243 |
| Ohio, Akron | 1138 | 1016 | 871 | 489 | 202 | 39 | 0 | 9 | 96 | 381 | 726 | 1070 | 6 037 |
| Cincinnati | 970 | 837 | 701 | 336 | 118 | 9 | 0 | 0 | 54 | 248 | 612 | 921 | 4 806 |
| Cleveland | 1159 | 1047 | 918 | 552 | 260 | 66 | 9 | 25 | 105 | 384 | 738 | 1088 | 6 351 |
| Columbus | 1088 | 949 | 809 | 426 | 171 | 27 | 0 | 6 | 84 | 347 | 714 | 1039 | 5 660 |
| Dayton | 1097 | 955 | 809 | 429 | 167 | 30 | 0 | 6 | 78 | 310 | 696 | 1045 | 5 622 |
| Mansfield | 1169 | 1042 | 924 | 543 | 245 | 60 | 9 | 22 | 114 | 397 | 768 | 1110 | 6 403 |
| Sandusky | 1107 | 991 | 868 | 495 | 198 | 36 | 0 | 6 | 66 | 313 | 684 | 1032 | 5 796 |
| Toledo | 1200 | 1056 | 924 | 543 | 242 | 60 | 0 | 16 | 117 | 406 | 792 | 1138 | 6 494 |
| Youngstown | 1169 | 1047 | 921 | 540 | 248 | 60 | 6 | 19 | 120 | 412 | 771 | 1104 | 6 417 |
| Okla., Oklahoma City | 868 | 664 | 527 | 189 | 34 | 0 | 0 | 0 | 15 | 164 | 498 | 766 | 3 725 |
| Tulsa | 893 | 683 | 539 | 213 | 47 | 0 | 0 | 0 | 18 | 158 | 522 | 787 | 3 860 |
| Oreg., Astoria | 753 | 622 | 636 | 480 | 363 | 231 | 146 | 130 | 210 | 375 | 561 | 679 | 5 186 |
| Burns | 1246 | 988 | 856 | 570 | 366 | 177 | 12 | 37 | 210 | 515 | 867 | 1113 | 6 957 |
| Eugene | 803 | 627 | 589 | 426 | 279 | 135 | 34 | 34 | 129 | 366 | 585 | 719 | 4 726 |
| Meacham | 1209 | 1005 | 983 | 726 | 527 | 339 | 84 | 124 | 288 | 580 | 918 | 1091 | 7 874 |
| Medford | 918 | 697 | 642 | 432 | 242 | 78 | 0 | 0 | 78 | 372 | 678 | 871 | 5 008 |
| Pendleton | 1017 | 773 | 617 | 396 | 205 | 63 | 0 | 0 | 111 | 350 | 711 | 884 | 5 127 |
| Portland | 825 | 644 | 586 | 396 | 245 | 105 | 25 | 28 | 114 | 335 | 597 | 735 | 4 635 |
| Roseburg | 766 | 608 | 570 | 405 | 267 | 123 | 22 | 16 | 105 | 329 | 567 | 713 | 4 491 |
| Salem | 822 | 647 | 611 | 417 | 273 | 144 | 37 | 31 | 111 | 338 | 594 | 729 | 4 754 |
| Sexton Summit | 958 | 809 | 818 | 609 | 465 | 279 | 81 | 81 | 171 | 443 | 666 | 874 | 6 254 |
| Pa., Allentown | 1116 | 1002 | 849 | 471 | 167 | 24 | 0 | 0 | 90 | 353 | 693 | 1045 | 5 810 |
| Erie | 1169 | 1081 | 973 | 585 | 288 | 60 | 0 | 25 | 102 | 391 | 714 | 1063 | 6 451 |
| Harrisburg | 1045 | 907 | 766 | 396 | 124 | 12 | 0 | 0 | 63 | 298 | 648 | 992 | 5 251 |
| Philadelphia | 1014 | 890 | 744 | 390 | 115 | 12 | 0 | 0 | 60 | 291 | 621 | 964 | 5 101 |
| Pittsburgh | 1119 | 1002 | 874 | 480 | 195 | 39 | 0 | 9 | 105 | 375 | 726 | 1063 | 5 987 |
| Reading | 1001 | 885 | 735 | 372 | 105 | 0 | 0 | 0 | 54 | 257 | 597 | 939 | 4 945 |
| Scranton | 1156 | 1028 | 893 | 498 | 195 | 33 | 0 | 19 | 132 | 434 | 762 | 1104 | 6 254 |
| Williamsport | 1122 | 1002 | 856 | 468 | 177 | 24 | 0 | 9 | 111 | 375 | 717 | 1073 | 5 934 |
| R.I., Block Is. | 1020 | 955 | 877 | 612 | 344 | 99 | 0 | 16 | 78 | 307 | 594 | 902 | 5 804 |
| Providence | 1110 | 988 | 868 | 534 | 236 | 51 | 0 | 16 | 96 | 372 | 660 | 1023 | 5 954 |
| S.C., Charleston | 487 | 389 | 291 | 54 | 0 | 0 | 0 | 0 | 0 | 59 | 282 | 471 | 2 033 |
| Columbia | 570 | 470 | 357 | 81 | 0 | 0 | 0 | 0 | 0 | 84 | 345 | 577 | 2 484 |
| Florence | 552 | 459 | 347 | 84 | 0 | 0 | 0 | 0 | 0 | 78 | 315 | 552 | 2 387 |
| Greenville | 648 | 535 | 434 | 120 | 12 | 0 | 0 | 0 | 0 | 112 | 387 | 636 | 2 884 |
| Spartanburg | 663 | 560 | 453 | 144 | 25 | 0 | 0 | 0 | 15 | 130 | 417 | 667 | 3 074 |
| S. Dak., Huron | 1628 | 1355 | 1125 | 600 | 288 | 87 | 9 | 12 | 165 | 508 | 1014 | 1432 | 8 223 |
| Rapid City | 1333 | 1145 | 1051 | 615 | 326 | 126 | 22 | 12 | 165 | 481 | 897 | 1172 | 7 345 |
| Sioux Falls | 1544 | 1285 | 1082 | 573 | 270 | 78 | 19 | 25 | 168 | 462 | 972 | 1361 | 7 839 |
| Tenn., Bristol | 828 | 700 | 598 | 261 | 68 | 0 | 0 | 0 | 51 | 236 | 573 | 828 | 4 143 |
| Chattanooga | 722 | 577 | 453 | 150 | 25 | 0 | 0 | 0 | 18 | 143 | 468 | 698 | 3 254 |
| Knoxville | 732 | 613 | 493 | 198 | 43 | 0 | 0 | 0 | 30 | 171 | 489 | 725 | 3 494 |
| Memphis | 729 | 585 | 456 | 147 | 22 | 0 | 0 | 0 | 18 | 130 | 447 | 698 | 3 232 |
| Nashville | 778 | 644 | 512 | 189 | 40 | 0 | 0 | 0 | 30 | 158 | 495 | 732 | 3 578 |
| Oak Ridge (Co.) | 778 | 669 | 552 | 228 | 56 | 0 | 0 | 0 | 39 | 192 | 531 | 772 | 3 817 |
| Tex., Abilene | 642 | 470 | 347 | 114 | 0 | 0 | 0 | 0 | 0 | 99 | 366 | 586 | 2 624 |
| Amarillo | 877 | 664 | 546 | 252 | 56 | 0 | 0 | 0 | 18 | 205 | 570 | 797 | 3 985 |
| Austin | 468 | 325 | 223 | 51 | 0 | 0 | 0 | 0 | 0 | 31 | 225 | 388 | 1 711 |
| Brownsville | 205 | 106 | 74 | 0 | 0 | 0 | 0 | 0 | 0 | 0 | 66 | 149 | 600 |
| Corpus Christi | 291 | 174 | 109 | 0 | 0 | 0 | 0 | 0 | 0 | 0 | 120 | 220 | 914 |
| Dallas | 601 | 440 | 319 | 90 | 6 | 0 | 0 | 0 | 0 | 62 | 321 | 524 | 2 363 |
| El Paso | 685 | 445 | 319 | 105 | 0 | 0 | 0 | 0 | 0 | 84 | 414 | 648 | 2 700 |
| Fort Worth | 614 | 448 | 319 | 99 | 0 | 0 | 0 | 0 | 0 | 65 | 324 | 536 | 2 405 |
| Galveston | 350 | 258 | 189 | 30 | 0 | 0 | 0 | 0 | 0 | 0 | 138 | 270 | 1 235 |
| Houston | 384 | 288 | 192 | 36 | 0 | 0 | 0 | 0 | 0 | 6 | 183 | 307 | 1 396 |
| Laredo | 267 | 134 | 74 | 0 | 0 | 0 | 0 | 0 | 0 | 0 | 105 | 217 | 797 |
| Lubbock | 800 | 613 | 484 | 201 | 31 | 0 | 0 | 0 | 18 | 174 | 513 | 744 | 3 578 |
| Midland | 651 | 468 | 322 | 90 | 0 | 0 | 0 | 0 | 0 | 87 | 381 | 592 | 2 591 |
| Port Arthur | 384 | 274 | 192 | 39 | 0 | 0 | 0 | 0 | 0 | 22 | 207 | 329 | 1 447 |

*(Continued)*

**TABLE 2-1.** *(Continued)*

| State and Station | Jan | Feb | Mar | Apr | May | June | July | Aug | Sep | Oct | Nov | Dec | Annual |
|---|---|---|---|---|---|---|---|---|---|---|---|---|---|
| San Angelo | 567 | 412 | 288 | 66 | 0 | 0 | 0 | 0 | 0 | 68 | 318 | 536 | 2 255 |
| San Antonio | 428 | 286 | 195 | 39 | 0 | 0 | 0 | 0 | 0 | 31 | 207 | 363 | 1 549 |
| Victoria | 344 | 230 | 152 | 21 | 0 | 0 | 0 | 0 | 0 | 6 | 150 | 270 | 1 173 |
| Waco | 596 | 389 | 270 | 66 | 0 | 0 | 0 | 0 | 0 | 43 | 270 | 456 | 2 030 |
| Wichita Falls | 698 | 518 | 378 | 120 | 6 | 0 | 0 | 0 | 0 | 99 | 381 | 632 | 2 832 |
| Utah, Milford | 1252 | 988 | 822 | 519 | 279 | 87 | 0 | 0 | 99 | 443 | 867 | 1141 | 6 497 |
| Salt Lake City | 1172 | 910 | 763 | 459 | 233 | 84 | 0 | 0 | 81 | 419 | 849 | 1082 | 6 052 |
| Wendover | 1178 | 902 | 729 | 408 | 177 | 51 | 0 | 0 | 48 | 372 | 822 | 1091 | 5 778 |
| Vt., Burlington | 1513 | 1333 | 1187 | 714 | 353 | 90 | 28 | 65 | 207 | 539 | 891 | 1349 | 8 269 |
| Va., Cape Henry | 694 | 633 | 536 | 246 | 53 | 0 | 0 | 0 | 0 | 112 | 360 | 645 | 3 279 |
| Lynchburg | 849 | 731 | 605 | 267 | 78 | 0 | 0 | 0 | 51 | 223 | 540 | 822 | 4 166 |
| Norfolk | 738 | 655 | 533 | 216 | 37 | 0 | 0 | 0 | 0 | 136 | 408 | 698 | 3 421 |
| Richmond | 815 | 703 | 546 | 219 | 53 | 0 | 0 | 0 | 36 | 214 | 495 | 784 | 3 865 |
| Roanoke | 834 | 722 | 614 | 261 | 65 | 0 | 0 | 0 | 51 | 229 | 549 | 825 | 4 150 |
| Wash., Nat'l. Ap. | 871 | 762 | 626 | 288 | 74 | 0 | 0 | 0 | 33 | 217 | 519 | 834 | 4 224 |
| Wash., Olympia | 834 | 675 | 645 | 450 | 307 | 177 | 68 | 71 | 198 | 422 | 636 | 753 | 5 236 |
| Seattle | 738 | 599 | 577 | 396 | 242 | 117 | 50 | 47 | 129 | 329 | 543 | 657 | 4 424 |
| Seattle Boeing | 831 | 655 | 608 | 411 | 242 | 99 | 34 | 40 | 147 | 384 | 624 | 763 | 4 838 |
| Seattle Tacoma | 828 | 678 | 657 | 474 | 295 | 159 | 56 | 62 | 162 | 391 | 633 | 750 | 5 145 |
| Spokane | 1231 | 980 | 834 | 531 | 288 | 135 | 9 | 25 | 168 | 493 | 879 | 1082 | 6 655 |
| Stampede Pass | 1287 | 1075 | 1085 | 855 | 654 | 483 | 273 | 291 | 393 | 701 | 1008 | 1178 | 92 83 |
| Tatoosh Is. | 713 | 613 | 645 | 525 | 431 | 333 | 295 | 279 | 306 | 406 | 534 | 639 | 5 719 |
| Walla Walla | 986 | 745 | 589 | 342 | 177 | 45 | 0 | 0 | 87 | 310 | 681 | 843 | 4 805 |
| Yakima | 1163 | 868 | 713 | 435 | 200 | 69 | 0 | 12 | 144 | 450 | 828 | 1039 | 5 941 |
| W. Va., Charleston | 880 | 770 | 648 | 300 | 96 | 9 | 0 | 0 | 63 | 254 | 591 | 865 | 4 476 |
| Elkins | 1008 | 896 | 791 | 444 | 198 | 48 | 9 | 25 | 135 | 400 | 729 | 992 | 5 675 |
| Huntington | 880 | 764 | 636 | 294 | 99 | 12 | 0 | 0 | 63 | 257 | 585 | 856 | 4 446 |
| Parkersburg | 942 | 826 | 691 | 339 | 115 | 6 | 0 | 0 | 60 | 264 | 606 | 905 | 4 754 |
| Wis., Green Bay | 1494 | 1313 | 1141 | 654 | 335 | 99 | 28 | 50 | 174 | 484 | 924 | 1333 | 8 029 |
| La Crosse | 1504 | 1277 | 1070 | 540 | 245 | 69 | 12 | 19 | 153 | 437 | 924 | 1339 | 7 589 |
| Madison | 1473 | 1274 | 1113 | 618 | 310 | 102 | 25 | 40 | 174 | 474 | 930 | 1330 | 7 863 |
| Milwaukee | 1376 | 1193 | 1054 | 642 | 372 | 135 | 43 | 47 | 174 | 471 | 876 | 1252 | 7 635 |
| Wyo., Casper | 1290 | 1084 | 1020 | 657 | 381 | 129 | 6 | 16 | 192 | 524 | 942 | 1169 | 7 410 |
| Cheyenne | 1228 | 1056 | 1011 | 672 | 381 | 102 | 19 | 31 | 210 | 543 | 924 | 1101 | 7 278 |
| Lander | 1417 | 1145 | 1017 | 654 | 381 | 153 | 6 | 19 | 204 | 555 | 1020 | 1299 | 7 870 |
| Sherian | 1355 | 1154 | 1054 | 642 | 366 | 150 | 25 | 31 | 219 | 539 | 948 | 1200 | 7 683 |

With the aid of Fig. 2-4 or eqs. 2-15, monthly degree-days $(D_m)$ can be predicted from the monthly average temperature, the 18.33 °C (65 °F) annual degree-day total $(D_a)$, and the new degree-day base; or known monthly degree-days can be converted to a new temperature base. The new monthly degree-day values can be summed and used to prepare an annual correction factor chart such as Fig. 2-5, which can be conveniently used to estimate fuel savings due to lower temperature settings at a given location.

**EXAMPLE 2-8**

Table 2-1 lists 414 °F-days for Atlanta in November and 2983 °F-days annually, using a 65 °F base. How many 55 °F degree-days are there in November?

For November, the actual daily degree-days are

$$\frac{414 \text{ °F-days}}{30 \text{ days}} = \frac{13.8 \text{ °F-days}}{\text{day}}$$

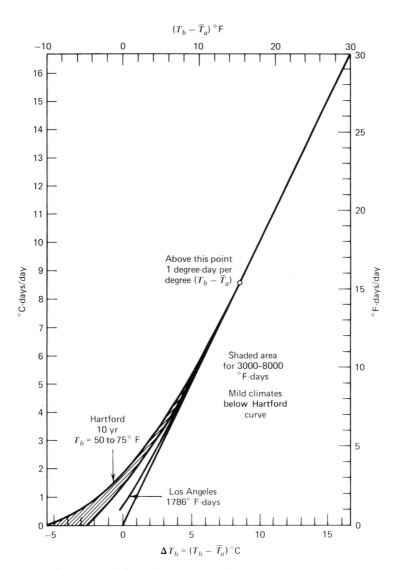

**FIGURE 2-4.** Relationship of monthly average temperature and degree-day base to degree-days.

From Fig. 2-4,

$$\Delta T_b = (T_b - \overline{T}_a) = 13.8\ °F$$
$$\overline{T}_a = 65\ °F - 13.8\ °F = 51.2\ °F$$

For a 55 °F base, using this average ambient temperature,

$$\Delta T_b = (T_b - \overline{T}_a) = (55 - 51.2)\ °F = 3.8\ °F$$

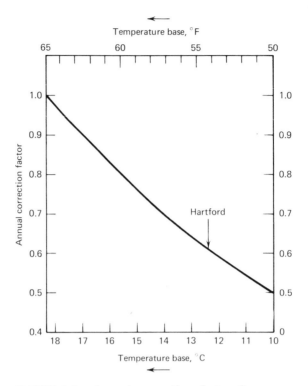

**FIGURE 2-5.** Annual correction factor for a new degree-day base for Hartford, Connecticut.

From Fig. 2-4 for this value of $\Delta T_b$ there are 5.5 °F-days per day—reading along the lower edge of the shaded area for the nearly 3000 °F-day location.

$$D_m = \frac{5.5\ °\text{F-days}}{\text{day}} \times 30\ \text{days} = 165\ °\text{F-days}$$

or

$$D_m = \frac{3.1\ °\text{C-days}}{\text{day}} \times 30\ \text{days} = 93\ °\text{C-days}$$

**EXAMPLE 2-8b**

Monthly °C-days for Hartford [65 °F base, appropriate for an inside ambient temperature of 75 °F (23.9 °C)] and monthly average temperatures have been taken from Appendix B and tabulated in Table 2-1a. What are the corresponding degree-days appropriate for a 20 °C (68 °F) indoor temperature?

The new temperature base is 68 °F − 10 °F = 58 °F or 14.44 °C. For May, $\overline{T}_a = 15.0\ °\text{C}$, while $D_a = 3527\ °\text{C-days}$

$$\Delta T_b = 14.44 - 15.0 = -0.56\ °\text{C}$$

**TABLE 2-1a.** Degree-C Days for Hartford, Connecticut (Example 2-8b)

| Month | Average Tempera-ture, °C (Appendix B) | 65 °F (18.3 °C) Base Degree-C Days (Appendix B) | 58 °F (14.44 °C) Base Degree-C Days (Calculated) | 50 °F (10 °C) Base Degree-C Days (Calculated) | 50 °F (10 °C) Base Degree-C Days (Weather Records) |
|---|---|---|---|---|---|
| Jan | − 3.8 | 686 | 565 | 428 | 428 |
| Feb | − 3.1 | 606 | 496 | 371 | 369 |
| Mar | 2.1 | 505 | 385 | 250 | 249 |
| Apr | 9.4 | 267 | 159 | 62 | 60 |
| May | 15.0 | 120 | 47 | 5 | 7 |
| Jun | 20.2 | 21 | 0 | 0 | 0 |
| Jul | 22.6 | 1 | 0 | 0 | 0 |
| Aug | 21.3 | 8 | 0 | 0 | 0 |
| Sep | 17.2 | 71 | 21 | 0 | 1 |
| Oct | 11.3 | 219 | 115 | 36 | 33 |
| Nov | 5.3 | 392 | 277 | 153 | 156 |
| Dec | − 2.0 | 631 | 510 | 373 | 373 |
| Total | | 3527 | 2575 | 1678 | 1676 |

by eq. 2-15a,

$$D_m = 31(-0.56) + \left[ 0.744 + 0.00387(3527) - \frac{0.5(3527)^2}{10^6} \right] 31$$

$$\times \exp - \left[ \frac{(-0.56 + 11.11)}{9.02} \right]^2$$

$$D_m = -17.4 + 64.5 = 47 \text{ °C-days}$$

Table 2-1a continues this example, listing 58 °F (14.44 °C) base degree-C days for all months, and also compares 50 °F (10 °C) degree-C days prepared by this method to those from actual weather records.

## Heating Demand

A normalized measure of the energy required to maintain comfortable temperatures in a given structure is characterized by solving eq. 2-14 for $UA$, the heating demand per degree-day.

$$\frac{Q_T}{D} = UA$$

In SI units it is conveniently expressed as MJ/°C-day, and in Engineering units as Btu/°F-day. $UA$ is a weather-independent, heat transfer constant that can be determined from analysis of the structure, as will be detailed below. Use of the

**TABLE 2-2.** Conductivities, Conductances and Resistances of Building and Insulating Materials at a Mean Temperature of 75 °F (24 °C)[a]

| Material | Description | Engineering Units | | | | | | SI Units | | | | | |
|---|---|---|---|---|---|---|---|---|---|---|---|---|---|
| | | Density, $\rho$ $\frac{lb}{ft^3}$ | Thermal Conductivity, $k$ $\frac{Btu\,in.}{hr\,°F\,ft^2}$ | Unit Conductance, $C$ $\frac{Btu}{ft^2\,°F\,hr}$ | Unit Resistance $\frac{1}{k}=r$ $\frac{hr\,°F\,ft^2}{Btu\,in.}$ | $\frac{1}{C}=R$ $\frac{ft^2\,hr\,°F}{Btu}$ | Specific Heat $\frac{Btu}{lb\,°F}$ | Density, $\rho$ $\frac{kg}{m^3}$ | Thermal Conductivity, $k$ $\frac{W}{°C\cdot m^2}$ | Unit Conductance, $C$ $\frac{W}{°C\cdot m^2}$ | Unit Resistance $\frac{1}{k}=r$ $\frac{°C\cdot m^2}{W}$ | $\frac{1}{c}=R$ $\frac{°C\cdot m^2}{W}$ | Specific Heat $\frac{kJ}{kg\cdot °C}$ |
| *Building board* | Asbestos-cement board ¼ in. or 6 mm | 120 | — | 16.5 | — | 0.07 | — | 1922 | — | 93.7 | — | 0.011 | — |
| Boards, panels, subflooring, sheathing, woodbased panel products | Gypsum or plasterboard ⅜ in. or 10 mm | 50 | — | 3.10 | — | 0.32 | — | 800 | — | 17.6 | — | 0.057 | — |
| | ½ in. or 13 mm | 50 | — | 2.25 | — | 0.45 | — | 800 | — | 12.8 | — | 0.078 | — |
| | Plywood | 34 | 0.80 | — | 1.25 | — | 0.29 | 545 | 0.12 | — | 8.70 | — | 1.21 |
| | ¼ in. or 6 mm | 34 | — | 3.20 | — | 0.31 | 0.29 | 545 | — | 18.2 | — | 0.055 | 1.21 |
| | ⅜ in. or 10 mm | 34 | — | 2.13 | — | 0.47 | 0.29 | 545 | — | 12.1 | — | 0.083 | 1.21 |
| | ½ in. or 13 mm | 34 | — | 1.60 | — | 0.62 | 0.29 | 545 | — | 9.09 | — | 0.110 | 1.21 |
| | ¾ in. or 20 mm | 34 | — | 1.07 | — | 0.93 | 0.29 | 545 | — | 6.08 | — | 0.165 | 1.21 |
| | Insulating board and sheathing ½ in. or 13 mm | 18 | — | 0.76 | — | 1.32 | 0.31 | 288 | — | 4.32 | — | 0.232 | 1.30 |
| | 25/32 in. or 20 mm | 18 | — | 0.49 | — | 2.06 | 0.31 | 288 | — | 2.78 | — | 0.359 | 1.30 |
| | Hardboard, high density, standard tempered | 63 | 1.00 | — | 1.00 | — | 0.33 | 1010 | 0.14 | — | 6.94 | — | 1.38 |
| | Particle board Medium density | 50 | 0.94 | — | 1.06 | — | 0.31 | 800 | 0.14 | — | 7.35 | — | 1.30 |
| | Underlayment ⅝ in. or 16 mm | 40 | — | 1.22 | — | 0.82 | 0.29 | 640 | — | 6.93 | — | 0.144 | 1.21 |
| | Wood subfloor ¾ in. or 20 mm | — | — | 1.06 | — | 0.94 | 0.34 | — | — | 6.02 | — | 0.166 | 1.42 |
| Building paper | Vapor—permeable felt | — | — | 16.7 | — | 0.06 | — | — | — | 94.8 | — | 0.011 | — |
| | Vapor—seal, two layers of mopped, 15-lb felt | — | — | 8.35 | — | 0.12 | — | — | — | 47.4 | — | 0.021 | — |
| Finish flooring materials | Carpet and fibrous pad | — | — | 0.48 | — | 2.08 | — | — | — | 2.73 | — | 0.367 | — |
| | Carpet and rubber pad | — | — | 0.81 | — | 1.23 | 0.34 | — | — | 4.60 | — | 0.217 | 1.42 |
| | Tile—asphalt, linoleum, vinyl, or rubber | — | — | 20.0 | — | 0.05 | 0.30 | — | — | 113.0 | — | 0.009 | 1.26 |

*Insulating materials*

| | | | | | | | | | | | | |
|---|---|---|---|---|---|---|---|---|---|---|---|---|
| **Blanket and batt** Mineral fiber—fibrous form processed from rock, slag, or glass | | | | | | | | | | | | |
|   Approximately 2 to 2¾ in. or 50 to 70 mm | — | — | 0.143 | — | 7 | 0.18 | — | — | 0.812 | — | 1.23 | 0.754 |
|   Approximately 3 to 3½ in. or 75 to 90 mm | — | — | 0.091 | — | 11 | 0.18 | — | — | 0.517 | — | 1.94 | 0.754 |
|   Approximately 5¼ to 6½ in. or 135 to 165 mm | — | — | 0.053 | — | 19 | 0.18 | — | — | 0.301 | — | 3.32 | 0.754 |
| **Board and slabs** | | | | | | | | | | | | |
|   Cellular glass | 9 | 0.40 | — | 2.50 | — | 0.24 | 144 | 0.058 | — | 17.2 | — | 1.0 |
|   Glass fiber, organic bonded | 4–9 | 0.25 | — | 4.00 | — | 0.19 | 64–144 | 0.036 | — | 27.8 | — | 8.0 |
|   Expanded polystyrene—molded beads | 1.0 | 0.28 | — | 3.57 | — | 0.29 | 16 | 0.040 | — | 25.0 | — | 1.2 |
|   Expanded polyurethane—R-11 expanded | 1.5 | 0.16 | — | 6.25 | — | 0.38 | 24 | 0.023 | — | 43.5 | — | 1.6 |
|   Mineral fiber with resin binder | 15 | 0.29 | — | 3.45 | — | 0.17 | 240 | 0.042 | — | — | — | 0.71 |
| **Loose fill** Mineral fiber—rock, slag, or glass | | | | | | | | | | | | |
|   Approximately 3 in. or 75 mm | — | — | 0.11 | — | 9 | 0.18 | — | — | 0.63 | — | 1.58 | 0.75 |
|   Approximately 4⅜ in. or 115 mm | — | — | 0.08 | — | 13 | 0.18 | — | — | 0.44 | — | 2.29 | 0.75 |
|   Approximately 6¼ in. or 160 mm | — | — | 0.05 | — | 19 | 0.18 | — | — | 0.30 | — | 3.35 | 0.75 |
|   Approximately 7¼ in. or 185 mm | — | — | 0.04 | — | 24 | 0.18 | — | — | 0.24 | — | 4.23 | 0.75 |
|   Silica aerogel | 7.6 | 0.17 | — | 5.88 | — | — | 122 | 0.025 | — | 40.8 | — | — |
|   Vermiculite (expanded) | 7–8 | 0.47 | — | 2.13 | — | — | 122 | 0.068 | — | 14.8 | — | — |
| **Roof insulation** Preformed, for use above deck | | | | | | | | | | | | |
|   Approximately ½ in. or 13 mm | — | — | 0.72 | — | 1.39 | — | — | — | 4.1 | — | 0.24 | 1.0 |
|   Approximately 1 in. or 25 mm | — | — | 0.36 | — | 2.78 | — | — | — | 2.0 | — | 0.49 | 2.1 |
|   Approximately 2 in. or 50 mm | — | — | 0.19 | — | 5.56 | — | — | — | 1.1 | — | 0.93 | 3.9 |
|   Cellular glass | 9 | 0.4 | — | 2.5 | — | 0.24 | 144 | 0.058 | — | 17.3 | — | 1.0 |
| ***Masonry materials*** **Concretes** | | | | | | | | | | | | |
|   Lightweight aggregates including expanded shale, clay, or slate; expanded slags; cinders; pumice; vermiculite; also cellular concretes | 200<br>100<br>80<br>40<br>20 | 5.2<br>3.6<br>2.5<br>1.15<br>0.70 | — | 0.19<br>0.28<br>0.40<br>0.86<br>1.43 | — | — | 3200<br>1600<br>1280<br>640<br>320 | 0.75<br>0.52<br>0.36<br>0.17<br>0.10 | — | 1.32<br>1.94<br>2.77<br>6.03<br>10.0 | — | — |
|   Sand and gravel or stone aggregate (not dried) | 140 | 12.0 | — | 0.08 | — | — | 2242 | 1.73 | — | 0.58 | — | — |

*(Continued)*

TABLE 2-2. *(Continued)*

| Material | Description | Engineering Units | | | | | | SI Units | | | | | |
|---|---|---|---|---|---|---|---|---|---|---|---|---|---|
| | | Density, $\rho$ $\frac{lb}{ft^3}$ | Thermal Conductivity, $k$ $\frac{Btu\,in.}{hr\,°F\,ft^2}$ | Unit Conductance, $C$ $\frac{Btu}{ft^2\,°F\,hr}$ | Unit Resistance $\frac{1}{k}=r$ $\frac{hr\,°F\,ft^2}{Btu\,in.}$ | $\frac{1}{C}=R$ $\frac{ft^2\,hr\,°F}{Btu}$ | Specific Heat $\frac{Btu}{lb\,°F}$ | Density, $\rho$ $\frac{kg}{m^3}$ | Thermal Conductivity, $k$ $\frac{W}{°C\cdot m^2}$ | Unit Conductance, $C$ $\frac{W}{°C\cdot m^2}$ | Unit Resistance $\frac{1}{k}=r$ $\frac{°C\cdot m^2}{W}$ | $\frac{1}{c}=R$ $\frac{°C\cdot m^2}{W}$ | Specific Heat $\frac{kJ}{kg\cdot°C}$ |
| Masonry units | Brick, common | 120 | 5.0 | — | 0.20 | — | — | 1922 | 0.72 | — | 1.39 | — | — |
| | Brick, face | 130 | 9.0 | — | 0.11 | — | — | 2082 | 1.30 | — | 0.77 | — | — |
| | Concrete blocks, three-oval core—sand and gravel aggregate | | | | | | | | | | | | |
| | 4 in. or 100 mm | — | — | 1.4 | — | 0.71 | — | — | — | 8.0 | — | 0.13 | — |
| | 8 in. or 200 mm | — | — | 0.9 | — | 1.11 | — | — | — | 5.1 | — | 0.20 | — |
| | 12 in. or 300 mm | — | — | 0.78 | — | 1.28 | — | — | — | 4.4 | — | 0.23 | — |
| | lightweight aggregate (expanded shale, clay slate or slag; pumice) | | | | | | | | | | | | |
| | 3 in. or 75 mm | — | — | 0.79 | — | 1.27 | — | — | — | 4.5 | — | 0.22 | — |
| | 4 in. or 100 mm | — | — | 0.67 | — | 1.50 | — | — | — | 3.8 | — | 0.26 | — |
| | 8 in. or 200 mm | — | — | 0.50 | — | 2.00 | — | — | — | 2.8 | — | 0.35 | — |
| | 12 in. or 300 mm | — | — | 0.44 | — | 2.27 | — | — | — | 2.5 | — | 0.40 | — |
| Plastering materials | Cement, plaster, sand, aggregate | 116 | 5.0 | — | 0.20 | — | — | 1858 | 0.72 | — | 1.39 | — | — |
| | Gypsum plaster: Lightweight aggregate | | | | | | | | | | | | |
| | ¼ in. or 13 mm | 45 | — | 3.12 | — | 0.32 | — | 721 | — | 17.7 | — | 0.056 | — |
| | ⅝ in. or 16 mm | 45 | — | 2.67 | — | 0.39 | — | 721 | — | 15.2 | — | 0.066 | — |

| Material | | | | | | | | | | | | |
|---|---|---|---|---|---|---|---|---|---|---|---|---|
| **Roofing** | | | | | | | | | | | | |
| Lightweight aggregate on metal lath ¾ in. or 20 mm | — | — | 2.13 | — | 0.47 | — | — | — | 12.1 | — | 0.083 | — |
| Asbestos-cement shingles | 120 | — | 4.76 | — | 0.21 | — | 1922 | — | 27.0 | — | 0.037 | — |
| Asphalt roll roofing | 70 | — | 6.50 | — | 0.15 | — | 1121 | — | 36.9 | — | 0.027 | — |
| Asphalt shingles | 70 | — | 2.27 | — | 0.44 | — | 1121 | — | 12.9 | — | 0.078 | — |
| Built-up roofing ⅜ in. or 10 mm | 70 | — | 3.00 | — | 0.33 | 0.35 | 1121 | — | 17.0 | — | 0.059 | — |
| Slate, ¼ in. or 13 mm | — | — | 20.00 | — | 0.05 | — | — | — | 113.6 | — | 0.009 | — |
| Wood shingles—plain or plastic film faced | — | — | 1.06 | — | 0.94 | 0.31 | — | — | 6.02 | — | 0.166 | — |
| **Siding materials (on flat surface)** | | | | | | | | | | | | |
| Shingles Asbestos-cement | 120 | — | 4.76 | — | 0.21 | — | 1922 | — | 27.0 | — | 3.70 | — |
| Siding Wood, drop, 1 in. or 25 mm | — | — | 1.27 | — | 0.79 | 0.31 | — | — | 7.21 | — | 0.139 | 1.30 |
| Wood, plywood, ⅜ in. or 10 mm. lapped | — | — | 1.59 | — | 0.59 | 0.29 | — | — | 9.03 | — | 0.111 | 1.21 |
| Aluminum or steel, over sheathing, hollowbacked | — | — | 1.61 | — | 0.61 | — | — | — | 9.14 | — | 0.109 | — |
| Insulating board—backed nominal, ⅜ in. or 10 mm | — | — | 0.55 | — | 1.82 | — | — | — | 3.12 | — | 0.320 | — |
| Insulating board—backed nominal, ⅜ in. or 10 mm foil-backed | — | — | 0.34 | — | 2.96 | — | — | — | 1.93 | — | 0.518 | — |
| Architectural glass | — | — | 10.00 | — | 0.10 | — | — | — | 56.8 | — | 0.018 | — |
| **Woods** | | | | | | | | | | | | |
| Maple, oak, and similar hardwoods | 45 | 1.10 | — | 0.91 | — | 0.30 | 721 | 0.159 | — | 6.3 | — | 1.26 |
| Fir, pine, and similar softwoods | 32 | 0.80 | — | 1.25 | — | 0.33 | 513 | 0.115 | — | 8.67 | — | 1.38 |
| **Metals** | | | | | | | | | | | | |
| Aluminum (1100) | 171 | 1536 | — | 0.000 65 | — | 0.214 | 2739 | 221.5 | — | 0.0045 | — | 0.896 |
| Steel, mild | 489 | 314 | — | 0.003 18 | — | 0.120 | 7833 | 45.3 | — | 0.022 | — | 0.502 |
| Steel, stainless | 494 | 108 | — | 0.009 26 | — | 0.109 | 7913 | 15.6 | — | 0.064 | — | 0.456 |

[a] Adapted by permission from ASHRAE Handbook of Fundamentals, 1972. Tables 2-2 to 2-5 have been reprinted with minor changes from McQuiston and Parker (3) © 1977 John Wiley & Sons, Inc., by permission of the publisher.

heating demand per degree-day gives the engineer a way to simply express heating demand and thus heat loss characteristics independent of the climatic conditions.

If $UA$ is given in W/°C (watts per degree Celsius), it can be multiplied by 0.0864 to give MJ/°C-day if degree-days are to be used later in calculations. Strictly speaking, degree-days are not proper SI units, but it seems certain that the practice of using them rather than degree-seconds will continue, making a conversion convenient at this point.

It is also customary to express $UA$ by giving the *design heating load* in watts or Btu/hr. To find $UA$ from this figure the engineer must know that the design temperature difference ($\Delta T_{design}$) is often *assumed* to be a 75 °F inside ambient less the $97\frac{1}{2}\%$ winter design temperature taken from *ASHRAE Fundamentals* (2) according to the location (see Example 2-12). Equation 2-2 may then be used to find $UA$ = design load/$\Delta T_{design}$. If the resultant value is in Btu/hr °F it can be multiplied by 24 to give Btu/°F-day.

## Heat Conduction Through Building Materials

The thermal conductivity of gases is quite low, but they transfer heat well if convection currents are allowed to form. Most insulating materials rely on interstitial air to limit conductive heat transfer and use the fibrous nature of the material to stop convective air currents. Since the fibers have a small cross section, their conduction is low, and the apparent conductivity approaches that of air, about 0.026 W/°C·m [0.18 Btu/hr ft² (°F/in.)] at room temperatures. Therefore, a wide range of insulating materials have about the same thermal properties regardless of their basic composition: Fibrous glass, rock wool, foam rubber, and polystyrene foam all have a thermal conductivity of about 0.035 W/°C·m [0.24 Btu/ft² (°F/in.)].

There are several techniques used to further reduce the thermal conductivity of insulating materials. One is the vacuum bottle concept, in which the convective space is evacuated. This is impractical with building materials in most circumstances. Another is the use of gas other than air. This is practical with certain polystyrene and polyurethane foams, which have a closed cell structure that can retain the gaseous foaming agent. When blown with a fluorocarbon, which has a thermal conductivity about one-half that of air, polyurethane foam has an overall conductivity of about 0.023 W/°C·m [0.16 Btu/hr ft² (°F/in.)]. Another technique is to use extremely fine powders, which can actually have better insulating properties than the air they contain. The usual gas-to-gas thermal conductivity is disrupted by molecular collisions with the walls of the powder, and the powder conductivity itself is low because the particles only have "point" contact with each other. Such materials are quite expensive but have seen applications in special situations where conductivities of 0.017 to 0.023 W/°C·m [0.12 to 0.16 Btu/hr ft² (°F/in.)] are important. Reflective coatings are also quite effective. When properly installed as a radiation barrier within a frame wall, several separated layers of aluminum foil can provide perhaps one-half the insulating value of a fibrous insulation in the same space.

Table 2-2 lists the thermal conductivity of many building materials. The conductivity of these materials varies somewhat with temperature. An increase in conductivity with temperature is typical for insulation, resulting directly from the increased velocity of the conductive gas molecules at a higher temperature.

## Overall Heat Transfer Coefficients

**Heat Transfer in Series Paths.**   For heat transfer coefficients $h_a$, $h_b$, ... and thermal conductivities $k_1$, $k_2$, ... in series, the overall heat transfer coefficient $U_T$ is found by noting that the overall $\Delta T$ in eq. 2-2 equals the sum of the individual $\Delta T$'s from eqs. 2-1 and 2-3:

$$\Delta T = \Delta T_1 + \Delta T_2 + \cdots + \Delta T_a + \Delta T_b + \cdots$$

Applying eq. 2-2 solved for $\Delta T$,

$$\frac{Q}{U_T A} = \frac{Q}{(k_1/L_1)A} + \frac{Q}{(k_2/L_2)A} + \cdots + \frac{Q}{h_a A} + \frac{Q}{h_b A} + \cdots$$

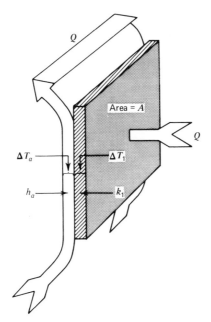

**FIGURE 2-6.**   Heat transfer in series paths.

From Fig. 2-6 it can be seen that with the heat transfer elements in series, the heat flows and areas are equal (if there are no transient effects), and therefore

$$\frac{1}{U_T} = \frac{L_1}{k_1} + \frac{L_2}{k_2} + \cdots + \frac{1}{h_a} + \frac{1}{h_b} + \cdots \qquad (2\text{-}16)$$

It is convenient to formulate a resistance form of the equation in which these series resistances are additive:

$$R_T = R_1 + R_2 + \cdots + R_a + R_b + \cdots$$

or $\qquad (2\text{-}17)$

$$R_T = r_1 L_1 + r_2 L_2 + \cdots + R_a + R_b$$

where the thermal resistivity is $r \equiv 1/k$, and thermal resistances are designated by $R_1 = r_1 L_1 = L_1/k_1$ and $R_a = 1/h_a$, etc. The previous table giving the thermal properties of building materials also includes their $r$ or $R$ value. If the thermal resistivity $r$ is tabulated, it is to be combined with the material thickness when using eq. 2-17. If the thermal resistance $R$ is tabulated, it already reflects the total resistance of the specified thickness of material.

**Heat Transfer in Parallel Paths.**   For thermal resistances in *parallel* the heat flows rather than the temperature differences are additive, as shown in Fig. 2-7.

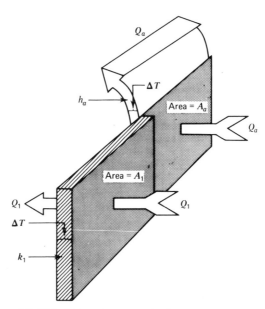

**FIGURE 2-7.**   Heat transfer in parallel paths.

Unlike resistances in series the areas need not be equal. The $\Delta T$'s, however, are identically equal, and eqs. 2-1 and 2-3 are then *summed* to express eq. 2-2. Therefore, since

$$Q_T = Q_1 + Q_2 + \cdots + Q_a + Q_b$$

then by eqs. 2-1, 2-2, and 2-3,

$$U_T A_T \,\Delta T = \frac{k_1}{L_1} A_1 \,\Delta T + \frac{k_2}{L_2} A_2 \,\Delta T + \cdots + h_a A_a \,\Delta T + h_b A_b \,\Delta T + \cdots$$

Dividing through by $A_T \,\Delta T$,

$$\frac{1}{R_T} = U_T = \frac{k_1}{L_1}\,(A_1/A_T) + \frac{k_2}{L_2}\,(A_2/A_T) + \cdots + h_a(A_a/A_T) + h_b(A_b/A_T)$$

$$(2\text{-}18)$$

If resistivities are preferred, they can also be used:

$$U_T A_T \,\Delta T = \frac{A_1 \,\Delta T}{r_1 L_1} + \frac{A_2 \,\Delta T}{r_2 L_2} + \cdots + \frac{A_a \,\Delta T}{R_a} + \frac{A_b \,\Delta T}{R_b} + \cdots$$

$$\frac{1}{R_T} = U_T = \frac{(A_1/A_T)}{r_1 L_1} + \frac{(A_2/A_T)}{r_2 L_2} + \cdots + \frac{(A_a/A_T)}{R_a} + \frac{(A_b/A_T)}{R_b} + \cdots$$

$$(2\text{-}19)$$

Parallel resistances occur in buildings when insulation is placed between studs, and on a larger scale when windows are put into walls. While each heat transfer path could be treated separately, it is often convenient to use eq. 2-2 only once for the total area with a single $U$-value calculated by eq. 2-18 or 2-19, since the temperature difference is usually the same for all the heat paths.

**Heat Transfer Through Air.**    The passage of heat through air is important in windows and in thin or highly conductive walls. In such situations the potential conduction through the solid surface is high because the material is thin (see eq. 2-1). When this is the case, the conductive heat transfer need not be quantified because the resistance of the convective air films is the controlling resistance to heat transfer.

There are two general cases for the convective transfer of heat through air. The first is the *air film*, a semi-infinite mass of air on one side of the conductive barrier. In the second there are two or more barriers, and heat transfer is through the enclosed *air space*, as well as through the air films on either side of the air space. Real situations are analyzed by combining appropriate thermal resistances in series corresponding to these configurations. A storm window, for instance, has a semi-infinite air mass on one side of the first conductive barrier, an air space between the two glass barriers, and a second air film on the other side of the second conductive glass barrier.

**TABLE 2-3.** Surface Heat Conductances and Unit Resistances for Air[a] Films

Surface Emissivities

| Position of Surface | Direction of Heat Flow | ε = 0.9 | | | | ε = 0.2 | | | | ε = 0.05 | | | |
|---|---|---|---|---|---|---|---|---|---|---|---|---|---|
| | | $h$ | | $R$ | | $h$ | | $R$ | | $h$ | | $R$ | |
| | | $\dfrac{Btu}{hr\,°F\,ft^2}$ | $\dfrac{W}{°C\cdot m^2}$ | $\dfrac{hr\,°F\,ft^2}{Btu}$ | $\dfrac{°C\cdot m^2}{W}$ | $\dfrac{Btu}{hr\,°F\,ft^2}$ | $\dfrac{W}{°C\cdot m^2}$ | $\dfrac{hr\,°F\,ft^2}{Btu}$ | $\dfrac{°C\cdot m^2}{W}$ | $\dfrac{Btu}{hr\,°F\,ft^2}$ | $\dfrac{W}{°C\cdot m^2}$ | $\dfrac{hr\,°F\,ft^2}{Btu}$ | $\dfrac{°C\cdot m^2}{W}$ |
| *Still air* | | | | | | | | | | | | | |
| Horizontal | Upward | 1.63 | 9.26 | 0.61 | 0.11 | 0.91 | 5.2 | 1.10 | 0.194 | 0.76 | 4.3 | 1.32 | 0.232 |
| Sloping—45° | Upward | 1.60 | 9.09 | 0.62 | 0.11 | 0.88 | 5.0 | 1.14 | 0.200 | 0.73 | 4.1 | 1.37 | 0.241 |
| Vertical | Horizontal | 1.46 | 8.29 | 0.68 | 0.12 | 0.74 | 4.2 | 1.35 | 0.238 | 0.59 | 3.4 | 1.70 | 0.298 |
| Sloping—45° | Downward | 1.32 | 7.50 | 0.76 | 0.13 | 0.60 | 3.4 | 1.67 | 0.294 | 0.45 | 2.6 | 2.22 | 0.391 |
| Horizontal | Downward | 1.08 | 6.13 | 0.92 | 0.16 | 0.37 | 2.1 | 2.70 | 0.476 | 0.22 | 1.3 | 4.55 | 0.800 |
| *Moving air* (Any Position) | | | | | | | | | | | | | |
| Wind is 15 mph or 6.7 m/s (for winter) | Any | 6.0 | 34.0 | 0.17 | 0.029 | | | | | | | | |
| Wind is 7½ mph or 3.4 m/s (for summer) | Any | 4.0 | 22.7 | 0.25 | 0.044 | | | | | | | | |

[a] Adapted by permission from *ASHRAE Handbook of Fundamentals*, 1972.

**TABLE 2-4.** Reflectance and Emittance of Various Surfaces and Effective Emittances of Air Space[a]

| Surface | Reflectance in Percent | Average Emit-tance, $\varepsilon$ | Effective Emittance $E$ of Air Space | |
|---|---|---|---|---|
| | | | With One Surface Having Emittance $\varepsilon$ and Other 0.90 | With Both Surfaces of Emittance $\varepsilon$ |
| Aluminum foil, bright | 92–97 | 0.05 | 0.05 | 0.03 |
| Aluminum sheet | 80–95 | 0.12 | 0.12 | 0.06 |
| Aluminum-coated paper, polished | 75–84 | 0.20 | 0.20 | 0.11 |
| Steel, galvanized, bright | 70–80 | 0.25 | 0.24 | 0.15 |
| Aluminum paint | 30–70 | 0.50 | 0.47 | 0.35 |
| Building materials—wood, paper, glass, masonry, nonmetallic paints | 5–15 | 0.90 | 0.82 | 0.82 |

[a] Adapted by permission from *ASHRAE Handbook of Fundamentals*, 1972.

A still air *film* has a low coefficient of heat transfer that is increased considerably by air movement due to wind (eqs. 2-8 and 2-9). The air *space* has two inner films that interact to form a convective air circulation cell, for which a single coefficient is assigned. The circulation is of course independent of the wind, but more heat is transmitted than with two still air films because there is definite air movement.

Table 2-3 shows thermal resistances (and related heat transfer coefficients) of the air film that forms between a mass of inside or outside air and a conductive barrier such as a pane of glass or metal plate. The values depend on the emissivity of the surface (Table 2-4), on whether the mass of air is moving, and whether the surface is horizontal, vertical, or somewhere in between. A reflective surface tends to increase the thermal resistance. Horizontal surfaces transfer heat more readily than do vertical surfaces. Since moving air tends to decrease the resistance, different values are recommended for winter and summer because of the higher winter winds.

Table 2-4 lists the emissivity of common structural materials and gives values for the combined emissivity $E$ to use when considering heat transfer across an air space.

Table 2-5 shows experimental values for thermal resistances of closed air spaces between conductive barriers spaced $\frac{3}{4}$ in. and 4 in. apart. The values vary widely with the average emissivity of the two surfaces (Table 2-4), somewhat with the temperature level, and very little with the air gap spacing. For ordinary glass surfaces, a winter heat transfer coefficient of about 6 W/°C·m² (1 Btu/hr °F ft²) is representative for air spaces.* For surfaces closer together than about $\frac{1}{2}$ in. the conductivity rises slightly because the thermal conductivity of the air becomes significant in such a thin layer. For this reason sealed panes of double glass are slightly less effective than the less convenient storm windows they replace. Such

* For simplicity these values can be used for all air spaces with little loss of accuracy. The corresponding $R$-values are 0.18 °C·m²/W and 1 hr °F ft²/Btu.

**TABLE 2-5a.** Unit Thermal Resistance of a Plane, ¾-in. (20 mm) Air Space[a]

| Position of Air Space | Direction of Heat Flow | Mean Air Temperature, °F | °C | Temperature Difference, °F | °C | $\varepsilon = 0.05$ $R \dfrac{\text{hr °F ft}^2}{\text{Btu}}$ | $R \dfrac{\text{°C·m}^2}{\text{W}}$ | $\varepsilon = 0.2$ $R \dfrac{\text{hr °F ft}^2}{\text{Btu}}$ | $R \dfrac{\text{°C·m}^2}{\text{W}}$ | $\varepsilon = 0.82$ $R \dfrac{\text{hr °F ft}^2}{\text{Btu}}$ | $R \dfrac{\text{°C·m}^2}{\text{W}}$ |
|---|---|---|---|---|---|---|---|---|---|---|---|
| Horizontal | Up | 90 | 32 | 10 | 6 | 2.26 | 0.398 | 1.63 | 0.287 | 0.76 | 0.134 |
| | | 50 | 10 | 30 | 17 | 1.67 | 0.294 | 1.37 | 0.241 | 0.78 | 0.137 |
| | | 50 | 10 | 10 | 6 | 2.23 | 0.393 | 1.71 | 0.301 | 0.87 | 0.153 |
| | | 0 | −18 | 20 | 11 | 1.79 | 0.315 | 1.52 | 0.268 | 0.93 | 0.164 |
| | | 0 | −18 | 10 | 6 | 2.16 | 0.380 | 1.78 | 0.313 | 1.02 | 0.180 |
| 45° Slope | Up | 90 | 32 | 10 | 6 | 2.81 | 0.495 | 1.9 | 0.335 | 0.81 | 0.143 |
| | | 50 | 10 | 30 | 17 | 1.95 | 0.343 | 1.54 | 0.271 | 0.83 | 0.146 |
| | | 50 | 10 | 10 | 6 | 2.78 | 0.490 | 2.02 | 0.356 | 0.94 | 0.166 |
| | | 0 | −18 | 20 | 11 | 2.27 | 0.400 | 1.74 | 0.306 | 1.01 | 0.178 |
| | | 0 | −18 | 10 | 6 | 2.71 | 0.477 | 2.13 | 0.375 | 1.13 | 0.199 |
| Vertical | Horizontal | 90 | 32 | 10 | 6 | 3.28 | 0.578 | 2.10 | 0.370 | 0.84 | 0.148 |
| | | 50 | 10 | 30 | 17 | 2.80 | 0.493 | 2.04 | 0.359 | 0.96 | 0.169 |
| | | 50 | 10 | 10 | 6 | 3.48 | 0.613 | 2.36 | 0.416 | 1.01 | 0.178 |
| | | 0 | −18 | 20 | 11 | 3.10 | 0.546 | 2.36 | 0.416 | 1.19 | 0.210 |
| | | 0 | −18 | 10 | 6 | 3.76 | 0.662 | 2.73 | 0.481 | 1.28 | 0.255 |
| 45° Slope | Down | 90 | 32 | 10 | 6 | 3.24 | 0.571 | 2.09 | 0.368 | 0.84 | 0.148 |
| | | 50 | 10 | 30 | 17 | 3.27 | 0.576 | 2.27 | 0.400 | 1.01 | 0.178 |
| | | 50 | 10 | 10 | 6 | 3.57 | 0.629 | 2.40 | 0.423 | 1.02 | 0.180 |
| | | 0 | −18 | 20 | 11 | 3.65 | 0.643 | 2.67 | 0.470 | 1.27 | 0.224 |
| | | 0 | −18 | 10 | 6 | 4.04 | 0.712 | 2.88 | 0.507 | 1.31 | 0.231 |

[a] Adapted by permission from *ASHRAE Handbook of Fundamentals*, 1972.


**TABLE 2-5b.** Unit Thermal Resistances of a Plane, 4-in. (100 mm) Air Space[a]

| Position of Air Space | Direction of Heat Flow | Mean Air Temperature, °F | °C | Temperature Difference, °F | °C | $\varepsilon = 0.05$ $R \dfrac{\text{hr °F ft}^2}{\text{Btu}}$ | $R \dfrac{\text{°C·m}^2}{\text{W}}$ | $\varepsilon = 0.02$ $R \dfrac{\text{hr °F ft}^2}{\text{Btu}}$ | $R \dfrac{\text{°C·m}^2}{\text{W}}$ | $\varepsilon = 0.82$ $R \dfrac{\text{hr °F ft}^2}{\text{Btu}}$ | $R \dfrac{\text{°C·m}^2}{\text{W}}$ |
|---|---|---|---|---|---|---|---|---|---|---|---|
| Horizontal | Up | 90 | 32 | 10 | 6 | 2.75 | 0.484 | 1.87 | 0.329 | 0.8 | 0.141 |
| | | 50 | 10 | 30 | 17 | 2.06 | 0.363 | 1.62 | 0.285 | 0.85 | 0.150 |
| | | 50 | 10 | 10 | 6 | 2.73 | 0.481 | 1.99 | 0.350 | 0.94 | 0.166 |
| | | 0 | −18 | 20 | 11 | 2.22 | 0.391 | 1.81 | 0.319 | 1.03 | 0.181 |
| | | 0 | −18 | 10 | 6 | 2.67 | 0.470 | 2.11 | 0.372 | 1.12 | 0.197 |
| 45° Slope | Up | 90 | 32 | 10 | 6 | 3.0 | 0.528 | 1.98 | 0.349 | 0.82 | 0.144 |
| | | 50 | 10 | 30 | 17 | 2.22 | 0.391 | 1.71 | 0.301 | 0.88 | 0.155 |
| | | 50 | 10 | 10 | 6 | 3.0 | 0.528 | 2.13 | 0.375 | 0.96 | 0.169 |
| | | 0 | −18 | 20 | 11 | 2.42 | 0.426 | 1.95 | 0.343 | 1.08 | 0.190 |
| | | 0 | −18 | 10 | 6 | 2.97 | 0.523 | 2.49 | 0.439 | 1.17 | 0.206 |
| Vertical | Horizontal | 90 | 32 | 10 | 6 | 3.44 | 0.606 | 2.16 | 0.380 | 0.91 | 0.160 |
| | | 50 | 10 | 30 | 17 | 2.62 | 0.461 | 1.94 | 0.342 | 0.94 | 0.166 |
| | | 50 | 10 | 10 | 6 | 3.45 | 0.608 | 2.34 | 0.412 | 1.01 | 0.178 |
| | | 0 | −18 | 20 | 11 | 2.86 | 0.504 | 2.22 | 0.391 | 1.16 | 0.204 |
| | | 0 | −18 | 10 | 6 | 3.42 | 0.602 | 2.55 | 0.449 | 1.24 | 0.218 |
| 45° Slope | Down | 90 | 32 | 10 | 6 | 4.36 | 0.768 | 2.50 | 0.440 | 0.9 | 0.158 |
| | | 50 | 10 | 30 | 17 | 3.39 | 0.597 | 2.33 | 0.410 | 1.02 | 0.180 |
| | | 50 | 10 | 10 | 6 | 4.41 | 0.777 | 2.75 | 0.484 | 1.08 | 0.190 |
| | | 0 | −18 | 20 | 11 | 3.73 | 0.657 | 2.71 | 0.477 | 1.27 | 0.224 |
| | | 0 | −18 | 10 | 6 | 4.39 | 0.773 | 3.05 | 0.537 | 1.35 | 0.238 |

[a] Adapted by permission from *ASHRAE Handbook of Fundamentals*, 1972.

**TABLE 2-5c.** Unit Thermal Resistances of Plane, Horizontal Air Spaces with Heat Flow Downward[a]

| Air Space Thickness, Inch mm | | Mean Temperature, °F °C | | $\varepsilon = 0.05$ | | $\varepsilon = 0.2$ | | $\varepsilon = 0.82$ | |
|---|---|---|---|---|---|---|---|---|---|
| | | | | $R\dfrac{\text{hr °F ft}^2}{\text{Btu}}$ | $R\dfrac{°\text{C}\cdot\text{m}^2}{\text{W}}$ | $R\dfrac{\text{hr °F ft}^2}{\text{Btu}}$ | $R\dfrac{°\text{C}\cdot\text{m}^2}{\text{W}}$ | $R\dfrac{\text{hr °F ft}^2}{\text{Btu}}$ | $R\dfrac{°\text{C}\cdot\text{m}^2}{\text{W}}$ |
| ¾ | 20 | 90 | 32 | 3.25 | 0.572 | 2.08 | 0.366 | 0.84 | 0.148 |
| | | 50 | 10 | 3.55 | 0.625 | 2.39 | 0.421 | 1.02 | 0.180 |
| | | 0 | −18 | 4.04 | 0.711 | 2.88 | 0.507 | 1.31 | 0.231 |
| 1½ | 38 | 90 | 32 | 5.24 | 0.923 | 2.76 | 0.486 | 0.93 | 0.164 |
| | | 50 | 10 | 5.74 | 1.01 | 3.21 | 0.565 | 1.14 | 0.201 |
| | | 0 | −18 | 6.59 | 1.16 | 3.97 | 0.699 | 1.50 | 0.264 |
| 4 | 100 | 90 | 32 | 8.08 | 1.42 | 3.38 | 0.595 | 0.99 | 0.174 |
| | | 50 | 10 | 8.94 | 1.57 | 4.02 | 0.708 | 1.23 | 0.217 |
| | | 0 | −18 | 10.9 | 1.92 | 5.20 | 0.916 | 1.65 | 0.291 |

[a] *Adapted by permission from ASHRAE Handbook of Fundamentals 1972.*

*insulating* glass is improved if a less conductive gas such as a fluorocarbon is sealed between the panes. This is often done, making such glass equivalent to the ordinary double-pane storm windows.

## EXAMPLE 2-9

Calculate the $R$ and $U$ values for a storm window in winter.

The heat transfer path is composed of series resistances, as shown in Fig. 2-8— the inside air film, the air space in the middle, and the outside air film:

$$R = R_{\text{inside film}} + R_{\text{air space}} + R_{\text{outside film}}$$

Using the tabulated values from Table 2-5b for a 4-in. air space having a 10 °C (50 °F) mean temperature value and a 6 °C (10 °F) temperature difference, and using air film values from Table 2-3,

$$R = 0.12 + 0.178 + 0.029 = 0.327 \text{ °C}\cdot\text{m}^2/\text{W}$$

$$U = 1/R = 3.06 \text{ W/°C}\cdot\text{m}^2$$

In Engineering units,

$$R = 0.68 + 1.01 + 0.17 = 1.86 \text{ hr °F ft}^2/\text{Btu}$$

$$U = 1/R = 0.54 \text{ Btu/hr °F ft}^2$$

The techniques just described deal with heat *loss* from windows, but not with the heat gains. In the ordinary structure—which has windows placed in most outside walls—modest gains from the sun will be taken into account by use of the below-ambient, degree-day temperature base. The wisely designed solar-heated structure, however, will minimize windows except in south-facing walls. These

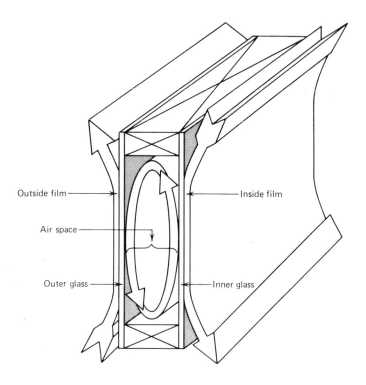

**FIGURE 2-8.** Heat transfer through a storm window in winter (sketch for Example 2-9).

windows will be shielded from the summer sun by an exterior overhang that permits free entry of the low winter sun—an inexpensive and desirable form of solar heating (see Chapter 13).

East and west windows are avoided because of the long exposure they have to the north-setting summer sun and the impossibility of using fixed blinds or overhang to eliminate this unwanted solar gain. North windows are no problem in summer (or winter) except for their thermal characteristics.

In most climates, expanses of south-facing windows can be considered as thermally neutral in the winter, with the losses approximately balanced by their heat gains due to winter sun. This is especially appropriate if the windows are equipped with drapes or other conservation devices, which can be closed to reduce losses at night and when the sun is not shining.

**Complex Walls.** A typical wall consists of a core material and inside and outside coverings. It may contain air spaces. The wall may not be of uniform cross section, because of supporting members such as studding or other framework.

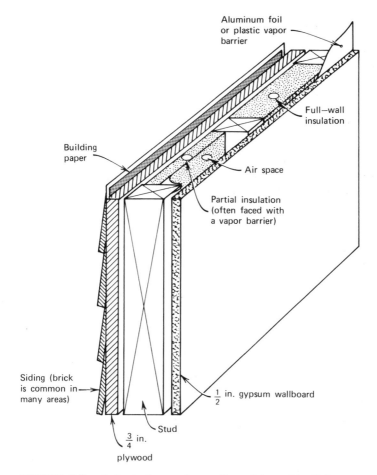

**FIGURE 2-9.** A typical complex wall used in residential construction.

The total $U$ value of a wall is determined by first adding the parallel resistances of studding, insulation, and air spaces together properly, then adding to that the series thermal resistances, and finally adding the applicable air film resistances outside both wall surfaces. Figure 2-9 shows the construction of a typical complex wall.

**EXAMPLE 2-10**

Calculate the $U$ value for an ordinary domestic frame wall (Fig. 2-9).

(a) The wall has 2.54 cm (1 in.) of fiberglass insulation.
(b) The wall has full-wall fiberglass insulation.

(a) The studs and insulation are parallel. If the insulation does not fill the air space, there is an air film on one or both sides, depending on the installation. Assuming one air space in the area between studs, which is 9 cm (3.5 in.) thick, by eq. 2-17,

$$R_{\text{between studs}} = R_{\text{air space}} + R_{\text{insulation}}$$

Using Table 2-2 thermal resistivities* and Tables 2-3 and 2-4 air space resistances, the thermal resistance in the area between studs is

$$R = 0.178 + 27.8 \times \frac{2.54}{100} = 0.884 \text{ °C} \cdot \text{m}^2/\text{W}$$

where the thickness of the insulation is expressed in *meters* by dividing the thickness in centimeters by 100. For the studs themselves,

$$R = 8.67 \times \frac{9}{100} = 0.78 \text{ °C} \cdot \text{m}^2/\text{W}$$

Since the studs and insulation are parallel heat paths, the areas devoted to each must be considered. The standard stud spacing is 16 in., and they are $1\frac{1}{2}$ in. wide. Thus, on the average, 1.5/16 or 9.4% of the wall is stud,† or 0.094 ft²/ft² of wall. The $R$ for studs and insulation is then calculated according to eq. 2-19 solved for $R_T$:

$$R_T = \frac{1}{\dfrac{0.906}{0.884} + \dfrac{0.094}{0.78}} = 0.873 \frac{\text{°C} \cdot \text{m}^2}{\text{W}}$$

The remaining resistances to heat transfer for the entire wall are in series and are simply added together with that of the insulation and studs to obtain the resistance of the total wall.

| | |
|---|---|
| Siding (1-in. overlap) | 0.139 |
| Vapor-permeable building paper | 0.011 |
| $\frac{3}{4}$ in. plywood ($\frac{3}{4}$ × 1.25) | 0.165 |
| Wallboard ($\frac{1}{2}$ in.) | 0.078 |
| Stud + insulation | 0.873 |
| Inside film | 0.120 |
| Outside film (winter) | 0.029 |
| Total thermal resistance | 1.415 $\dfrac{\text{°C} \cdot \text{m}^2}{\text{W}}$ |

(b) With full wall-thickness insulation, Table 2-2 lists a value of $R = 1.94$ °C·m²/W, and there are no air-films within the wall. Since the $R$ value for the wood is 0.78 °C·m²/W, eq. 2-19 gives

$$R_T = \frac{1}{\dfrac{0.906}{1.94} + \dfrac{0.094}{0.78}} = 1.70 \text{ °C} \cdot \text{m}^2/\text{W}$$

* Since 1-in. fiberglass batt is not listed, the r-value for fiberglass board has been used.
† This value will be higher if sills, beams, etc. are accounted for.

The remainder of the wall-components are just as they were before. Therefore, subtracting out the old internal wall section,

$$R = 1.412 - 0.873 + 1.70 = 2.24 \,°C \cdot m^2/W$$

In Engineering units,

(a) by eq. 2-17,*

$$R_{\text{between studs}} = 1.01 + 4 \times 1 = 5 \text{ hr }°F \text{ ft}^2/Btu$$

$$R_{\text{studs}} = 1.25 \times 3.5 = 4.38 \text{ hr }°F \text{ ft}^2/Btu$$

$$R_T = \frac{1}{\dfrac{0.906}{5} + \dfrac{0.094}{4.38}} = 4.95 \text{ hr }°F \text{ ft}^2/Btu$$

| | |
|---|---|
| Siding (1-in. overlap) | 0.79 |
| Vapor-permeable building paper | 0.06 |
| $\frac{3}{4}$ in. plywood | 0.93 |
| Wallboard ($\frac{1}{2}$ in.) | 0.45 |
| Stud + insulation | 4.95 |
| Inside film | 0.68 |
| Outside film (winter) | 0.17 |
| | ___ |
| Total thermal resistance | 8.03 hr °F ft²/Btu |

(b)

$$R_T = \frac{1}{\dfrac{0.906}{11} + \dfrac{0.094}{4.38}} = 9.63 \text{ hr }°F \text{ ft}^2/Btu$$

$$R = 8.03 - 4.95 + 9.63 = 12.8 \text{ hr }°F \text{ ft}^2/Btu$$

**Basement Walls Below Grade.** The conductivity of basement walls is usually rather high, and the heat lost through them can be significant when the temperature difference between the inside and outside is large. For this reason the well-insulated house will carry outside wall insulation down to and often below the ground level.

Below the grade, the outside ambient temperature is tempered by the ground temperature. The basic ground temperature is that of ground water, and varies from about 10 to 27 °C (50 to 80 °F) in the United States depending on the location. Using the ground temperature, ground losses can be estimated by using $U = 0.57 \text{ W}/°C \cdot m^2$ (0.1 Btu/hr °F ft²) for basement floors and $U = 1.14 \text{ W}/°C \cdot m^2$ (0.2 Btu/hr °F ft²) for the below-grade walls. These heat transfer coefficients vary considerably with the water content of the local soil because dry soil is a good insulator and saturated soil is a fair conductor. Because of uncertainties in soil water content and because temperature levels in unheated basements are

* See footnote, p. 44.

often held near groundwater temperatures anyway, many engineers omit them from heat load calculations. In the North, a solar-heated house should have 2 to 4 in. polystyrene foam under the basement floor and also along the walls if they are not otherwise insulated. Then the basement load can be ignored. In mild climates where concrete slabs on grade are used, losses are small if there is perimeter insulation to reduce direct losses to the ambient air. Such insulation is usually polystyrene foam, 1 to 2 in. thick, extending 1 to 2 ft inward or downward to keep heat from flowing from the slab to the air.

Methods for estimating losses from heated basements and concrete slabs are detailed in Chapter 24 of the *ASHRAE Fundamentals* (2).

## Vapor Barriers

Thermal insulations are usually permeable to water vapor. If the outside temperature falls below the dew point of the trapped air, there will be a depth to which condensation and even freezing occurs in the insulation. Water and ice will accumulate in this zone from the moisture leaving the building. This condensation is unfavorable thermally and disastrous when the ice melts during a warm spell or in the spring, since wet insulation can easily damage the wall or its protective coatings. For this reason a vapor barrier is always installed facing the heated area so that water vapor can never enter the insulation. The inexpensive barrier also serves to reduce the water necessary to restore normal humidity to the air if a humidifier is installed.

In parts of the southern United States where heating loads are small, the vapor barrier is reversed to face the outside to prevent condensation during the long air-conditioning season.

## Infiltration

The amount of air that infiltrates a building is highly variable, depending on the wind and the characteristic construction features. As a rule of thumb, one and one-half changes per hour are anticipated in an older residence, and one change per hour for newer residences, office buildings, and apartment rooms having outside walls. For the well-constructed, solar-heated house, one-half air change per hour should be anticipated.

The infiltration rate in houses may also be estimated by considering the wind velocity and the chimney effect of outside ambient temperature (5):

$$\text{Air changes/hr} = A + B(T_i - T_a) + Du \qquad (2\text{-}20)$$

where $T_i$ is the inside and $T_a$ the outside ambient temperature, and $u$ the wind velocity in m/s (mi/hr). For SI units $B = 0.011$ to $0.022$ and $D = 0.034$ to $0.067$; for Engineering units $B = 0.006$ to $0.012$ and $D = 0.015$ to $0.030$. The smaller values are for a tightly constructed house (c. 1970); the larger values for a loosely constructed house (c. 1950). The value for $A$ is 0.2 if combustion air for heating comes from inside the house, otherwise $A$ is 0.1.

Infiltrating air must be heated and represents a large and important part of the heating load. The load may be estimated from a dimensional equation generalized from Example 2-6:

$$Q_{in} = 0.335V(T_i - T_a) \tag{2-21a}$$

where $Q_{in}$ is the required heating rate in watts, $V$ the infiltration rate* in m$^3$/hr, and $T_i$ and $T_a$ are the inside and outside ambient temperature in degrees Celsius.

In Engineering units,

$$Q_{in} = 0.018V(T_i - T_a) \tag{2-21b}$$

where $Q_{in}$ is the heating rate in Btu/hr, $V$ the infiltration rate in ft$^3$/hr, and $T_i$ and $T_a$ are the inside and outside ambient temperatures in °F.

Note that infiltration losses, like heat losses, are proportional to the inside-outside temperature difference. The term $0.335V$ ($0.018V$) in the infiltration load equation is therefore expressed in the same units as an area-weighted heat transfer coefficient, W/°C (Btu/hr °F), and can be added directly to the other $UA$'s when determining the total heat losses and used in eq. 2-22 below for seasonal demand estimates.

Infiltration losses may also be estimated on the basis of the physical size of known cracks. Details are given in the ASHRAE fundamentals volume. The technique uses wind speed as input, which is changed to inches of water *stagnation* or velocity pressure. A graphical correlation is then used to predict the leakage rate as a function of crack width and pressure. A second effect, also explored in the fundamentals volume, is the chimney effect in which cold air comes in at the bottom of a vertical crack and hot air leaves at the top. Neither method is of much use where all known cracks are sealed and infiltration already minimized.

## Design Heat Load Calculations

The general plan for calculating the $UA$ of a structure is to first measure the areas of thermally distinctive portions of the structure: the windows, doors, ceilings, and walls. The heat transfer coefficient $U$ for each surface is then estimated using the previously developed techniques. Then the $UA$ product is determined for each portion of the structure. These amounts are summed with an infiltration $UA$ to produce the building $UA$. This is multiplied by the seasonal degree-days according to eq. 2-14 to get the annual heating load, and by the design temperature difference to get the design heating load. Units must be watched carefully because $D$ is given in degree-days, and $UA$ may not be in consistent units. Basement losses are determined separately, and since they are independent of weather, can be added to monthly hot water loads for further analysis.

* The volume times the number of air changes per hour.

Occasionally an average value for the heat transfer coefficient for walls, ceiling, or floor may be called for to evaluate the energy conservation features of the structure. This value is obtained by summing the $UA$ products for each element (for instance, an outside wall has elements of the basic wall, windows, and doors) and dividing by the total area (see eq. 2-18).

**EXAMPLE 2-11**

A house has 117.5 m² (1267 ft²) of attic floor with $R = 6.7$ °C·m²/W (38 hr °F ft²/Btu), 209 m² (2251 ft²) of walls with $R = 2.24$ °C·m²/W (12.8 hr °F ft²/Btu), and 28.2 m² (304 ft²) of windows and doors with $R = 0.327$ °C·m²/W (1.86 hr °F ft²/Btu). The basement is unheated.

(a) What is the heating demand per degree-day if the infiltration rate is one air change of 424 m³ (15 000 ft³) per hour?

(b) What is the average heat coefficient $U_0$ for the walls?

(a)

| | $R$ | $U$ | $A$ | $UA$ |
|---|---|---|---|---|
| Ceilings | 6.7 | 0.149 | 117.5 | 17.5 |
| Walls | 2.24 | 0.4464 | 209.0 | 93.3 |
| Windows | 0.327 | 3.06 | 28.2 | 86.3 |
| Infiltration | ($UA$ = 0.335 × 424 = 142.1) | | | 142.1 |
| Total $UA$ | | | | 339.2 W/°C |

$$UA = \frac{339.2 \text{ W}}{°C} \times \frac{J/s}{W} \times \frac{M}{10^6} \times \frac{3600 \text{ s}}{h} \times \frac{24 \text{ h}}{\text{day}} = 29.31 \text{ MJ/°C-day}$$

(b) $$U_0 = \frac{0.4464 \times 209 + 3.06 \times 28.2}{(209 + 28.2)} = 0.757 \text{ W/°C·m}^2$$

In Engineering units,

(a)

| | $R$ | $U$ | $A$ | $UA$ |
|---|---|---|---|---|
| Ceilings | 38 | 0.0263 | 1267 | 33.35 |
| Walls | 12.8 | 0.0781 | 2251 | 175.9 |
| Windows | 1.86 | 0.54 | 304 | 164.2 |
| Infiltration | ($UA$ = 0.018 × 15 000 = 270) | | | 270.0 |
| Total $UA$ | | | | 643.45 Btu/hr °F |

$$UA = \frac{643.45 \text{ Btu}}{\text{hr °F}} \times \frac{24 \text{ hr}}{\text{day}} = 15\,443 \text{ Btu/°F-day}$$

(b) $$U_0 = \frac{0.0781 \times 2251 + 0.54 \times 304}{(2251 + 304)} = 0.133 \text{ Btu/hr °F ft}^2$$

**EXAMPLE 2-12**

What is the design heating load for the house of Example 2-11, which has a basic heating demand of 29.31 MJ/°C-day (15 443 Btu/°F-day), if the winter design temperature is $-12.2$ °C (10 °F) outside and 23.9 °C (75 °C) inside?

$$Q_{design} = UA\ \Delta T = \frac{29.31\ MJ}{°C\text{-day}} \times \frac{1\ 000\ 000}{M} \times \frac{W \cdot s}{J} \times \frac{h}{3600\ s} \times \frac{day}{24\ h}$$

$$\times\ [23.9 - (-12.2)]\ °C = 12\ 246\ W$$

In Engineering units,

$$Q_{design} = UA\ \Delta T = \frac{15\ 443\ Btu}{°F} \times (75 - 10)\ °F \times \frac{day}{24\ hr} = 41\ 825\ Btu/hr$$

It is simpler to determine the design heating load by multiplying the $UA$ values expressed as W/°C·m² or Btu/hr °F ft² by the design temperature difference. The above procedure is demonstrated because $UA$ expressed as heating units per degree-day may often be the only parameter available.

## Seasonal Demand

Multiplying the structure heat loss coefficient $UA$ times the degree-day demand in a year gives the total seasonal heating load when units are appropriate (eq. 2-14). This can be converted into total seasonal fuel consumption by dividing by the efficiency and heating value of the fuel. Dimensional formulas are especially useful in this situation. The following equation can be used to convert loss coefficients into fuel usage in combination with the annual degree-day total $D_a$:

$$V_T = b\ UAD_a \tag{2-22}$$

Values of $b$ for SI and Engineering units for a number of fuels and units for $UA$ are given in Table 2-9 (p. 56). The reader should abstract values of $b$ that apply to his or her specific work and write them down in the form of eq. 2-22 to decrease the possibility of using the wrong constant through error.

**Calculations for Energy Conservation.** When considering steps to be taken to reduce heating load in a planned or existing structure, each $U$ and $A$ should be tabulated for each major source of energy loss. In a solar residence, major losses will most likely be, in order of magnitude, from doors and windows, infiltration, walls, ceilings, and basement. Then calculations should be made to show the $UA$ resulting from each loss, and the annual cost in fuel (or for solar, in capital) that results from heat losses through that area. When considering alternatives, differential costs may be quickly calculated by using eq. 2-22 in a convenient difference form:

$$\Delta V_T = b\ \Delta UAD_a \tag{2-23}$$

where $\Delta V_T$ is the incremental change in fuel usage resulting from a change in heat transfer coefficient $\Delta U$. The dimensional constant $b$ is unchanged and depends on the fuel and units used.

**Degree-Day Temperature Base.** The degree-days used in this book and elsewhere all have a base temperature of 65 °F (18.3 °C). This corresponds roughly to a 75 °F inside temperature. If other bases are used the seasonal fuel consumption can be predicted by using the proper degree-days in eqs. 2-22 and 2-23. The new degree-days can be calculated with the aid of Fig. 2-4 or eq. 2-15 as described earlier in this chapter. A yearly correction factor graph like that of Fig. 2-5 is convenient, for it permits continued use of the nominal 65 °F base degree-day figures.

The effect of thermostat night setback can be approximated by averaging the night and day correction factors, each weighted according to the time spent at the particular setting.

**EXAMPLE 2-13**

(a) How much fuel oil will be saved if 18.6 m² (200 ft²) of windows are changed from single to double glazing in Hartford where there are 3527 °C-days (6347 °F-days). How much electricity?

$$U_{\text{single}} = \frac{1}{R_T} = \frac{1}{0.12 + 0.029} = 6.71 \text{ W/°C·m}^2$$

$$U_{\text{double}} = \frac{1}{R_T} = \frac{1}{0.12 + 0.178 + 0.029} = 3.06 \text{ W/°C·m}^2$$

For fuel oil and electricity, respectively, $b = 0.000\ 726$ and $0.024$. Substituting in eq. 2-23,

$$\Delta V_{T(\text{oil})} = 0.000\ 726 \times (6.71 - 3.06) \times 18.6 \times 3527 = 174 \text{ gal}$$
$$\Delta V_{T(\text{elec})} = 0.024 \times (6.71 - 3.06) \times 18.6 \times 3527 = 5747 \text{ kWh}$$

(b) How much natural gas would it take to heat a house with a heat loss of 29.31 MJ/°C-day (15 443 Btu/°F-day)?

For natural gas, $b = 0.011\ 86$. Using eq. 2-22,

$$V_{T(\text{gas})} = 0.011\ 86 \times 29.31 \times 3527 = 1226 \text{ CCF (hundred cubic feet)}$$

(c) The degree-days of part (b) are based on an inside temperature of 23.9 °C (75 °F). What will the natural gas heating quantity be if the inside ambient temperature is lowered to 20 °C (68 °F)?

From Fig. 2-5, at Hartford, at a new temperature base of 14.4 °C (58 °F), the annual correction factor is 0.73.

$$V_{T(\text{gas})} = 0.73 \times 1226 = 895 \text{ CCF}$$

In Engineering units,

(a)
$$U_{single} = \frac{1}{0.68 + 0.17} = 1.176 \text{ Btu/hr ft}^2 \text{ °F}$$

$$U_{double} = \frac{1}{0.68 + 1.0 + 0.17} = 0.540 \text{ Btu/hr ft}^2 \text{ °F}$$

Eq. 2-23 gives

$$\Delta V_{T(oil)} = 0.000\,212\,6 \times (1.176 - 0.540) \times 200 \times 6347 = 172 \text{ gal}$$
$$\Delta V_{T(elec)} = 0.007\,032 \times (1.176 - 0.540) \times 200 \times 6347 = 5677 \text{ kWh}$$

(b)
$$V_{T(gas)} = 0.000\,012\,5 \times 15\,443 \times 6347 = 1225 \text{ CCF}$$

(c)
$$V_{T(gas)} = 0.73 \times 1225 \times 895 \text{ CCF}$$

**Humidity.** The heat required for humidification adds to the total load, but it is not large for new construction. In a well-sealed house, internal water generation will be adequate and many in fact require deliberate ventilation. Internal generation can be supplemented by clothes-dryer outlet air if more humidity is needed; this source is adequate to maintain humidity in almost any house that has a vapor barrier installed.

## Residential Cooling Load

Cooling loads are much smaller than heating loads (except in mild climates) because in summer the temperature difference between inside and outside ambients seldom exceeds 20 °F as opposed to the 70 °F difference common in the northern winters. The 20 °F reduction in temperature, however, is probably about as expensive to achieve as is the 70 °F heating load. Therefore, the occupancy and solar loads become a large portion of the total load and are more significant when cooling. As a further complication, "coolness" can be accumulated in the mass of the structure during the cooler nights and used to reduce the afternoon peak loads by allowing a few degrees rise in temperature. This method of operation also takes advantage of the characteristic *increase* in the capacity of an air-conditioning system as the inside temperature rises. Finally, water removal forms a very significant load in the air-conditioned structure; as much as one-third of the total load in a humid climate. Since it is impossible to achieve independent temperature and humidity control with a single air-cooling system, the system must be designed so that a proper temperature setting holds a comfortable humidity as well.

Because of these factors, the methods employed to predict peak and seasonal heating loads fail to predict the cooling load accurately. The practical approach recommended by ASHRAE and generally employed for residences uses the same general system as heating load prediction, but the temperature differences are modified to compensate for the heat transfer situations in which the air-conditioning heat gain inherently differs from the heating loss. The *latent* or humidity removal load is accounted for by a multiplier applied to the total load without analysis. The thermal loads due to occupancy and electrical consumption are estimated directly and added to the loads due to heat gain coming through the structure. The result of the analysis is not the seasonal cooling load, which is beyond simple analysis, but a recommended size for the cooling system. A description of the method is included here since a solar house may well incorporate summer cooling, and sizing that unit can easily be done at the same time as the determination of heating load.

The first step in calculating air-conditioning load is to establish effective temperature differences for the various surfaces that apply to the design outside ambient conditions. Since the temperature differences are usually only 15 to 25 °F, care is necessary since an error of a few degrees will appreciably affect the calculated cooling load. The basic method is to modify the actual temperature difference, as measured by dry-bulb temperatures, according to the sun load expected on the surface and the daily temperature range. This permits estimation of the higher load caused by the sun shining on exterior surfaces and allows for the *coolness* that can be stored in the structure as a result of lower night temperatures. The proper design temperature difference for calculating heat gains from a hot attic is therefore quite different from that appropriate for a north-facing wall. The proper temperature differences have been experimentally derived and are given in Table 2-6 as a function of the basic design temperature and the daily range of temperatures for a variety of domestic exterior walls and surfaces. For instance, the *ASHRAE Fundamentals* volume lists Hartford, Connecticut, as having a design dry-bulb temperature of 91 °F (1%), which means that the temperature exceeds that value during 1% of the hours during June, July, and August; about 22 hr/yr. The daily range is listed as 22 °F, which is M (Medium) according to Table 2-6 footnotes. High daily ranges are found inland, with low daily ranges in the mild seacoast climates. The values from the fourth column, therefore, apply but are for a 90 °F design and medium temperature range. They should be raised 1 °F, since the design temperature is 91 °F, not 90 °F. Interpolation between the L and M or the M and H values may also be called for if the daily range is far from the tabular values of 12, 20, and 30 °F.

Fenestration receives the very special treatment of Table 2-7. Since radiation load is very important for windows, the concept of temperature difference is not applicable at all, and the load is expressed directly as a heat gain in Btu/ft$^2$ which includes both radiation and convective heat gains. The proper value depends on the design temperature difference and the direction the windows face, as well as on the decorative coverage over the windows. North-facing windows add the least load, and east- and west-facing windows add the most load. The sun shines nearly perpendicularly into east and west windows when the sun is low, and it

## TABLE 2-6. Design Equivalent Temperature Differences

| Design Temperature, °F | 85 | 85 | 90 | 90 | 90 | 95 | 95 | 95 | 100 | 100 | 105 | 110 |
|---|---|---|---|---|---|---|---|---|---|---|---|---|
| Daily Temperature Range[a] | L | M | L | M | H | L | M | H | M | H | H | H |
| *Walls and doors* | | | | | | | | | | | | |
| 1. Frame and veneer-on-frame | 17.6 | 13.6 | 22.6 | 18.6 | 13.6 | 27.6 | 23.6 | 18.6 | 28.6 | 23.6 | 28.6 | 33.6 |
| 2. Masonry walls, 8-in. block or brick | 10.3 | 6.3 | 15.3 | 11.3 | 6.3 | 20.3 | 16.3 | 11.3 | 21.3 | 16.3 | 21.3 | 26.3 |
| 3. Partitions, frame | 9.0 | 5.0 | 14.0 | 10.0 | 5.0 | 19.0 | 15.0 | 10.0 | 20.0 | 15.0 | 20.0 | 25.0 |
|     masonry | 2.5 | 0 | 7.5 | 3.5 | 0 | 12.5 | 8.5 | 3.5 | 13.5 | 8.5 | 13.5 | 18.5 |
| 4. Wood doors | 17.6 | 13.6 | 22.6 | 18.6 | 13.6 | 27.6 | 23.6 | 18.6 | 28.6 | 23.6 | 28.6 | 33.6 |
| *Ceilings and Roofs[b]* | | | | | | | | | | | | |
| 1. Ceilings under naturally vented attic | | | | | | | | | | | | |
|     or vented flat roof—dark | 38.0 | 34.0 | 43.0 | 39.0 | 34.0 | 48.0 | 44.0 | 39.0 | 49.0 | 44.0 | 49.0 | 54.0 |
|     —light | 30.0 | 26.0 | 35.0 | 31.0 | 26.0 | 40.0 | 36.0 | 31.0 | 41.0 | 36.0 | 41.0 | 46.0 |
| 2. Built-up roof, no ceiling—dark | 38.0 | 34.0 | 43.0 | 39.0 | 34.0 | 48.0 | 44.0 | 39.0 | 49.0 | 44.0 | 49.0 | 54.0 |
|     —light | 30.0 | 26.0 | 35.0 | 31.0 | 26.0 | 40.0 | 36.0 | 31.0 | 41.0 | 36.0 | 41.0 | 46.0 |
| 3. Ceilings under unconditioned rooms | 9.0 | 5.0 | 14.0 | 10.0 | 5.0 | 19.0 | 15.0 | 10.0 | 20.0 | 15.0 | 20.0 | 25.0 |
| *Floors* | | | | | | | | | | | | |
| 1. Over unconditioned rooms | 9.0 | 5.0 | 14.0 | 10.0 | 5.0 | 19.0 | 15.0 | 10.0 | 20.0 | 15.0 | 20.0 | 25.0 |
| 2. Over basement, enclosed crawl space | 0 | 0 | 0 | 0 | 0 | 0 | 0 | 0 | 0 | 0 | 0 | 0 |
|     or concrete slab on ground | | | | | | | | | | | | |
| 3. Over open crawl space | 9.0 | 5.0 | 14.0 | 10.0 | 5.0 | 19.0 | 15.0 | 10.0 | 20.0 | 15.0 | 20.0 | 25.0 |

[a] Daily Temperature Range

| L (Low) Calculation Value: 12 °F. | M (Medium) Calculation Value: 20 °F. | H (High) Calculation Value: 30 °F. |
|---|---|---|
| *Applicable Range:* Less than 15 °F. | *Applicable Range:* 15 to 25 °F. | *Applicable Range:* More than 25 °F. |

[b] Ceilings and Roofs: For roofs in shade, 18-hr average = 11° temperature differential. At 90 °F design and medium daily range, equivalent temperature differential for light-colored roof equals 11 + (0.71)(39 − 11) = 31 °F.

## TABLE 2-7. Design Heat Transmission and Absorbed Solar Energy for Windows, Btu/hr ft²

| Outdoor | Regular Single Glass | | | | | | Regular Double Glass | | | | | | Heat-Absorbing Double Glass | | | | | |
|---|---|---|---|---|---|---|---|---|---|---|---|---|---|---|---|---|---|---|
| Design Temp. | 85 | 90 | 95 | 100 | 105 | 110 | 85 | 90 | 95 | 100 | 105 | 110 | 85 | 90 | 95 | 100 | 105 | 110 |
| | | | | | | No Awnings or inside Shading | | | | | | | | | | | | |
| North | 23 | 27 | 31 | 35 | 38 | 44 | 19 | 21 | 24 | 26 | 28 | 30 | 12 | 14 | 17 | 19 | 21 | 23 |
| NE and NW | 56 | 60 | 64 | 68 | 71 | 77 | 46 | 48 | 51 | 53 | 55 | 57 | 27 | 29 | 32 | 34 | 36 | 38 |
| East and West | 81 | 85 | 89 | 93 | 96 | 102 | 68 | 70 | 73 | 75 | 77 | 79 | 42 | 44 | 47 | 49 | 51 | 53 |
| SE and SW | 70 | 74 | 78 | 82 | 85 | 91 | 59 | 61 | 64 | 66 | 68 | 70 | 35 | 37 | 40 | 42 | 44 | 46 |
| South | 40 | 44 | 48 | 52 | 55 | 61 | 33 | 35 | 38 | 40 | 42 | 44 | 19 | 21 | 24 | 26 | 28 | 30 |
| | | | | | | Draperies or Venetian Blinds | | | | | | | | | | | | |
| North | 15 | 19 | 23 | 27 | 30 | 36 | 12 | 14 | 17 | 19 | 21 | 23 | 9 | 11 | 14 | 16 | 18 | 20 |
| NE and NW | 32 | 36 | 40 | 44 | 47 | 53 | 27 | 29 | 32 | 34 | 36 | 38 | 20 | 22 | 25 | 27 | 29 | 31 |
| East and West | 48 | 52 | 56 | 60 | 63 | 69 | 42 | 44 | 47 | 49 | 51 | 53 | 30 | 32 | 35 | 37 | 39 | 41 |
| SE and SW | 40 | 44 | 48 | 52 | 55 | 61 | 35 | 37 | 40 | 42 | 44 | 46 | 24 | 26 | 29 | 31 | 33 | 35 |
| South | 23 | 27 | 31 | 35 | 38 | 44 | 20 | 22 | 25 | 27 | 29 | 31 | 15 | 17 | 20 | 22 | 24 | 26 |
| | | | | | | Roller Shades Half-Drawn | | | | | | | | | | | | |
| North | 18 | 22 | 26 | 30 | 33 | 39 | 15 | 17 | 20 | 22 | 25 | 26 | 10 | 12 | 15 | 17 | 19 | 21 |
| NE and NW | 40 | 44 | 48 | 52 | 55 | 61 | 38 | 40 | 43 | 45 | 47 | 49 | 24 | 26 | 29 | 31 | 33 | 35 |
| East and West | 61 | 65 | 69 | 73 | 76 | 82 | 54 | 56 | 59 | 61 | 63 | 65 | 35 | 37 | 40 | 42 | 44 | 46 |
| SE and SW | 52 | 56 | 60 | 64 | 67 | 73 | 46 | 48 | 51 | 53 | 55 | 57 | 30 | 32 | 35 | 37 | 39 | 41 |
| South | 29 | 33 | 37 | 41 | 44 | 50 | 27 | 29 | 32 | 34 | 36 | 38 | 18 | 20 | 23 | 25 | 27 | 29 |
| | | | | | | Awnings | | | | | | | | | | | | |
| North | 20 | 24 | 28 | 32 | 35 | 41 | 13 | 15 | 18 | 20 | 22 | 24 | 10 | 12 | 15 | 17 | 19 | 21 |
| NE and NW | 21 | 25 | 29 | 33 | 36 | 42 | 14 | 16 | 19 | 21 | 23 | 25 | 11 | 13 | 16 | 18 | 20 | 22 |
| East and West | 22 | 26 | 30 | 34 | 37 | 43 | 14 | 16 | 19 | 21 | 23 | 25 | 12 | 14 | 17 | 19 | 21 | 23 |
| SE and SW | 21 | 25 | 29 | 33 | 36 | 42 | 14 | 16 | 19 | 21 | 23 | 25 | 11 | 13 | 16 | 18 | 20 | 22 |
| South | 21 | 24 | 28 | 32 | 35 | 41 | 13 | 15 | 18 | 20 | 22 | 24 | 11 | 13 | 16 | 18 | 20 | 22 |

53

does this for a long time, since in the summer the sun rises and sets to the north of the east-west line.

Awning-covered windows show the lowest heat gain, even less than a north-facing window, which receives more radiation from the warm surroundings. Roller shades, draperies, and clear windows show increasingly higher heat gains when they face the sun, as would be expected. Single glass has a significantly higher convective heat gain than double glass, but since even a north-facing window picks up appreciable radiation from the surroundings when the outside temperature is hot, the percentage difference is not large because the radiation gain, which is unaffected by the glazing, is a large portion of the load. Nevertheless, double glass is always worthwhile and is especially desirable in the South where the outdoor humidity often rises so high that condensation would occur on the *outside* of single glazing, wasting considerable cooling energy.

Heat-absorbing glass may be useful in some situations. This glass has a smoke color and typically absorbs 50% of the visible radiation. When direct rays strike the glass, it heats considerably but it transmits only one-half of the absorbed heat to the inside of the structure, causing a net reduction in cooling load. The reflecting glasses are a better solution because they reflect rather than absorb the light and heat that is not transmitted, but their appearance is so bizarre that they must be considered as a design feature and incorporated as an intimate part of the building design.

Low summer winds and the absence of the chimney effect reduce the importance of infiltration in the air-conditioning situation. Table 2-8 gives appropriate allowances in terms of cooling load per ft² of wall area for various design temperature differences.

The heat gain due to occupancy has been found to average about 225 Btu/hr per person, and a domestic kitchen contributes on the average about 1200 Btu/hr. Other electric loads are usually small, but they should be included if significant throughout the day.

The latent (humidity) load depends on the outside humidity and the moisture generated within the air-conditioned space, as well as on the infiltration rate. The design method does not take these conditions directly into account, however, but simply adds 30% to the sensible heat gains previously calculated.

The design technique that has just been described applies for small buildings

**TABLE 2-8.** Sensible Cooling Load Due to Infiltration and Ventilation

| Design Temperature, °F | 85 | 90 | 95 | 100 | 105 | 110 |
|---|---|---|---|---|---|---|
| Infiltration, Btu/hr ft² of gross exposed wall area | 0.7 | 1.1 | 1.5 | 1.9 | 2.2 | 2.6 |
| Mechanical ventilation, Btu/hr cfm | 11.0 | 16.0 | 22.0 | 27.0 | 32.0 | 38.0 |

of residential style only. The special tables used have been derived from careful measurements within experimental houses. They incorporate allowances for use of the *thermal flywheel* effect, which allows the house to heat up a few degrees on a hot day, taking advantage of coolness stored in the structure during the previous night. Correctly applied, this technique predicts the correct size cooling system that should be installed but gives no clue as to the average cooling load that will enable calculation of the seasonal cooling costs.

For larger buildings that do not lend themselves so readily to this simple analysis, ASHRAE has developed complex methods, described in the *Fundamentals* volume, for prediction of the air-conditioning load. These are beyond the scope of this book and are best implemented with the aid of a digital computer.

**Humidity Control.** Other things being equal, installation of an oversized cooling unit will provide less dehumidification than if a slightly undersized unit is installed. The smaller unit will remove the same amount of heat; however, it will run longer and condense more water from the air, and in so doing will produce a more comfortable environment. Never oversize an air-conditioning system.

**Cooling Degree-Days.** The *Climatic Atlas of the United States* lists cooling degree-days over the United States. These are calculated from basic weather information, using the difference between the average of a day's high and low and 65 °F, summed for positive values only over the cooling season. Figure 2-10 shows

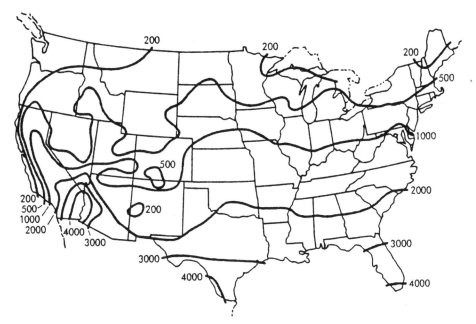

**FIGURE 2-10.** Regional distribution of annual degree-days of cooling.

the cooling degree-days plotted on a map of the United States. Cooling degree-days range from 500 in the cooler areas up to 3000 along the Gulf Coast and in southern desert areas, and as high as 4000 in a few places.

There has been no validation of the use of these degree-days to estimate seasonal cooling energy, but there is no other available method that can be simply applied.

### EXAMPLE 2-15

Estimate the design air-conditioning load for the house of Example 2-11, assuming conventional construction* and an even distribution of wall and window areas between N, E, S, and W directions. Locate the house in Hartford where the 1% design temperature is 91 °F and the daily range 22 °F.

(a) *Ceilings.* The $UA$ value is 33.35 Btu/hr °F, as before.* By interpolation of Table 2-6, the design $\Delta T$ under a light roof, for a design temperature of 91 °F at the medium temperature range, is 32 °F.

(b) *Walls.* The $UA$ value is 175.9 Btu/hr °F, as before. By interpolation of Table 2-6, the design $\Delta T$ for a 91 °F design temperature at the medium temperature range is 19.6 °F for frame construction.

(c) *Windows.* For double-pane glass with roller shades half drawn at 91 °F design temperature, Table 2-7 gives the heat gain as 17, 56, 29, and 56 Btu/ft² for N, E, S, and W windows, respectively. The area facing each direction is 304 ft²/4 = 76 ft².

(d) *Infiltration.* Table 2-8 estimates the infiltration rate as 1.1 Btu/hr ft² gross wall area, which is 2251 ft² + 304 ft² = 2555 ft².

(e) Occupation and kitchen loads are 225 Btu/hr for each person and 1200 Btu/hr, respectively.

### SUMMARY TABLE

|  | $UA$ (Btu/hr °F) | $\Delta T$ (°F) | Load (Btu/hr) |
|---|---|---|---|
| Ceilings | 33.35 | 32 | 1 067 |
| Walls | 175.9 | 19.6 | 3 448 |
| N Windows | 17 Btu/hr ft² × 76 ft² = | | 1 292 |
| E Windows | 56 Btu/hr ft² × 76 ft² = | | 4 246 |
| S Windows | 29 Btu/hr ft² × 76 ft² = | | 2 204 |
| W Windows | 56 Btu/hr ft² × 76 ft² = | | 4 256 |
| Infiltration | 1.1 × 2555 ft² = | | 2 810 |
| Occupation | 5 × 225 Btu/person = | | 1 125 |
| Kitchen | = | | 1 200 |
| TOTAL | | | 21 658 |

* There is some change in convection coefficients from winter to summer, but the overall effect is quite small except for windows, which are specially treated in the summer in any case. Therefore, the coefficients will be carried over from Example 2-10.

Multiplication by 1.3 to account for latent load gives 28 155 Btu/hr. Install the next smallest even size, probably 2¼ tons (27 000 Btu/hr).

**TABLE 2-9.** Values of $b$ for Use with Dimensional Eqs. 2-22 and 2-23

| | SI Units | | Engineering Units | |
| --- | --- | --- | --- | --- |
| Energy Source | $UA$ given in Degree-Day Units, MJ/°C-day | $U$ Given as a Heat Transfer Coefficient, W/°C·m² Area in m² | $UA$ given in Degree-Day Units, Btu/°F-day | $U$ Given as a Heat Transfer Coefficient, Btu/hr °F ft² Area in ft² |
| No. 2 fuel oil, gallons | 0.008 406 | 0.000 726 | $8.86 \times 10^{-6}$ | 0.000 212 6 |
| Natural gas, hundred cubic feet | 0.011 86 | 0.001 025 | $12.5 \times 10^{-6}$ | 0.000 300 |
| Propane, gallons | 0.012 88 | 0.001 113 | $13.6 \times 10^{-6}$ | 0.000 326 |
| Coal, tons | $60.81 \times 10^{-6}$ | $5.25 \times 10^{-6}$ | $0.064 \times 10^{-6}$ | $1.53 \times 10^{-6}$ |
| Electricity, kilowatt-hour | 0.2780 | 0.024 | 0.000 293 | $7.032 \times 10^{-3}$ |

No. 2 fuel oil: 80% combustion efficiency, 141 000 Btu/gal.
Natural gas: 80% combustion efficiency, 1000 Btu/ft³.
Propane: 80% combustion efficiency, 92 000 Btu/gal.
Coal: 60% combustion efficiency, 13 000 Btu/lb.
Electricity: 100% utilization efficiency, 3412 Btu/kWh.

# PROBLEMS

The Design House is a two-story, 189-m² (2035 ft²) suburban home with the following parameters:

| | |
| --- | --- |
| Projected roof area (ceilings) | 79.7 m² (1267 ft²) |
| Window area | 16.13 m² (256.5 ft²) |
| Door area | 2.64 m² (42 ft²) |
| Wall area (net) | 141.6 m² (2251.5 ft²) |
| Basement walls (above grade) area | 16.73 m² (266 ft²) |
| Basement walls (below grade) area | 41.83 m² (665 ft²) |
| Basement floor area | 61.89 m² (984 ft²) |
| Volume (basement excluded) | 437.4 m³ (15 451 ft³) |

Assume that the inside temperature is 23.9 °C (75 °F), the basement temperature 21.1 °C (70 °F), and the ground temperature 12.8 °C (55 °F). A winter design temperature of $-20$ °C ($-4$ °F) is to be used for calculating hourly losses to the outside air. Annually there are 3527 °C-days (6347 °F-days). Alternately, weather statistics may be used for your own location.

1.  (a)  Assuming 5 ft of soil with $k = 0.865$ W/°C·m (0.5 Btu/hr °F ft):
    i.  Calculate an equivalent $U$ coefficient in W/°C·m² (Btu/hr °F ft²).
    ii.  Find the heat loss in watts (Btu/hr) and MJ/day (Btu/day) through the basement floor area.
    (b)  Find the hourly and daily rate of heat loss through the below-grade basement walls.
    (c)  Sum the values above to determine the weather-independent heat losses on both hourly and daily bases, and find the annual loss in GJ (MBtu).

2.  If the $U$ loss coefficient for the windows is 3.06 W/°C·m² (0.54 Btu/hr °F ft²), estimate the window heat loss as $UA$ in watts (Btu/hr) and MJ/°C-day (Btu/°F-day) for the design house. This $U$ value is typical for a double-pane (storm) window installation.

3.  Infiltration losses are variable and difficult to estimate. A fair estimate for a tight house is one air change per hour based on the volume of the house, with the basement excluded since it has no direct communication with outside air. Considering that 24 °C (75 °F) air occupies 0.843 m³/kg (13.5 ft³/lb) and has a specific heat of 1.01 kJ/kg·C (0.242 Btu/lb °F),
    (a)  Calculate the infiltration losses in watts (Btu/hr) and MJ/°C-day (Btu/°F-day) for the design house.
    (b)  Find $UA$ for infiltration in W/°C (Btu/hr °F).
    (c)  Sum this value with $UA$ from the problem above and compare the window plus infiltration design heating loss computed from this total $UA$ value with the sum of the "Design" losses calculated in each problem. This technique is useful because of the ease of recalculation when thermal parameters are changed.

4.  Average hot water usage is 75 gal per family-day at 140 °F. Estimate the water heating requirements MJ/day (Btu/day) if the water supply is at ground temperature. How many GJ (MBtu) are used annually? Compare this load with the loss from the uninsulated basement [problem 1(c)] and comment.

5.  The losses from a solar collector depend on the outside temperature, the wind velocity, the sky temperature, and of course the temperature of the surface exposed to the weather. Although it is difficult to determine what the outside surface temperature will actually be, it is useful to explore the surface temperatures implied if a varying proportion of the incident energy is collected for use in heating the dwelling.
    At noon on December 21st the solar radiation on a 2.23 m² (24 ft²) solar collector is 934 W/m² (296 Btu/hr ft²). The outside temperature is 4.44 °C (40 °F). On a single graph, plot the losses due to convection caused by wind at 2.2 m/sec (5 mph) and radiation to a $-4$ °C (25 °F) sky ($\varepsilon, \alpha = 0.95$) versus the solar collector (outside) surface temperature for 10, 40, 70, 110, and 140 °C (50, 100, 150, 200, and 250 °F). Plot a third line showing total losses.
    (a)  What temperature will an unprotected collector assume with no heat removal (i.e., losses are equal to the solar radiation)?

(b) What cover glass temperatures are necessary if (i) 730 W (2500 Btu/hr) and (ii) 1460 W (5000 Btu/hr) are to be collected? (Here the heat removed plus the convection and radiation losses equal the solar radiation.)

(c) Calculate the overall heat transfer coefficient at each of the five temperatures. Do you think a constant value would serve well enough?

6. The heating load must be transferred from the heat source to the house using air or water as the heat transfer fluid. On the coldest day of the year, the heating load of the design house is 15 000 W (51 000 Btu/hr). Under this condition:

(a) Calculate the water flow in L/s (gpm) to a baseboard system assuming an 11.1 °C (20 °F) temperature loss.

(b) Calculate the air flow in cfm to an air circulation system assuming a 25 °C (45 °F) temperature loss.

7. Calculate U-values for the design house for use in determining present and potential fuel savings for the three following design situations:

(a) Circa 1959 insulation:
   i. Ceiling (8 in. studs, $\frac{1}{2}$ in. wallboard, 1 in. fiberglass between studs).
   ii. Walls as in Example 2-10(a).
   iii. Windows, single glazing.
   iv. Doors ($\frac{1}{2}$ in. wood).
   v. Basement walls above grade (8 in. cement mortar concrete).

(b) Circa 1970 insulation:
   i. Ceiling (8 in. studs, $\frac{1}{2}$ in. wallboard, 8 in. fiberglass).
   ii. Walls ["full wall" ($3\frac{1}{2}$ in.) fiberglass as in Example 2-10(b)].
   iii. Windows, double glazing.
   iv. Doors ($\frac{1}{2}$ in. wood) plus glass storm door.
   v. Basement walls above grade (8 in. cement mortar concrete) insulated with $3\frac{1}{2}$ in. fiberglass.

(c) Insulated to a practical maximum:
   i. Ceiling (8 in. studs, $\frac{1}{2}$ in. wallboard, 12 in. fiberglass).
   ii. Walls studded with 2 × 6's on 24 in. centers (actual dimensions $1\frac{1}{2}$ in. × $5\frac{1}{2}$ in.); full-wall fiberglass insulation $5\frac{1}{2}$ in. thick; otherwise the same wall as in Example 2-10(b).
   iii. Windows, triple glazing.
   iv. Doors, as in (b) above.
   v. Basement walls above grade as in (b) above.

8. The design heating loads in MJ/°C-day (Btu/°F-day) and watts (Btu/hr) (maximum) are best calculated methodically.

(a) Fill in the table below with A's (areas) and the values of U's for the three sets of insulation. Calculate UA's and enter the UA value for infiltration from problem 3 for the 1959 and 1970 house. For the maximum insulation house use half this infiltration rate on the assumption that this house will be more tightly constructed. Clearly, one air change per hour is unacceptable if heat losses are to be minimized.

(b) Sum the $UA$'s in each $UA$ column. Since the $UA$'s are directly proportional to the seasonal heating losses, perusal of these columns is worthwhile.

(c) Find the heating load in MJ/°C-day (Btu/°F-day) for each design situation.

(d) Find the annual fuel oil consumption for each of the cases above (assuming 80% combustion efficiency and 147.6 MJ/gal or 140 000 Btu/gal. If fuel oil costs 90¢ per gallon, calculate the corresponding annual fuel costs.

(e) Using the hot water thermal consumption from problem 4, calculate:
   i.  Fuel oil usage and cost to generate hot water annually using this fuel.
   ii. GJ (MBtu) per month for hot water.

Table for Use with Problems 7 and 8

| | | 1959 | | 1970 | | Maximum Insulation | |
|---|---|---|---|---|---|---|---|
| | $A$ | $U$ | $UA$ | $U$ | $UA$ | $U$ | $UA$ |
| Infiltration | NA | NA | | NA | | NA | |
| Ceilings | | | | | | | |
| Walls (net) | | | | | | | |
| Windows | | | | | | | |
| Doors | | | | | | | |
| Basement* | | | | | | | |
| $\Sigma\,UA$ | | | | | | | |
| $\Sigma\,UA \times 24$ (Btu/°F-day) | | | | | | | |

\* Walls above grade.

9. The annual heating demand tells nothing about the maximum heating demand. This occurs on the coldest day of the year at the winter design temperature.

(a) Find the maximum heating demand for the three design situations using the design temperature of $-20$ °C ($-4$ °F).

(b) Assuming 147.6 MJ or 140 000 Btu/gal and 80% combustion efficiency, size the auxiliary oil burner in gallons burned per hour.

(c) If a shower bath uses 3 gal/min of 115 °F hot water, what hourly, auxiliary oil-firing rate is needed?

10. Calculate the annual fuel savings in gallons of oil and in dollars (at 90¢ per gallon):

(a) Increasing the ceiling insulation to 8 in. from the 1 in. of the 1959 house. (The $U$-values have been tabulated in problem 8 for 1970 and 1959, respectively).

(b) Increasing the ceiling insulation to 12 in. from 8 in. (The $U$-values have been tabulated in problem 8 for "maximum" and 1970, respectively.

(c) Bringing the walls to full $3\frac{1}{2}$ in. insulation (1970) from the 1-in. thickness of 1959 would cost \$5.38/m² (50¢/ft²) if insulation is blown in. Estimate the annual savings per m² (ft²) of wall if the house is electrically heated at $3\frac{1}{2}$¢/kWh.

11. Estimate the annual return for triple glazing a presently double-glazed 4 ft × 8 ft window on the first floor, if gas heat is used at \$3.00/1000 ft$^3$. (Use the $U$-values that have already been calculated.)

12. For the 1970 house above, size an air-conditioning system for Hartford assuming:
    (a) The roofs are black.
    (b) Windows have roller shades half drawn:

    | | |
    |---|---|
    | S windows | 83.75 ft$^2$ |
    | N windows | 97.75 ft$^2$ |
    | E and W windows | 75.0 ft$^2$ |

    (c) 91 °F design temperature, medium daily range.

13. For your own location, prepare a monthly degree-day table for use with an inside ambient temperature of 20 °C (68 °F).

## REFERENCES

1. F. B. Rowley, A. B. Algren, and J. L. Blackshaw, ASHVE Research Report No. 869, "Surface conductances as affected by air velocity, temperature, and character of surface," *ASHVE Trans. 36*, 444 (1930).
2. *ASHRAE Handbook and Product Directory.* 1974 Applications Volume, 1975 Equipment Volume, 1976 Systems Volume, 1977 Fundamentals Volume. American Society of Heating, Refrigerating and Air-Conditioning Engineers, Inc., 345 E. 47th St., New York, 10017. These volumes are republished every four years.
3. F. C. McQuiston and J. D. Parker, *Heating, Ventilating and Air Conditioning*, New York: Wiley, 1977.
4. W. H. McAdams, *Heat Transmission*, Third edition, New York, McGraw-Hill, 1954.
5. J. E. Peterson, "Estimating Air Infiltration into Houses: An Analytical Approach,' *ASHRAE J.*, 60–62 (January 1979).
6. *Climatic Atlas of the United States*, U.S. Government Printing Office, 1968.

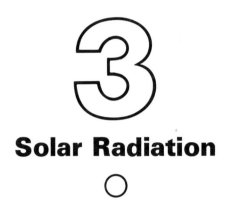

# Solar Radiation

The performance of any solar thermal system depends on the solar radiation available to it. Solar radiation is characterized by its variability. Even when abundant, it varies during the day, reaching a maximum at noon when the path length through the atmosphere is the shortest. Unless the collector is continuously turned to face the sun, the sun's changing altitude and azimuth will reduce the collected heat below the potential maximum. The hours of daylight also vary seasonally, being the shortest in winter when the need for heat is the greatest. Therefore, unlike most other power production equipment, solar collectors remain dormant for one-third to two-thirds of the day, increasing system cost considerably.

The local atmosphere, of course, appreciably affects the available sunlight. Ordinary pollution may reduce it 10% and, in problem locations such as the Los Angeles basin, there can be a much greater effect (1). Clouds are also a part of the climate in most localities, and the solar efficiency suffers accordingly.

In this chapter extraterrestrial radiation will be discussed and methods for calculating irradiance of sunlight will be described for clear days in the United States and Canada. Then solar radiation measurement will be introduced and methods described to obtain monthly values for total radiation on inclined surfaces from monthly total horizontal radiation. This monthly measurement, generally expressed as the daily average of the total radiation, is the only available solar radiation data for a great many locations.

When hourly radiation values are available from measurement or from cloud cover information, these data can be used more accurately for system design than average monthly information if they are preprocessed into detailed tables of radiation and weather data. Preparation of these tables will be discussed, and ways outlined to predict similar information for sites for which only monthly data are available.

## EXTRATERRESTRIAL RADIATION

The *irradiance* or intensity of the sun's radiant energy at the average earth-sun distance, measured normal to the earth–sun line but outside the earth's atmosphere, is called the solar constant, $I_{sc}$. The value of $I_{sc}$ has been determined within an estimated accuracy of $\pm 1.5\%$ as 1353 W/m². This value has been derived from numerous measurements of direct solar radiation flux made through the atmosphere at a number of solar zenith angles,* since the measured differences between them are directly translatable into an absolute measurement of the attenuation due to the atmosphere (2). These measurements have been confirmed by observations from high-flying aircraft, balloons, and space probes.

The extraterrestrial radiation from the sun is approximately the radiation of a *blackbody* at 5762 K, but shows certain peaks and valleys in the spectrum due to the radiative properties of the sun's incandescent gases. The blackbody and actual spectral irradiance are compared in Fig. 3-1.

The value of the solar constant $I_{sc}$ is believed to vary over a few percent from time to time due to not well-understood astronomical effects, but these variations have been averaged out by the use of a large number of measurements.†

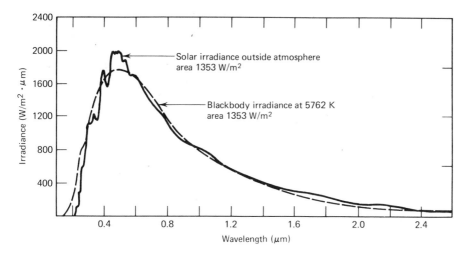

**FIGURE 3-1.** Solar spectral irradiance, standard curve, solar constant 1353 W/m² compared with blackbody radiation at 5762K (Ref. 2, 34).

* The sun's angle above the horizon is most conveniently expressed as the zenith angle ($\theta_z$). This is the angle between the sun's rays and the zenith, which is a line extending upward directly overhead.

† Satellite data from Nimbus 6 (1975–1976) show no such variance, the sun being constant $\pm 0.2\%$ over a year and one-half. Associated rocket flights indicate a value of about 1370 W/m² for the solar constant.

## Effect of Earth-Sun Distance

The apparent extraterrestrial solar irradiance $I_0$ varies over the year as the earth-sun distance changes seasonally, being about 3.5% higher than $I_{sc}$ in January and 3.5% lower in June. Table 3-1 gives monthly values of $I_0$ and tabulates the earth-sun distance in terms of the *astronomical unit*, the average earth-sun distance of $1.496 \times 10^8$ km or $92.956 \times 10^6$ miles. These values can be closely approximated by the empirical formula:

$$I_0 = I_{sc}\left[1 + 0.033 \cos\left(\frac{360N}{370}\right)\right] \tag{3-1}$$

where $N$ is the number of the day in the year.*

## Effect of Declination Angle

A much larger change in apparent extraterrestrial radiation is caused by the seasonally changing path of the sun through the sky. Associated changes in the sun's altitude and azimuth angles relative to the solar collector and changes in the length of time the sun spends above the horizon have an important effect on the total radiation received by the solar collector.

The seasonal change in the sun's path through the sky is due to the inclination of the earth's axis of daily rotation at 23.45° to a perhaps more logical axis perpendicular to the plane of the earth's orbit around the sun. Because of this all regions, but especially the areas nearest the poles, receive more light in summer and less light in winter than they otherwise would.

**TABLE 3-1.** $I_0$, $I_{sc}$, and Earth-Sun Distances Throughout the Year (Ref. 6, 4c)

| Dates | Jan 1 | Jan 4[a] | Feb 1 | Mar 1 | Apr 4[b] | May 1 | June 1 |
|---|---|---|---|---|---|---|---|
| $I_0$, Btu/hr ft² | 443.79 | 444.11 | 442.20 | 437.13 | 429.20 | 422.86 | 417.56 |
| $I_0$, W/m² | 1399 | 1400 | 1394 | 1378 | 1353 | 1333 | 1316 |
| Distance, A.U. | 0.9834 | 0.9831 | 0.9852 | 0.9909 | 1.0000 | 1.0075 | 1.0138 |

| Dates | July 1 | July 5[c] | Aug 1 | Sept 1 | Oct 5[b] | Nov 1 | Dec 1 |
|---|---|---|---|---|---|---|---|
| $I_0$, Btu/hr ft² | 415.24 | 415.24 | 416.51 | 421.27 | 429.10 | 435.54 | 441.56 |
| $I_0$, W/m² | 1309 | 1309 | 1313 | 1328 | 1353 | 1373 | 1392 |
| Distance, A.U. | 1.0167 | 1.0167 | 1.0151 | 1.0094 | 1.0000 | 0.9927 | 0.9859 |

[a] Perihelion, earth-sun distance is at its minimum.
[b] Earth-sun distance is 1.0000 astronomical unit: $I_0 = I_{sc}$.
[c] Aphelion, earth-sun distance is at its maximum.

* Equations 3-1 and 3-2 have been chosen to fit the year 1964 closely, which is the year selected for standardized tables published by ASHRAE (7). Other years are fit almost as well. The portions of the arguments in parentheses in eq. 3-1 and 3-2 are often given as $(360\ N/365)$ and $[(N + 284)/365]$ respectively (16). These circularize the earth's orbit, leading to persistent errors in winter of one to two percent in data elevated to a tilted surface.

**TABLE 3-2.** Parameters Used to Estimate Solar Radiation Intensity (Ref. 3, 6)

| Nominal Date | Day Number[a] | Declination Degrees | A Btu/hr ft² | A W/m² | B Air Mass⁻¹ | C Dimen-sionless | Equation of Time (min)[b] | Solar Noon |
|---|---|---|---|---|---|---|---|---|
| January 21 | 19.85 | − 20.0 | 390 | 1,230 | 0.142 | 0.058 | − 11.2 | Late |
| February 21 | 54.06 | − 10.0 | 385 | 1,215 | 0.144 | 0.060 | − 13.9 | Late |
| March 21 | 80.00 | 0.0 | 376 | 1,186 | 0.156 | 0.071 | − 7.5 | Late |
| April 21 | 110.47 | + 11.6 | 360 | 1,136 | 0.180 | 0.097 | + 1.1 | Early |
| May 21 | 140.15 | + 20.0 | 350 | 1,104 | 0.196 | 0.121 | + 3.3 | Early |
| June 21 | 172.50 | + 23.45 | 345 | 1,088 | 0.205 | 0.134 | − 1.4 | Late |
| July 21 | 201.84 | + 20.6 | 344 | 1,085 | 0.207 | 0.136 | − 6.2 | Late |
| August 21 | 232.49 | + 12.3 | 351 | 1,107 | 0.201 | 0.122 | − 2.4 | Late |
| September 21 | 265.00 | 0.0 | 365 | 1,151 | 0.177 | 0.092 | + 7.5 | Early |
| October 21 | 292.34 | − 10.5 | 378 | 1,192 | 0.160 | 0.073 | + 15.4 | Early |
| November 21 | 324.20 | − 19.8 | 387 | 1,221 | 0.149 | 0.063 | + 13.8 | Early |
| December 21 | 357.50 | − 23.45 | 391 | 1,233 | 0.142 | 0.057 | + 1.6 | Early |

[a] Selected to give the exact declination of the next column with eq. 3-2.
[b] Within one minute, this can be expressed mathematically by (28)

$$-\left[9\sin\left(\frac{N-1}{0.5}\right)\right] - 5 \quad (N < 100); \qquad \left[5\sin\left(\frac{N-100}{0.395}\right)\right] - 1 \quad (100 \le N \le 244);$$

$$\left[18.6\sin\left(\frac{N-242}{0.685}\right)\right] - 2.5 \quad (N > 242)$$

where the arguments of the angles are expressed as degrees.
*Source.* Except for day number,[a] this table has been taken from Ref. 6, but with correction of February declination.

The apparent inclination of the axis of the earth's rotation toward the sun with respect to a cylinder through and perpendicular to the earth's orbit is δ, the *declination* angle. The declination varies from + 23.45 to − 23.45° over the course of a year. Table 3-2 lists nominal values for the twenty-first day of each month

The declination of a given year differs from that of the previous year by about one-fourth of a day. Every four years there is a leap year correction, so that the declination repeats itself almost exactly in a four-year cycle.

The following empirical equation is recommended for calculation of declination:*

$$\delta_{\text{degrees}} = 23.45° \sin\left[\left(\frac{N - 80}{370}\right) \times 360\right]†  \qquad (3-2)$$

where $N$ is the day number in the year.‡

* See previous * footnote, p. 64.
† The argument of the sine function is in degrees.
‡ For the southern hemisphere, reverse the sign of the declination.

**TABLE 3-3.** Dates That Give the Monthly Average Declination Angle, for Use with Eqs. 3-1 and 3-2[a]

| Date[a] | Day Number[b] |
|---------|---------------|
| 17 Jan | 17 |
| 15 Feb | 46 |
| 17 Mar | 76 |
| 15 Apr | 105 |
| 15 May | 135 |
| 11 Jun | 162 |
| 18 Jul | 199 |
| 17 Aug | 229 |
| 15 Sep | 258 |
| 15 Oct | 288 |
| 14 Nov | 318 |
| 12 Dec | 346 |

[a] This table assumes 28 days in February.
[b] If using the alternate standard year given in the footnotes to eqs. 3-1 and 3-2, add one to February, subtract one from March and July, and subtract two from December (16).

Equation 3-2 is never more than a day away from the correct declination value. For monthly applications, the average monthly declination angle can be calculated almost exactly if the day numbers from Table 3-3 are used in eq. 3-2. These dates were selected by trial to most closely match the true four-year monthly average declination.

**The Seasons.** Figure 3-2 shows the annual astronomical situation as viewed from the top of the earth's orbit, looking downward at the north pole. Figure 3-3 shows the day–night hemispheres at the four seasonal extremes, as viewed from the orbital path just ahead of the earth's travel.

At the spring and fall equinox (about March and September 21, Fig. 3–3a and c), the sun's rays are perpendicular to the axis of daily rotation, and days and nights are everywhere of equal, 12 hour length.

At the summer solstice (winter in the southern hemisphere), about June 21 (Fig. 3-3b), the portion of the globe north of $90° - 23.5° = 66.5°N$ latitude is enjoying daylong radiation, while the portion south of 66.5°S has no direct sun at all. These zones are called the arctic zones. For the winter solstice, about December 21, their situation is reversed (Fig. 3-3d).

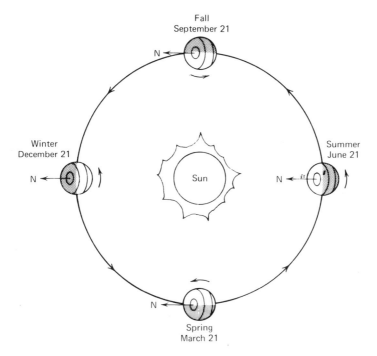

**FIGURE 3-2.** Seasonal changes in illumination of the earth, as viewed from well above the orbital plane.

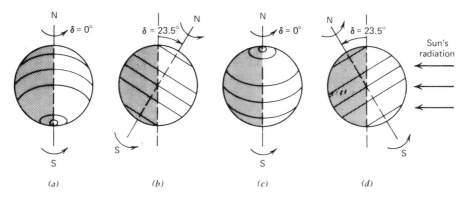

**FIGURE 3-3.** Seasonal changes in illumination of the earth, as viewed from the orbital path to be next travelled. (*a*) Spring equinox. (*b*) Summer solstice (winter in the southern hemisphere). (*c*) Fall equinox. (*d*) Winter solstice (summer in the southern hemisphere).

At the other extreme are the torrid zones, at latitudes less than 23.5°N or S, where the sun is directly overhead twice a year. The zones in between the arctic and the torrid zones are called the temperate zones. Here the sun is never directly overhead and there are no days when the sun does not rise and set.

These seasonal factors have a profound effect on fixed flat plate collectors. The sun's elevation at noon changes ±23.5°, a 47° range from summer to winter. Above latitudes of 40°N, the noon sun is less than 26.5° above the horizon at the winter solstice, seriously limiting available sunlight due to the short arc of travel between sunrise and sunset.

**The Sun's Path Through the Sky.** At either equinox the plane of the path that a given location travels during the daylight hours in Fig. 3-3a or c is perpendicular to the earth's axis and *parallel* with the sun's radiation, and the apparent path of the sun through the sky is therefore a straight line, with sunrise due east and sunset due west. The noontime elevation of the sun is 90° less the latitude.

At the winter solstice, since the plane of the path of the daylight travel of a given location (see Fig. 3-3d) remains perpendicular to the axis of the earth's rotation but is no longer parallel with the sun's radiation, the path of the sun's apparent travel through the sky is not straight, but traces an arc of a circle with sunrise to the south of east and sunset to the south of west in either hemisphere. The day is shortened below 12 h everywhere above the equator, and the sun reaches only to an angle of 90° less the latitude less 23.5° above the horizon at noon on the day of winter solstice.

At the summer solstice (southern hemisphere winter), about June 21 (Fig. 3-3b), the situation is reversed. The sun rises and sets to the north of the east–west line and also follows a circular path. Above the equator the days are everywhere longer than 12 hr, and the elevation of the sun at noon is (at the solstice) 90° less the latitude *plus* 23.5°.

## CLEAR SKY IRRADIANCE (5, 6, 7, 8)

### Atmospheric Effects

As the sun's radiation passes through the atmosphere it is attenuated in proportion to the length of its path according to an *extinction coefficient B*, to produce the *direct normal* irradiance at the earth's surface (3):

$$I_{DN} = A \exp\left(-\frac{P}{P_0} \times \frac{B}{\cos \theta_z}\right) \tag{3-3}$$

where $\theta_z$ is the zenith angle and $P/P_0$ is the pressure at the location concerned relative to a standard atmosphere, given by

$$\frac{P}{P_0} = \exp\left(-0.000\ 036\ 1 \times \text{altitude in feet above sea level}\right)$$

or

(3-3a)

$$\frac{P}{P_0} = \exp\left(-0.000\ 118\ 4 \times \text{altitude in meters above sea level}\right)$$

Monthly values for the apparent extraterrestrial solar intensity $A$ and the extinction coefficient $B$ for the United States and Canada are given in Table 3-2. The $A$ values are not derived from eq. 3-1 but were determined instead so that eq. 3-3 gives results consistent with Weather Bureau records for 70 locations in the United States and Canada.* The variations in $A$ are due to earth-sun distance changes and seasonal variations in the atmosphere. The extinction coefficient variations are mainly due to changes in atmospheric moisture from season to season.

The zenith angle term in eq. 3-3 accounts for the longer path length through the atmosphere when the sun is not perpendicular to the earth's surface. $I_{DN}$, the direct normal irradiance, is the energy of the direct solar beam falling on a unit area perpendicular to the beam at the earth's surface. To obtain the *global* irradiance the additional diffuse irradiance reflected from clouds and the clear sky must be included. This adds only to 5 to 10% when there is clear blue sky, but the diffuse component is a much higher proportion of the total when there are haze or clouds present. If the sun's disc is obscured, all the radiation flux is diffuse. The monthly average ratio of diffuse to direct normal irradiance for clear skies in the United States and Canada is given as parameter $C$ in Table 3-2 (31).

Much of the energy loss in passing through the atmosphere is due to absorption of infrared radiation by atmospheric water vapor. When the humidity is high the global irradiance may be reduced as much as 15% below that through the standard atmosphere due to additional absorption in the infrared portions of the spectrum. Figure 3-4 shows the change in the sun's spectrum due to one *standard air mass*, that is, perpendicular passage through a standard atmosphere. Water vapor attenuates the proportion of energy in the infrared ($\lambda > 1\ \mu$m) from its extraterrestrial value of 53% to about 38% of $I_{DN}$. The visible wavelengths from 0.4

---

* To estimate an $A$ value for the southern hemisphere temperate zones, divide an $A$ value from Table 3-2 by the extraterrestrial solar irradiance on the date indicated, then multiply it by the extraterrestrial solar irradiance on a date six months hence. This estimate for $A$ applies to the later date, and is surprisingly constant at 1155 W/m², except for September, October, and November, when it averages 1175 W/m². For constants $B$ and $C$ use the earlier values directly for the later date.

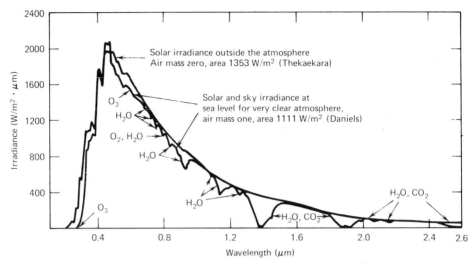

**FIGURE 3-4.** Spectral solar irradiance beyond the earth's atmosphere, air mass zero; and global irradiance at sea level, air mass = 1.0 (Ref. 2, 34).

to 0.7 $\mu$m are not greatly affected by passage through a clear atmosphere, except for the molecular scattering of blue light, which gives us the familiar blue sky.* The ultraviolet radiation from 0.3 to 0.4 $\mu$m is almost all scattered, reducing the portion of such energy in the direct radiation well under the 6.1% available extra-terrestrially. The far ultraviolet (below 0.3 $\mu$m), which contains 1.5% of the sun's radiant energy, is cut off completely by the ozone layer in the stratosphere. This loss is fortuitous, since these short wavelengths are too energetic to be withstood by many of earth's life forms.

The radiant beam is further reduced by atmospheric pollution, by seaside salt haze in coastal areas, and of course by fog and cloud as the weather changes. Interestingly, the decrease in effective transmissivity of the atmosphere due to the low winter sun is partly compensated for by a decrease in atmospheric water vapor and the closeness of the sun in the winter (northern hemisphere), so that in the United States there is only a 3% drop in noon global *normal* irradiance at 40°N latitude between the summer and winter extremes. On a *horizontal* flat surface though, the winter solstice noon, clear-sky irradiance is reduced to one-half of that on June 21. Combined with the effect of the shorter day, the winter total daily horizontal clear-sky radiation is reduced to only 30% of the summer

---

* Much of the scattered light is not lost but shows up as diffuse radiation. This is included in Fig. 3-4.

value. For this reason a tilted collector is generally necessary above latitudes of 23.5°, where the sun is never directly overhead.

### Irradiance on Plane Surfaces.

For any location, month, and time of day the direct terrestrial clear-sky radiation flux on any flat surface can be calculated using trigonometric relationships and the constants of Table 3-2 in combination with eq. 3-3 (4, 5, 6). The technique is first to calculate the direct normal radiation intensity, then the altitude of the sun above the horizon and the azimuth away from solar noon. The irradiance on a horizontal surface can then be calculated from simple trigonometric relationships. For tilted surfaces more complex trigonometric relationships establish the angle between the solar beam and a line normal to the collector surface as a function of the collector elevation angle and the fixed azimuth angle of the collector away from solar noon. The radiation flux on the collector is then determined as a simple function of the angle from the sun to the collector-normal line, the sun's altitude, and either the horizontal or the direct normal radiation flux.

### Irradiance on Fixed Flat Horizontal Surfaces.

Determination of the clear sky radiation flux on a flat horizontal surface is the first step toward finding the radiation on a tilted surface. The time of day is of course a most important variable. At solar noon the sun is most nearly overhead, and the *apparent solar time* (AST) is 12 hours. The local clock time (LCT) may be considerably different, since the same time is used over politically convenient geographical zones. The *time zone* (TZ) may be calculated by dividing the degrees longitude of the local Standard Time Meridian by 15, taking west longitudes as positive and east longitudes negative. This is the number of hours (+ or −) that must be added to the local clock time to obtain the time at 0° longitude. The time zones of the continental United States are EDT = +4, EST or CDT = +5, CST or MDT = +6, MST or PDT = +7, PST = +8.

For radiation calculations the apparent solar time is either assumed at a convenient value or found from the clock time by using the following dimensional equation, within which each term is expressed in *hours*:

$$AST = LCT + TZ - LONG/15 + EQT/60 \qquad (3\text{-}4)$$

LONG is the local longitude in *degrees*, taken as positive for west longitudes and negative for east longitudes. The "equation of time," EQT, is given in minutes in Table 3-2. It is an astronomical correction for irregularities in the earth's orbit.

Using the solar time the altitude and azimuth of the sun can be determined by the procedure in the following paragraphs.

Refer to Figs. 3-5 and 3-6 to visualize the various angles. The hour angle $\omega$ is the angle of the location's position away from solar noon caused by the earth's

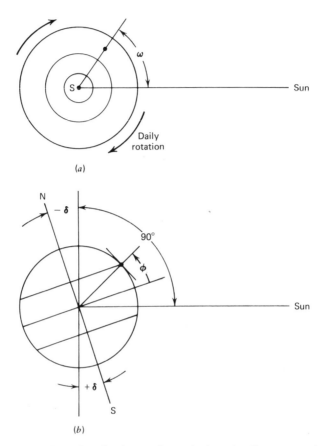

**FIGURE 3-5.** Angles used to calculate the direct normal radiation. (*a*) View from the South Pole showing hour angle ω. The paper is not necessarily in the plane of the earth's orbit. (*b*) View showing relationship of the latitude angle φ and declination angle δ.

rotation. To express ω in degrees, multiply the hours from solar noon by 360/24. In this text we define ω as negative in the morning and positive in the afternoon. The zenith angle $\theta_z$ and its complement the altitude α are then given by

$$\cos \theta_z = \sin \alpha = \cos \phi \cos \delta \cos \omega + \sin \phi \sin \delta \qquad (3\text{-}5)$$

where φ is the latitude (positive in either hemisphere) and δ is the solar declination.

The solar azimuth ψ, measured in degrees away from south (north in the southern hemisphere), is given by

$$\sin \psi = \frac{\cos \delta \sin \omega}{\cos \alpha} \qquad (3\text{-}6)$$

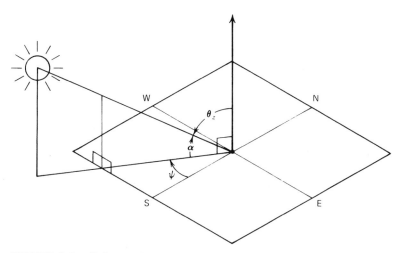

**FIGURE 3-6.** Solar zenith angle $\theta_z$, altitude $\alpha$, and azimuth $\psi$ shown in their relationship to a horizontal flat surface on the earth's surface.

where $\psi$ is negative in the morning and positive in the afternoon. Notice that this function is multivalued, and the choice of the two possible values for $\psi$ must be made correctly,* with $|\psi| \geq 90°$ when the solar altitude is less than it is when the azimuth of the sun is $\pm 90°$. That altitude is given by (26):

$$\sin \alpha = \frac{\sin \delta}{\sin \phi} \tag{3-6a}$$

Sin $\alpha$ will take absolute values greater than unity in the tropics if the sun stays in the northern or southern half of the sky all day.

After the solar zenith angle has been calculated from eq. 3-5, eq. 3-3 is then used with $A$ and $B$ values from Table 3-2 to find the direct normal irradiance $I_{DN}$, remembering that angle $\theta_z = (90 - \alpha)$ so that $\cos \theta_z = \sin \alpha$. The diffuse irradiance is given by parameter $C$ times $I_{DN}$, and the global irradiance on a flat horizontal surface is the sum of the direct and diffuse components:

$$H_{\text{global}} = I_{DN} \sin \alpha + C I_{DN} \tag{3-7}$$

The hour angle at sunset or sunrise, $\omega_s$, can be found by solving eq. 3-5 for $\omega$ when $\alpha = 0$:

$$\cos \omega_s = -\tan \phi \tan \delta \tag{3-8}$$

$\omega_s$ is negative for sunrise and positive for sunset. Absolute values of $\cos \omega_s$ over unity occur in the arctic zones when the sun does not rise or does not set.

---

* The obtuse angle is $\pm(180° - |\psi|)$, where $\psi$ is the acute angle and $\pm$ its sign, as determined with eq. 3-6.

To find $\omega_s$ in *hours* away from solar noon, use

$$\omega_s = \frac{\pm 24}{360} \times \arccos\left(-\tan\phi\tan\delta\right) \tag{3-8a}$$

Local time of sunrise and sunset (in hours) can then be calculated with eq. 3-4, solving for the local clock time when AST $= (12.00 + \omega_s)$ hours.

## EXAMPLE 3-1

Calculate the apparent solar time at Hartford, Connecticut (longitude 72.6°W), on November 21 at 9:50 A.M. EST.

From eq. 3-4, with all values in *hours*,

$$AST = 9\frac{50}{60} + 5 - \frac{72.6}{15} + \frac{13.8}{60} = 10.22 \text{ hr}$$

or $10 + 0.22 \times 60 = 10:13$ A.M. solar clock time, to the nearest minute.

## EXAMPLE 3-2

For a clear day at 10 A.M. solar time on November 21 (day 325) at 40°N latitude,*
(a) Find the direct normal irradiance at sea level and 1524 m (5000 ft).

First find the solar altitude and azimuth at $-2$ hours from noon:

$$\omega = -2 \times \frac{360}{24} = -30°$$

$$\phi = 40°$$

By eq. 3-2,

$$\delta = 23.45 \sin\left[\left(\frac{325 - 80}{370}\right) \times 360°\right] = -20.0°$$

using eq. 3-5

$$\sin\alpha = \cos 40° \cos(-20°)\cos(-30°) + \sin 40° \sin(-20°) = 0.4036$$
altitude $\alpha = \arcsin(0.4036) = 23.8°$

using eq. 3-6

$$\sin\psi = \frac{\cos(-20°)\sin(-30°)}{\cos 23.8°} = -0.5134$$

Azimuth angle $\psi = \arcsin(-0.5134) = -30.9°$ or $-149.1°$ of which the first value is correct, since eq. 3-6a gives a value of $\alpha = \arcsin[\sin(-20.0°)/(\sin 40°)]$ $= -32°$ (i.e., the sun has set) for the altitude when $\psi = \pm 90°$, which is indeed less than its current altitude of 23.8°.

* These integral values are chosen so that the results from Examples 3-2, 3-3, 3-4, and 3-6 can be compared with standard tables (Appendix A).

From eq. 3-3, with $A$ and $B$ from Table 3-2 for November 21, at 1524 m,

$$I_{DN} = 1221 \exp \left[ -\frac{0.149}{\cos (90 - 23.8)°} \times \exp (-0.000\ 118\ 4 \times 1524) \right]$$

$$I_{DN} = 897 \text{ W/m}^2$$

and at sea level,

$$I_{DN} = 1221 \exp [-0.149/\cos (90 - 23.8)°]$$
$$I_{DN} = 844 \text{ W/m}^2$$

In Engineering units, at 5000 ft,

$$I_{DN} = 387 \exp \left[ -\frac{0.149}{\cos (90 - 23.8)°} \times \exp (-0.000\ 036\ 1 \times 5000) \right]$$

$$I_{DN} = 284.3 \text{ Btu/hr ft}^2$$

and at sea level,

$$I_{DN} = 387 \exp [-0.149/\cos (90 - 23.8)°]$$
$$I_{DN} = 267.5 \text{ Btu/ft}^2$$

(b) Find the global irradiance at sea level.

The direct, diffuse, and global irradiance at sea level on the flat horizontal surface are found by using eq. 3-7, with $C = 0.063$ (Table 3-2):

$$H_{direct} = I_{DN} \sin \alpha = 844 \sin 23.8° = 340.5 \text{ W/m}^2$$
$$H_{diffuse} = I_{DN} \times C = 844 \times 0.063 = 53.2 \text{ W/m}^2$$
$$H_{global} = 340.5 + 53.2 = 393.7 \text{ W/m}^2$$

In Engineering units,

$$H_{direct} = 267.5 \times \sin 23.8° = 107.9 \text{ Btu/hr ft}^2$$
$$H_{diffuse} = 267.5 \times 0.063 = 16.9 \text{ Btu/hr ft}^2$$
$$H_{global} = 107.9 + 16.9 = 124.8 \text{ Btu/hr ft}^2$$

## EXAMPLE 3-3

Find the solar sunrise, sunset, and length of solar day at 40°N latitude on November 21.

Using eq. 3-8a,

$$\omega_s = \frac{\pm 24}{360} \arccos [-\tan 40° \tan (-20.0°)]$$

$$\omega_s = \pm 4.81 \text{ hr}$$

Sunrise $= 12 - 4.81 = 7.18$ hr or $7$ hr $+ 0.18 \times 60$ min
    $= 7$ hr, 11 min $= 0711$ solar clock time
Sunset $= 12 + 4.81 = 16.81$ hr or $16 + 0.81 \times 60 = 1649$ solar clock time
Length of solar day $= 4.81 \times 2 = 9.62$ hr

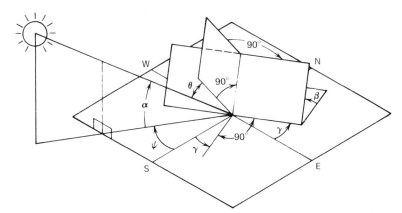

**FIGURE 3-7.** Angles used to calculate the irradiance on a tilted surface. The sun's altitude is $\alpha$ and its azimuth angle is $\psi$. The collector inclination is $\beta$, its azimuth angle $\gamma$, and the incident angle of the sun to the collector normal is $\theta$.

**Irradiance on a Tilted, Fixed Flat Surface.** Relationships among the angles involved with a tilted collector surface are shown in Fig. 3-7. The incident angle $\theta$ is the angle between the direct radiation beam from the sun and a line normal to a plane surface tilted upward at angle $\beta$ from the horizontal. It can be determined for a south*-facing collector with a variation of eq. 3-5:

$$\cos \theta = \cos (\phi - \beta) \cos \delta \cos \omega + \sin (\phi - \beta) \sin \delta \qquad (3\text{-}9)$$

The hour angle $\omega_s'$ of potential sunrise or sunset on the *collector* is given by

$$\cos \omega_s' = -\tan (\phi - \beta) \tan \delta \qquad (3\text{-}10)$$

Absolute values of $\cos \omega_s'$ greater than unity occur in the tropical zones when the sun does not rise and set on a south*-facing extraterrestrial surface.

In winter the hour angle of actual sunset often occurs first, while in summer the sun may go behind the collector before sunset on the horizon. Therefore, $\bar{\omega}_s$, the actual hour angle of sunset on the collector, is the minimum of $\omega_s$ from eq. 3-8 and $\omega_s'$ from eq. 3-10.

If the collector does not face south,* $\theta$ is found from the sun's altitude $\alpha$, azimuth $\psi$, and the positive or negative azimuth angle $\gamma$ that the flat surface faces with respect to due south.* $\gamma$ is negative when the surface faces east of south;* positive when it faces west of south.* Expressed analytically,

$$\cos \theta = \cos \alpha \cos (\psi - \gamma) \sin \beta + \sin \alpha \cos \beta \qquad (3\text{-}11)$$

Sunrise and sunset on a nonsouth*-facing collector can be found using eqs. 3-30 or 3-31.

* North in the southern hemisphere.

If the direct normal or the horizontal direct irradiance is known, the direct radiation flux on the tilted collector surface is given in terms of the incident angle and the sun's altitude by the following expression:

$$I_{direct} = H_{direct} \frac{\cos \theta}{\sin \alpha} = I_{DN} \cos \theta \qquad (3\text{-}12)$$

where $(\cos \theta)/(\sin \alpha)$ is sometimes called a magnification factor $M$. Since $I_{direct}$ cannot be negative, $\cos \theta$ is set to zero if eq. 3-9 or eq. 3-11 yields a value of $\theta$ greater than 90°, which indicates the sun is behind the collector.

The diffuse irradiance on the tilted collector is given by

$$I_{diffuse} = H_{diffuse} \frac{(1 + \cos \beta)}{2} \qquad (3\text{-}13)$$

This assumes that the surroundings are not reflective and that the diffuse light is distributed uniformly over the sky.

If the surroundings have a reflectivity $\rho$, then

$$I_{diffuse} = H_{diffuse} \frac{(1 + \cos \beta)}{2} + \rho(H_{direct} + H_{diffuse}) \frac{(1 - \cos \beta)}{2} \qquad (3\text{-}14)$$

where $\rho$ is about 0.2 for ordinary ground or vegetation, 0.8 for snow cover, and 0.15 for a gravel roof. Notice that the reflectivity of the surroundings does not affect either the direct or the diffuse irradiance on a horizontal surface.

To express the diffuse irradiance in terms of the direct normal radiation flux, eqs. 3-7 and 3-14 are combined to give

$$I_{diffuse} = I_{DN} \left[ C \frac{(1 + \cos \beta)}{2} + \rho(C + \sin \alpha) \frac{(1 - \cos \beta)}{2} \right] \qquad (3\text{-}14a)$$

## EXAMPLE 3-4

The altitude $\alpha$ of the sun in Example 3-2 was 23.8° and its azimuth $\psi$ was $-30.9°$. The sun is shining on a south-facing collector elevated 50° upward from the horizontal. Assume the reflectivity of the surroundings is zero.

(a) What is the global irradiance?

For the horizontal surface $H_{direct} = 340.5$ W/m² and $H_{diffuse} = 53.2$ W/m² as determined in Example 3-2(b).

Since the collector faces south, $\gamma = 0°$. Equation 3-11 therefore simplifies to

$$\cos \theta = \cos \alpha \cos \psi \sin \beta + \sin \alpha \cos \beta$$

and

$$\cos \theta = \cos 23.8° \cos (-30.9°) \sin 50° + \sin 23.8° \cos 50° = 0.8608$$
$$\text{incident angle } \theta = \text{arcos} (0.8608) = 30.59°$$

Using eq. 3-12,

$$M = 0.8608/\sin 23.8° = 2.133$$
$$I_{direct} = 340.5 \text{ W/m}^2 \times 2.133 = 726.6 \text{ W/m}^2$$

and from eq. 3-13

$$I_{diffuse} = 53.2 \text{ W/m}^2 \times \frac{(1 + \cos 50°)}{2} = 43.7 \text{ W/m}^2$$

$$I_{global} = 726.6 \text{ W/m}^2 + 43.7 \text{ W/m}^2 = 770.3 \text{ W/m}^2$$

Alternately eq. 3-9 gives the same value of $\theta$ in a single step:

$$\cos \theta = \cos (40° - 50°) \cos (-20.0) \cos (-30°)$$
$$+ \sin (40° - 50°) \sin (-20°) = 0.8608$$
$$\theta = 30.59°$$

In Engineering units, for the horizontal surface $H_{direct} = 107.9$ Btu/hr ft² and $H_{diffuse} = 16.9$ Btu/hr ft². Taking $M = 2.133$ as in the above solution,

$$I_{direct} = 107.9 \text{ Btu/hr ft}^2 \times 2.133 = 230.2 \text{ Btu/hr ft}^2$$

$$I_{diffuse} = 16.9 \text{ Btu/hr ft}^2 \times \frac{(1 + \cos 50°)}{2} = 13.9 \text{ Btu/hr ft}^2$$

$$I_{global} = 230.2 + 13.9 = 244.1 \text{ Btu/hr ft}^2$$

(b) If the collector is facing 20° east of due south what is the global irradiance on the tilted surface? (The solar time is 2 hours before noon.)

By eq. 3-11, with $\gamma = -20°$,

$$\cos \theta = \cos 23.8° \cos [30.9 - (-20)]° \sin 50° + \sin 23.8° \cos 50° = 0.9476$$
$$\theta = \text{arcos} (0.9476) = 18.62°$$

From eq. 3-12

$$I_{direct} = 340.5 \text{ W/m}^2 \times \frac{0.9476}{\sin 23.8°} = 800 \text{ W/m}^2$$

$$I_{global} = 800 \text{ W/m}^2 + 43.7 \text{ W/m}^2 = 843.5 \text{ W/m}^2$$

In Engineering units,

$$I_{direct} = 107.9 \text{ Btu/hr ft}^2 \times \frac{0.9476}{\sin 23.8°} = 253.5 \text{ Btu/hr ft}^2$$

$$I_{global} = 253.5 \text{ Btu/hr ft}^2 + 13.9 \text{ Btu/hr ft}^2 = 267.4 \text{ Btu/hr ft}^2$$

## Tabular Clear Sky Data

The preceding paragraphs have shown how to calculate clear sky irradiance for any collector tilt and azimuth at any time for any day of the year. Tables A-1 to A-6 in Appendix A have been prepared at hourly intervals in the same way, but with zero reflectivity, by Morrison (11), using Engineering units for six latitudes from 24 to 64°N. Tables A-7 to A-12 tabulate the corresponding incident angles. Monthly declinations were as listed in Table 3–2. The tables apply to south-facing surfaces inclined at angles 0°, 90°, the latitude plus 20°, 10°, 0°; and 10° less than the latitude. They are also available at collector azimuth angles of ±45° and ±90° in ASHRAE GRP 170 (4a). While such tables meet most needs for clear sky information, the techniques just introduced permit their preparation for exact latitudes, times, and tilt angles as well as for a collector azimuth other than due south.

Table A-13 has been prepared in the same fashion, but here monthly daily average SI total global radiation values are given at 2° latitude increments for 0°, 20°, 30°, 40°, 50°, 60°, 70°, and 90° collector tilt angles. The dates have been chosen from Table 3-3 to give accurate average monthly values. Table A-13 can be combined with percent of sunshine to estimate real radiation data, as will be shown in Examples 3-7(b) and 3-11.

Further corrections should be made to the tabular data to account for varying elevation and humidity throughout the United States. Figure 3-8 shows such corrections graphically. These "clearness numbers" range from 0.85 for the Gulf Coast winter to 1.15 for dry high-altitude summers, and are the multipliers for the Appendix A values.

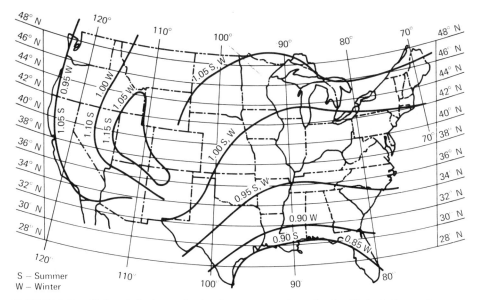

**FIGURE 3-8.** Estimated atmospheric clearness numbers in the United States for non-industrial localities. (Reprinted with permission from American Society of Heating, Refrigeration and Air Conditioning Engineers (Ref. 31).

## SOLAR RADIATION MEASUREMENT (4c, 9)

The accuracy of the analysis of many geophysical phenomena depends on knowledge of solar radiation. Its measurement has been an exacting subscience for nearly a hundred years.

There are two general types of instruments for measuring the sun's radiation:

1. *Pyrheliometer.* An instrument for measuring the intensity of direct solar radiation at normal incidence [i.e., pointing directly at and measuring the radiant beam plus a small portion (about 5.5°) of the nearby sky].

2. *Pyranometer.* An instrument for measuring the global (direct plus diffuse) irradiance from one hemisphere (a solid angle of 180°). It measures diffuse irradiance alone if the sensor is shaded from the sun with a disc. A fixed *shadow band* is often used for this purpose, but this itself shields some of the diffuse irradiance and requires corrections to the data.

These instruments can be either absolute or relative instruments. The absolute instruments are more expensive and serve as primary standards. They are almost exclusively pyrheliometers, which receive solar radiation on a blackened surface at the end of a blackened collimating tube.

The Ångström compensation pyrheliometer has two blackened metal strips, either of which can be exposed to sunlight. The dark strip is heated electrically to exactly the same temperature as the irradiated strip, establishing an exact value of the radiation received. A movable shutter allows each strip to be alternately illuminated or heated, increasing the accuracy.

The water-flow calorimetric pyrheliometer measures the temperature rise in a precisely controlled stream of water, removing the heat from a blackened surface at the bottom of a collimating tube. When a steady value is achieved, the radiation input is interrupted and an electrical current applied on the same surface to produce the same temperature rise. Measurement of the electrical energy again provides the absolute measurement.

Since these absolute instruments are difficult to use in the field, secondary standard instruments have been devised that are portable. The Abbot "silver disc" pyrheliometer has a blackened silver target that carefully incorporates a special mercury thermometer. The observed temperature rise in exactly 100 seconds exposure is the basis for a relative reading that is traceable to a primary standard. Many more modern high-accuracy instruments for field use have now been developed, and some of them claim absolute accuracies of $\pm 0.5\%$.

Very few pyrheliometers are actually used for regular radiation observation. Pyranometers, which do not need to track the sun, are much more generally applied. These devices are invariably thermoelectric, that is, they measure the temperature difference between a shaded and an irradiated thermopile (a thermopile is a group of thermocouples wired in series to increase their sensitivity). The

(a)

**FIGURE 3-9a.** The Eppley Precision Spectral Pyranometer. This instrument is of special quality and is often used for solar radiation research. (Courtesy of Eppley Laboratory.)

(b)

**FIGURE 3-9b.** The Eppley Black and White Pyranometer. A field instrument often used to monitor the solar radiation received by a collector array. (Courtesy of Eppley Laboratory.)

thermopiles are mounted flat, underneath one—sometimes two—glass hemispheres. With the aid of much electronics, these instruments produce an accurate and linear output, but must be calibrated periodically. *First-class* pyranometers, such as the Eppley Precision Spectral Pyranometer (Fig. 3-9a), are accurate to within $\pm 1\%$ and have a changing response as zenith angle changes within $\pm 3\%$ of the anticipated cosine relationship. *Second-class* instruments are generally used in the field. They have about twice the margin of error as the more expensive first-class instruments. One popular second-class instrument is the Eppley "Black and White" pyranometer (Fig. 3-9b). The white segments reflect almost 100% of the light and contain the "dark" thermopiles. The blackened sectors contain the irradiated thermopiles.

In a final category of instrument are the sunshine recorders. These attempt to record the presence or absence of the sun's disc, that is, the portion of time when a distinct shadow can be seen. Even though the data from these devices are difficult to interpret, there are so much sunshine data available (since operation is easy and trouble free) that there are occasional efforts to produce accurate radiation data from these readings (see Ref. 10 and *Sunshine Records*, p. 85).

### Measured Radiation Data.

So far in this chapter only clear sky radiation has been discussed. For most locations clear sky is an unusual situation and some cloud cover is normal. Figure 3-10 shows a continuous horizontal pyranometer record of a nominally clear spring

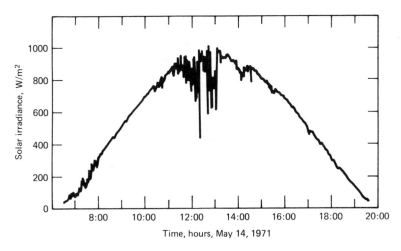

**FIGURE 3-10.** Global irradiance due to the sun and sky on a horizontal surface, measured at Greenbelt, Maryland, on 14 May, 1971. Total energy received during the day, 27.07 MJ/m² (Ref. 2).

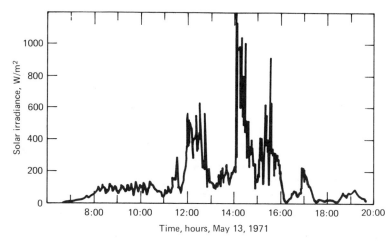

**FIGURE 3-11.** Global irradiance due to the sun and sky on a horizontal surface, measured at Greenbelt, Maryland, on 13 May, 1971. Total energy received during the day, 7.32 MJ/m² (Ref. 2).

day at Greenbelt, Maryland, a suburb of the District of Columbia. The peak irradiance is about 1000 W/m², but near noon there were occasional clouds, probably of the fair weather cumulus type.

A similar record from the preceding day, which was overcast, is shown in Fig. 3-11. The sun did come out briefly in midafternoon. Since flat plate collectors will operate above 200 to 400 W/m², both days would produce useful solar energy. It is interesting to note that the direct beam plus cloud cover produced occasional readings well over 1000 W/m² due to added reflected light, whereas the clear sky maximum at that time of day is about 900 W/m².

Given these two extremes of cloud cover plus all the intermediate situations, it is difficult to decide what information should be retained as a permanent record. Average hourly *global* irradiance on a horizontal surface or the *daily* global radiation total has been the standard format for National Weather Service (NWS) Data, which has been collected for many years. Hourly information is necessary for exact design but several approximate techniques have been devised to utilize the monthly average of the daily total radiation for system design. Since 1950 many stations have been involved (see Fig. 3-12) but by 1973 only 67 sites were providing daily totals and of these only 29 kept both hourly and daily totals. The data archived are not of good quality, however, because little maintenance was given most of the instruments after about 1960, and they drifted out of calibration, with errors estimated to be at least ±5% on up to ±30%. As a result there are probably a half-dozen or fewer stations in the United States with good hourly radiation data over a period of 10 years or longer.

**FIGURE 3-12.** Location of solar radiation stations with data archived at the National Climatic Center, Asheville, North Carolina. Numbers indicate more than one station in area. Data taken after 1960 have errors of from 5 to 30%. (Ref. 34)

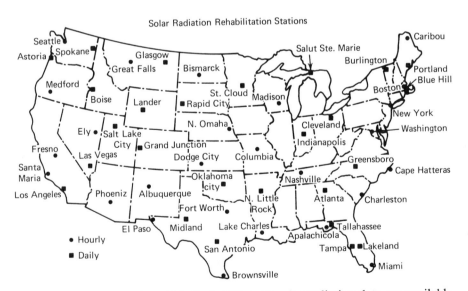

**FIGURE 3-13.** Stations for which "rehabilitated" solar radiation data are available. (Ref. 34)

To meet the new demand for radiation data created by the interest in alternate sources of energy, the NWS has "rehabilitated" the carelessly accumulated radiation data for the 52 stations shown in Fig. 3-13.* The instruments have been calibrated in hindsight by comparing recent clear day records with calculated clear day performance to estimate instrument drift. This method deals only superficially with the buildup of atmospheric pollutants that took place over most of the United States during the 1950–1970 period. Since the drift of the instruments

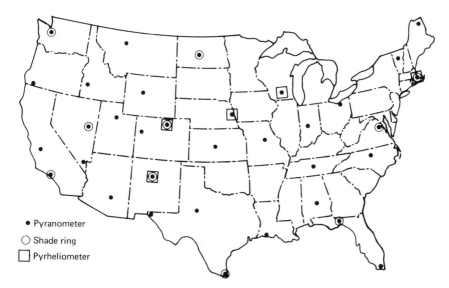

● Pyranometer
○ Shade ring
□ Pyrheliometer

**FIGURE 3-14.** New National Weather Service solar radiation station network (station at Fairbanks, Alaska, not indicated). (Ref. 34)

appears to have been not slow and steady but erratic, the accuracy of rehabilitated data is difficult to determine but has been estimated at $\pm 10\%$ (27).

A new well-maintained National Weather Service solar radiation station network has now been set up for the 35 U.S. locations shown in Fig. 3-14 plus Fairbanks, Alaska. It is anticipated that these data will be of highest quality, but since there is not even one station per state the network is rather sparse. Since it will be 1986 before 10 years data are accumulated, designers will have to look elsewhere for the 10-year data base desirable for solar system design.

**Sunshine Records.** The extensive U.S. network for measuring of the percentage of possible sunshine is shown in Fig. 3-15. Table 3-4 summarizes the long-term sunshine data (12).

* Twenty-six stations have been rehabilitated on an hourly basis, and a like number on a daily basis.

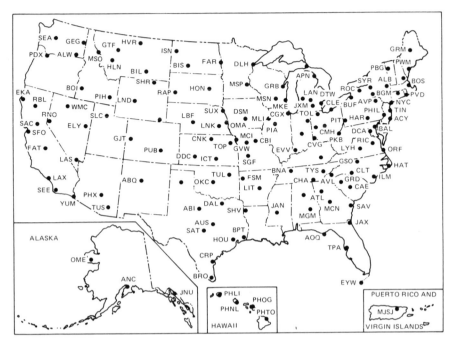

**FIGURE 3-15.** National Weather Service sunshine network 1973. (*Source:* From Edward Jessup, "A Brief History of the Solar Radiation Program," Ref. 9).

The utility of such sunshine data is controversial, but it is clear that there is no simple correlation with reported horizontal radiation data (13). If the *maximum* monthly clear sky radiation total (Table A-13) for a tilted collector at the proper latitude is multiplied by the percentage of sunshine (Table 3-4) and the clearness number (Fig. 3-8), the result approximates the total global radiation received above 15 W/m² on that surface, and the total global radiation on the *tilted* surface is about 1.08 times that value. The total global radiation on a horizontal surface may then be calculated from this figure. Validation of this technique must await a better data base, but from limited study I estimate that this method predicts winter radiation values within about 10%. For best summer accuracy, Table A-13 data on a collector tilt equal to the latitude is recommended [see Examples 3-7(*b*) and 3-11].

**Monthly Average Data.** In the form of average daily radiation on a horizontal surface, monthly records are available for a large number of U.S. and world locations. Löf, Duffie, and Smith surveyed these data in 1966 (14). Much of the

* This is the utilizable radiation, discussed later in this chapter.

**TABLE 3-4.** Mean Percentage of Possible Sunshine for Selected Locations[a] (Ref. 6)

| State and Station | Jan | Feb | Mar | Apr | May | Jun | Jul | Aug | Sept | Oct | Nov | Dec | Annual |
|---|---|---|---|---|---|---|---|---|---|---|---|---|---|
| Ala., Birmingham | 43 | 49 | 56 | 63 | 66 | 67 | 62 | 65 | 66 | 67 | 58 | 44 | 59 |
| Montgomery | 51 | 53 | 61 | 69 | 73 | 72 | 66 | 69 | 69 | 71 | 64 | 48 | 64 |
| Alaska, Anchorage | 39 | 46 | 56 | 58 | 50 | 51 | 45 | 39 | 35 | 32 | 33 | 29 | 45 |
| Fairbanks | 34 | 50 | 61 | 68 | 55 | 53 | 45 | 35 | 31 | 28 | 38 | 29 | 44 |
| Juneau | 30 | 32 | 39 | 37 | 34 | 35 | 28 | 30 | 25 | 18 | 21 | 18 | 30 |
| Nome | 44 | 46 | 48 | 53 | 51 | 48 | 32 | 26 | 34 | 35 | 36 | 30 | 41 |
| Ariz., Phoenix | 76 | 79 | 83 | 88 | 93 | 94 | 84 | 84 | 89 | 88 | 84 | 77 | 85 |
| Yuma | 83 | 87 | 91 | 94 | 97 | 98 | 92 | 91 | 93 | 93 | 90 | 83 | 91 |
| Ark., Little Rock | 44 | 53 | 57 | 62 | 67 | 72 | 71 | 74 | 71 | 74 | 58 | 47 | 62 |
| Calif., Eureka | 40 | 44 | 50 | 53 | 54 | 56 | 51 | 46 | 52 | 48 | 42 | 39 | 49 |
| Fresno | 46 | 63 | 72 | 83 | 89 | 94 | 97 | 97 | 93 | 87. | 73 | 47 | 78 |
| Los Angeles | 70 | 69 | 70 | 67 | 68 | 69 | 80 | 81 | 80 | 76 | 79 | 72 | 73 |
| Red Bluff | 50 | 60 | 65 | 75 | 79 | 86 | 95 | 94 | 89 | 77 | 64 | 50 | 75 |
| Sacramento | 44 | 57 | 67 | 76 | 82 | 90 | 96 | 95 | 92 | 82 | 65 | 44 | 77 |
| San Diego | 68 | 67 | 68 | 66 | 60 | 60 | 67 | 70 | 70 | 70 | 76 | 71 | 68 |
| San Francisco | 53 | 57 | 63 | 69 | 70 | 75 | 68 | 63 | 70 | 70 | 62 | 54 | 66 |
| Colo., Denver | 67 | 67 | 65 | 63 | 61 | 69 | 68 | 68 | 71 | 71 | 67 | 65 | 67 |
| Grand Junction | 58 | 62 | 64 | 67 | 71 | 79 | 76 | 72 | 77 | 74 | 67 | 58 | 69 |
| Conn., Hartford | 46 | 55 | 56 | 54 | 57 | 60 | 62 | 60 | 57 | 55 | 46 | 46 | 56 |
| D.C., Washington | 46 | 53 | 56 | 57 | 61 | 64 | 64 | 62 | 62 | 61 | 54 | 47 | 58 |
| Fla., Apalachicola | 59 | 62 | 62 | 71 | 77 | 70 | 64 | 63 | 62 | 74 | 66 | 53 | 65 |
| Jacksonville | 58 | 59 | 66 | 71 | 71 | 63 | 62 | 63 | 58 | 58 | 61 | 53 | 62 |
| Key West | 68 | 75 | 78 | 78 | 76 | 70 | 69 | 71 | 65 | 65 | 69 | 66 | 71 |
| Miami Beach | 56 | 72 | 73 | 73 | 68 | 62 | 65 | 67 | 62 | 62 | 65 | 65 | 67 |
| Tampa | 63 | 67 | 71 | 74 | 75 | 66 | 61 | 64 | 64 | 67 | 67 | 61 | 68 |
| Ga., Atlanta | 48 | 53 | 57 | 65 | 68 | 68 | 62 | 63 | 65 | 67 | 60 | 47 | 60 |
| Hawaii, Hilo | 48 | 42 | 41 | 34 | 31 | 41 | 44 | 38 | 42 | 41 | 34 | 36 | 39 |
| Honolulu | 62 | 64 | 60 | 62 | 64 | 66 | 67 | 70 | 70 | 68 | 63 | 60 | 65 |
| Lihue | 48 | 48 | 48 | 46 | 51 | 60 | 58 | 59 | 67 | 58 | 51 | 49 | 54 |
| Idaho, Boise | 40 | 48 | 59 | 67 | 68 | 75 | 89 | 86 | 81 | 66 | 46 | 37 | 66 |
| Pocatello | 37 | 47 | 58 | 64 | 66 | 72 | 82 | 81 | 78 | 66 | 48 | 36 | 64 |
| Ill., Cairo | 46 | 53 | 59 | 65 | 71 | 77 | 82 | 79 | 75 | 73 | 56 | 46 | 65 |
| Chicago | 44 | 49 | 53 | 56 | 63 | 69 | 73 | 70 | 65 | 61 | 47 | 41 | 59 |
| Springfield | 47 | 51 | 54 | 58 | 64 | 69 | 76 | 72 | 73 | 64 | 53 | 45 | 60 |
| Ind., Evansville | 42 | 49 | 55 | 61 | 67 | 73 | 78 | 76 | 73 | 67 | 52 | 42 | 64 |
| Ft. Wayne | 38 | 44 | 51 | 55 | 62 | 69 | 74 | 69 | 64 | 58 | 41 | 38 | 57 |
| Indianapolis | 41 | 47 | 49 | 55 | 62 | 68 | 74 | 70 | 68 | 64 | 48 | 39 | 59 |
| Iowa, Des Moines | 56 | 56 | 56 | 59 | 62 | 66 | 75 | 70 | 64 | 64 | 53 | 48 | 62 |
| Dubuque | 48 | 52 | 52 | 58 | 60 | 63 | 73 | 67 | 61 | 55 | 44 | 40 | 57 |
| Sioux City | 55 | 58 | 58 | 59 | 63 | 67 | 75 | 72 | 67 | 65 | 53 | 50 | 63 |
| Kans., Concordia | 60 | 60 | 62 | 63 | 65 | 73 | 79 | 76 | 72 | 70 | 64 | 58 | 67 |
| Dodge City | 67 | 66 | 68 | 68 | 68 | 74 | 78 | 78 | 76 | 75 | 70 | 67 | 71 |
| Wichita | 61 | 63 | 64 | 64 | 66 | 73 | 80 | 77 | 73 | 69 | 67 | 59 | 69 |
| Ky., Louisville | 41 | 47 | 52 | 57 | 64 | 68 | 72 | 69 | 68 | 64 | 51 | 39 | 59 |
| La., New Orleans | 49 | 50 | 57 | 63 | 66 | 64 | 58 | 60 | 64 | 70 | 60 | 46 | 59 |
| Shreveport | 48 | 54 | 58 | 60 | 69 | 78 | 79 | 80 | 79 | 77 | 65 | 60 | 69 |
| Maine, Eastport | 45 | 51 | 52 | 52 | 51 | 53 | 55 | 57 | 54 | 50 | 37 | 40 | 50 |
| Mass., Boston | 47 | 56 | 57 | 56 | 59 | 62 | 64 | 63 | 61 | 58 | 48 | 48 | 57 |
| Mich., Alpena | 29 | 43 | 52 | 56 | 59 | 64 | 70 | 64 | 52 | 44 | 24 | 22 | 51 |
| Detroit | 34 | 42 | 48 | 52 | 58 | 65 | 69 | 66 | 61 | 54 | 35 | 29 | 53 |
| Grand Rapids | 26 | 37 | 48 | 54 | 60 | 66 | 72 | 67 | 58 | 50 | 31 | 22 | 49 |
| Marquette | 31 | 40 | 47 | 52 | 53 | 56 | 63 | 57 | 47 | 38 | 24 | 24 | 47 |
| S. Ste. Marie | 28 | 44 | 50 | 54 | 54 | 59 | 63 | 58 | 45 | 36 | 21 | 22 | 47 |
| Minn., Duluth | 47 | 55 | 60 | 58 | 58 | 60 | 68 | 63 | 53 | 47 | 36 | 40 | 55 |
| Minneapolis | 49 | 54 | 55 | 57 | 60 | 64 | 72 | 69 | 60 | 54 | 40 | 40 | 56 |

*(Continued)*

**TABLE 3-4.** (*Continued*)

| State and Station | Jan | Feb | Mar | Apr | May | Jun | Jul | Aug | Sept | Oct | Nov | Dec | Annual |
|---|---|---|---|---|---|---|---|---|---|---|---|---|---|
| Miss., Vicksburg | 46 | 50 | 57 | 64 | 69 | 73 | 69 | 72 | 74 | 71 | 60 | 45 | 64 |
| Mo., Kansas City | 55 | 57 | 59 | 60 | 64 | 70 | 76 | 73 | 70 | 67 | 59 | 52 | 65 |
| St. Louis | 48 | 49 | 56 | 59 | 64 | 68 | 72 | 68 | 67 | 65 | 54 | 44 | 61 |
| Springfield | 48 | 54 | 57 | 60 | 63 | 69 | 77 | 72 | 71 | 65 | 58 | 48 | 63 |
| Mont. Havre | 49 | 58 | 61 | 63 | 63 | 65 | 78 | 75 | 64 | 57 | 48 | 46 | 62 |
| Helena | 46 | 55 | 58 | 60 | 59 | 63 | 77 | 74 | 63 | 57 | 48 | 48 | 60 |
| Kalispell | 28 | 40 | 49 | 57 | 58 | 60 | 77 | 73 | 61 | 60 | 28 | 20 | 53 |
| Nebr., Lincoln | 57 | 59 | 60 | 60 | 63 | 69 | 76 | 71 | 67 | 66 | 59 | 55 | 64 |
| North Platte | 63 | 63 | 64 | 62 | 63 | 72 | 78 | 74 | 72 | 70 | 62 | 58 | 68 |
| Nev., Ely | 61 | 64 | 68 | 65 | 67 | 79 | 79 | 81 | 81 | 73 | 67 | 62 | 72 |
| Las Vegas | 74 | 77 | 78 | 81 | 85 | 91 | 84 | 86 | 92 | 84 | 83 | 75 | 82 |
| Reno | 59 | 64 | 69 | 75 | 77 | 82 | 90 | 89 | 86 | 76 | 68 | 56 | 76 |
| Winnemucca | 52 | 60 | 64 | 70 | 76 | 83 | 90 | 90 | 86 | 75 | 62 | 53 | 74 |
| N.H., Concord | 48 | 53 | 55 | 53 | 51 | 56 | 57 | 58 | 55 | 50 | 43 | 43 | 52 |
| N.J., Atlantic City | 51 | 57 | 58 | 59 | 62 | 65 | 67 | 66 | 65 | 54 | 58 | 52 | 60 |
| N. Mex., Albuquerque | 70 | 72 | 72 | 76 | 79 | 84 | 76 | 75 | 81 | 80 | 79 | 70 | 76 |
| Roswell | 69 | 72 | 75 | 77 | 76 | 80 | 76 | 75 | 74 | 74 | 74 | 69 | 74 |
| N.Y., Albany | 43 | 51 | 53 | 53 | 57 | 62 | 63 | 61 | 58 | 54 | 39 | 38 | 53 |
| Binghamton | 31 | 39 | 41 | 44 | 50 | 56 | 54 | 51 | 47 | 43 | 29 | 26 | 44 |
| Buffalo | 32 | 41 | 49 | 51 | 59 | 67 | 70 | 67 | 60 | 51 | 31 | 28 | 53 |
| Canton | 37 | 47 | 50 | 48 | 54 | 61 | 63 | 61 | 54 | 45 | 40 | 31 | 49 |
| New York | 49 | 56 | 57 | 59 | 62 | 65 | 66 | 64 | 64 | 61 | 53 | 50 | 59 |
| Syracuse | 31 | 38 | 45 | 50 | 58 | 64 | 67 | 63 | 56 | 47 | 29 | 26 | 50 |
| N.C., Asheville | 48 | 53 | 56 | 61 | 64 | 63 | 59 | 59 | 62 | 64 | 59 | 48 | 58 |
| Raleigh | 50 | 56 | 59 | 64 | 67 | 65 | 62 | 62 | 63 | 64 | 62 | 52 | 61 |
| N. Dak., Bismarck | 52 | 58 | 56 | 57 | 58 | 61 | 73 | 69 | 62 | 59 | 49 | 48 | 59 |
| Devils Lake | 53 | 60 | 59 | 60 | 59 | 62 | 71 | 67 | 59 | 56 | 44 | 45 | 58 |
| Fargo | 47 | 55 | 56 | 58 | 62 | 63 | 73 | 69 | 60 | 57 | 39 | 46 | 59 |
| Williston | 51 | 59 | 60 | 63 | 66 | 66 | 78 | 75 | 65 | 60 | 48 | 48 | 63 |
| Ohio, Cincinnati | 41 | 46 | 52 | 56 | 62 | 69 | 72 | 68 | 68 | 60 | 46 | 39 | 57 |
| Cleveland | 29 | 36 | 45 | 52 | 61 | 67 | 71 | 68 | 62 | 54 | 32 | 25 | 50 |
| Columbus | 36 | 44 | 49 | 54 | 63 | 68 | 71 | 68 | 66 | 60 | 44 | 35 | 55 |
| Okla., Oklahoma City | 57 | 60 | 63 | 64 | 65 | 74 | 78 | 78 | 74 | 68 | 64 | 57 | 68 |
| Oreg., Baker | 41 | 49 | 56 | 61 | 63 | 67 | 83 | 81 | 74 | 62 | 46 | 37 | 60 |
| Portland | 27 | 34 | 41 | 49 | 52 | 55 | 70 | 65 | 55 | 42 | 28 | 23 | 48 |
| Roseburg | 24 | 32 | 40 | 51 | 57 | 59 | 79 | 77 | 68 | 42 | 28 | 18 | 51 |
| Pa., Harrisburg | 43 | 52 | 55 | 57 | 61 | 65 | 68 | 63 | 62 | 58 | 47 | 43 | 57 |
| Philadelphia | 45 | 56 | 57 | 58 | 61 | 62 | 64 | 61 | 62 | 61 | 53 | 49 | 57 |
| Pittsburg | 32 | 39 | 45 | 50 | 57 | 62 | 64 | 61 | 62 | 54 | 39 | 30 | 51 |
| R.I., Block Island | 45 | 54 | 47 | 56 | 58 | 60 | 62 | 62 | 60 | 59 | 50 | 44 | 56 |
| S.C., Charleston | 58 | 60 | 65 | 72 | 73 | 66 | 66 | 66 | 67 | 68 | 68 | 57 | 66 |
| Columbia | 53 | 57 | 62 | 68 | 69 | 68 | 63 | 65 | 64 | 68 | 64 | 51 | 63 |
| S. Dak., Huron | 55 | 62 | 60 | 62 | 65 | 68 | 76 | 72 | 66 | 61 | 52 | 49 | 63 |
| Rapid City | 58 | 62 | 63 | 62 | 61 | 66 | 73 | 73 | 69 | 66 | 58 | 54 | 64 |
| Tenn., Knoxville | 42 | 49 | 53 | 59 | 64 | 66 | 64 | 59 | 64 | 64 | 53 | 41 | 57 |
| Memphis | 44 | 51 | 57 | 64 | 68 | 74 | 73 | 74 | 70 | 69 | 58 | 45 | 64 |
| Nashville | 42 | 47 | 54 | 60 | 65 | 69 | 59 | 68 | 69 | 65 | 55 | 42 | 59 |
| Tex., Abilene | 64 | 68 | 73 | 66 | 73 | 86 | 83 | 85 | 73 | 71 | 72 | 66 | 73 |
| Amarillo | 71 | 71 | 75 | 75 | 75 | 82 | 81 | 81 | 79 | 76 | 76 | 70 | 76 |
| Austin | 46 | 50 | 57 | 60 | 62 | 72 | 76 | 79 | 70 | 70 | 57 | 49 | 63 |
| Brownsville | 44 | 49 | 51 | 57 | 65 | 73 | 78 | 78 | 67 | 70 | 54 | 44 | 61 |
| Del Rio | 53 | 55 | 61 | 63 | 60 | 66 | 75 | 80 | 69 | 66 | 58 | 52 | 63 |
| El Paso | 74 | 77 | 81 | 85 | 87 | 87 | 78 | 78 | 80 | 82 | 80 | 73 | 80 |
| Ft. Worth | 56 | 57 | 65 | 66 | 67 | 75 | 78 | 78 | 74 | 70 | 63 | 58 | 68 |
| Galveston | 50 | 50 | 55 | 61 | 69 | 76 | 72 | 71 | 70 | 74 | 62 | 49 | 63 |

88

(*Continued*)

**TABLE 3-4.** (*Continued*)

| State and Station | Jan | Feb | Mar | Apr | May | Jun | Jul | Aug | Sept | Oct | Nov | Dec | Annual |
|---|---|---|---|---|---|---|---|---|---|---|---|---|---|
| San Antonio | 48 | 51 | 56 | 58 | 60 | 69 | 74 | 75 | 69 | 67 | 55 | 49 | 62 |
| Utah, Salt Lake City | 48 | 53 | 61 | 68 | 73 | 78 | 82 | 82 | 84 | 73 | 56 | 49 | 69 |
| Vt., Burlington | 34 | 43 | 48 | 47 | 53 | 59 | 62 | 59 | 51 | 43 | 25 | 24 | 46 |
| Va., Norfolk | 50 | 57 | 60 | 63 | 67 | 66 | 66 | 66 | 63 | 64 | 60 | 51 | 62 |
| Richmond | 49 | 55 | 59 | 63 | 67 | 66 | 65 | 62 | 63 | 64 | 58 | 50 | 61 |
| Wash., North Head | 28 | 37 | 42 | 48 | 48 | 48 | 50 | 46 | 48 | 41 | 31 | 27 | 41 |
| Seattle | 27 | 34 | 42 | 48 | 53 | 48 | 62 | 56 | 53 | 36 | 28 | 24 | 45 |
| Spokane | 26 | 41 | 53 | 63 | 64 | 68 | 82 | 79 | 68 | 53 | 28 | 22 | 58 |
| Tatoosh Island | 26 | 36 | 39 | 45 | 47 | 46 | 48 | 44 | 47 | 38 | 26 | 23 | 40 |
| Walla Walla | 24 | 35 | 51 | 63 | 67 | 72 | 86 | 84 | 72 | 59 | 33 | 20 | 60 |
| Yakima | 34 | 49 | 62 | 70 | 72 | 74 | 86 | 86 | 74 | 61 | 38 | 29 | 65 |
| W. Va., Elkins | 33 | 37 | 42 | 47 | 55 | 55 | 56 | 53 | 55 | 51 | 41 | 33 | 48 |
| Parkersburg | 30 | 36 | 42 | 49 | 56 | 60 | 63 | 60 | 60 | 53 | 37 | 29 | 48 |
| Wis., Green Bay | 44 | 51 | 55 | 58 | 58 | 64 | 70 | 65 | 58 | 52 | 40 | 40 | 55 |
| Madison | 44 | 49 | 52 | 53 | 58 | 64 | 70 | 66 | 60 | 56 | 41 | 38 | 56 |
| Milwaukee | 44 | 48 | 53 | 56 | 60 | 65 | 73 | 67 | 62 | 56 | 44 | 39 | 57 |
| Wyo., Cheyenne | 65 | 66 | 64 | 61 | 59 | 68 | 70 | 68 | 69 | 69 | 65 | 63 | 66 |
| Lander | 66 | 70 | 71 | 66 | 65 | 74 | 76 | 75 | 72 | 67 | 61 | 62 | 69 |
| Sheridan | 56 | 61 | 62 | 61 | 61 | 67 | 76 | 74 | 67 | 60 | 53 | 52 | 64 |
| Yellowstone Park | 39 | 51 | 55 | 57 | 56 | 63 | 73 | 71 | 65 | 57 | 45 | 38 | 56 |
| P.R., San Juan | 64 | 69 | 71 | 66 | 59 | 62 | 65 | 67 | 61 | 63 | 63 | 65 | 65 |

NOTE. Based on period of record through December 1959, except in a few instances.
[a] These charts and tabulation derived from "Normals, Means, and Extremes" table in U.S. Weather Bureau publication *Local Climatological Data* (12).

U.S. data is available in the *Climatic Atlas of the United States* (12), which provides the basic data for most of the other monthly tabular data summaries such as those listed in *ASHRAE GRP 170* (4b). The data available are of varying accuracy, however, and contain occasional inconsistencies that can make city-by-city comparisons very misleading. Such monthly averaged horizontal radiation and climatic data were recently gathered by Klein for 171 North American cities (15). Readers are cautioned to verify the accuracy of chosen design data before making an actual application.

Under contract to the Department of Energy, the National Climatic Center in Asheville, North Carolina, has recently prepared the most exhaustive compilation of U.S. radiation data yet available (30). Based on the 26 rehabilitated data stations discussed earlier, data for 222 stations were calculated from a statistical model based on cloud cover (but not cloud type) using coefficients derived at the nearest of the rehabilitated sites. While the results are variable, depending on the consistency of the underlying data base and the similarity of the climate at the reported site to that of the rehabilitated station, there is generally 5 to 15% less radiation than listed in other sources. Figure 3-16 shows the locations of these stations, for which tabular data have been summarized in Table A-15.*

* References 15 and 30 list average daily temperatures. Average daytime temperatures for the winter months, used for solar calculations, can be estimated by adding 1 °C in coastal climates, 2 °C inland, and 3 °C in desert locations. References 4b and Table A-15 list daytime averages that have been calculated by summing 0.7 times the average temperature and 0.3 times the average daily maximum temperature.

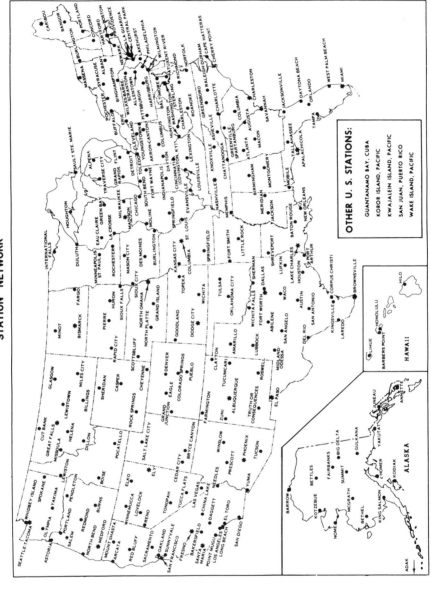

**STATION NETWORK**

**OTHER U. S. STATIONS:**

GUANTANAMO BAY, CUBA
KOROR ISLAND, PACIFIC
KWAJALEIN ISLAND, PACIFIC
SAN JUAN, PUERTO RICO
WAKE ISLAND, PACIFIC

**FIGURE 3-16.** Weather stations for which solar radiation data are presented in Ref. 30 and Appendix table A-15. Asterisks show rehabilitated data sites; data for the remainder have been calculated from a statistical correlation with cloud amount (but not type) in the meteorological records. (Ref. 35).

## MONTHLY AVERAGE RADIATION ON TILTED SURFACES

Practical use of monthly horizontal surface data requires the estimation of the corresponding radiation on the tilted surface of the solar collector. This is not a simple process because elevation of the direct and diffuse components involves a different procedure (eqs. 3-12 and 3-14), and generally global radiation data are all that are available. The procedure (16) requires calculation of the proportion of the potential extraterrestrial radiation available daily at the site, from which an empirical correlation then permits estimation of the diffuse component. To find the daily total extraterrestrial radiation, eq. 3-12 must be integrated over the daylight hours as the incident angle $\theta$ changes. The ratio of the extraterrestrial radiation on the tilted and horizontal surfaces is used to "elevate" the direct component, as described in the next section. Then the diffuse component is elevated, just as for clear sky data, and the two are summed to give the global radiation on the tilted surface.

### Extraterrestrial Radiation on Horizontal and Tilted Surfaces

**Horizontal Surfaces.** Equation 3-1 yields values of the extraterrestrial solar irradiance $I_0$ throughout the year. The extraterrestrial irradiance on a horizontal or tilted surface is found by combining eq. 3-1 with eq. 3-12 to give

$$I_{\text{extraterrestrial}} = I_0 \cos \theta \qquad (3\text{-}15)$$

where $\cos \theta$ is given by eq. 3-9, or by eq. 3-11 using the sun's altitude $\alpha$ and azimuth $\psi$ from eqs. 3-5 and 3-6, respectively. For a horizontal surface $\theta$ equals $\theta_z$, so the zenith angle and eq. 3-5 can be combined with eq. 3-12 to give the extraterrestrial horizontal irradiance,

$$H_0 = I_0[\cos \phi \cos \delta \cos \omega + \sin \phi \sin \delta] \qquad (3\text{-}16)$$

The total daily extraterrestrial radiation is determined by integrating eq. 3-16 over the solar day, giving

$$H_{0T} = \frac{I_{0(\text{daily sum})}}{\pi} [\cos \phi \cos \delta \sin \omega_s + (2\pi\omega_s/360) \sin \phi \sin \delta] \qquad (3\text{-}17)$$

where $I_{0(\text{daily sum})}$ is the total direct normal extraterrestrial radiation. If $I_0$ is expressed as radiation flux (e.g., Btu/hr ft$^2$ or W/m$^2$), then for a 24 hour period

$$I_{0(\text{daily sum})} = 24I_0 \qquad (3\text{-}17a)$$

for which the dimensions of $I_{0(\text{daily sum})}$ are in Btu/ft$^2$ or W·h/m$^2$. However, since a watt is one joule per second, the dimensions of W·h/m$^2$ are really (J/s) ×

hr/m², which is inconsistent with the SI system of units. Therefore, if $H_{0T}$ is to be expressed in SI energy units, then hours must be converted to seconds, giving

$$I_{0(\text{daily sum})} = \frac{24 \text{ hr}}{\text{day}} \times \frac{60 \text{ min}}{\text{hr}} \times \frac{60 \text{ sec}}{\text{min}} \times \frac{\text{MJ}}{1\,000\,000 \text{ J}} \times I_0$$

$$= 24 \times 0.0036 I_0 \tag{3-17b}$$

where $I_{0(\text{daily sum})}$ is now given in MJ/m², a conveniently sized unit for expressing the area normalized total energy.

In eq. 3-17, $\omega_s$ is the sunrise–sunset hour angle as derived from eq. 3-8, expressed as *degrees*. To use eq. 3-17 for accurate calculation of the monthly daily-average total extraterrestrial horizontal radiation $\bar{H}_{0T}$, a monthly average declination angle is necessary. This will be obtained if the dates listed in Table 3-3 are used in eqs. 3-1 and 3-2.

### EXAMPLE 3-5

On November 21 at 10 A.M. at 40°N latitude, problem 3-2(a) showed that the declination was $-20.0°$ and the hour angle $-30°$.

(a) What is the direct normal extraterrestrial irradiance?

From eq. 3-1,

$$I_0 = 1353 \left[ 1 + 0.033 \left( \cos \frac{360 \times 325}{370} \right) \right] = 1385 \text{ W/m}^2$$

(b) What is the extraterrestrial horizontal irradiance (i.e., on a surface parallel with the ground)?

Substituting in eq. 3-16,

$$H_0 = 1385[\cos 40° \cos(-20°) \cos(-30°) + \sin 40° \sin(-20°)]$$
$$= 559 \text{ W/m}^2$$

(c) What is the total extraterrestrial radiation received that day on the horizontal surface?

The sunset hour angle is given by eq. 3-8:

$$\omega_s = \arccos[-\tan 40° \tan(-20°)] = 72.2°$$

and the daily total radiation by eq. 3-17b and 3-17:

$$I_{0(\text{daily sum})} = 24 \times 1385 \times 0.0036 = 119.7 \text{ MJ/m}^2$$

$$H_{0T} = \frac{119.7}{\pi} \left[ \cos 40° \cos(-20.0°) \sin 72.2° \right.$$

$$\left. + \left( \frac{2 \times 72.2 \times \pi}{360} \right) \sin 40° \sin(-20°) \right] = 15.56 \text{ MJ/m}^2$$

**Tilted Surfaces.** If the total extraterrestrial radiation on a south-facing tilted surface is desired, eq. 3-9 can be substituted into eq. 3-12 and integrated over the solar day, giving

$$I_{0T} = \frac{I_{0(\text{daily sum})}}{\pi} [\cos (\phi - \beta) \cos \delta \sin \tilde{\omega}_s + (2\tilde{\omega}_s \pi / 360) \sin (\phi - \beta) \sin \delta]$$

$$(3\text{-}18)$$

where $\tilde{\omega}_s$ is the time of sunset on the tilted surface, that is, the minimum of $\omega_s$ (eq. 3-8) and $\omega'_s$ (eq. 3-10). The monthly daily-average total extraterrestrial radiation on the tilted surface $\bar{I}_{0T}$ can be calculated by using a proper monthly date from Table 3-3 for calculation of the declination angle $\delta$.

Equation 3-18 is written for values of $\tilde{\omega}_s$ expressed in *degrees*, not in units of time. This is consistent with the use of eq. 3-8 rather than 3-8a to find the "time" of sunrise and sunset.

### EXAMPLE 3-6

For a latitude of 40°N during the month of November,
   (a) Find the daily average extraterrestrial radiation on a horizontal surface.

Using day 318 from Table 3-3, eq. 3-1 gives

$$I_0 = 1353 \left[ 1 + 0.033 \cos \left( \frac{360 \times 318}{370} \right) \right] = 1381 \text{ W/m}^2$$

and by eq. 3-17b

$$I_{0(\text{daily sum})} = 1381 \times 24 \times 0.0036 = 119.3 \text{ MJ/m}^2$$

From eq. 3-2,

$$\delta = 23.45 \sin \left( \frac{318 - 80}{370} \times 360 \right) = -18.37°$$

From eq. 3-8,

$$\cos \omega_s = -\tan 40° \tan (-18.37°) = 0.2786$$
$$\omega_s = 73.82°$$

from eq. 3-17

$$\bar{H}_{0T} = \frac{119.3 \text{ MJ/m}^2}{\pi} \left[ \cos 40° \cos (-18.37°) \sin 73.82° \right.$$

$$\left. + \left( \frac{2 \times 73.82\pi}{360} \right) \sin 40° \sin (-18.37°) \right] = 16.61 \text{ MJ/m}^2$$

Notice that the *daily average* November radiation is 6% greater than the radiation calculated for November 21 in Example 3-5.

   (b) Find the daily average extraterrestrial radiation on a surface tilted upward 60° from the horizontal.

The sunset on the tilted collector occurs at the hour angle given by eq. 3-10 as

$$\cos \omega_s' = -\tan (40° - 60°) \tan (-18.37°) = -0.1209$$
$$\omega_s' = 96.94°$$

Since sunset on the horizon occurs earlier at $\omega_s = 73.82°$ hour angle, actual sunset for the collector occurs at $\tilde{\omega}_s = 73.82°$.

From eq. 3-18, the total daily radiation is

$$\bar{I}_{0T} = \frac{119.3 \text{ MJ/m}^2}{\pi} [\cos (40° - 60°) \cos (-18.37°) \sin 73.82°$$
$$+ (2 \times 73.82\pi/360) \sin (40° - 60°) \sin (-18.37°)]$$
$$= 37.81 \text{ MJ/m}^2$$

Table 3-5 lists the monthly daily average extraterrestrial radiation computed as in Example 3-6(a) for horizontal surfaces for a wide range of latitudes. Table A-14 lists the values for surfaces oriented 0°, 20°, 30°, 40°, 50°, 60°, 70°, and 90° to the horizontal for each degree of latitude. These tables are useful to estimate the ratio of the daily total direct or *beam* radiation on a tilted surface to that on a horizontal surface. This is the ratio of eq. 3-18 to eq. 3-17:

$$\bar{R}_b = \frac{\bar{I}_{0T}}{\bar{H}_{0T}} \approx \frac{\bar{I}_{Tb}}{\bar{H}_{Tb}} \tag{3-19}$$

$\bar{R}_b$ is an elevation factor used to elevate the direct beam component of the total radiation. Equation 3-19 is not exact for this purpose except at the equinoxes because late morning and afternoon in winter and summer can have a different proportion of direct beam radiation than when the sun is travelling a great circle through the sky. Overestimated direct radiation causes winter values of $\bar{R}_b$ to be 5 to 10% higher than the true values. This can be confirmed by hourly use of the techniques of Examples 3-2 to 3-4 to calculate total daily *direct* radiation on a tilted and a horizontal surface (i.e., with $C$ set to zero). The ratio of these totals is the true $\bar{R}_b$.

## Diffuse Radiation

Since most radiation data are global, it is necessary to determine the proportion of the radiation that is diffuse before the radiation on a tilted surface can be found. Obviously, the cloudier the climate, the higher the proportion of diffuse radiation. Page (13) and Liu and Jordan (17) have found that the proportions of diffuse radiation in horizontally collected monthly radiation data is a function of the clearness index $\bar{K}_T$, the proportion of the horizontal extraterrestrial radiation reaching the site. $\bar{K}_T$ is the ratio of the measured monthly total global radiation

**TABLE 3-5.** Monthly Average Daily Extraterrestrial Radiation on Horizontal Surfaces, MJ/m² (Northern Latitudes)

| Latitude Degrees North | Month Jan Day No. 17. | Feb 46. | Mar 76. | Apr 105. | May 135. | Jun 162. | Jul 199. | Aug 229. | Sept 258. | Oct 288. | Nov 318. | Dec 346. |
|---|---|---|---|---|---|---|---|---|---|---|---|---|
| 14° | 29.90 | 32.89 | 36.03 | 37.73 | 38.01 | 37.76 | 37.68 | 37.54 | 36.36 | 33.70 | 30.55 | 28.74 |
| 15° | 29.41 | 32.51 | 35.83 | 37.74 | 38.19 | 38.02 | 37.90 | 37.62 | 36.24 | 33.39 | 30.10 | 28.22 |
| 16° | 28.91 | 32.12 | 35.63 | 37.74 | 38.36 | 38.26 | 38.11 | 37.68 | 36.12 | 33.06 | 29.63 | 27.69 |
| 17° | 28.40 | 31.73 | 35.42 | 37.73 | 38.52 | 38.50 | 38.31 | 37.74 | 35.99 | 32.73 | 29.16 | 27.15 |
| 18° | 27.89 | 31.32 | 35.19 | 37.71 | 38.67 | 38.73 | 38.51 | 37.79 | 35.84 | 32.39 | 28.68 | 26.61 |
| 19° | 27.37 | 30.91 | 34.96 | 37.67 | 38.81 | 38.95 | 38.69 | 37.82 | 35.68 | 32.04 | 28.19 | 26.06 |
| 20° | 26.84 | 30.48 | 34.71 | 37.63 | 38.94 | 39.16 | 38.86 | 37.85 | 35.52 | 31.68 | 27.70 | 25.51 |
| 21° | 26.31 | 30.05 | 34.46 | 37.57 | 39.06 | 39.36 | 39.02 | 37.86 | 35.34 | 31.31 | 27.19 | 24.95 |
| 22° | 25.77 | 29.61 | 34.19 | 37.50 | 39.16 | 39.55 | 39.17 | 37.87 | 35.15 | 30.93 | 26.69 | 24.39 |
| 23° | 25.23 | 29.17 | 33.91 | 37.43 | 39.26 | 39.73 | 39.31 | 37.86 | 34.95 | 30.55 | 26.17 | 23.82 |
| 24° | 24.68 | 28.71 | 33.63 | 37.34 | 39.35 | 39.90 | 39.44 | 37.84 | 34.74 | 30.15 | 25.65 | 23.24 |
| 25° | 24.12 | 28.25 | 33.33 | 37.24 | 39.42 | 40.06 | 39.56 | 37.81 | 34.52 | 29.75 | 25.12 | 22.67 |
| 26° | 23.56 | 27.77 | 33.02 | 37.12 | 39.49 | 40.21 | 39.67 | 37.78 | 34.29 | 29.33 | 24.59 | 22.08 |
| 27° | 22.99 | 27.29 | 32.70 | 37.00 | 39.55 | 40.36 | 39.77 | 37.73 | 34.04 | 28.91 | 24.05 | 21.50 |
| 28° | 22.42 | 26.81 | 32.38 | 36.87 | 39.59 | 40.49 | 39.86 | 37.66 | 33.79 | 28.48 | 23.50 | 20.91 |
| 29° | 21.85 | 26.31 | 32.04 | 36.72 | 39.63 | 40.61 | 39.94 | 37.59 | 33.53 | 28.04 | 22.95 | 20.31 |
| 30° | 21.27 | 25.81 | 31.69 | 36.57 | 39.65 | 40.72 | 40.02 | 37.51 | 33.26 | 27.59 | 22.40 | 19.72 |
| 31° | 20.68 | 25.30 | 31.34 | 36.40 | 39.67 | 40.83 | 40.08 | 37.42 | 32.98 | 27.14 | 21.84 | 19.12 |
| 32° | 20.10 | 24.79 | 30.97 | 36.23 | 39.67 | 40.92 | 40.13 | 37.32 | 32.68 | 26.67 | 21.27 | 18.52 |
| 33° | 19.50 | 24.27 | 30.59 | 36.04 | 39.67 | 41.00 | 40.17 | 37.20 | 32.38 | 26.20 | 20.70 | 17.91 |
| 34° | 18.91 | 23.74 | 30.21 | 35.85 | 39.65 | 41.08 | 40.20 | 37.08 | 32.07 | 25.72 | 20.13 | 17.30 |
| 35° | 18.31 | 23.20 | 29.81 | 35.64 | 39.63 | 41.15 | 40.23 | 36.95 | 31.75 | 25.24 | 19.55 | 16.70 |
| 36° | 17.71 | 22.66 | 29.41 | 35.42 | 39.59 | 41.20 | 40.24 | 36.81 | 31.41 | 24.75 | 18.97 | 16.09 |
| 37° | 17.11 | 22.12 | 29.00 | 35.19 | 39.55 | 41.25 | 40.24 | 36.65 | 31.07 | 24.25 | 18.38 | 15.48 |
| 38° | 16.51 | 21.57 | 28.58 | 34.96 | 39.49 | 41.29 | 40.24 | 36.49 | 30.72 | 23.74 | 17.79 | 14.86 |
| 39° | 15.90 | 21.02 | 28.15 | 34.71 | 39.43 | 41.32 | 40.23 | 36.32 | 30.36 | 23.23 | 17.20 | 14.25 |
| 40° | 15.29 | 20.45 | 27.71 | 34.45 | 39.36 | 41.35 | 40.20 | 36.13 | 29.99 | 22.71 | 16.61 | 13.64 |
| 41° | 14.68 | 19.88 | 27.26 | 34.18 | 39.27 | 41.36 | 40.17 | 35.94 | 29.61 | 22.18 | 16.01 | 13.03 |
| 42° | 14.07 | 19.31 | 26.81 | 33.91 | 39.18 | 41.37 | 40.13 | 35.74 | 29.23 | 21.65 | 15.42 | 12.42 |
| 43° | 13.46 | 18.73 | 26.34 | 33.62 | 39.08 | 41.36 | 40.08 | 35.53 | 28.83 | 21.11 | 14.82 | 11.81 |
| 44° | 12.85 | 18.15 | 25.87 | 33.32 | 38.97 | 41.36 | 40.03 | 35.31 | 28.43 | 20.57 | 14.22 | 11.20 |
| 45° | 12.25 | 17.57 | 25.39 | 33.02 | 38.86 | 41.34 | 39.96 | 35.08 | 28.01 | 20.02 | 13.62 | 10.60 |
| 46° | 11.64 | 16.98 | 24.91 | 32.71 | 38.73 | 41.32 | 39.89 | 34.84 | 27.59 | 19.46 | 13.01 | 9.99 |
| 47° | 11.03 | 16.39 | 24.41 | 32.38 | 38.60 | 41.29 | 39.82 | 34.59 | 27.16 | 18.90 | 12.41 | 9.40 |
| 48° | 10.43 | 15.80 | 23.91 | 32.05 | 38.46 | 41.25 | 39.73 | 34.33 | 26.72 | 18.34 | 11.81 | 8.80 |
| 49° | 9.82 | 15.20 | 23.40 | 31.71 | 38.31 | 41.21 | 39.64 | 34.07 | 26.28 | 17.77 | 11.21 | 8.21 |
| 50° | 9.22 | 14.60 | 22.89 | 31.36 | 38.16 | 41.16 | 39.54 | 33.80 | 25.82 | 17.20 | 10.61 | 7.63 |
| 51° | 8.63 | 14.00 | 22.36 | 31.00 | 38.00 | 41.11 | 39.44 | 33.52 | 25.36 | 16.62 | 10.01 | 7.05 |
| 52° | 8.04 | 13.39 | 21.83 | 30.64 | 37.83 | 41.06 | 39.33 | 33.23 | 24.89 | 16.04 | 9.41 | 6.48 |
| 53° | 7.45 | 12.79 | 21.30 | 30.26 | 37.65 | 41.00 | 39.22 | 32.94 | 24.41 | 15.45 | 8.82 | 5.91 |
| 54° | 6.87 | 12.18 | 20.75 | 29.88 | 37.48 | 40.94 | 39.10 | 32.63 | 23.93 | 14.86 | 8.23 | 5.36 |
| 55° | 6.30 | 11.58 | 20.20 | 29.49 | 37.29 | 40.88 | 38.98 | 32.32 | 23.44 | 14.27 | 7.65 | 4.81 |
| 56° | 5.73 | 10.97 | 19.65 | 29.10 | 37.11 | 40.82 | 38.86 | 32.01 | 22.94 | 13.68 | 7.07 | 4.28 |
| 57° | 5.18 | 10.36 | 19.09 | 28.69 | 36.92 | 40.76 | 38.74 | 31.69 | 22.43 | 13.08 | 6.49 | 3.76 |
| 58° | 4.63 | 9.76 | 18.52 | 28.28 | 36.72 | 40.70 | 38.61 | 31.36 | 21.92 | 12.48 | 5.92 | 3.25 |
| 59° | 4.10 | 9.15 | 17.95 | 27.87 | 36.53 | 40.65 | 38.49 | 31.03 | 21.40 | 11.88 | 5.37 | 2.77 |
| 60° | 3.57 | 8.55 | 17.37 | 27.45 | 36.34 | 40.60 | 38.37 | 30.69 | 20.88 | 11.27 | 4.81 | 2.30 |
| 61° | 3.07 | 7.95 | 16.79 | 27.02 | 36.14 | 40.56 | 38.26 | 30.35 | 20.35 | 10.67 | 4.27 | 1.85 |
| 62° | 2.58 | 7.35 | 16.20 | 26.59 | 35.95 | 40.54 | 38.15 | 30.01 | 19.81 | 10.07 | 3.75 | 1.43 |
| 63° | 2.11 | 6.76 | 15.61 | 26.15 | 35.77 | 40.53 | 38.05 | 29.66 | 19.27 | 9.46 | 3.23 | 1.04 |
| 64° | 1.66 | 6.17 | 15.01 | 25.71 | 35.59 | 40.55 | 37.97 | 29.32 | 18.72 | 8.86 | 2.73 | 0.68 |

*(Continued)*

95

**TABLE 3-5.** (*Continued*)  Southern Hemisphere.

| Latitude Degrees South | Month Jan Day No. 17. | Feb 46. | Mar 76. | Apr 105. | May 135. | Jun 162. | Jul 199. | Aug 229. | Sept 258. | Oct 288. | Nov 318. | Dec 346. |
|---|---|---|---|---|---|---|---|---|---|---|---|---|
| 15° | 40.38 | 39.37 | 36.68 | 32.70 | 28.63 | 26.52 | 27.35 | 30.74 | 34.79 | 38.11 | 39.83 | 40.40 |
| 16° | 40.59 | 39.43 | 36.53 | 32.37 | 28.18 | 26.02 | 26.88 | 30.37 | 34.57 | 38.09 | 40.00 | 40.66 |
| 17° | 40.79 | 39.47 | 36.38 | 32.04 | 27.72 | 25.51 | 26.40 | 29.98 | 34.35 | 38.07 | 40.16 | 40.91 |
| 18° | 40.98 | 39.51 | 36.21 | 31.69 | 27.25 | 25.00 | 25.91 | 29.58 | 34.11 | 38.03 | 40.30 | 41.16 |
| 19° | 41.16 | 39.54 | 36.03 | 31.33 | 26.78 | 24.49 | 25.41 | 29.18 | 33.86 | 37.98 | 40.44 | 41.39 |
| 20° | 41.33 | 39.55 | 35.83 | 30.97 | 26.30 | 23.97 | 24.91 | 28.77 | 33.60 | 37.92 | 40.56 | 41.61 |
| 21° | 41.49 | 39.55 | 35.63 | 30.59 | 25.81 | 23.44 | 24.41 | 28.35 | 33.33 | 37.85 | 40.67 | 41.82 |
| 22° | 41.64 | 39.54 | 35.42 | 30.21 | 25.32 | 22.91 | 23.90 | 27.92 | 33.05 | 37.77 | 40.78 | 42.02 |
| 23° | 41.78 | 39.52 | 35.19 | 29.81 | 24.82 | 22.37 | 23.38 | 27.49 | 32.76 | 37.68 | 40.87 | 42.21 |
| 24° | 41.91 | 39.49 | 34.96 | 29.41 | 24.32 | 21.83 | 22.86 | 27.04 | 32.46 | 37.57 | 40.95 | 42.39 |
| 25° | 42.03 | 39.45 | 34.71 | 29.00 | 23.81 | 21.29 | 22.33 | 26.59 | 32.15 | 37.46 | 41.02 | 42.56 |
| 26° | 42.13 | 39.39 | 34.46 | 28.59 | 23.29 | 20.74 | 21.80 | 26.14 | 31.83 | 37.33 | 41.08 | 42.72 |
| 27° | 42.23 | 39.33 | 34.19 | 28.16 | 22.77 | 20.19 | 21.27 | 25.67 | 31.50 | 37.19 | 41.12 | 42.87 |
| 28° | 42.31 | 39.25 | 33.92 | 27.72 | 22.24 | 19.63 | 20.73 | 25.20 | 31.16 | 37.04 | 41.16 | 43.01 |
| 29° | 42.39 | 39.16 | 33.63 | 27.28 | 21.71 | 19.07 | 20.18 | 24.72 | 30.81 | 36.89 | 41.19 | 43.13 |
| 30° | 42.45 | 39.06 | 33.33 | 26.83 | 21.18 | 18.51 | 19.63 | 24.24 | 30.45 | 36.72 | 41.20 | 43.25 |
| 31° | 42.51 | 38.96 | 33.02 | 26.37 | 20.63 | 17.95 | 19.08 | 23.75 | 30.09 | 36.53 | 41.20 | 43.36 |
| 32° | 42.55 | 38.83 | 32.71 | 25.91 | 20.09 | 17.38 | 18.53 | 23.25 | 29.71 | 35.34 | 41.20 | 43.46 |
| 33° | 42.58 | 38.70 | 32.38 | 25.43 | 19.54 | 16.81 | 17.97 | 22.74 | 29.32 | 36.14 | 41.19 | 43.55 |
| 34° | 42.60 | 38.56 | 32.04 | 24.95 | 18.99 | 16.24 | 17.41 | 22.23 | 28.93 | 35.93 | 41.16 | 43.63 |
| 35° | 42.61 | 38.41 | 31.69 | 24.47 | 18.43 | 15.66 | 16.85 | 21.72 | 28.53 | 35.71 | 41.12 | 43.69 |
| 36° | 42.62 | 38.24 | 31.34 | 23.97 | 17.87 | 15.09 | 16.28 | 21.20 | 28.12 | 35.47 | 41.08 | 43.75 |
| 37° | 42.61 | 38.07 | 30.97 | 23.47 | 17.31 | 14.52 | 15.71 | 20.67 | 27.70 | 35.23 | 41.02 | 43.80 |
| 38° | 42.59 | 37.89 | 30.59 | 22.96 | 16.74 | 13.94 | 15.14 | 20.14 | 27.27 | 34.97 | 40.95 | 43.84 |
| 39° | 42.56 | 37.69 | 30.21 | 22.45 | 16.17 | 13.36 | 14.57 | 19.61 | 26.83 | 34.71 | 40.87 | 43.87 |
| 40° | 42.53 | 37.49 | 29.82 | 21.93 | 15.60 | 12.79 | 14.00 | 19.07 | 26.39 | 34.44 | 40.79 | 43.90 |
| 41° | 42.48 | 37.27 | 29.41 | 21.41 | 15.03 | 12.21 | 13.43 | 18.52 | 25.93 | 34.15 | 40.69 | 43.91 |
| 42° | 42.42 | 37.05 | 29.00 | 20.87 | 14.46 | 11.64 | 12.86 | 17.97 | 25.47 | 33.86 | 40.58 | 43.91 |
| 43° | 42.36 | 36.81 | 28.58 | 20.34 | 13.88 | 11.07 | 12.28 | 17.42 | 25.01 | 33.56 | 40.47 | 43.91 |
| 44° | 42.29 | 36.57 | 28.15 | 19.79 | 13.30 | 10.49 | 11.71 | 16.86 | 24.53 | 33.24 | 40.34 | 43.90 |
| 45° | 42.21 | 36.31 | 27.71 | 19.25 | 12.73 | 9.93 | 11.14 | 16.30 | 24.05 | 32.92 | 40.21 | 43.88 |

on a horizontal surface (usually expressed as the average daily total for the month) to the extraterrestrial quantity available, which is given by eq. 3-17, Table 3-5, or Table A-14:

$$\bar{K}_T = \frac{\bar{H}_T}{\bar{H}_{0T}} \tag{3-20}$$

Page (13) used measurements from 10 stations and produced a simple correlation that is in general agreement with the additional measurements of Choudhury (18), Stanhill (19), and Norris (20),

$$\left(\frac{\bar{H}_{Td}}{\bar{H}_T}\right) = 1.00 - 1.13\bar{K}_T \tag{3-21}$$

Liu and Jordan (17) relied heavily on Blue Hill, Massachusetts data to establish the proportion of diffuse energy in global monthly (or daily average) total solar

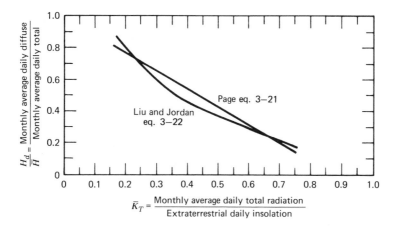

**FIGURE 3-17.** Comparison of the Page and Liu-Jordan equations for prediction of the proportion of diffuse radiation from monthly $\overline{K}_T$ data (Ref. 15).

radiation, but failed to apply a needed shade-ring correction factor. Using the current value of the solar constant, their relationship is expressed by (16),

$$\left(\frac{\overline{H}_{Td}}{\overline{H}_T}\right) = 1.390 - 4.027\overline{K}_T + 5.531\overline{K}_T{}^2 - 3.108\overline{K}_T{}^3 \qquad (3\text{-}22)$$

The data Liu and Jordan collected also give eq. 3-21 if the Blue Hill data are omitted. The two relationships are plotted in Fig. 3-17. They differ appreciably over the ordinary range of $\overline{K}_T$ from 0.4 to 0.5, which is found in winter over much of the United States, the ranges in which the Blue Hill data heavily influence the Liu-Jordan equation. The Liu-Jordan equation predicts a lesser proportion of diffuse radiation and 10 to 20% higher values for winter solar radiation. These values have not been confirmed by other data (16*, 29). It has seen wide application nonetheless, and it is incorporated into the data base used for virtually every simulation or performance prediction made for U.S. government agencies prior to 1978, and many thereafter.

Because the Page relationship can be shown to be consistent with measured and synthesized (see below) radiation data, it is recommended for general use.† After

---

* The measured data in Table 5 of this reference are incorrect and should be replaced by the corrected values published in the erratum.

† Reference 29 gives a detailed study of this and similar equations for Canadian stations, recommending $\overline{H}_{Td}/\overline{H}_T = 0.958 - 0.982\overline{K}_T$, which is interchangeable with eq. 3-23 at the relevant $\overline{K}_T$ values. Readers should periodically review current literature for new correlations to predict the proportion of diffuse radiation in monthly total global radiation data.

correction for the current value of the solar constant, the correct expression is

$$\left(\frac{\overline{H}_{Td}}{\overline{H}_T}\right) = 1.00 - 1.096\overline{K}_T \tag{3-23}$$

## Total Radiation on a Tilted Surface

Since the diffuse radiation is usually assumed to be isotropic, its value on the tilted collector is a fixed proportion of the diffuse radiation on the horizontal surface as given by eqs. 3-13 or 3-14. Using the monthly daily-average total radiation values, if eq. 3-14 is divided by $\overline{H}_T$,

$$\left(\frac{\bar{I}_{Td}}{\overline{H}_T}\right) = \left(\frac{\overline{H}_{Td}}{\overline{H}_T}\right)\frac{(1 + \cos\beta)}{2} + \rho\frac{(1 - \cos\beta)}{2} \tag{3-24}$$

Similarly, if eq. 3-19 is multiplied by $\overline{H}_{Tb}/\overline{H}_T$, the remaining proportion of the horizontal global radiation, which is direct or *beam*, then

$$\left(\frac{\bar{I}_{Tb}}{\overline{H}_T}\right) = \overline{R}_b\left(\frac{\overline{H}_{Tb}}{\overline{H}_T}\right) \tag{3-25}$$

Adding eq. 3-24 to 3-25,

$$\overline{R} = \frac{\bar{I}_T}{\overline{H}_T} = \overline{R}_b\left(\frac{\overline{H}_{Tb}}{\overline{H}_T}\right) + \frac{(1 + \cos\beta)}{2}\left(\frac{\overline{H}_{Td}}{\overline{H}_T}\right) + \rho\frac{(1 - \cos\beta)}{2} \tag{3-26}$$

where the factor $\overline{R}$ is an elevation factor for the monthly daily-average total horizontal global radiation to produce the radiation on the tilted collector:

$$\bar{I}_T = \overline{R}\overline{H}_T \tag{3-27}$$

Equation 3-26 is in final form for use in the elevation of horizontal total radiation data. The proportion of diffuse and direct beam radiation is found from the Page or an alternate relationship and substituted directly into the first and second terms. The elevation angle of the collector and the reflectivity of the surroundings are usually known for a given situation. The direct beam elevation factor $\overline{R}_b$ can be approximated from extraterrestrial radiation tables (or by direct calculation) as indicated in eq. 3-19, but as previously noted, the value will be several percent too high in winter.

If the Page equation is modified to

$$\left(\frac{\overline{H}_{Td}}{\overline{H}_T}\right) = 1.00 - 1.015\overline{K}_T \tag{3-28}$$

when used with eq. 3-26, it compensates for the inaccuracy caused by elevating the direct beam with the extraterrestrial radiation ratio (eq. 3-19). These two equations predict winter values of $\overline{R}$ within 2% of the Australian data given by

Klein (16)* and are also consistent with synthesized data presented in Appendix B† (see below).

The procedures for estimating $\bar{R}_b$, based on extraterrestrial radiation values, have been extended by Klein (16) to a nonsouth-facing tilted collector. The equations are quite complex, but they have been integrated into a closed form that appears quite practical for routine use using a computer or high-capacity programmable calculator. The direct beam elevation factor is given by

$$
\begin{aligned}
\bar{R}_b = \{ &[\cos\beta \sin\delta \sin\phi](\pi/180)[\tilde{\omega}_{ss} - \tilde{\omega}_{sr}] \\
&- [\sin\delta \cos\phi \sin\beta \cos\gamma](\pi/180)[\tilde{\omega}_{ss} - \tilde{\omega}_{sr}] \\
&+ [\cos\phi \cos\delta \cos\beta][\sin\tilde{\omega}_{ss} - \sin\tilde{\omega}_{sr}] \\
&+ [\cos\delta \cos\gamma \sin\phi \sin\beta][\sin\tilde{\omega}_{ss} - \sin\tilde{\omega}_{sr}] \\
&- [\cos\delta \sin\beta \sin\gamma][\cos\tilde{\omega}_{ss} - \cos\tilde{\omega}_{sr}]\}/ \\
&\{2[\cos\phi \cos\delta \sin\omega_s + (\pi/180)\omega_s \sin\phi \sin\delta]\}
\end{aligned}
\tag{3-29}
$$

Sunset and sunrise on the collector are different when the sun is shadowed by the collector. $\tilde{\omega}_{sr}$ and $\tilde{\omega}_{ss}$ are the sunrise and sunset hour angles on the tilted surface, given by:

if $\gamma < 0$

$$\tilde{\omega}_{sr} = -\min\{\omega_s, \text{arcos}\,[(AB + \sqrt{A^2 - B^2 + 1})/(A^2 + 1)]\} \tag{3-30a}$$

$$\tilde{\omega}_{ss} = \min\{\omega_s, \text{arcos}\,[(AB - \sqrt{A^2 - B^2 + 1})/(A^2 + 1)]\} \tag{3-30b}$$

if $\gamma > 0$

$$\tilde{\omega}_{sr} = -\min\{\omega_s, \text{arcos}\,[(AB - \sqrt{A^2 - B^2 + 1})/(A^2 + 1)]\} \tag{3-31a}$$

$$\tilde{\omega}_{ss} = \min\{\omega_s, \text{arcos}\,[(AB + \sqrt{A^2 - B^2 + 1})/(A^2 + 1)]\} \tag{3-31b}$$

where

$$A = \cos\phi/(\sin\gamma \tan\beta) + \sin\phi/\tan\gamma \tag{3-32}$$

$$B = \tan\delta[\cos\phi/\tan\gamma - \sin\phi/(\sin\gamma \tan\beta)] \tag{3-33}$$

and $\omega_s$ is given by eq. 3-8.

**EXAMPLE 3-7**

The south-facing collector of Examples 3-2 to 3-6 is again located at 40°N latitude and is inclined at 60° to the horizontal.

(a) If the measured horizontal global daily average total radiation in November is 6.91 MJ/m², estimate the daily-average radiation the collector receives.

---

* See footnote, p. 97.
† For high-altitude stations, $\bar{K}_T$ should be reduced before using eq. 3-28 so that it relates more accurately to actual cloudiness. For Denver the recommended reduction in $\bar{K}_T$ is 0.035.

Using $\bar{H}_{oT}$ from Example 3-6(a) or Table 3-5,

$$\bar{K}_T = \frac{\bar{H}_T}{\bar{H}_{oT}} = \frac{6.91 \text{ MJ}}{16.61 \text{ MJ}} = 0.4160$$

By the modified Page relationship (eq. 3-28) the diffuse portion is considered to be

$$\left(\frac{\bar{H}_{Td}}{\bar{H}_T}\right) = 1.00 - 1.015(0.4160) = 0.5777$$

The remaining portion of the radiation is the direct beam:

$$\left(\frac{\bar{H}_{Tb}}{\bar{H}_T}\right) = 1 - 0.5777 = 0.4223$$

The direct beam elevation factor $\bar{R}_b$ is found from eq. 3-19, using the extraterrestrial radiation totals from either Example 3-6 or Table A-14:

$$\bar{R}_b = \frac{\bar{I}_{ob}}{\bar{H}_{ob}} = \frac{37.81 \text{ MJ/m}^2}{16.61 \text{ MJ/m}^2} = 2.276$$

Substituting in eq. 3-26, with a value of 0.2 for ground reflectivity $\rho$,

$$\bar{R} = 2.276(0.4223) + \frac{(1 + \cos 60°)}{2}(0.5777) + 0.2\frac{(1 - \cos 60°)}{2} = 1.4457$$

and the total radiation on the tilted collector by eq. 3-27 is therefore

$$\bar{I}_T = 1.4457 \times 6.91 = 9.99 \text{ MJ/m}^2$$

(b) Weather service records indicate that the percentage of sunshine at this location is 46% in November. Estimate the monthly daily-average total global radiation received on the tilted collector, assuming a clearness number of 1.0.

From Table A-13 at 40°N latitude, the average total global radiation received daily at maximum on a tilted surface in November under clear sky conditions is 22.22 MJ/m², at a 60° tilt. The total radiation received above a 15 W/m² threshold is

$$\bar{I}_{T+} = 0.46 \times 22.22 = 10.22 \text{ MJ/m}^2$$

The total global radiation is 8% more than this (see *Sunshine Records* above):

$$\bar{I}_T = 10.22 \times 1.08 = 11.04 \text{ MJ/m}^2$$

Using the $\bar{R}$ value from part (a) the radiation on a horizontal surface is, by eq. 3-27,

$$\bar{H}_T = (11.04 \text{ MJ/m}^2)/1.4457 = 7.64 \text{ MJ/m}^2$$

which is within about 10% of the correct value. A new value of $\bar{R}$ appropriate for the $\bar{H}_T$ value may now be calculated and used for a better estimate of $\bar{H}_T$.

## SYNTHESIZED RADIATION DATA

The use of published monthly averaged radiation data, like the data in Reference 15, is not always recommended. It is much better if hourly radiation records of known high quality are available that can be analyzed by computer to give certain summary data for a site of interest. Since there is little high-quality data available it is recommended that hourly horizontal radiation data instead be predicted from ordinarily reported weather station information such as temperature, cloud cover, and relative humidity by the techniques discussed below. The resultant data have the advantage of consistency, and their accuracy is quite good. Since the diffuse component is separately predicted, elevation to a tilted surface can be done hourly and requires no correlation. The monthly horizontal data produced are consistent with eq. 3-23 and alternately they can be elevated accurately by using eqs. 3-19, 3-28, 3-26, and 3-27 in sequence.

The prediction of solar radiation from standard weather station information is a standard meteorological technique. Details regarding one method are available in technical papers and reports by M. A. Atwater, P. S. Brown, Jr., and J. T. Ball (21, 22, 23). Overall long-term accuracy is estimated at $\pm 5\%$, which is probably better than that of the rehabilitated U.S. Weather Service radiation data (27). A similar model has been used by the Canadian Atmospheric Environment Service to establish a 100 station 10-yr data base for Canada (33).

The radiation model predicts hourly radiation flux from local humidity and cloud cover information. The starting point of the model is the direct beam irradiance outside the atmosphere. This is attenuated by atmospheric absorption and scattering, with scattering producing diffuse irradiation. The attenuation of the radiant energy as it passes through the permanent atmospheric gases is described empirically using well-known meteorological relationships. The water vapor in the atmosphere is estimated from the surface dew point, and used to estimate further attenuation of the direct beam irradiance.

Transmission of visible radiant energy through clouds is a most important part of the technique. Coefficients available for nine reported types of clouds are applied in proportion to the portion of the sky reported as covered with cloud (23, 24). A recent improvement in the model also incorporates surface visibility, to include attenuation through the layer of pollution now common to most U.S. locations (32). Note carefully that this is a deterministic model, *not* a statistical correlation.

Obviously a method such as this will not be able to predict *hour-by-hour* data accurately, because there is no way to know the relative location of the sun and clouds with respect to each other and the observer. All that is needed, however, is radiation data that are representative in intensity and distribution and that total and average to the proper totals and averages. Cloud cover calculations have been shown to do this job very well and, since they have the effect of making radiation data available at about 250 U.S. stations, the radiation data problem could be solved if these data were generated and made generally available.

## Summary Radiation Tables

When real or synthesized *hourly* weather and radiation data are available, summary tables such as those in Appendix B can be prepared by computer. The tables

**TABLE 3-6a.** Radiation and Weather Information at 50° Inclination at Hartford

STATISTICS FOR HARTFORD, CCNN.                    1959-68

RADIATION UNITS ARE WATTS/SQ.M OR MJOULES/SQ.M
SLOPE = 50.0 DEG    AZIMUTH = 180.0 DEG
DEGREE-DAY BASE TEMP = 18.3 DEGREES C

**JAN**

| SOLAR RAD GROUP | 0 | 2-95 | 95-189 | 189-284 | 284-378 | 378-473 | 473-568 | 568-662 | 662-757 | 757-851 | 851-946 | OVER 946 |
|---|---|---|---|---|---|---|---|---|---|---|---|---|
| TIME AVG.TEMP | -4.9 | -2.5 | -1.1 | -2.9 | -2.0 | -0.3 | -1.9 | -2.7 | -3.3 | -3.7 | -3.6 | -5.2 |
| WATT/SQ.M | 0. | 36. | 139. | 239. | 325. | 427. | 54. | 611. | 714. | 694. | 906. | 923. |
| TOTAL TIME MSEC | 1.4598 | 0.4678 | 0.1717 | 0.1310 | 0.0856 | 0.0605 | 0.0547 | 0.0536 | 0.0494 | 0.0504 | 0.0371 | 0.0357 |
| GLOBAL RAD TOTAL | 0.0 | 17.4 | 23.7 | 31.3 | 28.8 | 25.8 | 23.7 | 32.8 | 33.2 | 49.5 | 33.6 | 36.4 |
| DIRECT RAD TOTAL | 0.0 | 3.1 | 6.3 | 16.5 | 16.7 | 16.7 | 22.0 | 27.1 | 28.8 | 35.4 | 30.6 | 34.0 |
| DEG-C DAYS | 394. | 117. | 39. | 32. | 21. | 13. | 13. | 13. | 12. | 13. | 9. | 10. |

**FEB**

| SOLAR RAD GROUP | 0 | 2-95 | 95-189 | 189-284 | 284-378 | 378-473 | 473-568 | 568-662 | 662-757 | 757-851 | 851-946 | OVER 946 |
|---|---|---|---|---|---|---|---|---|---|---|---|---|
| TIME AVG.TEMP | -4.5 | -1.7 | -1.2 | -2.0 | -2.0 | -1.2 | -0.4 | -2.2 | -1.9 | -1.4 | -2.9 | -2.5 |
| WATT/SQ.M | 0. | 40. | 139. | 237. | 331. | 423. | 522. | 620. | 706. | 810. | 901. | 1034. |
| TOTAL TIME MSEC | 1.2654 | 0.3751 | 0.1969 | 0.1170 | 0.1022 | 0.0713 | 0.0616 | 0.0530 | 0.0403 | 0.0511 | 0.0335 | 0.0727 |
| GLOBAL RAD TOTAL | 0.0 | 15.1 | 27.4 | 27.7 | 33.8 | 30.1 | 32.1 | 35.9 | 28.5 | 41.4 | 30.2 | 75.2 |
| DIRECT RAD TOTAL | 0.0 | 2.4 | 7.7 | 9.2 | 17.4 | 16.9 | 19.1 | 28.5 | 23.5 | 36.2 | 26.0 | 69.4 |
| DEG-C DAYS | 334. | 87. | 44. | 26. | 24. | 16. | 13. | 14. | 9. | 12. | 8. | 13. |

**MAR**

| SOLAR RAD GROUP | 0 | 2-95 | 95-189 | 189-284 | 284-378 | 378-473 | 473-563 | 568-662 | 662-757 | 757-851 | 851-946 | OVER 946 |
|---|---|---|---|---|---|---|---|---|---|---|---|---|
| TIME AVG.TEMP | 0.4 | 1.7 | 3.1 | 3.6 | 4.8 | 4.9 | 4.3 | 5.8 | 6.0 | 4.6 | 5.9 | 4.6 |
| WATT/SQ.M | 0. | 37. | 142. | 233. | 331. | 429. | 519. | 619. | 713. | 795. | 907. | 1042. |
| TOTAL TIME MSEC | 1.2877 | 0.4068 | 0.2405 | 0.1652 | 0.0029 | 0.0932 | 0.0958 | 0.0490 | 0.0594 | 0.0486 | 0.0425 | 0.0568 |
| GLOBAL RAD TOTAL | 0.0 | 14.9 | 34.1 | 38.5 | 30.7 | 40.0 | 49.7 | 30.3 | 42.3 | 39.6 | 39.5 | 100.9 |
| DIRECT RAD TOTAL | 0.0 | 2.2 | 4.9 | 12.8 | 9.9 | 19.5 | 28.9 | 17.6 | 30.7 | 31.1 | 32.6 | 90.6 |
| DEG-C DAYS | 267. | 78. | 43. | 28. | 14. | 14. | 15. | 7. | 8. | 8. | 6. | 16. |

**APR**

| SOLAR RAD GROUP | 0 | 2-95 | 95-189 | 189-284 | 284-378 | 378-473 | 473-568 | 568-662 | 662-757 | 757-851 | 851-946 | OVER 946 |
|---|---|---|---|---|---|---|---|---|---|---|---|---|
| TIME AVG.TEMP | 7.1 | 8.8 | 9.7 | 10.9 | 11.9 | 14.6 | 13.0 | 15.1 | 16.0 | 14.9 | 15.9 | 14.8 |
| WATT/SQ.M | 0. | 42. | 143. | 236. | 332. | 425. | 523. | 611. | 714. | 800. | 902. | 1033. |
| TOTAL TIME MSEC | 1.2989 | 0.2956 | 0.2070 | 0.1631 | 0.1433 | 0.0661 | 0.0832 | 0.0634 | 0.0677 | 0.0590 | 0.0463 | 0.0691 |
| GLOBAL RAD TOTAL | 0.0 | 12.5 | 29.6 | 38.4 | 47.5 | 40.9 | 43.5 | 38.7 | 43.3 | 46.4 | 42.2 | 71.4 |
| DIRECT RAD TOTAL | 0.0 | 2.4 | 1.9 | 5.4 | 15.8 | 14.7 | 22.4 | 18.8 | 29.8 | 32.1 | 33.1 | 63.4 |
| DEG-C DAYS | 167. | 32. | 21. | 14. | 11. | 5. | 5. | 3. | 3. | 3. | 2. | 4. |

**MAY**

| SOLAR RAD GROUP | 0 | 2-95 | 95-189 | 189-284 | 284-378 | 378-473 | 473-568 | 568-662 | 662-757 | 757-851 | 851-946 | OVER 946 |
|---|---|---|---|---|---|---|---|---|---|---|---|---|
| TIME AVG.TEMP | 12.1 | 15.1 | 13.3 | 16.2 | 17.9 | 20.2 | 20.2 | 21.7 | 21.4 | 21.3 | 21.2 | 21.2 |
| WATT/SQ.M | 0. | 46. | 140. | 239. | 328. | 425. | 54. | 613. | 711. | 801. | 900. | 1004. |
| TOTAL TIME MSEC | 1.3392 | 0.2203 | 0.1814 | 0.1674 | 0.1696 | 0.1328 | 0.1321 | 0.0990 | 0.0850 | 0.0623 | 0.0497 | 0.0336 |
| GLOBAL RAD TOTAL | 0.0 | 10.2 | 25.3 | 39.9 | 55.6 | 56.4 | 69.2 | 60.7 | 60.4 | 49.9 | 44.7 | 39.7 |
| DIRECT RAD TOTAL | 0.0 | 1.5 | 1.9 | 4.6 | 17.0 | 15.5 | 30.1 | 26.2 | 28.4 | 31.3 | 35.3 | 34.5 |
| DEG-C DAYS | 87. | 9. | 21. | 6. | 3. | 2. | 2. | 1. | 1. | 1. | 1. | 0. |

**JUN**

| SOLAR RAD GROUP | 0 | 2-95 | 95-189 | 189-284 | 284-378 | 378-473 | 473-568 | 568-662 | 662-757 | 757-851 | 851-946 | OVER 946 |
|---|---|---|---|---|---|---|---|---|---|---|---|---|
| TIME AVG.TEMP | 17.4 | 20.3 | 18.5 | 21.0 | 23.1 | 25.7 | 24.6 | 26.8 | 26.7 | 27.4 | 26.7 | 27.3 |
| WATT/SQ.M | 0. | 52. | 133. | 235. | 327. | 427. | 521. | 613. | 711. | 805. | 895. | 974. |
| TOTAL TIME MSEC | 1.2960 | 0.2203 | 0.1861 | 0.1652 | 0.1620 | 0.1177 | 0.1189 | 0.0918 | 0.1012 | 0.0598 | 0.0472 | 0.0153 |
| GLOBAL RAD TOTAL | 0.0 | 11.9 | 25.6 | 38.9 | 53.0 | 50.3 | 67.2 | 56.3 | 71.9 | 49.1 | 42.1 | 15.4 |
| DIRECT RAD TOTAL | 0.0 | 1.1 | 1.1 | 3.8 | 16.1 | 15.0 | 29.2 | 23.8 | 38.6 | 30.4 | 32.7 | 13.6 |
| DEG-C DAYS | 15. | 2. | 2. | 1. | 1. | 0. | 0. | 0. | 1. | 0. | 0. | 0. |

**JUL**

| | 0 | 2-95 | 95-189 | 189-284 | 284-378 | 378-473 | 473-568 | 568-662 | 662-757 | 757-851 | 851-946 | OVER 946 |
|---|---|---|---|---|---|---|---|---|---|---|---|---|
| SOLAR RAD GROUP | | | | | | | | | | | | |
| TIME AVG.TEMP | 19.8 | 22.8 | 21.8 | 24.3 | 25.5 | 27.3 | 26.9 | 28.2 | 28.1 | 28.9 | 29.1 | 23.2 |
| WATT/SQ.M | 0. | 51. | 139. | 242. | 329. | 428. | 520. | 614. | 706. | 805. | 897. | 976. |
| TOTAL TIME MSEC | 1.3392 | 0.2477 | 0.1483 | 0.1588 | 0.1883 | 0.1465 | 0.1325 | 0.1044 | 0.0997 | 0.0594 | 0.0410 | 0.0126 |
| GLOBAL RAD TOTAL | 0.0 | 12.8 | 20.6 | 38.4 | 62.0 | 62.7 | 68.9 | 64.1 | 70.4 | 47.8 | 36.8 | 12.3 |
| DIRECT RAD TOTAL | 0.0 | 2.4 | 1.1 | 6.0 | 18.4 | 19.2 | 30.2 | 29.8 | 41.3 | 33.9 | 29.4 | 10.9 |
| DEG-C DAYS | 1. | 0. | 0. | 0. | 0. | 0. | 0. | 0. | 0. | 0. | 0. | 0. |

**AUG**

| | 0 | 2-95 | 95-189 | 189-284 | 284-378 | 378-473 | 473-568 | 568-662 | 662-757 | 757-851 | 851-946 | OVER 946 |
|---|---|---|---|---|---|---|---|---|---|---|---|---|
| SOLAR RAD GROUP | | | | | | | | | | | | |
| TIME AVG.TEMP | 18.7 | 21.1 | 21.8 | 23.8 | 24.6 | 25.5 | 25.8 | 26.6 | 26.4 | 27.3 | 26.6 | 27.1 |
| WATT/SQ.M | 0. | 44. | 146. | 239. | 326. | 426. | 518. | 616. | 711. | 802. | 893. | 1001. |
| TOTAL TIME MSEC | 1.3392 | 0.2956 | 0.1638 | 0.1912 | 0.1634 | 0.1267 | 0.1156 | 0.0979 | 0.0668 | 0.0511 | 0.0407 | 0.0245 |
| GLOBAL RAD TOTAL | 0.0 | 13.2 | 23.9 | 45.6 | 53.2 | 54.0 | 59.8 | 60.3 | 48.9 | 41.0 | 36.3 | 24.5 |
| DIRECT RAD TOTAL | 0.0 | 2.7 | 1.3 | 8.7 | 15.7 | 20.2 | 29.1 | 31.4 | 30.9 | 28.7 | 29.0 | 21.5 |
| DEG-C DAYS | 7. | 1. | 1. | 2. | 0. | 0. | 0. | 0. | 0. | 0. | 0. | 0. |

**SEP**

| | 0 | 2-95 | 95-189 | 189-284 | 284-378 | 378-473 | 473-568 | 568-662 | 662-757 | 757-851 | 851-946 | OVER 946 |
|---|---|---|---|---|---|---|---|---|---|---|---|---|
| SOLAR RAD GROUP | | | | | | | | | | | | |
| TIME AVG.TEMP | 14.5 | 16.5 | 18.7 | 19.2 | 19.7 | 22.3 | 21.5 | 22.3 | 23.5 | 22.6 | 23.9 | 22.6 |
| WATT/SQ.M | 0. | 42. | 142. | 236. | 326. | 429. | 527. | 613. | 712. | 799. | 904. | 1008. |
| TOTAL TIME MSEC | 1.2780 | 0.3336 | 0.1746 | 0.1462 | 0.1289 | 0.0983 | 0.1055 | 0.0724 | 0.0745 | 0.0616 | 0.0518 | 0.0767 |
| GLOBAL RAD TOTAL | 0.0 | 13.5 | 24.8 | 34.5 | 42.1 | 42.2 | 55.5 | 44.3 | 53.1 | 49.2 | 46.9 | 78.9 |
| DIRECT RAD TOTAL | 0.0 | 3.6 | 9.0 | 19.8 | 23.2 | 33.7 | 28.6 | 39.3 | 39.3 | 40.5 | 40.5 | 71.2 |
| DEG-C DAYS | 53. | 8. | 3. | 2. | 2. | 1. | 1. | 1. | 0. | 1. | 0. | 1. |

**OCT**

| | 0 | 2-95 | 95-189 | 189-284 | 284-378 | 378-473 | 473-568 | 568-662 | 662-757 | 757-851 | 851-946 | OVER 946 |
|---|---|---|---|---|---|---|---|---|---|---|---|---|
| SOLAR RAD GROUP | | | | | | | | | | | | |
| TIME AVG.TEMP | 8.8 | 11.2 | 13.4 | 14.2 | 14.2 | 15.6 | 15.6 | 14.7 | 16.8 | 15.8 | 10.1 | 17.4 |
| WATT/SQ.M | 0. | 43. | 141. | 237. | 326. | 428. | 525. | 615. | 706. | 806. | 894. | 1006. |
| TOTAL TIME MSEC | 1.3291 | 0.3881 | 0.2225 | 0.1426 | 0.1170 | 0.0842 | 0.0770 | 0.0803 | 0.0515 | 0.0537 | 0.0432 | 0.0029 |
| GLOBAL RAD TOTAL | 0.0 | 16.7 | 31.4 | 33.8 | 38.2 | 36.1 | 40.4 | 49.3 | 36.3 | 47.3 | 33.6 | 9.3 |
| DIRECT RAD TOTAL | 0.0 | 5.3 | 8.3 | 13.6 | 20.4 | 22.7 | 26.8 | 36.9 | 23.6 | 40.7 | 33.9 | 67.2 |
| DEG-C DAYS | 142. | 31. | 13. | 8. | 6. | 3. | 3. | 4. | 2. | 2. | 1. | 3. |

**NOV**

| | 0 | 2-95 | 95-189 | 189-284 | 284-378 | 378-473 | 473-568 | 568-662 | 662-757 | 757-851 | 851-946 | OVER 946 |
|---|---|---|---|---|---|---|---|---|---|---|---|---|
| SOLAR RAD GROUP | | | | | | | | | | | | |
| TIME AVG.TEMP | 3.8 | 6.7 | 7.0 | 7.4 | 7.8 | 6.9 | 6.7 | 7.0 | 8.0 | 8.4 | 8.1 | 6.9 |
| WATT/SQ.M | 0. | 39. | 137. | 235. | 330. | 431. | 519. | 614. | 716. | 803. | 902. | 1006. |
| TOTAL TIME MSEC | 1.4180 | 0.4716 | 0.2052 | 0.1235 | 0.0310 | 0.0565 | 0.0515 | 0.0457 | 0.0428 | 0.0335 | 0.0302 | 0.0324 |
| GLOBAL RAD TOTAL | 0.1 | 18.5 | 28.2 | 29.0 | 26.7 | 24.3 | 26.7 | 28.1 | 30.7 | 26.0 | 27.3 | 32.6 |
| DIRECT RAD TOTAL | 0.0 | 2.5 | 9.0 | 14.0 | 13.7 | 16.5 | 20.9 | 22.8 | 26.0 | 23.5 | 24.6 | 30.3 |
| DEG-C DAYS | 239. | 63. | 27. | 16. | 10. | 8. | 7. | 6. | 5. | 4. | 4. | 3. |

**DEC**

| | 0 | 2-95 | 95-189 | 189-284 | 284-378 | 378-473 | 473-568 | 568-662 | 662-757 | 757-851 | 851-946 | OVER 946 |
|---|---|---|---|---|---|---|---|---|---|---|---|---|
| SOLAR RAD GROUP | | | | | | | | | | | | |
| TIME AVG.TEMP | -3.0 | -0.5 | -0.2 | -0.7 | -0.4 | 0.6 | -0.7 | 2.2 | 1.1 | -1.3 | -0.6 | -1.9 |
| WATT/SQ.M | 0. | 42. | 139. | 234. | 330. | 424. | 505. | 612. | 711. | 809. | 908. | 975. |
| TOTAL TIME MSEC | 1.5523 | 0.4687 | 0.1735 | 0.1145 | 0.0684 | 0.0522 | 0.0508 | 0.0536 | 0.1034 | 0.0472 | 0.0328 | 0.0031 |
| GLOBAL RAD TOTAL | 0.0 | 19.6 | 24.2 | 26.8 | 22.5 | 22.2 | 26.7 | 32.8 | 25.8 | 38.1 | 32.8 | 27.4 |
| DIRECT RAD TOTAL | 0.0 | 3.5 | 8.3 | 14.6 | 14.6 | 15.4 | 21.7 | 28.7 | 22.6 | 34.9 | 29.3 | 25.6 |
| DEG-C DAYS | 384. | 102. | 36. | 25. | 15. | 11. | 11. | 13. | 8. | 11. | 7. | 7. |

**TABLE 3-6b.** Summary Radiation and Weather Information at 50° Inclination at Hartford

STATISTICS FOR HARTFORD, CONN.                1959-68

RADIATION UNITS ARE WATTS/SQ.M OR MJOULES/SQ.M
SLOPE = 50.0 DEG   AZIMUTH = 180.0 DEG
DEGREE-DAY BASE TEMP = 18.3 DEGREES C

**JAN**

| SOLAR RAD GROUP | ALL | 0+ | 95+ | 189+ | 284+ | 378+ | 473+ | 568+ | 662+ | 757+ | 851+ | 946+ |
|---|---|---|---|---|---|---|---|---|---|---|---|---|
| TIME AVG.TEMP | -3.8 | -2.4 | -2.3 | -2.7 | -2.7 | -2.8 | -3.3 | -3.6 | -3.9 | -4.1 | -4.4 | -5.2 |
| TOTAL TIME MSEC | 2.6784 | 1.2186 | 0.7308 | 0.5591 | 0.4280 | 0.3355 | 0.2790 | 0.2243 | 0.1706 | 0.1242 | 0.0738 | 0.0357 |
| GLOBAL RAD TOTAL | 335.1 | 333.1 | 314.8 | 291.0 | 259.8 | 231.0 | 205.2 | 176.5 | 143.7 | 110.6 | 70.0 | 36.4 |
| DIRECT RAD TOTAL | 233.3 | 235.3 | 235.2 | 228.9 | 212.4 | 195.7 | 179.1 | 157.1 | 130.0 | 101.2 | 64.7 | 34.0 |
| RAD AV TIME MSEC | 0.0394 | 0.0396 | 0.0302 | 0.0281 | 0.0248 | 0.0223 | 0.0209 | 0.0198 | 0.0176 | 0.0158 | 0.0122 | 0.0090 |
| DEG-C DAYS | 688. | 292. | 175. | 136. | 104. | 83. | 70. | 57. | 44. | 32. | 19. | 10. |

**FEB**

| SOLAR RAD GROUP | ALL | 0+ | 95+ | 189+ | 284+ | 378+ | 473+ | 568+ | 662+ | 757+ | 851+ | 946+ |
|---|---|---|---|---|---|---|---|---|---|---|---|---|
| TIME AVG.TEMP | -3.1 | -1.6 | -1.5 | -1.6 | -1.8 | -1.7 | -1.8 | -2.2 | -2.2 | -2.2 | -2.6 | -2.5 |
| TOTAL TIME MSEC | 2.4451 | 1.1197 | 0.8046 | 0.6077 | 0.4907 | 0.3394 | 0.3172 | 0.2556 | 0.1976 | 0.1573 | 0.1062 | 0.0727 |
| GLOBAL RAD TOTAL | 377.5 | 377.4 | 352.3 | 334.9 | 307.2 | 273.4 | 243.3 | 211.2 | 175.2 | 146.8 | 105.4 | 75.2 |
| DIRECT RAD TOTAL | 257.1 | 257.1 | 254.7 | 247.0 | 237.8 | 220.5 | 203.6 | 184.5 | 156.0 | 132.5 | 95.3 | 69.4 |
| RAD AV TIME MSEC | 0.0364 | 0.0425 | 0.0342 | 0.0302 | 0.0284 | 0.0256 | 0.0234 | 0.0223 | 0.0198 | 0.0180 | 0.0148 | 0.0126 |
| DEG-C DAYS | 606. | 272. | 184. | 140. | 114. | 90. | 74. | 61. | 47. | 37. | 26. | 18. |

**MAR**

| SOLAR RAD GROUP | ALL | 0+ | 95+ | 189+ | 284+ | 378+ | 473+ | 568+ | 662+ | 757+ | 851+ | 946+ |
|---|---|---|---|---|---|---|---|---|---|---|---|---|
| TIME AVG.TEMP | 2.1 | 3.6 | 4.3 | 4.7 | 5.1 | 5.1 | 5.2 | 5.3 | 5.2 | 4.9 | 5.0 | 4.6 |
| TOTAL TIME MSEC | 2.6784 | 1.3907 | 0.9839 | 0.7434 | 0.5782 | 0.4853 | 0.3920 | 0.2963 | 0.2473 | 0.1879 | 0.1393 | 0.0923 |
| GLOBAL RAD TOTAL | 459.6 | 458.6 | 443.7 | 409.6 | 371.1 | 340.4 | 300.3 | 250.7 | 220.4 | 178.1 | 139.4 | 100.9 |
| DIRECT RAD TOTAL | 280.7 | 278.5 | 254.7 | 273.6 | 260.8 | 250.9 | 231.5 | 202.6 | 185.0 | 154.3 | 123.2 | 90.6 |
| RAD AV TIME MSEC | 0.0364 | 0.0450 | 0.0356 | 0.0331 | 0.0295 | 0.0284 | 0.0259 | 0.0223 | 0.0212 | 0.0187 | 0.0153 | 0.0137 |
| DEG-C DAYS | 505. | 238. | 160. | 117. | 69. | 74. | 60. | 45. | 38. | 29. | 22. | 16. |

**APR**

| SOLAR RAD GROUP | ALL | 0+ | 95+ | 189+ | 284+ | 378+ | 473+ | 568+ | 662+ | 757+ | 851+ | 946+ |
|---|---|---|---|---|---|---|---|---|---|---|---|---|
| TIME AVG.TEMP | 9.4 | 11.8 | 12.7 | 13.4 | 14.1 | 14.8 | 14.8 | 15.3 | 15.4 | 15.2 | 15.3 | 14.8 |
| TOTAL TIME MSEC | 2.5920 | 1.2931 | 0.9976 | 0.7906 | 0.6275 | 0.4842 | 0.3831 | 0.3049 | 0.2216 | 0.1739 | 0.1159 | 0.0691 |
| GLOBAL RAD TOTAL | 459.5 | 459.5 | 446.9 | 417.1 | 379.0 | 331.5 | 290.6 | 247.1 | 203.3 | 160.0 | 113.7 | 71.4 |
| DIRECT RAD TOTAL | 239.8 | 239.8 | 237.4 | 235.5 | 230.1 | 214.3 | 199.6 | 177.2 | 153.4 | 129.6 | 96.5 | 63.4 |
| RAD AV TIME MSEC | 0.0364 | 0.0432 | 0.0356 | 0.0331 | 0.0310 | 0.0277 | 0.0256 | 0.0223 | 0.0225 | 0.0176 | 0.0151 | 0.0126 |
| DEG-C DAYS | 267. | 101. | 69. | 48. | 34. | 34. | 19. | 14. | 11. | 8. | 6. | 4. |

**MAY**

| SOLAR RAD GROUP | ALL | 0+ | 95+ | 189+ | 284+ | 378+ | 473+ | 568+ | 662+ | 757+ | 851+ | 946+ |
|---|---|---|---|---|---|---|---|---|---|---|---|---|
| TIME AVG.TEMP | 15.0 | 17.9 | 18.5 | 19.5 | 20.2 | 20.9 | 21.1 | 21.4 | 21.3 | 21.2 | 21.2 | 21.2 |
| TOTAL TIME MSEC | 2.6784 | 1.3392 | 1.1189 | 0.9374 | 0.7700 | 0.6005 | 0.4676 | 0.3355 | 0.2365 | 0.1516 | 0.0693 | 0.0396 |
| GLOBAL RAD TOTAL | 512.3 | 512.3 | 502.1 | 476.7 | 436.8 | 391.2 | 324.7 | 255.5 | 194.8 | 134.4 | 84.5 | 39.7 |
| DIRECT RAD TOTAL | 226.5 | 226.5 | 224.9 | 223.0 | 218.4 | 201.5 | 185.9 | 155.7 | 129.5 | 101.1 | 69.9 | 34.5 |
| RAD AV TIME MSEC | 0.0364 | 0.0432 | 0.0356 | 0.0342 | 0.0317 | 0.0274 | 0.0241 | 0.0205 | 0.0160 | 0.0162 | 0.0133 | 0.0101 |
| DEG-C DAYS | 120. | 33. | 24. | 14. | 9. | 6. | 4. | 3. | 2. | 1. | 1. | 0. |

**JUN**

| SOLAR RAD GROUP | ALL | 0+ | 95+ | 189+ | 284+ | 378+ | 473+ | 568+ | 662+ | 757+ | 851+ | 946+ |
|---|---|---|---|---|---|---|---|---|---|---|---|---|
| TIME AVG.TEMP | 20.2 | 23.0 | 23.6 | 24.6 | 25.4 | 26.1 | 26.2 | 26.9 | 26.9 | 27.1 | 26.8 | 27.3 |
| TOTAL TIME MSEC | 2.5920 | 1.2960 | 1.0757 | 0.8956 | 0.7243 | 0.5623 | 0.4446 | 0.3157 | 0.2239 | 0.1228 | 0.0630 | 0.0155 |
| GLOBAL RAD TOTAL | 480.3 | 480.3 | 468.7 | 443.1 | 404.2 | 351.2 | 300.9 | 233.8 | 177.5 | 105.6 | 57.5 | 15.4 |
| DIRECT RAD TOTAL | 206.2 | 206.2 | 204.3 | 203.2 | 199.3 | 183.3 | 168.2 | 139.1 | 115.3 | 76.7 | 46.3 | 13.6 |
| RAD AV TIME MSEC | 0.0364 | 0.0432 | 0.0371 | 0.0342 | 0.0324 | 0.0277 | 0.0245 | 0.0205 | 0.0176 | 0.0137 | 0.0115 | 0.0072 |
| DEG-C DAYS | 120. | 33. | 24. | 14. | 9. | 6. | 4. | 3. | 2. | 0. | 0. | 0. |

## JUL

| SOLAR RAD GROUP | ALL | 0+ | 95+ | 189+ | 284+ | 378+ | 473+ | 568+ | 662+ | 757+ | 851+ | 946+ |
|---|---|---|---|---|---|---|---|---|---|---|---|---|
| TIME AVG.TEMP | 22.6 | 25.4 | 26.1 | 26.7 | 27.2 | 27.8 | 27.9 | 28.4 | 28.5 | 28.9 | 28.9 | 28.2 |
| TOTAL TIME MSEC | 2.6784 | 1.3392 | 1.0915 | 0.9432 | 0.7844 | 0.5962 | 0.4496 | 0.3172 | 0.2128 | 0.1130 | 0.0556 | 0.0126 |
| GLOBAL RAD TOTAL | 495.8 | 496.8 | 484.0 | 463.8 | 425.0 | 363.0 | 300.3 | 231.5 | 167.3 | 96.9 | 49.1 | 12.3 |
| DIRECT RAD TOTAL | 222.7 | 222.7 | 220.3 | 219.2 | 213.2 | 194.8 | 175.6 | 145.4 | 115.6 | 74.2 | 40.3 | 10.9 |
| RAD AV TIME MSEC | 0.0864 | 0.0432 | 0.0364 | 0.0342 | 0.0317 | 0.0270 | 0.0233 | 0.0198 | 0.0166 | 0.0133 | 0.0104 | 0.0072 |
| DEG-C DAYS | 1. | 0. | 0. | 1. | 0. | 0. | 0. | 0. | 0. | 0. | 0. | 0. |

## AUG

| SOLAR RAD GROUP | ALL | 0+ | 95+ | 189+ | 284+ | 378+ | 473+ | 568+ | 662+ | 757+ | 851+ | 946+ |
|---|---|---|---|---|---|---|---|---|---|---|---|---|
| TIME AVG.TEMP | 21.3 | 24.0 | 24.8 | 25.4 | 25.8 | 26.2 | 26.4 | 26.7 | 26.8 | 27.0 | 26.8 | 27.1 |
| TOTAL TIME MSEC | 2.6784 | 1.3392 | 1.0436 | 0.8738 | 0.6897 | 0.5252 | 0.3905 | 0.2830 | 0.1850 | 0.1163 | 0.0652 | 0.0045 |
| GLOBAL RAD TOTAL | 460.8 | 460.8 | 447.7 | 423.8 | 378.2 | 334.9 | 270.9 | 211.0 | 159.7 | 101.8 | 60.9 | 24.5 |
| DIRECT RAD TOTAL | 219.2 | 219.2 | 216.6 | 215.2 | 208.5 | 190.8 | 170.6 | 145.0 | 110.1 | 79.2 | 50.5 | 21.5 |
| RAD AV TIME MSEC | 0.0864 | 0.0432 | 0.0349 | 0.0331 | 0.0302 | 0.0266 | 0.0230 | 0.0194 | 0.0169 | 0.0144 | 0.0110 | 0.0036 |
| DEG-C DAYS | 8. | 2. | 1. | 1. | 0. | 0. | 0. | 0. | 0. | 0. | 0. | 0. |

## SEP

| SOLAR RAD GROUP | ALL | 0+ | 95+ | 189+ | 284+ | 378+ | 473+ | 568+ | 662+ | 757+ | 851+ | 946+ |
|---|---|---|---|---|---|---|---|---|---|---|---|---|
| TIME AVG.TEMP | 17.2 | 19.9 | 21.0 | 21.5 | 22.0 | 22.6 | 22.6 | 22.9 | 23.1 | 22.9 | 23.1 | 22.6 |
| TOTAL TIME MSEC | 2.5920 | 1.3140 | 0.9904 | 0.8158 | 0.6696 | 0.5407 | 0.4424 | 0.3370 | 0.2646 | 0.1901 | 0.1265 | 0.0767 |
| GLOBAL RAD TOTAL | 484.9 | 484.9 | 471.3 | 446.5 | 412.1 | 370.0 | 327.9 | 272.3 | 228.0 | 174.9 | 125.7 | 78.0 |
| DIRECT RAD TOTAL | 312.3 | 312.3 | 308.7 | 304.8 | 295.8 | 276.0 | 252.8 | 219.1 | 190.5 | 151.2 | 111.7 | 71.2 |
| RAD AV TIME MSEC | 0.0864 | 0.0439 | 0.0356 | 0.0331 | 0.0313 | 0.0277 | 0.0245 | 0.0216 | 0.0198 | 0.0173 | 0.0148 | 0.0119 |
| DEG-C DAYS | 71. | 18. | 9. | 7. | 4. | 3. | 2. | 2. | 1. | 1. | 1. | 1. |

## OCT

| SOLAR RAD GROUP | ALL | 0+ | 95+ | 189+ | 284+ | 378+ | 473+ | 568+ | 662+ | 757+ | 851+ | 946+ |
|---|---|---|---|---|---|---|---|---|---|---|---|---|
| TIME AVG.TEMP | 11.3 | 13.8 | 14.9 | 15.3 | 15.8 | 16.2 | 16.3 | 16.4 | 17.0 | 17.1 | 17.6 | 17.4 |
| TOTAL TIME MSEC | 2.6970 | 1.3579 | 0.9698 | 0.7474 | 0.6048 | 0.4878 | 0.4036 | 0.3265 | 0.2452 | 0.1948 | 0.1351 | 0.0929 |
| GLOBAL RAD TOTAL | 463.5 | 463.5 | 446.8 | 415.5 | 331.6 | 343.4 | 307.4 | 266.9 | 217.6 | 181.2 | 133.9 | 95.3 |
| DIRECT RAD TOTAL | 324.5 | 324.5 | 319.3 | 311.0 | 297.3 | 276.9 | 254.3 | 227.4 | 190.5 | 161.9 | 121.1 | 87.0 |
| RAD AV TIME MSEC | 0.0864 | 0.0439 | 0.0356 | 0.0324 | 0.0299 | 0.0270 | 0.0245 | 0.0223 | 0.0194 | 0.0180 | 0.0148 | 0.0122 |
| DEG-C DAYS | 219. | 77. | 47. | 34. | 26. | 19. | 16. | 13. | 8. | 7. | 4. | 3. |

## NOV

| SOLAR RAD GROUP | ALL | 0+ | 95+ | 189+ | 284+ | 378+ | 473+ | 568+ | 662+ | 757+ | 851+ | 946+ |
|---|---|---|---|---|---|---|---|---|---|---|---|---|
| TIME AVG.TEMP | 5.3 | 7.1 | 7.4 | 7.6 | 7.6 | 7.6 | 7.7 | 8.0 | 8.3 | 8.5 | 8.6 | 8.9 |
| TOTAL TIME MSEC | 2.5920 | 1.1740 | 0.7024 | 0.4972 | 0.3737 | 0.2927 | 0.2362 | 0.1847 | 0.1390 | 0.0961 | 0.0626 | 0.0324 |
| GLOBAL RAD TOTAL | 299.1 | 299.0 | 280.6 | 252.3 | 223.3 | 196.6 | 172.2 | 145.5 | 117.5 | 86.8 | 59.9 | 32.6 |
| DIRECT RAD TOTAL | 203.8 | 203.8 | 201.3 | 192.4 | 178.3 | 164.6 | 148.1 | 127.2 | 104.4 | 78.4 | 59.9 | 30.3 |
| RAD AV TIME MSEC | 0.0864 | 0.0396 | 0.0320 | 0.0284 | 0.0252 | 0.0238 | 0.0220 | 0.0194 | 0.0169 | 0.0151 | 0.0122 | 0.0101 |
| DEG-C DAYS | 392. | 152. | 89. | 62. | 47. | 37. | 29. | 22. | 16. | 11. | 7. | 3. |

## DEC

| SOLAR RAD GROUP | ALL | 0+ | 95+ | 189+ | 284+ | 378+ | 473+ | 568+ | 662+ | 757+ | 851+ | 946+ |
|---|---|---|---|---|---|---|---|---|---|---|---|---|
| TIME AVG.TEMP | -2.0 | -0.6 | -0.7 | -0.8 | -0.9 | -1.0 | -1.3 | -1.5 | -1.2 | -1.3 | -1.2 | -1.9 |
| TOTAL TIME MSEC | 2.6784 | 1.1261 | 0.6574 | 0.4838 | 0.3694 | 0.3010 | 0.2468 | 0.1980 | 0.1444 | 0.1080 | 0.0608 | 0.0281 |
| GLOBAL RAD TOTAL | 296.0 | 299.9 | 276.3 | 252.1 | 191.0 | 202.8 | 180.6 | 153.9 | 121.1 | 95.3 | 57.1 | 27.4 |
| DIRECT RAD TOTAL | 217.5 | 217.5 | 213.9 | 205.6 | 191.0 | 176.4 | 161.0 | 139.3 | 110.7 | 83.0 | 53.1 | 25.6 |
| RAD AV TIME MSEC | 0.0864 | 0.0354 | 0.0302 | 0.0274 | 0.0245 | 0.0227 | 0.0220 | 0.0198 | 0.0166 | 0.0158 | 0.0112 | 0.0079 |
| DEG-C DAYS | 631. | 247. | 145. | 107. | 82. | 67. | 57. | 46. | 33. | 24. | 14. | 7. |

are prepared specifically for use with techniques to be developed in later chapters, taking advantage of the ease with which a digital computer can make conveniently processed information available. The student may, therefore, wish to delay study of the remainder of this chapter until Chapter 4 has been read through the "Summary Radiation Tables" section on p. 149.

Table 3-6a shows the primary output produced by processing the hourly radiation data that was synthesized for Hartford with a south-facing collector tilted at 50° to the horizontal. To make such a table, global hourly radiation data are placed into one of 12 radiation groupings according to their radiation intensity (see the SOLAR RAD GROUP entries). Within each group the global and direct radiation in megajoules have been totalized (as GLOBAL RAD TOTAL and DIRECT RAD TOTAL), and the total time elapsed while the radiation was within the specific levels is also noted (TOTAL TIME MSEC). Time is given in megaseconds because that unit is consistent with the units used for the radiation totals. From this information the average global irradiance within the grouping, WATT/SQ. M, has been computed (this is the global total radiation divided by the total time).

The average ambient temperature when the irradiance was within the stated irradiance limits of each group is also kept (TIME AVG. TEMP) as well as degree-days (DEG-C DAYS).

Tabulation of this detailed radiation information is made so that later we may consider only that radiation occuring when the collector is actually activated by circulating the fluid to be heated. Activation occurs only when the radiation exceeds a *threshold* radiation flux, which is the irradiance necessary to balance the thermal losses from the collector with the solar gains. The threshold radiation flux for a particular collector depends on the ambient temperature at the radiation threshold (i.e., when it is about to go into operation) and a number of collector-related physical parameters.

Each irradiance grouping in Table 3-6a summarizes the monthly data, applying when the radiation was at the average threshold irradiance listed WATT/SQ. M. Now it is apparent that if a collector is activated at a given irradiance it will be activated for *all higher levels* as well. It will therefore be useful to sum all the radiation values and operating times occurring within *and above* a particular radiation grouping and to compute a new overall average temperature, while retaining the average radiation and temperature *at* the radiation threshold level for identification of the proper irradiance grouping. Table 3-6b has been prepared from Table 3-6a in this manner. The new tabular groupings are identified by the lower irradiance limit from Table 3-6a (SOLAR RAD GROUP). The time totals (TOTAL TIME MSEC), radiation totals (GLOBAL RAD TOTAL and DIRECT RAD TOTAL), and degree-day totals (DEG-C DAYS) are computed from Table 3-6a by adding to the value in the corresponding column all the values from columns to the right (i.e., from higher radiation levels). Thus the time, radiation, and degree-day totals are the totals accrued while the radiation level exceeded the level listed in the SOLAR RAD GROUP entry. The TIME AVG. TEMP entry has been calculated to give the average temperature corresponding to the radiation and time totals.

The Appendix B summary table on page 558 has been derived from both Tables 3-6a and 3-6b. The first seven radiation columns from Table 3-6b have been combined with the corresponding threshold irradiance (AV. THRSH. RAD. LEVL) and the average temperature at threshold (AV. THRESHOLD TEMP) from Table 3-6a.

The losses from a solar collector are independent of the radiation level, and indeed occur at a *rate* equal to the radiation threshold level (AV. THRSH. RAD. LEVL) during the entire total *time* of operation (TOTAL TIME MSEC). Subtracting the product of these two entries (i.e., the total loss from a collector with the listed threshold) from the total global radiation available gives the total radiation *utilizable*, entered in the Appendix B summary table as GL RAD OVER THRSH.

Solar collector performance can also be estimated from the utilizable radiation and the time-averaged temperature. As will be seen in the next section, the utilizable radiation can be predicted with reasonable accuracy from the total global radiation and the radiation threshold for locations at which detailed tabular information is unavailable.

**EXAMPLE 3-8**

For the month of February, assemble the 284+ W/m² radiation group information given in Table 3-6b and Appendix B from the data in Table 3-6a that applies to Hartford (1959–1968) at a collector slope of 50°.

The AV. THRSH. RAD. LEVL (331 W/m²) and AV. THRESHOLD TEMP (−2.0 °C) are taken directly from the WATT/SQ. M and TIME AVG. TEMP entries in the 284 to 378 W/m² SOLAR RAD GROUP category of Table 3-6a. TOTAL TIME MSEC (0.4907 Ms), GLOBAL RAD TOTAL, and DEG-C DAYS are the sum of the 284-378 W/m² values and those from the higher levels:

$$
\begin{aligned}
\text{TOTAL TIME MSEC} &= 0.1022 + 0.0713 + 0.0616 + 0.0580 \\
&\quad + 0.0403 + 0.0511 + 0.0335 + 0.0727 \\
&= 0.4907 \text{ Ms}
\end{aligned}
$$

$$
\begin{aligned}
\text{GLOBAL RAD TOTAL} &= 33.8 + 30.1 + 32.1 + 35.9 + 28.5 + \\
&\quad + 41.4 + 30.2 + 75.2 = 307.2 \text{ MJ/m}^2
\end{aligned}
$$

$$
\begin{aligned}
\text{DIRECT RAD TOTAL} &= 17.4 + 16.9 + 19.1 + 28.5 + 23.5 \\
&\quad + 36.2 + 26.8 + 69.4 = 237.8 \text{ MJ/m}^2
\end{aligned}
$$

$$
\begin{aligned}
\text{DEG-C DAYS} &= 24 + 16 + 13 + 14 + 9 + 12 + 8 + 18 \\
&= 114 \text{ °C-day}
\end{aligned}
$$

The TIME AVG. TEMP entry ($-1.8\,°C$) is computed by time averaging the time-averaged temperature from Table 3-6a that applies to the 284–378 W/m² solar radiation level with those from all the higher levels:

TIME AVG. TEMP = $[(-2.0)(0.1022) + (-1.2)(0.0713) + (-0.4)(0.0616)$
$+ (-2.2)(0.0580) + (-1.9)(0.0403) + (-1.4)(0.0511)$
$+ (-2.9)(0.0335) + (-2.5)(0.0727)]/(0.1022 + 0.0713$
$+ 0.0616 + 0.0580 + 0.0403 + 0.0511$
$+ 0.0335 + 0.0727) = -1.8\,°C$

The GL RAD OVER THRSH entry is computed from the just-generated GLOBAL RAD TOTAL (307.2 MJ/m²), TOTAL TIME MSEC (0.4907 Ms), and the AV. THRSH. RAD. LEVEL (331 W/m²):

GL RAD OVER THRSH = 307.2 MJ/m² $-$ (331 W/m²)(0.4907 Ms)
$= 144.8$ MJ/m²
$\simeq 145.0$ MJ/m²

The RAD AV TIME MSEC entry will be explained in Chapter 9. It can be regarded as the typical daily time period for solar collection above the stated radiation threshold level.

**Prediction of Utilizable Radiation.**  When no radiation information is available beyond monthly average data, the utilizable radiation can be predicted by correlation with other data. The simple correlation detailed below is based on an analysis of the data in Appendix B, which was computed from cloud cover information.*

The total radiation utilizable to the collector—the total of the global radiation that occurred beyond and in excess of the radiation threshold—can be empirically related, on a monthly basis, to the total global radiation on the tilted surface by the following equation:

$$I_{T+} = I_T e^{-kI_{th}} \tag{3-34}$$

$I_{T+}$ is the utilizable radiation above the threshold radiation level $I_{th}$, and $I_T$ is the total global radiation on the tilted collector. The utilizability constant $k$, which determines how quickly the utilizable radiation falls off as the threshold value increases, is essentially constant until the utilizable radiation is less than

---

* A complex graphical method for prediction of utilizable radiation from monthly average data has been given by Liu and Jordan (4b, 25). The correlations are no longer regarded as correct because shade ring corrections were not made, leading to higher estimates of direct radiation than reported elsewhere (13, 16). Appendix B data, however, are consistent in their direct/diffuse ratio with important Australian data (16). As real data of high quality become available, it is possible that eq. 3-36 will be updated, but it is likely that its form will persist.

**TABLE 3-7.** Values of $k$ to be used in eq. 3-34, m²/W

| $\bar{K}_T$ | $\tilde{\theta}$ 0 to ±10° | ±11 to ±25° | ±26 to ±40° |
|---|---|---|---|
| 0.35 | 0.00360 | 0.0038 | 0.0042 |
| 0.40 | 0.00335 | 0.0035 | 0.0039 |
| 0.45 | 0.00310 | 0.0033 | 0.0037 |
| 0.50 | 0.00285 | 0.0031 | 0.0035 |
| 0.55 | 0.00255 | 0.0028 | 0.0032 |
| 0.60 | 0.00230 | 0.0025 | 0.0029 |
| 0.65 | 0.00205 | 0.0023 | 0.0026 |

one-third of the original total. It is given in Table 3-7 as a function of the cloudiness (expressed as $\bar{K}_T$) and the incident angle of the sun at solar noon:

$$\tilde{\theta} = \beta - \phi + \delta \tag{3-35}$$

or it can be expressed analytically, where $\tilde{\theta}$ is given in degrees, as

$$k = 0.00542 - 0.0052\bar{K}_T + 0.59(\tilde{\theta})^2/10^6 \tag{3-36}$$

The final value of $k$ is sufficiently precise if rounded to two significant figures. The magnitude of $\tilde{\theta}$ is tabulated for each month in Tables A-7 to A-12, or it can be worked out directly by eq. 3-35 from the tilt angle $\beta$, latitude $\phi$, and declination $\delta$ (eq. 3-2). When $\tilde{\theta}$ is positive the collector is inclined at greater than the optimum tilt for a particular month, thereby tending to be negative in winter and positive in summer. If $\bar{K}_T$ is not directly available for the site of interest, it can be estimated from data at a nearby city with enough accuracy for the use of eq. 3-36 or Table 3-7. $\bar{K}_T$ should first be reduced by about 0.035 for high-altitude cities (e.g., Denver) to obtain a value more indicative of actual cloud cover with which to enter the table.

When tabular data like that in Appendix B are available, the value of the constant $k$ for any listed data grouping can be found by solving eq. 3-34 for $k$:

$$k = \frac{\ln (I_T/I_{T+}^*)}{I_{th}^*} \tag{3-37}$$

where $I_{th}^*$ and $I_{T+}^*$ are corresponding values of AV. THRSH. RAD. LEVEL and GL RAD OVER THRSH, respectively, from the data grouping selected, and $I_T$ is taken from the ALL column.

## EXAMPLE 3-9

Estimate the total global radiation available above 233 W/m² for Hartford in March with a collector tilt of 50°. Use as a basis the listed tabular value of 458.5

MJ/m² for total global radiation on the tilted collector. Compare the result with the listed value of 236.4 MJ/m² for the total global radiation above that threshold.

For March, Table A-9 gives $|\bar{\theta}| = 10°$. At 0° elevation, the March total global radiation is 407.4 MJ/m², while the extraterrestrial horizontal radiation $H_{0T}$ is 26.81 MJ/m² day at 42°N latitude (Table 3-5). Therefore,

$$\bar{K}_T = \frac{407.4}{26.81 \times 31} = 0.49$$

and from Table 3-7, $k = 0.0029$, or from eq. 3-36

$$k = 0.00542 - 0.0052(0.49) + 0.59(10)^2/10^6 = 0.0029$$

From eq. 3-34

$$I_{T+} = \frac{458.5 \text{ MJ}}{\text{m}^2} \exp\left(\frac{-0.0029 \text{ m}^2}{\text{W}} \times \frac{233 \text{ W}}{\text{m}^2}\right)$$

$$I_{T+} = 233 \text{ MJ/m}^2$$

which is quite close to the tabulated value.

**EXAMPLE 3-10**

For November at a 40°N location ($\bar{K}_T = 0.416$) the daily average total radiation on a collector inclined at 60° was estimated at 9.99 MJ/m² (Example 3-7). What is the total global radiation utilizable above a threshold of 175 W/m²?

From Table A-9, $|\bar{\theta}| = 0°$. Since $\bar{K}_T = 0.416$, from Table 3-6 or eq. 3-36 $k = 0.0033$ m²/W. From eq. 3-34,

$$I_{T+} = \frac{9.99 \text{ MJ}}{\text{m}^2} \exp\left(\frac{-0.0033 \text{ m}^2}{\text{W}} \times \frac{175 \text{ W}}{\text{m}^2}\right)$$

$$I_{T+} = 5.61 \text{ MJ/m}^2$$

The procedures introduced in this chapter permit calculation of the utilizable global radiation—that in excess of a threshold—from the total global radiation on an inclined surface, or by an elevation routine from data relating to a horizontal surface. When this alternate to use of the detailed radiation tables of Appendix B is planned, the radiation and weather data are ordinarily calculated and tabulated in advance as demonstrated in Example 3-11.

**EXAMPLE 3-11**

A south-facing solar collector is inclined at 40° to the horizontal at Hartford, Connecticut (latitude 41°50′). Using only the horizontal ALL data from Appendix B, for each month tabulate the total radiation, the constant $k$, 24-hour average temperature, daytime average temperature, and degree-days. Elevate the horizontal data using the modified Page equation (3-28), assuming a ground reflectivity of 0.2.

**TABLE 3-8.** Elevation of Horizontal Data to a 40° Tilt Angle Using the Modified Page equation (3-28) (Example 3-11)

| | Basic Data | | | Intermediate Values | | | | | | | | | Tilted Data | | | |
|---|---|---|---|---|---|---|---|---|---|---|---|---|---|---|---|---|
| | $\bar{T}_a$ °C (24 hr) | $D_m$ °C-day (Monthly) | $H_T$ MJ/m² (Monthly) | $\bar{H}_T$ MJ/m² (Daily) | $\bar{H}_{oT}$ MJ/m² (Daily) | $\bar{I}_{oT}$ MJ/m² (Daily) | $R_b$ (Ratio) | $\bar{K}_T$ (Ratio) | $\bar{H}_{Td}/\bar{H}_T$ (Ratio) | $\bar{R}$ (Ratio) | $\bar{I}_T$ MJ/m² (Daily) | $\tilde{\delta}$ degrees | $k$ m²/W | $I_T$ MJ/m² (Monthly) | $\bar{T}_a$ °C (daytime) | $D_m$ °C-days (Monthly) |
| Jan | -3.8 | 686 | 204.2 | 6.587 | 14.17 | 33.30 | 2.350 | 0.4647 | 0.5283 | 1.5982 | 10.53 | -22° | 0.0033 | 326.4 | -1.8 | 686 |
| Feb | -3.1 | 606 | 268.6 | 9.525 | 19.40 | 35.97 | 1.854 | 0.4908 | 0.5018 | 1.3901 | 13.24 | -14° | 0.0030 | 373.2 | -1.1 | 606 |
| Mar | 2.1 | 505 | 407.4 | 13.142 | 26.88 | 37.45 | 1.393 | 0.4889 | 0.5038 | 1.1596 | 15.24 | -3° | 0.0029 | 472.4 | 4.1 | 505 |
| Apr | 9.4 | 267 | 497.3 | 16.577 | 33.95 | 36.72 | 1.082 | 0.4882 | 0.5045 | 1.0048 | 16.66 | 7° | 0.0029 | 499.7 | 11.4 | 267 |
| May | 15.0 | 120 | 627 | 20.226 | 39.20 | 35.02 | 0.8935 | 0.5160 | 0.4763 | 0.9119 | 18.44 | 17° | 0.0029 | 571.6 | 17.0 | 120 |
| Jun | 20.2 | 21 | 622.9 | 20.763 | 41.36 | 33.88 | 0.8192 | 0.5020 | 0.4905 | 0.8739 | 18.14 | 21° | 0.0031 | 544.0 | 22.2 | 21 |
| Jul | 22.6 | 1 | 632.7 | 20.410 | 40.14 | 34.24 | 0.8529 | 0.5085 | 0.4839 | 0.8909 | 18.18 | 19° | 0.0030 | 563.6 | 24.6 | 1 |
| Aug | 21.3 | 8 | 529.7 | 17.087 | 35.77 | 35.72 | 0.9986 | 0.4777 | 0.5152 | 0.9624 | 16.45 | 11° | 0.0030 | 510.0 | 23.3 | 8 |
| Sep | 17.2 | 71 | 450.4 | 15.013 | 29.29 | 36.84 | 1.258 | 0.5125 | 0.4798 | 1.1013 | 15.53 | 0° | 0.0028 | 495.7 | 19.2 | 71 |
| Oct | 11.3 | 219 | 340.3 | 10.977 | 21.74 | 36.32 | 1.671 | 0.5050 | 0.4874 | 1.3103 | 14.38 | -10° | 0.0029 | 445.8 | 13.3 | 219 |
| Nov | 5.3 | 392 | 196.4 | 6.547 | 15.52 | 33.92 | 2.186 | 0.4219 | 0.5717 | 1.4645 | 9.59 | -20° | 0.0035 | 287.7 | 7.3 | 392 |
| Dec | -2.0 | 631 | 171.1 | 5.519 | 12.52 | 32.06 | 2.561 | 0.4409 | 0.5525 | 1.6571 | 9.15 | -24° | 0.0035 | 283.6 | 0 | 631 |

[a] It is equally accurate to retain the 24-hour average temperature and later lower the system temperature 2 °C, since only their difference is used in subsequent equations.

Notice that this minimal data input corresponds to that available for the many locations listed in Appendix A-15. (The Hartford data there, however, are much too conservative.)

Results from this example are summarized in Table 3-8. The elevation technique from Example 3-7 has been applied using exact values for the latitude with $\tilde{\theta}$ calculated from eq. 3-35 and $k$ from eq. 3-36. Daytime temperatures are calculated by adding 2°C to the 24-hour average ambient.

A comparison with the 40° slope Hartford Appendix B table shows that most of the new elevated values are within a percent or two of those tabulated in the ALL column. Values of "global radiation over threshold" calculated using eq. 3-34 with the $k$ values just determined will also be found to be accurate.

A complete sample calculation is given below for the month of June. From Table 3-3,

$$N = 162$$

From eq. 3-1, the solar irradiance is

$$I_0 = 1353\left[1 + \cos\left(162 \times \frac{360°}{370}\right)\right] = 1311.7 \text{ MJ/m}^2$$

From eq. 3-2, the declination is

$$\delta = 23.5 \sin\left[(162 - 80) \times \frac{360°}{370}\right] = 23.08°$$

The latitude in decimal degrees is

$$41\tfrac{50}{60} = 41.833°$$

The positive hour angle of sunrise/sunset on the horizon is given by eq. 3-8:

$$\omega_s = \arccos\left[-\tan(41.833°)\tan(23.08°)\right] = 112.42°$$

and the positive hour angle of sunrise/sunset on the collector by eq. 3-10:

$$\omega_s' = \arccos\left[-\tan(41.833° - 40°)\tan(23.08°)\right] = 90.78°$$

From eq. 3-17b the extraterrestrial daily radiation total is

$$I_{0(\text{daily sum})} = 24 \times 0.0036 \times 1311.7 = 113.33 \text{ MJ/m}^2$$

and by eq. 3-17 the total daily extraterrestrial horizontal radiation is

$$\bar{H}_{0T} = \frac{113.33}{\pi}\left[\cos(41.833°)\cos(23.08°)\sin(112.42°)\right.$$
$$\left. + \frac{2\pi(112.42)}{360}\sin(41.833°)\sin(23.08°)\right]$$
$$= 41.36 \text{ MJ/m}^2$$

The positive hour angle that the collector goes into shadow is

$$\tilde{\omega}_s = \min(\omega_s, \omega_s') = 90.78°$$

and by eq. 3-18, the total daily-average extraterrestrial radiation on the tilted surface is

$$\bar{I}_{0T} = \frac{113.33}{\pi} \left[ \cos(41.833° - 40°) \cos(23.08°) \sin(90.78°) \right.$$

$$\left. + \frac{2\pi(90.78)}{360} \sin(41.833° - 40°) \sin(23.08°) \right]$$

$$= 33.88 \text{ MJ/m}^2$$

so that by eq. 3-19

$$\bar{R}_b = \frac{33.88}{41.36} = 0.8192$$

$\bar{R}_b$ may also be estimated quickly and accurately by using Table A-14. For 42 °N latitude, the June extraterrestrial horizontal radiation is

$$\bar{H}_{0T} = 41.37 \text{ MJ/m}^2$$

and, on a 40° tilted surface, it is

$$\bar{I}_{0T} = 33.95 \text{ MJ/m}^2$$

so that

$$\bar{R}_b = \frac{33.95}{41.37} = 0.8206$$

essentially the same as calculated so laboriously above. Only with Table A-14 can data elevation be reliably attempted with an ordinary hand calculator.

The Appendix B "ALL" horizontal radiation for June for Hartford is 622.9 MJ/m². The daily average is therefore

$$\bar{H}_T = \frac{622.9}{30} = 20.76 \text{ MJ/m}^2$$

so by eq. 3-20, the clearness index is

$$\bar{K}_T = \frac{20.76}{41.36} = 0.5020$$

and using eq. 3-28, the proportion of diffuse radiation is

$$\bar{H}_{Td}/\bar{H}_T = 1.00 - 1.015 \times 0.5020 = 0.4905$$

Since $\rho = 0.2$, eq. 3-26 gives the overall elevating factor as

$$\bar{R} = 0.8192(1 - 0.4905) + \frac{(1 + \cos 40°)}{2} \times (0.4905) + 0.2 \times \frac{(1 - \cos 40°)}{2}$$

$$= 0.8739$$

Notice that $\bar{R}$ and $\bar{R}_b$ are not very different, unlike the situation in winter. This makes the summer calculations relatively insensitive to inaccuracies in the direct-diffuse relationship.

The elevated radiation on the tilted surface is found with eq. 3-27:

$$I_T = 0.8739 \times 622.9 = 544.4 \text{ MJ/m}^2$$

which compares with 542.6 MJ/m² in the Appendix B table for a 40° slope. The daily average radiation is, by eq. 3-27,

$$\bar{I}_T = 0.8739 \times 20.76 = 18.14 \text{ MJ/m}^2$$

By eq. 3-35, the incident angle of the sun on the collector at noon is

$$\tilde{\theta} = 40° - 41.833° + 23.08° = 21.25°$$

Using eq. 3-36, the empirical utilizability constant $k$ is

$$k = 0.00542 - 0.0052 \times 0.5020 + [0.59 \times (21.25)^2/10^6] = 0.0031$$

The accuracy of eq. 3-36 can be assessed by using eq. 3-37 to determine $k$ directly from the Appendix B Hartford data on a 40° slope. Using the data above a 189+ W/m² threshold,

$$k = \frac{\ln (542.6/269.9)}{233} = 0.0030$$

Finally, raising the average temperature 2 °C gives an estimated daytime temperature of

$$\bar{T}_a = 20.2° + 2.0° = 22.2 °C$$

which is reasonably close to the 21.9 °C given in the 0$^+$ threshold column of the 0° slope table, but well under the more applicable value of 23.4 °C given in the 95$^+$ threshold grouping of the 40° slope table.

The tilted radiation can also be estimated using 8% more than the product of the percent of sunshine (Table 3-4) and the ASHRAE clear sky data (Table A-13). In summer, use data from a 40° tilt to most closely match the latitude:

$$\bar{I}_T = 0.60 \times 25.77 \times 1.08 = 16.70$$

which is about 9% under the value just determined. By June, however, the collector is considerably off its optimum inclination. The average radiation determined this way for November-March is quite exact, even though the monthly errors are as high as 9%. There is some reason to speculate that this method might well be as accurate as the blind use of data like those in Appendix A-15.

## PROBLEMS

1. Plot the declination angle $\delta$ and the extraterrestrial solar intensity $I_0$ over a year on the same graph, using time for the abscissa.

2. Using eq. 3-2 and integral calculus, derive an expression for the average declination angle over arbitrary days $D_1$ to $D_2$. Check your values with Table 3-3.

3. Consider clear skies (clearness factor 1.0) at Albuquerque, New Mexico, Amarillo, Texas, and Santa Maria, California, at 3 P.M. on March 21. All three locations are at 35°N latitude; however, the altitudes are 1619 m, 1098 m, and nearly sea level, respectively.
   (a) For each location, what is the direct normal irradiance?
   (b) For Amarillo,
       i. Find sunrise, sunset, and the length of the solar day.
       ii. Find the global irradiance on a south-facing collector tilted at 43°.
       iii. Find the global irradiance on a 43° tilted collector facing 40° west of south.
   (c) Repeat your (b) assignment for Albuquerque.
   (d) Repeat your (b) assignment for Santa Maria.
   (e) Perform (a) and (b) for your own location, but using a 60° tilt for (ii) and (iii).

4. Find the direct normal and horizontal extraterrestrial irradiance at 3 P.M. on March 21:
   (a) For 35°N latitude locations.
   (b) For your own location.

5. What is the total extraterrestrial radiation received on March 21 at 35°N latitude?
   (a) On a flat surface.
   (b) On a surface tilted at 43°.
   (c) On flat and 60° tilted surfaces at your own location.

6. Consider solar collectors at 60°N latitude at noon and 2 P.M. on July 19 (day 199). For tilts of (i) 0°, (ii) 50°, (iii) 90°, find:
   (a) Solar declination.
   (b) $I_0$.
   (c) Extraterrestrial solar irradiance on the inclined surfaces.
   (d) Hour angle of sunset on the collectors in degrees, hours, and solar clock time.
   (e) Total daily extraterrestrial radiation.

7. Find the average daily extraterrestrial radiation in March:
   (a) On a horizontal surface at 35°N latitude.
   (b) On a surface tilted upward 43° at 35°N latitude.
   (c) On flat and 60° tilted surfaces at your own location.

8. For March, the measured daily average horizontal global radiation at 35°N latitude is 13.26 MJ/m² (1194 Btu/ft²).
   (a) Find $\bar{K}_T$.
   (b) Estimate the proportion of diffuse radiation using the Page, modified Page, and Liu-Jordan correlations.
   (c) Using the modified Page correlation, find the elevation factors to be applied to the daily average horizontal global radiation to produce for a 43° tilted surface:
       i. The beam radiation.

ii. The diffuse sky radiation.

iii. The reflected radiation from the surroundings, if the reflectivity is 0.2.

(d) Find $\bar{R}$ and the daily average total global radiation on the tilted surface.

(e) Find the monthly total global radiation on the tilted surface.

(f) Repeat (c) to (e) for snow cover having a surface reflectivity of 0.7.

(g) Repeat (c) to (e) for the Liu-Jordan and Page correlations.

(h) Perform (a) to (f) using data from Table A-15 for March at your location.

9. Elevate 12 months of total daily average horizontal radiation data to a 60° tilted surface using the modified Page correlation and Table A-15 for the following locations:

(a) Albuquerque.

(b) Amarillo.

(c) Santa Monica.

(d) Your own location.

10. Using ASHRAE style, clear-sky tables (Table A-13), use percent of sunshine to approximate the tilted data calculated in problem 9 for:

(a) Albuquerque.

(b) Your own location.

11. Plot on one graph the monthly daily average extraterrestrial radiation throughout the year at 24°, 44° and 64°N latitude:

(a) For a flat surface.

(b) For a 60° tilted surface.

(c) Superimpose the ASHRAE clear-sky data on the 44°N 60° tilt line and comment.

12. Derive eq. 3-16 as indicated in the text, showing each step.

13. Derive eq. 3-17 from 3-16. *Note.* The angles must be in radians for integration, and correction factors later applied to nontrigonometric values so that degrees can be used throughout.

14. Derive eq. 3-18 as described in the text.

15. What clock time is solar noon at Wichita, Kansas (37°39′N, 97°25′N) on July 17?

16. Using data from Table 3-6 for the month of December, calculate values listed in the corresponding $378+$ W/m² data column in Appendix B:

(a) Global radiation total, total time, and degree-days.

(b) Time-averaged temperature.

(c) Average threshold radiation level and average threshold temperature.

(d) Global radiation over threshold.

17. Using December total radiation data tabulated in Appendix B for a 50° tilt, compare the tabulated radiation above the 189 W/m² threshold listed with that calculated from eq. 3-34:

(a) For Hartford.

(b) For the Appendix B data closest to your own location.

(c)  Compare $k$ as calculated from eq. 3-36 (or Table 3-7) with eq. 3-37 for parts (a) and (b).

18.  Elevate 12 months of horizontal data to a 60° slope in the manner of Example 3-11 (Table 3-11). Find the proper $k$ value for each month.
(a)  For Hartford.
(b)  For your closest Appendix B location.
(c)  For your closest Table A-15 location.

## REFERENCES

1.  H. E. Landsberg, "Man-made climatic changes," *Science 170*, 1265–1274 (1970).

2.  M. P. Thekaekara, "Solar Radiation Techniques and Instrumentation," *Solar Energy 18*, 309–325 (1976).

3.  D. G. Stephenson, *Tables of Solar Altitudes, Azimuth, Intensity, and Heat Gain Factors for Latitudes from 43 to 45 Degrees North*, Division of Building Research Technical Paper 243 NRC 9528, Ottawa: National Research Council of Canada, 1967.

4.  *Applications of Solar Energy for Heating and Cooling of Buildings*, edited by R. C. Jordan and B. Y. H. Liu, ASHRAE GRP 170.

4a.  E. A. Farber, C. A. Morrison, "Clear Day Design Values."

4b.  B. Y. H. Liu and R. C. Jordan, "Availability of Solar Energy for Flat Plate Collectors."

4c.  J. I. Yellott, "Solar Radiation Measurement."

5.  J. I. Yellott, "Solar Energy Utilization for Heating and Cooling," *1974 ASHRAE Handbook of Applications*.

6.  J. I. Yellott, "Available Sources of Insolation Data," NAS Conference proceedings, *Solar Radiation Considerations in Building, Planning and Design*, 1976.

7.  *ASHRAE Handbook and Product Directory.* 1978 Applications Volume. American Society of Heating, Refrigerating and Air-Conditioning Engineers, Inc., 345 E. 47th St., New York, 10017.

8.  *ASHRAE Handbook and Product Directory.* 1977 Fundamentals Volume. American Society of Heating, Refrigerating and Air-Conditioning Engineers, Inc., 345 E. 47th St., New York, 10017.

9.  *Report and Recommendations of the Solar Energy Data Workshop*, held November 29–30, 1973. Available from the U.S. Government Printing Office as NSF-RA-N-74-062.

10.  R. L. Hulstrom, *An Accurate, Economical Solar Insolation Computer Model for the United States*, Proceedings of the 1977 Annual Meeting, American Section ISES, Orlando, Florida, CONF 770603-P2, pp. 14–17.

11.  C. A. Morrison and E. A. Farber, "Development and Use of Insolation Data for South Facing Surfaces in Northern Latitudes," paper presented at the ASHRAE Symposium, Solar Energy Applications, Montreal, Canada, June 22, 1974.

12. *Climatic Atlas of the United States*, U.S. Government Printing Office, 1968.

13. J. K. Page, "The estimation of monthly mean values of daily total short wave radiation on vertical and inclined surfaces from sunshine records for latitudes 40°N–40°S," *Proc. UN Conf. on New Sources of Energy*, Paper No. 35/5/98 (1961).

14. G. O. G. Löf, J. A. Duffie, and C. O. Smith, *World distribution of solar radiation*, Report No. 21, Solar Energy Lab., Univ. of Wisconsin, July 1966.

15. S. A. Klein, W. A. Beckman, and J. A. Duffie, *Monthly average solar radiation on inclined surfaces for 171 North American cities*, Report No. 44, Engineering Experiment Station, University of Wisconsin, Madison, 1977. Erratum (below) also applies to this work.

16. S. A. Klein, "Calculation of monthly average insolation on tilted surfaces," *Solar Energy 19*, 325–329 (1977). See also Erratum, *Solar Energy 20*, 441 (1978).

17. B. Y. H. Liu and R. C. Jordan, "The interrelationship and characteristic distribution of direct, diffuse, and total solar radiation," *Solar Energy 4* (3), 1–19 (1960).

18. N. K. O. Choudhury, "Solar radiation at New Delhi," *Solar Energy 7*, (2), 44–52 (1963).

19. G. Stanhill, "Diffuse sky and cloud radiation in Israel," *Solar Energy 10*, (2), 96–101 (1966).

20. D. J. Norris, "Solar radiation on inclined surfaces," *Solar Energy 10*, 72–77 (1966).

21. M. A. Atwater and P. S. Brown, Jr., "Numerical computations of the latitudinal variation of solar radiation for an atmosphere of varying opacity," *J. Appl. Meteorology 13* (2), 289–297 (1974).

22. M. A. Atwater and J. T. Ball, *Regional variations of solar radiation with application to solar energy system design*. Final Report under NSF Grant No. AER 75-14536, July 1976. The Center for the Environment and Man, Inc., 275 Windsor St., Hartford, CT 06120.

23. M. A. Atwater and J. T. Ball, "A numerical radiation model based on standard meteorological observations," *Solar Energy 21*, 163–170 (1978). *Note.* The lines in Fig. 1 of this reference represent daily rather than monthly data and the comparison is not valid.

24. G. Haurwitz, "Insolation in relation to cloud type," *J. Meteorology 5*, 110–113 (1948).

25. B. Y. H. Liu and R. C. Jordan, "A rational procedure for predicting the long term average performance of flat plate solar energy collectors," *Solar Energy 4*, 1–19 (1960).

26. R. Walraven, "Calculating the position of the sun," *Solar Energy 20*, 393–397 (1978).

27. W. C. Dickinson, "Annual available radiation for fixed and tracking collectors," *Solar Energy 21*, 249–251 (1978).

28. A. R. Singer, University of Puerto Rico (Mayuguez), personal communication.

29. M. Iqbal, "A study of Canadian diffuse and total solar radiation data—I. Monthly Average Daily Horizontal Radiation." *Solar Energy 22*, 81–86 (1979).

30. V. Cinquemani, J. R. Owenby, Jr., and R. G. Baldwin, *Input Data for Solar Systems*, a November 1978 report to the Department of Energy by the National Climatic Center, Asheville, N.C.

31. J. L. Threlkeld, "Solar Radiation of Surfaces on Clear Days," *ASHRAE Trans. 69*, 24 (1963).

32. M. A. Atwater, P. J. Lunde, and G. D. Robinson, *A cloud-cover radiation model producing results equivalent to measured radiation data*, ISES Silver Jubilee Congress, 28 May to 1 June 1979, Atlanta, Ga.

33. C. R. Attwater, F. C. Hooper, A. C. Brunger, J. A. Davies, J. E. Hay, D. C. McKay, and T. K. Won, "The Canadian Solar Radiation Base," paper presented at ASHRAE Annual Meeting, Detroit, Mich., June 25, 1979.

34. "Solar Radiation Data Sources, Applications and Network Design," HCT/T5362-01, Department of Energy, April 1978.

35. "Hourly Solar Radiation—Surface Meterological Observations," SOLMET Volume 2—Final Report, TD-9724, Department of Energy, February 1979.

# Flat Plate Collectors—Collecting the Heat

The flat plate collector is the simplest and most widely used means to convert the sun's radiation into useful heat. Typical flat plate collectors are shown in Figs. 4-1. Solar radiation passes through the transparent cover plates and strikes the blackened collector absorber surface where it is absorbed, changing to thermal or heat energy.

Thermal energy is removed from the absorber by a flow of air or liquid that then delivers the heat to be used or stored. But often over half the heat absorbed is lost from the absorber surface, principally by reradiation and convection to the exterior surroundings. The losses are minimized by the use of the transparent cover plates, which act like storm windows on a house, passing visible and near-infrared radiation from the sun while reducing losses from the interior.

Cover plates reduce convective losses from the absorber because of the resistance to convective heat transfer created by the air spaces above the absorber surface and between the cover plates. The outer cover plate will therefore have a lower temperature than that of the absorber surface, and convective losses will be correspondingly lower. Radiation losses are also minimized because the glass—which is transparent to the sun's radiation—is a good absorber of *thermal* energy, being nearly opaque to the long-wave thermal radiation "shining" from the absorber to the surroundings.

The cover plates are therefore "black" with respect to the reradiation from the absorber surface, and the reradiation is then *absorbed* rather than *transmitted* by the cover plates. This heats the cover plate to a temperature higher than the surroundings, but below that of the absorber. The rate of radiative heat loss from the absorber surface is thereby reduced, because it depends on the temperature of the surface to which it can radiate. The cooler outer surface of the cover plate in turn reradiates to the surroundings (or to the next cover plate), but at a reduced rate relative to the unprotected absorber.

The collector must also be properly insulated to prevent conduction losses from

120

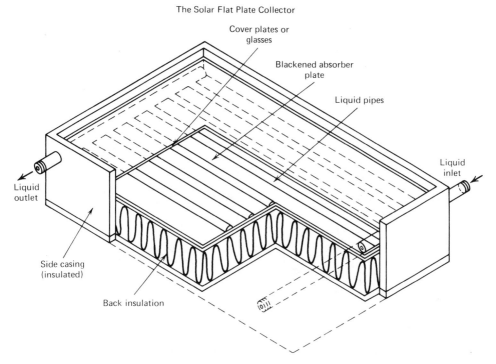

**FIGURE 4-1a.** A typical liquid-cooled fiat plate collector.

the back and sides. It is cheap and easy to do this but care is necessary to assure that all the possible conduction paths are blocked. The external framework and supports for the cover plates, for instance, must be insulated from the absorber plate and even from the air spaces, which are also well above the ambient temperature. Several inches of insulation are necessary under the plate (and at the sides) to reduce losses from these large areas to an economical minimum.

The convective and radiative losses from a flat plate collector tend to be large because the area for such losses is equal to the area for collection of the solar radiation. Proper design of a flat plate collector gives attention to reducing these losses by making it more difficult for the convection and radiation losses to occur. The heat transfer between the collecting fluid and the absorber surface must also be maximized so that the absorber surface is held at as low a temperature as possible consistent with the temperature at which the heat must be utilized.

The present chapter deals with the means for reducing the losses from the collector and the ways to calculate those losses, as the outside ambient and collector surface temperatures vary, so that the collector performance can be determined. The next chapter will discuss ways to improve the heat transfer to the collecting

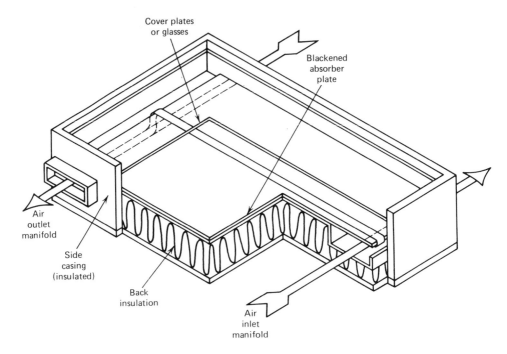

FIGURE 4-1*b*. A typical air-cooled flat plate collector.

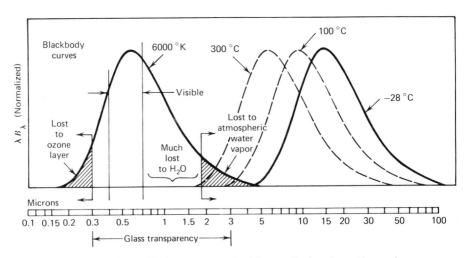

FIGURE 4-2. Solar radiation compared with reradiation from thermal sources.

fluid and to evaluate the effects of this heat transfer on the overall collector performance.

## Action of Materials on Light

When ordinary materials are exposed to light, they can *transmit, reflect,* or *absorb* it. If coefficients are defined equal to the proportion transmitted ($\tau$), reflected ($\rho$), and absorbed ($\alpha$), a simple equation can be written:

$$\tau + \rho + \alpha = 1 \qquad (4\text{-}1)$$

This equation is valid for any wavelength of light; it holds for the radiation from the sun and for the long-wave radiation from a solar collector's absorber plate. But the coefficients are not necessarily the same at different wavelengths, and in solar collector design, materials are of special interest if their transmissivity, reflectivity, or absorptivity changes favorably between the short wavelengths of visible and near-infrared light and the long wavelengths characteristic of thermal reradiation.

The two groupings of wavelength do not overlap. Figure 4-2 shows the frequency distribution of a given amount of radiated energy from a blackbody (a perfect radiator) at 5780 K—the sun's temperature—and thermal radiation at $-28$, 100, and 300 °C. Since wavelengths longer than two microns (a micron is a millionth of a meter) are for the most part absorbed by water in the earth's atmosphere, the two wavelength groupings are truly separate, with a 0.3- to 2-$\mu$ range for solar radiation, and 3 to 50 or 100 $\mu$ for long-wavelength reradiation.

## COVER PLATES

### Effect on Sunlight

A material must have a high transmissivity to be useful as a cover plate, and therefore the absorptivity and reflectivity must be minimized.

**Reflectivity.** The reflectivity of a transparent or translucent material depends on its refractive index and on the *angle of incidence* formed between the incoming radiation and a line perpendicular to the transmitting surface. If the incident angle is 0°, for a single surface,

$$\rho = \left(\frac{n-1}{n+1}\right)^2 \qquad (4\text{-}2)$$

where $n$ is the refractive index. In the case of glass, the most common cover plate material, $n = 1.53$ and

$$\rho = \left(\frac{1.53 - 1}{1.53 + 1}\right)^2 = 0.044$$

This means that 4.4% of the radiation is reflected at each surface, and 8.8% is reflected in passage through a single sheet of glass. This loss is serious when two or more cover plates are used and is the reason that in practice a maximum of two cover plates are installed. Reflective losses increase slowly as the incident angle increases, doubling to about 18% (both surfaces) when light strikes at an angle of 30° to the transmitting surface.

There are ways to reduce the reflectance loss from glass and other materials. If a clear film of a second *dielectric* (i.e., light-transmitting) material is coated on the first in a thickness of several microns or more, the reflectivity is reduced, and is given by

$$\rho = 1 - \frac{4n_1 n_2}{(n_2{}^2 + n_1)(n_1 + 1)} \tag{4-3}$$

(Notice that if $n_1 = n_2$ this expression reduces to eq. 4-2, as it must.) It can be shown that reflectivity is at a minimum for the situation when

$$n_2 = \sqrt{n_1} \tag{4-4}$$

and thus for glass,

$$n_2 = \sqrt{1.53} = 1.23$$

Consequently, according to eq. 4-3, the minimum reflectivity is then

$$\rho = 1 - \frac{4 \times 1.53 \times 1.23}{(1.23^2 + 1.53)(1.53 + 1)} = 0.022 = 2.2\%$$

which is one-half that of uncoated glass. The practical material with the most nearly ideal refractive index is one of the metallic fluorides familiar as camera lens coatings. For these materials $n = 1.38$, giving

$$\rho = 0.028 = 2.8\%$$

This is a significant reduction, yielding a reflectance of 5.6% for a single sheet of glass as opposed to 8.8% when the surface is uncoated. Unfortunately, such coatings are too expensive for use at present in solar collectors, since they must be vapor-deposited in a vacuum. Furthermore, the durability of such coatings is poor, and it is doubtful that the exterior surface coat would last the 20 to 40 years projected for the life of a solar collector.

A second approach to reflectivity reduction is the use of a thin-film, interference filter. This is the purpose of the thin film deposited on camera lenses. The thickness of this film should be one-fourth the wavelength of the dominant frequency of the transmitted light. The reflective losses with such a film are actually *zero* at a single wavelength, but extend over a broad range of wavelengths according to complex relationships beyond the scope of this book. Interestingly, the best refractive index for such a *thin film coating* is also given by eq. 4-4, and the fluoride

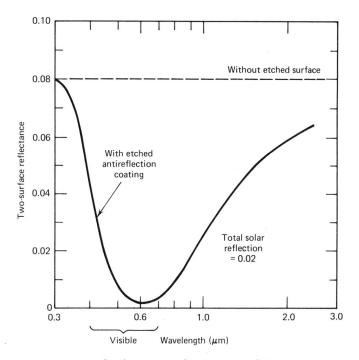

**FIGURE 4-3.** Reflection spectra for a sample of glass before and after etching (Ref. 1).

films are again the best practical choice. Since the durability of a deposited thin film would be even more of a problem than a thick one, it is fortunate that a proper thin film can be etched on the surface of glass with relative ease by immersion in a silica supersaturated fluorosilicic acid bath. The acid attacks the glass surface and leaves a skeletonized, porous silica layer having a near optimum refractive index (1, 2). Figure 4-3 shows the two-surface reflectivity of such a coating as a function of wavelength. Notice the sharp reduction in reflection to almost zero at 0.6 μ—the middle of the solar spectrum—and significant reductions over the entire range of interest (see Fig. 4-2) so that the overall reflection loss is reduced from 8% to 2% by the etching process.

Etched glass coatings are currently available commercially and would add little to the cost of a solar collector. Their durability has been a problem, but proof that such a coating can be very durable if properly done was recently given by Löf (3), who carefully tested etched glass from his 20-year-old solar house and found that the overall reflectance remained at only 2% per surface. Unfortunately, the manufacturer of the glass could not be located to discover what special techniques were responsible for this excellent durability. Mar, et al. (1) have found that a heat treatment makes the etched surface stable in high humidity, which otherwise quickly degrades the antireflective properties.

**Absorptivity.** There are many materials that transmit light very well, absorbing very little energy from the incident beam. Glass, air, water, and most clear plastics are among them. These materials generally absorb light slightly if any color at all can be observed through a thick layer. Glass, for instance, often has a green edge, and such ordinary glass absorbs about 5 to 10% of the incident light. Smoke-colored glass (often used for glazing commercial buildings) and heat-absorbing glass absorb much more. Fortunately it is an impurity (iron oxide) that controls the color and the absorptivity of glass. Low-iron glass, which has 0.05% iron oxide, will lose only about 2.5% of the transmitted light to absorption in the usual single-strength thickness of window glass. One very low-iron or "water-white" glass—0.01% iron oxide—is advertised to transmit about 91.4% of the sun's radiation.* Ordinary glass should never be used for solar collectors and, while the best solar grades now sell at a substantial premium, even they should eventually be available with little cost penalty as sufficient sales volume develops because low-iron raw materials are readily available.

Glass for solar collectors is usually tempered, to reduce the potential for breakage. The most common size is 34 in. × 76 in., in a $\frac{5}{32}$ in. thickness.

**Transmissivity.** Table 4-1 summarizes the properties of glass that are important to its use as a cover plate in solar collectors. Notice that the transmissivity of white glass is about 91%. This is consistent with the above discussion, for if eq. 4-1 is rearranged,

$$\tau = 1 - \rho - \alpha$$

and we have seen that $\rho = 0.088$ and $\alpha = 0$ to $0.01$.

**TABLE 4-1.** Properties of Glass

| Property | Type of Glass | | |
| --- | --- | --- | --- |
| | Ordinary Float | Sheet Lime | Water-White |
| Iron-oxide content, percent | 0.12 | 0.05 | 0.01 |
| Refractive index | 1.52 | 1.50 | 1.50 |
| Light transmittance (normal, percent) | 79–84 | 88–89 | 91.2–91.6 |
| Glass thickness, inches | 0.25–0.125 | 0.125–0.1875 | 0.125–0.21875 |
| Reflectance loss, percent | 8.2–8.0 | 8.1–8.0 | 8.0 |
| Absorption loss, percent | 8–13 | 3–4 | Under 1.0 |

* "Sunadex®," ASG Industries, Kingsport, Tenn.

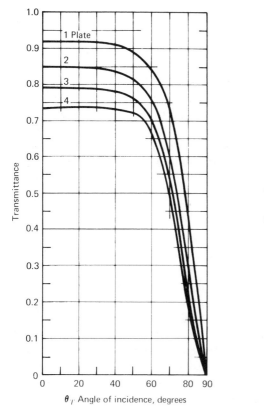

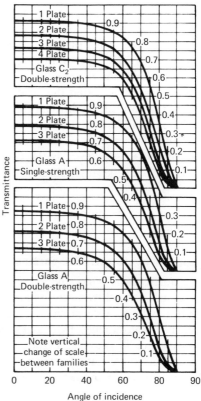

**FIGURE 4-4.** (*a*) Calculated transmittance of a system of glass plates, allowing for reflection losses only (refractive index = 1.526). (*b*) Calculated transmittance $\tau$ of systems of glass plates, allowing for reflection *and* absorption losses. Glass $C_2$ is water-white, 0.117 in. thick. Glass A is a window glass, 0.089 and 0.126 in. thick. Reproduced courtesy ASME (Ref. 4).

Maximum cover plate transmissivity is of major concern to the user of flat plate collectors, because losses here are directly "off the top," reducing the intensity of solar energy available at the absorber for conversion into useful heat.

The transmissivity of the more absorptive glasses can decrease considerably when the angle of incidence is more than 45° off the vertical. Reflectance changes only slightly with incident angle, but the absorptance increases in proportion to the path length of the light through the glass so that the proportion transmitted decreases with the cosine of the angle of incidence. If the glass is thick or not low in iron, the effect can be of importance in flat plate collectors not only because they collect a portion of their radiation in early morning and late afternoon but also because on cloudy days these collectors collect diffuse radiation that comes from all directions.

Figure 4-4a shows the effect of incident angle on transmissivity (without absorption losses) for one, two, three, and four sheets of low-iron plate glass. In Figure 4-4b the total losses are shown for this and two more absorbent grades of glass. These curves show that the incidence angle is not very important up to about 45° from the vertical. Additional sheets of glass, of course, reduce the overall transmission considerably. Even though the physics of several sheets is rather complex because of rereflections, for practical purposes the transmittance of several sheets of glass has been shown by Hottel and Woertz (4) to be the product of the transmissivities of the single sheets:

$$\tau = \tau_1 \cdot \tau_2 \cdot \ldots \cdot \tau_n \qquad (4\text{-}5)$$

Thus it is conventional to refer to the transmissivity of the complete cover plate assembly of a particular collector as $\tau$.

### Long-wavelength Properties

As noted above, since the frequencies involved in reradiation from the absorber plate are distinct from those in solar radiation, there is no special reason why the transmissivity, absorptivity, or reflectance should not change between the two frequency groupings. In actual fact, the change in the properties of glass with wavelength is extreme. Figure 4-5 shows that the transmissivity of glass rises sharply in the ultraviolet and remains high to about 2.7 $\mu$, where it drops rapidly to near zero. In the far infrared it becomes as well a reasonably good reflector (up to 30% at 10 $\mu$) of long-wavelength thermal radiation. This shift in properties is very desirable, because the glass then forms a radiation barrier between the heated absorber plate and the cooler surroundings, while still passing solar radiation.

Glass does not have ideal thermal radiation properties, however. A perfect material would *reflect* the thermal radiation completely and of course retain the high transmissivity of glass for incoming radiation. Attempts have been made to prepare such a material by vacuum deposition of special metallic oxides on a glass substrate. Indium oxide, for instance, has been deposited in very small samples by Tani (5) to give a transmissivity of 85% at 0.5 $\mu$ with a reflectivity of 97% at 4 $\mu$. It remains to be seen if the increase in thermal reflectivity is worth the decrease in solar transmissivity, and whether an inexpensive means for depositing the special layer can be developed.

### Alternate Cover Plate Materials

In the preceding paragraphs only glass has been discussed as a cover plate material. In practice there is only a very short list of alternative materials. Most clear plastics are too sensitive to ultraviolet degradation, and few can tolerate the tempera-

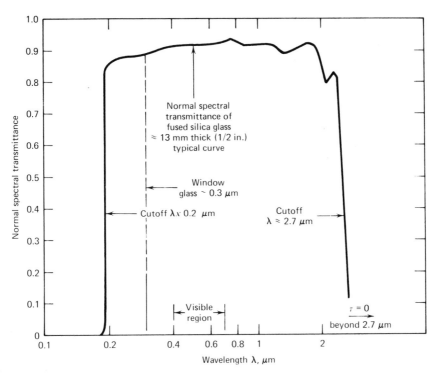

**FIGURE 4-5.** Transmittance of glass.

tures of 150 °C (300 °F) and higher that are easily reached if a collector is left idle in the summer. Polycarbonate (Lexan*) has seen solar application, as have certain acrylics (Plexiglass,† Lucite‡) but these materials have too low a melting point and are not as good as glass in terms of reradiative properties. Tedlar,‡ a polyfluorocarbon, quite good optically, lasts for several years and is easy to deal with since it is thermoplastic. Even so, Tedlar is not recommended for *interior* collector glazing because of the high temperatures likely to be reached there.

The optical properties of Tedlar are given in Fig. 4-6. While the reradiation properties are somewhat inferior to glass, the transparency is a little higher in the visible region because of its thinness and lower refractive index, which permits slightly less reflection—only 7% as compared to 8% for glass. Since Tedlar is basically very expensive it is used only in very thin sheets (about 4 mils). Tedlar film is superficially similar to polyethylene in having a milky but transparent

---

* Trademark of General Electric. Will appear this way throughout.
† Trademark of Rohm and Haas. Will appear this way throughout.
‡ Trademark of DuPont. Will appear this way throughout.

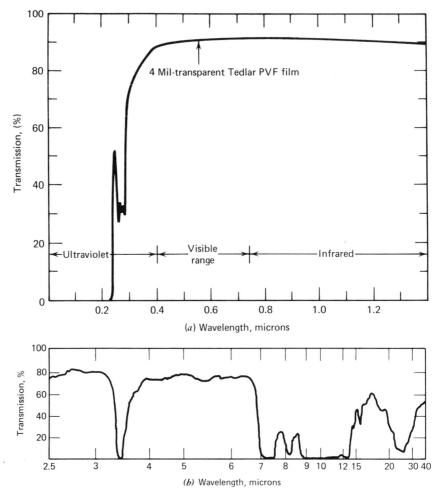

**FIGURE 4-6.** (*a*) Transmission of 4-mil (0.1 mm) Tedlar transparent polyvinyl fluoride film in the solar radiation wavelengths. (*b*) Transmission of Tedlar film at mid- and far-infrared wavelengths. (*Source:* DuPont promotional literature.)

appearance and in being far too flexible to be self-supporting. It must be installed under tension if it is to serve as a cover plate, and provision must be made to keep it taut as temperatures rise, since like all plastics the thermal expansion coefficient is several times that of glass, wood, or metal.

For inner glazings only, Teflon* has recently begun to see practical use. Teflon is an expensive, completely fluorinated polyfluorocarbon that is chemically inert to about 250 °C (480 °F). Exposure to sunlight for as long as 20 years has produced no detectable degredation. Its refractive index is 1.34, and consequently the two-

---
*Trademark of DuPont. Will appear this way throughout.

surface reflectivity is one-half that of glass—4.22%. Its day-long contribution to performance is even better than these values indicate, since not only is the basic reflectivity low but also there is less increase with increasing incident angle than with other materials. With an emissivity of 0.96, it is an excellent material for inner glazings, although its physical durability is insufficient for use as an outer cover. Crystal-clear, and as transparent as glass, it is used in 1-mil thickness (0.025 mm) for solar applications, at a material cost only one-fourth to one-half that of glass. It is so flimsy, however, that it must be mounted in a frame—like window screening—since it sags when heated.

A generally more practical alternative to glass—some would say the only alternative—is fiberglass sheet, a dense mat of glass fibers in a matrix of clear polyester plastic. This material is familiar as translucent, rigid corrugated plastic roofing, popular for sheds and decks and available at most building supply houses. In the clear grades, special attention is necessary to protect the material from long-term, ultraviolet degradation, which will interfere with transparency in a solar collector. One manufacturer's grade of this material suitable for long-term solar applications is called *Sun-lite*\*. Its transparency degrades only a few percent over a number of

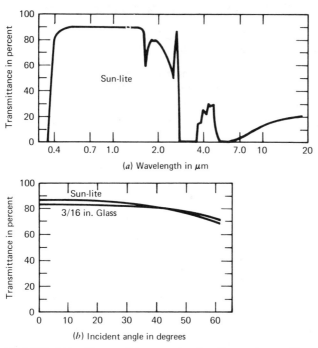

**FIGURE 4-7.** Transmittance of *Sun-lite* polyster-fiberglass sheet, (*a*) Versus wavelength. (*b*) Versus incident angle. (*Source*: Kalwall promotional literature.)

\* Trademark of Kalwall Corporation. Will appear this way throughout.

years, and its initial specification is almost the equal of ordinary glass. The transmittance of the material is shown and compared with ordinary window glass in Fig. 4-7. Since the material is about 95% glass, it resembles glass quite closely in optical characteristics, except that it is translucent rather than transparent. This causes no practical problem, but does make transmittance measurements difficult, accounting for often anomalous transmittance data.

Although fiberglass sheets are rigid when corrugated, they tend to sag in the flat sheets preferred for a solar collector application. They are usually installed under tension, or with spacers that keep the center of the material from sagging inward. With that exception, the material is easy to handle and has many advantages over glass because it is more resistant to fractures, lighter in weight, and can be cut from a roll on-site with heavy shearing equipment. It is also cheaper than glass in its basic cost per unit area, since it is used in 0.040-in. sheets rather than the $\frac{1}{8}$ to $\frac{3}{16}$ in. sheets ordinarily used for glass. It will not resist heat as well as glass, even though it will tolerate a few of the inevitable excursions to 150 to 200 °C (300–400 °F) that must be anticipated in any installation. It should not be used as an inner glazing or on a collector that is left uncooled in the summer, since such *stagnation* places a great stress on such thermally sensitive components.

## COLLECTOR PLATE SURFACES

The optical characteristics of the absorber plate are of great importance to the efficiency with which the light striking the collector can be collected as useful heat. The way the surface reradiates the thermal wavelengths determines the direct radiation losses, which are high in a solar collector even with the protective cover plates because the surface area is so large. The way the surface reflects visible and near-infrared light completely determines how much heat is absorbed, because the universally opaque absorber plate transmits no light.

For the absorber surface, eq. 4-1 therefore reduces to the simple

$$\alpha + \rho = 1 \tag{4-6}$$

because the transmissivity is zero. Observe again that there are only two effects possible when light strikes surfaces with which we are concerned: reflection and absorption. Since the absorptivity and the reflectivity sum to unity, only one effect need be discussed, and absorptivity is usually selected. But physicists also use another term, the *emissivity*, when they refer to the same optical property of the surface but at a temperature such that it emits rather than absorbs radiation.

In careful definition the emissivity and the absorptivity are interchangeable at any particular wavelength, but the two different words help keep track of the process involved. Solar scientists have found it useful to redefine these terms so that absorptivity ($\alpha$) is the absorptivity of a surface at the wavelengths of *solar* radiation, while emissivity ($\varepsilon$) is the same property but at the *thermal* radiation frequencies where reradiation is of interest. This distinction will be used henceforth and must be clearly understood.

It is also important to understand that the only alternative possibility to either emissivity or absorptivity on an opaque surface is reflectivity. There is only one word available here, and in ordinary usage it is applied only to incoming radiation. In solar work it must do double-duty and refer also to reradiation. A low emissivity therefore implies a high reflectance, meaning that the thermal energy *fails* to radiate outward because of the reflective character of the surface. This characteristic is generally unfamiliar to us because virtually all the usual hot radiating surfaces are nearly black optically. The ideal absorber plate, however, has a surface with a high absorptivity to absorb as much solar radiation as possible, but a low emissivity to reduce the reradiative losses to a minimum. Such a surface is referred to as a *selective* surface.

## Paints

Paints consist of a pigment material, an organic binder that polymerizes during drying, and solvents that permit easy application of the paint film. In drying, the solvent evaporates and the pigment and binder form a film to 1 to 3 mils thick (a mil is 0.025 millimeter or $\frac{1}{1000}$ inch). While many solar experimenters report good results using almost any flat, black paint on the absorber surfaces, it is wise to select a high-temperature variety or a special solar grade of paint* so that the most durable binder is used. Carbon black is almost universally used as the black pigment because it is very black, cheap, and quite durable. A carbon black paint is a good absorber but, since the paint film is not at all selective, it can be expected to have an absorptivity *and* emissivity of 0.95 to 0.98.

When paints are applied in a very thin film on the order of 0.05 to 0.1 mil thick, certain pigments become transparent to long-wave thermal radiation while keeping their high absorptivity for solar radiation. When coated on a reflecting substrate such as polished aluminum, the composite material can have the high absorptivity of the pigment with the low emissivity of the aluminum, thereby forming a selective surface. A number of such paints were formulated in recent studies by Mar and Lin (1, 2). The best of them were pigmented with varying mixtures of the calcined oxides of chromium, copper, iron, and manganese. The best optical properties were $\alpha = 0.92$, $\varepsilon = 0.1$ to $0.13$, which is certainly of practical interest. None of these materials, however, is yet available commercially and their durability has yet to be evaluated. Furthermore, since it is difficult to apply such a thin film uniformly, it is likely that such a paint must be factory coated just as are plated selective coatings (see below), but will not be quite as good in performance. A particular advantage, however, would be the ease of application

* Nextel 110-C10 (trademark of 3M Company) is one such product.

to materials such as aluminum on which it is difficult to form a good selective surface.

## Metallic Selective Surfaces

Nickel and chromium have for many years been electroplated on metallic substrates to serve as a protective and cosmetic coating in both metallic and "black" finishes. Both surfaces are the same basic material, but the black surface has been plated on a metallic substrate under conditions that give a rough surface structure to the plating, causing it to appear black. It so happens that such black coatings form an excellent selective surface when properly prepared. The reason for this is the *physical* character of the surface. The fine patina of the electroplated surface has a microstructure that is very fine when compared to thermal radiation, but very rough when compared to solar radiation. In other words, the surface is pitted, with dimensions of this microstructure on the order of 1 $\mu$.

The basic materials in the plated black selective coatings are not black, in spite of their appearance. They have a typical metallic or metallic oxide surface—shiny with a low emissivity and a high reflectance. The absorptivity is also low. These superficial properties are those actually observed at thermal wavelengths, because the surface roughness is very fine compared to long-wave radiation. In the visible and near-infrared range, however, the wavelengths of the radiation are small enough that the light can enter the pitted microstructure. When it does this, the reflection is likely to be to another part of the microstructure rather than back out of a pit. Since each reflection absorbs a bit of the light energy, after a number of reflections, the light can be nearly all absorbed and the shiny surface will appear to be, and in fact will be black. An analogy may help to understand the phenomena. If a hillside were densely pitted with deep tunnels the size of ping-pong balls, baseballs (thermal radiation) thrown at the surface would bounce off and the reflected balls would lose little energy. Half-inch ball bearings (solar radiation), however, would enter the tunnels and bounce around, losing their energy after several bounces from the hard surface.

**Practical Selective Surfaces.** Electroplated selective surfaces are not difficult to deposit for those engaged in the art, but they are expensive because of the materials involved and the care that is necessary in cleaning and controlling the process. Not all the coatings are durable, and their durability in high humidity is a special problem, since rainwater will often leak into the collector.

**Two-layer black nickel on nickel plated steel.** This has received much attention from Harry Tabor, an early solar pioneer (6, 7). His plating process is rather complex, for the coatings also contain zinc and sulfur. The selectivity of the resultant surface is the best yet reported, with $\alpha = 0.96$ and $\epsilon = 0.07$. This coating is commercially available and in use on some collectors. There are doubts

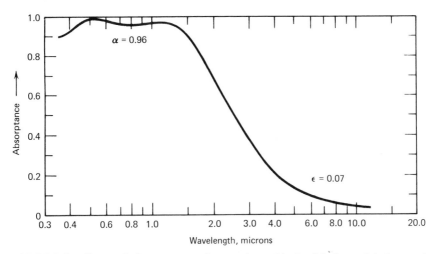

**FIGURE 4-8.** Spectral absorptance of a two-layer black nickel on nickel coated steel (Ref. 1).

about its humidity resistance, but in spite of failure in certain humidity tests black nickel performs well in practice in dry climates. Figure 4-8 shows spectral absorptance for a similar surface prepared by Mar and Lin (1, 2). Their surface degrades quickly in a hot and humid environment.

**Black chrome.**  Chromium can be plated by ordinary commercial processes on nickel-plated steel or copper to give a good selective surface. The nickel underlayer gives good humidity resistance and the plating process is simpler than for two-layer black nickel. Spectral absorptance of this material is shown in Fig. 4-9, and the integrated spectral properties are $\alpha = 0.95$, $\varepsilon = 0.1$. Commercial platers now routinely produce similar black chrome with properties better than black nickel.

Black chrome can also be deposited directly on copper. The spectral properties are only moderately good with $\alpha = 0.95$, $\varepsilon = 0.15$. Apparently a nickel undersurface is necessary to achieve emissivities below 0.1. Efforts to improve the performance of black chrome emphasize the importance of the surface roughness of the nickel underlayer.

Selective surfaces can also be applied chemically. Copper so treated has been used commercially in "Sunworks" brand solar collectors. Such coatings do not have as high an absorptivity or as low an emissivity as plated selective surfaces, although continuing improvement can be expected.

Applying a selective surface to aluminum is a special problem because aluminum reacts chemically with ordinary plating baths. The special techniques available to plate a nickel undercoat on aluminum themselves run to about $1 per square

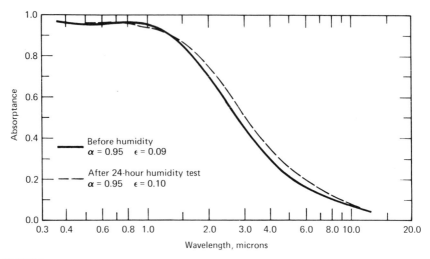

**FIGURE 4-9.** Black chrome on ½ mil nickel before and after humidity test (Ref. 1).

foot, probably too expensive to warrant serious consideration. Research recently reported (8) shows that moderately effective selective surfaces can be formed on aluminum by high-temperature, solid-state reactions with nickel. These materials have the best properties yet achieved directly on aluminum—$\alpha = 0.95$, $\varepsilon = 0.3$.

## Insulation

At least two inches (and preferably six inches or more) of good insulation must be used underneath and on exposed sides of the collector. Of the ordinary insulating materials, only fiberglass or rock wool batts are satisfactory. The material used must be free of an organic binder often used in ordinary applications, because high collector temperatures cause this material to deposit on the underside of the transparent cover, adversely affecting the collector transmissivity.

Most plastic materials such as polystyrene or polyurethane foam are damaged too easily by the heat of a stagnant collector, either melting or expanding destructively. One way to use such materials is to place a layer of fiberglass under the collector plate and install that assembly in a foamed or built-up plastic structure. Insulated by the fiberglass, the plastic will not normally overheat and can then serve structurally.

## Supportive Structure

The structure of a solar collector is generally fabricated from aluminum, steel, or wood. A few collectors use fiberglass construction. Any of these materials can be fashioned into an attractive durable structure. Exposed steel or wood, however, must be specially treated—perhaps with a plastic surface—if they are to withstand a 20- to 50-year, projected collector life.

If a metal supportive structure is used, heat paths from the warm sections of the collector to the basic structure must be eliminated. The supports for the cover plates, for instance, must be insulated not only from the absorber surface but also from glass and the air spaces, lest heat be lost by transfer to the supports.

If the collector is to be shipped with the cover glass installed, extra rigidity is required so that the glass is not fractured as the collector is handled. On-site installation of the cover plates is somewhat more complicated than with window glass, because the inclination angle increases exposure to water and because of the generally expressed need for simply removable plates to facilitate repair, cleaning, and replacement. Provision must be made for expansion and contraction of the cover plate material, because its expansion coefficient is likely to be quite different than the framework. All these requirements are generally met by using a U-shaped, extruded rubber gasket held in place within a metal trim strip (Fig. 4-10). A silicone rubber is a wise choice because this material is very weather resistant.

The market place has not yet determined whether a collector is best built on-site

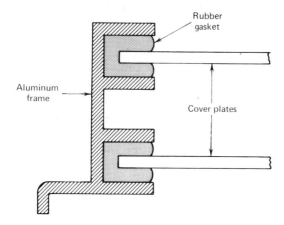

**FIGURE 4-10.** Provisions for support of cover plates, showing the soft rubber gasket used for sealing and to compensate for the difference in thermal expansion between the aluminum frame and the glass cover plates. For better performance, the exposed aluminum between the cover plates should be insulated.

or in a factory. The factory collector can be built with more quality control, efficiency, and by cheaper labor because of automation; while the on-site collector can be of less impressive but cheaper construction because it does not need to be built to withstand shipment. The on-site collector can rely on the building structure for support and serve as the watertight roofing, giving a further potential for cost reduction. At present, however, it is wise to install a manufactured collector because its performance and durability are more certain. The on-site collector may prove to be superior in the future because of lower cost and in many cases, improved appearance.

## COLLECTOR PERFORMANCE

A solar collector, like many thermal devices, is at first glance a very complex set of interacting thermal paths that defy analysis. Nevertheless, years of interacting theoretical and experimental work have shown ways to simplify the problem by combining complex parameters so that tractable and accurate prediction of solar performance can be obtained.

### Collector Heat Loss Coefficients

The single most important variable for solar collector performance is the heat loss from the collector. The heat loss is the main factor in fixing the time periods of collector operation—there is no collection when losses exceed solar gain—and it also determines to a large extent the collector efficiency when the collector is operating.

The practical choices that determine heat loss are the number of cover plates and the optical characteristics of the absorber plate. One or two can be considered (with more, the optical efficiency $\tau\alpha$ falls too much), and the highly absorptive collector surface can be selective or not. Other possibilities exist, of course, but these are the practical solutions available now.

Heat loss coefficients for these conditions are difficult to determine because they must include the effect of reradiation to the sky. The sky temperature depends to a great extent on the water vapor in the air and is an uncontrolled variable that is difficult to measure. Its effect on performance is not overwhelming, especially with selective surfaces or more than a single cover plate, but its uncontrolled variability during data collection makes precision difficult, especially when the data are sought as a function of temperature.

Perhaps the best exploration of the collector heat loss coefficient $U_L$ is given by Hottel and Woertz (4). Their data, basically analytical but calibrated with careful experimental data, are given in Fig. 4-11. Unfortunately they were not at that date able to test selective surfaces. Duffie and Beckman (9) made a more comprehensive analytical study, but do not have experimental confirmation.

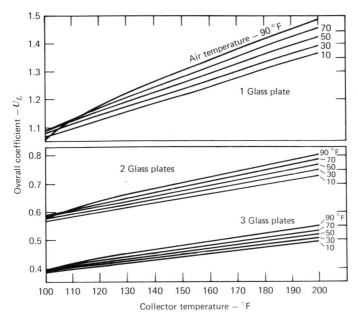

**FIGURE 4-11.** Overall heat transfer coefficient for use in calculation of heat losses from flat plate collectors (tilt from horizontal = 30°; collector-surface emissivity = 0.95; wind = 10 mph). Reprinted courtesy ASME (Ref. 4).

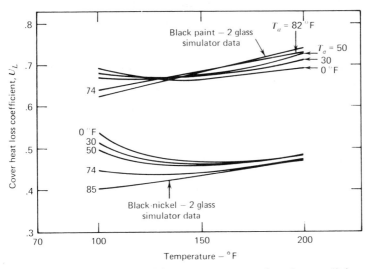

**FIGURE 4-12.** Effect of ambient temperature on heat loss coefficient (Ref. 10).

Measurements of collector loss $U_L$ made by the National Aeronautics and Space Administration (10) are given in Fig. 4-12. These data were collected indoors under controlled conditions, but with these conditions carefully adjusted to correlate with data taken outside. They corroborate the Hottel and Woertz data fairly closely, overlapping the two-glass, black-paint data, although the change is not so great with temperature—perhaps because the artificial "sky" did not have a low enough temperature. Neither Fig. 4-11 nor 4-12 treats the important case of one cover glass and a selective surface. In that case the heat loss coefficient can be expected to compare roughly with the two-glass, black-surface data (9).

The heat loss coefficient $U_L$ in Figs. 4-11 and 4-12 is seen to vary over a narrow range depending on both the collector absorber-plate temperature and the ambient temperature. In this book the analyses all assume an average heat loss coefficient. This has been shown to be a reasonable assumption, especially for two-glass collectors. Simulation studies by Klein, Beckman, and Duffie (11) found the effect of the variations in $U_L$ on system performance to be small if an average value of $U_L$ can be chosen at a representative collector plate temperature. This is usually the case using measured collector efficiency points statistically fitted to a straight line as the source of the loss coefficient information.

**Heat Balance.**    The starting point for analysis of a solar collector is a simple heat balance (4). Assuming that the collector is operating in the steady state—that is, that the system is not changing with time—the heat collected will equal the heat absorbed minus the losses to the environment:

$$Q = Q_a - Q_l \qquad (4\text{-}7)$$

The heat absorbed is equal to the product of the radiation flux $I$,* the collector area $A_c$, and the cover plate-absorber transmissivity-absorptivity product $\tau\alpha$, referred to as the *optical efficiency*.

$$Q_a = \tau\alpha I A_c \qquad (4\text{-}8)$$

The heat lost from the system is that lost from the collector absorber plate at temperature $T_c$ to the surroundings at the ambient temperature $T_a$. There are small losses from the collector to the sides and back via conduction and major losses by convection and radiation to the ambient temperature and by radiation to the sky and surroundings. To keep the relationship simple, a combined collector heat loss coefficient $U_L$ is defined in the customary way so that

$$Q_l = U_L A_c (T_c - T_a) \qquad (4\text{-}9)$$

---

* Flux is a time rate normalized for a unit area. Thus radiation flux is measured in W/m² or Btu/hr ft².

Combining terms according to eq. 4-7, the overall heat balance is

$$Q = \tau\alpha I A_c - U_L A_c(T_c - T_a) \tag{4-10}$$

Notice that the arriving irradiance $I$ is expressed as a flux in energy rate per unit area while the collected heat $Q$ is the net energy collection rate from all $A_c$. This can be confusing. The collector analysis is the most versatile if all the terms are expressed as fluxes, that is per unit of collector area. If equation 4-10 is divided by the area $A_c$, the resultant expression is

$$q = \tau\alpha I - U_L(T_c - T_a) \tag{4-11}$$

where $Q/A_c \equiv q$. Lowercase $q$ is defined to be the heat flux collected (energy collection rate per unit area), and should not be confused with uppercase $Q$, the energy collection rate for all the area. Equation 4-11, in which each term is expressed per unit of collector area, is a most important equation and will be used as the starting point for analysis of many collector situations. It is often called the Hottel-Whillier equation, after the solar researchers responsible for its formulation (13).

**Collector Efficiency.** It is sometimes convenient to consider the *collector efficiency*, $\eta$. This is the ratio of the rate of heat collected to that available

$$\eta = q/I \tag{4-12}$$

The value of the collector efficiency is usually between zero and one, but negative values result when the radiation flux cannot make up for the losses. If eq. 4-11 is divided by $I$, the efficiency is expressed as

$$\eta = \tau\alpha - U_L \frac{(T_c - T_a)}{I} \tag{4-13}$$

This equation is useful because it can be plotted to give a very informative curve. If $U_L$ is constant, a straight line results when $\eta$ is plotted on the ordinate with the collector *operating point* or efficiency function

$$f_c = (T_c - T_a)/I \tag{4-14}$$

on the abscissa. The $y$-axis intercept is then $\tau\alpha$ and the (negative) slope is $U_L$. The $x$-axis intercept is $\tau\alpha/U_L$.

A plot of eq. 4-14 is called a collector efficiency curve. An example of such a curve is given in Fig. 4-13. The optical efficiency $\tau\alpha$ is 0.8, a representative figure for either two glass cover plates and a flat black absorber or a selective surface with one cover, and the slope is 5.22 W/°C·m² (0.92 Btu/hr °F ft²), which is a reasonable but not exceptional loss coefficient.

**Use of the Efficiency Curve.** When the efficiency curve is given, collector performance can be predicted with no need to refer to or even be aware of the

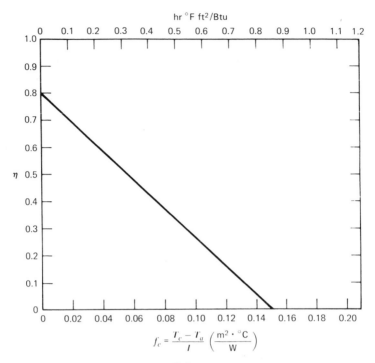

FIGURE 4-13. A collector efficiency curve. $\tau\alpha = 0.8$, $U_L = 5.22$ W/°C·m² (0.92 Btu/hr °F ft²).

values for $\tau\alpha$ or $U_L$. All that is needed is to evaluate the collector operating point $f_c = (T_c - T_a)/I$, read the efficiency from the plot, and apply a rearranged form of eq. 4-12:

$$q = \eta I \qquad\qquad (4\text{-}15)$$

**EXAMPLE 4-1**

The collector having the efficiency curve of Fig. 4-13 [$\tau\alpha = 0.8$, $U_L = 5.22$ W/m² (0.92 Btu/hr °F ft²)] is located at 40°N latitude and tilted at 50° to the horizontal on March 21. The sky is clear and the collector temperature is 48.9 °C (120 °F).

(a) What is the collector efficiency and the rate of heat collection at 11 A.M. if the ambient temperature is 1.7 °C (35 °F)?

From Table A-9 in Appendix A, $I = 307$ Btu/hr ft² or $3.152 \times 307 = 967.7$ W/m².

$$f_c = \frac{T_c - T_a}{I} = \frac{(48.9 - 1.7)\ °C}{967.7\ W/m^2} = 0.0488\ °C\cdot m^2/W$$

From Fig. 4-11 $\eta = 0.545$:

$$q = \eta I = 0.545 \times 967.7 \text{ W/m}^2 = 527.4 \text{ W/m}^2$$

Or using eqs. 4-11 and 4-12,

$$q = 0.8 \times 967.7 \text{ W/m}^2 - 5.22 \text{ W/°C} \cdot \text{m}^2 \times (48.9 - 1.7) \text{ °C}$$
$$= 527.8 \text{ W/m}^2$$

$$\eta = \frac{q}{I} = 527.8/967.7 = 0.545$$

or, using eq. 4-13,

$$\eta = 0.8 - \left[ 5.22 \text{ W/°C} \cdot \text{m}^2 \times \frac{(48.9 - 1.7) \text{ °C}}{967.7 \text{ W/m}^2} \right]$$
$$= 0.545$$
$$q = \eta I = 0.545 \times 967.7 \text{ W/m}^2 = 527.7 \text{ W/m}^2$$

In Engineering units,

$$f_c = \frac{T_c - T_a}{I} = \frac{(120 - 35) \text{ °F}}{307 \text{ Btu/hr °F ft}^2} = 0.277 \frac{\text{hr °F ft}^2}{\text{Btu}}$$

From Fig. 4-13 $\eta = 0.545$:

$$q = \eta I = 0.545 \times 307 \text{ Btu/hr ft}^2 = 167.32 \text{ Btu/hr ft}^2$$

Or using eqs. 4-11 and 4-12,

$$q = 0.8 \times 307 \text{ Btu/hr ft}^2 - 0.92 \text{ Btu/hr °F ft}^2 \times (120 - 35) \text{ °F}$$
$$= 167.4 \text{ Btu/hr ft}^2$$
$$\eta = \frac{q}{I} = \frac{167.4}{307} = 0.545$$

or using eq. 4-13,

$$\eta = 0.8 - \left[ 0.92 \text{ Btu/hr °F ft}^2 \times \frac{(120 - 35) \text{ °F}}{307 \text{ Btu/hr ft}^2} \right]$$
$$= 0.545$$
$$q = \eta I = 0.545 \times 307 \text{ Btu/hr ft}^2 = 167.32 \text{ Btu/hr ft}^2$$

(b) What is the lowest radiation level at which heat can be collected? This is called the *threshold* radiation level.

The collector will have zero efficiency as it begins to deliver heat. From Fig. 4-13, the efficiency is zero when the collector operating point is $0.1532 \text{ °C} \cdot \text{m}^2/\text{W}$ $(0.870 \text{ hr °F ft}^2/\text{Btu})$.

$$f_c = \frac{T_c - T_a}{I_{\text{th}}}$$

$$I_{\text{th}} = \frac{T_c - T_a}{f_c} = \frac{(48.9 - 1.7) \text{ °C}}{0.1532 \text{ °C} \cdot \text{m}^2/\text{W}} = 308.2 \text{ W/m}^2$$

or by solving eq. 4-11 or 4-13 for $I$ with $q$ or $\eta = 0$, respectively,

$$I_{th} = \frac{U_L(T_c - T_a)}{\tau\alpha} = \frac{5.22(48.9 - 1.7)}{0.8} = 308.0 \text{ W/m}^2$$

In Engineering units,

$$f_c = \frac{T_c - T_a}{I_{th}}$$

$$I_{th} = \frac{T_c - T_a}{f_c} = \frac{(120 - 35)\,^\circ\text{F}}{0.870 \text{ hr }^\circ\text{F ft}^2/\text{Btu}} = 97.70 \text{ Btu/hr ft}^2$$

or by solving eq. 4-11 or 4-13 for $I$ with $q$ or $\eta = 0$, respectively,

$$I_{th} = \frac{U_L(T_c - T_a)}{\tau\alpha} = 0.92 \times \frac{120 - 35}{0.8} = 97.75 \text{ Btu/hr ft}^2$$

(c) At 11 A.M. what is the stagnation temperature of the collector—the temperature that is reached if no heat is collected?

This problem is similar to the previous example in that the collector efficiency and heat collected are zero, and therefore the collector operating point is the same. Now, however, the radiation level is known but the collector temperature unknown. From eq. 4-14,

$$T_c = f_c I + T_a$$
$$T_c = 0.1532\,^\circ\text{C}\cdot\text{m}^2/\text{W} \times 967.7 \text{ W/m}^2 + 1.7\,^\circ\text{C} = 150.0\,^\circ\text{C}$$

or from eq. 4-11 or 4-13

$$T_c = \frac{\tau\alpha I}{U_L} + T_a = \frac{0.8 \times 967.7 \text{ W/m}^2}{5.22 \text{ W/m}^2} + 1.7\,^\circ\text{C} = 150.0\,^\circ\text{C}$$

In Engineering units, from eq. 4-14,

$$T_c = f_c I + T_a$$
$$T_c = 0.87 \text{ hr }^\circ\text{F ft}^2/\text{Btu} \times 307 \text{ Btu/hr ft}^2 + 35\,^\circ\text{F} = 302.1\,^\circ\text{F}$$

or from eq. 4-11 or 4-13

$$T_c = \frac{\tau\alpha I}{U_L} + T_a = \frac{0.8 \times 307 \text{ Btu/hr ft}^2}{0.92 \text{ Btu/hr }^\circ\text{F ft}^2} + 35\,^\circ\text{F} = 302\,^\circ\text{F}$$

Efficiency curves such as those in Fig. 4-13 provide an easy way to solve certain kinds of solar radiation problems. Equations 4-11, 4-12, and 4-13 treat the same information on a more analytical basis. Since the availability of inexpensive calculators makes the equations as quick to use as the efficiency curve, the analytical method is stressed here. It is more versatile, provides more insight, and is all but indispensable if a computer or programmable calculator is used for system calculations.

In Chapter 5 the concept of the efficiency curve will be extended to efficiency curves based on the inlet, outlet, and average fluid temperature. These are in practical use, since the average collector surface temperature is difficult to measure.

## Long-Term Collector Performance

The analysis introduced above is valid for any particular time, but does not tell much about long-term performance. The sun's irradiance and the outside ambient temperature (and even the collector temperature) change continually during the day making the instantaneous heat collection rate or efficiency difficult to interpret effectively. Furthermore, it is the total radiation collected that is of the most interest, not the collector efficiency or even an average collector efficiency. The following analysis develops an expression for the total radiation collected over the time period of operation as a direct function of weather, radiation, and collector parameters.

Consider the operation of a collector over the course of a time period $t_T$. Assume that the storage tank for the collected heat is quite large so that the collector temperature is always the temperature of the stored fluid and is thus unchanging with time. Rewriting eq. 4-11 to emphasize the time-varying quantities,

$$q(t) = \tau\alpha I(t) - U_L[T_c - T_a(t)]$$

For an instantaneous time differential $dt$, the equation takes the form

$$q(t)\,dt = \tau\alpha I(t)\,dt - U_L[T_c - T_a(t)]\,dt$$

which may be integrated term by term over a period of interest to give the total radiation collected:

$$q_T = \tau\alpha \int_0^{t_T} I(t)\,dt - U_L \int_0^{t_T} [T_c - T_a(t)]\,dt \tag{4-16}$$

Notice that these integrations can be carried out with relative ease. Compare this with the difficult mathematics that would result if integration of eq. 4-13 had been attempted instead.

Equation 4-16 can be evaluated by formulating the expressions for the time-averaged values of $I(t)$ and $T_a(t)$ over the time period $t_T$:

$$\bar{I} = \frac{\int_0^{t_T} I(t)\,dt}{t_T} \tag{4-17}$$

$$\bar{T}_a = \frac{\int_0^{t_T} T_a(t)\,dt}{t_T} \tag{4-18}$$

Each of these can be solved for its integral, and when both are substituted into eq. 4-16 they give a new form of the equation

$$q_T = \tau\alpha\bar{I}t_T - U_L(T_c - \bar{T}_a)t_T$$

Since the total radiation $I_T$ received in time $t_T$ is $\bar{I}t_T$, this equation can be written

$$q_T = \tau\alpha I_T - U_L(T_c - \bar{T}_a)t_T \tag{4-19}$$

This relationship tells us that the heat gains and the losses of a collector are separate and noninteracting over the entire time period of collector operation, as well as from moment to moment. A little reflection will confirm the logic. A collector at a given temperature loses heat to the environment at a rate determined only by the ambient temperature. Extra heat supplied to the absorber by an increase in radiation only results in more heat being carried off in the fluid removing heat from the collector; the conditions that determine heat loss from the collector do not change. Similarly, less radiation means that less net heat can be collected but, since the collector will remain at the same temperature—that of the collecting fluid—the heat losses will again remain the same.

The heat losses of collectors modeled by eqs. 4-11 and 4-19 are changed only by changing the time of collector operation or by changing the collector temperature. This is why circulation of the heat transfer fluid must be stopped when the collector efficiency falls below zero. Below the *threshold* radiation level, radiation received can no longer make up for the losses, and if fluid circulation were continued there would be a net heat loss from the collector. Conversely, the collector is 100% efficient in producing heat from radiation that reaches the absorber plate in excess of the fixed heat loss. Thus a fixed proportion of additional radiant energy is transformed into heat once the losses have been met.

Equation 4-19 also makes it clear that there is no need for concern about the time distribution of the radiation that falls on the collector. All that is needed is the *total* of the radiation occurring during the collector operating time. Even though the radiation changes during the course of the day—or if the sun goes behind a cloud—the character of the radiation will be adequately and accurately represented by the quantity $I_T$.

Similarly, there is no need for concern about the pattern of ambient temperature change, for the ambient temperature is represented in eq. 4-19 only by $\bar{T}_a$, the average ambient temperature. This time-averaged temperature is the ordinary average that is familiar to us all. For instance, if temperatures are available hourly over $n$ hours,

$$\bar{T}_a = \frac{\sum_{i=1}^{i=n} T_a(i)}{n} \tag{4-20}$$

which means that $\bar{T}_a$ is the sum of the hourly temperature values divided by the number of observations.

It can also be shown that if the collector temperature also changes with time, the time-averaged collector temperature appears in a more general form of eq. 4-19, the integrated collector equation (12):

$$q_T = \tau\alpha I_T - U_L(\bar{T}_c - \bar{T}_a)t_T \tag{4-21}$$

The proof of this is left as an exercise for the student. This relationship shows that no matter how the collector temperature varies there is an average temperature that can be used successfully in performance evaluation; and furthermore, that temperature is simply the time-averaged collector temperature.

### EXAMPLE 4-2

If the collector having the efficiency curve shown in Fig. 4-13 is exposed to clear sky radiation all day on December 21 at 40° N latitude, how much heat will be collected if the collector temperature is 48.9 °C (120 °F)?

Hourly radiation flux and a temperature* for each hour of the solar day have been given in the first two columns of Table 4-2. Radiation flux in the SI unit of $W/m^2$ are 3.152 times the tabulated hourly $Btu/ft^2$ values listed in Table A-3.† For each hour, the total radiation available is thus listed in $W \cdot h/m^2$.

The time for operation is determined by solving eq. 4-11 at each hour of the day, giving the net energy collected listed in the last column of Table 4-2. Negative values indicate that the collector should not be operated during that hour.

The hours of solar operation are 0830 to 1530 solar time, a total of seven hours. During this time the total available radiation is 5169 $W \cdot h/m^2$, and the average ambient temperature using eq. 4-20 is 2.52 °C. Since the collector temperature is given as 48.9 °C, eq. 4-21 can be applied:

$$q_T = \left\{ 0.8 \times 5169 \frac{W \cdot h}{m^2} - \left[ \frac{5.22 \, W}{°C \cdot m^2} (48.9 - 2.52) \, °C \times 7 \, h \right] \right\}$$

$$\times \frac{3600 \, s}{h} \times \frac{J/s}{W} \times \frac{M}{10^6} = 8.79 \, MJ/m^2$$

The conversion to $MJ/m^2$ is made because $W \cdot h$ is not acceptable as an SI unit, having the inconsistent dimensions of $J \cdot h/s$.

The use of eq. 4-21 gives exactly the same solution as the hourly approach of Table 4-2. This is shown by summing the positive values of $q$ from the last column, which gives 2441 $W \cdot h/m^2$, which is also 8.79 $MJ/m^2$.

In Engineering units the solution is simpler because the hourly radiation intervals and the radiation fluxes are both based on a single time unit, the hour. In the Engineering version of Table 4-2 the hourly heat collection total and rate are listed in the last column as calculated by eq. 4-11. There are 7 hours of positive solar heat collection when the collection would be operated. During this time the total available radiation is 1640 $Btu/ft^2$, and the average ambient temperature according to eq. 4-20 is 36.53 °F. Equation 4-21 is then applied using the constant 120 °F collector temperature:

$$q_T = 0.8 \times \frac{1640 \, Btu}{ft^2} - \left[ \frac{0.92 \, Btu}{hr \, °F \, ft^2} (120 - 36.53) \, °F \times 7 \, hr \right]$$

$$q_T = 774.45 \, Btu/ft^2$$

* These are monthly average temperatures for that hour at a Connecticut location.
† This conversion factor is based on the *thermochemical* Btu.

which is also the value produced by adding together the hourly values calculated by eq. 4-11, as shown in the last column of Table 4-2 (Eng.).

**TABLE 4-2 (SI).** Hourly Collection Evaluation (December 21, Table A-3, 50° tilt), $\tau\alpha = 0.8$, $U_L = 5.22$ W/°C·m²

| Time | $I_t$ Clear Sky Incident Irradiance W/m² (J/s·m²) | $T_a$ Ambient Temperature (10-yr hourly average), °C | $T_c$ Collector Temperature °C | $q$ Usable Energy Collected, W·h/m² |
|---|---|---|---|---|
| 8 | 157.6 | 0.20 | 48.89 | −128.07 |
| 9[a] | 516.9 | 0.91 | 48.89 | 163.08 |
| 10[a] | 740.7 | 1.68 | 48.89 | 346.12 |
| 11[a] | 870.0 | 2.27 | 48.89 | 452.63 |
| 12[a] | 914.1 | 2.86 | 48.89 | 491.02 |
| 13[a] | 870.0 | 3.35 | 48.89 | 458.27 |
| 14[a] | 740.7 | 3.33 | 48.89 | 354.76 |
| 15[a] | 516.9 | 3.23 | 48.89 | 175.22 |
| 16 | 157.6 | 3.04 | 48.89 | −113.22 |
| Average[a] | | 2.52[a] | 48.89 | |
| Total/7 hr[a] | 5169[a] | | | 2441.10[a] |

[a] Collector operating.

**TABLE 4-2 (Eng.).** Hourly Collector Evaluation (December 21, Table A-3, 50° tilt), $\tau\alpha = 0.8$, $U_L = 0.92$ Btu/hr °F ft²

| Time | $I_t$ Hourly Clear Sky Incident Irradiance Btu/hr ft² | $T_a$ Ambient Temperature (10-yr hourly average), °F | $T_c$ Collector Temperature °F | $q$ Usable Energy Collected Btu/ft² |
|---|---|---|---|---|
| 8 | 50 | 32.36 | 120 | −40.63 |
| 9[a] | 164 | 33.63 | 120 | 51.74 |
| 10[a] | 235 | 35.02 | 120 | 109.81 |
| 11[a] | 276 | 36.09 | 120 | 143.60 |
| 12[a] | 290 | 37.15 | 120 | 155.78 |
| 13[a] | 276 | 38.03 | 120 | 145.39 |
| 14[a] | 235 | 37.99 | 120 | 112.55 |
| 15[a] | 164 | 37.81 | 120 | 55.59 |
| 16 | 50 | 37.48 | 120 | −35.92 |
| Average[a] | | 36.53 | 120 | |
| Total/7 hr[a] | 1640[a] | | | 774.46[a] |

[a] Collector operating.

## Summary Radiation Tables

A special advantage in the use of eq. 4-19 or 4-21 becomes apparent when it is seen that the time periods making up the total time $t_T$ do not have to be consecutive. The time period involved could be an entire month, for instance, and, provided there were a way to determine the total operating hours in that month, the average ambient temperature during that time, and the total radiation over the same period, the total heat collected could be calculated. Since the sky is never clear for a month, such data cannot be calculated directly but are characterized as *weather data*. Once the minimum radiation level for collector operation is established such data can be taken from weather records *without further regard for the collector characteristics*. If the weather data are assembled beforehand for a number of radiation intervals, as described in Chapter 3, the proper set of radiation data for use with a particular collector can be selected at the time the data are applied.

Such monthly tables of weather data are given in Appendix B for several locations in proper form for use with this averaging technique. The average temperature while at the listed threshold radiation level is given to make it easy to estimate the actual radiation threshold for the collector. This is found by solving eq. 4-11 for the radiation flux when the heat collected is exactly zero:

$$I_{th} = \frac{U_L(T_c - \bar{T}_{th})}{\tau\alpha} \tag{4-22}$$

The lowest radiation grouping having the needed threshold radiation flux is used to provide the total collector operating time, total radiation, and average temperature during operation. These are needed to apply eq. 4-21.

At this point it is suggested that the student review the "Radiation Summary Tables" section from Chapter 3. The utility of the tables will become clearer now that the integrated collector equation and the concept of radiation threshold have been introduced.

### EXAMPLE 4-3

For the collector efficiency curve of Fig. 4-13, $\tau\alpha = 0.8$ and $U_L = 5.22$ W/°C·m². The collector is operated at $\bar{T}_c = 48.9$ °C inclined at 50° to the horizontal at Hartford, Connecticut, during the month of December.

(a) From which radiation grouping should Appendix B data be taken?
(b) What is the total radiation collected in December?

(a) From Appendix B, the 189+ W/m² radiation group threshold data are substituted into eq. 4-22 to give the threshold for this collector:

$$I_{th} = \frac{5.22[48.9 - (-0.7)]}{0.8} = 324 \text{ W/m}^2$$

and for 284+ W/m²,

$$I_{th} = \frac{5.22[48.9 - (-0.4)]}{0.8} = 322 \text{ W/m}^2$$

Since 324 W/m² is above the average threshold radiation level of 234 W/m² listed for the 189+ W/m² radiation grouping, this grouping is not applicable. For the 284+ W/m² grouping, however, the needed 322 W/m² are available, because the average threshold is listed as 330 W/m². The data from that group therefore apply.

(b) From Appendix B, for the 284+ W/m² radiation group,

$$I_T = 225.3 \text{ MJ/m}^2, \qquad t_T = 0.3694 \text{ Ms}, \quad \text{and} \quad \bar{T}_a = -0.9 \text{ °C}$$

Using eq. 4-21, and noting that 1 MJ = 1 MW·s,

$$q_T = \frac{0.8 \times 225.3 \text{ MJ}}{\text{m}^2} - \left\{ \frac{5.22 \text{ W}}{\text{°C·m}^2} [48.9 - (-0.9)] \text{ °C} \times 0.3694 \text{ Ms} \right\}$$

$$= 84.21 \text{ MJ/m}^2$$

The simplicity of Example 4-3 shows how powerful the integrated collector equation is when combined with properly tabulated weather data. By use of the proper Appendix B tables, eqs. 4-21 and 4-22 can predict on a monthly basis the heat collected by a given collector at the selected location and inclination angle. An important limitation of this approach is that the collector temperature must be fixed or the average collector temperature must be known. This limitation will be dealt with in subsequent chapters. For the present the collector temperature should be considered constant, as would be the case for a very large heat storage capacity and good collector plate-to-storage heat transfer.

## Utilizability

If the total radiation is expressed as

$$I_T = \bar{I} t_T = (\bar{I}_{th+} + I_{th}) t_T \tag{4-23}$$

in which $\bar{I}_{th+}$ is the average level by which the irradiance is in excess of the threshold irradiance $I_{th}$, then eq. 4-21 can be rewritten

$$q_T = \tau\alpha(\bar{I}_{th+} + I_{th}) t_T - U_L(\bar{T}_c - \bar{T}_a) t_T \tag{4-24}$$

If $I_{th}$ as given by eq. 4-22 is substituted into eq. 4-24, provided $\bar{T}_{th}$ is taken equal to $\bar{T}_a$ it yields

$$q_T = \tau\alpha \bar{I}_{th+} t_T$$

Solving eq. 4-23 for $\bar{I}_{th+} t_T$,

$$\bar{I}_{th+} t_T = I_T - I_{th} t_T \equiv I_{T+}$$

$I_{T+}$ is the total global radiation occurring in excess of the threshold irradiance during a particular time period—typically a month. $I_{T+}$ is thus the total radiation available to the collector after the constant losses have been subtracted out. It is a continuous function that is tabulated in Appendix B for the stated values of the threshold radiation under the entry GL RAD OVER THRSH. It can be termed the *utilizable* radiation (13).

Combining the two previous equations, the total heat collected is

$$q_T = \tau\alpha I_{T+} \qquad (4\text{-}25)$$

This equation can be used with eq. 4-22 in place of eq. 4-21. The procedure is more complex mathematically than using eqs. 4-21 and 4-22 because the tables must be interpolated to use the concept accurately.

If the utilizable radiation is available in a mathematical form, however, the application is simplified. For instance, using the analytical expression for utilizable radiation as a function of the radiation threshold given in Chapter 3 (eq. 3-34) in combination with eq. 4-25 and setting $T_{th} = \bar{T}_a$,

$$q_T = \tau\alpha I_T \exp\left[-kU_L(\bar{T}_c - \bar{T}_a)/\tau\alpha\right] \qquad (4\text{-}26)$$

$I_T$ is the total radiation available on the tilted surface, and $k$ is a constant dependent on the noontime angle of incidence and the cloud cover (see Chapter 3). Since the exponential relationship is accurate to only a percent or two in any case, eq. 4-26 is best used with a single average ambient temperature for all thresholds, avoiding the need for explicitly determining the threshold irradiance.

Equation 4-26 can also be used to interpolate the radiation tables if a representative value of $k$ is found from eq. 3-37 for use with a typical threshold. Since $k$ remains nearly constant with reasonable collector orientations and ordinary collector thresholds, this is a convenient way to store the information in the tables.

## EXAMPLE 4-4

Find the total radiation collected for the conditions of Example 4-3.

(a) Using the tabular utilizability method.
(b) Using the analytical utilizability method (a value for $k$ of 0.0032 m²/W can be calculated from the Hartford 50° table in December at $I_{th} = 325$ W/m²).*

(a) Since the threshold temperature and the average temperature are not the same, the average temperature will be used in eq. 4-22†, giving

$$I_{th} = \frac{5.22[48.9 - (-0.9)]}{0.8} = 325 \text{ W/m}^2$$

* Equation 3-36 gives $k = 0.003\ 24$ for this example.
† The average temperature has been taken from the 284+ W/m² radiation grouping for which the average threshold level is 330 W/m², which is closest to the calculated threshold.

To apply eq. 4-25 the radiation above the 325 W/m² threshold must be found by interpolation between the 234 and 330 W/m² values given in the Appendix B tables:

$$I_{T+} = 103.6 + \frac{(325 - 330)}{(234 - 330)} (138.8 - 103.6) = 105.4 \text{ MJ/m}^2$$

$$q_T = 0.8 \times 105.4 = 84.3 \text{ MJ/m}^2$$

(b) From Appendix B, $I_T = 296$ MJ/m² and at an average threshold radiation level of 330 W/m², $\bar{T}_a = -0.9$ °C. Using eq. 4-26,

$$q_T = 0.8 \times 296 \exp \{-0.0032 \times 5.22[48.9 - (-0.9)]/0.8\}$$
$$= 83.7 \text{ MJ/m}^2$$

## Efficiency of Collection

Even when eq. 4-21 or 4-25 is used to calculate the collected heat, measures of collector efficiency may be of interest. Efficiency may be determined by comparing the predicted performance with the maximum radiation available *over the time period of interest*, and applying eq. 4-12. When the time period is the same as the collection period, the *average* collector operating efficiency is calculated. If the *total* radiation available over the entire month (or other base time) is used, the *overall* efficiency of collection can be calculated based on the total radiation available during the time period whether the collector was operating or not.

**EXAMPLE 4-5**

For Example 4-3(b), calculate the average collector efficiency and the overall efficiency of collection.

The average collector efficiency relates to the radiation available above 284 + W/m²:

$$\bar{\eta} = \frac{84.21 \text{ MJ/m}^2}{225.3 \text{ MJ/m}^2} = 37.4\%$$

The overall collection efficiency is taken relative to the total radiation available at all levels, which is available in the "All" column in Appendix B:

$$\eta_0 = \frac{84.21 \text{ MJ/m}^2}{296.0 \text{ MJ/m}^2} = 28.5\%$$

The above example incidentally demonstrates the major problem in solar heating with flat plate collectors: the efficiency is low. Furthermore, the particular collector chosen (Fig. 4-13), being inoperative when it would lose more heat than it could collect, never had a chance to collect some 20% of the total radiation available. Of the remaining 80%, it collected only 37.37%, an unimpressive portion that is nonetheless in the range of current practice. There is clearly room for improvement here, and much of the research in solar collector technology is directed to that end. As a result, several high-efficiency collectors have been developed that have very promising performance. Unfortunately, these efforts are not yet cost effective because the user would be better off financially to install an increased area using ordinary collectors than to utilize an expensive but efficient collector. Although this may be the general case, if the high-efficiency collector is being used in a very cloudy area or on a building for which the area for the collector must be limited, costs can favor it even now.

High-efficiency collectors will probably be necessary for systems having solar-powered air conditioning, since these heat-operated systems perform well only as temperatures approach boiling—a temperature rise of some 70 °C (120 °F) above the ambient, compared with the 30 to 55 °C (50–100 °F) temperature rise needed for the average solar-heating system. These applications can be expected to encourage a volume of production that will reduce the cost of these collectors as soon as cost-effective solar air-conditioning units become available.

## COLLECTOR IMPROVEMENT

The simplest way to higher efficiency is to increase the radiation taken up by the absorber plate. The absorptivity of even common black paint, however, is so high that improved absorption will not have much effect. Consequently, water-white glass, with its high transmissivity, is very desirable, and most high-efficiency collectors—as well as some ordinary collectors—are incorporating it. Etched-surface, antireflective coatings are the next logical step, and we can soon expect them to be available commercially.

While improved cover glass can increase the heat that gets to the absorber plate, reducing the heat losses can put more of the absorbed heat to practical use. The most workable way to do this is to use a selective surface on the absorber plate, which reduces the radiation losses, thereby effectively reducing the heat transfer coefficient.

Convection losses also need reduction, but this is more difficult and more costly. The most logical approach is to increase the number of glass plates, but more than two plates are not seen in commercial practice because of the associated transmission loss with uncoated glass.

The convection coefficient can also be reduced by modifying the character of the convection between the absorber and cover plate. This can be done with a clear (or reflective) honeycomb having cells perpendicular to the absorber plate, that is inserted between the two surfaces. If the optical properties of the honeycomb are good it will absorb little light, and if the material is thin and nonmetallic little heat will be conducted to the cover plate. Reflective losses are at a minimum because the light reflected from the honeycomb is not lost but instead reflected into the collector toward the absorber.

The geometry of the honeycomb is critical (14). Good results have been obtained with cells about 1 cm across and 5 cm deep. Thinner integral cover plate–honeycomb laminates are also under development that have the added advantage of structural integrity. Heat loss characteristics somewhat better than double-glazed collectors are anticipated from honeycombs, with less loss in optical efficiency than a second cover plate.

Current materials proposed for honeycombs are transparent glass, Mylar,* and Lexan. Teflon, Tedlar, and polyester-fiberglass are under study as associated cover plates. Except for glass, however, the thermal environment is very difficult, since stagnation temperatures are above the service temperatures of the plastics.

Mylar accordian plaits between cover plate and absorber are a related approach that has seen commercialization. The plaits are deep and quite closely spaced to reduce reflective losses to a minimum. Temperature excursions are dealt with by a separate passive cooling system.

Comparative evaluations of any of these convection-suppressing configurations are difficult, but further development seems worthwhile.

### Evacuated Tubular Collectors

The idea of evacuating the space between the cover glass and the absorber plate is very attractive because this reduces convection losses to zero. With a selective surface the radiation losses could also be reduced to near zero, producing an ideal collector. Because of the pressure of the atmosphere, however, evacuating the space between flat plates is technically very difficult. Honeycombs could offer good physical support if the space were evacuated, but many sealing problems remain and this approach is so far impractical. On the other hand, there is much technology available in evacuating glass tubes for fluorescent lighting applications, and collectors based on using evacuated tubes have been successfully developed.

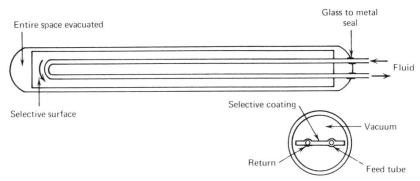

**FIGURE 4-14.** The Corning evacuated tube collector (not to scale).

* Trademark of DuPont. Will appear this way throughout.

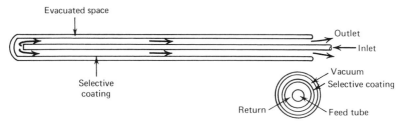

**FIGURE 4-15.** The Owens-Illinois evacuated tube collector (not to scale).

Figures 4-14 and 4-15 show two types of high-efficiency, evacuated-tube collectors. They were developed by two of the leading U.S. glass companies. The Corning collector uses a large-diameter, glass tube surrounding a narrow but conventional selectively surfaced flat plate collector. The Owens-Illinois approach is more dramatic, with the collector surface no longer a flat plate but a second tube within the first. The annular space between them is evacuated so that the entire glass assembly is like an over-long, wide-mouthed vacuum bottle, but with an absorptive rather than a reflective coating on the outside surface of the inner glass. Air or water is circulated to the closed end of the glass with ordinary tubing so that the returning stream collects heat by convection from the glass inner wall. The absorptive coating is selective and, because it is exposed only to vacuum, it should be quite durable. Its absorptivity of 0.85 is not as good as conventional selective surfaces, reducing the present performance of the collector below its potential. The reflection losses are specular, giving a futuristic appearance due to the grey-blue, mirror-like surface.

Each of the collectors can collect heat from radiation reaching the back as well as the front of the absorber surface. Utilization of this effect, pioneered by Owens-Illinois, has been achieved by spacing the tubes rather far apart and using a white matte surface reflector behind the assembly to illuminate the rear surface, as is shown in Fig. 4-16. While this arrangement does not increase the radiation that falls on a given surface area, it does increase the radiation that a given tube can collect, decreasing the cost of the heat collected. Spacing the collectors apart in this way also increases the efficiency of the tubular concept because the tubes do not then shade each other until later in the day. On a superficial area basis, however, the spreading of the tubes is bound to decrease the collector efficiency, albeit at less cost per unit of thermal energy collected.

Manufacturer's efficiency curves for the two collectors are presented in Fig. 4-17. The Corning collector curve is much higher, since it was rated in a close packed configuration, while the Owens-Illinois collector was arranged as shown in Fig. 4-16. Use of the Owens-Illinois curve is especially difficult, since it changes somewhat with the angle of the sun, with the reflector least effective at the time of curve measurement, which is solar noon. This situation is best handled by the

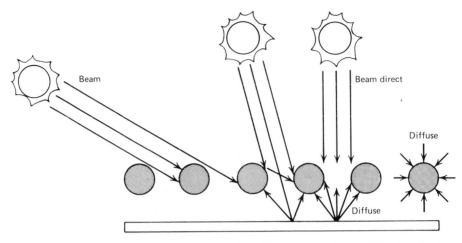

**FIGURE 4-16.** Diffuse reflector installed with widely spaced Owens-Illinois collectors. Notice that the circular absorber surface is illuminated by the direct rays of the sun in the same manner for most of the day. (*Source:* Owens-Illinois promotional literature.)

use of a properly averaged optical efficiency $\overline{\tau\alpha}$ in eq. 4-21. Moan (15) suggests determination of $\overline{\tau\alpha}$ by plotting daily average instead of instantaneous points in determining the collector efficiency curve. This value can be approximated by integration of the instantaneous incident angle modifier (discussed in the next section) over an entire day, using clear sky radiation data unless some sort of average day can be postulated.

Fluid coolant circuiting poses problems with each collector. The Corning collector uses a glass-to-metal seal through the vacuum, adding considerably to its expense and creating pressure drop problems, since the tube diameter is held to an absolute minimum. The Owens-Illinois collector uses no glass-to-metal seals but breakage of a single tube can cause loss of all the coolant fluid. This collector holds up a rather large quantity of collector liquid, making morning startup slow and inefficient. It is better suited for use with air than the Corning collector, however, since its pressure drop can be quite low.

A collector similar to the Owens-Illinois collector has been developed by General Electric. The fluid never leaves the metal tube, however, contacting the glass by means of a circular fin. With this arrangement a broken glass cannot cause fluid spillage, and less fluid is used to fill the collectors, resulting in faster startups and therefore greater collection efficiency when used with liquids.

Each collector has a good optical efficiency and a very low heat loss coefficient. Stagnation temperatures are therefore very high, and boilout is a problem with any conceivable liquid coolant whenever one or more tubes are not adequately cooled while exposed to the sun. Refilling the hot tubes with liquid is then unwise because of the danger of glass breakage due to thermal shock, and performance may not resume until the tubes have cooled naturally. Failure of the vacuum seal also presents peculiar problems. The failed tube envelope will heat excessively, causing a system heat loss that may go undetected for some time. Such a tube presents an invisible danger to maintenance personnel, since it is heated by the array to the coolant temperature, which in summer can be quite hot.

These collectors are too expensive at the present time to compete with ordinary flat plate collectors for home heating or in the solar hot water market. For air conditioning, where high temperatures are an inherent requirement, a collector capable of high temperatures is a virtual necessity. Each of these collectors has an important advantage over concentrating collectors (see Chapter 6) for such

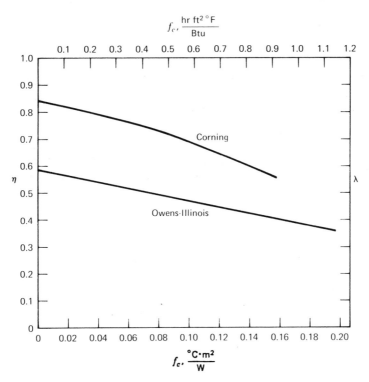

**FIGURE 4-17.** Manufacturer's efficiency curves for the Corning and Owens-Illinois evacuated tube collectors. The Corning curve is based on average temperature of the coolant, the Owens-Illinois on the inlet temperature, but with no flow specified. These curves are not sufficient for design. Note in comparing the collectors that the Owens-Illinois collector curve is based on the superficial area, thus including the empty spaces between tubes.

an application, because they can collect diffuse as well as direct radiation. This is a very important feature when use in the cloudy northeastern or midwestern areas of the United States is being considered.

## EFFECT OF INCIDENT ANGLE

During the day the angle of incidence of the sun on a solar collector changes constantly, changing the direct radiation flux in proportion to the cosine of the angle $\theta$ between the incoming radiation and a line normal to the collector surface (eq. 3-12). This effect has been incorporated into the tables of Appendixes A and B. The diffuse irradiance also reaches the collector from all incident angles, but it is always assumed isotropic over the separate earth and sky hemispheres so that its intensity is a simple function of the irradiance that is received on a horizontal plane and the collector tilt angle (eq. 3-13 or 3-14). This effect is also incorporated into Appendixes A and B.

The irradiance actually received at the absorber plate of course depends on the reflection and absorption through the cover plates and on the absorber surface. The effect is incorporated into eqs. 4-11 and 4-21 by use of the $\tau\alpha$ term, but since $\tau\alpha$ is ordinarily measured with irradiance perpendicular to the cover plates ($\theta = 0° = \theta_n$) the change in $\tau\alpha$ with incident angle has yet to be considered.

One way to do this is to determine the collector performance curve using points taken over an entire day instead of just at noon. To analyze such daylong data, the integrated collector equation (4-21) is divided through by $t_T$, and the average rate of heat flux collection defined as $\bar{q} = q_T/t_T$. Then if the average optical efficiency is $\overline{\tau\alpha}$, performance can be correlated by

$$\bar{q} = \overline{\tau\alpha}\bar{I} - U_L(\bar{T}_c - \bar{T}_a) \qquad (4\text{-}27)$$

which is a time-averaged form of eq. 4-11. The corresponding collector efficiency curve, having a slope $U_L$ and an intercept of $\overline{\tau\alpha}$, automatically includes the effect of incident angle. A curve determined in this manner will be characteristic of a particular tilt and test location, however, because the cloud cover affects the proportion of diffuse irradiance reaching the collector.

The effect of incident angle may be treated more generally by directly measuring its effect on performance and then analytically evaluating the effect of changing incident angle during daylong operation at the tilt and site involved. By using an *incident angle modifier* applied to the optical efficiency measured at normal incidence, a correctly averaged value of $\overline{\tau\alpha}$ can then be obtained for use in further performance analysis.

Figure 4-4b shows the effect of incident angle on $\tau$, the transmissivity through a system of glass plates with specified properties. The changes are a complex function of the incident angle $\theta_i$. Figure 4-18 shows the effect of incident angle on a col-

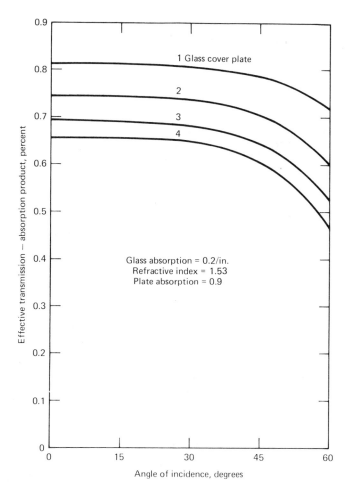

**FIGURE 4-18.** Effect of incident angle on the transmittance-absorptance product for flat black paint absorbers with a number of panes of glass. (Source: Colorado State University Report No. NSF/RANN/SE/GI-37815/PR/74/1, April 1974).

lector plate–absorber combination. Although the effect is small, this kind of plot is different from Fig. 4-4, since the absorptivity also changes with incident angle.

In practice an incident angle modifier for the normal incidence optical efficiency $(\tau\alpha)_n$ is measured as a part of the standard collector efficiency testing procedures and is given by

$$[K_{\tau\alpha}(\theta_i)]_{\text{direct}} = \frac{(\tau\alpha)_{\theta_i}}{(\tau\alpha)_n} \qquad (4\text{-}28)$$

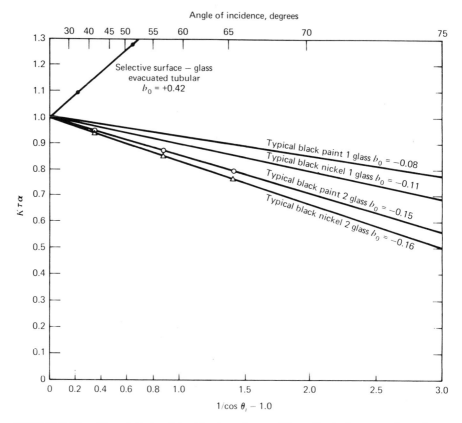

**FIGURE 4-19.** Correlation of incident angle modifier. Typical curves have been generalized from Ref. 10 and other sources.

where the normal incidence optical efficiency is measured by the usual testing procedures. Since these tests are carried out near solar noon on clear days when the proportion of diffuse light is low, this incident angle modifier is taken to apply to the direct irradiance.

Because the slope of an efficiency curve is not a function of $\tau\alpha$, it is only necessary to establish the $y$ intercept for a sampling of incident angles to determine the incident angle modifier values. This is done by tilting the collector and finding the efficiency when the collector operating point is zero (i.e., when the collector temperature equals the ambient temperature).

The resultant data can be neatly correlated (16) by use of the equation

$$(K_{\tau\alpha})_{\text{direct}} = 1 + b_0\left(\frac{1}{\cos\theta_i} - 1\right) \qquad (4\text{-}29)$$

where $b_0$ is the slope of a curve such as in Fig. 4-19, in which $K_{\tau\alpha}$ is plotted versus $1/\cos\theta_i - 1$ for several collectors. Note that even the Owens-Illinois evacuated tube collector plots linearly using this relationship, but with a positive slope because the reflector is least effective at solar noon.

For (isotropic) diffuse irradiance it can be shown (10) that

$$(K_{\tau\alpha})_{\text{diffuse}} = \frac{(\tau\alpha)_{\text{diffuse}}}{(\tau\alpha)_n} = 1 + b_0 \tag{4-30}$$

which evidently ignores the fact that tilted collectors do not "see" all the sky. The correct $\tau\alpha$ for a mixture of direct and diffuse irradiance is clearly the radiation-weighted, average optical efficiency:

$$(\tau\alpha)_{\theta i} = \frac{(\tau\alpha)_{\text{direct}}\, I_{\text{direct}} + (\tau\alpha)_{\text{diffuse}}\, I_{\text{diffuse}}}{I_{\text{global}}} \tag{4-31}$$

Dividing through by $(\tau\alpha)_n$, the optical efficiency at $\theta_i = 0°$,

$$\frac{(\tau\alpha)_{\theta_i}}{(\tau\alpha)_n} = (K_{\tau\alpha})_{\text{direct}} \frac{I_{\text{direct}}}{I_{\text{global}}} + (K_{\tau\alpha})_{\text{diffuse}} \frac{I_{\text{diffuse}}}{I_{\text{global}}} \tag{4-32}$$

Noting that $I_{\text{diffuse}} = I_{\text{global}} - I_{\text{direct}}$ and then incorporating eqs. 4-29 and 4-30, eq. 4-32 yields the *global* incident angle modifier as a function of incident angle and the proportion of direct irradiance:

$$(K_{\tau\alpha})_{\text{global}} = \frac{(\tau\alpha)_{\theta_i}}{(\tau\alpha)_n} = 1 + b_0\left[1 + \frac{I_{\text{direct}}}{I_{\text{global}}}\left(\frac{1}{\cos\theta_i} - 2\right)\right] \tag{4-33}$$

Equation 4-33 must be weighted by radiation intensity over an entire day, hour-by-hour, to find a correct average-value optical efficiency $\overline{\tau\alpha}$.

If the correct monthly value of $I_{\text{direct}}/I_{\text{global}}$ (taken above the proper collector-operating threshold) is applied over a clear sky day, the monthly $\overline{\tau\alpha}$ can be estimated, since the statistical angular distribution of the direct irradiance is adequately represented by clear sky data for most locations, where cloud cover occurs at random during the day. Expressed mathematically, if $\Delta t$ is the time increment, the average optical efficiency modifier is given by

$$\frac{\overline{\tau\alpha}}{(\tau\alpha)_n} = \frac{\displaystyle\sum_{\text{day}} I_{\text{global}}\left\{1 + b_0\left[1 + \frac{I_{\text{direct}}}{I_{\text{global}}}\left(\frac{1}{\cos\theta_i} - 2\right)\right]\right\}\Delta t}{I_{T\,\text{global}}} \tag{4-34}$$

Since $I_{\text{direct}}/I_{\text{global}}$ is a ratio, any convenient units may be used, and they need not be consistent with those used for hourly and total global radiation. The ratio can be calculated from the Appendix B tables (this equals $1 - I_{\text{direct}}/I_{\text{global}}$) or, alternately, the proportion of diffuse radiation can be estimated from $\overline{K}_T$ using the Page or other correlations as described in Chapter 3.

Integrating eq. 4-33 by the use of eq. 4-34 is cumbersome. With only a slight loss of accuracy, during the heating season, monthly average optical efficiency modifiers can be approximated by using the value of $\theta_i$ for $2\frac{1}{4}$ hours from solar noon in eq. 4-33. Simpler yet, a seasonal average can be estimated in a single step with the same equation by using the monthly average $\theta_i$ at $2\frac{1}{4}$ hours from noon with the monthly average proportion of direct radiation.

### EXAMPLE 4-6

Estimate the value of the incident angle optical efficiency modifier for the conditions of Examples 4-2 and 4-3. The collector is single glazed with a selective surface, and has a $b_0$ value of $-0.11$.

For a 50° tilt in December at 284+ $W/m^2$, $I_T = 225.3 \ MJ/m^2$ and $I_{T(\text{direct})} = 191.0 \ MJ/m^2$. Considering their ratio constant over the day,

$$\frac{I_{\text{direct}}}{I_{\text{global}}} = \frac{191.0}{225.3} = 0.8478$$

From Tables A-3 and A-9, the radiation level and incident angle of the sun on the collector are listed in Table 4-3 for the seven operating hours of the day.* For 9 A.M., eq. 4-33 gives

$$\frac{(\tau\alpha)_{\theta = 44.9°}}{(\tau\alpha)_n} = 1 + b_0\left[1 + \frac{I_{\text{direct}}}{I_{\text{global}}}\left(\frac{1}{\cos 44.9°} - 2\right)\right]$$

$$= 1 - 0.11\left[1 + 0.8478\left(\frac{1}{0.7083} - 2\right)\right]$$

$$= 0.9448$$

Application of eq. 4-34 is shown in the other entries of Table 4-3. The sum of the radiation incident on the collector is 1640 Btu/ft², and the sum of the hourly

**TABLE 4-3.** Determination of an Average Optical Efficiency Modifier for the Conditions of Examples 4-4 and 4-6

| Time | $\theta_i$ | $I$ (Btu/hr ft²) | $\left(\dfrac{(\tau\alpha)_{\theta i}}{(\tau\alpha)_n}\right)$ | $I_{\text{net}}$ (Btu/hr ft²) |
|------|-----------|------------------|---------------------------------------------------------------|-------------------------------|
| 0900 | 44.9° | 164 | 0.9448 | 154.9 |
| 1000 | 31.6° | 235 | 0.9670 | 227.3 |
| 1100 | 19.6° | 276 | 0.9775 | 269.8 |
| 1200 | 13.5° | 290 | 0.9806 | 284.4 |
| 1300 | 19.6° | 276 | 0.9775 | 269.8 |
| 1400 | 31.6° | 235 | 0.9670 | 227.2 |
| 1500 | 44.9° | 164 | 0.9448 | 154.9 |
|  |  | 1640 |  | 1588.4 |

* This time is for full sunshine. The average cloudiness for December might in the extreme reduce this to five hours, with an increase of 0.7% in the calculated value of the incident angle modifier. The radiation-averaged operating time, however, is consistent at $0.0245/0.0036 = 6.8$ hr (the radiation-averaged operating time will be discussed in Chapter 10).

totals weighted by the ratio of hourly to normal optical efficiencies is 1588 Btu/ft$^2$. Therefore, for December,

$$\frac{\overline{\tau\alpha}}{(\tau\alpha)_n} = \frac{1588}{1640} = 0.9685$$

Alternatively, using eq. 4-33, the monthly modifier can be estimated in a single step from the incident angle $2\frac{1}{4}$ hours from noon:

$$\frac{\overline{\tau\alpha}}{(\tau\alpha)_n} = 1 - 0.11\left[1 + 0.8478\left(\frac{1}{\cos 34.9°} - 2\right)\right]$$

$$= 0.9628$$

**TABLE 4-4.** Seasonal Tabulation of the Optical Efficiency Modifier $\overline{\tau\alpha}/(\tau\alpha)_n$ for the five winter months as in Example 4-6 [a]

| Month | Threshold Radiation Level | $I_{T(\text{direct})}/I_{T(\text{global})}$ | Daily Average Optical Efficiency Modifier by Eq. 4-34, $\overline{\tau\alpha}/(\tau\alpha)_n$ | | | Optical Efficiency Modifier by Eq. 4-33, $\theta_i$ taken $2\frac{1}{4}$ hr from Noon | | |
|---|---|---|---|---|---|---|---|---|
| | | | $b_0 = -0.08$ | $b_0 = -0.11$ | $b_0 = -0.15$ | $b_0 = -0.08$ | $b_0 = -0.11$ | $b_0 = -0.15$ |
| Nov | 189+ W/m$^2$ | 0.7623 | 0.9680 | 0.9560 | 0.9400 | 0.968 | 0.956 | 0.941 |
| Dec | 284+ W/m$^2$ | 0.8478 | 0.9771 | 0.9685 | 0.9571 | 0.973 | 0.963 | 0.949 |
| Jan | 284+ W/m$^2$ | 0.8176 | 0.9757 | 0.9665 | 0.9544 | 0.972 | 0.961 | 0.947 |
| Feb | 284+ W/m$^2$ | 0.7742 | 0.9674 | 0.9552 | 0.9389 | 0.970 | 0.958 | 0.944 |
| Mar | 189+ W/m$^2$ | 0.6679 | 0.9587 | 0.9432 | 0.9225 | 0.962 | 0.947 | 0.928 |
| Average optical efficiency modifier | | | 0.9694 | 0.9579 | 0.9426 | 0.969 | 0.958 | 0.943 |

[a] Direct/Global data is from Appendix B for Hartford, Connecticut. Other data is calculated from Table A-3 for 40°N at a 50° inclination angle.

The value of the average-incident-angle, optical efficiency modifier varies throughout the season, decreasing somewhat in the milder months as the azimuthal range of the sun increases and as the proportion of diffuse light rises because of higher humidity. Table 4-4 gives monthly averages for the modifier, continuing Example 4-3 for three different values of $b_0$ (see Fig. 4-19) using both methods of calculation. Overall averages, assuming an even solar contribution from each month, are also tabulated. Notice that the threshold radiation level, which affects the direct-to-global ratio, has been decreased in November and March because of rising ambient temperatures, also increasing the diffuse contribution.

A correction for dust on the collector plate can be introduced at this point. Hottel and Woertz (4) found that transmissivity was reduced by about 0.6% by the steady-state accumulation of dust on tilted collectors in the well-washed Boston climate. Accumulations of dust on the absorber plate can be much more serious, since they are not washed away by rain. Collectors must be well-sealed for this reason, especially if air is to be circulated as the heat transfer medium.

On balance, for a collector tilted to about the latitude plus 10° an average modifier of 0.935 is recommended for two-glass collectors ($b_0 = -0.15$), 0.95 for one-glass, selective surface collectors ($b_0 = -0.11$), and 0.96 for one-glass, black paint collectors ($b_0 = -0.08$). Such a correction can most easily be made by modifying the intercept of the experimental collector efficiency curve as it is being tabulated for use in a particular design situation. In almost all instances, it is simpler and just as accurate to use an average modifier throughout the season, since any error will be small considering the accuracy of most of the other data used in solar system analysis.

## PROBLEMS

For these problems use the Appendix B data closest to your own location.

1. A solar collector having $U_L = 4.656$ W/°C·m² (0.82 Btu/hr °F ft²) and $\tau\alpha = 0.76$ is tilted 50° from the horizontal on March 21. The sky is clear, and the average collector temperature is 50 °C (122 °F). Use the ASHRAE table (Appendix A) closest to your location.
   (a) What is the collector efficiency and rate of heat collection at 11:00 A.M. if the ambient temperature is −5 °C (23 °F)?
   (b) What is the lowest radiation level at which heat can be collected?
   (c) What is the stagnation temperature of the collector?

2. The collector from the above problem is exposed to clear sky radiation all day on December 21. Hourly ambient temperatures are those given in Table 4-2.
   (a) Calculate and sum over the day the hourly heat flux collected using the instantaneous collector performance equation 4-11.
   (b) Repeat the calculation using average and total values in the integrated collector equation 4-19.

3. For a slope of 40°, azimuth 180°, using Appendix B tables for March:
   (a) Find the total radiation falling when the radiation level exceeds 378 W/m² (120 Btu/hr ft²).
   (b) Find the total time the radiation level exceeded 378 W/m² (120 Btu/hr ft²).
   (c) Find the average temperature when the radiation level was over 378 W/m² (120 Btu/hr ft²).

4. Using the collector of problem 1 and continuing with a 50 °C (122 °F) collector temperature, use Appendix B tables and eq. 4-21 to calculate:
   (a) The average March radiation threshold.
   (b) The total heat collected in March.

5. Repeat problem 4 using eq. 4-25.

6. Repeat problem 4 using eq. 4-26. (You may wish to review the use of eq. 3-36.)

7. (a) For problem 4 (or 5 or 6) calculate the average and overall collection efficiency.
   (b) Draw an efficiency curve for the solar collector and plot the average collection efficiency determined in part (a) versus the collector function calculated using the basic data in Appendix B.

8. Assume that the collector of problem 1 has incident angle performance such that $b_0 = 0.12$. Find the monthly and average incident angle modifier for November to March:
   (a) Using hourly values (a programmable calculator will be needed).
   (b) Using the incidence angle $2\frac{1}{4}$ hours from solar noon.
   (c) Using the average seasonal incident angle modifier found by averaging the four monthly incidence angles $2\frac{1}{4}$ hours from solar noon during the winter months.

9. Derive eq. 4-33 as suggested in the text.

## REFERENCES

1. H. Y. B. Mar, J. H. Lin, P. B. Zimmer, R. E. Peterson, and J. S. Gross, *Optical Coatings for Flat Plate Solar Collection*. Final Report, 16 Sept 1974 to 16 Sept 1975. Available NTIS.
2. J. H. Lin, *Optimization of Coatings for Flat Plate Solar Collectors, Phase II*. January 1977. Available NTIS.
3. John C. Ward, personal communication.
4. H. C. Hottel and B. B. Woertz, "The Performance of Flat-Plate Solar-Heat Collectors," *Trans. ASME*, 91–104 (February 1942).
5. T. Tani, S. Sawata, T. Tanaka, and T. Horigome, *A Terrestrial Solar Thermal Energy Power System*, 1975 ISES meeting, Los Angeles.
6. H. Tabor, *Research on Optics of Selective Surfaces*, Final Report on Contract AF61 (052)-279, May 1963.
7. H. Tabor et al., *Further Studies on Selective Black Coatings*, Paper S/46, U.N. Conf. on New Sources of Energy, Rome; August 1961.
8. Teuvo Santala, *Selective Intermetallic Compound Surfaces*, 1975 ISES meeting, Los Angeles.
9. J. A. Duffie and W. A. Beckman, *Solar Energy Thermal Processes*, New York: Wiley, 1974.
10. F. F. Simon and E. H. Buyco, *Outdoor flat-plate collector performance prediction from solar simulator test data*. AIAA 10th Thermal Physics Conference, Paper No. 75-741, Denver, 1975.
11. S. A. Klein, W. A. Beckman, and J. A. Duffie, "A design procedure for solar heating systems." *Solar Energy 18*, 113–127 (1976).
12. P. J. Lunde, "Seasonal solar collector performance with maximum storage," *ASHRAE Journal* (November 1977).
13. H. C. Hottel and Austin Whillier, "Evaluation of flat-plate solar collector performance," Trans. of the Conf. on Use of Solar Energy II. Thermal Processes 74-104, University of Arizona, Tempe (1955).
14. H. Buchberg and D. K. Edwards, "Design considerations for solar collectors with cylindrical glass honeycombs," *Solar Energy 18*, 193–203 (1976).
15. K. L. Moan, *An Analysis of the low loss evacuated tubular collector using air as the heat transfer fluid*. Unpublished paper, Owens-Illinois, Inc., Toledo, Ohio.
16. A. F. Souka and H. H. Safwat, "Optimum orientations for the double-exposure, flat-plate collector and its reflectors," *Solar Energy 10*, 170 (1966).

# Flat Plate Collectors—Removing the Heat

Once the sun's heat has been absorbed on the surface of the collector plate it must be removed and delivered for use or storage. This is done by circulating a heat transfer fluid such as air or water to contact the absorber plate and be warmed by transfer of collected heat.

If the heat absorbed were not collected it would accumulate and raise the temperature of the collector until that temperature was high enough that the losses to the ambient equalled the collected heat. If the heat is inefficiently removed from the collector the same thing happens—the collector temperature rises until the losses to the heat transfer fluid plus those to the ambient just balance the solar input. The most efficient removal of the collected heat, therefore, requires that the collector surface be maintained at the lowest possible temperature, nearly that of the incoming heat transfer fluid. The success in keeping the collector surface temperature low is measured by a collector heat transfer efficiency factor that can be estimated analytically using the techniques described in this chapter.

### Sources of Inefficiency

**Flow Rate.** The fluid that has removed the heat from the collector necessarily becomes hotter as it receives the collected energy. If the flow rate is low, the temperature will rise appreciably in removing a given amount of heat. The average collector temperature is then raised even if the heat transfer is perfect. The flow rate of the heat transfer fluid therefore has an important effect on the efficiency of the collector. If the collector is performing a function such as heating hot water in a single pass, the flow rate must be limited because a certain temperature must be attained. But when the collector is used to collect heat for space heating, high temperatures are not necessary and circulation of air or water to the collector will usually be kept high so that the collector surface stays near the lowest possible temperature, that of the bulk of the stored fluid.

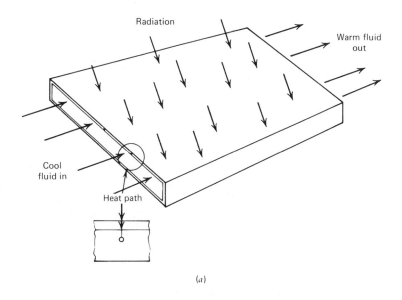

(a)

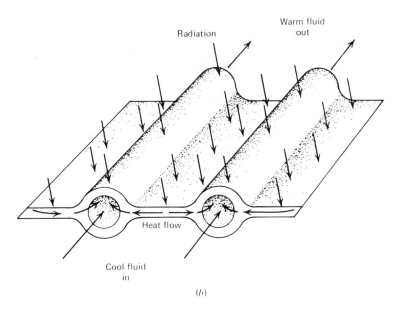

(b)

**FIGURE 5-1.** (a) General configuration of the absorber plate when the entire absorber surface is swept by the cooling fluid. Heat is conducted *perpendicular* to the absorber surface. (b) General configuration of the absorber plate when heat is conducted *parallel* to the absorber surface to fluid flowing in tubes.

167

**Fluid Heat Transfer.**   Maintaining an adequate flow rate, however, is not the only important consideration in fluid heat transfer. The heat must be transferred to the fluid, which is easy to do if the fluid is moving quickly but is somewhat harder with slow-moving fluids because they do not as effectively sweep heat from the sides of a duct or tube into the fluid. The ease of such heat transfer depends to a large degree on whether the fluid flow is *laminar* or *turbulent*. Laminar flow is smooth and gentle, with the fluid moving in slipping concentric layers along the tube walls. This flow is normal with slow-moving fluids. As the velocity is increased slight irregularities cause backward swirls and eddies to form, which tend to mix the fluid and improve the heat transfer. At a surprisingly definite point it becomes impossible to keep the eddies from forming throughout the fluid no matter how smooth the tube. Above this velocity the flow is said to be turbulent, and heat transfer coefficients increase markedly. Turbulent flow is therefore desirable for heat exchange. It is not always available because the energy to maintain the chaotic flow pattern must ultimately come from the pump or fan driving the fluid circulation. In general the flow can easily be kept turbulent with aqueous liquids, but with air or nonaqueous liquids the collector designer must work in or near the laminar flow range lest the circulation power consumption become prohibitive.

**Conductive Heat Transfer.**   The collected heat must be transferred from the absorber plate to the walls of the tube or duct carrying the heat transfer fluid. The heat is transferred either *through* the thickness of the absorber plate to the fluid circulating directly below it (Fig. 5-1a), or *along* a portion of the length (or width) of the absorber to reach widely spaced tubes or ducts (Fig. 5-1b). In the first case the heat must travel through only a thin plate and the thermal conductivity of the plate is not too important. In the second case, however, the absorber surface serves as an irradiated *fin* for the tube, and the collected heat must pass along the cross-sectional area of the plate from as far as half the distance to the next tube. Therefore, the conductivity and thickness of the collector plate become very important.

Design of the heat removal portions of a solar collector is fraught with opportunities for error, and these errors tend to be perpetuated by those who do not properly analyze the heat transfer situation. Many commercially marketed collectors have obvious design faults, and an understanding of the problems and trade-offs the designer faces is a help in avoiding them.

### Design of the Absorber

Figure 5-2 presents a number of absorber designs, showing the fluid passages used to remove the collected heat. Designs (*a*) to (*h*) are used with liquid coolants, while designs (*i*) to (*k*) are more practical with air cooling.

Design (*a*) in Fig. 5-2 is simply a serpentine connection of commercially available piping with no absorber plate as such. The pipes do not touch because available fittings do not permit extremely tight turns; therefore sunlight that falls between the tubes is wasted. Figures 5-1*b* and 5-1*c* show two ways to improve this situation. In design (*b*) a standard baseboard heating assembly using fins perpendicular to the tubes has been installed. This arrangement will generally intercept all the sunlight if the fins are closely spaced. The flow pattern between tubes can remain serpentine, or the tubes can be connected in parallel at entrance and exit to the collector. Such an absorber is simple to construct, although rather expensive.

In Fig. 5-1*c* large tubes have been flattened and joined together to form a continuous absorbing surface. Here the serpentine flow pattern is no longer practical; instead the tube ends must be soldered or welded into specially designed inlet and outlet headers. This design uses less material than design (*b*), but flow velocities are likely to be low and heat transfer somewhat inferior.

If the tubes are very small, as they could be if the absorber were extruded from a thermoplastic, then design (*d*) is practical. The small round tubes promote good heat transfer, and parallel flow provides a very low pressure drop.

Design (*e*) is used to reduce the material required in a metallic collector. The tubes are spaced far apart and the gap between them bridged with a "fin" soldered or welded to the tube. Fluid velocities will be high, promoting heat transfer. This design is easy to fabricate in a small shop because the fins are separately attached to each tube. When larger-scale manufacturing permits oven-soldering of the entire absorber, all the tubes can be attached at once to a continuous absorber plate as in Fig. 5-2(*f*), again using either a serpentine flow or parallel flow with headers. If the tubes are square, attachment will be more certain due to the large bond area, but the tubes may distort or even tear loose if they are overpressurized, especially if the bond is weak. Round tubes, however, will withstand pressures well in excess of any likely design without rupturing or distorting.

The flow passages in design (*g*) of Fig. 5-2 are made by using a special photographic process in which the piping network design is laid between two sheets of copper or aluminum and retained after the sheets are bonded by rolling them together. The tubes are then formed by overpressurizing the network so that the fluid circuit is *within* the fin. This very popular technique permits any number of flow configurations because a two-dimensional flow network can be custom-designed and headering done within the absorber plate.

If steel is used as an absorber plate, the flow circuit must be stamped in the metal before joining the plates together, since steel is not as ductile as copper or aluminum. The dimpled plate configuration in Fig. 5-1*h* is often used, because it minimizes the welded bond area holding the sheets together.

This configuration can be used with liquids, or if the passages are somewhat wider, with an air coolant. Most air collectors are simpler, however, often just a rectangular passage of sheet metal as in Fig. 5-1*a*. If the passageways are divided as in the design in Fig. 5-2*i*, support of the thin metal is easier and even flow distribution from the header is more likely.

Since air collectors suffer from poor heat transfer to the air stream, extended

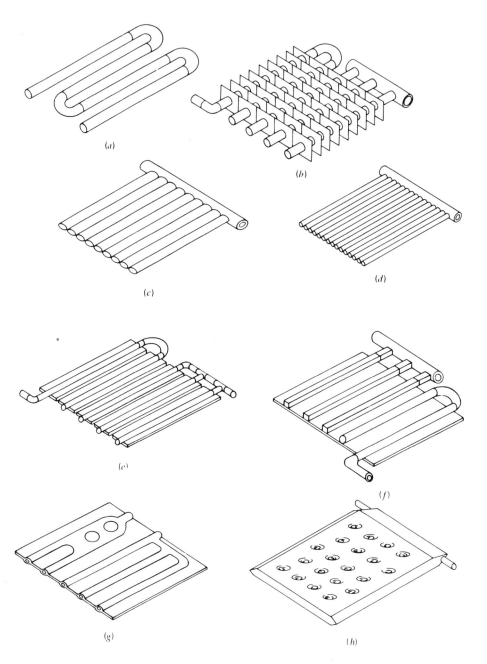

**FIGURE 5-2.** A selection of absorber designs, showing flow circuiting and headering. Each of these designs has been used in a commercial collector.

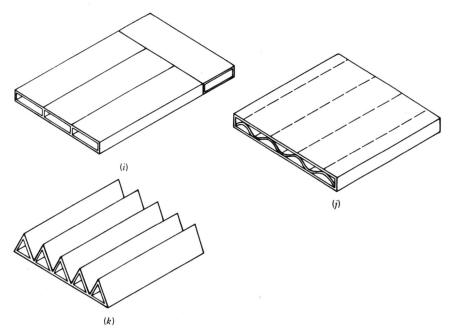

(i)

(j)

(k)

**FIGURE 5-2.** *(Continued)*

surface is sometimes installed. If the metal used in design (*i*) is thick enough, the entire channel will be heated by the absorber, doubling the available surface area. Absorber (*j*) is more elaborate, with corrugated metal quadrupling the area. Care must be taken, however, that the assembly be well bonded to assure good heat transfer.

There is no special reason to require a flat absorber plate. Some selective surfaces perform poorly at high incidence angles in which case design (*k*) will increase the average absorptivity. This design is most practical for an air-cooled collector, to take advantage of the extended surface inherent in the design.

**Collector Materials.** The practical materials for a flat plate collector are few. Metals are limited by cost to copper, aluminum, and steel. Brass might also be used if it were cheaper or its thermal conductivity higher. Corrosion is a problem with steel and especially with aluminum when an aqueous heat transfer fluid is used. Corrosion of steel can be overcome with proper inhibitors provided the system is sealed and kept free of air. Potentially, the same will be true for aluminum but, for the present, aluminum is a poor choice for collectors where the maximum life is desired unless less effective nonaqueous heat transfer fluids are used.

The application of plastics is limited by the fact that even with only one cover plate, temperatures of 150 °C (300 °F) can easily be reached if no heat is being

**TABLE 5-1a.** Physical Properties of Heat Transfer and Storage Media (SI Units)

| Fluid | Temperature, °C | Density, kg/m³ | Viscosity, m Pa·s [g/m·s] | Specific Heat kJ/kg·°C | Prandtl No. $(c\mu/k)$ | Thermal Conductivity W/°C·m | Coefficient of Thermal Expansion %/°C |
|---|---|---|---|---|---|---|---|
| Water | 38 | 993 | 0.684 | 4.166 [a] | 5 | 0.628 | 0.037 |
| | 66 | 980 | 0.432 | 4.187 [a] | 2.8 | 0.661 | 0.055 |
| | 93 | 963 | 0.305 | 4.208 [a] | 1.8 | 0.680 | 0.070 |
| Ethylene glycol–water | 38 | 1054 (1061) [f] | 2.3 | 3.43 | 19.83 | 0.398 | 0.057 |
| 50% by weight | 66 | 1035 (1041) [f] | 1.2 | 3.52 | 10.16 | 0.415 | 0.066 |
| (Dowtherm SR-1) [b] | 93 | 1016 (1024) [f] | 0.76 | 3.64 | 6.39 | 0.433 | 0.073 |
| Propylene glycol–water | 38 | 1025 (1027) [f] | 3.1 | 3.64 | 29.0 | 0.389 | 0.066 |
| 50% by weight | 66 | 1004 (1008) [f] | 1.5 | 3.73 | 14.6 | 0.384 | 0.082 |
| (Dowfrost) [b] | 93 | 985 (988) [f] | 0.9 | 3.83 | 9.06 | 0.381 | 0.080 |
| Silicone oil | 38 | 935 | 14.98 | 1.55 | 162 | 0.144 | |
| (Dow-Corning | 66 | 913 | 9.13 | 1.59 | 103 | 0.141 | 0.0928 |
| Syltherm 444) [c] | 93 | 889 | 6.40 | 1.63 | 75.5 | 0.138 | |
| Synthetic oil | 38 | 835 | 25.1 | 2.39 | 461 | 0.130 | 0.059 |
| H-30 [d] | 66 | 820 | 9.0 | 2.47 | 175 | 0.127 | |
| (to 175 °C) | 93 | 805 | 4.4 | 2.55 | 90 | 0.125 | |
| Mobiltherm light [e] | 38 | 966 | 4.4 | 1.83 | 67.1 | 0.120 | 0.063 |
| (maximum service | 66 | 950 | 2.5 | 1.93 | 40.9 | 0.118 | |
| temperature 205 °C) | 93 | 935 | 1.6 | 2.02 | 27.9 | 0.116 | |
| Mobiltherm 603 [e] | 38 | 855 | 17.1 | 1.89 | 241.7 | 0.134 | 0.063 |
| (maximum service | 66 | 838 | 7.54 | 1.99 | 112.8 | 0.133 | |
| temperature 315 °C) | 93 | 820 | 4.02 | 2.09 | 63.9 | 0.132 | |
| Air (50% rh at 70 °F) | 21 | 1.187 | 0.018 | 1.01 | 0.71 | 0.0260 | |
| | 66 | 1.033 | 0.021 | 1.02 | 0.70 | 0.0292 | |
| Rock (any common kind) | | 1600 (as packed) | | 0.837 | | | |
| Concrete (dense) | | 2043 | | 0.837 | | 1.731 | |
| Copper | | 8902 | | 0.393 | | 377 | |
| Aluminum | | 2722 | | 0.921 | | 206 | |
| Steel (not stainless steel) | | 7860 | | 0.448 | | 45 | |
| Stainless steel (300 series, typical) | | 8005 | | 0.502 | | 16 | |

[a] 4.19 is used for all examples, consistent with engineering practice.
[b] These are trademarks of Dow Chemical Company for their inhibited glycols. Properties courtesy Dow Chemical Company.
[c] Trademark Dow Corning Corporation. Properties courtesy Fluids and Lubricants Technical Service and Development Dept.
[d] Estimated from company literature, Mark Enterprises, Inc.
[e] Established using company literature, Mobil Oil Co.
[f] When inhibited. (Other properties are unaffected.)

removed. This condition might exist during construction or power failures and therefore the utmost care—too much, in the opinion of most people—is required to keep them below the 95 °C (200 °F) softening point of the most common plastics. Without a cover plate, however, there is no risk of overheating so that polyethylene and polypropylene in particular have seen wide application for solar swimming pool heaters, which are run so near the ambient temperature that convective losses are near zero even without a cover plate.

The only plastic promising for higher-temperature, structural use in collectors is molded fiberglass. This plastic is similar to the fiberglass used as a cover plate material but it can be more durable since it need not be clear. The collector surface can be blackened during manufacture if carbon black is added to the polyester

**TABLE 5-1*b*.** Physical Properties of Heat Transfer and Storage Media (Engineering Units)

| Fluid[c] | Temperature °F | Density, lb/ft³ | Viscosity, $\mu$, lb/min ft | Viscosity, $\mu$, cp | Specific Heat Btu/lb °F | Prandtl No. $(c\mu/k)$ | Thermal Conductivity Btu/hr° F ft | Coefficient of thermal expansion, %/°F |
|---|---|---|---|---|---|---|---|---|
| Water | 100 | 62.0 | 0.0276 | 0.684 | 0.9975[b] | 5 | 0.363 | 0.020 |
|  | 150 | 61.2 | 0.0174 | 0.432 | 1.000[b] | 2.8 | 0.382 | 0.031 |
|  | 200 | 60.1 | 0.0123 | 0.305 | 1.005[b] | 1.8 | 0.393 | 0.039 |
| Ethylene glycol–water | 100 | 65.8 (66.2)[d] | 0.0927 | 2.3 | 0.82 | 19.83 | 0.23 | 0.032 |
| 50% by weight | 150 | 64.6 (65.0)[d] | 0.0484 | 1.2 | 0.84 | 10.16 | 0.24 | 0.036 |
| (Dowtherm SR-1) | 200 | 63.4 (63.9)[d] | 0.0306 | 0.76 | 0.87 | 6.39 | 0.25 | 0.041 |
| Propylene glycol–water | 100 | 64.0 (64.1)[d] | 0.125 | 3.1 | 0.87 | 29.0 | 0.225 | 0.037 |
| 50% by weight | 150 | 62.7 (62.9)[d] | 0.605 | 1.5 | 0.89 | 14.6 | 0.222 | 0.045 |
| (Dowfrost) | 200 | 61.5 (61.7)[d] | 0.0363 | 0.9 | 0.915 | 9.06 | 0.220 | 0.044 |
| Silicone oil | 100 | 58.4 | 0.604 | 14.98 | 0.37 | 162 | 0.083 |  |
| (Dow-Corning | 150 | 57.0 | 0.368 | 9.13 | 0.38 | 103 | 0.0815 | 0.052 |
| Syltherm 444) | 200 | 55.5 | 0.258 | 6.40 | 0.39 | 75.5 | 0.0800 |  |
| Synthetic oil | 100 | 52.1 | 1.01 | 25.1 | 0.57 | 461 | 0.075 | 0.033 |
| H-30 (to | 150 | 51.2 | 0.364 | 9.0 | 0.59 | 175 | 0.0735 |  |
| 350 °F) | 200 | 50.2 | 0.178 | 4.4 | 0.61 | 90 | 0.072 |  |
| Mobiltherm light | 100 | 60.3 | 0.179 | 4.4 | 0.438 | 68.1 | 0.0691 | 0.035 |
| (maximum service | 150 | 59.3 | 0.100 | 2.5 | 0.460 | 40.5 | 0.0681 |  |
| temperature 400 °F) | 200 | 58.3 | 0.0641 | 1.6 | 0.483 | 27.7 | 0.0671 |  |
| Mobiltherm 603 | 100 | 53.4 | 0.690 | 17.1 | 0.452 | 241.4 | 0.0775 | 0.035 |
| (maximum service | 150 | 52.3 | 0.304 | 7.54 | 0.477 | 112.8 | 0.0768 |  |
| temperature 600 °F) | 200 | 51.2 | 0.162 | 4.02 | 0.500 | 63.9 | 0.0761 |  |
| Air (50% rh at 70 °F) | 70 | 13.5 ft³/lb | 0.000 726 | 0.018 | 0.242 | 0.71 | 0.0150 |  |
|  | 150[a] | 15.5 ft³/lb | 0.000 847 | 0.021 | 0.243 | 0.70 | 0.0169 |  |
| Rock (any common kind) |  | 100 (as packed) |  |  | 0.2 |  |  |  |
| Concrete (dense) |  | 150 |  |  | 0.2 |  | 1 |  |
| Copper | 212 | 556 |  |  | 0.094 |  | 218 |  |
| Aluminum | 212 | 170 |  |  | 0.22 |  | 119 |  |
| Steel (not stainless) | 212 | 491 |  |  | 0.107 |  | 26 |  |
| Stainless steel | 212 | 500 |  |  | 0.12 |  | 9 |  |
| (300 series, typical) |  |  |  |  |  |  |  |  |

[a] Same water content as 70 F air.
[b] 1.000 is used for all examples, consistent with engineering practice.
[c] Sources are listed in Table 5-1*a*.
[d] When inhibited. (Other properties are unaffected.)

plastic that forms the "glue" holding the matrix of glass fibers together. Such a material can be formulated to have long-term durability up to about 175 °C (350 °F).

## HEAT TRANSFER TO FLUIDS

The most important single consideration in heat transfer to a flowing fluid is the character of the flow regime, which has loosely been defined as either laminar or turbulent. Precise characterization is made by a dimensionless parameter known as the Reynolds number:

$$N_{Re} = \frac{4W}{\mu\pi D} \qquad (5\text{-}1a)$$

$D$ is the tube diameter, $W$ the mass flow rate, and $\mu$ is the viscosity of the fluid, all expressed in consistent units so that the Reynolds number will be dimensionless. Physical properties of the common solar heat transfer fluids are given in Table 5-1.

When the tube or duct is not circular in cross section, an equivalent diameter for use in eq. 5-1a is given by

$$D_e = \frac{4A_x}{P} \tag{5-2}$$

where $P$ is the perimeter of the tube or duct filled with flowing fluid, and $A_x$ is the cross-sectional area for fluid flow. The Reynolds number is then given more generally by

$$N_{Re} = \frac{4W}{\mu P} \tag{5-1b}$$

Reynolds numbers below 2100 mean that the flow is viscous, or *laminar*. Over 10 000 the flow is turbulent. In the range between 1000 and 10 000 the flow is said to be in the transition zone. Even though correlations are difficult, the transition zone is very important in solar work.

### EXAMPLE 5-1

A 1.22 m (4 ft) wide solar collector has flow of fluid averaging 65.55 °C (150 °F) flowing upward along its 2.44 m (8 ft) length. The irradiance is 788 W/m² (250 Btu/hr ft²), and the collector operates with an efficiency of 48% with the outdoor ambient at 10 °C (50 °F).

(a) What is the temperature rise if the collector is water cooled with a water flow of 25.2 $\mu$m³/s (24 gal/hr)?

$$q = 788 \text{ W/m}^2 \times 0.48 = 378 \text{ W/m}^2$$

$$W = \frac{25.2 \,\mu\text{m}^3}{\text{s}} \times \frac{10^{-6}}{\mu} \times \frac{980 \text{ kg}}{\text{m}^3} = 0.0247 \text{ kg/s}$$

$$c = \frac{4.19 \text{ kJ}}{\text{kg} \cdot {}^\circ\text{C}}$$

$$A = 1.22 \text{ m} \times 2.44 \text{ m} = 2.98 \text{ m}^2$$

$$\Delta T = \frac{Q}{Wc} = \frac{378 \text{ W}}{\text{m}^2} \times 2.98 \text{ m}^2 \times \frac{\text{s}}{0.0247 \text{ kg}} \times \frac{\text{kg} \cdot {}^\circ\text{C}}{4.19 \text{ kJ}} \times \frac{\text{k}}{10^3} = 10.88 \,^\circ\text{C}$$

(b) What is the Reynolds number if the water flow is distributed in parallel to eight 6.35-mm (¼ in.) inside-diameter tubes spaced with 152.4 mm (6 in.) between tubes?

$$W = \frac{0.0247 \text{ kg}}{\text{s}} \times \tfrac{1}{8} = 0.003 \ 09 \text{ kg/s}$$

Using eq. 5-1a

$$N_{Re} = \frac{4}{\pi} \times \frac{0.003\ 09\ kg}{s} \times \frac{m \cdot s}{0.432\ g} \times \frac{1}{6.35\ mm} \times \frac{m}{10^{-3}} \times \frac{10^3}{k}$$

$$= 1434$$

(c) What is the Reynolds number if the water flow is through the same tubes in series?

$$N_{Re} = \frac{4 \times 0.0247 \times 10^6}{\pi 6.35 \times 0.432} = 11\ 464$$

(d) What is the Reynolds number if the eight parallel tubes are 25.4 mm (1.0 in.) inside diameter?

$$N_{Re} = \frac{4W}{\mu \pi D} = \frac{4}{\pi} \times \frac{0.003\ 09\ kg}{s} \times \frac{m \cdot s}{0.432\ g} \times \frac{1}{25.4\ mm} \times \frac{m}{10^{-3}} \times \frac{10^3}{k}$$

$$= 359$$

(e) What is the temperature rise if air cooling is used at 0.037 76 m³/s (80 ft³/min)?

$$W = \frac{0.037\ 76\ m^3}{s} \times \frac{1.033\ kg}{m^3} = 0.039\ 02\ kg/s$$

$$\Delta T = \frac{Q}{Wc} = \frac{378\ W}{m^2} \times 2.98\ m^2 \times \frac{s}{0.039\ 02\ kg} \times \frac{kg \cdot {}^\circ C}{1.02\ kJ} \times \frac{k}{10^3}$$

$$= 28.27\ {}^\circ C$$

(f) What is the Reynolds number if the air flows up the collector in a 12.7 mm × 1.22 m (½ in. × 4 ft) channel behind the absorber plate?

From eq. 5-1b, if $P = 2[1.22 + (12.7/100)] = 2.465$ m,

$$N_{Re} = \frac{4W}{\mu P} = 4 \times \frac{0.039\ 02\ kg}{s} \times \frac{1}{2.465\ m} \times \frac{m \cdot s}{0.021\ g} \times \frac{10^3}{k}$$

$$= 3015$$

In Engineering units,

(a) $q = 250$ Btu/hr ft² × 0.48 = 120 Btu/hr ft²

$$W = \frac{24\ gal}{hr} \times \frac{ft^3}{7.48\ gal} \times \frac{61.2\ lb}{ft^3} = 196.36\ lb/hr$$

$$\Delta T = Q/Wc$$

$$= \frac{120\ Btu}{hr\ ft^2} \times 32\ ft^2 \times \frac{hr}{196.36\ lb} \times \frac{lb\ {}^\circ F}{1\ Btu}$$

$$= 19.55\ {}^\circ F$$

(b)
$$W = \frac{196.36 \text{ lb}}{\text{hr}} \times \frac{\text{hr}}{60 \text{ min}} = 3.273 \text{ lb/min}$$

(The flow through one tube is one-eighth of this.) Using eq. 5-1a

$$N_{Re} = \frac{4}{\pi} \times \frac{12 \text{ in.}}{\text{ft}} \times \frac{3.273 \text{ lb}}{8 \text{ min}} \times \frac{1}{0.25 \text{ in.}} \times \frac{\text{min ft}}{0.0174 \text{ lb}} = 1437$$

(c)
$$N_{Re} = \frac{4 \times 12 \times 3.273}{\pi \times 0.25 \times 0.0174} = 11\,496$$

(d)
$$N_{Re} = \frac{4W}{\mu \pi D} = \frac{4}{\pi} \times \frac{3.273 \text{ lb}}{8 \text{ min}} \times \frac{\text{min ft}}{0.0174 \text{ lb}} \times \frac{12 \text{ in.}}{\text{ft}} \times \frac{1}{1 \text{ in.}}$$
$$= 359$$

(e)
$$W = \frac{80 \text{ ft}^3}{\text{min}} \times \frac{60 \text{ min}}{\text{hr}} \times \frac{\text{lb}}{15.5 \text{ ft}^3} = 309.7 \text{ lb/hr}$$

$$\Delta T = \frac{Q}{Wc} = \frac{120 \text{ Btu}}{\text{hr ft}^2} \times 32 \text{ ft}^2 \times \frac{\text{hr}}{309.7 \text{ lb}} \times \frac{\text{lb }^\circ\text{F}}{0.243 \text{ Btu}} = 51.03 \text{ }^\circ\text{F}$$

(f) From eq. 5-1b, if $P = 2(4 + 0.5/12) = 8.083$ ft,
$$\frac{4W}{\mu P} = 4 \times \frac{5.161 \text{ lb/min}}{0.000\,847 \text{ lb/min ft}} \times \frac{1}{8.083 \text{ ft}} = 3015$$

The above examples demonstrate how readily the flows used in solar collectors can produce turbulent, laminar, and transitional flow of water or air.

**Turbulent Flow.** When the flow is turbulent the heat transfer coefficient is generally high and can be predicted with good accuracy by the following equation (3):

$$\frac{hD}{k} = 0.023(N_{Re})^{0.8}(N_{Pr})^{1/3} \tag{5-3}$$

where $hD/k$ is the dimensionless Nusselt number. The dimensionless Prandtl number,* $N_{Pr}$, is a function of fluid properties alone and is listed for common solar heat transfer fluids in Table 5-1. Each term in eq. 5-3 is dimensionless, and the equation can therefore be used easily with any consistent system of units. Extreme care with the units is necessary, however, and it is always recommended that they be carried throughout the calculations. For air and water simpler dimensional equations in Engineering units are available:

$$\text{For air, } h = \frac{0.5(u_s)^{0.8}}{(D')^{0.2}} \tag{5-3a}$$

$$\text{For water, } h = \frac{150(1 + 0.011T)(u_s)^{0.8}}{(D')^{0.2}} \tag{5-3b}$$

---

* $N_{Pr} = c\mu/k$ where $c$ is the specific heat of the fluid, $\mu$ its viscosity, and $k$ its thermal conductivity.

where $u_s$ is the fluid velocity in ft/sec, $D'$ the characteristic diameter in inches, $h$ the heat transfer coefficient in Btu/hr °F ft², and $T$ the temperature in °F. Like eq. 5-3, these equations hold only for Reynolds numbers above 10 000.

When the fluid is in turbulent flow within a helical or pancake coil, the heat transfer is increased due to additional agitation by a factor of $(1 + 3.5D_i/\bar{D}_c)$, where $D_i$ is the inside diameter of the coil and $\bar{D}_c$ is the arithmetic average diameter of the turns making up the coil.

**EXAMPLE 5-2**

Find the heat transfer coefficient for the tubes in Example 5-1(c).

$$N_{Re} = 11\ 464$$
$$N_{Pr} = 2.8 \quad \text{(from Table 5-1)}$$

Using eq. 5-3,

$$\frac{hD}{k} = 0.023(11\ 464)^{0.8}(2.8)^{1/3} = 57.31$$

Note that this intermediate result, called the Nusselt number, is also dimensionless.

$$h = N_{Nu} \times \frac{k}{D} = 57.31 \times \frac{0.661 \text{ W/°C·m}^2}{6.35 \text{ mm}} \times \frac{m}{10^{-3}} = \frac{5966 \text{ W}}{°C·m^2}$$

In Engineering units,

$$h = N_{Nu} \times \frac{k}{D} = 57.31 \times \frac{0.382 \text{ Btu/hr ft °F}}{0.25 \text{ in.} \times \text{ft/12 in.}}$$

$$h = 1051 \text{ Btu/hr °F ft}^2$$

or, using eq. 5-3b with Engineering units,

$$u_s = \frac{3.273 \text{ lb}}{\text{min}} \times \frac{\text{ft}^3}{61.2 \text{ lb}} \times \frac{\text{min}}{60 \text{ sec}} \times \frac{1}{0.000\ 340\ 88 \text{ ft}^2} = 2.615 \text{ ft/sec}$$

$$h = 150(1 + 0.011 \times 150)\frac{(2.615)^{0.8}}{(0.25)^{0.2}} = 1131 \frac{\text{Btu}}{\text{hr °F ft}^2}$$

**Transitional Flow.** In fast laminar and transitional flow, entrance effects become important. When the fluid first enters a tube from a large-diameter header, or when it has just taken a sharp turn, the slipping layers of laminar flow are upset and the fluid is well-mixed for a time just as it is during turbulent flow, and an improved heat transfer coefficient results. The effect only lasts for as long as it takes for the laminar flow to become reestablished, but in many common heat transfer applications—most air-cooling "coils," for example—the entrance effects provide the turbulence necessary for efficient heat transfer. This is the reason why air-conditioning heat exchangers are seldom over 2 in. thick—further flow would be in the laminar flow regime where heat transfer is very poor.

Entrance effects are also important in the flow circuits of most solar collectors. Clever designers use them to keep the heat transfer rates high even when the basic flow tends to be laminar, which usually means poor heat transfer. The entrance effects are significant for a certain number of diameters downstream of a disturbance, which can be estimated (1) by

$$\frac{L}{D} = 0.05 N_{Re} N_{Pr} \tag{5-4}$$

The variables used to correlate heat transfer in the laminar and transitional range ($N_{Re} < 10\,000$) are the Reynolds number and the length-of-tube to diameter-of-tube ratio $L/D$. These are related via the Colburn $j$ factor by the graph in Fig. 5-3a (2).

The $j$ factor includes the Prandtl number, the *mass velocity* $W/A_x$ (the fluid mass flow rate per unit area for flow), and the specific heat. Like many other correlating factors in fluid flow and heat transfer, $j$ is dimensionless.

$$j = \frac{h A_x}{c W} (N_{Pr})^{2/3} \tag{5-5}$$

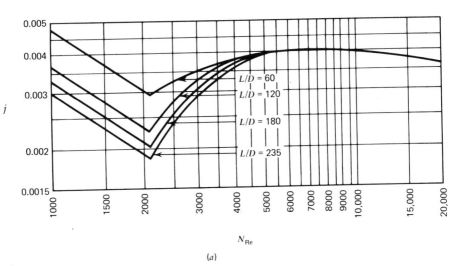

$(a)$

**FIGURE 5-3.** $(a)$ The Colburn $j$ factor for correlating heat transfer in the transitional flow range (Ref. 2). From Chemical Engineers Handbook edited by J. H. Perry, Third edition. Copyright © 1950 McGraw-Hill Book Company. Used with permission of McGraw-Hill Book Company.

We use it in transposed form:

$$h = \frac{jcW}{A_x(N_{Pr})^{2/3}} \tag{5-5a}$$

**EXAMPLE 5-3a**

Find the heat transfer coefficient for the laminar (but transitional) flow of water through the solar collector of Example 5-1(b).

The length of one tube is 2.44 m and the diameter is 6.35 mm.

$$L/D = \frac{2.44 \text{ m}}{6.35 \text{ mm}} \times \frac{\text{m}}{10^{-3}} = 384$$

$$A_x = \frac{\pi}{4} \times (6.35 \text{ mm})^2 = 31.67 \text{ (mm)}^2$$

From Fig. 5-3, $j = 0.0023$ at $N_{Re} = 1434$. Using eq. 5-5,

$$h = \frac{jcW}{A_x(N_{Pr})^{2/3}} = \frac{0.0023}{31.67 \text{ (mm)}^2} \times \frac{4.19 \text{ kJ}}{\text{kg °C}} \times \frac{(0.0247/8) \text{ kg}}{(2.8)^{2/3} \text{ s}}$$

$$\times \frac{\text{m}^2}{10^{-6}} \times \frac{10^3}{k} \times \frac{\text{Ws}}{\text{J}} = \frac{472.9 \text{ W}}{\text{°C·m}^2}$$

**EXAMPLE 5-3b**

Find the heat transfer coefficient for the laminar air flow in the collector of Example 5-1(f). By eq. 5-2:

$$D_e = 2 \times 12.7 \text{ mm} \times \frac{10^{-3}}{\text{m}} = 0.0254 \text{ m}$$

$$L/D = \frac{2.44 \text{ m}}{0.0254 \text{ m}} = 96$$

From Fig. 5-3, $j = 0.0033$ at $N_{Re} = 3015$.

$$h = \frac{0.0033 \, cW}{(N_{Pr})^{2/3} A_x} = \frac{0.0033}{(0.7)^{2/3}} \times \frac{1.02 \text{ kJ}}{\text{kg °C}} \times \frac{0.039 \, 02 \text{ kg}}{\text{s}}$$

$$\times \frac{1}{(1.22 \text{ m})(0.0127 \text{ m})} \times \frac{\text{Ws}}{\text{J}} \times \frac{10^3}{k} = \frac{10.75 \text{ W}}{\text{°C·m}^2}$$

Note the poor heat transfer coefficient, which is characteristic of slow-moving air.

**EXAMPLE 5-3c**

If the flow in Example 5-3b is broken up by baffles at 0.6096 m (2 ft) intervals, what is the new heat transfer coefficient?

$$L/D = \frac{2 \text{ ft} \times 12 \text{ in./ft}}{1.0 \text{ in.}} = \frac{0.6096 \text{ m}}{0.0254 \text{ m}} = 24$$

At $N_{Re} = 3015$ the $j$ factor can only be estimated from Fig. 5-3:

$$j = 0.004$$

$$h = \frac{0.004}{(0.7)^{2/3}} \times \frac{1.02 \text{ kJ}}{\text{kg} \cdot ^\circ\text{C}} \times \frac{0.039 \ 02 \text{ kg/s}}{1.22 \text{ m} \times 0.0127 \text{ m}} \times \frac{10^3}{\text{k}} \times \frac{\text{Ws}}{\text{J}}$$

$$= 13.03 \text{ W/}^\circ\text{C} \cdot \text{m}^2$$

In Engineering units,

(a)
$$A_x = \frac{\pi}{4} \times \left(\frac{0.25}{12}\right)^2 \text{ ft}^2 = 0.000 \ 340 \ 88 \text{ ft}^2$$

$$L/D = \frac{8 \text{ ft} \times 12 \text{ in./ft}}{0.25 \text{ in.}} = 384$$

From Fig. 5-3, $j = 0.0023$ at $N_{Re} = 1437$.

$$h = \frac{jcW}{A_x(N_{Pr})^{2/3}} = \frac{0.0023 \times 1 \text{ Btu/lb } ^\circ\text{F} \times (3.273/8) \text{ lb/min} \times 60 \text{ min/hr}}{0.000 \ 340 \ 88 \text{ ft}^2 \times (2.8)^{2/3}}$$

$$= 83.37 \text{ Btu/hr } ^\circ\text{F ft}^2$$

(b)
$$D_e = 4A_x/P \approx 2D = 2 \times 0.5 \text{ in.} = 1.0 \text{ in.}$$

$$L/D = \frac{8 \text{ ft} \times 12 \text{ in./ft}}{1.0 \text{ in.}} = 96$$

From Fig. 5-3, $j = 0.0033$ at $N_{Re} = 3015$.

$$h = \frac{0.0033cW}{A_x(N_{Pr})^{2/3}} = \frac{0.0033}{(0.7)^{2/3}} \times \frac{0.243 \text{ Btu}}{\text{lb } ^\circ\text{F}} \times \frac{1858 \text{ lb}}{\text{hr ft}^2} = 1.89 \text{ Btu/hr } ^\circ\text{F ft}^2$$

(c)
$$L/D = \frac{2 \text{ ft} \times 12 \text{ in./ft}}{1.0 \text{ in.}} = 24$$

From Fig. 5-3, $j \approx 0.004$ at $N_{Re} = 3015$.

$$h = \frac{0.004}{(0.7)^{2/3}} \times \frac{0.243 \text{ Btu}}{\text{lb } ^\circ\text{F}} \times \frac{1858 \text{ lb}}{\text{hr ft}^2} = 2.29 \text{ Btu/hr } ^\circ\text{F ft}^2$$

Example 5-3 shows that breaking up the laminar film is an effective way to increase laminar heat transfer coefficients. This can be done with baffles, abrupt changes of direction, sudden contraction and expansion, and in other ways. As the $L/D$ becomes shorter, heat transfer coefficients continue to improve. Using analytical techniques beyond the scope of this book, heat transfer coefficients in the $L/D$ range of 5 to 50 can be predicted closely enough to aid in the design of solar collectors (3). But even with such techniques the heat transfer coefficients in air collectors are so poor that extended surface—fins or corrugations—is often installed.

**Laminar Flow.** If the flow regime is fully developed laminar flow ($N_{\text{Re}} < 1000$), the heat transfer coefficient is a constant because the slipping layers do not mix and the fluid acts much like a solid, transferring heat by conduction. If heat is absorbed on the collector everywhere at nearly the same rate, the Nusselt number has a single value and the heat transfer coefficient is dependent on the diameter and the thermal conductivity alone (1, 3):

$$\frac{hD}{k} = 4.36 \tag{5-6}$$

The heat transfer rate is proportional to $U_f A_f$, which in a circular tube is given by

$$U_f A_f \approx h A_f = \left(\frac{4.36k}{D}\right) \pi DL = 4.36 \pi k L \tag{5-7}$$

Equation 5-7 leads to the interesting conclusion that the heat transfer rate for coolant flow in circular tubes is independent of the tube diameter or flow rate if the flow is laminar, all other factors, including the temperature driving force being equal.

### EXAMPLE 5-4

What is the heat transfer coefficient for the low Reynolds number flow of Example 5-1(d)?

Solving eq. 5-6 for $h$,

$$h = \frac{4.36k}{D} = \frac{4.36}{0.0254 \text{ m}} \times 0.661 \text{ W/}^\circ\text{C}\cdot\text{m}^2 = 113.5 \text{ W/}^\circ\text{C}\cdot\text{m}^2$$

In Engineering units

$$h = \frac{4.36k}{D} = 4.36 \times 0.382 \text{ Btu/hr }^\circ\text{F ft}^2 \times \frac{1}{1 \text{ in.}} \times \frac{12 \text{ in.}}{\text{ft}} = 20 \text{ Btu/hr }^\circ\text{F ft}^2$$

## Collector Design

The design of liquid-cooled collectors is less difficult than for those using air. It is relatively easy to keep liquid heat transfer coefficients up above 500 $\text{W/}^\circ\text{C}\cdot\text{m}^2$ (100 Btu/hr $^\circ$F ft$^2$), which will have little effect on collector performance when the tubes are reasonably sized. When the flow is laminar, as it will usually be in collectors when the tubes are in parallel, if there are enough tubes there will often be enough area to give adequate heat transfer even if the heat transfer coefficient is low (eq. 5-7). Otherwise such a collector will need an oversized water pump to provide the flow rates necessary to provide good heat transfer efficiency, and will be a poor performer for domestic hot water heating if the flow is kept low in order to gain temperature rise.

Air collectors, on the other hand, do not follow eq. 5-7 because changing the depth of the rectangular flow passage does not change the heat transfer area. The Reynolds number is also invariant with these changes (eq. 5-1b). The heat transfer coefficient, however, varies in accordance with eq. 5-5 or 5-6, therefore decreasing as the designer opens the flow passage to reduce the air flow pressure drop.

**Pressure Drop.**  Since the pressure drop through solar collectors follows the general quantitative relations for tubes and ducts, analysis will be delayed until Chapter 12 for discussion in conjunction with pumps, fans, piping, and duct work.

**Flow Distribution.**  The pressure drop in liquid collectors can actually be *too* low. If the pressure drop of the collector is less than the piping associated with it and the collectors are connected in the usual parallel configuration, the differing pipe runs to each collector will cause the flow to be less in the collectors furthest away from supply. The problem can be eased considerably by using the reverse-return configuration, in which all flow path lengths are equal (see Fig. 12-1).

A balancing valve in series with each collector is often incorporated in a collector array. This means extra labor for installation and adjustment, and there is substantial risk of clogging after years of use. The best solution includes design of a modest pressure drop into the collector, in which case the flows will be self balancing. Clogging can be minimized by increasing the pressure drop with long runs of smaller-diameter tubing or the use of turbulence-promoting configurations instead of small-diameter orifices.

The analogous flow distribution problem exists *within* the collector when the flow is in parallel through the individual collector tubes. For this reason a series flow is to be preferred, assuring equal flow over all the collector. When this is inconvenient reverse return can also be used within the collector to give an equal nominal fluid path length for each tube, reducing the tendency for poor flow distribution.

**Thermosyphoning.**  When parallel flow is used through large-diameter tubes on the absorber, exposure to sunlight can cause heated water to rise with enough buoyancy to overcome the minute pressure drop of such tubes. This self-pumping or *thermosyphoning* is of assistance in providing an even flow to all tubes, since poor flow distribution will be accompanied by a naturally occurring higher temperature rise through that tube.

**Air collectors.** Pressure drop is a major problem in air collectors. Pumping power must often be large to promote heat transfer and assure good flow distribution, and pumping power can result in considerable operating cost. Reduction of air pressure drop and better heat transfer—simultaneously—is a most important research goal for air collectors.

## HEAT TRANSFER FACTORS

### Fin Efficiency

Extended surface is often installed to help transfer heat from the absorber surface to the heat transfer fluid. In liquid-cooled collectors, the absorber surface serves as a fin on the tubes through which the liquid circulates so that the tubes may be made small, keeping costs down, simplifying distribution problems, and assuring a high liquid-side heat transfer coefficient (Fig. 5-1*b*).

In air-cooled collectors, the extended surface is in the air stream, but its analysis will be identical to that of the fin on the absorber tube. The heat transfer situation is shown in Fig. 5-4. Solar radiation uniformly irradiates the fin (i.e., absorber plate) and is conducted to the closest tube. There is no heat flow across the plane equidistant between the tubes. There is more and more heat flowing through the fin as the distance to the tube becomes less—a situation analogous to a stream becoming larger near its mouth. To conduct heat along the fin, temperatures must rise as distance from the root of the fin increases, with the temperature at the center point between the tubes being highest. These higher temperatures mean greater losses to the ambient through the cover glass (not shown).

If the net radiation collected is given by the basic collector equation (4-11), solution of appropriate partial differential equations (4) can take all factors into

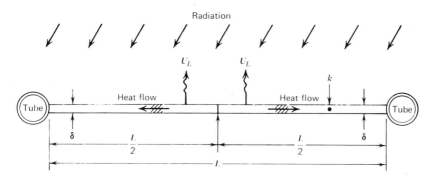

**FIGURE 5-4.** Heat transfer to the absorber coolant, showing nomenclature used to define fin efficiency.

account to give a new equation for the collected heat in terms of the fin efficiency, $\eta_F$:

$$\frac{Q}{A} = \eta_F[\tau\alpha I - U_L(T_c - T_a)] \tag{5-8}$$

Here $T_c$ is the temperature at the base of the fin (i.e., at the outside tube wall),

$$\eta_F = \frac{\tanh B}{B} \tag{5-9}$$

and

$$B = \frac{L}{2}\sqrt{U_L/k\,\delta} \tag{5-10}$$

where $L$ is the length of the fin—the distance between the tubes from *root* to *root*, not center-to-center. The thermal conductivity $k$, fin thickness $\delta$, heat transfer coefficient $U_L$, and distance $L$ must be expressed in proper units so that $B$ is dimensionless.

The hyperbolic tangent is defined as

$$\tanh x = \frac{e^x - e^{-x}}{e^x + e^{-x}}$$

Calculations of hyperbolic functions are laborious without a calculator programmed to handle hyperbolic functions directly. Therefore, Fig. 5-5 has been prepared from which the fin efficiency $\eta_F$ can be quickly determined once $B$ is known.

Equation 5-8 shows that fin efficiency is a constant factor acting on the efficiency the collector would have had if the entire absorber surface were at the temperature of the root of the fin. The fin efficiency factor, therefore, does *not* depend on either temperature level, the incident radiation, or the amount of heat conducted through the fin. It is only a function of the collector heat loss coefficient $U_L$ and physical nature of the collector plate. In the next section, it will be shown how the fin efficiency can be combined with the fluid heat transfer coefficient and the flow rate to produce a similar, constant overall heat-transfer efficiency factor that is multiplied by the heat collected when the entire absorber is at the *fluid* temperature to give the true collected heat.

**EXAMPLE 5-5a**

If the fins of Example 5-1b are made from 0.381 mm (0.015 in.) copper and the heat loss coefficient is 4.542 W/°C·m² (0.8 Btu/hr °F ft²), what is the fin efficiency?

$$B = \frac{L}{2}\sqrt{\frac{U_L}{k\delta}} = \frac{0.1524\text{ m}}{2}\sqrt{\frac{4.542\text{ W}}{°C \cdot m^2} \times \frac{°C \cdot m}{381\text{ W}} \times \frac{(m/10^{-3})}{0.381\text{ mm}}} = 0.4264$$

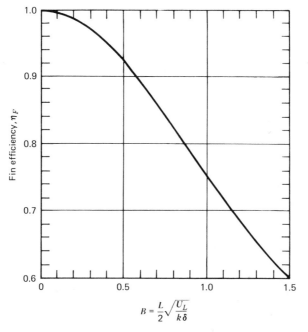

**FIGURE 5-5.** Graphical solution of eq. 5-9 to determine fin efficiency.

In Engineering units,

$$B = \frac{L}{2}\sqrt{\frac{U_L}{k\delta}} = \frac{6 \text{ in.}}{2} \times \frac{\text{ft}}{12 \text{ in.}} \sqrt{\frac{(0.8 \text{ Btu/hr } °F \text{ ft}^2)(12 \text{ in./ft})}{(220 \text{ Btu/hr } °F \text{ ft})(0.015 \text{ in.})}}$$

$$B = 0.4264$$

and then

$$\eta_F = \frac{\tanh (0.4264)}{0.4264} = 0.9435$$

**EXAMPLE 5-5b**

If the collector is made less efficient by removing one cover plate so that $U_L = 7.61 \text{ W/°C·m}^2$ (1.34 Btu/hr °F ft²), what is the fin efficiency?

$$B = \frac{0.1524}{2}\sqrt{\frac{7.61}{381} \times \frac{10^3}{0.381}} = 0.552$$

or, in Engineering units,

$$B = \frac{6}{2 \times 12}\sqrt{\frac{1.34 \times 12}{220 \times 0.015}} = 0.552$$

and then

$$\eta_F = \frac{\tanh (0.552)}{0.552} = 0.918$$

Notice that Example 5-5 shows that with a better (lower) heat loss coefficient, the fin efficiency rises. A careful analysis will show that the overall losses due to fin inefficiency are increased with the lower heat loss coefficient. There is more total heat lost because of fin inefficiency in Example 5-5a than in Example 5-5b because the term within the brackets of eq. 5-8 is much greater when the losses are low.

**Maximum Temperature.** The temperature rise along the fin from the collection tube to the hottest absorber surface depends on all the collector conditions, and is given by

$$\Delta T_{\max} = \left( \frac{\tau \alpha I}{U_L} + T_a - T_r \right) \left( 1 - \frac{1}{\cosh B} \right) \tag{5-10}$$

where $\cosh (x) = \frac{1}{2}(e^x + e^{-x})$, and $T_r$ is the temperature on the collection tube at the root of the fin, which in most practical collectors is approximately that of the coolant.

### EXAMPLE 5-6

Using parameters from Examples 5-5a and 5-1, what is the maximum temperature of the collector surface? Assume the ambient temperature to be 10 °C (50 °F) and the root of the fin to be 65.56 °C (150 °F). $\tau \alpha$ is 0.8633.

$$\Delta T_{\max} = \left( \frac{0.8633 \times 788 \text{ W/m}^2}{4.542 \text{ W/°C·m}^2} + 10 \text{ °C} - 65.56 \text{ °C} \right) \left( 1 - \frac{1}{\cosh (0.4264)} \right)$$

$$= 94.21 \text{ °C} \times 0.08450 = 7.96 \text{ °C}$$

$$T_{\max} = 65.56 \text{ °C} + 7.96 \text{ °C} = 73.52 \text{ °C}$$

or, in Engineering units,

$$\Delta T_{\max} = \left( \frac{0.8633 \times 250 \text{ Btu/hr ft}^2}{0.8 \text{ Btu/hr °F ft}^2} + 50 \text{ °F} - 150 \text{ °F} \right) \left( 1 - \frac{1}{\cosh (0.4264)} \right)$$

$$\Delta T_{\max} = 169.78 \text{ °F} \times 0.08450 = 14.3 \text{ °F}$$

$$T_{\max} = 150 \text{ °F} + 14.3 \text{ °F} = 164.3 \text{ °F}$$

## Effect of Flow Rate and Fluid Heat Transfer on Collector Performance

The losses anywhere on an absorber surface depend on the local temperature. When a fluid removes heat from the collector, the losses are therefore lowest near the fluid inlet and greatest near the outlet. However, the average losses can be closely approximated by using the average fluid temperature. When this assumption is made simple expressions can be derived for a heat transfer factor that incorporate the effect of flow rate and fluid heat transfer as well as fin efficiency.

Referring to the simple fluid-cooled collector of Fig. 5-6, let the running (i.e., continuously variable) temperature of the fluid as it passes through the collector be $T_f$, with inlet temperature $T_i$ and outlet temperature $T_o$. Let the effective absorber temperature corresponding to $T_f$ be $T_c$. Let the total area for fluid heat transfer inside the tube be $A_{fT}$, and the running area traversed from the inlet to a given distance from the inlet be $A_f$. Let the similarly traversed collector area, also a running variable (which may be different because of extended surface), be $A_c$. Let the fluid heat transfer coefficient be $U_f$ and the average fin efficiency be $\bar{\eta}_F$. The cumulative heat collected at a given distance from the inlet is $Q$, running from zero to $Q_T$, the total heat collected. The fluid flow rate is $W$, and its specific heat is $c$.

Assume that the heat collection rate and consequently the temperature difference between the fluid and the collector surface is uniform everywhere on the collector. This is necessarily the case if the average collector temperature is to be considered the average of the inlet and outlet fluid temperatures plus the average temperature difference for heat transfer to the fluid. The total heat transferred to the fluid is then

$$Q = U_f A_f (T_c - T_f) \qquad (5\text{-}12)$$

where $Q$, $A_f$, $T_c$, and $T_f$ are measured anywhere (but all at the same place) along the fluid path. The total heat absorbed by the fluid at that same point is

$$Q = Wc(T_f - T_{fi}) \qquad (5\text{-}13)$$

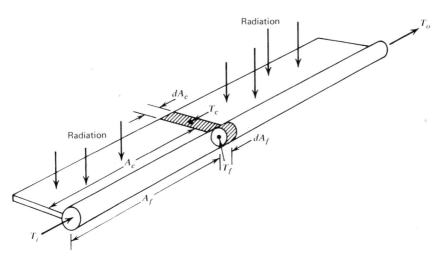

**FIGURE 5-6.** Nomenclature for derivation of the heat transfer factor equations.

Differentiating each equation with respect to traversed area $A_f$ gives

$$\frac{dQ}{dA_f} = Wc \frac{dT_f}{dA_f} \tag{5-14}$$

($T_{fi}$ is a constant) and, since ($T_c - T_f$) is constant,

$$\frac{dQ}{dA_f} = U_f(T_c - T_f) \tag{5-15}$$

Eliminating $dQ/dA_f$ between eqs. 5-14 and 5-15 and integrating to a point along the fluid path,

$$\int_{T_i}^{T_f} dT_f = \frac{U_f}{Wc}(T_c - T_f) \int_0^{A_f} dA_f$$

giving

$$T_f = T_i + \frac{U_f}{Wc}(T_c - T_f)A_f \tag{5-16}$$

If eq. 5-12 is written for the end of the fluid path, since ($T_c - T_f$) is constant,

$$Q_T = U_f(T_c - T_f)A_{fT} \tag{5-17}$$

Solving for $T_f$, eq. 5-17 becomes

$$T_f = T_c - \frac{Q_T}{U_f A_{fT}}$$

which is substituted into eq. 5-16 to give

$$T_c = T_i + \frac{Q_T}{A_{fT}} \left( \frac{A_f}{Wc} + \frac{1}{U_f} \right) \tag{5-18}$$

Writing the basic collector equation with the fin efficiency equation (5-8) for *collector* differential area and an average fin efficiency $\bar{\eta}_F$,

$$dQ = \bar{\eta}_F[\tau\alpha I \, dA_c - U_L(T_c - T_a) \, dA_c]$$

Substituting eq. 5-18 for $T_c$,

$$\frac{dQ}{\bar{\eta}_F} = \tau\alpha I \, dA_c - U_L\left[T_i + \frac{Q_T}{A_{fT}}\left(\frac{A_f}{Wc} + \frac{1}{U_f}\right) - T_a\right] dA_c$$

Integrating over the entire fluid path length,

$$\frac{1}{\bar{\eta}_F}\int_0^{Q_T} dQ = \tau\alpha I \int_0^{A_{cT}} dA_c - U_L(T_i - T_a)\int_0^{A_{cT}} dA_c - \frac{Q_T U_L}{A_{fT}}\int_0^{A_{cT}}\left(\frac{A_f}{Wc} + \frac{1}{U_f}\right) dA_c \tag{5-19}$$

If the fluid and collector area are related by

$$A_f = KA_c$$

then the last integral is

$$\int_0^{A_{cT}}\left(\frac{KA_c}{Wc} + \frac{1}{U_f}\right) dA_c = \left[\frac{KA_c^2}{2Wc} + \frac{A_c}{U_f}\right]_0^{A_{cT}} = \frac{KA_{cT}^2}{2Wc} + \frac{A_{cT}}{U_f} = \frac{A_{fT}A_{cT}}{2Wc} + \frac{A_{cT}}{U_f} \tag{5-20}$$

Integrating eq. 5-19 then gives

$$\frac{Q_T}{\bar{\eta}_F} = \tau\alpha I A_{cT} - U_L(T_i - T_a)A_{cT} - Q_T U_L\left(\frac{A_{cT}}{2Wc} + \frac{A_{cT}}{U_f A_{fT}}\right)$$

which can be solved for the total collected heat flux $q = Q_T/A_{cT}$.

Incorporating a *heat transfer* factor $F_i$, the heat flux collected is

$$q = F_i[\tau\alpha I - U_L(T_i - T_a)] \tag{5-21}$$

where $F_i$ is given by

$$F_i = \frac{1}{\dfrac{1}{\bar{\eta}_F} + \dfrac{U_L A_c}{U_f A_f} + \dfrac{U_L A_c}{2Wc}} \tag{5-22}$$

In eq. 5-22, $A_c$ and $A_f$ represent the total area considered—usually the entire collector.

Sometimes a significant portion of the collector surface exposed to radiation will be covered with tubes. That portion—being directly in contact with the fluid and so utilizing no extended surface—operates with a fin efficiency of unity. If the proportion of the surface covered with such tubes is given by $r$, the average fin efficiency is given by the following equation:

$$\bar{\eta}_F = \eta_F + r(1 - \eta_F) \tag{5-22a}$$

Through the application of eqs. 5-21 and 5-22, collector performance can be analyzed in terms of the coolant inlet temperature. There is no need for direct concern with collector surface temperatures, fluid temperatures, fin temperatures, and so on. These are all accounted for in these simple equations. Finding the correct values of $\eta_F$ and $U_f$ may not always be so simple, as we have seen above, but these are separable difficulties. The effect of heat transfer coefficients, fluid flow rate, and fin efficiency on collector performance—in terms of the fluid inlet temperature—are given simply by eqs. 5-21 and 5-22.

The heat transfer efficiency factor derivation need not have been oriented toward use of $T_i$ as the basis for eqs. 5-21 and 5-22. If the outlet temperature is used as a base the same sort of derivation gives

$$q = F_o[\tau\alpha I - U_L(T_o - T_a)] \tag{5-23}$$

$$F_o = \frac{1}{\dfrac{1}{\bar{\eta}_F} + \dfrac{U_L A_c}{U_f A_f} - \dfrac{U_L A_c}{2Wc}} \tag{5-24}$$

More widely used is a heat transfer efficiency factor based on average or *mean* fluid temperature:

$$q = F_m[\tau\alpha I - U_L(T_m - T_a)] \tag{5-25}$$

$$F_m = \frac{1}{\dfrac{1}{\bar{\eta}_F} + \dfrac{U_L A_c}{U_f A_f}} \tag{5-26}$$

Use of $F_m$ is interesting because the actual fluid flow need not be known. This is not to imply that predicted performance is *independent* of the flow, for the flow influences $T_m$, which will usually be found by

$$T_m = \frac{T_i + T_o}{2} \tag{5-27}$$

However, an efficiency curve based on eq. 5-25 (see eq. 5-31) will be independent of flow, except to the extent that flow affects $U_f$.*

### EXAMPLE 5-7

The water-cooled collector of Example 5-1, which has tubes in parallel, has been presented with the following parameters:

$$\eta_F = 0.9435$$
$$\tau\alpha = 0.8633$$
$$U_L = 4.542 \text{ W/}°C \cdot m^2$$
$$U_f = 472.9 \text{ W/}°C \cdot m^2$$
$$W = 0.0247 \text{ kg/s}$$
$$c = 4.19 \text{ kJ/kg } °C$$
$$A_c = 1.22 \text{ m} \times 2.44 \text{ m} = 2.98 \text{ m}^2$$
$$A_f = 8 \text{ tubes} \times \frac{2.44 \text{ m}}{\text{tube}} \times \frac{\pi 0.006 \ 35 \text{ m}^2}{\text{m}} = 0.3894 \text{ m}^2$$

(a) What is the rate of heat collection if the inlet temperature is 60 °C (140 °F)? What are the outlet and the mean water temperatures?

Using eq. 5-22,

$$\frac{U_L A_c}{U_f A_f} = \frac{4.542 \text{ W/}°C \cdot m^2}{472.9 \text{ W/}°C \cdot m^2} \times \frac{2.98 \text{ m}^2}{0.3894 \text{ m}^2} = 0.0735$$

$$\frac{U_L A_c}{2Wc} = \frac{4.542 \text{ W/}°C \cdot m^2}{2 \times 0.0247 \text{ kg/s}} \times \frac{2.98 \text{ m}^2}{4.19 \text{ kJ/kg } °C} \times \frac{J}{W \cdot s} \times \frac{k}{10^3} = 0.0654$$

$$F_i = \frac{1}{\dfrac{1}{0.9435} + 0.0735 + 0.0654} = 0.8342$$

* In other sources $F_i$ is often given as $F_R$ and $F_m$ as $F'$. Additional factors in the literature with still other meanings suggest the utility of the more consistent nomenclature used here. A more exact expression is available for the inlet heat transfer factor (4):

$$F_R = \frac{Wc}{U_L A_c} \left\{ 1 - \exp\left[ -\frac{U_L A_c}{Wc} \bigg/ \left( \frac{1}{\eta_F} + \frac{U_L A_c}{U_f A_f} \right) \right] \right\} \tag{5-28}$$

The simpler eq. 5-22 is equivalent in all practical collector situations, in which the outlet temperature is far from the stagnation temperature. It may be used accurately until the collector outlet has risen about half the potential rise toward stagnation, after which eq. 5-28 is preferred.

Equation 5-23 then gives

$$q = 0.8342[0.8633 \times 788 - 4.542(60 - 10)] = 378.04 \text{ W/m}^2$$

$$Q = 378.04 \text{ W/m}^2 \times 2.98 \text{ m}^2 = 1126.6 \text{ W}$$

$$T_o = T_i + \frac{Q}{Wc} = 60 \,°C + 1126.6 \text{ W} \times \frac{s}{0.0247 \text{ kg}} \times \frac{\text{kg} \,°C}{4.19 \text{ kJ}} \times \frac{J}{W \cdot s} \times \frac{k}{10^3}$$

$$= 70.89 \,°C$$

$$T_m = \frac{60 + 70.89}{2} = 65.44 \,°C$$

(b) If the outlet temperature is 70.89 °C (159.55 °F) calculate the heat collection flux using eqs. 5-24 and 5-23.

$$F_o = \frac{1}{\dfrac{1}{0.9435} + 0.0735 - 0.0654} = 0.9363$$

$$q = 0.9363[0.8633 \times 788 - 4.542(70.89 - 10)] = 378.00 \text{ W/m}^2$$

just as before.

(c) If the average fluid temperature is 65.44 °C (149.78 °F), find the collected heat flux using eqs. 5-26 and 5-25.

$$F_m = \frac{1}{\dfrac{1}{0.9435} + 0.0735} = 0.8823$$

$$q = 0.8823[0.8633 \times 788 - 4.542(65.44 - 10)] = 378.05 \text{ W/m}^2$$

Again, this answer is essentially just as before.

In Engineering units, the collector parameters are

$$\eta_F = 0.9435$$
$$\tau\alpha = 0.8633$$
$$U_L = 0.8 \text{ Btu/hr} \,°F \text{ ft}^2$$
$$U_f = 83.37 \text{ Btu/hr} \,°F \text{ ft}^2$$
$$W = \frac{0.053\,48 \text{ ft}^3}{\text{min}} \times \frac{61.2 \text{ lb}}{\text{ft}^3} = 3.273 \text{ lb/min}$$
$$c = 1 \text{ Btu/lb} \,°F$$
$$A_c = 32 \text{ ft}^2$$
$$A_f = 8 \text{ tubes} \times \frac{8 \text{ ft}}{\text{tube}} \times \frac{\pi 0.25 \text{ in.}^2}{\text{in.}} \times \frac{\text{ft}}{12 \text{ in.}} = 4.189 \text{ ft}^2$$

(a) Using eq. 5-22,

$$\frac{U_L A_c}{U_f A_f} = \frac{0.8 \text{ Btu/hr } ^\circ\text{F ft}^2}{83.37 \text{ Btu/hr } ^\circ\text{F ft}^2} \times \frac{32 \text{ ft}^2}{4.189 \text{ ft}^2} = 0.073\ 30$$

$$\frac{U_L A_c}{2Wc} = \frac{0.8 \text{ Btu/hr } ^\circ\text{F ft}^2 \times 32 \text{ ft}}{2 \times 3.273 \dfrac{\text{lb}}{\text{min}}} \times \frac{60 \text{ min}}{\text{hr}} \times \frac{1 \text{ Btu}}{\text{lb } ^\circ\text{F}} = 0.065\ 18$$

$$F_i = \frac{1}{\dfrac{1}{0.9435} + 0.0733 + 0.065\ 18} = 0.8344$$

Equation 5-21 then gives

$$q = 0.8344[0.8633 \times 250 - 0.8(140 - 50)] = 120.00 \text{ Btu/ft}^2$$

$$Q = 120.00 \frac{\text{Btu}}{\text{ft}^2} \times 32 \text{ ft}^2 = 3840 \text{ Btu/hr}$$

$$T_o = T_i + \frac{Q}{Wc} = 140 \ ^\circ\text{F} + \frac{3840 \text{ Btu/hr}}{3.273 \text{ lb/min} \times 60 \text{ min/hr}} = 159.55 \ ^\circ\text{F}$$

$$T_m = \frac{140 + 159.55}{2} = 149.78 \ ^\circ\text{F}$$

(b) Using eqs. 5-24 and 5-23, when the outlet temperature is 159.55 °F,

$$F_o = \frac{1}{\dfrac{1}{0.9435} + 0.073\ 30 - 0.065\ 18} = 0.9363$$

$$q = 0.9363[0.8633 \times 250 - 0.8(159.55 - 50)] = 120.02 \text{ Btu/ft}^2$$

(c) Using eqs. 5-26 and 5-25, when the mean temperature is 149.78 °F,

$$F_m = \frac{1}{\dfrac{1}{0.9435} + 0.073\ 30} = 0.8824$$

$$q = 0.8824[0.8633 \times 250 - 0.8(149.78 - 50)] = 120.01 \text{ Btu/ft}^2$$

Example 5-7 demonstrates that the user can work with $F_i$, $F_o$, or $F_m$ as desired and still get precisely the same analytical results.

## Efficiency Curves

If the respective heat transfer factors are multiplied through in eqs. 5-21, 5-23, and 5-25 and the efficiency found by dividing through both sides by $I$, then resulting equations

$$\eta_i = F_i \tau \alpha - \left[ \frac{F_i U_L (T_i - T_a)}{I} \right] \tag{5-29}$$

$$\eta_o = F_o\tau\alpha - \left[\frac{F_oU_L(T_o - T_a)}{I}\right] \tag{5-30}$$

$$\eta_m = F_m\tau\alpha - \left[\frac{F_mU_L(T_m - T_a)}{I}\right] \tag{5-31}$$

will all give the same efficiency for a given case in which the data are consistent. The $\eta$ subscripts serve here to identify the respective efficiency curves in the following discussion.

Each of these equations can be plotted to give a valid efficiency curve with $F_i\tau\alpha$, $F_m\tau\alpha$, or $F_o\tau\alpha$ as $y$ intercept, a slope of $F_iU_L$, $F_mU_L$, or $F_oU_L$, and an $x$-intercept of $\tau\alpha/U_L$.

Experimental data—which are the only practical source for collector parameters—should initially be plotted as an $\eta_m$ curve because minor (and inevitable) flow rate variations have no effect on $\eta_m$, while points for $\eta_i$ or $\eta_o$ curves must be corrected to a single flow rate if they are to be most accurate. Once the $\eta_m$ curve is available, $F_m\tau\alpha$, $F_mU_L$, and $\tau\alpha/U_L$ can be determined graphically from the $y$-intercept, slope, and $x$-intercept, respectively. From these, $\eta_i$ or $\eta_o$ curves can be determined as shown below for the original or a new rate of flow. However, for a new heat transfer fluid or for flow rate changes beyond perhaps $\pm 25\%$ of the experimental conditions, $\eta_m$ itself may not stay constant because of changes in the fluid heat transfer coefficient $U_f$. It is therefore advisable to use the experimental curves only near the experimental conditions of flow unless corrections are made for the change in $U_f$.

Since the $\eta_i$ curve is most often needed for system calculations the newer ASHRAE* testing method calls for reporting of experimental data based on the inlet fluid temperature instead of the mean fluid temperature used in the original NBS* test procedure. Although this has the disadvantage of needing a separate curve for each fluid flow rate, the method is popular because the needed design parameters are immediately available from the slope and intercept of the plotted curve. Because its intercept is the lowest of the three, the efficiency curve based on inlet temperature is not popular with the collector manufacturers, since it may put their product in a bad light if an unsophisticated customer compares an $\eta_i$ curve with a competitor's $\eta_m$ curve.

**Optical Efficiency Modifier.** When an efficiency curve has been obtained, it applies only when the sun is perpendicular to the solar collector. For system performance analysis, the optical efficiency incident angle modifier must be incorporated (see Chapter 4). This may be done before consideration of the heat

* The National Bureau of Standards (NBS) and the American Society of Heating, Refrigerating and Air Conditioning Engineers (ASHRAE) have both set standards for collector testing procedures that differ only in detail. These standards are continuously being revised, and care should be taken that the most current standards available are being applied.

transfer factor, perhaps by directly modifying the experimental $\eta_i$ or $\eta_m$ curve. More conventionally, this correction is made analytically after the desired efficiency curve has been obtained by simply multiplying the y-intercept value by the modifier.

## EXAMPLE 5-8

A liquid-cooled collector with a single cover glass and a selective surface is cooled with a glycol solution [$c = 3.517$ kJ/kg °C (0.84 Btu/lb °F)] flowing at 24.21 g/s·m² (17.85 lb/hr ft²). The experimental efficiency curve based on mean fluid temperature has a y-intercept of 0.8421 and a slope of 5.224 W/°C·m² (0.92 Btu/hr °F ft²).*

(a) Plot $\eta_m$ and $\eta_i$ curves.

The $\eta_m$ curve is plotted directly on Fig. 5-7 from the experimental data, with a y-intercept of $F_m \tau\alpha = 0.8421$ and an x-intercept of

$$x = \tau\alpha/U_L = F_m\tau\alpha/F_mU_L = 0.8421/5.224 = 0.1612 \ °C\cdot m^2/W$$

The *flow capacity rate* is

$$\frac{Wc}{A_c} = \frac{24.21 \text{ g}}{\text{s}\cdot\text{m}^2} \times \frac{3.517 \text{ kJ}}{\text{kg} \ °C} \times \frac{W\cdot s}{J} = \frac{85.15 \text{ W}}{°C\cdot m^2}$$

To plot the $\eta_i$ curve eqs. 5-22 and 5-26 are combined to give

$$F_i = \frac{1}{\dfrac{1}{F_m} + \dfrac{U_L A_c}{2Wc}}$$

Multiplying through by $\tau\alpha$ gives

$$F_i\tau\alpha = \frac{1}{\dfrac{1}{F_m\tau\alpha} + \dfrac{A_c}{2Wc(\tau\alpha/U_L)}}$$

$$= \frac{1}{\dfrac{1}{0.8421} + \dfrac{\text{s}\cdot\text{m}^2}{2 \times 24.21 \text{ g}} \times \dfrac{\text{kg} \ °C}{3.52 \text{ kJ}} \times \dfrac{W}{0.1612 \ °C\cdot m^2} \times \dfrac{J}{W\cdot s}}$$

$$= \frac{1}{\dfrac{1}{0.8421} + 0.0364} = 0.8171$$

Similarly, multiplying the same equation through by $U_L$ gives

$$F_iU_L = \frac{1}{\dfrac{1}{F_mU_L} + \dfrac{A_c}{2Wc}} = \frac{1}{\dfrac{°C\cdot m^2}{5.224 \text{ W}} + \dfrac{°C\cdot m^2}{2 \times 85.15 \text{ W}}} = 5.069 \ W/°C\cdot m^2$$

---

* Mathematically, the slope is negative. Throughout we refer only to its absolute magnitude, since that is numerically the same as the loss coefficient times the heat transfer factor.

The slope is unnecessary for plotting, however, since the $x$-intercept is the same for all three efficiency curves. The $\eta_i$ curve is therefore plotted from the $F_i \tau \alpha = 0.8171$ $y$-intercept to the same $x$-intercept, $\tau \alpha / U_L = 0.1612 \, °\text{C} \cdot \text{m}^2/\text{W}$.

(b) Find $F_i \overline{\tau \alpha}$ and $F_i U_L$ for system design use.

From Chapter 4, the incident angle modifier $\overline{\tau \alpha}/(\tau \alpha)_n$ for a one-glass selective surface collector is 0.95. Therefore,

$$F_i \overline{\tau \alpha} = [\overline{\tau \alpha}/(\tau \alpha)_n] F_i(\tau \alpha)_n = 0.8171 \times 0.95 = 0.7762$$

$F_i U_L = 5.069 \, \text{W}/°\text{C} \cdot \text{m}^2$   (as determined above—this value does not change)

In Engineering units

(a)      $\tau \alpha / U_L = 0.8421/0.92 = 0.9153 \, \text{hr} \, °\text{F} \, \text{ft}^2/\text{Btu}$

$$\frac{Wc}{A_c} = \frac{17.85 \, \text{lb}}{\text{hr ft}^2} \times \frac{0.8407 \, \text{Btu}}{\text{lb} \, °\text{F}} = 15.0 \, \text{Btu/hr} \, °\text{F} \, \text{ft}^2$$

$$F_i \tau \alpha = \cfrac{1}{\cfrac{1}{0.8421} + \cfrac{\text{hr} \, °\text{F} \, \text{ft}^2}{2 \times 15 \, \text{Btu}} \times \cfrac{\text{Btu}}{0.9153 \, \text{hr} \, °\text{F} \, \text{ft}^2}}$$

$$= \cfrac{1}{\cfrac{1}{0.8421} + 0.0363} = 0.8171$$

$$F_i U_L = \cfrac{1}{\cfrac{1}{0.92} + \cfrac{1}{2 \times 17.85 \times 0.84}} = 0.8926 \, \text{Btu/hr} \, °\text{F} \, \text{ft}^2$$

(b)      $F_i \overline{\tau \alpha} = 0.8171 \times 0.95 = 0.7762$

Techniques similar to those used in Example 5-8 can be used to transform any of the three efficiency curves into any of the others, either before or after application of the incident angle modifier—it makes no difference analytically when the reduction in $\tau \alpha$ is made to obtain the system values needed for long-term performance prediction (see Chapter 4). Appropriate transformation formulas have been derived and tabulated in Table 5-2 in terms of $y$-intercept and either slope or $x$-intercept. In that table use is made of the normalized flow rate

$$w = \frac{W}{A_c}$$

It is generally desirable to utilize flow rate in this fashion (as was done in the statement for the above example). The flow factors based on inlet and outlet temperature can then be written

$$F_i = \cfrac{1}{\cfrac{1}{F_m} + \cfrac{U_L}{2wc}} \tag{5-32a}$$

$$F_o = \frac{1}{\dfrac{1}{F_m} - \dfrac{U_L}{2wc}}$$

(5-32b)

with $F_m$ defined by eq. 5-26 as before.

The product of the fluid flow rate and the heat capacity $Wc$ is often called the *flow capacity rate*. When the flow rate is area-normalized the $wc$ term can be referred to as the *normalized* flow capacity rate.

When the normalized flow capacity rate is to be changed and other factors (especially $U_f$) are constant, the new $y$-intercept and slope of an $\eta_i$ curve can be obtained by multiplying by

$$\frac{(F_i)_{new}}{(F_i)_{old}} = \frac{1}{1 + \dfrac{(F_iU_L)_{old}}{2(wc)_{new}} - \dfrac{(F_iU_L)_{old}}{2(wc)_{old}}}$$

(5-32c)

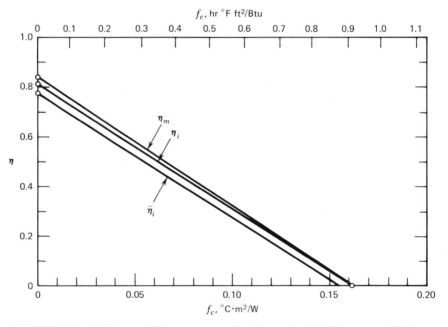

**FIGURE 5-7.** Collector efficiency curves for Example 5-8. The mean efficiency curve ($\eta_m$) was measured, transformed into an inlet based curve ($\eta_i$), which was modified for system use by the incident angle modifier ($\bar{\eta}_i$).

**TABLE 5-2.** Formulas for Converting One Efficiency Curve to Another. With appropriate subscripts, $y$-intercepts are designated $y$ and (positive) slopes $s$. The $x$-intercept needs no subscript and is designated $x$.

$\eta_i$ curve and $wc$ known: $y_i = F_i\tau\alpha$, $\quad x = \tau\alpha/U_L$, $\quad s_i = F_iU_L$

$$F_m\tau\alpha = y_m = \frac{1}{\left(\dfrac{1}{y_i} - \dfrac{s_i}{2y_iwc}\right)} = \frac{1}{\left(\dfrac{1}{y_i} - \dfrac{1}{2xwc}\right)}$$

$$F_mU_L = s_m = \frac{1}{\left(\dfrac{1}{s_i} - \dfrac{1}{2wc}\right)} = \frac{1}{\left(\dfrac{x}{y_i} - \dfrac{1}{2wc}\right)}$$

$$F_o\tau\alpha = y_o = \frac{1}{\left(\dfrac{1}{y_i} - \dfrac{s_i}{y_iwc}\right)} = \frac{1}{\left(\dfrac{1}{y_i} - \dfrac{1}{xwc}\right)}$$

$$F_oU_L = s_o = \frac{1}{\left(\dfrac{1}{s_i} - \dfrac{1}{wc}\right)} = \frac{1}{\left(\dfrac{x}{y_i} - \dfrac{1}{wc}\right)}$$

$\eta_m$ curve and $wc$ known: $y_m = F_m\tau\alpha$, $\quad x = \tau\alpha/U_L$, $\quad s_m = F_mU_L$

$$F_i\tau\alpha = y_i = \frac{1}{\left(\dfrac{1}{y_m} + \dfrac{s_m}{2y_mwc}\right)} = \frac{1}{\left(\dfrac{1}{y_m} + \dfrac{1}{2xwc}\right)}$$

$$F_iU_L = s_i = \frac{1}{\left(\dfrac{1}{s_m} + \dfrac{1}{2wc}\right)} = \frac{1}{\left(\dfrac{x}{y_m} + \dfrac{1}{2wc}\right)}$$

$$F_o\tau\alpha = y_o = \frac{1}{\left(\dfrac{1}{y_m} - \dfrac{s_m}{2y_mwc}\right)} = \frac{1}{\left(\dfrac{1}{y_m} - \dfrac{1}{2xwc}\right)}$$

$$F_oU_L = s_o = \frac{1}{\left(\dfrac{1}{s_m} - \dfrac{1}{2wc}\right)} = \frac{1}{\left(\dfrac{x}{y_m} - \dfrac{1}{2wc}\right)}$$

$\eta_o$ curve and $wc$ known: $y_o = F_o\tau\alpha$, $\quad x = \tau\alpha/U_L$, $\quad s_o = F_oU_L$

$$F_m\tau\alpha = y_m = \frac{1}{\left(\dfrac{1}{y_o} + \dfrac{s_o}{2y_owc}\right)} = \frac{1}{\left(\dfrac{1}{y_o} + \dfrac{1}{2xwc}\right)}$$

$$F_mU_L = s_m = \frac{1}{\left(\dfrac{1}{s_o} + \dfrac{1}{2wc}\right)} = \frac{1}{\left(\dfrac{x}{y_o} + \dfrac{1}{2wc}\right)}$$

$$F_i\tau\alpha = y_i = \frac{1}{\left(\dfrac{1}{y_o} + \dfrac{s_o}{y_owc}\right)} = \frac{1}{\left(\dfrac{1}{y_o} + \dfrac{1}{xwc}\right)}$$

$$F_iU_L = s_i = \frac{1}{\left(\dfrac{1}{s_o} + \dfrac{1}{wc}\right)} = \frac{1}{\left(\dfrac{x}{y_o} + \dfrac{1}{wc}\right)}$$

This equation can be derived from eqs. 5-26, 5-22, and the equation for $F_m U_L$ given on the left in Table 5-2.

If other factors (such as $U_f$) change during application of a solar collector, a more complete analysis must be made with the aid of laboratory $\tau\alpha$ information in order to estimate new flow factors.

**EXAMPLE 5-9**

(a) The flow rate for the collector used in Example 5-8 is to be halved. If $U_f$ is considered constant, what are the new $F_i\tau\alpha$ and $F_iU_L$?

Using eq. 5-32c,

$$\frac{(F_i)_{new}}{(F_i)_{old}} = \frac{1}{1 + \dfrac{5.069}{2 \times (85.15/2)} - \dfrac{5.069}{2 \times 85.15}} = 0.9711$$

$$F_i\tau\alpha = 0.9711 \times 0.8171 = 0.7935$$
$$F_iU_L = 0.9711 \times 5.069 = 4.923 \text{ W/}^\circ\text{C} \cdot \text{m}^2$$

(b) A silicone fluid is to be substituted for glycol at the same mass flow rate using the collector of Example 5-8. If both glycol and silicone flows are laminar, what are the proper values of $F_i\tau\alpha$ and $F_iU_L$? A calculation has been made to show that $\bar{\eta}_F = 0.98$. Manufacturer's data specifies the glass cover transmissivity to be 0.907, and the absorptivity of the selective surface is 0.97.

$$\tau\alpha = 0.907 \times 0.97 = 0.88$$

$$F_m = \frac{F_m\tau\alpha}{\tau\alpha} = \frac{0.8421}{0.88} = 0.9569$$

$$U_L = \frac{F_mU_L}{F_m} = \frac{5.224 \text{ W/}^\circ\text{C} \cdot \text{m}^2}{0.9569} = 5.459 \text{ W/}^\circ\text{C} \cdot \text{m}^2$$

Solving eq. 5-26 for the remaining unknown quantities,

$$\frac{U_fA_f}{A_c} = \frac{U_L}{\dfrac{1}{F_m} - \dfrac{1}{\bar{\eta}_F}} = \frac{5.459 \text{ W/}^\circ\text{C} \cdot \text{m}^2}{\dfrac{1}{0.9569} - \dfrac{1}{0.98}} = 221.9 \text{ W/}^\circ\text{C} \cdot \text{m}^2$$

When the silicone fluid is used, only the value of $U_f$ will change in the above expression. The ratio of the heat transfer coefficients for the old and new cases can be calculated using the techniques just introduced. In this case since the flow is designated laminar, by either eq. 5-6 or 5-7,

$$\frac{h_{new}}{h_{old}} = \frac{k_{new}}{k_{old}} = \frac{0.144}{0.415} = 0.3470$$

and

$$\left(\frac{U_f A_f}{A_c}\right)_{new} = \frac{221.9 \text{ W}}{°C \cdot m^2} \times 0.3470 = 77.01 \text{ W/}°C \cdot m^2$$

From eq. 5-22, retaining the value for $w$ from Example 5-8 of 24.21 g/s·m² but using the new value of $c = 1.55$ kJ/kg·°C,

$$F_i = \frac{1}{\dfrac{1}{0.98} + \dfrac{5.459}{77.01} + \dfrac{5.459}{2 \times 24.21 \times 1.55}} = \frac{1}{1.020 + 0.0709 + 0.0727}$$

$$F_i = 0.8591$$
$$F_i \tau\alpha = 0.8591 \times 0.88 = 0.7560$$
$$F_i U_L = 0.8591 \times 5.459 = 4.69 \text{ W/}°C \cdot m^2$$

In Engineering units,

(a)
$$\frac{(F_i)_{new}}{(F_i)_{old}} = \frac{1}{1 + \dfrac{0.8926}{2 \times (15/2)} - \dfrac{0.8926}{2 \times 15}} = 0.9711$$

$$F_i \tau\alpha = 0.9711 \times 0.8171 = 0.7934$$
$$F_i U_L = 0.9711 \times 0.8926 = 0.8668 \text{ Btu/hr }°F \text{ ft}^2$$

(b)
$$\tau\alpha = 0.907 \times 0.97 = 0.88$$
$$F_m = 0.8421/0.88 = 0.9569$$
$$U_L = 0.92/0.9569 = 0.9614 \text{ Btu/hr }°F \text{ ft}^2$$

$$\frac{U_f A_f}{A_c} = \frac{0.9614}{\dfrac{1}{0.9569} - \dfrac{1}{0.98}} = 39.03 \text{ Btu/hr }°F \text{ ft}^2$$

$$\frac{h_{new}}{h_{old}} = \frac{0.083}{0.24} = 0.3458$$

$$\left(\frac{U_f A_f}{A_c}\right)_{new} = 39.03 \times 0.3458 = 13.5 \text{ Btu/hr }°F \text{ ft}^2$$

$$F_i = \frac{1}{\dfrac{1}{0.98} + \dfrac{0.9614}{13.50} + \dfrac{0.9614}{2 \times 17.85 \times 0.37}} = \frac{1}{1.020 + 0.0712 + 0.0728}$$

$$= 0.8588$$
$$F_i \tau\alpha = 0.8588 \times 0.88 = 0.7557$$
$$F_i U_L = 0.8588 \times 0.9614 = 0.8257 \text{ Btu/hr }°F \text{ ft}^2$$

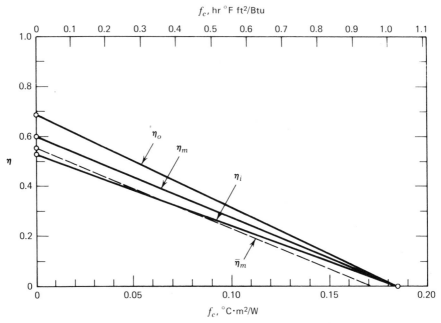

**FIGURE 5-8.** Collector efficiency curves for the air collector of Example 5-10. The normalized coolant flow capacity rate is smaller than for the liquid-cooled collector of Fig. 5-7.

**EXAMPLE 5-10**

An air collector having two glass cover plates ($b_o = -0.16$) is specified by its experimental $\eta_o$ curve, with a $y$-intercept of 0.6882 and slope of 3.727 W/°C·m² (0.6564 Btu/hr °F ft²). The air flow is 0.010 16 m³/s·m² (2 scfm/ft²). Use 21 °C (70 °F) fluid properties.

(a) Plot $\eta_i$, $\eta_m$, and $\eta_o$.

$$wc = \frac{0.01016 \text{ m}^3}{\text{s·m}^2} \times \frac{1.187 \text{ kg}}{\text{m}^3} \times \frac{1.01 \text{ kJ}}{\text{kg·°C}} \times \frac{\text{W·s}}{J} \times \frac{10^3}{k} = \frac{12.18 \text{ W}}{\text{°C·m}^2}$$

From Table 5-2,

$$F_m \tau \alpha = \frac{1}{\dfrac{1}{y_o} + \dfrac{s_o}{2y_o wc}} = \frac{1}{\dfrac{1}{0.6882} + \dfrac{1}{2} \times \dfrac{3.727 \text{ W}}{\text{°C·m}^2} \times \dfrac{1}{0.6882} \times \dfrac{\text{°C·m}^2}{12.18 \text{ W}}}$$

$$= 0.5969$$

$$F_m U_L = \frac{1}{\dfrac{1}{s_o} + \dfrac{1}{2wc}} = \frac{1}{\dfrac{\text{°C·m}^2}{3.727 \text{ W}} + \dfrac{\text{°C·m}^2}{2 \times 12.18 \text{ W}}} = 3.232 \frac{\text{W}}{\text{°C·m}^2}$$

$$x = \frac{\tau\alpha}{U_L} = \frac{F_o\tau\alpha}{F_o U_L} = \frac{0.6882 \; °C \cdot m^2}{3.727 \; W} = 0.1847 \frac{°C \cdot m^2}{W}$$

$$F_i\tau\alpha = \frac{1}{\dfrac{1}{y_o} + \dfrac{1}{xwc}} = \frac{1}{\dfrac{1}{0.6882} + \dfrac{1}{0.1847 \times 12.18}} = 0.5270$$

Using $x$ and the $y$-intercepts $F_m\tau\alpha$, $F_i\tau\alpha$, and $F_o\tau\alpha$, the three efficiency curves have been plotted in Fig. 5-8. An air collector such as this, operated with a relatively low flow rate of air, provides the greatest difference between the curves that is found in practice.

(b) Find $F_m\overline{\tau\alpha}$ and $F_m U_L$ for system design use.

From Chapter 4, the incident angle modifier is about 0.93 for this two-glass collector. Therefore,

$$F_m\overline{\tau\alpha} = 0.5969 \times 0.93 = 0.5551$$

and as before,

$$F_m U_L = 3.232 \; W/°C \cdot m^2$$

(c) If the collector cover plates are made from water-white glass with a transmissivity of 0.908, and a black paint absorber with an absorptivity of 0.992 is used, what are $U_L$ and $U_f$?

$$\tau\alpha = 0.908 \times 0.908 \times 0.992 = 0.8180$$

$$F_m = \frac{F_m\tau\alpha}{\tau\alpha} = \frac{0.5969}{0.8180} = 0.7297$$

$$U_L = \frac{F_m U_L}{F_m} = \frac{3.232 \; W/°C \cdot m^2}{0.7297} = 4.429 \; W/°C \cdot m^2$$

by eq. 5-26,

$$F_m = \frac{1}{\dfrac{1}{\overline{\eta}_F} + \dfrac{U_L A_c}{U_f A_f}}$$

For an air collector with no extended surface,

$$\overline{\eta}_F = 1 \quad \text{and} \quad A_c = A_f$$

so that

$$F_m = \frac{1}{1 + \dfrac{U_L}{U_f}}$$

$$U_f = \frac{U_L}{\dfrac{1}{F_m} - 1} = \frac{4.429 \; W/°C \cdot m^2}{\dfrac{1}{0.7297} - 1} = 11.96 \; W/°C \cdot m^2$$

This completes the characterization of this collector.

In Engineering units,

(a) $\quad wc = \dfrac{2 \text{ ft}^3}{\text{min ft}^2} \times \dfrac{60 \text{ min}}{\text{hr}} \times \dfrac{\text{lb}}{13.5 \text{ ft}^3} \times \dfrac{0.242 \text{ Btu}}{\text{lb °F}} = 2.151 \text{ Btu/hr °F ft}^2$

$$F_m \tau\alpha = \dfrac{1}{\dfrac{1}{0.6882} + \dfrac{1}{2} \times \dfrac{0.6564 \text{ Btu}}{\text{hr °F ft}^2} \times \dfrac{1}{0.6882} \times \dfrac{\text{hr °F ft}^2}{2.151 \text{ Btu}}} = 0.5971$$

$$F_m U_L = \dfrac{1}{\dfrac{\text{hr °F ft}^2}{0.6564 \text{ Btu}} + \dfrac{\text{hr °F ft}^2}{2 \times 2.151 \text{ Btu}}} = 0.5695 \text{ Btu/hr °F ft}^2$$

$$x = \dfrac{0.6882}{1} \times \dfrac{\text{hr °F ft}^2}{0.6564 \text{ Btu}} = 1.048$$

$$F_i \tau\alpha = \dfrac{1}{\dfrac{1}{0.6882} + \dfrac{1}{1.048 \times 2.151}} = 0.5272$$

(b) $\qquad\qquad F_m \overline{\tau\alpha} = 0.5971 \times 0.93 = 0.5553$

$\qquad\qquad F_m U_L = 0.5695 \text{ Btu/hr °F ft}^2$

(c) $\qquad\qquad \tau\alpha = 0.908 \times 0.908 \times 0.992 = 0.8180$

$$F_m = \dfrac{0.5971}{0.8180} = 0.7300$$

$$U_L = \dfrac{0.5695 \text{ Btu}}{0.7300 \text{ hr °F ft}^2} = 0.7801 \text{ Btu/hr °F ft}^2$$

$$U_f = \dfrac{0.7801 \text{ Btu/hr °F ft}^2}{\dfrac{1}{0.7300} - 1} = 2.109 \text{ Btu/hr °F ft}^2$$

**Utility.** In Example 5-10 only a single experimental curve and $\tau\alpha$ measurements were needed to determine $F_i$, $F_m$, $F_o$, $U_L$, and $U_f$. $\eta_i$, $\eta_m$, and $\eta_o$ can then be plotted, as well as $\eta$, the original efficiency curve of Chapter 4 defined by

$$\eta = \tau\alpha - \dfrac{(T_c - T_a)}{I} \qquad (4\text{-}13)$$

where $T_c$ is the average temperature of the collector absorber plate.

Each curve has its area of proper application. The $\eta_m$ curve is the most informative, since it is independent of flow and yet can simply produce the $\eta_i$ or $\eta_o$ curves. The degree to which the y-intercept $y_m$ falls below $\tau\alpha$ defines $F_m$, and tells at a casual glance the magnitude of heat transfer problems in that collector. The x-intercept value need only be mentally divided into $\tau\alpha$ to give the heat transfer coefficient just as quickly.

**Aperture-Area and Gross-Area Based Collector Efficiency Curves.**
The new ASHRAE test procedures call for efficiency curves to be based on the gross
area of the solar collector, that is, on the nominal exterior dimensions of the
collector frame. If this is used consistently all calculations will be correct even
though the slope ($U_L$) and $y$-intercept ($\tau\alpha$) of the curve will each be reduced in
value by the ratio of aperture to gross area. To convert a curve from this basis
to the scientifically more meaningful aperture area used for reporting the NBS test
results, merely increase the $y$-intercept and slope by the ratio of gross area to
aperture area thus retaining the $x$-intercept ($\tau\alpha/U_L$) unchanged.

**Flow Rate Selection**
Occasionally a particular temperature rise is desired from a solar collector.
If the inlet temperature is known, selecting the proper flow rate is usually a trial-
and-error problem. This can be avoided by combining eqs. 5-25, 5-26, and 2-10
to give

$$wc = F_m \left[ \frac{\tau\alpha I - U_L(T_m - T_a)}{T_o - T_i} \right] \tag{5-33}$$

Because of the basic assumption of equal temperature difference between the
absorbing fluid and the collector plate anywhere on the collector, this equation
should be used only when the outlet temperature is less than halfway between the
collector inlet and stagnation temperature, which was shown in Example 4-1 to
be given by

$$T_c = \frac{\tau\alpha I}{U_L} + T_a$$

**EXAMPLE 5-11**

What flow rate of water is needed to maintain a 82.22 °C (180 °F) outlet temperature when $T_i = 48.89$ °C (120 °F) and $I = 788$ W/m² (250 Btu/hr ft²) for
the collector of example 5-7 when $T_a = 10$ °C (50 °F)?

$$F_m = \frac{1}{\dfrac{1}{\overline{\eta}_F} + \dfrac{U_L A_c}{U_f A_f}} = \frac{1}{\dfrac{1}{0.9435} + 0.07330} = 0.8825$$

from eq. 5-33,

$$w = \frac{F_m}{c} \left[ \frac{\tau\alpha I - U_L(T_m - T_a)}{T_o - T_i} \right]$$

$$w = \frac{0.8825}{4.19\,\text{kJ/kg\,°C}} \left[ \frac{0.8633 \times 788\dfrac{W}{m^2} - \dfrac{4.542\,W}{°C\cdot m^2}\left(\dfrac{82.22 + 48.89}{2} - 10\right)°C}{(82.22 - 48.89)\,°C} \times \dfrac{J}{W\cdot s} \right]$$

$$w = 2.71\,\text{g/s}\cdot\text{m}^2$$

$$W = \frac{2.71\,\text{g}}{\text{s}\cdot\text{m}^2} \times 2.98\,\text{m}^2 = 8.06\,\text{g/s}$$

The stagnation temperature of the collector under these conditions is

$$T_c = \frac{0.8633 \times 788}{4.542} + 48.89 = 198.7\ ^\circ C$$

(see Example 4-1) and thus the equation applies for outlet temperatures of less than

$$\frac{198.7 + 48.9}{2} = 123.8\ ^\circ C$$

In Engineering units,

$$F_m = \frac{1}{\dfrac{1}{0.9435} + 0.073\,30} = 0.8825$$

$$w = \frac{0.8825}{1\ \text{Btu/lb}\ ^\circ F} \left[ \frac{0.8633 \times 250\,\dfrac{\text{Btu}}{\text{hr ft}^2} - \dfrac{0.8\ \text{Btu}}{\text{hr}\ ^\circ F\ \text{ft}^2}\left(\dfrac{180 + 120}{2} - 50\right)\,^\circ F}{(180 - 120)\ ^\circ F} \right]$$

$$w = 2.00\ \text{lb/hr ft}^2$$

$$W = \frac{2.00\ \text{lb}}{\text{hr ft}^2} \times 32\ \text{ft}^2 = 64\ \text{lb/hr}$$

The stagnation temperature of the collector under these conditions is

$$T_c = \frac{0.8633 \times 250}{0.8} + 120 = 390\ ^\circ F$$

and thus the equation applies for outlet temperatures of less than $(390 + 120)/2 = 255\ ^\circ F$.

## Experimental Efficiency Curves

If several efficiency curves are made available from different experimental tests, the relationships between them can be difficult to use because the data must be very consistent to obtain useful results.

As previously noted, it is best to do all curve fitting and analysis with an experimental $\eta_m$ curve, preparing the other curves when required, analytically, even when the flow data are in the range desired. When the flow rate desired is not within $\pm 25\%$ of the experimental data or when another heat transfer fluid is to be used, it will be worthwhile to estimate analytically the proportional change in $U_f$ between the two flow conditions (see Example 7-4) and to use this change to modify $\eta_m$ before calculating the $\eta_i$ parameters. Before a major investment is made in building a system, experimental data near the desired flow rate should be obtained from a source that is reliable and independent of the collector manufacturer.

Current U.S. test procedures call for a range of fluid temperatures up to nearly 90 °C (200 °F) to produce variations in the collector operating point so that the efficiency curve can be determined. This may give a gentle convex (downward) bend to the curve, since the radiation loss increases with the fourth power of the absolute temperature, rather than linearly. In such cases, the heat transfer coefficient and intercept should be taken from the smoothed curve at an operating point where the experimental collector fluid temperature was in the most important

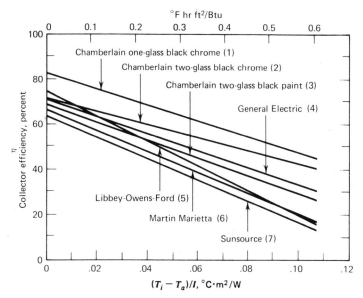

**FIGURE 5-9.** Solar collector efficiency curves based on inlet temperature for a number of commercial collectors. These curves were taken in the NASA Lewis solar simulator, an indoor test facility, prior to outdoor exposure (Ref. 5):

1. Single-glazed flat-plate collector, steel absorber, black-chrome selective surface.
2. Double-glazed flat-plate collector, steel absorber, black-chrome selective surface.
3. Double-glazed flat-plate collector, steel absorber, flat-black paint.
4. Double Lexan cover plates on flat plate collector, aluminum absorber, selective paint.
5. Double-glazed flat-plate collector, aluminum absorber, black paint.
6. Double-glazed flat-plate collector, aluminum absorber, optical black paint.
7. Single-glazed flat-plate collector, steel-tube absorber, black-nickel selective surface.

For all these collectors the flow capacity rate was 4.88 kg/h·m² (10 lb/hr ft²).

range for collector performance—generally about 40 to 50 °C (100 to 125 °F). This point varies from 0.012 °C·m²/W (0.07 hr °F ft²/Btu) for data taken during summer testing to perhaps 0.045 °C·m²/W (0.25 hr °F ft²/Btu) at near freezing winter conditions.

Figures 5-9 and 5-10 show measured $\eta_i$ curves for a number of commercial collectors taken by two U.S. agencies with long experience in solar collector

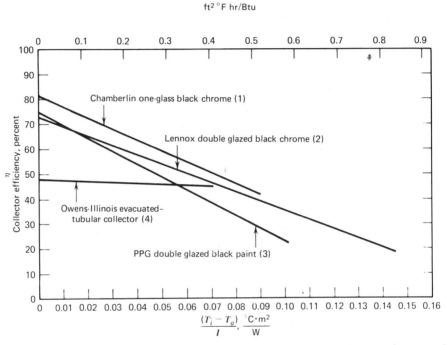

**FIGURE 5-10.** Solar collector efficiency curves based on inlet temperature for several commercial collectors. These curves were taken by the National Bureau of Standards in an outdoor test facility (Ref. 7).

1. Single-glazed flat-plate collector, steel absorber, black-chrome selective surface. $F_i\tau\alpha = 0.808$, $F_iU_L = 4.43$ W/°C·m².
2. Double-glazed flat-plate collector, antireflective coating on three surfaces, copper tubing, stell absorber plate with black-chrome selective surface. $F_i\tau\alpha = 0.726$, $F_iU_L = 3.62$ W/°C·m².
3. Double-glazed flat-plate collector, aluminum absorber, flat-black paint. $F_i\tau\alpha = 0.747$, $F_iU_L = 5.14$ W/°C·m².
4. Single-glazed evacuated-tubular collector, concentric selective absorber. $F_i\tau\alpha = 0.475$, $F_iU_L = 0.390$ W/°C·m².

Flow capacity rates of water for these collectors was 7.32 kg/m²·h (15 lb/hr ft²), except for number 4, which was about one-third of this.

performance testing (5, 7). Fig. 5-9 was made with a solar simulator and Fig. 5-10 data were taken outdoors using natural sunlight. The loss coefficients for the one collector common to the two sets of tests (labelled "1" in each figure) are rather different. Such inconsistency is not unusual, for a deliberate experiment in which identical collectors were tested by twenty-odd laboratories produced results having a standard deviation in $F_m\tau\alpha$ of 0.04 to 0.06 and $\pm 15$ to 25% standard deviation from the mean $F_m U_L$ value (6).

When a reliable $\eta_i$ curve is available at the flow conditions desired with the preferred heat transfer fluid, the $F_i\tau\alpha$ and $F_i U_L$ parameters can be taken directly from the curve and used for system design after applying the incident angle modifier to $F_i\tau\alpha$ as detailed in Chapter 4. Equations 5-21 and 4-21 then become

$$q = F_i\overline{\tau\alpha}I - F_i U_L(T_i - T_a) \tag{5-34}$$

$$q_T = F_i\overline{\tau\alpha}I_T - F_i U_L(\overline{T}_i - \overline{T}_a)t_T \tag{5-35}$$

These two equations retain the form of eqs. 4-11 and 4-21 but use intercept $F_i\overline{\tau\alpha}$ and slope $F_i U_L$ from the incident angle modified $\eta_i$ curve. The true collector temperature is no longer needed, and the collector inlet temperature is much more convenient in practical use.

## Application to Other Collector Fluid Flow Situations

Equations for $F_i$, $F_o$, and $F_m$ are generally applicable to collector heat transfer situations if the areas and coefficients are chosen with care even if the fluid flow situation differs in appearance from Fig. 5-6.

### EXAMPLE 5-12

A swimming pool collector is made from 2.54 mm (0.1 in.) polypropylene plastic with $\alpha = 0.9$, thermal conductivity 0.173 W/°C·m (0.1 Btu/hr °F ft). Water flows underneath the entire surface at a flow rate so high that the water temperature rise is negligible. What is $F_i$? What will the efficiency be when $T_i = 26.67$ °C (80 °F), $T_a = 15.55$ °C (60 °F), and $H = 788$ W/m² (250 Btu/hr ft²)? The loss coefficient on the unprotected surface is 14.2 W/°C·m² (2.5 Btu/hr °F ft²).

Since there is no finned surface, $\eta_F = 1$. Since there is no temperature rise, $U_L A_c/2wc$ is zero and eqs. 5-22, 5-24, and 5-26 all reduce to

$$F = \frac{1}{1 + \dfrac{U_L A_c}{U_f A_f}}$$

Since the areas for the fluid and the collector are equal, $F = 1/(1 + U_L/U_f)$. $U_f$ is the resistance to heat transfer from the collector to the fluid, and this results from the plastic, not the fast-moving fluid.

$$U_f = \frac{k}{L} = \frac{0.173 \text{ W/°C·m}}{0.00254 \text{ m}} = 68.11 \text{ W/°C·m}^2$$

Since $U_L = 14.2$ W/°C·m²

$$F = \frac{1}{1 + \frac{14.2}{68.11}} = 0.8275$$

From eqs. 5-29, 5-30, or 5-31,

$$\eta = 0.8275\left[0.9 - \frac{14.2(26.67 - 15.55)}{788}\right] = 0.5789$$

In Engineering units,

$$U_f = \frac{k}{L} = \frac{0.1 \text{ Btu/hr °F ft}}{0.1 \text{ in.} \times \text{ft/12 in.}} = \frac{12 \text{ Btu}}{\text{hr °F ft}^2}$$

Since $U = 2.5$ Btu/hr ft² °F,

$$F = \frac{1}{1 + 2.5/12} = 0.8276$$

From eqs. 5-29, 5-30, or 5-31,

$$\eta = 0.8276\left[0.9 - \frac{2.5(80 - 60)}{250}\right] = 0.5793$$

## EXAMPLE 5-13

A conventional liquid-cooled, flat plate collector with a loss coefficient of 3.975 W/°C·m² (0.7 Btu/hr °F ft²) has 19.05 mm (¾ in.) square tubes on centers 152.4 mm (6 in.) apart, bonded with 0.254 mm (0.010 in.) thick bonding material (on one face of the tube), which has a thermal conductivity of 0.173 W/°C·m (0.1 Btu/hr ft °F). What will be the effect on performance of the bond if the fluid heat transfer coefficient is 1700 W/°C·m² (300 Btu/hr °F ft²)?

First estimate the $h$ factor due to the bond:

$$h = \frac{k}{L} = \frac{0.173 \text{ W}}{°C \cdot m} \times \frac{1}{0.000 \, 254 \text{ m}} = 681.1 \text{ W/°C·m}^2$$

The fluid film coefficient of 1700 W/°C·m² is in series with the bond, but this coefficient extends over four times the area, that of the *entire* tube wall. Proper use of the series and parallel path equations for combinations of $h$'s given in Chapter 2 yields

$$\frac{1}{U_f} = \frac{1}{h_b} + \frac{1}{h_f} = \frac{1}{681} + \frac{1}{(1700)4}$$

$$U_f = 619 \text{ W/°C·m}^2$$

Using $F_m$ and assuming a thick and efficient collector surface so that $\eta_F = 1$,

$$F_m = \frac{1}{1 + \frac{U_L A_c}{U_f A_f}} = \frac{1}{1 + \frac{3.975}{619} \times \frac{152.4}{19.05}} = 0.9511$$

while without the bond,

$$F_m = \frac{1}{1 + \dfrac{3.975}{1700} \times \dfrac{152.4}{19.05 \times 4}} = 0.9953$$

a drop of about 5% in collector efficiency. Notice that the full inside area of the tube was used for heat transfer through the metal but only the bond area for the bond. This assumes that the heat is free to flow along the cross section of the tube, an assumption based on the fact that most tubes are made with rather thick walls to withstand handling.

In Engineering units,

$$\frac{1}{U_f} = \frac{1}{h_b} + \frac{1}{h_f} = \frac{1}{120} + \frac{1}{(300 \times 4)} = 109 \text{ Btu/hr } °F \text{ ft}^2$$

$$F_m = \frac{1}{1 + \dfrac{U_L A_c}{U_f A_f}} = \frac{1}{1 + \dfrac{0.7}{109} \times \dfrac{6}{0.75}} = 0.9511 \quad \text{with the bond}$$

while without the bond,

$$F_m = \frac{1}{1 + \dfrac{0.7}{300} \times \dfrac{6}{0.75 \times 4}} = 0.9954$$

### EXAMPLE 5-14

Fins are added to the air side of the air collector of Examples 5-1(f) and 5-3b, tripling its heat transfer area. What is the effect on $F_i$?

With no extended collector surfaces,

$$F_i = \frac{1}{1 + \dfrac{U_L A_c}{U_f A_f} + \dfrac{U_L A_c}{2Wc}}$$

With perfect fins on the air side,

$$\frac{U_L A_c}{U_f A_f} = \frac{4.542}{10.75} \times \frac{1}{3} = 0.1408$$

$$\frac{U_L A_c}{2Wc} = \frac{4.542 \text{ W}}{2 \, °C \cdot m^2} \times \frac{2.98 \text{ m}^2}{0.039\,02 \text{ kg/s}} \times \frac{kg \cdot °C}{1.02 \text{ kJ}} \times \frac{J}{W \cdot s} \times \frac{k}{10^3} = 0.1700$$

$$F_i = \frac{1}{1 + 0.1411 + 0.1701} = 0.7627$$

Without the fins installed,

$$\frac{U_L A_c}{U_f A_f} = \frac{4.542}{10.75} = 0.4225$$

$$F_i = \frac{1}{1 + 0.4225 + 0.1701} = 0.6279$$

so there is an improvement in $F_i$ and hence efficiency of about 20%.

In Engineering units,

$$\frac{U_L A_c}{U_f A_f} = \frac{0.8}{1.89} \times \frac{1}{3} = 0.1411$$

$$\frac{U_L A_c}{2Wc} = \frac{0.8 \text{ Btu}}{2 \text{ hr } °F \text{ ft}^2} \times \frac{32 \text{ ft}^2}{5.161 \text{ lb/min}} \times \frac{\text{lb } °F}{0.243 \text{ Btu}} \times \frac{\text{hr}}{60 \text{ min}} = 0.1701$$

$$F_i = \frac{1}{1 + 0.1411 + 0.1701} = 0.7626$$

Without the fins installed,

$$\frac{U_L A_c}{U_f A_f} = \frac{0.8}{1.89} = 0.4233$$

$$F_i = \frac{1}{1 + 0.4233 + 0.1701} = 0.6276$$

so there is an improvement in $F_i$ and hence efficiency of about 20%.

The solutions to the above examples assume that $U_f$ was unchanged when fins were added. More often $U_f$ will be reduced, and a careful fluid flow analysis may be required. It is also assumed that the fin efficiency on the air side is unity. Since the air side fins can be made from relatively cheap aluminum, this will almost always be the case.

## PROBLEMS

1. A collector measures 2 m × 1 m (6.56 ft × 3.28 ft). A flow of 40 g/s (317 lb/hr) of 50% ethylene glycol cools the collector. Assume 66 °C fluid properties (150 °F).
   (a) What is the normalized flow capacity rate $wc$?
   (b) What is the Reynolds number for flow through a single serpentine $\frac{1}{2}$ in. type M tube (see Table 12-2) having ten 0.8 m (2.62 ft) passages 20 cm (7.87 in.) apart, measured center-to-center?
   (c) Estimate the heat transfer coefficient $U_f$.
   (d) The absorber surface is 0.889 mm (0.035 in.) thick copper. Calculate the average fin efficiency if the collector heat loss coefficient $U_L$ is 4.656 W/°C·m² (0.82 Btu/hr °F ft²) and $\tau\alpha = 0.85$.
   (e) Calculate $F_i$, $F_m$, and $F_o$; and $F_i\tau\alpha$, $F_i U_L$, $F_m\tau\alpha$, $F_m U_L$.
   (f) Calculate the area-normalized rate of heat collection by the fluid if the irradiance is 788 W/m² (250 Btu/hr ft²) and the collector inlet temperature is 35 °C (95 °F) with the outside ambient 5 °C (41 °F).
   (g) Find the outlet and mean fluid temperatures and show that the heat collection rate can be calculated using them with $F_o$ and $F_m$, respectively.
   (h) Find the maximum temperature on the absorber. Ignore the small temperature rise from fluid to fin root.
   (i) Calculate $F_R$ using eq. 5-28 and compare with $F_i$.

2. Silicone oil is substituted for glycol-water and circulated at the same mass flow rate as in problem 1. What are $F_i$ and the heat collection rate?

3. Repeat problem 2 with the original value for the normalized flow capacity rate. By what factor must the volumetric flow be increased?

4. Repeat problem 1 with the flow through five 1.8 m (5.9 ft) tubes in parallel.

5. (a) Repeat problem 1 using an air flow rate of 0.025 m³/sec (53 ft³/min) under the absorber plate, parallel to the long dimension in a 1.5875 cm (⅝ in.) passage.
   (b) Plot $\eta_i$, $n_m$, and $\eta_o$ curves for part (a).

6. Using parameters from problem 1(d),
   (a) Plot the fin efficiency as the copper thickness is changed from 0.25 to 2.5 mm (0.01 to 0.100 in.).
   (b) For a metal thickness of 0.010 in. (0.254 mm), plot the *average* fin efficiency as the spacing *between* tubes is varied from 25 mm (1 in.) to 150 mm (6 in.).

7. Repeat problem 5, doubling the air flow rate.

8. Repeat problem 5, halving the air passage thickness.

9. Repeat problem 5, using fins to double the air side heat exchange area.

10. A commercial collector using ASHRAE test procedures gives $F_i \tau \alpha = 0.622$ and $F_i U_L = 3.657$ W/°C·m² (0.644 Btu/hr °F ft²). The cooling water flow rate was 0.0378 m³/s (5 lb/min) for the 2.338 m² (25.17 ft²) area.
    (a) What are $F_m \tau \alpha$ and $F_m U_L$?
    (b) The ambient temperature is 10 °C (50 °F) and the irradiance is 790 W/m² (250 Btu/hr ft²). What flow rate will give a 50 °C (90 °F) temperature rise from 40 °C (104 °F) at the inlet?
    (c) Suppose the flow rate is reduced to 60% of the test value. Assuming no change in $U_f$, what are the new $F_i \tau \alpha$ and $F_i U_L$? What are the new $F_m \tau \alpha$ and $F_m U_L$?
    (d) Assuming $\tau \alpha = 0.75$ and $\bar{\eta}_F = 0.96$, what is $U_f$ for the collector?
    (e) Ethylene glycol is to be used instead of water in an actual installation. Considering initial and final flows to be (1) turbulent and (2) laminar, estimate the proper values of $F_i \tau \alpha$ and $F_i U_L$ using the original flow capacity rate in each situation.

11. Derive eqs. 5-25 and 5-26.

12. Derive eq. 5-32c.

13. Derive the last two equations in the right-hand column of Table 5-2.

14. For the swimming pool situation in Example 5-12, plot collection efficiency versus plastic thickness by recalculating using several thicknesses.

**15.** The heat transfer coefficient in a helical coil is increased by a factor of $(1 + 3.5 D_i/D_c)$ due to constant turning, where $D_i$ is the coil diameter. Show that for a pancake coil the correct factor is $(1 + 3.5 D_i/\bar{D}_c)$, where $\bar{D}_c$ is the arithmetic average coil diameter.

## REFERENCES

1. W. H. Giedt, *Principles of Engineering Heat Transfer*, New York: Van Nostrand, 1957.
2. R. H. Perry and C. H. Chilton, *Chemical Engineers' Handbook*, Fifth Edition, New York: McGraw-Hill, 1973, p. **10**-14.
3. W. H. McAdams, *Heat Transmission*, Third Edition, New York: McGraw-Hill, 1954.
4. R. W. Bliss, Jr., "The derivations of several 'Plate-efficiency factors' useful in the design of flat-plate solar heat collectors," *Solar Energy 3*, 55 (1959).
5. R. H. Knoll and S. M. Johnson, "Baseline performance of solar collectors for NASA Langley solar building test facility," Proceedings 1977 Flat-plate Solar Collector Conference, edited by Delbert B. Ward, CONF-770253 UC-59 (FSEC 77-8), pp. 485–500 (March 1978).
6. E. R. Streed, J. E. Hill, et al., "Results and analysis of a round robin test program for liquid-heating flat-plate collectors," *Solar Energy 22*, 235–249 (1979).
7. J. E. Hill, J. P. Jenkins, and D. E. Jones, "Experimental Verification of a Standard Test Procedure for Solar Collectors," NBS Building Science Series 117, U.S. Department of Commerce, January 1979.

# Optically Concentrating Collectors and Reflectors

○

## CONCENTRATING COLLECTORS

The purpose of all solar collectors is to intercept the radiant energy of the sun and to transform as much of this energy as possible into useful heat. In an *optically concentrating collector* the radiant energy is concentrated *optically* before it is transformed into heat. Light entering a large *aperture* area is reflected or refracted to a relatively small *target* or *receiver* where it is transformed into heat energy that is then collected in a conventional way. A flat plate collector does the same job, but the transformation into heat takes place immediately, and heat rather than light is transported to the (usually) smaller area through which the collecting fluid flows.

Interest in concentrating collectors has developed not as a result of anything inherently more desirable in concentrating radiant energy, but rather because thermal losses can be lessened when the area that is heated to the operating temperature is reduced. This reduction in losses then makes it possible for the collector to heat fluid to a higher temperature with reasonable efficiency. We therefore associate concentrating collectors with high temperatures, but it should be understood that the reason for such performance is not the concentration itself but the reduced losses that the concentration makes possible.

There is a great variety in concentrators, much more so than with flat plate collectors. It would be difficult to describe them all. Instead the elements that make up concentrating collectors will be described in general, and a few representative types discussed.

### Parabolic Concentrators

The first active element that solar radiation reaches in a concentrator is either a transparent lens called a refractor, or a mirror. There may also be a transparent cover plate over the whole assembly to protect it from dirt and provide further thermal isolation from the ambient. The optics concentrate the incoming radiation into the receiver, which ideally is a point for a circular three-dimensional concentrator or a line for the trough-like, two-dimensional concentrator. A *parabolic reflector* is often used to intercept the radiant beam because rays parallel to its axis are concentrated at the focus of the parabola, as in Fig. 6-1a. In Fig. 6-1b the same effect is produced with refracting optics using a spherical or cylindrical lens to refract parallel light rays to a common focus. Each design can be made in

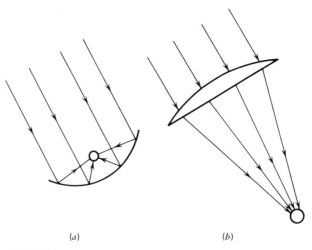

(a)                                              (b)

**FIGURE 6-1.**  Optical paths for imaging concentrating collectors. (a) Reflecting. (b) Refracting.

either two dimensions (Fig. 6-2) or three dimensions (Fig. 6-3) producing either a line or a point focus for the receiver. Figure 6-2b shows use of a *Fresnel* lens, which is optically equivalent to a cylindrical lens. It is often used because it can be molded easily in plastics and is much lighter than the lens it replaces. For reflecting mirrors, flat segments in the form of a small oblong mirror (for a three-dimensional reflector) or a thin strip (for a two-dimensional reflector) serve analogously. Glass mirrors (the best kind) cannot be economically made from curved stock, so that such *faceted* reflectors are as conventional as Fresnel lenses in solar concentrators.

The optical concentrators in Figs. 6-1 to 6-3 are *imaging* optics because an image of the source is available at the focus, even though it may be distorted. The source is the sun, but since the sun is not a point but a disc that subtends an arc of about 0.5°, its image cannot be focussed to a point (or a line for the two-dimensional concentrator) but must be a disc or long rectangle. The *concentration*

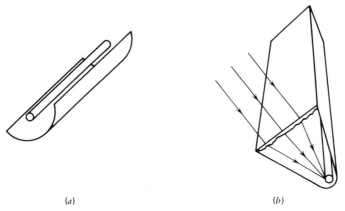

**FIGURE 6-2.** Two-dimensional concentrators. (*a*) Reflecting. (*b*) Refracting.

*ratio* is the ratio of the aperture area to the surface area of the absorber. The *maximum* concentration ratio is a theoretical measure of the optical energy concentration attainable by the concentrator. This is a definite number that can be worked out from the physical dimensions of the lens or reflector and the geometry of the target by simple optical principles.

Since the sun's image is a disc or long rectangle for the two- and three-dimensional imaging concentrators, a reasonable geometry for the receiver is a disc- or long rectangle-shaped flat plate collector. Orienting the collector plate so that

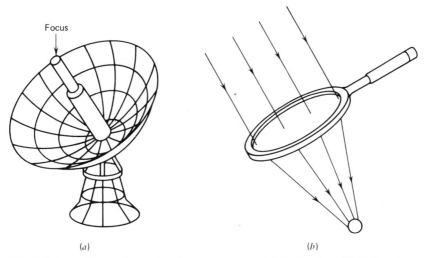

**FIGURE 6-3.** Three-dimensional concentrators. (*a*) Reflecting. (*b*) Refracting.

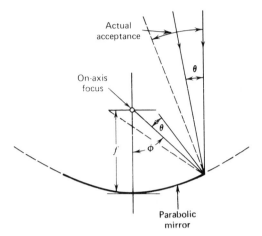

**FIGURE 6-4.** Parabolic mirror, showing relationships between acceptance half-angle $\theta$ and rim angle $\phi$.

only one side receives reflected radiation, as in Fig. 6-4,* the maximum concentration ratio for a parabolic reflector with a flat, one-sided receiver is expressed by

$$C_{3\text{-dim}} = \frac{\sin^2 \phi \cos^2 (\phi + \theta)}{\sin^2 \theta} \tag{6-1}$$

for the three-dimensional concentrator and

$$C_{2\text{-dim}} = \frac{\sin \phi \cos (\phi + \theta)}{\sin \theta} - 1 \tag{6-2}$$

for the two-dimensional case. Similar expressions are available for other configurations of reflector and receiver (1). In these equations, $\phi$ is the rim angle defined in Fig. 6-4, a convenient measure of the aperture size. $\theta$ is the acceptance angle of the collector. The actual acceptance is $2\theta$, the angle through which a point source can be moved and still focus on the receiver (see Fig. 6-4). The concentration ratio of a given collector is limited by the requirement that the actual acceptance be wide enough to admit the sun's entire disc. From eqs. 6-1 and 6-2 it can be shown that the concentration ratio not only depends on the aperture but that there is a maximum concentration ratio that can be obtained for a given acceptance angle. This occurs at concentration ratios of

$$C_{3\text{-dim}} = \frac{1}{4 \sin^2 \theta} - \frac{1}{2 \sin \theta} - \frac{3}{4} \tag{6-3}$$

* To show the geometry clearly, the size of $\theta$ and the target has been exaggerated so that it casts a significant shadow on the reflector. In practice at such a low concentration ratio a two-sided or a cylindrical receiver would be used so that any radiation lost by shadowing is absorbed directly.

and

$$C_{\text{2-dim}} = \frac{1}{2 \sin \theta} - \frac{3}{2} \tag{6-4}$$

when

$$\phi = \frac{1}{2}\left(\frac{\pi}{2} - \theta\right) \tag{6-5}$$

Therefore, once the acceptance angle is set, angle $\phi$ is determined, which fixes the ratios of focal length, receiver size, and aperture if the concentrator is to have the maximum concentration ratio. Consequently, the entire optimal design of a parabolic concentrator can be made from the choice of acceptance angle and aperture. Angle $\phi$ will always be somewhat less than the 45° shown in Fig. 6-4 (eq. 6-5). For concentrating the sun, the maximum concentration ratios for a parabolic reflector are about 110 and 13 000 for two- and three-dimensional concentrators, respectively—well above the ratios commonly used for solar thermal applications.

Nothing in the above argument indicates that parabolic reflectors (or the spherical or cylindrical refractors, which follow the same mathematics) are the best possible choice for reflector geometry, nor is there reason to presume that imaging optics are the only efficient possibility. Thermodynamically, it can be shown (1) that the ideal or maximum *possible* concentration ratio is much higher than that achieved by the parabolic reflector, and is given by

$$C_{\substack{\text{ideal} \\ \text{2-dim}}} = 1/\sin \theta \tag{6-6}$$

$$C_{\substack{\text{ideal} \\ \text{3-dim}}} = 1/\sin^2 \theta \tag{6-7}$$

which are about 230 and 50 000 for the two- and three-axis concentrators, respectively, using the sun as a source. We examine ideal concentrators not to produce high-concentration ratios directly, however, but to broaden the acceptance angle of the collector so that it need not track the sun accurately and in some cases need not be moved at all. A 60° actual acceptance with an ideal concentrator still permits a concentration ratio of 2, for instance, while a parabolic concentrator permits no concentration at all with an actual acceptance greater than 23°.

Ideal optical concentrations can be approached only with nonimaging concentrators, in which the image is not only distorted but also lacks symmetry with the source because of differing multiple reflection paths that depend on the angle of incidence.

## Nonimaging Concentrators

There are any number of nonimaging concentrators, ranging from simple but inefficient reflectors to the nearly ideal, but complexly curved, compound parabolic concentrator (CPC) shown in Fig. 6-5. We will consider only two-dimensional or

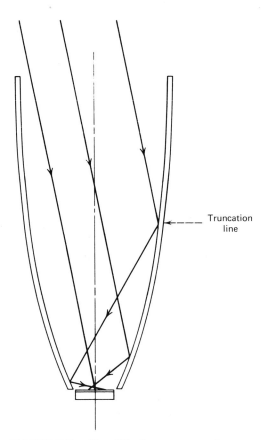

Truncation line

**FIGURE 6-5.** The Winston compound parabolic concentrator (CPC). Rays trace the path of off-axis radiation through one, two, and three reflections. The portion of the reflector above the truncation line may be removed with minimal effect on the performance.

line-focusing nonimaging collectors, because three-dimensional nonimaging geometry becomes too complex for ordinary solar thermal use. All two-dimensional CPCs, as we will note later, have a desirable acceptance angle along the trough, which often enables them to function during a full day without need for tracking when the long axis is oriented east-west.

The CPC shown in Fig. 6-5 is one of a general class discovered by Hinterberger and Winston (2) in the United States and independently by Baranov and Melnikov (3) in the Soviet Union. The concept is often referred to as the Winston collector, after Roland Winston, who holds the U.S. patent (which is assigned to the University of Chicago). It was originally devised for concentrating faint radiation developed in nuclear physics experiments.

The virtue of the CPC is that it truly approaches the thermodynamic limit for the concentration ratio given by eq. 6-6. It improves on the parabolic reflector in that it concentrates by using multiple reflections as well as direct interception and single reflection. Since its acceptance angle is as wide as is theoretically possible, it therefore sets the standard for a solar concentrator.

The wide acceptance angle of the Winston collector, given by eq. 6-6, has a second advantage that is of great importance for applications in cloudy areas such as the northeastern quarter of the United States. Diffuse light is intercepted and concentrated as well, with a proportion $1/C$ of the total diffuse light from the sky made available to the collector. Use of a low-concentration-ratio collector (say, $C = 1.5$), then makes two-thirds of the diffuse light available,* with an actual acceptance of 84°. Combined with a low-loss absorber—probably an evacuated tube with a low-emittance coating—a very effective collector results.

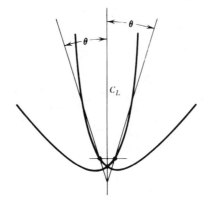

**FIGURE 6-6.** Construction of a compound parabolic concentrator. The acceptance half-angle $\theta$ is defined by the required concentration ratio. The focus of each parabola is placed on the intersection of the flat plate receiver and the acceptance angle line. The receiver is moved in or out until the (unused) arm of each parabola passes through the focus of the other parabola.

* This compares with 93% for a flat plate collector inclined at 30° and 82% at 50° (eq. 3-13). Such a CPC should be tilted so that the light reflected from the surroundings is also within the acceptance angle.

The advantage comes not from the concentration per se, but from the redesign of the absorber from a flat to a tubular shape, which can more easily be protected from heat loss.

The Winston collector is composed of sections of two parabolas as shown in Fig. 6-6. The focus of the parabolas is at either end of the receiver, and they are tilted about their own focus at the acceptance angle to form a funnel down which light will be accepted.

**FIGURE 6-7.** A solar collector incorporating three compound parabolic reflectors of the Winston design. The acceptance angle is so wide that the reflectors take the black coloration of the tubular absorbers that are within the evacuated glass tubes visible in the photograph. The concentration ratio is $1 \cdot 5$, and the aperture area is $1 \cdot 55$ m$^2$ ($16 \cdot 7$ ft$^2$). (Courtesy of First Solar Industries, Inc., Plymouth, Conn.)

Application of the CPC is restricted by its depth. This can be lessened by truncating the collector perhaps along the line shown in Fig. 6-5. Cutting the top half of the reflector away only decreases the concentration ratio by perhaps 10% in the average configuration. Actual losses will be somewhat less because of the decrease in the average number of reflections and the additional light that can spill down the abbreviated cone from outside the acceptance angle, plus the fact that more of the abbreviated troughs can be packed into a given area since they are somewhat narrower. Even so, a single trough is too deep for application in a useful size. The usual CPC has a number of troughs and in final form resembles a flat plate collector (Fig. 6-7).

## Other Forms of Concentrating Collectors

There are many other concentrators that can be regarded as either an imperfect parabolic collector (if they form an image) or an imperfect CPC (if they do not). Since they often have a wider but less sharply defined acceptance angle than the perfect shapes, the receiver must be larger if it is to intercept the concentrated beam over the full range of the acceptance angle. They can approach the performance of the ideal forms in a stationary application (more often they will not) because they can intercept more of the sunlight at the ends of the day, and because the flat mirror segments sometimes used can have a higher reflectivity than the CPC shape, which is so complex that of economic necessity it must be made from aluminum, which has a relatively low reflectivity.

Some of these shapes are shown in Fig. 6-8a to 6-8e. Figure 6-8a is a faceted, fixed mirror concentrator with a moving receiver. Figure 6-8b shows the same kind of approach applied to a *spherical* reflector. Ray tracings clearly show the utility of the moving absorber with either two- or three-dimensional concentrators using this easily manufactured reflector. Figures 6-8c and 6-8d both use flat plate collectors as receivers. Although the concentration ratio is only unity, using both sides of the absorber plate helps reduce costs. Figure 6-8e is actually a primitive CPC reflector, devised before that concept was well known.

CPC configurations can be approximated by other geometrical shapes with results ranging from fair to good. Approximations have been made by using a single ellipse for *both* sides of the parabolic form (5) and by the "V-groove" collector (6, 7), which uses a single mirror on each side of the receiver. When three flat segments are used on each side (7) the surface is a very good approximation to the parabolic shape required. Probably the expense is also a good approximation to the CPC as well.

The variations found in practice sometimes seem overwhelming in numbers but with due consideration they can usually be reduced to one of the elementary forms. For analysis, one wishes for information seldom available: the optical efficiency and theoretical concentration ratio as a function of the incident angle. For performance prediction, daylong averages are preferred, as explained in a following section, but even instantaneous efficiency curves are seldom available except at solar noon.

### Tracking

All three-dimensional concentrators must be moved or "tracked" to follow the sun. With their typically high-concentration ratios and correspondingly narrow acceptance angles, those collectors must be tracked along two axes if they are to keep the sun on their receiver.

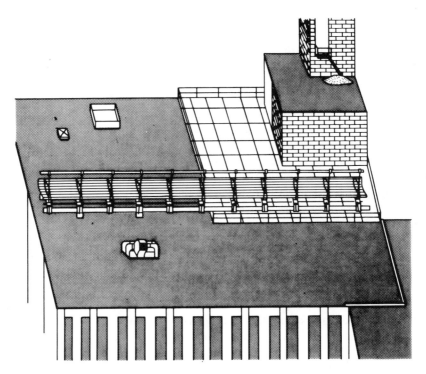

**FIGURE 6-8a.** Faceted fixed mirror concentrator developed at the Georgia Institute of Technology. The long rod at the top of the assembly is the absorber tube, which is moved during the day to track the reflections from the slatted mirrors below. The total collector area is 540 ft² (50 m²). [*Source:* W. F. Williams and S. F. Hutchins, *Development of a Solar Heat Supply System with Fixed Mirror Concentrators*, Workshop on Solar Collectors for Heating and Cooling of Buildings, New York, November 1974 (NSF-RA-N-75-019) (150–57).]

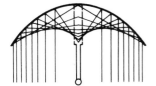

4c. Position of the absorber at
    12:00 noon.

4b. Position of the absorber at 10:00
    A.M. or 2:00 P.M. Shaded area is
    inactive.

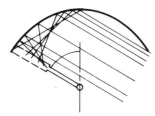

4a. Position of the absorber at 8:00
    A.M. or 4:00 P.M.

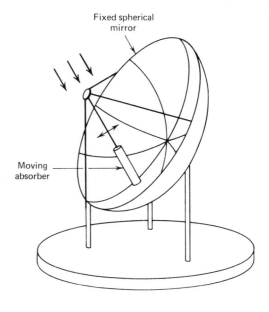

Fixed spherical
mirror

Moving
absorber

**FIGURE 6-8*b*.** Stationary reflector, tracking absorber concept using a spherical segment for a reflector. This is one of the very few three-dimensional concentrators proposed for solar thermal applications. The subsidiary figures show how the rays intercept the moving absorber at various times of day. The two-dimensional form of this concept, using a cylindrical trough, seldom has a moving absorber. The above ray tracings make it clear that such a collector needs a large absorber to be effective (see Fig. 6.8*c*). (*Source*: J. F. Kreider, "Modeling and construction of the stationary reflector/tracking absorber solar collector," 1974 Solar collector conference, p. 176 and p. 189, NSF-RA-N-75-019.)

Two-dimensional or line-focusing collectors can be oriented with the long axis facing either north-south or east-west. In either case the plane of the collector(s) should be perpendicular to the plane of the ecliptic, that is, tilted upward toward the sun's path of travel. When the troughs are oriented north-south, as in Fig. 6-9, the collectors must be tracked during the day to follow the sun. The orientation need not be adjusted seasonally, however, because the acceptance angle of

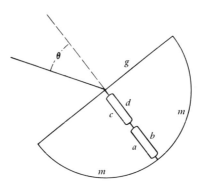

**FIGURE 6-8c.** Cross section of collector showing the absorber surfaces, the mirror *M*, and cover glass *G*. See the ray-tracing in Fig. 6-8*b* to understand why this collector needs such a large traget. [*Source:* H. S. Robertson, *A moderately focusing flat-plate solar collector*, 1977 Flat plate solar collector conference CONF-770263 (247-251).]

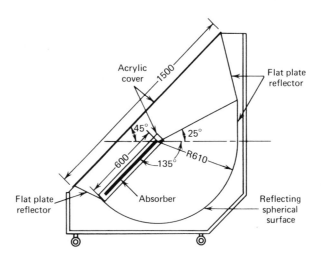

**FIGURE 6-8d.** Multiple reflectors used to irradiate both sides of a flat plate absorber. Notice how this empirical design approaches the cusp Winston collectors of Fig. 6-11. (*Source:* Y. Saito, *Design and test of a double exposure flat-plate collector using fixed spherical and flat reflectors*, ISES Abstracts 1975 Los Angeles meeting.

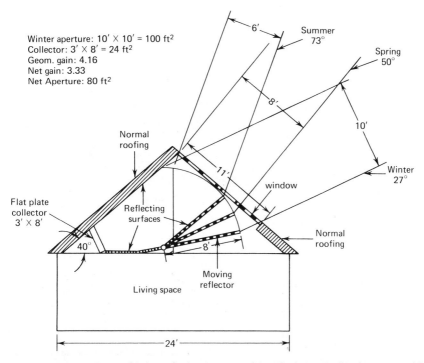

Winter aperture: 10' × 10' = 100 ft²
Collector: 3' × 8' = 24 ft²
Geom. gain: 4.16
Net gain: 3.33
Net Aperture: 80 ft²

**FIGURE 6-8e.** Pyramidal optical system—with window. Reflective pyramid optical concentrator. Wormser Scientific Company's unique utilization of the attic of a house as housing for a mirror concentrator. Flaps adjust seasonally according to the sun's angle. (Source: Promotional literature, Wormser Scientific Co., Stamford, Connecticut.)

a two-dimensional concentrator is broad along the long axis. Figure 6-9 also shows an interesting peculiarity of this configuration. The collectors must be separated from each other so that they do not shade their neighbors when the sun's angle is shallow during the morning and afternoon. The output from the separated collectors will be almost constant over the day until they do begin to shade one another, however, and this is a desirable pattern for many applications. It is important to realize that the advantage of a constant collection rate is not due to the effectiveness of the configuration, but instead is caused by the necessity for spacing. The daylong *optical efficiency** of such a spaced array, based on the overall area, will be inferior to the equivalent close packed stationary east-west array described below. Since the collector cost, performance, and heat losses are

* The optical efficiency is the proportion of the radiation incident on a collector that is transformed into heat on the absorber. It thus refers to the heat made available before consideration of convection, conduction, and radiation losses from the absorber.

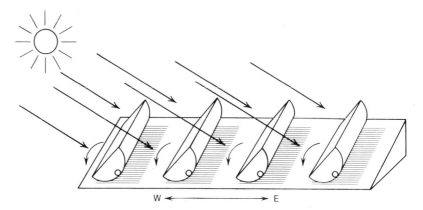

**FIGURE 6-9.** North-south oriented parabolic trough concentrator array, showing shadowing caused in late afternoon (and early morning) by finite collector spacing.

proportional to the actual, not the superficial area, the tracked array may be able to deliver more heat at less collector cost in high-temperature applications.

The east-west orientation is the most common for the two-dimensional (line focus) concentrating collectors. The sun's travel is then along the long axis of the collector for which there is a wide acceptance angle. The performance is similar to an ordinary flat plate collector, reaching a maximum at noon and otherwise falling off with the cosine of the sun's incident angle. Since the sun's travel approximates a straight line, there will ordinarily be no need for tracking in the north-south direction so that moderately concentrating collectors can be stationary, needing only seasonal north-south corrections that can be made by hand. If the concentration ratio is more than about 2 there will be some loss of performance with parabolic concentrators at the ends of the day in winter or summer, because the sun follows a circular path through the sky in those seasons. The appeal of a greater acceptance angle is especially strong in this situation, for not only will this permit better daily performance but the seasonal adjustments need be made less often. With a ×1.5 CPC (i.e., a concentration ratio of 1.5), a CPC will work the whole year without adjustment, and only four adjustments are needed for the ×3 concept.

A large acceptance angle also benefits performance because it allows collection of some of the diffuse energy from the sky. Parabolic-style concentrators do not intercept significant diffuse energy because the acceptance angle is so small.

The mechanical provisions for tracking, especially two-axis tracking, are troublesome. Even the relatively simple one-axis tracking can have trouble with ice and snow. If the acceptance angle is reasonably broad there is no need for the expensive feedback control devices in which the sun is sensed electronically and servomechanisms used to keep the tracking accurate. Instead devices resembling clockwork can be used to track the sun's movement. This avoids problems of "hunting" when the sun goes behind a cloud, and makes it easy to find the sun in the morning or when it emerges from cloud cover.

The optical element does not always have to be moved in a tracking concentrator. Many useful designs hold the reflector stationary and move the receiver into the focus of the reflector, which varies with the time of day. The off-center focus is not as good as is produced by moving the entire assembly, but the mechanical simplicity achieved is considerable.

### Receiver Shape and Orientation

The only receiver emphasized thus far has been the simple one-sided flat plate. When used at the focus of a parabolic reflector this has the disadvantage of shading the reflector without regaining the lost radiation, and when used at the bottom of a CPC or related concentrator there are losses out the back through the insulation, which are often significant because of the high temperatures at which these collectors are often run. Three other CPC designs are compared with the basic configuration in Fig. 6-10. They each collect heat from all sides of their receiver and need no opaque insulation. Their concentration ratio is the aperture area divided by the illuminated area, for example, $\pi DL$ for a tubular receiver of diameter $D$ and length $L$.

Figure 6-11 shows a nonsymmetrical CPC having a single cusp or "sea-shell" design. There are several commercial collectors that resemble this configuration. They do not use the careful CPC reflector design but evolved by trial and error before the appropriate theory became available.

Varied reflector shapes all have important effects on the efficiency, concentration ratio, and acceptance angle of their collectors. A few have been carefully analyzed, but most have not, relying instead on experimental measurements for evaluation.

**Receiver Configuration.** In solar electric "power tower" concentrators many nearly flat mirrors track the sun, each reflecting to a central receiver. The heat losses of such a configuration fall dramatically because of the small size of the receiver possible with the high-concentration ratio. With two-dimensional concentrators, however, pains must be taken that the advantage gained through optical concentration is not lost due to excessive receiver heat loss. Sometimes a simple black-painted flat plate receiver can be used as a target for parabolic concentrators or simple nonimaging concentrators such as the vee-trough, especially if the collector is used in a low-temperature application. Generally, however, a selective surface is mandatory, particularly with CPCs having a low-concentration ratio. With a broad acceptance angle, this kind of collector not only collects but also reradiates heat to the sky much more readily than a parabolic design.

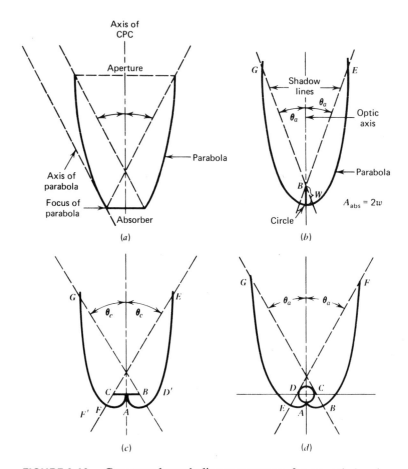

**FIGURE 6-10.** Compound parabolic concentrators for several absorber configurations. (*a*) Flat one-sided absorber. (*b*) Two-sided vertical fin absorber. (*c*) Two-sided horizontal fin absorber. (*d*) Tubular absorber. Reprinted by permission, *Solar Energy 18* (93-111), Copyright 1976 Pergamon Press.

A cover plate is often included, especially with complex mirror designs using aluminum reflectors, which should not be exposed to the weather. Recently, evacuated-tube receivers have been widely applied to eliminate convection losses in experimental concentrators, using both the flat-plate and circular vacuum bottle configurations described in Chapter 5. Since these also have a selective surface, total thermal losses are very low indeed. The marriage of concentrators and evacuated tubes is a good one, even though there is a semantics problem as to whether the evacuated tubes are being augmented by reflectors, or the concen-

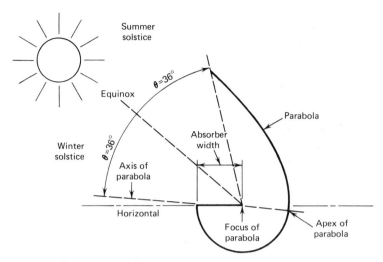

**FIGURE 6-11.** Nonsymmetrical compound parabolic concentrator. This is a "sea shell" collector having maximal output in winter. Reprinted by permission, *Solar Energy 18* (93-111), Copyright 1976 Pergamon Press.

trator is just using a new style target. In either case the performance is impressive, with heat losses negligible to as high as 100 °C (212 °F). There is a price, however, due to losses on necessarily imperfect reflective surfaces. Such losses in optical efficiency bring performance at lower temperatures down well below that of ordinary flat plate collectors. Unless these losses can be better controlled, the role of the concentrator will remain in the high-temperature applications until costs can be reduced well below those of ordinary flat plate collectors.

## Performance Analysis

By making a heat balance around a concentrating collector similar to that leading to eq. 4-11, the instantaneous performance of a generalized concentrating collector can be given as

$$q = \rho\gamma\tau\alpha I - \frac{U_L(T_c - T_a)}{(A_a/A_r)} \tag{6-8}$$

where $\rho$ is the specular reflectivity of the reflector, $\gamma$ is the proportion of the reflected beam intercepted by the receiver, $\tau$ the proportion of the incident beam reaching the reflectors, and $\alpha$ the proportion of the received beam absorbed by the receiver. $A_a/A_r$ is the concentration ratio of the collector: the ratio of the aperture area to the receiver *surface* area. The other terms are the same as in eq. 4-11 except that $U_L$ is based on the receiver surface area. If $U_L$ is based on the

**TABLE 6-1.** Specular Reflectivity of Mirror Surfaces

| | $\rho$ |
|---|---|
| Silver, fresh untarnished optical reflector | 0.93–0.95 |
| Back-silvered, low-iron glass[a] | 0.88 |
| Back-aluminized glass[a] | 0.76–0.80 |
| Electroplated silver | 0.96 |
| Aluminum lighting sheet | 0.82 |
| Aluminized Teflon (second-surface mirror)[a] | 0.77 |
| Silvered Teflon (second-surface mirror)[a] | 0.86 |
| Aluminized fiberglass | 0.92 |

[a] Plus the reflectivity of the front surface.

aperture area, the $A_a/A_r$ term is deleted from eq. 6-8. Values for $\rho$ are given in Table 6-1 for a number of reflecting materials.

Comparison of eqs. 6-8 and 4-11 makes it clear that while the area for heat loss is reduced by the concentration ratio, the heat reaching the absorber in the form of radiation is inherently less in a concentrator. The concentrator therefore may be inferior in performance to a flat plate collector when the collector temperature is low or the ambient temperature is high.

Equation 6-8 is not particularly useful because it is difficult to evaluate $\rho$, $\gamma$, and sometimes $\tau$ in practical collectors. Instead all these factors are often combined into an optical efficiency term called $\tau\alpha$ and the heat loss coefficient (can be) based on the aperture area. The performance can then be characterized just as flat plate collectors are, using eq. 4-11. The resultant efficiency curve also corresponds directly. Figure 6-12 shows performance curves from some concentrating collectors measured using flat plate collector techniques.

**Overall Performance.** Experimental collector efficiency curves are necessary to gauge the performance of concentrating collectors accurately. These curves are usually taken with the sun perpendicular to the collector, however, and since the off-normal performance is not considered it is not possible to predict the performance of concentrating collectors in an operating system without more information.

One important consideration is that all concentrators have a limited acceptance angle and therefore do not intercept as much diffuse energy as flat plate collectors. For parabolic reflectors the acceptance angle is so small that diffuse energy can be ignored. For other configurations the diffuse acceptance must be measured or known, as it is for the Winston collector where it is the reciprocal of the concentration ratio. The monthly total diffuse energy received can then be estimated using the Appendix B tables in which the diffuse energy is the total radiation less the direct radiation at the threshold radiation level involved.

Except for the single isolated tracking parabolic reflector, the direct energy

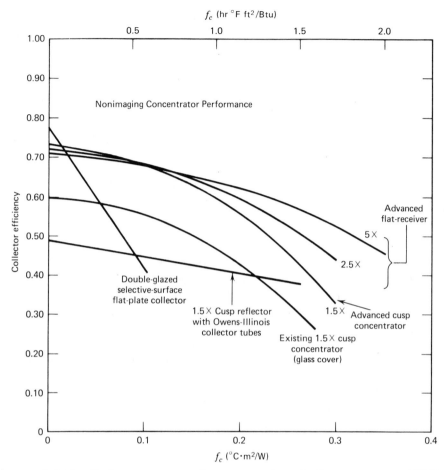

$f_c$ (hr °F ft²/Btu)

Nonimaging Concentrator Performance

Advanced
flat-receiver

5 X

2.5 X

Double-glazed
selective-surface
flat-plate collector

1.5 X Cusp reflector
with Owens-Illinois
collector tubes

1.5 X Advanced cusp
concentrator

Existing 1.5 X cusp
concentrator
(glass cover)

Collector efficiency

$f_c$ (°C·m²/W)

**FIGURE 6-12.** Performance curves for compound parabolic concentrators (CPC) compared with an excellent flat plate collector. (Source: Agonne National Laboratory Technical Reports.)

collected *over a day* by a concentrator is hard to estimate because of the shadowing effects with multiple tracked concentrators (see Fig. 6-9) or the angular variations in performance that characterize most of the fixed concentrators. These effects attenuate the collection day so that neither the monthly total of direct radiation nor the time above threshold applies. Since only the consistent diurnal pattern of *direct* radiation is involved, correction factors can be experimentally or analytically determined to account for these effects. The integrated collector equation (4-21) might be modified for the concentrating situation to give

$$q_T = \overline{\tau\alpha}(\Phi I_{T\ \text{direct}} + \Omega I_{T\ \text{diffuse}}) - U_L(\overline{T}_c - \overline{T}_a)\Psi t_T \qquad (6\text{-}9)$$

where $\tau\alpha$ is the average optical efficiency of the concentrator, $\Phi$ is a factor by which the total direct radiation falling on a fixed surface over a day above the threshold level can be multiplied to determine the direct radiation available to the collector, $\Omega$ is the proportion of the total diffuse radiation available to the collector, and $\Psi$ is the proportion of the time spent above the nominal radiation threshold that the collector can actually be operated. $I_{T \text{ direct}}$ can be taken directly from Appendix B for the threshold involved, and $I_{T \text{ diffuse}}$ is the difference between $I_T$ and $I_{T \text{ direct}}$. Threshold determination will probably not be much problem because these collectors have such low loss characteristics that a low and almost constant threshold will be characteristic of a given collector.

## REFLECTORS

When the target for a concentrating collector is an ordinary flat plate collector *array* and the reflectors are flat planes adjoining the edges of the collector array the assembly is regarded as a reflector-augmented, flat plate collector array, and the performance is characterized by an enhancement factor $\Phi_r$, which may apply to either the instantaneous or the average daily radiation level on the collectors. The enhancement factor is small unless the reflectors are specular and oriented to direct the reflected beam to the surface of the flat plate collector.

The mathematical description of an infinitely large specular reflector is not too difficult provided that both the collector and reflector are illuminated by the sun and do not shade one another. When the collectors are of reasonable dimensions the analysis becomes complex because the proportion of the reflected light that impinges on the collector changes as the sun's altitude and azimuth change. McDaniels et al. (8) summarize the applicable equations and have explored the relevant variables, summarizing their results in terms of instantaneous (8) and average (9) enhancement factors.

### Reflector Orientation

With an ordinary solar collector the optimal elevation angle is perpendicular to the radiation (Fig. 6-13a). If a reflector is used, the pair should be oriented so that the array is perpendicular to the incident radiation (assuming for the moment that the reflector reflectivity is unity), as in Fig. 6-13b. When the angle between collector and reflector is acute, all intercepted rays reach the absorber but, since the width of the array is unnecessarily small, less total energy is intercepted than if an obtuse angle is used (Fig. 6-13c). The obtuse angle, however, can permit much of the reflected energy to miss the collector at certain sun angles. The optimal angle between them is 90°, with the array oriented perpendicular to the incident sunlight (Fig. 6-13d).

The collector-reflector *system* should be inclined to give the greatest enhancement in November, December, January, and February when the incident energy

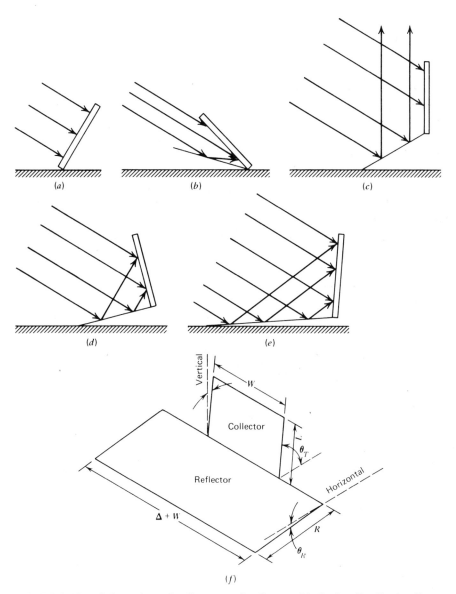

**FIGURE 6-13.** Orientation of collector and reflector. (*a*) Optimally tilted collector with no reflector. (*b*) Collector and reflector separated at an acute angle. The collector intercepts less energy than (*a*), but all the reflected energy strikes the collector. (*c*) Collector and reflector separated by an obtuse angle. The combination intercepts more radiation than (*a*), but not all the reflected radiation strikes the collector. (*d*) Collector and reflector separated at 90°. This is the optimal orientation for equal sized reflector and collector. (*e*) Optimal orientation for a collector and a reflector twice its size. For latitude 46 °N (Refs. 8, 9) the collector is tilted at about 85°, and the reflector is tilted downward by about 5°. (*f*) The optimal array of (*e*) showing nomenclature for Figs. 6-14 to 6-17.

is lowest. The optimum inclination is ordinarily about latitude plus 15°, but both the optimum orientation and the optimum angles between the collectors vary somewhat with reflector size and reflectivity. For the northern United States it is nearly optimum to use a near-vertical collector (inclination angle 80–90°) with an oversized reflector dropped slightly below the horizontal (inclined downward 0–10°). This combination is interesting and architecturally practical (Fig. 6-13*e* and *f*).

For analysis, it is convenient to characterize the effect of a reflector by the *enhancement* factor. The direct radiation on an *optimally oriented* flat plate collector is multiplied by this factor to predict the direct radiation on the actual collector as installed with a reflector. Figure 6-14 shows the variation with col-

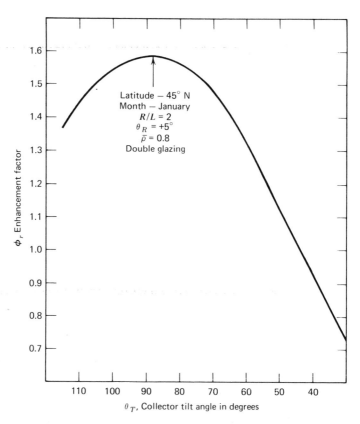

Latitude — 45° N
Month — January
$R/L = 2$
$\theta_R = +5°$
$\bar{\rho} = 0.8$
Double glazing

$\phi_r$, Enhancement factor

$\theta_T$, Collector tilt angle in degrees

**FIGURE 6-14.** Effect of collector angle on the otherwise optimal collector-reflector array. (Figures 6-14 to 6-17 reprinted by permission *Solar Energy 20*, (415-417) [Ref. 9], Copyright 1978, Pergamon Press.)

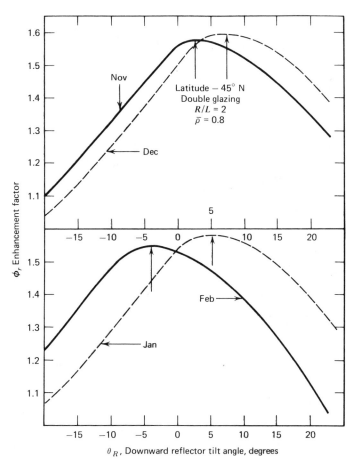

**FIGURE 6-15.** Effect of reflector angle on the otherwise optimal collector-reflector array. Positive values of reflector tilt are below horizontal.

found by numerical integration over a clear January day. This plot applies at 45°N latitude for a reflector dropped 5° below the horizontal, having a reflector dimension $R$ perpendicular to the common edge twice that of the collector height $L$. The optimum collector tilt is seen to be 88°.*

Figure 6-15 shows how the enhancement factor varies as the reflector angle is changed while the collector is held at the optimum tilt of about 90°. The optimum tilts in November, December, and January are close to +5° (downward), and by February the optimum tilt is about −5° (upward). These results are all for a

* Figures 6-14 to 6-17 include in the values of $\Phi_r$ the effect of incident angle on the reflectivity of two collector surfaces for both the direct and reflected beam.

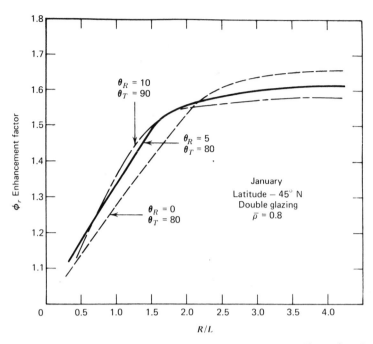

**FIGURE 6-16.** Effect of the (usually) shorter reflector dimension $L$ on the otherwise optimal collector-reflector array.

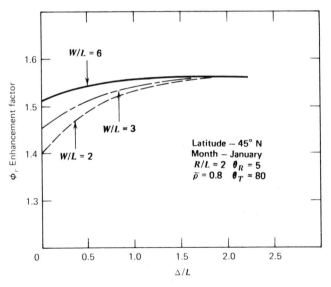

**FIGURE 6-17.** Effect of increasing the long reflector dimension on the otherwise optimal collector-reflector array. $\Delta$ is the increase beyond the long dimension $W$ of the collector array.

45°N latitude. For lower latitudes (down to as low as 35°N) the optimal values for reflector and collector tilt are algebraically less by the (positive) deviation in the actual latitude from 45°N.

The size of the reflector is of importance to the architect, even though the cost is minimal because of the roof area utilized. Figure 6-16 shows that for angles of 5 and 10° below the horizontal (at 45°N) in January, the reflector depth $R$ should be about twice as large as the collector dimension $L$ in the direction perpendicular to their common edge. For a horizontal reflector inclination the reflector should be deeper, making a strong case for tilting the reflector downward.

The reflector should also be somewhat longer than the collector's other dimension $W$, parallel to the common edge, so that enhancement will continue at times well away from solar noon. The enhancement factor varies with the ratio $W/L$ of the collector-reflector array. Figure 6-17 shows the enhancement as a function of the ratio of added reflector width $\Delta$ to the dimension $W$ (which is parallel to the common edge) for several $W/L$ values. Even with no added width the enhancement factor suffers little as long as $W/L > 6$.

## Performance Analysis

To analyze the performance of a collector-reflector system, the appropriate enhancement factor $\Phi_r$ should be chosen from Figs. 6-14 to 6-17 or synthesized rationally from a combination of them with reference to the actual situation. The factor is then applied to the direct component only by using a modified form of eq. 6-9 to find the total radiation collected:

$$q_T = \overline{\tau\alpha}(\Phi_r I_{T\ \text{direct}} + I_{T\ \text{diffuse}}) - U_L(\overline{T}_c - \overline{T}_a)t_T \qquad (6\text{-}10)$$

Reflector enhancement $\Phi_r$ is defined relative to latitude plus 15° inclination radiation information.

Proper values of $\Phi_r$ (8) can be used in a modification of eq. 4-11 to find the instantaneous rate of energy collection:

$$q = \overline{\tau\alpha}(\Phi_r I_{\text{direct}} + I_{\text{diffuse}}) - U_L(T_c - T_a) \qquad (6\text{-}11)$$

However, Appendix B has retained only values of total *global* radiation at threshold so that trial calculations with eq. 6-10 must be used to thus determine the operating threshold implicitly, as in the following example.

### EXAMPLE 6-1

Consider once again the collector of Fig. 4-13, which was evaluated in Examples 4-3 and 4-4 for December at Hartford when inclined at 50° to the horizontal. What monthly performance can be expected if it is replaced by an optimally oriented near-vertical collector of the same dimensions and a near-horizontal reflector twice the size of the collector? The collector temperature is 48.9 °C (120 °F). The array length is such that the $W/L$ ratio is 6 (see Fig. 6-13f).

Hartford is located at 42°N. The optimal orientation as determined from Figs. 6-14 and 6-15, respectively, when $R/L = 2$, is 88° collector tilt, with a 5° downward tilt of the reflector. For 42°N latitude, the corresponding angles are 85 and 2° downward. Figure 6-17 then shows the enhancement factor $\Phi_r$ to be 1.52 if $W/L = 6$. Since the optimum inclination angle for a collector without reflector at Hartford is $42° + 15° = 57°$, data for a 60° inclination angle will be used in eq. 6-10.

The proper solar radiation grouping will be determined by trial using eq. 6-10. For the $189 + $ W/m² grouping, total and direct radiation totals are 257.5 and 213.5 W/m², respectively. Diffuse radiation is their difference times a correction factor $R$ to elevate the diffuse radiation from 60° to the 85° of the actual configuration.

From eq. 3-24, ignoring ground reflection,

$$I_{T \text{ diffuse}} = H_{T \text{ diffuse}} \left[ \frac{1 + \cos \beta}{2} \right]$$

and therefore

$$R = \frac{\left( \dfrac{1 + \cos 85°}{2} \right)}{\left( \dfrac{1 + \cos 60°}{2} \right)} = 0.7248$$

and

$$I_{T \text{ diffuse}} = (257.5 - 213.5) \times 0.7248 = 31.89 \text{ MJ/m}^2$$

from eq. 6-10, using $\tau\alpha = 0.8$, $U_L = 5.22$ W/m² and $T_C = 48.89$ °C:

$$q_T = 0.8(1.52 \times 213.5 + 31.89) - 5.22[48.89 - (-0.8)] \times 0.4799$$
$$q_T = 285.13 - 124.48 = 161 \text{ MJ/m}^2$$

For the $284 + $ grouping

$$I_{T \text{ diffuse}} = (232.1 - 199.3) \times 0.7248 = 23.77 \text{ MJ/m}^2$$
$$q_T = 0.8(1.52 \times 199.3 + 23.77) - 5.22[48.89 - (-0.9)] \times 0.3704$$
$$q_T = 261.36 - 96.27 = 165 \text{ MJ/m}^2$$

which further trial will show to be the maximum collection rate.


Since Example 6-1 predicts twice as much energy collected as in Example 4-4, large reflectors should receive due consideration, particularly where winter sunshine is sparse. Even though the enhancement factor was only 1.52, the overall performance was doubled. This is because there are practically no additional losses from the collector when the reflector is added—only those due to a change in the incident angle—so that most of the added radiation appears as increased collector output.

## PROBLEMS

**1.** Plot maximum concentration ratio versus acceptance angle over a realistic range for
   (a) Parabolic reflector (two-dimensional).
   (b) Ideal reflector (two-dimensional).
   (c) For the ideal reflector, plot the percentage of the available diffuse radiation available versus the concentration ratio.

**2.** An optimally oriented reflector is added to the collector described in problem 4-4. Reestimate the performance assuming that the reflector dimension $R$ is twice the collector dimension $L$, the array is four times as wide ($W$) as it is long ($L$), and that the collector width ($W$) is equal to that of the reflector.

**3.** Design a parabolic reflector having a concentration ratio of 6 for a 4-in.-wide, flat-plate receiver as in Fig. 6-4.

## REFERENCES

1. Ari Rabl, "Comparison of Solar Concentrators," *Solar Energy 18*, 93–111 (1976).
2. H. Hinterberger and R. Winston, *Rev. Sci. Instr. 37*, 1094 (1966).
3. V. K. Baronov and G. K. Melnikov, *Soviet Journal of Optical Technology 33*, 408 (1966).
4. M. H. Cobble, "Theoretical concentrations for solar furnaces," *Solar Energy 5*, 61 (1961).
5. R. E. Jones, "Optical properties of cylindrical elliptic concentrators, *Proceedings* American section of ISES, 1977, p. 36.21.
6. See technical reports written for ERDA by Kudret Selcuk, Jet Propulsion Laboratory.
7. M. M. Shapiro, "Non-focussing solar concentrators of easy manufacture," *Solar Energy 19*, 212–213 (1977).
8. D. K. McDaniels et al., "Enhanced Solar Energy Collection Using Reflector Solar Thermal Collector Combinations," *Solar Energy 17*, 277–283 (1975).
9. S. Baker, D. K. McDaniels, H. D. Kaehn, and D. H. Lowndes, "Time integrated calculation of the insolation collected by a reflector-collector system," *Solar Energy 20*, 415–417 (1978).

# Transfer of the Collected Heat

Heat is removed from active solar collectors by the circulation of an air or a liquid stream to and from the collectors. In either case, one or more heat exchangers will usually be necessary to supply heated room air and domestic hot water to the building served by the solar-heating system.

Although the heated air from an air collector can be used directly for heating, if heat is to be stored the sensible heat in the air is exchanged with that in a rock or pebble storage bed because such a bed has a far higher heat capacity in a given volume than does the air. Domestic water heating will obviously also require heat exchange when air collectors are used.

Liquid-cooled collectors do not generally need heat exchange to permit heat to be stored, but some means must be used to deal with the problem of the liquid freezing in the collector. Most often a nonfreezing liquid is used for cooling the collector, which then requires heat exchange to heat the bulk of the storage that is invariably water because of its low cost and excellent thermal properties. A *terminal* heat exchanger will also be needed to use the heat stored in the water to heat the air in the building. Still another will probably be installed to heat domestic hot water.

In this chapter the sizing procedures for these heat exchangers will be set forth and the effect of their performance on the overall solar-heating system evaluated.

## FREEZE PROTECTION

Most liquid-cooled collectors will suffer serious damage if water is allowed to freeze in fluid passages. Expanding water can exert tremendous force, invariably causing leaks in metal piping. Plastic collectors, however, are flexible and can usually withstand freezing. Because of this their use is occasionally attempted in spite of their other shortcomings. Here of course we deal with those collectors that are in need of positive protection from freezing.

The simplest way to avoid the damage that results from a collector freeze-up is to use auxiliary heat to keep the collector a few degrees above freezing. This can be done with strip electrical heaters or by briefly circulating the stored water to the collector whenever the temperature drops to 2 °C (36 °F), when nightime radiation to the sky presents a danger of freezing. But, except in the mildest of climates, such a solution is not only inelegant, but expensive in energy costs. Furthermore, it is subject to electrical power failure. A more desirable treatment of the problem is to drain the fluid from the collector when it is not in use. This is a practical procedure but it is not without its problems. Design of a drain-down system can be tricky, since the water will sometimes vapor-lock and remain in lines as large as 15 mm (0.5 in.) in diameter, even if there is proper venting to allow air to replace the drained water. The vents themselves can freeze while the collectors are still warm. Collectors with large tubes tend to be less efficient because of low fluid velocities unless extra pumping power is used. The collectors are more prone to corrosion because of the continuous exposure to air, making aluminum or steel out of the question in the fluid circuit. Circulation pumps will be more expensive if the collectors are roof mounted, because the inexpensive, single-stage centrifugal pumps commonly available for hydronic heating systems only produce a few feet of head, not enough to raise fluid to an empty collector, although easily sufficient to circulate fluid to and from a full one.

A more trouble-free method for freeze protection is the use of an antifreeze solution in the collector with subsequent transfer of the collected heat to water for storage and use. Such systems are in common use and will continue in popularity until drain-down systems demonstrate high reliability and good performance over a period of years.

## Choice of Fluids

Water is a nearly ideal heat transfer fluid, being low in cost, nonflammable, nontoxic, and noncorrosive in plastics and some metals. Its heat transfer properties are good, with a high density, high specific heat, low viscosity, and a reasonably high boiling point. The problem, of course, is the freezing point.

Some alternate fluids having a low freezing point are listed in Table 7-1. Water itself can be made acceptable without affecting most of the other properties by adding inorganic salts. Unfortunately, most (if not all) of these produce critical corrosion problems in metal collectors and their use is seldom seriously considered. The traditional alcohol and glycol automobile antifreezes are another approach. They are used in high concentrations—often 50% by weight—and perform well. Methyl alcohol, however, is poisonous and lowers the boiling point too much. Ethyl alcohol would be a good choice if the system could be pressurized slightly, but of course it is an intoxicant under close government control and poisonous when denatured. Ethylene and propylene glycol are good choices and indeed are

**TABLE 7-1.** A Qualitative Evaluation of Heat Exchange Fluids for Solar Collectors. See Table 5-1 for Numerical Values of Important Thermal Properties

| | High Density[a] | High Specific Heat | Low Viscosity | Non-corrosive | Non-flammable | Toxicity | Non-freezing | Low Cost | High Boiling Point | Water— Miscibility |
|---|---|---|---|---|---|---|---|---|---|---|
| Salt water | + | + | + | × | + | + | + | + | − | + |
| Ethyl alcohol and water | − | − | + | + | + | + | + | − | × | + |
| Ethylene glycol and water | + | + | + | +[b] | + | × | + | − | + | + |
| Propylene glycol and water | + | + | − | +[b] | + | + | + | − | + | + |
| Kerosene | × | × | + | + | × | − | + | + | + | × |
| Mineral oil | − | × | × | + | × | + | + | − | + | × |
| Silicone oil | + | × | × | + | − | + | + | × | + | × |
| Air | × | × | + | + | + | + | + | + | + | NA |

[a] The symbol + = good, − = fair, × = poor.
[b] In copper or steel only, with proper inhibitors.

in general use; a 50% by weight mixture approaches water in most properties. Petroleum oils are also in use but, when high enough in molecular weight not to be an immediate fire hazard, they are viscous and will cause a loss of efficiency in collectors operated below perhaps 65 °C (150 °F). Silicone oils are somewhat better because they do not ignite as easily, but they are still quite viscous and very expensive. They are the best current choice for use in aluminum collectors.

For practical use in ordinary solar systems, ethylene glycol is the most popular antifreeze even though it is mildly poisonous. If potable water is to be heated, fear of leakage often causes its less toxic chemical relative, propylene glycol, to be specified. If the collectors are to be steel or copper, these liquids can perform well when proper corrosion inhibitors are used if proper care is taken to monitor them for acidity and to discard them periodically just as in automobile radiator usage, especially if they are allowed to stagnate above 300 °F in idle collectors. Adequate corrosion prevention to permit use of glycols in aluminum collectors is not yet available, although proper inhibitors will probably soon be formulated. In the meantime the use of aluminum collectors can be recommended only for systems in which silicone or petroleum oils are used for heat transfer. The less flammable silicone fluids are preferred for domestic applications; a proper design will minimize system holdup and make the quantity used small enough to be affordable. Special attention is needed to assure the highest possible heat transfer and, since the specific heat is low, increased pumping rates may be required.*

Air collectors, of course, do not have a freezing problem. But the penalty in

---

* This is an advantage on shutdown, however, because less heat is left in the collector to be lost overnight and replaced before startup the next day.

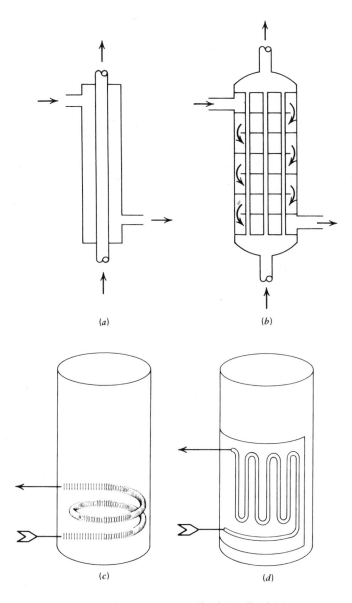

**FIGURE 7-1.** Some common liquid-to-liquid heat exchanger designs. (*a*) Concentric tube. (*b*) Shell and tube. (*c*) Within tank exchanger. (*d*) Traced tank heat exchanger.

system performance due to the poor heat transfer properties of air is about the same as that due to a heat exchanger, so that their advantage does not lie in performance or cost but rather in the convenience and simplicity of using inexhaustible and nontoxic air as a coolant.

### Heat Exchanger Design

There are many kinds of liquid-to-liquid heat exchangers. Some of the more likely designs for use in solar work are shown in Fig. 7-1. The concentric tube design, once reserved for textbook and laboratory, has been successful commercially for small systems when coiled for use. For larger systems, the shell-and-tube design is predominant. To reduce heat exchanger costs for hot water systems, heat exchangers are often built into the tank wall or made from coiled tubing placed inside the tank. Their effectiveness is often low because heat exchange on the water side must rely on natural convection for liquid circulation.

## HEAT EXCHANGER ANALYSIS

The most practical approach to heat exchanger design for solar work is the effectiveness concept. This is generally presented in a graphical form following the approach of Kays and London (1). With the advent of digital computers and programmable calculators the rather complex transcendental functions involved have become as easy to work with as the graphs, and they will be used in the following analysis (2).

The basic goal of heat exchanger analysis is to extend a knowledge of local heat transfer (where $Q = UA \, \Delta T$) to situations where the $\Delta T$ varies along the flow path. This is the case in a heat exchanger because each stream is either absorbing or losing heat. Mathematical analysis will show that the heat transfer relationship need not involve the unknown temperatures along each stream, but can instead be expressed in terms of the two *inlet* temperatures plus the flows and known overall heat transfer properties.

With a solar heat exchanger the inlet temperatures are not often known because they vary with weather patterns. Further analysis then permits evaluation of the performance of the exchanger in terms of the *storage* temperature, fluid flows, and heat transfer properties. This analysis takes into account the decrease in overall system performance that installation of the heat exchanger causes by elevating the collector temperature, leading to increased system heat losses. The final result is a *heat exchanger factor* by which the system performance is degraded by installation of a heat exchanger.

### Effectiveness

Consider the countercurrent heat exchanger of Fig. 7-2 within which two streams exchange heat. One of the streams is equal to or larger than the other in its flow

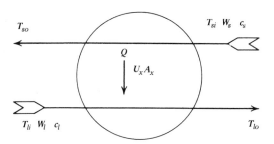

**FIGURE 7-2.** Nomenclature for the analysis of a countercurrent heat exchanger.

capacity rate $Wc$, the mass flow rate of the stream times its specific heat. This stream is known as the larger stream, and the other is the smaller. Flows are characterized by $W_l c_l$ and $W_s c_s$, respectively. The inlet temperatures to the heat exchanger $T_{li}$ and $T_{si}$ are considered known, and the corresponding outlet temperatures are $T_{lo}$ and $T_{so}$.

The maximum quantity of heat that could possibly be transferred is that bringing the small stream to the temperature of the incoming large stream:

$$Q_{max} = W_s c_s (T_{si} - T_{li}) \tag{7-1}$$

The heat exchange actually achieved can be expressed as a proportion $\varepsilon$ of this maximum, where $\varepsilon$ is called the *effectiveness* of the heat exchanger.

$$Q = \varepsilon W_s c_s (T_{si} - T_{li}) \tag{7-2}$$

This approach is useful because the effectiveness is independent of the temperature level, depending only on the exchanger heat transfer properties and the fluid flows. Once these are fixed, the effectiveness can be used to determine the performance of the heat exchanger in terms of the inlet temperatures.

Before proceeding further, the heat transfer rates and fluid flows will be redefined, normalized for a unit collector area. This will simplify later analyses and reduce the realistic range of the variables involved.

Dividing eq. 7-2 by the collector area, $A_c$, and designating $q = Q/A_c$ and $w = W/A_c$,

$$q = \varepsilon w_s c_s (T_{si} - T_{li}) \tag{7-3}$$

Equation 2-10 can also be rewritten in terms of area-normalized quantities to give

$$q = wc \, \Delta T \tag{7-4}$$

Applied directly to each stream in Fig. 7-2, this gives

$$q = w_s c_s (T_{si} - T_{so}) \tag{7-5}$$

$$q = w_l c_l (T_{lo} - T_{li}) \tag{7-6}$$

The $q$'s are equal since the heat lost from one stream is the heat gained by the other. If the small stream is to be cooled, the quantity $q$ will be positive; if it is to be heated $q$ will be negative. Since this $q$ is the same as given in eq. 7-3; eqs. 7-3, 7-5, and 7-6 can be combined in several ways to eliminate $q$. This leads to measures of the required effectiveness and flow capacity ratio expressed as ratios of temperature differences required for proper performance.

$$\varepsilon = \frac{T_{si} - T_{so}}{T_{si} - T_{li}} \tag{7-7}$$

$$\frac{w_s c_s}{w_l c_l} = \frac{T_{lo} - T_{li}}{T_{si} - T_{so}} \tag{7-8}$$

Equation 7-7 defines effectiveness as the ratio of the temperature change actually achieved in the smaller stream to the most that could possibly be achieved. Similarly, eq. 7-8 fixes the ratio of the flow capacity rates $wc$ (or for that matter, $Wc$) as the ratio of the temperature changes occurring in the two streams.

The effectiveness can be predicted directly from the heat transfer properties of the heat exchanger and the flow capacity rates. For countercurrent flow, heat exchanger effectiveness is given by

$$\varepsilon = \frac{1 - e^{-B}}{\left(1 - \dfrac{w_s c_s}{w_l c_l} e^{-B}\right)} \tag{7-9}$$

where, if $N_{TU} = U_x a_x / w_s c_s$,

$$B = U_x a_x \left(\frac{1}{w_s c_s} - \frac{1}{w_l c_l}\right) = \left(1 - \frac{w_s c_s}{w_l c_l}\right) N_{TU} \tag{7-10}$$

The ratio $U_x a_x / w_s c_s$ (or $U_x A_x / W_s c_s$, which is interchangeable) is a dimensionless measure of the adequacy of the heat exchanger area sometimes called the *number of transfer units*, which we will abbreviate $N_{TU}$. For proper heat exchange, $N_{TU}$ will generally be in the range of 1 to 10.

Equations 7-9 and 7-10 are written using the normalized heat exchanger area $a_x = A_x / A_c$ and the area-normalized flow rates. Without other changes these and subsequent equations can be modified for absolute areas and flows by simultaneously replacing $a_x$ with $A_x$ and $w$'s with $W$, since the equations always involve ratios of these quantities.

Equations 7-9 and 7-10 can be solved explicitly for $N_{TU}$ for use when the heat exchanger area must be found that provides a particular effectiveness, giving

$$N_{TU} = \frac{U_x a_x}{w_s c_s} = \frac{\ln\left(\dfrac{1 - \varepsilon \dfrac{w_s c_s}{w_l c_l}}{1 - \varepsilon}\right)}{\left(1 - \dfrac{w_s c_s}{w_l c_l}\right)} \tag{7-11}$$

Very often the flow capacity rate will be equal on both sides of the heat exchanger. With $w_l c_l = w_s c_s$, the above relationships become indeterminate and instead the following equations apply:

$$\varepsilon = \frac{N_{TU}}{1 + N_{TU}} \tag{7-9a}$$

$$B = \frac{U_x a_x}{wc} \equiv N_{TU} \tag{7-10a}$$

or, if the effectiveness is known,

$$N_{TU} = \frac{U_x a_x}{wc} = \left(\frac{\varepsilon}{1 - \varepsilon}\right) \tag{7-11a}$$

When one flow is very much larger than the other,* the equations simplify considerably to

$$\varepsilon = 1 - \exp\left(-U_x a_x / wc\right) = 1 - e^{-N_{TU}} \tag{7-9b}$$

$$N_{TU} = \frac{U_x a_x}{wc} = \ln\frac{1}{1 - \varepsilon} \tag{7-11b}$$

where $wc$ is the flow capacity of the smaller stream.

The last two equations may also be used for situations in which temperatures on one side of the exchanger are constant, as with a coil in or wrapped around a tank.

An outline of the method for derivation of eq. 7-9 and similar expressions for other heat exchanger configurations is given in Eckart and Drake (2). Table 7-2 lists some of these effectiveness equations, most of which cannot be solved explicitly for $U_x a_x / wc$ as can the simpler countercurrent relationships. Only use of the countercurrent effectiveness expressions are demonstrated in this book, on the premise that the higher efficiencies needed in solar work usually warrant true countercurrent exchange.

Readers already familiar with the "log-mean temperature difference" approach to heat exchanger analysis should be assured that the effectiveness technique gives precisely the same solutions. The difference is that the effectiveness method is oriented to work from $UA$, flows, and inlet temperatures to predict performance, while the older technique works well only when temperatures and flows are used to predict the $UA$ value. The effectiveness method is therefore much better at dealing with prediction of performance of a given heat exchanger when the inlet flows and temperatures are known. It provides an important step toward predicting the performance of the solar collector/heat exchanger combination, permitting calculation of effectiveness wherever heat exchanger $U_x A_x$ and flow

* A 15:1 flow capacity ratio gives an error of 5%; 30:1, 2%.

**TABLE 7-2.** Thermal Effectiveness of Heat Exchangers with Various Flow Arrangements. (Reference 2.)

Parallel flow:

$$\varepsilon = \frac{1 - \exp\left[- (UA/W_s c_s)(1 + W_s c_s/W_l c_l)\right]}{1 + \dfrac{W_s c_s}{W_l c_l}}$$

Counterflow:

$$\varepsilon = \frac{1 - \exp\left[- (UA/W_s c_s)(1 - W_s c_s/W_l c_l)\right]}{1 - \dfrac{W_s c_s}{W_l c_l}\exp\left[- (UA/W_s c_s)(1 - W_s c_s/W_l c_l)\right]}$$

Crossflow:

$$\varepsilon = \frac{\dfrac{UA}{W_s c_s}}{\left[\dfrac{\dfrac{UA}{W_s c_s}}{1 - \exp\left(-\dfrac{UA}{W_s c_s}\right)} + \dfrac{\dfrac{UA}{W_l c_l}}{1 - \exp\left(-\dfrac{UA}{W_l c_l}\right)} - 1\right]}$$

Reverse flow exchanger:

$$\varepsilon = \frac{2}{\left\{1 + \dfrac{W_s c_s}{W_l c_l} + \dfrac{1 + \exp\left[-\dfrac{UA}{W_s c_s}\sqrt{1 + \dfrac{(W_s c_s)^2}{(W_l c_l)^2}}\right]}{1 - \exp\left[-\dfrac{UA}{W_s c_s}\sqrt{1 + \dfrac{(W_s c_s)^2}{(W_l c_l)^2}}\right]}\sqrt{1 + \dfrac{(W_s c_s)^2}{(W_l c_l)^2}}\right\}}$$

capacity rates are known. These factors are independent of other system characteristics.

## EXAMPLE 7-1

Thirty 1.22 m × 2.44 m (4 ft × 8 ft) collectors are cooled by a heat exchanger rated at $U_x A_x = 5033$ W/°C (9541.5 Btu/hr °F). The temperature leaving the collectors is 65.55 °C (150 °F), while the well-mixed storage has a temperature of 60 °C (140 °F). 0.504 kg/s (4000 lb/hr) of water circulates in the storage loop, and 0.3629 kg/s (2880 lb/hr) of 50% ethylene glycol with a specific heat of 3.52 kJ/kg·°C (0.84 Btu/lb °F) circulates in the collector loop. What are temperatures of the return to storage and to the collector?

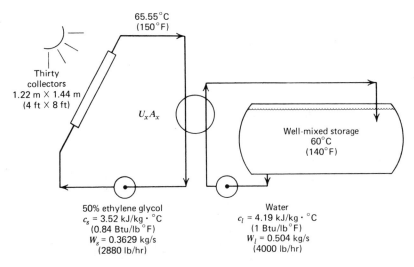

**FIGURE 7-3.** Collector, heat exchanger, and storage tank for Examples 7-1 to 7-4.

First calculate the basic normalized quantities.

$$A_c = 30 \times 1.22 \text{ m} \times 2.44 \text{ m} = 89.2 \text{ m}^2$$

$$w_s c_s = \frac{W_s c_s}{A_c} = \frac{0.3629 \text{ kg/s}}{89.2 \text{ m}^2} \times \frac{3.52 \text{ kJ}}{\text{kg} \cdot {}^\circ\text{C}} \times \frac{1000}{k} = 14.32 \text{ W/}{}^\circ\text{C} \cdot \text{m}^2$$

$$w_l c_l = \frac{W_l c_l}{A_c} = \frac{0.504 \text{ kg/s}}{89.2 \text{ m}^2} \times \frac{4.19 \text{ kJ}}{\text{kg} \cdot {}^\circ\text{C}} \times \frac{1000}{k} = 23.67 \text{ W/}{}^\circ\text{C} \cdot \text{m}^2$$

$$U_x a_x = \frac{5033 \text{ W/}{}^\circ\text{C}}{89.2 \text{ m}^2} = 56.42 \text{ W/}{}^\circ\text{C} \cdot \text{m}^2$$

Then calculate $\varepsilon$ using eqs. 7-9 and 7-10:

$$N_{TU} = \frac{U_x a_x}{w_s c_s} = \frac{56.42}{14.32} = 3.94$$

$$B = 3.940\left(1 - \frac{14.32}{23.67}\right) = 1.556$$

$$\varepsilon = \frac{1 - e^{-1.556}}{\left(1 - \dfrac{14.32}{23.67} e^{-1.556}\right)} = 0.9045$$

The heat transfer rate is given by eq. 7-3:

$$q = 0.9045 \times \frac{14.32 \text{ W}}{{}^\circ\text{C} \cdot \text{m}^2} (65.55 - 60) \, {}^\circ\text{C} = 71.89 \text{ W/m}^2$$

and the outlet temperatures are, by rearrangement of eqs. 7-5 and 7-6,

$$T_{so} = T_{si} - \frac{q}{w_s c_s} = 65.55\ °C - \frac{71.89\ W/m^2}{14.32\ W/°C \cdot m^2} = 60.53\ °C$$

$$T_{lo} = T_{li} + \frac{q}{w_l c_l} = 60\ °C + \frac{71.89}{23.67}\ °C = 63.04\ °C$$

which are temperatures of the return to the collector and return to storage, respectively.

In Engineering units,

$$A_c = 30 \times 4\ ft \times 8\ ft = 960\ ft^2$$

$$w_s c_s = \frac{W_s c_s}{A_c} = \frac{2880\ lb/hr}{960\ ft^2} \times \frac{0.84\ Btu}{lb\ °F} = 2.52\ Btu/hr\ °F\ ft^2$$

$$w_l c_l = \frac{W_l c_l}{A_c} = \frac{4000\ lb/hr}{960\ ft^2} \times \frac{1.0\ Btu}{lb\ °F} = 4.167\ Btu/hr\ °F\ ft^2$$

$$U_x a_x = \frac{9541.5\ Btu/hr\ °F}{960\ ft^2} = 9.939\ Btu/hr\ °F\ ft^2$$

$$\frac{U_x a_x}{w_s c_s} = \frac{9.939}{2.52} = 3.9440$$

Using eqs. 7-9 and 7-10,

$$B = 3.9440\left(1 - \frac{2.52}{4.167}\right) = 1.5589$$

$$\varepsilon = \frac{1 - e^{-1.5589}}{\left(1 - \dfrac{2.52}{4.167}\ e^{-1.5589}\right)} = 0.9047$$

by eq. 7-3,

$$q = 0.9047 \times \frac{2.52\ Btu}{hr\ °F\ ft^2}\ (150 - 140)\ °F = 22.80\ Btu/hr\ ft^2$$

and the outlet temperatures are, by rearrangement of eqs. 7-5 and 7-6,

$$T_{so} = T_{si} - \frac{q}{w_s c_s} = 150\ °F - \frac{22.80\ \dfrac{Btu}{hr\ ft^2}}{2.52\ \dfrac{Btu}{hr\ °F\ ft^2}} = 140.95\ °F$$

$$T_{lo} = T_{li} + \frac{q}{w_l c_l} = 140\ °F + \frac{22.80}{4.167}\ °F = 145.47\ °F$$

The above example assumed that the system was at a steady state, implicitly requiring that the solar flux be exactly that needed to maintain the finally calculated heat flux $q$. In the next section it is shown how the proper value for the solar heat collected and transferred can be determined in a single step using the usual collector equations plus a simple heat exchanger factor to correct for the effect of the heat exchanger on the system. In the new equations a collector temperature is no longer needed, but instead the more readily available storage temperature is used.

## THE HEAT EXCHANGER FACTOR

The solar collector, heat exchanger, and well-mixed storage tank shown in Fig. 7-4 will now be considered as an interacting system in order to derive a new collector performance equation that directly incorporates the effect of the heat exchanger (3).

Nomenclature used for heat exchanger analysis, as well as for solar collector performance, has been shown redundantly in Fig. 7-4. The basic collector performance equation (5-21) can be rewritten

$$q = F_i \tau \alpha I - F_i U_L (T_i - T_a) \tag{7-12}$$

If the collector flow capacity rate is smaller than that from storage the outlet temperature from the collector is

$$T_o = T_i + \frac{q}{w_s c_s}$$

The heat collected can also be expressed as

$$q = w_s c_s (T_o - T_i)$$

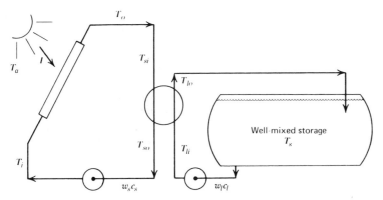

**FIGURE 7-4.** Nomenclature for the derivation of the heat exchange factor. Note that heat exchanger and collector terms are used redundantly.

which when solved for $T_i$ can be substituted into eq. 7-12 to give

$$q = F_i \tau \alpha I - F_i U_L \left( T_o - \frac{q}{w_s c_s} - T_a \right)$$

This can be solved for $q$ to give

$$q = \frac{F_i \tau \alpha I - F_i U_L (T_o - T_a)}{1 - \dfrac{F_i U_L}{w_s c_s}} \tag{7-13}$$

Equation 7-3 (which applies to the heat exchanger) is then solved for $T_{si}$ (which is identically the same temperature as $T_o$ in eq. 7-13) giving

$$T_o = T_{si} = T_{H} + \frac{q}{\varepsilon w_s c_s} \tag{7-14}$$

Substituting into eq. 7-13,

$$q = \frac{F_i \tau \alpha I - F_i U_L \left( T_{H} + \dfrac{q}{\varepsilon w_s c_s} - T_a \right)}{1 - \dfrac{F_i U_L}{w_s c_s}}$$

which can be solved for $q$ to give

$$q = \frac{F_i \tau \alpha I - F_i U_L (T_{H} - T_a)}{1 + \dfrac{F_i U_L}{w_s c_s} \left( \dfrac{1}{\varepsilon} - 1 \right)}$$

and, since $T_{H}$ is the same as the storage temperature $T_s$, the equation will be rewritten in terms of the storage temperature.

The performance of the system in Fig. 7-4 can therefore be described by

$$q = \frac{F_i \tau \alpha I - F_i U_L (T_s - T_a)}{1 + \dfrac{F_i U_L}{w_s c_s} \left( \dfrac{1}{\varepsilon} - 1 \right)} \tag{7-15}$$

In this equation the heat collected (and transferred) is given in terms of the storage temperature and effectiveness in combination with the usual collector and weather parameters. The relationship does much more, however. When compared with eq. 5-21 or 7-12, the denominator is the only new term, and that term is a *constant* since it contains none of the time-varying parameters. This constant is called the de Winter heat exchanger factor (3):

$$F_x = \frac{1}{1 + \dfrac{F_i U_L}{wc} \left( \dfrac{1}{\varepsilon} - 1 \right)} \tag{7-16}$$

where the subscripts have been dropped from the collector flow capacity rate $w_s c_s$ to be consistent with the nomenclature of Chapter 5.

For the above derivation it was assumed that the flow capacity rate from storage was larger than that circulating in the collector loop. A more general derivation can be made (4), where if $wc$ is the collector flow capacity and $w_s c_s$ is the smaller of the two rates, then

$$F_x = \frac{1}{1 + \dfrac{F_i U_L}{wc}\left(\dfrac{wc}{\varepsilon w_s c_s} - 1\right)} \qquad (7\text{-}16a)$$

This equation reduces to eq. 7-16 if the small flow is in the collector loop. The same result can be obtained with eq. 7-16 as written provided that $\varepsilon$ is always calculated using $w_s c_s$ for the *collector* flow (eqs. 7-9 and 7-10). The intermediate $\varepsilon$ value will not be according to previous definitions, however.

When the flow capacity rates are equal then eqs. 7-9a and 7-10a can be combined to give a simple expression for $\varepsilon$ that can be substituted into eq. 7-16 or 7-16a to give (3)

$$F_x = \frac{1}{\dfrac{F_i U_L}{U_x a_x} + 1} = \frac{1}{\dfrac{F_i U_L A_c}{U_x A_x} + 1} \qquad (7\text{-}16b)$$

in which case the heat exchanger factor can thus be calculated directly from $U_x a_x$ and collector characteristics.

For convenience in notation, eq. 7-15 can be rewritten as

$$q = F_x[F_i \tau \alpha I - F_i U_L(T_s - T_a)] \qquad (7\text{-}17)$$

and, if the $F_i$ and $F_x$ factors are combined into a single factor $F_i'$, where

$$F_i' = F_x F_i \qquad (7\text{-}18)$$

then the results can be stated even more simply as

$$q = F_i'[\tau \alpha I - U_L(T_s - T_a)] \qquad (7\text{-}19)$$

Equation 7-19 is the preferred form to record incorporation of the heat exchanger factor. In actual use and for a number of derivations, however, the factor $F_i'$ will be expanded into its component parts and applied in the form of eq. 7-17 because the factors $F_i \tau \alpha$ and $F_i U_L$ are directly available input from an $\eta_i$ efficiency curve.

In this section the heat exchanger factor has been derived, and formulas given for its application to equal and unequal flow capacity rates in collector and storage loops. This factor depends only on the system parameters and the effectiveness of the heat exchanger, and is independent of weather characteristics and system temperatures, which change during system operation.

**EXAMPLE 7-2**

Find the heat exchanger factor for the situation in Example 7-1 if $F_iU_L = 5.069$ W/°C·m² (0.8927 Btu/hr °F ft²).

Using eq. 7-16, and noting that $\varepsilon = 0.9047$ and $wc = 14.32$ W/°C·m² (2.52 Btu/hr °F ft²),

$$F_x = \frac{1}{1 + \dfrac{5.069}{14.32}\left(\dfrac{1}{0.9047} - 1\right)} = 0.9640$$

In Engineering units, with $\varepsilon = 0.9047$ and $wc = 2.52$ Btu/hr °F ft²,

$$F_x = \frac{1}{1 + \dfrac{0.8927}{2.52}\left(\dfrac{1}{0.9047} - 1\right)} = 0.9640$$

## Economical Heat Exchanger Area

The simplicity of the heat exchanger factor and its application is an important contribution to keeping the performance equations as simple as possible when analyzing the generally complex solar-operating systems. The relationship makes it clear that the effect of a given heat exchanger on the system is to reduce the performance of the system by a constant factor. This will have the direct consequence of increasing the required collector area in an exact inverse proportion to $F_x$ if an overall performance specification is to be met. Since the factor is constant as weather and temperature change, its value is under the control of the system designer when the heat exchanger is sized and the flows set.

The heat exchanger factor will typically have a value of about 0.95 when an economic evaluation of the cost of collector area versus the cost of the heat exchanger is made (3). Preliminary sizing can be made from such information if eq. 7-16 is solved for $\varepsilon$, giving

$$\varepsilon = \frac{1}{\left(\dfrac{1}{F_x} - 1\right)\dfrac{wc}{F_iU_L} + 1^*} \tag{7-20}$$

which can then be used in eq. 7-11 or 7-11a to determine $U_xa_x$.

For equal flow capacity rates an analytical expression can be derived to determine the optimum heat exchanger area (3). If the cost of the collector per unit collector area is $C_c$ and the cost of the heat exchanger is $C_x$ per unit exchanger area, then the *total* system cost is

$$C = \left(C_c + \frac{A_x}{A_c}C_x\right)\frac{A_0}{F_x}$$

---

* Replace this numeral 1 with $w_sc_s/wc$ if the storage flow is the smaller. This is unnecessary when the calculated $\varepsilon$ is to be used in eq. 7-11 provided that the *collector* flow is consistently substituted for $w_sc_s$ and the storage flow substituted for $w_ic_i$.

where $A_0$ is the area required to give the desired performance with perfect heat exchange ($F_x = 1$) and the factor $A_0/F_x$ is thus the area of a system having the same performance but another value of the $A_x/A_c$ ratio. If the minimum of this expression is taken with respect to the heat exchanger area (with $F_x$ given by eq. 7-16b) an optimum (minimum cost) $A_x/A_c$ ratio can be found as

$$\frac{A_x}{A_c} = a_x = \sqrt{\frac{F_i U_L C_c}{U_x C_x}} \qquad (7\text{-}21a)$$

or this can be expressed in terms of the normalized heat exchanger parameters

$$U_x a_x = \sqrt{\frac{F_i U_L C_c}{C_{UxAx}}} \qquad (7\text{-}21b)$$

where $C_{UxAx} = C_x/U_x = $ (cost of $Hx$)$/U_x A_x$; that is, the cost of the heat exchanger in terms of its often specified $UA$ parameter.

Equations 7-21 are very useful because they give area ratios that are near optimum even when the flows (expressed as $wc$ or $Wc$) differ by factors of up to about 3. Errors will always be on the safe side of the optimum (i.e., giving a larger than optimum heat exchanger) as long as the collector flow is the smaller, thereby affording a rapid method for heat exchanger sizing in practical systems. Proper use is demonstrated in Example 11-10.

The economical heat exchanger area has now been shown to be independent of all system parameters except $F_i U_L$ and the normalized collector and heat exchanger costs. Therefore, the heat exchanger size can be optimized *before* optimization of the collector area and storage tank size.

**EXAMPLE 7-3**

Determine the proper $U_x A_x$ value to make the heat exchanger factor 0.95 in Examples 7-1 and 7-2.

Using eq. 7-20, since the smaller flow is in the collector loop and therefore $w_s c_s/wc = 1$,

$$\varepsilon = \frac{1}{\left(\dfrac{1}{0.95} - 1\right)\dfrac{14.32}{5.069} + 1} = 0.8706$$

From eq. 7-11,

$$U_x a_x = 14.32 \times \frac{\ln\left(\dfrac{1 - 0.8706 \times \dfrac{14.32}{23.67}}{1 - 0.8706}\right)}{\left(1 - \dfrac{14.32}{23.67}\right)} = 47.00 \text{ W/}^\circ\text{C} \cdot \text{m}^2$$

$$U_x A_x = U_x a_x A_c = \frac{47.00 \text{ W}}{^\circ\text{C} \cdot \text{m}^2} \times 89.2 \text{ m}^2 = 4193 \text{ W/}^\circ\text{C}$$

In Engineering units,

$$\varepsilon = \frac{1}{\left(\dfrac{1}{0.95} - 1\right)\dfrac{2.52}{0.8927} + 1} = 0.8706$$

From eq. 7-11

$$U_xA_x = 2.52 \times \frac{\ln\left(\dfrac{1 - 0.8706\dfrac{2.52}{4.167}}{1 - 0.8706}\right)}{\left(1 - \dfrac{2.52}{4.167}\right)} = 8.273 \text{ Btu/hr } °\text{F ft}^2$$

$$U_xA_x = U_xa_xA_c = \frac{8.273 \text{ Btu}}{\text{hr }°\text{F ft}^2} \times 960 \text{ ft}^2 = 7942 \text{ Btu/hr }°\text{F}$$

### EXAMPLE 7-4

Find the proper size heat exchanger for $F_x = 0.95$ if the storage flow is very large. Other conditions from Example 7-3 apply.

From eq. 7-20, $\varepsilon = 0.8706$ as before. From eq. 7-11b,

$$N_{TU} = \frac{U_xa_x}{wc} = \ln\frac{1}{1 - \varepsilon} = \ln\frac{1}{1 - 0.8706} = 2.0448$$

$$U_xa_x = 14.32 \frac{\text{W}}{°\text{C}\cdot\text{m}^2} \times 2.0448 = 29.28 \text{ W/}°\text{C}\cdot\text{m}^2$$

$$U_xA_x = 29.28 \times 89.2 = 2612 \text{ W/}°\text{C}$$

In Engineering units, from eq. 7-20, $\varepsilon = 0.8706$ as before. From eq. 7-11b,

$$\frac{U_xa_x}{wc} = \ln\frac{1}{1 - \varepsilon} = \ln\frac{1}{1 - 0.8706} = 2.0448$$

$$U_xa_x = 2.52 \frac{\text{Btu}}{\text{hr }°\text{F ft}^2} \times 2.0448 = 5.150 \text{ Btu/hr }°\text{F ft}^2$$

$$U_xA_x = 5.150 \times 960 = 4947 \text{ Btu/hr }°\text{F}$$

The decrease in the required area (expressed as $U_xA_x$) relative to Example 7-3 is due to the large stream flowing so fast that its temperature is always at $T_{\text{inlet}}$. This situation can be approached when the heat exchanger is located near the storage tank so that the line losses are low.

### Effect of Flow Rates on the Heat Exchanger Factor

When the ratio of the flows is held constant, increasing the flow capacity rates in both collector and storage loops *decreases* $F_x$ if the collector flow is smaller, *increases* it when it is larger, and leaves $F_x$ *unaffected* when they are equal (eqs.

7-9 and 7-16a). Change in the heat transfer factor (eq. 5-22), however, assures that the overall system performance (indicated by $F_i'$) properly rises in each of these situations. This is shown concisely by combining eq. 5-26 with eq. 5-22 to express $F_i$ in terms of $F_m$ and the *collector* flow capacity rate ($wc$ or $Wc$) and then eliminating $F_i$ in eq. 7-16 or 7-16a. This gives

$$\frac{F_i'}{F_m} = \frac{1}{1 + \frac{F_m U_L}{wc}\left(\frac{1^*}{\varepsilon} - \frac{1}{2}\right)} = \frac{1}{1 + \frac{F_m U_L A_c}{Wc}\left(\frac{1^*}{\varepsilon} - \frac{1}{2}\right)} \tag{7-22}$$

This equation evaluates the performance of the collector–heat exchanger *system* in a single factor which can be applied to $F_m \tau \alpha$ and $F_m U_L$ to get $F_i' \tau \alpha$ and $F_i' U_L$ once the heat exchanger effectiveness is known. Equation 7-22 is a most convenient form for use with a collector curve based on the more desirable mean temperatures (see Chapter 5). It reduces to eq. 5-32a when no heat exchanger is used.

By maximizing the value of $F_i'/F_m$, one can determine the best split of a total flow between the collector and storage loops. This optimum depends on the heat exchanger area but surprisingly does not change with the value of $F_m U_L$. If the heat exchanger is small, perhaps one-third the flow should be in the collector loop. As the exchanger becomes grossly oversized, the optimum split approaches equal flow in each loop.

The strange behavior of the optimum occurs because a low flow in the storage loop elevates the *entire* collector to a higher temperature. Lower flow in the collector loop does not affect the collector inlet temperature at all, and therefore raises the average collector temperature less than an equivalent decrease in the storage flow rate.

When the flow capacity rates are the same, eq. 7-22 becomes indeterminate but can be restated without $\varepsilon$ in the manner of eq. 7-16b:

$$\frac{F_i'}{F_m} = \frac{1}{1 + \left(\frac{F_m U_L}{U_x a_x} + \frac{F_m U_L}{2 w_s c_s}\right)} = \frac{1}{1 + \left(\frac{F_m U_L A_c}{U_x A_x} + \frac{F_m U_L A_c}{2 W_s c_s}\right)} \tag{7-22a}$$

so that in the limiting case of very high flow rates in both circuits,

$$\frac{F_i'}{F_m} = \frac{1}{1 + \frac{F_m U_L}{U_x a_x}} = \frac{1}{1 + \frac{F_m U_L A_c}{U_x A_x}} \tag{7-22b}$$

The equations in this section relate collector and heat exchanger performance to the flow rate directly, *including* the effect of flow rate on performance of the collector, which has been only implicit through $F_i$ in the previous relationships. The next example shows that either technique yields the same final result.

* When the storage flow is the smaller, replace this numeral 1 with $wc/w_s c_s$ or consistently substitute the collector flow for $w_s c_s$ and the storage flow for $w_i c_i$ when calculating $\varepsilon$.

**EXAMPLE 7-5**

An economic analysis with eq. 7-21b indicates that the optimal $U_x a_x$ is 79.62 W/°C·m² (14.02 Btu/hr °F ft²) using the collector of Example 5-8, for which the ethylene glycol collector flow is 24.21 g/s·m² (17.85 lb/hr ft²), $F_m \tau\alpha$ is 0.8421, and $F_m U_L$ is 5.224 W/°C·m² (0.92 Btu/hr °F ft²). Find $F_i' \overline{\tau\alpha}$ and $F_i' U_L$, assuming as in Example 5-8 that $\overline{\tau\alpha}/(\tau\alpha)_n = 0.95$. Use the same ratio $w_s c_s/w_l c_l$ as in Example 7-1 to 7-3, and the same fluid properties.

First find the effectiveness of the exchanger:

$$U_x a_x = 79.62 \text{ W/°C·m}^2$$

$$wc = w_s c_s = \frac{24.21 \text{ g}}{\text{s·m}^2} \times \frac{3.517 \text{ kJ}}{\text{kg·°C}} \times \frac{\text{W}}{\text{J/s}} = 85.15 \text{ W/°C·m}^2$$

$$w_l c_l = 85.15 \text{ W/°C·m}^2 \times \frac{23.67}{14.32} = 140.75 \text{ W/°C·m}^2$$

$$N_{TU} = \frac{U_x a_x}{w_s c_s} = \frac{79.62}{85.15} = 0.9351$$

From eqs. 7-10 and 7-9,

$$B = 0.9351\left(1 - \frac{85.15}{140.75}\right) = 0.3694$$

$$\varepsilon = \frac{1 - e^{-0.3694}}{1 - \dfrac{85.15}{140.75} e^{-0.3694}} = 0.5308$$

Then, eq. 7-22 is used directly to find the collector system characteristics:

$$\frac{F_i'}{F_m} = \frac{1}{1 + \dfrac{5.224}{85.15}\left(\dfrac{1}{0.5308} - \dfrac{1}{2}\right)} = 0.9217$$

This is the system penalty due to flow in the collector and heat transfer in the heat exchanger. The system parameters are then

$$F_i' \overline{\tau\alpha} = F_m \tau\alpha \times \frac{F_i'}{F_m} \times \frac{\overline{\tau\alpha}}{(\tau\alpha)_n} = 0.8421 \times 0.9217 \times 0.95 = 0.7374$$

where 0.95 corrects for the effect of incident angle, and

$$F_i U_L = 0.9217 \times 5.224 \text{ W/°C·m}^2 = 4.816 \text{ W/°C·m}^2$$

The same answers can be reached less directly by using the heat exchanger factor with the Example 5-8 values of $F_i \tau\alpha = 0.8171$ and $F_i U_L = 5.069$ W/°C·m². From eq. 7-16,

$$F_x = \frac{1}{1 + \dfrac{5.069}{85.15}\left(\dfrac{1}{0.5308} - 1\right)} = 0.950$$

$$F_i' \overline{\tau\alpha} = F_i \times F_x \times \frac{\overline{\tau\alpha}}{(\tau\alpha)_n} = 0.871 \times 0.95 \times 0.95 = 0.7374$$

and
$$F_i'U_L = F_x \times F_iU_L = 0.950 \times 5.069 = 4.816 \text{ W/}°\text{C·m}^2$$

In Engineering units,
$$U_xa_x = 14.02 \text{ Btu/hr }°\text{F ft}^2$$

$$wc = w_sc_s = \frac{17.85 \text{ lb}}{\text{hr ft}^2} \times \frac{0.8407 \text{ Btu}}{\text{lb }°\text{F}} = 15 \text{ Btu/hr }°\text{F ft}^2$$

$$N_{TU} = \frac{14.02}{15} = 0.9347$$

$$w_ic_i = \frac{15 \text{ Btu}}{\text{hr }°\text{F ft}^2} \times \frac{4.167}{2.52} = 24.81 \text{ Btu/hr }°\text{F ft}^2$$

by eqs. 7-10 and 7-9,

$$B = 14.02\left(\frac{1}{15} - \frac{1}{24.81}\right) = 0.3694$$

$$\varepsilon = \frac{1 - e^{-0.3694}}{1 - \dfrac{15}{24.81} e^{-0.3694}} = 0.5308$$

and by eq. 7-22

$$\frac{F_i'}{F_m} = \frac{1}{1 + \dfrac{0.92}{15}\left(\dfrac{1}{0.5308} - \dfrac{1}{2}\right)} = 0.9217$$

so that

$$F_i'\overline{\tau\alpha} = 0.8421 \times 0.9217 \times 0.95 = 0.7374$$
$$F_i'U_L = 0.9217 \times 0.92 = 0.8480 \text{ Btu/hr }°\text{F ft}^2$$

## Finding Heat Exchanger Properties from Catalog Data

While engineers often size heat exchangers using $UA$ values, in the heating and air-conditioning trade, heat exchangers are most often specified from an extensive table detailing their performance according to the chosen design conditions. Since a solar system does not have a design point, heat exchangers must be sized in terms of $U_xA_x$ to use the techniques of this chapter. The next example is a typical case in which $U_xA_x$ is obtained from manufacturer's design data.

### EXAMPLE 7-6

A heat exchanger is specified by its performance as given by Fig. 7-5. What is $U_xA_x$? Use 38 °C (100 °F) properties for the large stream, and 66 °C (150 °F) properties for the small stream.

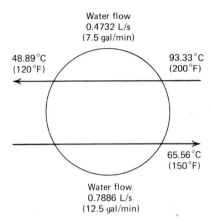

**FIGURE 7-5.** Diagram of a countercurrent heat exchanger showing performance data for use with Example 7-6.

In this example, eqs. 7-3 to 7-11 will not be used in normalized form because a collector area has not yet been associated with $U_x A_x$.

$$W_s c_s = \frac{0.4732\ \text{L}}{\text{s}} \times \frac{0.980\ \text{kg}}{\text{L}} \times \frac{4.19\ \text{kJ}}{\text{kg} \cdot {}^\circ\text{C}} \times \frac{\text{W}}{\text{J/s}} \times \frac{1000}{\text{k}} = 1943\ \text{W}/{}^\circ\text{C}$$

$$W_l c_l = \frac{0.7886\ \text{L}}{\text{s}} \times \frac{0.993\ \text{kg}}{\text{L}} \times \frac{4.19\ \text{kJ}}{\text{kg} \cdot \text{C}} \times \frac{\text{W}}{\text{J/s}} \times \frac{1000}{\text{k}} = 3281\ \text{W}/{}^\circ\text{C}$$

From eq. 7-8,

$$T_{lo} - T_{li} = \frac{W_s c_s}{W_l c_l}(T_{si} - T_{so}) = \frac{1943}{3281}(93.33 - 48.89) = 26.32\ {}^\circ\text{C}$$

$$T_{li} = T_{lo} - 26.32\ {}^\circ\text{C} = (65.56 - 26.32) = 39.24\ {}^\circ\text{C}$$

and from eq. 7-7,

$$\varepsilon = \frac{93.33 - 48.89}{93.33 - 39.24} = 0.8215$$

thus from eq. 7-11,

$$U_x A_x = 1943 \times \frac{\ln\left(\dfrac{1 - 0.8215 \times \dfrac{1943}{3281}}{1 - 0.8215}\right)}{\left(1 - \dfrac{1943}{3281}\right)} = 5033\ \text{W}/{}^\circ\text{C}$$

This is the heat exchanger used in Examples 7-1 and 7-2 (Fig. 7-3), yielding $F_x = 0.9640$.

In Engineering units,

$$W_s c_s = \frac{7.5 \text{ gal}}{\text{min}} \times \frac{61.2 \text{ lb}}{\text{ft}^3} \times \frac{\text{ft}^3}{7.48 \text{ gal}} \times \frac{60 \text{ min}}{\text{hr}} \times \frac{1 \text{ Btu}}{\text{lb }^\circ\text{F}} = 3682 \text{ Btu/hr }^\circ\text{F}$$

$$W_l c_l = \frac{12.5 \text{ gal}}{\text{min}} \times \frac{62 \text{ lb}}{\text{ft}^3} \times \frac{\text{ft}^3}{7.48 \text{ gal}} \times \frac{60 \text{ min}}{\text{hr}} \times \frac{1 \text{ Btu}}{\text{lb }^\circ\text{F}} = 6216.5 \text{ Btu/hr }^\circ\text{F}$$

From eq. 7-8,

$$T_{lo} - T_{li} = \frac{W_s c_s}{W_l c_l}(T_{si} - T_{so}) = \frac{3682}{6216.5}(200 - 120) \,^\circ\text{F} = 47.38 \,^\circ\text{F}$$

$$T_{li} = T_{lo} - 47.38 \,^\circ\text{F} = (150 - 47.38) \,^\circ\text{F} = 102.62 \,^\circ\text{F}$$

and from eq. 7-7,

$$\varepsilon = \frac{200 - 120}{200 - 102.62} = 0.8215$$

thus from eq. 7-11

$$U_x A_x = 3682 \times \frac{\ln\left(\dfrac{1 - 0.8215 \times \dfrac{3682}{6216.5}}{1 - 0.8215}\right)}{\left(1 - \dfrac{3682}{6216.5}\right)} = 9541.5 \text{ Btu/hr }^\circ\text{F}$$

## Natural Convection Heat Exchangers

For the important case of in-tank or traced-tank coils (Fig. 7-1c and 7-1d) the heat exchanger parameter $U_x A_x$ varies with the temperature difference between tube and tank circuits because this temperature difference determines the flow rate of the stored fluid in the circulating natural convection loop. For a tank heating water with a given temperature difference, a heat transfer coefficient can be estimated (7) using*

$$h = 142\left(\frac{\Delta T_x}{D_0}\right)^{0.25} \tag{7-23a}$$

where $h$ is in W/$^\circ$C·m$^2$, $\Delta T_x$ is the temperature difference in $^\circ$C, and $D_0$ the length of the characteristic path for natural convection in *meters*. $D_0$ is thus the height of the traced portion of a tank, half the perimeter of a bare horizontal tube or a finned tube, or the height of vertical tubes.

In Engineering units $h$ is given as Btu/hr $^\circ$F ft$^2$ by

$$h = 54\left(\frac{\Delta T_x}{D_0}\right)^{0.25} \tag{7-23b}$$

where $\Delta T_x$ is in degrees Fahrenheit and $D_0$ is in *inches*.

* Equations 7-23 apply when the average of the coil and tank temperature is 38 °C (100 °F). For 65 °C (150 °F), multiply by 1.15; for 21 °C (70 °F) divide by 1.25. Much below that natural convection is weak, ceasing entirely below 4 °C (39 °F), the maximum density of water.

Although this expression has solid theoretical roots, its applicability in solar work has not been verified, and its extension to finned surfaces is dubious. There is nothing published to indicate the proper way to determine a long-term value of $\varepsilon$ from $U_x$, for the distribution of collected heat flux is unknown. For the present, it is recommended that an average flux estimated from preliminary studies be used to determine $U_x$ for calculation of $\varepsilon$, using eqs. 7-23 and 7-9b, respectively.

In actual practice, $U_x$ may be well under that predicted by eq. 7-23 if there is significant resistance to heat transfer in the flowing fluid or across the metallic interface. These can be predicted by the methods of Chapter 4 and combined into an overall $U_x$ as explained in Chapter 2.

### EXAMPLE 7-7

A collector array having an area of 6 m² (64.6 ft²) is to be used with fluids and flows given in Example 5-8 to heat a 120-gal tank containing 443 kg (975 lb) water. The cylindrical tank is 55.88 cm (22 in.) in diameter and traced upward for 91.42 cm (36 in.) with an integral heat exchanger having perfect heat exchange to the wall inside the tank. What is the effectiveness of the heat exchanger and the heat exchanger factor at an average temperature of 100 °F?

$$A_x = \pi Dh = \pi \times 0.5588 \text{ m} \times 0.9142 \text{ m} = 1.605 \text{ m}^2$$

$$A_c/A_x = 6 \text{ m}^2/1.605 \text{ m}^2 = 3.74$$

from Example 5-8,

$$wc = 85.15 \text{ W/°C·m}^2$$

$$F_t U_L = 5.069 \text{ W/°C·m}^2$$

If $q$ is the heat flux collected per unit collector area, the heat transferred through the heat exchanger is

$$qA_c = U_x A_x \, \Delta T_x$$

Solving eq. 7-23a for $\Delta T_x$,

$$T_x = U_x^4 \, D_0/142^4$$

Substituting this into the preceding expression,

$$U_x^5 = \frac{142^4 q(A_c/A_x)}{D_0}$$

$$U_x = 52.7 \left[ \frac{q(A_c/A_x)}{D_0} \right]^{0.2}$$

An average collected heat flux can be estimated from Example 4-3 as

$$q = \frac{q_T}{t_T} = \frac{84.2 \text{ MJ/m}^2}{0.3694 \text{ Ms}} = 228 \text{ W/m}^2$$

so that the previous expression yields

$$U_x = 52.7 \left[ \frac{(228)(3.74)}{0.6095} \right]^{0.2} = 224 \text{ W/°C·m}^2$$

using eq. 7-9b,

$$N_{TU} = \frac{U_x a_x}{wc} = \frac{U_x(A_x/A_c)}{wc} = \frac{224 \text{ W}}{\text{C} \cdot \text{m}^2} \times \frac{(1/3.74)}{85.15 \text{ W}/^\circ\text{C} \cdot \text{m}^2} = 0.7034$$

$$\varepsilon = 1 - \exp(-0.7034) = 0.5051$$

and by eq. 7-16,

$$F_x = \frac{1}{1 + \dfrac{5.069 \text{ W}/^\circ\text{C} \cdot \text{m}^2}{85.15 \text{ W}/^\circ\text{C} \cdot \text{m}^2}\left(\dfrac{1}{0.5051} - 1\right)} = 0.9449$$

This applies at an average flux and should approximate the average value of $F_x$. It can be reestimated as better average flux values become available.

In Engineering units,

$$A_x = \pi \times 22 \text{ in.} \times 36 \text{ in.} \times \frac{\text{ft}^2}{144 \text{ in.}^2} = 17.27 \text{ ft}^2$$

$$A_c/A_x = 64.6 \text{ ft}^2/17.27 \text{ ft}^2 = 3.74$$

from Example 5-8

$$wc = 15 \text{ Btu/hr } ^\circ\text{F ft}^2$$

and

$$F_i U_L = 0.8926 \text{ Btu/hr } ^\circ\text{F ft}^2$$

The heat transferred through the exchanger is

$$qA_c = U_x A_x \, \Delta T_x$$

from eq. 7-23b,

$$\Delta T_x = U_x{}^4 D_0/54^4$$

and so

$$U_x = 24.32\left[\frac{q(A_c/A_x)}{D_0}\right]^{0.2}$$

An average flux collected from Example 4-3 is

$$q = \frac{84.21 \text{ MJ/m}^2}{0.369 \text{ Ms}} \times \frac{\text{W} \cdot \text{s}}{\text{J}} \times \frac{\text{Btu/hr } ^\circ\text{F ft}^2}{3.1456 \text{ W/m}^2} = 72.5 \text{ Btu/hr } ^\circ\text{F ft}^2$$

$$U_x = 24.32\left[\frac{(72.5)(3.74)}{24}\right]^{0.2} = 39.49 \text{ Btu/hr } ^\circ\text{F ft}^2$$

$$N_{TU} = \frac{U_x a_x}{wc} = \frac{U_x(A_x/A_c)}{wc} = \frac{(39.49 \text{ Btu/hr } ^\circ\text{F ft}^2)(1/3.74)}{15 \text{ Btu/hr } ^\circ\text{F ft}^2} = 0.7040$$

$$\varepsilon = 1 - \exp(-0.7040) = 0.5054$$

$$F_x = \frac{1}{1 + \dfrac{0.8926}{15}\left(\dfrac{1}{0.5040} - 1\right)} = 0.9450$$

## Variation in the Performance of Heat Exchangers

Because of changes in fluid heat transfer characteristics with respect to flow rate, temperature, and composition, the overall $U_x A_x$ may vary from its value at the rated conditions. The dependence is not strong if an exchanger is used within about 25% of the rated flow with the liquid for which it was rated. Beyond this, an estimate of new $U_x A_x$ is best made by adjusting the old value with the analytical *ratio* of the new to old overall heat transfer coefficients as estimated from the McAdams relationship (eq. 5-3) by techniques described in the next paragraph. It is simplest to assume that the two internal heat transfer coefficients are about equal on either side of the tube wall, that the flow is turbulent, and that the heat transfer resistance of the metal wall is unimportant. Such characteristics are often found in well-designed, liquid-to-liquid heat exchangers.

By eq. 5-3, for fluid on either side of the exchanger walls,

$$\frac{hD}{k} = 0.023 N_{\mathrm{Re}}^{0.8} N_{\mathrm{Pr}}^{1/3} \tag{5-3}$$

but the characteristic dimensions cannot be easily determined even if drawings are available because of the complexity in the flow pattern on at least one side of most heat exchangers.* Therefore, the *ratio* of $h$ (as used) to $h$ (as rated) will be found by dividing eq. 5-3 (as used) by eq. 5-3 (as rated), designated by subscripts $u$ and $r$, respectively.

$$\frac{\left(\dfrac{hD}{k}\right)_{\mathrm{used}}}{\left(\dfrac{hD}{k}\right)_{\mathrm{rated}}} = \frac{0.023\left(\dfrac{4W}{\mu P}\right)_{\mathrm{used}}^{0.8}[(N_{\mathrm{Pr}})_{\mathrm{used}}^{1/3}]}{0.023\left(\dfrac{4W}{\mu P}\right)_{\mathrm{rated}}^{0.8}[(N_{\mathrm{Pr}})_{\mathrm{rated}}^{1/3}]}$$

and cancelling like terms yields

$$R \equiv \frac{h_u}{h_r} = \frac{k_r}{k_u}\left(\frac{W_u \mu_r}{W_r \mu_u}\right)^{0.8}\left[\frac{(N_{\mathrm{Pr}})_u}{(N_{\mathrm{Pr}})_r}\right]^{1/3} \tag{7-24}$$

which applies to the fluid film on either side of the tube wall through which the heat is transferred.

Letting additional $s$ and $t$ subscripts represent shell and tube sides of the exchanger, then eq. 7-24 gives

$$h_{tu} = R_t h_{tr}$$

and

$$h_{su} = R_s h_{sr}$$

---

* $N_{\mathrm{Re}}^{0.6}$ has been found to describe shell side performance more accurately, but is not used here in the interest of simplicity and generality.

which are the new coefficients for each side of the tube. Combining these into the overall "as used" coefficient, by eq. 2-16,

$$\frac{1}{U_{xu}} = \frac{1}{R_t h_{tr}} + \frac{1}{R_s h_{sr}}$$

while for the "as rated" condition,

$$\frac{1}{U_{xr}} = \frac{1}{h_{tr}} + \frac{1}{h_{sr}}$$

Taking the ratio of these two equations,

$$\frac{U_{xu}}{U_{xr}} = \frac{\left(\dfrac{1}{h_{tr}} + \dfrac{1}{h_{sr}}\right)}{\left(\dfrac{1}{R_t h_{tr}} + \dfrac{1}{R_s h_{sr}}\right)}$$

To continue, an estimation of $h_{tr}/h_{sr}$ must be made. Assuming conveniently that the coefficients are equal,* as they will be in a well-designed exchanger, a simple estimation of the new overall coefficient "as used" can then be made if the $h$ terms are cancelled.

The overall as used coefficient is then

$$U_{xu} = \frac{2U_{xr}}{\dfrac{1}{R_t} + \dfrac{1}{R_s}} \qquad (7\text{-}24a)$$

Any estimate of the "as rated" ratio of the tube and shell coefficients $h_{tr}/h_{sr}$ may be used to derive an alternate statement of eq. 7-24a. This will in effect place different weighting factors on $R_t$ and $R_s$, the parameters that were estimated analytically using eq. 7-24.† The most accurate estimation of the $h_{tr}/h_{sr}$ ratio can be made if the heat transfer coefficient of one side can be determined analytically. This can often be done with the tube side of a shell and tube heat exchanger, allowing the shell side coefficient to be calculated directly from the known performance.

**EXAMPLE 7-8**

Estimate more carefully the performance of the heat exchanger of Example 7-6 for the situation of Example 7-1 considering the decrease in heat transfer coefficients resulting from the difference in flow rates and the change in fluid on the tube

---

* Note, however, that the as *used* values will *not* necessarily be equal.
† If $h_{tr}/h_{sr} \equiv k$, then

$$U_{xu} = \frac{1 + k}{\dfrac{1}{R_t} + \dfrac{k}{R_s}} U_{xr} \qquad (7\text{-}24b)$$

**TABLE 7-3.** Comparison of "Used" and "Rated" Thermal Properties and Flow Rates for Example 7-8

| | Tube-side Flow (Collector) | | Shell-side Flow (Storage) | |
|---|---|---|---|---|
| | As used | As rated | As used | As rated |
| Prandtl number, dimensionless | 10.16 | 2.8 | 2.8 | 2.8 |
| Viscosity, centipoise or m Pa·s | 1.2 | 0.432 | 0.432 | 0.432 |
| Thermal conductivity $k$ | | | | |
| W/m·°C | 0.415 | 0.661 | 0.661 | 0.661 |
| Btu/hr ft °F | 0.24 | 0.382 | 0.382 | 0.382 |
| Flow rate $W$ | | | | |
| kg/s | 0.3629 | 0.4637 | 0.504 | 0.7728 |
| lb/min | 48.0 | 61.36 | 66.67 | 102.2 |

(collector) side to 50% ethylene glycol. Consider all fluid properties at 66 °C (150 °F).

Relevant fluid properties have been taken from Table 5-1 to prepare Table 7-3, which compares rated and used flow conditions. Flow rates through the exchanger are as follows:

### Tube or Collector Flow

$$W_{\text{used} \atop \text{(glycol)}} = 0.3629 \text{ kg/s}$$

$$W_{\text{rated} \atop \text{(water)}} = \frac{0.4732 \text{ L}}{\text{s}} \times \frac{0.980 \text{ kg}}{\text{L}} = 0.4637 \text{ kg/s}$$

### Shell or Storage Flow

$$W_{\text{used} \atop \text{(water)}} = 0.504 \text{ kg/s}$$

$$W_{\text{rated} \atop \text{(water)}} = \frac{0.7886 \text{ L}}{\text{s}} \times \frac{0.98 \text{ kg}}{\text{L}} = 0.7728 \text{ kg/s}$$

For the shell (storage) side eq. 7-24 gives

$$R_s = \frac{h_u}{h_r} = \frac{0.661}{0.661}\left(\frac{0.504}{0.7728} \times \frac{0.432}{0.432}\right)^{0.8}\left(\frac{2.8}{2.8}\right)^{1/3} = \left(\frac{0.504}{0.7728}\right)^{0.8} = 0.7104$$

and for the tube (collector) side

$$R_t = \frac{h_u}{h_r} = \frac{0.661}{0.415}\left(\frac{0.3629}{0.4637} \times \frac{0.432}{1.2}\right)^{0.8}\left(\frac{10.16}{2.8}\right)^{1/3}$$

$$= 1.593 \times 0.3630 \times 1.537 = 0.8900$$

In the projected "as used" situation, by eq. 7-24a,

$$U_{xu} = \frac{2U_{xr}}{\dfrac{1}{0.8900} + \dfrac{1}{0.7104}} = 0.7902 U_{xr}$$

and the "as used" value for $U_x A_x$ is then

$$U_x A_x = \frac{5033 \text{ W}}{°C} \times 0.7902 = 3977 \text{ W/}°C$$

and

$$U_x a_x = \frac{3977 \text{ W}}{°C} \times \frac{1}{89.2 \text{ m}^2} = 44.58 \text{ W/}°C \cdot \text{m}^2$$

$\varepsilon$ can be found by eqs. 7-9 and 7-10:

$$B = 44.58\left(\frac{1}{14.32} - \frac{1}{23.67}\right) = 1.230$$

$$\varepsilon = \frac{1 - e^{-1.230}}{\left(1 - \dfrac{14.32}{23.67} e^{-1.230}\right)} = 0.8597$$

and $F_x$ as in Example 7-2:

$$F_x = \frac{1}{1 + \dfrac{5.069}{14.32}\left(\dfrac{1}{0.8597} - 1\right)} = 0.9454$$

In Engineering units,

### *Tube or Collector Flow*

$$W_{\substack{\text{used} \\ \text{(glycol)}}} = 2880 \text{ lb/hr} \times \frac{\text{hr}}{60 \text{ min}} = 48 \text{ lb/min}$$

$$W_{\substack{\text{rated} \\ \text{(water)}}} = \frac{7.5 \text{ gal}}{\text{min}} \times \frac{61.2 \text{ lb}}{\text{ft}^3} \times \frac{\text{ft}^3}{7.48 \text{ gal}} = 61.36 \text{ lb/min}$$

### *Shell or Storage Flow*

$$W_{\substack{\text{used} \\ \text{(water)}}} = \frac{4.167 \text{ lb}}{\text{hr ft}^2} \times 960 \text{ ft}^2 \times \frac{\text{hr}}{60 \text{ min}} = 66.67 \text{ lb/min}$$

$$W_{\substack{\text{rated} \\ \text{(water)}}} = \frac{12.5 \text{ gal}}{\text{min}} \times \frac{61.2 \text{ lb}}{\text{ft}^3} \times \frac{\text{ft}^3}{7.48 \text{ gal}} = 102.2 \text{ lb/min}$$

For the shell (storage) side eq. 7-24 gives

$$R_s = \frac{h_u}{h_r} = \frac{0.382}{0.382}\left(\frac{66.67}{102.2} \times \frac{0.432}{0.432}\right)^{0.8}\left(\frac{2.8}{2.8}\right)^{1/3} = \left(\frac{66.67}{102.2}\right)^{0.8} = 0.7105$$

and for the tube (collector) side

$$R_t = \frac{h_u}{h_r} = \frac{0.382}{0.24}\left(\frac{48.0}{61.36}\times\frac{0.432}{1.2}\right)^{0.8}\left(\frac{10.16}{2.8}\right)^{1/3}$$

$$= 1.592 \times 0.3628 \times 1.537 = 0.8877$$

In the projected "used" situation by eq. 7-24a,

$$U_{xu} = \frac{2U_{xr}}{\dfrac{1}{0.8877} + \dfrac{1}{0.7105}}$$

$$U_{xu} = 0.7892 U_{xr}$$

and the "as used" value for $U_x A_x$ is then

$$U_x A_x = \frac{9541.5\ \text{Btu}}{\text{hr }^\circ\text{F}} \times 0.7892 = 7530\ \frac{\text{Btu}}{\text{hr }^\circ\text{F ft}^2}$$

and

$$U_x a_x = 7530/960 = 7.844\ \text{Btu/hr }^\circ\text{F ft}^2$$

$\varepsilon$ can be found by eqs. 7-9 and 7-10:

$$B = 7.844\left(\frac{1}{2.52} - \frac{1}{4.167}\right) = 1.230$$

$$\varepsilon = \frac{1 - e^{-1.230}}{1 - \dfrac{2.52}{4.167}e^{-1.230}} = 0.8597$$

and $F_x$ as in Example 7-2:

$$F_x = \frac{1}{1 + \dfrac{0.8927}{2.52}\left(\dfrac{1}{0.8601} - 1\right)} = 0.9454$$

This example shows a typical case where the used flows were reasonably close to the rated flows. By slowing down the fluids and changing to ethylene glycol on one side, each heat transfer coefficient was reduced to about 80% of its original value, and the total system performance deteriorated nearly 2%. This situation is better handled by finding ratings having flows and fluids like those that will actually be used. When dealing with heat exchanger manufacturer's representatives, if fluids, flows, and a typical temperature pattern are specified (as in Example 7-2) that require the proper $U_x A_x$, in most instances the home office will be able to supply the proper heat exchanger by the use of computer programs incorporating detailed thermal analysis.

## TERMINAL HEAT EXCHANGE

Heat that has been placed in storage ultimately must be used to heat the air in a structure. With rock-bed storage, hot air is available simply by reversing the flow through the bed. The rocks or pebbles have so much surface area that heat transfer is no problem. When heat is to be transferred from a storage liquid—invariably water—to the room air the problem is more difficult. A liquid-to-forced-air heat exchanger and air ducting are most often used.

*Hydronic* baseboard convectors to which the water is pumped may initially seem feasible but, since heat exchange is carried out by natural convection, high fluid temperatures are required to transfer the heat in realizable lengths and such temperatures hurt the solar system performance. A more attractive method for heating the living space uses the *radiant* heating from pipes imbedded in ceiling or floor through which stored water is circulated.

In this section methods for the analysis of these kinds of terminal heat exchange will be outlined so that they can be used routinely for system studies in a later chapter when better information as to terminal heating loads and storage temperatures will be available.

### Liquid-to-Air Heat Exchangers

**Forced Circulation.** When air is being heated (or cooled) by a liquid the heat exchanger effectiveness analysis technique applies just as in liquid-liquid heat exchange. In the construction of the exchangers themselves, however, there is little similarity. Figure 7-6 shows a typical liquid-to-air heat exchanger. The heat transfer coefficient on the air side is always so much lower than on the liquid side that extended surface is always placed in the air stream. The Reynolds number in the small spaces between the fins is always low, making entrance effects predominant in heat exchange. If the heat exchanger fins are continuous, they are seldom more than about 50 mm (2 in.) in depth, for additional depth would not yield much more heat exchange in the developing laminar flow.* This is the reason that *finned-tube* exchangers (which have a circular fin integral with each tube) are better than *fin-tube* heat exchangers†—the air film is broken up periodically as it passes between tubes.

Either configuration is likely to be of the cross-flow design, that is, there is so little depth to the arrangement that the water cannot practically be made to flow

---

* Equation 5-4 gives the length that entrance effects persist as 1.5 in. when $N_{Re} = 2000$ with a 0.01 in. fin spacing ($D_e = 0.02$ in.).

† This is the traditional air-conditioning coil, which is not a coil at all but long, closely spaced fins with air flowing through the spaces between them, perforated so that tubes pass through the fins perpendicular to the air flow.

**FIGURE 7-6.** A liquid-to-air, finned-tube heat exchanger designed for hot water service. This exchanger is about 86 cm high (34 in.) with two rows of tubes. The tubes are 1.6 cm ($\frac{5}{8}$ in.) diameter, and the fins are 3.65 cm (1.44 in.) diameter, spaced as closely as 14 fins per inch. (Courtesy of Aerofin Corporation, Lynchburg, Va.)

countercurrent to the air stream. Unless the water flow (in terms of $Wc$) can be much greater than the air flow, cross-flow effectiveness is limited to about two-thirds of the counterflow values. When high effectiveness is needed, as is often the case in solar applications, it is suggested that two of the cross-flow exchangers be used in series. This will permit increased area without excessive duct size, proper exploitation of entrance effects, and a close approach to countercurrent flow (5). Such an arrangement is shown in Fig. 7-7a.

When two *counterflow* heat exchangers are used in series in the air stream, the effectiveness of the combination in counterflow can be quickly determined when the effectiveness of one exchanger is known, by (5):

$$\varepsilon_2 = \frac{2\varepsilon_1}{1 + \varepsilon_1} \qquad (7\text{-}25a)$$

where $\varepsilon_1$ is the effectiveness of one exchanger and $\varepsilon_2$ the effectiveness of the two together. For cross-flow exchangers in series the same relationship overpredicts performance, but by only a few percent.

When the desired effectiveness $\varepsilon_2$ is known, eq. 7-25a can be rearranged to give the effectiveness of one of the exchangers as

$$\varepsilon_1 = \frac{\varepsilon_2}{2 - \varepsilon_2} \qquad (7\text{-}25b)$$

These relationships permit proper exchangers to be chosen quickly from conventional catalogs even if they do not list exchangers having the high effectiveness often desired for the terminal heat exchanger.

Performance can be enhanced if the water flow can be made very large relative to the air flow, and the two exchangers can then be connected in parallel (Fig.

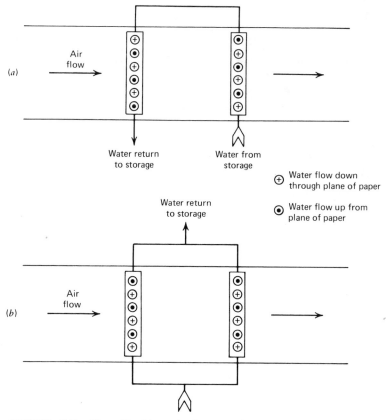

FIGURE 7-7. Two liquid-to-air heat exchangers arranged in (a) counterflow and, for very high water flows, (b) in parallel flow. Within one exchanger the fluids are in cross-flow.

7-7b). Equations 7-9b and 7-11b then apply to determine effectiveness if $U_xA_x$ (or $U_xA_x$) is known and vice versa, respectively.*

### EXAMPLE 7-9

377.6 $L^3$/s (800 ft³/min) of air at 23.89 °C (75 °F) is to be heated to 27.78 °C (82 °F) by a large flow of water at 29.44 °C (85 °F). Two heat exchangers arranged countercurrently in series are to be used. What are $\varepsilon$ and $U_xA_x$ for each?

$$Wc = \frac{377.6\ L^3}{s} \times \frac{1.187\ kg}{m^3} \times \frac{1.01\ kJ}{kg\ °C} \times \frac{m^3}{1000\ L^3} \times \frac{W}{J/s} \times \frac{10^3}{k} = 452.7\ W/°C$$

From eq. 7-7,

$$\varepsilon = \frac{27.78 - 23.89}{29.44 - 23.89} = 0.7$$

from eq. 7-11b,

$$N_{TU} = \frac{(U_xA_x)_{both}}{Wc} = \ln\frac{1}{1-\varepsilon} = \ln\frac{1}{1-0.7} = 1.2040$$

$$(U_xA_x)_{both} = 452.7\ W/°C \times 1.2040 = 545\ W/°C$$

$$(U_xA_x)_{each} = \frac{545}{2} = 272.5\ W/°C$$

from eq. 7-25b,

$$\varepsilon_1 = \frac{\varepsilon_2}{2 - \varepsilon_2} = \frac{0.7}{2 - 0.7} = 0.5385$$

In Engineering units,

$$Wc = \frac{800\ ft^3}{min} \times \frac{60\ min}{hr} \times \frac{lb}{13.5\ ft^2} \times \frac{0.24\ Btu}{lb\ °F} = 853.3\ Btu/hr\ °F$$

$$\varepsilon = \frac{82 - 75}{85 - 75} = 0.7$$

$$(U_xA_x)_{both} = Wc\ \ln\frac{1}{1-\varepsilon} = \frac{853.3\ Btu}{hr\ °F}\ \ln\frac{1}{1-0.7} = 1027.4\ Btu/hr\ °F$$

$$(U_xA_x)_{each} = \frac{1027.4}{2} = 513.7\ Btu/hr\ °F$$

$$\varepsilon_1 = \frac{\varepsilon_2}{2 - \varepsilon_2} = \frac{0.7}{2 - 0.7} = 0.5385$$

* The equations apply to either countercurrent or cross flow, for they both approach the same limit as water flow increases.

**EXAMPLE 7-10**

An air flow is being heated from 23.89 °C (75 °F) to 26.67 °C (80 °F) by a cross-flow exchanger using rapidly circulating water at 29.44 °C (85 °F). To what temperature will two such heat exchangers in counterflow heat the air?

$$\varepsilon_1 = \frac{26.67 - 23.89}{29.44 - 23.89} = 0.5$$

$$\varepsilon_2 = \frac{2\varepsilon_1}{1 + \varepsilon_1} = \frac{2 \times 0.5}{1 + 0.5} = \frac{1}{1.5} = 0.6667$$

$$\frac{T_0 - 23.89}{29.44 - 23.89} = 0.6667$$

$$T_0 = 0.6667 \times 5.55 + 23.89 = 27.59 \,°C$$

In Engineering units,

$$\varepsilon_1 = \frac{80 - 75}{85 - 75} = 0.5$$

$$\varepsilon_2 = \frac{2\varepsilon_1}{1 + \varepsilon_1} = \frac{2 + 0.5}{1 + 0.5} = \frac{1}{1.5} = 0.6667$$

$$\frac{T_0 - 75}{85 - 75} = 0.6667$$

$$T_0 = 0.6667 \times 10 + 75 = 81.67 \,°F$$

**Natural Circulation.**  Baseboard convectors such as those shown in Fig. 7-8 have been in common domestic and commercial use for years. The temperatures needed for ordinary application are much higher than available from solar heat—in the 65–95 °C (150–200 °F) range.

Performance of baseboard convectors is given as the heating rate per meter (foot) of length, with a different performance associated with each configuration.

Once performance is specified for one rated water temperature $T_r$, it can be changed to reflect the temperature $T$ of actual operation by multiplying by the factor (6):

$$\left[\frac{(T - T_a)}{(T_r - T_a)}\right]^{1.4} \tag{7-26}$$

where $T_a$ is the entering air temperature, commonly 18.3 °C (65 °F) at floor level.

**EXAMPLE 7-11**

A hydronic baseboard convector has a capacity of 1019 W/m$_{\text{baseboard}}$ (1060 Btu/hr ft$_{\text{baseboard}}$) when rated at 87.78 °C (190 °F). How much heat will 22.86 m

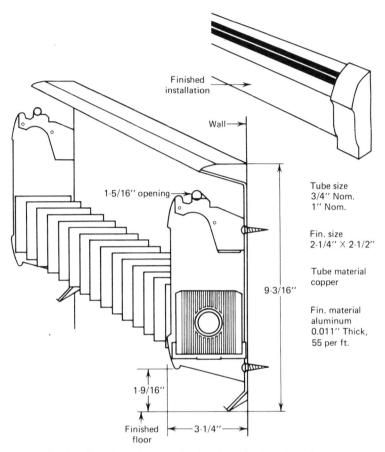

Finished installation

Wall

1-5/16" opening

9-3/16"

Tube size
3/4" Nom.
1" Nom.

Fin. size
2-1/4" X 2-1/2"

Tube material
copper

Fin. material
aluminum
0.011" Thick,
55 per ft.

1-9/16"

Finished floor

3-1/4"

**FIGURE 7-8.** Baseboard convector for terminal heat exchange.

(75 ft) of this baseboard put into a room if the water temperature is 37.78 °C (100 °F)?

The new rating (eq. 7-26) is

$$Q = \left(\frac{37.78 - 18.33}{87.78 - 18.33}\right)^{1.4} \times \frac{1019}{m_{\text{baseboard}}} = 171.4 \text{ W/m}_{\text{baseboard}}$$

Thus, for 22.86 m of baseboard

$$Q = \frac{171.4 \text{ W}}{m_{\text{baseboard}}} \times 22.86 \text{ m} = 3918 \text{ W}$$

In Engineering units,

$$Q = \left[\frac{(100 - 65)}{(190 - 65)}\right]^{1.4} \times 1060 \frac{\text{Btu}}{\text{hr ft}_{\text{baseboard}}} = 178 \text{ Btu/hr ft}_{\text{baseboard}}$$

Thus, for 75 ft of baseboard

$$Q = \frac{178 \text{ Btu}}{\text{hr ft}_{\text{baseboard}}} \times 75 \text{ ft} = 13\,380 \text{ Btu/hr}$$

The decrease to 20% of the nominal capacity is typical. It severely limits the application of baseboard terminal heat exchange in solar heating.

**Direct Radiation.** Contrary to intuition a warm ceiling (or floor) will transfer most of its heat directly to the floor (or ceiling) and furniture by radiation. Such hydronic terminal systems were common in the late 1940s and 1950s, but lost popularity due to cost and flow-balancing problems. Electric heat took over the concept in the 1960s and radiant electric ceilings are still popular. They generally operate at about 35 °C (95 °F) or lower, an ideal temperature range for solar heat. The heat flux is so low that the poor conductivity of the plaster ceiling (which serves as a fin) can be tolerated if the tubes are close enough together. To estimate the heat transfer rate, eq. 2-4 or 2-6 apply directly for SI and engineering units, respectively:

$$Q = \frac{5.673\varepsilon A}{10^8} (T_H{}^4 - T_L{}^4) \tag{2-4}$$

$$Q = \frac{0.1713\varepsilon A}{10^8} (T_H{}^4 - T_L{}^4) \tag{2-6}$$

**EXAMPLE 7-12**

What is the radiation heat transfer rate from a ceiling with an emissivity of 0.8 to a room kept at 21.11 °C (70 °F) if the ceiling is uniformly 37.78 °C (100 °F)?

$$Q = \frac{5.673 \times 0.8}{10^8} [(37.8 + 273)^4 - (21.1 + 273)^4] = 83.9 \text{ W/m}^2$$

This is sufficient for heating a well-insulated house. For instance, if a 28.47 MJ/°C-day (15 000 Btu/°F-day) house has 167.2 m² (1800 ft²) of floor area, when the temperature is −15 °C (5 °F) the demand will be

$$\frac{28.47 \text{ MJ}}{°\text{C-day}} \times \frac{\text{day}}{24 \text{ hr}} \times \frac{\text{hr}}{3600 \text{ s}} \times \frac{10^6}{\text{M}} \times \frac{\text{W}}{\text{J/s}} \times \frac{[18.33 - (-15)] °\text{C}}{167.2 \text{ m}^2}$$

$$= 65.68 \text{ W/m}^2_{\text{floor or ceiling}}$$

which is a very low heating flux.

In Engineering units,

$$Q = \frac{0.1713 \times 0.8}{10^8} [(100 + 460)^4 - (70 + 460)^4] = 26.64 \text{ Btu/hr ft}^2$$

## PROBLEMS

1. A heat exchanger heats one stream of water from 50 to 75 °C (122–167 °F) with an equal flow of 90 °C (194 °F) water.
   (a) What is the effectiveness?
   (b) What is the number of transfer units?
   (c) What is the unknown outlet temperature?
   (d) If the flow rate is 0.25 kg/s (1984 lb/hr), what is $U_x A_x$?

2. Repeat problem 1 for a heat exchanger that *cools* a stream of water from 75 to 50 °C (167–122 °F) with *twice* the flow of water at 35 °C (95 °F).

3. A heat exchanger is operating with an effectiveness of 0.7. The small flow capacity rate is equal to the larger. Consider $U_x A_x$ as constant while the small flow rate is reduced to one-third the original. Does $N_{TU}$ rise or fall?
   (a) What are the new values of $\varepsilon$ and $N_{TU}$?
   (b) How much is the rate of heat transfer changed?

4. The applications brochure for a heat exchanger recovering waste heat from household furnace flue gas states that the flue gas is cooled from 310 to 200 °C (590 to 392 °F) and that the unit produces air heated to 45 °C (113 °F) when the room temperature is 20 °C (68 °F). (Assume the specific heat of the flue gas equal to that of air to simplify this problem. It will actually be somewhat higher due to the water vapor content of these gases.)
   (a) Estimate the temperature of the exiting flue gas when this exchanger is applied to a furnace of the same size having a stack temperature of 240 °C (464 °F). By what factor is the recovered heat reduced?
   (b) If installed on twice as large a furnace, to what temperature can the 450 °F stack gas be cooled? By what factor is the recovered heat changed?
   (c) The cooling air flow is 0.29 kg/s (2300 lb/hr). What is $U_x A_x$?

5. A heat exchanger is cooling silicone oil from 80 to 50 °C (176–122 °F) using a volumetric water flow twice times that of the oil at 35 °C (95 °F). The water flow is increased to 100 times that of the oil.
   (a) What are initial and final $N_{TU}$ and effectiveness?
   (b) What is the new exiting oil temperature?

6. A solar collector having $U_L = 6.3$ W/°C·m² (1.11 Btu/hr °F ft²) circulates 50% ethylene glycol to a heat exchanger cooled by an equal flow capacity rate of storage water.
   (a) Plot values of $F_x$ and $F_t'/F_m$ versus $1/wc$ as the equal flow capacity rates each increase from 1 to 50 W/°C·m² (0.2 to 9 Btu/hr °F ft²) if $U_x a_x = 20$ W/°C·m² (3.52 Btu/hr °F ft²).
   (b) Repeat part (a) using half and twice the initial $U_x a_x$ value.

7. The area $U_x a_x$ of the collector-to-storage heat exchanger is 20 W/°C·m².
   (a) If the total flow capacity rate for the two circulating streams is 18 W/°C·m² (3.17 Btu/hr °F ft²), find the value of $F_t'/F_m$ if 20%, 30%, 40%, and 60%

of the total flow is in the collector loop. Plot these values and estimate the optimal flow split.

(b) Repeat part (a) for 10 and 40 W/°C·m² (1.73 and 6.92 Btu/hr °F ft²) combined flow rate.

**8.** The heat exchanger factor for a 200 m² (2153 ft²) collector array is to be 0.95. The flow capacity rate is 3.87 W/°C·m² (22 Btu/hr °F ft²). $F_i' U_L$ is 4.82 W/°C·m² (0.849 Btu/hr °F ft²).

(a) What is the needed heat exchanger effectiveness?

(b) What volumetric flow rate of 50% propylene glycol is required in the collector loop?

(c) If the storage flow capacity rate is twice that in the collector loop, size the heat exchanger in terms of both $U_x a_x$ and $U_x A_x$.

**9.** The cost of collectors for a 37.2 m² (400 ft²) proposed heating system is $215.29/m² ($20/ft²). A heat exchanger having $U_x A_x$ = 2500 W/°C (4739 Btu/hr °F) is available for $300. Assuming that the cost of the heat exchanger is proportional to the magnitude of $U_x A_x$, what multiple of the original heat exchanger should be installed if $F_i U_L$ = 4.82 W/°C·m² (0.849 Btu/hr °F ft²)?

**10.** A solar collector has a loss coefficient $F_m U_L$ of 5.128 W/°C·m² (0.903 Btu/hr °F ft²). If a heat exchanger having $U_x A_x$ = 26 000 W/°C (49 290 Btu/hr °F) is to be used with a 325 m² (3500 ft²) collector array, what is the *minimum* reduction in performance due to heat exchanger installation?

**11.** The heat exchanger of Example 7-6 is to be used in the situation of Example 7-1, but with a silicone oil heat transfer fluid in the collector loop.

(a) Estimate the new heat transfer factor if the flow capacity rates are held constant.

(b) What is the volumetric flow rate of the silicone oil? Compare with Example 7-1 and explain the change.

**12.** 1200 L³/s (2543 ft³/min) of air at 20 °C (68 °F) is heated by water at 30 °C (86 °F) in a terminal heat exchanger to 25 °C (77 °F). Assume countercurrent flow.

(a) What is the effectiveness?

(b) Find the required heat exchanger $U_x A_x$ if the water flow is very high.

(c) Find the required heat exchanger $U_x A_x$ if the flow capacity rates of water and air are equal. What are the outlet water temperature and water flow rate?

(d) Find the effectiveness with double the areas for cases (b) and (c) above. Compare with the use of eq. 7-25a fo find the effectiveness of two counterflow exchangers in series.

**13.** (a) Repeat problems 12(a) to 12(c) for a cross-flow heat exchanger using the proper formula from Table 7-2.

(b) Repeat problem 12(d) for a series countercurrent interconnection of two cross-flow heat exchangers as in Fig. 7.7a.

14. 900 L³/s (1907 ft³/min) of air is to be heated from 25 °C (77 °F) to 35 °C (95 °F) using water at 40 °C (104 °F). If the water flow is very high,
    (a) Find the required effectiveness.
    (b) If two smaller exchangers are used in counterflow, what is the effectiveness of each?
    (c) What is the needed $U_x A_x$?

15. A 185 m² (1911 ft²) house is to be baseboard-heated with natural convectors rated at 865 W/m (900 Btu/hr ft) for a 93.33 °C (200 °F) water temperature.
    (a) If the house has two stories and is twice as long as it is wide, how many feet of baseboard can be installed at maximum?
    (b) Estimate the heating loss (MJ/°C-day, W/°C, or Btu/°F-day) that can be met if the baseboard temperature is 40 °C (104 °F), the outside ambient is −15 °C (5 °F), and 20 °C (68 °F) is held inside the house.
    (c) Plot the heating output versus operating temperatures from 30 to 85 °C (85–185 °F).

16. Repeat problem 15 for radiant heating from the ceilings.

17. (a) Correct the plot of problem 16 for fin efficiency if the ceiling is heated by ¾ in., type K copper tubes set in 16 mm (⅝ in.) gypsum plaster on 20.3 cm (8 in.) center lines.
    (b) Add the effect of metal plaster lath that is made from 0.51 mm (0.020 in.) steel covering 10% of the superficial area.

18. (a) By formulating and integrating the proper differential equations, derive eq. 7-9b by making a balance between the rate of heat transfer and the resultant heat rise in the small stream over a differential area.
    (b) In a similar fashion derive eq. 7-9. (*Hint.* The temperature of one stream in terms of the other can be given by a generalized form of eq. 7-8, where $T_{li}$ and $T_{so}$ are running variables.)

19. Using L'Hopital's theorem, from eq. 7-9 derive
    (a) Equation 7-9a from eq. 7-9.
    (b) Equation 7-11a from 7-11.

20. Compare the results of using eqs. 7-9b and 7-9 when the flow capacity ratio is 20:1 for $N_{TU}$ values of 1, 3, and 10.

21. Derive eq. 7-16a using a derivation similar to that made for eq. 7-16.

22. Derive eq. 7-16b.

23. Derive eq. 7-20.

24. Derive eqs. 7-21a and 7-21b.

25. Derive eq. 7-22.

26. Derive eq. 7-22a.

27. Derive eq. 7-25a.

# REFERENCES

1. W. M. Kays and A. L. London, *Compact Heat Exchangers*, National Press, Palo Alto, Calif., 1955.

2. E. R. G. Eckert and R. M. Drake, Jr., *Heat and Mass Transfer*, Second Edition, New York: McGraw-Hill, 1959.

3. Francis DeWinter, "Heat exchanger penalties in double loop solar water heating systems," *Solar Energy 17*, 335–337 (1975).

4. S. A. Klein, W. A. Beckman, and J. A. Duffie, "A design procedure for solar heating systems," *Solar Energy 18*, 113–127 (1976).

5. P. J. Lunde, *Preliminary Design of a Solar-Powered Desiccant Air Conditioning System*, CEM Report 4186-555, October 1976 (Appendix A). Available from the National Technical Information Service.

6. Adapted from Sterling Radiator Co. sales bulletin SK859A, Westfield, Mass. The bulletin was prepared in consultation with the Hydronics Institute.

7. W. H. McAdams, *Heat Transmission*, Third Edition, New York: McGraw-Hill, 1954. The expressions have been developed from Ackerman's data in Fig. 7-10 using the general form of the Nusselt or Lorenz equation (7-2).

# Storage of the Collected Heat

There are two basic types of storage used in solar-heating systems. *Well-mixed* storage is most common with water storage in space-heating systems, while *stratified* storage is virtually mandatory in air-heating systems and often used for domestic hot water systems. In the following pages the two systems will be described analytically, and the ways in which they are equivalent will be explored.

## STRATIFIED STORAGE

When heated water or air returned from the collector does not mix with fluid near the collector inlet port, the storage is said to be *stratified*. The temperature of the fluid leaving the tank does not change as the storage tank heats, but stays constant. The collector therefore operates at a constant efficiency until the storage is "full," at which time the first-collected heat reaches the tank exit port and a marked rise in collector inlet temperature occurs, accompanied by marked decrease in collector efficiency. The process can then be repeated with the collector operating at a new constant efficiency until storage becomes full again.

With perfectly stratified water storage, heated water from the collector is returned to the top of the tank and will remain there because it is less dense than the water below. When there is hotter (less dense) water already at the top, the returned water will descend until it reaches a level in the tank having the same temperature. Imperfection in such storage exists because the descending warm water tends to entrain some of the hotter water as it descends, mixing the tank upper layers. This ordinarily happens only when the returned fluid is cooler than that at the top of the tank, a situation normally occurring only in the late afternoon as the sun's intensity is waning.

Stratified *discharge* is desirable and theoretically possible from all storage whether charged in the well-mixed or the stratified mode. This is done by returning the cooled fluid to the bottom of the storage tank, not letting it mix with warmer layers above. In practice this is easily done with domestic hot water, and the next day's collector operation is thereby conducted for as long as possible with cold water, reducing the collector heat losses.

Stratified storage for air is quite different. Although the heat transfer medium is air, the storage medium is a solid, typically stones a few centimeters (an inch or so) in diameter. Ideally the stones are uniform spheres that pack together closely in a pattern not unlike a crystal structure. They touch each other only at a single point, and therefore heat is transferred only with difficulty by conduction from one stone to another.

The beds are usually arranged with the heated air returning from the collectors at the top, and with the collector supply air taken from the bottom of the bed. This arrangement is not mandatory, as it is with the stratified water storage, and some rock beds are ported at the sides instead.

When hot air enters the bed, it heats only the first few layers of stone because the storage bed is quite large in cross section and the velocity of the air very low. By the time these few layers are heated, the air has been cooled to the original temperature of the stones, and it continues at that temperature until it exits at the bottom. When more air enters the top of the bed it tends to sweep the previously collected heat down the bed a short distance and replace it with heat at a new temperature, regardless of whether the new temperature is cooler or warmer than the old. Thus the temperature moves down the bed as a wave—a wave moving much more slowly than the actual air flow. In practice the stratification is not perfect, and there will be heat transfer between adjacent areas of the bed, moderating the peaks and valleys that the daily weather pattern will impose on the bed. Such broadening of the temperature stratification is of little consequence, since the heat is not lost, but still stored in almost the same volume. The only difficulty comes at the interface between the coldest portion of the bed and the stored heat— heat transfer past that interface reduces the capacity remaining for the bed to supply the cool air desirable for most efficient collector operation.

More serious imperfections in the stratified rock bed include channeling, where the flow is faster in one area of the bed than the other, and ineffective stratification because of too fast an air flow, too large stones, or too shallow a bed.

Discharge from a rock storage bed is simple and does not upset the stratification. It is accomplished by reversing the flow through the bed. Cool return air then enters the bottom and forces the stored heat upward. It is a "last in–first out" discharge technique, with the bottom of the bed automatically returning to room temperature to provide the collectors with cool air for as long as possible during the next day.

## Heat Collection with Stratified Storage

Consider the solar collector and stratified storage tank in Fig. 8-1 under constant operating conditions. Until the storage tank is full, the collector collects heat

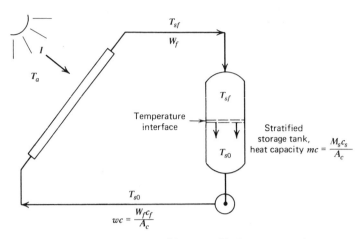

**FIGURE 8-1.** A collector with a stratified storage tank.

according to eq. 5-35 with $\overline{T}_i = T_{s0}$, and thus over a time period $t_T$:

$$q_T = F_i'\overline{\tau\alpha}I_T - F_i'U_L(T_{s0} - \overline{T}_a)t_T \tag{8-1}$$

where a heat exchanger factor has been included as $F_i' = F_iF_x$, even though it is not used in this particular illustration. The storage becomes full after a time $t_s$ sufficient for the incoming collector fluid flow $W_f$ to heat the entire mass $M_s$ of the contents of the storage tank to the final storage temperature $T_{sf}$. At this time the heat collected exactly equals the heat required to raise the entire tank to that outlet temperature and is also equal to the heat delivered by flow of fluid to the storage tank over time period $t_s$; thus

$$Q_T = M_sc_s(T_{sf} - T_{s0}) = W_fc_f(T_{sf} - T_{s0})t_s$$

In this equation $c_s$ is the heat capacity of the storage tank and $c_f$ the storage fluid heat capacity, which will be identical for water storage but differ with the rock-air storage system.

Solving for $t_s$, the storage turnover time, and dividing by the collector area,

$$t_s = \frac{M_sc_s}{W_fc_f} = \frac{m_sc_s}{w_fc_f} \tag{8-2}$$

where $m_sc_s$ and $w_fc_f$ are the storage capacity and fluid flow capacity rate, respectively, normalized for the collector area.

After the expiration of one turnover time period a new storage tank temperature is calculated:

$$T_{sf} = T_{s0} + \frac{q_T}{m_sc_s} \tag{8-3}$$

and the process repeated for the length of the next collection period. If the operating conditions are not constant, use of the customary totals and averages in eqs. 8-1 and 8-3 will give the mean storage temperature* so long as $t_T \leq t_s$.

Notice that stratified storage implies a sufficiently low flow to the collectors that there will be an appreciable temperature rise in the returned stream. This will cause the heat transfer factor $F'_i$ to become smaller, decreasing the collector efficiency. If the flow is kept high the turnover time will be so short that the storage will in reality become well-mixed.

**EXAMPLE 8-1**

A solar collector having $F_m\overline{\tau\alpha} = 0.8$ and $F_m U_L = 5.224$ W/°C·m² (0.92 Btu/hr °F ft²) is operated without a heat exchanger with a liquid flow of water of $wc = 45.43$ W/°C·m² (8 Btu/hr °F ft²) and a stratified storage capacity $mc = 0.3271$ MJ/°C·m² (16 Btu/°F ft²). Estimate the total heat collected and the final storage temperature after 6 hours of operation. Consider the solar input constant at 788.6 W/m² (250 Btu/hr ft²), the ambient temperature at $-1.11$ °C (30 °F), and an initial storage temperature $T_{so}$ of 15.55 °C (60 °F).

First, the operating parameters $F_i\overline{\tau\alpha}$ and $F_i U_L$ are determined, assuming that $U_f$ is constant. From Table 5-2,

$$F_i\overline{\tau\alpha} = \cfrac{1}{\cfrac{1}{F_m\overline{\tau\alpha}} + \cfrac{F_m U_L}{2F_m\overline{\tau\alpha}wc}} = \cfrac{1}{\cfrac{1}{0.8} + \cfrac{5.224}{2 \times 0.8 \times 45.43}} = 0.7565$$

$$F_i U_L = \cfrac{1}{\cfrac{1}{F_m U_L} + \cfrac{1}{2wc}} = \cfrac{1}{\cfrac{1}{5.224} + \cfrac{1}{2 \times 45.43}} = 4.940 \text{ W/°C·m}^2$$

Since there is no heat exchanger, $F'_i = F_i$.

The storage turnover time is given by eq. 8-2 as

$$t_s = \frac{mc}{wc} = \frac{0.3271 \text{ MJ/°C·m}^2}{45.43 \text{ W/°C·m}^2} \times \frac{\text{W}}{\text{J/s}} = 0.007\,20 \text{ Ms}$$

Equation 8-1 therefore applies for times up to 0.0072 Ms (2 hr). The total heat collected in the first 2 hr is therefore given by eq. 8-1 (remembering that $I_T = \overline{I}t_T$):

$$q_T = 0.7565 \times 788.6 \times 0.0072 - 4.940[15.55 - (-1.11)] \times 0.0072$$
$$= 3.703 \text{ MJ/m}^2$$

After one turnover time the entire storage tank will therefore be at the temperature given by eq. 8-3:

$$T_{sf} = T_{so} + \frac{q_T}{mc} = 15.55 \text{ °C} + \frac{3.703 \text{ MJ/m}^2}{0.3271 \text{ MJ/°C·m}^2} = 26.87 \text{ °C}$$

---

* The temperature the tank would have if its contents were mixed.

During the second turnover time collection will continue with the inlet temperature constant at 26.87 °C:

$$q_T = 0.7565 \times 788.6 \times 0.0072 - 4.940[26.87 - (-1.11)] \times 0.0072$$
$$= 3.300 \text{ MJ/m}^2$$

$$T_{sf} = 26.87 + \frac{3.300}{0.3271} = 36.96 \text{ °C}$$

and for the third turnover time period,

$$q_T = 0.7565 \times 788.6 \times 0.0072 - 4.940[36.96 - (-1.11)] \times 0.0072$$
$$= 2.941 \text{ MJ/m}^2$$

$$T_{sf} = 36.96 \text{ °C} + \frac{2.942 \text{ °C}}{0.3271} = 45.95 \text{ °C}$$

The total heat collected over the 6-hr period is

$$q_T = (3.703 + 3.301 + 2.942) = 9.946 \text{ MJ/m}^2$$

The rise in temperature of the storage tank in Example 8-1 is shown graphically in Fig. 8-2. Although the three segments seem to be almost a single line, the rate of temperature rise is 20% less in the last 2 hr than in the initial period.

In Engineering units,

$$F_i \overline{\tau\alpha} = \frac{1}{\dfrac{1}{F_m \tau\alpha} + \dfrac{F_m U_L}{2 F_m \overline{\tau\alpha} wc}} = \frac{1}{\dfrac{1}{0.8} + \dfrac{0.92}{2 \times 0.8 \times 8}} = 0.7565$$

$$F_i U_L = \frac{1}{\dfrac{1}{F_m U_L} + \dfrac{1}{2wc}} = \frac{1}{\dfrac{1}{0.92} + \dfrac{1}{2 \times 8}} = 0.8700 \text{ Btu/hr °F ft}^2$$

Since there is no heat exchanger, $F_i' = F_i$.

The storage turnover time is given by eq. 8-2 as

$$t_s = \frac{mc}{wc} = \frac{16 \text{ Btu/°F ft}^2}{8 \text{ Btu/hr °F ft}^2} = 2 \text{ hr}$$

For the first 2 hours,

$$q_T = 0.7565 \times 250 \times 2 - 0.87(60 - 30)2 = 326.05 \text{ Btu/ft}^2$$

$$T_{sf} = T_{so} + \frac{q_T}{mc} = 60 \text{ °F} + \frac{326.05 \text{ Btu/ft}^2}{16 \text{ Btu/°F ft}^2} = 80.38 \text{ °F}$$

with the inlet temperature constant at 80.38 °F:

$$q_T = 0.7565 \times 250 \times 2 - 0.87(80.38 - 30)2 = 290.59 \text{ Btu/ft}^2$$

$$T_{sf} = 80.38 \text{ °F} + \frac{290.59}{16} \text{ °F} = 98.54 \text{ °F}$$

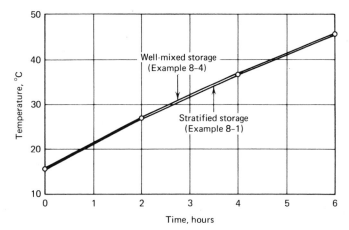

**FIGURE 8-2.** Performance of collector-storage systems having the same collector and storage capacity with well-mixed and stratified storage.

and, for the third turnover time period,

$$q_T = 0.7565 \times 250 \times 2 - 0.87(98.54 - 30)2 = 258.99 \text{ Btu/ft}^2$$

$$T_{sf} = 98.54 \text{ °F} + \frac{258.99}{16} \text{ °F} = 114.73 \text{ °F}$$

The total heat collected over the 6-hr period is

$$q_T = (326.05 + 290.59 + 258.99) = 875.63 \text{ Btu/ft}^2$$

## WELL-MIXED STORAGE

Perfectly well-mixed storage is easy to visualize. It is practical only with liquid storage. When a volume of water in storage is well-mixed, its temperature is uniform and tends to rise slowly as it is heated during the day. As the storage rises in temperature, the efficiency of the collector tends to drop, other factors being equal. Hence the stored water temperature tends to rise more rapidly early in the collection period. Well-mixed storage might seem to be inherently less efficient than stratified storage, but compensating for this is the high heat transfer factor $F_i$ attainable with the rapid flow of fluid through the collector, which makes the initial collector efficiency higher than with stratified storage.

When less than perfectly well-mixed, the circulating fluid can bypass from inlet to outlet, which has the effect of reducing short-term storage although the long-term capacity may be retained. This situation is often controlled by building baffles into the tank or by using the circulation pump to deliberately mix the contents of the tank.

A well-mixed storage tank is usually discharged the same way it is heated—by gradual cooling during a very rapid turnover of the contents to keep the tank well mixed. If the returned fluid is cold—as in the case of a domestic hot water tank—stratified discharge is desirable. Well-mixed operation the following day would then ideally involve only the cold water portion of storage. Collector supply and return would be only to the cool portion of the tank until it reached the temperature of the remainder of the tank. Such a strategy would operate the collector at the lowest possible temperature, but it is never practiced, due to plumbing inflexibility.* Instead, the water return is placed at the top of the tank and, since the collector flow must be high with well-mixed storage, the potential advantage of stratified discharge is quickly dissipated.

## Heat Collection with Well-Mixed Storage

In Chapter 4 the overall energy balance concept was applied to a solar collector and eqs. 4-11 and 4-21 were derived for prediction of instantaneous collector performance and the prediction of collector performance over a period of time. Each of these equations was based on the average collector surface temperature $T_c$. In Chapter 5 the average collector temperature was eliminated and eqs. 5-34 and 5-35 were derived based on the collector *inlet* temperature with the addition of the *flow factor* $F_i$. In Chapter 7 by means of a *heat exchanger factor* $F_x$ the

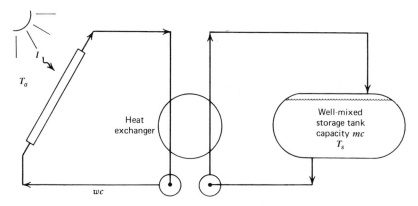

**FIGURE 8-3.** Schematic of a collection system comprising collector, heat exchanger, and well-mixed storage tank.

* It is approximated, however, in a vertical hot water tank with an immersed heat exchanger at the bottom. Natural convection will mix the contents of the tank that lie below the stratified hot layer.

instantaneous equation has been modified to a form that applies to the system in Fig. 8-3:

$$q = F_i'[\tau\alpha I - U_L(T_s - T_a)] \qquad (7\text{-}19)$$

and there is a corresponding form of the integrated equation

$$q_T = F_i'[\overline{\tau\alpha}I_T - U_L(\overline{T}_s - \overline{T}_a)t_T] \qquad (8\text{-}4)$$

In these two equations the system temperature involved is the storage temperature, and the collector inlet temperature is no longer needed.* The form of eq. 8-4 can still use improvement, however, for the average storage temperature $\overline{T}_s$ is a dependent variable, varying not only with the initial storage temperature but also with the progress of heat collection throughout the time period in question. We will therefore now derive modifications to permit use of the initial storage temperature to predict performance over a single collection period, typically a day, for which the initial storage temperature is known.

## The Integrated Storage Equation

Simulations have shown that for a short operating period characterized as is the solar day by a symmetrical radiation pattern about solar noon, the time-averaged temperature of the storage tank over the operating time period can be approximated by averaging the initial storage temperature $T_{s0}$ and the final storage temperature $T_{sf}$. Expressed mathematically,

$$\overline{T}_s = \frac{T_{sf} + T_{s0}}{2} \qquad (8\text{-}5)$$

The final temperature can also be characterized by making a heat balance over the period of collector operation:

$$\frac{\text{Heat collected}}{\text{by collector}} = \frac{\text{Heat added}}{\text{to storage}}$$

$$Q_T = Mc(T_{sf} - T_{s0})$$

---

* If heat loss from the tank and piping reaching the *outside ambient* is included, the proper coefficient for the heat loss term is $(F_i'U_L + U_s a_s + U_p a_p)$ where $U_s$ and $U_p$ are the storage and piping heat loss coefficients; and $a_s$ and $a_p$ are the ratio of storage and piping heat loss area to collector area. This correction need not be made if storage or piping heat losses are small, or provide heat to the structure and so are not net losses. When these or similar equations such as eqs. 9-2, 9-3, 9-4, 9-16, 9-29, and 9-31 are applied over a monthly time period $t_m$, the storage loss is continuous over the full month and its loss term should be modified to $U_s a_s t_m/t_T$, where $t_m/t_T$ can be estimated from Appendix B tables. This assumes storage losses over the month occur from the average *operating* storage temperature to the average *operating* ambient temperature, a convenient simplification.

where $M$ is the mass of the liquid in storage. If both sides of this equation are divided by the collector area $A_c$

$$q_T = mc(T_{sf} - T_{s0}) \tag{8-6}$$

where $m = M/A_c$. This normalization for collector area permits expression of the storage capacity as $mc$, the product of the mass in storage per unit collector area times its specific heat. $T_{sf}$ is then eliminated by combining eq. 8-5 with 8-6 giving

$$\bar{T}_s = T_{s0} + \frac{q_T}{2mc} \tag{8-7}$$

which can be substituted in eq. 8-4 and solved for $q_T$ to give the integrated storage equation:

$$q_T = \frac{F_i'[\overline{\tau\alpha}I_T - U_L(T_{s0} - \bar{T}_a)t_T]}{1 + \dfrac{F_i'U_Lt_T}{2mc}} \tag{8-8}$$

This equation expresses the performance of a solar collector, heat exchanger, and well-mixed storage tank over an operating period $t_T$. It takes into account the fact that the storage temperature rises throughout the collection period, continuously decreasing the collector performance because of the increased heat losses that accompany an increasing collector temperature. The numerator is in the form of eq. 8-4, and is in fact the performance with an infinite storage capacity having a constant temperature of $T_{s0}$. The denominator modifies this to account for the real storage.

The equation can be applied not only at the end of a day, but to any time period $t_T$ from the start of collection. Consider, for instance, the day of operation summarized in Table 4-2.

**EXAMPLE 8-2** (See diagram in Fig. 8-4)

A collector is oriented south with a 50° tilt at 40°N on December 21. Using ASHRAE tables (Table A-3) describe the hourly performance of the collector and well-mixed storage tank if $F_i' = 1$, $\overline{\tau\alpha} = 0.8$, $U_L = 5.224$ W/°C·m² (0.92 Btu/hr °F ft²), $T_{s0} = 46.11$ °C (115 °F), and $mc = 0.2044$ MJ/°C·m² (10 Btu/°F ft²). Hourly temperatures and radiation levels are the same as those given in Table 4-2.

The most accurate answer will be obtained by using eq. 8-8 to find $q_T$ for each hourly period and then finding a new $T_{s0}$ for the next hour by solving for $T_{sf}$ in eq. 8-6. This procedure is called a *simulation*.

$$T_{s0(\text{new})} = T_{sf} = T_{s0(\text{old})} + \frac{q_T}{mc}$$

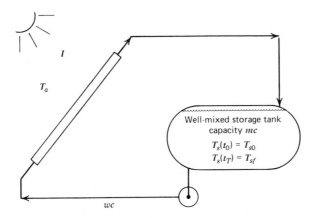

**FIGURE 8-4.** A collector with a well-mixed storage tank.

This hourly simulation will be shown below to be equivalent to using the exact eq. 8-4.

For the first hour (0.0036 Ms), $I$ is 519.6 W/m² and $T_a$ is 0.91 °C, so that

$$q_T = \frac{0.8 \times 516.9 \times 0.0036 - 5.224(46.11 - 0.91) \times 0.0036}{1 + \dfrac{5.22 \times 0.0036}{2 \times 0.2044}}$$

$$= \frac{0.6386}{1.046} = 0.6105 \text{ MJ/m}^2$$

$$T_{s0(new)} = 46.11 \text{ °C} + \frac{0.6105 \text{ MJ/m}^2}{0.2044 \text{ MJ/°C·m}^2} = 49.10 \text{ °C}$$

This process is repeated hourly, with results given as $q_T$ and $T_s$ in Table 8-1a.

The accuracy of the assumption of eq. 8-5 can now be checked for the hour-long steps. From 10:30 A.M. to 11:30 A.M., for instance,

$$T_{s0} = 54.90 \text{ °C} \quad \text{and} \quad T_{sf} = 61.99 \text{ °C}$$

therefore

$$\bar{T}_s = (54.90 + 61.99)/2 = 58.445 \text{ °C}$$

Using this in eq. 8-4 gives

$$q_T = 0.8 \times 870 \times 0.0036 - 5.224(58.445 - 2.27) \times 0.0036$$
$$= 1.4491 \text{ MJ/m}^2$$

exactly as in Table 8-1a, showing that eq. 8-8 is very accurate for 1-hour time steps, and is equivalent to use of eq. 8-4.

**TABLE 8-1a** (SI Units). Performance of a Flat Plate Collector with Storage at 40°N Tilted 50° with Horizontal on December 21, for Example 8-2

| Time | $I$ W/m² | $T_a$ °C | Using eq. 8-4[a] $q_T$ MJ/m² | Using eq. 8-4[a] $T_s$ °C | Using eq. 8-8 $T_s$ °C |
|------|------|------|------|------|------|
| 9 | 516.9 | 0.91 | 0.6105 | 46.11 | 46.11 |
| 10 | 740.7 | 1.68 | 1.1869 | 49.10 | 49.10 |
| 11 | 870.0 | 2.27 | 1.4491 | 54.90 | 54.78 |
| 12 | 914.1 | 2.86 | 1.4537 | 61.99 | 61.66 |
| 13 | 870.0 | 3.35 | 1.2132 | 69.10 | 68.57 |
| 14 | 740.7 | 3.33 | 0.7501 | 75.04 | 74.50 |
| 15 | 516.9 | 3.23 | 0.0661 | 78.71 | 78.57 |
| Final | | | | 79.03 | 79.97 |
| Average | | 2.52 | | | |
| Total | 5169.3 | | 6.7296 | | |

NOTE: System characteristics: $\overline{\tau\alpha} = 0.8$; $U_L = 5.224$ W/°C·m²; $mc = 0.2044$ MJ/°C·m², $T_{s0} = 46.11$ °C.

[a] As explained in the text, this is implemented by hourly use of eq. 8-8.


**TABLE 8-1b** (Engineering Units). Performance of a Flat Plate Collector with Storage at 40°N Tilted 50° with Horizontal on December 21, for Example 8-2

| Time | $I$ Btu/hr ft² | $T_a$ °F | Using eq. 8-4 $q_T$ Btu/ft² | Using eq. 8-4 $T_{s0}$ °F | Using eq. 8-8 $T_s$ °F |
|------|------|------|------|------|------|
| 9 | 164 | 33.63 | 53.86 | 115. | 115. |
| 10 | 235 | 35.02 | 104.65 | 120.39 | 120.39 |
| 11 | 276 | 36.09 | 127.74 | 130.85 | 130.63 |
| 12 | 290 | 37.15 | 128.15 | 143.63 | 143.03 |
| 13 | 276 | 38.03 | 106.94 | 156.44 | 155.48 |
| 14 | 235 | 37.99 | 66.14 | 167.12 | 166.16 |
| 15 | 164 | 37.81 | 5.87 | 173.75 | 173.50 |
| Final | | | | 174.34 | 176.02 |
| Average | | 36.53 | | | |
| Total | 1640 | | 593.36 | | |

NOTE: System characteristics: $\overline{\tau\alpha} = 0.8$; $U_L = 0.92$ Btu/hr °F ft²; $mc = 10$ Btu/°F ft²; $T_{s0} = 115$ °F.

Equation 8-8 can be used to predict the temperature with a longer time step but there may be a loss of accuracy. Over the first five hours, for instance,

$$I_T = 516.9 + 740.7 + 870.0 + 914.1 + 870.0 = 3912 \text{ W·h/m}^2$$
$$I_T = 3912 \text{ W·h/m}^2 \times 0.0036 \text{ Ms/h} = 14.083 \text{ MJ/m}^2$$
$$\bar{T}_a = (0.91 + 1.68 + 2.27 + 2.86 + 3.35)/5 = 2.214 \text{ °C}$$
$$t_T = 5 \text{ h} \times 0.0036 \text{ Ms/h} = 0.018 \text{ Ms}$$

$$q_T = \frac{0.8 \times 14.083 - 5.224(46.11 - 2.214)0.018}{\left(1 + \dfrac{5.22 \times 0.018}{2 \times 0.2044}\right)} = \frac{7.139}{1.2299} = 5.804 \text{ MJ/m}^2$$

$$T_{sf} = 46.11 + \frac{5.804}{0.2044} = 74.51 \text{ °C}$$

which compares with eq. 8-4 values of 75.04 °C and a total of 5.913 MJ/m² collected, showing that eq. 8-8 is not exact for a longer period. The difference of 2% is due to the approximation in eq. 8-5. It is not large enough to be of engineering significance.

Results using eq. 8-8 to determine each hourly temperature in a single step are tabulated in the last column of Table 8-1$a$. The entire day's performance is predicted by

$$q_T = \frac{[0.8 \times 5169 - 5.224(46.11 - 2.52)7]0.0036}{1 + \dfrac{5.224 \times 7 \times 0.0036}{2 \times 0.2044}} = \frac{9.148}{1.322}$$

$$= 6.92 \text{ MJ/m}^2$$

$$T_{sf} = 46.11 + \frac{6.92}{0.2044} = 79.97 \text{ °C}$$

which is some 3% higher in $q_T$ than the exact value computed by the hourly simulation, which has been shown to be equivalent to eq. 8-4. Values from eq. 8-8 compared to the exact simulation in the upper pair of curves in Fig. 8-5, which plot the last two columns of Table 8-1. While the deviations are not large, they are relatively high since this is an efficient collector and the storage volume is really too small.

In Engineering units, for the first hour,

$$q_T = \frac{0.8 \times 164 - 0.92(115 - 33.63) \times 1}{1 + \dfrac{0.92 \times 1}{2 \times 10}} = 53.86 \text{ Btu/ft}^2$$

$$T_{s0(new)} = 115 \text{ °F} + \frac{53.86 \text{ Btu/ft}^2}{10 \text{ Btu/°F ft}^2} = 120.39 \text{ °F}$$

and, from 10:30 A.M. to 11:30 A.M.,

$$T_{s0} = 130.85 \text{ °F} \quad \text{and} \quad T_{sf} = 143.63 \text{ °F}$$
$$\bar{T}_s = (130.85 + 143.63)/2 = 137.24 \text{ °F}$$

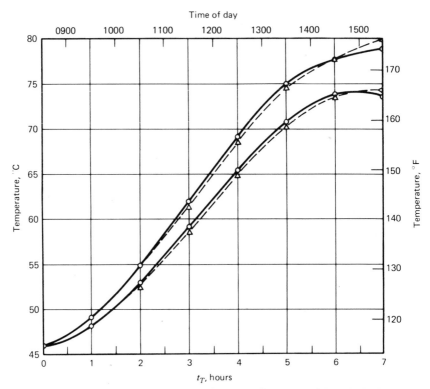

**FIGURE 8-5.** Plot of temperatures from Tables 8-1 and 8-2. Solid lines are the hour-by-hour simulation, dotted lines are the integrated equation applied from the time zero initial conditions.

Using this in eq. 8-4 gives

$$q_T = 0.8 \times 276 - 0.92(137.24 - 36.09) = 127.74 \text{ Btu/ft}^2$$

exactly as in Table 8-1$b$.

Equation 8-8 can be used to predict the temperature at any hour. At the end of the fifth hour, for instance,

$$I_T = 164 + 235 + 276 + 290 + 276 = 1241 \text{ Btu/ft}^2$$
$$\bar{T}_a = (33.63 + 35.02 + 36.09 + 37.15 + 38.03)/5 = 35.984 \text{ °F}$$
$$t_T = 5 \text{ hr}$$

$$q_T = \frac{0.8 \times 1241 - 0.92(115 - 35.984)5}{\left(1 + \dfrac{0.92 \times 5}{2 \times 10}\right)} = 511.65 \text{ Btu/ft}^2$$

$$T_{sf} = 115 + \frac{511.65}{10} = 166.16 \text{ °F}$$

which compares with Table 8-1b values of 167.12 °F and a total of 521.34 Btu/ft² collected, showing that eq. 8-4 is not exact for longer periods. At the end of the day's operation, using eqs. 8-4 and 8-6,

$$q_T = \frac{0.8 \times 1640 - 0.92(115 - 36.53)7}{1 + \dfrac{0.92 \times 7}{2 \times 10}} = 610.18 \text{ Btu/ft}^2$$

$$T_{sf} = 115 + \frac{610.18}{10} = 176.02 \text{ °F}$$

## The Integrated Storage Equation with Simultaneous Heat Load

Equation 8-8 does not represent a realistic situation for most solar applications because demand satisfied during the collector operation period is not considered. Such demand, shown in the system schematic of Fig. 8-6, will keep the temperature in storage down somewhat to permit a few percent more heat to be collected under given conditions.

Demand can be included by making an alternate derivation in which a new variable, the net heat collected, is given by

$$q_N = q_T - l_t \tag{8-9}$$

where $l_t$ is the demand expressed as the heat load per unit of collector area that occurs during the time $t_T$. When eq. 8-7 is rederived using net heat, it can be restated as

$$\overline{T}_s = T_{s0} + \frac{q_N}{2mc} \tag{8-7a}$$

which can be substituted into eq. 8-4. If the resultant expression for $q_T$ is substituted in eq. 8-9 and solved for $q_N$, the net heat collection is given by

$$q_N = \frac{F_i'[\overline{\tau\alpha}I_T - U_L(T_{s0} - \overline{T}_a)t_T] - l_t}{1 + \dfrac{F_i'U_Lt_T}{2mc}} \tag{8-10}$$

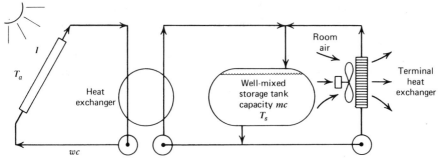

**FIGURE 8-6.** Schematic of a collection system including collector, heat exchanger, well-mixed storage tank, and terminal heat exchanger.

which can be substituted into eq. 8-9 to give the total heat collected as

$$q_T = \frac{F_i'[\overline{\tau\alpha}I_T - U_L(T_{s0} - \overline{T}_a)t_T]}{1 + \dfrac{F_i'U_Lt_T}{2mc}} + \frac{l_t}{1 + \dfrac{2mc}{F_i'U_Lt_T}} \qquad (8\text{-}11)$$

Equation 8-10 is the integrated net storage equation, which can be used to predict the rise in temperature of the storage during the day in combination with

$$T_{sf} = T_{s0} + \frac{q_N}{mc} \qquad (8\text{-}12)$$

Equation 8-11, the integrated total storage equation, predicts the total heat collected whether it was used or stored.*

The form of eqs. 8-8, 8-10, and 8-11 is interesting. The numerator in eq. 8-8 is the amount of heat that would be collected if the storage temperature were constant at the initial value. The denominator is a storage factor that corrects this value for the rising storage temperature during the day. This factor gets larger as $t_T$ increases, and $t_T$ is the only time variant variable in the denominator. Equation 8-10 is similar but the numerator gives the *net* heat that would go to storage if the storage temperature were constant. In eq. 8-11 there are two terms. One is associated with the heat collected, the other with the heating demand. The first term is exactly that of eq. 8-8, that is, the total heat that would be collected if there were no load. The second term is *added* to the first and represents the increased heat recovery due to having the demand simultaneous with the heat collection, keeping the storage temperature somewhat lower and thereby increasing the collector efficiency. Notice that either equation approaches eq. 8-4 in form as the storage capacity is increased.

When the storage capacity is very high these integrated equations are highly accurate for predicting daily performance. As the storage capacity is decreased to the practical range of 0.2 to 0.4 MJ/°C·m² (10 to 20 Btu/°F ft²), the daily errors become noticeable but never approach the ordinary tolerance of engineering equations, which is at best about ±5%. This accuracy (and usually much better) is held regardless of the imposed weather patterns because the storage installed is adequate to damp out such short-term variations. Therefore, even with realistic storage installed, we can continue to make daily calculations unconcerned about the time distribution of bright and cloudy weather, since the variation in performance due to this effect is not great.

**EXAMPLE 8-3**

Repeat Example 8-2 considering the effect of a simultaneous heating load.

* See footnote p. 287.

Table 8-2a is based on the same weather, collector, and storage information as Table 8-1a. Values of hourly load $l_t$ have been added for each hour that takes into account the outside temperature and a particular ratio of installed collector area to structure loss coefficient. Exact values of $q_T$ and $T_s$ have been calculated by an hour-by-hour simulation using eqs. 8-11, 8-9, and 8-12, which is equivalent to use of eq. 8-4.

The total collected heat $q_T$ is about 6% larger than in Table 8-1, because of the lower collector temperature with simultaneous load.

To calculate the temperature at any given time in a single step, the solutions of Example 8-2 using eq. 8-8 can be used as the first term in eq. 8-11. The second term of eq. 8-11 is then added to the first term and eq. 8-12 is applied. For instance, for the entire day of Table 8-2a, the second term in eq. 8-11 is

$$\frac{l_t}{1 + \dfrac{2mc}{F_t' U_L t_T}} = \frac{1.5349 \text{ MJ/m}^2}{1 + \dfrac{2 \times 0.2044}{5.224 \times 7 \times 0.0036}} = 0.3739 \text{ MJ/m}^2$$

$$q_T = 6.9200 + 0.3739 = 7.2939 \text{ MJ/m}^2$$
$$\text{(from Example 8-2)}$$

and

$$T_{sf} = \frac{(7.2939 - 1.5349)}{0.2044} + 46.11 = 74.29 \, ^\circ\text{C}$$

The exact hourly storage temperatures and those calculated in a single step from the integrated equations have been plotted as the lower pair of lines in Fig. 8-5. Once again it is seen that the errors are small and the integrated equations are nearly exact for 2 to 3 hours.

**TABLE 8-2a** (SI Units).  Performance for the Situation in Table 8-1 with Consideration of a Simultaneous Heat Load (Example 8-3)

| Time | $I$ W/m² | $T_a$ °C | $l_t^b$ MJ/m² | Using eq. 8-4[a] $q_T$ MJ/m² | Using eq. 8-4[a] $T_{s0}$ °C | Using eq. 8-10 $T_s$ °C |
|---|---|---|---|---|---|---|
| 9 | 516.9 | 0.91 | 0.2416 | 0.6212 | 46.11 | 46.11 |
| 10 | 740.7 | 1.68 | 0.2310 | 1.2173 | 47.97 | 47.97 |
| 11 | 870.0 | 2.27 | 0.2228 | 1.4968 | 52.79 | 52.67 |
| 12 | 914.1 | 2.86 | 0.2146 | 1.5164 | 59.03 | 58.67 |
| 13 | 870.0 | 3.35 | 0.2078 | 1.2890 | 65.39 | 64.81 |
| 14 | 740.7 | 3.33 | 0.2081 | 0.8376 | 70.68 | 70.05 |
| 15 | 516.9 | 3.23 | 0.2095 | 0.1643 | 73.76 | 73.49 |
| Final | | | | | 73.54 | 74.29 |
| Average | | 2.52 | | | | |
| Total | 5169.3 | | 1.5353 | 7.1426 | | |

[a] As explained in the text, this is implemented by hourly use of eq. 8-10.

[b] $l_t = (T_b - T_a)^+ / [D_a (A_c/L_a)] = (T_b - T_a)^+ / [(UA)_{\text{bldg}}/A_c]$ where the quantity in brackets is arbitrarily but realistically 72.11 hr °C·m²/MJ (1.4741 °F hr ft²/M Btu).

**TABLE 8-2b** (Engineering Units). Performance for the Situation in Table 8-1 with Consideration of a Simultaneous Heat Load (Example 8-3)

| Time | $I$ Btu/hr ft² | $T_a$ °F | $l_t$ Btu/ft² | Using eq. 8-4 $q_T$ Btu/ft² | Using eq. 8-4 $T_s$ °F | Using eq. 8-10 $T_s$ °F |
|------|------|------|------|------|------|------|
| 9 | 164 | 33.63 | 21.28 | 54.80 | 115. | 115. |
| 10 | 235 | 35.02 | 20.34 | 107.33 | 118.35 | 118.35 |
| 11 | 276 | 36.09 | 19.61 | 131.95 | 127.05 | 126.83 |
| 12 | 290 | 37.15 | 18.89 | 133.67 | 138.28 | 137.65 |
| 13 | 276 | 38.03 | 18.30 | 113.62 | 149.76 | 148.72 |
| 14 | 235 | 37.99 | 18.32 | 73.84 | 159.30 | 158.16 |
| 15 | 164 | 37.81 | 18.45 | 14.51 | 164.85 | 164.35 |
| Final | | | | | 164.45 | 165.79 |
| Average | | 36.53 | | | | |
| Total | 1640 | | 135.20 | 629.73 | | |

In Engineering units, for the entire day, the second term of eq. 8-11 gives

$$\frac{l_t}{1 + \dfrac{2mc}{F_i' U_L t_T}} = \frac{135.20 \text{ Btu/ft}^2}{1 + \dfrac{2 \times 10}{0.92 \times 7}} = 32.93 \text{ Btu/ft}^2$$

$$q_T = 610.18 + 32.93 = 643.11 \text{ Btu/ft}^2$$
$$\text{(from Example 8-2)}$$

and

$$T_{sf} = \frac{643.11 - 135.20}{10} + 115 = 165.79 \text{ °F}$$

which compares with 164.45 °F using the exact solution in which eq. 8-10 was applied hourly, as shown in Table 8-2b.

## COMPARISON OF STRATIFIED STORAGE WITH WELL-MIXED STORAGE

If the storage and the collector are considered together as a system, analysis can be simplified. In the following paragraphs, the *system* consisting of a collector plus stratified storage is compared with the corresponding collector and well-mixed storage system. Their performance, surprisingly, is comparable because the disadvantages of a low collector flow are compensated by the advantages of stratified storage.

Equation 8-1 gave the performance of stratified storage as

$$q_T = F_i'[\overline{\tau\alpha}I_T - U_L(T_{s0} - \overline{T}_a)t_T] \tag{8-1}$$

which can be rewritten considering that

$$F_i = \cfrac{1}{\cfrac{1}{\bar{\eta}_F} + \cfrac{U_i A_c}{U_f A_f} + \cfrac{U_L}{2wc}} = \cfrac{1}{\cfrac{1}{F_m} + \cfrac{U_L}{2wc}} \tag{5-32a}$$

to give

$$q_T = \cfrac{F_x[\overline{\tau\alpha}I_T - U_L(T_{s0} - \bar{T}_a)t_T]}{\cfrac{1}{F_m} + \cfrac{U_L}{2wc}} \tag{8-13}$$

while for well-mixed storage eq. 8-8 can similarly be rewritten as

$$q_T = \cfrac{F_x[\overline{\tau\alpha}I_T - U_L(T_{s0} - \bar{T}_a)t_T]}{\cfrac{1}{F_m} + \cfrac{U_L}{2wc} + \cfrac{F_x U_L t_T}{2mc}} \tag{8-14}$$

after dividing numerator and denominator by $F_i$ and applying eq. 5-32a.

Now if we choose, we may compare stratified storage at an ordinary rate of flow with well-mixed storage at a very high rate of flow—so high that $F_i = F_m$. In this case eq. 8-14 becomes

$$q_T = \cfrac{F_x[\overline{\tau\alpha}I_T - U_L(T_{s0} - \bar{T}_a)t_T]}{\cfrac{1}{F_m} + \cfrac{F_x U_L t_T}{2mc}} \tag{8-15}$$

Now, since the turnover time $t_s = mc/wc$, $wc = mc/t_s$ can be substituted into eq. 8-13 to give, for stratified storage

$$q_T = \cfrac{F_x[\overline{\tau\alpha}I_T - U_L(T_{s0} - \bar{T}_a)t_T]}{\cfrac{1}{F_m} + \cfrac{U_L t_s}{2mc}} \tag{8-16}$$

If we choose to consider performance at the end of one turnover time when $t_s = t_T$, eqs. 8-15 and 8-16 are identical except for the $F_x$ factor in the denominator of eq. 8-15. If $F_x = 1.0$, the equations are exactly equal, and it is thus shown that the combination of collector with well-mixed storage at a rapid flow rate with no heat exchanger (Fig. 8-4) produces the same performance as a collector having a lower flow combined with stratified storage (Fig. 8-1), provided that the systems are evaluated at one turnover time.* For the second turnover time period, the new $T_{s0}$'s are the same (since the heat collected was the same) and the same reasoning holds, leading to the general conclusion that for an integral number of turnover times the performance of stratified and well-mixed systems are equivalent.

---

* If $F_x < 1$ the well-mixed storage is superior. The effect is insignificant at $F_x = 0.95$.

After a few turnover times the cumulative performance of the systems are virtually equivalent at any time.

### EXAMPLE 8-4

For the situation of Example 8-1 and Fig. 8-4 find the heat collected and storage temperature after 2, 4, and 6 hours if the storage is now well-mixed and the flow so fast that $F_i = F_m$.

For the first 2-hour period use of eq. 8-8 with $F_m$ values gives

$$q_T = \frac{0.8 \times 788.6 \times 0.0072 - 5.224[15.55 - (-1.11)]0.0072}{1 + \dfrac{5.224 \times 0.0072}{2 \times 0.3271}}$$

$$= \frac{3.9157}{1.0575} = 3.7028 \text{ MJ/m}^2$$

$$T_{sf} = 15.55 + \frac{3.7028}{0.3271} = 26.87 \text{ °C}$$

and, for the second 2-hour period,

$$q_T = \frac{0.8 \times 788.6 \times 0.0072 - 5.224[26.87 - (-1.11)]0.0072}{1.0575}$$

$$= \frac{3.4899}{1.0575} = 3.3002 \text{ MJ/m}^2$$

$$T_{sf} = 26.87 + \frac{3.3002}{0.3271} = 36.96 \text{ °C}$$

and the third,

$$q_T = \frac{0.8 \times 788.6 \times 0.0072 - 5.224[36.96 - (-1.11)]0.0072}{1.0575}$$

$$= \frac{3.1104}{1.0575} = 2.9413 \text{ MJ/m}^2$$

$$T_{sf} = 36.96 + \frac{2.9413}{0.3271} = 45.95 \text{ °C}$$

Comparison with Example 8-1 shows these results to be identical with those produced using $F_i$ values and eq. 8-1. The results of each are plotted in Fig. 8-2.

The integrated storage equation can also be used to determine performance in a single step over the entire period for stratified storage provided $F_m$ is used instead of $F_i'$, giving for this example

$$q_T = \frac{0.8 \times 788.6 \times 6 \times 0.0036 - 5.224[15.55 - (-1.11)]6 \times 0.0036}{1 + \dfrac{5.224 \times 6 \times 0.0036}{2 \times 0.3271}}$$

$$= \frac{11.7471}{1.1725} = 10.02 \text{ MJ/m}^2$$

$$T_{sf} = 15.55 + \frac{10.02}{0.3271} = 46.18 \, °C$$

The small error (less than 1% in $q_T$) when compared to the analysis above or to Example 8-1 is due to using the integrated storage equation over a relatively long time period.

In Engineering units, for the first 2-hour period, eq. 8-8 gives

$$q_T = \frac{0.8 \times 250 \times 2 - 0.92(60 - 30)2}{1 + \dfrac{0.92 \times 2}{2 \times 16}} = \frac{3448}{1.0575} = 326.05 \, Btu/ft^2$$

$$T_{sf} = 60 + \frac{326.05}{16} = 80.38 \, °F$$

and for the second 2-hour period,

$$q_T = \frac{0.8 \times 250 \times 2 - 0.92(80.38 - 30)2}{1.0575} = \frac{307.30}{1.0575} = 290.59 \, Btu/ft^2$$

$$T_{sf} = 80.38 + \frac{290.59}{16} = 98.54 \, °F$$

and the third,

$$q_T = \frac{0.8 \times 250 \times 2 - 0.92(98.54 - 30)2}{1.0575} = \frac{273.89}{1.0575} = 258.99 \, Btu/ft^2$$

$$T_{sf} = 98.54 + \frac{258.99}{16} = 114.72 \, °F$$

Using the same equation for the entire 6-hour period produces only a small error:

$$q_T = \frac{0.8 \times 250 \times 6 - 0.92(60 - 30)6}{1 + \dfrac{0.92 \times 6}{2 \times 16}} = \frac{1034.4}{1.1725} = 882.22 \, Btu/ft^2$$

$$T_{sf} = 60 + \frac{88.22}{16} = 115.14 \, °F$$

## Choosing the Storage System

When the effects of simultaneous load are considered, the equality in performance described previously shifts slightly to favor the well-mixed case. This is because only the well-mixed storage benefits immediately from simultaneous load through a lowering of collector temperature.

The preceding arguments make it clear that a slower collector flow plus stratified storage is almost, but not quite as good, as rapidly circulating well-mixed storage. The higher the collector flow the shorter the turnover time and the more closely the system performance approaches the ideal case. In practice, as long as the turnover time is less than the normal operating day,* the two storage techniques are equivalent. Stratified storage is useful and often mandatory whenever the flow rate of the cooling fluid is reduced so much that there is a significant temperature rise across the collector. In such cases the detrimental effect of that temperature rise can be balanced out of the system by storing the heat in a stratified bed.

There is little appeal in such an approach when hot water is being stored for space heating; it is easy and cheap to provide high flow rates of fluid to the collector. With air collectors, however, it is impossible to provide enough air flow to the collectors to provide good performance with well-mixed storage because of the inherently high pressure drops and resultant high fan power in the overall system. Stratified storage is thereby the single factor most important in making such air systems competitive with liquid systems.

Systems that circulate the collector fluid slowly have another disadvantage that needs emphasis. As the flow rate of the collector cooling fluid is lowered, the heat transfer coefficient between the fluid and the tube wall or absorber plate decreases until the flow finally becomes laminar. The general relationship follows the form of eq. 5-3 with the coefficient $U_f$ decreasing as the 0.8 power of the ratio of the flow rates. This can have a great effect on the system performance, especially if the flow rate has been slowed considerably in order to obtain high temperatures for special uses, such as hot water for process heat.

In such cases the heat transfer coefficient is under control of the collector designer who can maintain a particular flow rate and still increase velocity as is necessary— with liquid collectors at least—without undue pumping penalties. With air collectors the design situation is more difficult, and a low heat transfer coefficient is a system characteristic. The compensating factor keeping air systems competitive is the availability of low-temperature collector inlet air, a consequence of stratified storage, and the virtually infinite heat exchange area in the storage bed of rock or pebbles.

## HOT WATER SYSTEMS

The previous sections apply to solar heat storage in general, but assume that the load is independent of the stored water temperature. This is true for space heating but not for heating of hot water. Unless the storage tank is above the hot water use temperature the feed water can only be *preheated*, with elevation to the final temperature accomplished with auxiliary fuel or electricity.

It is possible to maintain the entire storage tank at the use temperature by reducing the water flow or by adding auxiliary energy directly, but this would be inefficient

* About 6 to 8 hours.

because higher-temperature coolant would ultimately be pumped through the collectors, increasing losses to the outside ambient temperature. The most efficient water-heating systems devote a special storage volume to *solar*-heated hot water. Cold makeup water enters at the bottom, and preheated water is taken from the top at the same rate. Within a second storage volume, auxiliary heat is added to bring a small quantity of stored water to the final temperature. The second volume can be in the same storage tank at the level above application of the auxiliary heat, or in a separate tank. Since the surface area for heat losses is larger with two tanks, the single tank is preferred.

For most efficient operation, the second storage volume should be minimized. From the user's standpoint, however, a large stored volume of hot water is desirable. Current domestic solar hot water tanks devote about one-quarter of their volume to use-temperature storage, but there has been little formal study of the situation. The *best* approach is to store no fully heated water at all, adding auxiliary heat only as actually needed using a "tankless coil" with a hot water "boiler" or a very high-capacity, in-line electric heater. The tankless coil should be bypassed to provide auxiliary heat only as needed, since it is typically held at 70 to 95 °C (160–200 °F) for space heating, a level that would almost always add heat to the solar-heated water.

Storage in the main storage volume is nearly always well-mixed during solar heating. The well-mixed volume initially will be only that of the cold water that has entered the bottom of the tank during stratified *discharge*, if heat is added by a submerged heat exchanger on the bottom of the tank. As the lower portion heats, the warmer upper layers will join the well-mixed volume, but the cold start increases collector efficiency. For a traced tank or if an external heat exchanger is used, the entire solar tank volume is always active. Stratified *storage* is seldom attempted except for the natural circulation systems used in mild climates, which need no heat exchanger.

## Hot Water Heating with Well-Mixed Storage

In the following analyses we will consider the entire solar storage to be well-mixed. The actual hot water use pattern is unknown and variable; a constant demand is therefore assumed, placing about a third of the demand within the solar day.

If a mass flow rate of water $W_{HW}$ is to be heated from $T_{CW}$ to $T_{HW}$, then the rate of heat withdrawal from a storage tank that is at temperature $T_s$ is

$$L = (Wc)_{HW}(T_s - T_{CW}) \qquad (8\text{-}17)$$

when $T_{CW} \leq T_s \leq T_{HW}$. Normalizing for collector area,

$$l = (wc)_{HW}(T_s - T_{CW}) \qquad (8\text{-}18)$$

Integrating over a time period $t_T$ it can easily be shown that the total load is proportional to the difference in the average temperatures:

$$l_t = (wc)_{HW}(\bar{T}_s - \bar{T}_{CW})t_T \tag{8-19}$$

Assuming again that $\bar{T}_s = (T_{so} + T_{sf})/2$, as in the derivation of eqs. 8-8 and 8-10, combining eqs. 8-4, 8-7a, and 8-9 yields

$$q_N = \frac{F_i'[\bar{\tau\alpha}I_T - U_L(T_{so} - \bar{T}_a)t_T] - (wc)_{HW}(T_{so} - \bar{T}_{CW})t_T}{1 + \dfrac{[F_i'U_L + (wc)_{HW}]t_T}{2mc}} \tag{8-20}$$

This is the integrated net storage equation for hot water. In a single step it predicts the net solar heat added to the storage tank over a time period of solar heating $t_T$, using only the physical parameters of the system, weather data summed or averaged over $t_T$, and the inlet cold water temperature and starting tank temperature. The solar heat collected is automatically decreased as the tank temperature rises while the solar heat delivered to the exiting hot water increases.

The accuracy of eq. 8-20 depends on the validity of eq. 8-5, in which the average storage temperature is taken as the mean between initial and final storage temperatures. This will be true for periods of a few hours, but is less accurate over an entire day. For 1 hour, eq. 8-20 is equivalent to a simulation with an infinitely small time step. For longer periods it will not be accurate if the temperature rises above the hot water use temperature or falls below the supply temperature, since eq. 8-17 no longer holds. The integrated net storage equation (8-10) then applies, where if $T_s > T_{HW}$,* $l_t$ holds at a constant value,

$$l_t = (wc)_{HW}(T_{HW} - \bar{T}_{CW})t_T \tag{8-21}$$

or if $T_s < T_{CW}$,

$$l_t = 0$$

Once $q_N$ has been determined, the temperature rise in the storage tank is given by eq. 8-12. The net heat supplied to the hot water discharged is given by combining eqs. 8-7a and 8-19:

$$l_t = (wc)_{HW}\left(\frac{q_N}{2mc} + T_{so} - \bar{T}_{CW}\right)t_T \tag{8-22}$$

and the total heat collected is

$$q_T = q_N + l_t \tag{8-23}$$

---

* Actually, the water is delivered at $T_s$ even when $T_s > T_{HW}$. An automatic (or manual) mixing control reduces the heated water flow rate in order to continue to deliver the heat quantity expressed in eq. 8-21 by later adding cold water, which reduces the final temperature to $T_{HW}$.

## EXAMPLE 8-5

The system and one-day weather data of Example 8-2 are used to preheat a steady water flow of 0.3867 g/s·m² (0.2854 lb/hr ft²) from its supply temperature of 15.55 °C (60 °F). The final hot water temperature is to be 60 °C (140 °F). If the initial storage temperature is 25 °C (77 °F) find the daily solar contribution to the hot water load, the total heat collected, and the final collector temperature (a) by hourly simulation, and (b) by a single application of the integrated net storage equation for hot water.

(a) First the water flow capacity rate is found.

$$(wc)_{HW} = \frac{0.3867 \text{ g}}{\text{s} \cdot \text{m}^2} \times \frac{4.19 \text{ kJ}}{\text{kg} \cdot °\text{C}} \times \frac{\text{W} \cdot \text{s}}{\text{J}} = 1.6203 \text{ W}/°\text{C} \cdot \text{m}^2$$

Equation 8-20 will now be applied hourly to give results equivalent to a simulation. Refer to Table 8-3 for hourly values for $I$ and $T_a$. When using eq. 8-20, remember that 1 hr is 0.0036 Ms. For the first hour of simulation,

$$q_N = \frac{0.8 \times 516.9 \times 0.0036 - 5.224(25 - 0.91)(0.0036) - 1.6203(25 - 15.55)(0.0036)}{1 + \dfrac{(5.224 + 1.6203)(0.0036)}{2 \times 0.2044}}$$

$$= \frac{0.9805}{1.060} = 0.92476 \text{ MJ/m}^2$$

By eq. 8-12,

$$T_{sf} = \frac{0.92476 \text{ MJ}}{\text{m}^2} \times \frac{°\text{C} \cdot \text{m}^2}{0.2044 \text{ MJ}} + 25 °\text{C} = 29.52 °\text{C}$$

**TABLE 8-3a** (SI Units). Performance for the Situation of Example 8-5 in which there is a Simultaneous Hot Water Load

| | | | Simulated[a] | | | | Eq. 8-20 | |
|---|---|---|---|---|---|---|---|---|
| Time | $I$ W/m² | $T_a$ °C | $l_t$ MJ/m² | $l_t$ (total) MJ/m² | $q_T$ MJ/m² | $T_s$ °C | $l_t$ (total) MJ/m² | $T_{so}$ °C |
| 9 (start) | 516.9 | 0.91 | 0.0683 | 0.0683 | 0.9931 | 25.00 | 0.0683 | 25.00 |
| 10 | 740.7 | 1.68 | 0.1021 | 0.1704 | 1.543 | 29.52 | 0.1770 | 29.52 |
| 11 | 870 | 2.27 | 0.1460 | 0.3164 | 1.785 | 36.58 | 0.3337 | 36.44 |
| 12 | 914.1 | 2.86 | 0.1920 | 0.5084 | 1.775 | 44.59 | 0.5334 | 44.24 |
| 13 | 870 | 3.35 | 0.2330 | 0.7414 | 1.525 | 52.34 | 0.7607 | 51.82 |
| 14 | 740.7 | 3.33 | 0.2628 | 1.0042 | 1.056 | 58.66 | 0.9921 | 58.26 |
| 15[b] | 516.9 | 3.23 | 0.2593 | 1.2635 | 0.368 | 62.54 | 1.1967 | 62.79 |
| Final | | | | | | 63.07 | | 64.71 |
| Average | | 2.52 | | | | | | |
| Total | 5169.3 | | 1.2635 | | 9.045 | | | |

[a] As explained in the text, this is implemented by the hourly use of eq. 8-20.
[b] Since $T_{sf} > T_{HW}$, the integrated storage equation was used for the last hour with a constant load from eq. 8-21. This does not affect the final temperature in this case.

and by eq. 8-22

$$l_t = \frac{1.6203 \text{ W}}{°C \cdot m^2} \left( \frac{0.924\,76 \text{ MJ/m}^2}{2 \times 0.2044 \text{ MJ/}°C \cdot m^2} + 25\,°C - 15.55\,°C \right) \times 0.0036 \text{ Ms} \times \frac{J}{W \cdot s}$$

$$= 0.06832 \text{ MJ/m}^2$$

The second hour is calculated similarly using new weather data and $T_{s0} = 29.52\,°C$. Table 8-3a lists values for all seven operating hours determined in this way.

(b) Equation 8-20 can also be used to determine the performance in a single calculation from total and average parameters. For the entire 7-day hour, $I_T = 5169.3$ W·h/m² and $\bar{T}_a = 2.52\,°C$ so that

$$q_N = \frac{\begin{array}{c} 0.8 \times 5169.3 \times 0.0036 - 5.224(25 - 2.52)(0.0036)(7) \\ - 1.6203(25 - 15.55)(0.0036)(7) \end{array}}{1 + \dfrac{(5.224 + 1.6203)(0.0036)(7)}{2 \times 0.2044}}$$

$$= \frac{11.542}{1.4219} = 8.117 \text{ MJ/m}^2$$

which is within about 4% of the prediction by simulation. The final temperature is

$$T_{sf} = \frac{8.117}{0.2044} + 25 = 64.71\,°C$$

and the solar contribution to the hot water load is

$$l_t = 1.6203 \left( \frac{8.117}{2 \times 0.2044} + 25 - 15.55 \right)(0.0036)(7) = 1.1966 \text{ MJ/m}^2$$

which compares with 1.26 MJ/m² in Table 8-3.

**TABLE 8-3b** (Engineering Units). Performance for the Situation of Example 8-5 in which there is a Simultaneous Hot Water Load

| Time | $I$ Btu/hr ft² | $T_a$ °F | $l_t$ Btu/ft² | $l_t$ (total) Btu/ft² | $q_T$ Btu/ft² | $T_s$ °F | $l_t$ (total) Btu/ft² | $T_{s0}$ °F |
|------|------|------|------|------|------|------|------|------|
| | | | Simulated[a] | | | | Eq. 8-20 | |
| 9 | 164 | 33.63 | 6.01 | 6.01 | 87.55 | 77.00 | 6.01 | 77.00 |
| 10 | 235 | 35.02 | 8.99 | 15.00 | 136.03 | 85.15 | 15.59 | 85.15 |
| 11 | 276 | 36.09 | 12.86 | 27.86 | 157.33 | 97.86 | 29.39 | 97.61 |
| 12 | 290 | 37.15 | 16.92 | 44.78 | 156.44 | 112.30 | 46.99 | 111.66 |
| 13 | 276 | 38.03 | 20.53 | 65.31 | 134.39 | 126.26 | 67.01 | 125.33 |
| 14 | 235 | 37.99 | 23.16 | 88.47 | 93.10 | 137.64 | 87.41 | 136.93 |
| 15[b] | 164 | 37.81 | 22.83 | 111.30 | 32.48 | 144.63 | 105.43 | 145.09 |
| Final | — | | — | | — | 145.59 | | 148.55 |
| Average | | 36.53 | | | | | | |
| Total | 1640 | | 111.3 | | 797.33 | | | |

[a] As explained in the text, this is implemented by the hourly use of eq. 8-20.
[b] Since $T_{sf} > T_{Hw}$, the integrated storage equation was used for the last hour with a constant load from eq. 8-21. This does not affect the final temperature in this case.

In Engineering units,

(a) $\qquad (wc)_{HW} = \dfrac{0.2854 \text{ lb}}{\text{hr ft}^2} \times \dfrac{1 \text{ Btu}}{\text{lb °F}} = 0.2854 \text{ Btu/hr °F ft}^2$

By eq. 8-20, using data given in Table 8-3,

$$q_N = \frac{0.8 \times 164 - 0.92(77 - 33.63)1 - (77 - 60)(0.2854)1}{1 + \dfrac{(0.92 + 0.2854)1}{2 \times 10}}$$

$$= \frac{86.45}{1.0603} = 81.53 \text{ Btu/ft}^2$$

By eq. 8-12,

$$T_{sf} = \frac{81.53}{10} + 77 = 85.15 \text{ °F}$$

(This is the starting temperature for the next hour.)
By eq. 8-22,

$$l_t = 0.2854\left(\frac{81.53}{2 \times 10} + 77 - 60\right) \times 1 = 6.01 \text{ Btu/ft}^2$$

Values for all hours are given in Table 8-3.

(b) For the seventh hour, a single calculation with eq. 8-20 gives

$$q_N = \frac{0.8 \times 1640 - 0.92(77 - 36.53)7 - (77 - 60)(0.2854)7}{1 + \dfrac{(0.92 + 0.2854)7}{2 \times 10}}$$

$$= \frac{1017.4}{1.4219} = 715.54 \text{ Btu/ft}^2$$

$$T_{sf} = \frac{715.54}{10} + 77 = 148.55 \text{ °F}$$

$$l_t = 0.2854\left(\frac{715.54}{2 \times 10} + 77 - 60\right)7 = 105.43 \text{ Btu/ft}^2$$

which compares with 111.3 Btu/ft² in Table 8-3.

Table 8-3 also lists values calculated using eq. 8-20 to determine the performance at other times of the day in a single step. Resultant storage temperatures by both methods are plotted by the upper lines of Fig. 8-7. The results look much like Fig. 8-5. The integrated equation is quite accurate until the final full-day calculation, but even then the result is within engineering accuracy.

The lower lines in Fig. 8-7 compare the simulation with the integrated equation for another situation with five times the hot water flow. Here half the collected heat is removed during collector operation. This overloaded system is also well predicted for any hour but the last using the one-step integrated equation. Even then the error in total heat collected is only about 6%, and less than 3% in the heat delivered to the hot water flow. Understand, however, that these calculations have been made using the minimum practical storage capacity. The one-time use of eq. 8-20 will be more accurate when more storage capacity is installed.

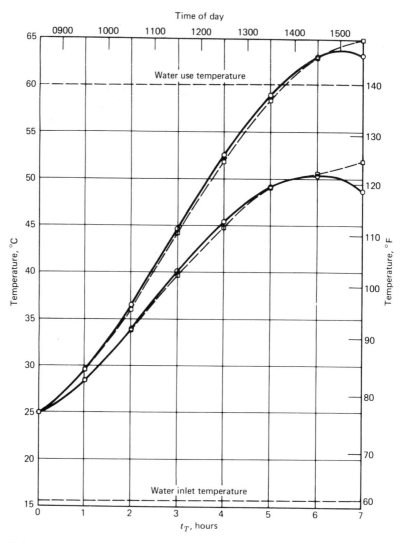

**FIGURE 8-7.** Plot of storage temperatures for hot water heating as detailed in Table 8-3. Solid lines are the hour-by-hour simulation, dotted lines are the hot water integrated equation applied from the time zero initial conditions to each plotted point.

## PRACTICAL CONSIDERATIONS IN THE DESIGN OF STORAGE

It is relatively easy to obtain well-mixed or stratified storage approximating the theoretical requirements if care is taken in the construction of the storage device.

Water tanks will stratify automatically if they are constructed like ordinary hot water tanks with a height greater than the tank diameter. Flow should enter the tank in such a way that the velocity of the entering water does not tend to mix the tank. As water storage tanks get large enough for space heating a vertical configuration becomes less practical and horizontal tanks are most often installed. These tanks may have to be baffled so that the entire contents are turned over by the circulating pump. If the baffling is done well the tank can serve as a stratified tank at low flow rates* and as a well-mixed tank when the circulating flow is very high. A series of smaller tanks provides the same effect. In lieu of baffles, deliberate mixing by the kinetic energy of the pumped return stream will keep the tank well-mixed in most circumstances if the pumped flow rate is kept high.

Water storage tanks can be made from steel, concrete, or fiberglass. The fiberglass tanks are probably best but they are expensive in the conservative wall thicknesses now recommended by the manufacturers in lieu of actual long-term, high-temperature operating experience. Steel tanks are the most commonly used, but they must be lined or coated to retard corrosion, since the tank will usually be operated either vented to the air or with well-oxygenated makeup for domestic hot water. Concrete tanks are inexpensive but must be coated to reduce water permeation. They will not corrode but there is a risk of fracture with resultant loss of water from the system. If temperatures are kept low, a wooden or metal box and a vinyl plastic liner can serve as inexpensive storage.

Rock- or pebble-bed storage is always stratified, of course, and the flow through the bed is always at a low superficial velocity. Almost any configuration will stratify sufficiently if the flow distribution is good and the bed pressure drop within an acceptable range. The "cold" manifold generally covers the entire bottom of the bed, and free space at the top provides a "hot" manifold.

Successful beds have been made in depths from 1.3 meters (4 ft) on up to two stories high. The taller beds must have larger rocks to keep the pressure drop down, while the more squat beds use smaller rocks to keep the pressure drop high enough to avoid air distribution problems. The net result is that most beds are designed for a pressure drop of about 3 to 12 mm (0.1–0.5 in.) of water pressure drop. Stones 20 to 40 mm (0.75–1.5 in.) in diameter are recommended and a 1.5 to 2 m (5–7 ft) depth is suitable for most applications. Stream rounded gravel, if available, is much more desirable than crushed rock, since the pressure drop will be lower. In either case the material should be graded carefully because undersized particles will fill in the spaces that are to serve as air passages.

---

* This has often been proposed, but data are lacking. The preceding analysis suggests that for space heating with liquid storage the advantages of stratified storage are negligible.

Rock-bed storage takes about three times the volume of a water tank. This is because the specific heat of the rocks is low, about one-fifth that of water, while the density as usually packed is 1.5 times that of water. The size of rock or pebbles does not affect the density; as long as they are uniform and spherical the density will be the same. The type of rock does not matter either, for there is little variance except when specific minerals such as iron or aluminum oxide are used. Although these materials in particular have more desirable thermal properties, the bulk of the rock bed is seldom a serious enough problem to warrant additional investment and they are seldom if ever used.

Phase-change materials are also occasionally used for heat storage. Relying on the heat of crystallization from an aqueous solution to form hydrates, *eutectic salts* have seen many years of experimentation and use. They can be handled analytically with ease; the base temperature is the "melting" temperature and they store heat at exactly that temperature until their capacity is used up. There have been recurring problems with eutectic salts, however, usually related to their tendency to form supersaturated solutions instead of crystallizing properly as they age. A recent material is discussed and the state of the art reviewed in publications of Maria Telkes (1, 2), who has developed much of this technology.

All storage containers must be well insulated. A thermal resistance or $R$ value of at least 2 °C·m²/W (11 hr °F ft²/Btu) is desirable in all cases. The cost of insulation can be minimized by using inexpensive polystyrene foam over a thinner layer of fiberglass insulation, which can withstand the high temperature next to the tank. Cubical storage tanks are often recommended, since they have a minimum insulation cost for a given heat loss. If the tank must be outside the structure or buried in the ground, or if the system is for hot water heating only, losses from the tank are losses from the system and two to three times as much insulation should be used.

## PROBLEMS

NOTE: Unless a programmable calculator is available, stop simulations after 4 hours.

1.  The collector of problem 5-1 has $F_t \tau\alpha = 0.7938$, $F_m \tau\alpha = 0.8092$, $F_i U_L = 4.348$ W/°C·m² (0.7657 Btu/hr °F ft²), and $F_m U_L = 4.433$ W/°C·m² (0.7807 Btu/hr °F ft²) at a normalized flow capacity rate of 114.27 W/°C·m² (21.65 Btu/hr °F ft²). Assume that $F_m$ stays constant as the flow rate is changed, and that water is used as the collector coolant.

    (a)  Find $F_i \overline{\tau\alpha}$ and $F_i U_L$ for a flow capacity rate reduced to one-fourth the original value. Use an incident angle modifier of 0.96.

    (b)  Find the turnover time for a tank having a storage capacity of 98.13 kg (215.9 lb) water (about 100 L or 26.4 gal) for the 2 m² (21.52 ft²) collector.

    (c)  Calculate the total heat collected using the appropriate integrated storage equation.

(d) For December 21 at 48°N, use the hourly ASHRAE table for a 68° collector tilt as the basis for a simulation of water heating using stratified storage. For ambient temperatures use the hourly sequence $-15$, $-10$, $-5$, 0, 0, $-5$, $-5$ °C (5, 14, 23, 32, 32, 23, 23 °F).

2. Repeat problems 1(c) and (d) with well-mixed storage and an *infinite* flow rate.

3. Repeat the problems 1(c) and (d) with well-mixed storage and the original flow rate.

4. For a constant hourly heat withdrawal of 0.454 MJ (431 Btu) from the storage tank:
   (a) Repeat problem 1(c) and (d).
   (b) Repeat problem 2.
   (c) Repeat problem 3.

5. The collector of problem 1 is to be applied to hot water heating. The cold water temperature is 15 °C (59 °F) and the hot water use temperature is 45 °C (113 °F). The storage capacity is 120 gal (453 kg or 1000 lb), and four of the collectors are installed. Using the specified flow rate, for a hot water load of 4 gal/hr (15.11 kg/h or 33.32 lb/hr) predict the proportion of demand satisfied for the solar day given in problem 1c if the starting temperature is 30 °C (86 °F).
   (a) Using the integrated storage equation.
   (b) Using an hourly simulation.
   (c) Continue the simulation for 24 hours total and calculate the portion of demand satisfied that day.

6. Repeat problem 5 if there is a hazy cloud cover reducing the sun's intensity by 30% during each hour of the day.

7. The hot water storage tank of problem 5 is 0.559 m (22 in.) in diameter, covered on its ends and its cylindrical portion with 7.62 cm (3 in.) fiberglass insulation ($R = 1.937$ °C·m²/W or 11 °F hr ft²/Btu).
   (a) Incorporating losses to a 20 °C (68 °F) room ambient, repeat problem 5(b).
   (b) Rederive eq. 8-20 to include the heat loss from the storage tank.
   (c) Apply your new equation for the 24 hour simulation of problem 5(c).

8. Instead of using eq. 8-7, let $\bar{T}_s = T_{s0} + (\delta q_T/2mc)$ where $\delta$ is to be an empirically determined factor approximating unity that will make the integrated storage equations exact.
   (a) Derive a form of the integrated storage equation 8.8 utilizing $\delta$.
   (b) For Example 8-2, find a daily average value of $\delta$.
   (c) Repeat (b) for problems 1 to 4, as available.

9. (a) Find the average collector temperature in Example 8-2.
   (b) Equate eq. 8-4 and the new $\delta$ form of eq. 8-8 from problem 8a and solve for $\bar{T}_c$.
   (c) Test this method of finding $\bar{T}_c$ using Example 8-2 and values of $\delta$ from problem 8, comparing your answers with part (a).

**10.** Repeat problem 8 for Example 8-3, using the integrated storage equations *with load*.

**11.** Repeat problem 8 for the hot water situation with load, and find δ for problems 5 and 6.

## REFERENCES

### General

P. J. Lunde, "Prediction of the Performance of Solar Heating Systems Utilizing Annual Annual Storage," *Solar Energy 22*, 69–75 (1979).

### Cited

1. Maria Telkes, R. P. Mozzer, "Thermal storage in salt-hydrate eutectics," Proceedings of the 1978 Annual Meeting, American Section ISES, Denver, Colorado.
2. Maria Telkes, "Solar Energy Storage," Chapter 14 of *Critical Materials in Energy Production*, Academic Press, New York (1976).

# Long-Term System Performance

The subject of this chapter is the performance of the overall solar system—collector, heat exchanger, storage, and load—over monthly and longer time periods.

When real weather is considered, there will generally be less solar heat collected than would be the case if a uniform radiation pattern existed. If the sun were to appear regularly for a few hours each day, the temperature in storage would need to be elevated only slightly to provide for overnight heating needs. Instead there are sunny days and cloudy days, and often a series of each. When there are a series of sunny days, the storage media must save the excess heat collected until such a time as it can be used. In the process of saving the heat the storage becomes hot and subsequent solar collection therefore becomes less efficient. In this way the pattern of the sunny days (and to a lesser extent, the heating demand) has an important impact on system performance.

There is in fact a definite relationship between the energy that is collected with real weather patterns and energy that might ideally be collected if the radiation distribution were uniform. Determination of that relationship permits calculation of the actual performance from that predicted using the equations previously developed.

The amount of storage installed further complicates the situation. If there is only a little storage capacity the temperatures elevate quickly and the collector efficiency decreases so much that little solar heat can be collected at the end of a series of bright days. In such cases the exact weather pattern becomes important. On the other hand, if there is a great deal of storage, the pattern of sunny days is of little consequence, for the temperature will stay near its starting or *base* temperature during the entire collection period. The performance for this *base* or *maximum storage* situation will be determined *analytically* using the techniques from Chapter 4 and subsequently will be modified to consider the weather pattern and the storage installed so that the actual performance can be predicted over a given time.

The statistical relationship between weather and the many variables in a solar-heating system does not become apparent on studying the *heat* collected from the collector at the base temperature. What is important is how the amount of heat that can be collected compares with the demand during the same time period—usually one month. If the ratio of these quantities is small, then performance near that of the base conditions can be expected. This is because the little heat that is collected will be used almost immediately, so that even if the storage is relatively small there will not be enough excess heat to raise its temperature appreciably. If the heat collected at base conditions is comparable to the total heat load then there will be an increasing loss in performance as the quantity of storage is decreased, since considerable heat will need to be stored during the sunny periods when there is too much solar heat to be utilized by the daily load. Finally, if the ratio of solar heat available at base conditions to the load is high enough (well over unity), a certain minimum amount of storage will be adequate to permit supply of all the heat needed during the time period of reference. The storage temperature will then rise gradually throughout the time period, with the increased losses from the collector dissipating the excess available heat to the atmosphere.

## THRESHOLD RADIATION DETERMINATION

A solar collector is operated only when the solar radiation is above the radiation threshold at which the heat collected is exactly zero. Therefore, only radiation data taken when the radiation flux is above the threshold level are useful for prediction of the performance of the system.

In any Appendix B table there are several columns of monthly radiation summary data, each of which applies to a different radiation threshold. The monthly radiation threshold level of the system must be below the "Average Threshold Radiation Level" if the data listed for a particular column are to be valid. The threshold for a particular month is determined using a form of eq. 4-22 that incorporates the incident angle modified optical efficiency:*

$$I_{th} = \frac{U_L}{\overline{\tau\alpha}} (\overline{T}_s - \overline{T}_{th}) \qquad (9\text{-}1)$$

The average ambient temperature at the radiation threshold $\overline{T}_{th}$ is taken from the proper column of Appendix B (AV. THRESHOLD TEMP) to be consistent with the final threshold value. Values of $U_L/\overline{\tau\alpha}$, $F_i'U_L/F_i'\overline{\tau\alpha}$ or $F_mU_L/F_m\overline{\tau\alpha}$, may be used interchangeably in eq. 9-1; the heat transfer factors cancel because at the threshold, the collected radiation is zero no matter what the value of the heat transfer factor happens to be. At the threshold, the heat collected just compensates

---

* See p. 158 for a discussion of the incident angle modifier and p. 201 for its application.

for the losses so that the temperatures in storage, everywhere on the collector, and at the heat exchanger are all equal and interchangeable with $T_s$.

**EXAMPLE 9-1\***

If $F_t \overline{\tau\alpha} = 0.7762$ and $F_m U_L = 5.069 \text{ W}/°\text{C} \cdot \text{m}^2$ (0.8927 Btu/hr $°\text{F}$ ft²) as in Example 5-8(b), find the threshold radiation level for a collector inclined at 60° to the horizontal in Hartford, Connecticut, in January if the storage is at 35 °C (95 °F). Use Appendix B data.

Using the Hartford January data at a 60° tilt angle, the 189 + W/m² group lists $\overline{T}_{th} = -2.9$ °C and $I_{th} = 238$ W/m². From eq. 9-1, the threshold radiation to activate the collector is

$$I_{th} = \frac{5.069 \text{ W}/°\text{C} \cdot \text{m}^2}{0.7762} [35 - (-2.9)] \, °\text{C} = 247 \text{ W/m}^2$$

which is too high to use the 189 + W/m² data. The next column gives the proper data. There $\overline{T}_{th} = -2.0$ °C and $I_{th} = 325$ W/m². For the collector,

$$I_{th} = \frac{5.069 \text{ W}/°\text{C} \cdot \text{m}^2}{0.7762} [35 - (-2.0)] \, °\text{C} = 242 \text{ W/m}^2$$

which is below the threshold for that data column, and the 284 + W/m² column therefore applies.

Data from any *higher* column also apply, but of course the collector would be operated to use the maximum available radiation in most cases.

**Graphical Procedure.** The process illustrated in Example 9-1 can be simplified considerably by noting that for a given $T_s$, eq. 9-1 is a linear equation expressing $I_{th}$ as a function of $\overline{T}_{th}$. When plotted as in Fig. 9-1, the equation has a slope that is independent of the value of $T_s$. A zero value for $I_{th}$ occurs when $T_s = \overline{T}_{th}$, fixing one point on a *threshold line*. When the ambient temperature is zero, the threshold radiation level is $(U_L/\overline{\tau\alpha})T_s$, which fixes a second point on the line. If these points are plotted on abscissa and ordinate, respectively, one $T_s$ value gives a threshold line from which other lines corresponding to different $T_s$ values may easily be plotted, as shown in the following example.

The radiation threshold plot applies to a given collector under any operating conditions. It is independent of location, tilt angle, or time of year. It can be prepared and filed for later use with any analysis involving that collector or any other collector with the same value of $\overline{\tau\alpha}/U_L$. It gives a very rapid method for selecting the proper data to be taken from the Appendix B tables.

---

\* In Chapters 9, 10, and 11, solutions to the examples are given in SI units only. This has been made necessary because space has not permitted a double set of Appendix B tables.

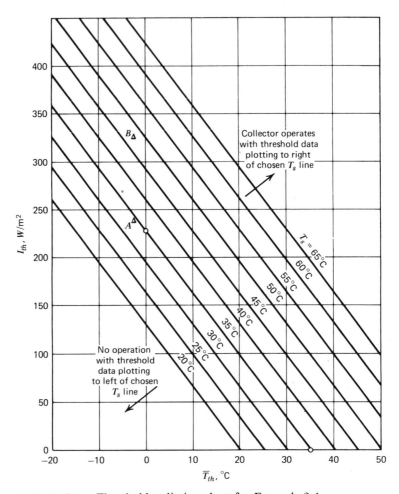

**FIGURE 9-1.** Threshold radiation chart for Example 9-1.

## EXAMPLE 9-2

Construct threshold lines for the collector of Example 9-1 and again find the January threshold radiation for a 60° inclination at Hartford with $T_s = 35$ °C.

Figure 9-1 is a plot of $I_{th}$ (ordinate) versus $\overline{T}_{th}$ (abscissa). A single line is first constructed for $T_s = 35$ °C. The value of $I_{th}$ when $\overline{T}_{th}$ is zero is, by eq. 9-1,

$$I_{th} = \frac{U_L}{\tau\alpha} T_s = \frac{5.069 \text{ W/°C·m}^2}{0.7762} \times 35 \text{ °C} = 228.6 \text{ W/m}^2$$

The bottom (x-axis) intercept is $T_s = 35$ °C, when $T_s = \overline{T}_{th}$. The threshold line is drawn between these points as shown. Other threshold lines are parallel to the first, with $x$-intercepts equal to their respective $T_s$ values. A number of threshold lines have been plotted at 5 °C intervals so that this threshold chart can be used in later work.

For the indicated January condition, for the 189+ W/m² radiation group the threshold temperature is $-2.9$ °C, and the average radiation threshold level is 238 W/m². This point is plotted at point $A$. Since it is to the left of the 35 °C line, there is no operation. The 284+ W/m² level plots at point $B$. This point is above and to the right of the 35 °C line, and operation at 35 °C is therefore practical with 284+ W/m² data.

## PERFORMANCE WITH MAXIMUM STORAGE

If the storage temperature is considered constant for a given month the performance predicted is for the limiting case of *maximum* or infinite storage. This places an upper bound of the monthly heat *collectible* on $q_T$ which can be exactly determined *analytically* using one of the three techniques developed in Chapter 4.

These techniques are summarized in the next few paragraphs. The relevant equations have been modified as necessary to incorporate the incidence angle modifier, the inlet heat transfer factor, and the collector-to-storage heat exchanger.

For systems use, eq. 4-25 becomes

$$q_T = F_i' \overline{\tau\alpha} I_{T+} \tag{9-2}$$

where $I_{T+}$ is the global radiation in excess of the threshold, interpolated from Appendix B values as shown in Example 4-4, using the threshold radiation given by eq. 9-1:

$$I_{th} = \frac{F_i' U_L}{F_i' \overline{\tau\alpha}} (\overline{T}_s - \overline{T}_{th}) \tag{9-1a}$$

Using eq. 4-26, the collectible heat can be determined without explicit determination of the radiation threshold by

$$q_T = F_i' \overline{\tau\alpha} I_T \exp\left[-k \frac{F_i' U_L}{F_i' \overline{\tau\alpha}} (\overline{T}_s - \overline{T}_a)\right] \tag{9-3}$$

where $k$ relates $I_T$, the total global radiation above a zero threshold, with the utilizable global radiation—that in excess of the threshold as described in Chapter 4. The utilizability constant $k$ can be determined by a correlation such as is outlined in Chapter 3, or it can be calculated from actual data like that in Appendix B by use of eq. 3-37.

Finally, if tabulated data is to be used directly in the integrated collector equation (4-21),

$$q_T = F_i' \overline{\tau\alpha} I_T - F_i' U_L (\overline{T}_s - \overline{T}_a) t_T \tag{9-4}$$

where the proper values for $\overline{T}_a$, $t_T$, and $I_T$ are chosen by graphical use of eq. 9-1 as explained above.

Users can choose one of the three techniques according to their own particular needs. Equation 9-2 is inconvenient for hand calculations but requires less data storage for machine use, especially if the threshold and average temperature are taken as equal. Equation 9-3 is the only choice where only the *total* global radiation is available, although reduced accuracy must be anticipated if a correlation is used to determine values of $k$.

For hand use eq. 9-4 is especially convenient, since it requires manipulation of only three tabular values. Equally excellent accuracy is achieved by hand or using a computer. The large data base that is retained permits solutions to related applications such as annual storage, as will be described later in this chapter.

**Performance as a Proportion of Monthly Demand.**   For certain applications it is convenient to relate the monthly heat available to the monthly demand. Taking the temperature of the storage as constant at a minimum or *base* temperature $T_{sB}$ below which auxiliary heat will be provided, the potential monthly solar participation (expressed as a proportion of the monthly load) is $f'_B$, given by

$$f'_B = \frac{q_T A_c}{L_m} \tag{9-5}$$

where $L_m$ is the monthly load and $q_T$ is the heat collectible at the base temperature.

**Normalized Area and Demand.**   If numerator and denominator of eq. 9-5 are divided by $L_a$, the total annual load, and the ratio of the area to annual load $A_c/L_a$ is defined as $S$, then

$$f'_B = \frac{q_T S}{L_m/L_a} \tag{9-6}$$

where $q_T$ has been evaluated at the base temperature. This approach removes one variable from the systems analysis by identifying the system size with the *ratio* of collector area to annual demand, rather than using both the area and demand as independent variables.

When the parameter $S$ is used as a measure of collector area, the actual area is then *normalized* to the annual load. A given $S$ describes the collector area in a much more significant way than the area $A_c$ itself.

For any given situation, $S$ can be determined by

$$S = \frac{A_c}{L_a} = \frac{A_c}{(L_{aH} + L_{aW})} \tag{9-7}$$

where $L_{aH}$ is the annual heating load and $L_{aw}$ the annual hot water load.* The annual heating load can be determined by any method, but is most often calculated by

$$L_{aH} = (UA)_{\text{loss}} D_a \qquad (9\text{-}8)$$

that is, as the product of the heat loss coefficient of the structure to be heated (expressed as heat units per degree-day) and the total degree-days annually at the location involved. The annual hot water load is generally determined by assuming a constant daily hot water load,† but this may also be estimated by any desired means.

When the ratio $L_m/L_a$ is used with $S$ in eq. 9-6, it is the only measure needed for monthly load. Determining the ratio $L_m/L_a$ is trivial in a heating only situation, where it is simply the ratio of the monthly to annual degree-days $D_m/D_a$. When heating and hot water loads are combined, assuming that hot water load is uniform for each month,‡ the ratio depends only on the monthly degree-day ratio and the annual ratio of hot water to total demand, $L_{aw}/L_a$.

$$L_m/L_a = [1 - L_{aw}/L_a] \frac{D_m}{D_a} + \frac{1}{12} \frac{L_{aw}}{L_a} \qquad (9\text{-}9)$$

The use of the *normalized collector area* $S$, the *monthly demand ratio* $L_m/L_a$, and eq. 9-6 will be stressed in this book. For analysis of a system having a fixed area, eq. 9-5 may be preferred and is equally valid.

Once the $S$ value is known an average flow capacity rate for the water that will be heated can be determined easily with just the temperature rise of the heated water. Integrating eq. 2-10 annually over $t_a$, the number of time units in a year,

$$Q_{aT} \times (L_{aw}/L_a) = (Wc)_{HW} \Delta T t_a$$

Dividing by the collector area and solving for the flow capacity rate,

$$(wc)_{HW} = \frac{q_{aT}}{\Delta T t_a} (L_{aw}/L_a)$$

Since $q_{aT}$ is the annual load per unit collector area, $S = 1/q_{aT}$, and therefore

$$(wc)_{HW} = \frac{L_{aw}/L_a}{S \Delta T t_a} \qquad (9\text{-}10)$$

---

* Other constant loads that can be estimated monthly, such as basement floor and below-ground losses, can also be included in $L_{aw}$ if desired.

† At a supply temperature of 60 °C (140 °F), 75 gal/day for each family (50 gal/day without a clothes washer) is widely used. If family composition is known, use 20 gal/day for each of the first two people and 15 gal/day for each additional person.

‡ Because of the irregularity of hot water use, consideration of the number of days in each month hardly seems worthwhile. To do this replace $\frac{1}{12}$ in eq. 9-9 with the days in a particular month divided by 365.25, the average number of days in a year.

**EXAMPLE 9-3**

A house has a heat loss of $UA = 17.72$ MJ/°C-day (9331 Btu/°F-day). The collector area is 34.2 m² (368 ft²). What are $S$ and $L_m/L_a$ for the heating-only situation in Hartford, Connecticut?

From Appendix B, $D_a = 3527$ °C-day:

$$S = \frac{A_c}{L_a} = \frac{34.2 \text{ m}^2}{\dfrac{17.72 \text{ MJ}}{\text{°C-day}} \times 3527 \text{ °C-day}} = \frac{0.000\ 547\ 2 \text{ m}^2}{\text{MJ}} = \frac{0.5472 \text{ m}^2}{\text{GJ}}$$

$L_m/L_a$ values are calculated directly from Appendix B values for monthly degree-days (from the left-hand column, under "all").* They are tabulated for this problem in Table 9-1. For January,

$$L_m/L_a = 686/3527 = 0.1945$$

**EXAMPLE 9-4**

If the hot water load in the previous example is 283 L/day (75 gal/day) heated from 12.8 to 60 °C (55 to 140 °F), what are $S$ and $L_m/L_a$?

$$L_{aW} = \frac{283 \text{ L}}{\text{day}} \times \frac{1 \text{ kg}}{\text{L}} \times \frac{4.19 \text{ kJ}}{\text{kg °C}} \times (60 - 12.8) \text{ °C} \times \frac{\text{G}}{10^6 \text{ k}} \times \frac{365 \text{ days}}{\text{yr}}$$

$$= 20.43 \text{ GJ/yr}$$

$$L_{aH} = \frac{17.72 \text{ MJ}}{\text{°C-day}} \times 3527 \text{ °C-day} = 62.50 \text{ GJ/yr}$$

From eq. 9-2b,

$$S = \frac{A_c}{(L_{aH} + L_{aW})} = \frac{34.2 \text{ m}^2}{(62.5 + 20.43) \text{ GJ}} = 0.4124 \text{ m}^2/\text{GJ}$$

To calculate $L_m/L_a$, first find $L_{aW}/L_a$:

$$L_{aW}/L_a = \frac{20.43}{20.43 + 62.5} = 0.2464$$

from eq. 9-9

$$L_m/L_a = (1 - 0.2464)\frac{D_m}{3527} + \frac{1}{12}(0.2464) = \frac{D_m}{4680} + 0.0205$$

for January

$$L_m/L_a = 686/4680 + 0.0205 = 0.1672$$

This example is completed in Table 9-1.

* Due to computer roundoff the Appendix B total is 3527, while Table 9-1 values total 3526.

**TABLE 9-1.** $L_m/L_a$ Determined for Examples 9-1 to 9-3

| Month | $D_m$ °F-day | $D_m$ °C-day | Heating Only $L_m/L_a$ | Heating and Hot Water $L_m/L_a$ | Hot Water Only $L_m/L_a$ |
|-------|--------|--------|--------|--------|--------|
| Jan   | 1235 | 686 | 0.1946 | 0.1672 | 0.084 87 |
| Feb   | 1090 | 606 | 0.1717 | 0.1500 | 0.077 34 |
| Mar   | 909  | 505 | 0.1432 | 0.1285 | 0.084 87 |
| Apr   | 481  | 267 | 0.0758 | 0.0776 | 0.082 14 |
| May   | 216  | 120 | 0.0340 | 0.0462 | 0.084 87 |
| Jun   | 38   | 21  | 0.0060 | 0.0250 | 0.082 14 |
| Jul   | 2    | 1   | 0.0003 | 0.0208 | 0.084 87 |
| Aug   | 15   | 8   | 0.0024 | 0.0223 | 0.084 87 |
| Sep   | 128  | 71  | 0.0202 | 0.0357 | 0.082 14 |
| Oct   | 395  | 219 | 0.0622 | 0.0674 | 0.084 87 |
| Nov   | 705  | 392 | 0.1111 | 0.1042 | 0.082 14 |
| Dec   | 1135 | 631 | 0.1788 | 0.1553 | 0.084 87 |
| Total | 6347 | 3526 | 1.0003 | 1.0002 | 1.000 00 |

## EXAMPLE 9-5

Assume that a system has only the hot water load from Example 9-4 and that the collector area is 4.41 m² (95 ft²).

(a) What are $S$ and $L_m/L_a$?
(b) What is $(wc)_{HW}$?

(a) $$S = A_c/L_a = 4.41/20.43 = 0.2159 \text{ m}^2/\text{GJ}$$

assuming equal monthly loads:

$$L_m/L_a = 1/12 = 0.083\ 33$$

For any hot water only system, if the *monthly* loads are taken as equal, $L_m/L_a$ is 0.083 33 for any month. If the *daily* hot water loads are assumed equal, $L_m/L_a$ is then in proportion to the number of days in the month. For January, for instance,

$$L_m/L_a = 31/365.25 = 0.084\ 87$$

Annual performance will be about the same on either basis, but the monthly values —particularly February—may be noticeably different. $L_m/L_a$ values based on equal daily loads are listed in Table 9-1 and will be used for hot water only situations.

(b) From eq. 9-10, since $L_{aW}/L_a = 1.0$,

$$(wc)_{HW} = \frac{GJ}{0.2159 \text{ m}^2} \times \frac{1}{(60 - 12.8)\ ^\circ\text{C}} \times \frac{1}{365 \text{ day}} \times \frac{\text{day}}{24 \text{ h}} \times \frac{\text{h}}{3600 \text{ s}} \times \frac{10^9}{G} \times \frac{\text{W} \cdot \text{s}}{J}$$

$$= 3.11 \text{ W}/^\circ\text{C} \cdot \text{m}^2$$

The previous examples make it clear that the collector size and thermal load aspects of a particular solar application can be completely described in terms of only two values—$S$ and $L_{aw}/L_a$.* Since $A_c$ is often unknown at the start of a study, it is often convenient to determine the performance at several convenient values of $S$ (which is easier to estimate than $A_c$), finding the actual desired value of $A_c$ only at the end of the study by multiplying the selected $S$ times $L_a$. The advantage in this approach is that the relatively small practical range of $S$, about 0.2 to 2 m²/GJ (2–20 ft²/MBtu), gives an intuitive feel for the performance expected that is not otherwise available. Furthermore, the results of one study can often be applied directly to another structure.

## MONTHLY SYSTEM PERFORMANCE WITH CONVENTIONAL STORAGE

The integrated storage equations (8-10 and 8-11) apply to each day's operation provided that the starting temperature $T_{so}$ is known. Equation 8-11 describes the total heat collected for each operating day by†

$$q_d = \frac{F_i'[\overline{\tau\alpha}I_d - U_L(T_{sod} - \overline{T}_{ad})t_d]}{1 + \dfrac{F_i'U_Lt_d}{2mc}} + \frac{l_d}{1 + \dfrac{2mc}{F_i'U_Lt_d}} \tag{9-11}$$

For the entire month of $N$ operating days, the total heat collected is

$$q_T = q_1 + q_2 + q_3 + \cdots + q_N$$

Substituting eq. 9-11 into this equation for each operating day, when like terms are added and averages taken as in deriving eq. 4-21, there results

$$q_T = \frac{F_i'\{\overline{\tau\alpha}[I_1 + I_2 + \cdots + I_N] - U_L(\overline{T}_{so} - \overline{T}_a)[t_1 + t_2 + \cdots + t_N]\}}{1 + \dfrac{F_i'U_L\bar{t}}{2mc}}$$

$$+ \frac{[l_1 + l_2 + \cdots + l_N]}{1 + \dfrac{2mc}{F_i'U_L\bar{t}}} \tag{9-12}$$

This consolidation of terms is mathematically possible only if there is an average operating time $\bar{t}$. This is the characteristic time of solar collection. Studies beyond the scope of this book‡ have shown that the characteristic operating time $\bar{t}$ can be closely approximated by averaging the lengths of time when the radiation

---

* $(wc)_{HW}$ is not needed for seasonal analysis.

† $d$ subscripts are used to indicate daily total or average values. Note that the numerator of the first term is $q_T$, which may be determined from eq. 9-2 or 9-3 or as shown here, from eq. 9-4.

‡ By the author.

level is above the application threshold. This is done through a radiation-weighing procedure that produces $\bar{t}$ by

$$\bar{t} = \frac{\sum_{d=1}^{N} t_d I_d}{I_T} \tag{9-13}$$

where

$$I_T = \sum_{d=1}^{N} I_d$$

and $I_d$ is the total daily radiation at the desired tilt and azimuth accumulated during the times while the instantaneous radiation is above the appropriate threshold.

The unusual characteristic of $\bar{t}$ (eq. 9-13) is that it is a *meterological* quantity that can be mechanically calculated and tabulated along with $\bar{T}_a$, $t_T$, and $I_T$. The "radiation averaged time" entries in Appendix B have been calculated by eq. 9-13 for use with equations in which the terms of eq. 9-12 have been further consolidated to give the form

$$q_T = \frac{F'_i[\overline{\tau\alpha}I_T - U_L(T_{so} - \bar{T}_a)t_T]}{1 + \dfrac{F'_i U_L \bar{t}}{2mc}} + \frac{l_t}{1 + \dfrac{2mc}{F'_i U_L \bar{t}}} \tag{9-14}$$

Defining the proportion of monthly demand actually met by solar energy as $f$, since $q_T$ is the heat flux actually collected, then

$$f = \frac{q_T A_c}{L_m} = \frac{q_T S}{L_m/L_a} \tag{9-15}$$

Multiplying both sides of eq. 9-14 by $A_c/L_m$,

$$f = \frac{F'_i[\overline{\tau\alpha}I_T - U_L(\bar{T}_{so} - \bar{T}_a)t_T]A_c/L_m}{\left(1 + \dfrac{F'_i U_L \bar{t}}{2mc}\right)} + \frac{L_t/L_m}{\left(1 + \dfrac{2mc}{F'_i U_L \bar{t}}\right)} \tag{9-16a}$$

or incorporating eq. 9-7,

$$f = \frac{F'_i[\overline{\tau\alpha}I_T - U_L(\bar{T}_{so} - \bar{T}_a)t_T]}{\left(1 + \dfrac{F'_i U_L \bar{t}}{2mc}\right)} \left(\frac{S}{L_m/L_a}\right) + \frac{L_t/L_m}{\left(1 + \dfrac{2mc}{F'_i U_L \bar{t}}\right)} \tag{9-16b}$$

where $L_t/L_m$ is the proportion of the monthly load occurring during $t_T$. $L_t$ and $l_t$ are the load and the area-normalized load, respectively, occurring during $t_T$, the time of collector operation. The terms $L_T$ and $l_T$ are used for the total loads occurring over a period of time, which includes both periods of collector operation and the time between such periods.

Equation 9-16 cannot usually be evaluated because $\bar{T}_{so}$, the average initial storage temperature, is unknown. It can be determined only after the fact and is an operating time-weighted average given by

$$\overline{T}_{s0} = \frac{\sum\limits_{d=1}^{N} (T_{s0d} \times t_d)}{\sum\limits_{d=1}^{N} t_d} \tag{9-16c}$$

When the monthly solar participation $f$ is less than about 50%, simulations have shown that the system will return to the base temperature $T_{sB}$ before operation on most days. Therefore, if $T_{sB}$ is used in place of $\overline{T}_{s0}$, eq. 9-16 can be written as

$$f = \frac{f_B'}{1 + \dfrac{F_i'U_L\bar{t}}{2mc}} + \frac{L_t/L_m}{1 + \dfrac{2mc}{F_i'U_L\bar{t}}} \tag{9-17}$$

This equation analytically predicts real monthly performance with storage. It predicts too high a value for $f$ if there are days in the month when the initial storage temperature is greater than the base temperature. Because of this, eq. 9-17 is recommended for use only below $f = 0.5$. Even so, the equation has special value in understanding system operation and will be used as a basis for correlations of performance on up to $f = 1.0$.

The numerator of the first term of eq. 9-17 is the monthly $f_B'$ given by eq. 9-5 or 9-6, that is, the portion of the monthly load that *could* be collected by solar energy if the storage were infinite so that it would stay at the base temperature. This is divided by a factor correcting for the periodic *daily* rise in storage temperature caused by the smaller storage installed

The second term is much smaller (on the order of 0.01 to 0.04), quantifying the positive effect of simultaneous heat load on system $f$. It is an additive term *independent* of the area of collector and the weather parameters used to evaluate $f_B'$. $L_t/L_m$ is approximated by $D_t/D_a$ for heating systems ($D_t$, the degree-days simultaneous with $t_T$, are tabulated in Appendix B tables), or by $t_T/t_m$ for hot water systems, where $t_m$ is the clock time in a month.

$Q_T$ and $q_T$, the heats supplied monthly by solar energy, can be determined as needed by

$$q_T = \frac{f \times L_m}{A_c} = \frac{f \times (L_m/L_a)}{S} \tag{9-18}$$

and

$$Q_T = f \times L_m = \frac{f \times (L_m/L_a) \times A_c}{S} \tag{9-19}$$

The proportion of the *annual* demand met monthly by solar energy is also of interest. Monthly values for annual solar participation are given by

$$f_a = \frac{Q_T}{L_a} = \frac{q_T}{l_a} = q_T S = f \times (L_m/L_a) \tag{9-20}$$

If $f_a$ is calculated monthly* and monthly values summed over the entire year, the total $\sum f_a$ is the annual solar participation—the proportion of the annual load carried annually by solar energy. This is a most important ratio—the one single performance number by which a system will often be judged.

## EXAMPLE 9-6

For the load situation of Example 9-3, with $S = 0.5472 \text{ m}^2/\text{GJ}$ ($6.521 \text{ ft}^2/\text{M Btu}$) and 2839 L (750 gal) of water storage, and with the collector of Example 9-1 ($F_i\overline{\tau\alpha} = 0.7762$, $F_iU_L = 5.069 \text{ W/°C·m}^2$) and $F_x = 0.95$, what are January, February, and March performance in terms of $f$? The collector is now to be inclined at 40° to the horizontal, and the base temperature is 30 °C (86 °F).

$$A_c = 34.2 \text{ m}^2$$

$$mc = \frac{Mc}{A_c} = 2839 \text{ L} \times \frac{1 \text{ kg}}{\text{L}} \times \frac{4.19 \text{ kJ}}{\text{°C}} \times \frac{1}{34.2 \text{ m}^2} \times \frac{M}{1000 \text{ k}}$$

$$= 0.3478 \text{ MJ/°C·m}^2$$

$$F_i'\overline{\tau\alpha} = 0.7762 \times 0.95 = 0.7374$$

$$F_i'U_L = 5.069 \text{ W/°C·m}^2 \times 0.95 = 4.816 \text{ W/°C·m}^2$$

$$T_{sB} = 30.0 \text{ °C}$$

$$S = 0.4124 \text{ m}^2/\text{GJ}$$

The threshold chart in Fig. 9-1 still applies. For January, with $\overline{T}_{s0} = 30$ °C, operation occurs with $\bar{I}_{th} = 237 \text{ W/m}^2$ in the 189+ W/m² radiation group, and $\overline{T}_a = -2.7$ °C. From the same column,

$$\overline{T}_a = -2.7 \text{ °C} \qquad \bar{i} = 0.0277 \text{ Ms}$$
$$t_T = 0.5638 \text{ Ms} \qquad D_t = 137 \text{ °C-days}$$
$$I_T = 280.3 \text{ MJ/m}^2 \qquad D_m = 686 \text{ °C-days}$$

$$L_m/L_a = 0.1672 \quad \text{(Table 9-1)}$$

Using eq. 9-24

$$f = \frac{\{0.7374 \times 280.3 - 4.816[30 - (-2.7)]0.5638]\} \dfrac{0.4124 \times 10^{-3}}{0.1672}}{1 + \dfrac{4.816 \times 0.0277}{2 \times 0.3478}}$$

$$+ \frac{\dfrac{137/686}{2 \times 0.3478}}{1 + \dfrac{4.816 \times 0.0277}{4.816 \times 0.0277}}$$

$$f = \frac{0.2908}{1.1918} + \frac{0.1997}{6.214} = 0.2440 + 0.0321 = 0.2762$$

* When following this procedure $f_a$ should never be derived from any $f$ greater than unity. To do so would credit the system annual performance with excess heat always potentially available in fall and late spring.

So that in January about 27% of the monthly load is carried by solar energy. The monthly heat collected is therefore (9-18)

$$q_T = \frac{0.2762 \times 0.1672}{0.4124 \times 10^{-3}} = 112.0 \text{ MJ/m}^2$$

$$Q_T = \frac{112.0 \text{ MJ}}{\text{m}^2} \, 34.2 \text{ m}^2 = 3.83 \text{ GJ}$$

and the annual solar participation in January is (eq. 9-20)

$$f_a = 0.2762 \times 0.1672 = 0.0462$$

For February, the proper solar radiation group is still $189+$ W/m², and $L_m/L_a = 0.1500$. Equation 9-17 gives

$$f = \frac{0.4102}{1.2022} + \frac{0.2277}{5.946} = 0.3412 + 0.0383 = 0.3795$$

For March, the threshold is the same, $L_m/L_a = 0.1285$ and

$$f = \frac{0.6925}{1.2271} + \frac{0.2317}{5.4035} = 0.5644 + 0.0429 = 0.6072$$

Since the value of $f$ for March is over 0.5, it is a few percent higher than the true value of about 0.5635, which will be calculated later. January and February values are more accurate, but also high, relative to the values of 0.2678 and 0.3643 that will be estimated later.

## The Base-temperature Correlation for $f$

In eq. 9-17 there are three dimensionless groups—$f_B'$, $F_i'U_L\bar{t}/2mc$, and $L_t/L_m$. These groups can be used to determine the actual performance of the system expressed as monthly solar participation $f$ in a way analogous to that in which the dimensionless Reynolds number, Prandtl number, and Nusselt number describe heat transfer in a flowing fluid (5-3). An advantage in this case is that eq. 9-17 gives an idea of the proper form of the correlation needed to describe an actual system in which $T_{sB}$ varies from day to day.

The more variables involved in a correlation, the better the potential accuracy. For solar-heating systems, a simple correlation using only two variables produces good accuracy. This correlation has also been modified to apply to hot water systems, as will be shown later in this chapter. The primary variable is $f_B'$, the potential monthly solar participation with maximum storage and a constant temperature $T_{sB}$.

The secondary dimensionless group is $F_i'U_L\bar{t}/2mc$. For the correlation, only the ratio $F_i'U_L/mc$ was used—since it is desirable to avoid excessive use of the tables, neither the radiation averaged time $\bar{t}$ nor the $L_t/L_m$ ratio (which will be reflected in the overall fit) were directly involved. The $F_i'U_L/mc$ ratio is not dimensionless, but has dimensions of $(Ms)^{-1}$ ($hr^{-1}$ for Engineering units). The correlation using the potential solar contribution and the *storage function* $F_i' U_L/mc$ can be expressed as

$$f = f\left(f_B', \frac{F_i' U_L}{mc}\right) \tag{9-21}$$

The form of eq. 9-17 (realizing that the second term is quite small and rises slowly with $f_B'$)* suggests a simple form for eq. 9-21:

$$f = f_B' \times \varepsilon\left(f_B', \frac{F_i' U_L}{mc}\right) \tag{9-22}$$

where $\varepsilon$ is a multiplier that reduces maximum system performance $f_B'$ to true performance $f$. Statistical correlation of 1500 months of simulated performance produced a good fit from eq. 9-22 for $3.5\,(Ms)^{-1} \leq F_i' U_L/mc \leq 35\,(Ms)^{-1}$ using the form

$$f = f_B'\left[1.138 - 0.271f_B' + 0.006(f_B')^3 - 0.01214\left(\frac{F_i' U_L}{mc}\right) + 0.000\,152\left(\frac{F_i' U_L}{mc}\right)^2\right] \tag{9-23}$$

This equation predicts monthly performance within about 5%, and yearly data will generally be within 1% of the correct values determined from a simulation made sharing the same data base and using the hour-by-hour techniques described in Chapter 8.†

Since predictions using the correlation are better than those from eq. 9-17 at values of $f$ over about 0.2 (and as good below $f = 0.2$),‡ eq. 9-23 is preferred for systems analysis, especially since less input data is required.

In Fig. 9-2, eq. 9-23 is plotted so that $f$ can be predicted from $f_B'$ using $F_i' U_L/mc$ as an intermediate parameter. The figure makes quite clear the effect of limited storage on system performance. Unless $F_i' U_L/mc$ is greater than $14\,(Ms)^{-1}$ (0.05 $hr^{-1}$) it is very difficult to achieve $f = 1$ even at values of $f_B'$ up to 2.2. This is because there will always be a few successive days without sun and, if the storage is too small, auxiliary heat may be needed even when *monthly* sunlight is plentiful. Values of $f_B'$ over 2.2 are beyond the range of the correlation, and a value of $f = 1$ is then arbitrarily recommended. In such cases there is so much excess sun that storage installed is likely to be adequate.

It is significant that $F_i' U_L$ is just as important as the actual storage capacity $mc$ in determining the effect of storage. A high-loss collector cannot collect heat

* This effect may be seen in Example 9-6.

† Predictions using eq. 9-23 are slightly better than given in reference 1 for a more complex correlation. Accuracy is best in the northeastern and north central United States. In the sun belt $\Sigma f_a$ was predicted about 0.02 too high (see Fig. 10-16).

‡ Being a mathematical "fit," the statistical eq. 9-23 happens to predict $f$ values slightly higher than $f_B'$ for certain $f_B'$ under 0.3. The smaller value should be used, since $f_B'$ is the maximum possible performance.

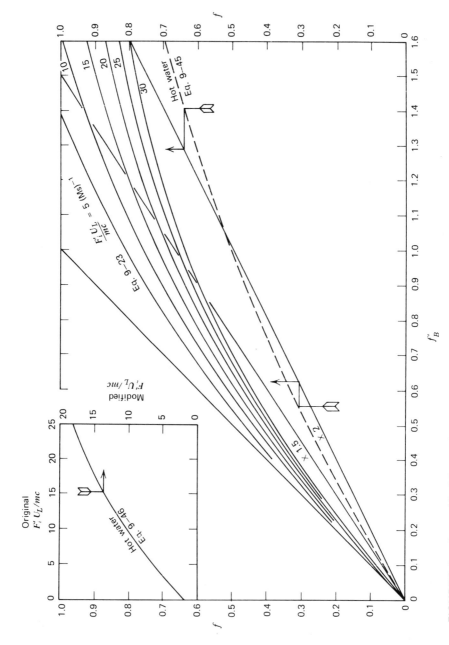

**FIGURE 9-2.** A graphical form of eq. 9-23 that may be used to find performance with finite storage when the parameter $f'_B$ and the storage capacity are known. An enlarged Figure 9-2 is located between pages 328-329.

326

efficiently when the storage is hot and it therefore needs more storage to provide a given performance.

**EXAMPLE 9-7**

Continuing with the parameters from Example 9-6, predict the monthly and annual performance using eq. 9-23 over the entire year.

The threshold chart in Fig. 9-1 still applies. For January, the correct radiation grouping is still 189+ W/m². Using eqs. 9-4 and 9-6,

$$f_B' = \{0.7374 \times 280.3 - 4.816[30 - (-2.7)]0.5638\} \frac{0.4124 \times 10^{-3}}{0.1672}$$

$$= 0.2908$$

From either eq. 9-23 or Fig. 9-2, with

$$\frac{F_i' U_L}{mc} = \frac{4.816 \text{ W/°C} \cdot \text{m}^2}{0.3478 \text{ MJ/°C} \cdot \text{m}^2} \times \frac{\text{J}}{\text{W} \cdot \text{s}} = 13.85 \text{ (Ms)}^{-1}$$

the monthly performance is

$$f = 0.2678$$

From eq. 9-20

$$f_a = 0.2678 \times 0.1672 = 0.0447$$

Table 9-2a lists weather data for the other months and certain other parameters. Of these, $q_T$, $L_m/L_a$, and $f_B'$, plus $f$ and $f_a$ are the only parameters that need to be tabulated to define the system performance.

Notice the broad range over which $f = 1.0$. When $f_B' > 2.2$ it is obvious that $f = 1.0$ and then $f_a = L_m/L_a$, and therefore the detailed calculations need not be made. A final step is the summing of monthly $f_a$ values to give the overall performance as percent of annual demand supplied annually. In this case $\Sigma f_a = 0.5386$.

It is also interesting to examine the results using the elevated data and $k$ values developed in Example 3-11 (Table 3-8) instead of the tabular data. Equation 9-3 is then to be used for calculating $q_T$ with $\bar{T}_s$ taken equal to $T_{sB}$.

$$q_T = 0.7374 \times \frac{323.3 \text{ MJ}}{\text{m}^2} \times \exp\left\{\frac{-0.0033 \text{ m}^2}{\text{W}} \times \frac{4.8160 \text{ W}}{\text{°C} \cdot \text{m}^2} \times \frac{[30 - (-1.8)] \text{°C}}{0.7374}\right\}$$

$$= \frac{120.1 \text{ MJ}}{\text{m}^2}$$

by eq. 9-6,

$$f_B' = \frac{120.1 \text{ MJ}}{\text{m}^2} \times \frac{0.4124 \times 10^{-3}}{0.1672} \frac{\text{m}^2}{\text{MJ}} = 0.2963$$

**TABLE 9-2a.** Monthly Performance of a Solar-Heating System for Example 9-7 Using Tabular Data from Appendix B. $F'_i\overline{\tau\alpha} = 0.7374$, $F'_iU_L = 4.8160\ \text{W}/°\text{C}\cdot\text{m}^2$, $T_{sB} = 30.0\ °\text{C}$, $mc = 0.3478\ \text{MJ}/°\text{C}\cdot\text{m}^2$, $L_{aw}/L_a = 0.2462$, and $S = 0.4124\ \text{m}^2/\text{GJ}$.

| | $D_m$ °C-days | $\overline{T}_a$ °C | $t_T$ Ms | $I_T$ MJ/ m² | $q_T$ MJ/ m² | $S$ m²/GJ | $f'_{aB}$ | $L_m/L_a$ | $f'_B$ | $f$ | $f_a$ | $f_B$ | $f_{aB}$ |
|---|---|---|---|---|---|---|---|---|---|---|---|---|---|
| Jan | 686. | −2.7 | 0.5638 | 280.3 | 117.9 | 0.4124 | 0.0486 | 0.1671 | 0.2909 | 0.2678 | 0.0447 | 0.2909 | 0.0486 |
| Feb | 606. | −1.5 | 0.6030 | 326.4 | 149.2 | 0.4124 | 0.0615 | 0.1500 | 0.4101 | 0.3643 | 0.0546 | 0.4101 | 0.0615 |
| Mar | 505. | 4.8 | 0.7466 | 415.5 | 215.7 | 0.4124 | 0.0889 | 0.1284 | 0.6928 | 0.5635 | 0.0723 | 0.6928 | 0.0889 |
| Apr | 267. | 12.5 | 1.0994 | 482.9 | 263.4 | 0.4124 | 0.1086 | 0.0775 | 1.4003 | 0.8907 | 0.0691 | 1.0000 | 0.0775 |
| May | 120. | 18.4 | 1.2618 | 559.5 | 342.0 | 0.4124 | 0.1410 | 0.0461 | 3.0553 | 1.0000 | 0.0461 | 1.0000 | 0.0461 |
| Jun | 21. | 23.4 | 1.2287 | 531.4 | 352.7 | 0.4124 | 0.1454 | 0.0250 | 5.8150 | 1.0000 | 0.0250 | 1.0000 | 0.0250 |
| Jul | 1. | 25.2 | 1.4836 | 559.8 | 378.5 | 0.4124 | 0.1560 | 0.0207 | 7.5236 | 1.0000 | 0.0207 | 1.0000 | 0.0207 |
| Aug | 8. | 24.5 | 1.1995 | 494.4 | 332.7 | 0.4124 | 0.1372 | 0.0222 | 6.1703 | 1.0000 | 0.0222 | 1.0000 | 0.0222 |
| Sep | 71. | 20.9 | 1.0408 | 489.5 | 315.3 | 0.4124 | 0.1300 | 0.0357 | 3.6424 | 1.0000 | 0.0357 | 1.0000 | 0.0357 |
| Oct | 219. | 14.9 | 0.9475 | 438.9 | 254.7 | 0.4124 | 0.1050 | 0.0673 | 1.5603 | 0.9346 | 0.0629 | 1.0000 | 0.0673 |
| Nov | 392. | 7.6 | 0.4964 | 243.4 | 125.9 | 0.4124 | 0.0519 | 0.1042 | 0.4979 | 0.4306 | 0.0449 | 0.4979 | 0.0519 |
| Dec | 631. | −0.8 | 0.4864 | 240.4 | 105.1 | 0.4124 | 0.0433 | 0.1553 | 0.2790 | 0.2577 | 0.0400 | 0.2790 | 0.0433 |
| Total | 3527. | | | | | | | | | 0.5386 | | | 0.5892 |

**TABLE 9-2b.** Monthly Performance of a Solar-Heating System for Example 9-7 Using Data Generated from Horizontal Radiation in Example 3-11. Compare to Table 9-2a.

| | $D_m$ °C-days | $\overline{T}_a$ °C | $I_T$ MJ/ m² | $k$ m²/W | $q_T$ MJ/ m² | $S$ m²/GJ | $f'_{aB}$ | $L_m/L_a$ | $f'_a$ | $f$ | $f_a$ | $f_B$ | $f_{aB}$ |
|---|---|---|---|---|---|---|---|---|---|---|---|---|---|
| Jan | 686 | −1.8 | 326.4 | 0.0033 | 121.3 | 0.4124 | 0.0500 | 0.1671 | 0.2993 | 0.2748 | 0.0459 | 0.2993 | 0.0500 |
| Feb | 606 | −1.1 | 373.2 | 0.0030 | 149.6 | 0.4124 | 0.0617 | 0.1500 | 0.4114 | 0.3653 | 0.0548 | 0.4114 | 0.0617 |
| Mar | 505 | 4.1 | 472.4 | 0.0029 | 213.3 | 0.4124 | 0.0879 | 0.1284 | 0.6849 | 0.5585 | 0.0717 | 0.6849 | 0.0880 |
| Apr | 267 | 11.4 | 499.7 | 0.0029 | 259.1 | 0.4124 | 0.1068 | 0.0776 | 1.3772 | 0.8836 | 0.0686 | 1.0 | 0.0776 |
| May | 120 | 17. | 571.6 | 0.0029 | 329.5 | 0.4124 | 0.1360 | 0.0462 | 2.9432 | 1.0 | 0.0462 | 1.0 | 0.0462 |
| Jun | 21 | 22.2 | 544.0 | 0.0031 | 342.6 | 0.4124 | 0.1412 | 0.0250 | 5.6466 | 1.0 | 0.0250 | 1.0 | 0.0250 |
| Jul | 1 | 24.6 | 563.6 | 0.0030 | 373.9 | 0.4124 | 0.1538 | 0.0207 | 7.4314 | 1.0 | 0.0207 | 1.0 | 0.0207 |
| Aug | 8 | 23.3 | 510.0 | 0.0030 | 329.8 | 0.4124 | 0.1357 | 0.0222 | 6.1144 | 1.0 | 0.0222 | 1.0 | 0.0222 |
| Sep | 71 | 19.2 | 495.7 | 0.0028 | 300.0 | 0.4124 | 0.1237 | 0.0357 | 3.4653 | 1.0 | 0.0357 | 1.0 | 0.0357 |
| Oct | 219 | 13.3 | 445.8 | 0.0029 | 239.6 | 0.4124 | 0.0988 | 0.0673 | 1.4675 | 0.9105 | 0.0613 | 1.0 | 0.0673 |
| Nov | 392 | 7.3 | 287.7 | 0.0035 | 126.3 | 0.4124 | 0.0520 | 0.1043 | 0.4993 | 0.4317 | 0.0450 | 0.4993 | 0.0521 |
| Dec | 631 | 0.0 | 283.6 | 0.0035 | 105.4 | 0.4124 | 0.0434 | 0.1554 | 0.2797 | 0.2583 | 0.0401 | 0.2797 | 0.0434 |
| Total | 3527 | | | | | | | | | 0.5373 | | | 0.5900 |

and with

$$F'_iU_L/mc = 13.85\ (\text{Ms})^{-1}$$

eq. 9-23 or Fig. 9-2 gives

$$f = 0.2724$$

Table 9-2b continues this analysis for the entire year. Compare the results with Table 9-2a. While these virtually identical results are not to be expected in every case, the simple data correlation techniques of Chapter 3 in general perform well when only the total global radiation on a horizontal (or tilted) surface is available.

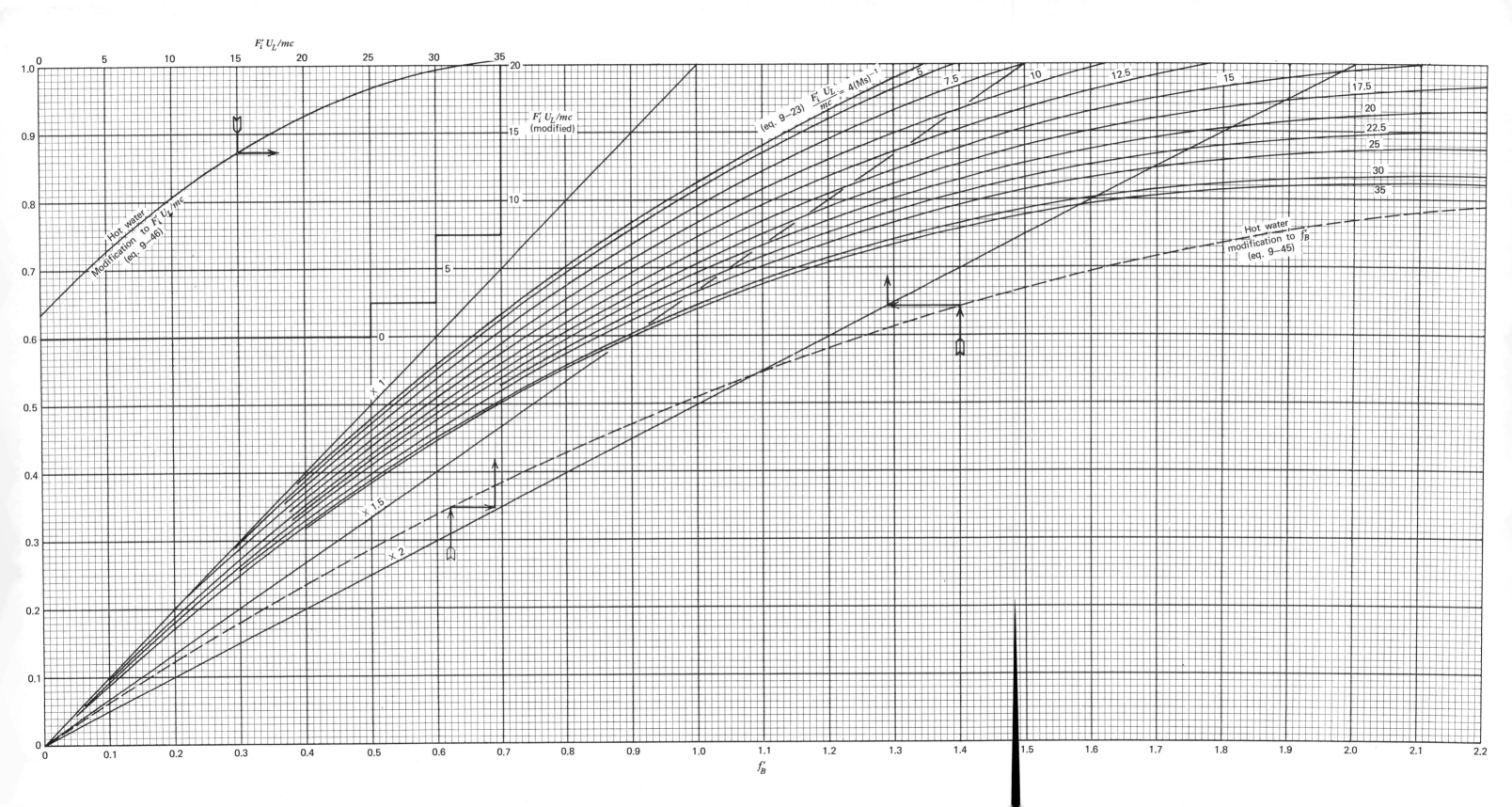

**Limitations.** Equation 9-23 and Fig. 9-2 have certain limitations. Most important is that the $F_i'U_L/mc$ term be in the range of 3.5 to 35 $(Ms)^{-1}$ (0.0125 to 0.125 $hr^{-1}$). Extrapolations beyond the range fail. For lower values (oversized storage) eq. 9-17 will give better (although perhaps too high) values.

The correlation was prepared without consideration of "boil-off," which occurs in water storage systems when the calculated storage temperature exceeds the boiling point of water. This limits the effectiveness of storage if $mc$ is in the range of 0.2 $MJ/°C·m^2$ (10 $Btu/°F ft^2$), but analytically it can generally be disregarded.*

## ANNUAL HEAT STORAGE

It is very attractive to develop a solar system based on storing heat collected in the summer for use in winter heating. In practice there are several difficulties. The storage volume must be 20 to 50 times that used in the usual system where storage of only a few days' sunshine is anticipated. A large storage vessel makes the area for losses from the storage tank quite large, and therefore there must be a great deal of insulation installed. Since so much storage is needed, it should be used to maximum efficiency, with summer heat raising storage temperatures as high as possible. Such high temperatures cause high collector losses and low collection efficiency unless expensive low-loss collectors are used. Yet to be attractive, annual storage systems must use a *smaller* collector area than do ordinary systems for a given performance.

Annual storage systems have so much storage installed that the integrated storage eqs. (8-10 and 8-11) can be applied over an entire month with near-absolute accuracy, provided as usual that the collector properties are known and hold constant during the period of collector operation, and that the losses from the storage media are negligible. Hooper (2) reports that in his analyses of economically optimum systems, losses run from 10% of the heat collected (for an apartment building) to 25% (for a typical house). Therefore, a more complex version of the integrated storage equations is derived below that takes into account such losses.

### The Integrated Annual Storage Equations

System calculations with annual storage are best made at monthly intervals so that the threshold radiation level can be recalculated monthly as the storage temperature rises or falls. The loss from the storage tanks during the month is

$$L_s = U_s A_s (\overline{T}_s - T_g) t_m \qquad (9\text{-}24)$$

---

* At worst, the last month to use auxiliary heat in the spring will have its performance reduced 5 to 10%. Even this is not likely unless the collector has a high $F_i'\tau\alpha$, a low $F_i'U_L$, or if the system has a very low $S$, extending the low $f$ months well into the spring when demand is irregular.

where $U_s$ is the storage tank heat loss coefficient, $A_s$ the area for such loss, $T_g$ the ground temperature, and $t_m$ the time of the heat loss, which is the total clock time in a month. Considering an entire system similar to that in Fig. 8-6, the normalized net heat to storage during the entire monthly time period $t_m$ (which includes time period of solar collection totalling $t_T$) is

$$q_N = q_T - l_m - l_s \qquad (9\text{-}25)$$

where the normalized total monthly load is $l_m = L_m/A_c$ and the normalized storage tank heat loss is $l_s = L_s/A_c$, thereby normalizing the system loads to the collector area. Let $U_s a_s$ be the collector area-normalized, storage tank heat loss coefficient defined by

$$U_s a_s = \frac{U_s A_s}{A_c} \qquad (9\text{-}26)$$

Dividing eq. 9-24 by $A_c$ gives

$$l_s = U_s a_s (\overline{T}_s - T_g) t_m \qquad (9\text{-}27)$$

which is substituted into eq. 9-25 along with the usual expression for the total heat collected given by eq. 9-4 to give

$$q_N = F_i' \overline{\tau \alpha} I_T - F_i' U_L (\overline{T}_s - \overline{T}_a) t_T - l_m - U_s a_s (\overline{T}_s - T_g) t_m \qquad (9\text{-}28)$$

$\overline{T}_s$ is now eliminated by use of eq. 8-7a, and the resultant expression solved for $q_N$. This gives the net integrated annual storage equation:

$$q_N = \frac{F_i'[\overline{\tau \alpha} I_T - U_L(T_{s0} - \overline{T}_a)t_T] - l_m - U_s a_s(T_{s0} - T_g)t_m}{1 + \dfrac{F_i U_L t_T + U_s a_s t_m}{2mc}} \qquad (9\text{-}29)$$

which is combined with eq. 8-12 to give the change in temperature of the storage media over time $t_m$:

$$T_{sf} = T_{s0} + \frac{q_N}{mc} \qquad (8\text{-}12)$$

The total *useful* heat collected (i.e., excluding the storage losses) is found by summing the heat to the load and the net heat to storage, redefining $q_T$ for this case as the useful heat:

$$q_T = q_N + l_m \qquad (9\text{-}30)$$

which in combination with eq. 9-29 gives

$$q_T = \frac{F_R'[\overline{\tau \alpha} I_T - U_L(T_{s0} - T_a)t_T] - U_s a_s(T_{s0} - T_g)t_m}{1 + \dfrac{F_R' U_L t_T + U_s a_s t_m}{2mc}} + \frac{l_m}{1 + \dfrac{2mc}{F_R' U_L t_T + U_s a_s t_m}} \qquad (9\text{-}31)$$

Equations 9-29, 8-12, and either 9-30 or 9-31 are used monthly to predict performance for the annual storage situation. They are almost exact because the storage capacity must be so large with annual storage that eq. 8-5 is followed quite closely (3).

When the integrated annual storage equations are used, it is best to begin with March when the system will be at its lowest usable storage temperature, the base temperature $T_{sB}$. The storage temperature will increase, month by month, until October or November. If auxiliary heat is to be needed, the final storage temperature will fall below $T_{sB}$ and the calculation of $q_N$ must be repeated using eq. 9-28 with $\overline{T}_s = T_{sB}$. This permits calculation of the required auxiliary heat by difference assuming that the solar heat continues to be delivered into the seasonal storage media and the losses from the tank continue. Other possibilities, which will not be explored here, are to use a smaller auxiliary storage to smooth the daily heat collection once the annual storage is exhausted, letting the annual storage volume cool to ground temperature, or to dump the collected heat directly into the terminal heat exchanger with no storage at all.

Use of eqs. 9-29 or 9-31 requires determination of $l_m$, the normalized monthly heat demand. Since these equations are designed for use with tables from Appendix B, it is desirable to use degree-days or the tabulated ratio $L_m/L_a$ directly. The basic definition of $l_m$ is

$$l_m = \frac{L_m}{A_c} \tag{9-32}$$

which can be used directly when $L_m$ and $A_c$ are known. When the installed collector area is expressed as $S$ and the monthly load ratio $L_m/L_a$ is available, then dividing eq. 9-32 by $L_a$ produces the monthly demand per unit collector area

$$l_m = \frac{L_m/L_a}{S} \tag{9-33}$$

which is a great convenience when using eqs. 9-29 or 9-31. The form of eq. 9-33 may be more easily kept in mind if it is noted that $S$ is the reciprocal of the normalized *annual* load

$$l_a = \frac{L_a}{A_c} = \frac{1}{S} \tag{9-34}$$

**EXAMPLE 9-8**

For the load situation in Example 9-3, 170 m³ (44 909 gal) of water storage are installed in a square, 3 m deep tank insulated on all sides with 15 cm (5.906 in.) of foam having $k = 0.036$ W/m·°C (0.25 Btu in./hr °F ft²). If the collector area is one-half again as large at 51.3 m² and, as in Example 9-1 and 9-6, $F_i\overline{\tau\alpha} = 0.7762$, $F_iU_L = 5.069$ W/°C·m² (0.8927 Btu/hr °F ft²), $F_x = 0.95$, and $T_{sB} = 30$ °C

(86 °F), what is the annual performance at Hartford, Connecticut with the collector inclined at 40° to the horizontal? The ground water temperature $T_g$ is 10 °C (50 °F).

$$mc = \frac{Mc}{A_c} = 170 \text{ m}^3 \times \frac{980 \text{ kg}}{\text{m}^3} \times \frac{4.19 \text{ kJ}}{\text{kg} \cdot °\text{C}} \times \frac{1}{68.4 \text{ m}^2} \times \frac{M}{1000 \text{ k}}$$

$$= 10.20 \text{ MJ/}°\text{C} \cdot \text{m}^2$$

If $v = l^2 h$ ($v$ = volume, $l$ = length, $h$ = height),

$$l = \sqrt{\frac{v}{h}} = \sqrt{\frac{170 \text{ m}^3}{3 \text{ m}}} = 7.528 \text{ m}$$

Area of top and bottom $= 2 \times 7.528^2 \quad = 113.3 \text{ m}^2$

Area of sides $\quad = 4 \times 7.528 \times 3 = \quad 90.3 \text{ m}^2$

Total tank area for heat loss $\quad = 203.6 \text{ m}^2$

the tank heat loss coefficient is

$$\text{Tank } U_s = \frac{k}{l} = \frac{0.036 \text{ W}}{\text{m} \cdot °\text{C}} \times \frac{1}{15 \text{ cm}} \times \frac{100 \text{ cm}}{\text{m}} = 0.24 \text{ W/}°\text{C} \cdot \text{m}^2$$

and the normalized $UA$ for the tank is

$$U_s a_s = \frac{U_s A_s}{A_c} = \frac{0.24 \text{ W}}{°\text{C} \cdot \text{m}^2} \times \frac{203.6 \text{ m}^2 \text{(storage)}}{68.4 \text{ m}^2 \text{(collector)}} = 0.7144 \text{ W/}°\text{C} \cdot \text{m}^2$$

From Example 9-3, $S = 1.5 \times 0.5472 \text{ m}^2/\text{GJ} = 0.8208 \text{ m}^2/\text{GJ}$. Values for $L_m/L_a$ are taken from the fourth column of Table 9-1. The collector properties are

$$F_i'\overline{\tau\alpha} = F_x F_i\overline{\tau\alpha} = 0.95 \times 0.7762 = 0.7374$$

$$F_i'U_L = F_x F_i U_L = 0.95 \times 5.069 = 4.816 \text{ W/}°\text{C} \cdot \text{m}^2$$

The threshold plot of Fig. 9-1 applies here because its slope $F_i U_L / F_i\overline{\tau\alpha} = 5.069/0.7762 = 6.53 \text{ W/}°\text{C} \cdot \text{m}^2$, as in Example 9-2. Assuming that the storage is at a minimum temperature of $T_{sB} = 30$ °C on March 1, by Fig. 9-1 the proper radiation group for Hartford at a 40° tilt is 189+ W/m², using Appendix B values. Then $\overline{T}_a = 4.8$ °C, $t_T = 0.7466$ Ms, $I_T = 41.55$ MJ/m², $t_m = 2.6784$ Ms and from Table 9-1, $L_m/L_a = 0.1285$. Substituting in eq. 9-33,

$$l_m = 0.1285/(0.8208 \times 10^{-3} \text{ m}^2/\text{MJ}) = 156.55 \text{ MJ/m}^2$$

Equation 9-29 gives

$$q_N = \frac{0.7374 \times 415.5 - 4.816(30 - 4.8)0.7466 - 156.55 - 0.7144(30 - 10)2.6784}{1 + \dfrac{4.816 \times 0.7466 + 0.7144 \times 2.6784}{2 \times 10.20}}$$

$$q_N = \frac{20.96}{1.270} = 16.50 \text{ MJ/m}^2$$

and the final temperature is then given by eq. 8-12 as

$$T_{sf} = \frac{16.50}{10.20} + 30 = 31.62\ °C$$

The average storage temperature is then $(30 + 31.62)/2 = 30.81\ °C$. This value can be used to evaluate the storage loss $l_s$ by eq. 9-27:

$$l_s = 0.7144(30.81 - 10)2.6784 = 39.82\ MJ/m^2$$

and by eq. 8-4 the total heat collected is

$$q_T = 0.7374 \times 415.5 - 4.816(30.81 - 4.8)0.7466 = 212.87\ MJ/m^2$$

Cross-checking with eq. 9-25,

$$q_N = 212.87 - 156.55 - 39.82 = 16.50\ MJ/m^2$$

and also with eq. 9-31, the useful heat collected is

$$q_T = \frac{20.96 + 156.55}{1.270} + \frac{156.55}{1 + \dfrac{1}{0.270}} = 139.77 + 33.28 = 173.05\ MJ/m^2$$

the same as $q_N + l_m = 16.50 + 156.55 = 173.05\ MJ/m^2$.

Table 9-3 gives the results of the complete analysis of this example. The net heat is positive until September, when the storage temperature starts to fall. At the end of December the storage is at $33.22\ °C$, nearly at the base temperature, and January is therefore calculated at a constant $\overline{T}_s = T_{sB} = 30\ °C$ (86 °F) using eq. 9-10. The January drop to $30\ °C$ provides $3.22\ °C \times 10.2\ MJ/°C \cdot m^2 = 32.84\ MJ/m^2$ which will appear in the auxiliary heat that is found by summing the net heat for all months. The minus value of $(-163.05\ MJ/m^2)$ in Table 9-3 indicates the

TABLE 9-3.  Monthly Performance Using Annual Storage (Example 9-8).

| | Thresh-old Level | $T_{s0}$ | $T_a$ | $t_T$ | $I_T$ | $L_m/L_a$ | $t_{all}$ | $q_N$ | $t_{sf}$ | $q_T$ | $l_m$ | $l_s$ | $q_{aux}$ |
|---|---|---|---|---|---|---|---|---|---|---|---|---|---|
| Mar | 189+ | 30.00 | 4.8 | 0.7466 | 415.5 | 0.1285 | 2.6784 | 16.50 | 31.62 | 173.05 | 156.55 | 38.27 | |
| Apr | 189+ | 31.62 | 13.4 | 0.8388 | 446.2 | 0.0776 | 2.5920 | 93.78 | 40.81 | 188.32 | 94.54 | 40.03 | |
| May | 189+ | 40.81 | 19.4 | 0.9961 | 521.3 | 0.0462 | 2.6784 | 125.25 | 53.09 | 181.53 | 56.29 | 58.96 | |
| Jun | 189+ | 53.09 | 24.6 | 0.9526 | 491.5 | 0.0250 | 2.5920 | 92.33 | 62.14 | 122.79 | 30.46 | 79.79 | |
| Jul | 284+ | 62.14 | 27.1 | 0.8294 | 469.7 | 0.0208 | 2.6784 | 63.02 | 68.32 | 88.36 | 25.34 | 99.77 | |
| Aug | 284+ | 68.32 | 25.7 | 0.7232 | 408.8 | 0.0223 | 2.6784 | 11.26 | 69.43 | 38.43 | 27.17 | 111.59 | |
| Sep | 378+ | 69.43 | 22.5 | 0.5537 | 382.2 | 0.0357 | 2.592 | 2.59 | 69.68 | 46.09 | 43.49 | 110.04 | |
| Oct | 378+ | 69.68 | 16.2 | 0.4900 | 338.3 | 0.0674 | 2.687 | -60.69 | 63.73 | 21.43 | 82.12 | 114.56 | |
| Nov | 378+ | 63.73 | 7.6 | 0.2898 | 186.9 | 0.1042 | 2.592 | -144.03 | 49.61 | -17.08 | 126.95 | 99.49 | |
| Dec | 284+ | 49.61 | -0.8 | 0.3661 | 212.3 | 0.1553 | 2.6784 | -167.19 | 33.22 | 22.01 | 189.21 | 75.79 | |
| Jan | 189+ | 30.0 | -2.7 | 0.5638 | 280.3 | 0.1672 | 2.6784 | -124.07 | 30.00 | 79.64 | 203.70 | 38.27 | 91.22[a] |
| Feb | 189+ | 30.0 | -1.5 | 0.6030 | 326.4 | 0.1500 | 2.6784 | -71.81 | 30.00 | 110.94 | 182.75 | 110.94 | 71.81 |
| Total | | | | | | | | -163.05 | | 1055.51 | 1218.57 | 977.50 | 163.03 |

[a] $(124.07 - 3.22 \times 10.2)$.

need for auxiliary heat; a positive value, indicating over 100% of demand supplied by solar energy, would be accompanied by $T_s$ rising over the 12-month period.

The auxiliary heat can also be calculated as a cross-check by summing $q_T$ and $l_s$ separately and applying eq. 9-30 solved for $q_N$:

$$q_N = 1055.51 - 1218.57 = -163.06 \text{ MJ/m}^2$$

The portion of energy collected that is lost to the ground is quite large:

$$\frac{977}{1056 + 977} = 48.06\%$$

indicating a need for better insulation (see problem 9-6).

The overall portion of annual demand supplied by solar energy on an annual basis is designated $\sum f_a$, and is given in this case by

$$\sum f_a = \frac{l_a - q_{\text{aux}}}{l_a} = \frac{1218.57 - 163.03}{1218.6} = 86.62\%$$

Notice that in the previous analysis it was unnecessary to work out any thermal values other than those normalized, that is, per unit of collector area. This is generally the best way to proceed, for conversion to absolute values is simple enough when the need arises by multiplying by $A_c$.

## TERMINAL HEAT EXCHANGE

After determination of system performance, the terminal heat exchanger that is required for liquid systems can be sized. Its performance will previously have been chosen by the selection of the base temperature, which will usually be 5 to 10 °C (10–20 °F) above the planned inside ambient, less about 3 °C (5 °F) if hot water preheat is also installed.* A larger exchanger can produce a lower temperature, but the cost to do this will generally not be worthwhile, and furthermore air circulating at less than 32 °C (90 °F) will often cause discomfort due to drafts.† A smaller heat exchanger will require a higher base temperature, which will affect system performance adversely.

The designer can, however, select any desired base temperature to use with the techniques below to size the terminal heat exchanger.

It is usually possible to increase the water flow to the terminal heat exchanger so that the temperature drop in that liquid is very low. If the flow (in terms of

---

* Heat exchanger sizing, however, is carried out using the base temperature without this correction.

† Note that because of stratification, pebble bed storage can be cooled to room temperature without this problem. However, if air from the collectors was originally delivered to storage at less than 90 °F, discomfort may result. This may be avoided by raising the threshold for operation (and performance prediction) from zero to that collected heat flux necessary to produce the required temperature rise in the circulating air stream.

*wc* or *Wc*) of the water is at least 15 times that of the air, eq. 7-11b can be used to determine the heat exchanger area. Otherwise eq. 7-11 is called for if the flow is countercurrent, as will be the case if two cross-flow heat exchangers in series are used (see the section on liquid-to-air heat exchangers in Chapter 7).

The size of the terminal heat exchanger depends on the quantity of solar heat to be delivered. It is sized to remove the largest monthly solar load from liquid supplied at the base temperature for 24 hours a day.

The effectiveness of the heat exchanger determines how much cooler than the base temperature the air delivered for heating will be. The cooler the air, the greater the air flow required and in general the less comfortable will be the occupants. In practical application the effectiveness will be such that the air is heated to within 1 to 5 °C (2 to 10 °F) of the stored fluid temperature. The optimum depends on the temperature the occupants will accept, the airflow for which the circulation system can be economically designed, and the cost of the heat exchanger, which will rise rapidly as the closer approaches are desired.

The system must be designed so that auxiliary heat can be added to the conditioned space without interfering with removal of heat from the storage tank, because economics dictate that the solar heat exchanger must work around-the-clock when at its design conditions even when auxiliary heat is also required. Auxiliary heat should therefore be added to the air downstream of the solar heat exchanger with a separate heater, or even added to the building with a separate system. It should *never* be added to the storage tank or to a liquid stream leaving the tank. Such operation will increase the collector heat losses and the size of the solar heat exchanger unnecessarily by failing to take advantage of the easy heat exchange always available with high-temperature, auxiliary heat.

## EXAMPLE 9-9

Size terminal heat exchangers for the conditions of Example 9-7 (Table 9-2a) if the water flow is very large and the temperature approach is 1 °C (1.8 °F) and 3 °C (5.4 °F), as shown in Fig. 9-3.

The March solar heating duty is $f_a = 0.0723$. Therefore,

$$q_{March} = 0.0723 \times \frac{GJ}{0.4124 \text{ m}^2} \times \frac{1000 \text{ M}}{G} = 175.3 \text{ MJ/m}^2$$

If the heat is to be released at a constant rate, the rate of heating is

$$q_{March} = 175.3 \text{ MJ/m}^2 \times \frac{month}{2.6784 \text{ Ms}} = 65.46 \text{ W/m}^2$$

Figure 9-3 shows the system conditions for an approach of 1 and 3 °C.

$$\text{For 1 °C} \quad \varepsilon = \frac{32 - 23.89}{33 - 23.89} = 0.8902$$

$$\text{For 3 °C} \quad \varepsilon = \frac{30 - 23.89}{33 - 23.89} = 0.6707$$

since $q = wc\,\Delta T$, the air flow capacity rate is,

$$\text{For 1 °C} \qquad wc = \frac{q}{\Delta T} = \frac{65.46}{(32 - 23.89)} = 8.071 \text{ W/°C·m}^2$$

$$\text{For 3 °C} \qquad wc = \frac{q}{\Delta T} = \frac{65.46}{(30 - 23.89)} = 10.71 \text{ W/°C·m}^2$$

From eq. 7-11b,

$$\text{For 1 °C} \qquad U_x a_x = w_s c_s \ln \frac{1}{1 - \varepsilon} = 8.071 \ln \frac{1}{1 - 0.8902} = 17.83 \text{ W/°C·m}^2$$

$$\text{For 3 °C} \qquad U_x a_x = 10.71 \ln \frac{1}{1 - 0.6707} = 11.90 \text{ W/°C·m}^2$$

and thus to hold the smaller differential the heat exchanger must be 50% larger.

To evaluate the base temperature choice, the performance criteria of Klein, Duffie, and Beckman (1) suggest that a system should have a heat exchanger sized in proportion to the loss from the structure (heating load) so that $1 < \lambda < 3$ where $\lambda$ is a dimensionless ratio defined by

$$\lambda = \varepsilon W_s c_s D_a / L_{aH} \tag{9-35}$$

$\lambda$ is not a function of the approach. It depends on the base temperature and the *total* heating load, not the solar load, and therefore applies more directly to

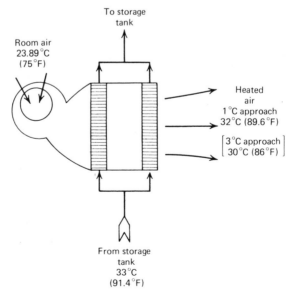

**FIGURE 9-3.** Terminal heat exchanger conditions for Example 9-9.

systems in which the heat exchanger also transfers the auxiliary heat. With the aid of an integrated eq. 7-3 and eq. 9-20, $\lambda$ can be redefined as

$$\lambda = \frac{\varepsilon W_s c_s D_a S}{(1 - L_{aW}/L_a)} = \frac{f_a D_a}{\Delta T(1 - L_{aW}/L_a)t_m} \tag{9-36}$$

where $\Delta T$ is the difference between the base storage fluid temperature and the room ambient. The criteria and either eq. 9-35 or 9-36, or a constant $\Delta T$, can be used for preliminary sizing. A definitive economic evaluation, however, requires a comparison of the cost increment for a change in heat exchanger size with the change in system performance due to the consequent shift in base temperature.

**EXAMPLE 9-10**

Evaluate the choice of 33 °C (91.4 °F) as base temperature in the preceding analyses.

From eq. 9-35, for 1 °C,

$$\lambda = \frac{\varepsilon W_s c_s D_a}{L_{aH}} = \frac{0.8902 \times (8.071 \text{ W/°C·m}^2) \times 3527 \text{ °C-day}}{62.5 \text{ GJ}} \times 34.2 \text{ m}^2$$

$$\times \frac{24 \text{ hr}}{\text{day}} \times \frac{3600 \text{ sec}}{\text{hr}} \times \frac{\text{J/s}}{\text{W}} \times \frac{\text{G}}{10^9} = 1.2$$

and for 3 °C, the criteria is the same, since $\lambda$ is independent of the approach:

$$\lambda = \frac{\varepsilon W_s c_s D_a}{L_{aH}} = \frac{0.6707 \times 10.71 \times 3527 \times 34.2 \times 24 \times 3600}{62.5 \times 10^9} = 1.2$$

or from eq. 9-36,

$$\lambda = \frac{0.0723/\text{mo} \times 3527 \text{ °C-day}}{(33 - 23.89) \text{ °C} \times (1 - 0.2464) \times 31 \text{ day/mo}} = 1.2$$

Since the criteria is at the low end of the 1 to 3 range these measures indicate that from an economic standpoint a 33 °C (91.4 °F) base storage temperature is somewhat high.*

In this case the conclusion is probably correct; the base temperature was not chosen on an economic basis, but rather to assure that there would be no discomfort due to drafts.

## ANNUAL PERFORMANCE PREDICTION WITH SIMPLE DATA

### Base Temperature Method

When the annual but not the monthly performance of conventional systems is of importance, eq. 9-3 can be used more simply than was demonstrated in Table

---

* Because colder domestic water is available and preheated, the base temperature used for the calculation of performance is 3 °C (5 °F) lower, 30 °C (86 °F).

**TABLE 9-4.** Recalculation of the Performance of the Solar-Heating System for Example 9-7 Using Data Generated from Horizontal Radiation from Example 3-11, but with $k$ Equal to 1/1000 the Days in the Month (in m²/W) and $T_{sB} = 28$ °C. Compare with Tables 9-2a and 9-2b. $F_i'\tau\alpha = 0.7374$, $F_i'U_L = 4.8160$ W/°C·m², $mc = 0.3478$ MJ/°C·m², $L_{aW}/L_a = 0.2464$, and $S = 0.4124$ m²/GJ.

| | $D_m$ °C-days | $\bar{T}_a$ °C | $I_T$ MJ/m² | $k$ m²/W | $q_T$ MJ/m² | $S$ m²/GJ | $f'_{aB}$ | $L_m/L_a$ | $f'_B$ | $f$ | $f_a$ | $f_B$ | $f_{aB}$ |
|------|------|-------|-------|--------|-------|--------|--------|--------|--------|--------|--------|--------|--------|
| Jan | 686 | −3.8 | 326.4 | 0.0031 | 126.4 | 0.4124 | 0.0521 | 0.1671 | 0.3120 | 0.2854 | 0.0477 | 0.3120 | 0.0521 |
| Feb | 606 | −3.1 | 373.2 | 0.0028 | 155.8 | 0.4124 | 0.0643 | 0.1500 | 0.4284 | 0.3785 | 0.0568 | 0.4284 | 0.0643 |
| Mar | 505 | 2.1 | 472.4 | 0.0031 | 206.2 | 0.4124 | 0.0850 | 0.1284 | 0.6621 | 0.5439 | 0.0699 | 0.6621 | 0.0850 |
| Apr | 267 | 9.4 | 499.7 | 0.0030 | 256.0 | 0.4124 | 0.1056 | 0.0776 | 1.3606 | 0.8783 | 0.0681 | 1.0 | 0.0776 |
| May | 120 | 15.0 | 571.6 | 0.0031 | 324.0 | 0.4124 | 0.1337 | 0.0462 | 2.8937 | 1.0 | 0.0462 | 1.0 | 0.0462 |
| Jun | 21 | 20.2 | 544.0 | 0.0030 | 344.3 | 0.4124 | 0.1419 | 0.0250 | 5.6754 | 1.0 | 0.0250 | 1.0 | 0.0250 |
| Jul | 1 | 22.6 | 563.6 | 0.0031 | 372.5 | 0.4124 | 0.1533 | 0.0207 | 7.4053 | 1.0 | 0.0207 | 1.0 | 0.0207 |
| Aug | 8 | 21.3 | 510.0 | 0.0031 | 328.3 | 0.4124 | 0.1351 | 0.0222 | 6.0877 | 1.0 | 0.0222 | 1.0 | 0.0222 |
| Sep | 71 | 17.2 | 495.7 | 0.0030 | 295.8 | 0.4124 | 0.1220 | 0.0357 | 3.4168 | 1.0 | 0.0357 | 1.0 | 0.0357 |
| Oct | 219 | 11.3 | 445.8 | 0.0031 | 234.4 | 0.4124 | 0.0966 | 0.0673 | 1.4359 | 0.9014 | 0.0607 | 1.0 | 0.0673 |
| Nov | 392 | 5.3 | 287.7 | 0.0030 | 136.0 | 0.4124 | 0.0561 | 0.1043 | 0.5377 | 0.4594 | 0.0479 | 0.5377 | 0.0561 |
| Dec | 631 | −2.0 | 283.6 | 0.0031 | 113.9 | 0.4124 | 0.0470 | 0.1554 | 0.3025 | 0.2775 | 0.0431 | 0.3025 | 0.0470 |
| Total | 3527 | | | | | | | | | | 0.5441 | | 0.5993 |

9-2b. Notice that the value of $k$ there does not vary widely. It can in fact be approximated empirically for most U.S. locations as 1/1000 the number of days in the month (in m²/W) with little error in the final annual performance, so long as the collector is tilted to within 15° of the optimum tilt of the latitude plus about 12°.

Table 9-4 shows the analysis of Table 9-2b repeated with these new values for $k$. Average 24-hour temperatures from Table 3-5 have been used, and the base temperature correspondingly has been lowered 2 °C to compensate. Notice that the errors in performance for the winter months tend to be compensated for by opposite errors in the spring and fall months. Using this technique, annual performance can be calculated directly from monthly values for the tilted total global radiation, degree-days, and 24-hour average temperature. The method will give accurate results for collector tilts of 15 to 25° away from optimum if the base temperature is raised an additional 2 °C.

## F-Chart Method

The F-chart method of annual performance prediction (4) also utilizes monthly input values of tilted global radiation, degree-days, and 24-hour average temperature. It is good for collector tilts within 15 to 20° of the optimum. System input is the same as the annual base temperature method described above except that there is no entry for system temperature. For this reason a different F-chart correlation is required for each situation to be analyzed. Thus far, validated F-charts have been published for air and liquid heating systems where there is an indoor ambient temperature of 20 °C (68 °F). (Both correlations use inlet-based collector parameters.) The correlations are both commonly used with 18.3 °C (65 °F) base degree-days, an improper application that overestimates the annual solar contribution $\Sigma f_a$.

The F-chart correlation first requires determination of two dimensionless parameters. In the nomenclature of this book,

$$Y' = \frac{F_i' \overline{\tau\alpha} I_T A_c}{L_m} = \frac{F_i' \overline{\tau\alpha} I_T S}{L_m/L_a} \tag{9-37}$$

$$X' = \frac{F_i' U_L (T_{\text{ref}} - \overline{T}_a) t_m A_c}{L_m} = \frac{F_i' U_L (T_{\text{ref}} - \overline{T}_a) t_m S}{L_m/L_a} \tag{9-38}$$

where $I_T$ is the total global radiation on the tilted surface, $\overline{T}_a$ the monthly 24-hour average temperature, and $t_m$ the total time in the month. $T_{\text{ref}}$ is a reference temperature, the boiling point of water (100 °C). Correction factors are applied to each term to account for terminal heat exchanger size and storage capacity, respectively:

$$Y = [0.394 + 0.65 \exp(-0.139/\lambda)] Y' \tag{9-39}$$

where $\lambda$ is defined by eq. 9-35 ($Y = Y'$ when $\lambda = 2$), and

$$X = \left(\frac{mc}{0.307}\right)^{-0.25} X' \qquad (0.15 < mc < 1.2) \tag{9-40}$$

where $mc$ is the usual area-normalized storage capacity, given in MJ/°C·m². 

The correlation equation for liquid systems is

$$f = 1.029 \, Y - 0.065 \, X - 0.245 \, Y^2 + 0.0018 \, X^2 + 0.0215 \, Y^3 \tag{9-41}$$
$$\text{(for } 0 < Y < 3 \text{ and } 0 < X < 18)$$

and for air systems

$$f = 1.040 \, Y - 0.065 \, X - 0.159 \, Y^2 + 0.00187 \, X^2 - 0.0095 \, Y^3 \tag{9-42}$$
$$\text{(for } 0 < Y < 3 \text{ and } 0 < X < 18)$$

For air systems $Y$ always equals $Y'$ but, instead of eq. 9-40, the following expression is used:

$$X = X' \left(\frac{mc}{0.409}\right)^{-0.3} \left(\frac{v}{10.1}\right)^{0.28} \tag{9-43}$$

where $v$ is the air flow rate in L/sec·m².

The annual performance predicted by F-chart (with $\lambda = 2$) is closely approximated by the monthly or the annual version of the base-temperature method using a base temperature of 22 °C (71.6 °F), or 24 °C (75.2 °F) if daytime temperatures are used.

The 24 °C base temperature corresponds to the conditions used for development of the F-chart correlation—20 °C inside ambient temperature, plus about 7 °C for heat exchange, less 3 °C for concurrent hot water usage. When it is used differences in $\sum f_a$ predicted by the two methods are seldom over 0.02 and are usually less than 0.015. Monthly predictions may not correspond very closely because the F-chart method cannot account for seasonally changing threshold values.

## EXAMPLE 9-10

Using simple data from Table 3-8, predict the annual performance of the heating system described in Example 9-7 using the F-chart and annual base-temperature techniques.

Tables 9-5a and 9-5b list the input data for each technique. Calculation of $L_m/L_a$ is as in Example 9-4.

For January:

For F-chart, using the second expression of eq. 9-37, and letting $\lambda = 2$,

$$Y = 0.7374 \times \frac{326.4 \text{ MJ}}{\text{m}^2} \times \frac{0.4124 \text{ m}^2}{\text{GJ}} \times \frac{\text{GJ}}{1000 \text{ MJ}} \times \frac{1}{0.1671} = 0.5940$$

and by eq. 9-38,

$$X' = \frac{4.816 \text{ W}}{\text{°C} \cdot \text{m}^2} \times [100 - (-3.8)] \text{ °C} \times 2.6784 \text{ Ms} \times \frac{0.4124 \text{ m}^2}{\text{GJ}} \times \frac{\text{GJ}}{1000 \text{ MJ}}$$

$$\times \frac{1}{0.1671} \times \frac{\text{J}}{\text{W} \cdot \text{s}} = 3.304$$

**TABLE 9-5a.** Performance Prediction by F-Chart for Example 9-7 (Table 9-4). $F_t'\overline{\tau\alpha} = 0.7374$, $F_t'U_L = 4.8160 \text{ W/°C} \cdot \text{m}^2$, $mc = 0.3478 \text{ MJ/°C} \cdot \text{m}^2$, $L_{aw}/L_a = 0.2464$, and $S = 0.4124 \text{ m}^2/\text{GJ}$.

| | $D_m$ °C-days | $\overline{T}_a$ °C | $I_T$ MJ/m² | $t_m$ Ms | $Y$ | $X$ | $q_T$ MJ/m² (delivered) | $S$ | $L_m/L_a$ | $f$ | $f_a$ |
|-----|------|------|-------|--------|--------|-------|--------|--------|--------|--------|--------|
| Jan | 686 | −3.8 | 326.4 | 2.6784 | 0.5940 | 3.203 | 137.6 | 0.4124 | 0.1671 | 0.3396 | 0.0567 |
| Feb | 606 | −3.1 | 373.2 | 2.4451 | 0.7566 | 3.201 | 166.6 | 0.4124 | 0.1500 | 0.4580 | 0.0687 |
| Mar | 505 | 2.1 | 472.4 | 2.6784 | 1.119 | 3.930 | 201.5 | 0.4124 | 0.1284 | 0.6469 | 0.0831 |
| Apr | 267 | 9.4 | 499.7 | 2.5920 | 1.958 | 5.827 | 173.0 | 0.4124 | 0.0776 | 0.9195 | 0.0713 |
| May | 120 | 15.0 | 571.6 | 2.6784 | 3.762 | 9.492 | 112.0 | 0.4124 | 0.0462 | 1.0 | 0.0462 |
| Jun | 21 | 20.2 | 544 | 2.5920 | 6.617 | 15.91 | 60.67 | 0.4124 | 0.0250 | 1.0 | 0.0250 |
| Jul | 1 | 22.6 | 563.6 | 2.6784 | 8.279 | 19.24 | 50.31 | 0.4124 | 0.0207 | 1.0 | 0.0207 |
| Aug | 8 | 21.3 | 510.0 | 2.6784 | 6.986 | 18.24 | 53.93 | 0.4124 | 0.0222 | 1.0 | 0.0222 |
| Sep | 71 | 17.2 | 495.7 | 2.5920 | 4.222 | 11.57 | 86.58 | 0.4124 | 0.0357 | 1.0 | 0.0357 |
| Oct | 219 | 11.3 | 445.8 | 2.6784 | 2.014 | 6.793 | 146.2 | 0.4124 | 0.0673 | 0.8956 | 0.0603 |
| Nov | 392 | 5.3 | 287.7 | 2.5920 | 0.8388 | 4.531 | 112.8 | 0.4124 | 0.1043 | 0.4459 | 0.0465 |
| Dec | 631 | −2.0 | 283.6 | 2.6784 | 0.5550 | 3.386 | 113.0 | 0.4124 | 0.1554 | 0.3001 | 0.0466 |
| Total | 3527 | | | | | | | | | | 0.5832 |

by eq. 9-40,

$$X = 3.304 \times \left(\frac{0.3478}{0.307}\right)^{-0.25} = 3.203$$

by eq. 9-41,

$$f(3.203, 0.5940) = 0.3396$$

using eq. 9-15 solved for $q_T$, the *actual* heat collected is

$$q_T = 0.3396 \times 0.1671 \times \frac{GJ}{0.4124 \text{ m}^2} \times \frac{1000 \text{ MJ}}{GJ} = 137.6 \text{ MJ/m}^2$$

and by eq. 9-20,

$$f_a = 0.3396 \times 0.1671 = 0.0567$$

For the annual base-temperature method, using $T_{sB} = 22$ °C, by eq. 9-3 at maximum storage

$$q_{T(\text{max storage})} = 0.7374 \times \frac{326.4 \text{ MJ}}{\text{m}^2}$$

$$\times \exp\left\{-\frac{0.0031 \text{ W}}{\text{m}^2} \times \frac{4.8160 \frac{\text{W}}{\text{C} \cdot \text{m}^2}}{0.7374} \times [22 - (-3.8)] \text{ °C}\right\}$$

$$= 142.8 \text{ MJ/m}^2$$

by eq. 9-6,

$$f'_B = \frac{142.8 \text{ MJ}}{\text{m}^2} \times \frac{0.4124 \text{ m}^2}{GJ} \times \frac{GJ}{1000 \text{ MJ}} \times \frac{1}{0.1671} = 0.3523$$

$$\frac{F'_i U_L}{mc} = \frac{4.8160 \text{ W}}{\text{m}^2} \times \frac{\text{°C} \cdot \text{m}^2}{0.3478 \text{ MJ}} \times \frac{J}{\text{W} \cdot \text{s}} = 13.85 \text{ (Ms)}^{-1}$$

**TABLE 9-5b.** Performance Prediction Using the Annual Base Temperature Method with $T_{sB} = 22$ °C to Compare with Table 9-5a. $F'_i \overline{\tau\alpha} = 0.7374$, $F'_i U_L = 4.8160 \text{ W/°C} \cdot \text{m}^2$, $T_{sB} = 22$ °C, $mc = 0.3478 \text{ MJ/°C} \cdot \text{m}^2$, $L_{aW}/L_a = 0.2464$, and $S = 0.4124 \text{ m}^2/\text{GJ}$.

| | $D_m$ °C-days | $\overline{T}_a$ °C | $I_T$ MJ/m² | $k$ m²/W | $q_T$ MJ/m² (at max. storage) | $S$ m²/GJ | $L_m/L_a$ | $f$ | $f_a$ |
|---|---|---|---|---|---|---|---|---|---|
| Jan | 686 | −3.8 | 326.4 | 0.0031 | 142.8 | 0.4124 | 0.1671 | 0.3185 | 0.0532 |
| Feb | 606 | −3.1 | 373.2 | 0.0028 | 173.9 | 0.4124 | 0.1500 | 0.4161 | 0.0624 |
| Mar | 505 | 2.1 | 472.4 | 0.0031 | 232.8 | 0.4124 | 0.1284 | 0.5974 | 0.0767 |
| Apr | 267 | 9.4 | 499.7 | 0.0030 | 287.9 | 0.4124 | 0.0776 | 0.9272 | 0.0719 |
| May | 120 | 15.0 | 571.6 | 0.0031 | 365.8 | 0.4124 | 0.0462 | 1.0 | 0.0462 |
| Jun | 21 | 20.2 | 544 | 0.0030 | 387.3 | 0.4124 | 0.0250 | 1.0 | 0.0250 |
| Jul | 1 | 22.6 | 563.6 | 0.0031 | 420.7 | 0.4124 | 0.0207 | 1.0 | 0.0207 |
| Aug | 8 | 21.3 | 510.0 | 0.0031 | 370.7 | 0.4124 | 0.0222 | 1.0 | 0.0222 |
| Sep | 71 | 17.2 | 495.7 | 0.0030 | 332.7 | 0.4124 | 0.0357 | 1.0 | 0.0357 |
| Oct | 219 | 11.3 | 445.8 | 0.0031 | 264.7 | 0.4124 | 0.0673 | 0.9490 | 0.0639 |
| Nov | 392 | 5.3 | 287.7 | 0.0030 | 152.9 | 0.4124 | 0.1043 | 0.5059 | 0.0528 |
| Dec | 631 | −2.0 | 283.6 | 0.0031 | 128.7 | 0.4124 | 0.1554 | 0.3097 | 0.0481 |
| Total | 3527 | | | | | | | | 0.5790 |

by eq. 9-23 or Fig. 9-2,

$$f[0.3523, 13.85 \, (Ms)^{-1}] = 0.3185$$

by eq. 9-20,

$$f_a = 0.3185 \times 0.1671 = 0.0532$$

Tables 9-5a and 9-5b show monthly details of performance predictions by F-chart and the base-temperature method using the input data for Example 9-7 (Table 9-4) and a 22 °C base temperature. Notice that the F-chart, which is specifically not intended for monthly use, does not agree too closely with the base-temperature technique month-by-month but that the annual $\sum f_a$ values are within 0.01. This is the typical situation. Also typical are the high values of $\sum f_a$—about 0.58—relative to example 9-7, which predicted a $\sum f_a$ of 0.5373 because of the realistically high 30 °C base temperature. The overprediction is also due to the inconsistent use of 65 °F degree-days with a 68 °F indoor ambient. The error is not too great if applied to an annual load estimated using the proper degree-days (based on about 58 °F), but if 65 °F base degree-days are used, the absolute magnitude of the solar heat delivered by a particular system easily can be overestimated by 30 to 50%.

## MONTHLY SYSTEM PERFORMANCE FOR HEATING HOT WATER

The techniques presented thus far do not apply directly for heating hot water. Hot water demand is generally regarded as constant from day-to-day, while heating demand varies with the statistical swings of outside temperature.* Most important, water can be *preheated* when solar energy is scarce, giving higher collector efficiency due to operation at a low temperature.

The stored water temperature is not under designer control. It will seek an equilibrium value such that the solar energy collected will match the heat actually delivered. When more solar energy is available, the storage temperature rises, making it possible to deliver more heat to the water flow being preheated. Since this also hurts collector efficiency, the thermal return per unit area for a hot water system falls off quite rapidly as the collector area is increased. Conversely, a small amount of preheat is economical almost anywhere in the United States, provided that the demand is large enough to be worth the expenditure of fixed costs for engineering, pumps, and so on.

Since hot water demand is steady from day-to-day, the penalty for undersized storage is less with hot water systems. Nevertheless, storage capacity $mc$ of less than 0.2 MJ/°C·m² (10 Btu/°F ft²) (including the accessible auxiliary-heated storage volume) is still not recommended because of the potential for boilover in the summer. The tank should be very well insulated (R = 5 °C·m²/W or

---

* Demand patterns within a day differ, but it should be clear by now that the system performance is insensitive to this factor.

30 hr °F ft²/Btu); recent tests have shown the usual tank insulation to be inadequate.

The base-temperature method for space heating can bracket the probable system performance. A design using a base temperature equal to the supply temperature $T_{CW}$ will surely be too optimistic. A design using a base temperature at the use temperature $T_{HW}$ will be too conservative, for no credit is taken for the preheat collector efficiency improvement. A logical choice might be an average between $T_{CW}$ and $T_{HW}$. This is surprisingly good. It predicts monthly performance at about $\sum f_a = 0.8$ and is conservative by about 0.05 in $\sum f_a$ at $\sum f_a = 0.4$. Therefore, a complete new correlation is not necessary to predict the performance of hot water systems. The next section details corrections to be made to $f_B'$ and the storage function $F_i' U_L / mc$ prior to the use of eq. 9-23. These will allow direct application of the predictive techniques already developed to the heating of hot water, provided losses from the storage tank are not excessive.

### Application of the Base Temperature Method to Hot Water Heating

**Base Temperature Selection.** The base temperature for hot water applications is defined as

$$T_{sB} = (T_{CW} + T_{HW})/2 \tag{9-44}$$

Table 9-6 shows simulation results justifying this assumption. The two identical systems performed almost identically even though one was preheating water from 15.55 to 60 °C (60 to 140 °F) and the other from 26.67 to 48.89 °C (80 to 120 °F). The average water temperature is the same: 27.78 °C (100 °F)—and this is taken as the base temperature.

**Modification of $f_B'$.** The use of eq. 9-44 for the base temperature produces too high a value for $f_B'$ when the monthly solar participation is high, and too low when it is low. Before use of eq. 9-23, $f_B'$ is corrected using

$$(f_B')_{\text{effective}} = 1.27 f_B' - 0.25(f_B')^2 \tag{9-45}$$
$$\text{(valid for } 0 < f_B' < 2.2)$$

This correction is plotted in Fig. 9-4. A reduction is made in $f_B'$ values above 1.08, while lesser values are increased up to 26% to compensate for lower storage temperatures.

The dashed line on Fig. 9-2 that roughly follows the ×2 line also represents eq. 9-45. When using Fig. 9-2 for hot water, enter with $f_B'$ as usual, on the abscissa. Then proceed upward to the dotted line. From that intersection go horizontally to the ×2 line. The corrected $f_B'$ value can now be read from the abscissa. Then proceed vertically to the effective value of the storage function $F_i' U_L / mc$, which is determined as explained below.

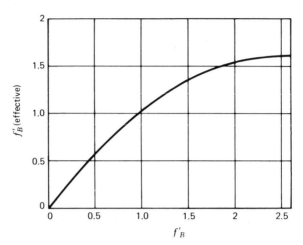

**FIGURE 9-4.** Correction to be made to $f_B'$ for hot water correlations (eq. 9-45).

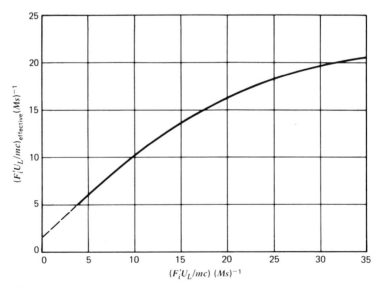

**FIGURE 9-5.** Correction to be made to the storage function for hot water correlations (eq. 9-46).

**TABLE 9-6.** Hot Water Simulations at Hartford (Calendar 1959). $F_i'\overline{\tau\alpha} = 0.73$, $F_i'U_L = 4.316$ W/°C·m², $mc = 0.4088$ MJ/°C·m², and $S = 0.2642$ m²/GJ.

|  | Case 1 | Case 2 |
|---|---|---|
| Cold Water Temperature | 15.55 °C (60 °F) | 26.67 °C (80 °F) |
| Hot Water Temperature | 60 °C (140 °F) | 48.89 °C (120 °F) |
|  | $\Sigma f_a$ | $\Sigma f_a$ |
| Jan | 0.3202 | 0.3079 |
| Feb | 0.4816 | 0.4728 |
| Mar | 0.6708 | 0.6810 |
| Apr | 0.6884 | 0.6941 |
| May | 0.8217 | 0.8602 |
| Jun | 0.6307 | 0.6402 |
| Jul | 0.7608 | 0.7883 |
| Aug | 0.8067 | 0.8225 |
| Sep | 0.7820 | 0.8008 |
| Oct | 0.5379 | 0.5349 |
| Nov | 0.2902 | 0.2553 |
| Dec | 0.3002 | 0.2751 |
| $\Sigma f_a$ | 0.5917 | 0.5953 |

**Modification of $F_i'U_L/mc$.** Since hot water usage is more regular than space heating, smaller storage volumes are more effective when heating hot water.* The corrected storage function is given by

$$\left(\frac{F_i'U_L}{mc}\right)_{\text{effective}} = 1.667 + 0.974\left(\frac{F_i'U_L}{mc}\right) - 0.01246\left(\frac{F_i'U_L}{mc}\right)^2 \qquad (9\text{-}46)$$

where the units of $F_i'U_L/mc$ are (Ms)$^{-1}$. Equation 9-46 is plotted in Fig. 9-5 and also on Fig. 9-2. Values of the storage function are decreased (leading to better performance) when $(F_i'U_L/mc)$ is below 10.56 (Ms)$^{-1}$, and increased when above that value.

**Limitations.** The simulations that produced eqs. 9-45 and 9-46 were run over the same range of parameters as those producing the basic correlation equation (9-23). As before, $F_i'U_L/mc$ must be 3.5 to 35 (Ms)$^{-1}$ (0.0125 to 0.125 hr$^{-1}$). Since eq. 9-45 applies only for $f_B' < 2.2$, a monthly solar participation of 1.0 is arbitrarily recommended for higher values. Storage capacities below 0.2 MJ/°C·m² (10 Btu/°F ft²) are not recommended because of boilover problems.

It is difficult to meet 100% of any month's hot water requirements using solar energy, because then the tank temperature must always remain above the hot

* Remember, however, to consider only that portion of the storage tank that can be solar heated as solar storage.

water use temperature. Simulations and eq. 9-45 confirm this difficulty. Values of 90 to 95% are much easier, however, and if the consumer can modify his day-to-day use pattern it is likely that 100% can be achieved.

The simulations producing the hot water correlation equations used a steady hot water demand. This means that one-quarter to one-half of the demand occurs during the day, depending on the season. Heavy daytime users should expect a few percent more annual solar participation (unless they concentrate usage so early in the morning that they must use stored energy); nighttime users a little less. Applications with periodic demands (high school locker rooms, for instance) should install extra storage so that solar heat can be accumulated efficiently during periods of lessened hot water usage.*

**EXAMPLE 9-11**

Consider once again a solar system with parameters from Examples 9-6 and 9-7 (Table 9-2). What performance can be expected for heating water from 10 to 50 °C (50 to 122 °F)?

The base temperature by eq. 9-44 is

$$T_{sB} = \frac{10 + 50}{2} = 30 \,°C$$

which is the same as in the earlier examples. Since the threshold levels depend only on the collector parameters and the base temperature, they are also unchanged. Therefore, the solar data and the collector performance at maximum storage $q_T$ are the same as in Example 9-7. However, $f'_B$ is different because $L_m/L_a$ is now in proportion to the days in the month (Example 9-5 and Table 9-1).

Table 9-7a repeats the weather data abstracted from Appendix B in Example 9-7. For January, using eqs. 9-4 and 9-6,

$$f'_B = \{0.7374 \times 280.3 - 4.816[30 - (-2.7)]0.5638\} \frac{0.4124 \times 10^{-3}}{0.08487}$$

$$= 0.5729$$

by eq. 9-45,

$$(f'_B)_{\text{effective}} = 1.27(0.5729) - 0.25(0.5729)^2 = 0.6455$$

by eq. 9-46,

$$\left(\frac{F'_t U_L}{mc}\right)_{\text{effective}} = 1.667 + 0.974(13.85) - 0.01246(13.85)^2 = 12.77 \,(Ms)^{-1}$$

by eq. 9-23 or Fig. 9-2, the monthly performance is

$$f(0.6455, 12.77) = 0.5388$$

---

* Equation 8-20 can be helpful when making such designs.

**TABLE 9-7a.** Monthly Performance of a Solar Hot Water System for Example 9-11 Using Tabular Data from Appendix B. $F_i'\overline{\tau\alpha} = 0.7374$, $F_i'U_L = 4.8160$ W/°C·m², $mc = 0.3478$ MJ/°C·m², $L_{aw}/L_a = 1$, $S = 0.4124$ m²/GJ, and $T_{sB} = 30$ °C.

| | Data Col- umn | $\overline{T}_a$ °C | $t_T$ Ms | $I_T$ MJ/m² | $q_T$ MJ/m² | $S$ m²/GJ | $L_m/L_a$ | $f_B'$ | $f_B'$ (effective) $f$ | $f_a$ |
|---|---|---|---|---|---|---|---|---|---|---|
| Jan | 4 | −2.7 | 0.5638 | 280.3 | 117.9 | 0.4124 | 0.084 87 | 0.5729 | 0.6455 | 0.5388 | 0.0457 |
| Feb | 4 | −1.5 | 0.6030 | 326.4 | 149.2 | 0.4124 | 0.077 34 | 0.7955 | 0.8521 | 0.6653 | 0.0515 |
| Mar | 4 | 4.8 | 1.7466 | 415.5 | 215.7 | 0.4124 | 0.084 87 | 1.0481 | 1.0564 | 0.7699 | 0.0653 |
| Apr | 3 | 12.5 | 1.0994 | 482.9 | 263.4 | 0.4124 | 0.082 14 | 1.3225 | 1.2423 | 0.8483 | 0.0697 |
| May | 3 | 18.4 | 1.2618 | 559.5 | 342.0 | 0.4124 | 0.084 87 | 1.6618 | 1.4201 | 0.9093 | 0.0772 |
| Jun | 3 | 23.4 | 1.2287 | 531.4 | 352.7 | 0.4124 | 0.082 14 | 1.7709 | 1.4650 | 0.9227 | 0.0758 |
| Jul | 2 | 25.2 | 1.4836 | 559.8 | 378.5 | 0.4124 | 0.084 87 | 1.8391 | 1.4901 | 0.9298 | 0.0789 |
| Aug | 3 | 24.5 | 1.1995 | 494.4 | 332.7 | 0.4124 | 0.084 87 | 1.6166 | 1.3997 | 0.9029 | 0.0766 |
| Sep | 3 | 20.9 | 1.0408 | 489.5 | 315.3 | 0.4124 | 0.082 14 | 1.5831 | 1.3840 | 0.8979 | 0.0738 |
| Oct | 3 | 14.9 | 0.9475 | 438.9 | 254.7 | 0.4124 | 0.084 87 | 1.2376 | 1.1888 | 0.8273 | 0.0702 |
| Nov | 4 | 7.6 | 0.4964 | 243.4 | 125.9 | 0.4124 | 0.082 14 | 0.6321 | 0.7029 | 0.5761 | 0.0473 |
| Dec | 4 | −0.8 | 0.4864 | 240.4 | 105.1 | 0.4124 | 0.084 87 | 0.5107 | 0.5834 | 0.4965 | 0.0421 |
| Total | | | | | | | | | | | 0.7741 |

**TABLE 9-7b.** Monthly Performance of a Solar Hot Water System for Example 9-11 Using Data Generated from Horizontal Radiation from Example 3-11. Compare with Table 9-6a. $F_i'\overline{\tau\alpha} = 0.7374$, $F_i'U_L = 4.8160$ W/°C·m², $mc = 0.3478$ MJ/°C·m², $L_{aw}/L_a = 1$, $S = 0.4124$ m²/GJ, and $T_{sB} = 30$ °C.

| | $\overline{T}_a$ °C | $I_T$ MJ/m² | $k$ m²/W | $q_T$ MJ/m² | $S$ m²/GJ | $L_m/L_a$ | $f_B'$ | $f_B'$ (effective) $f$ | $f_a$ |
|---|---|---|---|---|---|---|---|---|---|---|
| Jan | −1.8 | 326.4 | 0.0033 | 121.3 | 0.4124 | 0.084 87 | 0.5894 | 0.6617 | 0.5494 | 0.0466 |
| Feb | −1.1 | 373.2 | 0.0030 | 149.6 | 0.4124 | 0.077 34 | 0.7979 | 0.8542 | 0.6664 | 0.0515 |
| Mar | 4.1 | 472.4 | 0.0029 | 213.3 | 0.4124 | 0.084 87 | 1.0365 | 1.0477 | 0.7658 | 0.0650 |
| Apr | 11.4 | 499.7 | 0.0029 | 259.1 | 0.4124 | 0.082 14 | 1.3008 | 1.2290 | 0.8431 | 0.0693 |
| May | 17. | 571.6 | 0.0029 | 329.5 | 0.4124 | 0.084 87 | 1.6012 | 1.3926 | 0.9007 | 0.0764 |
| Jun | 22.2 | 544 | 0.0031 | 342.6 | 0.4124 | 0.082 14 | 1.7201 | 1.4448 | 0.9167 | 0.0753 |
| Jul | 24.6 | 563.6 | 0.0030 | 373.9 | 0.4124 | 0.084 87 | 1.8166 | 1.4821 | 0.9275 | 0.0787 |
| Aug | 23.3 | 510.0 | 0.0030 | 329.8 | 0.4124 | 0.084 87 | 1.6024 | 1.3931 | 0.9008 | 0.0765 |
| Sep | 19.2 | 495.7 | 0.0028 | 300.0 | 0.4124 | 0.082 14 | 1.5064 | 1.3458 | 0.8854 | 0.0727 |
| Oct | 13.3 | 445.8 | 0.0029 | 239.6 | 0.4124 | 0.084 87 | 1.1614 | 1.1397 | 0.8069 | 0.0685 |
| Nov | 7.3 | 287.7 | 0.0035 | 126.3 | 0.4124 | 0.082 14 | 0.6339 | 0.7046 | 0.5772 | 0.0474 |
| Dec | 0.0 | 283.6 | 0.0035 | 105.4 | 0.4124 | 0.084 87 | 0.5119 | 0.5846 | 0.4974 | 0.0422 |
| Total | | | | | | | | | | 0.7702 |

and from eq. 9-20

$$f_a = 0.5388 \times 0.08487 = 0.0457$$

Table 9-7a continues this example for a year. $\sum f_a$ is 0.7741, much higher than in Example 9-7. This is partly due to the lower temperatures maintained when preheating hot water, and partly due to the shifting of load into the summer months.

The method works equally well with elevated data and $k$ values from Example

3-11 and Table 3-8 as is shown in Table 9-7*b*. Using eq. 9-3, $q_T$ is unchanged from Example 9-7:

$$q_T = 0.7374 \times \frac{326.4 \text{ MJ}}{\text{m}^2}$$

$$\times \exp \left\{ \frac{-0.0033 \text{ m}^2}{\text{W}} \times \frac{4.8160 \text{ W}}{{}^\circ\text{C} \cdot \text{m}^2} \times \frac{[30 - (-1.8)] \,{}^\circ\text{C}}{0.7374} \right\}$$

$$= 121.3 \text{ MJ/m}^2$$

by eq. 9-6,

$$f_B' = \frac{121.3 \text{ MJ}}{\text{m}^2} \times \frac{0.4124 \times 10^{-3} \text{ m}^2}{0.084 \, 87 \text{ MJ}} = 0.5894$$

$$(f_B')_{\text{effective}} = 1.27 \times 0.5894 - 0.25 \times (0.5894)^2 = 0.6617$$

The effective storage function remains as previously determined. Using eq. 9-23 or Fig. 9-2,

$$f(0.6617, 12.77) = 0.5494$$

and from eq. 9-20,

$$f_a = 0.5494 \times 0.084 \, 87 = 0.0466$$

Table 9-7*b* continues this analysis for the entire year. Results are quite comparable to Table 9-7*a*. The annual performance technique (Table 9-5*b*) also works well with hot water when the corrections are made to $f_B'$ and the storage function.

**Variable Water Temperature.**   When the flow or the cold or hot water temperature varies from month to month, $L_m/L_a$ must be determined according to the load:

$$L_m/L_a = \frac{L_m}{\sum\limits_{\text{yr}} L_m} = \frac{(W_{HW})_m \, \Delta T_m t_m}{\sum\limits_{\text{yr}} [(W_{HW})_m \, \Delta T_m t_m]} \tag{9-47}$$

where $\Delta T_m$ is the temperature rise and $t_m$ is the monthly water usage time at flow rate $(W_{HW})_m$. Since $L_m/L_a$ is a ratio, these values can be expressed by any convenient measures in using eq. 9-47.

When only annual results are needed, the calculations can be simplified if a load-weighted monthly average base temperature is used throughout the year. This is given by

$$\overline{T}_{sB} = \sum\limits_{\text{yr}} [(L_m/L_a) \times (T_{sB})_m] \tag{9-48}$$

The final $\sum f_a$ determined in this way will be virtually identical to that using the individual monthly base temperatures $(T_{sB})_m$.

**EXAMPLE 9-12**

Doctors at a very fictitious nursing home have ordered patients to take 4 baths a day in winter, 3 in spring and fall, and 2 in summer. The incoming feed water

**TABLE 9-8.** Predicted Performance of a Hot Water System Having Uneven Seasonal Demands and Supply Temperatures. One Calculation Was Made with Monthly Base Temperatures, the Other with an Average Base Temperature Using Eq. 9-48 (Example 9-12)

| | Monthly Base Temperature, °C | Loads | $L_m/L_a$ | $f$ Using Monthly Base Temperatures | $f$ Using Weighted Average Base Temperature (33.27 °C) |
|---|---|---|---|---|---|
| Jan | 30. | 160. | 0.1368 | 0.207 | 0.194 |
| Feb | 30. | 160. | 0.1368 | 0.253 | 0.239 |
| Mar | 35. | 90. | 0.0769 | 0.507 | 0.518 |
| Apr | 35. | 90. | 0.0769 | 0.583 | 0.596 |
| May | 35. | 90. | 0.0769 | 0.679 | 0.692 |
| Jun | 37.5 | 50. | 0.0427 | 0.873 | 0.896 |
| Jul | 37.5 | 50. | 0.0427 | 0.898 | 0.918 |
| Aug | 37.5 | 50. | 0.0427 | 0.861 | 0.886 |
| Sep | 35. | 90. | 0.0769 | 0.657 | 0.669 |
| Oct | 35. | 90. | 0.0769 | 0.567 | 0.579 |
| Nov | 35. | 90. | 0.0769 | 0.329 | 0.340 |
| Dec | 30. | 160. | 0.1368 | 0.187 | 0.175 |
| Sum | | 1170. | | | |
| $\Sigma f_a$ | | | | 0.456 | 0.460 |

NOTE: Calculated using $F_i'\overline{\tau\alpha} = 0.7374$, $F_i'U_L = 4.8160$ W/°C·m², $mc = 0.3478$ MJ/°C·m², and $S = 0.2062$ m²/GJ, with tabular data from Table 9-2a.

runs 10 °C (50 °F) in winter, 20 °C (68 °F) in spring and fall, and 25 °C (77 °F) in summer. Hot water for baths is to be provided by solar panels.

(a) What are monthly $L_m/L_a$ values? Assume the months of equal length.
(b) What is the proper average base temperature if the hot water supply is held at 50 °C (122 °F)?

(a) The monthly loads are the product of the flow rate and the temperature rise. For January, the numerator of eq. 9-47 is

$$\text{Load} = 4 \times (50 - 10) = 160 \text{ units}$$

The total load for the entire year will be found to be 1170 units. By eq. 9-47,

$$L_m/L_a = 160/1170 = 0.1368$$

The other monthly values are listed in Table 9-8.

(b) By eq. 9-44, the average monthly base temperature for January is

$$T_{sB} = (50 + 10)/2 = 30 \text{ °C}$$

Other monthly values are listed in Table 9-8. The weighted average base temperature determined by eq. 9-48 using Table 9-8 values is

$$\overline{T}_{sB} = 3 \times 0.1368 \times 30 + 6 \times 0.0769 \times 35 + 3 \times 0.0427 \times 37.5$$
$$= 33.27 \text{ °C}$$

Using Appendix B data with the same parameters as Example 9-11, but half the area $S$, Table 9-8 shows that the performance using $\overline{T}_{sB}$ for the base temperature is about the same on an annual basis as when the monthly $T_{sB}$ values are used. Even the monthly values are almost interchangeable.

If the monthly loads are weighted by the days in the month when eq. 9-47 is applied, the monthly $L_m/L_a$ values differ somewhat, and the base temperature rises 0.02 °C. The complication is unnecessary, for final $\sum f_a$ values are within 0.001 of those determined above.

## PROBLEMS

For collector $A$, $F_i\overline{\tau\alpha} = 0.808$, $F_iU_L = 4.43$ W/°C·m² (0.766 Btu/hr °F ft²). For collector $B$, $F_i\tau\alpha = 0.726$, $F_iU_L = 3.62$ W/°C·m² (0.626 Btu/hr °F ft²). The liquid flow capacity rates in the collector circuit are 82.3 W/°C·m² (14.5 Btu/hr °F ft²).

1. (a) Draw the collector efficiency curves, using an incident angle modifier of 0.96.
   (b) For collector $A$, what are the threshold radiation levels in Atlanta, Boston, and Denver in January? Use a 30 °C (86 °F) base temperature and the closest Appendix B table for a latitude $+10°$ inclination. Assume a drain-down system that requires no heat exchanger.
   (c) Using the table corresponding to that threshold, estimate the collectible heat $(q_T)$ for each location with eq. 9-4. Locate the operating point on the efficiency curve.
   (d) Estimate $q_T$ for each location using eq. 9-2, interpolating between the columns for the exact threshold radiation.
   (e) Estimate $q_T$ for each location using eq. 9-3, determining the constant $k$ from Table 3-7.
   (f) List the advantages and disadvantages of each method for finding the collectable heat $q_T$.
   (g) Using answers from part (c) estimate the maximum solar contribution $f'_B$ at the three locations if the structure heat loss is 24 MJ/°C-day (12 646 Btu/°F-day) and the collector area is 40 m² (430 ft²). Use eqs. 9-5 and 9-6 alternately for the three locations, assuming heating load only.
   (i) Repeat (h) if there is an additional hot water load of 300 L/day (79.26 gal/day) from 20 °C (68 °F) to 60 °C (140 °F).
   (j) Find the monthly solar participation $f$ for part (i) at the three locations if the storage medium is 4000 L (1057 gal) of water, using eq. 9-17. What portion is due to simultaneous load?
   (k) Recalculate part (j) using eq. 9-23 or Fig. 9-2.
   (l) What portions of the *annual* thermal demand are met in part (k)?
   (m) Express the answers to part (k) in terms of thermal units (MJ or M Btu).
   (n) Summarize the relevant system parameters $F_i\overline{\tau\alpha}$, $F_iU_L$, $T_{sB}$, $mc$, $S$, and $L_{aw}/L_a$ at the three locations for combined hot water and space heating.

2. Repeat problem 1 for the entire year at Boston, and find $\sum f_a$ for space heating only using the method of 1(c).

3. Repeat problem 1 for the entire year at Atlanta, and find $\sum f_a$ for hot water and space heating using the method of 1(d).

4. Repeat problem 1 for the entire year at Denver, and find $\sum f_a$ for hot water heating only using the method of 1(e) and 1/6 the stated collector area.

5. Calculate problem 1 for collector $B$.

6. The system of problem 1(n) is modified for annual storage by installation of a 200 000 L (52 910 gal) water storage tank insulated with 25.4 cm (10 in.) poly-urethane foam expanded with $R$-11. The tank is square and 2 m (6.56 ft) deep.
   (a)  Find the solar participation during March and April at Boston, Atlanta, and Denver, using ground temperatures of 55, 70, and 60°, respectively.

7. For one city estimate the annual solar contribution of the annual storage system in problem 6.

8. Find the proportion of the heat collected in problem 7 that is lost to the ground.

9. Using values of $f_a$ available from problems 3, 4, or 5, size the terminal heat exchanger. Remember to size with 3 °C (5 °F) higher than the designated base temperature because of hot water load. Assume that the water flow is very large and that the approach is 2 °C (3.6 °F).
   (a)  What are values of $U_x a_x$?
   (b)  What are values of $U_x A_x$?
   (c)  What is the minimum air temperature entering the living space?
   (d)  What is the required air flow rate in m³/s (ft³/min)?

10. Repeat Examples 9-9 and 9-10 continuing with the parameters stated but assuming a base temperature (before consideration of hot water load) of 30 °C (86 °F).

11. Find monthly values of $L_m/L_a$ and the average base temperature for a car wash having the following load characteristics.

| | Cars per Day | Water Inlet Temperature | | Water Use Temperature | |
|---|---|---|---|---|---|
| | | C° | F° | C° | F° |
| Jan | 30 | 8.89 | 48.0 | 65 | 149 |
| Feb | 51 | 10.83 | 51.5 | 60 | 140 |
| Mar | 64 | 12.78 | 55.0 | 58 | 136.4 |
| Apr | 81 | 14.44 | 58.0 | 55 | 131 |
| May | 125 | 16.83 | 62.3 | 55 | 131 |
| Jun | 186 | 18.55 | 65.4 | 50 | 122 |
| Jul | 138 | 20.0 | 68.0 | 50 | 122 |
| Aug | 152 | 22.22 | 72.0 | 50 | 122 |
| Sep | 122 | 22.78 | 73.0 | 50 | 122 |
| Oct | 75 | 20.56 | 69.0 | 55 | 131 |
| Nov | 58 | 16.67 | 62.0 | 60 | 140 |
| Dec | 41 | 12.78 | 55.0 | 65 | 149 |

12. Derive eq. 9-2 from eqs. 9-4 and 9-1a. What simplifying assumption is necessary?

**13.** Modify eq. 9-4 to include
   (a) Losses from outside piping, which has $a_p$ times the collector heat loss area and a loss coefficient $U_p$.
   (b) Losses from an outside storage tank having $a_s$ times the collector heat loss area and a loss coefficient $U_s$.

**14.** Derive eq. 9-9.

## REFERENCES

1. P. J. Lunde, "Prediction of the performance of solar heating systems over a range of storage capacities." *Solar Energy 23*, 115–121 (1979).
2. F. C. Hooper, University of Toronto, personal communication.
3. P. J. Lunde, "Prediction of the performance of solar heating systems utilizing annual storage," *Solar Energy 22*, 69–75 (1979).
4. S. A. Klein, W. A. Beckman, and J. A. Duffie. "A design procedure for solar heating systems," *Solar Energy 18*, 113–127 (1976).

# Parametric Studies

A basic characteristic of solar-heating and hot water systems is the large number of system variables acting in combination with ever-changing patterns of demand and solar radiation. The designer must make rational choices between parameters such as collector area, fluid flow rates, heat exchanger size and storage volume, as well as determining the particular collector to be used. It is not enough to know that one condition or another is superior; the effect on system performance must be judged quantitatively so that the return may be compared with the cost.

The analysis techniques already introduced are sufficient to permit the performance evaluation of entire systems. The calculations tend to be tedious— mostly because each month must be evaluated separately and the results combined to determine the annual return in the form of solar heat.

In this chapter we will learn ways to make multiple calculations by working from a given case to other cases of interest without complete recalculation. This can be done because only the location, collector angle, base operating temperature, and the nature of the collector determine the threshold radiation values. Once the threshold values are known the weather parameters taken from Appendix B do not change; thus the choice of weather parameters depends only on the collector, its orientation, location, and the type of service expected. Since these are likely to be fixed for a given application, a single set of weather data can serve throughout for the investigation of performance parameters such as flow rate variations, installed area, heat exchanger size, and storage capacity.

Efficient shortcut methods for working through needed parametric studies are given in full detail. Although they are complex to explain and difficult to understand, once learned, the reader will find them convenient for reducing the labor of repetitive calculations. Even when programming a general-purpose computer, these methods can help reduce running time and hence the cost of parametric studies.

In conjunction with the methods for making parametric comparisons, representative results from a single system will be presented so that the reader can gain some insight into the effects of the variables. There are so many possibilities, however, that the studies are necessarily incomplete.

## SOLAR COLLECTION SYSTEMS

Let us briefly review the elements and variables in a solar heating or hot water system. The solar collector, which is described by a heat transfer modified optical efficiency and heat loss coefficient calculated from test results. The optical efficiency will be further modified for the drop in collector performance occurring because the incident angle becomes greater near morning and evening. The collector can be air or liquid cooled. The area of collectors installed is expressed as $S$, the ratio of area to annual load, which will be in the range of 0.2 to 2 m²/GJ (2–20 ft²/MBtu). The monthly load on the system will depend on the load ratio $L_{aW}/L_a$ and the monthly degree-day pattern. The collectors will be operated at as low a temperature as possible unless they are of a high-efficiency, low-loss design.

For space-heating systems the base temperature of storage—the temperature below which auxiliary heat will be used—is estimated by adding a temperature differential for terminal heat exchange to the desired inside ambient temperature (see Chapter 9). For liquid systems, 5 to 10 °C is recommended. The higher differential will be more comfortable for the occupants, since air circulating below about 32 °C (90 °F) may feel drafty. If hot water preheat is included, the effective base temperature is then lowered about 3 °C. For air systems the rock-bed heat exchange is always good and the base temperature is taken at the desired room ambient,* with no adjustment for hot water preheat.

For systems providing only hot water preheat, the base temperature used is the average between the supply and the use temperature. The resultant value of $f_B'$, the potential monthly solar participation, is then corrected to account for lower or higher base temperatures actually anticipated depending on the initial value of $f_B'$. The effective storage is also adjusted to account for the more regular hot water use pattern.

The fluid flow to the collector should be as high as possible, limited in the case of an air collector by the fan power requirement, and occasionally limited for hot water only systems by a need for a high temperature rise. The effect of the flow rate on collector efficiency is expressed as the collector heat transfer factor $F_i$, which will be as high as 0.95 to 0.98 for a liquid system and as low as 0.6 for an air system.

Storage capacity is installed with operating solar systems so that excess heat from sunny periods can be put to use at night or during periods of cloudiness.

---

* However, if air is to be delivered to the heated space at a minimum temperature, that temperature is used as the threshold temperature in eq. 9-1.

Economical storage capacity generally ranges from 0.2 to 0.4 MJ/°C·m² (10–20 Btu/°F ft²).

In liquid-cooled systems an antifreeze loop will heat the water in storage through a heat exchanger, unless a drain-down collector is used. The effect of the heat exchanger on the system is expressed as $F_x$, which is multiplied by $F_i$ to give an overall heat transfer factor $F_i'$. $F_x$ typically has a value of about 0.95.

In air systems there is no heat exchanger as such, and heat storage utilizes a rock bed. The rock bed stratifies the stored heat, compensating for the limited air flow to the collector by making the heat transfer factor for the storage and collector combination equal to $F_m$, which is higher than $F_i$—perhaps about 0.75.

For space heating the water storage tank will be well-mixed, since stratification yields little advantage in liquid systems unless high temperatures are needed. Domestic hot water systems can utilize stratified storage in their vertical tanks provided hot water is delivered to the top of the tank and cold water fed to the collectors from the bottom. Again, when stratified storage is used, its effect on system performance is compensated for by making system calculations with $F_i = F_m$. Ordinary hot water systems with a natural convection heat exchanger coil at the bottom of the tanks benefit from stratified discharge, but do not stratify storage in the usual sense. Hot water systems perform well because they can utilize the plentiful summer sun, and because they operate at low temperatures when the solar participation is low, using auxiliary fuel to attain the high temperature required for actual use.

Heat is delivered from the rock storage medium for space heating by reversing the air flow. Since no heat exchanger is needed, the process is cheap and efficient. A liquid storage medium must heat room air with the aid of a terminal heat exchanger, which means that the base temperature cannot be as low as that with an air system.

## MONTHLY PERFORMANCE CURVES

Once the performance of the system has been determined for several values of the important parameters, some convenient presentation of the results is in order. Graphical presentations can make the system effects of the variables clear at a glance. They can be useful for the student and engineer as well as to the customer for solar energy whose first emotion—to supply all the heating demand with solar energy—must be dealt with rationally.

Figure 10-1 shows a seasonal plot of the monthly demand and performance from Example 9-7, which is very informative regarding the time relationship of collected energy to demand. The first hump-shaped curve expresses the demand as monthly demand ratio $L_m/L_a$. The potential performance curve is a plot of $f_{aB}'$, the proportion of the annual demand *available* monthly at base conditions (i.e.,

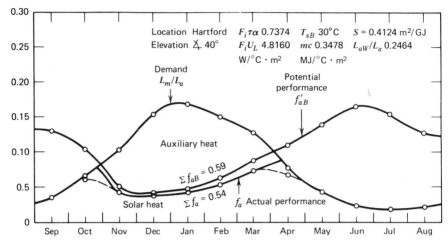

**FIGURE 10-1.** Monthly performance curves for the system analyzed in Example 9-7.

at infinite storage capacity), taken from Table 9-2. Thus, while $f_B'$ is the potential monthly solar participation, $f_{aB}'$ is the potential *annual* solar participation for a particular month.

The new variable $f_{aB}'$ is calculated by converting $f_B'$, the potential monthly solar participation at base conditions, from a monthly to an annual basis:

$$f_{aB}' = f_B' \times (L_m/L_a) \tag{10-1}$$

It may be most conveniently calculated routinely by following the new procedure in the two equations below. Combining eqs. 9-6 and 10-1, where $q_T$ is the heat collectible monthly at base conditions determined by either eq. 9-2, 9-3, or 9-4,

$$f_{aB}' = q_T S = \frac{q_T A_c}{L_a} \tag{10-2}$$

and then $f_B'$ is found by

$$f_B' = f_{aB}'/(L_m/L_a) \tag{10-3}$$

The lower performance curve in Fig. 10-1 is a plot of $f_a$, the annual solar participation for each month. $f_a$ is derived from $f$, which has previously been given by

$$f = f_B'\left[1.138 - 0.271f_B' + 0.006(f_B')^3 - 0.012\,14\frac{F_i'U_L}{mc} + 0.000\,152\left(\frac{F_i'U_L}{mc}\right)^2\right] \tag{9-23}$$

by combination with $L_m/L_a$ to give

$$f_a = f \times (L_m/L_a) \tag{10-4}$$

which has been taken from eq. 9-20.

In one glance the performance curves show how much demand there was, how much energy there was available to the system at base conditions, and how much that energy was reduced as a result of rising storage temperatures with finite storage.

The total demand is proportional to the area beneath the $L_m/L_a$ curve. The auxiliary energy needed is represented by the area between the demand curve and the $f'_{aB}$ curve (base conditions) or the $f_a$ curve (finite storage). The demand actually met by solar energy is the area below both the demand curve and the actual system performance curve.

The total annual solar participation has been noted numerically on each performance curve. These were determined by

$$\Sigma f_a = \sum_{\substack{\text{all} \\ \text{months}}} f_a \qquad (10\text{-}5)$$

and

$$\Sigma f_{aB} = \sum_{\substack{\text{all} \\ \text{months}}} \min \left[ f'_{aB}, (L_m/L_a) \right] \qquad (10\text{-}6)$$

$\Sigma f_a$ is the most important systems number, indicating the annual performance of the actual system. $\Sigma f_{aB}$ is a useful benchmark for establishing the losses due to finite storage, since performance at base conditions is equivalent to performance with infinite storage.

The pattern of Fig. 10-1 is typical of an economically optimized solar heating and hot water system. The available solar energy is excessive during much of the year—except for the dark winter days when demand is high. Some demand continues throughout the year, but further increase in the collector area will contribute to demand only during the November to March heating season and can therefore never return as much energy per unit of collector area as the area now installed, much of which is used year-round.

Some care is necessary in construction of the performance curves because the data points are so few. The October $f_a$ curve, for instance, is not plotted to the logical point (dotted line), but instead to the end of October (solid line) because otherwise some auxiliary energy would be indicated for over half of the October period.

## CHANGING THE COLLECTOR AREA

Increasing the area of collectors in a particular installation will obviously increase performance unless the system was already meeting 100% of the demand. When area or any other variable is changed, however, it is important that the size of other related variables be changed also. It would be misleading, for instance, to double the collector area and then evaluate the new performance using the old total coolant flow rate, which clearly should be doubled as well. For this reason

**TABLE 10-1.** Performance with Fixed Storage and Several Areas as Determined for Example 10-1. Notice that the Calculations Through the $q_T$ Column are Exactly as in Table 9-2a. Continuing with Table 9-2b Gives Similar Results (Corresponding $\sum f_a$ values 0.5293, 0.3597, 0.7291, and 0.8513).

| | Appendix B Column Number | $D_m$ °C-days | $\bar{T}_a$ °C | $t_T$ Ms | $I_T$ MJ/m² | $q_T$ MJ/m² | $f_{aB}$ | $L_m/L_a$ | $f_B$ | $S = 0.4$ m²/GJ | | $S = 0.2$ m²/GJ | | $S = 0.8$ m²/GJ | | $S = 1.2$ m²/GJ | |
|---|---|---|---|---|---|---|---|---|---|---|---|---|---|---|---|---|---|
| | | | | | | | | | | $f$ | $f_a$ | $f$ | $f_a$ | $f$ | $f_a$ | $f$ | $f_a$ |
| Jan | 4 | 686. | −2.7 | 0.5638 | 280.3 | 117.9 | 0.0471 | 0.1671 | 0.2822 | 0.2604 | 0.0435 | 0.1355 | 0.0226 | 0.4781 | 0.0799 | 0.6547 | 0.1094 |
| Feb | 4 | 606. | −1.5 | 0.6030 | 326.4 | 149.2 | 0.0596 | 0.1500 | 0.3978 | 0.3547 | 0.0532 | 0.1880 | 0.0282 | 0.6257 | 0.0938 | 0.8185 | 0.1227 |
| Mar | 3 | 505. | 4.8 | 0.7466 | 415.5 | 215.7 | 0.0863 | 0.1284 | 0.6720 | 0.5502 | 0.0706 | 0.3051 | 0.0391 | 0.8728 | 0.1121 | 1.0000 | 0.1284 |
| Apr | 3 | 267. | 12.5 | 1.0994 | 482.9 | 263.4 | 0.1053 | 0.0775 | 1.3582 | 0.8774 | 0.0680 | 0.5547 | 0.0430 | 1.0000 | 0.0775 | 1.0000 | 0.0775 |
| May | 3 | 120. | 18.4 | 1.2618 | 559.5 | 342.0 | 0.1368 | 0.0461 | 2.9634 | 1.0000 | 0.0461 | 0.9143 | 0.0422 | 1.0000 | 0.0461 | 1.0000 | 0.0461 |
| Jun | 3 | 21. | 23.4 | 1.2287 | 531.4 | 352.7 | 0.1411 | 0.0250 | 5.6402 | 1.0000 | 0.0250 | 1.0000 | 0.0250 | 1.0000 | 0.0250 | 1.0000 | 0.0250 |
| Jul | 2 | 1. | 25.2 | 1.4836 | 559.8 | 378.5 | 0.1514 | 0.0207 | 7.2974 | 1.0000 | 0.0207 | 1.0000 | 0.0207 | 1.0000 | 0.0207 | 1.0000 | 0.0207 |
| Aug | 3 | 8. | 24.5 | 1.1995 | 494.4 | 332.7 | 0.1331 | 0.0222 | 5.9848 | 1.0000 | 0.0222 | 1.0000 | 0.0222 | 1.0000 | 0.0222 | 1.0000 | 0.0222 |
| Sep | 3 | 71. | 20.9 | 1.0408 | 489.5 | 315.3 | 0.1261 | 0.0357 | 3.5329 | 1.0000 | 0.0357 | 0.9776 | 0.0349 | 1.0000 | 0.0357 | 1.0000 | 0.0357 |
| Oct | 3 | 219. | 14.9 | 0.9475 | 438.9 | 254.7 | 0.1018 | 0.0673 | 1.5134 | 0.9228 | 0.0621 | 0.6028 | 0.0405 | 1.0000 | 0.0673 | 1.0000 | 0.0673 |
| Nov | 4 | 392. | 7.6 | 0.4964 | 243.4 | 125.9 | 0.0503 | 0.1042 | 0.4830 | 0.4196 | 0.0437 | 0.2254 | 0.0235 | 0.7174 | 0.0748 | 0.9051 | 0.0943 |
| Dec | 4 | 631. | −0.8 | 0.4864 | 240.4 | 105.1 | 0.0420 | 0.1553 | 0.2706 | 0.2505 | 0.0389 | 0.1302 | 0.0202 | 0.4619 | 0.0717 | 0.6351 | 0.0986 |
| Total | | 3527. | | | | | | | | | 0.5301 | | 0.3625 | | 0.7272 | | 0.8485 |

NOTE: *System parameters:* Data from Appendix B for Hartford, Connecticut (1959–1968) on a 40° slope. $F_i'\tau\alpha = 0.7374$, $F_i'U_L = 4.8160$ W/°C·m², $T_{sB} = 30$ °C, $mc = 0.3478$ MJ/°C·m², $L_{aw}/L_a = 0.2464$.

this book throughout emphasizes area-normalized quantities for coolant flow rate, heat exchanger area, and storage capacity.

In the practical design of a solar system the engineer generally proceeds immediately to the question of how much area to install. It is really unnecessary to select a trial area suiting a particular structure as we have done so far. All that is needed is to select a few convenient values of $S$; performance from these are plotted to determine an optimum; then this optimum $S$ value can be used for all similar applications and the actual area found with eq. 9-7 when it is needed. If values for $S$ are chosen in ratios of 3:2:1 or 4:2:1 the results will be more easily managed.

Recalculation of performance with a new area or $S$ value is most conveniently done by modifying the original potential solar participation $f'_B$ from eq. 10-3 to calculate a new value

$$(f'_B)_{new} = (f'_B)_{old} \times \frac{S_{new}}{S_{old}} \qquad (10\text{-}7)$$

This procedure avoids repeating the tedious calculations of $q_T$ for use in eqs. 10-2 and 10-3. If the potential monthly solar participation is needed on an annual basis (most often it will not be), then

$$(f'_{aB})_{new} = (f'_{aB})_{old} \times \frac{S_{new}}{S_{old}} \qquad (10\text{-}8)$$

**EXAMPLE 10-1**

Repeat the monthly performance calculations of Example 9-7 (Table 9-2a) using a starting normalized area of $S = 0.2$ m²/GJ and other areas in the ratio of 4:2:1. Assume that the storage size is changed in proportion to the collector area considered. (Using Engineering units, suitable values are $S = 10$, 5, and 20 ft²/M Btu.)

The appropriate areas are $S = 0.4$, 0.2, and 0.8 m²/GJ (approaching them in this order will be seen to be convenient).

The threshold levels and basic weather data from Table 9-2a are listed again in Table 10-1, unchanged. Using eqs. 10-2, 10-3, 9-23, and 10-4, in sequence, for January, for $S = 0.4$ m²/GJ:

$$f'_{aB} = \{0.7374 \times 280.3 - 4.8160[30 - (-2.7)]0.5638\}(0.4 \times 10^{-3}) = 0.047\,16$$

$$f'_B = \frac{0.047\,16}{0.1672} = 0.2822$$

$$\frac{F'_i U_L}{mc} = \frac{4.8160}{0.3478} = 13.85 \ (Ms)^{-1}$$

$$f\left(f'_B = 0.2822, \frac{F'_i U_L}{mc} = 13.85\right) = 0.2604 \qquad \text{(eq. 9-23 or Fig. 9-2)}$$

$$f_a = 0.2604 \times 0.1672 = 0.0435$$

For $S = 0.2$ m$^2$/GJ, $q_T$ need not be recalculated. Using eq. 10-2, $f_B'$ changes in direct proportion to the area:

$$f_B' = 0.2822/2 = 0.1411$$
$$f = 0.1355 \text{ by eq. 9-23 or Fig. 9-2}$$
$$f_a = 0.1355 \times 0.1672 = 0.0226$$

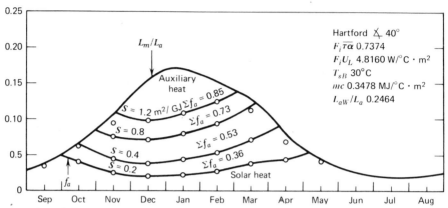

**FIGURE 10-2.** Monthly performance curves for Example 10-1. Here the area related parameter $S$ is varied.

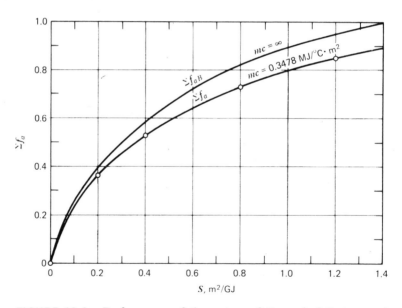

**FIGURE 10-3.** Performance of the system of Example 9-7 at several values of the collector area (expressed as $S$, the ratio of area to annual system demand), using data from Example 10-1 and Fig. 10-2.

For $S = 0.8$ m²/GJ,

$$f'_B = 0.2822 \times 2 = 0.5644$$
$$f = 0.4781$$
$$f_a = 0.4781 \times 0.1672 = 0.0799$$

Only the values of $f$ and $f_a$ are listed in Table 10-1 to reduce its bulk. After all the January cases are done, the other months are calculated in succession.* For completeness an $S = 1.2$ m²/GJ case has also been calculated for Table 10-1. $f'_B$ is not entered (except for the initial case), since it can be reconstructed easily if needed. The minimum data record of Table 10-1 still permits any desired information to be calculated, and the most relevant monthly performance information to be plotted as in Fig. 10-2. $f'_{aB}$ and $f_{aB}$ can be ratioed from the value for $S = 0.4$ m²/GJ according to the area if needed.

Figure 10-3 is a plot of the normalized area $S$ versus $\sum f_a$, the annual solar participation. The lower line is the case just calculated, while the upper line is a plot showing $\sum f_{aB}$, the potential annual solar participation (which was calculated from eq. 10-6, but is not shown in Table 10-1). This type of presentation is quite important because an economic analysis can be made from it, as we will see in Chapter 11. Such *annual performance curves* demonstrate the way the law of diminishing returns operates with increasing solar collector area.

## CHANGING THE STORAGE CAPACITY

When the effect of varying storage capacity is to be explored, the starting point is the same as for a change in area, with $f'_B$, the potential monthly solar participation. The calculated procedure is the same, but a new value for the storage function $F'_i U_L/mc$ is used when $f$ is calculated. As with $S$, since there is no reason to select particular $mc$ values when making a preliminary design, integral values are usually most convenient.

### EXAMPLE 10-2

Repeat the calculations of Example 10-1 for new storage values of $mc = 0.2$, 0.4, and 0.8 MJ/°C·m². (With Engineering units, use 10, 20, and 40 Btu/°F ft².)

$$\frac{F'_i U_L}{mc} = \frac{4.8160}{0.2} = 24.08/\text{Ms} \qquad \text{(for } mc = 0.2 \text{ MJ/°C·m²)}$$

---

* In the interest of brevity, summer entries need not be made ($f$ always is unity in a heating or heating plus hot water case in the summer) if values of $L_m/L_a$ (0.0223 + 0.0208 + 0.0250 = 0.0681) are added to $\sum f_{aB}$ and $\sum f_a$ for the other nine months to give credit for summer hot water heating.

$$\frac{F_i'U_L}{mc} = \frac{4.8160}{0.4} = 12.04/\text{Ms} \qquad \text{(for } mc = 0.4 \text{ MJ/°C·m}^2)$$

$$\frac{F_i'U_L}{mc} = \frac{4.8160}{0.8} = 6.02/\text{Ms} \qquad \text{(for } mc = 0.8 \text{ MJ/°C·m}^2)$$

The calculations in this example are within the capability of a hand electronic calculator, except for eq. 9-23, which is best done graphically or with a programmable calculator. To speed use of the graph, use the straight $\times 2$ line to divide graphically the entered $f_B'$ by 2 (for $S = 0.2$ m²/GJ) and multiply it by 2 (for $S = 0.8$ m²/GJ) as illustrated for January in Fig. 10-4. Values of $f$ are then read off the vertical lines at needed values of $F_i'U_L/mc$.

For January with $S = 0.4$, 0.2, and 0.8 m²/GJ, as before, for $f_B' = 0.2822$, $0.2822/2 = 0.1411$; and $0.2822 \times 2 = 0.5644$, respectively,

$$f = 0.2420, 0.1264, 0.4414 \qquad \text{(for } mc = 0.2 \text{ MJ/°C·m}^2)$$
$$f = 0.2645, 0.1376, 0.4865 \qquad \text{(for } mc = 0.4 \text{ MJ/°C·m}^2)$$
$$f = 0.2805, 0.1411, 0.5184 \qquad \text{(for } mc = 0.8 \text{ MJ/°C·m}^2)$$

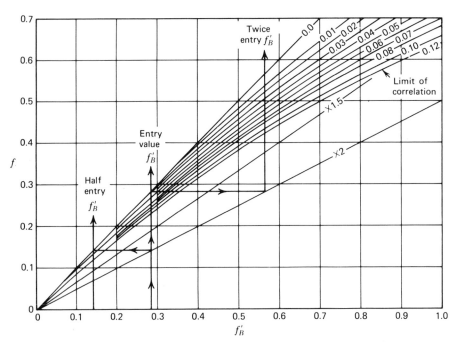

**FIGURE 10-4.** Use of the "$\times 2$" line in Fig. 9-2 to graphically divide and multiply entered $f_B'$ values by two.

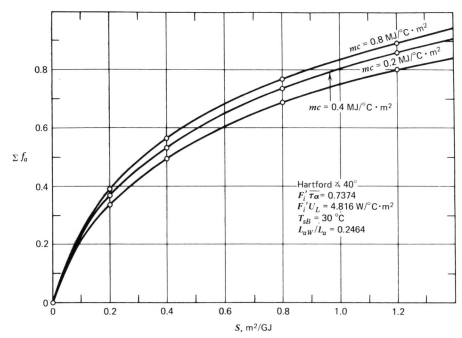

**FIGURE 10-5.** Annual performance curves for Example 10-2 showing the effect of storage capacity.

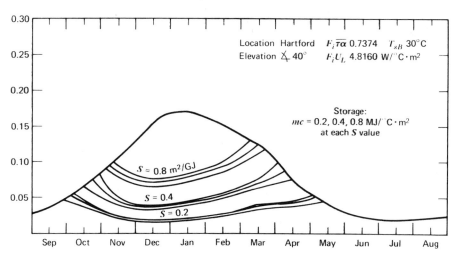

**FIGURE 10-6.** Monthly performance of the system of Figs. 10-1 and 10-3 for integral values of storage capacity (data from Table 10-2, as determined in Example 10-2).

**TABLE 10-2.** Performance with Several Areas and Storage Capacities from Example 10-2. $F_i'\overline{\tau\alpha} = 0.7374$, $F_i'U_L = 4.8160$ W/°C·m², $T_{sB} = 30$ °C, $L_m/L_a = 0.2464$.

| $mc = $ 0.2000 MJ/ °C·m² | $S = $ 0.4000 m²/GJ | $S = $ 0.2000 m²/GJ | | $S = $ 0.4000 m²/GJ | | $S = $ 0.8000 m²/GJ | |
|---|---|---|---|---|---|---|---|
| | $f_B'$ | $f$ | $f_a$ | $f$ | $f_a$ | $f$ | $f_a$ |
| Jan | 0.2822 | 0.1264 | 0.0211 | 0.2420 | 0.0404 | 0.4414 | 0.0737 |
| Feb | 0.3978 | 0.1750 | 0.0262 | 0.3288 | 0.0493 | 0.5739 | 0.0861 |
| Mar | 0.6720 | 0.2833 | 0.0363 | 0.5064 | 0.0650 | 0.7853 | 0.1008 |
| Apr | 1.3582 | 0.5105 | 0.0396 | 0.7890 | 0.0612 | 1.0000 | 0.0775 |
| May | 2.9634 | 0.8178 | 0.0377 | 1.0000 | 0.0461 | 1.0000 | 0.0461 |
| Jun | 5.6402 | 1.0000 | 0.0250 | 1.0000 | 0.0250 | 1.0000 | 0.0250 |
| Jul | 7.2974 | 1.0000 | 0.0207 | 1.0000 | 0.0207 | 1.0000 | 0.0207 |
| Aug | 5.9848 | 1.0000 | 0.0222 | 1.0000 | 0.0222 | 1.0000 | 0.0222 |
| Sep | 3.5329 | 0.8626 | 0.0307 | 1.0000 | 0.0357 | 1.0000 | 0.0357 |
| Oct | 1.5134 | 0.5535 | 0.0372 | 0.8242 | 0.0554 | 1.0000 | 0.0673 |
| Nov | 0.4830 | 0.2097 | 0.0218 | 0.3882 | 0.0404 | 0.6545 | 0.0682 |
| Dec | 0.2706 | 0.1214 | 0.0188 | 0.2329 | 0.0361 | 0.4266 | 0.0662 |
| Total | | 0.3379 | | 0.4981 | | 0.6900 | |

$mc = 0.4000$ MJ/ C·m²

| | $f_B'$ | $f$ | $f_a$ | $f$ | $f_a$ | $f$ | $f_a$ |
|---|---|---|---|---|---|---|---|
| Jan | 0.2822 | 0.1376 | 0.0230 | 0.2645 | 0.0442 | 0.4865 | 0.0813 |
| Feb | 0.3978 | 0.1909 | 0.0286 | 0.3606 | 0.0541 | 0.6375 | 0.0956 |
| Mar | 0.6720 | 0.3101 | 0.0398 | 0.5602 | 0.0719 | 0.8927 | 0.1146 |
| Apr | 1.3582 | 0.5648 | 0.0438 | 0.8975 | 0.0696 | 1.0000 | 0.0775 |
| May | 2.9634 | 0.9362 | 0.0432 | 1.0000 | 0.0461 | 1.0000 | 0.0461 |
| Jun | 5.6402 | 1.0000 | 0.0250 | 1.0000 | 0.0250 | 1.0000 | 0.0250 |
| Jul | 7.2974 | 1.0000 | 0.0207 | 1.0000 | 0.0207 | 1.0000 | 0.0207 |
| Aug | 5.9848 | 1.0000 | 0.0222 | 1.0000 | 0.0222 | 1.0000 | 0.0222 |
| Sep | 3.5329 | 1.0000 | 0.0357 | 1.0000 | 0.0357 | 1.0000 | 0.0357 |
| Oct | 1.5134 | 0.6140 | 0.0413 | 0.9452 | 0.0636 | 1.0000 | 0.0673 |
| Nov | 0.4830 | 0.2290 | 0.0238 | 0.4268 | 0.0445 | 0.7317 | 0.0763 |
| Dec | 0.2706 | 0.1322 | 0.0205 | 0.2545 | 0.0395 | 0.4699 | 0.0730 |
| Total | | 0.3680 | | 0.5374 | | 0.7357 | |

$mc = 0.8000$ MJ/ C·m²

| | $f_B'$ | $f$ | $f_a$ | $f$ | $f_a$ | $f$ | $f_a$ |
|---|---|---|---|---|---|---|---|
| Jan | 0.2822 | 0.1411 | 0.0235 | 0.2805 | 0.0468 | 0.5184 | 0.0866 |
| Feb | 0.3978 | 0.1989 | 0.0298 | 0.3831 | 0.0574 | 0.6825 | 0.1023 |
| Mar | 0.6720 | 0.3291 | 0.0422 | 0.5981 | 0.0768 | 0.9687 | 0.1244 |
| Apr | 1.3582 | 0.6032 | 0.0467 | 0.9743 | 0.0755 | 1.0000 | 0.0775 |
| May | 2.9634 | 1.0000 | 0.0461 | 1.0000 | 0.0461 | 1.0000 | 0.0461 |
| Jun | 5.6402 | 1.0000 | 0.0250 | 1.0000 | 0.0250 | 1.0000 | 0.0250 |
| Jul | 7.2974 | 1.0000 | 0.0207 | 1.0000 | 0.0207 | 1.0000 | 0.0207 |
| Aug | 5.9848 | 1.0000 | 0.0222 | 1.0000 | 0.0222 | 1.0000 | 0.0222 |
| Sep | 3.5329 | 1.0000 | 0.0357 | 1.0000 | 0.0357 | 1.0000 | 0.0357 |
| Oct | 1.5134 | 0.6568 | 0.0442 | 1.0000 | 0.0673 | 1.0000 | 0.0673 |
| Nov | 0.4830 | 0.2415 | 0.0251 | 0.4541 | 0.0473 | 0.7863 | 0.0820 |
| Dec | 0.2706 | 0.1353 | 0.0210 | 0.2698 | 0.0419 | 0.5005 | 0.0777 |
| Total | | 0.3828 | | 0.5632 | | 0.7680 | |

Corresponding $f_a$'s are the above $f$ values times $L_m/L_a$, which is 0.1672 for this month. Table 10-2 gives complete values for all months. Tabulation of $f_B'$ is superfluous; this is directly proportional to area and only the $S = 0.4 \text{ m}^2/\text{GJ}$ values are shown, just as tabulated in Table 10-1. The annual performance $\sum f_a$ for the three storage quantities is plotted versus normalized area in Fig. 10-5, and the corresponding monthly performance curves are plotted in Fig. 10-6. These plots clearly show relationships between areas and performance under different conditions. For instance, if $\sum f_a$ is 0.6, Fig. 10-5 indicates that 0.59 units of area are required if $mc = 0.2 \text{ MJ}/°\text{C·m}^2$, while only 0.45 units of area are necessary if $mc = 0.8 \text{ MJ}/°\text{C·m}^2$.

## CHANGING $L_m/L_a$ OR $L_{aW}/L_a$

When the monthly demand ratio changes, as it might when considering a heating system without hot water, or for another structure having a different value of $L_{aW}/L_a$ (hot water to space heating annual demand ratio), the potential annual solar participation ($f_{aB}'$) is unchanged if the base temperature is the same. That is, the same net energy is *available* monthly for a given $S$. Therefore, new potential monthly solar participation values $f_B'$ can be calculated starting with eq. 10-3, using old $f_{aB}'$ values but a new $L_m/L_a$ for each month.

When considering a single structure with a fixed collector area, however, $S$ will be different from case to case when $L_m/L_a$ changes, because the annual demand will be different. It is then necessary to recalculate $S$ (or $L_a$) and start the analysis over using eq. 10-2 or 10-8, but even this avoids recalculation of $q_T$.

**EXAMPLE 10-3**

Use the parameters of Example 10-1 to recalculate the monthly and annual performance of systems for:
(a) Space heating only.
(b) Hot water only.

(a) For space heating only, Table 9-1 gives $L_m/L_a = 0.1946$ for January. In Example 10-1, $f_{aB}'$ was 0.047 16 for $S = 0.4 \text{ m}^2/\text{GJ}$. Therefore, for the same $S$ value,

$$f_B' = \frac{0.047\ 16}{0.1946} = 0.2425$$

$(F_i'U_L/mc)$ is as before, 13.85 $(\text{Ms})^{-1}$. From eq. 9-23 or Fig. 9-2,

$$f(0.2423, 13.85) = 0.2263$$
$$f_a = 0.2263 \times 0.1946 = 0.0440$$

See Table 10-3 for values for the other months. Annual values of $\sum f_a$ are plotted on Fig. 10-7a.

(b) For hot water only, Table 9-1 gives $L_m/L_a = 0.084\,87$ for January. Since in Example 10-1 $f'_{aB}$ was 0.047 16 for $S = 0.4\ \text{m}^2/\text{GJ}$,

$$f'_B = \frac{0.047\,16}{0.084\,87} = 0.5556$$

Using eqs. 9-45, 9-46, and 9-23 (or just Fig. 9-2),

$$(f'_B)_{\text{effective}} = 0.6285$$

$$\left(\frac{F'_i U_L}{mc}\right)_{\text{effective}} = 12.77\ (\text{Ms})^{-1}$$

$$f(0.6285, 12.77) = 0.5274$$
$$f_a = 0.5274 \times 0.084\,87 = 0.0448$$

The remainder of the calculations are summarized in Table 10-4, and the annual values are also plotted on Fig. 10-7 along with values from the combined space heating and hot water case of Example 10-1.

For comparison, Fig. 10-7 also has plotted results from the hot water system analysis using eq. 9-23 without correcting $f'_B$ and $(F'_i U_L/mc)$ for hot water service. The curves are quite similar, but there is a significant difference in $\sum f_a$ values in the economically important range of 0.4 to 0.7.

Observe that in Fig. 10-7 the most uniform load (hot water) always gives the highest annual solar participation $\sum f_a$ (if the area is sized in proportion to the annual load, i.e., at a given $S$), and that the seasonally most concentrated heating load gives the least return. The economics will, therefore, always be better for hot water when the temperature range is similar regardless of the particular application.

**TABLE 10-3.** Performance of the System of Example 10-1 (Table 10-1) for Heating only at Several Collector Areas. The Calculations in Table 10-1 Apply Directly up to the $L_m/L_a$ Column with Which This Table Begins (Example 10-3).

| | $S = 0.4\ \text{m}^2/\text{GJ}$ | | | | $S = \dfrac{0.2\ \text{m}^2}{\text{GJ}}$ | $S = \dfrac{0.8\ \text{m}^2}{\text{GJ}}$ | $S = \dfrac{1.2\ \text{m}^2}{\text{GJ}}$ |
|---|---|---|---|---|---|---|---|
| | $L_m/L_a$ | $f'_B$ | $f$ | $f_a$ | $f$ | $f$ | $f$ |
| Jan | 0.1945 | 0.2425 | 0.2263 | 0.0440 | 0.1171 | 0.4211 | 0.5850 |
| Feb | 0.1718 | 0.3473 | 0.3144 | 0.0540 | 0.1653 | 0.5647 | 0.7539 |
| Mar | 0.1432 | 0.6028 | 0.5046 | 0.0722 | 0.2766 | 0.8233 | 0.9847 |
| Apr | 0.0757 | 1.3919 | 0.8882 | 0.0672 | 0.5655 | 1.0000 | 1.0000 |
| May | 0.0340 | 4.0217 | 1.0000 | 0.0340 | 1.0000 | 1.0000 | 1.0000 |
| Jun | 0.0060 | 23.7014 | 1.0000 | 0.0060 | 1.0000 | 1.0000 | 1.0000 |
| Jul | 0.0003 | 533.9848 | 1.0000 | 0.0003 | 1.0000 | 1.0000 | 1.0000 |
| Aug | 0.0023 | 58.6883 | 1.0000 | 0.0023 | 1.0000 | 1.0000 | 1.0000 |
| Sep | 0.0201 | 6.2660 | 1.0000 | 0.0201 | 1.0000 | 1.0000 | 1.0000 |
| Oct | 0.0621 | 1.6410 | 0.9533 | 0.0592 | 0.6401 | 1.0000 | 1.0000 |
| Nov | 0.1111 | 0.4532 | 0.3974 | 0.0442 | 0.2125 | 0.6870 | 0.8780 |
| Dec | 0.1789 | 0.2350 | 0.2198 | 0.0393 | 0.1136 | 0.4100 | 0.5711 |
| $\sum f_a$ | | | | 0.4429 | 0.2800 | 0.6470 | 0.7845 |

**TABLE 10-4.** Performance of the System of Example 10-1 (Table 10-1) for Hot Water Heating Only at Several Collector Areas. The Calculations in Table 10-1 Apply Directly up to the $L_m/L_a$ Column with Which This Table Begins (Example 10-3).

| | $S = 0.4 \ \text{m}^2/\text{GJ}$ | | | | | $S = \dfrac{0.2 \ \text{m}^2}{\text{GJ}}$ | $S = \dfrac{0.8 \ \text{m}^2}{\text{GJ}}$ | $S = \dfrac{1.2 \ \text{m}^2}{\text{GJ}}$ |
|---|---|---|---|---|---|---|---|---|
| | $L_m/L_a$ | $f'_B$ | $(f'_B)_{\text{effective}}$ | $f$ | $f_a$ | $f$ | $f$ | $f$ |
| Jan | 0.084 87 | 0.5556 | 0.6285 | 0.5274 | 0.0448 | 0.3061 | 0.7908 | 0.9100 |
| Feb | 0.077 34 | 0.7716 | 0.8311 | 0.6534 | 0.0505 | 0.4010 | 0.8916 | 0.9583 |
| Mar | 0.084 87 | 1.0169 | 1.0329 | 0.7588 | 0.0644 | 0.4949 | 0.9456 | 1.0000 |
| Apr | 0.082 14 | 1.2829 | 1.2178 | 0.8388 | 0.0689 | 0.5817 | 0.9613 | 1.0000 |
| May | 0.084 87 | 1.6122 | 1.3977 | 0.9023 | 0.0766 | 0.6704 | 1.0000 | 1.0000 |
| Jun | 0.082 14 | 1.7181 | 1.4440 | 0.9165 | 0.0753 | 0.6951 | 1.0000 | 1.0000 |
| Jul | 0.084 87 | 1.7838 | 1.4699 | 0.9241 | 0.0784 | 0.7095 | 1.0000 | 1.0000 |
| Aug | 0.084 87 | 1.5684 | 1.3769 | 0.8956 | 0.0760 | 0.6597 | 1.0000 | 1.0000 |
| Sep | 0.082 14 | 1.5357 | 1.3608 | 0.8904 | 0.0731 | 0.6515 | 1.0000 | 1.0000 |
| Oct | 0.084 87 | 1.2005 | 1.1643 | 0.8172 | 0.0694 | 0.5564 | 0.9602 | 1.0000 |
| Nov | 0.082 14 | 0.6133 | 0.6848 | 0.5645 | 0.0464 | 0.3327 | 0.8243 | 0.9299 |
| Dec | 0.084 87 | 0.4953 | 0.5677 | 0.4855 | 0.0412 | 0.2774 | 0.7492 | 0.8817 |
| $\Sigma f_a$ | | | | | 0.7650 | 0.5286 | 0.9270 | 0.9733 |

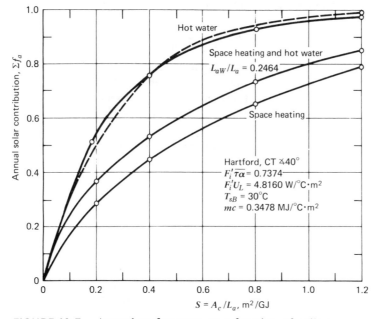

**FIGURE 10-7.** Annual performance as a function of collector area for heating, heating and hot water, and hot water (only). The expression of area as a ratio to the annual load ($S$) is necessary to make this comparison useful. Dotted line shows hot water case calculated without hot water correction, eqs. 9-45 and 9-46.

## CHANGES IN $F_i'$

Equation 10-3 is a good starting point for system performance analysis. In combination with eq. 9-4 (or similarly from eq. 9-2 or 9-3) and eq. 10-2, it can be written as

$$f_B' = \left[ \frac{\overline{\tau\alpha}I_T - U_L(T_{sB} - \overline{T}_a)t_T}{L_m/L_a} \right] F_i'S \tag{10-10}$$

The potential monthly solar participation $f_B'$ (and $f_{aB}'$), once calculated, holds for all combinations of $F_i'$ and $S$ for which the product $F_i'S$ is constant. Therefore, the *same* values of $f_{aB}'$ and $f_B'$ can be used after a change in $F_i'$, but they apply for another $S$, given by

$$S_{new} = S_{old} \times \frac{(F_i')_{old}}{(F_i')_{new}} \tag{10-11}$$

Recalculations for a new $F_i'$ can therefore begin with eq. 9-23, but apply to new $S$ values. The old values for actual monthly solar participation $f$ also remain valid; they hold for the new $S$ *and* the $mc$ value that retains the original value of the storage function $F_i'U_L/mc$, given by

$$(mc)_{new} = \frac{(mc)_{old}}{[(F_i')_{old}/(F_i')_{new}]} \tag{10-12}$$

Thus, when eq. 10-12 is used, no recalculation at all is necessary for a change in $F_i'$; it is only necessary to find the values of $S$ and $mc$ for which the old $f$ applies. The procedure is especially useful when there has been a change in $F_i'$ if $S$ and $mc$ must be reoptimized, because the old values of $S$ and $mc$ were often arbitrarily chosen anyway.

Alternately, if the same $S$ and $mc$ values must be retained, new calculations can be made by starting with new $f_{aB}'$ and $f_B'$ values given by

$$(f_{aB}')_{new} = (f_{aB}')_{old} \times \frac{(F_i')_{new}}{(F_i')_{old}} \tag{10-13}$$

$$(f_B')_{new} = (f_B')_{old} \times \frac{(F_i')_{new}}{(F_i')_{old}} \tag{10-14}$$

which still avoids direct recalculation of $q_T$ but requires use of eq. 9-23.

In eqs. 10-11 to 10-14, if only $F_x$ or $F_i$ has been changed, use of $F_i$ or $F_x$ instead of $F_i'$ in those relationships is of course equally accurate.

### EXAMPLE 10-4

In the preceding problems, $F_i\overline{\tau\alpha} = 0.7762$, $F_iU_L = 5.069$ W/°C·m² (0.8927 Btu/hr °F ft²), and $F_x = 0.95$ (see Example 9-6). As a part of a parametric study, the flow rate to the collector is decreased so that $F_i\overline{\tau\alpha} = 0.6986$, $F_iU_L = 4.5625$ (0.8035 Btu/hr

°F ft²), and $F_x = 0.92$. Prepare a new $\sum f_a$ versus $S$ plot working from the Example 10-2 results presented in Table 10-3.

(a) New $S$ values. From eq. 10-11,

$$\frac{(S)_{\text{new}}}{(S)_{\text{old}}} = \frac{(F_i'\tau\alpha)_{\text{old}}}{(F_i'\tau\alpha)_{\text{new}}} = \frac{(F_i\tau\alpha)_{\text{old}}(F_x)_{\text{old}}}{(F_i\tau\alpha)_{\text{new}}(F_x)_{\text{new}}} = \frac{0.8927}{0.8034} \times \frac{0.95}{0.92} = 1.1474$$

The three new $S$ values are

i)   $S = 1.1474 \times 0.2 = 0.2295$ m²/GJ
ii)   $S = 1.1474 \times 0.4 = 0.4590$ m²/GJ
iii)   $S = 1.1474 \times 0.8 = 0.9179$ m²/GJ

(b) New storage quantities. Using eq. 10-12, the new storage values are

i)   $mc = 0.2/1.1474 = 0.1743$ MJ/°C·m²
ii)   $mc = 0.4/1.1474 = 0.3486$ MJ/°C·m²
iii)   $mc = 0.8/1.1474 = 0.6972$ MJ/°C·m²

Figure 10-8 is a plot of the new annual system performance. Here the *same* $\sum f_a$ numbers from Table 10-2 have been plotted against the new $S$ values (solid lines). Fully recalculated values of $\sum f_a$ for the new $F_i'$ and the original $mc$ values of 0.2, 0.4, and 0.8 MJ/°C·m² have also been plotted (dotted lines). They are so close to those just calculated that for casual use the correction for $mc$ given in eq. 10-12 is unnecessary even in this example of an unusually large shift in $F_i'$.

## OTHER CHANGES

Changes in collector orientation, base temperature, or collector type require complete recalculations. Fortunately, these choices can be fixed early in a particular study.

The collector should face south* and be at an architecturally convenient angle of at least the latitude, or 10° less for hot water heating. The base temperature should be as low as possible, as outlined earlier in this chapter. The initial collector selection can then be made by comparing the annual solar participation $\sum f_a$ under the same conditions; it is recommended that an efficiency curve be obtained from an independent source and evaluated with $F_x = 0.95$, $mc = 0.3$ MJ/°C·m² (15 Btu/°F ft²), and an $S$ value of 0.6 m²/GJ (7 ft²/M Btu). An economic choice can then be made on a cost per thermal unit basis once the cost of the

---

* For a few locations with consistently more cloud cover in the morning or evening, due South may not be optimum. For most locations, however, within 20° of due South is close enough to use the tables in Appendix A or B. Beyond that the available energy falls off rapidly in the winter months. Appendix B includes tables for a 40° collector slope at Hartford that is oriented 30° and 45° east of south (azimuth angles 150° and 135°). These can be used only for performance prediction with the monthly base temperature method, for the annual methods do not allow for the increased values of the utilizability constant $k$ characteristic of the off-south orientation.

collector is known. If results are close, the final decision may await final calculations or be made on the basis of less tangible characteristics.

## Effect of Base Temperature

The performance of a solar system is strongly influenced by the base temperature. Were it not for an inherently low base temperature, air collectors would not be economically viable. With hydronic baseboard heat the base temperature must be too high for practical use. Because the base temperature can be kept low, hot water preheat of a small part of the hot water load is economical in many commercial and industrial installations.

For moderate-temperature commercial or industrial applications, a higher base temperature may be used with the techniques of this chapter, which have been validated to base temperatures of 55 °C (130 °F) and should apply at higher temperatures as well. Equation 9-16 will apply directly for many of these situations and, if more than 50% solar contribution is expected, correlation equation (9-23) (with eqs. 9-45 and 9-46 for hot water) can be used.*

---

\* This gives slightly less performance than a uniform day-to-day usage. This assumption is not at all critical relative to the 1 to 2% accuracy (relative to simulation) of the correlation, since the entire second term in eq. 9-16 seldom exceeds 0.05.

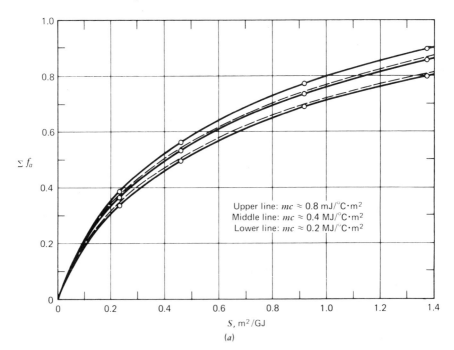

$$\Sigma f_a$$

Upper line: $mc \approx 0.8\ \text{mJ/}^\circ\text{C·m}^2$
Middle line: $mc \approx 0.4\ \text{MJ/}^\circ\text{C·m}^2$
Lower line: $mc \approx 0.2\ \text{MJ/}^\circ\text{C·m}^2$

$S$, m²/GJ

(a)

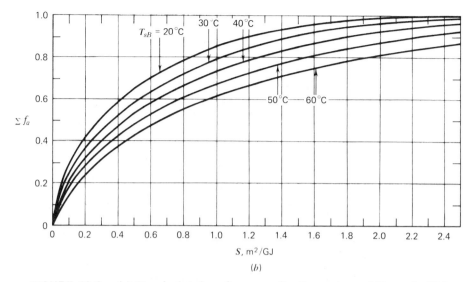

**FIGURE 10-8.** (a) Recalculated performance for the system of Example 10-2 (Fig. 10-5) when $F_i'$ changes as detailed in Example 10-4. Dotted lines show exact values for the nominal storage capacities. The solid lines are simply a replot of the Example 10-2 data (Table 10-2) and strictly speaking apply to slightly less storage, as noted in Example 10-4. (b) Effect of changes in base temperature on the performance of the system in Example 10-1 (Fig. 10-3).

Figure 10-8b shows the effect of changes in base temperature on the system of Example 10-1 (Table 10-1 and Fig. 10-2). Since this performance is typical of an efficient collector, the student will find it worthwhile to give this figure careful examination.

## Effect of Collector Orientation

Figure 10-9a, b, and c shows how inclination angle affects performance of three systems—the heating-only system, the hot water system, and the heating plus hot water system—all at the same base temperature [30 °C (86 °F)]. The same collector properties and location are used as in Examples 10-1 to 10-3.

The effect of inclination on annual performance is seen to be small in any case, but more extreme with the heating-only system, which cannot use the high summer sun. As the normalized area $S$ rises, there are minor changes in optimum inclination angle in all three applications. At low $S$ the inclination is less important, but as $S$ rises, annual performance can be improved by further tilting of the collector to pick up more of the winter sun. The differences are slight overall.

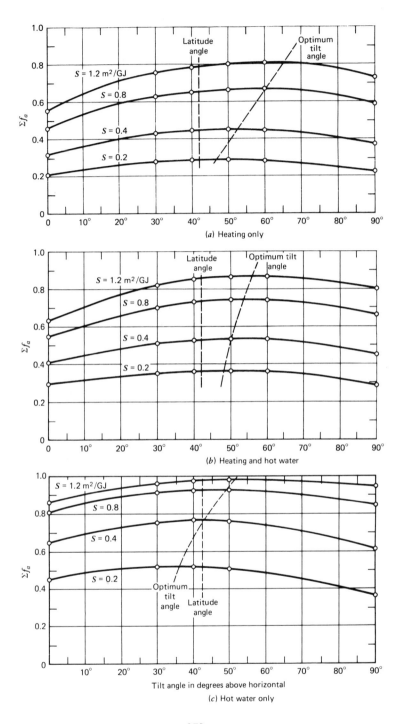

(a) Heating only

(b) Heating and hot water

Tilt angle in degrees above horizontal

(c) Hot water only

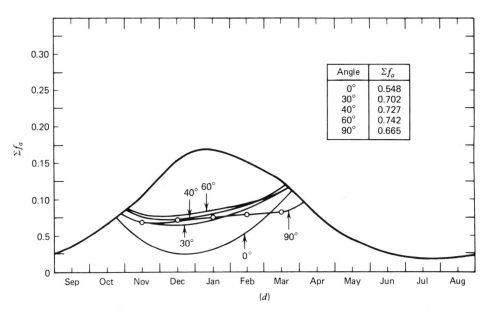

| Angle | $\Sigma f_a$ |
|-------|--------------|
| 0° | 0.548 |
| 30° | 0.702 |
| 40° | 0.727 |
| 60° | 0.742 |
| 90° | 0.665 |

FIGURE 10-9. The effect of the collector tilt angle on annual performance. (*a*) For heating only. (*b*) For heating and hot water. (*c*) For hot water only. (*d*) Monthly for the heating and hot water system only with $S = 0.8$ m²/GJ. These cases are all analyzed using $F_i'\overline{\tau\alpha} = 0.7374$, $F_i'U_L = 4.816$ W/°C·m², $T_{sB} = 30$ °C, $mc = 0.3478$ MJ/°C·m², and for the heating and hot water case, $L_{aW}/L_a = 0.2464$.

For this location (Hartford), in general, angles of 40 to 50° (about equal to or several degrees greater than the latitude) are a good choice.

The reason that collector inclination is not critical is shown by Fig. 10-9*d*. The broad-base solar coverage in spring and fall is constant, and the performance in the winter months is somewhat better with a high inclination angle. For a typical solar system that supplies only a portion of the load, however, optimizing tilt for December will reduce the March contribution by about the same amount.

For best performance the collector should face directly south but, within 20° of south, the loss in performance is minimal. For the conditions of Fig. 10-9*a* to *c*, $\Sigma f_a$ drops about 0.02 when the azimuth is 30° off due south, and 0.05 for 45° (0.04 for water heating). The effect is important only in the winter months, because during the summer half of the year the sun's azimuthal range exceeds 180° and the off-south collector remains in sunlight most of the day.

## Effect of the Collector Characteristics

In the last two chapters, for consistency, the discussion has been restricted to a collector with particular values of $F_i'\overline{\tau\alpha}$ and $F_i'U_L$. There are of course many other

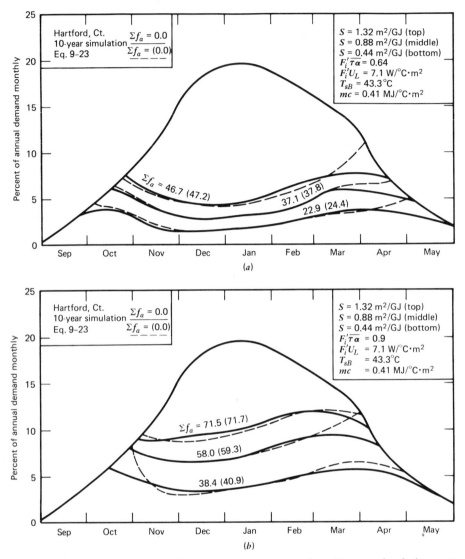

**FIGURE 10-10.** Monthly performance curves comparing 10-year simulations at Hartford with performance predictions using eq. 9-23 for three $S$ values over a wide range of collector characteristics.

combinations of these variables. A collector will tend to have a low $\tau\alpha$ if there are two or more cover plates, or if there is a selective surface that has a relatively low $\alpha$. The loss coefficient $U_L$ will tend to be low if there are two or more cover plates or a selective surface; a high coefficient can result from mistakes in design that

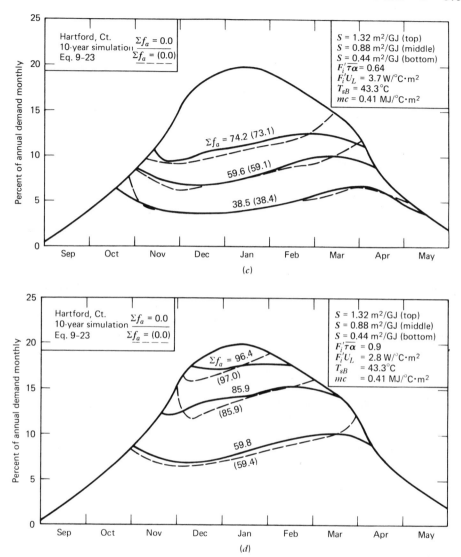

**FIGURE 10-10.** *(Continued)*

permit excessive heat loss. $F_i'$ will invariably be low in air collectors—values of $F_m$, used instead of $F_i'$ with stratified storage, will be 0.6 to 0.8. Poor heat transfer or low flow rates can also make $F_i'$ low in liquid collectors, especially with collectors heating hot water, when the liquid flow can be laminar if a high-temperature rise is taken in a single pass.

Figure 10-10a, b, c, and d show monthly performance curves of several $F_i'\overline{\tau\alpha}$ and $F_i'U_L$ combinations with $\overline{\tau\alpha}/U_L$ ranging from 0.09 to 0.32 °C·m²/MJ (0.5 to 1.8 °F hr ft²/Btu). The dotted lines show performance predictions using the weather tables in Appendix B with the correlation equation (9-23) plus the other methods of this and the previous chapter. The solid lines are results of simulations using the simple methods outlined in Chapter 8 with hourly data for 10 years. The correlation is seen to work quite well, having trouble only at either end of the season when the heat collected may come at the wrong time to be effectively utilized (e.g., in March the sunny period is often at the end of the month while the demand is at the beginning).

Figure 10-11a shows the effect of changing only one variable—$F_i'U_L$—by a factor of 2. The system performance has increased from $\sum f_a = 0.47$ to 0.74. This comparison is somewhat misleading, however. It is more informative to find the collector area required to achieve the same performance. The new area is smaller by a factor of 2 (dotted line). Thus for this set of conditions the required area is linear with $F_i'U_L$.

Figure 10-11b shows the effect of varying $F_i'\overline{\tau\alpha}$ in a similar way. It is seen that by changing $F_i'\overline{\tau\alpha}$ from 0.64 to 0.9 the required area for a given performance can be halved. This system, then, is more than linearly dependent on $F_i'\overline{\tau\alpha}$; proportionally speaking, its value is therefore more important than $F_i'U_L$ to the system designer. In practice, however, the two characteristics usually interact. Reducing the number of glazings to increase $\tau$ will also increase $U_L$. Use of a selective coating to reduce $U_L$ will also reduce $\alpha$, since these coatings are not quite as absorptive as black paint.

Figure 10-12 explores the effect of base temperature and storage on monthly performance by simulation for calendar 1959 with the same collector used in Fig. 10-10c. Performance is seen to be reduced by about 5% in $f_a$ for every 10 °C (20 °F) rise in base temperature, and drops about 3% when $mc$ is halved.

## Seasonal and Annual Variations

The pattern of the previous monthly performance curves has been consistent—generally a roughly constant monthly $f_a$, which decreases somewhat in November and December and reaches a high in March. This pattern varies with the locality and with the years used for the weather data.

Figure 10-13a to i compares the results of simulated performance using the same collector for nine successive winter seasons at Hartford using the same set of solar operating parameters. The yearly simulated performance has been plotted (solid lines) and compared with 10-year simulation averages (dotted lines). The upper curve is the demand, and the season's degree-days are noted along with the percentage of the 10-year average that this represents. The bottom curves show the solar heat collected, and the annual performance $\sum f_a$ has been

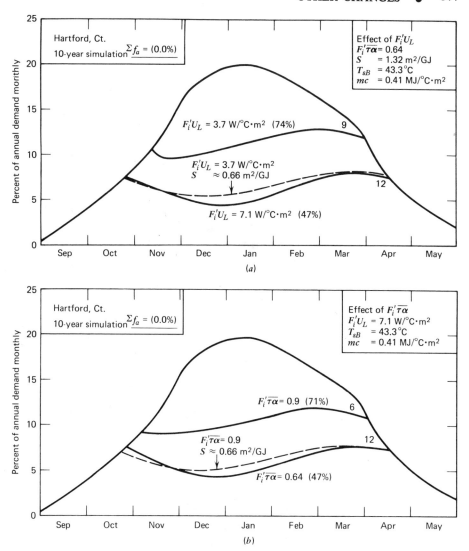

**FIGURE 10-11.** Monthly performance curves showing how a change in collector parameter $F_i' \overline{\tau\alpha}$ from 0.64 to 0.9 is equivalent to a change in loss coefficient $F_i' U_L$ from 7.1 to 3.7 W/°C·m². These are the result of 10-year simulations at Hartford.

similarly noted. There is only one season (1961–1962) that mimics the 10-year situation closely. The demand in that year was 99% of normal, and the annual performance of the system $\sum f_a$ was 59.4% of demand, which compares with 59.6% for the 10-year average. The performance in other years randomly varies within a 5 to 10% range in demand and performance.

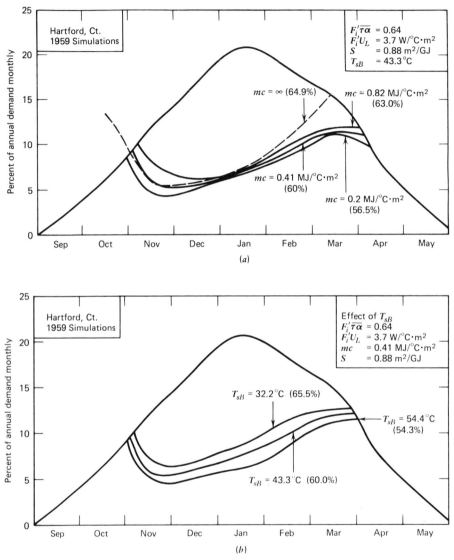

**FIGURE 10-12.** Annual performance curves showing the effect of changes in (*a*) storage capacity (*mc*), and (*b*) base temperature on the performance of one system. These curves are from one-year simulations at Hartford.

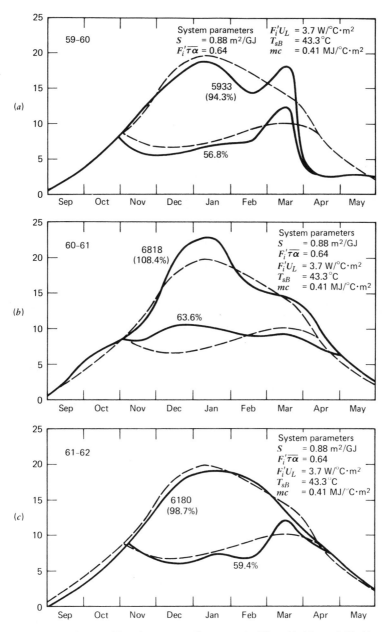

**FIGURE 10-13.** For the same collector as in Figs. 10-10*c* and 10-12, the simulated annual performance is shown for 10 successive winter seasons. The dotted lines show the 10-year averages. Annual degree-days are listed on the upper (demand) curves with the percentage of

379

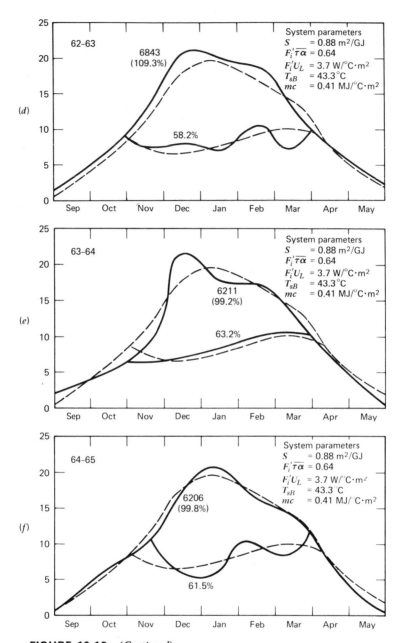

**FIGURE 10-13.** (*Continued*)
10-year demand given in parentheses. Percentage of the particular year's demand met by solar energy ($\sum f_a$) is noted on lower (supply) curves.

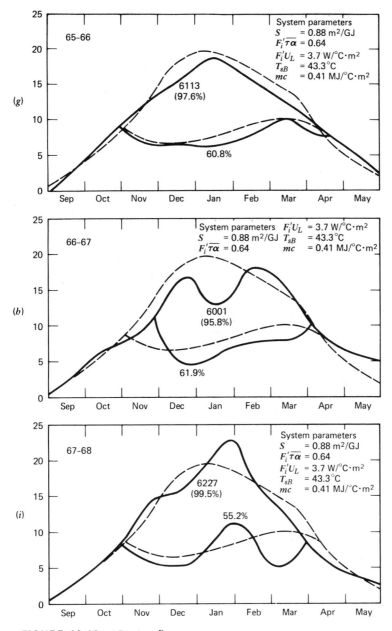

**FIGURE 10-13.** (*Continued*)

For Hartford it has been found that *calendar* 1959* is also a good predictor of 10-year performance. Even though 1959 has a somewhat darker November and a brighter March than the 10-year average, annual performance has been shown to be virtually identical. Once located, such a year can be very useful for simulations because the cost of a simulation is proportional to the time period that must be analyzed.

Unlike simulations, the base temperature correlation technique predicts performance at a cost independent of the length of the data base incorporated into the analysis. Cost for preparation of the data base, of course, does depend on its length. A 10-year time period is widely regarded as long enough to gain representative solar-operating information,† and the Hartford data in Appendix B have been prepared from weather data from the 1959–1968 time period. Data for the other locations are given for the 2-year calendar 1971–1972 period. While such data predict performance for that period accurately, comparisons have shown that systems designed with these data will be somewhat oversized.

## EFFECT OF STRATIFIED STORAGE

With air systems having rock storage, the system characteristics change. As shown in Chapter 8, as the turnover time is reduced, the performance of a collector with stratified storage approaches that of a well-mixed storage system having $F_i = F_m$. Figures 10-14 and 10-15 are the results of system simulations that demonstrate this point. Figure 10-14 parameters are from a commercial collector, conservatively designed with two cover plates to have a low-loss coefficient with a black-paint absorber surface. Figure 10-15 represents a hypothetical one-glass, black-paint collector with a very high value of $F_m$ but the high losses that might be typical of manufactured-in-place units. Interestingly, the performance of these collectors is almost identical.

The upper line in either set of curves is a performance prediction for 1959 at Hartford with $F_i = F_m$ using well-mixed storage. In Fig. 10-14 this line is from a simulation. In Fig. 10-15 it is from the base temperature correlation using eq. 9-23. The other lines represent simulations *using ideally stratified storage* with a properly calculated $F_i$ at turnover times from 4 to 10 hours, the latter a typical manufacturer's recommendation. The stratified lines indeed approach the well-mixed $F_m$ performance line as the turnover time is reduced. On the basis of this graph, to account for a turnover time of 10 hours, about 0.03 should be subtracted from $\sum f_a$ as computed using $F_m$ with the equation 9-23 correlation. For a

---

* That is, fall 1959 followed by spring 1959. Incongruous, perhaps, but no trouble in analysis since the storage is always at the base temperature ($T_{sB}$) on January 1.
† NBSIR 76-1187, "Interim Performance Criteria for Solar Heating and Cooling Systems in Commercial Buildings," November 1976, specifies use of a 10-year data base (p. 4).

turnover time of 4 hours, only 0.005 need be subtracted. For a turnover time of 7 hours, subtract 0.02.

The longer turnover times for air systems hurt performance because there are then many days in most months in which the operating time is less than the turnover time. The storage on these days is not fully utilized before the bed is discharged. When stratified storage systems are used for hot water, the turnover times are shorter and $\sum f_a$ values calculated with $F_m$ can be used directly.

## OTHER LOCATIONS

The variability of solar radiation and demand with location is of course great. The variations between nearby sites can also be great, especially on the West Coast of the United States or where there are differences in elevation that can cause variations in load and radiation over short distances. Nevertheless, the designer often has no choice but to use the nearest location for which proper data are available. As a first approximation it is reasonable to resize a solar installation for construction at a different location by changing the area installed in proportion to the new annual load. This is equivalent to using the same value for $S$ in both places. Presentation of results as $\sum f_a$ versus $S$ is therefore a convenient way to compare performance of a given system at different locations.

Figure 10-16 shows the simulated annual performance (solid lines) in terms of $S$ for a number of locations using the same 1971–1972 data base in Appendix B for a particular heating-only case. The angle of the collector tilt was held to 8° plus the latitude to compensate for the effect of latitude on performance. For a given normalized area $S$, annual performance was identical for DCA (Washington, D.C.), Boston, Concord (NH), and Hartford.* Madison, Wisconsin was evidently less sunny relative to its degree-day load (producing the lowest performance), Atlanta and Denver were somewhat brighter, and Los Angeles was decidedly more sunny. The differences (Los Angeles excluded) are significant but not startling. An $S = 1.135 \ \text{m}^2/\text{GJ}$ (10 ft²/M Btu) system in Denver produces 70% of annual demand, while in Madison that size system produces only 51% of the annual demand. However, it takes nearly twice as large a system (in terms of $S$) to achieve 70% of the demand with solar energy in Madison as in Denver.

The base temperature correlation is seen to perform well at most of these locations (compare the solid and dotted lines). For the Northeast and Madison, the correlation is excellent, with $\sum f_a$ within $\pm 0.01$. For Denver and Atlanta it is high by about 0.02, probably as a result of longer solar days that elevate average collector temperatures. For Los Angeles, which has a very low heating demand

* The reason for such good correspondence needs further study. After all, not only the load and radiation levels but also the heating season and length of solar day vary as well.

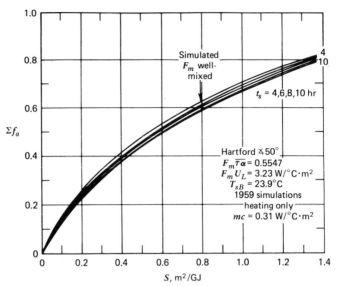

**FIGURE 10-14.** Simulated performance for a system with well-mixed storage using $F_m$ parameters (upper line) compared with stratified storage simulations conventionally made using $F_t$ parameters for various turnover times (i.e., collector flow rates).

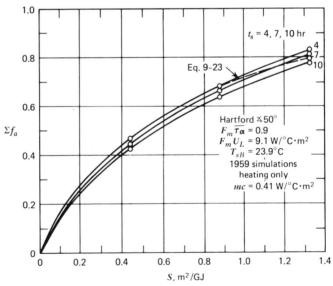

**FIGURE 10-15.** For another system, results like those of Fig. 10-14 are shown. The well-mixed performance has been *predicted* using eq. 9-23 with $F_m$ parameters.

384

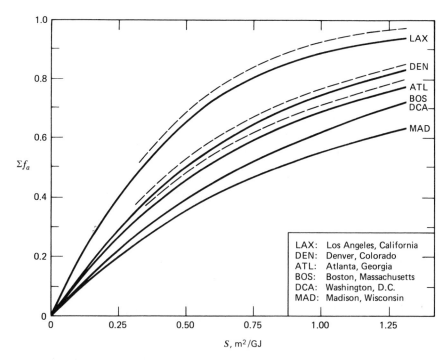

**FIGURE 10-16.** Performance prediction using eq. 9-23 (dotted lines) at six locations compared to results from two-year simulations (solid lines). Dotted lines for the two lower curves overlap the solid lines closely and are not shown. For each location $F_i' \overline{\tau\alpha} = 0.8$, $F_i'U_L = 5.224$ W/°C·m², $T_{sB} = 48.9$ °C (120 °F), $mc = 0.4088$ MJ/°C·m², and the inclination angle was 8° plus the local latitude.

and long solar days* due to very clear skies, the overprediction is about 0.04, still within ordinary engineering accuracy.

The general similarity of the curves in Fig. 10-16 indicates that if solar systems are sized according to $S$ values, data from only a few cities would suffice for many more, provided the pattern of the similarities were known. This is an area for further research.

## PROBLEMS

1. For one location assemble the input data [as in problem 9-1(n)] and the performance data up to $q_T$ [as in problem 9-1(c), (d), or (e)] in a table like Table 10-1.

* This suggests that a correlation retaining $\bar{t}$ in the storage function might be useful (eq. 9-17).

    (a)  Using convenient $S$ values (0.8, 0.4, and 1.2 GJ/m² or 10, 5, and 15 ft² M Btu) assemble performance information as in Table 10-1, using methods of Example 10-1.

    (b)  Plot the results as in Fig. 10-2.

    (c)  Plot the results in terms of $\sum f_a$ as in Fig. 10-3.

**2.** Repeat problem 1 using integral values of 0.2, 0.4, and 0.8 MJ/°C·m² (10, 20, and 40 Btu/°F ft²) for the storage quantity, using methods of Example 10-2 to produce results that are to be presented as in Table 10-2, Fig. 10-5, and Fig. 10-6.

**3.** Recalculate the base case of problem 1(a) for

    (a)  heating only

    (b)  hot water only, assuming a uniform daily hot water load with a supply temperature of 15 °C (59 °F) and a use temperature of 50 °C (122 °F).

**4.** The collector efficiency curve used to calculate problem 1 derives from problem 9-1. For collector $A$, the water flow capacity rates are given at the start of the Chapter 9 problems.

    (a)  Recalculate problem 1(a) if the flow is halved, using methods of Example 10-4.

    (b)  What is the exact storage capacity value for which these results apply?

**5.** Predict and plot the annual performance of the collector of problem 4 for hot water only, assuming a supply temperature of 15 °C (59 °F) and a use temperature of 50 °C (122 °F).

**6.** The collector of problem 4 is now to be modified to be an air collector. Predict and plot the annual performance if the fluid film coefficient is 11 W/°C·m² (1.94 Btu/hr ft² °F) and the flow rate of air is 0.010 16 m³/s per m² collector area (2 ft³/min ft²). The inside ambient temperature is to be 24 °C (75 °F). Correct the final $\sum f_a$ values for the proper turnover time.

# Economic Evaluation

All solar systems require a large initial investment to realize a modest annual rate of return in terms of fuel savings. To be economical, the solar system must perform for many years, returning its initial cost plus interest over a long lifetime.

It is of no consequence whether the money invested is borrowed or not. No matter who is the owner, capital should return proper interest, for there are always other profitable investments to be made.

The value of a solar system investment is very dependent on the economic criteria selected. If an aggressive 12% interest is required and no allowance taken for advancing fuel prices, probably no existing system is economical. On the other hand, if a 6% interest is satisfactory and a 10% annual increase in fuel costs is assumed, almost any system will be very profitable (and why not—the fuel will increase in price by a factor of 10 in 25 years!)

In this chapter the methods for making economic evaluations will be summarized, and then it will be shown how the previous performance analyses can be quickly adapted for use in determining the optimal collector area, storage capacity, and heat exchanger size. Optimization of installed collector area is the most important step, for it is seldom economical for a solar system to supply the last several percent of the load by installation of enough area to catch extra sun for a few days in December and January. Since the added area would be worthless during the rest of the year, it is more economical to supply the heat from auxiliary sources. The more expensive the auxiliary fuel, the higher the proportion of annual demand that can be economically supplied by solar energy.

## FUNDAMENTALS OF ECONOMIC ANALYSIS

### Net Present Value and Return on Investment (1)

The starting point for all economic analyses is the time value of money. If a sum of money is in hand and has a present value of $P$, it will be worth at a future time a value $F$ given by

$$F = P(1 + ni) \qquad (11\text{-}1)$$

where $i$ is the fractional interest rate, and $n$ is the number of years into the future at which $F$ is to be evaluated. If the interest is compounded annually (and this is routinely the basis for economic evaluations), then repeated application of eq. 11-1 at one-year intervals results in the annual compound interest relationship:

$$F = P(1 + i)^n \qquad (11\text{-}2)$$

The equivalence expressed in eq. 11-2 between the value of money now and in the future is fundamental. Modern investment analyses refer all future returns of money to the present time by use of a rearrangement of eq. 11-2:

$$P = \frac{F}{(1 + i)^n} \qquad (11\text{-}3)$$

where $i$ has been renamed the discount rate. Thus future money is not as valuable as money at the present, and must be "discounted" by the factor $1/(1 + i)^n$.

The difference between the present value of all future returns $P$ and the present capital $C$ required to make an investment is the net present worth or net present value (NPV) of the investment:*

$$\text{NPV} = P - C \qquad (11\text{-}4)$$

The NPV is similar in concept to profit, but with all monies discounted to the present time so that a dollar returned in 10 years is not as profitable as a dollar returned in 1 year. Only investments with positive NPVs need be considered.

The rate of return on investment (ROI) is the maximum discount rate one could afford to pay, that is, the discount rate which makes NPV = 0. It is often found by trial, although in many simple cases graphical solutions are practical.

**EXAMPLE 11-1**

A solar collector costs $15 and returns $5 annually for 5 years, after which it is to be discarded. The interest rate is 8%. What is the present value of all the annual returns?

Case 1 in Table 11-1 details the situation. For each year the present (i.e., discounted) value of the $5 return is listed and the total is summed in the total present value row. For instance, at the end of the third year, the $5 return is presently worth (by eq. 11-3)

$$P = \$5/(1 + 0.08)^3 = \$3.969$$

Since the total present worth of all the returns is $19.96, the net present value of the investment by eq. 11-4 is

$$\text{NPV} = \$19.96 - \$15 = \$4.96$$

* This is often called the *net worth*.

This means that purchasing the collector with money borrowed at 8% is financially equivalent to being given—right now—$4.96 in cash. Investing *that* amount at 8% interest will produce by the end of the term the same bank balance as borrowing $15 and receiving $5 income for 5 years while simultaneously paying off the $15 mortgage annually.

The return on investment in this case is 19.85%, determined by trial so that the NPV = 0 in Case 2 of Table 11-1.

### Annual Return

Equal annual payments of amount $A$ deposited over $n$ years at the end of each year can produce a future sum of money. The present value of these payments is given by summing values from eq. 11-3 for these payments over $n$ years. If the resultant equation is multiplied through by $(1 + i)^n$ and that result by $(1 + i)$ and then the last two equations subtracted, solving for $P$ gives

$$P = A\left[\frac{(1 + i)^n - 1}{i(1 + i)^n}\right] = A\left[\frac{1 - (1 + i)^{-n}}{i}\right] \qquad (11\text{-}5)$$

This equation can be rewritten in functional form as

$$P = AR(i, n) \qquad (11\text{-}6)$$

where $R$ is the payback ratio, a function of interest rate $i$ and time period $n$. $R$ is the ratio of capital invested at time zero to the annual return which that capital produces.* It is given by

$$R(i, n) = \frac{1 - (1 + i)^{-n}}{i} \qquad (11\text{-}7)$$

The payback ratio can be thought of as the capital necessary to return a dollar annually for $n$ years if the discount rate is $i$. When that ratio in an actual investment is less than $R$, the investment is *profitable*. When it is equal to $R$, it is *marginal*. When it is greater than $R$, it is unprofitable and the investment should be avoided.

$R$ is also used to convert $n$ annual payments to their present value. It therefore describes the mortgage or annuity situation. Figure 11-1 is a graphical representation of eq. 11-7 in which $R$ has been plotted versus $n$ for a number of integral values of the discount rate. This graph permits determination of $i$ when $R$ and $n$ are known, which cannot be done directly with eq. 11-7.

Example 11-1 may be solved more conveniently by means of eq. 11-7 or Fig.

---

* The term *payback time* should be assiduously avoided here. It is guaranteed to create confusion. The payback time is the value of $R$ in years for the *actual* investment, the time at which the total dollar return equals the initial investment. It continues in wide use as a measure of economic desirability in spite of the unanimous protests from economists, who point out that it is independent of the interest rate and does not take into account the additional length of time beyond the breakeven point for which profits may continue to be made.

**TABLE 11-1.** Yearly Economic Analyses for Examples 11-1 and 11-5

| Case | Initial Annual Return, $ | Inflation Rate | Interest or Discount Rate, $i$ | Total Present Value, $ | Annual Return/Present Value | | | | |
|------|------|------|------|------|------|------|------|------|------|
| | | | | | Year 1, $ | Year 2, $ | Year 3, $ | Year 4, $ | Year 5, $ |
| 1. Base case | 5.00 | 0.0 | 0.08 | — | 5.000 | 5.000 | 5.000 | 5.000 | 5.000 |
| | | | | 19.96 | 4.630 | 4.287 | 3.969 | 3.675 | 3.403 |
| 2. ROI: Equivalent annual return; present value $15 | 5.00 | 0.0 | 0.1985 (by trial) | — | 5.00 | 5.00 | 5.00 | 5.00 | 5.00 |
| | | | | 15.001 | 4.172 | 3.481 | 2.904 | 2.423 | 2.021 |
| 3. Actual situation with inflated return | 5.00 | 0.06 | 0.08 | — | 5.300 | 5.618 | 5.955 | 6.312 | 6.691 |
| | | | | 23.645 | 4.907 | 4.816 | 4.727 | 4.640 | 4.555 |
| 4. Equivalent; constant return: $i$ from eq. 11-8 | 5.00 | 0.0 | 0.018 87 | — | 5.000 | 5.000 | 5.000 | 5.000 | 5.000 |
| | | | | 23.645 | 4.907 | 4.817 | 4.727 | 4.640 | 4.554 |
| 5. ROI: Equivalent annual return; present value $15 | 5.00 | 0.06 | 0.2704 (by trial) | — | 5.300 | 5.618 | 5.955 | 6.312 | 6.691 |
| | | | | 15.00 | 4.172 | 3.481 | 2.904 | 2.423 | 2.022 |

11-1, shortcutting the year-by-year calculations of Table 11-1. The technique is valid in those cases for which the annual return and discount rate are constant.

### EXAMPLE 11-2

Recalculate Example 11-1 using the payback ratio.

From the definition of $R$,

$$R = \$15/\$5 = 3$$

When $R = 3$ and $n = 5$, from eq. 11-7 by trial or directly from Fig. 11-1, $i = 0.1985$. This is the return on investment (ROI).*

The present value of all the annual returns is based on $R$ calculated using eq. 11-7 or Fig. 11-1, with the discount rate equal to the interest rate:

$$R(0.08, \text{5 yr}) = \frac{1 - 1.08^{-5}}{0.08} = 3.993$$

The $5 annual return is then presently worth, by eq. 11-6,

$$P = \$5 \times 3.993 = \$19.96$$

and by eq. 11-4,

$$\text{NPV} = \$19.96 - \$15.00 = \$4.96$$

which are the same results as before.

## Effect of Federal Income Tax and Solar Incentives

For commercial investments, the tax situation is complex and the effect of taxes must often be calculated year by year as in Example 11-1, unless it is a regular

---

* The payback time for this investment, if it *must* be calculated, is the value of $R$ in years for the actual situation. In this case it is 3 years.

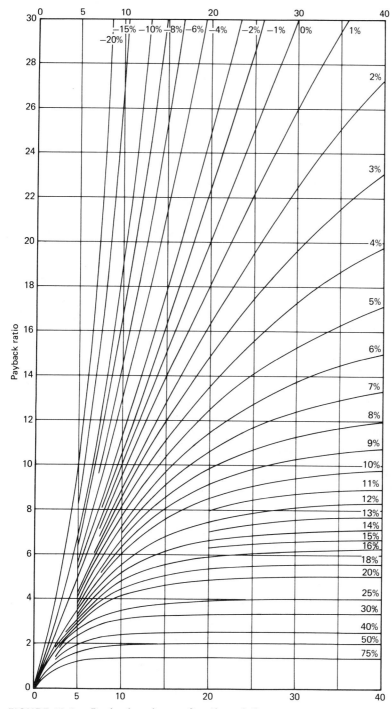

**FIGURE 11-1.** Payback ratio as a function of discount rate and length of investment.

payment that can be discounted to the present in a single step. For the home-owner, money paid as interest or received as a solar incentive is not taxable, and the value of such a gift depends on the individual's tax bracket.

**Interest Payments.** If additional income of the investor is ordinarily taxed at a rate $t$, then while a dollar of income can pay a dollar's interest, it is worth only $(1 - t)$ dollars for other uses. The value of a dollar paid as interest is thus effectively multiplied by $1/(1 - t)$ and the interest rate thereby reduced by the factor $(1 - t)$.

**Incentives.** Nontaxable solar incentives such as tax credits and subsidies are worth more than their face value to an investor by a factor $1/(1 - t)$. This higher value reflects the actual incremental income necessary for that investor to other-wise accumulate a dollar amount of capital equal to the solar incentive. In an economic analysis, however, the capital cost of the investment is reduced by only the face amount and the remaining calculations are made in the ordinary manner. This is consistent with the original computation of capital cost using after-tax dollars.

**EXAMPLE 11-3**

The income of the investor of Example 11-1 is taxed at 30%.

(a) What are the new NPV and ROI?
(b) If there is a 20% tax credit available what are NPV and ROI?

(a) The effective interest rate is

$$i = 0.08 \times (1 - 0.3) = 0.056$$

The payback ratio for the new discount rate is, by eq. 11-7,

$$R(0.056, 5 \text{ yr}) = \frac{1 - 1.056^{-5}}{0.056} = 4.259$$

and the present value of the returns is thereby (eq. 11-6)

$$P = \$5 \times 4.259 = \$21.29$$

which gives a NPV of the investment by eq. 11-4 of

$$\text{NPV} = \$21.29 - \$15.00 = \$6.29$$

The ROI on the amount actually invested is as before determined from

$$R(\text{ROI}, 5 \text{ yr}) = 15/5 = 3$$

remaining at 19.85%. Considering that the interest is not taxable, the ROI is $19.85/0.7 = 28.36\%$.*

* The payback time here is $R(i = 0.2836, n = 5) = 2.51$ yr, considering the tax situation.

(b) The tax credit is

$$\$15 \times 0.2 = \$3.00$$

Its value to this investor is

$$\frac{\$3}{(1 - 0.3)} = \$4.29$$

The net investment is

$$\$15.00 - \$3.00 = \$12.00$$
$$NPV = \$21.29 - \$12.00 = \$9.29$$
$$R(ROI, 5 \text{ yr}) = \$12.00/\$5.00 = 2.4$$
$$ROI = 30.7\%$$

Considering his income tax bracket, for this investor the ROI is equivalent to 0.307/0.7 or 43.9%.

## Other Costs

**Scrap Value.** The investment required should be discounted by the present value of the investment recoverable at the end of the time period before proceeding with the other calculations.

**Effect of Maintenance or Operating Costs.** No matter when scheduled or what their amount, maintenance and operating costs can be treated by making up a table like Table 11-1, discounting all these costs to the present time, and increasing the capital investment by that amount. If the costs can be estimated as a fixed annual percentage of either the original investment or of the inflated return, they may be handled more simply with eq. 11-5.

**EXAMPLE 11-4**

(a) If the collector in Example 11-2 is worth $3 at the end of year 5 as scrap, recalculate the ROI and NPV.

The present value as scrap is given by eq. 11-3 as

$$P = \$3/(1 + 0.08)^5 = \$2.04$$

The net investment is therefore $15.00 - $2.04 = $12.96. By definition, the payback ratio* is

$$R(ROI, n) = \frac{\$12.96}{\$5.00} = 2.592$$

and for $n = 5$ yr, ROI $= 0.268 = 26.8\%$ by Fig. 11-1 or by trial with eq. 11-7.

The NPV is computed using the present value of the return, which is the same as before, with the new capital value.

$$NPV = P - C = \$19.96 - \$12.96 = \$7.00$$

* In this example this is also the payback time in years.

(b) If maintenance costs of $1.50 per year are anticipated for each of the first 2 years, modify the values of ROI and NPV accordingly.

By eq. 11-5, the present value of the maintenance cost is

$$P = \$1.50\left[\frac{1 - (1 + 0.08)^{-2}}{0.08}\right] = \$2.67$$

$$\text{NPV} = \$7.00 - 2.67 = \$4.33$$

The net investment is now $12.96 + $2.67 = $15.63. Therefore, the payback ratio* is

$$R(\text{ROI}, n) = \frac{\$15.63}{\$5.00} = 3.126$$

$$\text{ROI} = 18.0\%$$

## Effect of Inflation

Inflation does not affect the sunken cost of a solar installation, but it does affect the return. Most analyses include a constant inflation rate in the value of the return to account for the rising value of the fuel saved.

The year-by-year approach of Table 11-1 (cases 1 and 2) handles this situation easily, as indeed it does most cases. The simpler payback ratio method can be adapted by defining a new effective discount rate

$$i_d = \frac{1 + i_c}{1 + i_i} - 1 \approx (i_c - i_i) \tag{11-8}$$

where $i_c$ is the interest rate on capital and $i_i$ the inflation rate. The new discount rate gives identically the same annual and total present value for the annual returns as would a yearly analysis. The NPV can then be determined from the investment.

The ROI is calculated by first determining the effective discount rate $i$ that reduces the NPV to zero. Then eq. 11-8 is solved for $i_c$, giving

$$i_c = (1 + i_d)(1 + i_i) - 1 \approx i_d + i_i \tag{11-9}$$

where $i_c$ is then the ROI with inflation accounted for.

Use of the approximate forms of eqs. 11-8 and 11-9 will not produce significant error in making an analysis, since the few tenths of a percent difference is small compared to the uncertainty in predicting the inflation rate.

### EXAMPLE 11-5

The collector described in Example 11-1 is to be applied where fuel costs and hence annual returns are increasing at 6% annually. What are NPV and ROI?

---

* In this example this is also the payback time in years.

By eq. 11-8, the effective discount rate is

$$i_d = \frac{1 + 0.08}{1 + 0.06} - 1 = 0.018\ 87$$

The present value of returns on investment is estimated using the marginal payback ratio determined from Fig. 11-1 or eq. 11-7* as

$$R(0.018\ 87,\ 5\ \mathrm{yr}) = \frac{1 - (1 + 0.018\ 87)^{-5}}{0.018\ 87} = 4.729$$

By eq. 11-6, the present value of the return is

$$P = \$5 \times 4.729 = \$23.65$$

so by eq. 11-4

$$\mathrm{NPV} = \$23.65 - \$15.00 = \$8.65$$

Cases 3 and 4 in Table 11-1 compare the true year-by-year analysis with that using eq. 11-8 followed by eq. 11-5. Notice that the net present value of each year's annual return as well as their total is identical by either method.

The value for effective discount rate that reduces the NPV to $15 is unchanged and has been found by trial to be 0.1985. It can also be found directly from Fig. 11-1 from $R(i,\ 5\ \mathrm{yr}) = \$15/\$3 = 5$. Applying eq. 11-9,

$$\mathrm{ROI} = i_c = (1 + 0.1985)(1 + 0.06) - 1 = 0.2704$$

or 27.04%.† The true year-by-year situation, given as case 5 in Table 11-1, produces the same ROI by trial when the present value of the returns is reduced to $15.

## SYSTEM OPTIMIZATIONS

### Costs

The optimization methods described below are mathematically exact, but their utility depends on the accuracy of the incoming data. Costs are notoriously difficult to find and tricky to assign to one part of the system or another. Engineering judgement is therefore often called for in determining the costs to be used in economic analysis.

Throughout this book, apart from fixed costs, all costs are considered to be *linear* with the installed area. This is, of course, not always true. Derivation of accurate system costs must therefore start with an initial cost estimate for a system near the final size. Fixed costs for pumps, controls, piping, and electrical

---

* The approximate form of eq. 11-8 yields $R = 4.713$.
† The payback time in this case is $R(i = 0.2704,\ n = 5\ \mathrm{yr})$, or 2.58 yr. The inflated total return after 2.58 yr indeed equals $15 (plot the annual values from case 3, Table 11-1 to see this).

work are estimated first, and then purchased costs for collectors, heat exchangers, and storage tanks are added. Installation costs must be included and distributed among the components. The total installed cost is then divided by the total installed area to produce the unit system cost ($/m² or $/ft²). Individual costs for collector, storage, and heat exchangers are simultaneously calculated on a similar basis—per unit *collector* area—to simplify the economic analyses.

Alternatively, any component can be considered to have both variable and fixed costs associated with it. While the total cost is eventually needed for evaluating the investment, optimization can be done with either the variable costs or the total costs as is most convenient.

When all the system optimizing is done, it is time once again to estimate the costs of the system and components, based on optimal values. If the unit costs are changed, then another "pass" through the calculations will be warranted. Thus solar-heating systems, like so many other systems, must be optimized and reoptimized as better and better cost and performance data become available.

The techniques in this book are designed to be applied repetitively and have been organized to minimize the work involved. An analysis is finally complete only when all the assumptions made to solve the design problem are confirmed in its solution.

## Collector Area Optimization

The lower performance line on Fig. 10-3 is the typical result of a system design for a solar-heating system. The area to be installed has not yet been chosen, but it is clear that there are diminishing performance returns as $S$ increases. This performance line has been replotted in Fig. 11-2. There are data for $S = 0.2$, 0.4, 0.8, and 1.2 m²/GJ.

The abscissa of Fig. 11-2 is $S$, the ratio of collector area to thermal units of annual demand. The ordinate is $\sum f_a$, the proportion of annual demand met by solar energy. This can be restated as the ratio of the thermal units of annual demand supplied by solar energy to the thermal units of annual demand. The ratio of the coordinates at any point on the curve ($y/x$) is the annual thermal return

$$\frac{y}{x} = \frac{\sum f_a}{S} = \frac{Q_a/L_a}{A_c/L_a} = \frac{Q_a}{A_c} \tag{11-10}$$

where $Q_a$ is the annual heat supplied by solar energy and $A_c$ is the total collector area.

The slope of a line from the origin through any point on the curve is also equal to $y/x$, which eq. 11-10 shows to be the thermal units of demand met per unit of collector area at that $S$ value. This slope is therefore the annual thermal return on an investment of a unit of collector area.

In the upper curve of Fig. 11-3 values of the slopes of such lines have been

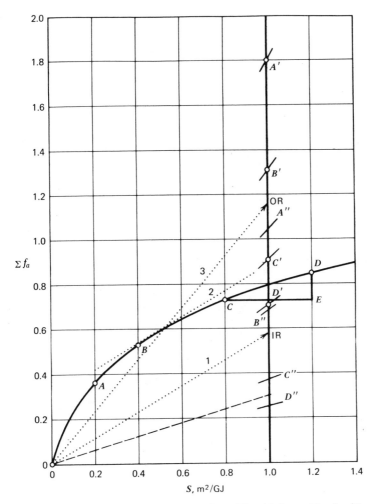

**FIGURE 11-2.** Performance data from Fig. 10-3 combined with points on the $S = 1$ m²/GJ line showing how the values plotted in Fig. 11-3 were obtained.

plotted against $S$. Points were plotted mechanically from Fig. 11-2 by projecting the $S = 1$ m²/GJ (or 10 ft²/M Btu) line vertically and drawing lines from the origin through the point considered and on to the integrally valued $S$ line. The slope of such a line is numerically equal to the ordinate value at the $S = 1$ m²/GJ line (or $\frac{1}{10}$ that of the 10 ft²/M Btu line) and is shown by points $A'$, $B'$, $C'$, and $D'$ for the four plotted points. The upper line in Fig. 11-3 expresses these points as the GJ delivered per m² (M Btu/ft²).

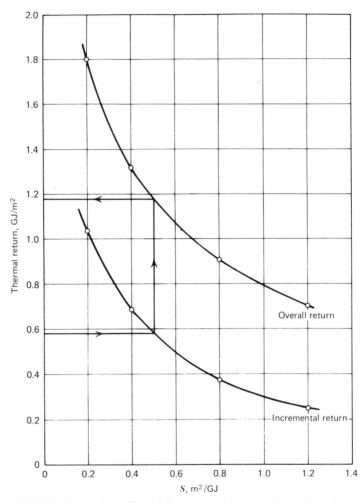

**FIGURE 11-3.** Overall and incremental return from the solar-heating system of Fig. 11-2 (Example 11-7).

The overall thermal return per unit area is of primary importance for calculating ROI, but it is not of much use in optimization. Installation of enough area so that the ROI becomes marginal means that the thermal return would be such that the present value of the total capital invested *equalled* the present value of the annual returns in the form of fuel savings. Since it is obvious that the performance of any lesser area will return at a greater rate than *that* investment, it follows that the last increment of collector area added returns *less* than the value of the capital necessary to provide it. Therefore, why install that incremental area at all? What is really needed is a measure not of what the return is on the entire collector area, but of the return on the *last* increment of area. Then $S$ can be decreased until *that* return is equal to the marginal return. Automatically the remainder of the array will then perform at above this return to generate NPV for the investment.

The slope of the performance curve in Fig. 11-2 itself can provide such information. The slope is a measure of the incremental return at the indicated $S$ value, as can be shown by considering the $\sum f_a$ values at $S = 0.8$ m²/GJ and $1.2$ m²/GJ, which are $0.7272$ and $0.8485$, respectively. The incremental performance return attained by increasing the area from the former value of $S$ to the latter is therefore $(0.8485 - 0.7272) = 0.1213$ units of $\sum f_a$, and the incremental thermal return is $0.1213/(1.2 - 0.8) = 0.303\ 25$ GJ/m². The ratio of these values is exactly the slope of the line connecting points $C$ and $D$, or considering triangle $CDE$ in terms of line segments, slope $= \overline{DE}/\overline{CE}$. To measure such a slope simply, use two drafting triangles to transfer the slope running between $C$ and $D$ to a parallel line passing through the origin and the $S = 1$ m²/GJ line. This line is shown dashed, and its slope of $0.303$ can be read directly from the $\sum f_a$ scale at its $S = 1$ m²/GJ intercept. That slope really applies at about $S = 1$ m²/GJ, and could more easily have been established by taking a tangent to the $\sum f_a$ versus $S$ curve at that point. In this way points $A''$, $B''$, $C''$, and $D''$ have been determined, giving values of slopes at $A$, $B$, $C$, and $D$, respectively. The lower line in Fig. 11-3 is a plot of these points, which shows the rate of incremental return as $S$ changes.

Figure 11-3 is all that is needed to optimize the collector area and estimate the overall thermal return of investment. First, however, the dollar return expected from the capital investment must be transformed into an equivalent thermal return. To do this the marginal payback ratio is established from the discount rate and the system lifetime using eq. 11-7. In combination with fuel cost, the marginal capital to collect a unit quantity of energy annually can be determined. Then, given the collector cost, the marginal thermal return per unit collector area can be calculated.

## EXAMPLE 11-6

The cost of electricity is 5¢/kWh, and solar collectors are available at a system cost of \$200/m² (\$18.58/ft²). If fuel escalation is estimated at 5% annually, and capital is available at 9%, what is the marginal annual thermal return from a unit of collector area if the system life is estimated at 30 years? The tax bracket of the investor is 30%.

The effective interest rate is $0.09 \times (1 - 0.3) = 0.063$.
The effective discount rate is

$$i_d = \frac{1 + 0.063}{1 + 0.05} - 1 = 0.012\ 38$$

The marginal payback ratio is $R(0.012\,38, 30\text{ yr})$:

$$R = \frac{1 - 1.012\,38^{-30}}{0.012\,38} = 24.93$$

The cost of electricity in thermal units is

$$\frac{\$0.05}{kWh} \times \frac{kWh}{3.6\ MJ} \times \frac{1000\ MJ}{GJ} = \frac{\$13.89}{GJ}$$

The capital that is a marginal investment in returning 1 GJ annually is therefore

$$\frac{\$13.89}{GJ} \times 24.93 = \frac{\$346.25}{GJ}$$

In terms of collector area this is equivalent to

$$\frac{1\ GJ}{\$346.25} \times \frac{\$200}{m^2} = 0.5776\ GJ/m^2$$

When the inflation rate is higher than the effective interest rate the discount rate becomes negative. The calculations proceed just as outlined above, but the economic advantage of the solar system improves markedly.

**Area Optimization Procedure.** Given the curves of Fig. 11-3, selection of the optimum collector area and overall thermal return is easy and the optimization can be repeated quickly and consistently for a variety of economic assumptions. First the incremental return curve is used to determine the optimum (normalized) area $S$ from the marginal thermal return calculated as shown in Example 11-6. Then the same optimal $S$ value is used with the overall return curve to determine the overall thermal return. Note once more that the *overall return* on investment is *not* optimized, since the return on investment will continue to rise as collector area falls.* Instead the collector area is optimized so that the *incremental* return on any additional collector area is exactly the marginal rate of return.

This procedure allows the purchaser to put only as much money as is economical into the solar system, thus maximizing the net present value (NPV) of the investment.

The NPV is the monetary value of the total annual thermal returns (expressed as present value) less the investment. Both returns and investment are best kept normalized per unit collector area, and the NPV found by multiplying their difference by either the installed collector area or by $S$, the area per unit annual thermal demand.

### EXAMPLE 11-7

Optimize the collector area for the system plotted in Figs. 11-2 and 11-3 using the marginal thermal return of $0.5776\ GJ/m^2$ ($0.049\,28$ M Btu/ft²) from Example 11-6. What are ROI and NPV?

---

* Until the fixed costs such as pumps and valves predominate in the cost per unit collector area figure.

The arrows in Fig. 11-3 show the optimization procedure. The incremental return of 0.5776 GJ/m² occurs at about $S = 0.50$ m²/GJ, sizing the system to carry 59% of the annual demand. The overall annual return at that same $S$ is 1.18 GJ/m², as read from the upper curve.* The payback ratio from this return is

$$R(i, 30) = \frac{\$200/m^2}{1.18 \text{ GJ}/m^2} \times \frac{GJ}{\$13.89} = 12.20$$

which, using Fig. 11-1, is seen equivalent to a discount rate of 0.0717. The effective ROI discount rate is then, by eq. 11-9,

$$i_c = (1 + 0.0717)(1 + 0.05) - 1 = 0.1253$$

which for the investor is equivalent to an interest rate of

$$i = \frac{0.1253}{0.7} = 0.1790$$

in an ordinary interest-bearing taxable account.†

The NPV of the investment is the present value of the annual returns less the capital cost. On a per unit area basis,

$$NPV = \left(\frac{1.18 \text{ GJ}}{m^2} \times \frac{\$346.25}{GJ} - \frac{\$200}{m^2}\right) = \$208.60/m^2$$

Per unit of annual demand, the NPV is

$$NPV = \left(\frac{1.18 \text{ GJ}}{m^2} \times \frac{\$346.25}{GJ} - \frac{\$200}{m^2}\right) \frac{0.50 \text{ m}^2}{GJ} = \$104.30/GJ$$

The actual optimized collector area is

$$A_c = \frac{0.50 \text{ m}^2}{GJ} \times 82.93 \text{ GJ} = 41.46 \text{ m}^2$$

or, in Engineering units, 446 ft². Therefore, the absolute value of the NPV is

$$NPV = \frac{\$208.60}{m^2} \times 41.46 \text{ m}^2 = \$8649$$

or, using the NPV per unit of annual demand,

$$NPV = \frac{\$104.30}{GJ} \times 82.93 \text{ GJ} = \$8649$$

**Other ways to optimize.** The previous example demonstrates a quick way to find the economic optimum. Direct maximization of the net present value gives

---

* Notice that the overall return times the collector area $S$ equals the annual $\sum f_a$. This is useful for checking construction of the overall return curve, since $\sum f_a$ can also be taken directly from the $\sum f_a$ versus $S$ performance curve.

† The payback time here is $R(i = 0.1790, n = 30) = 5.55$ yr. This number can be shown to be independent of the interest rate and project life span, and it is therefore not a suitable measure of economic worth.

the same result with somewhat more effort by using only the overall return line from Fig. 11-3. When plotted, the NPV versus $S$ curve gives not only the optimum but also suboptimal values showing how closely the optimum need be approached to gain reasonable financial benefits.

### EXAMPLE 11-8

Find the economic optimum area using the net present value (NPV) of the investment at several values of $S$.

For $S = 0.4 \text{ m}^2/\text{GJ}$, the overall return curve of Fig. 11-3 gives a return of $1.32 \text{ GJ/m}^2$.

$$\text{NPV} = \left(\frac{1.32 \text{ GJ}}{\text{m}^2} \times \frac{\$13.89}{\text{GJ}} \times 24.93 - \frac{\$200}{\text{m}^2}\right)\frac{0.4 \text{ m}^2}{\text{GJ}} = \$102.83/\text{GJ}$$

For $S = 0.8 \text{ m}^2/\text{GJ}$, return is $0.91 \text{ GJ/m}^2$:

$$\text{NPV} = (0.91 \times \$13.89 \times 24.93 - \$200)0.8 = \$92.09/\text{GJ}$$

Other values, including that for the optimum from Example 11-7, have been plotted in Fig. 11-4. The quick graphical procedure of Example 11-7 is quite accurate, and in good agreement with Fig. 11-4 at $S = 0.50 \text{ m}^2/\text{GJ}$.

The optimization method used in Fig. 11-3 can also be implemented directly on Fig. 11-2 using just two triangles. Draw a line from the origin to the $S = 1 \text{ m}^2/\text{GJ}$ line intersecting at a value on the ordinate axis equal to the incremental return in $\text{GJ/m}^2$. Using triangles, transfer that slope upward until it is tangent to the $\sum f_a$ versus $S$ curve. Read the optimum $S$ at the tangent point. Draw a second line from the origin through that point to the $S = 1 \text{ m}^2/\text{GJ}$ line, where the overall return in $\text{GJ/m}^2$ is read.

The three lines needed for this procedure are lightly dotted on Fig. 11-2, numbered 1, 2, and 3. While this method is faster for a given case it is not as consistent and convenient for repetitive use as the overall and incremental return plots of Fig. 11-3.

## Optimization of the Storage Capacity

Figure 10-5 shows the effect of storage on system performance as presented by three $\sum f_a$ versus $S$ curves. To optimize the storage capacity the $S$ optimization procedure just described is applied for each of the three storage capacities, producing three cost-optimized systems each with its own optimal $S$ and optimal performance $\sum f_a$. Each is then evaluated for net present value, and an intermediate storage value is then found graphically that will give the best NPV of all. Then the corresponding optimal $S$ is found for the new storage value from a plot of optimal $S$ versus $mc$ prepared from the first three cases.

The procedure used for drawing the thermal return curves in Example 11-7 can be simplified further by preparing all graphs on a single piece of paper. Figure

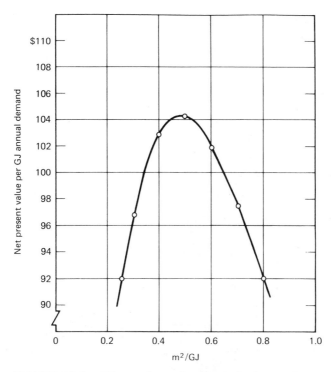

**FIGURE 11-4.** Direct determination of the optimum collector area (Example 11-8).

11-5 has been thus prepared from Fig. 10-5 (Table 10-2) data by first plotting $S$ versus $\sum f_a$ for each storage quantity. Overall and incremental return lines are then constructed as in Fig. 11-3, but this time on the same graph so that the need to refer to actual numbers is completely avoided. The graphical work must be done consistently using carefully smoothed curves because the optimum is usually quite shallow.

**EXAMPLE 11-9**

Find the optimal storage capacity using data from the three sets of return curves in Fig. 11-5 that have been derived from the data of Example 10-2 (Table 10-2). The cost of the storage is $1500 for a 1000 gal (3.785 m³) tank, and other system costs are $163.74/m².

First the storage costs are put on a collector area basis: The cost of water storage is

$$\frac{\$1500}{1000 \text{ gal}} \times \frac{\text{gal}}{8.33 \text{ lb}} \times \frac{\text{lb}}{0.454 \text{ kg}} = \$0.3966/\text{kg}$$

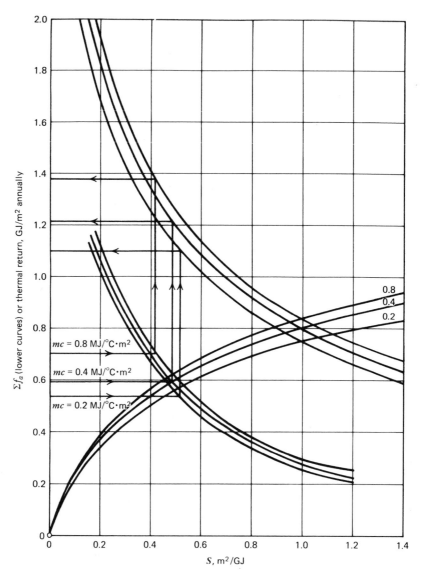

**FIGURE 11-5.** Graphical performance, incremental, and overall return for the data of Example 10-2 (Table 10-2) that will be used to perform an optimization for storage capacity (Example 11-9).

Calculating the area-normalized cost of storage for the three storage quantities, and assuming cost is proportional to capacity,

for $mc = 0.2$ MJ/$°C \cdot m^2$     $\dfrac{0.2 \text{ MJ}}{°C \cdot m^2} \times \dfrac{kg \cdot °C}{4.19 \text{ kJ}} \times \dfrac{0.3966}{kg} \times \dfrac{10^3 \text{ k}}{M} = \$18.93/m^2$

for $mc = 0.4$ MJ/$°C \cdot m^2$     $0.4 \times \$0.3966 \times 10^3/4.19 = \$37.86/m^2$

for $mc = 0.8$ MJ/$°C \cdot m^2$     $0.8 \times \$0.3966 \times 10^3/4.19 = \$75.72/m^2$

so that the total cost is

$$\$167.08 + \$18.93 = \$186.01/m^2 \qquad \text{for } mc = 0.2 \text{ MJ/}°C \cdot m^2$$
$$\$167.08 + \$37.86 = \$204.94/m^2 \qquad \text{for } mc = 0.4 \text{ MJ/}°C \cdot m^2$$
$$\$167.08 + \$75.72 = \$242.80/m^2 \qquad \text{for } mc = 0.8 \text{ MJ/}°C \cdot m^2$$

Then the marginal thermal returns of a unit collector area are calculated. The marginal capital cost of 1 GJ annually has already been determined as \$346.25 (Example 11-6). Therefore, the marginal *annual thermal* returns per unit collector area (instead of per dollar) are, respectively,

$$\frac{\$186.01}{m^2} \times \frac{GJ}{\$346.25} = 0.5372 \text{ GJ}/m^2$$

$$\frac{\$204.94}{m^2} \times \frac{GJ}{\$346.25} = 0.5919 \text{ GJ}/m^2$$

$$\frac{\$242.80}{m^2} \times \frac{GJ}{\$346.25} = 0.7012 \text{ GJ}/m^2$$

To compare the three investments the marginal returns calculated above are entered as three horizontal lines on Fig. 11-5. The intersection with their incremental return lines determines the optimum $S$ for each case. Projecting the points of intersection upward to their respective overall return lines gives three values for the overall return, each of which is now financially optimized so that further increase in area results in an incremental rate of return below the marginal value. The net present value is then calculated for each storage capacity case. A plot of NPV versus $mc$ is made and the optimum storage taken at the maximum NPV. The corresponding $S$ is then found for the optimum $mc$ by a plot of optimum $S$ versus $mc$.

For $mc = 0.2$ MJ/$°C \cdot m^2$ the total thermal return corresponding to a marginal return of 0.5372 GJ/$m^2$ is graphically determined by the above procedure to be 1.10 MJ/$m^2$ annually, and the optimum $S$ is 0.52 $m^2$/GJ demand. Since the present value of 1 GJ annually is \$346.25, with the system cost \$186.01/$m^2$, the net present value of that investment is

$$\text{NPV}_{(mc = 0.2 \text{ MJ/}°C \cdot m^2)} = \left( \frac{1.10}{m^2} \times \frac{\$346.25}{GJ} - \frac{\$186.01}{m^2} \right) \frac{0.52 \text{ m}^2}{GJ}$$

$$= \$101.33/\text{GJ demand}$$

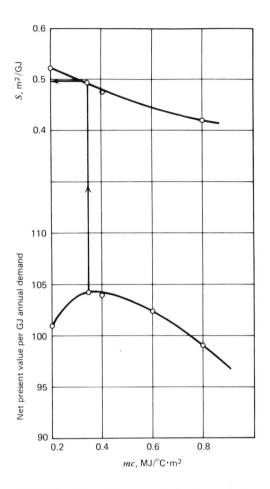

**FIGURE 11-6.** Direct optimization of the storage capacity and the corresponding collec-tor area (Example 11-9).

Similarly, the other storage capacities give

$$\text{NPV}_{(mc = 0.4 \text{ MJ/°C·m}^2)} = (1.215 \times \$346.25 - \$204.94)0.482 = \$103.99/\text{GJ}$$

$$\text{NPV}_{(mc = 0.8 \text{ MJ/°C·m}^2)} = (1.381 \times \$346.25 - \$242.80)0.422 = \$99.33/\text{GJ}$$

Figure 11-6 is a plot of storage capacity versus NPV including values for $mc = 0.6$ MJ/°C·m², which have not been detailed. The optimum storage lies between 0.3 and 0.4 MJ/°C·m². The case in Figs. 11-2 and 11-3 for which

$mc = 0.3478$ MJ/$°C \cdot m^2$ is therefore essentially at the optimum. The cost of the system is therefore as in Example 11-7:

$$\$167.08 + \frac{0.3478 \times 0.3966}{4.19} \times 1000 = \$200/m^2$$

and the NPV and ROI are therefore as in Example 11-7.

The actual storage mass is

$$M = \frac{(mc)A_c}{c} = \frac{0.3478 \text{ MJ/°C} \cdot m^2}{4.19 \text{ kJ/kg} \cdot °C} \times 41.46 \text{ m}^2 \times \frac{1000 \text{ k}}{M} = 3441 \text{ kg}$$

or 3441 L. In Engineering units the optimum storage volume is

$$3441 \text{ L} \times \frac{\text{gal}}{3.785 \text{ L}} = 909 \text{ gal}$$

## Optimization of the Heat Exchanger Area

The size of heat exchanger installed between the collector and storage tank depends on the cost of the exchanger. If the heat exchanger adds little to system costs on a dollars per unit collector area basis, a high heat exchanger factor with a large heat exchange area will be economical. Conversely, when the heat exchanger is expensive, smaller $F_x$ values and heat exchanger area will be optimal.

**Costs.** The basis for heat exchanger costs is the unit *collector* area. As in previous work, there is little need (once costs are available) to be concerned with a given structure because normalized values provide a much better "feel" of the system and permit many simplifications.

**Analytical Technique.** Equation 7-21 gives the exact optimized heat exchanger area $U_x a_x$ or $U_x A_x$ if there are equal flow capacity rates in the storage and collector loops. If the ratio of flow capacity rates is less than about 3:1, the expression still predicts the optimized area with good accuracy over the normal range of collector characteristics. Ordinarily this calculation will be made early in the design of a system and will be the basis for determination of $F_x$.

### EXAMPLE 11-10

The cost of the heat exchanger for the system of Example 10-2 as optimized in Example 11-7 is $414.27 for the building proposed in Example 9-3. Using $F_x = 0.95$ for that system, $U_x a_x = 79.62$ W/$°C \cdot m^2$ (14.02 Btu/hr °F ft²) (Example 7-5). If the other system costs are $190.01/m² ($17.65/ft²) estimate the optimum heat exchanger size using eq. 7-21.

From Example 11-7, the optimal $S$ is 0.50 m²/GJ.

$$A_c = \frac{0.50 \text{ m}^2}{\text{GJ}} \times 82.93 \text{ GJ} = 41.46 \text{ m}^2$$

$$U_x A_x = U_x a_x A_c = \frac{79.62 \text{ W}}{°C \cdot m^2} \times 41.46 \text{ m}^2 = 3301 \text{ W/°C}$$

$$C_{U_x A_x} = \frac{\$414.27}{3301 \text{ W/°C}} = \frac{\$0.1255}{\text{W/°C}}$$

The ratio of flow capacity rates is $w_l c_l / w_s c_s = 23.67/14.32 = 1.65$ (see Example 7-5), which is well under 3:1. From Example 5-8, $F_i U_L = 5.069$ W/°C·m². From eq. 7-21b,

$$U_x a_x = \sqrt{\frac{F_i U_L C_c}{C_{U_x A_x}}} = \sqrt{\frac{5.069 \text{ W}}{°C \cdot m^2} \times \frac{\$190.01}{m^2} \times \frac{\text{W/°C}}{\$0.1255}} = 87.60 \text{ W/°C} \cdot m^2$$

When costs are available eq. 7-21 is a much better method for heat exchanger sizing than is sizing to an arbitrary value of $F_x$ as was done in Chapter 7.

**Optimization of Heat Exchanger Area by Performance.** The technique of the previous example is simple and ordinarily recommended. By a more careful analysis of performance, a system can be optimized for other changing parameters as the need arises. This will be demonstrated in the next example by optimizing the heat exchanger area. This will serve to develop faith in eq. 7-21, which has been presented without proof, as well as to demonstrate a general technique.

The starting point is the analysis already made, and costs already developed. Then the size of the component to be optimized will be changed to several arbitrary values. Using techniques from Chapter 10, the new annual performance generally can be determined without extensive recalculations. Then a new optimum area (expressed as $S$) is determined for each new situation. Reoptimization for storage capacity is usually not necessary, since that optimum is generally quite broad. Net present value is then calculated for each situation from a plot of the component size versus net present value, an optimum can be chosen, and this will have a corresponding optimum $S$.

**EXAMPLE 11-11**

Optimize the heat exchanger from Example 11-10 by consideration of the performance of exchangers having 0.5, 1.5, and 2.0 times the nominal $U_x a_x$.

(a) Find the marginal annual return for all four systems, expressed as GJ$_{annually}$/ m²$_{collector}$ (M Btu/ft²).

Using the heat exchanger cost from Example 11-10 of $0.1255/(W/°C), and $190.01/m² for collector and other system costs, the normalized cost of the entire system can be expressed per unit collector area as

$$C_c = \$190.01 + \$0.1255(U_x a_x)$$

or for the four systems,

$$\times 1.0: \quad \$190.01 + \$0.1255 \times 79.62 \times 1.0 = \$200/m^2$$
$$\times 1.5: \quad \$190.01 + \$0.1255 \times 79.62 \times 1.5 = \$205/m^2$$
$$\times 0.5: \quad \$190.01 + \$0.1255 \times 79.62 \times 0.5 = \$195/m^2$$
$$\times 2.0: \quad \$190.01 + \$0.1255 \times 79.62 \times 2.0 = \$210/m^2$$

The marginal return is calculated by dividing the system cost by the present value of 1 GJ annually delivered for the life of the system, which has already been established as \$346.25/GJ (Example 11-6).

$$\times 1.0: \quad \$200.00/m^2 \times GJ/\$346.25 = 0.5776 \ GJ_{annually}/m^2$$
$$\times 1.5: \quad \$205.00/m^2 \times GJ/\$346.25 = 0.5921 \ GJ_{annually}/m^2$$
$$\times 0.5: \quad \$195.00/m^2 \times GJ/\$346.25 = 0.5632 \ GJ_{annually}/m^2$$
$$\times 2.0: \quad \$210.00/m^2 \times GJ/\$346.25 = 0.6065 \ GJ_{annually}/m^2$$

## Performance Parameter Changes

In this case and many others of practical interest, the component-size change directly affects only the heat transfer factor $F_i'$. The new factor is analytically evaluated using appropriate techniques that have already been established. In the case of the heat exchanger, new $\varepsilon$ and $F_x$ values are calculated.

### EXAMPLE 11-11$b$

For all four heat exchangers calculate new $F_x$ values. Values from Example 7-5 are

$$(wc)_t = 85.15 \ W/^{\circ}C \cdot m^2 \ (15.00 \ Btu/hr \ ^{\circ}F \ ft^2)$$
$$(wc)_s = 140.75 \ W/^{\circ}C \cdot m^2 \ (24.79 \ Btu/hr \ ^{\circ}F \ ft^2)$$
$$F_i U_L = 5.069 \ W/^{\circ}C \cdot m^2 \ (0.8927 \ Btu/hr \ ^{\circ}F \ ft^2)$$

For $\times 1.5$:

$$U_x a_x = 79.62 \times 1.5 = 119.43 \ W/^{\circ}C \cdot m^2$$

$$B = 119.43\left(\frac{1}{85.15} - \frac{1}{140.75}\right) = 0.5541$$

$$\varepsilon = \frac{1 - e^{-0.5441}}{\left(1 - \dfrac{85.15}{140.75} e^{-0.5441}\right)} = 0.6521$$

$$F_x = \frac{1}{1 + \dfrac{5.069}{85.15}\left(\dfrac{1}{0.6521} - 1\right)} = 0.9692$$

For other values see Table 11-2.

**TABLE 11-2.** Heat Exchanger Optimization (Example 11-11)

| HX Case | $C_e$ | Marginal Return | $U_x a_x$ | $B$ | $\varepsilon$ | $F_x$ | $F = \dfrac{F_{x(old)}}{F_{x(new)}}$ | Entering Marginal Return Value | Leaving Overall Return Value | Correct Overall Return Value | Leaving S Value | Correct S Value | $\Sigma f_a$ |
|---|---|---|---|---|---|---|---|---|---|---|---|---|---|
| ×1.0 | 200 | 0.5776 | 79.62 | 0.3694 | 0.5308 | 0.9500 | 1.0000 | 0.5776 | 1.180 | 1.180 | 0.500 | 0.500 | 0.590 |
| ×1.5 | 205 | 0.5921 | 119.40 | 0.5541 | 0.6521 | 0.9692 | 0.9802 | 0.5804 | 1.183 | 1.207 | 0.497 | 0.487 | 0.588 |
| ×0.5 | 195 | 0.5632 | 39.81 | 0.1847 | 0.3393 | 0.8961 | 1.0601 | 0.5971 | 1.200 | 1.139 | 0.480 | 0.509 | 0.576 |
| ×2.0 | 210 | 0.6065 | 159.24 | 0.7387 | 0.7346 | 0.9789 | 0.9704 | 0.5886 | 1.190 | 1.226 | 0.490 | 0.4755 | 0.583 |

**Optimized System Performance.** Changes in the area- and storage-optimized performance curve of Fig. 11-2 as a result of changes in $F_i'$ or $F_x$ are determined by eq. 10-11, which gives new $S$ values for use with the *same* $\Sigma f_a$ values, but for the new $F_i'$ or $F_x$; and by eq. 10-12 which gives new $mc$ values. This technique permits use of the existing system calculations simply by substitution of new values for $S$ and $mc$. For the modest changes in $F_x$ given in Table 11-2, the slightly different value of $mc$ to which these figures apply can be neglected.

New plots like those of Figs. 11-2 and 11-3 can now be prepared for each of the three new heat exchanger factors by using eq. 10-12 with the new $F_x$ values in Table 11-2. It is much simpler, however, to note that it is equivalent to multiply both of the coordinates in Fig. 11-3 and the abscissa in Fig. 11-2 by the factor $F = F_{x(old)}/F_{x(new)}$ derived from eq. 10-11.* This factor is tabulated in Table 11-2. The overall and incremental curves can then be used exactly as before, provided only that this factor $F$ be properly applied to values entering and leaving the graph in order to read it correctly.

### EXAMPLE 11-11c

For Table 11-2 values, find optimal $S$, overall return, and $\Sigma f_a$ for the four heat exchanger areas using the plots in Fig. 11-3 developed for $mc = 0.3478$ MJ/ °C·m² (17 Btu/°F ft²).

For the ×1.5 case, the factor $F$ is 0.9802 (Table 11-2).
Entering Fig. 11-3 with a marginal return of

$$0.5921 \times 0.9802 = 0.5804 \text{ GJ/m}^2$$

gives an $S$ value from Fig. 11-3 of 0.497 m²/GJ and an overall return that is read as 1.183 GJ/m². The correct overall return is

$$1.183/0.9802 = 1.207 \text{ GJ/m}^2$$

The correct $S$ value is

$$0.497 \times 0.9802 = 0.487 \text{ m}^2/\text{GJ}$$

* Changes in $F_i$ are treated similarly.

The product of these values is $\sum f_a$:

$$0.487 \times 1.207 = 0.588$$

which checks with the Fig. 11-2 entry for $S = 0.497$ m²/GJ, as it should. Values for other cases are calculated analogously and are entered in Table 11-2.

Net present values of investment for all four cases follow as before from the values in Table 11-2:

$$\times 1.0: \quad (1.18 \times 346.25 - 200.00)0.50 = \$104.29/GJ_{\text{annual load}}$$
$$\times 1.5: \quad (1.207 \times 346.25 - 205.00)0.487 = \$103.69/GJ_{\text{annual load}}$$
$$\times 0.5: \quad (1.139 \times 346.25 - 195.00)0.509 = \$101.48/GJ_{\text{annual load}}$$
$$\times 2.0: \quad [1.226 \times 346.25 - 210.00]0.4755 = \$102.00/GJ_{\text{annual load}}$$

Net present value, optimized area, and $\sum f_a$ are plotted versus the proportional change in exchanger area in Fig. 11-7. The optimum heat exchanger area is seen as $1.1 \times 79.62 = 87.6$ W/°C·m², just as in Example 11-10. That is so little change that the original heat exchanger area is retained, and with it the system area

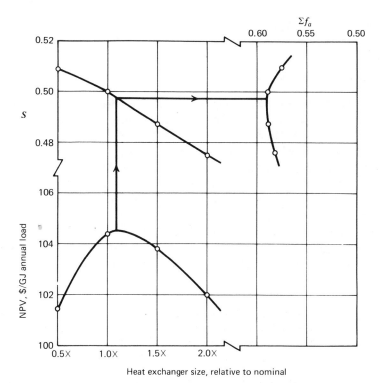

**FIGURE 11-7.** Direct optimization of the heat exchanger area and the corresponding normalized collector area and annual solar contribution (Example 11-11).

$S = 0.5$ m²/GJ. Since chances are very small that the new heat exchanger area would change system economics sufficiently to affect the storage optimization, this example concludes the design problem.

The actual heat exchanger $U_x A_x$ is the product of the nominal $U_x a_x$, the area, and the optimal proportion of that area installed. In this case

$$U_x A_x = \frac{79.62 \text{ W}}{°C \cdot m^2} \times 41.46 \text{ m}^2 \times 1.0 = 3301 \text{ W/°C}$$

and, from the operating information in Example 7-5, the "small" flow of anti-freeze is

$$W_s = \frac{(wc)_s}{c} A_c = \frac{85.15 \text{ W}}{°C \cdot m^2} \times \frac{kg \cdot °C}{3.52 \text{ kJ}} \times 41.46 \text{ m}^2 \times \frac{J}{W \cdot s} \times \frac{k}{1000} = 1.00 \text{ kg/s}$$

which is about

$$\frac{1.00 \text{ kg}}{s} \times \frac{m^3}{1035 \text{ kg}} \times \frac{1000 \text{ L}}{m^3} = 0.9662 \text{ L/s}$$

or

$$\frac{0.9662 \text{ L}}{s} \times \frac{60 \text{ s}}{\min} \times \frac{gal}{3.785 \text{ L}} = 15.3 \text{ gal/min}$$

## Linearization of Costs

In some cases the costs of components may not be linear with their size. There may be a large fixed cost, as with the pumps and controls, or the equipment may cost proportionally more in a small size.

Costs often vary with the 0.6 power of the size, a common empirical relationship observed with capital equipment such as heat exchangers and storage tanks. Such costs may be linearized for the range of application by splitting them into a fixed plus a variable component applying within the anticipated range of usage. This will speed the optimization process by reducing the changes in unit costs necessary for successive reoptimizations, but it will not change the final answer.

The economical range of the optimum will be larger when using incremental costs, and this will more accurately represent the true situation.

### EXAMPLE 11-12

Figure 11-8 shows a cost curve for the storage tanks of Example 11-9. It rises as the 0.6 power of the tank capacity. If the anticipated range of usage is 1000 to 2000 gal, find the fixed and variable costs per unit collector area as a function of $mc$, the area-normalized storage capacity.

A straight line is drawn in Fig. 11-8 along the average slope of the cost line at 1500 gal and extended to the $y$-axis, which it intersects at $750. This is the fixed portion of the cost. The variable portion is given by the slope of the line; in this case the triangle drawn gives a slope of $1162/1500 gal = $0.7747/gal.

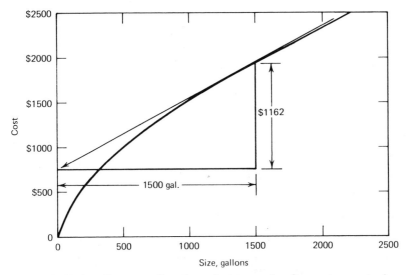

**FIGURE 11-8.** Cost as a function of storage size for a storage tank, showing the splitting of costs in the 1000 to 2000 gal range into fixed and variable portions. For this curve the cost rises as the 0.6 power of the size (Example 11-12).

The cost may be converted to \$/kg by dividing by 3.785 kg/gal giving

$$C_{\text{total}} = \$750 + \$0.2047M$$

where $M$ is the mass of the storage in kilograms. Since $M = (mc)A_c/c$, then in dimensional form,

$$M = \frac{(mc)A_c}{4.19/1000} = 238.7(mc)(A_c)$$

where 4.19/1000 is the specific heat of water in MJ/kg·°C, $mc$ is in MJ/°C·m², and $A_c$ is in m². The total cost is then

$$C_{\text{total}} = \$750 + \$0.2047[238.7(mc)(A_c)]$$
$$C_{\text{total}} = \$750 + \$48.86(mc)(A_c)$$

and if $mc = 0.3478$ MJ/°C·m², as in Example 11-9,

$$C_{\text{total}} = \$750 + \$16.99A_c$$

and therefore the variable portion of the storage cost is \$16.99/m², and the fixed portion \$750.

Splitting costs into fixed and variable components is optional, and can be done for any or all the system components. Once determined the fixed costs are summed and retained but only the variable costs are used in the optimization analyses. The fixed costs do not need to be incorporated, since by definition they do not affect the optimizations—fixed costs are unchanging as the installed normalized area $S$, area $A_c$, storage capacity $mc$, or heat exchanger area $U_x a_x$ are varied. The fixed costs do affect the monetary value of the NPV and ROI, but they can be conveniently ignored as an unnecessary complication until the very last step in analysis, after all the optimizations are finished. Then final record values of NPV and ROI are calculated. These may be expressed as the actual fixed costs for a particular structure, or the fixed costs per unit annual load for a more generalized study.

If a tax credit of a fixed amount for one or more years is available, its effect on the investment can also be evaluated at this point.

**EXAMPLE 11-13**

The system optimized in Example 11-7 has been shown to be optimized not only for normalized area but also for storage capacity and heat exchanger area. Assume that the $200/m^2$ collector costs represent variable costs only, and that there is an additional $1592 fixed cost for this installation, which involves an annual load of 82.93 GJ. The total collector area for the optimal $S$ of 0.50 m²/GJ has been determined (Example 11-7) as 41.46 m². The NPV, now considered to apply to variable costs only, is $104.30/GJ (Example 11-7). The variable system cost is $200/m² and the annual return is 1.18 GJ/m². What are the new NPV and ROI?

*Method A* preserves the basis of previous work that was normalized per unit annual load:

$$L_a = 82.93 \text{ GJ}$$

The fixed cost per unit annual load is then

$$\frac{C_{\text{fixed}}}{L_a} = \frac{\$1592}{82.93 \text{ GJ}} = \frac{\$19.20}{\text{GJ}}$$

in which case

$$\text{NPV} = \frac{\$104.30}{\text{GJ}} - \frac{\$19.20}{\text{GJ}} = \$85.10/\text{GJ}$$

or, on the basis of the total system installed,

$$\text{NPV} = \frac{\$85.10}{\text{GJ}} \times 82.93 \text{ GJ} = \$7057$$

For the optimized system,

$$\text{Unit cost} = \frac{\$200}{m^2} + \frac{\$19.20}{\text{GJ}} \times \frac{\text{GJ}}{0.50 \text{ m}^2} = \$238.40/m^2$$

The return can be computed from

$$R(i, 30) = \frac{\$238.40/m^2}{1.18 \text{ GJ}/m^2} \times \frac{\text{GJ}}{\$13.89} = 14.55$$

*Method B* is simpler and more direct.

$$\text{NPV} = \frac{\$104.30}{\text{GJ}} \times 82.93 \text{ GJ} - \$1592 = \$7058$$

$$\text{System cost} = \frac{\$200}{\text{m}^2} \times 41.46 \text{ m}^2 + \$1592 = \$9884$$

$$R(i, 30) = \frac{\$9884}{1.18 \text{ GJ/m}^2} \times \frac{\text{GJ}}{\$13.89} \times \frac{1}{41.46 \text{ m}^2} = 14.55$$

By either method, using Fig. 11-1 the discount rate is 5.49%, and the effective ROI discount rate is then

$$i_c = (1 + 0.0549)(1 + 0.05) - 1 = 0.1077$$

which is equivalent for this investor to

$$i = \frac{0.1077}{0.7} = 0.1538$$

or 15.38%, still a desirable return.*

### EXAMPLE 11-14

There is a tax credit of $500 per year available for three years and $175 for a fourth year for the system in Example 11-13.
Evaluate the effect on the investment.

Since the investor's tax bracket is 30%, in terms of incremental income, the tax credit is worth $500/(1 − 0.3) = $714.29 per year for three years, and $175/0.7 = $250 for the fourth year. The investment is reduced by the face amounts discounted to their present value at the effective interest rate (Example 11-6), using eqs. 11-5 and 11-3 to give the discounted value of the tax credits as

$$\$500\left[\frac{1 - (1 + 0.063)^{-3}}{0.063}\right] + \frac{\$175}{(1 + 0.063)^4} = \$1329 + \$137 = \$1466$$

$$\text{System cost} = \$9884 - \$1466 = \$8418$$

$$R(i, 30) = \frac{8418}{1.18} \times \frac{1}{13.89} \times \frac{1}{41.46} = 12.39$$

$$i_d = 0.0702$$
$$i_c = (1 + 0.0702)(1 + 0.05) - 1 = 0.1237$$

which for this investor is equivalent to

$$i = \frac{0.1237}{0.7} = 0.1767$$

or 17.7%.†

* The payback time here is $R(i = 0.1538, n = 30 \text{ yr}) = 6.41$ yr.
† The payback time is reduced to $R(i = 0.1767, n = 30 \text{ yr}) = 5.62$ yr.

## Optimization of Conservation Measures

In Example 11-6 the dollars invested to save a unit of heat annually were calculated from the cost of energy and the payback ratio. This same *marginal investment* is appropriate for conservation measures as well.*

The thermal return from conservation measures can usually be explicitly calculated in a few steps. First, one of the techniques of Chapter 2 is applied to calculate the heat loss coefficient, and eq. 2-14 is used to determine the annual heat loss flux $Q_a/A$. This is plotted versus the extent of the variable that is to be optimized, showing, for instance, the effect of insulation thickness or a number of panes of glass on resultant annual heat loss. This plot is similar in function to Fig. 11-2, but it has a negative slope. It may be regarded as a plot of annual losses in $GJ/m^2$ (on the ordinate) versus *extent*$/m^2$ (on the abscissa). The optimal extent of the conservation measure can then be determined by exactly the same graphical techniques used for determination of the optimal collector area in Example 11-7, matching the incremental return with the slope of the curve.

When the relationship between the annual heat loss and the extent of the conservation measure can be expressed analytically, the optimization can be carried out analytically with very little effort. For instance, consider the problem of optimal insulation for the solar house. The annual heat loss is given by combining eqs. 2-16 and 2-14 to give

$$\frac{Q_a}{A} = \frac{D_a}{\dfrac{L_1}{k_1} + B} \tag{11-11}$$

where $L_1$ and $k_1$ are the thickness and thermal conductivity of the insulation and $B$ is the sum of all the other appropriate $R$ values for the wall.

Differentiating eq. 11-11 with respect to the thickness $L_1$ gives

$$-\frac{d(Q_a/A)}{dL_1} \equiv s = \frac{D_a}{k_1\left(\dfrac{L_1}{k_1} + B\right)^2} \equiv \frac{D_a r_1}{(L_1 r_1 + B)^2} \tag{11-12}$$

where $r_1$ is the thermal resistivity, which is either tabulated directly in Table 2-2 or derivable from the thickness and the listed $R$ value ($r = R/L$). The slope $s$, which is the annual thermal return per unit thickness of insulation, is to be set equal to the marginal thermal return. It is calculated from the marginal investment (previously determined) and the cost of the insulation. Ideally this is the cost of additional insulation at the optimum thickness, but insulation costs can usually be approximated as being directly proportional to thickness.

---

* It is in fact conservative in assuming a finite lifetime for them.

**EXAMPLE 11-15**

Installed fiberglass insulation is available at $45/m³ in a 12-in. thickness. What is the economic thickness for use at Hartford ($D_a$ = 3527 °C-days) if the system is optimized as in Examples 11-6 to 11-14, for which the marginal thermal return was $346.25/GJ annual return?

From Table 2-2, the $R$ value for 6-in. fiberglass batt is 3.32 m² · °C/W.

$$r = \frac{R}{L} = \frac{3.32 \text{ m}^2 \cdot \text{C}}{\text{W}} \times \frac{1}{6 \text{ in.}} \times \frac{39.37 \text{ in.}}{\text{m}}$$

$$r = 21.78 \text{ m} \cdot °\text{C/W}$$

The marginal investment permits an annual *incremental* thermal return of

$$s = \frac{\text{GJ}}{\$346.25} \times \frac{\$45}{\text{m}^3} = 0.13 \text{ GJ/m}^3 \quad \text{or} \quad 0.13 \frac{\text{GJ/m}^2}{\text{m}_{\text{thickness}}}$$

When eq. 11-12 is solved for $L_1$, if $B \ll L_1/k_1$, this gives

$$L_1 = \sqrt{\frac{D_a}{r_1 s}}$$

so that the optimal insulation thickness is

$$L_1 = \sqrt{3527 \text{ °C-days} \times \frac{24 \text{ h}}{\text{day}} \times \frac{3600 \text{ s}}{\text{h}} \times \frac{\text{W}}{21.78 \text{ m} \cdot °\text{C}} \times \frac{\text{GJ}}{10^9 \text{ W} \cdot \text{s}} \times \frac{\text{m}^3}{0.13 \text{ GJ}}}$$

$$= 0.328 \text{ m or about } 12.91 \text{ in.}$$

For such thick insulation, the other series resistances $B$ can indeed be neglected.

As of this writing, insulation is available even more cheaply than demonstrated in Example 11-15. Even thicker insulation may now be economical, as a result of the high value placed on thermal energy when a uniform energy cost escalation is presumed.

For the general case, it is easy to use the basic method of Example 11-15 to derive a dimensional equation for the economical thickness of insulation:

$$L_1 = L_i \sqrt{\frac{D_a C_v R b}{R_i C_i}} - L_i \times \frac{R_{\text{existing}}}{R_i} \tag{11-13}$$

where $L_1$ = Economical thickness of added insulation, same units as $L_i$.

$L_i$ = Thickness of insulation that has a thermal resistance of $R_i$.

$D_a$ = Degree-days at the location, °C-days or °F-days, consistent with the units of $R_i$.

$C_v$ = Cost of the commercial unit of fuel, consistent with Table 2-9.

$R$ = Required payback ratio (or simple payback time, years).

$R_i$ = Thermal resistance of a thickness $L_i$ of insulation, measured in °C·m²/W or hr °F ft²/Btu.

$R_{existing}$ = Thermal resistance of the existing structure, in the same thermal units as $R_i$.

$C_i$ = Cost of a thickness $L_i$ of added insulation per m² or ft², consistent with units of $R_i$.

$b$ = Dimensional constant from Table 2-9, consistent with the units of $R_i$ and depending on the fuel used.*

### EXAMPLE 11-16

The partially insulated frame wall of Example 2-10 has an $R$ value of 1.412 °C·m²/W (8.03 hr °F ft²/Btu). Insulation is available in a 15.24 cm (6 in.) thickness for $3.23/m² (30¢/ft²) that has an $R$ value of 3.32 °C·m²/W (19hr °Fft²/Btu). Fuel cost at Burlington, Vermont is 95¢ per gallon of #2 fuel oil. What added thickness of insulation will return the original investment cost in 5 years?

Annual degree days at Burlington are given in Table 2-1 as 8269 °F-days, or 4594 °C-days.

For oil fuel with SI units $b$ is 0.000 726, or in Engineering units it is 0.000 212 6. By eq. 11-13,

$$L_1 = 15.24 \text{ cm} \sqrt{\frac{4594 \times \$0.95 \times 5 \times 0.000\,726}{3.32 \times \$3.23}} - 15.24 \text{ cm} \times \frac{1.412}{3.32}$$

$$L_1 = 18.52 - 6.48 = 12.04 \text{ cm}$$

Since there is not room for this much insulation in the wall, consideration may be given to adding more wall structure.

In Engineering units, by eq. 11-13,

$$L_1 = 6 \text{ in.} \sqrt{\frac{8269 \times 95¢ \times 5 \times 0.000\,212\,6}{19 \times 30¢}} - 6 \text{ in.} \times \frac{8.03}{19}$$

$$L_1 = 7.26 - 2.54 = 4.73 \text{ in.}$$

## PROBLEMS

1. Consider a solar collector that costs $250/m² ($23.22/ft²). Its life is 6 years and there is no scrap value. If the interest rate is 10% and the collector returns $75/m² ($6.97/ft²) annually,
   (a) What is the present value of the annual returns using the payback ratio approach?
   (b) Prepare a table like Table 11-1 (Base Case) showing the annual returns and their present values. What are the total NPV and the ROI?

2. Repeat problem 1 if there is also an inflation rate of 4.5% annually.

* Units of $b$ are (Ms/day)/(MJ/fuel unit) or (hr/day)/(Btu/fuel unit).

3. Repeat problem 2 if there is an investment tax credit of $10/m$^2$ ($0.929/ft$^2$) and a scrap value of $20/m$^2$ ($1.858/ft$^2$).

4. Repeat problem 1 if the investor is in a 25% tax bracket and can deduct interest costs. Make an additional row showing the investor's tax savings, and discount their value to the present.

5. Repeat problem 1 if there are maintenance costs of $5/m$^2$ ($0.4645/ft$^2$) each year.

6. Use the plot from problem 10-1(c) to prepare a plot like Fig. 11-3 where the overall and incremental thermal return is plotted against $S$.

7. Assume that the cost of electric auxiliary heat is 4.5¢/kWh and that the collector costs $250/m$^2$ ($23.22/ft$^2$) and has a 25-year life. The interest rate is 10% and the fuel escalation rate is 4.5% annually.
   (a) What is the marginal thermal return?
   (b) What is the optimal system $S$ using the plot of problem 6?
   (c) What is the optimal annual solar participation?
   (d) What is the overall thermal return?
   (e) What are the NPV and ROI (per unit collector area)?
   (f) Summarize the final system sizes.

8. Confirm the methodology of problem 7 by plotting NPV versus $S$ (as in Fig. 11-4) for several $S$ values.

9. Using plots from problem 10-2, optimize the storage quantity if the tank costs $750/1000 gal.

10. The plot of problem 10-1(b) has been prepared for a drain-down–type solar collector [problem 9-1(b)]. The water flow capacity rate for collector $A$ is given at the start of the Chapter 9 problems. For a heat exchanger having $U_x A_x = 2100$ W/°C (3981 Btu/hr °F), the cost is $200. Assuming the cost for other sizes is proportional:
    (a) What is the optimal heat exchanger size (expressed as $U_x a_x$) if costs are as in problem 7? Use equal flow capacity rates in collector and storage loops.
    (b) Plot a new $S$ versus $\sum f_a$ performance curve.
    (c) Reoptimize the collector area ($S$) as in problem 7 using the curve generated in part (b), and compare your results with the analytical technique of Example 11-11.
    (d) Quantify all the parameters of the optimized system.

11. A storage tank costs $1800 for 1800 gal. Assuming the costs rise as the 0.6 power of the size, find the expression for fixed plus variable costs using the slope at 2000 gal.

12. Modify the ROI and NPV determined in problem 7 if additional fixed costs of $2000 are now considered.

**13.** Further modify the ROI and NPV if a tax credit of $1200 is available 1 year after construction.

**14.** Derive eq. 11-5 as suggested in the text.

**15.** If electricity costs 5¢/kWh, and insulation costs $25/m³ (70.8¢/ft³), what is the economical insulation thickness for the preferred fuel in your area if the fuel escalation rate is
(a) 0%,  (b) 5%,  (c) 10%

**16.** Oil costing $1.20 per gallon delivers 118 MJ (112 000 Btu) per gallon. How many panes of glass are economical if glass costs $12/m² ($1.11/ft²) for each additional pane? Use your own location, a 25 yr life, and fuel escalation rates of
(a) 5%,  (b) 10%,  (c) 20%.

## REFERENCES

1. F. A. Holland, F. A. Watson, J. K. Wilkinson, *An Introduction to Process Economics*, New York: Wiley, 1974.

# Solar Systems Design

○

The systems that provide solar heating and hot water have many more components than the collectors, heat exchangers, and storage tanks that have been the subject of the analytical designs explored thus far. All the components are finally assembled into a working system that varies considerably from application to application. In this chapter a number of these systems will be detailed and their rationale explained. Design procedures for the more conventional systems equipment such as fans, tubes, and ductwork will also be set forth, and auxiliary heating systems discussed.

## SWIMMING POOL HEATERS

Solar heaters for swimming pools are easily the most practical application for solar heating today. A swimming pool requires much low-temperature heat to maintain the water temperature at a minimum of 24 °C (75 °F) and preferably at 27 °C (80 °F). Supplying the needed heat with fossil fuels or electric heat is prohibitively expensive except where the price of natural gas has been held below market price by regulation. Even in these areas legislation is likely to prevent the use of such a valuable resource for such a pedestrian purpose and, when the resulting laws are finally enforced, the consumer will have no choice but to limit the swimming season according to the natural climate or install a solar-heating system.

A typical outdoor solar pool-heating system is detailed in Fig. 12-1. The pool's own pump is utilized to circulate pool water to the heaters, a function easily achieved since these pumps are always oversized so as to conveniently handle the pool-cleaning chore. The solar panels are installed in series with the filter, but downstream of it so that the normal buildup of pressure as the filter loads will not pressurize the panels. A check valve is installed in the line to the panels so that they will not drain into the filter and backflush the accumulated dirt back into the pool when the pump is not operating.

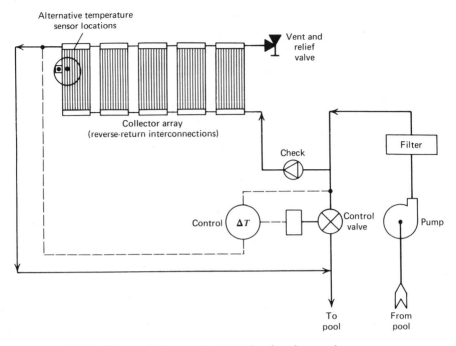

**FIGURE 12-1.** Schematic for a solar-heated swimming pool.

The pool heaters are actuated when the temperature of a sensor near the solar collector is higher than that of the pool, provided the temperature of the pool is lower than the user-set maximum for heating. Closing of the control valve then forces water to pass through the collector array. A vent valve is often provided to permit the removal of trapped air, although a properly designed array will have sufficient water flow to flush out accumulated air with little trouble. If the collectors cannot withstand the full pump-operating pressure, an elevated open vent with a standpipe or a pressure relief valve should also be installed.

The solar panels are often made from plastic (melt-blackened polypropylene) and generally have no cover plate, since the losses are small with the collector operating near-ambient temperatures. With no cover plates the collector stagnation temperature is low and the plastic is safe from melting. A plastic collector is also freeze-tolerant, expanding slightly to accommodate ice with no structural damage. Finally, of course, a plastic collector is cheap, perhaps half what a metal collector might be. Metal swimming pool collectors are used where the durability of the plastic collectors is in doubt* or where the collectors must be covered with glazing. This is necessary for spa or ski-center type operation in which higher-

* Plastic collectors are currently guaranteed for up to 10 years but that much actual operating experience has not yet been accumulated. The physical properties of the plastic would indicate a nearly infinite life.

than-normal pool temperatures are desired or colder-than-typical outdoor temperatures can be expected. Only copper collectors should be used, since aluminum quickly corrodes in the highly oxygenated swimming pool water. Since scale will precipitate from some kinds of water on heated metal, a positive shutoff of the water circulated to the collectors is desirable on shutdown.

Water flow to pool-heating solar collectors should always be as high as possible to keep the collector temperature low and the heat transfer factor high. If the flow is so high that the collectors run within a degree or so of the pool temperature, however, a control system operating on water temperature differential may call for a premature system shutdown. In that case, either the flow should be restricted somewhat or preferably the control sensor should be located in an uncooled portion of the collector or in a dummy collector section located nearby.

Since pool-heating collectors must be designed to handle as much flow as possible, flows within the collector and the array are almost always in parallel. Series flow is used only when the collectors are stacked two or three high in a billboard-type array. Figure 12-1 shows the "reverse-return" connections used for parallel flow.

Solar heating for indoor pools is more difficult because the air above the pool rather than the water is heated, to keep the relative humidity of the air low enough to avoid condensation within the roof structure.

Serpentine pipes (Fig. 5-2a) are sometimes used as a solar collector for pool heating but they require more installation area unless they are placed very closely together. When spaced apart, the pipes should be oriented up and down (with pipe runs north-south) to take advantage of the ability of round surfaces to intercept the same amount of radiant energy regardless of sun angle. The thinnest wall tubing available should be used to keep the heat transfer factor as high as possible, and air must be carefully vented from the system.

## System Sizing

It is very difficult to size a pool-heating collector array analytically because of the extreme effect of weather conditions on the pool heat losses. Wind can cause enormous loss, acting through both convection and evaporation of water from the pool surface. Nighttime radiative losses are also quite variable—large where the nights are cool and dry, but small where they are warm or humid. The direct solar gain to the pool is also difficult to estimate. All the ultraviolet and much of the infrared energy is absorbed in the first 5 cm or so of depth (within a few inches), and if the water surface is dormant this layer of heated water may lose energy to the surroundings at an increased rate. Even so, estimates of the effective $\tau\alpha$ for the pool with a light-colored bottom range up to 0.8.

An analysis of the pool heat loss situation has been made by Löf (1) for the midwestern United States with the wind at 0.35 m/s (8 mi/h), but this overestimates losses considerably for less windy locations. His tables for pool losses with

**TABLE 12-1a.** Calculated Equilibrium Swimming Pool Temperatures and Heat Required to Maintain 80 °F when Continuously Covered with Floating Transparent Plastic Film (1). Basis: 80 Percent Absorption of Solar Radiation, 8 mi/hr (0.35 m/s) Wind Velocity.[a]

| Mean Air Temperature, °F | Solar Radiation: 5.7 MJ/m²·day (500 Btu/ft²·day) | | 11.4 MJ/m²·day (1000 Btu/ft² day) | | 17.0 MJ/m²·day (1500 Btu/ft² day) | | 22.7 MJ/m²·day (2000 Btu/ft² day) | | 28.4 MJ/m²·day (2500 Btu/ft² day) | |
|---|---|---|---|---|---|---|---|---|---|---|
| | Eq'lb $T$, °F | Heat Req. Btu/ft² day | Eq'lb $T$, °F | Heat Req. Btu/ft² day | Eq'lb $T$, °F | Heat Req. Btu/ft² day | Eq'lb $T$, °F | Heat Req. Btu/ft² day | Eq'lb $T$, °F | Heat Req. Btu/ft² day |
| 40 | 46 | 2402 | 51 | 2002 | 57 | 1602 | 62 | 1202 | 67 | 802 |
| 50 | 56 | 1717 | 61 | 1317 | 67 | 917 | 72 | 517 | 77 | 117 |
| 60 | 66 | 1025 | 71 | 625 | 77 | 225 | 81 | 0 | 87 | 0 |
| 70 | 76 | 318 | 81 | 0 | 87 | 0 | 92 | 0 | 97 | 0 |
| 80 | 86 | 0 | 91 | 0 | 97 | 0 | 100+ | 0 | 100+ | 0 |

**TABLE 12-1b.** Calculated Equilibrium Swimming Pool Temperatures and Heat Required to Maintain 80 °F (for open pools). Basis: 80 Percent Absorption of Solar Radiation, 50 Percent Relative Humidity, 8 mi/hr (0.35 m/s) Wind Velocity.[a]

| Mean Air Temperature, °F | Solar Radiation: 5.7 MJ/m²·day (500 Btu/ft² day) | | 11.4 MJ/m²·day (1000 Btu/ft² day) | | 17.0 MJ/m²·day (1500 Btu/ft² day) | | 22.7 MJ/m²·day (2000 Btu/ft² day) | | 28.4 MJ/m²·day (2500 Btu/ft² day) | |
|---|---|---|---|---|---|---|---|---|---|---|
| | Eq'lb $T$, °F | Heat Req. Btu/ft² day | Eq'lb $T$, °F | Heat Req. Btu/ft² day | Eq'lb $T$, °F | Heat Req. Btu/ft² day | Eq'lb $T$, °F | Heat Req. Btu/ft² day | Eq'lb $T$, °F | Heat Req. Btu/ft² day |
| 40 | 41 | 4749 | 45 | 4349 | 49 | 3949 | 53 | 3549 | 56 | 3149 |
| 50 | 48 | 3910 | 51 | 3510 | 55 | 3110 | 58 | 2710 | 62 | 2310 |
| 60 | 54 | 3015 | 58 | 2615 | 62 | 2215 | 65 | 1815 | 68 | 1415 |
| 70 | 61 | 2027 | 64 | 1627 | 67 | 1227 | 70 | 827 | 73 | 427 |
| 80 | 67 | 820 | 69 | 420 | 72 | 20 | 75 | 0 | 78 | 0 |

[a] Reprinted by permission of the American Section of the International Solar Energy Society, Inc., the University of Delaware, Newark, Delaware, copyright © 1977.

and without a clear plastic cover to retard evaporation are given as Tables 12-1a and 12-1b. (Daily solar radiation needed to enter these tables can be estimated from Appendix B data at 0° inclination angle.)

A useful rule of thumb for sizing collectors for pool use suggests installation of at least one-half the pool area in solar collectors inclined at the optimum angle, which is about 35° for the northern United States, 25 to 30° below the Mason-Dixon line, and 40° in south Texas and Florida where temperatures permit outdoor winter swimming. If the collectors must be laid flat, half again as much area should be installed.

This sizing method works in any location because the climatic conditions during the critical month or so for which heating extends the swimming season are roughly the same anywhere in the United States.

## HOT WATER HEATING

### Antifreeze System

A complete schematic for a hot water system that uses an antifreeze loop and a separate auxiliary heater is shown in Fig. 12-2. Dual thermostatic tempering valves (with the downstream valve set somewhat hotter) assure a safe water temperature and maximum usage of solar-preheated water while permitting the lowest possible solar collection temperatures. In many installations a somewhat larger solar storage tank is used alone with an electric auxiliary heater (shown) at the *top* of the solar tank, thus minimizing the area for heat losses from storage.

Some states permit the use of a nontoxic antifreeze in a potable water system such as this—propylene glycol or silicone oil is usually recommended. Most states permit use of poisonous ethylene glycol if there is no possibility of a glycol leak with a single material failure. For this reason "traced" tanks are often used in which the heat exchanger is flat sheet, containing fluid passages, which is wrapped around the storage tank. Figure 12-2 shows another popular design, a finned coil at the bottom of the tank that is often made safe by expanding the fluid-carrying tube within the finned tube to provide a double seal against leakage.

The antifreeze loop is protected against thermal expansion by an expansion tank—a long-standard item for hydronic-heating systems. A relief valve guards against the danger of overheating and overpressurization. The circulation pump is actuated when a large enough temperature differential is sensed between the storage tank temperature and the temperature near or under the collector cover plates. In liquid systems the "turnon" differential is usually quite high—about 10 °C (20 °F) so that enough heat will be accumulated in the collector to keep the fluid temperature at the sensor above the storage temperature even when the cold fluid laying in the connecting lines is pumped through the collector. The "turnoff" differential is ideally zero, and is usually set at about 1 °C (2 °F) due to electronics limitations. Increased efficiency can be had by using a low temperature differential for "turnon" as well, if the consequent cycling of the circulating pump is reduced or eliminated by reducing the flow rate at first so that the cold liquid does not flood the collector; or by requiring several minutes of flow once the system has turned on. The problem can be done away with altogether if the collector manufacturer locates a proper well for the sensor in a dormant section of the collectors.

A check valve is installed just downstream of the circulating pump so that cool water cannot thermosyphon into the heat exchanger when the collectors have

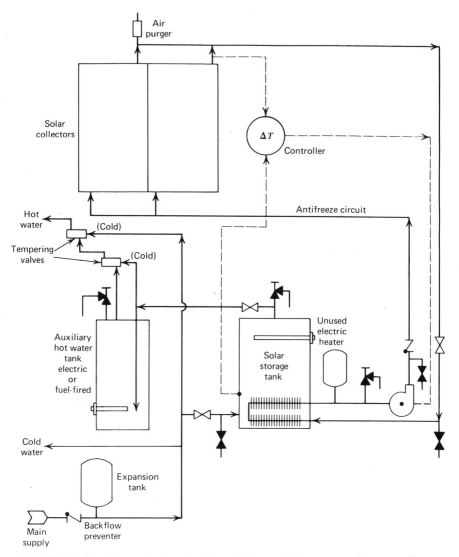

**FIGURE 12-2.** Schematic for solar-heated domestic hot water using an antifreeze fluid for freeze protection and an auxiliary hot water tank.

cooled after the day's operation. Fluid to the collectors is fed from the bottom so that air will more easily flow upward for release through the purge valve located at the highest point in the system.

Cold water enters the system through a backflow preventer, which is required by many codes when connecting into the water mains. A surge tank prevents overpressurization by heating if the backflow preventer has isolated the system and there is no demand for hot water. Cold water is used to supply makeup and tempering water to the solar hot water system. Hot water, withdrawn for use from either of the water tanks, is replaced with water from the *top* of the solar storage tank and that water is in turn replaced with cold water fed to the bottom of the tank. At night this system permits stratified discharge, but during the day the tank is solar-heated from the bottom, and the contents are well-mixed by the resulting convection currents. Heat exchanger efficiency is often quite unknown, a situation that is of course intolerable when trying to make a rational design. Furthermore the tanks usually have only three inches of insulation, leading to losses recently estimated as 30% of the collected heat.

During periods of excess sunshine a solar water heater can overheat considerably, and the stored water may actually boil, activating a relief valve. The controller is often set to stop fluid circulation in such cases, so that the collectors stagnate. A glycol-water antifreeze fluid will boil out of them, of course, but that will be the end of the problem provided that the collector and fluid can withstand the stagnation temperatures safely. If additional protection is desired, cold water can be flushed through the storage tanks by deliberate dumping of the hot water, possibly by an automatic control. The tempering valve that mixes cold water with heated water to keep the delivered hot water temperature to a safe 65 °C (150 °F) is particularly necessary in solar systems because of the probability of such temperature excursions.

## Drain-Down Systems

The use of glycol antifreeze in solar applications is troublesome. It must be monitored for acidity and changed every few years just as is automobile antifreeze. Furthermore, the heat exchanger factor causes a loss in system performance. Silicone-oil antifreeze fluid is maintenance-free but expensive, and causes an even greater reduction in heat transfer.

A drain-down system bypasses these problems by draining the collector whenever the circulating pump is not energized. Such a system is shown in Fig. 12-3. This is a "drainback" system which returns the water to the storage tank when the collector is drained. Since the system pressure is at that of the water mains, the venting arrangements are critical. It is first necessary to have an absolutely fail-safe drain-down collector. These often have a simple parallel tube layout such as that of Fig. 4-1a with tubes up to 2 cm ($\frac{3}{4}$ in.) in diameter. Such large collector passages reduce collector efficiency unless provided with heroic water flows to keep the fluid velocity high, so that the performance advantage of eliminating the heat exchanger is not fully realized from the very start. More troublesome, perhaps, is the additional pumping requirement when the collectors are located high above the storage tank, which is often in the basement. In this pressurized system,

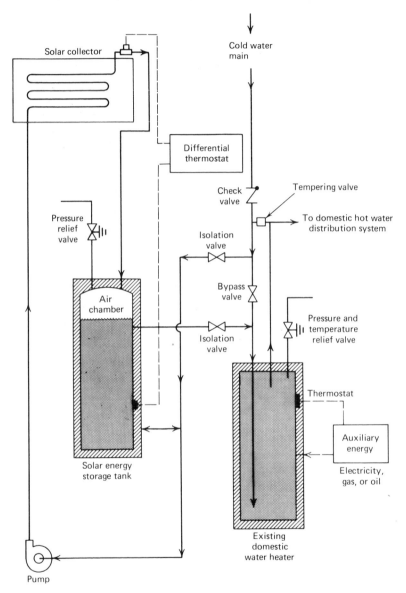

**FIGURE 12-3.** Schematic for solar-heated domestic hot water using drain-down for protection from freezing. (*Source:* From T. P. Mastronarde, M.S. thesis, University of Connecticut.)

the circulating pump must continuously lift the fluid that entire height during operation, because the return line will not fill since it must be large enough to provide effective collector venting.* This not only increases the pumping power, but it forces the use of a more expensive pump than the customary single-stage, centrifugal pump, which is capable of only a few feet of pressure head.

The drain-down system of Fig. 12-3 drains to a pressurized solar storage tank having air trapped above the liquid. When the collector circulation pump stops, air from the chamber rises up the collector fluid return pipe, and most of the water drains from the collector through the dormant pump and returns to the solar storage tank. The trapped air volume must be maintained by deliberate introduction of air, or by air dissolved in the cold water (not shown), because if the air volume is dissolved away into the domestic water, the collectors cannot drain and they will be ruined at the first freeze. Alternatively, there may be a diaphragm installed to retain the air, but then any collector leakage will possibly cause loss of all the air and maintenance will be necessary to replace it.

The remainder of the system is very similar to that of Fig. 12-2, but hot water is taken from the auxiliary storage tank and replaced with water from the solar storage tank, and that is in turn replaced with cold water from the mains, a less efficient arrangement. Because of the air chamber, no expansion tank is needed.

## Hot Air System

A third solution to the freezing problem is to use the safest fluid of all in a heat exchanger loop—air. The disadvantages are those related to the poor density and heat transfer characteristics of air, which provide a substantial penalty in the form of pumping power and a low heat transfer factor. The advantages are simple: Air is nontoxic, need never be replaced, and is spill-proof and noncorrosive.

An air water-heating system is shown in Fig. 12-4. While the system is very simple, there must be a water circulation pump as well as a fan to drive air to the solar collectors and a backdraft damper to prevent freezing through natural convection when the system is shut down on cold nights. The water circulates to an inside-mounted, air-to-liquid heat exchanger of customary fin-tube design. The water pump should be located as close to the storage tank as possible so that even if air gets into the circuit a simple pump without much lift can begin pumping.

## Stratification

None of the systems shown above use stratified storage, although they all probably achieve stratified discharge. For stratified storage the collector flow must be

* If a separate vent line from the air space to the top of the collectors is installed, the return line can be smaller. It will fill with liquid during operation if the fluid velocity is maintained above about 0.7 m/s (2 ft/sec). Such a vent line is opened on shutdown by a solenoid valve at the bottom of the line and protected against water entry by a check valve at the top.

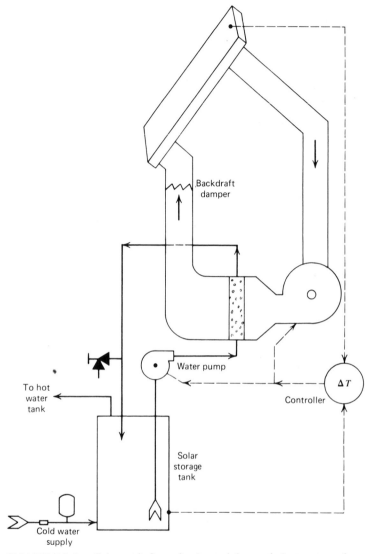

**FIGURE 12-4.** Schematic for solar-heated domestic hot water using air as an antifreeze fluid.

reduced so that solar-heated hot water is made available at the desired usage temperature. The user then has the option of adjusting the rate of water usage to match the available supply of solar-heated water. In areas in which freezing is a problem (and this is virtually all the United States except for the southern tips of Florida and Texas) stratified storage can be attained only with a heat exchanger mounted out-of-tank through which the water and antifreeze fluids flow with equal flow capacity rate. A drain-down collector is not well suited for this application because it is difficult to make an efficient low-flow solar collector drain reliably.

## Natural Circulation Water Heaters

In areas where freezing is not a frequent problem such as Israel and Florida, natural circulation or *thermosyphon* solar water heaters provide a simple and elegant system. The plumbing arrangement is shown in Fig. 12-5. One important limitation is that the entire water storage capacity must be located above the solar water heater upper outlet,* so that roof-mounted storage tanks are generally installed. Aesthetically unattractive, they can be improved considerably if decorated to resemble a chimney.

In colder areas, natural circulation systems have been installed with in-tank heat exchangers serving a glycol loop. These systems will be less efficient, since they cannot stratify the storage tank.

**Operation.** When sun shines on the dormant collector, it heats the water present in the tubes. This water becomes less dense than the colder water in the

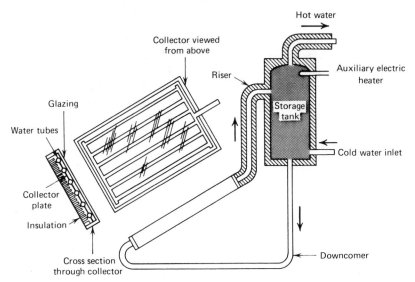

**FIGURE 12-5.** Thermosyphon domestic hot water heater. Courtesy Heating/Piping/Air Condition (October 1966).

* The tank can be placed at the same level as the collectors if there is a check valve in the thermosyphon line to prevent reverse action at night. The circulation rate may drop, however.

tank and thus it rises to the highest point in the assembly, the top of the storage tank. It is replaced in the collector by cold water, which flows from the bottom of the tank and is in turn heated. The process repeats continuously forming what is termed a *thermosyphon* water circulation pattern.

Collectors designed for natural circulation have large-diameter tubes to reduce their pressure drop and so enhance the natural circulation flow rate. This also facilitates draining, which is their only method of freeze protection.* The flow pattern is often serpentine. Excessive storage capacity must be avoided so that the water temperature will reach a usable temperature (50 °C or 120 °F) on most days. For most locations 0.2 to 0.4 MJ/°C·m² (10–15 Btu/°F ft²) will be about right.

Systems like those in Fig. 12-5 were installed in California and south Florida by the thousands in the 1920 to 1940 period (2). They were manually drained when freezing temperatures threatened, and served well for many years. Some current systems used in these areas use electric heaters or (if installed) the circulation pump to provide heat to protect the collector on the few nights when freezing threatens. This solution is not only wasteful of heat or electricity, but it fails in case of power outage, which occasionally accompanies freezing temperatures.

Natural convection systems would seem to be difficult to analyze, and indeed they are difficult to *design* analytically. Once properly designed, however, adequate natural circulation is assured and analysis is easy using techniques from Chapters 5, 8, and 9.

The flow through the collector will always be laminar, or just barely turbulent. The heat transfer coefficient is therefore independent of flow rate (which will vary widely during the collection day), and $F_m$ can be accurately estimated from the Nusselt number (see eq. 5-6 and Example 5-4) and eq. 5-26 once the usual thermal and physical collector parameters are known. Natural circulation storage tanks are almost perfectly stratified for much of the day, so that system performance is also independent of the flow rate. The tank turnover time is typically short—two or three hours—so that the well-mixed integrated storage equations of Chapter 8 apply for the daily situation, and eq. 9-23 can be used to predict monthly performance.

The solar water heaters installed in Florida had storage tank leakage problems starting about 15 years after installation. This has been attributed to electrochemical action between the steel tank or its galvanized coating and the copper in the collector and its piping. New installations should therefore use more durable linings to avoid such corrosion, and install electrically isolating (dielectric) pipe fittings between collectors and storage tank if they are made from different metals.

* Because of high domestic water pressures, draindown to atmospheric pressure after each collection period is impractical in most areas. Drainback to the pressurized storage tank is impossible due to its location. For a few days per year, however, manual drain*out* is practical when outside temperatures approach freezing.

# HEATING AND HOT WATER SYSTEMS

## Drain-Down Systems

As with domestic hot water systems, there is no clear superiority of either drain-down or antifreeze for freeze protection when space heating is included. The drain-down technique is more practical with a heating system, however, because the water is usually stored at atmospheric pressure and therefore there is no cause for concern over the supply of air for venting the collectors.

**Hot Water Preheat.** Figure 12-6 shows a typical drain-down system for heating and hot water preheat. Notice that the hot water is heated by the main storage tank. There is never an advantage to deliberately heating domestic hot water to its final temperature with solar energy, because auxiliary energy will always be needed in a system if the main storage temperature is below the minimum hot water temperature of perhaps 50 °C (120 °F). Since auxiliary energy costs virtually the same no matter what the temperature of use, a proper place to apply the auxiliary energy—which must be supplied somewhere in the system—is in raising hot water to the final temperature. The collectors then operate at the lowest system temperature and hence are more efficient. During the months when solar energy is in plentiful supply, the entire storage tank will heat well above the minimum hot water temperature and auxiliary energy will seldom be needed.

There are a number of techniques for supplying preheat to the pressurized hot water. The preheat tank can be submerged in the main storage tank so that heat will transfer by natural convection through its walls, eliminating the need for a circulating pump. The water will automatically cool, too, when the main storage cools, and there is added risk of corrosion since the tank will usually be of steel. Alternately, a standard solar-heated hot water preheat tank having a finned coil in the bottom can be installed nearby; the coil is supplied with water pumped from the main storage tank. The preheat tank can then be held above the main storage temperature when that temperature drops by stopping the pump. Sometimes the preheat tank is left out altogether, and the "tankless coil" concept, popular in hydronic oil-fired heating systems, is used. With such a system there will be little preheat when the storage is cool, but when the storage temperature is above about 65 °C (150 °F) a tankless coil will have adequate heat transfer to heat a limited domestic hot water flow. Auxiliary heat, however, must be available upon demand since no hot water is stored.

In Fig. 12-6 a submerged coil is used in the main storage tank to preheat pressurized hot water recirculated from the preheat storage tank. The pump is turned off when the preheat water approaches either the main storage tank temperature or the desired hot water temperature. Cold water makeup enters the system through a backflow preventer, and a vacuum breaker protects the tanks from accidental collapse during malfunction or servicing.

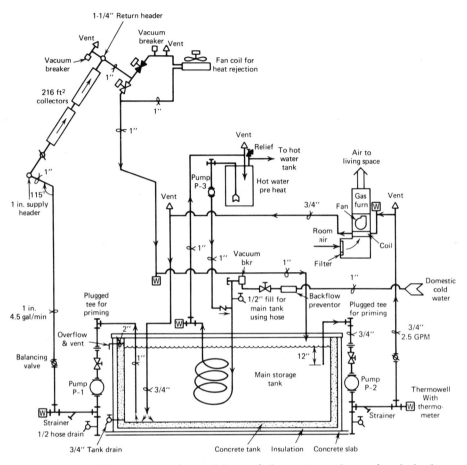

**FIGURE 12-6.** Solar space heating and domestic hot water preheat using drain-down for freeze protection. (*Source:* Adapted from U.S. Navy drawings.)

**Space Heating.** Since the circulating loop for the space-heating coil in Fig. 12-6 need not be pressurized, the stored fluid is pumped directly. The circulating pump is located out of the tank, arranged so that it can be primed with water to facilitate the first start-up (the ordinary centrifugal pump will not pump air). Water is taken from near the top of the tank, where it is warmest. Notice that care has been taken that the coil not be located at the highest point in the loop, where air will accumulate, and that vents are located at high points to remove air as necessary.

Solar panels heat the main storage tank by means of the loop located at the left of the drawing. Stored water circulates through the collectors, is heated, and returns to the tank. If the stored water is nearing the boiling point, the water is circuited to a cooling coil mounted near the collectors to protect the system from boil-off.

Since this is a drain-down system the circulating pump must lift water to the collectors during each start-up.* Plug or ball type balancing valves are located in the main circulation line and in the flow circuit to each collector. Unless the collectors have been designed to have a high-pressure drop (and this is difficult to combine with drain-down capability) the flow to each collector must be adjusted in the field to assure that the collectors furthest from the source of supply receive adequate waterflow. Reverse return (Fig. 12-1) helps balance the collector flows, but is not completely effective on a large array.

The drain-down provisions are actuated when the circulating pump P-1 is stopped. The pressure at the pump inlet is then equivalent to perhaps 70 cm (2 ft) of water above the atmospheric pressure at the tank surface. With one or two stories of water located in the pipe above the pump outlet the water will drop through the pump and on into the tank by gravity, until the pressure is equal on both sides of the pump. If the fluid lines are not vented, however, a vacuum will form in the upper lines and the water will be held in place by the atmospheric pressure. The vacuum breaker prevents this, venting atmospheric air into the system at its highest point. The water in the upper part of the system then gravity-drains fully into the storage tank, provided that the lines are carefully inclined to prevent possible pockets of accumulated water from blocking complete drainage. Lines that are to drain must be at least 2 cm ($\frac{3}{4}$ in.) in diameter to prevent vapor lock. Lines smaller than this cannot be trusted to drain completely.

Notice that there are no penetrations of the storage tank below the water line for the circulating fluids. This is a wise precaution with concrete tanks, which are subject to cracking. The only penetration is the small tank drain located at the bottom of the tank. The risk in this procedure is that one or more of the circulating pumps could lose its prime. This is a particular possibility with pump P-1, which can have all its water siphoned out on draindown as a result of the inertia of water rapidly flowing into the tank. The balancing valve in that line should be adjusted to limit the drain-down rate and so prevent this.

### Antifreeze Systems

If antifreeze is used in the system, freeze protection is virtually assured but the system becomes more complex. Figure 12-7 shows a typical antifreeze system, with the antifreeze loop at the left of the diagram. Pump P-1 circulates antifreeze within the collector–heat-exchanger loop. This loop is sealed, having its own expansion tank, and generally will be pressurized to about 200 kPa (15 psig). Corrosion inhibitors are necessary with glycol solutions although there is no fresh

* Lines should be sized for at least 0.7 m/s (0.2 ft/sec) velocity to assure air purging.

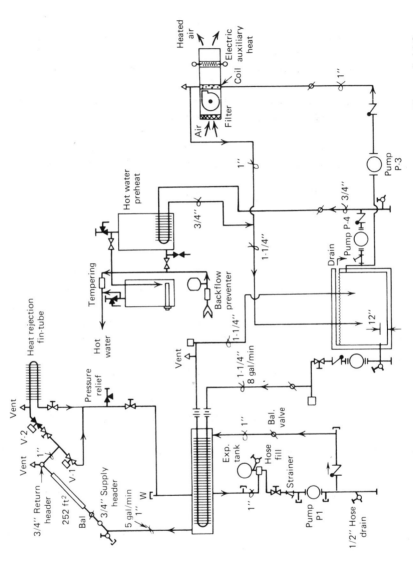

**FIGURE 12-7.** Solar space heating and domestic hot water preheat using an antifreeze fluid for freeze protection. (*Source:* Adapted from U.S. Navy drawings.)

source of oxygen because of the gradual breakdown of ethylene or propylene glycol. If a mineral, synthetic, or silicone oil is used aluminum collectors are permissible, but otherwise steel or copper collectors are mandatory.

The collector arrangement is much as in Fig. 12-6, although the fan on the heat rejection unit has wisely been replaced with natural convection finned tubing. The antifreeze heat exchanger is a two-pass type, which is somewhat less efficient than a countercurrent arrangement but simpler to install and maintain. The choice of the shell-side passages for antifreeze is not optimal, since it increases the volume of antifreeze needed to charge the system. The placing of the heat exchanger in the flow path upstream of the collectors permits the use of a single vent for the residual air of both units.

The storage tank for this system happens to be steel or fiberglass, so tank penetrations are made for the pump inlets to insure a trouble-free prime.

The circulating loops to the air-heating coil and the hot water-preheating tank are similar to that of the air-heating coil in Fig. 12-4. The hot water preheat tank is under domestic water pressure, and a finned coil is used in the hot water tank to transfer stored heat to the tank. The remainder of the hot water system is conventional.

## Controls

Controls for liquid systems should be as simple as possible. The basic control is the differential thermostat. As previously described, the thermostat turns on the pump, which circulates fluid to the collectors whenever the temperature on the collector absorber surface under the cover glass is warmer than the stored fluid.* The fluid flow rate may be modulated just after start-up to avoid premature shutdown due to cold liquid held up in the interconnecting lines, but full flow should be assured after the first few minutes of operation.

The hot water preheat pump is energized whenever the stored preheated water is cooler than the main storage. The pump supplying the room-air heating coil should be energized by a two-stage thermostat. As the temperature in the living space drops, this pump is first energized. If the temperature continues to fall, the auxiliary heater is also energized. When the water nears the base temperature, pump circulation is stopped and the system then uses only auxiliary heat until more solar heat is available.

## Air Systems

The air solar system in Fig. 12-8 is much simpler than any of the liquid systems. There is little possibility of freeze-up, no chance for vapor locks, no worry over

* A check valve must be installed in this circulation line to prevent thermosyphon cooling when the system is shut down, unless the line is drained.

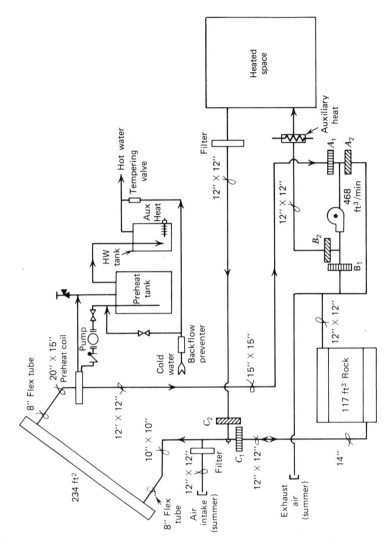

**FIGURE 12-8.** Solar space heating and domestic hot water preheat with an air circulation system and rock storage. (*Source:* adapted from U.S. Navy drawings.)

spillage, and a single fan takes the place of almost all the pumps used in a liquid system. In actual practice all the flow switching can be accomplished with only three flapper valves, conveniently mounted in an air-handling unit also housing the fan.

There are drawbacks, however. The pressure loss through the ductwork and components is often high enough that fan-operating costs are significant. While the system may be simple, implementing it is not. It is far easier (and cheaper) to run a one-inch copper tube than a one-foot duct (and twice that cross section will be required in a cold climate). Air leakage plays havoc with efficiency and all joints must be carefully taped. The physical size of the storage container must be three times as large as with water storage for the same heat capacity, and retrofit is almost impossible.

**Operation.** In Fig. 12-8 shutoff louvers $A_1$, $B_1$, and $C_1$ are shown opened, and $A_2$, $B_2$, and $C_2$ are closed; the proper positions for collection of heat for delivery to storage. Cold air leaves the bottom of the rock bed and passes directly to the collectors. After being heated, it passes through the hot water preheat coil and then returns through the fan to the rock bed, where it gives up its heat. Although the hot water is preheated only during operation, the high outlet temperature from the collector permits adequate heat exchange.

If space heating is desired at the same time as solar collection, the rock bed is bypassed. Louvers $B_1$ and $C_1$ are closed and $B_2$ and $C_2$ opened so that the fan pushes air from the collectors into the heated space, withdrawing cool room air through louver $C_2$ and on to the collectors for reheating.

If no solar collection is possible and no heat is required, the fan is turned off. If heat is required, louvers $A_2$, $B_2$, $C_1$, and $C_2$ are each opened and $A_1$ and $B_1$ closed to reverse the flow through the bed. The fan then takes in hot air from the top of the bed and delivers it to the heated space. Auxiliary heat, if needed, can be added downstream of the fan. Cool air leaving the room passes to the bottom of the bed and is heated as it rises through the bed.

Summer cooling of the collectors, desirable in any system, is simple in this air system. The summer air intake and exhaust vents on either side of the rock bed are opened, short-circuiting the usual flow of air through the bed. Although there will still be some flow through the bed, most of the air will come from and be returned to the outside. This mode of operation permits summer water heating without the use of rock storage, a convenience since heat stored in the rock bed cannot be made available to the hot water system. Leakage through these vents would be thermally disastrous in winter, and therefore positive manual sealing is strongly recommended.

The popularity of air solar systems is rising as their advantages become appreciated. It remains to be seen whether they can compete effectively with the more compact and inherently more efficient liquid systems. In an air system, the heat transfer between the circulating air and the solar absorber surface is inherently poor, flow through the collectors is inherently too low, and fan power costs are inevitably high. A low inlet temperature, however, is almost constantly available at the exit of the rock bed to compensate for the first problem. Stratification of the storage bed compensates for the second, as we have seen. There is every indication that fan power can be reduced considerably, and when this is done air systems may well displace liquid systems from their present commanding place on the solar scene.

## PUMPS AND FANS

In any active solar heating/hot water system, electrical power must be supplied to circulate liquid or air for heat transfer between the collectors, storage, and point of use. The fans or pumps called on for these tasks are likely to be operating almost continuously for several months of the year so that more than ordinary attention to their efficiency is warranted.

### Pumps

For loops circulating liquid, only centrifugal pumps need be considered. A typical centrifugal pump is shown in Fig. 12-9.

These pumps are available for a wide variety of flow and head requirements. In the larger sizes (hundreds of gal/min or 0.005 to 0.05 m³/s) they are efficient, translating 50 to 80% of the applied electricity into fluid energy. In residential applications—and they have been used in hydronic heating systems for years—they are absurdly inefficient, with only a few percent of the applied power converted into fluid energy. Most of the lost energy is used to turn the shaft against the shaft seal. Recently the shaft seal has been completely eliminated in newer pump designs by the use of a magnetic coupling or by "canning" the motor so that its moving parts can be wet by the circulating fluid without problems. Such pumps are gradually displacing the shaft-sealed pumps in hydronic-heating systems, dropping the shaft power requirement from 1/8 hp (90 W) to 1/20 hp (40 W). Such pumps should be used exclusively in solar applications. They are not only more efficient but also more reliable and no more expensive.

Figure 12-10 shows a performance curve for a commercial magnet drive pump. Because of the low velocities used in hydronic-heating systems to reduce noise, the available head is quite low—2.5 m (8 ft) water maximum—but it is usually economical to design the system for such operation. Drain-down systems, however, need much higher heads to lift water to the drained collectors on system start-up, and need a more sophisticated pump.

**FIGURE 12-9.** A centrifugal pump suitable for solar applications. This "canned motor" design requires no shaft seal, reducing power consumption. (Photo courtesy of Grundfos Pumps Corporation.)

The theoretical energy required to recirculate in the usual closed low-pressure–drop loops is very small, and is given by

$$\text{Work} = W \, \Delta h (g/g_c) \tag{12-1}$$

where $\Delta h$ is the pressure drop of the system expressed in terms of the head of liquid flowing, sometimes referred to as the *lost head*, measured in m (ft). In SI units, $g/g_c$ is 9.80 N/kg or 9.80 W/(kg·m/s), while in the Engineering system of units $g/g_c = 1$. The term $g/g_c$ is a conversion factor necessary in SI units because

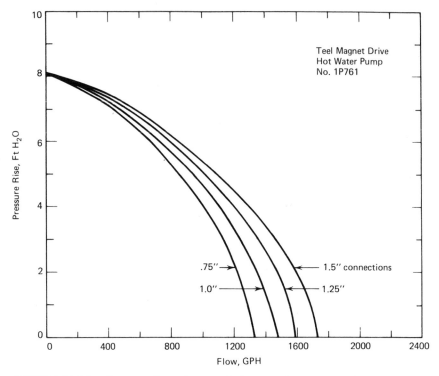

**FIGURE 12-10.** Pressure-flow characteristic curve for a magnetic drive centrifugal pump. (*Source:* from W. W. Grainger #343 catalog data, p. 429.)

the flow rate $W$ is expressed in units of mass (kg/s) instead of force (lb/sec), which happens to incorporate the needed conversion factor automatically.

### EXAMPLE 12-1

The flow in a solar collector loop is 0.5 kg/s (66.2 lb/min) of 50% ethylene glycol. What is the work required to circulate this fluid if the lost heat is 1.2 m (3.94 ft) glycol?

By eq. 12-1,

$$\text{Work} = \frac{0.5 \text{ kg}}{\text{s}} \times 1.2 \text{ m} \times \frac{9.8 \text{ W} \cdot \text{s}}{\text{kg} \cdot \text{m}} = 5.88 \text{ W}$$

or in Engineering units

$$\text{Work} = \frac{66.2 \text{ lb}}{\text{min}} \times 3.94 \text{ ft} \times \frac{\text{min hp}}{33\,000 \text{ ft-lb}} = 0.0079 \text{ hp}$$

This calculation quickly shows that even with magnet drive the pumping efficiency is very low and therefore the theoretical power requirements are not usually worthwhile calculating. Further attention by pump manufacturers to improvements in efficiency can increase the attractiveness of liquid-based solar systems, which often use several circulating loops consuming appreciable total power.

## Fans

Centrifugal fans are in almost exclusive use to circulate air within air-conditioning and solar air systems. Unlike liquid loops, the theoretical power requirement is significant. It is given by

$$\text{Work} = W\nu \, \Delta h \rho_w (g/g_c) \tag{12-2}$$

where $\Delta h$ is in mm (in.) of *water*, reflecting Engineering practice. $W\nu$ is the volumetric flow rate of the air, which is often used instead of the weight flow. The density of air here is expressed as the specific volume, $\nu$(m³/kg or ft³/lb).

## EXAMPLE 12-2

An air solar system has a flow of 0.45 m³/s (953 ft³/min) and an overall pressure drop of 22 mm $H_2O$ (0.906 in. $H_2O$). What is the theoretical power required for this air circulation?

By eq. 12-2,

$$\text{Work} = \frac{0.45 \text{ m}^3}{s} \times 23 \text{ mm} \times \frac{m}{1000 \text{ mm}} \times \frac{1000 \text{ kg}}{m^3} \times \frac{9.8 \text{ W}}{kg \cdot m/s} = 101.4 \text{ W}$$

or

$$\text{Work} = \frac{953 \text{ ft}^2}{\text{min}} \times 0.906 \text{ in.} \times \frac{ft}{12 \text{ in.}} \times \frac{62.4 \text{ lb}}{ft^3} \times \frac{\text{min hp}}{33\,000 \text{ ft-lb}}$$

$$= 0.1361 \text{ hp } (\sim 1/8 \text{ hp})$$

Just as with pumps, there has in the past been little motivation to approach theoretical power requirements with commercial fans. Consequently, the actual power consumed to circulate the air to air collectors can be $\frac{1}{2}$ or even 1 hp (400–800 W) in an ordinary residence. The usual flow rate has been 2 ft³/min for each ft² of collector, but recent installations having low pressure drop have been designed for twice as much flow.

Two kinds of centrifugal fans are used in solar air circulation service. The forward-curved or squirrel cage design (Fig. 12-11) is most popular in residential and light commercial use because of its low initial cost. The speed required for a given pressure rise is low, permitting a directly driven motor drive. The fan is compact and reasonably efficient, but is relatively noisy.

Figure 12-13 contains typical application data for a forward-curved fan. Notice that if the air flow is restricted, or if the motor is underpowered, the fan will

**FIGURE 12-11.** Impeller (top) and finished assembly of a forward-curved or "squirrel-cage" centrifugal blower. Note the direct-drive motor within the shroud (bottom). The impeller rotates counterclockwise. (Courtesy of Lau Division, Philips Industries, Inc.)

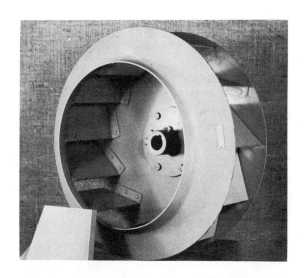

**FIGURE 12-12.** Impeller (top) and finished assembly of a backward-curved centrifugal blower. The impeller is belt-driven by a motor within the motor housing to the side of the shroud (bottom) because this fan must be driven at high speed. The impeller rotates counterclockwise. (Courtesy of Lau Division, Philips Industries, Inc.)

445

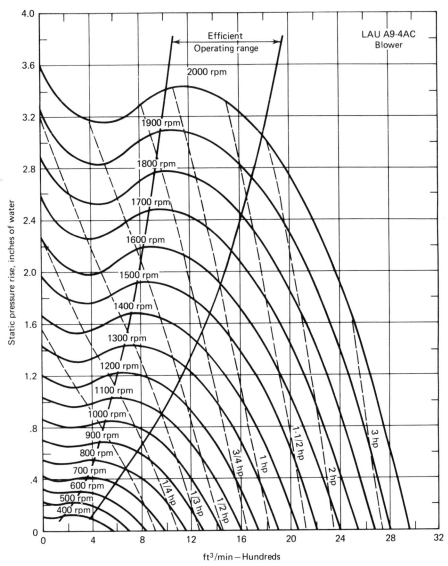

**FIGURE 12-13.** Pressure rise characteristic for a Lau forward-curved blower.

operate in the inefficient and noisy range in the valley of the pressure-low curve. Conversely, if the system pressure drop is too low, the fan becomes inefficient, demanding excessive motor power but moving only a little more air. Efficient operation requires careful selection of motor, fan, and fan speed to match the system.

Backward-curved centrifugal fans (Fig. 12-12) are quieter, more efficient, and not as sensitive to underloading or overloading as the forward-curved fans. Physically larger, they are also more expensive due to close running clearances. They generally must be belt-driven since they operate with twice the speed of the forward-curved type. They are the best choice for the air collector loop where the theoretical power required is high and good efficiency is therefore important.

**EXAMPLE 12-3**

Find the operating speed and efficiency of the fan of Fig. 12-13 for the application in Example 12-2.

The correct speed of 1100 rev/min is taken at the intersection of the 950 ft³/min and 0.9 in. H₂O lines on the chart (0.4483 m³/s and 2.286 cm H₂O). The power required is $\frac{1}{3}$ hp or 248.7 W. Efficiency is therefore

$$\eta = \frac{101.4 \text{ W}}{248.7 \text{ W}} = 40.8\%$$

or

$$\eta = \frac{0.1361 \text{ hp}}{\frac{1}{3} \text{ hp}} = 40.8\%$$

This efficiency includes drive losses, but not motor losses. Since fractional horse-power motors operate with an efficiency of about 65%, the overall efficiency is about 0.408 × 0.65 = 26.5%, and the electrical input 248.7 W/0.65 = 383 W.

The motor should be sized for the *minimum* pressure drop that is anticipated. At 1100 rpm, for instance, free flow delivery is 1600 ft³/min, and the nearly $\frac{3}{4}$ hp required would quickly burn out the small motor indicated above

The application of backward-curved fans is similar, except that less care is required in sizing since the pressure-rise curve falls off smoothly, having no hump as does the forward-curved fan; and the horsepower required does not increase at free delivery.

## SIZING PIPE AND DUCTWORK

Characterization of friction in fluid flow in straight tube or duct runs in the collector and associated piping is the starting point for sizing tubes and ductwork. An important fundamental relationship is the Fanning equation, given here in a modified form as

$$\Delta p = \frac{4fW^2L}{2g_c A_x^2 \rho D} \tag{12-1}$$

where $\Delta p$ is the frictional loss or pressure drop, $W$ the mass flow rate, $L$ the length of the tube or duct, $A_x$ its cross-sectional area, $g_c$ a constant (1 kg·m/N·s² or 32.17 ft/sec²), $\rho$ the fluid density, $D$ the equivalent diameter of the duct or tube, and $f$ a friction factor.

In the case of noncircular cross sections, $D$ is given by eq. 5-2 as

$$D_e = \frac{4A_x}{P} \tag{5-2}$$

where $P$ is the perimeter of the duct or tube. The friction factor $f$ is an empirical function of the relative roughness of the tube or duct, and the Reynolds number which is given by

$$N_{Re} = \frac{4W}{\mu\pi D} \tag{5-1a}$$

or if the cross section is not circular

$$N_{Re} = \frac{4W}{\mu P} \tag{5-1b}$$

For smooth tubes and ducts, when the flow is laminar ($N_{Re} < 2100$) the friction factor is given by

$$f = \frac{16}{N_{Re}} \tag{12-2a}$$

and when it is turbulent, for Reynolds numbers up to about $10^6$, by (3)

$$f = 0.001\ 40 + \frac{0.125}{N_{Re}^{0.32}} \tag{12-2b}$$

For circular ducts, pipes, or tubes, eq. 12-1 can be restated as

$$\Delta p = \frac{32fW^2L}{\pi^2 g_c D^5 \rho} \tag{12-3}$$

and for rectangular ducts as

$$\Delta p = \frac{fW^2LP\nu}{2g_c A_x{}^3} = \frac{fW^2LP}{2g_c \rho A_x{}^3} \tag{12-4}$$

For rectangular air ducts the density is also expressed as the specific volume $\nu = 1/\rho$, a convenience since air density in Engineering units is often given this way, as in Table 5-1.

In many systems pressure drop is most often expressed as the equivalent fluid head loss $\Delta h$ in terms of meters or feet of water, a convenient fluid for measurement of pressure drop. In terms of "lost head" of a measuring fluid such as water, eq. 12-3 and 12-4 become, for round tubes,

$$\Delta h = \frac{32fW^2L}{\pi^2 g D^5 \rho \rho_m} \tag{12-5}$$

where $\rho_m$ is the density of the measuring fluid and $g$ is the acceleration due to gravity (9.8 m/s² or 32.2 ft/sec²). For rectangular ducts,

$$\Delta h = \frac{fW^2LP\nu}{2gA_x^3\rho_m} = \frac{fW^2LP}{2g\rho\rho_mA_x^3} \tag{12-6}$$

Equations 12-3 to 12-6 should be examined carefully. Pressure drop or lost head is directly proportional to the length of tubing, proportional to the *square* of the flow rate, and proportional to the *fifth* power of the tubing or duct size. Therefore, the piping or ductwork designer can be relatively unconcerned about the length of the run, only moderately concerned with the circulation rate, but must be extremely sensitive that the size of the tube or duct is appropriate for the flow rate. Consequently, the pressure drop through the collectors is of highest interest since that is where the minimum dimensions are most likely to be found.

## Pressure Drop Through Collectors

### EXAMPLE 12-4a

Find the pressure drop for the water flow in the two different flow paths through the solar collector in Example 5-1(b) and 5-1(c).

For

$$N_{Re} = 1437, \text{ eq. 12-2a applies:}$$

$$f = \frac{16}{1437} = 0.0111$$

For

$$N_{Re} = 11\ 496, \text{ eq. 12-2b applies}$$

$$f = 0.001\ 40 + \frac{0.125}{(11\ 496)^{0.32}} = 0.007\ 67$$

Using eq. 12-5 for Example 5-1(b) and 5-1(c), respectively,

$$\Delta h = \frac{32fW^2L}{\pi^2 g D^5 \rho \rho_m}$$

$$= \frac{32 \times 0.0111}{\pi^2} \times \left(\frac{0.0247}{8}\frac{\text{kg}}{\text{s}}\right)^2 \times 2.44\text{ m} \times \frac{\text{s}^2}{9.8\text{ m}} \times \frac{1}{\left(\frac{6.35}{1000}\text{ m}\right)^5} \times \left(\frac{\text{m}^3}{980\text{ kg}}\right)^2$$

$$= 0.008\ 61\text{ m H}_2\text{O} \quad \text{or} \quad 8.61\text{ mm H}_2\text{O}$$

or using eq. 12-3

$$\Delta p = \frac{32fW^2L}{\pi^2 g_c D^5 \rho} = \frac{32 \times 0.0111}{\pi^2} \times \left(\frac{0.0247}{8}\frac{\text{kg}}{\text{s}}\right)^2 \times 2.44\text{ m} \times \frac{\text{N}\cdot\text{s}^2}{1\text{ kg}\cdot\text{m}}$$

$$\times \frac{1}{\left(\frac{6.35}{1000}\text{ m}\right)^5} \times \frac{\text{m}^3}{980\text{ kg}} \times \frac{\text{Pa}}{\text{N/m}^2} = 82.7\text{ Pa}$$

and for the higher flow rate,

$$\Delta h = \frac{32 \times 0.007\ 67 \times 0.0247^2 \times 2.44}{\pi^2 \times 9.8 \times (6.35/1000)^5 \times (980)^2} = 0.381 \text{ m } H_2O$$

or

$$\Delta p = \frac{32 \times 0.007\ 67 \times 0.0247^2 \times 2.44}{\pi^2 \times (6.35/1000)^5 \times 980} = 3658 \text{ Pa or } 3.66 \text{ kPa}$$

In Engineering units, from eq. 12-5,

$$\Delta h = \frac{32fW^2L}{\pi^2 g D^5 \rho_w{}^2} = \frac{32 \times 0.0111}{\pi^2} \times \left(\frac{3.273}{8}\right)^2 \frac{lb^2}{min^2} \times 8 \text{ ft} \times \frac{sec^2}{32.2 \text{ ft}}$$

$$\times \frac{1}{\left(\frac{0.25}{12} \text{ ft}\right)^5} \times \left(\frac{ft^3}{61.2 \text{ lb}}\right)^2 \times \left(\frac{min}{60 \text{ sec}}\right)^2 = 0.0283 \text{ ft } H_2O \\ \text{or } 0.34 \text{ in. } H_2O$$

or using eq. 12-3

$$\Delta p = \frac{32 \times 0.0111}{\pi^2} \times \left(\frac{3.273}{8}\right)^2 \frac{lb^2}{min^2} \times 8 \text{ ft} \times \frac{sec^2}{32.2 \text{ ft}} \times \frac{1}{\left(\frac{0.25}{12} \text{ ft}\right)^5} \times \frac{ft^3}{61.2 \text{ lb}}$$

$$\times \left(\frac{min}{60 \text{ sec}}\right)^2 \times \frac{ft^2}{144 \text{ in.}^2} = 0.0120 \text{ lb/in.}^2$$

and for the higher flow rate,

$$\Delta h = \frac{32 \times 0.007\ 67 \times 3.273^2 \times 8}{\pi^2 \times 32.2 \times (0.25/12)^5 \times (61.2)^2 \times (60)^2} = 1.25 \text{ ft } H_2O$$

or

$$\Delta p = \frac{32 \times 0.007\ 67 \times 3.273^2 \times 8}{\pi^2 \times 32.2(0.25/12)^5 \times 61.2 \times (60)^2 \times 144} = 0.532 \text{ lb/in.}^2$$

Example 12-4a shows that the actual collector pressure drop can be quite small in a parallel flow arrangement. This is not always an advantage, for it can lead to difficulty in distributing the flow evenly to a bank of collectors, since the total collector pressure drop may then depend on the details of the collector manifolding. More pressure drop must be provided or balancing valves introduced if the flow rate is to be the same for each collector. This is done automatically with the series flow path, although that pressure drop will be too high with the somewhat higher flow properly used with this collector unless the next larger tube size is used.

**EXAMPLE 12-4b**

What is the pressure drop of the air through the collector in Example 5-1(f)?

For

$$N_{\text{Re}} = 3015, \text{ by eq. 12-2b*}$$

$$f = 0.001\,40 + \frac{0.125}{(3015)^{0.32}} = 0.011$$

$$A_x = 1.22 \text{ m} \times 12.7 \text{ mm} \times 1 \text{ m}/1000 \text{ mm} = 0.015\,49 \text{ m}^2$$

from eq. 12-6a

$$\Delta h = \frac{fW^2LP}{2g\rho\rho_m A_x^3} = 0.011 \times \left(\frac{0.039\,02 \text{ kg}}{\text{s}}\right)^2 \times \frac{2.44 \text{ m}}{2 \times 9.8 \text{ m/s}^2} \times \frac{2.465 \text{ m}}{(0.015\,49 \text{ m}^2)^3}$$

$$\times \frac{\text{m}^3}{1.033 \text{ kg}} \times \frac{\text{m}^3}{1000 \text{ kg}}$$

$$= 0.001\,34 \text{ m or } 1.34 \text{ mm}$$

or using eq. 12-4

$$\Delta p = \frac{0.011 \times (0.039\,02)^2 \times 2.44 \times 2.465}{2 \times 1.033 \times (0.015\,49)^3} = 13.11 \text{ Pa}$$

In Engineering units,

$$A_x = \tfrac{1}{2} \text{ in.} \times 4 \text{ ft} \times \frac{\text{ft}}{12 \text{ in.}} = 0.1666 \text{ ft}^2$$

Using eq. 12-6a,

$$\Delta h = \frac{fW^2LP v}{2g A_x^3 \rho_m} = 0.011 \times \left(\frac{5.161 \text{ lb}}{\text{min}}\right)^2 \times \frac{8 \text{ ft}}{2 \times 32.2 \text{ ft/sec}^2} \times \frac{8.033 \text{ ft}}{(0.1666)^3 \text{ ft}^6}$$

$$\times \frac{15.5 \text{ ft}^3}{\text{lb}} \times \frac{\text{ft}^3}{62.4 \text{ lb}} \times \left(\frac{\text{min}}{60 \text{ sec}}\right)^2 \times \frac{12 \text{ in.}}{\text{ft}}$$

$$= 0.0524 \text{ in. H}_2\text{O}$$

and using eq. 12-4,

$$\Delta p = \frac{0.011(5.161)^2 \times 8 \times 8.033 \times 15.5}{2 \times 32.2 \times 0.1667^3 \times 60^2 \times 144} = 0.001\,89 \text{ lb/in.}^2$$

The pressure drop through the air collector in Example 12-4b is desirably low, but may permit distribution problems between collectors in parallel unless the pressure drop in the ductwork is identical for each collector. Deliberate increase of this pressure drop may be desirable so that the bulk of the pressure drop in

---

* For flows in the transition zone, pressure drop is variable and may depend on arbitrary factors or even be unstable.

the collector bank will be through the collectors, tending to distribute the flow equally to each collector.

## Pressure Drop Through Piping

Table 12-2 gives dimensions of copper tubing of types K, L, and M, which are in common use in the United States. Type K is extra heavy wall, and type L is

**TABLE 12-2.** Types K, L, and M copper tube dimensions in Engineering and SI units (Courtesy of Copper Development Association, Inc.)

| Type K Nominal Pipe Size (in.) | Diameter | | | | Wall Thickness | | Inside Cross-Sectional Area | |
|---|---|---|---|---|---|---|---|---|
| | OD | | ID | | | | | |
| | in. | mm | in. | mm | in. | mm | in.² | m² × 10³ |
| ¼ | 0.375 | 9.53 | 0.305 | 7.75 | 0.035 | 0.889 | 0.0731 | 0.0471 |
| ⅜ | 0.500 | 12.70 | 0.402 | 10.21 | 0.049 | 1.245 | 0.127 | 0.082 |
| ½ | 0.625 | 15.88 | 0.527 | 13.39 | 0.049 | 1.245 | 0.218 | 0.141 |
| ⅝ | 0.750 | 19.05 | 0.652 | 16.56 | 0.049 | 1.245 | 0.334 | 0.215 |
| ¾ | 0.875 | 22.23 | 0.745 | 18.92 | 0.065 | 1.651 | 0.436 | 0.281 |
| 1 | 1.125 | 28.58 | 0.995 | 25.27 | 0.065 | 1.651 | 0.778 | 0.502 |
| 1¼ | 1.375 | 34.93 | 1.245 | 31.62 | 0.065 | 1.651 | 1.217 | 0.785 |
| 1½ | 1.625 | 41.28 | 1.481 | 37.62 | 0.072 | 1.829 | 1.723 | 1.111 |
| 2 | 2.125 | 53.98 | 1.959 | 49.76 | 0.083 | 2.108 | 3.014 | 1.945 |
| 2½ | 2.625 | 66.68 | 2.435 | 61.85 | 0.095 | 2.413 | 4.657 | 3.004 |
| 3 | 3.125 | 79.38 | 2.907 | 73.84 | 0.109 | 2.769 | 6.637 | 4.282 |
| 3½ | 3.625 | 92.08 | 3.385 | 85.98 | 0.120 | 3.048 | 8.999 | 5.806 |
| 4 | 4.125 | 104.78 | 3.857 | 97.97 | 0.134 | 3.404 | 11.684 | 7.538 |
| Type L | | | | | | | | |
| ¼ | 0.375 | 9.53 | 0.315 | 8.00 | 0.030 | 0.762 | 0.0779 | 0.0503 |
| ⅜ | 0.500 | 12.70 | 0.430 | 10.92 | 0.035 | 0.889 | 0.145 | 0.094 |
| ½ | 0.625 | 15.88 | 0.545 | 13.84 | 0.040 | 1.016 | 0.233 | 0.151 |
| ⅝ | 0.750 | 19.05 | 0.666 | 16.92 | 0.042 | 1.067 | 0.348 | 0.225 |
| ¾ | 0.875 | 22.23 | 0.785 | 19.94 | 0.045 | 1.143 | 0.484 | 0.312 |
| 1 | 1.125 | 28.58 | 1.025 | 26.04 | 0.050 | 1.270 | 0.825 | 0.532 |
| 1¼ | 1.375 | 34.93 | 1.265 | 32.13 | 0.055 | 1.397 | 1.257 | 0.811 |
| 1½ | 1.625 | 41.28 | 1.505 | 38.23 | 0.060 | 1.524 | 1.779 | 1.148 |
| 2 | 2.125 | 53.98 | 1.985 | 50.42 | 0.070 | 1.778 | 3.095 | 1.997 |
| 2½ | 2.625 | 66.68 | 2.465 | 62.61 | 0.080 | 2.032 | 4.772 | 3.079 |
| 3 | 3.125 | 79.38 | 2.945 | 74.80 | 0.090 | 2.286 | 6.812 | 4.395 |
| 3½ | 3.625 | 92.08 | 3.425 | 87.00 | 0.100 | 2.540 | 9.213 | 5.944 |
| 4 | 4.125 | 104.78 | 3.905 | 99.19 | 0.110 | 2.794 | 11.977 | 7.727 |
| Type M | | | | | | | | |
| ¼ᵃ | 0.375 | 9.53 | | | | | | |
| ⅜ | 0.500 | 12.70 | 0.450 | 11.43 | 0.025 | 0.635 | 0.159 | 0.103 |
| ½ | 0.625 | 15.88 | 0.569 | 14.45 | 0.028 | 0.711 | 0.254 | 0.164 |
| ⅝ᵃ | 0.750 | 19.05 | | | | | | |
| ¾ | 0.875 | 22.23 | 0.811 | 20.60 | 0.032 | 0.813 | 0.517 | 0.333 |
| 1 | 1.125 | 28.58 | 1.055 | 26.80 | 0.035 | 0.889 | 0.874 | 0.564 |
| 1¼ | 1.375 | 34.93 | 1.291 | 32.79 | 0.042 | 1.067 | 1.309 | 0.845 |
| 1½ | 1.625 | 41.28 | 1.527 | 38.79 | 0.049 | 1.245 | 1.831 | 1.182 |
| 2 | 2.125 | 53.98 | 2.009 | 51.03 | 0.058 | 1.473 | 3.170 | 2.045 |
| 2½ | 2.625 | 66.68 | 2.495 | 63.37 | 0.065 | 1.651 | 4.889 | 3.154 |
| 3 | 3.125 | 79.38 | 2.981 | 75.72 | 0.072 | 1.829 | 6.979 | 4.503 |
| 3½ | 3.625 | 92.08 | 3.459 | 87.86 | 0.083 | 2.108 | 9.397 | 6.063 |
| 4 | 4.125 | 104.78 | 3.935 | 99.95 | 0.095 | 2.413 | 12.161 | 7.846 |

ᵃ Not manufactured in this size.

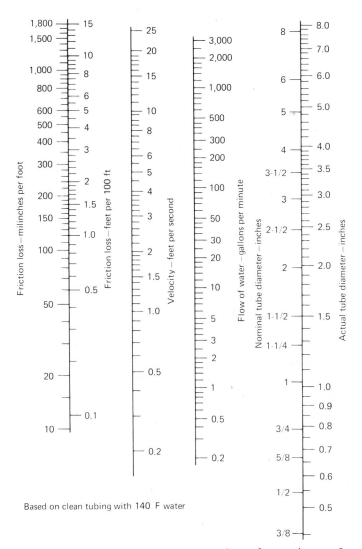

**FIGURE 12-14.** Friction loss due to flow of water in type L copper tube.

generally used for indoor plumbing. Type M is suitable for heating systems and is somewhat cheaper because of the thin wall. None of these wall thicknesses are subject to failure because of the pressures found in ordinary heating and hot water systems, but the thicker wall tubing is less subject to mechanical damage through abuse.

The nominal size of copper tubing relates not to the tube itself, but is $\frac{1}{8}$ in. larger than its outside diameter to refer instead to the size of iron pipe which is equivalent in its pressure drop characteristics. The first four sizes turn out to have an actual outside diameter exactly equal to the *next largest* nominal size, which is incongruous enough; furthermore, tubing of those same four sizes is sold and referred to by its *actual* outside diameter in the air-conditioning and automotive fields. Thus ¾ in. copper water pipe sold in a plumbing store has a 0.875 in. outside diameter while ¾ in. copper tube for air conditioning will be only 0.750 in. in diameter. The air-conditioning tube will usually be "soft"—supplied in coils— and the plumbing tube "hard"—supplied in lengths.

The pressure drop through tube or ductwork is calculated just as in Examples 12-4a and 12-4b, of course using the total fluid flows to all the collectors when appropriate. To speed pressure drop calculations in water tube, the nomograph in Fig. 12-14 can be used instead of eqs. 12-2b and 12-5. Perfect agreement should not be expected because the viscosity of water varies rapidly with temperature.

Fittings cause additional pressure drop, which can be roughly accounted for by adding 50% to the length of piping. When more exact estimates are needed, Table 12-3 gives the length of pipe to add for each fitting according to its nominal size.

Water velocity in tubes should not exceed 1.2 m/s (4 ft/sec) because of the high noise characteristic of high-velocity flow. Above 1.8 m/s (6 ft/sec) erosion

**TABLE 12-3.** Allowance for Friction Loss in Valves and Fittings Expressed as Equivalent Length of Tube (Courtesy of Copper Development Association, Inc.)

| Fitting Size, Inches | Standard Ells 90° | Standard Ells 45° | 90° Tee Side Branch | 90° Tee Straight Run | Coupling | Gate Valve | Globe Valve |
|---|---|---|---|---|---|---|---|
| ⅜ | 0.5 | 0.3 | 0.75 | 0.15 | 0.15 | 0.1 | 4 |
| ½ | 1 | 0.6 | 1.5 | 0.3 | 0.3 | 0.2 | 7.5 |
| ¾ | 1.25 | 0.75 | 2 | 0.4 | 0.4 | 0.25 | 10 |
| 1 | 1.5 | 1.0 | 2.5 | 0.45 | 0.45 | 0.3 | 12.5 |
| 1¼ | 2 | 1.2 | 3 | 0.6 | 0.6 | 0.4 | 18 |
| 1½ | 2.5 | 1.5 | 3.5 | 0.8 | 0.8 | 0.5 | 23 |
| 2 | 3.5 | 2 | 5 | 1 | 1 | 0.7 | 26 |
| 2½ | 4 | 2.5 | 6 | 1.3 | 1.3 | 0.8 | 33 |
| 3 | 5 | 3 | 7.5 | 1.5 | 1.5 | 1 | 40 |
| 3½ | 6 | 3.5 | 9 | 1.8 | 1.8 | 1.2 | 50 |
| 4 | 7 | 4 | 10.5 | 2 | 2 | 1.4 | 63 |
| 5 | 9 | 5 | 13 | 2.5 | 2.5 | 1.7 | 70 |
| 6 | 10 | 6 | 15 | 3 | 3 | 2 | 84 |

Equivalent Length of Tube, Feet (for meters, multiply by 0.3)

NOTE: Allowances are for streamlined soldered fittings and recessed threaded fittings. For threaded fittings, double the allowances shown in the table.

of the metal becomes serious, especially at elbows. Velocity in piping or ductwork is given by

$$u = \frac{W}{\rho A_x} \qquad (12\text{-}7)$$

To decrease noise in piping it is also desirable to control flow rates by the pressure drop through restrictive piping rather than with valves, which are often noisy.

**EXAMPLE 12-5**

Pressure drop through the parallel-tube collectors of Example 12-4a has been increased to 0.304 m (1 ft) water by doubling the flow rate to the collectors and installing deliberate pressure loss. What size copper water tube should be used for the 12.19 m (40 ft) run to 44.59 m² (480 ft²) of these 1.22 m × 2.43 m (4 × 8 ft) collectors? Use the water pump characteristic in Fig. 12-10.

The piping size is first estimated and then changed as necessary.

Number of collectors:

$$\frac{44.59 \text{ m}^2}{1.22 \text{ m} \times 2.43 \text{ m}} = 15$$

$$W = 0.0247 \text{ kg/s} \times 2 \times 15 = 0.741 \text{ kg/s}$$

Assume $1\frac{1}{4}$ in. copper tube is installed, and consult Table 12-2 for dimensions and area. The pipe run is (12.19 × 2) m both ways, or 36.6 m including an extra 50% for fittings. By eq. 5-1a,

$$N_{\text{Re}} = \frac{4 \times \dfrac{0.741 \text{ kg}}{\text{s}} \times \dfrac{1000}{\text{k}}}{\dfrac{0.432 \text{ g}}{\text{m}\cdot\text{s}} \times \pi \times \dfrac{32.13}{1000}} = 67\,970$$

$$f = 0.001\,40 + \frac{0.125}{(67\,970)^{0.32}} = 0.004\,95$$

From eq. 12-5,

$$\Delta h = \frac{32 \times 0.004\,95}{\pi^2} \times \frac{\text{s}^2}{9.8 \text{ m}} \times \left(\frac{0.741 \text{ kg}}{\text{s}}\right)^2 \times \frac{36.6 \text{ m}}{\left(\dfrac{32.13}{1000} \text{ m}\right)^5} \times \left(\frac{\text{m}^3}{980 \text{ kg}}\right)^2$$

$$= 1.0 \text{ m H}_2\text{O}$$

or from eq. 12-3,

$$\Delta p = \frac{32 \times 0.004\,95}{\pi^2} \times \left(\frac{0.741 \text{ kg}}{\text{s}}\right)^2 \times \frac{36.6 \text{ m}}{\left(\dfrac{32.13}{1000} \text{ m}\right)^5} \times \frac{\text{m}^3}{980 \text{ kg}} \times \frac{\text{N}\cdot\text{s}^2}{1 \text{ kg}\cdot\text{m}} \times \frac{\text{Pa}}{\text{N/m}^2}$$

$$= 9611 \text{ Pa}$$

The proper choice of tube can be confirmed by recalculations using 1 in. and $1\frac{1}{2}$ in. tube. The fifth power of diameter in eqs. 12-5 and 12-3 assures that only one of the three sizes is appropriate for the pressure rise available in Fig. 12-10.

In Engineering units,

Number of collectors installed:

$$480 \text{ ft}^2/(4 \text{ ft} \times 8 \text{ ft}) = 15$$

New flow rate:

$$W = 3.273 \text{ lb/min} \times 2 \times 15 = 98.19 \text{ lb/min}$$

or

$$98.19 \text{ lb/min}/(8.33 \text{ lb/gal}) = 11.79 \text{ gal/min}$$

If $1\frac{1}{4}$ in. copper tube is installed, the pressure drop as estimated from Fig. 12-14 is 3.5 ft/100 ft of tube. The velocity is about 3 ft/sec. For the 80-ft pipe length another 50% is added to account for fittings, giving an equivalent length of 120 ft, which is about 4.7 ft of lost heat, including the 1-ft head lost in the collectors. The pump in Fig. 12-10 can move about 970 gal/hr (16.2 gal/min) at this head, which is more than adequate. With one size smaller tubing the loss is 8 ft/100 ft of tube—too high for further consideration. Confirming these calculations with eqs. 5-1a, 12-2b, and 12-5,

$$N_{\text{Re}} = \frac{4W}{\mu\pi D} = \frac{4 \times 98.19 \text{ lb/min}}{0.0174 \text{ lb/min ft} \times \pi \times 1.265 \text{ in.}} \times \frac{12 \text{ in.}}{\text{ft}} = 68\,160$$

$$f = 0.001\,40 + \frac{0.125}{(68\,160)^{0.32}} = 0.004\,95$$

$$\Delta h = \frac{32 \times 0.004\,95}{\pi^2} \times \frac{\sec^2}{32.2 \text{ ft}} \times \left(\frac{98.19 \text{ lb}}{\text{min}}\right)^2 \times \frac{120 \text{ ft}}{\left(\frac{1.265}{12}\right)^5 \text{ft}^5} \times \left(\frac{\text{ft}^3}{61.2 \text{ lb}}\right)^2$$

$$\times \left(\frac{\text{min}}{60 \text{ sec}}\right)^2 = 3.29 \text{ ft H}_2\text{O}$$

and the total is 4.29 ft $\text{H}_2\text{O}$. This provides $1050/60 = 17.5$ gal/min flow with the pump specified, which is in good agreement with the 16.2 gal/min determined using Fig. 12-14.

**Effect of Antifreeze on Pump Characteristics.** For most centrifugal pumps the characteristic curve for water can be used for antifreeze mixtures if the curve is expressed in meters or feet of head as tested but interpreted in terms of the meters or feet of fluid pumped.* The nomograph in Fig. 12-14 will not

---

* This is the first approximation to make. The secondary effect of viscosity on the performance curve makes it desirable to obtain curves made with the exact fluid to be used. This is especially true when changing to a markedly different fluid, such as silicone oil.

work, of course. Calculations like those ending the previous example—but with antifreeze properties—will show an increased pressure drop (in terms of meters or feet of glycol) of 20 to 50% depending on the temperature.

## Pressure Drop Through Ductwork

Equations 12-5 and 12-6 express the lost head for air in round and rectangular ducts, respectively. Equivalent lengths of some common duct fittings are shown in Fig. 12-15.

To keep friction loss and noise to a minimum, a maximum velocity of 5 m/s (1000 ft/min) is recommended. Using a rearrangement of eq. 12-7 the required cross-sectional duct area can then be determined:

$$A_x = \frac{Wv}{u} \tag{12-8}$$

where $Wv$ is the volumetric flow rate (m³/s or ft³/min) and $u$, the velocity (m/s or ft/min).

### EXAMPLE 12-6

An air solar system has a circulation rate of 0.448 m³/s (950 ft³/min). What is the duct size and lost head to provide this flow through ducting with an equivalent length of 61 m (200 ft) at a velocity of 5.08 m/s (1000 ft/min)?

From eq. 12-8,

$$A_x = \frac{Wv}{u} = \frac{0.448 \text{ m}^3}{\text{s}} \times \frac{\text{s}}{5.08 \text{ m}} = 0.0882 \text{ m}^2$$

A 10 in. × 14 in. duct (0.254 m × 0.356 m) has an area of 0.254 m × 0.356 m = 0.0904 m², and a perimeter of 2(0.254 + 0.356) = 1.22 m. The weight flow is

$$W = \frac{0.448 \text{ m}^3}{\text{s}} \times \frac{1.187 \text{ kg}}{\text{m}^3} = 0.532 \text{ kg/s}$$

By eq. 5-1b,

$$N_{\text{Re}} = \frac{4 \times 0.532 \text{ kg/s} \times 1000/\text{k}}{(0.018 \text{ g/m·s}) \times 1.22 \text{ m}} = 96\,900$$

By eq. 12-2b,

$$f = 0.001\,40 + \frac{0.125}{(96\,900)^{0.32}} = 0.004\,57$$

By eq. 12-6,

$$\Delta h = \frac{0.004\,57 \times (0.532 \text{ kg/s})^2 \times 61 \text{ m} \times 1.22 \text{ m}}{2 \times 9.8 \text{ m/s}^2 \times (0.0904 \text{ m}^2)^3 \times 1000 \text{ kg/m}^3 \times 1.187 \text{ kg/m}^3}$$
$$= 0.0056 \text{ m H}_2\text{O}$$

or 5.6 mm H₂O.

Or by eq. 12-4,

$$\Delta p = \frac{0.004\ 57 \times (0.532\ \text{kg/s})^2 \times 61\ \text{m} \times 1.22\ \text{m}}{2 \times \dfrac{1\ \text{kg} \cdot \text{m}}{\text{N} \cdot \text{s}^2} \times \dfrac{\text{N/m}^2}{\text{Pa}} \times (0.0904\ \text{m}^2)^3 \times \dfrac{1.187\ \text{kg}}{\text{m}^3}} = 54.9\ \text{Pa}$$

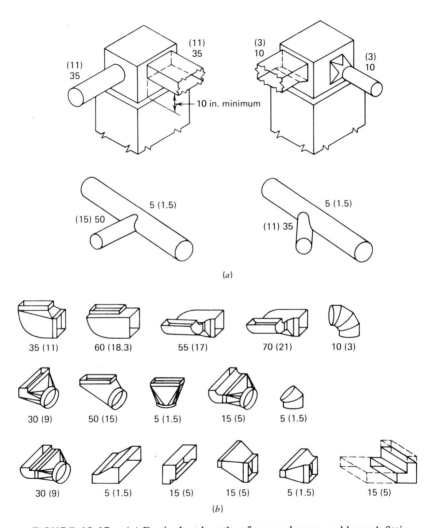

**FIGURE 12-15.** (*a*) Equivalent lengths of some plenum and branch fittings in feet and meters in parentheses. (*b*) Equivalent lengths of common duct fittings in feet and meters in parentheses. (Courtesy of Air Conditioning Contractors of America, 1228 17th St., N.W., Washington, D.C. 20036.)

In Engineering units,

$$A_x = \frac{Wv}{u} = \frac{950 \text{ ft}^3}{\text{min}} \times \frac{\text{min}}{1000 \text{ ft}} = 0.95 \text{ ft}^2$$

A duct 10 in. × 14 in. has an area of $10 \times 14/144 = 0.972 \text{ ft}^2$. This has a perimeter $P$ of $2(10 + 14) = 48$ in. The weight flow is

$$W = \frac{950 \text{ ft}^3}{\text{min}} \times \frac{\text{lb}}{13.5 \text{ ft}^3} = 70.37 \text{ lb/min}$$

By eq. 5-1b

$$N_{\text{Re}} = \frac{4 \times 70.37 \text{ lb}}{\text{min}} \times \frac{\text{min ft}}{0.000 \, 726 \text{ lb}} \times \frac{1}{48 \text{ in.}} \times \frac{12 \text{ in.}}{\text{ft}} = 96 \, 928$$

By eq. 12-2b,

$$f = 0.001 \, 40 + \frac{0.125}{(96 \, 928)^{0.32}} = 0.004 \, 57$$

By eq. 12-6,

$$\Delta h = \frac{0.004 \, 57 \times (70.37 \text{ lb/min})^2 \times 200 \text{ ft} \times 48 \text{ in.} \times 13.5 \text{ ft}^3/\text{lb}}{2 \times 32.2 \text{ ft/sec}^2 \times (0.972 \text{ ft}^2)^3 \times 62.4 \text{ lb/ft}^3 \times (60 \text{ sec/min})^2}$$

$$= 0.221 \text{ in. H}_2\text{O}$$

Or by eq. 12-4,

$$\Delta p = \frac{0.004 \, 57 \times (70.37 \text{ lb/min})^2 \times 200 \text{ ft} \times 48 \text{ in.} \times 13.5 \text{ ft}^3/\text{lb}}{2 \times 32.2 \text{ ft/sec}^2 \times (0.972 \text{ ft}^2)^3 \times 12 \text{ in./ft} \times 144 \text{ in.}^2/\text{ft}^2 \times (60 \text{ sec/min})^2}$$

$$= 0.007 \, 97 \text{ lb/in.}^2$$

The pressure drop of air in ductwork can also be quickly determined graphically. For round ducts, Fig. 12-16 is used directly. For rectangular ducts, Table 12-4 is used to establish a round-duct equivalent diameter with which to enter Fig. 12-16.

The complete design of ductwork is beyond the scope of this book. For further information, consult an air-conditioning textbook such as McQuiston and Parker (4) or the ASHRAE handbooks (5).

## AUXILIARY HEATING

Solar heat will be economical for only a fraction of the building heat and hot water load in most cases. This will continue to be true as long as auxiliary heating sources are available that require little capital investment and yet deliver heat at a cost comparable to the main solar system.

Most of the common auxiliary sources deliver auxiliary heat with *zero* net investment because a complete conventional heating system must be installed in a solar-heated building as a precaution against breakdowns or unseasonably cold

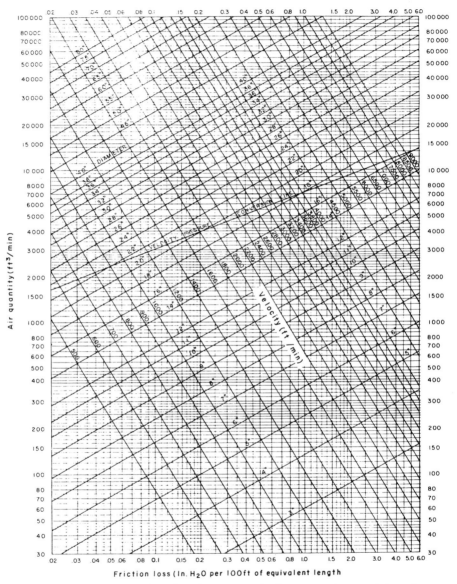

**FIGURE 12-16.** Pressure drop for air at atmospheric pressure in round duct. (*Source: Handbook of Air Conditioning Systems Design*, Carrier Air Conditioning Company, © 1965, McGraw-Hill. Used with permission of McGraw-Hill Book Company.)

**TABLE 12-4.** Air Duct Dimensions, Area, and Circular Equivalent Diameter [a]

| Side dimension | 6 Area sq ft | 6 Diam. in. | 8 Area sq ft | 8 Diam. in. | 10 Area sq ft | 10 Diam. in. | 12 Area sq ft | 12 Diam. in. | 14 Area sq ft | 14 Diam. in. | 16 Area sq ft | 16 Diam. in. | 18 Area sq ft | 18 Diam. in. | 20 Area sq ft | 20 Diam. in. | 22 Area sq ft | 22 Diam. in. |
|---|---|---|---|---|---|---|---|---|---|---|---|---|---|---|---|---|---|---|
| 10 | 0.39 | 8.4 | 0.52 | 9.8 | 0.65 | 10.9 | | | | | | | | | | | | |
| 12 | 0.45 | 9.1 | 0.62 | 10.7 | 0.77 | 11.9 | 0.94 | 13.1 | | | | | | | | | | |
| 14 | 0.52 | 9.8 | 0.72 | 11.5 | 0.91 | 12.9 | 1.09 | 14.2 | 1.28 | 15.3 | | | | | | | | |
| 16 | 0.59 | 10.4 | 0.81 | 12.2 | 1.02 | 13.7 | 1.24 | 15.1 | 1.45 | 16.3 | 1.67 | 17.5 | | | | | | |
| 18 | 0.66 | 11.0 | 0.91 | 12.9 | 1.15 | 14.5 | 1.40 | 16.0 | 1.63 | 17.3 | 1.87 | 18.5 | 2.12 | 19.7 | | | | |
| 20 | 0.72 | 11.5 | 0.99 | 13.5 | 1.26 | 15.2 | 1.54 | 16.8 | 1.81 | 18.2 | 2.07 | 19.5 | 2.34 | 20.7 | 2.61 | 21.9 | | |
| 22 | 0.78 | 12.0 | 1.08 | 14.1 | 1.38 | 15.9 | 1.69 | 17.6 | 1.99 | 19.1 | 2.27 | 20.4 | 2.57 | 21.7 | 2.86 | 22.9 | 3.17 | 24.1 |
| 24 | 0.84 | 12.4 | 1.16 | 14.6 | 1.50 | 16.6 | 1.83 | 18.3 | 2.14 | 19.8 | 2.47 | 21.3 | 2.78 | 22.6 | 3.11 | 23.9 | 3.43 | 25.1 |
| 26 | 0.89 | 12.8 | 1.26 | 15.2 | 1.61 | 17.2 | 1.97 | 19.0 | 2.31 | 20.6 | 2.66 | 22.1 | 3.01 | 23.5 | 3.35 | 24.8 | 3.71 | 26.1 |
| 28 | 0.95 | 13.2 | 1.33 | 15.6 | 1.71 | 17.7 | 2.09 | 19.6 | 2.47 | 21.3 | 2.86 | 22.9 | 3.25 | 24.4 | 3.60 | 25.7 | 4.00 | 27.1 |
| 30 | 1.01 | 13.6 | 1.41 | 16.1 | 1.82 | 18.3 | 2.22 | 20.2 | 2.64 | 22.0 | 3.06 | 23.7 | 3.46 | 25.2 | 3.89 | 26.7 | 4.27 | 28.0 |
| 32 | 1.07 | 14.0 | 1.48 | 16.5 | 1.93 | 18.8 | 2.36 | 20.8 | 2.81 | 22.7 | 3.25 | 24.4 | 3.68 | 26.0 | 4.12 | 27.5 | 4.55 | 28.9 |
| 34 | 1.13 | 14.4 | 1.58 | 17.0 | 2.03 | 19.3 | 2.49 | 21.4 | 2.96 | 23.3 | 3.43 | 25.1 | 3.89 | 26.7 | 4.37 | 28.3 | 4.81 | 29.7 |
| 36 | 1.18 | 14.7 | 1.65 | 17.4 | 2.14 | 19.8 | 2.61 | 21.9 | 3.11 | 23.9 | 3.63 | 25.8 | 4.09 | 27.4 | 4.58 | 29.0 | 5.07 | 30.5 |
| 38 | 1.23 | 15.0 | 1.73 | 17.8 | 2.25 | 20.3 | 2.76 | 22.5 | 3.27 | 24.5 | 3.80 | 26.4 | 4.30 | 28.1 | 4.84 | 29.8 | 5.37 | 31.4 |
| 40 | 1.28 | 15.3 | 1.81 | 18.2 | 2.33 | 20.7 | 2.88 | 23.0 | 3.43 | 25.1 | 3.97 | 27.0 | 4.52 | 28.8 | 5.07 | 30.5 | 5.62 | 32.1 |
| 42 | 1.33 | 15.6 | 1.86 | 18.5 | 2.43 | 21.1 | 2.98 | 23.4 | 3.57 | 25.6 | 4.15 | 27.6 | 4.71 | 29.4 | 5.31 | 31.2 | 5.86 | 32.8 |
| 44 | 1.38 | 15.9 | 1.95 | 18.9 | 2.52 | 21.5 | 3.11 | 23.9 | 3.71 | 26.1 | 4.33 | 28.2 | 4.90 | 30.0 | 5.55 | 31.9 | 6.12 | 33.5 |
| 46 | 1.43 | 16.2 | 2.01 | 19.2 | 2.61 | 21.9 | 3.22 | 24.3 | 3.88 | 26.7 | 4.49 | 28.7 | 5.10 | 30.6 | 5.76 | 32.5 | 6.37 | 34.2 |
| 48 | 1.48 | 16.5 | 2.09 | 19.6 | 2.71 | 22.3 | 3.35 | 24.8 | 4.03 | 27.2 | 4.65 | 29.2 | 5.30 | 31.2 | 5.97 | 33.1 | 6.64 | 34.9 |
| 50 | | | 2.16 | 19.9 | 2.81 | 22.7 | 3.46 | 25.2 | 4.15 | 27.6 | 4.84 | 29.8 | 5.51 | 31.8 | 6.19 | 33.7 | 6.87 | 35.5 |

[a] *Handbook of Air Conditioning Systems Design*, Carrier Air Conditioning Company, © 1965, McGraw-Hill. Used with permission of McGraw-Hill Book Company.

or cloudy weather. The single exception is an oversized annual storage system, which will have hot water on hand at the beginning of the season and therefore needs no backup.

Solar heating and hot water systems are most competitive when electrical energy is the alternative source of heat, and electricity will therefore be the most common auxiliary heat source. In spite of its cost, electricity has many advantages. It is quite economical if used to power a mechanical "heat pump" that operates independently without integration into the solar system (7). Electric power seems assured—for a price—because it can be made with coal or nuclear fuels, neither of which is in short supply (at least in the United States). Electricity is so essential to all heating systems (including solar) that it is not interruptible in a practical sense by boycott or strike.

Oil fuel has the major advantage that it can be stored. Once an adequate supply is on hand, it is completely uninterruptible except for the electricity to run its associated equipment. The price of this international commodity is currently rising above that of other fuels, and as time goes on and supplies run shorter there may be no limit to the price that oil can reach in a free market, because there is no replacement for it as a motor fuel in the quantities that are required.

Natural gas is an ideal fuel, burning cleanly in inexpensive furnaces. Its reliability is the poorest of all, with storage impossible, supplies limited, and free-market price levels constrained by law. It would be imprudent to rely on natural gas for an auxiliary fuel in a solar installation, although a gas system convertible to another fuel would be suitable.

Coal and wood are solid fuels with inherent disadvantages for automatic operation in residences and commercial buildings. Coal is almost out of the question because the low-pollution grades are scarce and expensive. Wood is enjoying new popularity, especially in the northeastern United States, where farmland abandoned 50 years ago is now yielding a harvest of hardwood trees. Given the reduced heating load in a well-designed solar house, an efficient wood stove is an attractive alternative energy source. The fuel is easily stored, although bulky. It is free from possible constraints on supply. For many homeowners it is free for the labor of cutting and splitting the supply growing on their own property. Wood is only for the enthusiast, however, since its use requires constant attention and labor, and its comfort level is debatable since it is not well suited for central heating.

## Combustion Efficiency

Fuel burners and furnace design are beyond the scope of this book. Interested readers are referred to Burkhardt's book (6), which is by far the best treatment of the subject. Over the next few pages we will explore the general characteristics of combustion systems, generalizing the current and likely future practice.

Since up to 50% of the energy supplied to a solar-heated structure may be supplied by traditional fuels, it is very worthwhile to make their use as efficient as possible. Engineering effort in the auxiliary system design and installation is therefore just as important as in the main solar-heating system.

Electrical systems are the easiest to make efficient. Inherently, of course, electricity is converted 100% into heat energy. With a little carelessness, however, this need not be true. Baseboard heaters can be recessed into outside walls so that some of the expensive energy is wasted by leakage under the higher temperature gradient through the thin insulation installed behind them. Attic and cellar ducting can be underinsulated, and heat can be wasted through installation of an *electric furnace* out of doors where the heat loss does not eventually heat the structure.

When burning common fuels, two areas must receive special attention to achieve the best energy yields:

1. The air that is used for combustion must be held to a minimum, so that the fuel energy is not used to heat an oversupply of air that passes wastefully up the chimney.
2. As much heat from the combustion process must be transferred from the exhaust gases to the heating system as is practical.

The thermal performance of a furnace is reflected in the *combustion efficiency* of the system, which can be determined from the carbon dioxide (or oxygen) content of the exhaust gases and their temperature, using simple but not inexpensive equipment.

**Limiting Combustion Air.** Natural gas, some light grades of fuel oil (kerosene) and wood can be burned effectively with air supplied by aspiration from the natural draft created by the chimney. The chimney is not just a vent. It is a tall column of hot air that tends to rise, aspirating fresh air into the system through the furnace or stove in rough proportion to its height and temperature. Wood must have more excess air than the other fuels so that the heavy liquids volatilized during the combustion process can be efficiently burned. For this reason wood is inherently less efficient in a furnace or stove than the other fuels.

Ordinary (No. 2) fuel oil is not as volatile as natural gas or kerosene, nor as well-structured as split wood for air contact, so that it must be atomized into fine droplets before burning and provided with forced air to promote combustion. No. 6 fuel oil, a commercial fuel, is so viscous that it must be heated in order to be pumped and atomized. It is used because it is less expensive than No. 2 fuel oil, a consequence of its formation as a natural waste product of oil refining.

The practical auxiliary fuels all burn cleanly in a proper furnace. Failure to provide enough air for combustion creates smoke with any carbonaceous fuel in most combustion systems; in extreme situations carbon monoxide formation can result with often deadly consequences.

Recent developments in both oil and wood combustion can be expected to increase combustion efficiency considerably. New oil burners are becoming available in which the flame is a vortex in which fuel particles recirculate, making soot formation *impossible*. Excess air can therefore be easily set at any desired level (including less than zero, in which case carbon monoxide is formed!). In newer wood stoves the excess air is reduced by providing air flow to the wood from the *top* of the stack to the bottom. In this way the volatiles driven off the fresh wood (on the top) are combusted thoroughly as they pass through the hot coals.

**Air Supply.** In home-heating systems, air for combustion is generally drawn from the heated living space. It is replaced through leakage induced by the partial vacuum created by the chimney. If a structure is well sealed, or the chimney ineffective, the exhaust gases may not be able to leave the structure. For this reason alone there should be separate provision for outside combustion air to be vented to the furnace in a tightly constructed solar house. The use of outside air rather than heated air also improves the overall combustion efficiency a percent or so but, except in the case of woodstoves and fireplaces, which use much excess air, this effect is not too important.

Fireplaces should receive special attention in any energy efficient structure. Not only do they use prodigious amounts of air, but they draw it all from infiltration through windows and other leaky areas into the heated space. In colder climates the additional heating load may require more heat than the fire produces for the house. The drafts created can cause discomfort except where the pleasant radiation is immediately available. Fireplaces therefore *must* be provided with their own source of outside air if they are to be used for space heating in any but the mildest climate. They can be completely isolated from the room air by the installation of a glass "screen" through which the radiant energy can pass, provided that the vents provided are closed and air supplied from elsewhere.

Excessive air is naturally provided to furnaces in most two-story structures through overgenerous chimney sizing. This situation can be nullified by the installation of a deliberate obstruction in the flue pipe. Because the penalty for a faulty installation may be carbon monoxide poisoning, this should only be done by specially trained personnel. The more common solution to this problem is to vent fresh air to the chimney through a *damper*. Damper installations are common, but unless they vent outside air to the chimney they waste a great deal of energy. Excessive draft can also be treated by restricting the flow of combustion air to the furnace. This is a safer procedure than restricting the flue but, since the furnace will be operating with more "draft" or vacuum than its original design called for, leaks may develop that will make it difficult to achieve good efficiency without smoke or carbon monoxide formation.

**Heat Transfer from the Combustion Gases.** Immediately after combustion, the flue gases are at 1100 to 1900 °C (2000–3500 °F), depending on the amount of excess air. They are cooled in a rudimentary heat exchanger designed for low pressure drop rather than for efficient cooling. This is appropriate because the gases should not be cooled below 120 to 175 °C (250–350 °F); they must be left hot enough to create a proper draft in the chimney.

The heat exchanger may heat circulating room air directly through sheet metal or it may heat water in what is called a *hot water boiler*. That arrangement is similar to heating a pot on a stove, but the water is never allowed to boil. The air heater is a little more sophisticated, with a larger area for heat exchange.

In traditional fuel oil furnaces the combustion gases leave at 200 to 300 °C (400–600 °F) with perhaps 60% excess air. This will give 77 to 82% combustion efficiency, which is reasonably good considering that about 90% is the maximum possible without condensing the water produced by combustion. The trouble is that the furnaces are difficult to adjust to these conditions and do not stay well-adjusted for long. Therefore, they are usually set with much more excess air to avoid the possibility of soot formation. Average combustion efficiency has been found to be about 60% in occupied houses, a consequence of the difficulty of furnace adjustments, the incompetence of servicemen, and the homeowner's own inattention to routine maintenance. It also should be noted that energy shortage or no, the fuel oil dealer can hardly be expected to have a compelling interest in reducing the homeowner's oil consumption, while he does have a great interest in reducing the number of service calls, on which he seldom makes money.

Temperatures leaving the furnace are lowered if a better heat exchanger is installed, of course. This increases the capital cost of the furnace and does not sell well because most furnaces are bought by contractors, not homeowners. Efficiency of ordinary furnaces can be improved by after-market heat exchangers installed in the flue, which heat a flow of air to be used for basement heating or ducted elsewhere. These devices are a good investment if flue gas leaves the furnace at more than about 230 °C (450 °F).

Contractor-installed furnaces tend to be oversized because increased capacity is cheaper than the engineering effort to size the furnace properly. Such furnaces can be reduced in capacity by changing orifices (for gas) or nozzles (for oil). Increased efficiency will often result because the heat flux will be less through the heat exchanger surfaces because of the longer duty cycle.

**Standby Heat Losses.**    When a furnace is in a standby condition, it is usually hot and some heat is being wasted up the chimney if natural convection occurs. In the case of a gas hot water heater, the losses can be extreme, with natural circulation of hot water quickly providing more heat to the heat exchanger so that it continues to heat the air passing through. In oil-fired hot water and hydronic heating systems the situation can also exist, although the more restricted inlet air passages of the oil burner fan reduce the circulation rate of the air considerably. In a hot air heating system, when combustion ceases the residual heat in the heat exchanger is likely to be swept up the chimney instead of into the house because of the same effect. These losses are all compounded if the air lost is interior air already heated by the furnace, which must be replaced by infiltration of outside air.

These problems can be dealt with by closing the flue when there is no combustion taking place. This must be done reliably, because if the flue is accidentally left closed when the flame restarts, fire and carbon monoxide poisoning are likely consequences. Devices to do this safely have been marketed in Europe for many years and are slowly reaching a conservative market in the United States. Their safe installation is a just cause for concern, but fail-safe devices to protect the homeowner are mandatory and seem likely to control the hazard.

## PROBLEMS

1. Air collectors having an area of 70 m² (753.5 ft²), $F_i \tau \alpha = 0.61$, and $F_i U_L = 3.2$ W/°C·m² (0.5635 Btu/hr ft² °F) are cooled with 0.7 m³/s (1483 ft³/min) of air provided by a fan drawing 1.3 hp (1119 W).
   (a) If the inlet air is 20 °C (68 °F) at what inlet-to-outlet temperature differential is the heat collected equal to the fan power?
   (b) Repeat for 50 °C (122 °F) and 80 °C (176 °F) inlet temperatures.
   (c) What is the efficiency of the fan-motor combination if the overall pressure drop is 21.84 mm H₂O (0.86 in. H₂O)?
   (d) If the fan and drive characteristics are given in Fig. 12-13, what is the motor efficiency?
   (e) There are an equivalent of 30 m (98.43 ft) of rectangular 22 in. wide ductwork to and from the collectors. Size the ducts, leaving 6.5 mm H₂O (0.256 in. H₂O) for the collectors and 10 mm H₂O (0.394 in. H₂O) for the rock storage.

2. The flow capacity rate of silicone oil through a 0.914 m × 2.13 m (3 ft × 7 ft) collector is 100 W/°C·m² (17.6 Btu/hr °F ft²) at 66 °C (150 °F).
   (a) What is the pressure drop in series through sixteen 0.9 m (2.95 ft) long serpentine passages of ⅜ in. type L copper tube?
   (b) What is the pressure drop in parallel flow through eight 2 m (6.56 ft) passages of ⅜ in. type M copper tube?
   (c) What size type L copper tube should be used for a total length run of 45 m (148 ft) to 20 collectors and back? The permissible pressure drop is 1 m H₂O (3.28 ft H₂O). What is the flow rate in L/s (gal/hr)?

3. Air flows underneath the collector plate in a 6.7 m (22 ft) long air collector built into a roof between joists. What is the pressure drop if the flow passage is 20 mm (0.787 in.) deep and the air flow at 21 °C (70 °F) is 2.54 L/s per m² (5 ft³/min ft²)?

4. The pump of Fig. 12-10 is to be used for silicone oil. What pressure rise is available at 1200 gal/hr (1.5 in. connections)?

# REFERENCES

1. G. O. G. Löf and G. A. Löf, "Performance of solar swimming pool heater—transparent cover type," Proceedings of the 1977 Annual Meeting, American Section of the International Solar Energy Society, Orlando, Fla.

2. J. E. Scott, "The solar water heater industry in south Florida: history and projections," *Solar Energy 18*, 387–393 (1976).

3. T. B. Drew, E. C. Koo, and W. M. McAdams, *Trans AIChE 28:* 56 (1932).

4. F. C. McQuiston and J. P. Parker, *Heating, Ventilating, and Air Conditioning*, Wiley: New York, 1977.

5. *ASHRAE Handbook & Product Directory, 1977 Fundamentals*, American Society of Heating, Refrigerating and Air-Conditioning Engineers, Inc., 345 East 47th Street, New York, N.Y. 10017.

6. C. H. Burkhardt, *Domestic and Commercial Oil Burners—Installation and Servicing*, 3rd Edition, McGraw-Hill, 1969.

7. T. L. Freeman, J. W. Mitchell, and T. E. Audit, "Performance of combined solar-heat pump systems," *Solar Energy 22*, 125–135 (1979).

# Passive Heating Systems

Thus far in this book we have dealt with *active* solar-heating and hot water systems. In an active system mechanical means are used to circulate a heat transfer fluid to the collector, to storage, and for distribution to the point of use. *Passive* systems use natural elements to distribute the collected heat. Often a passive system becomes a "live-in" solar collector with profound impact on the architectural design of the structure and the life-style of the occupants.

Since temperature differences are used to distribute heat, most passive heating systems cannot and do not attempt to maintain a uniform indoor temperature. Many achieve a degree of control with adjustable drapes or louvers on glazed portions of the structure, or even use circulation fans. Such systems are called *hybrid* systems if they mechanically circulate to a storage mass, but these are also true passive systems if they can function adequately without electrical power.

Passive systems are traditionally designed by architects because of the importance of the structure to their design, and because engineers have not known enough about the thermal processes involved to be of much help. The operating results from completed structures, as measured by auxiliary fuel usage, have not been generally useful to the analytical designer because the occupants have tended to be solar enthusiasts having a good tolerance for temperature extremes. They are also likely to use wood as an auxiliary fuel, which does not lend itself to accurate measurement of thermal energy consumption.

This impasse has recently been broken by J. D. Balcomb and co-workers at Los Alamos Scientific Laboratory (1). Using their engineering backgrounds and the considerable resources of that institution, they have quantified the study of passive structures and provided the basis for the first portion of this chapter.

## PASSIVE STRUCTURES

### Windows

The simplest passive element is the window. A south-facing, double-glazed window will admit more thermal energy as radiation than it loses through heat loss during any month in most U.S. locations. For energy conservation purposes then, as much glass as is desired can usually be added to south walls, so long as the sun is free to shine to the interior in the winter months. In a properly designed passive structure the excess heat from windows is collected and stored for use in making up other losses.

**EXAMPLE 13-1**

A double-glazed window is located on the south side of a house in Hartford. Low-iron glass (0.05% $Fe_2O_3$) is used for glazing. What are the potential losses, gains, and the net heat gain for January and March?

**Gains:** The transmittance of the glass is 88.5% (Table 4-1). From Appendix B at a 90° slope, $I_T = 300.1$ and 337.8 MJ/m² for January and March, respectively.

$$\text{January potential gain} = (0.885)^2 \times 300.1 = 235.0 \text{ MJ/m}^2$$
$$\text{March potential gain} = (0.885)^2 \times 337.8 = 264.6 \text{ MJ/m}^2$$

**Losses:** The average temperatures for losses are $-3.8$ and 2.1 °C, respectively. They continue over the entire month, for 2.6784 Ms. The heat loss coefficient is the same as from a storm window in winter (Example 2-9). Assuming an average inside temperature of 21 °C,

$$\text{January losses} = UA\, \Delta Tt_T$$
$$= \frac{3.06 \text{ W}}{°C \cdot m^2} \times 1 \text{ m}^2 \times [21 - (-3.8)] °C \times 2.6784 \text{ Ms} \times \frac{J/s}{W}$$
$$= 203.3 \text{ MJ}$$
$$\text{March losses} = 3.06 \times (21 - 2.1) \times 2.6784 = 154.9 \text{ MJ}$$
$$\text{January net gain} = 235.0 - 203.3 = 31.7 \text{ MJ/m}^2$$
$$\text{March net gain} = 264.6 - 154.9 = 109.7 \text{ MJ/m}^2$$

Notice that the gains and losses during January for the above example are about equal, and therefore the excess heat is limited. In practice the radiation will be reduced even further by incident angle effects and less than perfect absorption. The tables of Appendix B were computed without snow cover and another 10 to 20% may be anticipated in months when snow cover is persistent, favorably affecting this balance. Other data bases often show even more radiation on a 90° slope. These have inevitably been calculated from measured total global horizontal radiation (there being very little 90° data available) using the optimistic Liu-Jordan relationship (3-22) and the approximation of eq. 3-19 to calculate the direct beam radiation, which will overstate the available solar energy on a 90° slope in January at Hartford, for instance, by about 18%.

## Direct-Gain Passive System

As the windows in a south wall are enlarged to take advantage of the solar gain, the temperatures in the sunlit space will tend to rise to uncomfortable levels in the

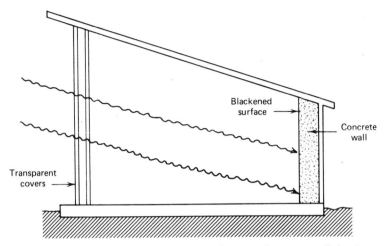

**FIGURE 13-1.** Direct gain structure using massive rear wall for heat storage.

daytime. If the sun shines on a deliberately massive wall or floor, much of the heat will instead be stored within the structure of the building and furnishings in what is called the *direct-gain* passive structure (Fig. 13-1). The wall will usually be at the back of the room, exposed directly to sunlight and painted a dark color. A wall out of the sun will not function well, since the room temperature must be well above that of the wall to deliver heat for storage to it. In fact, contrary to popular opinion, massive inside walls do not even help moderate winter temperatures—the heat transfer is too slow relative to the losses.* Vertical skylights or *clerestory* windows are a good application for the direct gain concept, since the occupants and furnishings are then out of the direct sun.

Large windows should be shaded by an overhang to reduce the incoming radiation in the warmer months. This is imperative in any passive structure.

## The Trombe Wall

The direct gain structure, quite literally, is a "live-in" collector. However, its occupants and furnishings do not thrive with so much sun—ordinarily only a portion of the living area can be so treated. Felix Trombe, a Frenchman, has perfected and popularized a second passive concept that deals quite effectively with this problem.

The Trombe wall (Fig. 13-2) is erected directly behind a large expanse of double

* However, for summer such walls add much to comfort through moderation. Summer cooling loads are much smaller than winter heating loads.

window glazing. It is a massive floor-to-ceiling wall of concrete, adobe, or even water (in drums), painted black or at least a dark color. Like the direct gain structure, it must have overhang to keep out the summer sun. Openings can be made in the wall to provide the desired window area. The wall intercepts most of the sunlight before it can come into the living area, converting it to heat which warms the massive wall. By the end of the day the wall can be tens of degrees above the room temperature with the rear surface providing indirect heat to compensate for the thermal losses of the structure.

If the Trombe wall is too thin, less than about 20 cm (8 in.) thick, it will heat up rapidly, increasing heat losses to the outside and causing premature heating of the inside ambient. If it is too thick—over about 50 cm (1.5 ft)—it will heat too slowly, and will not make its heat available during the same day to its rear surface unless the wall is of a very conductive material. The accumulated heat will then tend to be lost overnight through the exterior glazing, since the wall has not been adequately cooled.

Vents are usually provided at the top and bottom of the Trombe wall. They should be half the horizontal cross-sectional area of the air space behind the glass (2). They are intended to let natural convection speed the heat transfer to the living space from the front of the wall, reducing the wall temperature and thus decreasing exterior losses by bypassing storage for the immediate heating needs. The circulation in most installations has not been adequate, however, and a circulating fan is often installed to reduce the turnover time. Too long a turnover time will cause noticeable stratification of the air in the living space. The vents must, however, be closed at night lest a reverse thermosyphon effect *cool* the room

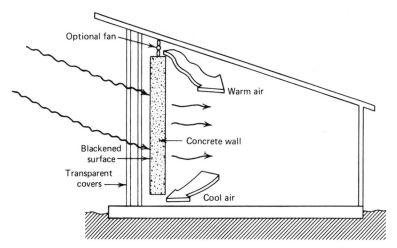

**FIGURE 13-2.** Trombe wall for passive solar heating. (From the Solar Home Book, B. Anderson and M. Riordan, Cheshire Books.)

through the exterior glazing. This can be done automatically with a check valve made from thin plastic, taped along its upper edge to the inside of the upper vent or the outside of the lower vent.


### Optimization

The effectiveness of the direct gain structure and the Trombe wall is dependent on the properties of the storage media and the transparent glazing. Balcomb has characterized the performance of the direct gain structure and the Trombe wall by simulation using Los Alamos, New Mexico weather conditions, and presented the results in a way that simplifies the optimal design of a passive structure (1).


**Thermal Coupling.** If the storage wall cannot transmit its heat to the room adequately, it will rise in temperature until it does finally lose that heat. In the Trombe wall the heat loss will be split to the inside, where it is useful, and to the outside, where it is lost. If there is too high a heat transfer coefficient between the storage wall and the room ambient, however, there will be no storage of heat and nights will require auxiliary heat while days may need deliberate cooling. Too low a coefficient will cause excessive wall temperatures and too much heat will be lost to the outside. There is, therefore, an optimum *coupling coefficient* for heat transfer to the room.

Figure 13-3 shows the effect of changes in this heat transfer coefficient on the thermal performance of both direct gain and Trombe wall structures, as determined by simulation with Los Alamos weather and the otherwise optimal system that will be developed in the next few pages. The Trombe wall has maximum annual performance with a coupling coefficient of $U_c = 8.5$ W/$°C·m^2$ (1.5 Btu/hr $°F$ ft$^2$). This is about what will be obtained with a rough-surfaced wall, including heat transfer by radiation (see Examples 2-2 to 2-4 and eqs. 2-4 to 2-9). This fortuitous circumstance accounts for much of the success of the Trombe wall. Because coupling coefficients from about 8–16 W/$°C·m^2$ (1.5–3 Btu/hr $°F$ ft$^2$) give near-optimal performance, it is unnecessary to be covering and uncovering the backside of the wall or to use a fan to remove heat from it.

In the case of the direct gain structure (1) the smaller the coupling coefficient the better, because losses only occur from the room itself. Nevertheless, values below about 6 W/$°C·m^2$ (1 Btu/hr $°F$ ft$^2$) are unrealistic, implying added transparent inside insulation and high wall temperatures.

If the coupling coefficient is high, the Trombe wall and the direct gain structure both stay at room temperature, discharging all collected heat immediately whether it is needed or not. They both, therefore, require additional auxiliary heat, each approaching the same solar contribution—in the case of Fig. 13-3, about 44% relative to 75% and 85% for the performance with realistic coupling coefficients.

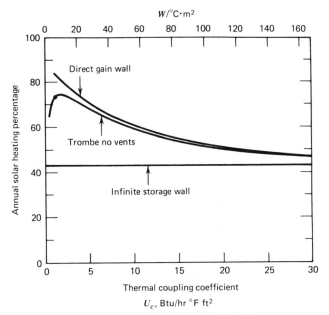

**FIGURE 13-3.** Effect of thermal coupling between room and storage, on annual performance. [Figs. 13-3 to 13-8 are taken from 1976 Passive Workshop Proceedings (LA-6637-C) p. 238–246, and Ref. 1.]

**Glazing.** The size of the collector is the size of the window, and this can be expressed as $S$, the load-normalized collector area. The annual solar-heating fraction $\sum f_a$ at Los Alamos is given in Fig. 13-4 for three glazing situations as a function of $S$ values calculated using 18.3 °C (65 °F) base degree-days. The Los Alamos area, however, while it has a high heating demand,* may have twice the sun of midwestern and northeastern U.S. locations.

It is easy to see that no matter how much glass is installed, single glazing is inadequate, permitting only a 32% annual solar contribution. Double glazing is good, with $\sum f_a = 0.7$ at $S = 0.5$ m²/GJ. Night insulation is better yet. Clearly desirable anywhere, it is almost mandatory in a more cloudy location at which $S$ must be increased to intercept adequate sunlight.

**Room Temperature Range.** Obviously, the greater the range of acceptable room temperatures, the better the solar contribution $\sum f_a$. Simulated Los Alamos results for different temperature ranges are shown in Fig. 13-5. There is clearly much to be gained. A tolerance of $\pm 3$ °C ($\pm 5$ °F) is probably what a disinterested family would accept. The greatly increased $\sum f_a$ available with greater tolerances

* 4080 °C-days (7300 °F-days) for Figs. 13-3 to 13-7.

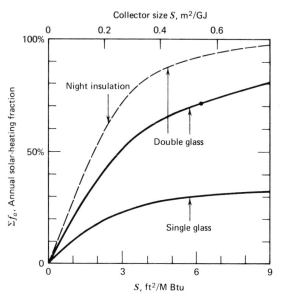

**FIGURE 13-4.** Effect of the area of the Trombe wall on annual performance at Los Alamos.

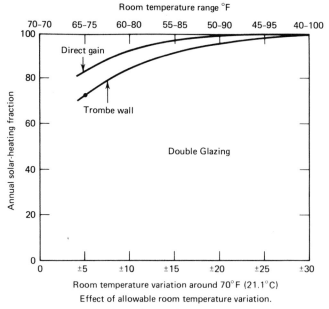

Effect of allowable room temperature variation.

**FIGURE 13-5.** Effect of the allowable temperature variation on annual performance at Los Alamos.

474

explains why solar enthusiasts are usually satisfied with their passive houses. If an installation falls short of its anticipated performance, an increase in the temperature range acceptable to the occupants will tend to compensate.

**Effect of Storage Capacity.** Figure 13-6 shows the effect of area-normalized storage capacity $mc$ on the performance of the Trombe wall with single* and

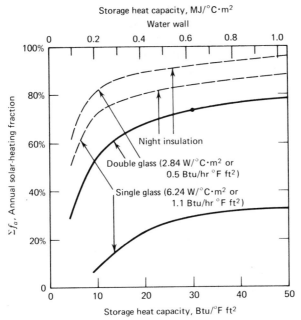

**FIGURE 13-6.** Effect of the heat storage capacity on annual performance for single and double glass with and without night insulation, at Los Alamos.

double† glazing, with and without night insulation. With night insulation‡ single glazing becomes acceptable and is even better than simple double glazing alone at Los Alamos.

The annual performance $\sum f_a$ improves as $mc$ increases with the high-conductivity storage wall used to prepare Fig. 13-6. Most of the benefit, however, is obtained at $mc = 0.6$ MJ/°C·m² (30 Btu/°F ft²). For concrete, this requires a 0.3 m (1 ft) thick wall behind the glass area, giving 1 ft³ of storage per ft² of wall

* $U_L = 6.24$ W/°C·m² (1.1 Btu/hr °F ft²).
† $U_L = 2.84$ W/°C·m² (0.5 Btu/hr °F ft²).
‡ $U_L = 0.57$ W/°C·m² (0.1 Btu/hr °F ft²).

area (0.3 m³/m² wall area). Concrete, adobe, and a "water wall" of water-filled containers are the only materials that can reasonably be considered, since other masonry materials have too little thermal conductivity.

Clearly the storage capacity usually optimum for active systems—about 0.3 MJ/°C·m² (15 Btu/°F ft²)—is inadequate for passive applications. So small a storage capacity means high wall temperatures, which cause an unacceptable heat loss because it continues for 24 hours a day, or uncomfortable excursions of room ambient in a direct gain structure.

For a direct gain wall the same amount of storage (0.6 MJ/°C·m² or 30 Btu/°F ft²) should be distributed on the sunlit interior surfaces—usually 6 to 8 in. of concrete on wall or floor. If the coupling coefficient can be reduced by smoothing the walls, the storage capacity used can be reduced proportionally.

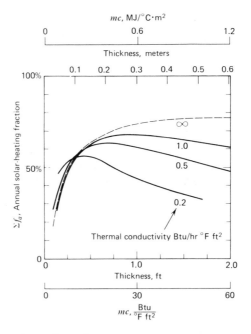

**FIGURE 13-7.** Effect of the thermal conductivity of a Trombe wall on annual performance as a function of wall thickness (storage capacity) at Los Alamos.

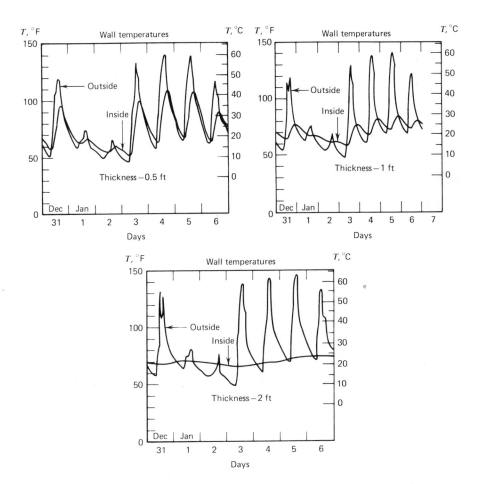

**FIGURE 13-8.** Inside and outside wall temperatures for a Trombe wall structure over a week of changing weather for three wall thicknesses. Results are simulations from Los Alamos (Ref. 1).

**Wall Conductivity.** If the conductivity of the wall is low, the heat will be slow to conduct to the interior wall surface. If the storage capacity is also too large, heat may not appear on the inside of the wall until the next day. This will reduce the system performance and, as shown in Fig. 13-7, there will therefore be an optimum storage capacity for each wall conductivity. If the conductivity is infinite, that maximum lies at infinity, but for a concrete wall, with a conductivity of 1.7 W/m·°C (1 Btu/hr ft °F), the optimum is about 0.6 MJ/°C·m² (30 Btu/°F ft²).

The wall temperature records from several days' simulation given in Fig. 13-8 make the situation clear. With the thin 0.15 m (6 in.) wall, inside and outside wall temperatures are not very different, and yet they undergo considerable excursions. With the 0.3 m (1 ft) wall [$mc$ = 0.6 MJ/°C·m² (30 Btu/°F ft²)] the inside wall temperatures are moderated considerably, almost to within the permitted 18 to 24 °C (65–75 °F) room temperature range. With the thick 0.6 m (2 ft) wall, the inside wall temperature is very steady, and yet Fig. 13-7 shows that the annual performance is somewhat less. The 0.3 m wall inside temperature peaks just before midnight, giving more heat to the occupants when it is most needed.

## PERFORMANCE PREDICTION

**Simulation.** The performance of a window or a direct gain structure can be simulated hourly using the techniques developed for Example 4-2 (Table 4-2) except that the losses and load continue to draw on stored or auxiliary heat for 24 hours per day—the collector is always "on." This method assumes that there is enough storage capacity to keep the room temperature nearly constant, which will require 0.6 MJ/°C·m² (30 Btu/°F ft²) distributed over wall or floor (see the Trombe wall discussions above). The same sort of simulation can be made for a water wall and extended to a Trombe wall and, since the heated storage wall loses heat in proportion to its temperature, the situation is quite similar to that of Example 8-3 (Table 8-2). There are more exact analyses that include interactions of the wall and room, but these require the use of complex differential equations that are beyond the scope of this text. Such *mathematical models* (1) are an important area for solar research, but at this writing neither they nor the simpler techniques presented in this chapter have been validated in the cloudy climates of the eastern half of the United States.

Several assumptions will be made to extend the simulation techniques of previous chapters to passive systems:

1. The heating load will be calculated using a lower degree-day temperature base appropriate for the low end of the interior ambient temperature range.
2. The lowest anticipated ambient temperature is taken as the base temperature. The collector and room are held at the base temperature by auxiliary heat when necessary.
3. Losses continue from the building and collector for 24 hours per day. The loss from the collector wall is not included in the loss from the building. Night insulation values are used when the solar flux drops to zero (i.e., beginning at sunset).
4. The room mass and the storage mass are decoupled by assuming that heat is removed from the storage mass only as necessary to maintain the room at the *minimum* ambient temperature. This overstates the losses from a water wall or a Trombe wall but underestimates the losses from the heated structure.

5. The passive wall has infinite thermal conductivity so that there are no thermal gradients to be considered. In practice this is approached by a water wall.

## EXAMPLE 13-2

Repeat the simulation of Example 8-2 for a passive wall with simultaneous heating load (as in Example 8-3 and Table 8-2) using clear sky data at 40°N on a 90° surface on December 21. For collector parameters use $\overline{\tau\alpha} = 0.7441$ and $U_L = 3.06$ W/°C·m$^2$; at night 0.9767 W/°C·m$^2$. The heat capacity of the storage wall is 0.613 26 MJ/°C·m$^2$ (30 Btu/°F ft$^2$). The permissible temperature swing is 18.33 to 23.89 °C (65–75 °F). The temperature of the wall at the start of the day is 19 °C (66.2 °F). As in Example 8-3, $S = 0.880\ 55$ m$^2$/GJ (10 ft$^2$/M Btu), based on 3412 °C-days (6142 °F-days) annually [temperature base 65 °F (18.33 °C)].

The hourly heating load calculation is based on maintaining 18.33 °C (65 °F), which implies a temperature base for degree-days of 12.78 °C (55 °F). From eq. 2-11 and basic definitions, the load in MJ/h·m$^2$ is

$$l = \frac{(T_b - T_a)^+}{D_a S} = \frac{(12.78 - T_a)^+}{3412 \text{ °C days}} \times \frac{\text{day}}{24 \text{ h}} \times \frac{\text{GJ}}{0.880\ 55 \text{ m}^2} \times \frac{1000 \text{ M}}{\text{G}}$$

$$l = \frac{(12.78 - T_a)^+}{72.11}$$

where $T_a$ is the hourly ambient temperature.

Table 13-1 lists hourly ambient temperatures and irradiance values from Table A-3, continuing the temperature values of Table 8-2 for the after-sunset hours. For the first hour of operation, eq. 8-10 gives

$$q_N = \frac{[0.7441 \times 176.5 - 3.06(19 - 0.2)]0.0036 - \dfrac{(12.78 - 0.2)}{72.11}}{1 + \dfrac{3.06 \times 0.0036}{2 \times 0.613\ 26}}$$

$$q_N = \frac{0.2657 - 0.1745}{1.009} = 0.0904 \text{ MJ/m}^2$$

and, by eq. 8-12,

$$T_{sf} = 19 \text{ °C} + \frac{0.0904}{0.613\ 26} = 19.15 \text{ °C}$$

This procedure has been repeated hourly in Table 13-1 both with and without night insulation, but only to midnight in its entirety. The passive wall collects enough heat so that if the next day is overcast, no auxiliary heat is needed until 1600 hours even if night insulation is not used.

A direct gain simulation is run in much the same way except that the storage temperature used in eq. 8-10 is always the base temperature of 18.33 °C (65 °F). If overheating is anticipated, 21 °C (70 °F) or higher may be more appropriate. It is convenient to continue to store the excess heat as in Table 13-1, using a suitable storage mass.

**TABLE 13–1.** Passive Simulations with and without Night Insulation (Example 13–2)[a]

| Time | $I$ W/m² | $T_a$ °C | $l_T$ MJ/m² | $U_L$ MJ/°C·m² | $T_c^b$ °C | $q_T$ MJ/m² | $U_L$ MJ/°C·m² | $T_c^b$ °C | $q_T$ MJ/m² |
|---|---|---|---|---|---|---|---|---|---|
| 0800 | 176.5 | 0.20 | 0.1744 | 3.06 | 19.00 | 0.2649 | 3.06 | 19.00 | 0.2649 |
| 0900 | 513.8 | 0.91 | 0.1646 | 3.06 | 19.15 | 1.1664 | 3.06 | 19.15 | 1.1664 |
| 1000 | 696.6 | 1.68 | 0.1539 | 3.06 | 20.78 | 1.6422 | 3.06 | 20.78 | 1.6422 |
| 1100 | 794.3 | 2.27 | 0.1457 | 3.06 | 23.21 | 1.8815 | 3.06 | 23.21 | 1.8815 |
| 1200 | 829.0 | 2.86 | 0.1375 | 3.06 | 26.04 | 1.9491 | 3.06 | 26.04 | 1.9491 |
| 1300 | 794.3 | 3.35 | 0.1307 | 3.06 | 28.99 | 1.8300 | 3.06 | 28.99 | 1.8300 |
| 1400 | 696.6 | 3.33 | 0.1310 | 3.06 | 31.76 | 1.5410 | 3.06 | 31.76 | 1.5410 |
| 1500 | 513.8 | 3.23 | 0.1324 | 3.06 | 34.06 | 1.0287 | 3.06 | 34.06 | 1.0287 |
| 1600 | 176.5 | 3.04 | 0.1350 | 3.06 | 35.52 | 0.1152 | 3.06 | 35.52 | 0.1152 |
| 1700 | 0 | 2.24 | 0.1461 | 0.9767 | 35.49 | −0.1146 | 3.06 | 35.49 | −0.3617 |
| 1800 | 0 | 1.26 | 0.1597 | 0.9767 | 35.06 | −0.1164 | 3.06 | 34.66 | −0.3633 |
| 1900 | 0 | 0.63 | 0.1685 | 0.9767 | 34.62 | −0.1169 | 3.06 | 33.81 | −0.3607 |
| 2000 | 0 | 0.38 | 0.1719 | 0.9767 | 34.15 | −0.1161 | 3.06 | 32.95 | −0.3540 |
| 2100 | 0 | −0.18 | 0.1797 | 0.9767 | 33.68 | −0.1164 | 3.06 | 32.09 | −0.3507 |
| 2200 | 0 | −0.59 | 0.1854 | 0.9767 | 33.20 | −0.1161 | 3.06 | 31.22 | −0.3457 |
| 2300 | 0 | −1.01 | 0.1912 | 0.9767 | 32.71 | −0.1158 | 3.06 | 30.36 | −0.3408 |
| 2400 | 0 | −1.45 | 0.1973 | 0.9767 | 32.20 | −0.1155 | 3.06 | 29.49 | −0.3360 |
| 0400 | 0 | −2.62 | 0.2135 | 0.9767 | 30.13 | −0.1122 | 3.06 | 25.17 | −0.3109 |
| 0800 | 0 | 0.20 | 0.1744 | 0.9767 | 28.15 | −0.0958 | 3.06 | 22.19 | −0.2460 |

[a] $\overline{\tau\alpha} = 0.7441$, $mc = 0.613\,26$ MJ/°C·m², $S = 0.880\,55$ m²/GJ ($D_a = 3412$ °C-days).
[b] Initial temperature for the hour given, and final temperature for the previous hour.

Table 13-1 should be compared carefully with Table 8-2, which uses a conventional collector of the same aperture. The passive collector collects much more heat because of lower temperatures for heat loss and a low heat-loss coefficient. Exactly why ordinary manufactured collectors cannot approach the loss coefficients claimed for passive walls is unclear, but edge and back losses are a likely source. Another factor is the slow upward drift in heat transfer coefficients with rising temperature (see Fig. 4-11), an effect not taken into account in passive simulations, while active collectors utilize data taken at somewhat higher temperatures. Further research is needed in this important area.

The advantage of the passive collector tends to be lost if the sun does not shine, since losses continue unabated. The potential for passive systems can be better evaluated through longer-term simulations, and better yet through careful monitoring of full-size structures.

**Long-Term Performance.** The monthly performance of a window or a direct gain structure is given by eq. 9-4 and eq. 9-5 or 9-6. The use of the maximum storage relationships assumes that the temperature for heat loss through the window is properly moderated to the specified room temperature by the interior storage mass, and that the direct gain storage wall is sunlit all day. To account for insulation added to the windows at night, the loss term in eq. 9-4 is broken into separate day and night expressions. Thus the monthly performance of a direct gain window with night insulation is given by

$$f = f_B' - \frac{U_n(\overline{T}_n - \overline{T}_{an})t_{Tn}S}{L_m/L_a} = f_B' - \frac{U_n(\overline{T}_n - \overline{T}_{an})t_{Tn}A_c}{L_m} \tag{13-1}$$

where $f_B'$ is evaluated with eq. 9-4 in combination with eq. 9-5 or 9-6 using $F_i' = 1$, the minimum room ambient as the base temperature, and the *daytime* weather and radiation data from the $0+$ column of a $90°$ slope Appendix B table. The load term for the structure that is used to calculate $f$ and $f_B'$ should not include either day or night losses from the collector wall. Values of $f$ over 1.0 indicate that solar heat is available in excess of that needed in a particular month.

The second term in eq. 13-1 corrects $f_B'$ for night losses, which will occur in proportion to the room temperature. $U_n$ is the night heat loss coefficient through the window, and $t_{Tn}$ is the time of night losses ($t_T$ from the ALL column minus $t_T$ from the $0+$ column). $\overline{T}_{an}$ is the average nighttime outside ambient temperature, which is numerically equal to the average threshold temperature listed in the ALL column. The true average temperature $\overline{T}_n$ for losses from the collector at night is unknown. Use

$$\overline{T}_n = T_{sB} \qquad [f_B' < 0.5] \tag{13-2a}$$

or

$$\overline{T}_n = T_{sB} + 3\,°\text{C}\,[5\,°\text{F}] \qquad [f_B' > 0.5] \tag{13-2b}$$

the latter equation accounting for a somewhat warmer structure when the solar participation is high.

For a passive wall directly behind a window, the potential daytime collector performance depends on the unknown wall temperature, and the situation is similar to that for which we have developed eq. 9-23.

If the wall is simulated as detailed in Example 13-2, the monthly performance can be estimated by replacing $f_B'$ in eq. 13-1 by the potential solar participation with daytime losses from storage considered, which can be calculated from eq. 9-23:

$$f = f_1\left(f_B', \frac{U_L}{mc}\right) - \frac{U_n(\overline{T}_n - \overline{T}_{an})t_n S}{L_m/L_a} = f_1\left(f_B', \frac{U_L}{mc}\right) - \frac{U_n(\overline{T}_n - \overline{T}_{an})t_n A_c}{L_m} \tag{13-3}$$

Here all terms are as just defined for eq. 13-1, except that $\overline{T}_n$ is selected on the basis of $f_1$ rather than $f_B'$ when using eq. 13-2. The function $f_1$, the predicted performance of the wall before consideration of night losses, is permitted values in excess of unity. When $f_B'$ exceeds 2.2, $f$ is taken as 1.0.

When the night loss term in eq. 13-3 exceeds perhaps half that of the final solar participation, the correlation should not be relied on, since it depends too much on the difference between two large numbers. This occurs when the system is in desperate need of the night insulation that is almost mandatory in cloudy areas.

Equation 13-3 applies directly for prediction of the performance of a highy conductive massive wall ($mc$ = 0.6 to 0.9 MJ/°C·m² or 30 to 45 Btu/°F ft²). To correct for a Trombe wall, refer to Fig. 13-7, which indicates a decrease of about 0.04 in $\sum f_a$ for a concrete wall with the optimum storage capacity when $\sum f_a \approx 0.7$. Equation 13-3 can be corrected for the higher surface temperatures reached with a Trombe wall by using half the true storage capacity $mc$ in eq. 9-23 (or Fig. 9-2).

Most other passive structures can be classified as either direct gain structures or as a modified Trombe wall. These can be analyzed using the techniques already introduced.

One important case that has yet to be satisfactorily treated is the solar greenhouse. The outline is clear: Water storage under benches, only a south wall glazed (and that with double glass or perhaps fiberglass), and aluminum foil interior wall covering. Too little field data are available to be sure of an analysis technique, but the general direct gain technique with a 0 °C (32 °F) base temperature would be a good starting point.

**EXAMPLE 13-3**

A south-facing double glazed passive window is installed at Hartford with a water wall directly behind it. The minimum acceptable room ambient is 18 °C, and 13 °C (55 °F) degree-days have been made available using the techniques of Chapter 2. The storage capacity is that of 1 ft (30.5 cm) of concrete, 30 Btu/°F ft² or 0.61326 MJ/°C·m². Low-iron glass (0.05% $Fe_2O_3$) is used for the glazing, and $R$ = 0.6971 °C·m²/W (4 hr °F ft²/Btu) night insulation is to be added to the windows at night. The collector area is 1.6 m²/GJ (18.17 ft²/M Btu) annual load (collection wall excepted). What is the annual solar participation?

Table 13-2 lists $L_m/L_a$ for heating only at Hartford using base 12.8 °C (55 °F) degree-days (see Chapter 2). There are 2210 °C-days altogether.

$\overline{T}_a$, $t_T$, and $I_T$ are taken from the 0+ column of the Hartford Appendix B table at 90° slope. $\overline{T}_{an}$ and $t_n$ are taken from "ALL" and "0+" data. For January,

$$\overline{T}_{an} = -4.9 \text{ °C}$$
$$t_n = 2.6784 - 1.2038 = 1.4746 \text{ Ms}$$

Assuming a perfect black absorber,

$$\overline{\tau\alpha} = 0.885^2 \times 0.95 = 0.7441$$

where 0.95 is the incidence angle modifier. ($F_i'$ = 1; the heat is formed in storage.) As in Example 13-1 $U_L$ = 3.06 W/°C·m². At night (see eq. 2-16),

$$U_n = \cfrac{1}{\cfrac{1}{3.06} + 0.6971} = 0.9767 \text{ W/°C·m}^2$$

The base temperature is 18 °C, and $S$ is based on losses from that inside ambient.

**TABLE 13-2.** Passive System Calculations

*Location:* Hartford, Connecticut, with collector slope 90°. $\overline{\tau\alpha} = 0.7441$, $U_L = 3.06$ W/°C·m², $mc = 0.613\,26$ MJ/°C·m², $T_{sB} = 18$ °C, $U_n = 0.9767$ W/°C·m². 13 °C (55 °F) base degree-days have been used.

| | Solar Rad Group | $T_a$ °C | $t_T$ Ms | $I_T$ MJ/m² | $q_T$ MJ/m² | $L_m/L_a$ | $S$ m²/ GJ | $f_B'$ | $f_1$ | $T_{an}$ °C | $t_n$ Ms | $\Delta f$ | $f$ | $f_a$ |
|---|---|---|---|---|---|---|---|---|---|---|---|---|---|---|
| Jan | 0+ | −2.4 | 1.2038 | 300.1 | 148.2 | 0.2325 | 1.6 | 1.0196 | 0.8271 | −4.9 | 1.4746 | −0.2567 | 0.5704 | 0.1326 |
| Feb | 0+ | −1.6 | 1.1462 | 317.4 | 167.4 | 0.2028 | | 1.3210 | 0.9736 | −4.3 | 1.2989 | −0.2532 | 0.7203 | 0.1461 |
| Mar | 0+ | 3.6 | 1.3864 | 337.8 | 190.3 | 0.1508 | | 2.0187 | 1.1779 | 0.4 | 1.2920 | −0.2758 | 0.9021 | 0.1360 |
| Apr | 0+ | 11.8 | 1.2481 | 293.6 | 194.8 | 0.0533 | | 5.8473 | 1.0 | 7.2 | 1.3439 | −0.5437 | 1. | 0.0533 |
| May | | | | | | | 0.0106 | | | | | | 1. | 0.0106 |
| Jun | | | | | | | 0. | | | | | | | 0. |
| Jul | | | | | | | 0. | | | | | | | 0. |
| Aug | | | | | | | 0. | | | | | | | 0. |
| Sep | | | | | | | 0.0040 | | | | | | 1. | 0.0040 |
| Oct | 0+ | 13.8 | 1.3406 | 378.7 | 264.6 | 0.0349 | | 12.1289 | 1. | 8.8 | 1.3374 | −0.7306 | 1. | 0.0349 |
| Nov | 0+ | 7.2 | 1.1462 | 263.8 | 158.41 | 0.1038 | | 2.4418 | 1. | 3.8 | 1.4458 | −0.3744 | 1. | 0.1038 |
| Dec | 0+ | −0.6 | 1.1142 | 273.7 | 140.24 | 0.2073 | | 1.0824 | 0.8611 | −3.0 | 1.5642 | −0.2830 | 0.5781 | 0.1198 |
| | | | | | | | | | | | | | $\sum f_a$ | 0.7412 |

By eqs. 9-4 and 9-6,

$$q_T = 0.7441 \times 300.1 - 3.06[18 - (-2.4)]1.2038 = 148.2 \text{ MJ/m}^2$$

$$f_B' = \frac{\dfrac{148.2 \text{ MJ}}{\text{m}^2} \times \dfrac{1.6 \text{ m}^2}{\text{GJ}} \times \dfrac{\text{G}}{1000 \text{ M}}}{0.2325} = 1.019$$

$$\frac{U_L}{mc} = \frac{3.06 \text{ W}}{°\text{C} \cdot \text{m}^2} \times \frac{°\text{C} \cdot \text{m}^2}{0.613\,26 \text{ MJ}} \times \frac{\text{J/s}}{\text{W}} = 4.99/\text{Ms}$$

From eq. 9-23 or Fig. 9-2,

$$f_1(1.019, 4.99/\text{Ms}) = 0.827$$

By eq. 13-3, using $T_n = 18 + 3 = 21$ °C (eq. 13-2b),

$$f = 0.827 - \frac{\{0.9767[21 - (-4.9)]1.4746\}\dfrac{1.6}{1000}}{0.2325} = 0.827 - 0.257 = 0.570$$

$$f_a = 0.570 \times 0.2325 = 0.1326$$

For all months, the solar participation $\sum f_a$ is 0.7412.

If the wall were constructed instead to the rear of the room or the mass were a part of the floor, performance of the resultant direct gain structure would be considerably better, presuming that all the heat received could be stored. In January, for instance, the solar contribution using eq. 13-1 instead of eq. 13-3 would be

$$f = 1.019 - 0.257 = 0.7620$$

Year-round solar participation would be about 0.9.

On the other hand, if a Trombe wall is considered, the January value of $f_1$ drops to 0.7765 because the effective storage function $U_L/mc$ increases to 9.99 MJ/°C·m², and

$$f = 0.7765 - 0.257 = 0.5195$$

For the entire year the solar contribution is about 0.69.

Figure 13-9 extends the case of Example 13-3 to other areas and collector configurations, producing annual performance curves like those made for active systems in Fig. 10-7. Although not strictly comparable because of differences in the inside ambient temperature that determines $L_m/L_a$, the active systems are clearly superior. The contest becomes more interesting if Fig. 10-7 and Fig. 13-9 are prepared for the same inside temperatures using radiation data incorporating snow cover, which can be expected for nearly three months of the year in the Hartford area. Snow cover reduces the collector area required for a given performance by one-third for passive systems! Passive systems are much more sensitive to radiation data and elevation techniques than are active systems. Many published simulation results and predictive correlations are inaccurate because they were prepared using dubious radiation information and do not accept radiation data as an input.

The analyses that led to Fig. 13-9 are very site-specific. At Hartford, only modest

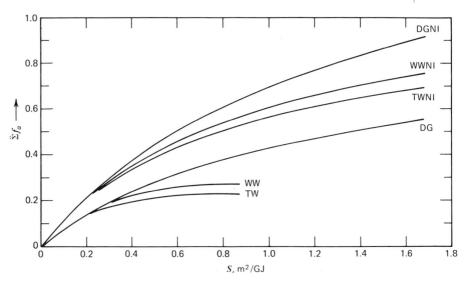

**FIGURE 13-9.** Performance predictions at Hartford for passive systems using eqs. 13-1 to 13-3. Direct gain (DG), water wall (WW), and Trombe wall (TW); with and without night insulation (NI). $\tau\alpha = 0.7741$, $U_L = 3.06$ W/°C·m², $U_n = 0.9767$ W/°C·m², $T_{sB} = 18$ °C, $mc = 0.6133$ MJ/°C·m². Monthly loads calculated with 12.78 °C (55 °F) base degree-days (Example 13-3).

help from the sun can be expected without night insulation. In the southwest, where there is twice the winter radiation, night insulation is proportionally *less* effective, but a high contribution system can work well without it.

In practice, it is a problem to supply night insulation when collector areas become large enough for house heating. It is physically difficult to move and to store foam insulation in the daytime, and fold-up curtains or blinds will not generally have enough insulation value. Mechanical systems have been devised to blow "beads" of foam to and from the space between glass windows, but the practical advantages of passive systems—low cost and simplicity—may not be retained.

**EXAMPLE 13-4**

If the house of Example 9-3 is to be heated passively using the passive wall of Example 13-3, what are the dimensions of the south wall?

The annual heating load for that house, which has a rather low heating requirement, is

$$L_a = \frac{17.72 \text{ MJ}}{\text{°C-day}} \times 2210 \text{ °C-day} \times \frac{G}{1000 \text{ M}} = 39.16 \text{ GJ}$$

If $S = 1.6 \text{ m}^2/\text{GJ}$,

$$A_c = S \times L_a = \frac{1.6 \text{ m}^2}{\text{GJ}} \times 39.16 \text{ GJ} = 62.66 \text{ m}^2 \text{ (or 674 ft}^2)$$

and if the wall is 5 m (16.4 ft) high, the required length is $62.66/5 = 12.53$ m or 41.1 ft long. If the house is a rectangular box two stories high and 8 m (26.2 ft) deep, the superficial area of the house is 200 m² (2156 ft²), which is an ordinary good-sized residence.

## ECONOMICS

It is extremely difficult to assign costs to a passive collection system, since it replaces part of the ordinary structure and has an impact (be it positive or negative) on the architectural aesthetics of the house and the life-style of the occupants.

If a credit is given for the wall structure that the collector replaces, passive installations are estimated to cost as little as half that of active systems per unit of installed area. If this actually turns out to be the case, such systems may well become competitive with active systems for space heating in the eastern half of the United States as well as in the sunny southwest. Careful analysis is needed, however, for almost every situation. To do this accurately, more field data and operating experience are needed, and better radiation data on a 90° slope will be required than are likely to result from current government programs. In spite of its many uncertainties, the passive concept is extremely attractive to many people, however, since it provides a low-technology, low-maintenance approach to the problem of higher fuel prices.

## PROBLEMS

These need be performed for only January, February, and March unless a programmable calculator is available.

1. Repeat Example 13-3 for Washington, Atlanta, and Denver.

2. Repeat Example 13-3 using a 24 °C (or 75 °F) base temperature and 18.33 °C (65 °F) base degree-days.

3. Repeat problem 1 without night insulation for the nearest city.

4. Repeat problem 1 with one-half the collector area for the nearest city.

5. Repeat problem 1 with triple glazing for the nearest city.

6. Repeat problem 1 with one half the collector area and a direct gain wall for the nearest city.

## REFERENCES

1. J. D. Balcomb, H. C. Hedstrom, and R. D. McFarland, "Simulation Analysis of Passive Solar Heated Buildings—Preliminary Results," *Solar Energy 19*, 277–282 (1977).
2. D. Kelbaugh, "The Kelbaugh House," *Solar Age*, 18–23 (July 1976).

# Solar Radiation Tables

24°N. Latitude[a]

| Date | Solar Time AM | PM | Solar Position Alt | Azm | Direct normal | Global Irradiance, Btu/hr ft² Horiz. | South-Facing Elevation Angle 14 | 24 | 34 | 44 | 90 |
|------|------|------|------|------|------|------|------|------|------|------|------|
| Jan 21 | 7 | 5 | 4.8 | 65.6 | 71 | 10 | 17 | 21 | 25 | 28 | 31 |
|  | 8 | 4 | 16.9 | 58.3 | 239 | 83 | 110 | 126 | 137 | 145 | 127 |
|  | 9 | 3 | 27.9 | 48.8 | 288 | 151 | 188 | 207 | 221 | 228 | 176 |
|  | 10 | 2 | 37.2 | 36.1 | 308 | 204 | 246 | 268 | 282 | 287 | 207 |
|  | 11 | 1 | 43.6 | 19.6 | 317 | 237 | 283 | 306 | 319 | 324 | 226 |
|  | 12 |  | 46.0 | 0.0 | 320 | 249 | 296 | 319 | 332 | 336 | 232 |
|  | Surface Daily Totals |  |  |  | 2766 | 1622 | 1984 | 2174 | 2300 | 2360 | 1766 |
| Feb 21 | 7 | 5 | 9.3 | 74.6 | 158 | 35 | 44 | 49 | 53 | 56 | 46 |
|  | 8 | 4 | 22.3 | 67.2 | 263 | 116 | 135 | 145 | 150 | 151 | 102 |
|  | 9 | 3 | 34.4 | 57.6 | 298 | 187 | 213 | 225 | 230 | 228 | 141 |
|  | 10 | 2 | 45.1 | 44.2 | 314 | 241 | 273 | 286 | 291 | 287 | 168 |
|  | 11 | 1 | 53.0 | 25.0 | 321 | 276 | 310 | 324 | 328 | 323 | 185 |
|  | 12 |  | 56.0 | 0.0 | 324 | 288 | 323 | 337 | 341 | 335 | 191 |
|  | Surface Daily Totals |  |  |  | 3036 | 1998 | 2276 | 2396 | 2446 | 2424 | 1476 |
| Mar 21 | 7 | 5 | 13.7 | 83.8 | 194 | 60 | 63 | 64 | 62 | 59 | 27 |
|  | 8 | 4 | 27.2 | 76.8 | 267 | 141 | 150 | 152 | 149 | 142 | 64 |
|  | 9 | 3 | 40.2 | 67.9 | 295 | 212 | 226 | 229 | 225 | 214 | 95 |
|  | 10 | 2 | 52.3 | 54.8 | 309 | 266 | 285 | 288 | 283 | 270 | 120 |
|  | 11 | 1 | 61.9 | 33.4 | 315 | 300 | 322 | 326 | 320 | 305 | 135 |
|  | *12 |  | 66.0 | 0.0 | 317 | 312 | 334 | 339 | 333 | 317 | 140 |
|  | Surface Daily Totals |  |  |  | 3078 | 2270 | 2428 | 2456 | 2412 | 2298 | 1022 |
| Apr 21 | 6 | 6 | 4.7 | 100.6 | 40 | 7 | 5 | 4 | 4 | 3 | 2 |
|  | 7 | 5 | 18.3 | 94.9 | 203 | 83 | 77 | 70 | 62 | 51 | 10 |
|  | 8 | 4 | 32.0 | 89.0 | 256 | 160 | 157 | 149 | 137 | 122 | 16 |
|  | 9 | 3 | 45.6 | 81.9 | 280 | 227 | 227 | 220 | 206 | 186 | 41 |
|  | 10 | 2 | 59.0 | 71.8 | 292 | 278 | 282 | 275 | 259 | 237 | 61 |
|  | 11 | 1 | 71.1 | 51.6 | 298 | 310 | 316 | 309 | 293 | 269 | 74 |
|  | 12 |  | 77.6 | 0.0 | 299 | 321 | 328 | 321 | 305 | 280 | 79 |
|  | Surface Daily Totals |  |  |  | 3036 | 2454 | 2458 | 2374 | 2228 | 2016 | 488 |
| May 21 | 6 | 6 | 8.0 | 108.4 | 86 | 22 | 15 | 10 | 9 | 9 | 5 |
|  | 7 | 5 | 21.2 | 103.2 | 203 | 98 | 85 | 73 | 59 | 44 | 12 |
|  | 8 | 4 | 34.6 | 98.5 | 248 | 171 | 159 | 145 | 127 | 106 | 15 |
|  | 9 | 3 | 48.3 | 93.6 | 269 | 233 | 224 | 210 | 190 | 165 | 16 |
|  | 10 | 2 | 62.0 | 87.7 | 280 | 281 | 275 | 261 | 239 | 211 | 22 |
|  | 11 | 1 | 75.5 | 76.9 | 286 | 311 | 307 | 293 | 270 | 240 | 34 |
|  | 12 |  | 86.0 | 0.0 | 288 | 322 | 317 | 304 | 281 | 250 | 37 |
|  | Surface Daily Totals |  |  |  | 3032 | 2556 | 2447 | 2286 | 2072 | 1800 | 246 |
| Jun 21 | 6 | 6 | 9.3 | 111.6 | 97 | 29 | 20 | 12 | 12 | 11 | 7 |
|  | 7 | 5 | 22.3 | 106.8 | 201 | 103 | 87 | 73 | 58 | 41 | 13 |
|  | 8 | 4 | 35.5 | 102.6 | 242 | 173 | 158 | 142 | 122 | 99 | 16 |
|  | 9 | 3 | 49.0 | 98.7 | 263 | 234 | 221 | 204 | 182 | 155 | 18 |
|  | 10 | 2 | 62.6 | 95.0 | 274 | 280 | 269 | 253 | 229 | 199 | 18 |
|  | 11 | 1 | 76.3 | 90.8 | 279 | 309 | 300 | 283 | 259 | 227 | 19 |
|  | 12 |  | 89.4 | 0.0 | 281 | 319 | 310 | 294 | 269 | 236 | 22 |
|  | Surface Daily Totals |  |  |  | 2994 | 2574 | 2422 | 2230 | 1992 | 1700 | 204 |

| Date | Solar Time AM | Solar Time PM | Solar Position Alt | Solar Position Azm | Direct normal | Global Irradiance, Btu/hr ft² Horiz. | South-Facing Elevation Angle 14 | 24 | 34 | 44 | 90 |
|---|---|---|---|---|---|---|---|---|---|---|---|
| Jul 21 | 6 | 6 | 8.2 | 109.0 | 81 | 23 | 16 | 11 | 10 | 9 | 6 |
| | 7 | 5 | 21.4 | 103.8 | 195 | 98 | 85 | 73 | 59 | 44 | 13 |
| | 8 | 4 | 34.8 | 99.2 | 239 | 169 | 157 | 143 | 125 | 104 | 16 |
| | 9 | 3 | 48.4 | 94.5 | 261 | 231 | 221 | 207 | 187 | 161 | 18 |
| | 10 | 2 | 62.1 | 89.0 | 272 | 278 | 270 | 256 | 235 | 206 | 21 |
| | 11 | 1 | 75.7 | 79.2 | 278 | 307 | 302 | 287 | 265 | 234 | 32 |
| | 12 | | 86.6 | 0.0 | 280 | 317 | 312 | 298 | 275 | 245 | 36 |
| | Surface Daily Totals | | | | 2932 | 2526 | 2412 | 2250 | 2036 | 1766 | 246 |
| Aug 21 | 6 | 6 | 5.0 | 101.3 | 35 | 7 | 5 | 4 | 4 | 4 | 2 |
| | 7 | 5 | 18.5 | 95.6 | 186 | 82 | 76 | 69 | 60 | 50 | 11 |
| | 8 | 4 | 32.2 | 89.7 | 241 | 158 | 154 | 146 | 134 | 118 | 16 |
| | 9 | 3 | 45.9 | 82.9 | 265 | 223 | 222 | 214 | 200 | 181 | 39 |
| | 10 | 2 | 59.3 | 73.0 | 278 | 273 | 275 | 268 | 252 | 230 | 59 |
| | 11 | 1 | 71.6 | 53.2 | 284 | 304 | 309 | 301 | 285 | 261 | 71 |
| | 12 | | 78.3 | 0.0 | 286 | 315 | 320 | 313 | 296 | 272 | 75 |
| | Surface Daily Totals | | | | 2864 | 2408 | 2402 | 2316 | 2168 | 1958 | 470 |
| Sep 21 | 7 | 5 | 13.7 | 83.8 | 173 | 57 | 60 | 60 | 59 | 56 | 26 |
| | 8 | 4 | 27.2 | 76.8 | 248 | 136 | 144 | 146 | 143 | 136 | 62 |
| | 9 | 3 | 40.2 | 67.9 | 278 | 205 | 218 | 221 | 217 | 206 | 93 |
| | 10 | 2 | 52.3 | 54.8 | 292 | 258 | 275 | 278 | 273 | 261 | 116 |
| | 11 | 1 | 61.9 | 33.4 | 299 | 291 | 311 | 315 | 309 | 295 | 131 |
| | 12 | | 66.0 | 0.0 | 301 | 302 | 323 | 327 | 321 | 306 | 136 |
| | Surface Daily Totals | | | | 2878 | 2194 | 2342 | 2366 | 2322 | 2212 | 992 |
| Oct 21 | 7 | 5 | 9.1 | 74.1 | 138 | 32 | 40 | 45 | 48 | 50 | 42 |
| | 8 | 4 | 22.0 | 66.7 | 247 | 111 | 129 | 139 | 144 | 145 | 99 |
| | 9 | 3 | 34.1 | 57.1 | 284 | 180 | 206 | 217 | 223 | 221 | 138 |
| | 10 | 2 | 44.7 | 43.8 | 301 | 234 | 265 | 277 | 282 | 279 | 165 |
| | 11 | 1 | 52.5 | 24.7 | 309 | 268 | 301 | 315 | 319 | 314 | 182 |
| | 12 | | 55.5 | 0.0 | 311 | 279 | 314 | 328 | 332 | 327 | 188 |
| | Surface Daily Totals | | | | 2868 | 1928 | 2198 | 2314 | 2364 | 2346 | 1442 |
| Nov 21 | 7 | 5 | 4.9 | 65.8 | 67 | 10 | 16 | 20 | 24 | 27 | 29 |
| | 8 | 4 | 17.0 | 58.4 | 232 | 82 | 108 | 123 | 135 | 142 | 124 |
| | 9 | 3 | 28.0 | 48.9 | 282 | 150 | 186 | 205 | 217 | 224 | 172 |
| | 10 | 2 | 37.3 | 36.3 | 303 | 203 | 244 | 265 | 278 | 283 | 204 |
| | 11 | 1 | 43.8 | 19.7 | 312 | 236 | 280 | 302 | 316 | 320 | 222 |
| | 12 | | 46.2 | 0.0 | 315 | 247 | 293 | 315 | 328 | 332 | 228 |
| | Surface Daily Totals | | | | 2706 | 1610 | 1962 | 2146 | 2268 | 2324 | 1730 |
| Dec 21 | 7 | 5 | 3.2 | 62.6 | 30 | 3 | 7 | 9 | 11 | 12 | 14 |
| | 8 | 4 | 14.9 | 55.3 | 225 | 71 | 99 | 116 | 129 | 139 | 130 |
| | 9 | 3 | 25.5 | 46.0 | 281 | 137 | 176 | 198 | 214 | 223 | 184 |
| | 10 | 2 | 34.3 | 33.7 | 304 | 189 | 234 | 258 | 275 | 283 | 217 |
| | 11 | 1 | 40.4 | 18.2 | 314 | 221 | 270 | 295 | 312 | 320 | 236 |
| | 12 | | 42.6 | 0.0 | 317 | 232 | 282 | 308 | 325 | 332 | 243 |
| | Surface Daily Totals | | | | 2624 | 1474 | 1852 | 2058 | 2204 | 2286 | 1808 |

[a] C. A. Morrison and E. A. Farber, ASHRAE Solar Energy Applications Symposium, Montreal, June 22, 1974. (A number of typographical errors have been corrected.)

NOTE: 1 Btu/hr ft² = 3.152 W/m².

## 32°N. Latitude

| Date | Solar Time AM | PM | Solar Position Alt | Azm | Direct normal | Global Irradiance, Btu/hr ft² | | | | | |
|------|-----|-----|------|------|------|--------|------|------|------|------|------|
| | | | | | | South-Facing Elevation Angle | | | | | |
| | | | | | | Horiz. | 22 | 32 | 42 | 52 | 90 |
| Jan 21 | 7 | 5 | 1.4 | 65.2 | 1 | 0 | 0 | 0 | 0 | 1 | 1 |
| | 8 | 4 | 12.5 | 56.5 | 203 | 56 | 93 | 106 | 116 | 123 | 115 |
| | 9 | 3 | 22.5 | 46.0 | 269 | 118 | 175 | 193 | 206 | 212 | 181 |
| | 10 | 2 | 30.6 | 33.1 | 295 | 167 | 235 | 256 | 269 | 274 | 221 |
| | 11 | 1 | 36.1 | 17.5 | 306 | 198 | 273 | 295 | 308 | 312 | 245 |
| | 12 | | 38.0 | 0.0 | 310 | 209 | 285 | 308 | 321 | 324 | 253 |
| | Surface Daily Totals | | | | 2458 | 1288 | 1839 | 2008 | 2118 | 2166 | 1779 |
| Feb 21 | 7 | 5 | 7.1 | 73.5 | 121 | 22 | 34 | 37 | 40 | 42 | 38 |
| | 8 | 4 | 19.0 | 64.4 | 247 | 95 | 127 | 136 | 140 | 141 | 108 |
| | 9 | 3 | 29.9 | 53.4 | 288 | 161 | 206 | 217 | 222 | 220 | 158 |
| | 10 | 2 | 39.1 | 39.4 | 306 | 212 | 266 | 278 | 283 | 279 | 193 |
| | 11 | 1 | 45.6 | 21.4 | 315 | 244 | 304 | 317 | 321 | 315 | 214 |
| | 12 | | 48.0 | 0.0 | 317 | 255 | 316 | 330 | 334 | 328 | 222 |
| | Surface Daily Totals | | | | 2872 | 1724 | 2188 | 2300 | 2345 | 2322 | 1644 |
| Mar 21 | 7 | 5 | 12.7 | 81.9 | 185 | 54 | 60 | 60 | 59 | 56 | 32 |
| | 8 | 4 | 25.1 | 73.0 | 260 | 129 | 146 | 147 | 144 | 137 | 78 |
| | 9 | 3 | 36.8 | 62.1 | 290 | 194 | 222 | 224 | 220 | 209 | 119 |
| | 10 | 2 | 47.3 | 47.5 | 304 | 245 | 280 | 283 | 278 | 265 | 150 |
| | 11 | 1 | 55.0 | 26.8 | 311 | 277 | 317 | 321 | 315 | 300 | 170 |
| | 12 | | 58.0 | 0.0 | 313 | 287 | 329 | 333 | 327 | 312 | 177 |
| | Surface Daily Totals | | | | 3012 | 2084 | 2378 | 2403 | 2358 | 2246 | 1276 |
| Apr 21 | 6 | 6 | 6.1 | 99.9 | 66 | 14 | 9 | 6 | 6 | 5 | 3 |
| | 7 | 5 | 18.8 | 92.2 | 206 | 86 | 78 | 71 | 62 | 51 | 10 |
| | 8 | 4 | 31.5 | 84.0 | 255 | 158 | 156 | 148 | 136 | 120 | 35 |
| | 9 | 3 | 43.9 | 74.2 | 278 | 220 | 225 | 217 | 203 | 183 | 68 |
| | 10 | 2 | 55.7 | 60.3 | 290 | 267 | 279 | 272 | 256 | 234 | 95 |
| | 11 | 1 | 65.4 | 37.5 | 295 | 297 | 313 | 306 | 290 | 265 | 112 |
| | 12 | | 69.6 | 0.0 | 297 | 307 | 325 | 318 | 301 | 276 | 118 |
| | Surface Daily Totals | | | | 3076 | 2390 | 2444 | 2356 | 2206 | 1994 | 764 |
| May 21 | 6 | 6 | 10.4 | 107.2 | 119 | 36 | 21 | 13 | 13 | 12 | 7 |
| | 7 | 5 | 22.8 | 100.1 | 211 | 107 | 88 | 75 | 60 | 44 | 13 |
| | 8 | 4 | 35.4 | 92.9 | 250 | 175 | 159 | 145 | 127 | 105 | 15 |
| | 9 | 3 | 48.1 | 84.7 | 269 | 233 | 223 | 209 | 188 | 163 | 33 |
| | 10 | 2 | 60.6 | 73.3 | 280 | 277 | 273 | 259 | 237 | 208 | 56 |
| | 11 | 1 | 72.0 | 51.9 | 285 | 305 | 305 | 290 | 268 | 237 | 72 |
| | 12 | | 78.0 | 0.0 | 286 | 315 | 315 | 301 | 278 | 247 | 77 |
| | Surface Daily Totals | | | | 3112 | 2582 | 2454 | 2284 | 2064 | 1788 | 469 |
| Jun 21 | 6 | 6 | 12.2 | 110.2 | 131 | 45 | 26 | 16 | 15 | 14 | 9 |
| | 7 | 5 | 24.3 | 103.4 | 210 | 115 | 91 | 76 | 59 | 41 | 14 |
| | 8 | 4 | 36.9 | 96.8 | 245 | 180 | 159 | 143 | 122 | 99 | 16 |
| | 9 | 3 | 49.6 | 89.4 | 264 | 236 | 221 | 204 | 181 | 153 | 19 |
| | 10 | 2 | 62.2 | 79.7 | 274 | 279 | 268 | 251 | 227 | 197 | 41 |
| | 11 | 1 | 74.2 | 60.9 | 279 | 306 | 299 | 282 | 257 | 224 | 56 |
| | 12 | | 81.5 | 0.0 | 280 | 315 | 309 | 292 | 267 | 234 | 60 |
| | Surface Daily Totals | | | | 3084 | 2634 | 2436 | 2234 | 1990 | 1690 | 370 |

| Date | Solar Time AM | PM | Solar Position Alt | Azm | Direct nor-mal | Global Irradiance, Btu/hr ft² Horiz. | South-Facing Elevation Angle 22 | 32 | 42 | 52 | 90 |
|---|---|---|---|---|---|---|---|---|---|---|---|
| Jul 21 | 6 | 6 | 10.7 | 107.7 | 113 | 37 | 22 | 14 | 13 | 12 | 8 |
|  | 7 | 5 | 23.1 | 100.6 | 203 | 107 | 87 | 75 | 60 | 44 | 14 |
|  | 8 | 4 | 35.7 | 93.6 | 241 | 174 | 158 | 143 | 125 | 104 | 16 |
|  | 9 | 3 | 48.4 | 85.5 | 261 | 230 | 220 | 205 | 185 | 159 | 31 |
|  | 10 | 2 | 60.9 | 74.3 | 271 | 274 | 269 | 254 | 232 | 204 | 54 |
|  | 11 | 1 | 72.4 | 53.3 | 277 | 302 | 300 | 285 | 262 | 232 | 69 |
|  | 12 |  | 78.6 | 0.0 | 279 | 311 | 310 | 296 | 273 | 242 | 74 |
|  | Surface Daily Totals |  |  |  | 3012 | 2558 | 2422 | 2250 | 2030 | 1754 | 458 |
| Aug 21 | 6 | 6 | 6.5 | 100.5 | 59 | 14 | 9 | 7 | 6 | 6 | 4 |
|  | 7 | 5 | 19.1 | 92.8 | 190 | 85 | 77 | 69 | 60 | 50 | 12 |
|  | 8 | 4 | 31.8 | 84.7 | 240 | 156 | 152 | 144 | 132 | 116 | 33 |
|  | 9 | 3 | 44.3 | 75.0 | 263 | 216 | 220 | 212 | 197 | 178 | 65 |
|  | 10 | 2 | 56.1 | 61.3 | 276 | 262 | 272 | 264 | 249 | 226 | 91 |
|  | 11 | 1 | 66.0 | 38.4 | 282 | 292 | 305 | 298 | 281 | 257 | 107 |
|  | 12 |  | 70.3 | 0.0 | 284 | 302 | 317 | 309 | 292 | 268 | 113 |
|  | Surface Daily Totals |  |  |  | 2902 | 2352 | 2388 | 2296 | 2144 | 1934 | 736 |
| Sep 21 | 7 | 5 | 12.7 | 81.9 | 163 | 51 | 56 | 56 | 55 | 52 | 30 |
|  | 8 | 4 | 25.1 | 73.0 | 240 | 124 | 140 | 141 | 138 | 131 | 75 |
|  | 9 | 3 | 36.8 | 62.1 | 272 | 188 | 213 | 215 | 211 | 201 | 114 |
|  | 10 | 2 | 47.3 | 47.5 | 287 | 237 | 270 | 273 | 268 | 255 | 145 |
|  | 11 | 1 | 55.0 | 26.8 | 294 | 268 | 306 | 309 | 303 | 289 | 164 |
|  | 12 |  | 58.0 | 0.0 | 296 | 278 | 318 | 321 | 315 | 300 | 171 |
|  | Surface Daily Totals |  |  |  | 2808 | 2014 | 2288 | 2308 | 2264 | 2154 | 1226 |
| Oct 21 | 7 | 5 | 6.8 | 73.1 | 99 | 19 | 29 | 32 | 34 | 36 | 32 |
|  | 8 | 4 | 18.7 | 64.0 | 229 | 90 | 120 | 128 | 133 | 134 | 104 |
|  | 9 | 3 | 29.5 | 53.0 | 273 | 155 | 198 | 208 | 213 | 212 | 153 |
|  | 10 | 2 | 38.7 | 39.1 | 293 | 204 | 257 | 269 | 273 | 270 | 188 |
|  | 11 | 1 | 45.1 | 21.1 | 302 | 236 | 294 | 307 | 311 | 306 | 209 |
|  | 12 |  | 47.5 | 0.0 | 304 | 247 | 306 | 320 | 324 | 318 | 217 |
|  | Surface Daily Totals |  |  |  | 2696 | 1654 | 2100 | 2208 | 2252 | 2232 | 1588 |
| Nov 21 | 7 | 5 | 1.5 | 65.4 | 2 | 0 | 0 | 0 | 1 | 1 | 1 |
|  | 8 | 4 | 12.7 | 56.6 | 196 | 55 | 91 | 104 | 113 | 119 | 111 |
|  | 9 | 3 | 22.6 | 46.1 | 263 | 118 | 173 | 190 | 202 | 208 | 176 |
|  | 10 | 2 | 30.8 | 33.2 | 289 | 166 | 233 | 252 | 265 | 270 | 217 |
|  | 11 | 1 | 36.2 | 17.6 | 301 | 197 | 270 | 291 | 303 | 307 | 241 |
|  | 12 |  | 38.2 | 0.0 | 304 | 207 | 282 | 304 | 316 | 320 | 249 |
|  | Surface Daily Totals |  |  |  | 2406 | 1280 | 1816 | 1980 | 2084 | 2130 | 1742 |
| Dec 21 | 8 | 4 | 10.3 | 53.8 | 176 | 41 | 77 | 90 | 101 | 108 | 107 |
|  | 9 | 3 | 19.8 | 43.6 | 257 | 102 | 161 | 180 | 195 | 204 | 183 |
|  | 10 | 2 | 27.6 | 31.2 | 288 | 150 | 221 | 244 | 259 | 267 | 226 |
|  | 11 | 1 | 32.7 | 16.4 | 301 | 180 | 258 | 282 | 298 | 305 | 251 |
|  | 12 |  | 34.6 | 0.0 | 304 | 190 | 271 | 295 | 311 | 318 | 259 |
|  | Surface Daily Totals |  |  |  | 2348 | 1136 | 1704 | 1888 | 2016 | 2086 | 1794 |

NOTE: 1 Btu/hr ft² = 3.152 W/m².

## 40°N. Latitude

| Date | Solar Time AM | PM | Solar Position Alt | Azm | Direct nor-mal | Global Irradiance, Btu/hr ft² Horiz. | South-Facing Elevation Angle 30 | 40 | 50 | 60 | 90 |
|------|------|------|------|------|------|------|------|------|------|------|------|
| Jan 21 | 8 | 4 | 8.1 | 55.3 | 142 | 28 | 65 | 74 | 81 | 85 | 84 |
|  | 9 | 3 | 16.8 | 44.0 | 239 | 83 | 155 | 171 | 182 | 187 | 171 |
|  | 10 | 2 | 23.8 | 30.9 | 274 | 127 | 218 | 237 | 249 | 254 | 223 |
|  | 11 | 1 | 28.4 | 16.0 | 289 | 154 | 257 | 277 | 290 | 293 | 253 |
|  | 12 |  | 30.0 | 0.0 | 294 | 164 | 270 | 291 | 303 | 306 | 263 |
|  | Surface Daily Totals |  |  |  | 2182 | 948 | 1660 | 1810 | 1906 | 1944 | 1726 |
| Feb 21 | 7 | 5 | 4.8 | 72.7 | 69 | 10 | 19 | 21 | 23 | 24 | 22 |
|  | 8 | 4 | 15.4 | 62.2 | 224 | 73 | 114 | 122 | 126 | 127 | 107 |
|  | 9 | 3 | 25.0 | 50.2 | 274 | 132 | 195 | 205 | 209 | 208 | 167 |
|  | 10 | 2 | 32.8 | 35.9 | 295 | 178 | 256 | 267 | 271 | 267 | 210 |
|  | 11 | 1 | 38.1 | 18.9 | 305 | 206 | 293 | 306 | 310 | 304 | 236 |
|  | 12 |  | 40.0 | 0.0 | 308 | 216 | 306 | 319 | 323 | 317 | 245 |
|  | Surface Daily Totals |  |  |  | 2640 | 1414 | 2060 | 2162 | 2202 | 2176 | 1730 |
| Mar 21 | 7 | 5 | 11.4 | 80.2 | 171 | 46 | 55 | 55 | 54 | 51 | 35 |
|  | 8 | 4 | 22.5 | 69.6 | 250 | 114 | 140 | 141 | 138 | 131 | 89 |
|  | 9 | 3 | 32.8 | 57.3 | 282 | 173 | 215 | 217 | 213 | 202 | 138 |
|  | 10 | 2 | 41.6 | 41.9 | 297 | 218 | 273 | 276 | 271 | 258 | 176 |
|  | 11 | 1 | 47.7 | 22.6 | 305 | 247 | 310 | 313 | 307 | 293 | 200 |
|  | 12 |  | 50.0 | 0.0 | 307 | 257 | 322 | 326 | 320 | 305 | 208 |
|  | Surface Daily Totals |  |  |  | 2916 | 1852 | 2308 | 2330 | 2284 | 2174 | 1484 |
| Apr 21 | 6 | 6 | 7.4 | 98.9 | 89 | 20 | 11 | 8 | 7 | 7 | 4 |
|  | 7 | 5 | 18.9 | 89.5 | 206 | 87 | 77 | 70 | 61 | 50 | 12 |
|  | 8 | 4 | 30.3 | 79.3 | 252 | 152 | 153 | 145 | 133 | 117 | 53 |
|  | 9 | 3 | 41.3 | 67.2 | 274 | 207 | 221 | 213 | 199 | 179 | 93 |
|  | 10 | 2 | 51.2 | 51.4 | 286 | 250 | 275 | 267 | 252 | 229 | 126 |
|  | 11 | 1 | 58.7 | 29.2 | 292 | 277 | 308 | 301 | 285 | 260 | 147 |
|  | 12 |  | 61.6 | 0.0 | 293 | 287 | 320 | 313 | 296 | 271 | 154 |
|  | Surface Daily Totals |  |  |  | 3092 | 2274 | 2412 | 2320 | 2168 | 1956 | 1022 |
| May 21 | 5 | 7 | 1.9 | 114.7 | 1 | 0 | 0 | 0 | 0 | 0 | 0 |
|  | 6 | 6 | 12.7 | 105.6 | 144 | 49 | 25 | 15 | 14 | 13 | 9 |
|  | 7 | 5 | 24.0 | 96.6 | 216 | 114 | 89 | 76 | 60 | 44 | 13 |
|  | 8 | 4 | 35.4 | 87.2 | 250 | 175 | 158 | 144 | 125 | 104 | 25 |
|  | 9 | 3 | 46.8 | 76.0 | 267 | 227 | 221 | 206 | 186 | 160 | 60 |
|  | 10 | 2 | 57.5 | 60.9 | 277 | 267 | 270 | 255 | 233 | 205 | 89 |
|  | 11 | 1 | 66.2 | 37.1 | 283 | 293 | 301 | 287 | 264 | 234 | 108 |
|  | 12 |  | 70.0 | 0.0 | 284 | 301 | 312 | 297 | 274 | 243 | 114 |
|  | Surface Daily Totals |  |  |  | 3160 | 2552 | 2442 | 2264 | 2040 | 1760 | 724 |
| Jun 21 | 5 | 7 | 4.2 | 117.3 | 22 | 4 | 3 | 3 | 2 | 2 | 1 |
|  | 6 | 6 | 14.8 | 108.4 | 155 | 60 | 30 | 18 | 17 | 16 | 10 |
|  | 7 | 5 | 26.0 | 99.7 | 216 | 123 | 92 | 77 | 59 | 40 | 14 |
|  | 8 | 4 | 37.4 | 90.7 | 246 | 182 | 159 | 142 | 121 | 97 | 16 |
|  | 9 | 3 | 48.8 | 80.2 | 263 | 233 | 219 | 202 | 179 | 151 | 47 |
|  | 10 | 2 | 59.8 | 65.8 | 272 | 272 | 266 | 248 | 224 | 193 | 74 |
|  | 11 | 1 | 69.2 | 41.9 | 277 | 296 | 296 | 278 | 253 | 221 | 92 |
|  | 12 |  | 73.5 | 0.0 | 279 | 304 | 306 | 289 | 263 | 230 | 98 |
|  | Surface Daily Totals |  |  |  | 3180 | 2648 | 2434 | 2224 | 1974 | 1670 | 610 |

| Date | Solar Time AM | PM | Solar Position Alt | Azm | Direct nor-mal | Global Irradiance, Btu/hr ft² Horiz. | South-Facing Elevation Angle 30 | 40 | 50 | 60 | 90 |
|------|------|------|------|------|------|------|------|------|------|------|------|
| Jul 21 | 5 | 7 | 2.3 | 115.2 | 2 | 0 | 0 | 0 | 0 | 0 | 0 |
|  | 6 | 6 | 13.1 | 106.1 | 138 | 50 | 26 | 17 | 15 | 14 | 9 |
|  | 7 | 5 | 24.3 | 97.2 | 208 | 114 | 89 | 75 | 60 | 44 | 14 |
|  | 8 | 4 | 35.8 | 87.8 | 241 | 174 | 157 | 142 | 124 | 102 | 24 |
|  | 9 | 3 | 47.2 | 76.7 | 259 | 225 | 218 | 203 | 182 | 157 | 58 |
|  | 10 | 2 | 57.9 | 61.7 | 269 | 265 | 266 | 251 | 229 | 200 | 86 |
|  | 11 | 1 | 66.7 | 37.9 | 275 | 290 | 296 | 281 | 258 | 228 | 104 |
|  | 12 |  | 70.6 | 0.0 | 276 | 298 | 307 | 292 | 269 | 238 | 111 |
|  | Surface Daily Totals |  |  |  | 3062 | 2534 | 2409 | 2230 | 2006 | 1728 | 702 |
| Aug 21 | 6 | 6 | 7.9 | 99.5 | 81 | 21 | 12 | 9 | 8 | 7 | 5 |
|  | 7 | 5 | 19.3 | 90.0 | 191 | 87 | 76 | 69 | 60 | 49 | 12 |
|  | 8 | 4 | 30.7 | 79.9 | 237 | 150 | 150 | 141 | 129 | 113 | 50 |
|  | 9 | 3 | 41.8 | 67.9 | 260 | 205 | 216 | 207 | 193 | 173 | 89 |
|  | 10 | 2 | 51.7 | 52.1 | 272 | 246 | 267 | 259 | 244 | 221 | 120 |
|  | 11 | 1 | 59.3 | 29.7 | 278 | 273 | 300 | 292 | 276 | 252 | 140 |
|  | 12 |  | 62.3 | 0.0 | 280 | 282 | 311 | 303 | 287 | 262 | 147 |
|  | Surface Daily Totals |  |  |  | 2916 | 2244 | 2354 | 2258 | 2104 | 1894 | 978 |
| Sep 21 | 7 | 5 | 11.4 | 80.2 | 149 | 43 | 51 | 51 | 49 | 47 | 32 |
|  | 8 | 4 | 22.5 | 69.6 | 230 | 109 | 133 | 134 | 131 | 124 | 84 |
|  | 9 | 3 | 32.8 | 57.3 | 263 | 167 | 206 | 208 | 203 | 193 | 132 |
|  | 10 | 2 | 41.6 | 41.9 | 280 | 211 | 262 | 265 | 260 | 247 | 168 |
|  | 11 | 1 | 47.7 | 22.6 | 287 | 239 | 298 | 301 | 295 | 281 | 192 |
|  | 12 |  | 50.0 | 0.0 | 290 | 249 | 310 | 313 | 307 | 292 | 200 |
|  | Surface Daily Totals |  |  |  | 2708 | 1788 | 2210 | 2228 | 2182 | 2074 | 1416 |
| Oct 21 | 7 | 5 | 4.5 | 72.3 | 48 | 7 | 14 | 15 | 17 | 17 | 16 |
|  | 8 | 4 | 15.0 | 61.9 | 204 | 68 | 106 | 113 | 117 | 118 | 100 |
|  | 9 | 3 | 24.5 | 49.8 | 257 | 126 | 185 | 195 | 200 | 198 | 160 |
|  | 10 | 2 | 32.4 | 35.6 | 280 | 170 | 245 | 257 | 261 | 257 | 203 |
|  | 11 | 1 | 37.6 | 18.7 | 291 | 199 | 283 | 295 | 299 | 294 | 229 |
|  | 12 |  | 39.5 | 0.0 | 294 | 208 | 295 | 308 | 312 | 306 | 238 |
|  | Surface Daily Totals |  |  |  | 2454 | 1348 | 1962 | 2060 | 2098 | 2074 | 1654 |
| Nov 21 | 8 | 4 | 8.2 | 55.4 | 136 | 28 | 63 | 72 | 78 | 82 | 81 |
|  | 9 | 3 | 17.0 | 44.1 | 232 | 82 | 152 | 167 | 178 | 183 | 167 |
|  | 10 | 2 | 24.0 | 31.0 | 268 | 126 | 215 | 233 | 245 | 249 | 219 |
|  | 11 | 1 | 28.6 | 16.1 | 283 | 153 | 254 | 273 | 285 | 288 | 248 |
|  | 12 |  | 30.2 | 0.0 | 288 | 163 | 267 | 287 | 298 | 301 | 258 |
|  | Surface Daily Totals |  |  |  | 2128 | 942 | 1636 | 1778 | 1870 | 1908 | 1686 |
| Dec 21 | 8 | 4 | 5.5 | 53.0 | 89 | 14 | 39 | 45 | 50 | 54 | 56 |
|  | 9 | 3 | 14.0 | 41.9 | 217 | 65 | 135 | 152 | 164 | 171 | 163 |
|  | 10 | 2 | 20.7 | 29.4 | 261 | 107 | 200 | 221 | 235 | 242 | 221 |
|  | 11 | 1 | 25.0 | 15.2 | 280 | 134 | 239 | 262 | 276 | 283 | 252 |
|  | 12 |  | 26.6 | 0.0 | 285 | 143 | 253 | 275 | 290 | 296 | 263 |
|  | Surface Daily Totals |  |  |  | 1978 | 782 | 1480 | 1634 | 1740 | 1796 | 1646 |

NOTE: 1 Btu/hr ft² = 3.152 W/m².

48°N. Latitude

| Date | Solar Time AM | PM | Solar Position Alt | Azm | Direct normal | Global Irradiance, Btu/hr ft² South-Facing Elevation Angle Horiz. | 38 | 48 | 58 | 68 | 90 |
|------|------|------|------|------|------|------|------|------|------|------|------|
| Jan 21 | 8 | 4 | 3.5 | 54.6 | 37 | 4 | 17 | 19 | 21 | 22 | 22 |
| | 9 | 3 | 11.0 | 42.6 | 185 | 46 | 120 | 132 | 140 | 145 | 139 |
| | 10 | 2 | 16.9 | 29.4 | 239 | 83 | 190 | 206 | 216 | 220 | 206 |
| | 11 | 1 | 20.7 | 15.1 | 261 | 107 | 231 | 249 | 260 | 263 | 243 |
| | 12 | | 22.0 | 0.0 | 267 | 115 | 245 | 264 | 275 | 278 | 255 |
| | Surface Daily Totals | | | | 1710 | 596 | 1360 | 1478 | 1550 | 1578 | 1478 |
| Feb 21 | 7 | 5 | 2.4 | 72.2 | 12 | 1 | 3 | 4 | 4 | 4 | 4 |
| | 8 | 4 | 11.6 | 60.5 | 188 | 49 | 95 | 102 | 105 | 106 | 96 |
| | 9 | 3 | 19.7 | 47.7 | 251 | 100 | 178 | 187 | 191 | 190 | 167 |
| | 10 | 2 | 26.2 | 33.3 | 278 | 139 | 240 | 251 | 255 | 251 | 217 |
| | 11 | 1 | 30.5 | 17.2 | 290 | 165 | 278 | 290 | 294 | 288 | 247 |
| | 12 | | 32.0 | 0.0 | 293 | 173 | 291 | 304 | 307 | 301 | 258 |
| | Surface Daily Totals | | | | 2330 | 1080 | 1880 | 1972 | 2024 | 1978 | 1720 |
| Mar 21 | 7 | 5 | 10.0 | 78.7 | 153 | 37 | 49 | 49 | 47 | 45 | 35 |
| | 8 | 4 | 19.5 | 66.8 | 236 | 96 | 131 | 132 | 129 | 122 | 96 |
| | 9 | 3 | 28.2 | 53.4 | 270 | 147 | 205 | 207 | 203 | 193 | 152 |
| | 10 | 2 | 35.4 | 37.8 | 287 | 187 | 263 | 266 | 261 | 248 | 195 |
| | 11 | 1 | 40.3 | 19.8 | 295 | 212 | 300 | 303 | 297 | 283 | 223 |
| | 12 | | 42.0 | 0.0 | 298 | 220 | 312 | 315 | 309 | 294 | 232 |
| | Surface Daily Totals | | | | 2780 | 1578 | 2208 | 2228 | 2182 | 2074 | 1632 |
| Apr 21 | 6 | 6 | 8.6 | 97.8 | 108 | 27 | 13 | 9 | 8 | 7 | 5 |
| | 7 | 5 | 18.6 | 86.7 | 205 | 85 | 76 | 68 | 59 | 48 | 21 |
| | 8 | 4 | 28.5 | 74.9 | 247 | 142 | 149 | 141 | 129 | 113 | 69 |
| | 9 | 3 | 37.8 | 61.2 | 268 | 191 | 216 | 208 | 194 | 174 | 115 |
| | 10 | 2 | 45.8 | 44.6 | 280 | 228 | 268 | 260 | 245 | 223 | 152 |
| | 11 | 1 | 51.5 | 24.0 | 286 | 252 | 301 | 294 | 278 | 254 | 177 |
| | 12 | | 53.6 | 0.0 | 288 | 260 | 313 | 305 | 289 | 264 | 185 |
| | Surface Daily Totals | | | | 3076 | 2106 | 2358 | 2266 | 2114 | 1902 | 1262 |
| May 21 | 5 | 7 | 5.2 | 114.3 | 41 | 9 | 4 | 4 | 4 | 3 | 2 |
| | 6 | 6 | 14.7 | 103.7 | 162 | 61 | 27 | 16 | 15 | 13 | 10 |
| | 7 | 5 | 24.6 | 93.0 | 219 | 118 | 89 | 75 | 60 | 43 | 13 |
| | 8 | 4 | 34.7 | 81.6 | 248 | 171 | 156 | 142 | 123 | 101 | 45 |
| | 9 | 3 | 44.3 | 68.3 | 264 | 217 | 217 | 202 | 182 | 156 | 86 |
| | 10 | 2 | 53.0 | 51.3 | 274 | 252 | 265 | 251 | 229 | 200 | 120 |
| | 11 | 1 | 59.5 | 28.6 | 279 | 274 | 296 | 281 | 258 | 228 | 141 |
| | 12 | | 62.0 | 0.0 | 280 | 281 | 306 | 292 | 269 | 238 | 149 |
| | Surface Daily Totals | | | | 3254 | 2482 | 2418 | 2234 | 2010 | 1728 | 982 |
| Jun 21 | 5 | 7 | 7.9 | 116.5 | 77 | 21 | 9 | 9 | 8 | 7 | 5 |
| | 6 | 6 | 17.2 | 106.2 | 172 | 74 | 33 | 19 | 18 | 16 | 12 |
| | 7 | 5 | 27.0 | 95.8 | 220 | 129 | 93 | 77 | 59 | 39 | 15 |
| | 8 | 4 | 37.1 | 84.6 | 246 | 181 | 157 | 140 | 119 | 95 | 35 |
| | 9 | 3 | 46.9 | 71.6 | 261 | 225 | 216 | 198 | 175 | 147 | 74 |
| | 10 | 2 | 55.8 | 54.8 | 269 | 259 | 262 | 244 | 220 | 189 | 105 |
| | 11 | 1 | 62.7 | 31.2 | 274 | 280 | 291 | 273 | 248 | 216 | 126 |
| | 12 | | 65.5 | 0.0 | 275 | 287 | 301 | 283 | 258 | 225 | 133 |
| | Surface Daily Totals | | | | 3312 | 2626 | 2420 | 2204 | 1950 | 1644 | 874 |

| Date | Solar Time AM | PM | Solar Position Alt | Azm | Direct nor-mal | Global Irradiance, Btu/hr ft$^2$ Horiz. | South-Facing Elevation Angle 38 | 48 | 58 | 68 | 90 |
|------|------|------|------|------|------|------|------|------|------|------|------|
| Jul 21 | 5 | 7 | 5.7 | 114.7 | 43 | 10 | 5 | 5 | 4 | 4 | 3 |
|  | 6 | 6 | 15.2 | 104.1 | 156 | 62 | 28 | 18 | 16 | 15 | 11 |
|  | 7 | 5 | 25.1 | 93.5 | 211 | 118 | 89 | 75 | 59 | 42 | 14 |
|  | 8 | 4 | 35.1 | 82.1 | 240 | 171 | 154 | 140 | 121 | 99 | 43 |
|  | 9 | 3 | 44.8 | 68.8 | 256 | 215 | 214 | 199 | 178 | 153 | 83 |
|  | 10 | 2 | 53.5 | 51.9 | 266 | 250 | 261 | 246 | 224 | 195 | 116 |
|  | 11 | 1 | 60.1 | 29.0 | 271 | 272 | 291 | 276 | 253 | 223 | 137 |
|  | 12 |  | 62.6 | 0.0 | 272 | 279 | 301 | 286 | 263 | 232 | 144 |
|  | Surface Daily Totals |  |  |  | 3158 | 2474 | 2386 | 2200 | 1974 | 1694 | 956 |
| Aug 21 | 6 | 6 | 9.1 | 98.3 | 99 | 28 | 14 | 10 | 9 | 8 | 6 |
|  | 7 | 5 | 19.1 | 87.2 | 190 | 85 | 75 | 67 | 58 | 47 | 20 |
|  | 8 | 4 | 29.0 | 75.4 | 232 | 141 | 145 | 137 | 125 | 109 | 65 |
|  | 9 | 3 | 38.4 | 61.8 | 254 | 189 | 210 | 201 | 187 | 168 | 110 |
|  | 10 | 2 | 46.4 | 45.1 | 266 | 225 | 260 | 252 | 237 | 214 | 146 |
|  | 11 | 1 | 52.2 | 24.3 | 272 | 248 | 293 | 285 | 268 | 244 | 169 |
|  | 12 |  | 54.3 | 0.0 | 274 | 256 | 304 | 296 | 279 | 255 | 177 |
|  | Surface Daily Totals |  |  |  | 2898 | 2086 | 2300 | 2200 | 2046 | 1836 | 1208 |
| Sep 21 | 7 | 5 | 10.0 | 78.7 | 131 | 35 | 44 | 44 | 43 | 40 | 31 |
|  | 8 | 4 | 19.5 | 66.8 | 215 | 92 | 124 | 124 | 121 | 115 | 90 |
|  | 9 | 3 | 28.2 | 53.4 | 251 | 142 | 196 | 197 | 193 | 183 | 143 |
|  | 10 | 2 | 35.4 | 37.8 | 269 | 181 | 251 | 254 | 248 | 236 | 185 |
|  | 11 | 1 | 40.3 | 19.8 | 278 | 205 | 287 | 289 | 284 | 269 | 212 |
|  | 12 |  | 42.0 | 0.0 | 280 | 213 | 299 | 302 | 296 | 281 | 221 |
|  | Surface Daily Totals |  |  |  | 2568 | 1522 | 2102 | 2118 | 2070 | 1966 | 1546 |
| Oct 21 | 7 | 5 | 2.0 | 71.9 | 4 | 0 | 1 | 1 | 1 | 1 | 1 |
|  | 8 | 4 | 11.2 | 60.2 | 165 | 44 | 86 | 91 | 95 | 95 | 87 |
|  | 9 | 3 | 19.3 | 47.4 | 233 | 94 | 167 | 176 | 180 | 178 | 157 |
|  | 10 | 2 | 25.7 | 33.1 | 262 | 133 | 228 | 239 | 242 | 239 | 207 |
|  | 11 | 1 | 30.0 | 17.1 | 274 | 157 | 266 | 277 | 281 | 276 | 237 |
|  | 12 |  | 31.5 | 0.0 | 278 | 166 | 279 | 291 | 294 | 288 | 247 |
|  | Surface Daily Totals |  |  |  | 2154 | 1022 | 1774 | 1860 | 1890 | 1866 | 1626 |
| Nov 21 | 8 | 4 | 3.6 | 54.7 | 36 | 5 | 17 | 19 | 21 | 22 | 22 |
|  | 9 | 3 | 11.2 | 42.7 | 179 | 46 | 117 | 129 | 137 | 141 | 135 |
|  | 10 | 2 | 17.1 | 29.5 | 233 | 83 | 186 | 202 | 212 | 215 | 201 |
|  | 11 | 1 | 20.9 | 15.1 | 255 | 107 | 227 | 245 | 255 | 258 | 238 |
|  | 12 |  | 22.2 | 0.0 | 261 | 115 | 241 | 259 | 270 | 272 | 250 |
|  | Surface Daily Totals |  |  |  | 1668 | 596 | 1336 | 1448 | 1518 | 1544 | 1442 |
| Dec 21 | 9 | 3 | 8.0 | 40.9 | 140 | 27 | 87 | 98 | 105 | 110 | 109 |
|  | 10 | 2 | 13.6 | 28.2 | 214 | 63 | 164 | 180 | 192 | 197 | 190 |
|  | 11 | 1 | 17.3 | 14.4 | 242 | 86 | 207 | 226 | 239 | 244 | 231 |
|  | 12 |  | 18.6 | 0.0 | 250 | 94 | 222 | 241 | 254 | 260 | 244 |
|  | Surface Daily Totals |  |  |  | 1444 | 446 | 1136 | 1250 | 1326 | 1364 | 1304 |

NOTE: 1 Btu/hr ft$^2$ = 3.152 W/m$^2$.

## 56°N. Latitude

| Date | Solar Time AM | PM | Solar Position Alt | Azm | Direct nor-mal | Global Irradiance, Btu/hr ft² Horiz. | South-Facing Elevation Angle 46 | 56 | 66 | 76 | 90 |
|------|------|------|------|------|------|------|------|------|------|------|------|
| Jan 21 | 9 | 3 | 5.0 | 41.8 | 78 | 11 | 50 | 55 | 59 | 60 | 60 |
| | 10 | 2 | 9.9 | 28.5 | 170 | 39 | 135 | 146 | 154 | 156 | 153 |
| | 11 | 1 | 12.9 | 14.5 | 207 | 58 | 183 | 197 | 206 | 208 | 201 |
| | 12 | | 14.0 | 0.0 | 217 | 65 | 198 | 214 | 222 | 225 | 217 |
| | Surface Daily Totals | | | | 1126 | 282 | 934 | 1010 | 1058 | 1074 | 1044 |
| Feb 21 | 8 | 4 | 7.6 | 59.4 | 129 | 25 | 65 | 69 | 72 | 72 | 69 |
| | 9 | 3 | 14.2 | 45.9 | 214 | 65 | 151 | 159 | 162 | 161 | 151 |
| | 10 | 2 | 19.4 | 31.5 | 250 | 98 | 215 | 225 | 228 | 224 | 208 |
| | 11 | 1 | 22.8 | 16.1 | 266 | 119 | 254 | 265 | 268 | 263 | 243 |
| | 12 | | 24.0 | 0.0 | 270 | 126 | 268 | 279 | 282 | 276 | 250 |
| | Surface Daily Totals | | | | 1986 | 740 | 1640 | 1716 | 1742 | 1716 | 1598 |
| Mar 21 | 7 | 5 | 8.3 | 77.5 | 128 | 28 | 40 | 40 | 39 | 37 | 32 |
| | 8 | 4 | 16.2 | 64.4 | 215 | 75 | 119 | 120 | 117 | 111 | 97 |
| | 9 | 3 | 23.3 | 50.3 | 253 | 118 | 192 | 193 | 189 | 180 | 158 |
| | 10 | 2 | 29.0 | 34.9 | 272 | 151 | 249 | 251 | 246 | 234 | 205 |
| | 11 | 1 | 32.7 | 17.9 | 282 | 172 | 285 | 288 | 282 | 268 | 236 |
| | 12 | | 34.0 | 0.0 | 284 | 179 | 297 | 300 | 294 | 280 | 246 |
| | Surface Daily Totals | | | | 2586 | 1268 | 2066 | 2084 | 2040 | 1938 | 1700 |
| Apr 21 | 5 | 7 | 1.4 | 108.8 | 0 | 0 | 0 | 0 | 0 | 0 | 0 |
| | 6 | 6 | 9.6 | 96.5 | 122 | 32 | 14 | 9 | 8 | 7 | 6 |
| | 7 | 5 | 18.0 | 84.1 | 201 | 81 | 74 | 66 | 57 | 46 | 29 |
| | 8 | 4 | 26.1 | 70.9 | 239 | 129 | 143 | 135 | 123 | 108 | 82 |
| | 9 | 3 | 33.6 | 56.3 | 260 | 169 | 208 | 200 | 186 | 167 | 133 |
| | 10 | 2 | 39.9 | 39.7 | 272 | 201 | 259 | 251 | 236 | 214 | 174 |
| | 11 | 1 | 44.1 | 20.7 | 278 | 220 | 292 | 284 | 268 | 245 | 200 |
| | 12 | | 45.6 | 0.0 | 280 | 227 | 303 | 295 | 279 | 255 | 209 |
| | Surface Daily Totals | | | | 3024 | 1892 | 2282 | 2186 | 2038 | 1830 | 1458 |
| May 21 | 4 | 8 | 1.2 | 125.5 | 0 | 0 | 0 | 0 | 0 | 0 | 0 |
| | 5 | 7 | 8.5 | 113.4 | 93 | 25 | 10 | 9 | 8 | 7 | 6 |
| | 6 | 6 | 16.5 | 101.5 | 175 | 71 | 28 | 17 | 15 | 13 | 11 |
| | 7 | 5 | 24.8 | 89.3 | 219 | 119 | 88 | 74 | 58 | 41 | 16 |
| | 8 | 4 | 33.1 | 76.3 | 244 | 163 | 153 | 138 | 119 | 98 | 63 |
| | 9 | 3 | 40.9 | 61.6 | 259 | 201 | 212 | 197 | 176 | 151 | 109 |
| | 10 | 2 | 47.6 | 44.2 | 268 | 231 | 259 | 244 | 222 | 194 | 146 |
| | 11 | 1 | 52.3 | 23.4 | 273 | 249 | 288 | 274 | 251 | 222 | 170 |
| | 12 | | 54.0 | 0.0 | 275 | 255 | 299 | 284 | 261 | 231 | 178 |
| | Surface Daily Totals | | | | 3340 | 2374 | 2374 | 2188 | 1962 | 1682 | 1218 |
| Jun 21 | 4 | 8 | 4.2 | 127.2 | 21 | 4 | 2 | 2 | 2 | 2 | 1 |
| | 5 | 7 | 11.4 | 115.3 | 122 | 40 | 14 | 13 | 11 | 10 | 8 |
| | 6 | 6 | 19.3 | 103.6 | 185 | 86 | 34 | 19 | 17 | 15 | 12 |
| | 7 | 5 | 27.6 | 91.7 | 222 | 132 | 92 | 76 | 57 | 38 | 15 |
| | 8 | 4 | 35.9 | 78.8 | 243 | 175 | 154 | 137 | 116 | 92 | 55 |
| | 9 | 3 | 43.8 | 64.1 | 257 | 212 | 211 | 193 | 170 | 143 | 98 |
| | 10 | 2 | 50.7 | 46.4 | 265 | 240 | 255 | 238 | 214 | 184 | 133 |
| | 11 | 1 | 55.6 | 24.9 | 269 | 258 | 284 | 267 | 242 | 210 | 156 |
| | 12 | | 57.5 | 0.0 | 271 | 264 | 294 | 276 | 251 | 219 | 164 |
| | Surface Daily Totals | | | | 3438 | 2562 | 2388 | 2166 | 1910 | 1606 | 1120 |

| Date | Solar Time AM | PM | Solar Position Alt | Azm | Direct normal | Global Irradiance, Btu/hr ft² Horiz. | South-Facing Elevation Angle 46 | 56 | 66 | 76 | 90 |
|------|------|------|------|------|------|------|------|------|------|------|------|
| Jul 21 | 4 | 8 | 1.7 | 125.8 | 0 | 0 | 0 | 0 | 0 | 0 | 0 |
| | 5 | 7 | 9.0 | 113.7 | 91 | 27 | 11 | 10 | 9 | 8 | 6 |
| | 6 | 6 | 17.0 | 101.9 | 169 | 72 | 30 | 18 | 16 | 14 | 12 |
| | 7 | 5 | 25.3 | 89.7 | 212 | 119 | 88 | 74 | 58 | 41 | 15 |
| | 8 | 4 | 33.6 | 76.7 | 237 | 163 | 151 | 136 | 117 | 96 | 61 |
| | 9 | 3 | 41.4 | 62.0 | 252 | 201 | 208 | 193 | 173 | 147 | 106 |
| | 10 | 2 | 48.2 | 44.6 | 261 | 230 | 254 | 239 | 217 | 189 | 142 |
| | 11 | 1 | 52.9 | 23.7 | 265 | 248 | 283 | 268 | 245 | 216 | 165 |
| | 12 | | 54.6 | 0.0 | 267 | 254 | 293 | 278 | 255 | 225 | 173 |
| | Surface Daily Totals | | | | 3240 | 2372 | 2342 | 2152 | 1926 | 1646 | 1186 |
| Aug 21 | 5 | 7 | 2.0 | 109.2 | 1 | 0 | 0 | 0 | 0 | 0 | 0 |
| | 6 | 6 | 10.2 | 97.0 | 112 | 34 | 16 | 11 | 10 | 9 | 7 |
| | 7 | 5 | 18.5 | 84.5 | 187 | 82 | 73 | 65 | 56 | 45 | 28 |
| | 8 | 4 | 26.7 | 71.3 | 225 | 128 | 140 | 131 | 119 | 104 | 78 |
| | 9 | 3 | 34.3 | 56.7 | 246 | 168 | 202 | 193 | 179 | 160 | 126 |
| | 10 | 2 | 40.5 | 40.0 | 258 | 199 | 251 | 242 | 227 | 206 | 166 |
| | 11 | 1 | 44.8 | 20.9 | 264 | 218 | 282 | 274 | 258 | 235 | 191 |
| | 12 | | 46.3 | 0.0 | 266 | 225 | 293 | 285 | 269 | 245 | 200 |
| | Surface Daily Totals | | | | 2850 | 1884 | 2218 | 2118 | 1966 | 1760 | 1392 |
| Sep 21 | 7 | 5 | 8.3 | 77.5 | 107 | 25 | 36 | 36 | 34 | 32 | 28 |
| | 8 | 4 | 16.2 | 64.4 | 194 | 72 | 111 | 111 | 108 | 102 | 89 |
| | 9 | 3 | 23.3 | 50.3 | 233 | 114 | 181 | 182 | 178 | 168 | 147 |
| | 10 | 2 | 29.0 | 34.9 | 253 | 146 | 236 | 237 | 232 | 221 | 193 |
| | 11 | 1 | 32.7 | 17.9 | 263 | 166 | 271 | 273 | 267 | 254 | 223 |
| | 12 | | 34.0 | 0.0 | 266 | 173 | 283 | 285 | 279 | 265 | 233 |
| | Surface Daily Totals | | | | 2368 | 1220 | 1950 | 1962 | 1918 | 1820 | 1594 |
| Oct 21 | 8 | 4 | 7.1 | 59.1 | 104 | 20 | 53 | 57 | 59 | 59 | 57 |
| | 9 | 3 | 13.8 | 45.7 | 193 | 60 | 138 | 145 | 148 | 147 | 138 |
| | 10 | 2 | 19.0 | 31.3 | 231 | 92 | 201 | 210 | 213 | 210 | 195 |
| | 11 | 1 | 22.3 | 16.0 | 248 | 112 | 240 | 250 | 253 | 248 | 230 |
| | 12 | | 23.5 | 0.0 | 253 | 119 | 253 | 263 | 266 | 261 | 241 |
| | Surface Daily Totals | | | | 1804 | 688 | 1516 | 1586 | 1612 | 1588 | 1480 |
| Nov 21 | 9 | 3 | 5.2 | 41.9 | 76 | 12 | 49 | 54 | 57 | 59 | 58 |
| | 10 | 2 | 10.0 | 28.5 | 165 | 39 | 132 | 143 | 149 | 152 | 148 |
| | 11 | 1 | 13.1 | 14.5 | 201 | 58 | 179 | 193 | 201 | 203 | 196 |
| | 12 | | 14.2 | 0.0 | 211 | 65 | 194 | 209 | 217 | 219 | 211 |
| | Surface Daily Totals | | | | 1094 | 284 | 914 | 986 | 1032 | 1046 | 1016 |
| Dec 21 | 9 | 3 | 1.9 | 40.5 | 5 | 0 | 3 | 4 | 4 | 4 | 4 |
| | 10 | 2 | 6.6 | 27.5 | 113 | 19 | 86 | 95 | 101 | 104 | 103 |
| | 11 | 1 | 9.5 | 13.9 | 166 | 37 | 141 | 154 | 163 | 167 | 164 |
| | 12 | | 10.6 | 0.0 | 180 | 43 | 159 | 173 | 182 | 186 | 182 |
| | Surface Daily Totals | | | | 748 | 156 | 620 | 678 | 716 | 734 | 722 |

NOTE: 1 Btu/hr ft² = 3.152 W/m².

64°N. Latitude

| Date | Solar Time AM | PM | Solar Position Alt | Azm | Direct normal | Global Irradiance, Btu/hr ft² Horiz. | South-Facing Elevation Angle 54 | 64 | 74 | 84 | 90 |
|------|------|------|------|------|------|------|------|------|------|------|------|
| Jan 21 | 10 | 2 | 2.8 | 28.1 | 22 | 2 | 17 | 19 | 20 | 20 | 20 |
| | 11 | 1 | 5.2 | 14.1 | 81 | 12 | 72 | 77 | 80 | 81 | 81 |
| | 12 | | 6.0 | 0.0 | 100 | 16 | 91 | 98 | 102 | 103 | 103 |
| | Surface Daily Totals | | | | 306 | 45 | 268 | 290 | 302 | 306 | 304 |
| Feb 21 | 8 | 4 | 3.4 | 58.7 | 35 | 4 | 17 | 19 | 19 | 19 | 19 |
| | 9 | 3 | 8.6 | 44.8 | 147 | 31 | 103 | 108 | 111 | 110 | 107 |
| | 10 | 2 | 12.6 | 30.3 | 199 | 55 | 170 | 178 | 181 | 178 | 173 |
| | 11 | 1 | 15.1 | 15.3 | 222 | 71 | 212 | 220 | 223 | 219 | 213 |
| | 12 | | 16.0 | 0.0 | 228 | 77 | 225 | 235 | 237 | 232 | 226 |
| | Surface Daily Totals | | | | 1432 | 400 | 1230 | 1286 | 1302 | 1282 | 1252 |
| Mar 21 | 7 | 5 | 6.5 | 76.5 | 95 | 18 | 30 | 29 | 29 | 27 | 25 |
| | 8 | 4 | 12.7 | 62.6 | 185 | 54 | 101 | 102 | 99 | 94 | 89 |
| | 9 | 3 | 18.1 | 48.1 | 227 | 87 | 171 | 172 | 169 | 160 | 153 |
| | 10 | 2 | 22.3 | 32.7 | 249 | 112 | 227 | 229 | 224 | 213 | 203 |
| | 11 | 1 | 25.1 | 16.6 | 260 | 129 | 262 | 265 | 259 | 246 | 235 |
| | 12 | | 26.0 | 0.0 | 263 | 134 | 274 | 277 | 271 | 258 | 246 |
| | Surface Daily Totals | | | | 2296 | 932 | 1856 | 1870 | 1830 | 1736 | 1656 |
| Apr 21 | 5 | 7 | 4.0 | 108.5 | 27 | 5 | 2 | 2 | 2 | 1 | 1 |
| | 6 | 6 | 10.4 | 95.1 | 133 | 37 | 15 | 9 | 8 | 7 | 6 |
| | 7 | 5 | 17.0 | 81.6 | 194 | 76 | 70 | 63 | 54 | 43 | 37 |
| | 8 | 4 | 23.3 | 67.5 | 228 | 112 | 136 | 128 | 116 | 102 | 91 |
| | 9 | 3 | 29.0 | 52.3 | 248 | 144 | 197 | 189 | 176 | 158 | 145 |
| | 10 | 2 | 33.5 | 36.0 | 260 | 169 | 246 | 239 | 224 | 203 | 188 |
| | 11 | 1 | 36.5 | 18.4 | 266 | 184 | 278 | 270 | 255 | 233 | 216 |
| | 12 | | 37.6 | 0.0 | 268 | 190 | 289 | 281 | 266 | 243 | 225 |
| | Surface Daily Totals | | | | 2982 | 1644 | 2176 | 2082 | 1936 | 1736 | 1594 |
| May 21 | 4 | 8 | 5.8 | 125.1 | 51 | 11 | 5 | 4 | 4 | 3 | 3 |
| | 5 | 7 | 11.6 | 112.1 | 132 | 42 | 13 | 11 | 10 | 9 | 8 |
| | 6 | 6 | 17.9 | 99.1 | 185 | 79 | 29 | 16 | 14 | 12 | 11 |
| | 7 | 5 | 24.5 | 85.7 | 218 | 117 | 86 | 72 | 56 | 39 | 28 |
| | 8 | 4 | 30.9 | 71.5 | 239 | 152 | 148 | 133 | 115 | 94 | 80 |
| | 9 | 3 | 36.8 | 56.1 | 252 | 182 | 204 | 190 | 170 | 145 | 128 |
| | 10 | 2 | 41.6 | 38.9 | 261 | 205 | 249 | 235 | 213 | 186 | 167 |
| | 11 | 1 | 44.9 | 20.1 | 265 | 219 | 278 | 264 | 242 | 213 | 193 |
| | 12 | | 46.0 | 0.0 | 267 | 224 | 288 | 274 | 251 | 222 | 201 |
| | Surface Daily Totals | | | | 3470 | 2236 | 2312 | 2124 | 1898 | 1624 | 1436 |
| Jun 21 | 3 | 9 | 4.2 | 139.4 | 21 | 4 | 2 | 2 | 2 | 2 | 1 |
| | 4 | 8 | 9.0 | 126.4 | 93 | 27 | 10 | 9 | 8 | 7 | 6 |
| | 5 | 7 | 14.7 | 113.6 | 154 | 60 | 16 | 15 | 13 | 11 | 10 |
| | 6 | 6 | 21.0 | 100.8 | 194 | 96 | 34 | 19 | 17 | 14 | 13 |
| | 7 | 5 | 27.5 | 87.5 | 221 | 132 | 91 | 74 | 55 | 36 | 23 |
| | 8 | 4 | 34.0 | 73.3 | 239 | 166 | 150 | 133 | 112 | 88 | 73 |
| | 9 | 3 | 39.9 | 57.8 | 251 | 195 | 204 | 187 | 164 | 137 | 119 |
| | 10 | 2 | 44.9 | 40.4 | 258 | 217 | 247 | 230 | 206 | 177 | 157 |
| | 11 | 1 | 48.3 | 20.9 | 262 | 231 | 275 | 258 | 233 | 202 | 181 |
| | 12 | | 49.5 | 0.0 | 263 | 235 | 284 | 267 | 242 | 211 | 189 |
| | Surface Daily Totals | | | | 3650 | 2488 | 2342 | 2118 | 1862 | 1558 | 1356 |

| Date | Solar Time AM | PM | Solar Position Alt | Azm | Direct nor-mal | Global Irradiance, Btu/hr ft² Horiz. | South-Facing Elevation Angle 54 | 64 | 74 | 84 | 90 |
|------|------|------|------|------|------|------|------|------|------|------|------|
| Jul 21 | 4 | 8 | 6.4 | 125.3 | 53 | 13 | 6 | 5 | 5 | 4 | 4 |
|  | 5 | 7 | 12.1 | 112.4 | 128 | 44 | 14 | 13 | 11 | 10 | 9 |
|  | 6 | 6 | 18.4 | 99.4 | 179 | 81 | 30 | 17 | 16 | 13 | 12 |
|  | 7 | 5 | 25.0 | 86.0 | 211 | 118 | 86 | 72 | 56 | 38 | 28 |
|  | 8 | 4 | 31.4 | 71.8 | 231 | 152 | 146 | 131 | 113 | 91 | 77 |
|  | 9 | 3 | 37.3 | 56.3 | 245 | 182 | 201 | 186 | 166 | 141 | 124 |
|  | 10 | 2 | 42.2 | 39.2 | 253 | 204 | 245 | 230 | 208 | 181 | 162 |
|  | 11 | 1 | 45.4 | 20.2 | 257 | 218 | 273 | 258 | 236 | 207 | 187 |
|  | 12 | | 46.6 | 0.0 | 259 | 223 | 282 | 267 | 245 | 216 | 195 |
|  | Surface Daily Totals | | | | 3372 | 2248 | 2280 | 2090 | 1864 | 1588 | 1400 |
| Aug 21 | 5 | 7 | 4.6 | 108.8 | 29 | 6 | 3 | 3 | 2 | 2 | 2 |
|  | 6 | 6 | 11.0 | 95.5 | 123 | 39 | 16 | 11 | 10 | 8 | 7 |
|  | 7 | 5 | 17.6 | 81.9 | 181 | 77 | 69 | 61 | 52 | 42 | 35 |
|  | 8 | 4 | 23.9 | 67.8 | 214 | 113 | 131 | 123 | 112 | 97 | 87 |
|  | 9 | 3 | 29.6 | 52.6 | 234 | 144 | 190 | 182 | 169 | 150 | 138 |
|  | 10 | 2 | 34.2 | 36.2 | 246 | 168 | 237 | 229 | 215 | 194 | 179 |
|  | 11 | 1 | 37.2 | 18.5 | 252 | 183 | 268 | 260 | 244 | 222 | 205 |
|  | 12 | | 38.3 | 0.0 | 254 | 188 | 278 | 270 | 255 | 232 | 215 |
|  | Surface Daily Totals | | | | 2808 | 1646 | 2108 | 2008 | 1860 | 1662 | 1522 |
| Sep 21 | 7 | 5 | 6.5 | 76.5 | 77 | 16 | 25 | 24 | 24 | 23 | 21 |
|  | 8 | 4 | 12.7 | 62.6 | 163 | 51 | 92 | 92 | 90 | 85 | 81 |
|  | 9 | 3 | 18.1 | 48.1 | 206 | 83 | 159 | 159 | 156 | 147 | 141 |
|  | 10 | 2 | 22.3 | 32.7 | 229 | 108 | 212 | 213 | 209 | 198 | 189 |
|  | 11 | 1 | 25.1 | 16.6 | 240 | 124 | 246 | 248 | 243 | 230 | 220 |
|  | 12 | | 26.0 | 0.0 | 244 | 129 | 258 | 260 | 254 | 241 | 230 |
|  | Surface Daily Totals | | | | 2074 | 892 | 1726 | 1736 | 1696 | 1608 | 1532 |
| Oct 21 | 8 | 4 | 3.0 | 58.5 | 17 | 2 | 9 | 9 | 10 | 10 | 10 |
|  | 9 | 3 | 8.1 | 44.6 | 122 | 26 | 86 | 91 | 93 | 92 | 90 |
|  | 10 | 2 | 12.1 | 30.2 | 176 | 50 | 152 | 159 | 161 | 159 | 155 |
|  | 11 | 1 | 14.6 | 15.2 | 201 | 65 | 193 | 201 | 203 | 200 | 195 |
|  | 12 | | 15.5 | 0.0 | 208 | 71 | 207 | 215 | 217 | 213 | 208 |
|  | Surface Daily Totals | | | | 1238 | 358 | 1088 | 1136 | 1152 | 1134 | 1106 |
| Nov 21 | 10 | 2 | 3.0 | 28.1 | 23 | 3 | 18 | 20 | 21 | 21 | 21 |
|  | 11 | 1 | 5.4 | 14.2 | 79 | 12 | 70 | 76 | 79 | 80 | 79 |
|  | 12 | | 6.2 | 0.0 | 97 | 17 | 89 | 96 | 100 | 101 | 100 |
|  | Surface Daily Totals | | | | 302 | 46 | 266 | 286 | 298 | 302 | 300 |
| Dec 21 | 11 | 1 | 1.8 | 13.7 | 4 | 0 | 3 | 4 | 4 | 4 | 4 |
|  | 12 | | 2.6 | 0.0 | 16 | 2 | 14 | 15 | 16 | 17 | 17 |
|  | Surface Daily Totals | | | | 24 | 2 | 20 | 22 | 24 | 24 | 24 |

NOTE: 1 Btu/hr ft² = 3.152 W/m².

**TABLE A-7.** Latitude 24°N. Incident Angles for Horizontal and South-Facing Tilted Surfaces

| | Tilt Angle: | Horiz. | Lat. −10 | Lat. | Lat. +10 | Lat. +20 | Vert. |
|---|---|---|---|---|---|---|---|
| | 7  5 | 86.8 | 80.5 | 76.3 | 72.4 | 68.9 | 62.6 |
| Dates: (Decl) | 8  4 | 75.1 | 67.5 | 62.7 | 58.6 | 55.4 | 56.6 |
| Dec 21 (−23.45) | 9  3 | 64.5 | 55.3 | 49.6 | 44.9 | 41.8 | 51.1 |
| | 10  2 | 55.7 | 44.5 | 37.4 | 31.6 | 28.0 | 46.6 |
| | 11  1 | 49.5 | 36.5 | 27.6 | 19.6 | 14.3 | 43.6 |
| | 12 | 7.4 | 33.4 | 23.5 | 13.5 | 3.4 | 42.5 |

| | Tilt Angle: | Horiz. | Lat. −10 | Lat. | Lat. +10 | Lat. +20 | Vert. |
|---|---|---|---|---|---|---|---|
| | 7  5 | 85.2 | 79.6 | 75.9 | 72.6 | 69.8 | 65.7 |
| Jan 21 (−19.9) | 8  4 | 73.1 | 66.2 | 62.0 | 58.5 | 56.0 | 59.8 |
| Nov 21 (−19.9) | 9  3 | 62.1 | 53.5 | 48.4 | 44.5 | 42.2 | 54.4 |
| | 10  2 | 52.8 | 42.1 | 35.5 | 30.6 | 28.2 | 50.0 |
| | 11  1 | 46.4 | 33.4 | 24.8 | 17.6 | 14.1 | 47.0 |
| | 12 | 44.0 | 30.0 | 20.0 | 10.0 | 0.0 | 46.0 |

| | Tilt Angle: | Horiz. | Lat. −10 | Lat. | Lat. +10 | Lat. +20 | Vert. |
|---|---|---|---|---|---|---|---|
| | 7  5 | 80.7 | 77.2 | 75.2 | 73.7 | 72.6 | 74.8 |
| Feb 21 (−10.6) | 8  4 | 67.7 | 62.9 | 60.5 | 59.0 | 58.5 | 69.0 |
| Oct 21 (−10.7) | 9  3 | 55.6 | 49.0 | 45.9 | 44.3 | 44.5 | 63.8 |
| | 10  2 | 44.9 | 35.9 | 31.5 | 29.5 | 30.6 | 59.6 |
| | 11  1 | 37.0 | 25.0 | 18.0 | 14.8 | 17.6 | 56.9 |
| | 12 | 34.0 | 20.0 | 10.0 | 0.0 | 10.0 | 56.0 |

| | Tilt Angle: | Horiz. | Lat. −10 | Lat. | Lat. +10 | Lat. +20 | Vert. |
|---|---|---|---|---|---|---|---|
| | 7  5 | 76.3 | 75.2 | 75.0 | 75.2 | 75.9 | 84.0 |
| Mar 21 (0.0) | 8  4 | 62.8 | 60.5 | 60.0 | 60.5 | 62.0 | 78.3 |
| Sep 21 (0.0) | 9  3 | 49.8 | 45.9 | 45.0 | 45.9 | 48.4 | 73.3 |
| | 10  2 | 37.7 | 31.5 | 30.0 | 31.5 | 35.5 | 69.4 |
| | 11  1 | 28.1 | 18.0 | 15.0 | 18.0 | 24.8 | 66.9 |
| | 12 | 24.0 | 10.0 | 0.0 | 10.0 | 20.0 | 66.0 |

| | Tilt Angle: | Horiz. | Lat. −10 | Lat. | Lat. +10 | Lat. +20 | Vert. |
|---|---|---|---|---|---|---|---|
| | 6  6 | 85.3 | 88.0 | 90.0 | 92.0 | 93.9 | 100.6 |
| | 7  5 | 71.7 | 73.5 | 75.3 | 77.6 | 80.2 | 94.6 |
| Apr 21 (+11.9) | 8  4 | 58.0 | 58.9 | 60.7 | 63.4 | 67.0 | 89.1 |
| Aug 21 (+12.1) | 9  3 | 44.4 | 44.2 | 46.2 | 49.7 | 54.4 | 84.4 |
| | 10  2 | 31.0 | 29.5 | 32.0 | 36.8 | 43.2 | 80.7 |
| | 11  1 | 18.9 | 14.8 | 18.9 | 26.2 | 34.9 | 78.4 |
| | 12 | 12.4 | 1.6 | 11.6 | 21.6 | 31.6 | 77.6 |

| | Tilt Angle: | Horiz. | Lat. −10 | Lat. | Lat. +10 | Lat. +20 | Vert. |
|---|---|---|---|---|---|---|---|
| | 6  6 | 82.0 | 86.0 | 90.0 | 93.4 | 96.7 | 108.2 |
| | 7  5 | 68.8 | 72.6 | 75.9 | 79.6 | 83.6 | 102.3 |
| May 21 (+20.3) | 8  4 | 55.4 | 58.5 | 62.0 | 66.2 | 71.1 | 97.0 |
| Jul 21 (+20.5) | 9  3 | 41.7 | 44.5 | 48.4 | 53.5 | 59.5 | 92.4 |
| | 10  2 | 28.0 | 30.6 | 35.5 | 42.1 | 49.6 | 88.9 |
| | 11  1 | 14.5 | 17.6 | 24.8 | 33.4 | 42.6 | 86.7 |
| | 12 | 4.0 | 10.0 | 20.0 | 30.0 | 40.0 | 86.0 |

| | Tilt Angle: | Horiz. | Lat. −10 | Lat. | Lat. +10 | Lat. +20 | Vert. |
|---|---|---|---|---|---|---|---|
| | 6  6 | 80.7 | 86.0 | 90.0 | 94.0 | 97.8 | 111.3 |
| | 7  5 | 67.7 | 72.4 | 76.3 | 80.5 | 85.0 | 105.5 |
| Jun 21 (+23.45) | 8  4 | 54.5 | 58.6 | 62.7 | 67.5 | 72.8 | 100.2 |
| | 9  3 | 41.0 | 44.9 | 49.6 | 55.8 | 61.7 | 95.7 |
| | 10  2 | 27.4 | 31.6 | 37.4 | 44.5 | 52.4 | 92.3 |
| | 11  1 | 13.7 | 19.7 | 27.6 | 36.5 | 45.9 | 90.2 |
| | 12 | 0.6 | 13.4 | 23.4 | 33.4 | 43.4 | 89.4 |

**TABLE A-8.** Latitude 32°N. Incident Angles for Horizontal and South-Facing Tilted Surfaces

| | Tilt Angle: | Horiz. | Lat. −10 | Lat. | Lat. +10 | Lat. +20 | Vert. |
|---|---|---|---|---|---|---|---|
| | 8    4 | 79.7 | 67.5 | 62.7 | 58.6 | 55.4 | 54.5 |
| Dates: (Decl.) | 9    3 | 70.2 | 55.3 | 49.6 | 44.9 | 41.8 | 47.1 |
| Dec 21 (−23.45) | 10   2 | 62.4 | 44.5 | 37.4 | 31.6 | 28.0 | 40.7 |
| | 11   1 | 57.3 | 36.5 | 27.6 | 19.6 | 14.3 | 36.2 |
| | 12 | 55.4 | 33.4 | 23.4 | 13.5 | 3.5 | 34.5 |

| | Tilt Angle: | Horiz. | Lat. −10 | Lat. | Lat. +10 | Lat. +20 | Vert. |
|---|---|---|---|---|---|---|---|
| | 7    5 | 88.6 | 79.6 | 75.9 | 72.6 | 69.8 | 65.2 |
| | 8    4 | 77.5 | 66.2 | 62.0 | 58.5 | 56.0 | 57.4 |
| Jan 21 (−19.9) | 9    3 | 66.7 | 53.5 | 48.4 | 44.5 | 42.2 | 50.0 |
| Nov 21 (−19.9) | 10   2 | 59.4 | 42.1 | 35.5 | 30.6 | 28.2 | 43.8 |
| | 11   1 | 53.9 | 33.4 | 24.8 | 17.6 | 14.1 | 39.6 |
| | 12 | 52.0 | 30.0 | 20.0 | 10.0 | 0.0 | 38.0 |

| | Tilt Angle: | Horiz. | Lat. −10 | Lat. | Lat. +10 | Lat. +20 | Vert. |
|---|---|---|---|---|---|---|---|
| | 7    5 | 82.9 | 77.2 | 75.2 | 73.7 | 72.6 | 73.6 |
| | 8    4 | 71.0 | 62.9 | 60.5 | 59.0 | 58.5 | 65.9 |
| Feb 21 (−10.6) | 9    3 | 60.1 | 49.0 | 45.9 | 44.3 | 44.5 | 58.9 |
| Oct 21 (−10.7) | 10   2 | 50.9 | 35.9 | 31.5 | 29.5 | 30.6 | 53.2 |
| | 11   1 | 44.4 | 25.0 | 18.0 | 14.8 | 17.6 | 49.4 |
| | 12 | 42.0 | 20.0 | 10.0 | 0.0 | 10.0 | 48.0 |

| | Tilt Angle: | Horiz. | Lat. −10 | Lat. | Lat. +10 | Lat. +20 | Vert. |
|---|---|---|---|---|---|---|---|
| | 7    5 | 77.3 | 75.2 | 75.0 | 75.2 | 75.9 | 82.1 |
| | 8    4 | 64.9 | 60.5 | 60.0 | 60.5 | 62.0 | 74.6 |
| Mar 21 (0.0) | 9    3 | 53.2 | 45.9 | 45.0 | 45.9 | 48.4 | 68.0 |
| Sep 21 (0.0) | 10   2 | 42.7 | 31.5 | 30.0 | 31.5 | 35.5 | 62.7 |
| | 11   1 | 35.0 | 18.0 | 15.0 | 18.0 | 24.8 | 59.2 |
| | 12 | 32.0 | 10.0 | 0.0 | 10.0 | 20.0 | 58.0 |

| | Tilt Angle: | Horiz. | Lat. −10 | Lat. | Lat. +10 | Lat. +20 | Vert. |
|---|---|---|---|---|---|---|---|
| | 6    6 | 83.9 | 88.0 | 90.0 | 92.0 | 93.9 | 99.8 |
| | 7    5 | 71.2 | 73.5 | 75.3 | 77.6 | 80.2 | 92.1 |
| Apr 21 (+11.9) | 8    4 | 58.5 | 58.9 | 60.7 | 63.4 | 67.0 | 84.9 |
| Aug 21 (+12.1) | 9    3 | 46.1 | 44.2 | 46.2 | 49.7 | 54.4 | 78.7 |
| | 10   2 | 34.3 | 29.5 | 32.0 | 36.8 | 43.2 | 3.8 |
| | 11   1 | 24.6 | 14.8 | 18.9 | 26.2 | 34.9 | 70.7 |
| | 12 | 20.4 | 1.6 | 11.6 | 21.6 | 31.6 | 69.6 |

| | Tilt Angle: | Horiz. | Lat. −10 | Lat. | Lat. +10 | Lat. +20 | Vert. |
|---|---|---|---|---|---|---|---|
| | 6    6 | 79.6 | 86.6 | 90.0 | 93.4 | 96.7 | 106.9 |
| | 7    5 | 67.2 | 72.6 | 75.9 | 79.6 | 83.6 | 99.3 |
| May 21 (+20.3) | 8    4 | 54.6 | 58.5 | 62.0 | 66.2 | 71.1 | 92.4 |
| Jul 21 (+20.5) | 9    3 | 41.9 | 44.5 | 48.4 | 53.5 | 59.5 | 86.4 |
| | 10   2 | 29.4 | 30.6 | 35.5 | 42.1 | 49.6 | 81.9 |
| | 11   1 | 18.0 | 17.6 | 24.8 | 33.4 | 42.6 | 79.0 |
| | 12 | 12.0 | 10.0 | 20.0 | 30.0 | 40.0 | 78.0 |

| | Tilt Angle: | Horiz. | Lat. −10 | Lat. | Lat. +10 | Lat. +20 | Vert. |
|---|---|---|---|---|---|---|---|
| | 6    6 | 77.8 | 86.0 | 90.0 | 94.0 | 97.8 | 109.7 |
| | 7    5 | 65.7 | 72.4 | 76.3 | 80.5 | 85.0 | 102.2 |
| | 8    4 | 53.1 | 58.6 | 62.7 | 67.5 | 72.8 | 95.4 |
| Jun 21 (+23.45) | 9    3 | 40.4 | 44.9 | 49.6 | 55.3 | 61.7 | 89.6 |
| | 10   2 | 27.8 | 31.6 | 37.4 | 44.5 | 52.4 | 85.2 |
| | 11   1 | 15.8 | 19.6 | 27.6 | 36.5 | 45.8 | 82.4 |
| | 12 | 8.6 | 13.4 | 23.4 | 33.4 | 43.4 | 81.4 |

**TABLE A-9.** Latitude 40°N. Incident Angles for Horizontal and South-Facing Tilted Surfaces

|  | Tilt Angle: | Horiz. | Lat. −10 | Lat. | Lat. +10 | Lat. +20 | Vert. |
|---|---|---|---|---|---|---|---|
|  | 8    4 | 84.5 | 67.5 | 62.7 | 58.6 | 55.4 | 53.2 |
| Dates: (Decl.) | 9    3 | 76.0 | 55.3 | 49.6 | 44.9 | 41.8 | 43.8 |
| Dec 21 (−23.45) | 10    2 | 69.3 | 44.5 | 37.4 | 31.6 | 28.0 | 35.4 |
|  | 11    1 | 65.0 | 36.5 | 27.6 | 19.6 | 14.3 | 29.0 |
|  | 12 | 63.4 | 33.4 | 23.4 | 13.5 | 3.5 | 26.6 |

|  | Tilt Angle: | Horiz. | Lat. −10 | Lat. | Lat. +10 | Lat. +20 | Vert. |
|---|---|---|---|---|---|---|---|
|  | 8    4 | 81.9 | 66.2 | 62.0 | 58.5 | 56.0 | 55.7 |
| Jan 21 (−19.9) | 9    3 | 73.2 | 53.5 | 48.4 | 44.5 | 42.2 | 46.4 |
| Nov 21 (−19.9) | 10    2 | 66.2 | 42.1 | 35.5 | 30.6 | 28.2 | 38.3 |
|  | 11    1 | 61.6 | 33.4 | 24.8 | 17.6 | 14.1 | 32.3 |
|  | 12 | 60.0 | 30.0 | 20.0 | 10.0 | 0.0 | 30.0 |

|  | Tilt Angle: | Horiz. | Lat. −10 | Lat. | Lat. +10 | Lat. +20 | Vert. |
|---|---|---|---|---|---|---|---|
|  | 7    5 | 85.2 | 77.2 | 75.2 | 73.7 | 72.6 | 72.7 |
|  | 8    4 | 74.6 | 62.9 | 60.5 | 59.0 | 58.5 | 63.3 |
| Feb 21 (−10.6) | 9    3 | 65.0 | 49.0 | 45.9 | 44.3 | 44.5 | 59.5 |
| Oct  21 (−10.7) | 10    2 | 57.2 | 35.9 | 31.5 | 29.5 | 30.6 | 47.1 |
|  | 11    1 | 51.9 | 25.0 | 18.0 | 14.8 | 17.6 | 41.9 |
|  | 12 | 50.0 | 20.0 | 10.0 | 0.0 | 10.0 | 40.0 |

|  | Tilt Angle: | Horiz. | Lat. −10 | Lat. | Lat. +10 | Lat. +20 | Vert. |
|---|---|---|---|---|---|---|---|
|  | 7    5 | 78.6 | 75.2 | 75.0 | 75.2 | 75.9 | 80.4 |
|  | 8    4 | 67.5 | 60.5 | 60.0 | 60.5 | 62.0 | 71.3 |
| Mar 21 (0.0) | 9    3 | 57.2 | 45.9 | 45.0 | 45.9 | 48.4 | 63.0 |
| Sep 21 (0.0) | 10    2 | 48.4 | 31.5 | 30.0 | 31.5 | 35.5 | 56.2 |
|  | 11    1 | 42.3 | 18.0 | 15.0 | 18.0 | 24.8 | 51.6 |
|  | 12 | 40.0 | 10.0 | 0.0 | 10.0 | 20.0 | 50.0 |

|  | Tilt Angle: | Horiz. | Lat. −10 | Lat. | Lat. +10 | Lat. +20 | Vert. |
|---|---|---|---|---|---|---|---|
|  | 6    6 | 82.6 | 88.0 | 90.0 | 92.0 | 93.9 | 98.9 |
|  | 7    5 | 71.1 | 73.5 | 75.3 | 77.6 | 80.2 | 89.5 |
| Apr 21 (+11.9) | 8    4 | 59.7 | 58.9 | 60.7 | 63.4 | 67.0 | 80.7 |
| Aug 21 (+12.1) | 9    3 | 48.7 | 44.2 | 46.2 | 49.7 | 54.4 | 73.1 |
|  | 10    2 | 38.8 | 29.5 | 32.0 | 36.8 | 43.2 | 67.0 |
|  | 11    1 | 31.3 | 14.8 | 18.9 | 26.2 | 34.9 | 63.0 |
|  | 12 | 28.4 | 1.6 | 11.6 | 21.6 | 31.6 | 61.6 |

|  | Tilt Angle: | Horiz. | Lat. −10 | Lat. | Lat. +10 | Lat. +20 | Vert. |
|---|---|---|---|---|---|---|---|
|  | 5    7 | 88.1 | 100.4 | 104.1 | 107.4 | 110.2 | 114.7 |
|  | 6    6 | 77.3 | 86.6 | 90.0 | 93.4 | 96.7 | 105.2 |
|  | 7    5 | 66.0 | 72.6 | 75.9 | 79.6 | 83.6 | 96.1 |
| May 21 (+20.3) | 8    4 | 54.6 | 58.5 | 62.0 | 66.2 | 71.1 | 87.7 |
| Jul 21 (+20.5) | 9    3 | 43.2 | 44.5 | 48.4 | 53.5 | 59.5 | 80.5 |
|  | 10    2 | 32.5 | 30.6 | 35.5 | 42.1 | 49.6 | 74.9 |
|  | 11    1 | 23.8 | 17.6 | 24.8 | 33.4 | 42.6 | 71.2 |
|  | 12 | 20.0 | 10.0 | 20.0 | 30.0 | 40.0 | 70.0 |

|  | Tilt Angle: | Horiz. | Lat. −10 | Lat. | Lat. +10 | Lat. +20 | Vert. |
|---|---|---|---|---|---|---|---|
|  | 5    7 | 85.8 | 99.5 | 103.7 | 107.6 | 111.1 | 117.2 |
|  | 6    6 | 75.2 | 86.0 | 90.0 | 94.0 | 97.8 | 107.7 |
|  | 7    5 | 64.0 | 72.4 | 76.3 | 80.5 | 85.0 | 98.8 |
| Jun 21 (+23.45) | 8    4 | 52.6 | 58.6 | 62.7 | 67.5 | 72.8 | 90.6 |
|  | 9    3 | 41.2 | 44.9 | 49.6 | 55.3 | 61.7 | 83.6 |
|  | 10    2 | 30.2 | 31.6 | 37.4 | 44.5 | 52.4 | 78.1 |
|  | 11    1 | 20.8 | 19.6 | 27.6 | 36.5 | 45.8 | 74.6 |
|  | 12 | 16.6 | 13.4 | 23.4 | 33.4 | 43.4 | 73.4 |

**TABLE A-10.** Latitude 48°N. Incident Angles for Horizontal and South-Facing Tilted Surfaces

| Dates (Decl.) | Tilt Angle: | | Horiz. | Lat. −10 | Lat. | Lat. +10 | Lat. +20 | Vert. |
|---|---|---|---|---|---|---|---|---|
| Dec 21 (−23.45) | 9 | 3 | 82.0 | 55.3 | 49.6 | 44.9 | 41.8 | 41.6 |
| | 10 | 2 | 76.4 | 44.5 | 37.4 | 31.6 | 28.0 | 31.1 |
| | 11 | 1 | 72.7 | 36.5 | 27.6 | 19.6 | 14.3 | 22.4 |
| | 12 | | 71.4 | 33.4 | 23.4 | 13.5 | 3.5 | 18.5 |

| | Tilt Angle: | | Horiz. | Lat. −10 | Lat. | Lat. +10 | Lat. +20 | Vert. |
|---|---|---|---|---|---|---|---|---|
| | 8 | 4 | 86.5 | 66.2 | 62.0 | 58.5 | 56.0 | 54.7 |
| Jan 21 (−19.9) | 9 | 3 | 79.0 | 53.5 | 48.4 | 44.5 | 42.2 | 43.7 |
| Nov 21 (−19.9) | 10 | 2 | 73.1 | 42.1 | 35.5 | 30.6 | 28.2 | 33.5 |
| | 11 | 1 | 69.3 | 33.4 | 24.8 | 17.6 | 14.1 | 25.4 |
| | 12 | | 68.0 | 30.0 | 20.0 | 10.0 | 0.0 | 22.0 |

| | Tilt Angle: | | Horiz. | Lat. −10 | Lat. | Lat. +10 | Lat. +20 | Vert. |
|---|---|---|---|---|---|---|---|---|
| | 7 | 5 | 87.6 | 72.2 | 75.2 | 73.7 | 72.6 | 72.2 |
| Feb 21 (−10.6) | 8 | 4 | 78.4 | 62.9 | 60.5 | 59.0 | 58.5 | 61.2 |
| Oct 21 (−10.7) | 9 | 3 | 70.3 | 49.0 | 45.9 | 44.3 | 44.5 | 50.7 |
| | 10 | 2 | 63.8 | 35.9 | 31.5 | 29.5 | 30.6 | 41.4 |
| | 11 | 1 | 59.5 | 25.0 | 18.0 | 14.8 | 17.5 | 34.6 |
| | 12 | | 58.0 | 20.0 | 10.0 | 0.0 | 10.0 | 32.0 |

| | Tilt Angle: | | Horiz. | Lat. −10 | Lat. | Lat. +10 | Lat. +20 | Vert. |
|---|---|---|---|---|---|---|---|---|
| | 7 | 5 | 80.0 | 75.2 | 75.0 | 73.2 | 75.9 | 78.9 |
| Mar 21 (0.0) | 8 | 4 | 70.5 | 60.5 | 60.0 | 60.5 | 62.0 | 68.2 |
| Sep 21 (0.0) | 9 | 3 | 61.8 | 45.9 | 45.0 | 45.9 | 48.4 | 58.3 |
| | 10 | 2 | 54.6 | 31.5 | 30.0 | 31.5 | 35.5 | 49.9 |
| | 11 | 1 | 49.7 | 18.0 | 15.0 | 18.0 | 24.8 | 44.1 |
| | 12 | | 48.0 | 10.0 | 0.0 | 10.0 | 20.0 | 42.0 |

| | Tilt Angle: | | Horiz. | Lat. −10 | Lat. | Lat. +10 | Lat. +20 | Vert. |
|---|---|---|---|---|---|---|---|---|
| | 6 | 6 | 81.4 | 88.0 | 90.0 | 92.0 | 93.9 | 97.7 |
| | 7 | 5 | 71.4 | 73.5 | 75.3 | 77.6 | 80.2 | 86.9 |
| Apr 21 (+11.9) | 8 | 4 | 61.5 | 58.9 | 60.7 | 63.4 | 67.0 | 76.7 |
| Aug 21 (+12.1) | 9 | 3 | 52.2 | 44.2 | 46.2 | 49.7 | 54.4 | 67.7 |
| | 10 | 2 | 44.2 | 29.5 | 32.0 | 36.8 | 43.2 | 60.3 |
| | 11 | 1 | 38.5 | 14.8 | 18.9 | 26.2 | 34.9 | 55.3 |
| | 12 | | 36.4 | 1.6 | 11.6 | 21.6 | 31.6 | 53.6 |

| | Tilt Angle: | | Horiz. | Lat. −10 | Lat. | Lat. +10 | Lat. +20 | Vert. |
|---|---|---|---|---|---|---|---|---|
| | 5 | 7 | 84.8 | 100.4 | 104.1 | 107.4 | 110.2 | 114.2 |
| | 6 | 6 | 75.3 | 86.6 | 90.0 | 93.4 | 96.7 | 103.2 |
| | 7 | 5 | 65.4 | 72.8 | 75.9 | 79.6 | 83.6 | 92.8 |
| May 21 (+20.3) | 8 | 4 | 55.4 | 58.5 | 62.0 | 66.2 | 71.1 | 83.1 |
| Jul 21 (+20.5) | 9 | 3 | 45.7 | 44.5 | 48.4 | 53.5 | 59.5 | 74.6 |
| | 10 | 2 | 37.0 | 30.6 | 35.5 | 42.1 | 49.6 | 67.9 |
| | 11 | 1 | 30.5 | 17.6 | 24.8 | 33.4 | 42.6 | 63.5 |
| | 12 | | 28.0 | 10.0 | 20.0 | 30.0 | 40.0 | 62.0 |

| | Tilt Angle: | | Horiz. | Lat. −10 | Lat. | Lat. +10 | Lat. +20 | Vert. |
|---|---|---|---|---|---|---|---|---|
| | 5 | 7 | 82.1 | 99.5 | 103.7 | 107.6 | 111.1 | 116.3 |
| | 6 | 6 | 72.8 | 86.0 | 90.0 | 94.0 | 97.3 | 105.4 |
| | 7 | 5 | 63.0 | 72.4 | 76.3 | 80.5 | 85.0 | 95.2 |
| Jun 21 (+23.45) | 8 | 4 | 52.9 | 58.6 | 62.7 | 67.5 | 72.8 | 85.7 |
| | 9 | 3 | 43.1 | 44.9 | 49.6 | 55.3 | 61.7 | 77.5 |
| | 10 | 2 | 34.2 | 31.6 | 37.4 | 44.5 | 52.4 | 71.1 |
| | 11 | 1 | 27.3 | 19.6 | 27.6 | 36.5 | 45.8 | 66.9 |
| | 12 | | 24.6 | 13.4 | 23.4 | 33.4 | 43.4 | 65.4 |

**TABLE A-11.** Latitude 56°N. Incident Angles for Horizontal and South-Facing Tilted Surfaces

| Dates (Decl.) | Tilt Angle: | | Horiz. | Lat. −10 | Lat. | Lat. +10 | Lat. +20 | Vert. |
|---|---|---|---|---|---|---|---|---|
| Dec 21 (−23.45) | 9 | 3 | 83.1 | 55.3 | 49.6 | 44.9 | 41.8 | 40.5 |
| | 10 | 2 | 83.4 | 44.5 | 37.4 | 31.6 | 28.0 | 28.2 |
| | 11 | 1 | 80.5 | 36.5 | 27.6 | 19.6 | 14.3 | 16.8 |
| | 12 | | 79.4 | 33.4 | 23.4 | 13.5 | 3.4 | 10.5 |

| | Tilt Angle: | | Horiz. | Lat. −10 | Lat. | Lat. +10 | Lat. +20 | Vert. |
|---|---|---|---|---|---|---|---|---|
| Jan 21 (−19.9) | 9 | 3 | 85.0 | 53.5 | 48.4 | 44.5 | 42.2 | 42.1 |
| Nov 21 (−19.9) | 10 | 2 | 80.1 | 42.1 | 35.5 | 30.6 | 28.2 | 30.0 |
| | 11 | 1 | 77.1 | 33.4 | 24.8 | 17.6 | 14.1 | 19.3 |
| | 12 | | 76.0 | 30.0 | 20.0 | 10.0 | 0.0 | 14.0 |

| | Tilt Angle: | | Horiz. | Lat. −10 | Lat. | Lat. +10 | Lat. +20 | Vert. |
|---|---|---|---|---|---|---|---|---|
| Feb 21 (−10.6) | 8 | 4 | 82.5 | 62.9 | 60.5 | 59.0 | 58.5 | 59.6 |
| Oct 21 (−10.7) | 9 | 3 | 75.8 | 49.0 | 45.9 | 44.3 | 44.5 | 47.6 |
| | 10 | 2 | 70.6 | 35.9 | 31.5 | 29.5 | 30.6 | 36.5 |
| | 11 | 1 | 67.2 | 25.0 | 18.0 | 14.8 | 17.6 | 27.7 |
| | 12 | | 66.0 | 20.0 | 10.0 | 0.0 | 10.0 | 24.0 |

| | Tilt Angle: | | Horiz. | Lat. −10 | Lat. | Lat. +10 | Lat. +20 | Vert. |
|---|---|---|---|---|---|---|---|---|
| | 7 | 5 | 81.7 | 75.2 | 75.0 | 75.2 | 75.9 | 77.6 |
| Mar 21 (0.0) | 8 | 4 | 73.9 | 60.5 | 60.0 | 60.5 | 62.0 | 65.0 |
| Sep 21 (0.0) | 9 | 3 | 66.7 | 45.9 | 45.0 | 45.9 | 48.4 | 54.1 |
| | 10 | 2 | 61.0 | 31.5 | 30.0 | 31.5 | 35.5 | 44.1 |
| | 11 | 1 | 57.3 | 18.0 | 15.0 | 18.0 | 24.8 | 36.8 |
| | 12 | | 56.0 | 10.0 | 0.0 | 10.0 | 20.0 | 34.0 |

| | Tilt Angle: | | Horiz. | Lat. −10 | Lat. | Lat. +10 | Lat. +20 | Vert. |
|---|---|---|---|---|---|---|---|---|
| | 5 | 7 | 88.6 | 102.4 | 104.7 | 106.5 | 107.9 | 108.8 |
| | 6 | 6 | 80.4 | 88.0 | 90.0 | 92.0 | 93.9 | 96.5 |
| Apr 21 (+11.9) | 7 | 5 | 72.0 | 73.5 | 75.3 | 77.6 | 80.2 | 84.4 |
| Aug 21 (+12.1) | 8 | 4 | 63.9 | 58.9 | 60.7 | 63.4 | 67.0 | 72.9 |
| | 9 | 3 | 56.4 | 44.2 | 46.2 | 49.7 | 54.4 | 62.5 |
| | 10 | 2 | 50.1 | 29.5 | 32.0 | 36.8 | 43.2 | 53.8 |
| | 11 | 1 | 45.9 | 14.8 | 18.9 | 26.2 | 34.9 | 47.8 |
| | 12 | | 44.4 | 1.6 | 11.6 | 21.6 | 31.6 | 45.6 |

| | Tilt Angle: | | Horiz. | Lat. −10 | Lat. | Lat. +10 | Lat. +20 | Vert. |
|---|---|---|---|---|---|---|---|---|
| | 4 | 8 | 88.8 | 113.8 | 118.0 | 121.5 | 124.0 | 125.5 |
| | 5 | 7 | 81.5 | 100.4 | 104.1 | 107.4 | 110.2 | 113.1 |
| | 6 | 6 | 73.5 | 86.6 | 90.0 | 93.4 | 96.7 | 101.0 |
| May 21 (+20.3) | 7 | 5 | 65.2 | 72.6 | 75.9 | 79.6 | 83.6 | 89.4 |
| Jul 21 (+20.5) | 8 | 4 | 56.9 | 58.5 | 62.0 | 66.2 | 71.1 | 78.6 |
| | 9 | 3 | 49.1 | 44.5 | 48.4 | 53.5 | 59.5 | 68.9 |
| | 10 | 2 | 42.4 | 30.6 | 35.5 | 42.1 | 49.6 | 61.1 |
| | 11 | 1 | 37.7 | 17.6 | 24.8 | 33.4 | 42.6 | 55.9 |
| | 12 | | 36.0 | 10.0 | 20.0 | 30.0 | 40.0 | 54.0 |

| | Tilt Angle: | | Horiz. | Lat. −10 | Lat. | Lat. +10 | Lat. +20 | Vert. |
|---|---|---|---|---|---|---|---|---|
| | 4 | 8 | 85.8 | 112.5 | 117.3 | 121.4 | 124.6 | 1271. |
| | 5 | 7 | 78.6 | 99.5 | 103.7 | 107.6 | 111.1 | 114.8 |
| | 6 | 6 | 70.7 | 86.0 | 90.0 | 94.0 | 97.8 | 102.9 |
| Jun 21 (+23.45) | 7 | 5 | 62.4 | 72.4 | 76.3 | 80.5 | 85.0 | 91.5 |
| | 8 | 4 | 54.1 | 58.6 | 62.7 | 67.5 | 72.8 | 80.9 |
| | 9 | 3 | 46.2 | 44.9 | 49.6 | 55.3 | 61.7 | 71.6 |
| | 10 | 2 | 39.3 | 31.6 | 37.4 | 41.5 | 52.4 | 64.1 |
| | 11 | 1 | 34.4 | 19.6 | 27.6 | 36.5 | 45.8 | 59.2 |
| | 12 | | 32.6 | 13.4 | 23.4 | 33.4 | 43.4 | 57.4 |

**TABLE A-12.** Latitude 64°N. Incident Angles for Horizontal and South-Facing Tilted Surfaces

| Tilt Angle: | | Horiz. | Lat. −10 | Lat. | Lat. +10 | Lat. +20 | Vert. |
|---|---|---|---|---|---|---|---|
| Dates (Decl.) | 11 1 | 88.2 | 36.5 | 27.6 | 19.6 | 14.3 | 13.9 |
| Dec 21 (−23.45) | 12 | 87.4 | 33.4 | 23.4 | 13.5 | 3.4 | 2.5 |

| Tilt Angle: | | Horiz. | Lat. −10 | Lat. | Lat. +10 | Lat. +20 | Vert. |
|---|---|---|---|---|---|---|---|
| Jan 21 (−19.9) | 10 2 | 87.2 | 42.1 | 35.5 | 30.6 | 26.2 | 28.2 |
| Nov 21 (−19.9) | 11 1 | 84.8 | 33.4 | 24.8 | 17.6 | 14.1 | 15.0 |
| | 12 | 84.0 | 30.0 | 20.0 | 10.0 | 0.0 | 6.0 |

| Tilt Angle: | | Horiz. | Lat. −10 | Lat. | Lat. +10 | Lat. +20 | Vert. |
|---|---|---|---|---|---|---|---|
| | 8 4 | 86.6 | 62.9 | 60.5 | 59.0 | 58.5 | 58.8 |
| Feb 21 (−10.6) | 9 3 | 81.4 | 49.0 | 45.9 | 44.3 | 44.5 | 45.4 |
| Oct 21 (−10.7) | 10 2 | 77.4 | 35.9 | 31.5 | 29.5 | 30.6 | 32.6 |
| | 11 1 | 74.9 | 25.0 | 18.0 | 14.8 | 17.6 | 21.4 |
| | 12 | 74.0 | 20.0 | 10.0 | 0.0 | 10.0 | 16.0 |

| Tilt Angle: | | Horiz. | Lat. −10 | Lat. | Lat. +10 | Lat. +20 | Vert. |
|---|---|---|---|---|---|---|---|
| | 7 5 | 83.5 | 75.2 | 75.0 | 75.2 | 75.9 | 76.5 |
| | 8 4 | 77.3 | 60.5 | 60.0 | 60.5 | 62.0 | 63.3 |
| Mar 21 (0.0) | 9 3 | 71.9 | 45.9 | 45.0 | 45.9 | 48.4 | 50.5 |
| Sep 21 (0.0) | 10 2 | 67.7 | 31.5 | 30.0 | 31.5 | 35.5 | 38.9 |
| | 11 1 | 64.9 | 18.0 | 15.0 | 18.0 | 24.8 | 29.8 |
| | 12 | 64.0 | 10.0 | 0.0 | 10.0 | 20.0 | 26.0 |

| Tilt Angle: | | Horiz. | Lat. −10 | Lat. | Lat. +10 | Lat. +20 | Vert. |
|---|---|---|---|---|---|---|---|
| | 5 7 | 86.0 | 102.4 | 104.7 | 106.5 | 107.9 | 108.4 |
| | 6 6 | 79.6 | 88.0 | 90.0 | 92.0 | 93.9 | 95.1 |
| | 7 5 | 73.0 | 73.5 | 75.3 | 77.6 | 80.2 | 82.0 |
| Apr 21 (+11.9) | 8 4 | 66.7 | 58.9 | 60.7 | 63.4 | 67.0 | 69.4 |
| Aug 21 (+12.1) | 9 3 | 61.0 | 44.2 | 46.2 | 49.7 | 54.4 | 57.7 |
| | 10 2 | 56.5 | 29.5 | 32.0 | 36.8 | 43.2 | 47.6 |
| | 11 1 | 53.5 | 14.8 | 18.9 | 26.2 | 34.9 | 40.3 |
| | 12 | 52.4 | 1.6 | 11.6 | 21.6 | 31.6 | 37.6 |

| Tilt Angle: | | Horiz. | Lat. −10 | Lat. | Lat. +10 | Lat. +20 | Vert. |
|---|---|---|---|---|---|---|---|
| | 4 8 | 84.2 | 113.8 | 118.0 | 121.5 | 124.0 | 124.9 |
| | 5 7 | 78.4 | 100.4 | 104.1 | 107.4 | 110.2 | 111.6 |
| | 6 6 | 72.1 | 86.6 | 90.0 | 93.4 | 96.7 | 98.6 |
| May 21 (+20.3) | 7 5 | 65.5 | 72.6 | 75.9 | 79.6 | 83.6 | 86.1 |
| Jul 21 (+20.5) | 8 4 | 59.1 | 58.5 | 62.0 | 66.2 | 71.1 | 74.2 |
| | 9 3 | 53.2 | 44.5 | 48.4 | 53.5 | 59.5 | 63.4 |
| | 10 2 | 48.4 | 30.6 | 35.5 | 42.1 | 49.6 | 54.4 |
| | 11 1 | 45.1 | 17.6 | 24.8 | 33.4 | 42.6 | 48.3 |
| | 12 | 44.0 | 10.0 | 20.0 | 30.0 | 40.0 | 46.0 |

| Tilt Angle: | | Horiz. | Lat. −10 | Lat. | Lat. +10 | Lat. +20 | Vert. |
|---|---|---|---|---|---|---|---|
| | 3 9 | 85.8 | 124.7 | 130.4 | 135.1 | 138.2 | 139.2 |
| | 4 8 | 81.0 | 112.5 | 117.3 | 121.4 | 124.6 | 125.9 |
| | 5 7 | 75.3 | 99.5 | 103.7 | 107.6 | 111.1 | 112.8 |
| | 6 6 | 69.0 | 86.0 | 90.0 | 94.0 | 97.8 | 100.0 |
| Jun 21 (+23.45) | 7 5 | 62.5 | 72.4 | 76.3 | 80.5 | 85.0 | 87.8 |
| | 8 4 | 56.0 | 58.8 | 62.7 | 67.5 | 72.8 | 76.2 |
| | 9 3 | 50.1 | 44.9 | 49.6 | 55.6 | 61.7 | 65.9 |
| | 10 2 | 45.1 | 31.6 | 37.4 | 44.5 | 52.4 | 57.3 |
| | 11 1 | 41.7 | 19.6 | 27.6 | 36.5 | 45.8 | 51.5 |
| | 12 | 40.6 | 13.5 | 23.4 | 33.4 | 43.4 | 49.4 |

**TABLE A-13.** Monthly Daily-Average Total Global Radiation on South-Facing Surfaces for Clear Skies at Sea Level with a Clearness Number of 1.0[a], MJ/m²

| | Jan | Feb | Mar | Apr | May | Jun | Jul | Aug | Sep | Oct | Nov | Dec |
|---|---|---|---|---|---|---|---|---|---|---|---|---|
| A | 1230. | 1215. | 1186. | 1136. | 1104. | 1088. | 1085. | 1107. | 1151. | 1192. | 1221. | 1223. |
| B | 0.142 | 0.144 | 0.156 | 0.18 | 0.196 | 0.205 | 0.207 | 0.201 | 0.177 | 0.16 | 0.149 | 0.142 |
| C | 0.058 | 0.06 | 0.071 | 0.097 | 0.121 | 0.134 | 0.136 | 0.122 | 0.092 | 0.073 | 0.063 | 0.057 |
| Day No. | 17. | 46. | 76. | 105. | 135. | 162. | 199. | 229. | 258. | 288. | 318. | 346. |

0°=HORIZ
90°=VERT

| | Jan | Feb | Mar | Apr | May | Jun | Jul | Aug | Sep | Oct | Nov | Dec |
|---|---|---|---|---|---|---|---|---|---|---|---|---|
| Lat | 24° | 24° | 24° | 24° | 24° | 24° | 24° | 24° | 24° | 24° | 24° | 24° |
| Tilt 0° | 18.13 | 21.49 | 25.24 | 27.39 | 28.82 | 29.16 | 28.76 | 27.61 | 25.78 | 22.53 | 18.96 | 16.82 |
| 20° | 23.68 | 26.01 | 27.62 | 27.37 | 26.83 | 26.31 | 26.35 | 26.80 | 27.17 | 26.22 | 24.16 | 22.52 |
| 30° | 25.45 | 27.16 | 27.62 | 26.20 | 24.74 | 23.85 | 24.10 | 25.28 | 26.71 | 26.95 | 25.74 | 24.42 |
| 40° | 26.48 | 27.52 | 26.82 | 24.28 | 22.01 | 20.79 | 21.23 | 23.06 | 25.48 | 26.90 | 26.57 | 25.60 |
| 50° | 26.72 | 27.07 | 25.24 | 21.68 | 18.67 | 17.17 | 17.78 | 20.21 | 23.52 | 26.07 | 26.62 | 26.03 |
| 60° | 26.19 | 25.83 | 22.93 | 18.47 | 14.83 | 13.27 | 13.93 | 16.80 | 20.89 | 24.48 | 25.89 | 25.70 |
| 70° | 24.88 | 23.83 | 19.97 | 14.74 | 10.88 | 9.23 | 10.00 | 12.94 | 17.67 | 22.18 | 24.41 | 24.61 |
| 90° | 20.15 | 17.84 | 12.43 | 6.45 | 3.10 | 2.35 | 2.67 | 4.91 | 9.88 | 15.77 | 19.37 | 20.32 |
| | | | | | | | | | | | | |
| Lat | 26° | 26° | 26° | 26° | 26° | 26° | 26° | 26° | 26° | 26° | 26° | 26° |
| Tilt 0° | 17.19 | 20.69 | 24.73 | 27.20 | 28.92 | 29.38 | 28.93 | 27.54 | 25.38 | 21.83 | 18.05 | 15.86 |
| 20° | 22.82 | 25.42 | 27.39 | 27.47 | 27.19 | 26.78 | 26.77 | 27.00 | 27.06 | 25.76 | 23.37 | 21.62 |
| 30° | 24.67 | 26.70 | 27.54 | 26.44 | 25.21 | 24.40 | 24.61 | 25.60 | 26.75 | 26.63 | 25.05 | 23.58 |
| 40° | 25.81 | 27.20 | 26.90 | 24.66 | 22.59 | 21.45 | 21.84 | 23.52 | 25.66 | 26.72 | 26.00 | 24.86 |
| 50° | 26.18 | 26.90 | 25.47 | 22.18 | 19.35 | 17.92 | 18.49 | 20.77 | 23.84 | 26.04 | 26.18 | 25.40 |
| 60° | 25.79 | 25.82 | 23.31 | 19.08 | 15.59 | 14.01 | 14.64 | 17.46 | 21.35 | 24.60 | 25.60 | 25.20 |
| 70° | 24.63 | 23.98 | 20.48 | 15.45 | 11.64 | 10.01 | 10.76 | 13.68 | 18.25 | 22.46 | 24.27 | 24.26 |
| 90° | 20.22 | 18.29 | 13.16 | 7.23 | 3.66 | 2.69 | 3.20 | 5.61 | 10.64 | 16.31 | 19.55 | 20.29 |
| | | | | | | | | | | | | |
| Lat | 28° | 28° | 28° | 28° | 28° | 28° | 28° | 28° | 28° | 28° | 28° | 28° |
| Tilt 0° | 16.23 | 19.87 | 24.18 | 26.98 | 28.98 | 29.57 | 29.05 | 27.43 | 24.95 | 21.10 | 17.12 | 14.91 |
| 20° | 21.94 | 24.78 | 27.12 | 27.54 | 27.52 | 27.22 | 27.16 | 27.17 | 26.91 | 25.25 | 22.55 | 20.72 |
| 30° | 23.87 | 26.19 | 27.42 | 26.64 | 25.64 | 24.93 | 25.08 | 25.89 | 26.74 | 26.25 | 24.31 | 22.76 |
| 40° | 25.09 | 26.83 | 26.93 | 25.00 | 23.15 | 22.08 | 22.43 | 23.94 | 25.81 | 26.49 | 25.37 | 24.13 |
| 50° | 25.59 | 26.68 | 25.66 | 22.66 | 20.01 | 18.64 | 19.17 | 21.31 | 24.13 | 25.96 | 25.68 | 24.79 |
| 60° | 25.33 | 25.75 | 23.64 | 19.67 | 16.33 | 14.73 | 15.40 | 18.10 | 21.77 | 24.68 | 25.24 | 24.72 |
| 70° | 24.33 | 24.07 | 20.95 | 16.14 | 12.38 | 10.78 | 11.50 | 14.40 | 18.79 | 22.68 | 24.06 | 23.92 |
| 90° | 20.23 | 18.68 | 13.86 | 8.00 | 4.36 | 3.21 | 3.72 | 6.37 | 11.37 | 16.81 | 19.66 | 20.27 |
| | | | | | | | | | | | | |
| Lat | 30° | 30° | 30° | 30° | 30° | 30° | 30° | 30° | 30° | 30° | 30° | 30° |
| Tilt 0° | 15.28 | 19.02 | 23.59 | 26.72 | 29.00 | 29.71 | 29.14 | 27.29 | 24.49 | 20.35 | 16.19 | 13.97 |
| 20° | 21.05 | 24.10 | 26.80 | 27.56 | 27.82 | 27.63 | 27.51 | 27.30 | 26.72 | 24.70 | 21.70 | 19.84 |
| 30° | 23.05 | 25.62 | 27.25 | 26.80 | 26.04 | 25.42 | 25.52 | 26.15 | 26.70 | 25.83 | 23.54 | 21.95 |
| 40° | 24.37 | 26.40 | 26.92 | 25.31 | 23.67 | 22.69 | 22.99 | 24.32 | 25.91 | 26.21 | 24.70 | 23.42 |
| 50° | 24.98 | 26.40 | 25.80 | 23.09 | 20.64 | 19.34 | 19.83 | 21.82 | 24.38 | 25.83 | 25.13 | 24.19 |
| 60° | 24.86 | 25.62 | 23.93 | 20.23 | 17.05 | 15.48 | 16.14 | 18.71 | 22.15 | 24.70 | 24.83 | 24.26 |
| 70° | 24.00 | 24.10 | 21.38 | 16.80 | 13.10 | 11.54 | 12.24 | 15.10 | 19.30 | 22.85 | 23.80 | 23.61 |
| 90° | 20.23 | 19.01 | 14.53 | 8.76 | 5.04 | 3.74 | 4.36 | 7.13 | 12.08 | 17.26 | 19.70 | 20.28 |
| | | | | | | | | | | | | |
| Lat | 32° | 32° | 32° | 32° | 32° | 32° | 32° | 32° | 32° | 32° | 32° | 32° |
| Tilt 0° | 14.34 | 18.14 | 22.98 | 26.42 | 28.98 | 29.81 | 29.19 | 27.11 | 24.00 | 19.57 | 15.25 | 13.01 |
| 20° | 20.17 | 23.38 | 26.44 | 27.54 | 28.08 | 28.01 | 27.82 | 27.39 | 26.49 | 24.11 | 20.83 | 18.92 |
| 30° | 22.24 | 25.01 | 27.04 | 26.93 | 26.43 | 25.91 | 25.96 | 26.38 | 26.61 | 25.36 | 22.74 | 21.09 |
| 40° | 23.66 | 25.91 | 26.86 | 25.57 | 24.16 | 23.32 | 23.52 | 24.67 | 25.97 | 25.88 | 23.99 | 22.65 |
| 50° | 24.39 | 26.05 | 25.90 | 23.50 | 21.24 | 20.01 | 20.46 | 22.29 | 24.59 | 25.64 | 24.54 | 23.53 |
| 60° | 24.40 | 25.43 | 24.18 | 20.76 | 17.75 | 16.23 | 16.85 | 19.29 | 22.50 | 24.66 | 24.37 | 23.73 |
| 70° | 23.69 | 24.06 | 21.77 | 17.44 | 13.81 | 12.28 | 12.95 | 15.77 | 19.78 | 22.97 | 23.49 | 23.23 |
| 90° | 20.23 | 19.28 | 15.16 | 9.50 | 5.73 | 4.34 | 5.03 | 7.88 | 12.76 | 17.66 | 19.71 | 20.21 |

[a] This table was prepared in the same way as Tables A-1 to A-6 but with dates from Table 3-3 and units of MJ/m².

$$\frac{MJ/m^2}{0.011357} = Btu/S.F.$$

| | Jan | Feb | Mar | Apr | May | Jun | Jul | Aug | Sep | Oct | Nov | Dec |
|---|---|---|---|---|---|---|---|---|---|---|---|---|
| Lat | 34° | 34° | 34° | 34° | 34° | 34° | 34° | 34° | 34° | 34° | 34° | 34° |
| Tilt 0° | 13.39 | 17.25 | 22.33 | 26.09 | 28.92 | 29.88 | 29.19 | 26.89 | 23.46 | 18.76 | 14.32 | 12.03 |
| 20° | 19.27 | 22.61 | 26.04 | 27.49 | 28.30 | 28.34 | 28.10 | 27.45 | 26.22 | 23.47 | 19.95 | 17.95 |
| 30° | 21.41 | 24.35 | 26.78 | 27.02 | 26.78 | 26.38 | 26.37 | 26.57 | 26.48 | 24.84 | 21.94 | 20.17 |
| 40° | 22.93 | 25.36 | 26.75 | 25.80 | 24.61 | 23.80 | 24.01 | 24.98 | 25.99 | 25.49 | 23.28 | 21.79 |
| 50° | 23.77 | 25.64 | 25.94 | 23.86 | 21.82 | 20.66 | 21.06 | 22.73 | 24.75 | 25.40 | 23.95 | 22.78 |
| 60° | 23.91 | 25.16 | 24.38 | 21.25 | 18.42 | 16.97 | 17.55 | 19.84 | 22.81 | 24.57 | 23.91 | 23.10 |
| 70° | 23.35 | 23.95 | 22.12 | 18.04 | 14.53 | 13.01 | 13.65 | 16.42 | 20.21 | 23.03 | 23.17 | 22.73 |
| 90° | 20.21 | 19.49 | 15.77 | 10.25 | 6.40 | 5.02 | 5.70 | 8.61 | 13.42 | 18.01 | 19.70 | 20.03 |
| Lat | 36° | 36° | 36° | 36° | 36° | 36° | 36° | 36° | 36° | 36° | 36° | 36° |
| Tilt 0° | 12.42 | 16.34 | 21.65 | 25.72 | 28.82 | 29.90 | 29.16 | 26.64 | 22.90 | 17.92 | 13.39 | 11.05 |
| 20° | 18.33 | 21.81 | 25.60 | 27.40 | 28.48 | 28.65 | 28.35 | 27.47 | 25.91 | 22.79 | 19.07 | 16.93 |
| 30° | 20.52 | 23.63 | 26.48 | 27.07 | 27.10 | 26.81 | 26.74 | 26.72 | 26.31 | 24.27 | 21.12 | 19.17 |
| 40° | 22.12 | 24.76 | 26.60 | 25.99 | 25.03 | 24.32 | 24.47 | 25.26 | 25.96 | 25.05 | 22.56 | 20.86 |
| 50° | 23.06 | 25.17 | 25.94 | 24.19 | 22.36 | 21.29 | 21.64 | 23.13 | 24.87 | 25.10 | 23.34 | 21.93 |
| 60° | 23.33 | 24.83 | 24.54 | 21.71 | 19.07 | 17.68 | 18.22 | 20.37 | 23.07 | 24.41 | 23.43 | 22.35 |
| 70° | 22.91 | 23.77 | 22.42 | 18.62 | 15.27 | 13.72 | 14.33 | 17.04 | 20.61 | 23.02 | 22.83 | 22.12 |
| 90° | 20.09 | 19.62 | 16.33 | 11.00 | 7.16 | 5.70 | 6.36 | 9.32 | 14.04 | 18.31 | 19.68 | 19.73 |
| Lat | 38° | 38° | 38° | 38° | 38° | 38° | 38° | 38° | 38° | 38° | 38° | 38° |
| Tilt 0° | 11.44 | 15.41 | 20.94 | 25.31 | 28.68 | 29.91 | 29.09 | 26.35 | 22.30 | 17.07 | 12.44 | 10.05 |
| 20° | 17.33 | 20.97 | 25.11 | 27.27 | 28.63 | 28.93 | 28.55 | 27.45 | 25.55 | 22.07 | 18.15 | 15.85 |
| 30° | 19.56 | 22.87 | 26.13 | 27.08 | 27.39 | 27.22 | 27.08 | 26.84 | 26.10 | 23.66 | 20.26 | 18.10 |
| 40° | 21.22 | 24.11 | 26.40 | 26.14 | 25.41 | 24.81 | 24.90 | 25.49 | 25.89 | 24.55 | 21.78 | 19.82 |
| 50° | 22.26 | 24.63 | 25.90 | 24.48 | 22.86 | 21.89 | 22.18 | 23.50 | 24.95 | 24.73 | 22.66 | 20.96 |
| 60° | 22.64 | 24.44 | 24.64 | 22.13 | 19.69 | 18.38 | 18.87 | 20.85 | 23.29 | 24.20 | 22.87 | 21.48 |
| 70° | 22.36 | 23.53 | 22.67 | 19.16 | 15.98 | 14.42 | 15.05 | 17.64 | 20.97 | 22.96 | 22.42 | 21.37 |
| 90° | 19.85 | 19.69 | 16.86 | 11.73 | 7.92 | 6.37 | 7.02 | 10.02 | 14.64 | 18.54 | 19.58 | 19.29 |
| Lat | 40° | 40° | 40° | 40° | 40° | 40° | 40° | 40° | 40° | 40° | 40° | 40° |
| Tilt 0° | 10.45 | 14.47 | 20.20 | 24.87 | 28.51 | 29.90 | 28.99 | 26.02 | 21.68 | 16.19 | 11.47 | 9.05 |
| 20° | 16.28 | 20.10 | 24.58 | 27.09 | 28.74 | 29.19 | 28.73 | 27.39 | 25.16 | 21.31 | 17.18 | 14.72 |
| 30° | 18.53 | 22.07 | 25.74 | 27.06 | 27.63 | 27.62 | 27.39 | 26.92 | 25.84 | 22.99 | 19.33 | 16.95 |
| 40° | 20.24 | 23.40 | 26.15 | 26.25 | 25.76 | 25.28 | 25.30 | 25.69 | 25.78 | 24.00 | 20.92 | 18.68 |
| 50° | 21.35 | 24.04 | 25.80 | 24.73 | 23.34 | 22.48 | 22.70 | 23.83 | 24.98 | 24.31 | 21.89 | 19.87 |
| 60° | 21.84 | 23.98 | 24.70 | 22.52 | 20.28 | 19.07 | 19.49 | 21.31 | 23.47 | 23.91 | 22.22 | 20.48 |
| 70° | 21.69 | 23.22 | 22.88 | 19.67 | 16.66 | 15.16 | 15.76 | 18.20 | 21.28 | 22.83 | 21.90 | 20.48 |
| 90° | 19.49 | 19.69 | 17.34 | 12.44 | 8.67 | 7.04 | 7.78 | 10.70 | 15.20 | 18.71 | 19.37 | 18.69 |
| Lat | 42° | 42° | 42° | 42° | 42° | 42° | 42° | 42° | 42° | 42° | 42° | 42° |
| Tilt 0° | 9.45 | 13.53 | 19.43 | 24.40 | 28.29 | 29.87 | 28.87 | 25.66 | 21.02 | 15.29 | 10.50 | 8.05 |
| 20° | 15.18 | 19.20 | 24.02 | 26.88 | 28.81 | 29.41 | 28.88 | 27.28 | 24.72 | 20.51 | 16.16 | 13.53 |
| 30° | 17.42 | 21.24 | 25.30 | 26.98 | 27.85 | 27.98 | 27.68 | 26.95 | 25.54 | 22.27 | 18.33 | 15.71 |
| 40° | 19.16 | 22.65 | 25.85 | 26.32 | 26.10 | 25.77 | 25.71 | 25.86 | 25.62 | 23.39 | 19.97 | 17.44 |
| 50° | 20.33 | 23.41 | 25.64 | 24.94 | 23.78 | 23.04 | 23.20 | 24.12 | 24.96 | 23.82 | 21.02 | 18.66 |
| 60° | 20.91 | 23.47 | 24.70 | 22.87 | 20.84 | 19.74 | 20.10 | 21.73 | 23.60 | 23.56 | 21.46 | 19.32 |
| 70° | 20.87 | 22.85 | 23.03 | 20.15 | 17.32 | 15.91 | 16.46 | 18.74 | 21.56 | 22.62 | 21.27 | 19.42 |
| 90° | 18.97 | 19.64 | 17.78 | 13.12 | 9.41 | 7.80 | 8.53 | 11.36 | 15.72 | 18.82 | 19.04 | 17.92 |
| Lat | 44° | 44° | 44° | 44° | 44° | 44° | 44° | 44° | 44° | 44° | 44° | 44° |
| Tilt 0° | 8.45 | 12.59 | 18.63 | 23.89 | 28.06 | 29.81 | 28.74 | 25.26 | 20.33 | 14.37 | 9.52 | 7.07 |
| 20° | 14.02 | 18.29 | 23.41 | 26.63 | 28.84 | 29.60 | 29.00 | 27.14 | 24.25 | 19.67 | 15.09 | 12.29 |
| 30° | 16.23 | 20.38 | 24.81 | 26.87 | 28.03 | 28.31 | 27.94 | 26.95 | 25.19 | 21.51 | 17.26 | 14.41 |
| 40° | 17.97 | 21.88 | 25.49 | 26.35 | 26.42 | 26.22 | 26.10 | 26.00 | 25.41 | 22.72 | 18.93 | 16.10 |
| 50° | 19.18 | 22.74 | 25.43 | 25.11 | 24.19 | 23.57 | 23.67 | 24.37 | 24.90 | 23.27 | 20.05 | 17.32 |
| 60° | 19.84 | 22.93 | 24.64 | 23.17 | 21.37 | 20.37 | 20.68 | 22.11 | 23.68 | 23.15 | 20.58 | 18.03 |
| 70° | 19.90 | 22.44 | 23.13 | 20.58 | 17.96 | 16.64 | 17.14 | 19.24 | 21.78 | 22.35 | 20.50 | 18.21 |
| 90° | 18.30 | 19.54 | 18.17 | 13.78 | 10.13 | 8.57 | 9.27 | 12.02 | 16.21 | 18.85 | 18.57 | 16.98 |

*(Continued)*

|       | Jan | Feb | Mar | Apr | May | Jun | Jul | Aug | Sep | Oct | Nov | Dec |
|-------|-----|-----|-----|-----|-----|-----|-----|-----|-----|-----|-----|-----|
| Lat | 46° | 46° | 46° | 46° | 46° | 46° | 46° | 46° | 46° | 46° | 46° | 46° |
| Tilt 0° | 7.45 | 11.64 | 17.81 | 23.35 | 27.80 | 29.73 | 28.58 | 24.83 | 19.62 | 13.45 | 8.53 | 6.12 |
| 20° | 12.80 | 17.36 | 22.75 | 26.33 | 28.85 | 29.75 | 29.09 | 26.96 | 23.73 | 18.80 | 13.96 | 11.06 |
| 30° | 14.96 | 19.50 | 24.27 | 26.71 | 28.18 | 28.59 | 28.16 | 26.90 | 24.79 | 20.69 | 16.11 | 13.08 |
| 40° | 16.68 | 21.07 | 25.09 | 26.33 | 26.71 | 26.64 | 26.46 | 26.09 | 25.15 | 21.99 | 17.79 | 14.72 |
| 50° | 17.91 | 22.03 | 25.17 | 25.23 | 24.58 | 24.06 | 24.11 | 24.58 | 24.78 | 22.65 | 18.95 | 15.93 |
| 60° | 18.61 | 22.34 | 24.53 | 23.43 | 21.88 | 20.98 | 21.24 | 22.45 | 23.71 | 22.66 | 19.56 | 16.67 |
| 70° | 18.77 | 21.99 | 23.17 | 20.98 | 18.58 | 17.34 | 17.80 | 19.70 | 21.95 | 22.00 | 19.59 | 16.92 |
| 90° | 17.44 | 19.39 | 18.51 | 14.40 | 10.84 | 9.32 | 9.99 | 12.69 | 16.65 | 18.81 | 17.95 | 15.93 |
|       |     |     |     |     |     |     |     |     |     |     |     |     |
| Lat | 48° | 48° | 48° | 48° | 48° | 48° | 48° | 48° | 48° | 48° | 48° | 48° |
| Tilt 0° | 6.48 | 10.69 | 16.96 | 22.78 | 27.52 | 29.60 | 28.39 | 24.36 | 18.87 | 12.51 | 7.54 | 5.23 |
| 20° | 11.55 | 16.39 | 22.06 | 26.00 | 28.83 | 29.86 | 29.13 | 26.74 | 23.17 | 17.88 | 12.78 | 9.88 |
| 30° | 13.62 | 18.56 | 23.68 | 26.51 | 28.30 | 28.84 | 28.35 | 26.82 | 24.35 | 19.83 | 14.88 | 11.81 |
| 40° | 15.30 | 20.20 | 24.62 | 26.28 | 26.97 | 27.03 | 26.78 | 26.14 | 24.84 | 21.21 | 16.55 | 13.40 |
| 50° | 16.52 | 21.24 | 24.85 | 25.30 | 24.93 | 24.52 | 24.52 | 24.75 | 24.61 | 21.97 | 17.74 | 14.59 |
| 60° | 17.27 | 21.66 | 24.35 | 23.65 | 22.36 | 21.56 | 21.76 | 22.75 | 23.68 | 22.10 | 18.40 | 15.36 |
| 70° | 17.50 | 21.45 | 23.15 | 21.34 | 19.18 | 18.01 | 18.42 | 20.13 | 22.08 | 21.57 | 18.53 | 15.67 |
| 90° | 16.43 | 19.16 | 18.78 | 15.00 | 11.54 | 10.06 | 10.70 | 13.32 | 17.04 | 18.69 | 17.16 | 14.92 |
|       |     |     |     |     |     |     |     |     |     |     |     |     |
| Lat | 50° | 50° | 50° | 50° | 50° | 50° | 50° | 50° | 50° | 50° | 50° | 50° |
| Tilt 0° | 5.55 | 9.72 | 16.09 | 22.17 | 27.22 | 29.44 | 28.17 | 23.86 | 18.10 | 11.57 | 6.57 | 4.36 |
| 20° | 10.32 | 15.36 | 21.32 | 25.62 | 28.77 | 29.92 | 29.14 | 26.48 | 22.57 | 16.94 | 11.55 | 8.69 |
| 30° | 12.29 | 17.56 | 23.05 | 26.27 | 28.38 | 29.05 | 28.49 | 26.69 | 23.86 | 18.93 | 13.58 | 10.51 |
| 40° | 13.90 | 19.24 | 24.10 | 26.17 | 27.19 | 27.37 | 27.06 | 26.15 | 24.48 | 20.38 | 15.21 | 12.03 |
| 50° | 15.11 | 20.36 | 24.46 | 25.33 | 25.25 | 24.94 | 24.88 | 24.88 | 24.39 | 21.23 | 16.40 | 13.19 |
| 60° | 15.87 | 20.88 | 24.11 | 23.83 | 22.80 | 22.10 | 22.25 | 23.01 | 23.61 | 21.47 | 17.10 | 13.97 |
| 70° | 16.17 | 20.79 | 23.06 | 21.65 | 19.74 | 18.66 | 19.02 | 20.52 | 22.14 | 21.08 | 17.31 | 14.33 |
| 90° | 15.33 | 18.80 | 19.00 | 15.55 | 12.21 | 10.78 | 11.39 | 13.93 | 17.39 | 18.49 | 16.21 | 13.79 |
|       |     |     |     |     |     |     |     |     |     |     |     |     |
| Lat | 52° | 52° | 52° | 52° | 52° | 52° | 52° | 52° | 52° | 52° | 52° | 52° |
| Tilt 0° | 4.66 | 8.75 | 15.20 | 21.53 | 26.89 | 29.25 | 27.91 | 23.32 | 17.30 | 10.62 | 5.63 | 3.50 |
| 20° | 9.12 | 14.29 | 20.54 | 25.20 | 28.66 | 29.94 | 29.10 | 26.17 | 21.92 | 15.97 | 10.31 | 7.42 |
| 30° | 10.99 | 16.48 | 22.36 | 25.98 | 28.41 | 29.21 | 28.60 | 26.51 | 23.33 | 17.99 | 12.23 | 9.08 |
| 40° | 12.54 | 18.19 | 23.53 | 26.02 | 27.37 | 27.67 | 27.30 | 26.11 | 24.07 | 19.49 | 13.81 | 10.48 |
| 50° | 13.72 | 19.36 | 24.01 | 25.32 | 25.56 | 25.37 | 25.25 | 24.98 | 24.11 | 20.43 | 14.98 | 11.57 |
| 60° | 14.50 | 19.98 | 23.80 | 23.95 | 23.21 | 22.60 | 22.70 | 23.23 | 23.47 | 20.71 | 15.71 | 12.33 |
| 70° | 14.85 | 20.00 | 22.90 | 21.91 | 20.26 | 19.27 | 19.58 | 20.86 | 22.15 | 20.51 | 15.98 | 12.72 |
| 90° | 14.25 | 18.30 | 19.15 | 16.07 | 12.86 | 11.48 | 12.05 | 14.50 | 17.69 | 18.22 | 15.11 | 12.37 |
|       |     |     |     |     |     |     |     |     |     |     |     |     |
| Lat | 54° | 54° | 54° | 54° | 54° | 54° | 54° | 54° | 54° | 54° | 54° | 54° |
| Tilt 0° | 3.80 | 7.77 | 14.29 | 20.87 | 26.53 | 29.03 | 27.62 | 22.76 | 16.48 | 9.67 | 4.73 | 2.68 |
| 20° | 7.88 | 13.15 | 19.72 | 24.74 | 28.51 | 29.97 | 29.02 | 25.83 | 21.24 | 14.96 | 9.09 | 6.06 |
| 30° | 9.61 | 15.31 | 21.61 | 25.65 | 28.41 | 29.33 | 28.65 | 26.30 | 22.74 | 17.00 | 10.90 | 7.51 |
| 40° | 11.06 | 17.03 | 22.89 | 25.82 | 27.51 | 27.94 | 27.50 | 26.03 | 23.60 | 18.55 | 12.41 | 8.75 |
| 50° | 12.19 | 18.25 | 23.50 | 25.26 | 25.84 | 25.77 | 25.58 | 25.04 | 23.78 | 19.56 | 13.54 | 9.73 |
| 60° | 12.95 | 18.93 | 23.42 | 24.03 | 23.58 | 23.07 | 23.11 | 23.41 | 23.27 | 20.00 | 14.29 | 10.42 |
| 70° | 13.34 | 19.06 | 22.67 | 22.13 | 20.75 | 19.86 | 20.11 | 21.17 | 22.10 | 19.86 | 14.61 | 10.81 |
| 90° | 12.94 | 17.65 | 19.23 | 16.55 | 13.52 | 12.16 | 12.69 | 15.04 | 17.93 | 17.86 | 13.97 | 10.62 |
|       |     |     |     |     |     |     |     |     |     |     |     |     |
| Lat | 56° | 56° | 56° | 56° | 56° | 56° | 56° | 56° | 56° | 56° | 56° | 56° |
| Tilt 0° | 2.95 | 6.79 | 13.36 | 20.17 | 26.13 | 28.81 | 27.30 | 22.17 | 15.63 | 8.72 | 3.87 | 1.93 |
| 20° | 6.54 | 11.96 | 18.85 | 24.23 | 28.32 | 29.98 | 28.92 | 25.45 | 20.51 | 13.92 | 7.87 | 4.69 |
| 30° | 8.08 | 14.07 | 20.82 | 25.26 | 28.35 | 29.44 | 28.67 | 26.05 | 22.11 | 15.96 | 9.56 | 5.88 |
| 40° | 9.39 | 15.76 | 22.18 | 25.57 | 27.60 | 28.18 | 27.65 | 25.92 | 23.07 | 17.54 | 10.97 | 6.91 |
| 50° | 10.42 | 17.00 | 22.91 | 25.16 | 26.07 | 26.15 | 25.87 | 25.06 | 23.38 | 18.61 | 12.06 | 7.74 |
| 60° | 11.14 | 17.74 | 22.97 | 24.05 | 23.90 | 23.52 | 23.48 | 23.54 | 23.01 | 19.14 | 12.79 | 8.33 |
| 70° | 11.54 | 17.95 | 22.36 | 22.30 | 21.21 | 20.43 | 20.60 | 21.44 | 21.99 | 19.11 | 13.16 | 8.69 |
| 90° | 11.31 | 16.82 | 19.23 | 16.99 | 14.19 | 12.83 | 13.31 | 15.55 | 18.10 | 17.40 | 12.72 | 8.62 |

|  | Jan | Feb | Mar | Apr | May | Jun | Jul | Aug | Sep | Oct | Nov | Dec |
|---|---|---|---|---|---|---|---|---|---|---|---|---|
| Lat | 58° | 58° | 58° | 58° | 58° | 58° | 58° | 58° | 58° | 58° | 58° | 58° |
| Tilt 0° | 2.16 | 5.82 | 12.41 | 19.44 | 25.71 | 28.60 | 26.98 | 21.57 | 14.77 | 7.77 | 3.03 | 1.28 |
| 20° | 5.15 | 10.71 | 17.93 | 23.69 | 28.09 | 29.96 | 28.81 | 25.04 | 19.74 | 12.83 | 6.57 | 3.42 |
| 30° | 6.44 | 12.73 | 19.96 | 24.83 | 28.26 | 29.51 | 28.65 | 25.76 | 21.42 | 14.85 | 8.08 | 4.36 |
| 40° | 7.55 | 14.38 | 21.41 | 25.27 | 27.64 | 28.40 | 27.78 | 25.76 | 22.49 | 16.44 | 9.36 | 5.17 |
| 50° | 8.43 | 15.61 | 22.25 | 24.99 | 26.25 | 26.50 | 26.14 | 25.03 | 22.91 | 17.55 | 10.36 | 5.83 |
| 60° | 9.07 | 16.38 | 22.43 | 24.02 | 24.18 | 23.94 | 23.83 | 23.63 | 22.68 | 18.16 | 11.06 | 6.32 |
| 70° | 9.44 | 16.67 | 21.96 | 22.41 | 21.61 | 20.97 | 21.07 | 21.67 | 21.80 | 18.23 | 11.44 | 6.63 |
| 90° | 9.35 | 15.78 | 19.15 | 17.37 | 14.82 | 13.48 | 13.97 | 16.03 | 18.22 | 16.80 | 11.18 | 6.65 |
| Lat | 60° | 60° | 60° | 60° | 60° | 60° | 60° | 60° | 60° | 60° | 60° | 60° |
| Tilt 0° | 1.46 | 4.87 | 11.44 | 18.68 | 25.26 | 28.36 | 26.66 | 20.95 | 13.88 | 6.81 | 2.24 | 0.72 |
| 20° | 3.79 | 9.42 | 16.97 | 23.10 | 27.85 | 29.90 | 28.68 | 24.59 | 18.93 | 11.68 | 5.20 | 2.16 |
| 30° | 4.81 | 11.32 | 19.05 | 24.36 | 28.12 | 29.54 | 28.61 | 25.43 | 20.68 | 13.65 | 6.48 | 2.80 |
| 40° | 5.69 | 12.88 | 20.57 | 24.92 | 27.64 | 28.57 | 27.88 | 25.55 | 21.84 | 15.23 | 7.57 | 3.36 |
| 50° | 6.40 | 14.08 | 21.50 | 24.78 | 26.40 | 26.81 | 26.37 | 24.96 | 22.38 | 16.37 | 8.44 | 3.82 |
| 60° | 6.93 | 14.86 | 21.80 | 23.93 | 24.42 | 24.32 | 24.14 | 23.67 | 22.28 | 17.03 | 9.07 | 4.17 |
| 70° | 7.25 | 15.20 | 21.47 | 22.47 | 21.98 | 21.47 | 21.51 | 21.85 | 21.54 | 17.20 | 9.43 | 4.40 |
| 90° | 7.25 | 14.55 | 18.97 | 17.71 | 15.42 | 14.15 | 14.62 | 16.46 | 18.25 | 16.03 | 9.31 | 4.46 |
| Lat | 62° | 62° | 62° | 62° | 62° | 62° | 62° | 62° | 62° | 62° | 62° | 62° |
| Tilt 0° | 0.87 | 3.97 | 10.46 | 17.90 | 24.81 | 28.11 | 26.33 | 20.31 | 12.97 | 5.85 | 1.52 | 0.29 |
| 20° | 2.53 | 8.11 | 15.96 | 22.47 | 27.60 | 29.81 | 28.52 | 24.10 | 18.07 | 10.47 | 3.82 | 1.00 |
| 30° | 3.26 | 9.86 | 18.07 | 23.83 | 27.95 | 29.54 | 28.53 | 25.05 | 19.88 | 12.37 | 4.83 | 1.32 |
| 40° | 3.90 | 11.32 | 19.65 | 24.52 | 27.61 | 28.70 | 27.93 | 25.30 | 21.13 | 13.91 | 5.70 | 1.60 |
| 50° | 4.42 | 12.45 | 20.67 | 24.50 | 26.51 | 27.08 | 26.56 | 24.84 | 21.78 | 15.05 | 6.40 | 1.83 |
| 60° | 4.81 | 13.22 | 21.08 | 23.79 | 24.67 | 24.72 | 24.47 | 23.69 | 21.80 | 15.75 | 6.92 | 2.01 |
| 70° | 5.07 | 13.60 | 20.88 | 22.47 | 22.32 | 21.93 | 21.91 | 21.97 | 21.20 | 15.99 | 7.23 | 2.13 |
| 90° | 5.12 | 13.16 | 18.68 | 17.99 | 15.99 | 14.82 | 15.24 | 16.85 | 18.21 | 15.08 | 7.21 | 2.18 |
| Lat | 64° | 64° | 64° | 64° | 64° | 64° | 64° | 64° | 64° | 64° | 64° | 64° |
| Tilt 0° | 0.39 | 3.12 | 9.47 | 17.10 | 24.37 | 27.89 | 25.98 | 19.64 | 12.04 | 4.91 | 0.91 | 0.05 |
| 20° | 1.28 | 6.80 | 14.91 | 21.80 | 27.31 | 29.71 | 28.32 | 23.57 | 17.16 | 9.21 | 2.56 | 0.22 |
| 30° | 1.68 | 8.38 | 17.02 | 23.26 | 27.75 | 29.56 | 28.41 | 24.62 | 19.03 | 10.99 | 3.28 | 0.30 |
| 40° | 2.03 | 9.71 | 18.65 | 24.06 | 27.55 | 28.80 | 27.94 | 24.99 | 20.36 | 12.47 | 3.91 | 0.37 |
| 50° | 2.33 | 10.76 | 19.73 | 24.17 | 26.58 | 27.33 | 26.71 | 24.66 | 21.10 | 13.58 | 4.43 | 0.43 |
| 60° | 2.55 | 11.50 | 20.25 | 23.60 | 24.88 | 25.10 | 24.75 | 23.65 | 21.24 | 14.29 | 4.82 | 0.47 |
| 70° | 2.70 | 11.90 | 20.17 | 22.40 | 22.62 | 22.38 | 22.27 | 22.05 | 20.77 | 14.59 | 5.06 | 0.50 |
| 90° | 2.75 | 11.64 | 18.28 | 18.22 | 16.53 | 15.48 | 15.82 | 17.18 | 18.09 | 13.92 | 5.11 | 0.52 |

**TABLE A-14.** Monthly Daily-Average Total Extraterrestrial Radiation on Tilted Surfaces Facing South, MJ/m² (Northern Latitudes)

| Day No. | Jan 17. | Feb 46. | Mar 76. | Apr 105. | May 135. | Jun 162. | Jul 199. | Aug 229. | Sep 258. | Oct 288. | Nov 318. | Dec 346. |
|---|---|---|---|---|---|---|---|---|---|---|---|---|
| Lat | 14° | 14° | 14° | 14° | 14° | 14° | 14° | 14° | 14° | 14° | 14° | 14° |
| Tilt 0° | 29.90 | 32.89 | 36.03 | 37.73 | 38.01 | 37.76 | 37.68 | 37.54 | 36.36 | 33.70 | 30.55 | 28.74 |
| 20° | 37.67 | 38.21 | 37.50 | 35.22 | 32.35 | 30.72 | 31.31 | 33.74 | 36.27 | 37.67 | 37.60 | 37.19 |
| 30° | 39.89 | 39.16 | 36.53 | 32.37 | 28.18 | 26.02 | 26.88 | 30.37 | 34.57 | 37.97 | 39.45 | 39.78 |
| 40° | 40.90 | 38.92 | 34.45 | 28.59 | 23.29 | 20.74 | 21.80 | 26.14 | 31.83 | 37.11 | 40.10 | 41.16 |
| 50° | 40.66 | 37.50 | 31.33 | 23.97 | 17.87 | 15.09 | 16.28 | 21.20 | 28.12 | 35.12 | 39.54 | 41.29 |
| 60° | 39.19 | 34.95 | 27.25 | 18.69 | 12.15 | 9.36 | 10.57 | 15.74 | 23.56 | 32.07 | 37.77 | 40.16 |
| 70° | 36.53 | 31.32 | 22.34 | 12.95 | 6.47 | 3.99 | 5.06 | 10.00 | 18.29 | 28.04 | 34.86 | 37.82 |
| 90° | 27.99 | 21.37 | 10.66 | 1.51 | 0.0 | 0.0 | 0.0 | 0.06 | 6.34 | 17.57 | 25.98 | 29.78 |
| Lat | 15° | 15° | 15° | 15° | 15° | 15° | 15° | 15° | 15° | 15° | 15° | 15° |
| Tilt 0° | 29.41 | 32.51 | 35.83 | 37.74 | 38.19 | 38.02 | 37.90 | 37.62 | 36.24 | 33.39 | 30.10 | 28.22 |
| 20° | 37.35 | 38.04 | 37.53 | 35.45 | 32.72 | 31.15 | 31.71 | 34.03 | 36.38 | 37.58 | 37.32 | 36.83 |
| 30° | 39.67 | 39.10 | 36.68 | 32.70 | 28.63 | 26.52 | 27.35 | 30.74 | 34.79 | 37.98 | 39.28 | 39.52 |
| 40° | 40.79 | 38.98 | 34.71 | 29.00 | 23.81 | 21.29 | 22.33 | 26.59 | 32.15 | 37.23 | 40.04 | 41.00 |
| 50° | 40.66 | 37.67 | 31.68 | 24.47 | 18.43 | 15.66 | 16.85 | 21.72 | 28.53 | 35.35 | 39.59 | 41.24 |
| 60° | 39.31 | 35.22 | 27.69 | 19.25 | 12.73 | 9.93 | 11.14 | 16.30 | 24.05 | 32.40 | 37.94 | 40.23 |
| 70° | 36.75 | 31.70 | 22.87 | 13.54 | 7.02 | 4.49 | 5.59 | 10.58 | 18.84 | 28.46 | 35.13 | 37.99 |
| 90° | 28.41 | 21.90 | 11.29 | 2.00 | 0.0 | 0.0 | 0.0 | 0.30 | 6.96 | 18.14 | 26.43 | 30.16 |
| Lat | 16° | 16° | 16° | 16° | 16° | 16° | 16° | 16° | 16° | 16° | 16° | 16° |
| Tilt 0° | 28.91 | 32.12 | 35.63 | 37.74 | 38.36 | 38.26 | 38.11 | 37.68 | 36.12 | 33.06 | 29.63 | 27.69 |
| 20° | 37.02 | 37.85 | 37.56 | 35.66 | 33.08 | 31.57 | 32.11 | 34.31 | 36.48 | 37.47 | 37.04 | 36.46 |
| 30° | 39.44 | 39.03 | 36.81 | 33.02 | 29.07 | 27.01 | 27.82 | 31.11 | 35.00 | 37.98 | 39.10 | 39.24 |
| 40° | 40.67 | 39.02 | 34.95 | 29.41 | 24.32 | 21.83 | 22.86 | 27.04 | 32.46 | 37.35 | 39.97 | 40.83 |
| 50° | 40.66 | 37.83 | 32.03 | 24.95 | 18.99 | 16.24 | 17.41 | 22.23 | 28.93 | 35.58 | 39.63 | 41.18 |
| 60° | 39.41 | 35.48 | 28.13 | 19.79 | 13.30 | 10.49 | 11.71 | 16.86 | 24.53 | 32.73 | 38.09 | 40.27 |
| 70° | 36.97 | 32.06 | 23.38 | 14.12 | 7.58 | 5.00 | 6.12 | 11.15 | 19.39 | 28.88 | 35.39 | 38.15 |
| 90° | 28.82 | 22.43 | 11.91 | 2.51 | 0.0 | 0.0 | 0.0 | 0.62 | 7.58 | 18.70 | 26.87 | 30.52 |
| Lat | 17° | 17° | 17° | 17° | 17° | 17° | 17° | 17° | 17° | 17° | 17° | 17° |
| Tilt 0° | 28.40 | 31.73 | 35.42 | 37.73 | 38.52 | 38.50 | 38.31 | 37.74 | 35.99 | 32.73 | 29.16 | 27.15 |
| 20° | 36.68 | 37.66 | 37.57 | 35.87 | 33.44 | 31.99 | 32.49 | 34.57 | 36.57 | 37.35 | 36.74 | 36.07 |
| 30° | 39.21 | 38.95 | 36.94 | 33.33 | 29.51 | 27.50 | 28.28 | 31.47 | 35.20 | 37.98 | 38.91 | 38.95 |
| 40° | 40.54 | 39.05 | 35.19 | 29.81 | 24.82 | 22.37 | 23.38 | 27.49 | 32.76 | 37.45 | 39.89 | 40.64 |
| 50° | 40.64 | 37.97 | 32.37 | 25.43 | 19.54 | 16.81 | 17.97 | 22.74 | 29.32 | 35.79 | 39.66 | 41.10 |
| 60° | 39.50 | 35.73 | 28.56 | 20.34 | 13.88 | 11.07 | 12.28 | 17.42 | 25.01 | 33.04 | 38.23 | 40.31 |
| 70° | 37.17 | 32.41 | 23.89 | 14.71 | 8.14 | 5.52 | 6.66 | 11.73 | 19.93 | 29.29 | 35.63 | 38.30 |
| 90° | 29.21 | 22.94 | 12.53 | 3.03 | 0.0 | 0.0 | 0.0 | 1.00 | 8.20 | 19.26 | 27.31 | 30.87 |
| Lat | 18° | 18° | 18° | 18° | 18° | 18° | 18° | 18° | 18° | 18° | 18° | 18° |
| Tilt 0° | 27.89 | 31.32 | 35.19 | 37.71 | 38.67 | 38.73 | 38.51 | 37.79 | 35.84 | 32.39 | 28.68 | 26.61 |
| 20° | 36.33 | 37.45 | 37.57 | 36.07 | 33.78 | 32.40 | 32.87 | 34.83 | 36.65 | 37.21 | 36.43 | 35.68 |
| 30° | 38.96 | 38.85 | 37.05 | 33.63 | 29.94 | 27.98 | 28.74 | 31.83 | 35.38 | 37.96 | 38.70 | 38.65 |
| 40° | 40.39 | 39.07 | 35.41 | 30.21 | 25.32 | 22.91 | 23.90 | 27.92 | 33.05 | 37.55 | 39.79 | 40.45 |
| 50° | 40.61 | 38.10 | 32.69 | 25.91 | 20.09 | 17.38 | 18.53 | 23.25 | 29.71 | 35.99 | 39.68 | 41.01 |
| 60° | 39.58 | 35.97 | 28.98 | 20.87 | 14.46 | 11.64 | 12.86 | 17.97 | 25.47 | 33.35 | 38.36 | 40.34 |
| 70° | 37.36 | 32.75 | 24.39 | 15.29 | 8.71 | 6.05 | 7.21 | 12.31 | 20.47 | 29.69 | 35.87 | 38.43 |
| 90° | 29.60 | 23.46 | 13.15 | 3.58 | 0.0 | 0.0 | 0.0 | 1.41 | 8.82 | 19.81 | 27.73 | 31.21 |
| Lat | 19° | 19° | 19° | 19° | 19° | 19° | 19° | 19° | 19° | 19° | 19° | 19° |
| Tilt 0° | 27.37 | 30.91 | 34.96 | 37.67 | 38.81 | 38.95 | 38.69 | 37.82 | 35.68 | 32.04 | 28.19 | 26.06 |
| 20° | 35.97 | 37.24 | 37.56 | 36.25 | 34.12 | 32.80 | 33.24 | 35.08 | 36.71 | 37.07 | 36.12 | 35.28 |
| 30° | 38.69 | 38.75 | 37.16 | 33.92 | 30.36 | 28.45 | 29.19 | 32.17 | 35.56 | 37.92 | 38.49 | 38.34 |
| 40° | 40.24 | 39.08 | 35.63 | 30.59 | 25.81 | 23.44 | 24.41 | 28.35 | 33.33 | 37.63 | 39.69 | 40.24 |
| 50° | 40.56 | 38.22 | 33.01 | 26.37 | 20.63 | 17.95 | 19.08 | 23.75 | 30.09 | 36.19 | 39.68 | 40.91 |
| 60° | 39.65 | 36.20 | 29.40 | 21.41 | 15.03 | 12.21 | 13.43 | 18.52 | 25.93 | 33.64 | 38.47 | 40.35 |
| 70° | 37.53 | 33.09 | 24.89 | 15.86 | 9.28 | 6.59 | 7.76 | 12.88 | 21.00 | 30.08 | 36.09 | 38.55 |
| 90° | 29.98 | 23.96 | 13.76 | 4.13 | 0.01 | 0.0 | 0.0 | 1.85 | 9.44 | 20.35 | 28.15 | 31.54 |

| | Jan | Feb | Mar | Apr | May | Jun | Jul | Aug | Sep | Oct | Nov | Dec |
|---|---|---|---|---|---|---|---|---|---|---|---|---|
| Lat | 20° | 20° | 20° | 20° | 20° | 20° | 20° | 20° | 20° | 20° | 20° | 20° |
| Tilt 0° | 26.84 | 30.48 | 34.71 | 37.63 | 38.94 | 39.16 | 38.86 | 37.85 | 35.52 | 31.68 | 27.70 | 25.51 |
| 20° | 35.60 | 37.01 | 37.53 | 36.43 | 34.45 | 33.19 | 33.60 | 35.31 | 36.77 | 36.92 | 35.79 | 34.87 |
| 30° | 38.42 | 38.63 | 37.25 | 34.20 | 30.78 | 28.92 | 29.63 | 32.50 | 35.72 | 37.88 | 38.26 | 38.02 |
| 40° | 40.07 | 39.07 | 35.83 | 30.97 | 26.30 | 23.97 | 24.91 | 28.77 | 33.60 | 37.70 | 39.57 | 40.02 |
| 50° | 50.50 | 38.33 | 33.32 | 26.83 | 21.18 | 18.51 | 19.63 | 24.24 | 30.45 | 36.37 | 39.67 | 40.80 |
| 60° | 39.70 | 36.42 | 29.80 | 21.93 | 15.60 | 12.79 | 14.00 | 19.07 | 26.39 | 33.93 | 38.57 | 40.34 |
| 70° | 37.70 | 33.41 | 25.37 | 16.44 | 9.85 | 7.13 | 8.31 | 13.46 | 21.52 | 30.46 | 36.30 | 38.66 |
| 90° | 30.35 | 24.46 | 14.37 | 4.70 | 0.17 | 0.0 | 0.0 | 2.32 | 10.05 | 20.89 | 28.55 | 31.86 |
| | | | | | | | | | | | | |
| Lat | 21° | 21° | 21° | 21° | 21° | 21° | 21° | 21° | 21° | 21° | 21° | 21° |
| Tilt 0° | 26.31 | 30.05 | 34.46 | 37.57 | 39.06 | 39.36 | 39.02 | 37.86 | 35.34 | 31.31 | 27.19 | 24.95 |
| 20° | 35.23 | 36.77 | 37.50 | 36.59 | 34.76 | 33.57 | 33.95 | 35.54 | 36.81 | 36.75 | 35.45 | 34.44 |
| 30° | 38.14 | 38.50 | 37.33 | 34.47 | 31.18 | 29.38 | 30.06 | 32.83 | 35.88 | 37.83 | 38.02 | 37.69 |
| 40° | 39.89 | 39.05 | 36.02 | 31.33 | 26.78 | 24.49 | 25.41 | 29.18 | 33.86 | 37.75 | 39.44 | 39.78 |
| 50° | 40.43 | 38.42 | 33.62 | 27.28 | 21.71 | 19.07 | 20.18 | 24.72 | 30.81 | 36.53 | 39.65 | 40.67 |
| 60° | 39.75 | 36.63 | 30.19 | 22.45 | 16.17 | 13.36 | 14.57 | 19.61 | 26.83 | 34.20 | 38.66 | 40.33 |
| 70° | 37.85 | 33.72 | 25.85 | 17.01 | 10.42 | 7.68 | 8.87 | 14.03 | 22.04 | 30.84 | 36.50 | 38.75 |
| 90° | 30.70 | 24.94 | 14.97 | 5.27 | 0.43 | 0.0 | 0.00 | 2.81 | 10.66 | 21.42 | 28.94 | 32.16 |
| | | | | | | | | | | | | |
| Lat | 22° | 22° | 22° | 22° | 22° | 22° | 22° | 22° | 22° | 22° | 22° | 22° |
| Tilt 0° | 25.77 | 29.61 | 34.19 | 37.50 | 39.16 | 39.55 | 39.17 | 37.87 | 35.15 | 30.93 | 26.69 | 24.39 |
| 20° | 34.84 | 36.52 | 37.45 | 36.75 | 35.07 | 33.95 | 34.29 | 35.76 | 36.84 | 36.57 | 35.11 | 34.01 |
| 30° | 37.84 | 38.36 | 37.40 | 34.73 | 31.58 | 29.83 | 30.48 | 33.14 | 36.02 | 37.76 | 37.77 | 37.34 |
| 40° | 39.70 | 39.03 | 36.20 | 31.69 | 27.25 | 25.00 | 25.91 | 29.58 | 34.11 | 37.80 | 39.29 | 39.54 |
| 50° | 40.35 | 38.51 | 33.90 | 27.72 | 22.24 | 19.63 | 20.73 | 25.20 | 31.16 | 36.69 | 39.62 | 40.53 |
| 60° | 39.77 | 36.82 | 30.58 | 22.96 | 16.74 | 13.94 | 15.14 | 20.14 | 27.27 | 34.47 | 38.74 | 40.30 |
| 70° | 37.99 | 34.02 | 26.32 | 17.57 | 11.00 | 8.24 | 9.44 | 14.60 | 22.55 | 31.20 | 36.69 | 38.83 |
| 90° | 31.05 | 25.42 | 15.57 | 5.85 | 0.75 | 0.0 | 0.11 | 3.32 | 11.27 | 21.94 | 29.33 | 32.45 |
| | | | | | | | | | | | | |
| Lat | 23° | 23° | 23° | 23° | 23° | 23° | 23° | 23° | 23° | 23° | 23° | 23° |
| Tilt 0° | 25.23 | 29.17 | 33.91 | 37.43 | 39.26 | 39.73 | 39.31 | 37.86 | 34.95 | 30.55 | 26.17 | 23.82 |
| 20° | 34.44 | 36.26 | 37.40 | 36.89 | 35.37 | 34.31 | 34.63 | 35.96 | 36.87 | 36.39 | 34.75 | 33.57 |
| 30° | 37.54 | 38.20 | 37.45 | 34.98 | 31.97 | 30.28 | 30.90 | 33.45 | 36.15 | 37.68 | 37.51 | 36.99 |
| 40° | 39.50 | 38.99 | 36.37 | 32.04 | 27.72 | 25.51 | 26.40 | 29.98 | 34.35 | 37.84 | 39.14 | 39.28 |
| 50° | 40.25 | 38.58 | 34.18 | 28.16 | 22.77 | 20.19 | 21.27 | 25.67 | 31.50 | 36.84 | 39.57 | 40.38 |
| 60° | 39.79 | 37.01 | 30.95 | 23.47 | 17.31 | 14.52 | 15.71 | 20.67 | 27.70 | 34.72 | 38.81 | 40.25 |
| 70° | 38.11 | 34.31 | 26.79 | 18.14 | 11.57 | 8.80 | 10.00 | 15.17 | 23.06 | 31.55 | 36.86 | 38.90 |
| 90° | 31.38 | 25.89 | 16.16 | 6.43 | 1.11 | 0.00 | 0.34 | 3.84 | 11.88 | 22.46 | 29.70 | 32.73 |
| | | | | | | | | | | | | |
| Lat | 24° | 24° | 24° | 24° | 24° | 24° | 24° | 24° | 24° | 24° | 24° | 24° |
| Tilt 0° | 24.68 | 28.71 | 33.63 | 37.34 | 39.35 | 39.90 | 39.44 | 37.84 | 34.74 | 30.15 | 25.65 | 23.24 |
| 20° | 34.03 | 35.99 | 37.33 | 37.02 | 35.66 | 34.67 | 34.95 | 36.16 | 36.87 | 36.19 | 34.38 | 33.13 |
| 30° | 37.22 | 38.04 | 37.50 | 35.22 | 32.35 | 30.72 | 31.31 | 33.74 | 36.27 | 37.59 | 37.24 | 36.63 |
| 40° | 39.28 | 38.93 | 36.53 | 32.37 | 28.18 | 26.02 | 26.88 | 30.37 | 34.57 | 37.86 | 38.97 | 39.01 |
| 50° | 40.14 | 38.64 | 34.45 | 28.59 | 23.29 | 20.74 | 21.80 | 26.14 | 31.83 | 36.97 | 39.52 | 40.21 |
| 60° | 39.79 | 37.18 | 31.32 | 23.97 | 17.87 | 15.09 | 16.28 | 21.20 | 28.12 | 34.96 | 38.86 | 40.19 |
| 70° | 38.23 | 34.58 | 27.24 | 18.69 | 12.15 | 9.36 | 10.57 | 15.74 | 23.56 | 31.89 | 37.02 | 38.95 |
| 90° | 31.70 | 26.35 | 16.75 | 7.02 | 1.50 | 0.11 | 0.63 | 4.37 | 12.48 | 22.97 | 30.06 | 32.99 |
| | | | | | | | | | | | | |
| Lat | 25° | 25° | 25° | 25° | 25° | 25° | 25° | 25° | 25° | 25° | 25° | 25° |
| Tilt 0° | 24.12 | 28.25 | 33.33 | 37.24 | 39.42 | 40.06 | 39.56 | 37.81 | 34.52 | 29.75 | 25.12 | 22.67 |
| 20° | 33.61 | 35.71 | 37.25 | 37.14 | 35.94 | 35.02 | 35.27 | 36.34 | 36.87 | 35.98 | 34.01 | 32.67 |
| 30° | 36.89 | 37.86 | 37.53 | 35.45 | 32.72 | 31.15 | 31.71 | 34.03 | 36.38 | 37.49 | 36.96 | 36.25 |
| 40° | 39.05 | 38.87 | 36.67 | 32.70 | 28.63 | 26.52 | 27.35 | 30.74 | 34.79 | 37.87 | 38.79 | 38.73 |
| 50° | 40.02 | 38.69 | 34.70 | 29.00 | 23.81 | 21.29 | 22.33 | 26.59 | 32.15 | 37.10 | 39.44 | 40.03 |
| 60° | 39.78 | 37.34 | 31.68 | 24.47 | 18.43 | 15.66 | 16.85 | 21.72 | 28.53 | 35.20 | 38.90 | 40.12 |
| 70° | 38.32 | 34.85 | 27.69 | 19.25 | 12.73 | 9.93 | 11.14 | 16.30 | 24.05 | 32.23 | 37.17 | 38.99 |
| 90° | 32.01 | 26.81 | 17.33 | 7.61 | 1.93 | 0.34 | 0.96 | 4.90 | 13.08 | 23.47 | 30.41 | 33.24 |

*(Continued)*

|  | Jan | Feb | Mar | Apr | May | Jun | Jul | Aug | Sep | Oct | Nov | Dec |
|---|---|---|---|---|---|---|---|---|---|---|---|---|
| Lat | 26° | 26° | 26° | 26° | 26° | 26° | 26° | 26° | 26° | 26° | 26° | 26° |
| Tilt 0° | 23.56 | 27.77 | 33.02 | 37.12 | 39.49 | 40.21 | 39.67 | 37.78 | 34.29 | 29.33 | 24.59 | 22.08 |
| 20° | 33.19 | 35.42 | 37.16 | 37.25 | 36.21 | 35.36 | 35.57 | 36.52 | 36.86 | 35.76 | 33.62 | 32.20 |
| 30° | 36.55 | 37.68 | 37.55 | 35.66 | 33.08 | 31.57 | 32.11 | 34.31 | 36.48 | 37.38 | 36.67 | 35.86 |
| 40° | 38.81 | 38.79 | 36.81 | 33.02 | 29.07 | 27.01 | 27.82 | 31.11 | 35.00 | 37.87 | 38.60 | 38.44 |
| 50° | 39.89 | 38.73 | 34.95 | 29.41 | 24.32 | 21.83 | 22.86 | 27.04 | 32.46 | 37.21 | 39.36 | 39.84 |
| 60° | 39.75 | 37.49 | 32.02 | 24.95 | 18.99 | 16.24 | 17.41 | 22.23 | 28.93 | 35.42 | 38.92 | 40.04 |
| 70° | 38.41 | 35.10 | 28.13 | 19.79 | 13.30 | 10.49 | 11.71 | 16.86 | 24.53 | 32.55 | 37.30 | 39.01 |
| 90° | 32.30 | 27.25 | 17.91 | 8.20 | 2.37 | 0.62 | 1.33 | 5.45 | 13.68 | 23.96 | 30.75 | 33.48 |
| Lat | 27° | 27° | 27° | 27° | 27° | 27° | 27° | 27° | 27° | 27° | 27° | 27° |
| Tilt 0° | 22.99 | 27.29 | 32.70 | 37.00 | 39.55 | 40.36 | 39.77 | 37.73 | 34.04 | 28.91 | 24.05 | 21.50 |
| 20° | 32.75 | 35.12 | 37.05 | 37.35 | 36.47 | 35.69 | 35.87 | 36.68 | 36.84 | 35.53 | 33.23 | 31.73 |
| 30° | 36.20 | 37.48 | 37.57 | 35.87 | 33.44 | 31.99 | 32.49 | 34.57 | 36.57 | 37.26 | 36.36 | 35.47 |
| 40° | 38.56 | 38.70 | 36.94 | 33.33 | 29.51 | 27.50 | 28.28 | 31.47 | 35.20 | 37.86 | 38.40 | 38.13 |
| 50° | 39.74 | 38.75 | 35.18 | 29.81 | 24.82 | 22.37 | 23.38 | 27.49 | 32.76 | 37.31 | 39.26 | 39.63 |
| 60° | 39.71 | 37.62 | 32.36 | 25.43 | 19.54 | 16.81 | 17.97 | 22.74 | 29.32 | 35.63 | 38.94 | 39.93 |
| 70° | 38.48 | 35.35 | 28.56 | 20.34 | 13.88 | 11.07 | 12.28 | 17.42 | 25.01 | 32.86 | 37.43 | 39.02 |
| 90° | 32.58 | 27.68 | 18.49 | 8.79 | 2.84 | 0.95 | 1.73 | 6.01 | 14.27 | 24.45 | 31.08 | 33.70 |
| Lat | 28° | 28° | 28° | 28° | 28° | 28° | 28° | 28° | 28° | 28° | 28° | 28° |
| Tilt 0° | 22.42 | 26.81 | 32.38 | 36.87 | 39.59 | 40.49 | 39.86 | 37.66 | 33.79 | 28.48 | 23.50 | 20.91 |
| 20° | 32.30 | 34.80 | 36.94 | 37.44 | 36.72 | 36.02 | 36.16 | 36.84 | 36.80 | 35.29 | 32.82 | 31.24 |
| 30° | 35.84 | 37.27 | 37.57 | 36.07 | 33.78 | 32.40 | 32.87 | 34.83 | 36.65 | 37.13 | 36.05 | 35.06 |
| 40° | 38.29 | 38.60 | 37.05 | 33.63 | 29.94 | 27.98 | 28.74 | 31.83 | 35.38 | 37.84 | 38.18 | 37.81 |
| 50° | 39.58 | 38.76 | 35.41 | 30.21 | 25.32 | 22.91 | 23.90 | 27.92 | 33.05 | 37.40 | 39.15 | 39.41 |
| 60° | 39.66 | 37.74 | 32.69 | 25.91 | 20.09 | 17.38 | 18.53 | 23.25 | 29.71 | 35.83 | 38.94 | 39.82 |
| 70° | 38.54 | 35.58 | 28.98 | 20.87 | 14.46 | 11.64 | 12.86 | 17.97 | 25.47 | 33.16 | 37.54 | 39.01 |
| 90° | 32.85 | 28.11 | 19.05 | 9.39 | 3.32 | 1.31 | 2.16 | 6.57 | 14.86 | 24.93 | 31.39 | 33.91 |
| Lat | 29° | 29° | 29° | 29° | 29° | 29° | 29° | 29° | 29° | 29° | 29° | 29° |
| Tilt 0° | 21.85 | 26.31 | 32.04 | 36.72 | 39.63 | 40.61 | 39.94 | 37.59 | 33.53 | 28.04 | 22.95 | 20.31 |
| 20° | 31.85 | 34.48 | 36.82 | 37.52 | 36.96 | 36.33 | 36.44 | 36.98 | 36.76 | 35.03 | 32.41 | 30.75 |
| 30° | 35.47 | 37.05 | 37.55 | 36.25 | 34.12 | 32.80 | 33.24 | 35.08 | 36.71 | 36.98 | 35.72 | 34.64 |
| 40° | 38.01 | 38.49 | 37.15 | 33.92 | 30.36 | 28.45 | 29.19 | 32.17 | 35.56 | 37.80 | 37.95 | 37.48 |
| 50° | 39.40 | 38.76 | 35.62 | 30.59 | 25.81 | 23.44 | 24.41 | 28.35 | 33.33 | 37.48 | 39.03 | 39.18 |
| 60° | 39.59 | 37.86 | 33.01 | 26.37 | 20.63 | 17.95 | 19.08 | 23.75 | 30.09 | 36.01 | 38.92 | 39.69 |
| 70° | 38.58 | 35.80 | 29.39 | 21.41 | 15.03 | 12.21 | 13.43 | 18.52 | 25.93 | 33.45 | 37.63 | 38.99 |
| 90° | 33.10 | 28.52 | 19.61 | 9.98 | 3.82 | 1.71 | 2.61 | 7.13 | 15.44 | 25.40 | 31.69 | 34.10 |
| Lat | 30° | 30° | 30° | 30° | 30° | 30° | 30° | 30° | 30° | 30° | 30° | 30° |
| Tilt 0° | 21.27 | 25.81 | 31.69 | 36.57 | 39.65 | 40.72 | 40.02 | 37.51 | 33.26 | 27.59 | 22.40 | 19.72 |
| 20° | 31.39 | 34.15 | 36.68 | 37.58 | 37.19 | 36.63 | 36.70 | 37.11 | 36.70 | 34.77 | 31.98 | 30.25 |
| 30° | 35.09 | 36.82 | 37.53 | 36.43 | 34.45 | 33.19 | 33.60 | 35.31 | 36.77 | 36.82 | 35.39 | 34.21 |
| 40° | 37.72 | 38.37 | 37.24 | 34.20 | 30.78 | 28.92 | 29.63 | 32.50 | 35.72 | 37.76 | 37.71 | 37.14 |
| 50° | 39.21 | 38.75 | 35.82 | 30.97 | 26.30 | 23.97 | 24.91 | 28.77 | 33.60 | 37.54 | 38.89 | 38.93 |
| 60° | 39.51 | 37.96 | 33.31 | 26.83 | 21.18 | 18.51 | 19.63 | 24.24 | 30.45 | 36.19 | 38.89 | 39.54 |
| 70° | 38.60 | 36.01 | 29.79 | 21.93 | 15.60 | 12.79 | 14.00 | 19.07 | 26.39 | 33.73 | 37.71 | 38.95 |
| 90° | 33.34 | 28.92 | 20.17 | 10.58 | 4.33 | 2.13 | 3.07 | 7.70 | 16.02 | 25.86 | 31.98 | 34.28 |
| Lat | 31° | 31° | 31° | 31° | 31° | 31° | 31° | 31° | 31° | 31° | 31° | 31° |
| Tilt 0° | 20.68 | 25.30 | 31.34 | 36.40 | 39.67 | 40.83 | 40.08 | 37.42 | 32.98 | 27.14 | 21.84 | 19.12 |
| 20° | 30.92 | 33.80 | 36.53 | 37.64 | 37.41 | 36.93 | 36.96 | 37.23 | 36.63 | 34.50 | 31.55 | 29.74 |
| 30° | 34.69 | 36.57 | 37.50 | 36.59 | 34.76 | 33.57 | 33.95 | 35.54 | 36.81 | 36.66 | 35.04 | 33.77 |
| 40° | 37.42 | 38.23 | 37.32 | 34.47 | 31.18 | 29.38 | 30.06 | 32.83 | 35.88 | 37.70 | 37.46 | 36.78 |
| 50° | 39.01 | 38.72 | 36.01 | 31.33 | 26.78 | 24.49 | 25.41 | 29.18 | 33.86 | 37.60 | 38.74 | 38.67 |
| 60° | 39.41 | 38.04 | 33.61 | 27.28 | 21.71 | 19.07 | 20.18 | 24.72 | 30.81 | 36.35 | 38.85 | 39.38 |
| 70° | 38.61 | 36.20 | 30.19 | 22.45 | 16.17 | 13.36 | 14.57 | 19.61 | 26.83 | 34.00 | 37.78 | 38.90 |
| 90° | 33.56 | 29.32 | 20.72 | 11.17 | 4.85 | 2.57 | 3.55 | 8.27 | 16.59 | 26.31 | 32.26 | 34.43 |

|  | Jan | Feb | Mar | Apr | May | Jun | Jul | Aug | Sep | Oct | Nov | Dec |
|---|---|---|---|---|---|---|---|---|---|---|---|---|
| Lat | 32° | 32° | 32° | 32° | 32° | 32° | 32° | 32° | 32° | 32° | 32° | 32° |
| Tilt 0° | 20.10 | 24.79 | 30.97 | 36.23 | 39.67 | 40.92 | 40.13 | 37.32 | 32.68 | 26.67 | 21.27 | 18.52 |
| 20° | 30.43 | 33.45 | 36.37 | 37.68 | 37.62 | 37.21 | 37.21 | 37.35 | 36.55 | 34.21 | 31.11 | 29.22 |
| 30° | 34.29 | 36.32 | 37.45 | 36.75 | 35.07 | 33.95 | 34.29 | 35.76 | 36.84 | 36.48 | 34.68 | 33.32 |
| 40° | 37.10 | 38.08 | 37.39 | 34.73 | 31.58 | 29.83 | 30.48 | 33.14 | 36.02 | 37.63 | 37.20 | 36.41 |
| 50° | 38.79 | 38.69 | 36.19 | 31.69 | 27.25 | 25.00 | 25.91 | 29.58 | 34.11 | 37.64 | 38.58 | 38.39 |
| 60° | 39.30 | 38.12 | 33.90 | 27.72 | 22.24 | 19.63 | 20.73 | 25.20 | 31.16 | 36.51 | 38.80 | 39.20 |
| 70° | 38.61 | 36.39 | 30.57 | 22.96 | 16.74 | 13.94 | 15.14 | 20.14 | 27.27 | 34.26 | 37.83 | 38.82 |
| 90° | 33.77 | 29.70 | 21.26 | 11.77 | 5.39 | 3.03 | 4.04 | 8.84 | 17.16 | 26.75 | 32.52 | 34.58 |
| Lat | 33° | 33° | 33° | 33° | 33° | 33° | 33° | 33° | 33° | 33° | 33° | 33° |
| Tilt 0° | 19.50 | 24.27 | 30.59 | 36.04 | 39.67 | 41.00 | 40.17 | 37.20 | 32.38 | 26.20 | 20.70 | 17.91 |
| 20° | 29.95 | 33.09 | 36.20 | 37.71 | 37.82 | 37.49 | 37.45 | 37.45 | 36.46 | 33.92 | 30.66 | 28.70 |
| 30° | 33.87 | 36.05 | 37.39 | 36.89 | 35.37 | 34.31 | 34.63 | 35.96 | 36.87 | 36.29 | 34.31 | 32.86 |
| 40° | 36.78 | 37.92 | 37.45 | 34.98 | 31.97 | 30.28 | 30.90 | 33.45 | 36.15 | 37.55 | 36.92 | 36.03 |
| 50° | 38.56 | 38.64 | 36.36 | 32.04 | 27.72 | 25.51 | 26.40 | 29.98 | 34.35 | 37.67 | 38.41 | 38.10 |
| 60° | 39.17 | 38.18 | 34.17 | 28.16 | 22.77 | 20.19 | 21.27 | 25.67 | 31.50 | 36.65 | 38.73 | 39.01 |
| 70° | 38.59 | 36.56 | 30.95 | 23.47 | 17.31 | 14.52 | 15.71 | 20.67 | 27.70 | 34.51 | 37.87 | 38.74 |
| 90° | 33.97 | 30.07 | 21.80 | 12.36 | 5.92 | 3.50 | 4.55 | 9.42 | 17.73 | 27.19 | 32.76 | 34.70 |
| Lat | 34° | 34° | 34° | 34° | 34° | 34° | 34° | 34° | 34° | 34° | 34° | 34° |
| Tilt 0° | 18.91 | 23.74 | 30.21 | 35.85 | 39.65 | 41.08 | 40.20 | 37.08 | 32.07 | 25.72 | 20.13 | 17.30 |
| 20° | 29.45 | 32.72 | 36.02 | 37.73 | 38.01 | 37.76 | 37.68 | 37.54 | 36.36 | 33.62 | 30.20 | 28.17 |
| 30° | 33.45 | 35.77 | 37.33 | 37.02 | 35.66 | 34.67 | 34.95 | 36.16 | 36.87 | 36.08 | 33.93 | 32.39 |
| 40° | 36.43 | 37.75 | 37.49 | 35.22 | 32.35 | 30.72 | 31.31 | 33.74 | 36.27 | 37.46 | 36.63 | 35.63 |
| 50° | 38.31 | 38.57 | 36.52 | 32.37 | 28.18 | 26.02 | 26.88 | 30.37 | 34.57 | 37.69 | 38.22 | 37.79 |
| 60° | 39.03 | 38.22 | 34.44 | 28.59 | 23.29 | 20.74 | 21.80 | 26.14 | 31.83 | 36.78 | 38.64 | 38.80 |
| 70° | 38.55 | 36.72 | 31.31 | 23.97 | 17.87 | 15.09 | 16.28 | 21.20 | 28.12 | 34.75 | 37.89 | 38.63 |
| 90° | 34.15 | 30.43 | 22.33 | 12.95 | 6.47 | 3.99 | 5.06 | 10.00 | 18.29 | 27.61 | 33.00 | 34.81 |
| Lat | 35° | 35° | 35° | 35° | 35° | 35° | 35° | 35° | 35° | 35° | 35° | 35° |
| Tilt 0° | 18.31 | 23.20 | 29.81 | 35.64 | 39.63 | 41.15 | 40.23 | 36.95 | 31.75 | 25.24 | 19.55 | 16.70 |
| 20° | 28.94 | 32.33 | 35.83 | 37.74 | 38.19 | 38.02 | 37.90 | 37.62 | 36.24 | 33.30 | 29.73 | 27.63 |
| 30° | 33.01 | 35.49 | 37.25 | 37.14 | 35.94 | 35.02 | 35.27 | 36.34 | 36.87 | 35.87 | 33.54 | 31.91 |
| 40° | 36.08 | 37.56 | 37.53 | 35.45 | 32.72 | 31.15 | 31.71 | 34.03 | 36.38 | 37.35 | 36.33 | 35.22 |
| 50° | 38.05 | 38.50 | 36.67 | 32.70 | 28.63 | 26.52 | 27.35 | 30.74 | 34.79 | 37.69 | 38.01 | 37.47 |
| 60° | 38.87 | 38.26 | 34.70 | 29.00 | 23.81 | 21.29 | 22.33 | 26.59 | 32.15 | 36.89 | 38.54 | 38.57 |
| 70° | 38.50 | 36.86 | 31.67 | 24.47 | 18.43 | 15.66 | 16.85 | 21.72 | 28.53 | 34.97 | 37.90 | 38.51 |
| 90° | 34.31 | 30.78 | 22.85 | 13.54 | 7.02 | 4.49 | 5.59 | 10.58 | 18.84 | 28.03 | 33.22 | 34.90 |
| Lat | 36° | 36° | 36° | 36° | 36° | 36° | 36° | 36° | 36° | 36° | 36° | 36° |
| Tilt 0° | 17.71 | 22.66 | 29.41 | 35.42 | 39.59 | 41.20 | 40.24 | 36.81 | 31.41 | 24.75 | 18.97 | 16.09 |
| 20° | 28.43 | 31.94 | 35.63 | 37.74 | 38.36 | 38.26 | 38.11 | 37.68 | 36.12 | 32.98 | 29.25 | 27.08 |
| 30° | 32.57 | 35.19 | 37.15 | 37.25 | 36.21 | 35.36 | 35.57 | 36.52 | 36.86 | 35.65 | 33.14 | 31.42 |
| 40° | 35.71 | 37.37 | 37.55 | 35.66 | 33.08 | 31.57 | 32.11 | 34.31 | 36.48 | 37.23 | 36.01 | 34.80 |
| 50° | 37.78 | 38.41 | 36.81 | 33.02 | 29.07 | 27.01 | 27.82 | 31.11 | 35.00 | 37.69 | 37.79 | 37.13 |
| 60° | 38.69 | 38.28 | 34.94 | 29.41 | 24.32 | 21.83 | 22.86 | 27.04 | 32.46 | 37.00 | 38.43 | 38.33 |
| 70° | 38.43 | 36.99 | 32.02 | 24.95 | 18.99 | 16.24 | 17.41 | 22.23 | 28.93 | 35.18 | 37.89 | 38.36 |
| 90° | 34.45 | 31.12 | 23.37 | 14.12 | 7.58 | 5.00 | 6.12 | 11.15 | 19.39 | 28.44 | 33.42 | 34.97 |
| Lat | 37° | 37° | 37° | 37° | 37° | 37° | 37° | 37° | 37° | 37° | 37° | 37° |
| Tilt 0° | 17.11 | 22.12 | 29.00 | 35.19 | 39.55 | 41.25 | 40.24 | 36.65 | 31.07 | 24.25 | 18.38 | 15.48 |
| 20° | 27.90 | 31.54 | 35.42 | 37.73 | 38.52 | 38.50 | 38.31 | 37.74 | 35.99 | 32.65 | 28.77 | 26.52 |
| 30° | 32.11 | 34.88 | 37.05 | 37.35 | 36.47 | 35.69 | 35.87 | 36.68 | 36.84 | 35.41 | 32.72 | 30.91 |
| 40° | 35.33 | 37.16 | 37.56 | 35.87 | 33.44 | 31.99 | 32.49 | 34.57 | 36.57 | 37.11 | 35.69 | 34.37 |
| 50° | 37.49 | 38.31 | 36.93 | 33.33 | 29.51 | 27.50 | 28.28 | 31.47 | 35.20 | 37.67 | 37.56 | 36.77 |
| 60° | 38.50 | 38.29 | 35.18 | 29.81 | 24.82 | 22.37 | 23.38 | 27.49 | 32.76 | 37.09 | 38.30 | 38.07 |
| 70° | 38.35 | 37.11 | 32.35 | 25.43 | 19.54 | 16.81 | 17.97 | 22.74 | 29.32 | 35.39 | 37.87 | 38.20 |
| 90° | 34.58 | 31.44 | 23.88 | 14.71 | 8.14 | 5.52 | 6.66 | 11.73 | 19.93 | 28.83 | 33.61 | 35.02 |

*(Continued)*

|        | Jan   | Feb   | Mar   | Apr   | May   | Jun   | Jul   | Aug   | Sep   | Oct   | Nov   | Dec   |
|--------|-------|-------|-------|-------|-------|-------|-------|-------|-------|-------|-------|-------|
| Lat    | 38°   | 38°   | 38°   | 38°   | 38°   | 38°   | 38°   | 38°   | 38°   | 38°   | 38°   | 38°   |
| Tilt 0° | 16.51 | 21.57 | 28.58 | 34.96 | 39.49 | 41.29 | 40.24 | 36.49 | 30.72 | 23.74 | 17.79 | 14.86 |
| 20°    | 27.37 | 31.13 | 35.19 | 37.71 | 38.67 | 38.73 | 38.51 | 37.79 | 35.84 | 32.30 | 28.28 | 25.96 |
| 30°    | 31.64 | 34.56 | 36.94 | 37.44 | 36.72 | 36.02 | 36.16 | 36.84 | 36.80 | 35.17 | 32.30 | 30.40 |
| 40°    | 34.94 | 36.93 | 37.56 | 36.07 | 33.78 | 32.40 | 32.87 | 34.83 | 36.65 | 36.97 | 35.35 | 33.92 |
| 50°    | 37.18 | 38.19 | 37.04 | 33.63 | 29.94 | 27.98 | 28.74 | 31.83 | 35.38 | 37.64 | 37.32 | 36.40 |
| 60°    | 38.29 | 38.29 | 35.40 | 30.21 | 25.32 | 22.91 | 23.90 | 27.92 | 33.05 | 37.17 | 38.15 | 37.79 |
| 70°    | 38.24 | 37.22 | 32.68 | 25.91 | 20.09 | 17.38 | 18.53 | 23.25 | 29.71 | 35.58 | 37.83 | 38.02 |
| 90°    | 34.69 | 31.76 | 24.38 | 15.29 | 8.71  | 6.05  | 7.21  | 12.31 | 20.47 | 29.22 | 33.78 | 35.05 |
| Lat    | 39°   | 39°   | 39°   | 39°   | 39°   | 39°   | 39°   | 39°   | 39°   | 39°   | 39°   | 39°   |
| Tilt 0° | 15.90 | 21.01 | 28.15 | 34.71 | 39.43 | 41.32 | 40.23 | 36.32 | 30.36 | 23.23 | 17.20 | 14.25 |
| 20°    | 26.83 | 30.71 | 34.96 | 37.67 | 38.81 | 38.95 | 38.69 | 37.82 | 35.68 | 31.95 | 27.77 | 25.38 |
| 30°    | 31.16 | 34.22 | 36.81 | 37.52 | 36.96 | 36.33 | 36.44 | 36.98 | 36.76 | 34.91 | 31.87 | 29.87 |
| 40°    | 34.53 | 36.70 | 37.55 | 36.25 | 34.12 | 32.80 | 33.24 | 35.08 | 36.71 | 36.82 | 34.99 | 33.45 |
| 50°    | 36.86 | 38.06 | 37.15 | 33.92 | 30.36 | 28.45 | 29.19 | 32.17 | 35.56 | 37.60 | 37.06 | 36.02 |
| 60°    | 38.07 | 38.27 | 35.61 | 30.59 | 25.81 | 23.44 | 24.41 | 28.35 | 33.33 | 37.24 | 37.99 | 37.49 |
| 70°    | 38.12 | 37.31 | 33.00 | 26.37 | 20.63 | 17.95 | 19.08 | 23.75 | 30.09 | 35.75 | 37.77 | 37.82 |
| 90°    | 34.78 | 32.06 | 24.87 | 15.86 | 9.28  | 6.59  | 7.76  | 12.88 | 21.00 | 29.60 | 33.94 | 35.06 |
| Lat    | 40°   | 40°   | 40°   | 40°   | 40°   | 40°   | 40°   | 40°   | 40°   | 40°   | 40°   | 40°   |
| Tilt 0° | 15.29 | 20.45 | 27.71 | 34.45 | 39.36 | 41.35 | 40.20 | 36.13 | 29.99 | 22.71 | 16.61 | 13.64 |
| 20°    | 26.29 | 30.28 | 34.71 | 37.63 | 38.94 | 39.16 | 38.86 | 37.85 | 35.52 | 31.58 | 27.26 | 24.80 |
| 30°    | 30.67 | 33.88 | 36.68 | 37.58 | 37.19 | 36.63 | 36.70 | 37.11 | 36.70 | 34.64 | 31.42 | 29.33 |
| 40°    | 34.11 | 36.46 | 37.53 | 36.43 | 34.45 | 33.19 | 33.60 | 35.31 | 36.77 | 36.65 | 34.63 | 32.97 |
| 50°    | 36.52 | 37.92 | 37.24 | 34.20 | 30.78 | 28.92 | 29.63 | 32.50 | 35.72 | 37.55 | 36.78 | 35.61 |
| 60°    | 37.82 | 38.24 | 35.82 | 30.97 | 26.30 | 23.97 | 24.91 | 28.77 | 33.60 | 37.30 | 37.81 | 37.17 |
| 70°    | 37.98 | 37.39 | 33.31 | 26.83 | 21.18 | 18.51 | 19.63 | 24.24 | 30.45 | 35.92 | 37.70 | 37.60 |
| 90°    | 34.85 | 32.35 | 25.36 | 16.44 | 9.85  | 7.13  | 8.31  | 13.46 | 21.52 | 29.96 | 34.07 | 35.04 |
| Lat    | 41°   | 41°   | 41°   | 41°   | 41°   | 41°   | 41°   | 41°   | 41°   | 41°   | 41°   | 41°   |
| Tilt 0° | 14.68 | 19.88 | 27.26 | 34.18 | 39.27 | 41.36 | 40.17 | 35.94 | 29.61 | 22.18 | 16.01 | 13.03 |
| 20°    | 25.73 | 29.84 | 34.45 | 37.57 | 39.06 | 39.36 | 39.02 | 37.86 | 35.34 | 31.21 | 26.74 | 24.22 |
| 30°    | 30.16 | 33.53 | 36.53 | 37.64 | 37.41 | 36.93 | 36.96 | 37.23 | 36.63 | 34.37 | 30.97 | 28.79 |
| 40°    | 33.68 | 36.20 | 37.49 | 36.59 | 34.76 | 33.57 | 33.95 | 35.54 | 36.81 | 36.48 | 34.25 | 32.48 |
| 50°    | 36.17 | 37.77 | 37.32 | 34.47 | 31.18 | 29.38 | 30.06 | 32.83 | 35.88 | 37.48 | 36.49 | 35.19 |
| 60°    | 37.56 | 38.19 | 36.01 | 31.33 | 26.78 | 24.49 | 25.41 | 29.18 | 33.86 | 37.35 | 37.62 | 36.83 |
| 70°    | 37.81 | 37.45 | 33.60 | 27.28 | 21.71 | 19.07 | 20.18 | 24.72 | 30.81 | 36.07 | 37.61 | 37.35 |
| 90°    | 34.90 | 32.62 | 25.84 | 17.01 | 10.42 | 7.68  | 8.87  | 14.03 | 22.04 | 30.32 | 34.19 | 35.01 |
| Lat    | 42°   | 42°   | 42°   | 42°   | 42°   | 42°   | 42°   | 42°   | 42°   | 42°   | 42°   | 42°   |
| Tilt 0° | 14.07 | 19.31 | 26.81 | 33.91 | 39.18 | 41.37 | 40.13 | 35.74 | 29.23 | 21.65 | 15.42 | 12.42 |
| 20°    | 25.17 | 29.39 | 34.19 | 37.50 | 39.16 | 39.55 | 39.17 | 37.87 | 35.15 | 30.83 | 26.22 | 23.62 |
| 30°    | 29.65 | 33.16 | 36.37 | 37.68 | 37.62 | 37.21 | 37.21 | 37.35 | 36.55 | 34.08 | 30.50 | 28.23 |
| 40°    | 33.23 | 35.93 | 37.45 | 36.75 | 35.07 | 33.95 | 34.29 | 35.76 | 36.84 | 36.29 | 33.86 | 31.97 |
| 50°    | 35.80 | 37.60 | 37.38 | 34.73 | 31.58 | 29.83 | 30.48 | 33.14 | 36.02 | 37.40 | 36.18 | 34.75 |
| 60°    | 37.28 | 38.13 | 36.19 | 31.69 | 27.25 | 25.00 | 25.91 | 29.58 | 34.11 | 37.38 | 37.41 | 36.47 |
| 70°    | 37.63 | 37.50 | 33.89 | 27.72 | 22.24 | 19.63 | 20.73 | 25.20 | 31.16 | 36.22 | 37.50 | 37.08 |
| 90°    | 34.92 | 32.88 | 26.31 | 17.57 | 11.00 | 8.24  | 9.44  | 14.60 | 22.55 | 30.66 | 34.30 | 34.94 |
| Lat    | 43°   | 43°   | 43°   | 43°   | 43°   | 43°   | 43°   | 43°   | 43°   | 43°   | 43°   | 43°   |
| Tilt 0° | 13.46 | 18.73 | 26.34 | 33.62 | 39.08 | 41.36 | 40.08 | 35.53 | 28.83 | 21.11 | 14.82 | 11.81 |
| 20°    | 24.60 | 28.93 | 33.91 | 37.43 | 39.26 | 39.73 | 39.31 | 37.86 | 34.95 | 30.44 | 25.68 | 23.02 |
| 30°    | 29.12 | 32.79 | 36.20 | 37.71 | 37.82 | 37.49 | 37.45 | 37.45 | 36.46 | 33.78 | 30.02 | 27.65 |
| 40°    | 32.77 | 35.64 | 37.39 | 36.89 | 35.37 | 34.31 | 34.63 | 35.96 | 36.87 | 36.09 | 33.45 | 31.45 |
| 50°    | 35.41 | 37.42 | 37.44 | 34.98 | 31.97 | 30.28 | 30.90 | 33.45 | 36.15 | 37.31 | 35.86 | 34.29 |
| 60°    | 36.98 | 38.06 | 36.36 | 32.04 | 27.72 | 25.51 | 26.40 | 29.98 | 34.35 | 37.40 | 37.18 | 36.09 |
| 70°    | 37.43 | 37.54 | 34.17 | 28.16 | 22.77 | 20.19 | 21.27 | 25.67 | 31.50 | 36.34 | 37.37 | 36.79 |
| 90°    | 34.93 | 33.13 | 26.77 | 18.14 | 11.57 | 8.80  | 10.00 | 15.17 | 23.06 | 30.99 | 34.38 | 34.85 |

|  |  | Jan | Feb | Mar | Apr | May | Jun | Jul | Aug | Sep | Oct | Nov | Dec |
|---|---|---|---|---|---|---|---|---|---|---|---|---|---|
| Lat |  | 44° | 44° | 44° | 44° | 44° | 44° | 44° | 44° | 44° | 44° | 44° | 44° |
| Tilt | 0° | 12.85 | 18.15 | 25.87 | 33.32 | 38.97 | 41.36 | 40.03 | 35.31 | 28.43 | 20.57 | 14.22 | 11.20 |
|  | 20° | 24.02 | 28.47 | 33.62 | 37.34 | 39.35 | 39.90 | 39.44 | 37.84 | 34.74 | 30.04 | 25.14 | 22.41 |
|  | 30° | 28.59 | 32.40 | 36.02 | 37.73 | 38.01 | 37.76 | 37.68 | 37.54 | 36.36 | 33.47 | 29.53 | 27.07 |
|  | 40° | 32.29 | 35.35 | 37.32 | 37.02 | 35.66 | 34.67 | 34.95 | 36.16 | 36.87 | 35.88 | 33.03 | 30.91 |
|  | 50° | 35.01 | 37.22 | 37.49 | 35.22 | 32.35 | 30.72 | 31.31 | 33.74 | 36.27 | 37.21 | 35.52 | 33.81 |
|  | 60° | 36.66 | 37.97 | 36.51 | 32.37 | 28.18 | 26.02 | 26.88 | 30.37 | 34.57 | 37.40 | 36.93 | 35.68 |
|  | 70° | 37.20 | 37.56 | 34.43 | 28.59 | 23.29 | 20.74 | 21.80 | 26.14 | 31.83 | 36.46 | 37.23 | 36.47 |
|  | 90° | 34.91 | 33.36 | 27.22 | 18.69 | 12.15 | 9.36 | 10.57 | 15.74 | 23.56 | 31.32 | 34.44 | 34.74 |
| Lat |  | 45° | 45° | 45° | 45° | 45° | 45° | 45° | 45° | 45° | 45° | 45° | 45° |
| Tilt | 0° | 12.25 | 17.57 | 25.39 | 33.02 | 38.86 | 41.34 | 39.96 | 35.08 | 28.01 | 20.02 | 13.62 | 10.60 |
|  | 20° | 23.43 | 27.99 | 33.33 | 37.24 | 39.42 | 40.06 | 39.56 | 37.81 | 34.52 | 29.63 | 24.59 | 21.79 |
|  | 30° | 28.04 | 32.00 | 35.83 | 37.74 | 38.19 | 38.02 | 37.90 | 37.62 | 36.24 | 33.15 | 29.03 | 26.47 |
|  | 40° | 31.79 | 35.04 | 37.24 | 37.14 | 35.94 | 35.02 | 35.27 | 36.34 | 36.87 | 35.66 | 32.59 | 30.35 |
|  | 50° | 34.58 | 37.01 | 37.52 | 35.45 | 32.72 | 31.15 | 31.71 | 34.03 | 36.38 | 37.09 | 35.17 | 33.31 |
|  | 60° | 36.32 | 37.86 | 36.66 | 32.70 | 28.63 | 26.52 | 27.35 | 30.74 | 34.79 | 37.40 | 36.67 | 35.25 |
|  | 70° | 36.95 | 37.56 | 34.69 | 29.00 | 23.81 | 21.29 | 22.33 | 26.59 | 32.15 | 35.56 | 37.06 | 36.13 |
|  | 90° | 34.87 | 33.58 | 27.67 | 19.25 | 12.73 | 9.93 | 11.14 | 16.30 | 24.05 | 31.62 | 34.48 | 34.59 |
| Lat |  | 46° | 46° | 46° | 46° | 46° | 46° | 46° | 46° | 46° | 46° | 46° | 46° |
| Tilt | 0° | 11.64 | 16.98 | 24.91 | 32.71 | 38.73 | 41.32 | 39.89 | 34.84 | 27.59 | 19.46 | 13.01 | 9.99 |
|  | 20° | 22.84 | 27.51 | 33.02 | 37.12 | 39.49 | 40.21 | 39.67 | 37.78 | 34.29 | 29.21 | 24.03 | 21.16 |
|  | 30° | 27.48 | 31.60 | 35.62 | 37.74 | 38.36 | 38.26 | 38.11 | 37.68 | 36.12 | 32.82 | 28.52 | 25.86 |
|  | 40° | 31.28 | 34.72 | 37.15 | 37.25 | 36.21 | 35.36 | 35.57 | 36.52 | 36.86 | 35.43 | 32.14 | 29.78 |
|  | 50° | 34.14 | 36.79 | 37.54 | 35.66 | 33.08 | 31.57 | 32.11 | 34.31 | 36.48 | 36.96 | 34.79 | 32.78 |
|  | 60° | 35.96 | 37.74 | 36.80 | 33.02 | 29.07 | 27.01 | 27.82 | 31.11 | 35.00 | 37.38 | 36.38 | 34.80 |
|  | 70° | 36.68 | 37.55 | 34.93 | 29.41 | 24.32 | 21.83 | 22.86 | 27.04 | 32.46 | 36.65 | 36.87 | 35.75 |
|  | 90° | 34.80 | 33.78 | 28.11 | 19.79 | 13.30 | 10.49 | 11.71 | 16.86 | 24.53 | 31.92 | 34.50 | 34.41 |
| Lat |  | 47° | 47° | 47° | 47° | 47° | 47° | 47° | 47° | 47° | 47° | 47° | 47° |
| Tilt | 0° | 11.03 | 16.39 | 24.41 | 32.38 | 38.60 | 41.29 | 39.82 | 34.59 | 27.16 | 18.90 | 12.41 | 9.40 |
|  | 20° | 22.23 | 27.02 | 32.70 | 37.00 | 39.55 | 40.36 | 39.77 | 37.73 | 34.04 | 28.78 | 23.46 | 20.53 |
|  | 30° | 26.90 | 31.18 | 35.41 | 37.73 | 38.52 | 38.50 | 38.31 | 37.74 | 35.99 | 32.47 | 28.00 | 25.24 |
|  | 40° | 30.76 | 34.39 | 37.04 | 37.35 | 36.47 | 35.69 | 35.87 | 36.68 | 36.84 | 35.18 | 31.68 | 29.18 |
|  | 50° | 33.67 | 36.55 | 37.55 | 35.87 | 33.44 | 31.99 | 32.49 | 34.57 | 36.57 | 36.82 | 34.40 | 32.24 |
|  | 60° | 35.57 | 37.61 | 36.92 | 33.33 | 29.51 | 27.50 | 28.28 | 31.47 | 35.20 | 37.34 | 36.08 | 34.32 |
|  | 70° | 36.38 | 37.52 | 35.17 | 29.81 | 24.82 | 22.37 | 23.38 | 27.49 | 32.76 | 36.73 | 36.66 | 35.35 |
|  | 90° | 34.70 | 33.96 | 28.54 | 20.34 | 13.88 | 11.07 | 12.28 | 17.42 | 25.01 | 32.21 | 34.49 | 34.20 |
| Lat |  | 48° | 48° | 48° | 48° | 48° | 48° | 48° | 48° | 48° | 48° | 48° | 48° |
| Tilt | 0° | 10.43 | 15.80 | 23.91 | 32.05 | 38.46 | 41.25 | 39.73 | 34.33 | 26.72 | 18.34 | 11.81 | 8.80 |
|  | 20° | 21.62 | 26.52 | 32.37 | 36.87 | 39.59 | 40.49 | 39.86 | 37.66 | 33.79 | 28.34 | 22.88 | 19.88 |
|  | 30° | 26.32 | 30.75 | 35.19 | 37.71 | 38.67 | 38.73 | 38.51 | 37.79 | 35.84 | 32.12 | 27.46 | 24.60 |
|  | 40° | 30.21 | 34.04 | 36.93 | 37.44 | 36.72 | 36.02 | 36.16 | 36.84 | 36.80 | 34.93 | 31.20 | 28.57 |
|  | 50° | 33.19 | 36.30 | 37.55 | 36.07 | 33.78 | 32.40 | 32.87 | 34.83 | 36.65 | 36.67 | 33.99 | 31.67 |
|  | 60° | 35.16 | 37.46 | 37.03 | 33.63 | 29.94 | 27.98 | 28.74 | 31.83 | 35.38 | 37.30 | 35.75 | 33.80 |
|  | 70° | 36.06 | 37.48 | 35.39 | 30.21 | 25.32 | 22.91 | 23.90 | 27.92 | 33.05 | 36.79 | 36.42 | 34.91 |
|  | 90° | 34.58 | 34.13 | 28.96 | 20.87 | 14.46 | 11.64 | 12.86 | 17.97 | 25.47 | 32.48 | 34.46 | 33.95 |
| Lat |  | 49° | 49° | 49° | 49° | 49° | 49° | 49° | 49° | 49° | 49° | 49° | 49° |
| Tilt | 0° | 9.82 | 15.20 | 23.40 | 31.71 | 38.31 | 41.21 | 39.64 | 34.07 | 26.28 | 17.77 | 11.21 | 8.21 |
|  | 20° | 21.00 | 26.01 | 32.04 | 36.72 | 39.63 | 40.61 | 39.94 | 37.59 | 33.53 | 27.90 | 22.30 | 19.23 |
|  | 30° | 25.72 | 30.30 | 34.95 | 37.67 | 38.81 | 38.95 | 38.69 | 37.82 | 35.68 | 31.76 | 26.91 | 23.94 |
|  | 40° | 29.65 | 33.68 | 36.80 | 37.52 | 36.96 | 36.33 | 36.44 | 36.98 | 36.76 | 34.66 | 30.70 | 27.93 |
|  | 50° | 32.68 | 36.03 | 37.54 | 36.25 | 34.12 | 32.80 | 33.24 | 35.08 | 36.71 | 36.50 | 33.56 | 31.07 |
|  | 60° | 34.72 | 37.29 | 37.13 | 33.92 | 30.36 | 28.45 | 29.19 | 32.17 | 35.56 | 37.24 | 35.40 | 33.26 |
|  | 70° | 35.70 | 37.42 | 35.60 | 30.59 | 25.81 | 23.44 | 24.41 | 28.35 | 33.33 | 36.84 | 36.17 | 34.44 |
|  | 90° | 34.42 | 34.28 | 29.37 | 21.41 | 15.03 | 12.21 | 13.43 | 18.52 | 25.93 | 32.74 | 34.41 | 33.67 |

(Continued)

|      | Jan | Feb | Mar | Apr | May | Jun | Jul | Aug | Sep | Oct | Oct | Dec |
|------|-----|-----|-----|-----|-----|-----|-----|-----|-----|-----|-----|-----|
| Lat  | 50° | 50° | 50° | 50° | 50° | 50° | 50° | 50° | 50° | 50° | 50° | 50° |
| Tilt 0°  | 9.22  | 14.60 | 22.89 | 31.36 | 38.16 | 41.16 | 39.54 | 33.80 | 25.82 | 17.20 | 10.61 | 7.63  |
| 20°  | 20.37 | 25.49 | 31.69 | 36.57 | 39.65 | 40.72 | 40.02 | 37.51 | 33.26 | 27.44 | 21.71 | 18.57 |
| 30°  | 25.10 | 29.85 | 34.70 | 37.63 | 38.94 | 39.16 | 38.86 | 37.85 | 35.52 | 31.38 | 26.35 | 23.27 |
| 40°  | 29.07 | 33.31 | 36.67 | 37.58 | 37.19 | 36.63 | 36.70 | 37.11 | 36.70 | 34.38 | 30.19 | 27.27 |
| 50°  | 32.15 | 35.75 | 37.52 | 36.43 | 34.45 | 33.19 | 33.60 | 35.31 | 36.77 | 36.32 | 33.11 | 30.44 |
| 60°  | 34.25 | 37.11 | 37.22 | 34.20 | 30.78 | 28.92 | 29.63 | 32.50 | 35.72 | 37.16 | 35.03 | 32.68 |
| 70°  | 35.32 | 37.34 | 35.80 | 30.97 | 26.30 | 23.97 | 24.91 | 28.77 | 33.60 | 36.88 | 35.88 | 33.93 |
| 90°  | 34.22 | 34.42 | 29.77 | 21.93 | 15.60 | 12.79 | 14.00 | 19.07 | 26.39 | 32.99 | 34.32 | 33.34 |
| Lat  | 51° | 51° | 51° | 51° | 51° | 51° | 51° | 51° | 51° | 51° | 51° | 51° |
| Tilt 0°  | 8.63  | 14.00 | 22.36 | 31.00 | 38.00 | 41.11 | 39.44 | 33.52 | 25.36 | 16.62 | 10.01 | 7.05  |
| 20°  | 19.74 | 24.96 | 31.33 | 36.40 | 39.67 | 40.83 | 40.08 | 37.42 | 32.98 | 26.98 | 21.11 | 17.90 |
| 30°  | 24.47 | 29.39 | 34.45 | 37.57 | 39.06 | 39.36 | 39.02 | 37.86 | 35.34 | 31.00 | 25.77 | 22.58 |
| 40°  | 28.46 | 32.92 | 36.52 | 37.64 | 37.41 | 36.93 | 36.96 | 37.23 | 36.63 | 34.08 | 29.66 | 26.59 |
| 50°  | 31.59 | 35.45 | 37.48 | 36.59 | 34.76 | 33.57 | 33.95 | 35.54 | 36.81 | 36.13 | 32.64 | 29.78 |
| 60°  | 33.75 | 36.90 | 37.30 | 34.47 | 31.18 | 29.38 | 30.06 | 32.83 | 35.88 | 37.08 | 34.63 | 32.07 |
| 70°  | 34.90 | 37.24 | 35.99 | 31.33 | 26.78 | 24.49 | 25.41 | 29.18 | 33.86 | 36.90 | 35.57 | 33.38 |
| 90°  | 33.99 | 34.53 | 30.16 | 22.45 | 16.17 | 13.36 | 14.57 | 19.61 | 26.83 | 33.22 | 34.21 | 32.96 |
| Lat  | 52° | 52° | 52° | 52° | 52° | 52° | 52° | 52° | 52° | 52° | 52° | 52° |
| Tilt 0°  | 8.04  | 13.39 | 21.83 | 30.64 | 37.83 | 41.06 | 39.33 | 33.23 | 24.89 | 16.04 | 9.41  | 6.48  |
| 20°  | 19.09 | 24.43 | 30.96 | 36.23 | 39.67 | 40.92 | 40.13 | 37.32 | 32.68 | 26.50 | 20.50 | 17.21 |
| 30°  | 23.82 | 28.91 | 34.18 | 37.50 | 39.16 | 39.55 | 39.17 | 37.87 | 35.15 | 30.61 | 25.18 | 21.88 |
| 40°  | 27.84 | 32.52 | 36.36 | 37.68 | 37.62 | 37.21 | 37.21 | 37.35 | 36.55 | 33.78 | 29.11 | 25.87 |
| 50°  | 31.00 | 35.13 | 37.43 | 36.75 | 35.07 | 33.95 | 34.29 | 35.76 | 36.84 | 35.92 | 32.14 | 29.09 |
| 60°  | 33.23 | 36.68 | 37.37 | 34.73 | 31.58 | 29.83 | 30.48 | 33.14 | 36.02 | 36.97 | 34.20 | 31.41 |
| 70°  | 34.44 | 37.12 | 36.17 | 31.69 | 27.25 | 25.00 | 25.91 | 29.58 | 34.11 | 36.90 | 35.23 | 32.79 |
| 90°  | 33.72 | 34.63 | 30.55 | 22.96 | 16.74 | 13.94 | 15.14 | 20.14 | 27.27 | 33.44 | 34.06 | 32.53 |
| Lat  | 53° | 53° | 53° | 53° | 53° | 53° | 53° | 53° | 53° | 53° | 53° | 53° |
| Tilt 0°  | 7.45  | 12.79 | 21.30 | 30.26 | 37.65 | 41.00 | 39.22 | 32.94 | 24.41 | 15.45 | 8.82  | 5.91  |
| 20°  | 18.43 | 23.89 | 30.59 | 36.04 | 39.67 | 41.00 | 40.17 | 37.20 | 32.38 | 26.02 | 19.88 | 16.52 |
| 30°  | 23.16 | 28.43 | 33.90 | 37.43 | 39.26 | 39.73 | 39.31 | 37.86 | 34.95 | 30.20 | 24.58 | 21.15 |
| 40°  | 27.18 | 32.10 | 36.19 | 37.71 | 37.82 | 37.49 | 37.45 | 37.45 | 36.46 | 33.46 | 28.53 | 25.13 |
| 50°  | 30.38 | 34.80 | 37.38 | 36.89 | 35.37 | 34.31 | 34.63 | 35.96 | 36.87 | 35.70 | 31.62 | 28.35 |
| 60°  | 32.66 | 36.45 | 37.43 | 34.98 | 31.97 | 30.28 | 30.90 | 33.45 | 36.15 | 36.86 | 33.75 | 30.71 |
| 70°  | 33.94 | 36.98 | 36.34 | 32.04 | 27.72 | 25.51 | 26.40 | 29.98 | 34.35 | 36.89 | 34.85 | 32.14 |
| 90°  | 33.41 | 34.70 | 30.92 | 23.47 | 17.31 | 14.52 | 15.71 | 20.67 | 27.70 | 33.64 | 33.88 | 32.05 |
| Lat  | 54° | 54° | 54° | 54° | 54° | 54° | 54° | 54° | 54° | 54° | 54° | 54° |
| Tilt 0°  | 6.87  | 12.18 | 20.75 | 29.88 | 37.48 | 40.94 | 39.10 | 32.63 | 23.93 | 14.86 | 8.23  | 5.36  |
| 20°  | 17.76 | 23.33 | 30.20 | 35.85 | 39.65 | 41.08 | 40.20 | 37.08 | 32.07 | 25.53 | 19.25 | 15.81 |
| 30°  | 22.48 | 27.93 | 33.62 | 37.34 | 39.35 | 39.90 | 39.40 | 37.84 | 34.74 | 29.78 | 23.96 | 20.39 |
| 40°  | 26.51 | 31.67 | 36.01 | 37.73 | 38.01 | 37.76 | 37.68 | 37.54 | 36.36 | 33.13 | 27.94 | 24.36 |
| 50°  | 29.73 | 34.45 | 37.31 | 37.02 | 35.66 | 34.67 | 34.95 | 36.16 | 36.87 | 35.46 | 31.08 | 27.58 |
| 60°  | 32.06 | 36.19 | 37.47 | 35.22 | 32.35 | 30.72 | 31.31 | 33.74 | 36.27 | 36.72 | 33.26 | 29.96 |
| 70°  | 33.41 | 36.82 | 36.50 | 32.37 | 28.18 | 26.02 | 26.88 | 30.37 | 34.57 | 36.87 | 34.44 | 31.44 |
| 90°  | 33.05 | 34.75 | 31.28 | 23.97 | 17.87 | 15.09 | 16.28 | 21.20 | 28.12 | 33.83 | 33.66 | 31.51 |
| Lat  | 55° | 55° | 55° | 55° | 55° | 55° | 55° | 55° | 55° | 55° | 55° | 55° |
| Tilt 0°  | 6.30  | 11.58 | 20.20 | 29.49 | 37.29 | 40.88 | 38.98 | 32.32 | 23.44 | 14.27 | 7.65  | 4.81  |
| 20°  | 17.08 | 22.77 | 29.81 | 35.64 | 39.63 | 41.15 | 40.23 | 36.95 | 31.75 | 25.04 | 18.61 | 15.09 |
| 30°  | 21.77 | 27.41 | 33.32 | 37.24 | 39.42 | 40.06 | 39.56 | 37.81 | 34.52 | 29.36 | 23.32 | 19.61 |
| 40°  | 25.80 | 31.22 | 35.82 | 37.74 | 38.19 | 38.02 | 37.90 | 37.62 | 36.24 | 32.78 | 27.32 | 23.54 |
| 50°  | 29.05 | 34.08 | 37.22 | 37.14 | 35.94 | 35.02 | 35.27 | 36.34 | 36.87 | 35.22 | 30.50 | 26.76 |
| 60°  | 31.41 | 35.91 | 37.50 | 35.45 | 32.72 | 31.15 | 31.71 | 34.03 | 36.38 | 36.58 | 32.75 | 29.16 |
| 70°  | 32.82 | 36.64 | 36.64 | 32.70 | 28.63 | 26.52 | 27.35 | 30.74 | 34.79 | 36.83 | 34.00 | 30.67 |
| 90°  | 32.63 | 34.78 | 31.64 | 24.47 | 18.43 | 15.66 | 16.85 | 21.72 | 28.53 | 34.00 | 33.40 | 30.89 |

| | Jan | Feb | Mar | Apr | May | Jun | Jul | Aug | Sep | Oct | Nov | Dec |
|---|---|---|---|---|---|---|---|---|---|---|---|---|
| Lat | 56° | 56° | 56° | 56° | 56° | 56° | 56° | 56° | 56° | 56° | 56° | 56° |
| Tilt 0° | 5.73 | 10.97 | 19.65 | 29.10 | 37.11 | 40.82 | 38.86 | 32.01 | 22.94 | 13.68 | 7.07 | 4.28 |
| 20° | 16.39 | 22.20 | 29.40 | 35.42 | 39.59 | 41.20 | 40.24 | 36.81 | 31.41 | 24.53 | 17.96 | 14.35 |
| 30° | 21.04 | 26.89 | 33.01 | 37.12 | 39.49 | 40.21 | 39.67 | 37.78 | 34.29 | 28.92 | 22.66 | 18.80 |
| 40° | 25.06 | 30.76 | 35.61 | 37.74 | 38.36 | 38.26 | 38.11 | 37.68 | 36.12 | 32.43 | 26.68 | 22.68 |
| 50° | 28.32 | 33.69 | 37.13 | 37.25 | 36.21 | 35.36 | 35.57 | 36.52 | 36.86 | 34.95 | 29.89 | 25.88 |
| 60° | 30.72 | 35.60 | 37.52 | 35.66 | 33.08 | 31.57 | 32.11 | 34.31 | 36.48 | 36.41 | 32.19 | 28.29 |
| 70° | 32.18 | 36.43 | 36.78 | 33.02 | 29.07 | 27.01 | 27.82 | 31.11 | 35.00 | 36.77 | 33.51 | 29.83 |
| 90° | 32.16 | 34.78 | 31.98 | 24.95 | 18.99 | 16.24 | 17.41 | 22.23 | 28.93 | 34.15 | 33.09 | 30.19 |
| Lat | 57° | 57° | 57° | 57° | 57° | 57° | 57° | 57° | 57° | 57° | 57° | 57° |
| Tilt 0° | 5.18 | 10.36 | 19.09 | 28.69 | 36.92 | 40.76 | 38.74 | 31.69 | 22.43 | 13.08 | 6.49 | 3.76 |
| 20° | 15.68 | 21.62 | 28.99 | 35.19 | 39.55 | 41.25 | 40.24 | 36.65 | 31.07 | 24.02 | 17.30 | 13.59 |
| 30° | 20.29 | 26.35 | 32.69 | 37.00 | 39.55 | 40.36 | 39.77 | 37.73 | 34.04 | 28.47 | 21.99 | 17.96 |
| 40° | 24.29 | 30.28 | 35.40 | 37.73 | 38.52 | 38.50 | 38.31 | 37.74 | 35.99 | 32.06 | 26.01 | 21.78 |
| 50° | 27.55 | 33.28 | 37.03 | 37.35 | 36.47 | 35.69 | 35.87 | 36.68 | 36.84 | 34.67 | 29.25 | 24.94 |
| 60° | 29.97 | 35.28 | 37.53 | 35.87 | 33.44 | 31.99 | 32.49 | 34.57 | 36.57 | 36.23 | 31.59 | 27.34 |
| 70° | 31.48 | 36.20 | 36.90 | 33.33 | 29.51 | 27.50 | 28.28 | 31.47 | 35.20 | 36.69 | 32.98 | 28.91 |
| 90° | 31.61 | 34.75 | 32.32 | 25.43 | 19.54 | 16.81 | 17.97 | 22.74 | 29.32 | 34.29 | 32.73 | 29.40 |
| Lat | 58° | 58° | 58° | 58° | 58° | 58° | 58° | 58° | 58° | 58° | 58° | 58° |
| Tilt 0° | 4.63 | 9.76 | 18.52 | 28.28 | 36.72 | 40.70 | 38.61 | 31.36 | 21.92 | 12.48 | 5.92 | 3.25 |
| 20° | 14.95 | 21.04 | 28.57 | 34.96 | 39.49 | 41.29 | 40.24 | 36.49 | 30.72 | 23.49 | 16.62 | 12.81 |
| 30° | 19.51 | 25.80 | 32.36 | 36.87 | 39.59 | 40.49 | 39.86 | 37.66 | 33.79 | 28.01 | 21.29 | 17.07 |
| 40° | 23.47 | 29.78 | 35.17 | 37.71 | 38.67 | 38.73 | 38.51 | 37.79 | 35.84 | 31.67 | 25.31 | 20.81 |
| 50° | 26.72 | 32.85 | 36.91 | 37.44 | 36.72 | 36.02 | 36.16 | 36.84 | 36.80 | 34.38 | 28.56 | 23.92 |
| 60° | 29.16 | 34.93 | 37.53 | 36.07 | 33.78 | 32.40 | 32.87 | 34.83 | 36.65 | 36.03 | 30.95 | 26.31 |
| 70° | 30.71 | 35.94 | 37.01 | 33.63 | 29.94 | 27.98 | 28.74 | 31.83 | 35.38 | 36.60 | 32.39 | 27.89 |
| 90° | 30.99 | 34.70 | 32.64 | 25.91 | 20.09 | 17.38 | 18.53 | 23.25 | 29.71 | 34.40 | 32.31 | 28.50 |
| Lat | 59° | 59° | 59° | 59° | 59° | 59° | 59° | 59° | 59° | 59° | 59° | 59° |
| Tilt 0° | 4.10 | 9.15 | 17.95 | 27.87 | 36.53 | 40.65 | 38.49 | 31.03 | 21.40 | 11.88 | 5.37 | 2.77 |
| 20° | 14.21 | 20.44 | 28.14 | 34.71 | 39.43 | 41.32 | 40.23 | 36.32 | 30.36 | 22.96 | 15.93 | 11.99 |
| 30° | 18.69 | 25.23 | 32.02 | 36.72 | 39.63 | 40.61 | 39.94 | 37.59 | 33.53 | 27.54 | 20.56 | 16.13 |
| 40° | 22.61 | 29.26 | 34.94 | 37.67 | 38.81 | 38.95 | 38.69 | 37.82 | 35.68 | 31.27 | 24.57 | 19.78 |
| 50° | 25.83 | 32.40 | 36.79 | 37.52 | 36.96 | 36.33 | 36.44 | 36.98 | 36.76 | 34.06 | 27.83 | 22.82 |
| 60° | 28.28 | 34.55 | 37.52 | 36.25 | 34.12 | 32.80 | 33.24 | 35.08 | 36.71 | 35.82 | 30.25 | 25.17 |
| 70° | 29.86 | 35.65 | 37.11 | 33.92 | 30.36 | 28.45 | 29.19 | 32.17 | 35.56 | 36.48 | 31.75 | 26.76 |
| 90° | 30.29 | 34.61 | 32.96 | 26.37 | 20.63 | 17.95 | 19.08 | 23.75 | 30.09 | 34.50 | 31.83 | 27.47 |
| Lat | 60° | 60° | 60° | 60° | 60° | 60° | 60° | 60° | 60° | 60° | 60° | 60° |
| Tilt 0° | 3.57 | 8.55 | 17.37 | 27.45 | 36.34 | 40.60 | 38.37 | 30.69 | 20.88 | 11.27 | 4.81 | 2.30 |
| 20° | 13.44 | 19.83 | 27.70 | 34.45 | 39.36 | 41.35 | 40.20 | 36.13 | 29.99 | 22.42 | 15.22 | 11.15 |
| 30° | 17.83 | 24.65 | 31.68 | 36.57 | 39.65 | 40.72 | 40.02 | 37.51 | 33.26 | 27.05 | 19.81 | 15.13 |
| 40° | 21.69 | 28.72 | 34.69 | 37.63 | 38.94 | 39.16 | 38.86 | 37.85 | 35.52 | 30.86 | 23.79 | 18.66 |
| 50° | 24.88 | 31.91 | 36.65 | 37.58 | 37.19 | 36.63 | 36.70 | 37.11 | 36.70 | 33.73 | 27.05 | 21.61 |
| 60° | 27.31 | 34.14 | 37.49 | 36.43 | 34.45 | 33.19 | 33.60 | 35.31 | 36.77 | 35.58 | 29.49 | 23.91 |
| 70° | 28.92 | 35.33 | 37.20 | 34.20 | 30.78 | 28.92 | 29.63 | 32.50 | 35.72 | 36.35 | 31.04 | 25.49 |
| 90° | 29.48 | 34.49 | 33.26 | 26.83 | 21.18 | 18.51 | 19.63 | 24.24 | 30.45 | 34.58 | 31.28 | 26.29 |
| Lat | 61° | 61° | 61° | 61° | 61° | 61° | 61° | 61° | 61° | 61° | 61° | 61° |
| Tilt 0° | 3.07 | 7.95 | 16.79 | 27.02 | 36.14 | 40.56 | 38.26 | 30.35 | 20.35 | 10.67 | 4.27 | 1.85 |
| 20° | 12.65 | 19.21 | 27.25 | 34.18 | 39.27 | 41.36 | 40.17 | 35.94 | 29.61 | 21.87 | 14.50 | 10.26 |
| 30° | 16.93 | 24.05 | 31.32 | 36.40 | 39.67 | 40.83 | 40.08 | 37.42 | 32.98 | 26.56 | 19.02 | 14.06 |
| 40° | 20.70 | 28.15 | 34.43 | 37.57 | 39.06 | 39.36 | 39.02 | 37.86 | 35.34 | 30.43 | 22.97 | 17.43 |
| 50° | 23.84 | 31.40 | 36.50 | 37.64 | 37.41 | 36.93 | 36.96 | 37.23 | 36.63 | 33.39 | 26.22 | 20.28 |
| 60° | 26.25 | 33.70 | 37.46 | 36.59 | 34.76 | 33.57 | 33.95 | 35.54 | 36.81 | 35.32 | 28.67 | 22.50 |
| 70° | 27.87 | 34.97 | 37.28 | 34.47 | 31.18 | 29.38 | 30.06 | 32.83 | 35.88 | 36.19 | 30.25 | 24.05 |
| 90° | 28.54 | 34.32 | 33.56 | 27.28 | 21.71 | 19.07 | 20.18 | 24.72 | 30.81 | 34.63 | 30.64 | 24.92 |

(Continued)

|        | Jan | Feb | Mar | Apr | May | Jun | Jul | Aug | Sep | Oct | Nov | Dec |
|--------|-----|-----|-----|-----|-----|-----|-----|-----|-----|-----|-----|-----|
| Lat    | 62° | 62° | 62° | 62° | 62° | 62° | 62° | 62° | 62° | 62° | 62° | 62° |
| Tilt 0° | 2.58 | 7.35 | 16.20 | 26.59 | 35.95 | 40.54 | 38.15 | 30.01 | 19.81 | 10.07 | 3.75 | 1.43 |
| 20°    | 11.82 | 18.58 | 26.80 | 33.91 | 39.18 | 41.37 | 40.13 | 35.74 | 29.23 | 21.31 | 13.75 | 9.31 |
| 30°    | 15.97 | 23.43 | 30.95 | 36.23 | 39.67 | 40.92 | 40.13 | 37.32 | 32.68 | 26.05 | 18.19 | 12.89 |
| 40°    | 19.63 | 27.56 | 34.16 | 37.50 | 39.16 | 39.55 | 39.17 | 37.87 | 35.15 | 29.99 | 22.09 | 16.07 |
| 50°    | 22.70 | 30.86 | 36.34 | 37.68 | 37.62 | 37.21 | 37.21 | 37.35 | 36.55 | 33.02 | 25.31 | 18.77 |
| 60°    | 25.08 | 33.22 | 37.41 | 36.75 | 35.07 | 33.95 | 34.29 | 35.76 | 36.84 | 35.05 | 27.77 | 20.89 |
| 70°    | 26.69 | 34.58 | 37.34 | 34.73 | 31.58 | 29.83 | 30.48 | 33.14 | 36.02 | 36.01 | 29.38 | 22.39 |
| 90°    | 27.47 | 34.12 | 33.84 | 27.72 | 22.24 | 19.63 | 20.73 | 25.20 | 31.16 | 34.66 | 29.90 | 23.30 |
|        |     |     |     |     |     |     |     |     |     |     |     |     |
| Lat    | 63° | 63° | 63° | 63° | 63° | 63° | 63° | 63° | 63° | 63° | 63° | 63° |
| Tilt 0° | 2.11 | 6.76 | 15.61 | 26.15 | 35.77 | 40.53 | 38.05 | 29.66 | 19.27 | 9.46 | 3.23 | 1.04 |
| 20°    | 10.95 | 17.93 | 26.33 | 33.62 | 39.08 | 41.36 | 40.08 | 35.53 | 28.83 | 20.74 | 12.97 | 8.28 |
| 30°    | 14.93 | 22.78 | 30.57 | 36.04 | 39.67 | 41.00 | 40.17 | 37.20 | 32.38 | 25.52 | 17.32 | 11.58 |
| 40°    | 18.47 | 26.94 | 33.80 | 37.43 | 39.26 | 39.73 | 39.31 | 37.86 | 34.95 | 29.52 | 21.15 | 14.53 |
| 50°    | 21.44 | 30.28 | 36.16 | 37.71 | 37.82 | 37.49 | 37.45 | 37.45 | 36.46 | 32.63 | 24.33 | 17.04 |
| 60°    | 23.76 | 32.71 | 37.35 | 36.89 | 35.37 | 34.31 | 34.63 | 35.96 | 36.87 | 34.74 | 26.77 | 19.03 |
| 70°    | 25.35 | 34.13 | 37.39 | 34.98 | 31.97 | 30.28 | 30.90 | 33.45 | 36.15 | 35.80 | 28.40 | 20.44 |
| 90°    | 26.21 | 33.86 | 34.11 | 28.16 | 22.77 | 20.19 | 21.27 | 25.67 | 31.50 | 34.66 | 29.05 | 21.37 |
|        |     |     |     |     |     |     |     |     |     |     |     |     |
| Lat    | 64° | 64° | 64° | 64° | 64° | 64° | 64° | 64° | 64° | 64° | 64° | 64° |
| Tilt 0° | 1.66 | 6.17 | 15.01 | 25.71 | 35.59 | 40.55 | 37.97 | 29.32 | 18.72 | 8.86 | 2.73 | 0.68 |
| 20°    | 10.02 | 17.27 | 25.86 | 33.32 | 38.97 | 41.36 | 40.03 | 35.31 | 28.43 | 20.17 | 12.16 | 7.14 |
| 30°    | 13.81 | 22.12 | 30.19 | 35.85 | 39.65 | 41.08 | 40.20 | 37.08 | 32.07 | 24.98 | 16.39 | 10.09 |
| 40°    | 17.17 | 26.29 | 33.59 | 37.34 | 39.35 | 39.90 | 39.44 | 37.84 | 34.74 | 29.04 | 20.12 | 12.73 |
| 50°    | 20.02 | 29.67 | 35.98 | 37.73 | 38.01 | 37.76 | 37.68 | 37.54 | 36.36 | 32.22 | 23.24 | 14.99 |
| 60°    | 22.25 | 32.14 | 37.27 | 37.02 | 35.66 | 34.67 | 34.95 | 36.16 | 36.87 | 34.42 | 25.66 | 16.79 |
| 70°    | 23.81 | 33.64 | 37.44 | 35.22 | 32.35 | 30.72 | 31.31 | 33.74 | 36.27 | 35.57 | 27.29 | 18.08 |
| 90°    | 24.73 | 33.55 | 34.37 | 28.59 | 23.29 | 20.74 | 21.80 | 26.14 | 31.83 | 34.63 | 28.05 | 18.99 |

Daily Average Total Extraterrestrial Radiation on Tilted Surfaces Facing North, MJ/m² (Southern Latitudes)

| Month<br>Day No. | Jan<br>17. | Feb<br>46. | Mar<br>76. | Apr<br>105. | May<br>135. | Jun<br>162. | Jul<br>199. | Aug<br>229. | Sep<br>258. | Oct<br>288. | Nov<br>318. | Dec<br>346. |
|---|---|---|---|---|---|---|---|---|---|---|---|---|
| S. Lat | 30° | 30° | 30° | 30° | 30° | 30° | 30° | 30° | 30° | 30° | 30° | 30° |
| Tilt 0° | 42.45 | 39.06 | 33.33 | 26.83 | 21.18 | 18.51 | 19.63 | 24.24 | 30.45 | 36.72 | 41.20 | 43.25 |
| 20° | 39.15 | 38.90 | 37.25 | 34.12 | 30.45 | 28.43 | 29.22 | 32.34 | 35.72 | 38.01 | 38.83 | 38.94 |
| 30° | 35.94 | 37.14 | 37.54 | 36.25 | 33.77 | 32.17 | 32.75 | 34.97 | 36.75 | 36.98 | 36.05 | 35.29 |
| 40° | 31.79 | 34.30 | 36.68 | 37.29 | 36.06 | 34.93 | 35.29 | 36.54 | 36.67 | 34.84 | 32.30 | 30.76 |
| 50° | 26.84 | 30.48 | 34.71 | 37.19 | 37.26 | 36.62 | 36.75 | 37.00 | 35.48 | 31.68 | 27.70 | 25.51 |
| 60° | 21.27 | 25.81 | 31.69 | 35.96 | 37.32 | 37.21 | 37.10 | 36.34 | 33.21 | 27.59 | 22.40 | 19.72 |
| 70° | 15.29 | 20.45 | 27.71 | 33.64 | 36.25 | 36.66 | 36.32 | 34.57 | 29.93 | 22.71 | 16.61 | 13.64 |
| 90° | 3.57 | 8.55 | 17.37 | 26.03 | 30.87 | 32.28 | 31.50 | 27.97 | 20.76 | 11.27 | 4.81 | 2.30 |
| S. Lat | 32° | 32° | 32° | 32° | 32° | 32° | 32° | 32° | 32° | 32° | 32° | 32° |
| Tilt 0° | 42.55 | 38.83 | 32.71 | 25.91 | 20.09 | 17.38 | 18.53 | 23.25 | 29.71 | 36.34 | 41.20 | 43.46 |
| 20° | 39.67 | 39.12 | 37.06 | 33.54 | 29.60 | 27.47 | 28.31 | 31.66 | 35.38 | 38.08 | 39.26 | 39.55 |
| 30° | 36.66 | 37.58 | 37.57 | 35.88 | 33.08 | 31.33 | 31.99 | 34.48 | 36.63 | 37.27 | 36.69 | 36.09 |
| 40° | 32.69 | 34.95 | 36.94 | 37.13 | 35.55 | 34.24 | 34.69 | 36.25 | 36.78 | 35.35 | 33.13 | 31.73 |
| 50° | 27.89 | 31.32 | 35.19 | 37.25 | 36.93 | 36.11 | 36.33 | 36.92 | 35.80 | 32.39 | 28.68 | 26.61 |
| 60° | 22.42 | 26.81 | 32.38 | 36.24 | 37.20 | 36.89 | 36.87 | 36.47 | 33.74 | 28.48 | 23.50 | 20.91 |
| 70° | 16.51 | 21.57 | 28.58 | 34.13 | 36.34 | 36.54 | 36.30 | 34.91 | 30.65 | 23.74 | 17.79 | 14.86 |
| 90° | 4.63 | 9.76 | 18.52 | 26.89 | 31.36 | 32.56 | 31.88 | 28.69 | 21.81 | 12.48 | 5.92 | 3.25 |
| S. Lat | 34° | 34° | 34° | 34° | 34° | 34° | 34° | 34° | 34° | 34° | 34° | 34° |
| Tilt 0° | 42.60 | 38.56 | 32.04 | 24.95 | 18.99 | 16.24 | 17.41 | 22.23 | 28.93 | 35.93 | 41.16 | 43.63 |
| 20° | 40.15 | 39.30 | 36.82 | 32.93 | 28.72 | 26.47 | 27.37 | 30.93 | 34.99 | 38.11 | 39.65 | 40.13 |
| 30° | 37.34 | 37.98 | 37.56 | 35.47 | 32.34 | 30.45 | 31.18 | 33.93 | 36.47 | 37.52 | 37.29 | 36.86 |
| 40° | 33.56 | 35.56 | 37.16 | 36.93 | 34.98 | 33.51 | 34.03 | 35.90 | 36.83 | 35.82 | 33.91 | 32.67 |
| 50° | 28.91 | 32.12 | 35.63 | 37.27 | 36.56 | 35.54 | 35.86 | 36.78 | 36.08 | 33.06 | 29.63 | 27.69 |
| 60° | 23.56 | 27.77 | 33.02 | 36.48 | 37.03 | 36.50 | 36.59 | 36.54 | 34.23 | 29.33 | 24.59 | 22.08 |
| 70° | 17.71 | 22.66 | 29.41 | 34.58 | 36.37 | 36.35 | 36.21 | 35.19 | 31.35 | 24.75 | 18.97 | 16.09 |
| 90° | 5.73 | 10.97 | 19.65 | 27.72 | 31.80 | 32.77 | 32.20 | 29.36 | 22.83 | 13.68 | 7.07 | 4.28 |
| S. Lat | 36° | 36° | 36° | 36° | 36° | 36° | 36° | 36° | 36° | 36° | 36° | 36° |
| Tilt 0° | 42.62 | 38.24 | 31.34 | 23.97 | 17.87 | 15.09 | 16.28 | 21.20 | 28.12 | 35.47 | 41.08 | 43.75 |
| 20° | 40.59 | 39.43 | 36.53 | 32.28 | 27.80 | 25.44 | 26.40 | 30.18 | 34.57 | 38.09 | 40.00 | 40.66 |
| 30° | 37.99 | 38.33 | 37.50 | 35.01 | 31.56 | 29.53 | 30.33 | 33.35 | 36.26 | 37.73 | 37.85 | 37.59 |
| 40° | 34.39 | 36.13 | 37.33 | 36.69 | 34.37 | 32.72 | 33.34 | 35.51 | 36.85 | 36.25 | 34.66 | 33.58 |
| 50° | 29.90 | 32.89 | 36.03 | 37.25 | 36.14 | 34.92 | 35.33 | 36.60 | 36.32 | 33.70 | 30.55 | 28.74 |
| 60° | 24.68 | 28.71 | 33.63 | 36.70 | 36.80 | 36.25 | 36.25 | 36.57 | 34.68 | 30.15 | 25.65 | 23.24 |
| 70° | 18.91 | 23.74 | 30.21 | 34.99 | 36.35 | 36.10 | 36.07 | 35.43 | 32.00 | 25.72 | 20.13 | 17.30 |
| 90° | 6.87 | 12.18 | 20.75 | 28.51 | 32.18 | 32.92 | 32.46 | 29.99 | 23.82 | 14.86 | 8.23 | 5.36 |
| S. Lat | 38° | 38° | 38° | 38° | 38° | 38° | 38° | 38° | 38° | 38° | 38° | 38° |
| Tilt 0° | 42.59 | 37.89 | 30.59 | 22.96 | 16.74 | 13.94 | 15.14 | 20.14 | 27.27 | 34.97 | 40.95 | 43.84 |
| 20° | 40.98 | 39.51 | 36.21 | 31.58 | 26.85 | 24.38 | 25.40 | 29.38 | 34.10 | 38.03 | 40.30 | 41.16 |
| 30° | 38.59 | 38.64 | 37.40 | 34.51 | 30.75 | 28.57 | 29.44 | 32.73 | 36.00 | 37.89 | 38.36 | 38.28 |
| 40° | 35.18 | 36.65 | 37.46 | 36.39 | 33.71 | 31.88 | 32.59 | 35.08 | 36.82 | 36.64 | 35.38 | 34.45 |
| 50° | 30.86 | 33.61 | 36.38 | 37.17 | 35.65 | 34.23 | 34.74 | 36.36 | 36.51 | 34.29 | 31.44 | 29.77 |
| 60° | 25.77 | 29.61 | 34.19 | 36.81 | 36.51 | 35.54 | 35.85 | 36.54 | 35.09 | 30.93 | 26.69 | 24.39 |
| 70° | 20.10 | 24.79 | 30.97 | 35.34 | 36.26 | 35.77 | 35.86 | 35.61 | 32.61 | 26.67 | 21.27 | 18.52 |
| 90° | 8.04 | 13.39 | 21.83 | 29.25 | 32.50 | 32.99 | 32.65 | 30.57 | 24.78 | 16.04 | 9.41 | 6.48 |

## APPENDIX A

**TABLE A-15.** Input Data for Solar Systems.

Table A-15 has been prepared especially for this book under a contract with the National Climatic Center under the direction of Vincent Cinquemani. Day temperatures have been estimated as 0.3 times the average daily maximum plus 0.7 times the average daily temperature, and temperatures converted to SI units. Latitude and longitude are expressed as degrees and minutes, and station heights above sea level are given in meters.

The data base for these tables is a report titled "Input Data for Solar Systems" by V. Cinquemani, J. R. Owenby, Jr. and R. G. Baldwin, of the National Climatic Center, Asheville, North Carolina. That document was prepared by the Department of Commerce for the Department of Energy in November 1978 under Interagency Agreement No. E(49-26)-1041. Radiation data from 26 rehabilitated data stations (see Chapter 3) was extrapolated by cloud cover measurements and statistical correlations to 222 more sites. The result is the most comprehensive weather and radiation information ever available for the United States. At this writing government personnel in the Department of Commerce and Department of Energy contractors are evaluating the probable errors in the radiation estimates, which are generally about ten percent lower than previously available data.

**TABLE A-15. Radiation and Weather Data**

| | | JAN | FEB | MAR | APR | MAY | JUN | JUL | AUG | SEP | OCT | NOV | DEC | ANNUAL |
|---|---|---|---|---|---|---|---|---|---|---|---|---|---|---|
| **AK HOMER** | | | 5938N | 15130W | | 22 M | | | | | | | | |
| DAILY RAD | MJ/SQ M | 1.38 | 3.79 | 8.62 | 14.17 | 17.96 | 19.87 | 18.14 | 13.49 | 8.98 | 4.96 | 1.99 | .73 | 9.51 |
| DAY TEMP | DEG C | -4.8 | -2.8 | -1.2 | 2.9 | 7.1 | 10.6 | 12.6 | 12.6 | 9.6 | 4.2 | -1.1 | -4.9 | 3.7 |
| DEG DAY | C-DAY | 751 | 624 | 644 | 500 | 391 | 272 | 219 | 217 | 300 | 476 | 613 | 751 | 5758 |
| AVG TEMP | DEG C | -5.9 | -3.9 | -2.4 | 1.7 | 5.7 | 9.3 | 11.3 | 11.3 | 8.3 | 3.0 | -2.1 | -5.9 | 2.5 |
| **AL BIRMINGHAM** | | | 3334N | 8645W | | 192 M | | | | | | | | |
| DAILY RAD | MJ/SQ M | 8.02 | 10.97 | 14.71 | 18.99 | 21.07 | 21.77 | 20.54 | 19.56 | 16.51 | 13.74 | 9.74 | 7.51 | 15.26 |
| DAY TEMP | DEG C | 8.5 | 10.1 | 13.7 | 19.3 | 23.4 | 27.1 | 28.3 | 28.0 | 25.1 | 19.5 | 13.1 | 9.0 | 18.8 |
| DEG DAY | C-DAY | 363 | 287 | 216 | 64 | 11 | 0 | 0 | 0 | 3 | 76 | 217 | 341 | 1578 |
| AVG TEMP | DEG C | 6.8 | 8.3 | 11.8 | 17.3 | 21.4 | 25.2 | 26.6 | 26.2 | 23.3 | 17.4 | 11.2 | 7.3 | 16.9 |
| **AL MOBILE** | | | 3041N | 8815N | | 67 M | | | | | | | | |
| DAILY RAD | MJ/SQ M | 9.40 | 12.48 | 15.97 | 19.54 | 21.25 | 21.20 | 19.47 | 18.63 | 16.45 | 14.74 | 10.84 | 8.62 | 15.72 |
| DAY TEMP | DEG C | 12.3 | 13.9 | 16.9 | 21.6 | 25.5 | 28.4 | 29.0 | 29.0 | 26.8 | 22.3 | 16.6 | 13.3 | 21.3 |
| DEG DAY | C-DAY | 251 | 187 | 123 | 22 | 0 | 0 | 0 | 0 | 0 | 22 | 117 | 214 | 936 |
| AVG TEMP | DEG C | 10.7 | 12.2 | 15.2 | 19.9 | 23.8 | 26.8 | 27.6 | 27.5 | 25.3 | 20.5 | 14.7 | 11.6 | 19.7 |
| **AL MONTGOMERY** | | | 3218N | 8624W | | 62 M | | | | | | | | |
| DAILY RAD | MJ/SQ M | 8.53 | 11.50 | 15.21 | 19.62 | 21.53 | 22.38 | 20.89 | 19.81 | 16.66 | 14.32 | 10.39 | 8.16 | 15.75 |
| DAY TEMP | DEG C | 10.3 | 12.1 | 15.5 | 20.4 | 24.3 | 27.8 | 28.8 | 28.7 | 26.2 | 20.8 | 14.8 | 11.0 | 20.1 |
| DEG DAY | C-DAY | 309 | 233 | 166 | 42 | 0 | 0 | 0 | 0 | 0 | 52 | 170 | 284 | 1260 |
| AVG TEMP | DEG C | 8.6 | 10.3 | 13.6 | 18.4 | 22.4 | 26.1 | 27.2 | 27.1 | 24.4 | 18.8 | 12.8 | 9.2 | 18.2 |
| **AR FQRT SMITH** | | | 3520N | 9422W | | 141 M | | | | | | | | |
| DAILY RAD | MJ/SQ M | 8.44 | 11.34 | 14.89 | 18.34 | 21.70 | 23.71 | 23.44 | 21.31 | 17.04 | 13.63 | 9.66 | 7.73 | 15.94 |
| DAY TEMP | DEG C | 5.7 | 8.2 | 12.1 | 18.8 | 23.0 | 27.4 | 29.8 | 29.5 | 25.4 | 19.5 | 12.3 | 7.1 | 18.2 |
| DEG DAY | C-DAY | 448 | 338 | 262 | 73 | 9 | 0 | 0 | 0 | 3 | 75 | 243 | 405 | 1853 |
| AVG TEMP | DEG C | 3.9 | 6.3 | 10.2 | 16.8 | 21.2 | 25.6 | 27.9 | 27.4 | 23.3 | 17.3 | 10.2 | 5.3 | 16.3 |
| **AR LITTLE ROCK** | | | 3444N | 9214W | | 81 M | | | | | | | | |
| DAILY RAD | MJ/SQ M | 8.30 | 11.38 | 14.90 | 18.28 | 21.89 | 23.91 | 23.06 | 21.11 | 17.23 | 13.94 | 9.61 | 7.65 | 15.94 |
| DAY TEMP | DEG C | 5.9 | 7.9 | 12.1 | 18.5 | 22.9 | 27.5 | 29.3 | 29.0 | 25.0 | 19.2 | 12.2 | 7.1 | 18.0 |
| DEG DAY | C-DAY | 439 | 344 | 261 | 77 | 9 | 0 | 0 | 0 | 3 | 79 | 245 | 403 | 1863 |
| AVG TEMP | DEG C | 4.2 | 6.1 | 10.2 | 16.5 | 21.0 | 25.6 | 27.4 | 27.0 | 22.9 | 16.9 | 10.2 | 5.3 | 16.1 |
| **AZ PHOENIX** | | | 3326N | 11201W | | 339 M | | | | | | | | |
| DAILY RAD | MJ/SQ M | 11.59 | 15.59 | 20.59 | 26.72 | 30.37 | 31.09 | 28.22 | 26.02 | 22.87 | 17.89 | 13.06 | 10.58 | 21.22 |
| DAY TEMP | DEG C | 12.9 | 15.2 | 17.9 | 22.5 | 27.4 | 32.0 | 35.2 | 33.9 | 31.2 | 24.9 | 17.9 | 13.7 | 23.7 |
| DEG DAY | C-DAY | 238 | 162 | 103 | 33 | 0 | 0 | 0 | 0 | 0 | 0 | 101 | 216 | 862 |
| AVG TEMP | DEG C | 10.7 | 12.8 | 15.4 | 19.8 | 24.6 | 29.2 | 32.9 | 31.7 | 28.8 | 22.3 | 15.4 | 11.4 | 21.3 |
| **AZ PRESCOTT** | | | 3439N | 11226W | | 1531 M | | | | | | | | |
| DAILY RAD | MJ/SQ M | 11.53 | 15.15 | 20.17 | 25.82 | 29.84 | 31.34 | 26.21 | 23.74 | 22.18 | 17.51 | 12.94 | 10.52 | 20.58 |
| DAY TEMP | DEG C | | | | | | | | | | | | | |
| DEG DAY | C-DAY | 481 | 381 | 357 | 219 | 92 | 18 | 0 | 0 | 13 | 141 | 320 | 454 | 2476 |
| AVG TEMP | DEG C | 2.8 | 4.7 | 6.8 | 11.1 | 15.8 | 20.6 | 24.2 | 22.8 | 20.1 | 14.0 | 7.7 | 3.7 | 12.9 |

(Continued)

## TABLE A-15. (Continued)

Coordinates shown for each station: latitude (FEB column), longitude (MAR column), elevation in M (MAY column).

| | | JAN | FEB | MAR | APP | MAY | JUN | JUL | AUG | SEP | OCT | NOV | DEC | ANNUAL |
|---|---|---|---|---|---|---|---|---|---|---|---|---|---|---|
| **AZ TUCSON** | | | 3207N | 11056W | | 779 M | | | | | | | | |
| DAILY RAD | MJ/SQ M | 12.47 | 16.25 | 21.16 | 26.82 | 30.32 | 30.98 | 26.57 | 24.77 | 22.46 | 18.18 | 13.71 | 11.30 | 21.25 |
| DAY TEMP | DEG C | 12.6 | 14.2 | 16.5 | 21.1 | 25.8 | 30.5 | 32.2 | 30.7 | 28.9 | 23.5 | 17.0 | 13.2 | 22.2 |
| DEG DAY | C-DAY | 246 | 185 | 135 | 45 | 0 | 0 | 0 | 0 | 0 | 16 | 123 | 224 | 974 |
| AVG TEMP | DEG C | 10.5 | 11.9 | 14.2 | 18.6 | 23.1 | 27.8 | 30.2 | 28.8 | 26.7 | 21.2 | 14.7 | 11.1 | 19.9 |
| **AZ WINSLOW** | | | 3501N | 11044W | | 1488 M | | | | | | | | |
| DAILY RAD | MJ/SQ M | 11.17 | 15.06 | 20.07 | 25.91 | 29.45 | 30.77 | 26.63 | 24.29 | 21.68 | 17.17 | 12.70 | 10.15 | 20.45 |
| DAY TEMP | DEG C | 2.5 | 6.3 | 9.7 | 14.6 | 19.9 | 25.1 | 28.3 | 26.9 | 23.5 | 16.7 | 8.7 | 3.2 | 15.5 |
| DEG DAY | C-DAY | 558 | 403 | 348 | 193 | 69 | 8 | 0 | 0 | 11 | 140 | 363 | 537 | 2630 |
| AVG TEMP | DEG C | .3 | 3.9 | 7.1 | 12.1 | 17.1 | 22.1 | 25.7 | 24.5 | 20.8 | 14.1 | 6.2 | 1.0 | 12.9 |
| **AZ YUMA** | | | 3240N | 11436W | | 63 M | | | | | | | | |
| DAILY RAD | MJ/SQ M | 12.44 | 16.38 | 21.78 | 27.36 | 30.96 | 31.93 | 27.84 | 26.43 | 23.28 | 18.42 | 13.79 | 11.35 | 21.83 |
| DAY TEMP | DEG C | 15.0 | 17.4 | 20.0 | 24.2 | 28.4 | 32.4 | 36.3 | 35.7 | 32.8 | 26.7 | 19.7 | 15.5 | 25.3 |
| DEG DAY | C-DAY | 171 | 107 | 54 | 13 | 0 | 0 | 0 | 0 | 3 | 0 | 60 | 153 | 561 |
| AVG TEMP | DEG C | 13.0 | 15.2 | 17.7 | 21.8 | 25.9 | 29.9 | 34.3 | 33.8 | 30.6 | 24.4 | 17.5 | 13.5 | 23.1 |
| **CA ARCATA** | | | 4059N | 12406W | | 69 M | | | | | | | | |
| DAILY RAD | MJ/SQ M | 6.00 | 9.00 | 12.86 | 18.01 | 20.91 | 22.26 | 20.52 | 17.92 | 15.23 | 10.62 | 6.73 | 5.33 | 13.78 |
| DAY TEMP | DEG C | | | | | | | | | | | | | |
| DEG DAY | C-DAY | | | | | | | | | | | | | |
| AVG TEMP | DEG C | | | | | | | | | | | | | |
| **CA BAKERSFIELD** | | | 3525N | 11903W | | 150 M | | | | | | | | |
| DAILY RAD | MJ/SQ M | 8.70 | 12.50 | 18.10 | 23.77 | 28.48 | 31.20 | 30.45 | 27.47 | 22.60 | 16.55 | 10.69 | 7.69 | 19.85 |
| DAY TEMP | DEG C | 10.3 | 13.1 | 15.7 | 19.2 | 23.1 | 27.4 | 31.4 | 30.0 | 27.0 | 21.7 | 15.3 | 10.4 | 20.4 |
| DEG DAY | C-DAY | 302 | 196 | 148 | 78 | 12 | 0 | 0 | 0 | 0 | 31 | 153 | 294 | 1214 |
| AVG TEMP | DEG C | 8.6 | 11.3 | 13.7 | 17.1 | 21.0 | 24.9 | 28.8 | 27.6 | 24.8 | 19.4 | 13.3 | 8.8 | 18.3 |
| **CA CHINA LAKE** | | | 3541N | 11741W | | 681 M | | | | | | | | |
| DAILY RAD | MJ/SQ M | 10.32 | 13.95 | 19.69 | 25.35 | 28.92 | 31.17 | 29.65 | 29.69 | 22.47 | 16.71 | 11.73 | 9.54 | 20.77 |
| DAY TEMP | DEG C | | | | | | | | | | | | | |
| DEG DAY | C-DAY | | | | | | | | | | | | | |
| AVG TEMP | DEG C | | | | | | | | | | | | | |
| **CA DAGGETT** | | | 3452N | 11647W | | 568 M | | | | | | | | |
| DAILY RAD | MJ/SQ M | 10.87 | 14.53 | 20.11 | 25.81 | 29.41 | 31.39 | 29.55 | 27.04 | 22.79 | 17.20 | 12.31 | 9.94 | 20.91 |
| DAY TEMP | DEG C | | | | | | | | | | | | | |
| DEG DAY | C-DAY | 305 | 206 | 151 | 66 | 8 | 0 | 0 | 0 | 0 | 32 | 164 | 293 | 1225 |
| AVG TEMP | DEG C | 8.5 | 11.1 | 13.7 | 17.9 | 22.4 | 26.7 | 30.7 | 29.7 | 26.2 | 20.1 | 13.1 | 8.9 | 19.1 |
| **CA EL TORO** | | | 3390N | 11744W | | 116 M | | | | | | | | |
| DAILY RAD | MJ/SQ M | 10.75 | 14.03 | 18.27 | 21.89 | 23.49 | 24.90 | 26.82 | 24.46 | 19.72 | 15.40 | 11.65 | 9.86 | 18.44 |
| DAY TEMP | DEG C | | | | | | | | | | | | | |
| DEG DAY | C-DAY | | | | | | | | | | | | | |
| AVG TEMP | DEG C | | | | | | | | | | | | | |

| | | JAN | FEB | MAR | APR | MAY | JUN | JUL | AUG | SEP | OCT | NOV | DEC | ANNUAL |
|---|---|---|---|---|---|---|---|---|---|---|---|---|---|---|
| **CA FRESNO** | | | 3646N | 11943W | | 100 M | | | | | | | | |
| DAILY RAD | MJ/SQ M | 7.45 | 11.49 | 17.77 | 23.75 | 28.19 | 31.01 | 30.47 | 27.50 | 22.53 | 16.22 | 10.08 | 6.52 | 19.42 |
| DAY TEMP | DEG C | 9.0 | 11.8 | 14.3 | 18.1 | 22.2 | 26.0 | 29.9 | 28.7 | 26.1 | 20.5 | 14.0 | 9.1 | 19.1 |
| DEG DAY | C-DAY | 339 | 235 | 191 | 101 | 28 | 5 | 0 | 0 | 0 | 50 | 192 | 331 | 1472 |
| AVG TEMP | DEG C | 7.4 | 9.9 | 12.2 | 15.7 | 19.7 | 23.3 | 27.0 | 25.7 | 23.2 | 17.9 | 11.9 | 7.7 | 16.8 |
| **CA LONG BEACH** | | | 3349N | 11809W | | 17 M | | | | | | | | |
| DAILY RAD | MJ/SQ M | 10.53 | 13.79 | 18.27 | 21.99 | 23.43 | 24.29 | 26.10 | 23.83 | 19.30 | 15.05 | 11.39 | 9.61 | 18.13 |
| DAY TEMP | DEG C | 14.2 | 14.8 | 15.7 | 17.5 | 19.5 | 21.2 | 24.0 | 24.7 | 24.0 | 21.3 | 17.9 | 15.0 | 19.2 |
| DEG DAY | C-DAY | 188 | 152 | 137 | 82 | 39 | 13 | 0 | 4 | 4 | 27 | 86 | 164 | 892 |
| AVG TEMP | DEG C | 12.3 | 13.1 | 14.0 | 15.9 | 17.8 | 19.6 | 22.3 | 22.9 | 22.1 | 19.4 | 15.9 | 13.1 | 17.4 |
| **CA LOS ANGELES** | | | 3356N | 11824W | | 32 M | | | | | | | | |
| DAILY RAD | MJ/SQ M | 10.51 | 13.78 | 18.37 | 22.14 | 23.37 | 24.05 | 26.19 | 23.60 | 19.08 | 14.95 | 11.39 | 9.63 | 18.09 |
| DAY TEMP | DEG C | 14.0 | 14.5 | 14.8 | 16.1 | 17.7 | 19.0 | 21.3 | 21.9 | 21.6 | 19.7 | 17.3 | 15.4 | 17.8 |
| DEG DAY | C-DAY | 184 | 150 | 148 | 108 | 63 | 39 | 11 | 8 | 13 | 43 | 88 | 155 | 1010 |
| AVG TEMP | DEG C | 12.5 | 13.1 | 13.6 | 14.9 | 16.6 | 18.1 | 20.3 | 20.8 | 20.4 | 18.4 | 15.8 | 13.8 | 16.5 |
| **CA MOUNT SHASTA** | | | 4119N | 12219W | | 1093 M | | | | | | | | |
| DAILY RAD | MJ/SQ M | 6.36 | 9.73 | 14.19 | 19.93 | 24.81 | 27.65 | 29.25 | 25.11 | 19.69 | 13.11 | 7.48 | 5.73 | 16.92 |
| DAY TEMP | DEG C | 2.3 | 4.8 | 6.4 | 10.0 | 14.0 | 17.9 | 22.7 | 21.8 | 19.0 | 13.0 | 7.0 | 3.3 | 11.9 |
| DEG DAY | C-DAY | 541 | 423 | 424 | 312 | 206 | 99 | 21 | 36 | 81 | 234 | 388 | 508 | 3273 |
| AVG TEMP | DEG C | .9 | 3.2 | 4.7 | 7.9 | 11.8 | 15.6 | 19.9 | 18.9 | 16.2 | 10.8 | 5.4 | 1.9 | 9.8 |
| **CA NEEDLES** | | | 3446N | 11437W | | 270 M | | | | | | | | |
| DAILY RAD | MJ/SQ M | 11.18 | 15.36 | 20.71 | 26.29 | 30.09 | 31.68 | 28.84 | 25.85 | 22.86 | 17.45 | 12.75 | 10.36 | 21.12 |
| DAY TEMP | DEG C | | | | | | | | | | | | | |
| DEG DAY | C-DAY | 234 | 145 | 63 | 23 | 0 | 0 | 0 | 0 | 0 | 6 | 91 | 212 | 794 |
| AVG TEMP | DEG C | 10.9 | 13.6 | 16.4 | 21.3 | 26.4 | 31.3 | 35.2 | 34.1 | 30.5 | 23.5 | 15.9 | 11.5 | 22.6 |
| **CA OAKLAND** | | | 3744N | 12212W | | 2 M | | | | | | | | |
| DAILY RAD | MJ/SQ M | 8.03 | 11.55 | 16.53 | 21.81 | 25.10 | 26.67 | 26.36 | 23.29 | 19.31 | 13.75 | 9.33 | 7.34 | 17.42 |
| DAY TEMP | DEG C | 10.2 | 12.1 | 13.1 | 14.5 | 16.0 | 17.7 | 18.4 | 18.6 | 19.4 | 17.4 | 14.1 | 10.9 | 15.2 |
| DEG DAY | C-DAY | 282 | 204 | 194 | 150 | 107 | 63 | 44 | 41 | 33 | 75 | 162 | 260 | 1615 |
| AVG TEMP | DEG C | 9.2 | 11.1 | 12.1 | 13.4 | 14.9 | 16.6 | 17.3 | 17.5 | 18.1 | 16.2 | 12.9 | 9.9 | 14.1 |
| **CA POINT MUGU** | | | 3407N | 11907W | | 4 M | | | | | | | | |
| DAILY RAD | MJ/SQ M | 10.52 | 13.84 | 18.56 | 22.14 | 22.90 | 23.32 | 24.04 | 21.96 | 18.25 | 14.71 | 11.42 | 9.72 | 17.62 |
| DAY TEMP | DEG C | | | | | | | | | | | | | |
| DEG DAY | C-DAY | | | | | | | | | | | | | |
| AVG TEMP | DEG C | | | | | | | | | | | | | |
| **CA RED BLUFF** | | | 4009N | 12215W | | 108 M | | | | | | | | |
| DAILY RAD | MJ/SQ M | 6.47 | 10.13 | 15.37 | 21.68 | 26.95 | 29.50 | 30.32 | 26.22 | 20.94 | 13.93 | 8.02 | 5.80 | 17.94 |
| DAY TEMP | DEG C | 6.7 | 11.6 | 13.5 | 17.3 | 21.9 | 26.5 | 30.6 | 29.2 | 26.6 | 20.5 | 13.8 | 9.4 | 19.1 |
| DEG DAY | C-DAY | 341 | 233 | 203 | 121 | 36 | 4 | 0 | 0 | 0 | 46 | 188 | 321 | 1493 |
| AVG TEMP | DEG C | 7.3 | 10.0 | 11.8 | 15.3 | 19.7 | 24.2 | 27.9 | 26.6 | 24.1 | 18.3 | 12.1 | 8.0 | 17.1 |

*(Continued)*

**TABLE A-15.** *(Continued)*

| | | JAN | FEB | MAR | APR | MAY | JUN | JUL | AUG | SEP | OCT | NOV | DEC | ANNUAL |
|---|---|---|---|---|---|---|---|---|---|---|---|---|---|---|
| **CA SACRAMENTO** | | | 3831N | 12130W | | 8 M | | | | | | | | |
| DAILY RAD | MJ/SQ M | 6.77 | 10.66 | 16.55 | 22.74 | 27.63 | 30.46 | 30.51 | 26.88 | 21.64 | 14.92 | 8.87 | 6.11 | 18.65 |
| DAY TEMP | DEG C | 8.6 | 11.4 | 13.5 | 16.8 | 20.4 | 24.0 | 26.9 | 26.3 | 24.6 | 19.7 | 13.4 | 8.9 | 17.9 |
| DEG DAY | C-DAY | 343 | 237 | 207 | 126 | 67 | 11 | | | 3 | 56 | 200 | 331 | 1581 |
| AVG TEMP | DEG C. | 7.3 | 9.9 | 11.7 | 14.6 | 17.9 | 21.4 | 24.0 | 23.4 | 21.9 | 17.4 | 11.7 | 7.7 | 15.7 |
| **CA SAN DIEGO** | | | 3244N | 11710W | | 9 M | | | | | | | | |
| DAILY RAD | MJ/SQ M | 11.07 | 14.37 | 18.52 | 21.98 | 22.73 | 23.40 | 24.81 | 23.35 | 19.49 | 15.59 | 12.06 | 10.26 | 18.14 |
| DAY TEMP | DEG C | 14.5 | 15.2 | | 17.1 | 18.4 | 19.5 | 21.8 | 22.0 | 22.2 | 20.2 | 17.5 | 15.3 | 18.4 |
| DEG DAY | C-DAY | 174 | 132 | 122 | 80 | 44 | 29 | | | 9 | 24 | 78 | 143 | 838 |
| AVG TEMP | DEG C | 12.9 | 13.7 | 14.4 | 15.9 | 17.4 | 18.6 | 20.9 | 21.9 | 21.1 | 18.9 | 16.0 | 13.7 | 17.1 |
| **CA SAN FRANCISCO** | | | 3737N | 12223W | | 5 M | | | | | | | | |
| DAILY RAD | MJ/SQ M | 8.03 | 11.45 | 16.51 | 21.79 | 25.26 | 26.97 | 27.14 | 24.02 | 19.77 | 13.91 | 9.32 | 7.29 | 17.62 |
| DAY TEMP | DEG C | 10.2 | 11.9 | 13.0 | 14.3 | 16.0 | 17.9 | 18.3 | 18.7 | 19.4 | 17.7 | 14.3 | 10.9 | 15.2 |
| DEG DAY | C-DAY | 288 | 214 | 207 | 162 | 117 | 67 | 52 | 47 | 37 | 76 | 162 | 263 | 1692 |
| AVG TEMP | DEG C | 9.1 | 10.7 | 11.7 | 12.9 | 14.6 | 16.4 | 16.9 | 17.2 | 17.8 | 16.1 | 12.9 | 9.8 | 13.9 |
| **CA SANTA MARIA** | | | 3454N | 12027W | | 72 M | | | | | | | | |
| DAILY RAD | MJ/SQ M | 9.69 | 12.95 | 17.95 | 21.80 | 24.29 | 26.65 | 26.57 | 23.90 | 19.64 | 15.36 | 11.05 | 9.12 | 18.25 |
| DAY TEMP | DEG C | 12.3 | 13.0 | 13.5 | 14.6 | 15.7 | 17.0 | 18.3 | 18.5 | 18.9 | 17.9 | 15.6 | 13.1 | 15.7 |
| DEG DAY | C-DAY | 250 | 202 | 210 | 168 | 136 | 93 | 62 | 57 | 52 | 88 | 150 | 227 | 1695 |
| AVG TEMP | DEG C | 10.3 | 11.1 | 11.6 | 12.7 | 13.9 | 15.3 | 16.7 | 16.8 | 17.0 | 15.8 | 13.4 | 11.0 | 13.8 |
| **CA SUNNYVALE** | | | 3725N | 12204W | | 12 M | | | | | | | | |
| DAILY RAD | MJ/SQ M | 8.37 | 11.77 | 16.86 | 22.06 | 25.84 | 27.84 | 27.71 | 24.59 | 19.97 | 14.17 | 9.57 | 7.49 | 18.02 |
| DAY TEMP | DEG C | | | | | | | | | | | | | |
| DEG DAY | C-DAY | | | | | | | | | | | | | |
| AVG TEMP | DEG C | | | | | | | | | | | | | |
| **CO COLORADO SPRINGS** | | | 3849N | 10443W | | 1881 M | | | | | | | | |
| DAILY RAD | MJ/SQ M | 10.11 | 13.37 | 17.59 | 21.92 | 24.16 | 26.88 | 25.10 | 22.99 | 19.96 | 15.42 | 10.72 | 8.87 | 18.09 |
| DAY TEMP | DEG C | .2 | 1.7 | 3.9 | 10.1 | 15.2 | 20.4 | 23.8 | 22.8 | 18.4 | 12.6 | 5.1 | 1.5 | 11.3 |
| DEG DAY | C-DAY | 627 | 524 | 512 | 313 | 167 | 57 | 5 | | 86 | 253 | 458 | 586 | 3595 |
| AVG TEMP | DEG C | -1.9 | -.4 | 1.8 | 7.9 | 13.1 | 18.1 | 21.5 | 20.6 | 16.1 | 10.3 | 3.1 | -.6 | 9.1 |
| **CO DENVER** | | | 3945N | 10452W | | 1625 M | | | | | | | | |
| DAILY RAD | MJ/SQ M | 9.53 | 12.79 | 17.37 | 21.33 | 24.23 | 26.68 | 25.79 | 23.20 | 19.60 | 14.76 | 10.03 | 8.30 | 17.80 |
| DAY TEMP | DEG C | 1.1 | 2.7 | 5.0 | 10.9 | 16.1 | 21.2 | 25.2 | 24.0 | 19.6 | 13.6 | 6.4 | 2.6 | 12.4 |
| DEG DAY | C-DAY | 604 | 501 | 482 | 292 | 141 | 44 | | | 67 | 227 | 427 | 558 | 3343 |
| AVG TEMP | DEG C | -1.2 | .4 | 2.8 | 8.6 | 13.9 | 18.9 | 22.8 | 22.0 | 17.1 | 11.1 | 4.1 | .3 | 10.1 |
| **CO EAGLE** | | | 3939N | 10655W | | 1985 M | | | | | | | | |
| DAILY RAD | MJ/SQ M | 8.56 | 12.23 | 17.04 | 21.93 | 25.59 | 28.47 | 27.06 | 23.65 | 20.05 | 14.83 | 9.86 | 7.84 | 18.09 |
| DAY TEMP | DEG C | | | | | | | | | | | | | |
| DEG DAY | C-DAY | 809 | 649 | 584 | 385 | 236 | 106 | 24 | 44 | 158 | 348 | 568 | 770 | 4681 |
| AVG TEMP | DEG C | -7.8 | -4.8 | -.5 | 5.5 | 10.7 | 14.9 | 18.8 | 17.6 | 13.1 | 7.1 | -.6 | -6.5 | 5.6 |

Station coordinates (Latitude, Longitude, Elevation):
- CO GRAND JUNCTION — 3907N, 10832W, 1475 M
- CO PUEBLO — 3817N, 10431W, 1439 M
- CT HARTFORD — 4156N, 7241W, 55 M
- CU GUANTANAMO BAY — 1954N, 7509W, 16 M
- DC WASHINGTON-STERLING — 3857N, 7727W, 88 M
- DE WILMINGTON — 3940N, 7536W, 24 M
- FL APALACHICOLA — 2944N, 8502W, 6 M
- FL DAYTONA BEACH — 2911N, 8103W, 12 M

| State | Station | Parameter | JAN | FEB | MAR | APR | MAY | JUN | JUL | AUG | SEP | OCT | NOV | DEC | ANNUAL |
|---|---|---|---|---|---|---|---|---|---|---|---|---|---|---|---|
| CO | GRAND JUNCTION | DAILY RAD MJ/SQ M | 8.98 | 12.70 | 17.63 | 22.54 | 27.01 | 29.49 | 27.98 | 24.76 | 20.82 | 15.26 | 10.42 | 8.30 | 18.82 |
| | | DAY TEMP DEG C | -1.3 | 2.6 | 7.0 | 13.1 | 19.0 | 24.3 | 28.3 | 26.0 | 21.9 | 14.9 | 6.2 | .3 | 13.6 |
| | | DEG DAY C-DAY | 661 | 488 | 410 | 224 | 74 | 11 | 0 | 0 | 33 | 180 | 420 | 612 | 3113 |
| | | AVG TEMP DEG C | -3.0 | .9 | 5.1 | 10.9 | 16.8 | 21.8 | 25.9 | 24.1 | 19.6 | 12.7 | 4.3 | -1.4 | 11.5 |
| CO | PUEBLO | DAILY RAD MJ/SQ M | 10.15 | 13.30 | 17.75 | 22.20 | 24.54 | 27.63 | 26.23 | 23.85 | 20.19 | 15.44 | 10.82 | 8.88 | 18.42 |
| | | DAY TEMP DEG C | 1.5 | 4.0 | 6.9 | 13.4 | 18.6 | 24.0 | 27.1 | 26.0 | 21.5 | 15.2 | 7.5 | 3.1 | 14.1 |
| | | DEG DAY C-DAY | 601 | 471 | 431 | 225 | 82 | 16 | 0 | 0 | 31 | 186 | 403 | 551 | 2997 |
| | | AVG TEMP DEG C | -1.1 | 1.5 | 4.4 | 10.9 | 16.2 | 21.5 | 24.7 | 23.6 | 19.0 | 12.5 | 4.9 | .6 | 11.6 |
| CT | HARTFORD | DAILY RAD MJ/SQ M | 5.42 | 8.11 | 11.10 | 14.92 | 17.80 | 19.13 | 18.71 | 16.13 | 13.10 | 9.68 | 5.64 | 4.37 | 12.01 |
| | | DAY TEMP DEG C | -2.6 | -1.4 | 3.5 | 10.6 | 16.6 | 21.8 | 24.5 | 23.2 | 19.1 | 13.4 | 6.7 | -.7 | 11.2 |
| | | DEG DAY C-DAY | 692 | 594 | 506 | 288 | 126 | 13 | 0 | 0 | 59 | 213 | 395 | 634 | 3527 |
| | | AVG TEMP DEG C | -4.0 | -2.9 | 2.0 | 8.7 | 14.6 | 19.9 | 22.6 | 21.3 | 17.1 | 11.4 | 5.2 | -2.1 | 9.5 |
| CU | GUANTANAMO BAY | DAILY RAD MJ/SQ M | 15.92 | 18.70 | 21.66 | 24.06 | 23.12 | 22.25 | 23.63 | 22.73 | 20.70 | 17.98 | 16.25 | 14.90 | 20.18 |
| DC | WASHINGTON-STERLING | DAILY RAD MJ/SQ M | 6.49 | 9.25 | 12.77 | 16.56 | 19.50 | 21.57 | 20.63 | 18.36 | 15.21 | 11.39 | 7.39 | 5.46 | 13.71 |
| | | DAY TEMP DEG C | 1.6 | 2.6 | 7.3 | 13.7 | 19.3 | 23.7 | 25.9 | 25.0 | 21.4 | 15.3 | 8.9 | 2.7 | 13.9 |
| | | DEG DAY C-DAY | 567 | 486 | 399 | 196 | 73 | 3 | 0 | 0 | 24 | 162 | 338 | 534 | 2784 |
| | | AVG TEMP DEG C | .1 | 1.0 | 5.4 | 11.7 | 17.0 | 21.7 | 24.1 | 23.1 | 19.4 | 13.3 | 7.1 | 1.1 | 12.1 |
| DE | WILMINGTON | DAILY RAD MJ/SQ M | 6.48 | 9.39 | 13.00 | 16.80 | 19.41 | 21.36 | 20.69 | 18.32 | 14.95 | 11.17 | 7.31 | 5.54 | 13.71 |
| | | DAY TEMP DEG C | 1.4 | 2.3 | 6.9 | 13.1 | 18.7 | 23.6 | 25.9 | 25.0 | 21.7 | 15.8 | 9.2 | 2.9 | 13.9 |
| | | DEG DAY C-DAY | 568 | 488 | 403 | 212 | 71 | 0 | 0 | 0 | 18 | 141 | 322 | 522 | 2745 |
| | | AVG TEMP DEG C | .0 | .9 | 5.3 | 11.3 | 16.9 | 21.9 | 24.3 | 23.4 | 19.9 | 14.0 | 7.6 | 1.5 | 12.3 |
| FL | APALACHICOLA | DAILY RAD MJ/SQ M | 9.68 | 12.78 | 16.73 | 21.32 | 23.73 | 22.68 | 20.58 | 19.15 | 17.42 | 15.56 | 11.80 | 9.28 | 16.73 |
| | | DAY TEMP DEG C | 13.3 | 14.4 | 17.1 | 21.3 | 25.0 | 27.7 | 28.5 | 28.5 | 26.9 | 22.8 | 17.5 | 14.1 | 21.4 |
| | | DEG DAY C-DAY | 204 | 161 | 97 | 17 | 0 | 0 | 0 | 0 | 0 | 12 | 88 | 177 | 756 |
| | | AVG TEMP DEG C | 12.1 | 13.2 | 15.9 | 20.2 | 23.8 | 26.7 | 27.4 | 27.5 | 25.9 | 21.6 | 16.2 | 12.9 | 20.3 |
| FL | DAYTONA BEACH | DAILY RAD MJ/SQ M | 10.88 | 13.77 | 17.57 | 21.36 | 22.34 | 20.72 | 20.25 | 19.09 | 16.77 | 14.20 | 11.75 | 9.88 | 16.55 |
| | | DAY TEMP DEG C | 16.4 | 17.1 | 19.5 | 22.7 | 25.6 | 27.0 | 28.7 | 28.7 | 27.6 | 24.3 | 20.1 | 17.1 | 23.0 |
| | | DEG DAY C-DAY | 134 | 117 | 67 | | 0 | 0 | 0 | 0 | 0 | 12 | 54 | 118 | 502 |
| | | AVG TEMP DEG C | 14.7 | 15.3 | 17.7 | 20.9 | 23.9 | 26.3 | 27.2 | 27.3 | 26.4 | 22.9 | 18.4 | 15.3 | 21.4 |

*(Continued)*

**TABLE A-15.** (Continued)

| | | JAN | FEB | MAR | APR | MAY | JUN | JUL | AUG | SEP | OCT | NOV | DEC | ANNUAL |
|---|---|---|---|---|---|---|---|---|---|---|---|---|---|---|
| FL | JACKSONVILLE | | 3030N | 6142W | | 9 M | | | | | | | | |
| | DAILY RAD MJ/SQ M | 10.21 | 13.21 | 17.27 | 21.06 | 22.20 | 21.39 | 20.45 | 19.23 | 16.37 | 13.88 | 11.30 | 9.28 | 16.32 |
| | DAY TEMP DEG C | 14.2 | 15.3 | 18.1 | 21.9 | 25.2 | 27.7 | 28.7 | 28.7 | 27.0 | 22.8 | 17.9 | 14.7 | 21.8 |
| | DEG DAY C-DAY | 193 | 157 | 96 | 13 | | | | | | 11 | 89 | 176 | 737 |
| | AVG TEMP DEG C | 12.6 | 13.5 | 16.2 | 20.1 | 23.5 | 26.2 | 27.2 | 27.2 | 25.7 | 21.4 | 16.2 | 13.0 | 20.2 |
| FL | MIAMI | | 2548N | 8016W | | 2 M | | | | | | | | |
| | DAILY RAD MJ/SQ M | 12.00 | 14.91 | 18.20 | 21.10 | 20.92 | 19.38 | 20.01 | 18.50 | 16.53 | 14.78 | 12.69 | 11.57 | 16.72 |
| | DAY TEMP DEG C | 21.0 | 21.4 | 23.2 | 25.2 | 26.0 | 28.0 | 29.0 | 29.1 | 28.7 | 26.6 | 23.6 | 21.5 | 25.4 |
| | DEG DAY C-DAY | 29 | 37 | 9 | 0 | | | | | | | 7 | 31 | 113 |
| | AVG TEMP DEG C | 19.6 | 19.9 | 21.6 | 23.9 | 25.6 | 27.2 | 27.9 | 28.3 | 27.6 | 25.4 | 22.3 | 20.2 | 24.1 |
| FL | ORLANDO | | 2833N | 8120W | | 36 M | | | | | | | | |
| | DAILY RAD MJ/SQ M | 11.34 | 14.11 | 17.96 | 21.54 | 22.57 | 20.78 | 20.44 | 18.99 | 16.98 | 14.80 | 12.44 | 10.51 | 16.87 |
| | DAY TEMP DEG C | 17.4 | 18.1 | 20.5 | 23.5 | 26.4 | 28.3 | 28.8 | 29.0 | 28.0 | 24.9 | 20.8 | 18.1 | 23.7 |
| | DEG DAY C-DAY | 109 | 102 | 52 | 7 | | | | | | 0 | 42 | 94 | 406 |
| | AVG TEMP DEG C | 15.7 | 16.4 | 18.8 | 21.6 | 24.7 | 26.8 | 27.4 | 27.7 | 26.7 | 23.5 | 19.2 | 16.4 | 22.1 |
| FL | TALLAHASSEE | | 3023N | 8422W | | 21 M | | | | | | | | |
| | DAILY RAD MJ/SQ M | 9.95 | 12.91 | 16.79 | 20.69 | 21.97 | 21.37 | 19.84 | 19.01 | 16.94 | 14.95 | 11.44 | 9.22 | 16.26 |
| | DAY TEMP DEG C | 13.4 | 14.6 | 17.7 | 22.0 | 25.8 | 28.4 | 28.9 | 28.8 | 27.2 | 22.6 | 17.0 | 13.8 | 21.7 |
| | DEG DAY C-DAY | 227 | 179 | 104 | 19 | | | | | | 17 | 113 | 209 | 868 |
| | AVG TEMP DEG C | 11.4 | 12.7 | 15.1 | 19.9 | 23.8 | 26.7 | 27.3 | 27.3 | 25.6 | 20.7 | 14.9 | 11.8 | 19.8 |
| FL | TAMPA | | 2758N | 8232W | | 3 M | | | | | | | | |
| | DAILY RAD MJ/SQ M | 11.47 | 14.29 | 18.09 | 21.66 | 22.68 | 20.97 | 19.89 | 18.76 | 16.93 | 15.28 | 12.57 | 10.62 | 16.93 |
| | DAY TEMP DEG C | 17.5 | 18.2 | 20.6 | 24.0 | 26.8 | 28.7 | 29.1 | 29.3 | 28.5 | 25.3 | 21.0 | 18.2 | 23.9 |
| | DEG DAY C-DAY | 113 | 98 | 50 | 6 | | | | | | 0 | 39 | 94 | 399 |
| | AVG TEMP DEG C | 15.8 | 16.6 | 18.9 | 22.2 | 25.1 | 27.2 | 27.7 | 27.9 | 27.1 | 23.7 | 19.3 | 16.4 | 22.3 |
| FL | WEST PALM BEACH | | 2641N | 8006W | | 6 M | | | | | | | | |
| | DAILY RAD MJ/SQ M | 11.35 | 13.99 | 17.66 | 20.59 | 20.93 | 19.36 | 20.19 | 18.88 | 16.10 | 13.89 | 12.03 | 10.87 | 16.32 |
| | DAY TEMP DEG C | 20.2 | 20.6 | 22.6 | 24.8 | 26.7 | 28.2 | 29.0 | 29.3 | 28.6 | 26.3 | 23.1 | 20.9 | 25.0 |
| | DEG DAY C-DAY | 46 | 51 | 14 | 0 | | | | | | 0 | 12 | 43 | 166 |
| | AVG TEMP DEG C | 18.6 | 18.9 | 21.0 | 23.3 | 25.3 | 26.9 | 27.7 | 27.9 | 27.5 | 25.1 | 21.7 | 19.3 | 23.6 |
| GA | ATLANTA | | 3339N | 8426W | | 315 M | | | | | | | | |
| | DAILY RAD MJ/SQ M | 8.14 | 11.00 | 14.79 | 19.14 | 21.04 | 21.72 | 20.57 | 19.39 | 16.14 | 13.62 | 10.02 | 7.65 | 15.27 |
| | DAY TEMP DEG C | 7.3 | 8.8 | 12.3 | 17.9 | 22.3 | 25.7 | 27.0 | 26.8 | 23.9 | 18.6 | 12.5 | 7.9 | 17.6 |
| | DEG DAY C-DAY | 389 | 311 | 246 | 80 | 15 | | | | 4 | 76 | 227 | 371 | 1719 |
| | AVG TEMP DEG C | 5.8 | 7.2 | 10.6 | 16.2 | 20.6 | 24.2 | 25.6 | 25.3 | 22.4 | 16.9 | 10.8 | 6.4 | 16.0 |
| GA | AUGUSTA | | 3322N | 8158W | | 45 M | | | | | | | | |
| | DAILY RAD MJ/SQ M | 8.52 | 11.52 | 15.19 | 19.62 | 21.17 | 21.60 | 20.47 | 18.92 | 16.00 | 13.84 | 10.40 | 8.18 | 15.45 |
| | DAY TEMP DEG C | 9.6 | 11.1 | 14.6 | 19.8 | 24.1 | 27.6 | 28.6 | 28.2 | 25.3 | 20.0 | 14.3 | 10.0 | 19.4 |
| | DEG DAY C-DAY | 334 | 264 | 192 | 50 | 6 | | | | 58 | 191 | 321 | 1416 |
| | AVG TEMP DEG C | 7.7 | 9.1 | 12.6 | 17.7 | 22.1 | 25.7 | 26.9 | 26.4 | 23.4 | 17.8 | 12.1 | 8.0 | 17.4 |

**GA MACON** (3242N, 8339W, 110 M)

| | JAN | FEB | MAR | APR | MAY | JUN | JUL | AUG | SEP | OCT | NOV | DEC | ANNUAL |
|---|---|---|---|---|---|---|---|---|---|---|---|---|---|
| DAILY RAD (MJ/SQ M) | 8.73 | 11.57 | 15.47 | 19.70 | 21.39 | 21.78 | 20.26 | 19.49 | 16.33 | 14.15 | 10.66 | 8.27 | 15.65 |
| DAY TEMP (DEG C) | 10.6 | 12.1 | 15.6 | 20.9 | 25.1 | 28.0 | 29.2 | 29.0 | 26.1 | 20.8 | 15.0 | 10.9 | 20.3 |
| DEG DAY (C-DAY) | 302 | 235 | 166 | 37 | 3 |  |  |  | 0 | 46 | 169 | 288 | 1246 |
| AVG TEMP (DEG C) | 8.8 | 10.2 | 13.6 | 18.8 | 23.1 | 26.4 | 27.4 | 27.2 | 24.3 | 18.7 | 12.9 | 9.1 | 18.4 |

**GA SAVANNAH** (3208N, 8112W, 16 M)

| | JAN | FEB | MAR | APR | MAY | JUN | JUL | AUG | SEP | OCT | NOV | DEC | ANNUAL |
|---|---|---|---|---|---|---|---|---|---|---|---|---|---|
| DAILY RAD (MJ/SQ M) | 9.02 | 11.89 | 15.87 | 19.99 | 21.02 | 20.93 | 20.24 | 18.40 | 15.48 | 13.81 | 10.68 | 8.55 | 15.49 |
| DAY TEMP (DEG C) | 11.8 | 13.1 | 16.4 | 20.9 | 24.9 | 27.9 | 28.9 | 28.6 | 26.0 | 21.4 | 16.0 | 12.2 | 20.7 |
| DEG DAY (C-DAY) | 268 | 211 | 142 | 35 | 0 |  |  |  | 0 | 33 | 141 | 254 | 1084 |
| AVG TEMP (DEG C) | 9.9 | 11.2 | 14.4 | 18.9 | 22.9 | 26.2 | 27.3 | 27.0 | 24.6 | 19.5 | 13.9 | 10.2 | 18.8 |

**HI BARBERS POINT** (2119N, 15804W, 10 M)

| | JAN | FEB | MAR | APR | MAY | JUN | JUL | AUG | SEP | OCT | NOV | DEC | ANNUAL |
|---|---|---|---|---|---|---|---|---|---|---|---|---|---|
| DAILY RAD (MJ/SQ M) | 13.71 | 16.35 | 18.67 | 20.81 | 22.39 | 22.98 | 22.90 | 22.37 | 20.59 | 17.64 | 14.75 | 13.23 | 18.87 |
| DAY TEMP (DEG C) |  |  |  |  |  |  |  |  |  |  |  |  |  |
| DEG DAY (C-DAY) |  |  |  |  |  |  |  |  |  |  |  |  |  |
| AVG TEMP (DEG C) |  |  |  |  |  |  |  |  |  |  |  |  |  |

**HI HILO** (1943N, 15504W, 11 M)

| | JAN | FEB | MAR | APR | MAY | JUN | JUL | AUG | SEP | OCT | NOV | DEC | ANNUAL |
|---|---|---|---|---|---|---|---|---|---|---|---|---|---|
| DAILY RAD (MJ/SQ M) | 12.71 | 14.14 | 15.30 | 16.28 | 17.62 | 18.82 | 18.44 | 18.07 | 17.55 | 15.57 | 12.54 | 11.57 | 15.72 |
| DAY TEMP (DEG C) | 23.2 | 23.1 | 23.0 | 23.6 | 24.4 | 25.0 | 25.3 | 25.7 | 25.6 | 25.3 | 24.4 | 23.3 | 24.3 |
| DEG DAY (C-DAY) |  |  |  |  |  |  |  |  |  |  |  |  |  |
| AVG TEMP (DEG C) | 21.8 | 21.7 | 21.7 | 22.3 | 23.1 | 23.7 | 24.1 | 24.4 | 24.2 | 23.9 | 23.1 | 22.0 | 23.0 |

**HI HONOLULU** (2120N, 15755W, 5 M)

| | JAN | FEB | MAR | APR | MAY | JUN | JUL | AUG | SEP | OCT | NOV | DEC | ANNUAL |
|---|---|---|---|---|---|---|---|---|---|---|---|---|---|
| DAILY RAD (MJ/SQ M) | 13.39 | 15.85 | 18.40 | 20.38 | 22.12 | 22.75 | 22.72 | 22.32 | 20.54 | 17.48 | 14.37 | 12.85 | 18.60 |
| DAY TEMP (DEG C) | 23.6 | 23.5 | 23.9 | 24.9 | 26.1 | 27.2 | 27.8 | 28.0 | 28.1 | 27.2 | 25.8 | 24.3 | 25.9 |
| DEG DAY (C-DAY) |  |  |  |  |  |  |  |  |  |  |  |  |  |
| AVG TEMP (DEG C) | 22.4 | 22.4 | 22.8 | 23.8 | 24.9 | 26.1 | 26.7 | 27.1 | 26.9 | 26.1 | 24.7 | 23.2 | 24.7 |

**HI LIHUE** (2159N, 15921W, 45 M)

| | JAN | FEB | MAR | APR | MAY | JUN | JUL | AUG | SEP | OCT | NOV | DEC | ANNUAL |
|---|---|---|---|---|---|---|---|---|---|---|---|---|---|
| DAILY RAD (MJ/SQ M) | 12.52 | 14.75 | 16.75 | 18.62 | 20.70 | 21.20 | 21.14 | 20.63 | 19.77 | 16.45 | 13.10 | 11.95 | 17.30 |
| DAY TEMP (DEG C) | 22.9 | 22.9 | 23.1 | 23.9 | 25.1 | 26.2 | 26.7 | 27.1 | 27.0 | 26.2 | 24.9 | 23.4 | 25.0 |
| DEG DAY (C-DAY) |  |  |  |  |  |  |  |  |  |  |  |  |  |
| AVG TEMP (DEG C) | 21.8 | 21.8 | 22.1 | 22.9 | 24.2 | 25.3 | 25.8 | 26.2 | 26.0 | 25.2 | 24.0 | 22.5 | 24.0 |

**IA BURLINGTON** (4047N, 9107W, 214 M)

| | JAN | FEB | MAR | APR | MAY | JUN | JUL | AUG | SEP | OCT | NOV | DEC | ANNUAL |
|---|---|---|---|---|---|---|---|---|---|---|---|---|---|
| DAILY RAD (MJ/SQ M) | 6.57 | 9.74 | 13.22 | 17.45 | 21.29 | 24.07 | 23.66 | 20.75 | 16.08 | 12.04 | 7.53 | 5.45 | 14.82 |
| DAY TEMP (DEG C) | -3.6 | -1.1 | 4.3 | 12.5 | 18.4 | 23.6 | 25.9 | 25.1 | 20.4 | 14.8 | 5.9 | -1.1 | 12.1 |
| DEG DAY (C-DAY) | 725 | 587 | 484 | 231 | 96 |  |  | 4 | 39 | 178 | 420 | 644 | 3417 |
| AVG TEMP (DEG C) | -5.1 | -2.6 | 2.5 | 10.7 | 16.6 | 21.9 | 24.1 | 23.3 | 18.6 | 12.9 | 4.3 | -2.4 | 10.4 |

**IA DES MOINES** (4132N, 9339W, 294 M)

| | JAN | FEB | MAR | APR | MAY | JUN | JUL | AUG | SEP | OCT | NOV | DEC | ANNUAL |
|---|---|---|---|---|---|---|---|---|---|---|---|---|---|
| DAILY RAD (MJ/SQ M) | 6.59 | 9.77 | 13.40 | 17.67 | 21.19 | 24.11 | 23.80 | 20.74 | 16.27 | 12.12 | 7.47 | 5.53 | 14.89 |
| DAY TEMP (DEG C) | -5.6 | -3.0 | 2.5 | 11.4 | 17.7 | 22.9 | 25.6 | 24.6 | 19.7 | 14.2 | 4.7 | -2.6 | 11.0 |
| DEG DAY (C-DAY) | 786 | 634 | 536 | 258 | 103 | 14 |  | 4 | 52 | 194 | 453 | 689 | 3726 |
| AVG TEMP (DEG C) | -7.0 | -4.3 | 1.1 | 9.7 | 16.1 | 21.4 | 23.9 | 22.9 | 17.9 | 12.4 | 3.2 | -3.9 | 9.5 |

*(Continued)*

## TABLE A-15. (Continued)

Latitude, longitude, and elevation (M) appear in the FEB, MAR, and MAY columns of the DAILY RAD line for each station.

| | | JAN | FEB | MAR | APR | MAY | JUN | JUL | AUG | SEP | OCT | NOV | DEC | ANNUAL |
|---|---|---|---|---|---|---|---|---|---|---|---|---|---|---|
| **IA** | **MASON CITY** | | 4309N | 9320W | | 373 M | | | | | | | | |
| | DAILY RAD | MJ/SQ M | 6.28 | 9.49 | 13.25 | 17.23 | 21.51 | 23.99 | 23.65 | 20.80 | 15.95 | 11.47 | 6.81 | 5.03 | 14.62 |
| | DAY TEMP | DEG C | | | | | | | | | | | | | |
| | DEG DAY | C-DAY | 875 | 723 | 620 | 322 | 147 | 36 | 7 | 17 | 92 | 254 | 523 | 773 | 4389 |
| | AVG TEMP | DEG C | -9.9 | -7.5 | -1.7 | 7.6 | 14.1 | 19.6 | 21.8 | 21.1 | 15.7 | 10.3 | .9 | -6.6 | 7.1 |
| **IA** | **SIOUX CITY** | | 4224N | 9623W | | 336 M | | | | | | | | |
| | DAILY RAD | MJ/SQ M | 6.45 | 9.55 | 13.28 | 17.91 | 21.58 | 24.10 | 24.08 | 20.94 | 16.13 | 11.78 | 7.29 | 5.33 | 14.87 |
| | DAY TEMP | DEG C | -6.1 | -3.1 | 2.3 | 11.6 | 18.0 | 23.1 | 26.0 | 24.9 | 19.4 | 13.8 | 4.2 | -3.1 | 10.9 |
| | DEG DAY | C-DAY | 809 | 647 | 548 | 263 | 105 | 18 | 0 | 6 | 63 | 210 | 478 | 715 | 3862 |
| | AVG TEMP | DEG C | -7.8 | -4.8 | .7 | 9.7 | 16.1 | 21.3 | 24.1 | 23.1 | 17.4 | 11.7 | 2.4 | -4.7 | 9.1 |
| **ID** | **BOISE** | | 4334N | 11613W | | 874 M | | | | | | | | |
| | DAILY RAD | MJ/SQ M | 5.51 | 9.53 | 14.80 | 20.73 | 25.84 | 27.95 | 29.65 | 24.93 | 19.71 | 12.91 | 7.13 | 4.96 | 16.97 |
| | DAY TEMP | DEG C | -.4 | 3.3 | 6.8 | 11.5 | 16.3 | 20.5 | 26.3 | 24.9 | 19.7 | 13.3 | 5.9 | 1.2 | 12.4 |
| | DEG DAY | C-DAY | 620 | 459 | 412 | 267 | 140 | 54 | 0 | 7 | 71 | 226 | 420 | 567 | 3243 |
| | AVG TEMP | DEG C | -1.7 | 1.9 | 5.1 | 9.4 | 14.1 | 18.2 | 23.6 | 22.3 | 17.3 | 11.2 | 4.3 | .1 | 10.5 |
| **ID** | **LEWISTON** | | 4623N | 11701W | | 438 M | | | | | | | | |
| | DAILY RAD | MJ/SQ M | 3.85 | 6.91 | 11.57 | 16.29 | 20.91 | 22.87 | 26.51 | 21.92 | 16.28 | 9.76 | 4.68 | 3.25 | 13.73 |
| | DAY TEMP | DEG C | .7 | 4.7 | 7.7 | 12.1 | 16.6 | 20.5 | 25.6 | 24.5 | 19.8 | 12.9 | 6.0 | 2.6 | 12.8 |
| | DEG DAY | C-DAY | 582 | 418 | 381 | 245 | 129 | 47 | 0 | 11 | 69 | 227 | 408 | 520 | 3035 |
| | AVG TEMP | DEG C | -.4 | 3.4 | 6.1 | 10.2 | 14.5 | 18.3 | 23.0 | 21.9 | 17.4 | 11.0 | 4.7 | 1.6 | 11.0 |
| **ID** | **POCATELLO** | | 4255N | 11236W | | 1365 M | | | | | | | | |
| | DAILY RAD | MJ/SQ M | 6.12 | 10.01 | 15.56 | 20.66 | 25.88 | 28.14 | 29.50 | 25.41 | 20.08 | 13.65 | 7.82 | 5.41 | 17.35 |
| | DAY TEMP | DEG C | -3.4 | .1 | 3.6 | 9.5 | 14.7 | 19.0 | 24.8 | 23.6 | 17.9 | 11.5 | 3.8 | -1.4 | 10.3 |
| | DEG DAY | C-DAY | 720 | 554 | 510 | 328 | 187 | 77 | 0 | 11 | 107 | 286 | 488 | 656 | 3924 |
| | AVG TEMP | DEG C | -4.9 | -1.4 | 1.9 | 7.4 | 12.4 | 16.6 | 21.9 | 20.8 | 15.2 | 9.1 | 2.1 | -2.8 | 8.2 |
| **IL** | **CHICAGO** | | 4147N | 8745W | | 190 M | | | | | | | | |
| | DAILY RAD | MJ/SQ M | 5.75 | 8.62 | 12.56 | 16.56 | 20.30 | 22.78 | 22.06 | 19.51 | 15.36 | 11.00 | 6.42 | 4.56 | 13.79 |
| | DAY TEMP | DEG C | -3.1 | -1.4 | 4.0 | 11.5 | 17.3 | 23.1 | 25.3 | 24.8 | 20.5 | 14.6 | 5.9 | -.8 | 11.8 |
| | DEG DAY | C-DAY | 701 | 585 | 486 | 252 | 116 | 14 | 0 | 4 | 32 | 176 | 410 | 629 | 3405 |
| | AVG TEMP | DEG C | -4.3 | -2.6 | 2.7 | 9.9 | 15.6 | 21.4 | 23.7 | 23.2 | 18.8 | 13.0 | 4.7 | -1.9 | 10.3 |
| **IL** | **MOLINE** | | 4127N | 9031W | | 181 M | | | | | | | | |
| | DAILY RAD | MJ/SQ M | 6.07 | 9.21 | 12.69 | 16.56 | 19.90 | 22.35 | 22.00 | 19.46 | 15.40 | 11.30 | 6.75 | 4.91 | 13.89 |
| | DAY TEMP | DEG C | -4.4 | -2.1 | 3.6 | 12.1 | 18.0 | 23.3 | 25.4 | 24.5 | 20.0 | 14.4 | 5.5 | -1.7 | 11.6 |
| | DEG DAY | C-DAY | 749 | 611 | 504 | 242 | 102 | 11 | 0 | 6 | 44 | 191 | 430 | 661 | 3551 |
| | AVG TEMP | DEG C | -5.8 | -3.5 | 2.1 | 10.3 | 16.2 | 21.6 | 23.6 | 22.7 | 18.1 | 12.4 | 4.0 | -3.0 | 9.9 |
| **IL** | **SPRINGFIELD** | | 3950N | 8940W | | 187 M | | | | | | | | |
| | DAILY RAD | MJ/SQ M | 6.64 | 9.77 | 12.97 | 17.19 | 21.17 | 23.79 | 23.36 | 20.49 | 16.50 | 12.12 | 7.68 | 5.56 | 14.77 |
| | DAY TEMP | DEG C | -1.6 | .5 | 5.7 | 13.5 | 19.2 | 24.5 | 26.2 | 25.3 | 21.5 | 15.6 | 7.0 | .4 | 13.2 |
| | DEG DAY | C-DAY | 659 | 538 | 441 | 202 | 73 | 7 | 0 | 6 | 27 | 157 | 385 | 594 | 3087 |
| | AVG TEMP | DEG C | -2.9 | -.9 | 4.1 | 11.7 | 17.4 | 22.7 | 24.5 | 23.6 | 19.6 | 13.7 | 5.5 | -.8 | 11.5 |

| | | JAN | FEB | MAR | APR | MAY | JUN | JUL | AUG | SEP | OCT | NOV | DEC | ANNUAL |
|---|---|---|---|---|---|---|---|---|---|---|---|---|---|---|
| IN | EVANSVILLE | | 3803N | 8732W | | 118 M | | | | | | | | |
| | DAILY RAD MJ/SQ M | 6.51 | 9.43 | 13.06 | 17.03 | 20.23 | 22.50 | 21.79 | 19.69 | 15.93 | 12.34 | 7.75 | 5.66 | 14.32 |
| | DAY TEMP DEG C | 1.8 | 3.8 | 8.5 | 15.6 | 20.6 | 25.6 | 27.3 | 26.5 | 22.7 | 16.7 | 8.9 | 3.3 | 15.1 |
| | DEG DAY C-DAY | 558 | 453 | 363 | 146 | 53 | 3 | 0 | 0 | 19 | 131 | 335 | 512 | 2573 |
| | AVG TEMP DEG C | .3 | 2.2 | 6.8 | 13.7 | 18.7 | 23.7 | 25.4 | 24.6 | 20.6 | 14.6 | 7.2 | 1.8 | 13.3 |
| IN | FORT WAYNE | | 4100N | 8512W | | 252 M | | | | | | | | |
| | DAILY RAD MJ/SQ M | 5.17 | 7.92 | 11.44 | 15.44 | 18.97 | 20.90 | 20.28 | 18.09 | 14.45 | 10.49 | 5.86 | 4.19 | 12.74 |
| | DAY TEMP DEG C | -2.5 | -1.1 | 3.9 | 11.3 | 17.1 | 22.6 | 24.5 | 23.6 | 20.0 | 13.8 | 5.9 | -.7 | 11.5 |
| | DEG DAY C-DAY | 684 | 582 | 491 | 262 | 120 | 13 | 0 | 0 | 50 | 202 | 413 | 627 | 3451 |
| | AVG TEMP DEG C | -3.7 | -2.4 | 2.5 | 9.6 | 15.3 | 20.8 | 22.8 | 21.8 | 18.1 | 12.0 | 4.6 | -1.9 | 10.0 |
| IN | INDIANAPOLIS | | 3944N | 8617W | | 246 M | | | | | | | | |
| | DAILY RAD MJ/SQ M | 5.62 | 8.48 | 11.77 | 15.87 | 19.16 | 21.20 | 20.50 | 18.65 | 15.03 | 11.09 | 6.57 | 4.73 | 13.22 |
| | DAY TEMP DEG C | -.9 | .7 | 5.8 | 13.0 | 18.6 | 23.8 | 25.6 | 24.7 | 21.0 | 15.0 | 6.9 | .6 | 12.9 |
| | DEG DAY C-DAY | 639 | 533 | 436 | 215 | 88 | 6 | 0 | 3 | 35 | 168 | 388 | 587 | 3098 |
| | AVG TEMP DEG C | -2.3 | -.7 | 4.3 | 11.3 | 16.8 | 22.1 | 23.9 | 22.9 | 19.1 | 13.2 | 5.4 | -.6 | 11.3 |
| IN | SOUTH BEND | | 4142N | 8619W | | 236 M | | | | | | | | |
| | DAILY RAD MJ/SQ M | 4.72 | 7.49 | 11.26 | 15.74 | 19.55 | 21.81 | 21.02 | 18.91 | 14.65 | 10.32 | 5.64 | 3.86 | 12.92 |
| | DAY TEMP DEG C | -3.2 | -1.9 | 3.3 | 10.7 | 16.5 | 22.1 | 24.1 | 23.5 | 19.5 | 13.6 | 5.5 | -1.0 | 11.1 |
| | DEG DAY C-DAY | 706 | 602 | 512 | 282 | 136 | 19 | 0 | 13 | 54 | 204 | 423 | 634 | 3588 |
| | AVG TEMP DEG C | -4.4 | -3.2 | 1.8 | 8.9 | 14.7 | 20.3 | 22.4 | 21.7 | 17.7 | 11.9 | 4.2 | -2.1 | 9.5 |
| KS | DODGE CITY | | 3746N | 9958W | | 787 M | | | | | | | | |
| | DAILY RAD MJ/SQ M | 9.38 | 12.73 | 16.76 | 21.40 | 23.72 | 26.76 | 26.05 | 23.33 | 19.14 | 14.76 | 10.14 | 8.31 | 17.71 |
| | DAY TEMP DEG C | 1.3 | 3.8 | 7.2 | 14.4 | 19.8 | 25.2 | 28.3 | 27.7 | 22.6 | 16.5 | 8.1 | 2.6 | 14.8 |
| | DEG DAY C-DAY | 589 | 463 | 410 | 191 | 64 | 12 | 0 | 0 | 23 | 137 | 370 | 544 | 2803 |
| | AVG TEMP DEG C | -.7 | 1.8 | 5.1 | 12.2 | 17.8 | 23.2 | 26.2 | 25.6 | 20.5 | 14.4 | 6.0 | .8 | 12.7 |
| KS | GOODLAND | | 3922N | 10142W | | 1124 M | | | | | | | | |
| | DAILY RAD MJ/SQ M | 8.96 | 11.98 | 16.16 | 20.76 | 23.40 | 26.75 | 26.32 | 23.21 | 18.64 | 14.39 | 9.72 | 7.88 | 17.35 |
| | DAY TEMP DEG C | -.1 | 2.0 | 4.7 | 11.7 | 17.3 | 23.0 | 26.8 | 25.8 | 20.5 | 14.1 | 6.0 | 1.2 | 12.7 |
| | DEG DAY C-DAY | 644 | 521 | 494 | 272 | 120 | 31 | 0 | 0 | 60 | 215 | 442 | 601 | 3400 |
| | AVG TEMP DEG C | -2.4 | -.3 | 2.4 | 9.3 | 14.9 | 20.6 | 24.3 | 23.4 | 17.9 | 11.6 | 3.6 | -1.1 | 10.4 |
| KS | TOPEKA | | 3904N | 9538W | | 270 M | | | | | | | | |
| | DAILY RAD MJ/SQ M | 7.73 | 10.68 | 14.26 | 18.63 | 21.74 | 24.13 | 24.15 | 21.68 | 17.21 | 13.01 | 8.76 | 6.62 | 15.72 |
| | DAY TEMP DEG C | -.5 | 2.6 | 7.0 | 14.5 | 19.9 | 24.7 | 27.5 | 27.0 | 22.1 | 16.3 | 8.0 | 1.6 | 14.2 |
| | DEG DAY C-DAY | 637 | 492 | 414 | 183 | 61 | 7 | 0 | 0 | 31 | 144 | 368 | 572 | 2914 |
| | AVG TEMP DEG C | -2.2 | .8 | 5.1 | 12.5 | 18.1 | 23.1 | 25.7 | 25.1 | 20.1 | 14.2 | 6.1 | -.1 | 12.4 |
| KS | WICHITA | | 3739N | 9725W | | 408 M | | | | | | | | |
| | DAILY RAD MJ/SQ M | 8.90 | 12.01 | 15.95 | 20.23 | 23.10 | 25.70 | 25.40 | 23.06 | 18.34 | 14.18 | 9.88 | 7.83 | 17.05 |
| | DAY TEMP DEG C | 1.3 | 4.2 | 8.3 | 15.6 | 20.8 | 26.1 | 28.0 | 28.0 | 23.1 | 17.3 | 8.9 | 3.0 | 15.5 |
| | DEG DAY C-DAY | 581 | 447 | 373 | 153 | 50 | 4 | 0 | 0 | 18 | 117 | 337 | 526 | 2606 |
| | AVG TEMP DEG C | -.4 | 2.4 | 6.4 | 13.7 | 18.9 | 24.3 | 27.1 | 26.5 | 21.4 | 15.3 | 7.1 | 1.4 | 13.7 |

(Continued)

**TABLE A-15.** (Continued)

| | | JAN | FEB | MAR | APR | MAY | JUN | JUL | AUG | SEP | OCT | NOV | DEC | ANNUAL |
|---|---|---|---|---|---|---|---|---|---|---|---|---|---|---|
| KY | **LEXINGTON** | | 3802N | 8436W | | 301 M | | | | | | | | |
| | DAILY RAD MJ/SQ M | 6.19 | 8.85 | 12.48 | 16.79 | 19.83 | 21.53 | 21.00 | 19.13 | 15.46 | 11.85 | 7.46 | 5.51 | 13.84 |
| | DAY TEMP DEG C | 1.9 | 3.3 | 8.1 | 14.7 | 20.0 | 24.4 | 26.3 | 25.0 | 22.2 | 16.2 | 8.5 | 3.3 | 14.6 |
| | DEG DAY C-DAY | 553 | 462 | 374 | 168 | 59 | | | | 22 | 137 | 340 | 508 | 2627 |
| | AVG TEMP DEG C | .5 | 1.8 | 6.4 | 12.9 | 16.2 | 22.8 | 24.6 | 23.9 | 20.3 | 14.3 | 7.0 | 1.9 | 12.9 |
| KY | **LOUISVILLE** | | 3811N | 8544W | | 149 M | | | | | | | | |
| | DAILY RAD MJ/SQ M | 6.19 | 8.96 | 12.51 | 16.65 | 19.52 | 21.60 | 20.85 | 19.07 | 15.45 | 11.83 | 7.41 | 5.54 | 13.80 |
| | DAY TEMP DEG C | 2.2 | 3.6 | 8.3 | 15.1 | 20.0 | 24.7 | 26.7 | 26.2 | 22.5 | 16.5 | 8.9 | 3.4 | 14.8 |
| | DEG DAY C-DAY | 546 | 454 | 367 | 159 | 58 | | | | 19 | 134 | 333 | 506 | 2579 |
| | AVG TEMP DEG C | .7 | 2.1 | 6.7 | 13.3 | 18.2 | 22.9 | 24.9 | 24.4 | 20.6 | 14.5 | 7.2 | 2.0 | 13.1 |
| LA | **BATON ROUGE** | | 3032N | 9109W | | 23 M | | | | | | | | |
| | DAILY RAD MJ/SQ M | 8.91 | 11.96 | 15.65 | 19.08 | 21.24 | 21.86 | 19.81 | 19.03 | 16.62 | 14.77 | 10.44 | 8.36 | 15.64 |
| | DAY TEMP DEG C | 12.3 | 13.9 | 17.2 | 22.0 | 25.5 | 28.0 | 29.3 | 29.1 | 26.9 | 22.3 | 16.7 | 13.4 | 21.4 |
| | DEG DAY C-DAY | 251 | 186 | 116 | 18 | 0 | | | | 0 | 30 | 116 | 212 | 929 |
| | AVG TEMP DEG C | 10.6 | 12.2 | 15.4 | 20.2 | 23.8 | 26.8 | 27.8 | 27.6 | 25.3 | 20.3 | 14.8 | 11.6 | 19.7 |
| LA | **LAKE CHARLES** | | 3007N | 9313W | | 3 M | | | | | | | | |
| | DAILY RAD MJ/SQ M | 8.27 | 11.46 | 14.91 | 17.82 | 20.99 | 22.36 | 20.29 | 18.81 | 16.85 | 15.67 | 10.40 | 8.01 | 15.49 |
| | DAY TEMP DEG C | 12.8 | 14.1 | 17.3 | 22.0 | 25.5 | 28.0 | 29.5 | 29.4 | 27.0 | 23.0 | 17.5 | 14.0 | 21.8 |
| | DEG DAY C-DAY | 231 | 170 | 111 | 14 | 0 | | | | 0 | 20 | 99 | 188 | 832 |
| | AVG TEMP DEG C | 11.3 | 12.8 | 15.7 | 20.5 | 24.0 | 27.1 | 28.0 | 27.9 | 25.8 | 21.1 | 15.7 | 12.4 | 20.2 |
| LA | **NEW ORLEANS** | | 2959N | 9015W | | 3 M | | | | | | | | |
| | DAILY RAD MJ/SQ M | 9.47 | 12.62 | 16.06 | 20.20 | 22.33 | 22.74 | 20.58 | 19.48 | 17.18 | 15.15 | 11.04 | 8.84 | 16.31 |
| | DAY TEMP DEG C | 13.2 | 14.7 | 17.6 | 22.0 | 25.6 | 28.4 | 29.1 | 29.2 | 27.1 | 22.7 | 17.3 | 14.2 | 21.8 |
| | DEG DAY C-DAY | 224 | 166 | 104 | 16 | 0 | | | | 0 | 22 | 99 | 182 | 813 |
| | AVG TEMP DEG C | 11.6 | 13.1 | 15.9 | 20.3 | 23.9 | 26.9 | 27.7 | 27.7 | 25.7 | 21.0 | 15.6 | 12.7 | 20.2 |
| LA | **SHREVEPORT** | | 3228N | 9349W | | 79 M | | | | | | | | |
| | DAILY RAD MJ/SQ M | 8.65 | 11.78 | 15.22 | 18.30 | 21.41 | 23.43 | 22.86 | 21.30 | 17.63 | 14.79 | 10.54 | 8.29 | 16.19 |
| | DAY TEMP DEG C | 10.0 | 11.9 | 15.5 | 20.9 | 24.7 | 28.4 | 30.2 | 30.2 | 27.0 | 21.7 | 15.3 | 11.2 | 20.6 |
| | DEG DAY C-DAY | 307 | 231 | 162 | 36 | 3 | | | | 2 | 39 | 154 | 272 | 1204 |
| | AVG TEMP DEG C | 8.4 | 10.3 | 13.8 | 19.1 | 23.0 | 26.8 | 28.4 | 28.4 | 25.2 | 19.7 | 13.4 | 9.6 | 18.9 |
| MA | **BOSTON** | | 4222N | 7102N | | 5 M | | | | | | | | |
| | DAILY RAD MJ/SQ M | 5.40 | 8.05 | 11.53 | 15.05 | 18.39 | 20.62 | 19.85 | 16.87 | 14.30 | 10.10 | 5.71 | 4.57 | 12.54 |
| | DAY TEMP DEG C | -.4 | .3 | 4.5 | 10.5 | 16.2 | 21.4 | 24.3 | 23.2 | 19.3 | 14.3 | 8.4 | 1.6 | 12.0 |
| | DEG DAY C-DAY | 617 | 538 | 463 | 273 | 121 | 15 | | 4 | 42 | 167 | 330 | 551 | 3121 |
| | AVG TEMP DEG C | -1.6 | -.9 | 3.4 | 9.2 | 14.8 | 20.0 | 22.9 | 21.8 | 18.1 | 13.0 | 7.3 | .6 | 10.7 |
| MD | **BALTIMORE** | | 3911N | 7640W | | 47 M | | | | | | | | |
| | DAILY RAD MJ/SQ M | 6.66 | 9.53 | 13.19 | 16.89 | 19.45 | 21.33 | 20.69 | 18.15 | 15.10 | 11.32 | 7.49 | 5.67 | 13.79 |
| | DAY TEMP DEG C | 2.2 | 3.1 | 7.7 | 14.0 | 19.5 | 24.2 | 26.5 | 25.5 | 22.0 | 15.9 | 9.5 | 3.3 | 14.4 |
| | DEG DAY C-DAY | 544 | 470 | 382 | 189 | 61 | | | | 15 | 139 | 315 | 512 | 2627 |
| | AVG TEMP DEG C | .8 | 1.6 | 6.0 | 12.1 | 17.6 | 22.4 | 24.8 | 23.8 | 20.3 | 14.1 | 7.8 | 1.8 | 12.8 |

| | | JAN | FEB | MAR | APR | MAY | JUN | JUL | AUG | SEP | OCT | NOV | DEC | ANNUAL |
|---|---|---|---|---|---|---|---|---|---|---|---|---|---|---|
| **MD PATUXENT RIVER** | | | 3817N | 7625W | | 14 M | | | | | | | | |
| | DAILY RAD MJ/SQ M | 6.90 | 9.78 | 13.40 | 17.46 | 20.01 | 21.48 | 20.62 | 18.47 | 15.40 | 11.59 | 8.02 | 6.09 | 14.10 |
| | DAY TEMP DEG C | | | | | | | | | | | | | |
| | DEG DAY C-DAY | | | | | | | | | | | | | |
| | AVG TEMP DEG C | | | | | | | | | | | | | |
| **ME BANGOR** | | | 4448N | 6849W | | 62 M | | | | | | | | |
| | DAILY RAD MJ/SQ M | 5.16 | 8.23 | 12.41 | 16.35 | 19.62 | 21.07 | 21.10 | 18.29 | 14.24 | 9.52 | 5.35 | 4.30 | 12.97 |
| | DAY TEMP DEG C | | | | | | | | | | | | | |
| | DEG DAY C-DAY | | | | | | | | | | | | | |
| | AVG TEMP DEG C | | | | | | | | | | | | | |
| **ME CARIBOU** | | | 4652N | 6801W | | 190 M | | | | | | | | |
| | DAILY RAD MJ/SQ M | 4.76 | 8.22 | 12.86 | 16.05 | 17.91 | 19.94 | 20.00 | 17.03 | 12.51 | 7.81 | 4.16 | 3.52 | 12.06 |
| | DAY TEMP DEG C | -10.3 | -8.9 | -3.1 | 4.1 | 11.6 | 17.1 | 20.1 | 18.6 | 14.0 | 8.1 | .8 | -7.5 | 5.4 |
| | DEG DAY C-DAY | 935 | 811 | 713 | 472 | 263 | 94 | 47 | 68 | 182 | 365 | 560 | 842 | 5352 |
| | AVG TEMP DEG C | -11.8 | -10.6 | -4.7 | 2.6 | 9.8 | 15.3 | 18.3 | 16.8 | 12.3 | 6.6 | -.3 | -8.8 | 3.8 |
| **ME PORTLAND** | | | 4339N | 7019W | | 19 M | | | | | | | | |
| | DAILY RAD MJ/SQ M | 5.11 | 7.74 | 11.00 | 14.80 | 17.79 | 19.42 | 18.83 | 16.58 | 13.14 | 9.33 | 5.21 | 4.12 | 11.92 |
| | DAY TEMP DEG C | -4.2 | -3.3 | 1.4 | 7.6 | 13.3 | 18.6 | 21.8 | 21.0 | 16.7 | 11.3 | 5.1 | -2.0 | 9.0 |
| | DEG DAY C-DAY | 749 | 655 | 572 | 372 | 212 | 59 | 18 | 31 | 111 | 274 | 440 | 677 | 4167 |
| | AVG TEMP DEG C | -5.8 | -5.1 | -.1 | 5.9 | 11.5 | 16.8 | 20.0 | 19.1 | 14.8 | 9.5 | 3.7 | -3.5 | 7.2 |
| **MI ALPENA** | | | 4504N | 8334W | | 210 M | | | | | | | | |
| | DAILY RAD MJ/SQ M | 4.11 | 7.00 | 11.67 | 15.97 | 19.52 | 21.32 | 21.39 | 17.97 | 13.12 | 8.43 | 4.34 | 3.07 | 12.33 |
| | DAY TEMP DEG C | -6.4 | -5.9 | -1.5 | 6.4 | 12.5 | 18.3 | 20.9 | 20.1 | 15.5 | 10.4 | 2.9 | -3.5 | 7.5 |
| | DEG DAY C-DAY | 813 | 727 | 668 | 415 | 253 | 83 | 42 | 61 | 147 | 305 | 502 | 717 | 4733 |
| | AVG TEMP DEG C | -7.9 | -7.6 | -3.2 | 4.5 | 10.3 | 16.1 | 18.6 | 17.9 | 13.5 | 8.5 | 1.6 | -4.8 | 5.6 |
| **MI DETROIT** | | | 4225N | 8301W | | 191 M | | | | | | | | |
| | DAILY RAD MJ/SQ M | 4.74 | 7.72 | 11.35 | 15.88 | 19.47 | 21.18 | 20.83 | 17.88 | 14.22 | 9.94 | 5.42 | 3.90 | 12.71 |
| | DAY TEMP DEG C | -2.6 | -1.7 | 3.2 | 10.5 | 16.3 | 22.3 | 24.6 | 23.6 | 19.7 | 13.9 | 6.2 | -.4 | 11.3 |
| | DEG DAY C-DAY | 681 | 593 | 510 | 282 | 132 | 14 | 0 | 6 | 44 | 190 | 398 | 609 | 3459 |
| | AVG TEMP DEG C | -3.6 | -2.8 | 1.9 | 8.9 | 14.7 | 20.6 | 22.9 | 22.2 | 18.1 | 12.4 | 5.1 | -1.3 | 9.9 |
| **MI FLINT** | | | 4258N | 8344W | | 233 M | | | | | | | | |
| | DAILY RAD MJ/SQ M | 4.35 | 7.22 | 10.86 | 15.20 | 18.82 | 20.58 | 20.39 | 17.65 | 13.57 | 9.41 | 4.87 | 3.51 | 12.20 |
| | DAY TEMP DEG C | -4.1 | -3.2 | 1.8 | 9.5 | 15.1 | 20.6 | 22.9 | 22.0 | 18.0 | 12.4 | 4.8 | -1.7 | 9.8 |
| | DEG DAY C-DAY | 736 | 641 | 558 | 318 | 170 | 36 | 8 | 20 | 82 | 241 | 445 | 658 | 3913 |
| | AVG TEMP DEG C | -5.4 | -4.6 | .3 | 7.7 | 13.2 | 18.8 | 20.9 | 20.1 | 16.1 | 10.7 | 3.5 | -2.9 | 8.2 |
| **MI GRAND RAPIDS** | | | 4253N | 8531W | | 245 M | | | | | | | | |
| | DAILY RAD MJ/SQ M | 4.19 | 7.34 | 11.51 | 16.02 | 19.92 | 22.20 | 21.73 | 19.02 | 14.32 | 9.73 | 5.06 | 3.53 | 12.88 |
| | DAY TEMP DEG C | -3.7 | -2.8 | 2.1 | 9.9 | 15.9 | 21.6 | 23.9 | 23.1 | 18.8 | 13.0 | 5.0 | -1.5 | 10.4 |
| | DEG DAY C-DAY | 720 | 630 | 549 | 308 | 150 | 24 | 4 | 15 | 63 | 227 | 438 | 648 | 3776 |
| | AVG TEMP DEG C | -4.9 | -4.2 | .6 | 8.1 | 13.9 | 19.7 | 21.9 | 21.1 | 16.9 | 11.1 | 3.7 | -2.6 | 8.8 |

(*Continued*)

**TABLE A-15.** *(Continued)*

| | | | JAN | FEB | MAR | APR | MAY | JUN | JUL | AUG | SEP | OCT | NOV | DEC | ANNUAL |
|---|---|---|---|---|---|---|---|---|---|---|---|---|---|---|---|
| MI | HOUGHTON | | | 4710N | 8830W | | 329 M | | | | | | | | |
| | DAILY RAD | MJ/SQ M | 2.77 | 5.49 | 10.59 | 15.50 | 18.84 | 20.86 | 20.86 | 17.27 | 11.46 | 7.61 | 3.30 | 2.18 | 11.39 |
| | DAY TEMP | DEG C | | | | | | | | | | | | | |
| | DEG DAY | C-DAY | | | | | | | | | | | | | |
| | AVG TEMP | DEG C | | | | | | | | | | | | | |
| MI | SAULT STE. MARIE | | | 4628N | 8422W | | 221 M | | | | | | | | |
| | DAILY RAD | MJ/SQ M | 3.69 | 6.85 | 11.67 | 15.70 | 19.16 | 20.55 | 20.83 | 17.28 | 11.91 | 7.64 | 3.76 | 2.87 | 11.82 |
| | DAY TEMP | DEG C | -8.6 | -7.9 | -3.0 | 4.9 | 11.2 | 16.7 | 19.5 | 19.0 | 14.5 | 9.3 | 1.5 | -5.5 | 6.0 |
| | DEG DAY | C-DAY | 875 | 774 | 706 | 447 | 276 | 111 | 53 | 69 | 162 | 324 | 537 | 773 | 5107 |
| | AVG TEMP | DEG C | -9.9 | -9.3 | -4.4 | 3.4 | 9.4 | 14.8 | 17.7 | 17.3 | 12.9 | 7.9 | .4 | -6.6 | 4.5 |
| MI | TRAVERSE CITY | | | 4444N | 8535W | | 192 M | | | | | | | | |
| | DAILY RAD | MJ/SQ M | 3.53 | 6.44 | 11.36 | 15.95 | 19.62 | 21.70 | 21.67 | 18.26 | 13.22 | 8.56 | 4.28 | 2.91 | 12.29 |
| | DAY TEMP | DEG C | | | | | | | | | | | | | |
| | DEG DAY | C-DAY | 761 | 689 | 625 | 372 | 215 | 58 | 18 | 37 | 99 | 262 | 468 | 673 | 4277 |
| | AVG TEMP | DEG C | -6.2 | -6.3 | -1.8 | 5.9 | 11.6 | 17.6 | 20.4 | 19.7 | 15.2 | 9.9 | 2.7 | -3.4 | 7.1 |
| MN | DULUTH | | | 4650N | 9211W | | 432 M | | | | | | | | |
| | DAILY RAD | MJ/SQ M | 4.41 | 7.64 | 11.74 | 15.58 | 18.64 | 20.06 | 21.04 | 17.56 | 12.43 | 8.23 | 4.32 | 3.31 | 12.08 |
| | DAY TEMP | DEG C | -11.5 | -9.4 | -3.2 | 5.2 | 11.4 | 16.8 | 20.5 | 19.5 | 14.0 | 8.9 | -.8 | -8.4 | 5.2 |
| | DEG DAY | C-DAY | 973 | 823 | 715 | 440 | 269 | 108 | 37 | 58 | 177 | 339 | 610 | 872 | 5421 |
| | AVG TEMP | DEG C | -13.1 | -11.1 | -4.7 | 3.7 | 9.7 | 15.0 | 18.7 | 17.8 | 12.4 | 7.4 | -2.0 | -9.8 | 3.7 |
| MN | INTERNATIONAL FALLS | | | 4834N | 9323W | | 361 M | | | | | | | | |
| | DAILY RAD | MJ/SQ M | 4.04 | 7.52 | 11.87 | 16.38 | 19.48 | 21.03 | 21.80 | 18.37 | 12.72 | 7.99 | 3.92 | 3.08 | 12.35 |
| | DAY TEMP | DEG C | -14.9 | -11.8 | -4.4 | 5.3 | 12.1 | 17.8 | 20.8 | 19.4 | 13.5 | 8.1 | -2.7 | -11.4 | 4.3 |
| | DEG DAY | C-DAY | 1087 | 902 | 764 | 447 | 257 | 93 | 37 | 62 | 202 | 371 | 668 | 969 | 5859 |
| | AVG TEMP | DEG C | -16.7 | -13.9 | -6.3 | 3.4 | 10.1 | 15.8 | 18.8 | 17.3 | 11.7 | 6.4 | -3.9 | -12.9 | 2.5 |
| MN | MINNEAPOLIS-ST. PAUL | | | 4453N | 9313W | | 255 M | | | | | | | | |
| | DAILY RAD | MJ/SQ M | 5.27 | 8.67 | 12.52 | 16.36 | 19.72 | 21.88 | 22.36 | 19.15 | 14.24 | 9.76 | 5.45 | 4.01 | 13.28 |
| | DAY TEMP | DEG C | -9.5 | -7.0 | -.6 | 9.0 | 15.7 | 21.1 | 23.9 | 23.0 | 17.3 | 11.8 | 1.6 | -6.1 | 8.3 |
| | DEG DAY | C-DAY | 909 | 754 | 632 | 332 | 151 | 36 | 6 | 12 | 96 | 262 | 543 | 799 | 4532 |
| | AVG TEMP | DEG C | -11.0 | -8.6 | -2.1 | 7.3 | 13.9 | 19.4 | 22.2 | 21.2 | 15.6 | 10.0 | .2 | -7.4 | 6.7 |
| MN | ROCHESTER | | | 4355N | 9230W | | 402 M | | | | | | | | |
| | DAILY RAD | MJ/SQ M | 5.41 | 8.54 | 12.28 | 16.00 | 19.24 | 21.58 | 21.66 | 18.86 | 14.19 | 9.87 | 5.61 | 4.20 | 13.12 |
| | DAY TEMP | DEG C | -9.1 | -6.8 | -.9 | 8.7 | 15.3 | 20.6 | 22.9 | 22.1 | 17.0 | 11.6 | 1.7 | -5.9 | 8.1 |
| | DEG DAY | C-DAY | 897 | 748 | 641 | 342 | 162 | 43 | 12 | 19 | 103 | 269 | 540 | 794 | 4570 |
| | AVG TEMP | DEG C | -10.6 | -8.4 | -2.3 | 6.9 | 13.4 | 18.9 | 21.2 | 20.3 | 15.2 | 9.8 | .3 | -7.3 | 6.5 |
| MO | COLUMBIA | | | 3849N | 9213W | | 270 M | | | | | | | | |
| | DAILY RAD | MJ/SQ M | 6.94 | 9.93 | 13.38 | 17.32 | 21.33 | 23.71 | 24.01 | 21.31 | 16.46 | 12.49 | 7.97 | 5.93 | 15.07 |
| | DAY TEMP | DEG C | -.1 | 2.4 | 7.0 | 14.5 | 19.7 | 24.4 | 26.8 | 26.2 | 22.0 | 16.3 | 8.2 | 1.8 | 14.1 |
| | DEG DAY | C-DAY | 615 | 488 | 406 | 174 | 65 | 6 | 0 | 3 | 23 | 137 | 352 | 554 | 2823 |
| | AVG TEMP | DEG C | -1.5 | .9 | 5.4 | 12.8 | 18.0 | 22.8 | 25.2 | 24.4 | 20.2 | 14.4 | 6.6 | .4 | 12.5 |

| | | JAN | FEB | MAR | APR | MAY | JUN | JUL | AUG | SEP | OCT | NOV | DEC | ANNUAL |
|---|---|---|---|---|---|---|---|---|---|---|---|---|---|---|
| MO | KANSAS CITY | | 3918N | 9443W | | 315 M | | | | | | | | |
| | DAILY RAD | MJ/SQ M | 7.35 | 10.15 | 13.65 | 17.87 | 21.25 | 23.60 | 23.86 | 21.14 | 16.48 | 12.40 | 8.37 | 6.37 | 15.21 |
| | DAY TEMP | DEG C | -1.3 | 1.7 | 6.5 | 14.1 | 19.5 | 24.4 | 27.0 | 26.5 | 21.8 | 16.0 | 7.2 | .9 | 13.7 |
| | DEG DAY | C-DAY | 653 | 509 | 418 | 187 | 71 | | | | 28 | 144 | 378 | 581 | 2977 |
| | AVG TEMP | DEG C | -2.7 | .2 | 4.8 | 12.3 | 17.8 | 22.8 | 25.3 | 24.7 | 20.0 | 14.2 | 5.7 | -.4 | 12.1 |
| MO | SPRINGFIELD | | 3714N | 9323W | | 387 M | | | | | | | | |
| | DAILY RAD | MJ/SQ M | 7.76 | 10.51 | 14.02 | 18.21 | 21.36 | 23.55 | 23.41 | 21.26 | 16.80 | 12.98 | 8.80 | 6.84 | 15.46 |
| | DAY TEMP | DEG C | 2.2 | 4.5 | 8.5 | 15.5 | 20.2 | 24.9 | 27.3 | 27.0 | 22.7 | 17.0 | 9.3 | 3.8 | 15.3 |
| | DEG DAY | C-DAY | 553 | 436 | 367 | 153 | 52 | 6 | | 3 | 19 | 126 | 325 | 499 | 2539 |
| | AVG TEMP | DEG C | .5 | 2.8 | 6.7 | 13.6 | 18.4 | 23.1 | 25.4 | 25.1 | 20.7 | 15.0 | 7.5 | 2.2 | 13.4 |
| MO | ST. LOUIS | | 3845N | 9023W | | 172 M | | | | | | | | |
| | DAILY RAD | MJ/SQ M | 7.12 | 10.05 | 13.67 | 17.75 | 21.24 | 23.75 | 23.26 | 20.61 | 16.56 | 12.48 | 8.15 | 6.02 | 15.06 |
| | DAY TEMP | DEG C | 1.0 | 3.2 | 7.9 | 15.4 | 20.5 | 25.5 | 27.5 | 26.8 | 22.6 | 16.8 | 8.7 | 2.8 | 14.9 |
| | DEG DAY | C-DAY | 581 | 465 | 379 | 151 | 57 | | | 0 | 19 | 124 | 333 | 523 | 2638 |
| | AVG TEMP | DEG C | -.4 | 1.7 | 6.3 | 13.6 | 18.8 | 23.8 | 25.9 | 25.1 | 20.9 | 15.1 | 7.2 | 1.4 | 13.3 |
| MS | JACKSON | | 3219N | 9005W | | 101 M | | | | | | | | |
| | DAILY RAD | MJ/SQ M | 8.55 | 11.65 | 15.54 | 19.39 | 22.03 | 22.97 | 21.66 | 20.21 | 17.13 | 14.43 | 10.23 | 8.04 | 15.99 |
| | DAY TEMP | DEG C | 10.3 | 11.9 | 15.5 | 20.8 | 24.7 | 28.3 | 29.4 | 29.2 | 26.4 | 21.2 | 15.1 | 11.3 | 20.3 |
| | DEG DAY | C-DAY | 316 | 246 | 174 | 41 | 3 | | | 0 | 0 | 51 | 167 | 280 | 1278 |
| | AVG TEMP | DEG C | 8.4 | 9.9 | 13.4 | 18.7 | 22.6 | 26.3 | 27.6 | 27.3 | 24.4 | 18.8 | 12.9 | 9.4 | 18.3 |
| MS | MERIDIAN | | 3220N | 8845W | | 94 M | | | | | | | | |
| | DAILY RAD | MJ/SQ M | 8.44 | 11.49 | 15.07 | 18.86 | 21.10 | 22.27 | 20.69 | 19.74 | 16.50 | 14.27 | 10.18 | 7.94 | 15.55 |
| | DAY TEMP | DEG C | 10.2 | 11.9 | 15.5 | 20.7 | 24.6 | 28.2 | 29.2 | 29.0 | 26.0 | 20.6 | 14.6 | 10.8 | 20.1 |
| | DEG DAY | C-DAY | 319 | 246 | 173 | 44 | 4 | | | 0 | 0 | 62 | 184 | 294 | 1326 |
| | AVG TEMP | DEG C | 8.3 | 9.9 | 13.4 | 18.6 | 22.4 | 26.2 | 27.3 | 27.1 | 24.1 | 18.2 | 12.3 | 8.8 | 18.1 |
| MT | BILLINGS | | 4548N | 10832W | | 1088 M | | | | | | | | |
| | DAILY RAD | MJ/SQ M | 5.52 | 8.66 | 13.50 | 17.32 | 21.71 | 24.67 | 27.05 | 22.95 | 16.68 | 11.20 | 6.37 | 4.78 | 15.03 |
| | DAY TEMP | DEG C | -4.1 | -.9 | 1.9 | 8.9 | 14.4 | 18.8 | 24.4 | 23.5 | 17.0 | 11.6 | 3.6 | -1.4 | 9.8 |
| | DEG DAY | C-DAY | 742 | 585 | 558 | 340 | 185 | 73 | 6 | 8 | 123 | 271 | 488 | 658 | 4037 |
| | AVG TEMP | DEG C | -5.6 | -2.6 | .3 | 7.0 | 12.5 | 17.0 | 22.1 | 21.2 | 14.9 | 9.6 | 2.1 | -2.9 | 8.0 |
| MT | CUT BANK | | 4836N | 11222W | | 1170 M | | | | | | | | |
| | DAILY RAD | MJ/SQ M | 4.56 | 7.81 | 12.80 | 16.85 | 21.36 | 23.21 | 25.95 | 21.52 | 15.34 | 9.89 | 5.45 | 3.79 | 14.05 |
| | DEG DAY | C-DAY | 841 | 663 | 658 | 425 | 265 | 148 | 46 | 69 | 204 | 360 | 588 | 751 | 5018 |
| | AVG TEMP | DEG C | -8.8 | -5.3 | -2.9 | 4.2 | 9.8 | 13.6 | 18.0 | 17.0 | 11.8 | 6.7 | -1.3 | -5.9 | 4.7 |
| MT | DILLON | | 4515N | 11233W | | 1588 M | | | | | | | | |
| | DAILY RAD | MJ/SQ M | 5.97 | 9.60 | 14.52 | 18.60 | 22.57 | 24.33 | 27.15 | 22.96 | 17.26 | 11.61 | 6.83 | 5.11 | 15.54 |
| | DEG DAY | C-DAY | 772 | 614 | 609 | 398 | 252 | 132 | 30 | 47 | 181 | 344 | 553 | 708 | 4640 |
| | AVG TEMP | DEG C | -6.6 | -3.6 | -1.3 | 5.1 | 10.2 | 14.2 | 19.1 | 18.1 | 12.6 | 7.2 | -.1 | -4.5 | 5.9 |

(Continued)

## TABLE A-15. (Continued)

| | | JAN | FEB | MAR | APR | MAY | JUN | JUL | AUG | SEP | OCT | NOV | DEC | ANNUAL |
|---|---|---|---|---|---|---|---|---|---|---|---|---|---|---|
| **MT GLASGOW** | DAILY RAD MJ/SQ M | 4.40 | 4813N 7.62 | 10637N 12.54 | 700 M 16.89 | 20.74 | 23.23 | 24.89 | 21.14 | 15.21 | 9.96 | 5.43 | 3.79 | 13.82 |
| | DAY TEMP DEG C | -11.0 | -7.6 | -2.0 | 8.0 | 14.4 | 18.6 | 23.7 | 22.9 | 16.2 | 10.1 | .0 | -6.7 | 7.2 |
| | DEG DAY C-DAY | 961 | 774 | 686 | 370 | 191 | 84 | 8 | 17 | 146 | 321 | 600 | 825 | 4983 |
| | AVG TEMP DEG C | -12.7 | -9.3 | -3.8 | 6.0 | 12.3 | 16.7 | 21.4 | 20.6 | 14.0 | 8.0 | -1.7 | -8.3 | 5.3 |
| **MT GREAT FALLS** | DAILY RAD MJ/SQ M | 4.77 | 4729N 8.17 | 11122W 13.28 | 1116 M 16.89 | 20.97 | 23.85 | 26.43 | 21.94 | 15.64 | 10.49 | 5.65 | 3.81 | 14.33 |
| | DAY TEMP DEG C | -4.9 | -1.5 | .8 | 8.2 | 13.8 | 17.9 | 23.1 | 22.1 | 16.2 | 10.9 | 2.9 | -1.7 | 9.0 |
| | DEG DAY C-DAY | 767 | 597 | 594 | 360 | 204 | 90 | 10 | 23 | 144 | 291 | 507 | 663 | 4250 |
| | AVG TEMP DEG C | -6.4 | -3.0 | -1.8 | 6.3 | 11.8 | 16.0 | 20.7 | 19.7 | 14.1 | 9.1 | 1.4 | -3.1 | 7.2 |
| **MT HELENA** | DAILY RAD MJ/SQ M | 4.76 | 4636N 8.04 | 11200W 13.00 | 1188 M 16.87 | 21.11 | 23.15 | 26.49 | 21.91 | 16.03 | 10.51 | 5.91 | 4.13 | 14.33 |
| | DAY TEMP DEG C | -6.0 | -1.9 | 1.1 | 8.0 | 13.3 | 17.2 | 22.6 | 21.6 | 15.5 | 9.6 | 1.6 | -3.2 | 8.3 |
| | DEG DAY C-DAY | 808 | 616 | 592 | 372 | 223 | 108 | 18 | 32 | 169 | 339 | 555 | 718 | 4550 |
| | AVG TEMP DEG C | -7.7 | -3.7 | -.8 | 5.9 | 11.2 | 15.1 | 19.9 | 19.0 | 13.1 | 7.4 | -.2 | -4.8 | 6.2 |
| **MT LEWISTOWN** | DAILY RAD MJ/SQ M | 4.77 | 4703N 7.86 | 10927W 12.81 | 1264 M 16.39 | 20.51 | 23.37 | 25.96 | 21.58 | 15.57 | 10.27 | 5.70 | 4.12 | 14.08 |
| | DEG DAY C-DAY | 791 | 641 | 646 | 415 | 265 | 147 | 39 | 52 | 193 | 336 | 547 | 698 | 4770 |
| | AVG TEMP DEG C | -7.2 | -4.6 | -2.5 | 4.5 | 9.8 | 13.7 | 18.6 | 18.0 | 12.2 | 7.5 | .1 | -4.2 | 5.5 |
| **MT MILES CITY** | DAILY RAD MJ/SQ M | 5.19 | 4626N 8.46 | 10552W 13.45 | 803 M 17.50 | 21.51 | 24.35 | 26.02 | 22.44 | 16.39 | 10.90 | 6.25 | 4.53 | 14.75 |
| | DAY TEMP DEG C | -7.4 | -3.9 | .9 | 9.5 | 15.6 | 20.4 | 26.0 | 24.9 | 17.8 | 11.5 | 2.0 | -3.9 | 9.4 |
| | DEG DAY C-DAY | 854 | 675 | 599 | 328 | 160 | 65 | 5 | 9 | 121 | 282 | 543 | 741 | 4382 |
| | AVG TEMP DEG C | -9.2 | -5.8 | -1.0 | 7.4 | 13.5 | 18.3 | 23.6 | 22.5 | 15.5 | 9.3 | .2 | -5.6 | 7.4 |
| **MT MISSOULA** | DAILY RAD MJ/SQ M | 3.54 | 4655N 6.52 | 11405W 11.14 | 972 M 15.69 | 20.23 | 21.94 | 26.41 | 21.35 | 15.41 | 9.22 | 4.65 | 3.03 | 13.26 |
| | DAY TEMP DEG C | -4.9 | -1.2 | 2.5 | 8.7 | 13.5 | 17.2 | 22.2 | 21.2 | 15.5 | 8.8 | 1.6 | -2.9 | 8.5 |
| | DEG DAY C-DAY | 761 | 588 | 546 | 352 | 221 | 112 | 22 | 39 | 167 | 360 | 545 | 694 | 4407 |
| | AVG TEMP DEG C | -6.2 | -2.7 | .7 | 6.6 | 11.2 | 14.9 | 19.2 | 18.3 | 12.9 | 6.7 | .2 | -4.1 | 6.5 |
| **NC ASHEVILLE** | DAILY RAD MJ/SQ M | 8.19 | 3526N 11.02 | 8232W 14.82 | 661 M 18.92 | 20.48 | 21.05 | 20.16 | 18.46 | 15.44 | 13.02 | 9.63 | 7.46 | 14.89 |
| | DAY TEMP DEG C | 5.0 | 6.0 | 9.8 | 15.5 | 19.8 | 23.4 | 24.9 | 24.5 | 21.2 | 15.8 | 9.9 | 5.5 | 15.1 |
| | DEG DAY C-DAY | 467 | 398 | 329 | 155 | 56 | | | | 28 | 149 | 312 | 453 | 2355 |
| | AVG TEMP DEG C | 3.3 | 4.1 | 7.7 | 13.3 | 17.6 | 21.4 | 23.1 | 22.7 | 19.3 | 13.8 | 7.9 | 3.7 | 13.2 |
| **NC CAPE HATTERAS** | DAILY RAD MJ/SQ M | 7.78 | 3516N 10.81 | 7533W 15.05 | 2 M 20.13 | 22.26 | 23.10 | 21.80 | 19.35 | 16.69 | 12.90 | 9.91 | 7.47 | 15.61 |
| | DAY TEMP DEG C | 8.6 | 8.9 | 11.5 | 16.2 | 20.6 | 24.0 | 26.5 | 26.3 | 24.1 | 19.5 | 14.5 | 9.9 | 17.6 |
| | DEG DAY C-DAY | 339 | 299 | 254 | 104 | 26 | | | | | 42 | 154 | 298 | 1516 |
| | AVG TEMP DEG C | 7.4 | 7.7 | 10.3 | 14.9 | 19.4 | 23.5 | 25.6 | 25.3 | 23.2 | 18.4 | 13.3 | 8.7 | 16.5 |

**NC CHARLOTTE** — 3513N 8056W 234 M

| | Units | JAN | FEB | MAR | APR | MAY | JUN | JUL | AUG | SEP | OCT | NOV | DEC | ANNUAL |
|---|---|---|---|---|---|---|---|---|---|---|---|---|---|---|
| DAILY RAD | MJ/SQ M | 8.16 | 11.02 | 14.95 | 19.24 | 21.06 | 21.80 | 20.78 | 19.24 | 16.06 | 13.32 | 9.82 | 7.63 | 15.26 |
| DAY TEMP | DEG C | 7.3 | 8.5 | 12.3 | 18.0 | 22.3 | 26.1 | 27.5 | 27.0 | 23.9 | 18.4 | 12.5 | 7.5 | 17.6 |
| DEG DAY | C-DAY | 394 | 327 | 256 | 81 | 19 | 0 | 0 | 0 | 6 | 84 | 233 | 388 | 1788 |
| AVG TEMP | DEG C | 5.6 | 6.7 | 10.3 | 16.0 | 20.4 | 24.4 | 25.8 | 25.4 | 22.2 | 16.5 | 10.6 | 5.8 | 15.8 |

**NC CHERRY POINT** — 3454N 7653W 11 M

| | Units | JAN | FEB | MAR | APR | MAY | JUN | JUL | AUG | SEP | OCT | NOV | DEC | ANNUAL |
|---|---|---|---|---|---|---|---|---|---|---|---|---|---|---|
| DAILY RAD | MJ/SQ M | 8.59 | 11.63 | 15.73 | 20.36 | 21.84 | 22.00 | 20.76 | 18.55 | 16.20 | 13.27 | 10.29 | 8.15 | 15.62 |
| DAY TEMP | DEG C | | | | | | | | | | | | | |
| DEG DAY | C-DAY | | | | | | | | | | | | | |
| AVG TEMP | DEG C | | | | | | | | | | | | | |

**NC GREENSBORO** — 3605N 7957W 270 M

| | Units | JAN | FEB | MAR | APR | MAY | JUN | JUL | AUG | SEP | OCT | NOV | DEC | ANNUAL |
|---|---|---|---|---|---|---|---|---|---|---|---|---|---|---|
| DAILY RAD | MJ/SQ M | 8.12 | 11.01 | 14.90 | 19.10 | 21.20 | 22.17 | 21.15 | 19.25 | 16.09 | 12.95 | 9.52 | 7.47 | 15.25 |
| DAY TEMP | DEG C | 5.4 | 6.6 | 10.7 | 16.8 | 21.5 | 25.4 | 26.8 | 26.1 | 22.7 | 17.1 | 11.0 | 5.9 | 16.3 |
| DEG DAY | C-DAY | 453 | 379 | 302 | 113 | 33 | 0 | 0 | 0 | 13 | 116 | 278 | 437 | 2124 |
| AVG TEMP | DEG C | 3.7 | 4.8 | 8.8 | 14.8 | 19.5 | 23.6 | 25.1 | 24.4 | 20.9 | 15.1 | 9.1 | 4.2 | 14.5 |

**NC RALEIGH-DURHAM** — 3552N 7847W 134 M

| | Units | JAN | FEB | MAR | APR | MAY | JUN | JUL | AUG | SEP | OCT | NOV | DEC | ANNUAL |
|---|---|---|---|---|---|---|---|---|---|---|---|---|---|---|
| DAILY RAD | MJ/SQ M | 7.87 | 10.48 | 14.48 | 18.66 | 20.52 | 21.16 | 20.15 | 18.29 | 15.63 | 12.54 | 9.22 | 7.21 | 14.70 |
| DAY TEMP | DEG C | 6.5 | 7.5 | 11.5 | 17.4 | 21.7 | 25.4 | 27.0 | 26.4 | 23.3 | 17.7 | 12.0 | 6.9 | 16.9 |
| DEG DAY | C-DAY | 422 | 354 | 279 | 100 | 27 | 0 | 0 | 0 | 7 | 103 | 250 | 410 | 1952 |
| AVG TEMP | DEG C | 4.7 | 5.7 | 9.6 | 15.3 | 19.7 | 23.6 | 25.3 | 24.7 | 21.4 | 15.7 | 10.0 | 5.1 | 15.1 |

**ND BISMARCK** — 4646N 10045W 502 M

| | Units | JAN | FEB | MAR | APR | MAY | JUN | JUL | AUG | SEP | OCT | NOV | DEC | ANNUAL |
|---|---|---|---|---|---|---|---|---|---|---|---|---|---|---|
| DAILY RAD | MJ/SQ M | 5.30 | 8.80 | 13.26 | 16.56 | 20.97 | 23.38 | 24.78 | 21.30 | 15.37 | 10.30 | 5.76 | 4.23 | 14.17 |
| DAY TEMP | DEG C | -11.4 | 8.4 | -2.1 | 8.1 | 14.6 | 19.7 | 23.8 | 23.0 | 16.5 | 10.5 | .0 | -7.4 | 7.2 |
| DEG DAY | C-DAY | 978 | 801 | 687 | 367 | 188 | 68 | 10 | 19 | 140 | 313 | 602 | 851 | 5024 |
| AVG TEMP | DEG C | -13.2 | -10.3 | -3.8 | 6.1 | 12.4 | 17.7 | 21.6 | 20.7 | 14.2 | 8.2 | -1.7 | -9.1 | 5.2 |

**ND FARGO** — 4654N 9648W 274 M

| | Units | JAN | FEB | MAR | APR | MAY | JUN | JUL | AUG | SEP | OCT | NOV | DEC | ANNUAL |
|---|---|---|---|---|---|---|---|---|---|---|---|---|---|---|
| DAILY RAD | MJ/SQ M | 4.71 | 8.01 | 12.46 | 16.75 | 20.82 | 22.63 | 24.06 | 20.71 | 14.80 | 9.92 | 5.19 | 3.83 | 13.66 |
| DAY TEMP | DEG C | -12.9 | 10.2 | -2.8 | 7.4 | 14.6 | 20.0 | 23.5 | 22.7 | 16.3 | 10.2 | -.5 | -9.1 | 6.6 |
| DEG DAY | C-DAY | 1018 | 844 | 703 | 378 | 186 | 54 | 7 | 18 | 130 | 310 | 607 | 896 | 5151 |
| AVG TEMP | DEG C | -14.5 | -11.8 | -4.3 | 5.7 | 12.6 | 18.2 | 21.5 | 20.7 | 14.4 | 8.3 | -1.9 | -10.6 | 4.9 |

**ND MINOT** — 4816N 10117W 522 M

| | Units | JAN | FEB | MAR | APR | MAY | JUN | JUL | AUG | SEP | OCT | NOV | DEC | ANNUAL |
|---|---|---|---|---|---|---|---|---|---|---|---|---|---|---|
| DAILY RAD | MJ/SQ M | 4.35 | 7.44 | 11.85 | 16.58 | 20.95 | 22.41 | 23.80 | 20.43 | 14.49 | 9.64 | 4.98 | 3.52 | 13.37 |
| DAY TEMP | DEG C | | | | | | | | | | | | | |
| DEG DAY | C-DAY | 983 | 812 | 713 | 398 | 213 | 83 | 15 | 39 | 159 | 326 | 618 | 866 | 5225 |
| AVG TEMP | DEG C | -13.4 | -10.7 | -4.7 | 5.1 | 11.6 | 16.7 | 20.4 | 19.6 | 13.4 | 7.8 | -2.3 | -9.6 | 4.5 |

**NE GRAND ISLAND** — 4058N 9819W 566 M

| | Units | JAN | FEB | MAR | APR | MAY | JUN | JUL | AUG | SEP | OCT | NOV | DEC | ANNUAL |
|---|---|---|---|---|---|---|---|---|---|---|---|---|---|---|
| DAILY RAD | MJ/SQ M | 7.50 | 10.41 | 14.36 | 19.21 | 22.38 | 25.45 | 25.14 | 22.01 | 17.13 | 12.91 | 8.38 | 6.46 | 15.95 |
| DAY TEMP | DEG C | -3.6 | -.6 | 3.8 | 12.0 | 17.9 | 23.4 | 26.7 | 26.0 | 20.1 | 14.3 | 5.5 | -1.0 | 12.1 |
| DEG DAY | C-DAY | 736 | 580 | 508 | 256 | 102 | 19 | 3 | 3 | 59 | 201 | 447 | 654 | 3568 |
| AVG TEMP | DEG C | -5.4 | -2.4 | 1.9 | 9.9 | 15.9 | 21.5 | 24.6 | 23.9 | 18.0 | 12.1 | 3.4 | -2.8 | 10.1 |

*(Continued)*

**TABLE A-15.** *(Continued)*

|  |  | JAN | FEB | MAR | APR | MAY | JUN | JUL | AUG | SEP | OCT | NOV | DEC | ANNUAL |
|---|---|---|---|---|---|---|---|---|---|---|---|---|---|---|
| **NE** | **NORTH OMAHA** |  | 4122N | 9601W |  | 404 M |  |  |  |  |  |  |  |  |
|  | DAILY RAD MJ/SQ M | 7.19 | 10.12 | 13.87 | 17.69 | 21.25 | 24.09 | 23.91 | 21.09 | 15.58 | 11.91 | 7.31 | 5.80 | 14.99 |
|  | DAY TEMP DEG C | -5.1 | -2.1 | 3.0 | 11.8 | 17.8 | 22.9 | 25.7 | 24.9 | 19.8 | 14.4 | 4.9 | -2.1 | 11.3 |
|  | DEG DAY C-DAY | 772 | 614 | 523 | 253 | 103 | 18 | 4 | 6 | 55 | 190 | 452 | 677 | 3667 |
|  | AVG TEMP DEG C | -6.6 | -3.6 | 1.4 | 10.0 | 16.1 | 21.2 | 23.9 | 23.2 | 18.0 | 12.4 | 3.3 | -3.5 | 9.7 |
| **NE** | **NORTH PLATTE** |  | 4108N | 10041W |  | 849 M |  |  |  |  |  |  |  |  |
|  | DAILY RAD MJ/SQ M | 7.86 | 10.88 | 15.13 | 19.56 | 22.56 | 25.72 | 25.84 | 22.58 | 17.76 | 13.36 | 8.62 | 6.87 | 16.39 |
|  | DAY TEMP DEG C | -2.6 | -1.0 | 3.4 | 11.0 | 16.8 | 22.1 | 25.7 | 25.1 | 19.3 | 13.2 | 4.7 | -.7 | 11.5 |
|  | DEG DAY C-DAY | 717 | 574 | 529 | 290 | 132 | 36 | 4 | 4 | 78 | 244 | 480 | 658 | 3746 |
|  | AVG TEMP DEG C | -4.8 | -2.2 | 1.3 | 8.8 | 14.6 | 20.0 | 23.5 | 22.8 | 16.8 | 10.6 | 2.3 | -2.9 | 9.2 |
| **NE** | **SCOTTSBLUFF** |  | 4152N | 10336W |  | 1206 M |  |  |  |  |  |  |  |  |
|  | DAILY RAD MJ/SQ M | 7.67 | 10.79 | 14.84 | 18.93 | 21.94 | 25.38 | 25.92 | 22.69 | 18.15 | 12.99 | 8.21 | 6.53 | 16.17 |
|  | DAY TEMP DEG C | -1.7 | .9 | 3.6 | 10.3 | 15.9 | 21.1 | 25.7 | 24.5 | 18.9 | 12.7 | 4.7 | -.2 | 11.4 |
|  | DEG DAY C-DAY | 691 | 552 | 529 | 313 | 156 | 51 | 0 | 4 | 89 | 255 | 480 | 644 | 3764 |
|  | AVG TEMP DEG C | -3.9 | -1.4 | 1.3 | 7.9 | 13.6 | 18.8 | 23.2 | 22.0 | 16.2 | 10.1 | 2.3 | -2.4 | 9.0 |
| **NH** | **CONCORD** |  | 4312N | 7130W |  | 105 M |  |  |  |  |  |  |  |  |
|  | DAILY RAD MJ/SQ M | 5.21 | 7.79 | 11.05 | 14.95 | 17.96 | 19.34 | 19.00 | 16.52 | 12.94 | 9.27 | 5.25 | 4.11 | 11.95 |
|  | DAY TEMP DEG C | -4.6 | -3.4 | 1.9 | 8.9 | 15.1 | 20.3 | 23.1 | 21.7 | 17.4 | 11.8 | 5.0 | -2.4 | 9.6 |
|  | DEG DAY C-DAY | 764 | 659 | 563 | 347 | 175 | 32 | 9 | 25 | 101 | 271 | 450 | 692 | 4088 |
|  | AVG TEMP DEG C | -6.3 | -5.2 | .2 | 6.8 | 12.8 | 18.2 | 20.9 | 19.6 | 15.3 | 9.6 | 3.3 | -4.0 | 7.6 |
| **NJ** | **LAKEHURST** |  | 4002N | 7420W |  | 37 M |  |  |  |  |  |  |  |  |
|  | DAILY RAD MJ/SQ M | 6.35 | 9.04 | 12.58 | 16.52 | 18.98 | 20.14 | 19.33 | 17.39 | 14.31 | 10.85 | 7.05 | 5.39 | 13.16 |
|  | DAY TEMP DEG C |  |  |  |  |  |  |  |  |  |  |  |  |  |
|  | DEG DAY C-DAY |  |  |  |  |  |  |  |  |  |  |  |  |  |
|  | AVG TEMP DEG C |  |  |  |  |  |  |  |  |  |  |  |  |  |
| **NJ** | **NEWARK** |  | 4042N | 7410W |  | 9 M |  |  |  |  |  |  |  |  |
|  | DAILY RAD MJ/SQ M | 6.26 | 9.00 | 12.58 | 16.44 | 19.15 | 20.37 | 19.97 | 17.76 | 14.45 | 10.79 | 6.77 | 5.16 | 13.22 |
|  | DAY TEMP DEG C | .9 | 1.6 | 6.1 | 12.5 | 18.2 | 23.5 | 26.2 | 25.2 | 21.4 | 15.7 | 9.2 | 2.6 | 13.6 |
|  | DEG DAY C-DAY | 579 | 504 | 420 | 222 | 79 | 0 | 0 | 0 | 19 | 135 | 313 | 526 | 2797 |
|  | AVG TEMP DEG C | -.3 | .3 | 4.8 | 10.9 | 16.6 | 21.9 | 24.7 | 23.7 | 19.9 | 14.2 | 7.9 | 1.4 | 12.2 |
| **NM** | **ALBUQUERQUE** |  | 3503N | 10637W |  | 1619 M |  |  |  |  |  |  |  |  |
|  | DAILY RAD MJ/SQ M | 11.54 | 15.23 | 20.06 | 25.29 | 28.80 | 30.40 | 28.24 | 25.99 | 22.38 | 17.55 | 12.87 | 10.53 | 20.74 |
|  | DAY TEMP DEG C | 3.7 | 6.5 | 9.9 | 15.6 | 20.9 | 26.1 | 28.2 | 27.0 | 23.4 | 16.8 | 9.0 | 4.2 | 16.0 |
|  | DEG DAY C-DAY | 513 | 389 | 331 | 157 | 32 | 0 | 0 | 0 | 4 | 121 | 342 | 496 | 2385 |
|  | AVG TEMP DEG C | 1.8 | 4.4 | 8.2 | 13.2 | 18.5 | 23.7 | 25.9 | 24.8 | 21.2 | 14.6 | 6.9 | 2.3 | 13.7 |
| **NM** | **CLAYTON** |  | 3627N | 10309W |  | 1515 M |  |  |  |  |  |  |  |  |
|  | DAILY RAD MJ/SQ M | 10.91 | 14.08 | 18.74 | 23.15 | 25.22 | 27.44 | 25.92 | 23.80 | 20.45 | 16.27 | 11.66 | 9.77 | 18.95 |
|  | DAY TEMP DEG C | 3.0 | 4.6 | 7.2 | 12.9 | 18.0 | 23.0 | 25.4 | 24.7 | 20.6 | 15.1 | 8.2 | 4.1 | 13.9 |
|  | DEG DAY C-DAY | 549 | 449 | 424 | 239 | 96 | 21 | 0 | 3 | 41 | 180 | 378 | 515 | 2895 |
|  | AVG TEMP DEG C | .6 | 2.3 | 4.7 | 10.4 | 15.6 | 20.7 | 23.1 | 22.4 | 18.3 | 12.7 | 5.7 | 1.7 | 11.5 |

| | JAN | FEB | MAR | APR | MAY | JUN | JUL | AUG | SEP | OCT | NOV | DEC | ANNUAL |
|---|---|---|---|---|---|---|---|---|---|---|---|---|---|
| **NM FARMINGTON** (3645N, 10814W, 1677 M) | | | | | | | | | | | | | |
| DAILY RAD MJ/SQ M | 10.72 | 14.54 | 19.22 | 24.21 | 27.82 | 30.25 | 28.13 | 25.56 | 21.95 | 16.78 | 11.88 | 9.50 | 20.05 |
| DAY TEMP DEG C | | | | | | | | | | | | | |
| DEG DAY C-DAY | 627 | 467 | 420 | 258 | 102 | 20 | 0 | 3 | 37 | 208 | 430 | 601 | 3173 |
| AVG TEMP DEG C | -1.9 | 1.7 | 4.8 | 9.6 | 15.3 | 19.9 | 23.9 | 22.6 | 18.1 | 11.6 | 4.0 | -1.1 | 10.7 |
| **NM ROSWELL** (3324N, 10432W, 1103 M) | | | | | | | | | | | | | |
| DAILY RAD MJ/SQ M | 11.88 | 15.58 | 20.51 | 25.17 | 27.91 | 29.62 | 27.70 | 25.44 | 21.71 | 17.33 | 12.84 | 10.80 | 20.54 |
| DAY TEMP DEG C | 6.3 | 9.1 | 12.7 | 18.5 | 23.3 | 27.9 | 28.8 | 28.1 | 24.0 | 18.2 | 11.3 | 7.0 | 17.9 |
| DEG DAY C-DAY | 463 | 344 | 271 | 103 | 11 | 0 | 0 | 3 | 9 | 106 | 302 | 443 | 2054 |
| AVG TEMP DEG C | 3.4 | 6.1 | 9.6 | 15.4 | 20.3 | 25.0 | 26.2 | 25.5 | 21.3 | 15.3 | 8.3 | 4.1 | 15.0 |
| **NM TRUTH OR CONSEQUENCES** (3314N, 10716W, 1481 M) | | | | | | | | | | | | | |
| DAILY RAD MJ/SQ M | 12.68 | 16.47 | 21.41 | 26.53 | 29.02 | 30.07 | 26.84 | 25.15 | 22.02 | 17.92 | 13.81 | 11.38 | 21.11 |
| DAY TEMP DEG C | | | | | | | | | | | | | |
| DEG DAY C-DAY | 431 | 313 | 255 | 104 | 11 | 0 | 0 | 0 | 3 | 80 | 272 | 417 | 1886 |
| AVG TEMP DEG C | 4.4 | 7.2 | 10.1 | 15.3 | 20.3 | 24.9 | 26.3 | 25.2 | 22.0 | 16.3 | 9.3 | 4.9 | 15.1 |
| **NM TUCUMCARI** (3511N, 10336W, 1231 M) | | | | | | | | | | | | | |
| DAILY RAD MJ/SQ M | 11.45 | 14.71 | 19.43 | 23.81 | 26.26 | 28.19 | 26.66 | 24.56 | 20.76 | 16.38 | 12.17 | 10.33 | 19.56 |
| DAY TEMP DEG C | | | | | | | | | | | | | |
| DEG DAY C-DAY | 482 | 372 | 315 | 144 | 32 | 4 | 0 | 0 | 11 | 121 | 313 | 454 | 2248 |
| AVG TEMP DEG C | 2.8 | 5.1 | 8.2 | 13.8 | 18.7 | 23.9 | 25.8 | 24.6 | 20.9 | 14.8 | 7.9 | 3.7 | 14.2 |
| **NM ZUNI** (3506N, 10848W, 1965 M) | | | | | | | | | | | | | |
| DAILY RAD MJ/SQ M | 11.19 | 14.72 | 19.15 | 24.59 | 28.07 | 29.53 | 25.70 | 23.59 | 21.50 | 16.98 | 12.35 | 10.13 | 19.79 |
| DAY TEMP DEG C | | | | | | | | | | | | | |
| DEG DAY C-DAY | 598 | 473 | 437 | 282 | 147 | 38 | 0 | 7 | 51 | 216 | 415 | 568 | 3232 |
| AVG TEMP DEG C | -0.9 | 1.4 | 4.2 | 8.9 | 13.7 | 18.6 | 21.9 | 20.8 | 17.4 | 11.4 | 4.5 | .0 | 10.2 |
| **NV ELKO** (4050N, 11547W, 1547 M) | | | | | | | | | | | | | |
| DAILY RAD MJ/SQ M | 7.82 | 11.74 | 16.60 | 21.56 | 26.14 | 28.75 | 29.77 | 26.28 | 21.48 | 15.01 | 9.22 | 7.00 | 18.45 |
| DAY TEMP DEG C | -2.8 | .5 | 3.9 | 8.9 | 13.8 | 18.3 | 24.3 | 23.0 | 17.8 | 11.4 | 3.9 | -1.3 | 10.1 |
| DEG DAY C-DAY | 720 | 557 | 517 | 358 | 226 | 106 | 15 | 33 | 138 | 312 | 503 | 673 | 4158 |
| AVG TEMP DEG C | -4.9 | -1.6 | 1.7 | 6.4 | 11.1 | 15.3 | 20.8 | 19.4 | 14.2 | 8.3 | 1.6 | -3.4 | 7.4 |
| **NV ELY** (3917N, 11451W, 1906 M) | | | | | | | | | | | | | |
| DAILY RAD MJ/SQ M | 9.30 | 12.95 | 18.23 | 22.80 | 26.22 | 28.52 | 27.77 | 25.31 | 21.96 | 15.97 | 10.51 | 8.20 | 18.98 |
| DAY TEMP DEG C | -2.3 | .0 | 2.7 | 7.7 | 12.7 | 17.3 | 22.7 | 21.7 | 17.0 | 10.7 | 3.6 | -.9 | 9.4 |
| DEG DAY C-DAY | 713 | 577 | 554 | 395 | 261 | 134 | 13 | 34 | 147 | 327 | 517 | 668 | 4340 |
| AVG TEMP DEG C | -4.7 | -2.3 | .4 | 5.2 | 10.0 | 14.3 | 19.6 | 18.6 | 13.7 | 7.8 | 1.1 | -3.2 | 6.7 |
| **NV LAS VEGAS** (3605N, 11510W, 664 M) | | | | | | | | | | | | | |
| DAILY RAD MJ/SQ M | 11.10 | 15.20 | 20.69 | 26.32 | 30.03 | 31.52 | 29.38 | 26.72 | 23.12 | 17.47 | 12.32 | 9.99 | 21.16 |
| DAY TEMP DEG C | 8.7 | 11.5 | 14.8 | 19.9 | 25.6 | 30.4 | 34.0 | 33.1 | 29.2 | 21.8 | 13.9 | 9.2 | 21.0 |
| DEG DAY C-DAY | 358 | 251 | 180 | 70 | 6 | 0 | 0 | 0 | 0 | 41 | 198 | 341 | 1445 |
| AVG TEMP DEG C | 6.8 | 9.5 | 12.7 | 17.7 | 22.9 | 27.9 | 32.0 | 30.8 | 26.7 | 19.5 | 11.8 | 7.3 | 18.8 |

*(Continued)*

## TABLE A-15. (Continued)

### NV LOVELOCK (4004N, 11833W, 1190 M)

| Measure | Units | JAN | FEB | MAR | APR | MAY | JUN | JUL | AUG | SEP | OCT | NOV | DEC | ANNUAL |
|---|---|---|---|---|---|---|---|---|---|---|---|---|---|---|
| DAILY RAD | MJ/SQ M | 9.12 | 13.23 | 18.80 | 24.57 | 28.99 | 31.20 | 31.59 | 28.19 | 23.01 | 16.47 | 10.55 | 8.11 | 20.32 |
| DAY TEMP | DEG C | | | | | | | | | | | | | |
| DEG DAY | C-DAY | 622 | 463 | 429 | 275 | 142 | 48 | 0 | 9 | 70 | 238 | 443 | 589 | 3328 |
| AVG TEMP | DEG C | -1.7 | 1.8 | 4.5 | 9.2 | 14.2 | 18.7 | 22.8 | 21.8 | 17.1 | 10.7 | 3.6 | -1.7 | 7.4 |

### NV RENO (3930N, 11947W, 1341 M)

| Measure | Units | JAN | FEB | MAR | APR | MAY | JUN | JUL | AUG | SEP | OCT | NOV | DEC | ANNUAL |
|---|---|---|---|---|---|---|---|---|---|---|---|---|---|---|
| DAILY RAD | MJ/SQ M | 9.08 | 13.05 | 18.72 | 24.51 | 28.63 | 30.66 | 30.55 | 27.30 | 22.67 | 16.24 | 10.35 | 8.01 | 19.98 |
| DAY TEMP | DEG C | 2.2 | 5.2 | 7.2 | 11.1 | 15.5 | 19.5 | 24.4 | 23.1 | 19.3 | 13.4 | 7.2 | 2.8 | 12.6 |
| DEG DAY | C-DAY | 570 | 434 | 426 | 303 | 182 | 81 | 9 | 28 | 93 | 253 | 415 | 551 | 3345 |
| AVG TEMP | DEG C | -.1 | 2.8 | 4.6 | 8.2 | 12.6 | 16.4 | 20.7 | 19.4 | 15.7 | 10.2 | 4.5 | .6 | 9.6 |

### NV TONOPAH (3804N, 11708W, 1653 M)

| Measure | Units | JAN | FEB | MAR | APR | MAY | JUN | JUL | AUG | SEP | OCT | NOV | DEC | ANNUAL |
|---|---|---|---|---|---|---|---|---|---|---|---|---|---|---|
| DAILY RAD | MJ/SQ M | 10.42 | 14.46 | 20.17 | 25.54 | 29.25 | 31.64 | 30.67 | 27.67 | 23.18 | 17.26 | 11.70 | 9.38 | 20.95 |
| DAY TEMP | DEG C | | | | | | | | | | | | | |
| DEG DAY | C-DAY | 599 | 473 | 437 | 284 | 149 | 51 | 0 | 7 | 60 | 226 | 420 | 570 | 3276 |
| AVG TEMP | DEG C | -1.0 | 1.4 | 4.2 | 8.9 | 13.8 | 18.5 | 22.8 | 21.5 | 17.5 | 11.2 | 4.3 | -.1 | 10.3 |

### NV WINNEMUCCA (4054N, 11748W, 1323 M)

| Measure | Units | JAN | FEB | MAR | APR | MAY | JUN | JUL | AUG | SEP | OCT | NOV | DEC | ANNUAL |
|---|---|---|---|---|---|---|---|---|---|---|---|---|---|---|
| DAILY RAD | MJ/SQ M | 7.84 | 11.66 | 16.71 | 22.33 | 26.80 | 29.16 | 30.39 | 26.65 | 21.64 | 15.00 | 9.19 | 7.02 | 18.70 |
| DAY TEMP | DEG C | -.0 | 3.3 | 5.5 | 9.9 | 14.9 | 19.3 | 25.0 | 23.4 | 18.6 | 12.2 | 5.4 | 1.2 | 11.6 |
| DEG DAY | C-DAY | 634 | 481 | 472 | 332 | 199 | 83 | 3 | 23 | 111 | 288 | 462 | 596 | 3684 |
| AVG TEMP | DEG C | -2.1 | 1.2 | 3.1 | 7.3 | 12.1 | 16.5 | 21.7 | 19.9 | 15.1 | 9.1 | 2.9 | -.9 | 8.8 |

### NV YUCCA FLATS (3657N, 11603W, 1197 M)

| Measure | Units | JAN | FEB | MAR | APR | MAY | JUN | JUL | AUG | SEP | OCT | NOV | DEC | ANNUAL |
|---|---|---|---|---|---|---|---|---|---|---|---|---|---|---|
| DAILY RAD | MJ/SQ M | 10.82 | 14.45 | 20.02 | 25.49 | 29.25 | 31.02 | 30.11 | 27.03 | 22.94 | 17.21 | 11.82 | 9.68 | 20.82 |
| DAY TEMP | DEG C | | | | | | | | | | | | | |
| DEG DAY | C-DAY | | | | | | | | | | | | | |
| AVG TEMP | DEG C | | | | | | | | | | | | | |

### NY ALBANY (4245N, 7348W, 89 M)

| Measure | Units | JAN | FEB | MAR | APR | MAY | JUN | JUL | AUG | SEP | OCT | NOV | DEC | ANNUAL |
|---|---|---|---|---|---|---|---|---|---|---|---|---|---|---|
| DAILY RAD | MJ/SQ M | 5.18 | 7.81 | 11.19 | 15.15 | 17.82 | 19.63 | 19.58 | 17.01 | 13.28 | 9.27 | 5.19 | 4.04 | 12.10 |
| DAY TEMP | DEG C | -4.3 | -3.2 | 2.3 | 10.1 | 16.3 | 21.7 | 24.2 | 22.9 | 18.6 | 12.7 | 5.6 | -2.0 | 10.4 |
| DEG DAY | C-DAY | 749 | 646 | 544 | 302 | 141 | 22 | 5 | 12 | 75 | 234 | 423 | 673 | 3826 |
| AVG TEMP | DEG C | -5.8 | -4.7 | .8 | 8.3 | 14.3 | 19.7 | 22.2 | 20.9 | 16.6 | 10.8 | 4.2 | -3.4 | 8.7 |

### NY BINGHAMTON (4213N, 7559W, 499 M)

| Measure | Units | JAN | FEB | MAR | APR | MAY | JUN | JUL | AUG | SEP | OCT | NOV | DEC | ANNUAL |
|---|---|---|---|---|---|---|---|---|---|---|---|---|---|---|
| DAILY RAD | MJ/SQ M | 4.38 | 6.53 | 9.77 | 14.09 | 16.98 | 19.08 | 18.83 | 16.17 | 12.84 | 8.84 | 4.70 | 3.37 | 11.30 |
| DAY TEMP | DEG C | -3.9 | -3.9 | .9 | 8.6 | 14.5 | 19.8 | 22.2 | 21.2 | 17.2 | 11.6 | 4.5 | -2.7 | 9.1 |
| DEG DAY | C-DAY | 741 | 657 | 581 | 338 | 178 | 42 | 12 | 22 | 96 | 253 | 447 | 682 | 4049 |
| AVG TEMP | DEG C | -5.6 | -5.1 | -.4 | 7.1 | 12.8 | 18.2 | 20.6 | 19.6 | 15.7 | 10.2 | 3.4 | -3.7 | 7.7 |

### NY BUFFALO (4256N, 7844W, 215 M)

| Measure | Units | JAN | FEB | MAR | APR | MAY | JUN | JUL | AUG | SEP | OCT | NOV | DEC | ANNUAL |
|---|---|---|---|---|---|---|---|---|---|---|---|---|---|---|
| DAILY RAD | MJ/SQ M | 3.96 | 6.20 | 10.08 | 14.92 | 18.12 | 20.47 | 20.16 | 17.17 | 13.07 | 8.90 | 4.58 | 3.21 | 11.74 |
| DAY TEMP | DEG C | -3.6 | -3.1 | 1.2 | 8.6 | 14.4 | 20.3 | 22.7 | 21.8 | 18.0 | 12.3 | 5.4 | -1.3 | 9.7 |
| DEG DAY | C-DAY | 711 | 632 | 567 | 335 | 178 | 32 | 7 | 18 | 77 | 233 | 420 | 639 | 3849 |
| AVG TEMP | DEG C | -4.6 | -4.2 | .1 | 7.2 | 12.8 | 18.7 | 21.2 | 20.2 | 16.4 | 10.8 | 4.3 | -2.3 | 8.4 |

| State | Station | Variable | Units | JAN | FEB | MAR | APR | MAY | JUN | JUL | AUG | SEP | OCT | NOV | DEC | ANNUAL |
|---|---|---|---|---|---|---|---|---|---|---|---|---|---|---|---|---|
| NY | MASSENA (4456N 7451W 63M) | DAILY RAD | MJ/SQ M | 4.44 | 7.04 | 11.09 | 15.24 | 18.31 | 20.19 | 19.87 | 16.84 | 12.75 | 8.36 | 4.40 | 3.34 | 11.82 |
| | | DAY TEMP | DEG C | | | | | | | | | | | | | |
| | | DEG DAY | C-DAY | 870 | 751 | 644 | 380 | 194 | 43 | 12 | 32 | 107 | 284 | 485 | 773 | 4575 |
| | | AVG TEMP | DEG C | -9.7 | -8.5 | -2.4 | 5.7 | 12.3 | 16.9 | 19.8 | 18.0 | 15.1 | 9.2 | 2.2 | -6.6 | 6.0 |
| NY | NEW YORK CITY (CENTRAL PARK) (4047N 7358W 57M) | DAILY RAD | MJ/SQ M | 5.68 | 8.18 | 11.77 | 15.48 | 18.57 | 19.41 | 19.15 | 16.83 | 13.77 | 10.16 | 6.05 | 4.58 | 12.47 |
| | | DAY TEMP | DEG C | 1.2 | 1.9 | 6.3 | 12.6 | 18.3 | 23.5 | 26.0 | 25.2 | 21.6 | 16.2 | 9.7 | 2.9 | 13.8 |
| | | DEG DAY | C-DAY | 565 | 492 | 412 | 215 | 76 | 0 | 0 | 0 | 16 | 116 | 293 | 508 | 2693 |
| | | AVG TEMP | DEG C | .1 | .8 | 5.1 | 11.2 | 16.8 | 22.0 | 24.8 | 23.8 | 20.2 | 14.8 | 8.6 | 1.9 | 12.5 |
| NY | NEW YORK CITY (LA GUARDIA) (4046N 7354W 16M) | DAILY RAD | MJ/SQ M | 6.21 | 9.02 | 12.68 | 16.53 | 19.18 | 20.45 | 20.25 | 17.97 | 14.53 | 10.79 | 6.73 | 5.18 | 13.29 |
| | | DAY TEMP | DEG C | 1.0 | 1.6 | 5.9 | 12.1 | 17.9 | 23.3 | 26.1 | 25.0 | 21.2 | 15.7 | 9.5 | 2.9 | 13.5 |
| | | DEG DAY | C-DAY | 567 | 496 | 420 | 222 | 81 | 0 | 0 | 0 | 17 | 124 | 295 | 506 | 2728 |
| | | AVG TEMP | DEG C | .1 | .6 | 4.8 | 10.9 | 16.6 | 21.9 | 24.8 | 23.8 | 20.1 | 14.5 | 8.5 | 2.0 | 12.4 |
| NY | ROCHESTER (4307N 7740W 169M) | DAILY RAD | MJ/SQ M | 4.13 | 6.35 | 10.25 | 15.20 | 18.23 | 20.62 | 20.21 | 17.24 | 13.16 | 8.87 | 4.58 | 3.19 | 11.84 |
| | | DAY TEMP | DEG C | -3.2 | -2.7 | 1.9 | 9.5 | 15.4 | 21.2 | 23.6 | 22.5 | 18.6 | 13.0 | 6.0 | -1.0 | 10.4 |
| | | DEG DAY | C-DAY | 706 | 626 | 551 | 315 | 158 | 26 | 5 | 14 | 70 | 221 | 408 | 632 | 3732 |
| | | AVG TEMP | DEG C | -4.4 | -4.0 | .6 | 7.8 | 13.6 | 19.4 | 21.8 | 20.7 | 16.8 | 11.3 | 4.7 | -2.1 | 8.9 |
| NY | SYRACUSE (4307N 7607W 124M) | DAILY RAD | MJ/SQ M | 4.37 | 6.48 | 10.10 | 15.02 | 17.91 | 20.18 | 19.95 | 17.06 | 13.22 | 8.82 | 4.52 | 3.24 | 11.74 |
| | | DAY TEMP | DEG C | -3.4 | -2.8 | 2.0 | 9.7 | 15.6 | 21.2 | 23.7 | 22.7 | 18.9 | 13.0 | 6.2 | -1.0 | 10.5 |
| | | DEG DAY | C-DAY | 713 | 628 | 548 | 308 | 151 | 26 | 6 | 10 | 67 | 218 | 400 | 636 | 3711 |
| | | AVG TEMP | DEG C | -4.7 | -4.1 | .7 | 8.1 | 13.8 | 19.4 | 21.9 | 20.9 | 17.1 | 11.4 | 5.0 | -2.2 | 8.9 |
| OH | AKRON-CANTON (4055N 8126W 377M) | DAILY RAD | MJ/SQ M | 4.86 | 7.37 | 10.94 | 15.40 | 18.93 | 20.87 | 20.28 | 18.11 | 14.43 | 10.30 | 5.73 | 4.01 | 12.60 |
| | | DAY TEMP | DEG C | -1.9 | -1.0 | 3.9 | 11.0 | 16.7 | 22.0 | 23.9 | 23.1 | 19.4 | 13.6 | 6.1 | -.3 | 11.4 |
| | | DEG DAY | C-DAY | 667 | 580 | 496 | 275 | 128 | 18 | 6 | 9 | 56 | 205 | 405 | 613 | 3457 |
| | | AVG TEMP | DEG C | -3.2 | -2.4 | 2.3 | 9.2 | 14.8 | 20.2 | 22.1 | 21.3 | 17.6 | 11.8 | 4.8 | -1.4 | 9.8 |
| OH | CINCINNATI (COVINGTON, KY) (3904N 8440W 271M) | DAILY RAD | MJ/SQ M | 5.68 | 8.38 | 11.66 | 15.87 | 18.98 | 20.85 | 20.10 | 18.55 | 14.88 | 11.23 | 6.68 | 4.91 | 13.15 |
| | | DAY TEMP | DEG C | .9 | 2.3 | 7.1 | 14.0 | 19.2 | 24.1 | 26.0 | 25.0 | 21.9 | 15.7 | 8.1 | 2.3 | 13.9 |
| | | DEG DAY | C-DAY | 584 | 493 | 401 | 189 | 77 | 5 | 0 | 0 | 24 | 151 | 353 | 539 | 2816 |
| | | AVG TEMP | DEG C | -.5 | .7 | 5.4 | 12.2 | 17.3 | 22.3 | 24.2 | 23.6 | 19.9 | 13.8 | 6.6 | .9 | 12.2 |
| OH | CLEVELAND (4124N 8151W 245M) | DAILY RAD | MJ/SQ M | 4.41 | 6.82 | 10.47 | 15.31 | 19.08 | 20.92 | 20.74 | 17.96 | 14.07 | 9.84 | 5.29 | 3.61 | 12.38 |
| | | DAY TEMP | DEG C | -1.7 | -1.1 | 3.6 | 10.7 | 16.3 | 21.7 | 23.6 | 22.8 | 19.4 | 13.7 | 6.5 | .1 | 11.3 |
| | | DEG DAY | C-DAY | 656 | 577 | 498 | 278 | 136 | 22 | 5 | 9 | 53 | 197 | 390 | 598 | 3419 |
| | | AVG TEMP | DEG C | -2.8 | -2.3 | 2.3 | 9.1 | 14.6 | 19.9 | 21.9 | 21.1 | 17.7 | 12.1 | 5.3 | -.9 | 9.8 |

*(Continued)*

**TABLE A-15.** *(Continued)*

Location data (from header row): latitude under FEB, longitude under MAR, elevation (M) under MAY.

| | | | JAN | FEB | MAR | APR | MAY | JUN | JUL | AUG | SEP | OCT | NOV | DEC | ANNUAL |
|---|---|---|---|---|---|---|---|---|---|---|---|---|---|---|---|
| **OH COLUMBUS** | | | | 4000N | 8253W | | 254 M | | | | | | | | |
| | DAILY RAD | MJ/SQ M | 5.21 | 7.68 | 11.12 | 15.35 | 18.69 | 20.57 | 19.92 | 18.62 | 14.54 | 10.73 | 6.10 | 4.39 | 12.74 |
| | DAY TEMP | DEG C | -.7 | .5 | 5.7 | 12.6 | 18.1 | 23.2 | 25.0 | 24.1 | 20.5 | 14.4 | 6.9 | .6 | 12.6 |
| | DEG DAY | C-DAY | 631 | 540 | 444 | 232 | 98 | 7 | | 4 | 42 | 190 | 388 | 591 | 3167 |
| | AVG TEMP | DEG C | -2.0 | -.9 | 4.0 | 10.7 | 16.2 | 21.3 | 23.1 | 22.2 | 18.4 | 12.3 | 5.4 | -.7 | 10.8 |
| **OH DAYTON** | | | | 3954N | 8413W | | 306 M | | | | | | | | |
| | DAILY RAD | MJ/SQ M | 5.55 | 8.23 | 11.64 | 15.92 | 19.29 | 21.26 | 20.54 | 18.67 | 14.96 | 11.00 | 6.40 | 4.62 | 13.17 |
| | DAY TEMP | DEG C | -.9 | .6 | 5.4 | 12.5 | 18.2 | 23.5 | 25.3 | 24.5 | 20.8 | 14.8 | 6.8 | .6 | 12.7 |
| | DEG DAY | C-DAY | 636 | 538 | 448 | 229 | 92 | | | | 35 | 171 | 387 | 587 | 3134 |
| | AVG TEMP | DEG C | -2.2 | -.9 | 3.9 | 10.8 | 16.4 | 21.8 | 23.7 | 22.8 | 19.1 | 13.1 | 5.4 | -.6 | 11.1 |
| **OH TOLEDO** | | | | 4136N | 8348W | | 211 M | | | | | | | | |
| | DAILY RAD | MJ/SQ M | 4.93 | 7.72 | 11.31 | 15.71 | 19.48 | 21.32 | 20.99 | 18.34 | 14.48 | 10.34 | 5.65 | 4.03 | 12.86 |
| | DAY TEMP | DEG C | -2.7 | -1.4 | 3.6 | 11.0 | 16.8 | 22.4 | 24.3 | 23.5 | 19.6 | 13.6 | 5.6 | -1.0 | 11.3 |
| | DEG DAY | C-DAY | 692 | 589 | 503 | 277 | 127 | 18 | 10 | | 55 | 211 | 423 | 637 | 3545 |
| | AVG TEMP | DEG C | -4.0 | -2.7 | 2.1 | 9.1 | 14.9 | 20.5 | 22.4 | 21.6 | 17.7 | 11.7 | 4.2 | -2.2 | 9.6 |
| **OH YOUNGSTOWN** | | | | 4116N | 8040W | | 361 M | | | | | | | | |
| | DAILY RAD | MJ/SQ M | 4.37 | 6.66 | 10.10 | 14.51 | 18.00 | 19.96 | 19.68 | 17.09 | 13.55 | 9.66 | 5.18 | 3.58 | 11.86 |
| | DAY TEMP | DEG C | -2.3 | -1.6 | 3.3 | 10.5 | 16.1 | 21.3 | 23.3 | 22.5 | 18.9 | 13.2 | 5.9 | -.6 | 10.9 |
| | DEG DAY | C-DAY | 677 | 596 | 512 | 288 | 143 | 23 | 12 | | 66 | 213 | 412 | 623 | 3570 |
| | AVG TEMP | DEG C | -3.5 | -2.9 | 1.8 | 8.7 | 14.2 | 19.4 | 21.5 | 20.7 | 17.1 | 11.4 | 4.6 | -1.8 | 9.3 |
| **OK OKLAHOMA CITY** | | | | 3524N | 9736W | | 397 M | | | | | | | | |
| | DAILY RAD | MJ/SQ M | 9.09 | 11.97 | 15.89 | 19.56 | 21.77 | 24.33 | 24.15 | 22.13 | 17.64 | 13.99 | 10.22 | 8.23 | 16.58 |
| | DAY TEMP | DEG C | 4.5 | 7.1 | 10.9 | 17.0 | 21.9 | 26.6 | 29.3 | 29.2 | 24.7 | 18.9 | 11.5 | 6.2 | 17.4 |
| | DEG DAY | C-DAY | 486 | 369 | 296 | 100 | 20 | | | | | 82 | 263 | 431 | 2054 |
| | AVG TEMP | DEG C | 2.7 | 5.2 | 9.0 | 15.8 | 20.2 | 24.9 | 27.5 | 27.3 | 22.8 | 16.9 | 9.6 | 4.4 | 15.5 |
| **OK TULSA** | | | | 3612N | 9554W | | 206 M | | | | | | | | |
| | DAILY RAD | MJ/SQ M | 8.30 | 11.10 | 14.82 | 18.19 | 20.68 | 22.93 | 23.04 | 21.17 | 16.71 | 13.21 | 9.39 | 7.48 | 15.59 |
| | DAY TEMP | DEG C | 4.3 | 6.9 | 11.0 | 17.8 | 22.2 | 26.8 | 29.6 | 29.3 | 24.9 | 19.2 | 11.6 | 6.1 | 17.5 |
| | DEG DAY | C-DAY | 489 | 370 | 293 | 98 | 16 | | | | | 79 | 260 | 434 | 2045 |
| | AVG TEMP | DEG C | 2.6 | 5.1 | 9.1 | 16.0 | 20.4 | 25.2 | 27.8 | 27.4 | 22.9 | 17.2 | 9.7 | 4.3 | 15.6 |
| **OR ASTORIA** | | | | 4609N | 12353W | | 7 M | | | | | | | | |
| | DAILY RAD | MJ/SQ M | 3.57 | 6.19 | 9.83 | 14.22 | 18.25 | 18.45 | 19.82 | 17.01 | 13.43 | 8.09 | 4.39 | 2.96 | 11.35 |
| | DAY TEMP | DEG C | 5.8 | 7.6 | 8.2 | 10.1 | 12.6 | 14.8 | 16.8 | 17.1 | 16.2 | 12.9 | 9.2 | 7.0 | 11.5 |
| | DEG DAY | C-DAY | 420 | 333 | 355 | 287 | 219 | 142 | 91 | 84 | 112 | 210 | 308 | 382 | 2943 |
| | AVG TEMP | DEG C | 4.8 | 6.4 | 6.9 | 8.8 | 11.3 | 13.6 | 15.6 | 15.7 | 14.7 | 11.6 | 8.1 | 6.0 | 10.3 |
| **OR BURNS** | | | | 4335N | 11903W | | 1271 M | | | | | | | | |
| | DAILY RAD | MJ/SQ M | 5.56 | 8.99 | 13.47 | 18.71 | 23.29 | 25.87 | 27.92 | 23.64 | 18.39 | 11.83 | 6.74 | 4.89 | 15.77 |
| | DAY TEMP | DEG C | -2.1 | 1.2 | 4.2 | 9.1 | 13.6 | 17.5 | 23.1 | 21.8 | 17.4 | 11.0 | 4.0 | -.7 | 10.0 |
| | DEG DAY | C-DAY | 686 | 529 | 498 | 347 | 223 | 114 | 17 | | 126 | 305 | 487 | 639 | 4009 |
| | AVG TEMP | DEG C | -3.8 | -.6 | 2.3 | 6.6 | 11.2 | 15.0 | 20.2 | 18.9 | 14.6 | 8.5 | 2.1 | -2.3 | 7.7 |

| State | Location | Measure | Units | JAN | FEB | MAR | APR | MAY | JUN | JUL | AUG | SEP | OCT | NOV | DEC | ANNUAL |
|---|---|---|---|---|---|---|---|---|---|---|---|---|---|---|---|---|
| OR | MEDFORD (4222N, 12252W, 396M) | DAILY RAD | MJ/SQ M | 4.62 | 8.37 | 12.85 | 16.60 | 23.08 | 25.85 | 28.09 | 24.07 | 18.03 | 11.14 | 5.72 | 3.82 | 15.35 |
| | | DAY TEMP | DEG C | 3.8 | 6.9 | 9.1 | 12.4 | 16.5 | 20.5 | 25.2 | 24.2 | 20.9 | 14.2 | 7.9 | 4.2 | 13.8 |
| | | DEG DAY | C-DAY | 489 | 369 | 348 | 247 | 139 | 52 | 6 | 12 | 49 | 200 | 358 | 470 | 2739 |
| | | AVG TEMP | DEG C | 2.6 | 5.2 | 7.1 | 10.1 | 14.1 | 17.9 | 22.1 | 21.3 | 18.0 | 11.9 | 6.4 | 3.2 | 11.6 |
| OR | NORTH BEND (4325N, 12415W, 5M) | DAILY RAD | MJ/SQ M | 4.98 | 7.99 | 12.01 | 17.13 | 21.08 | 22.63 | 23.92 | 20.27 | 15.63 | 10.13 | 5.95 | 4.32 | 13.84 |
| | | DAY TEMP | DEG C | | | | | | | | | | | | | |
| | | DEG DAY | C-DAY | 351 | 286 | 312 | 265 | 205 | 135 | 104 | 93 | 112 | 174 | 248 | 319 | 2604 |
| | | AVG TEMP | DEG C | 7.0 | 8.1 | 8.3 | 9.5 | 11.7 | 13.8 | 15.0 | 15.4 | 14.7 | 12.7 | 10.1 | 8.1 | 11.2 |
| OR | PENDLETON (4541N, 11851W, 456M) | DAILY RAD | MJ/SQ M | 3.95 | 6.96 | 11.84 | 17.05 | 21.85 | 24.33 | 27.19 | 22.63 | 17.05 | 10.31 | 4.97 | 3.32 | 14.29 |
| | | DAY TEMP | DEG C | 1.1 | 5.1 | 8.1 | 12.3 | 16.7 | 20.8 | 25.5 | 24.3 | 19.9 | 13.2 | 6.5 | 3.1 | 13.1 |
| | | DEG DAY | C-DAY | 568 | 406 | 365 | 235 | 122 | 39 | 3 | 7 | 54 | 213 | 393 | 504 | 2909 |
| | | AVG TEMP | DEG C | .0 | 3.8 | 6.6 | 10.5 | 14.7 | 18.7 | 23.1 | 21.9 | 17.8 | 11.4 | 5.2 | 2.1 | 11.3 |
| OR | PORTLAND (4536N, 12236W, 12M) | DAILY RAD | MJ/SQ M | 3.52 | 6.29 | 10.16 | 14.84 | 18.88 | 20.12 | 23.12 | 18.99 | 13.81 | 8.21 | 4.40 | 2.95 | 12.11 |
| | | DAY TEMP | DEG C | 4.3 | 7.2 | 9.0 | 11.9 | 15.4 | 18.3 | 21.5 | 21.1 | 18.7 | 13.6 | 8.5 | 5.7 | 13.0 |
| | | DEG DAY | C-DAY | 463 | 346 | 332 | 240 | 147 | 71 | 27 | 31 | 66 | 193 | 328 | 418 | 2662 |
| | | AVG TEMP | DEG C | 3.4 | 6.0 | 7.6 | 10.3 | 13.7 | 16.7 | 19.5 | 19.2 | 16.8 | 12.1 | 7.4 | 4.8 | 11.5 |
| OR | REDMOND (4416N, 12109W, 940M) | DAILY RAD | MJ/SQ M | 5.57 | 8.79 | 13.51 | 19.10 | 23.60 | 25.96 | 27.76 | 23.48 | 17.98 | 11.34 | 6.49 | 4.82 | 15.70 |
| | | DAY TEMP | DEG C | | | | | | | | | | | | | |
| | | DEG DAY | C-DAY | 599 | 454 | 454 | 343 | 236 | 122 | 31 | 57 | 129 | 286 | 433 | 544 | 3688 |
| | | AVG TEMP | DEG C | -1.0 | 2.1 | 3.7 | 7.7 | 10.7 | 14.6 | 18.7 | 17.7 | 14.3 | 9.1 | 3.9 | .8 | 8.2 |
| OR | SALEM (4455N, 12301W, 61M) | DAILY RAD | MJ/SQ M | 3.77 | 6.67 | 10.75 | 15.55 | 19.72 | 20.98 | 24.31 | 20.14 | 15.07 | 8.73 | 4.66 | 3.14 | 12.79 |
| | | DAY TEMP | DEG C | 4.9 | 7.5 | 8.9 | 11.8 | 15.2 | 18.4 | 21.9 | 21.5 | 19.0 | 13.6 | 8.6 | 6.0 | 13.1 |
| | | DEG DAY | C-DAY | 451 | 344 | 341 | 253 | 164 | 74 | 24 | 29 | 67 | 203 | 330 | 415 | 2695 |
| | | AVG TEMP | DEG C | 3.8 | 6.1 | 7.3 | 9.9 | 13.2 | 16.2 | 19.2 | 18.9 | 16.6 | 11.8 | 7.3 | 4.9 | 11.3 |
| PA | ALLENTOWN (4039N, 7526W, 117M) | DAILY RAD | MJ/SQ M | 5.99 | 8.66 | 12.24 | 16.00 | 18.58 | 20.17 | 20.03 | 17.55 | 14.05 | 10.51 | 6.45 | 4.88 | 12.92 |
| | | DAY TEMP | DEG C | -1.0 | -.0 | 5.0 | 11.8 | 17.5 | 22.7 | 25.3 | 23.9 | 20.0 | 14.2 | 7.3 | .6 | 12.3 |
| | | DEG DAY | C-DAY | 641 | 554 | 463 | 252 | 106 | 12 | 0 | 3 | 47 | 191 | 378 | 591 | 3238 |
| | | AVG TEMP | DEG C | -2.3 | -1.4 | 3.4 | 9.9 | 15.6 | 20.8 | 23.4 | 22.1 | 18.2 | 12.3 | 5.7 | -.7 | 10.6 |
| PA | ERIE (4205N, 8011W, 225M) | DAILY RAD | MJ/SQ M | 3.92 | 6.55 | 10.44 | 15.42 | 18.68 | 20.96 | 20.80 | 16.51 | 13.63 | 9.39 | 4.72 | 3.15 | 12.02 |
| | | DAY TEMP | DEG C | -2.7 | -1.2 | 1.8 | 8.6 | 14.1 | 19.6 | 21.8 | 21.1 | 17.8 | 12.3 | 5.6 | -.7 | 9.7 |
| | | DEG DAY | C-DAY | 687 | 619 | 553 | 337 | 187 | 44 | 13 | 24 | 78 | 231 | 415 | 618 | 3806 |
| | | AVG TEMP | DEG C | -3.8 | -3.8 | .5 | 7.1 | 12.6 | 18.1 | 20.4 | 19.7 | 16.3 | 10.9 | 4.5 | -1.6 | 8.4 |

*(Continued)*

**TABLE A-15.** *(Continued)*

| | | JAN | FEB | MAR | APR | MAY | JUN | JUL | AUG | SEP | OCT | NOV | DEC | ANNUAL |
|---|---|---|---|---|---|---|---|---|---|---|---|---|---|---|
| **PA HARRISBURG** | | | 4013N | 7651W | | 106 M | | | | | | | | |
| DAILY RAD | MJ/SQ M | 6.08 | 8.75 | 12.29 | 16.01 | 18.75 | 20.48 | 20.01 | 17.60 | 14.37 | 10.60 | 6.57 | 5.08 | 13.05 |
| DAY TEMP | DEG C | .2 | 1.5 | 6.6 | 13.4 | 19.2 | 24.1 | 26.3 | 25.1 | 21.3 | 15.1 | 8.1 | 1.6 | 13.5 |
| DEG DAY | C-DAY | 601 | 509 | 413 | 206 | 71 | 14 | 0 | 0 | 28 | 163 | 353 | 558 | 2902 |
| AVG TEMP | DEG C | -1.1 | .2 | 5.0 | 11.6 | 17.3 | 22.2 | 24.5 | 23.3 | 19.4 | 13.2 | 6.6 | .3 | 11.9 |
| **PA PHILADELPHIA** | | | 3953N | 7515W | | 9 M | | | | | | | | |
| DAILY RAD | MJ/SQ M | 6.30 | 9.02 | 12.58 | 16.27 | 18.84 | 20.55 | 19.95 | 17.87 | 14.54 | 10.88 | 7.03 | 5.34 | 13.26 |
| DAY TEMP | DEG C | 1.5 | 2.4 | 7.1 | 13.4 | 19.1 | 24.2 | 26.6 | 25.4 | 21.8 | 15.9 | 9.4 | 3.1 | 14.2 |
| DEG DAY | C-DAY | 563 | 484 | 398 | 204 | 68 | 14 | 0 | 0 | 21 | 138 | 313 | 513 | 2702 |
| AVG TEMP | DEG C | .2 | 1.1 | 5.5 | 11.6 | 17.3 | 22.4 | 24.9 | 23.8 | 20.1 | 14.1 | 7.9 | 1.8 | 12.5 |
| **PA PITTSBURGH** | | | 4030N | 8013W | | 373 M | | | | | | | | |
| DAILY RAD | MJ/SQ M | 4.82 | 7.10 | 10.70 | 14.94 | 18.18 | 19.99 | 19.17 | 17.14 | 13.72 | 10.16 | 5.73 | 3.94 | 12.13 |
| DAY TEMP | DEG C | -1.0 | -.2 | 4.9 | 11.9 | 17.3 | 22.2 | 23.9 | 23.0 | 19.5 | 13.6 | 6.5 | .3 | 11.8 |
| DEG DAY | C-DAY | 636 | 556 | 463 | 247 | 116 | 14 | 0 | 10 | 54 | 207 | 395 | 594 | 3295 |
| AVG TEMP | DEG C | -2.2 | -1.5 | 3.4 | 10.1 | 15.4 | 20.3 | 22.2 | 21.2 | 17.7 | 11.8 | 5.2 | -.8 | 10.2 |
| **PA WILKES-BARRE-SCRANTON** | | | 4120N | 7544W | | 289 M | | | | | | | | |
| DAILY RAD | MJ/SQ M | 5.16 | 7.81 | 11.25 | 15.20 | 18.05 | 19.97 | 19.81 | 17.17 | 13.61 | 10.17 | 5.56 | 4.17 | 12.33 |
| DAY TEMP | DEG C | -2.1 | -1.3 | 3.7 | 10.9 | 16.8 | 21.8 | 24.1 | 22.9 | 19.0 | 13.2 | 6.2 | -.4 | 11.2 |
| DEG DAY | C-DAY | 672 | 587 | 499 | 275 | 122 | 16 | 0 | 10 | 64 | 217 | 403 | 618 | 3487 |
| AVG TEMP | DEG C | -3.3 | -2.6 | 2.2 | 9.2 | 14.9 | 19.9 | 22.3 | 21.1 | 17.2 | 11.4 | 4.9 | -1.6 | 9.6 |
| **PN KOROR ISLAND** | | | 720N | 13429E | | 33 M | | | | | | | | |
| DAILY RAD | MJ/SQ M | 15.85 | 17.69 | 18.51 | 19.20 | 17.79 | 16.52 | 16.37 | 16.83 | 17.33 | 17.10 | 16.44 | 15.19 | 17.07 |
| DAY TEMP | DEG C | 28.2 | 28.0 | 28.3 | 28.8 | 28.8 | 28.6 | 28.3 | 28.3 | 28.5 | 28.7 | 28.7 | 28.4 | 28.5 |
| DEG DAY | C-DAY | | | | | | | | | | | | | |
| AVG TEMP | DEG C | 27.3 | 27.1 | 27.7 | 27.7 | 27.8 | 27.6 | 27.3 | 27.3 | 27.6 | 27.7 | 27.7 | 27.4 | 27.5 |
| **PN KWAJALEIN ISLAND** | | | 844N | 16744E | | 8 M | | | | | | | | |
| DAILY RAD | MJ/SQ M | 17.86 | 19.88 | 20.39 | 19.75 | 18.46 | 18.28 | 18.27 | 19.15 | 18.33 | 17.31 | 16.55 | 16.46 | 18.39 |
| DAY TEMP | DEG C | 28.0 | 28.1 | 28.3 | 28.3 | 28.4 | 28.4 | 28.6 | 28.9 | 28.9 | 28.0 | 28.3 | 28.2 | 28.4 |
| DEG DAY | C-DAY | | | | | | | | | | | | | |
| AVG TEMP | DEG C | 27.3 | 27.4 | 27.6 | 27.7 | 27.7 | 27.7 | 27.8 | 28.1 | 28.1 | 28.0 | 27.6 | 27.6 | 27.7 |
| **PN WAKE ISLAND** | | | 1917N | 16639E | | 4 M | | | | | | | | |
| DAILY RAD | MJ/SQ M | 15.31 | 17.84 | 20.56 | 22.18 | 23.32 | 23.22 | 21.84 | 21.24 | 19.73 | 17.83 | 16.31 | 14.88 | 19.52 |
| DAY TEMP | DEG C | 25.8 | 25.8 | 26.2 | 26.5 | 27.4 | 28.3 | 28.6 | 29.0 | 28.9 | 28.4 | 27.6 | 26.6 | 27.4 |
| DEG DAY | C-DAY | | | | | | | | | | | | | |
| AVG TEMP | DEG C | 25.0 | 25.0 | 25.4 | 25.7 | 26.6 | 27.5 | 27.8 | 28.1 | 28.1 | 27.6 | 26.9 | 25.9 | 26.6 |
| **PR SAN JUAN** | | | 1826N | 6600W | | 19 M | | | | | | | | |
| DAILY RAD | MJ/SQ M | 15.04 | 17.43 | 20.29 | 21.46 | 20.57 | 20.62 | 21.26 | 20.86 | 19.00 | 17.20 | 15.52 | 14.02 | 18.61 |
| DAY TEMP | DEG C | 25.2 | 25.2 | 25.8 | 26.4 | 27.3 | 28.0 | 28.2 | 28.4 | 28.4 | 28.1 | 27.0 | 25.9 | 27.0 |
| DEG DAY | C-DAY | | | | | | | | | | | | | |
| AVG TEMP | DEG C | 24.1 | 24.1 | 24.6 | 25.3 | 26.2 | 26.9 | 27.2 | 27.4 | 27.3 | 27.0 | 25.9 | 24.9 | 25.9 |

| | | | | JAN | FEB | MAR | APP | MAY | JUN | JUL | AUG | SEP | OCT | NOV | DEC | ANNUAL |
|---|---|---|---|---|---|---|---|---|---|---|---|---|---|---|---|---|
| RI | PROVIDENCE | DAILY RAD | MJ/SQ M | 5.74 | 4144N 8.38 | 7126W 11.71 | 15.59 | 19 M 18.78 | 20.15 | 19.24 | 17.01 | 13.72 | 10.29 | 6.10 | 4.75 | 12.62 |
| | | DAY TEMP | DEG C | -.7 | -1.1 | 4.0 | 10.1 | 15.5 | 20.8 | 23.8 | 22.9 | 19.1 | 13.8 | 7.7 | 1.1 | 11.5 |
| | | DEG DAY | C-DAY | 631 | 554 | 484 | 295 | 144 | 6 | | | 52 | 194 | 362 | 577 | 3319 |
| | | AVG TEMP | DEG C | -2.0 | -1.4 | 2.7 | 8.5 | 13.8 | 19.1 | 22.3 | 21.3 | 17.4 | 12.1 | 6.3 | -.3 | 10.0 |
| SC | CHARLESTON | DAILY RAD | MJ/SQ M | 8.45 | 3254N 11.30 | 6002W 15.19 | 19.66 | 12 M 21.11 | 20.93 | 20.42 | 17.99 | 15.82 | 13.54 | 10.60 | 8.18 | 15.26 |
| | | DAY TEMP | DEG C | 11.1 | 12.2 | 15.5 | 20.0 | 24.1 | 27.0 | 28.3 | 27.9 | 25.5 | 20.8 | 15.5 | 11.5 | 20.0 |
| | | DEG DAY | C-DAY | 289 | 233 | 167 | 38 | | | | | | 41 | 151 | 271 | 1193 |
| | | AVG TEMP | DEG C | 9.2 | 10.3 | 13.6 | 18.1 | 22.3 | 25.5 | 26.8 | 26.4 | 24.0 | 18.9 | 13.5 | 9.6 | 18.2 |
| SC | COLUMBIA | DAILY RAD | MJ/SQ M | 8.64 | 3357N 11.58 | 8107W 15.38 | 19.82 | 69 M 21.50 | 22.09 | 20.90 | 19.32 | 16.33 | 13.75 | 10.45 | 8.19 | 15.67 |
| | | DAY TEMP | DEG C | 9.4 | 10.7 | 14.4 | 20.0 | 24.3 | 27.0 | 29.1 | 28.6 | 25.0 | 20.0 | 14.3 | 9.8 | 19.5 |
| | | DEG DAY | C-DAY | 338 | 274 | 200 | 46 | | | | | | 62 | 189 | 327 | 1443 |
| | | AVG TEMP | DEG C | 7.4 | 8.7 | 12.3 | 17.8 | 22.3 | 26.0 | 27.3 | 26.8 | 23.6 | 17.9 | 12.1 | 7.8 | 17.5 |
| SC | GREENVILLE-SPARTANBURG | DAILY RAD | MJ/SQ M | 8.28 | 3454N 11.14 | 8213W 15.08 | 19.26 | 296 M 20.87 | 21.77 | 20.77 | 19.28 | 15.95 | 13.39 | 9.99 | 7.61 | 15.28 |
| | | DAY TEMP | DEG C | 7.3 | 8.5 | 12.3 | 17.9 | 22.4 | 26.1 | 27.3 | 26.8 | 23.6 | 18.3 | 12.4 | 7.6 | 17.5 |
| | | DEG DAY | C-DAY | 391 | 321 | 250 | 80 | 16 | | | | | 81 | 233 | 381 | 1758 |
| | | AVG TEMP | DEG C | 5.7 | 6.9 | 10.5 | 16.1 | 20.6 | 24.4 | 25.7 | 25.3 | 22.1 | 16.5 | 10.6 | 6.1 | 15.9 |
| SD | HURON | DAILY RAD | MJ/SQ M | 5.54 | 4423N 8.45 | 9813W 12.64 | 17.36 | 393 M 21.24 | 23.84 | 24.77 | 21.48 | 16.09 | 11.22 | 6.55 | 4.60 | 14.48 |
| | | DAY TEMP | DEG C | -9.0 | -6.0 | .0 | 9.7 | 16.0 | 21.5 | 25.3 | 24.5 | 18.2 | 12.0 | 2.1 | -5.4 | 9.1 |
| | | DEG DAY | C-DAY | 904 | 733 | 620 | 320 | 152 | | | | 94 | 268 | 543 | 789 | 4475 |
| | | AVG TEMP | DEG C | -10.8 | -7.8 | -1.7 | 7.7 | 13.9 | 19.5 | 23.2 | 22.3 | 15.9 | 9.8 | .2 | -7.1 | 7.1 |
| SD | PIERRE | DAILY RAD | MJ/SQ M | 6.01 | 4423N 9.02 | 10017W 13.69 | 18.32 | 526 M 22.32 | 24.91 | 25.85 | 22.61 | 16.98 | 11.94 | 7.07 | 5.02 | 15.31 |
| | | DAY TEMP | DEG C | | | | | | | | | | | | | |
| | | DEG DAY | C-DAY | 851 | 694 | 606 | 312 | 148 | 41 | 3 | 6 | 84 | 251 | 520 | 749 | 4265 |
| | | AVG TEMP | DEG C | -9.1 | -6.4 | -1.2 | 7.9 | 14.1 | 19.7 | 24.0 | 23.3 | 16.7 | 16.7 | 1.0 | -5.8 | 8.4 |
| SD | RAPID CITY | DAILY RAD | MJ/SQ M | 6.15 | 4403N 9.38 | 10304W 13.95 | 18.03 | 966 M 21.41 | 24.19 | 25.23 | 22.27 | 17.23 | 12.07 | 7.34 | 5.41 | 15.22 |
| | | DAY TEMP | DEG C | -3.6 | -1.5 | .0 | 9.1 | 14.9 | 19.9 | 24.8 | 24.4 | 18.2 | 12.3 | 3.9 | -1.1 | 10.2 |
| | | DEG DAY | C-DAY | 742 | 610 | 582 | 340 | 177 | 74 | | | 106 | 263 | 493 | 663 | 4066 |
| | | AVG TEMP | DEG C | -5.6 | -3.4 | -.4 | 7.0 | 12.9 | 17.9 | 22.6 | 22.0 | 15.8 | 10.0 | 1.9 | -3.1 | 8.1 |
| SD | SIOUX FALLS | DAILY RAD | MJ/SQ M | 6.04 | 4334N 9.10 | 9644W 13.08 | 17.51 | 435 M 21.49 | 23.83 | 24.40 | 20.93 | 16.00 | 11.41 | 6.89 | 5.01 | 14.64 |
| | | DAY TEMP | DEG C | -8.2 | -5.3 | .5 | 9.8 | 16.3 | 21.7 | 24.9 | 24.1 | 18.1 | 12.2 | 2.3 | -5.1 | 9.3 |
| | | DEG DAY | C-DAY | 875 | 709 | 603 | 315 | 144 | 36 | | | 92 | 258 | 532 | 775 | 4355 |
| | | AVG TEMP | DEG C | -9.9 | -7.0 | -1.1 | 7.8 | 14.3 | 19.8 | 22.9 | 22.1 | 16.1 | 10.1 | .6 | -6.7 | 7.4 |

*(Continued)*

## TABLE A-15. (Continued)

| | | JAN | FEB | MAR | APR | MAY | JUN | JUL | AUG | SEP | OCT | NOV | DEC | ANNUAL |
|---|---|---|---|---|---|---|---|---|---|---|---|---|---|---|
| TN | CHATTANOOGA | | 3502N | 8512W | | 210 M | | | | | | | | |
| | DAILY RAD MJ/SQ M | 7.16 | 9.74 | 13.35 | 17.59 | 19.65 M | 20.78 | 19.69 | 18.50 | 15.16 | 12.58 | 8.77 | 6.59 | 14.13 |
| | DAY TEMP DEG C | 6.2 | 7.8 | 11.8 | 17.9 | 22.4 | 26.4 | 27.8 | 27.4 | 24.1 | 18.1 | 11.4 | 6.7 | 17.3 |
| | DEG DAY C-DAY | 427 | 347 | 268 | 92 | 28 | 0 | 0 | 0 | 5 | 101 | 268 | 410 | 1946 |
| | AVG TEMP DEG C | 4.6 | 6.1 | 9.9 | 15.8 | 20.3 | 24.4 | 26.0 | 25.6 | 22.2 | 16.0 | 9.4 | 5.1 | 15.4 |
| TN | KNOXVILLE | | 3549N | 8359W | | 299 M | | | | | | | | |
| | DAILY RAD MJ/SQ M | 7.04 | 9.80 | 13.51 | 18.15 | 20.46 M | 21.58 | 20.48 | 18.91 | 15.70 | 12.72 | 8.61 | 6.46 | 14.45 |
| | DAY TEMP DEG C | 6.2 | 7.5 | 11.7 | 17.7 | 22.1 | 25.9 | 27.3 | 26.8 | 23.7 | 17.9 | 11.2 | 6.7 | 17.1 |
| | DEG DAY C-DAY | 420 | 350 | 269 | 96 | 26 | 0 | 0 | 0 | 6 | 97 | 263 | 405 | 1932 |
| | AVG TEMP DEG C | 4.8 | 6.0 | 9.9 | 15.7 | 20.2 | 24.2 | 25.7 | 25.2 | 22.0 | 16.1 | 9.6 | 5.3 | 15.4 |
| TN | MEMPHIS | | 3503N | 8959W | | 87 M | | | | | | | | |
| | DAILY RAD MJ/SQ M | 7.75 | 10.72 | 14.50 | 18.60 | 21.39 | 23.20 | 22.38 | 20.70 | 16.69 | 13.67 | 9.27 | 7.13 | 15.50 |
| | DAY TEMP DEG C | 6.2 | 8.1 | 12.2 | 18.6 | 23.3 | 27.6 | 29.2 | 28.6 | 24.9 | 19.2 | 12.3 | 7.4 | 18.1 |
| | DEG DAY C-DAY | 422 | 330 | 254 | 73 | 12 | 0 | 0 | 0 | 4 | 79 | 235 | 384 | 1793 |
| | AVG TEMP DEG C | 4.7 | 6.6 | 10.6 | 16.9 | 21.6 | 25.9 | 27.6 | 26.9 | 23.1 | 17.2 | 10.5 | 5.9 | 16.5 |
| TN | NASHVILLE | | 3607N | 8641W | | 180 M | | | | | | | | |
| | DAILY RAD MJ/SQ M | 6.58 | 9.35 | 12.82 | 17.52 | 20.71 M | 22.28 | 21.46 | 19.71 | 15.86 | 12.64 | 8.07 | 5.91 | 14.41 |
| | DAY TEMP DEG C | 5.1 | 6.6 | 11.0 | 17.5 | 22.2 | 26.0 | 28.2 | 27.6 | 24.1 | 18.1 | 10.9 | 6.2 | 17.0 |
| | DEG DAY C-DAY | 460 | 373 | 291 | 98 | 25 | 0 | 0 | 0 | 6 | 100 | 277 | 424 | 2054 |
| | AVG TEMP DEG C | 3.5 | 5.0 | 9.3 | 15.6 | 20.3 | 24.8 | 26.4 | 25.8 | 22.2 | 16.1 | 9.1 | 4.7 | 15.2 |
| TX | ABILENE | | 3226N | 9941W | | 534 M | | | | | | | | |
| | DAILY RAD MJ/SQ M | 10.48 | 13.42 | 17.89 | 20.92 | 23.12 M | 25.07 | 24.28 | 22.20 | 18.13 | 14.93 | 11.44 | 9.80 | 17.64 |
| | DAY TEMP DEG C | 8.5 | 10.8 | 14.6 | 20.5 | 24.4 | 28.7 | 30.7 | 30.6 | 26.4 | 20.9 | 14.3 | 10.0 | 20.0 |
| | DEG DAY C-DAY | 367 | 266 | 197 | 58 | 6 | 0 | 0 | 0 | 4 | 49 | 187 | 321 | 1451 |
| | AVG TEMP DEG C | 6.5 | 8.8 | 12.5 | 18.4 | 22.4 | 26.8 | 28.8 | 28.7 | 24.5 | 18.9 | 12.3 | 8.0 | 18.1 |
| TX | AMARILLO | | 3514N | 10142W | | 1098 M | | | | | | | | |
| | DAILY RAD MJ/SQ M | 10.90 | 14.11 | 18.51 | 22.91 | 25.10 M | 27.16 | 25.88 | 23.87 | 19.98 | 15.93 | 11.72 | 9.89 | 18.83 |
| | DAY TEMP DEG C | 4.5 | 6.5 | 10.0 | 16.0 | 20.6 | 25.9 | 28.1 | 27.5 | 23.3 | 17.5 | 10.2 | 5.8 | 16.3 |
| | DEG DAY C-DAY | 499 | 393 | 334 | 153 | 45 | 6 | 0 | 0 | 11 | 114 | 312 | 457 | 2324 |
| | AVG TEMP DEG C | 2.2 | 4.3 | 7.6 | 13.6 | 18.7 | 23.7 | 25.9 | 25.3 | 21.0 | 15.3 | 7.9 | 3.6 | 14.1 |
| TX | AUSTIN | | 3018N | 9742W | | 189 M | | | | | | | | |
| | DAILY RAD MJ/SQ M | 9.81 | 12.76 | 16.22 | 18.22 | 20.81 | 23.51 | 23.89 | 21.92 | 18.23 | 15.13 | 11.20 | 9.36 | 16.76 |
| | DAY TEMP DEG C | 11.5 | 13.6 | 17.1 | 22.1 | 25.7 | 29.2 | 31.0 | 31.1 | 27.8 | 23.0 | 16.9 | 13.1 | 21.9 |
| | DEG DAY C-DAY | 268 | 191 | 124 | 24 | 0 | 0 | 0 | 0 | 2 | 22 | 114 | 222 | 965 |
| | AVG TEMP DEG C | 9.8 | 11.8 | 15.3 | 20.3 | 24.0 | 27.6 | 29.2 | 29.3 | 26.1 | 21.2 | 15.1 | 11.3 | 20.1 |
| TX | BROWNSVILLE | | 2554N | 9726W | | 6 M | | | | | | | | |
| | DAILY RAD MJ/SQ M | 10.36 | 12.89 | 16.54 | 19.71 | 21.87 M | 24.01 | 25.11 | 23.01 | 19.22 | 16.33 | 11.97 | 9.79 | 17.57 |
| | DAY TEMP DEG C | 17.3 | 19.0 | 21.3 | 25.2 | 27.6 | 29.5 | 30.5 | 30.5 | 28.0 | 25.8 | 21.6 | 18.7 | 24.7 |
| | DEG DAY C-DAY | 125 | 84 | 49 | 4 | 0 | 0 | 0 | 0 | 0 | 3 | 19 | 81 | 361 |
| | AVG TEMP DEG C | 15.7 | 17.4 | 19.8 | 23.8 | 26.3 | 28.2 | 29.1 | 29.1 | 27.6 | 24.3 | 20.1 | 17.1 | 23.2 |

| | | JAN | FEB | MAR | APR | MAY | JUN | JUL | AUG | SEP | OCT | NOV | DEC | ANNUAL |
|---|---|---|---|---|---|---|---|---|---|---|---|---|---|---|
| **TX CORPUS CHRISTI** | | | 27 46 N | 97 30 W | | 13 M | | | | | | | | |
| DAILY RAD | MJ/SQ M | 10.19 | 13.02 | 16.23 | 18.64 | 21.18 | 23.76 | 24.81 | 22.59 | 19.15 | 16.07 | 11.83 | 9.59 | 17.26 |
| DAY TEMP | DEG C | 15.2 | 17.0 | 20.0 | 24.2 | 26.9 | 29.5 | 30.9 | 31.0 | 28.7 | 25.0 | 20.6 | 16.8 | 23.8 |
| DEG DAY | C-DAY | 169 | 111 | 67 | 0 | 0 | 0 | 0 | 0 | 0 | 4 | 45 | 122 | 518 |
| AVG TEMP | DEG C | 13.5 | 15.3 | 18.3 | 22.7 | 25.5 | 28.0 | 29.3 | 29.5 | 27.2 | 23.3 | 18.3 | 15.1 | 22.2 |
| **TX DALLAS** | | | 32 51 N | 96 51 W | | 149 M | | | | | | | | |
| DAILY RAD | MJ/SQ M | 9.32 | 12.16 | 16.14 | 18.46 | 21.43 | 24.23 | 24.08 | 22.13 | 18.01 | 14.48 | 10.63 | 8.85 | 16.66 |
| DAY TEMP | DEG C | 9.1 | 11.3 | 15.0 | 20.8 | 24.8 | 29.1 | 31.4 | 31.6 | 27.3 | 21.7 | 15.0 | 10.6 | 20.6 |
| DEG DAY | C-DAY | 338 | 243 | 174 | 39 | 0 | 0 | 0 | 0 | 0 | 31 | 158 | 289 | 1272 |
| AVG TEMP | DEG C | 7.4 | 9.7 | 13.2 | 19.1 | 23.2 | 27.6 | 29.8 | 29.9 | 25.7 | 20.0 | 13.3 | 9.0 | 19.0 |
| **TX DEL RIO** | | | 29 22 N | 100 55 W | | 313 M | | | | | | | | |
| DAILY RAD | MJ/SQ M | 10.88 | 13.68 | 17.93 | 19.29 | 20.74 | 22.97 | 23.31 | 21.98 | 17.98 | 15.43 | 12.02 | 10.24 | 17.20 |
| DAY TEMP | DEG C | 12.5 | 15.3 | 19.3 | 24.4 | 27.7 | 31.0 | 32.5 | 32.1 | 28.7 | 23.8 | 17.4 | 13.4 | 23.2 |
| DEG DAY | C-DAY | 249 | 157 | 91 | 9 | 0 | 0 | 0 | 0 | 0 | 19 | 102 | 219 | 846 |
| AVG TEMP | DEG C | 10.4 | 13.2 | 17.0 | 22.2 | 25.7 | 29.1 | 30.4 | 30.1 | 26.8 | 21.8 | 15.3 | 11.3 | 21.1 |
| **TX EL PASO** | | | 31 48 N | 106 24 W | | 1194 M | | | | | | | | |
| DAILY RAD | MJ/SQ M | 12.77 | 16.80 | 21.67 | 26.82 | 29.51 | 30.44 | 27.81 | 25.93 | 22.55 | 18.60 | 14.11 | 11.70 | 21.56 |
| DAY TEMP | DEG C | 8.7 | 11.5 | 14.9 | 20.2 | 24.8 | 29.3 | 30.0 | 29.0 | 25.6 | 20.2 | 13.3 | 9.1 | 20.4 |
| DEG DAY | C-DAY | 368 | 258 | 182 | 49 | 0 | 0 | 0 | 0 | 0 | 51 | 223 | 355 | 1486 |
| AVG TEMP | DEG C | 6.4 | 9.1 | 12.6 | 17.7 | 22.3 | 26.8 | 27.9 | 26.9 | 23.4 | 17.8 | 10.9 | 6.9 | 17.4 |
| **TX FORT WORTH** | | | 32 50 N | 97 03 W | | 164 M | | | | | | | | |
| DAILY RAD | MJ/SQ M | 9.14 | 12.14 | 15.99 | 18.34 | 21.45 | 24.43 | 24.46 | 22.50 | 18.40 | 14.67 | 10.64 | 8.69 | 16.74 |
| DAY TEMP | DEG C | 8.9 | 11.1 | 14.7 | 20.3 | 24.2 | 28.7 | 31.1 | 31.3 | 27.2 | 21.7 | 15.2 | 10.6 | 20.4 |
| DEG DAY | C-DAY | 348 | 253 | 186 | 51 | 0 | 0 | 0 | 0 | 0 | 33 | 159 | 294 | 1322 |
| AVG TEMP | DEG C | 7.1 | 9.3 | 12.8 | 18.4 | 22.5 | 27.0 | 29.3 | 29.4 | 25.4 | 19.8 | 13.2 | 8.8 | 18.6 |
| **TX HOUSTON** | | | 29 59 N | 95 22 W | | 33 M | | | | | | | | |
| DAILY RAD | MJ/SQ M | 8.77 | 11.74 | 14.72 | 17.28 | 20.14 | 21.54 | 20.75 | 19.14 | 16.69 | 14.48 | 10.49 | 8.28 | 15.33 |
| DAY TEMP | DEG C | 12.9 | 14.7 | 17.8 | 22.4 | 26.0 | 29.0 | 30.2 | 30.4 | 28.0 | 23.7 | 18.1 | 14.4 | 22.3 |
| DEG DAY | C-DAY | 231 | 163 | 105 | 13 | 0 | 0 | 0 | 0 | 0 | 13 | 86 | 185 | 796 |
| AVG TEMP | DEG C | 11.2 | 12.9 | 16.0 | 20.8 | 24.3 | 27.3 | 28.5 | 28.6 | 26.2 | 21.6 | 16.2 | 12.6 | 20.5 |
| **TX KINGSVILLE** | | | 27 31 N | 97 49 W | | 17 M | | | | | | | | |
| DAILY RAD | MJ/SQ M | 10.35 | 13.18 | 16.28 | 18.87 | 21.16 | 23.10 | 23.96 | 21.81 | 18.44 | 15.77 | 11.74 | 9.64 | 17.03 |
| DAY TEMP | DEG C | | | | | | | | | | | | | |
| DEG DAY | C-DAY | | | | | | | | | | | | | |
| AVG TEMP | DEG C | | | | | | | | | | | | | |
| **TX LAREDO** | | | 27 32 N | 99 28 W | | 158 M | | | | | | | | |
| DAILY RAD | MJ/SQ M | 10.88 | 13.57 | 17.20 | 19.60 | 22.15 | 23.53 | 24.19 | 22.80 | 19.35 | 15.98 | 11.81 | 10.09 | 17.60 |
| DAY TEMP | DEG C | | | | | | | | | | | | | |
| DEG DAY | C-DAY | 166 | 98 | 48 | 0 | 0 | 0 | 0 | 0 | 0 | 4 | 41 | 128 | 485 |
| AVG TEMP | DEG C | 13.6 | 16.1 | 19.8 | 24.6 | 27.4 | 30.0 | 31.1 | 30.9 | 28.3 | 24.2 | 18.4 | 14.8 | 23.3 |

(Continued)

**TABLE A-15.** (Continued)

| TX Station / Measure | JAN | FEB | MAR | APR | MAY | JUN | JUL | AUG | SEP | OCT | NOV | DEC | ANNUAL |
|---|---|---|---|---|---|---|---|---|---|---|---|---|---|
| **LUBBOCK** (3339N, 10149W, 988 M) | | | | | | | | | | | | | |
| DAILY RAD MJ/SQ M | 11.70 | 15.11 | 20.00 | 24.60 | 27.19 | 28.88 | 27.37 | 25.06 | 20.66 | 16.66 | 12.67 | 10.61 | 20.04 |
| DAY TEMP DEG C | 6.3 | 8.3 | 11.9 | 18.0 | 22.6 | 27.0 | 28.6 | 27.9 | 23.8 | 18.4 | 11.7 | 7.5 | 17.7 |
| DEG DAY C-DAY | 446 | 347 | 282 | 106 | 16 | | | | 4 | 90 | 270 | 408 | 1969 |
| AVG TEMP DEG C | 3.9 | 5.9 | 9.4 | 15.6 | 20.3 | 25.1 | 26.5 | 25.8 | 21.7 | 16.1 | 9.3 | 5.2 | 15.4 |
| **LUFKIN** (3114N, 9445W, 96 M) | | | | | | | | | | | | | |
| DAILY RAD MJ/SQ M | 9.01 | 12.13 | 15.62 | 18.43 | 21.18 | 23.33 | 22.77 | 21.15 | 17.37 | 15.31 | 10.93 | 8.71 | 16.33 |
| DAY TEMP DEG C | | | | | | | | | | | | | |
| DEG DAY C-DAY | 283 | 206 | 142 | 31 | 0 | | | 0 | | 29 | 142 | 244 | 1077 |
| AVG TEMP DEG C | 9.3 | 11.2 | 14.4 | 19.6 | 23.4 | 26.8 | 28.3 | 28.4 | 25.3 | 20.1 | 14.0 | 10.4 | 19.3 |
| **MIDLAND-ODESSA** (3156N, 10212W, 871 M) | | | | | | | | | | | | | |
| DAILY RAD MJ/SQ M | 12.27 | 15.69 | 20.87 | 24.88 | 27.58 | 29.08 | 27.12 | 25.08 | 20.93 | 17.27 | 13.35 | 11.35 | 20.45 |
| DAY TEMP DEG C | 8.8 | 11.2 | 14.9 | 20.4 | 24.8 | 28.8 | 30.1 | 29.8 | 26.2 | 21.0 | 14.2 | 10.1 | 20.0 |
| DEG DAY C-DAY | 368 | 268 | 194 | 54 | 18 | | | | 0 | 45 | 198 | 329 | 1456 |
| AVG TEMP DEG C | 6.4 | 8.8 | 12.4 | 17.9 | 22.4 | 26.6 | 27.9 | 27.7 | 24.1 | 18.8 | 11.8 | 7.7 | 17.7 |
| **PORT ARTHUR** (2957N, 9401W, 7 M) | | | | | | | | | | | | | |
| DAILY RAD MJ/SQ M | 9.08 | 12.15 | 15.36 | 18.27 | 21.23 | 22.82 | 20.95 | 19.70 | 17.33 | 15.00 | 10.81 | 8.56 | 15.94 |
| DAY TEMP DEG C | 12.7 | 14.5 | 17.3 | 22.1 | 25.4 | 28.6 | 29.8 | 30.0 | 27.7 | 23.0 | 17.4 | 14.0 | 21.9 |
| DEG DAY C-DAY | 233 | 168 | 112 | 18 | | | | | 0 | 19 | 102 | 190 | 842 |
| AVG TEMP DEG C | 11.1 | 12.8 | 15.6 | 20.5 | 23.9 | 27.1 | 28.3 | 28.4 | 26.1 | 21.1 | 15.7 | 12.3 | 20.2 |
| **SAN ANGELO** (3122N, 10030W, 582 M) | | | | | | | | | | | | | |
| DAILY RAD MJ/SQ M | 10.91 | 13.71 | 18.23 | 21.00 | 23.04 | 24.81 | 24.09 | 22.31 | 18.24 | 15.17 | 11.85 | 10.15 | 17.79 |
| DAY TEMP DEG C | 10.1 | 12.4 | 15.9 | 21.8 | 25.6 | 29.5 | 31.3 | 31.2 | 26.8 | 21.6 | 15.2 | 11.2 | 21.1 |
| DEG DAY C-DAY | 321 | 229 | 159 | 41 | 7 | | | | 0 | 41 | 166 | 288 | 1245 |
| AVG TEMP DEG C | 8.0 | 10.2 | 13.9 | 19.6 | 23.6 | 27.6 | 29.3 | 29.2 | 24.9 | 19.6 | 13.1 | 9.1 | 19.0 |
| **SAN ANTONIO** (2932N, 9828W, 242 M) | | | | | | | | | | | | | |
| DAILY RAD MJ/SQ M | 10.16 | 13.10 | 16.46 | 18.30 | 21.50 | 23.48 | 24.07 | 22.10 | 18.59 | 15.32 | 11.45 | 9.61 | 17.01 |
| DAY TEMP DEG C | 12.2 | 14.4 | 18.0 | 22.7 | 26.1 | 29.6 | 31.1 | 31.1 | 28.0 | 23.3 | 17.3 | 13.7 | 22.3 |
| DEG DAY C-DAY | 251 | 172 | 108 | 17 | | | | | 0 | 18 | 99 | 207 | 872 |
| AVG TEMP DEG C | 10.4 | 12.5 | 16.0 | 20.9 | 24.4 | 27.9 | 29.3 | 29.3 | 26.3 | 21.4 | 15.4 | 11.8 | 20.5 |
| **SHERMAN** (3343N, 9640W, 233 M) | | | | | | | | | | | | | |
| DAILY RAD MJ/SQ M | 9.01 | 11.77 | 15.50 | 18.28 | 21.02 | 23.99 | 23.57 | 21.92 | 17.93 | 14.39 | 10.43 | 8.44 | 16.35 |
| DAY TEMP DEG C | | | | | | | | | | | | | |
| DEG DAY C-DAY | 401 | 297 | 228 | 63 | 7 | | | 0 | 0 | 50 | 196 | 348 | 1590 |
| AVG TEMP DEG C | 5.4 | 7.7 | 11.3 | 17.6 | 21.8 | 26.3 | 28.7 | 28.7 | 24.4 | 18.8 | 11.9 | 7.1 | 17.5 |
| **WACO** (3137N, 9713W, 155 M) | | | | | | | | | | | | | |
| DAILY RAD MJ/SQ M | 9.45 | 12.44 | 16.20 | 18.30 | 20.13 | 23.97 | 24.18 | 22.22 | 18.17 | 14.77 | 10.86 | 9.11 | 16.65 |
| DAY TEMP DEG C | 10.1 | 12.3 | 15.9 | 21.4 | 25.3 | 29.4 | 31.5 | 31.7 | 27.8 | 22.5 | 16.0 | 11.7 | 21.3 |
| DEG DAY C-DAY | 310 | 223 | 156 | 31 | 0 | | | | 0 | 28 | 134 | 262 | 1144 |
| AVG TEMP DEG C | 8.3 | 10.5 | 14.0 | 19.6 | 23.6 | 27.7 | 29.8 | 29.8 | 26.1 | 20.6 | 14.2 | 9.9 | 19.5 |

| | JAN | FEB | MAR | APR | MAY | JUN | JUL | AUG | SEP | OCT | NOV | DEC | ANNUAL |
|---|---|---|---|---|---|---|---|---|---|---|---|---|---|
| **TX  WICHITA FALLS** (3358N, 9829W, 314 M) | | | | | | | | | | | | | |
| DAILY RAD MJ/SQ M | 9.78 | 12.74 | 16.70 | 20.01 | 22.89 | 25.21 | 24.59 | 22.35 | 18.18 | 14.66 | 10.86 | 9.07 | 17.25 |
| DAY TEMP DEG C | 7.3 | 9.8 | 13.6 | 20.1 | 24.5 | 29.5 | 32.1 | 32.0 | 27.0 | 21.1 | 13.8 | 8.8 | 20.0 |
| DEG DAY C-DAY | 405 | 297 | 227 | 62 | 7 | | | | | 51 | 205 | 358 | 1612 |
| AVG TEMP DEG C | 5.3 | 7.7 | 11.4 | 17.9 | 22.4 | 27.4 | 29.9 | 29.7 | 25.0 | 18.9 | 11.6 | 6.8 | 17.8 |
| **UT  BRYCE CANYON** (3742N, 11209W, 2313 M) | | | | | | | | | | | | | |
| DAILY RAD MJ/SQ M | 10.37 | 14.02 | 19.12 | 24.21 | 27.85 | 30.13 | 27.51 | 24.48 | 21.79 | 16.62 | 11.52 | 9.29 | 19.74 |
| DAY TEMP DEG C | | | | | | | | | | | | | |
| DEG DAY C-DAY | 778 | 650 | 625 | 455 | 324 | 183 | 71 | 98 | 202 | 382 | 572 | 734 | 5074 |
| AVG TEMP DEG C | -6.8 | -4.9 | -1.8 | 3.2 | 7.9 | 13.4 | 16.4 | 15.5 | 11.6 | 6.0 | -.7 | -5.3 | 4.1 |
| **UT  CEDAR CITY** (3742N, 11306W, 1712 M) | | | | | | | | | | | | | |
| DAILY RAD MJ/SQ M | 10.01 | 13.39 | 18.56 | 23.75 | 28.00 | 30.71 | 28.41 | 25.44 | 22.34 | 16.57 | 11.26 | 8.91 | 19.78 |
| DAY TEMP DEG C | | | | | | | | | | | | | |
| DEG DAY C-DAY | 625 | 496 | 458 | 298 | 156 | 48 | 0 | 3 | 63 | 236 | 437 | 589 | 3409 |
| AVG TEMP DEG C | -1.8 | .6 | 3.6 | 8.4 | 13.4 | 18.3 | 22.9 | 21.8 | 17.3 | 10.8 | 3.8 | -.7 | 9.9 |
| **UT  SALT LAKE CITY** (4046N, 11158W, 1288 M) | | | | | | | | | | | | | |
| DAILY RAD MJ/SQ M | 7.25 | 11.22 | 16.50 | 21.50 | 26.81 | 29.06 | 29.39 | 25.58 | 20.92 | 14.68 | 8.94 | 6.47 | 18.19 |
| DAY TEMP DEG C | -.7 | 2.4 | 6.1 | 11.7 | 17.0 | 21.5 | 27.5 | 26.2 | 20.8 | 13.7 | 5.8 | .5 | 12.7 |
| DEG DAY C-DAY | 637 | 492 | 437 | 263 | 132 | 49 | 0 | 3 | 58 | 223 | 432 | 598 | 3324 |
| AVG TEMP DEG C | -2.2 | .8 | 4.2 | 9.6 | 14.6 | 19.0 | 24.8 | 23.6 | 18.2 | 11.3 | 3.9 | -.9 | 10.6 |
| **VA  NORFOLK** (3654N, 7612W, 9 M) | | | | | | | | | | | | | |
| DAILY RAD MJ/SQ M | 7.70 | 10.58 | 14.54 | 19.03 | 21.42 | 22.70 | 21.03 | 19.07 | 15.84 | 12.29 | 9.21 | 7.08 | 15.04 |
| DAY TEMP DEG C | 6.1 | 6.7 | 10.5 | 16.0 | 20.9 | 25.1 | 27.1 | 26.3 | 23.4 | 17.9 | 12.4 | 7.1 | 16.6 |
| DEG DAY C-DAY | 422 | 367 | 296 | 126 | 29 | | | | | 78 | 223 | 391 | 1937 |
| AVG TEMP DEG C | 4.7 | 5.2 | 8.9 | 14.3 | 19.3 | 23.6 | 25.7 | 24.9 | 22.1 | 16.5 | 10.9 | 5.7 | 15.2 |
| **VA  RICHMOND** (3730N, 7720W, 50 M) | | | | | | | | | | | | | |
| DAILY RAD MJ/SQ M | 7.17 | 9.95 | 13.74 | 17.77 | 20.00 | 21.25 | 20.14 | 18.16 | 15.30 | 11.72 | 8.32 | 6.43 | 14.16 |
| DAY TEMP DEG C | 4.7 | 5.9 | 10.2 | 16.4 | 21.1 | 25.3 | 27.2 | 26.3 | 22.9 | 17.1 | 11.4 | 5.6 | 16.2 |
| DEG DAY C-DAY | 474 | 398 | 316 | 126 | 36 | | | | 12 | 113 | 267 | 448 | 2190 |
| AVG TEMP DEG C | 3.1 | 4.1 | 8.3 | 14.3 | 19.2 | 23.4 | 25.5 | 24.6 | 21.1 | 15.2 | 9.4 | 3.9 | 14.3 |
| **VA  ROANOKE** (3719N, 7958W, 358 M) | | | | | | | | | | | | | |
| DAILY RAD MJ/SQ M | 7.50 | 10.21 | 14.03 | 17.95 | 20.02 | 21.36 | 20.38 | 18.39 | 15.41 | 12.26 | 8.68 | 6.70 | 14.41 |
| DAY TEMP DEG C | 4.0 | 5.0 | 9.2 | 15.3 | 19.9 | 23.9 | 25.8 | 25.2 | 21.9 | 16.3 | 10.0 | 4.5 | 15.1 |
| DEG DAY C-DAY | 493 | 418 | 339 | 157 | 56 | | | | 18 | 131 | 305 | 476 | 2393 |
| AVG TEMP DEG C | 2.4 | 3.4 | 7.4 | 13.3 | 18.0 | 22.1 | 24.0 | 23.4 | 20.0 | 14.3 | 8.2 | 3.0 | 13.3 |
| **VT  BURLINGTON** (4428N, 7309W, 104 M) | | | | | | | | | | | | | |
| DAILY RAD MJ/SQ M | 4.37 | 6.89 | 10.47 | 14.71 | 17.86 | 19.62 | 19.53 | 16.74 | 12.74 | 8.40 | 4.25 | 3.21 | 11.58 |
| DAY TEMP DEG C | -6.9 | -5.8 | -.1 | 7.8 | 14.6 | 20.3 | 22.9 | 21.5 | 16.9 | 11.0 | 4.0 | -3.9 | 8.5 |
| DEG DAY C-DAY | 830 | 722 | 618 | 367 | 184 | 35 | 11 | 27 | 106 | 279 | 467 | 730 | 4376 |
| AVG TEMP DEG C | -8.4 | -7.4 | -1.6 | 6.1 | 12.7 | 18.4 | 21.0 | 19.7 | 15.2 | 9.3 | 2.8 | -5.2 | 6.9 |

*(Continued)*

**TABLE A-15.** *(Continued)*

### WA OLYMPIA — 46°58′N, 122°25′W, 61 m

| | JAN | FEB | MAR | APR | MAY | JUN | JUL | AUG | SEP | OCT | NOV | DEC | ANNUAL |
|---|---|---|---|---|---|---|---|---|---|---|---|---|---|
| DAILY RAD MJ/SQ M | 3.05 | 5.71 | 9.59 | 14.24 | 18.52 | 19.22 | 21.71 | 17.58 | 13.13 | 7.22 | 3.85 | 2.51 | 11.36 |
| DAY TEMP DEG C | 4.0 | 6.4 | 8.0 | 10.9 | 14.8 | 17.1 | 20.0 | 19.5 | 17.0 | 12.1 | 7.6 | 5.2 | 11.9 |
| DEG DAY C-DAY | 479 | 373 | 376 | 280 | 189 | 109 | 49 | 57 | 110 | 246 | 362 | 439 | 3071 |
| AVG TEMP DEG C | 2.9 | 5.0 | 6.2 | 9.0 | 12.2 | 14.9 | 17.6 | 17.1 | 14.8 | 10.3 | 6.3 | 4.2 | 10.0 |

### WA SEATTLE-TACOMA — 47°27′N, 122°18′W, 122 m

| | JAN | FEB | MAR | APR | MAY | JUN | JUL | AUG | SEP | OCT | NOV | DEC | ANNUAL |
|---|---|---|---|---|---|---|---|---|---|---|---|---|---|
| DAILY RAD MJ/SQ M | 2.97 | 5.62 | 9.60 | 14.68 | 19.45 | 20.45 | 25.51 | 18.34 | 13.02 | 7.45 | 3.83 | 2.40 | 11.95 |
| DAY TEMP DEG C | 4.3 | 6.8 | 8.0 | 10.7 | 14.3 | 17.0 | 19.8 | 19.3 | 16.8 | 12.4 | 8.0 | 5.5 | 11.9 |
| DEG DAY C-DAY | 462 | 353 | 360 | 272 | 174 | 93 | 44 | 46 | 94 | 221 | 340 | 422 | 2881 |
| AVG TEMP DEG C | 3.4 | 5.7 | 6.7 | 9.3 | 12.7 | 15.4 | 18.1 | 17.7 | 15.3 | 11.2 | 7.0 | 4.7 | 10.6 |

### WA SPOKANE — 47°38′N, 117°32′W, 721 m

| | JAN | FEB | MAR | APR | MAY | JUN | JUL | AUG | SEP | OCT | NOV | DEC | ANNUAL |
|---|---|---|---|---|---|---|---|---|---|---|---|---|---|
| DAILY RAD MJ/SQ M | 3.57 | 6.88 | 11.81 | 16.97 | 21.77 | 23.64 | 26.75 | 22.04 | 16.29 | 9.54 | 4.51 | 2.90 | 13.89 |
| DAY TEMP DEG C | -2.7 | 1.2 | 4.5 | 9.6 | 14.6 | 18.4 | 23.4 | 22.3 | 17.5 | 10.5 | 3.0 | -.8 | 10.1 |
| DEG DAY C-DAY | 682 | 510 | 474 | 315 | 182 | 80 | 12 | 26 | 109 | 296 | 492 | 620 | 3798 |
| AVG TEMP DEG C | -3.7 | .1 | 3.1 | 7.8 | 12.6 | 16.4 | 20.9 | 20.0 | 15.3 | 8.8 | 1.9 | -1.7 | 8.5 |

### WA WHIDBEY ISLAND — 48°21′N, 122°40′W, 17 m

| | JAN | FEB | MAR | APR | MAY | JUN | JUL | AUG | SEP | OCT | NOV | DEC | ANNUAL |
|---|---|---|---|---|---|---|---|---|---|---|---|---|---|
| DAILY RAD MJ/SQ M | 3.21 | 6.04 | 10.42 | 15.26 | 19.98 | 20.66 | 22.48 | 18.07 | 13.32 | 7.43 | 4.05 | 2.64 | 11.96 |
| DAY TEMP DEG C | | | | | | | | | | | | | |
| DEG DAY C-DAY | | | | | | | | | | | | | |
| AVG TEMP DEG C | | | | | | | | | | | | | |

### WA YAKIMA — 46°34′N, 120°32′W, 325 m

| | JAN | FEB | MAR | APR | MAY | JUN | JUL | AUG | SEP | OCT | NOV | DEC | ANNUAL |
|---|---|---|---|---|---|---|---|---|---|---|---|---|---|
| DAILY RAD MJ/SQ M | 4.14 | 7.56 | 12.74 | 18.13 | 22.79 | 24.61 | 26.76 | 22.41 | 16.83 | 10.11 | 5.04 | 3.35 | 14.54 |
| DAY TEMP DEG C | -1.0 | 3.8 | 7.6 | 12.2 | 16.9 | 20.6 | 24.4 | 23.2 | 19.1 | 12.5 | 5.2 | .9 | 12.1 |
| DEG DAY C-DAY | 646 | 456 | 399 | 256 | 133 | 52 | 11 | 21 | 82 | 257 | 443 | 581 | 3339 |
| AVG TEMP DEG C | -2.5 | 2.1 | 5.4 | 9.7 | 14.4 | 18.1 | 21.5 | 20.3 | 16.3 | 10.1 | 3.6 | -.4 | 9.9 |

### WI EAU CLAIRE — 44°52′N, 91°29′W, 273 m

| | JAN | FEB | MAR | APR | MAY | JUN | JUL | AUG | SEP | OCT | NOV | DEC | ANNUAL |
|---|---|---|---|---|---|---|---|---|---|---|---|---|---|
| DAILY RAD MJ/SQ M | 5.13 | 8.47 | 12.37 | 16.18 | 19.07 | 21.24 | 21.41 | 16.39 | 13.57 | 9.38 | 5.11 | 3.87 | 12.85 |
| DAY TEMP DEG C | | | | | | | | | | | | | |
| DEG DAY C-DAY | 918 | 772 | 649 | 342 | 163 | 36 | 8 | 21 | 112 | 281 | 550 | 809 | 4661 |
| AVG TEMP DEG C | -11.3 | -9.2 | -2.6 | 6.9 | 13.4 | 18.9 | 21.4 | 20.2 | 14.8 | 9.3 | .0 | -7.8 | 6.2 |

### WI GREEN BAY — 44°29′N, 88°08′W, 214 m

| | JAN | FEB | MAR | APR | MAY | JUN | JUL | AUG | SEP | OCT | NOV | DEC | ANNUAL |
|---|---|---|---|---|---|---|---|---|---|---|---|---|---|
| DAILY RAD MJ/SQ M | 5.12 | 8.23 | 12.53 | 16.33 | 19.51 | 21.65 | 21.43 | 18.41 | 13.82 | 9.31 | 5.28 | 3.97 | 12.97 |
| DAY TEMP DEG C | -7.8 | -6.2 | -.5 | 8.3 | 14.4 | 19.9 | 22.6 | 21.7 | 16.8 | 11.3 | 2.4 | -4.9 | 8.2 |
| DEG DAY C-DAY | 854 | 731 | 627 | 353 | 188 | 51 | 12 | 30 | 106 | 272 | 515 | 759 | 4498 |
| AVG TEMP DEG C | -9.2 | -7.8 | -1.9 | 6.6 | 12.5 | 18.1 | 20.7 | 19.8 | 14.9 | 9.6 | 1.2 | -6.2 | 6.5 |

### WI LA CROSSE — 43°52′N, 91°15′W, 205 m

| | JAN | FEB | MAR | APR | MAY | JUN | JUL | AUG | SEP | OCT | NOV | DEC | ANNUAL |
|---|---|---|---|---|---|---|---|---|---|---|---|---|---|
| DAILY RAD MJ/SQ M | 5.46 | 8.68 | 12.49 | 16.19 | 19.44 | 21.62 | 21.57 | 16.91 | 14.09 | 9.80 | 5.60 | 4.19 | 13.17 |
| DAY TEMP DEG C | -7.3 | -5.0 | 1.0 | 10.4 | 16.7 | 21.9 | 24.4 | 23.6 | 18.2 | 12.7 | 3.2 | -4.4 | 9.6 |
| DEG DAY C-DAY | 842 | 700 | 584 | 290 | 124 | 22 | | 22 | 72 | 234 | 493 | 754 | 4120 |
| AVG TEMP DEG C | -8.8 | -6.7 | -.5 | 8.7 | 15.0 | 20.3 | 22.7 | 21.9 | 16.6 | 11.0 | 1.9 | -5.7 | 8.0 |

**WI — MADISON** (43 08N, 89 20W, 262 M)

| | Units | JAN | FEB | MAR | APR | MAY | JUN | JUL | AUG | SEP | OCT | NOV | DEC | ANNUAL |
|---|---|---|---|---|---|---|---|---|---|---|---|---|---|---|
| DAILY RAD | MJ/SQ M | 5.85 | 9.12 | 12.89 | 15.87 | 19.78 | 22.11 | 21.95 | 19.38 | 14.75 | 10.34 | 5.72 | 4.41 | 13.52 |
| DAY TEMP | DEG C | -7.0 | -5.0 | .5 | 9.2 | 15.2 | 20.6 | 23.0 | 22.3 | 17.3 | 11.8 | 2.9 | -4.3 | 8.9 |
| DEG DAY | C-DAY | 830 | 696 | 599 | 328 | 165 | 40 | 8 | 22 | 96 | 263 | 505 | 742 | 4294 |
| AVG TEMP | DEG C | -8.4 | -6.5 | -1.0 | 7.4 | 13.3 | 18.8 | 21.2 | 20.4 | 15.4 | 9.9 | 1.5 | -5.6 | 7.2 |

**WI — MILWAUKEE** (42 57N, 87 54W, 211 M)

| | Units | JAN | FEB | MAR | APR | MAY | JUN | JUL | AUG | SEP | OCT | NOV | DEC | ANNUAL |
|---|---|---|---|---|---|---|---|---|---|---|---|---|---|---|
| DAILY RAD | MJ/SQ M | 5.44 | 8.36 | 12.36 | 16.37 | 20.07 | 22.44 | 22.26 | 19.51 | 14.87 | 10.30 | 5.95 | 4.29 | 13.52 |
| DAY TEMP | DEG C | -5.7 | -4.0 | 1.0 | 8.7 | 14.1 | 19.9 | 22.8 | 22.4 | 17.9 | 12.3 | 3.8 | -3.1 | 9.2 |
| DEG DAY | C-DAY | 786 | 661 | 579 | 336 | 193 | 50 | 8 | 20 | 78 | 244 | 475 | 703 | 4135 |
| AVG TEMP | DEG C | -7.0 | -5.3 | -.3 | 7.1 | 12.3 | 18.1 | 21.1 | 20.7 | 16.2 | 10.6 | 2.5 | -4.3 | 7.6 |

**WV — CHARLESTON** (38 22N, 81 36W, 290 M)

| | Units | JAN | FEB | MAR | APR | MAY | JUN | JUL | AUG | SEP | OCT | NOV | DEC | ANNUAL |
|---|---|---|---|---|---|---|---|---|---|---|---|---|---|---|
| DAILY RAD | MJ/SQ M | 5.66 | 8.02 | 11.46 | 15.39 | 18.61 | 20.15 | 19.09 | 17.19 | 14.44 | 11.03 | 6.96 | 4.99 | 12.75 |
| DAY TEMP | DEG C | 2.9 | 4.1 | 8.7 | 15.3 | 20.1 | 24.1 | 25.7 | 24.9 | 21.6 | 15.9 | 9.2 | 3.8 | 14.7 |
| DEG DAY | C-DAY | 526 | 443 | 357 | 159 | 63 | 6 | | | 26 | 148 | 327 | 496 | 2551 |
| AVG TEMP | DEG C | 1.4 | 2.5 | 6.9 | 13.3 | 18.1 | 22.2 | 23.9 | 23.1 | 19.7 | 13.9 | 7.4 | 2.3 | 12.9 |

**WV — HUNTINGTON** (38 22N, 82 33W, 255 M)

| | Units | JAN | FEB | MAR | APR | MAY | JUN | JUL | AUG | SEP | OCT | NOV | DEC | ANNUAL |
|---|---|---|---|---|---|---|---|---|---|---|---|---|---|---|
| DAILY RAD | MJ/SQ M | 5.97 | 8.59 | 12.11 | 16.43 | 19.41 | 20.92 | 20.07 | 17.93 | 14.82 | 11.39 | 7.24 | 5.30 | 13.35 |
| DAY TEMP | DEG C | 2.7 | 3.8 | 8.6 | 15.1 | 20.0 | 24.3 | 25.8 | 25.0 | 21.7 | 15.9 | 9.1 | 3.7 | 14.6 |
| DEG DAY | C-DAY | 529 | 449 | 361 | 163 | 64 | 6 | | | 26 | 147 | 325 | 499 | 2569 |
| AVG TEMP | DEG C | 1.3 | 2.3 | 6.8 | 13.2 | 18.1 | 22.4 | 24.1 | 23.3 | 19.8 | 13.9 | 7.5 | 2.2 | 12.9 |

**WY — CASPER** (42 55N, 106 28W, 1612 M)

| | Units | JAN | FEB | MAR | APR | MAY | JUN | JUL | AUG | SEP | OCT | NOV | DEC | ANNUAL |
|---|---|---|---|---|---|---|---|---|---|---|---|---|---|---|
| DAILY RAD | MJ/SQ M | 7.75 | 11.50 | 16.35 | 20.96 | 25.01 | 28.39 | 28.76 | 25.26 | 19.85 | 13.83 | 8.68 | 6.74 | 17.76 |
| DAY TEMP | DEG C | -3.2 | -1.1 | 1.4 | 8.1 | 13.7 | 19.0 | 24.3 | 23.6 | 17.4 | 11.0 | 2.9 | -1.6 | 9.6 |
| DEG DAY | C-DAY | 720 | 594 | 586 | 372 | 216 | 82 | 7 | 9 | 127 | 298 | 518 | 668 | 4197 |
| AVG TEMP | DEG C | -4.9 | -2.9 | -.6 | 5.9 | 11.5 | 16.6 | 21.7 | 20.9 | 14.8 | 8.7 | 1.1 | -3.2 | 7.5 |

**WY — CHEYENNE** (41 09N, 104 49W, 1872 M)

| | Units | JAN | FEB | MAR | APR | MAY | JUN | JUL | AUG | SEP | OCT | NOV | DEC | ANNUAL |
|---|---|---|---|---|---|---|---|---|---|---|---|---|---|---|
| DAILY RAD | MJ/SQ M | 8.69 | 12.12 | 16.26 | 20.09 | 22.64 | 25.63 | 25.31 | 22.31 | 18.92 | 14.09 | 9.34 | 7.61 | 16.92 |
| DAY TEMP | DEG C | -1.1 | .3 | 1.8 | 8.1 | 13.5 | 18.5 | 23.2 | 22.2 | 17.0 | 11.1 | 3.9 | .3 | 9.9 |
| DEG DAY | C-DAY | 661 | 560 | 575 | 372 | 219 | 87 | 12 | 17 | 125 | 294 | 492 | 617 | 4031 |
| AVG TEMP | DEG C | -3.0 | -1.7 | -.2 | 5.9 | 11.3 | 16.3 | 20.6 | 19.8 | 14.6 | 8.8 | 1.9 | -1.6 | 7.7 |

**WY — ROCK SPRINGS** (41 36N, 109 04W, 2056 M)

| | Units | JAN | FEB | MAR | APR | MAY | JUN | JUL | AUG | SEP | OCT | NOV | DEC | ANNUAL |
|---|---|---|---|---|---|---|---|---|---|---|---|---|---|---|
| DAILY RAD | MJ/SQ M | 8.34 | 12.36 | 17.36 | 22.06 | 26.60 | 29.22 | 28.91 | 25.42 | 20.80 | 14.82 | 9.38 | 7.38 | 18.56 |
| DEG DAY | C-DAY | 789 | 647 | 622 | 415 | 252 | 110 | 10 | 27 | 149 | 349 | 572 | 730 | 4672 |
| AVG TEMP | DEG C | -7.1 | -4.8 | -1.7 | 4.5 | 10.2 | 14.9 | 20.1 | 18.9 | 13.6 | 7.1 | -.7 | -5.2 | 5.8 |

**WY — SHERIDAN** (44 46N, 106 58W, 1209 M)

| | Units | JAN | FEB | MAR | APR | MAY | JUN | JUL | AUG | SEP | OCT | NOV | DEC | ANNUAL |
|---|---|---|---|---|---|---|---|---|---|---|---|---|---|---|
| DAILY RAD | MJ/SQ M | 5.87 | 8.94 | 13.67 | 17.45 | 21.37 | 24.47 | 26.43 | 22.77 | 17.04 | 11.41 | 6.70 | 5.01 | 15.09 |
| DAY TEMP | DEG C | -4.0 | -1.4 | 1.5 | 8.6 | 13.9 | 18.4 | 23.9 | 23.3 | 16.9 | 11.2 | 2.9 | -1.6 | 9.5 |
| DEG DAY | C-DAY | 758 | 608 | 586 | 357 | 208 | 93 | 16 | 17 | 136 | 296 | 527 | 681 | 4283 |
| AVG TEMP | DEG C | -6.1 | -3.4 | -.6 | 6.4 | 11.7 | 16.2 | 21.3 | 20.7 | 14.4 | 8.8 | .8 | -3.6 | 7.2 |

# Solar Radiation Data on Inclined Surfaces for Selected U.S. Locations
## for Use with the Base-Temperature Methods for Performance Prediction

O

Calculated from data tabulated in: *Solar Radiation Data on Inclined Surfaces for Selected U.S. Locations*, The Center for the Environment and Man, Inc., 275 Windsor St., Hartford, CT 06120. (Compiled by P. J. Lunde and P. S. Brown, Jr., from a model developed by M. A. Atwater.)

STATISTICS FOR HARTFORD, CONN.    1959-68

RADIATION UNITS ARE WATTS/SQ.M OR MJOULES/SQ.M
SLOPE = 0.0 DEG    AZIMUTH = 0.0 DEG
DEGREE-DAY BASE TEMP = 18.3 DEGREES C

**JAN**

| SOLAR RAD GROUP | ALL | 0+ | 95+ | 189+ | 284+ | 378+ | 473+ |
|---|---|---|---|---|---|---|---|
| AV.THRSH.RAD.LEVL | 0. | 40. | 143. | 237. | 330. | 424. | 503. |
| AV.THRESHOLD TEMP | -4.8 | -2.5 | -1.7 | -1.7 | -2.6 | -3.2 | -4.8 |
| GL RAD OVER THRSH | 204.2 | 159.2 | 87.6 | 45.9 | 13.0 | 3.2 | 0.0 |
| TIME AVG.TEMP | -3.8 | -2.4 | -2.3 | -2.7 | -3.1 | -3.6 | -4.8 |
| TOTAL TIME MSEC | 2.6784 | 1.1138 | 0.6955 | 0.4450 | 0.3013 | 0.1562 | 0.0407 |
| GLOBAL RAD TOTAL | 204.2 | 204.2 | 187.3 | 151.4 | 117.3 | 69.5 | 20.4 |
| DIRECT RAD TOTAL | 98.8 | 98.8 | 96.1 | 85.9 | 74.1 | 48.9 | 16.9 |
| RAD AV TIME MSEC | 0.0864 | 0.0360 | 0.0266 | 0.0212 | 0.0180 | 0.0126 | 0.0086 |
| DEG-C DAYS | 686. | 267. | 166. | 108. | 75. | 39. | 11. |

**FEB**

| SOLAR RAD GROUP | ALL | 0+ | 95+ | 189+ | 284+ | 378+ | 473+ |
|---|---|---|---|---|---|---|---|
| AV.THRSH.RAD.LEVL | 0. | 42. | 138. | 239. | 331. | 427. | 521. |
| AV.THRESHOLD TEMP | -4.3 | -2.1 | -0.8 | -1.3 | -1.6 | -1.4 | -1.7 |
| GL RAD OVER THRSH | 268.6 | 222.1 | 146.5 | 90.1 | 52.6 | 25.2 | 8.2 |
| TIME AVG.TEMP | -3.1 | -1.6 | -1.4 | -1.6 | -1.7 | -1.8 | -2.0 |
| TOTAL TIME MSEC | 2.4451 | 1.1182 | 0.7841 | 0.5591 | 0.4072 | 0.2862 | 0.1793 |
| GLOBAL RAD TOTAL | 268.6 | 268.6 | 254.7 | 223.6 | 187.4 | 147.3 | 101.7 |
| DIRECT RAD TOTAL | 133.7 | 133.7 | 131.4 | 124.7 | 113.8 | 97.2 | 73.0 |
| RAD AV TIME MSEC | 0.0864 | 0.0400 | 0.0313 | 0.0259 | 0.0223 | 0.0187 | 0.0155 |
| DEG-C DAYS | 606. | 258. | 179. | 129. | 94. | 67. | 42. |

**MAR**

| SOLAR RAD GROUP | ALL | 0+ | 95+ | 189+ | 284+ | 378+ | 473+ |
|---|---|---|---|---|---|---|---|
| AV.THRSH.RAD.LEVL | 0. | 44. | 141. | 236. | 329. | 428. | 519. |
| AV.THRESHOLD TEMP | 0.4 | 2.0 | 2.8 | 4.1 | 3.6 | 4.4 | 5.2 |
| GL RAD OVER THRSH | 407.3 | 347.8 | 249.2 | 175.7 | 122.5 | 77.6 | 46.4 |
| TIME AVG.TEMP | 2.1 | 3.6 | 4.2 | 4.6 | 4.8 | 5.1 | 5.3 |
| TOTAL TIME MSEC | 2.6784 | 1.3608 | 1.0159 | 0.7697 | 0.5771 | 0.4504 | 0.3420 |
| GLOBAL RAD TOTAL | 407.4 | 407.3 | 392.3 | 357.6 | 312.1 | 270.4 | 224.0 |
| DIRECT RAD TOTAL | 204.8 | 204.8 | 201.9 | 197.3 | 185.6 | 170.8 | 149.8 |
| RAD AV TIME MSEC | 0.0864 | 0.0443 | 0.0356 | 0.0320 | 0.0284 | 0.0252 | 0.0220 |
| DEG-C DAYS | 505. | 232. | 167. | 122. | 91. | 69. | 52. |

**JUL**

| SOLAR RAD GROUP | ALL | 0+ | 95+ | 189+ | 284+ | 378+ | 473+ |
|---|---|---|---|---|---|---|---|
| AV.THRSH.RAD.LEVL | 0. | 33. | 140. | 240. | 331. | 429. | 520. |
| AV.THRESHOLD TEMP | 19.4 | 20.7 | 21.8 | 24.1 | 24.2 | 26.5 | 26.3 |
| GL RAD OVER THRSH | 632.5 | 575.2 | 428.6 | 317.5 | 232.6 | 159.9 | 105.6 |
| TIME AVG.TEMP | 22.6 | 24.3 | 25.3 | 26.2 | 26.6 | 27.2 | 27.3 |
| TOTAL TIME MSEC | 2.6784 | 1.7525 | 1.3705 | 1.1088 | 0.9346 | 0.7434 | 0.5962 |
| GLOBAL RAD TOTAL | 632.7 | 632.6 | 620.1 | 583.5 | 541.7 | 478.5 | 415.4 |
| DIRECT RAD TOTAL | 295.0 | 295.0 | 292.9 | 283.9 | 274.8 | 254.1 | 234.6 |
| RAD AV TIME MSEC | 0.0864 | 0.0569 | 0.0461 | 0.0396 | 0.0360 | 0.0313 | 0.0284 |
| DEG-C DAYS | 1. | 0. | 0. | 0. | 0. | 0. | 0. |

**AUG**

| SOLAR RAD GROUP | ALL | 0+ | 95+ | 189+ | 284+ | 378+ | 473+ |
|---|---|---|---|---|---|---|---|
| AV.THRSH.RAD.LEVL | 0. | 44. | 144. | 236. | 333. | 425. | 521. |
| AV.THRESHOLD TEMP | 18.4 | 20.2 | 21.7 | 23.0 | 24.6 | 24.4 | 25.5 |
| GL RAD OVER THRSH | 529.7 | 460.3 | 338.7 | 246.2 | 171.3 | 115.2 | 71.1 |
| TIME AVG.TEMP | 21.3 | 23.4 | 24.3 | 24.8 | 25.4 | 25.6 | 26.0 |
| TOTAL TIME MSEC | 2.6784 | 1.5656 | 1.2229 | 1.0051 | 0.7729 | 0.6073 | 0.4583 |
| GLOBAL RAD TOTAL | 529.7 | 529.7 | 514.5 | 483.2 | 428.4 | 373.3 | 310.0 |
| DIRECT RAD TOTAL | 237.9 | 237.9 | 234.6 | 229.3 | 215.5 | 201.6 | 177.3 |
| RAD AV TIME MSEC | 0.0864 | 0.0504 | 0.0414 | 0.0371 | 0.0317 | 0.0288 | 0.0245 |
| DEG-C DAYS | 8. | 2. | 1. | 1. | 1. | 0. | 0. |

**SEP**

| SOLAR RAD GROUP | ALL | 0+ | 95+ | 189+ | 284+ | 378+ | 473+ |
|---|---|---|---|---|---|---|---|
| AV.THRSH.RAD.LEVL | 0. | 37. | 143. | 236. | 333. | 427. | 523. |
| AV.THRESHOLD TEMP | 14.5 | 16.3 | 17.6 | 19.6 | 19.8 | 21.9 | 21.5 |
| GL RAD OVER THRSH | 450.4 | 399.0 | 284.4 | 206.3 | 141.7 | 92.9 | 54.2 |
| TIME AVG.TEMP | 17.2 | 19.5 | 20.5 | 21.3 | 21.7 | 22.3 | 22.4 |
| TOTAL TIME MSEC | 2.5920 | 1.4663 | 1.0714 | 0.8417 | 0.6685 | 0.5202 | 0.4025 |
| GLOBAL RAD TOTAL | 450.4 | 450.4 | 438.1 | 405.2 | 364.3 | 314.9 | 264.6 |
| DIRECT RAD TOTAL | 253.2 | 253.2 | 250.9 | 242.2 | 232.1 | 210.1 | 186.1 |
| RAD AV TIME MSEC | 0.0864 | 0.0475 | 0.0382 | 0.0328 | 0.0299 | 0.0256 | 0.0227 |
| DEG-C DAYS | 71. | 21. | 12. | 7. | 5. | 3. | 2. |

### APR

| SOLAR RAD GROUP | ALL | 0+ | 95+ | 189+ | 284+ | 378+ | 473+ |
|---|---|---|---|---|---|---|---|
| AV.THRSH.RAD.LEVL | 0. | 38. | 144. | 234. | 333. | 423. | 519. |
| AV.THRESHOLD TEMP | 6.9 | 8.4 | 9.6 | 10.5 | 11.7 | 12.7 | 13.3 |
| GL RAD OVER THRSH | 497.3 | 440.0 | 320.5 | 239.7 | 171.8 | 122.8 | 82.5 |
| TIME AVG.TEMP | 9.4 | 11.3 | 12.2 | 12.9 | 13.7 | 14.2 | 14.6 |
| TOTAL TIME MSEC | 2.5920 | 1.5037 | 1.1326 | 0.8892 | 0.6858 | 0.5472 | 0.4187 |
| GLOBAL RAD TOTAL | 497.3 | 497.3 | 483.1 | 448.2 | 400.5 | 354.3 | 299.9 |
| DIRECT RAD TOTAL | 236.4 | 236.4 | 234.1 | 228.1 | 219.1 | 206.5 | 166.6 |
| RAD AV TIME MSEC | 0.0864 | 0.0504 | 0.0407 | 0.0356 | 0.0320 | 0.0288 | 0.0256 |
| DEG-C DAYS | 267. | 125. | 83. | 59. | 41. | 29. | 21. |

### MAY

| SOLAR RAD GROUP | ALL | 0+ | 95+ | 189+ | 284+ | 378+ | 473+ |
|---|---|---|---|---|---|---|---|
| AV.THRSH.RAD.LEVL | 0. | 39. | 142. | 238. | 331. | 429. | 523. |
| AV.THRESHOLD TEMP | 11.8 | 13.1 | 13.8 | 16.4 | 16.3 | 19.4 | 18.7 |
| GL RAD OVER THRSH | 626.9 | 562.4 | 424.8 | 320.2 | 238.3 | 168.6 | 113.7 |
| TIME AVG.TEMP | 15.0 | 16.9 | 17.9 | 18.8 | 19.4 | 20.2 | 20.3 |
| TOTAL TIME MSEC | 2.6784 | 1.6682 | 1.3262 | 1.0912 | 0.8881 | 0.7106 | 0.5818 |
| GLOBAL RAD TOTAL | 627.0 | 627.0 | 613.7 | 580.3 | 531.9 | 473.2 | 418.0 |
| DIRECT RAD TOTAL | 279.0 | 279.0 | 276.7 | 270.1 | 261.1 | 244.8 | 227.6 |
| RAD AV TIME MSEC | 0.0864 | 0.0540 | 0.0446 | 0.0389 | 0.0349 | 0.0310 | 0.0274 |
| DEG-C DAYS | 120. | 49. | 32. | 21. | 14. | 8. | 7. |

### JUN

| SOLAR RAD GROUP | ALL | 0+ | 95+ | 189+ | 284+ | 378+ | 473+ |
|---|---|---|---|---|---|---|---|
| AV.THRSH.RAD.LEVL | 0. | 33. | 142. | 237. | 333. | 428. | 520. |
| AV.THRESHOLD TEMP | 16.8 | 18.3 | 19.3 | 20.7 | 22.1 | 24.5 | 24.5 |
| GL RAD OVER THPSH | 622.9 | 565.4 | 420.3 | 319.6 | 236.3 | 170.7 | 113.7 |
| TIME AVG.TEMP | 20.2 | 21.9 | 23.0 | 23.9 | 24.7 | 25.3 | 25.5 |
| TOTAL TIME MSEC | 2.5920 | 1.7190 | 1.3342 | 1.0624 | 0.8676 | 0.6901 | 0.5800 |
| GLOBAL RAD TOTAL | 622.9 | 622.9 | 610.0 | 571.4 | 525.2 | 466.1 | 418.9 |
| DIRECT RAD TOTAL | 284.2 | 284.2 | 282.4 | 273.7 | 267.3 | 248.9 | 233.5 |
| RAD AV TIME MSEC | 0.0864 | 0.0576 | 0.0468 | 0.0403 | 0.0374 | 0.0324 | 0.0299 |
| DEG-C DAYS | 21. | 9. | 5. | 3. | 1. | 1. | 0. |

### OCT

| SOLAR RAD GROUP | ALL | 0+ | 95+ | 189+ | 284+ | 378+ | 473+ |
|---|---|---|---|---|---|---|---|
| AV.THRSH.RAD.LEVL | 0. | 43. | 142. | 236. | 331. | 427. | 524. |
| AV.THRESHOLD TEMP | 9.0 | 11.6 | 13.0 | 13.5 | 14.9 | 15.3 | 16.0 |
| GL RAD OVER THRSH | 340.3 | 285.7 | 193.6 | 128.2 | 79.6 | 43.3 | 18.2 |
| TIME AVG.TEMP | 11.3 | 13.9 | 14.7 | 15.3 | 15.9 | 16.3 | 16.8 |
| TOTAL TIME MSEC | 2.6870 | 1.2730 | 0.9299 | 0.6995 | 0.5116 | 0.3758 | 0.2592 |
| GLOBAL RAD TOTAL | 340.3 | 340.3 | 325.6 | 292.9 | 248.6 | 203.8 | 154.0 |
| DIRECT RAD TOTAL | 185.8 | 185.8 | 182.2 | 175.3 | 160.2 | 140.5 | 112.6 |
| RAD AV TIME MSEC | 0.0364 | 0.0410 | 0.0324 | 0.0284 | 0.0241 | 0.0209 | 0.0173 |
| DEG-C DAYS | 219. | 71. | 46. | 32. | 21. | 14. | 9. |

### NOV

| SOLAR RAD GROUP | ALL | 0+ | 95+ | 189+ | 284+ | 378+ | 473+ |
|---|---|---|---|---|---|---|---|
| AV.THRSH.RAD.LEVL | 0. | 40. | 138. | 236. | 331. | 426. | 513. |
| AV.THRESHOLD TEMP | 3.9 | 6.4 | 7.0 | 7.3 | 7.5 | 8.2 | 8.3 |
| GL RAD OVER THRSH | 196.4 | 150.9 | 86.2 | 45.0 | 20.1 | 5.8 | 0.5 |
| TIME AVG.TEMP | 5.3 | 7.0 | 7.4 | 7.7 | 7.9 | 8.3 | 8.4 |
| TOTAL TIME MSEC | 2.5920 | 1.1297 | 0.6624 | 0.4198 | 0.2632 | 0.1498 | 0.0616 |
| GLOBAL RAD TOTAL | 195.4 | 196.4 | 177.6 | 144.1 | 107.1 | 69.6 | 32.0 |
| DIRECT RAD TOTAL | 83.9 | 88.9 | 86.6 | 79.0 | 65.9 | 47.7 | 25.1 |
| RAD AV TIME MSEC | 0.0864 | 0.0382 | 0.0277 | 0.0223 | 0.0184 | 0.0148 | 0.0104 |
| DEG-C DAYS | 392. | 148. | 84. | 52. | 32. | 17. | 7. |

### DEC

| SOLAR RAD GROUP | ALL | 0+ | 95+ | 189+ | 284+ | 378+ | 473+ |
|---|---|---|---|---|---|---|---|
| AV.THRSH.RAD.LEVL | 0. | 40. | 136. | 239. | 334. | 422. | 482. |
| AV.THRESHOLD TEMP | -3.0 | -0.5 | -0.8 | 0.2 | -0.7 | -1.8 | 1.2 |
| GL RAD OVER THRSH | 171.1 | 127.8 | 67.2 | 29.8 | 9.1 | 0.2 | 0.0 |
| TIME AVG.TEMP | -2.0 | -0.6 | -0.7 | -0.6 | -1.2 | -1.7 | 1.2 |
| TOTAL TIME MSEC | 2.6784 | 1.0940 | 0.6156 | 0.3715 | 0.2178 | 0.1015 | 0.0025 |
| GLOBAL RAD TOTAL | 171.1 | 171.1 | 152.1 | 118.5 | 81.8 | 43.0 | 1.2 |
| DIRECT RAD TOTAL | 82.9 | 82.9 | 80.5 | 71.1 | 55.0 | 32.9 | 1.1 |
| RAD AV TIME MSEC | 0.0864 | 0.0356 | 0.0256 | 0.0202 | 0.0158 | 0.0112 | 0.0047 |
| DEG-C DAYS | 631. | 240. | 136. | 82. | 49. | 23. | 1. |

### ALL MONTHS

| SOLAR RAD GROUP | ALL | 0+ | 95+ | 189+ | 284+ | 378+ | 473+ |
|---|---|---|---|---|---|---|---|
| TIME AVG.TEMP | 9.7 | 12.9 | 14.3 | 15.6 | 16.6 | 17.9 | 19.4 |
| TOTAL TIME MSEC | 31.5705 | 16.7069 | 12.1612 | 9.2628 | 7.0956 | 5.3388 | 3.9226 |
| GLOBAL RAD TOTAL | 4947.9 | 4947.7 | 4769.1 | 4359.7 | 3846.3 | 3264.0 | 2660.3 |
| DIRECT RAD TOTAL | 2380.5 | 2380.5 | 2350.5 | 2260.6 | 2124.5 | 1904.1 | 1624.1 |
| RAD AV TIME MSEC | 0.0864 | 0.0490 | 0.0395 | 0.0345 | 0.0311 | 0.0276 | 0.0251 |
| DEG-C DAYS | 3527. | 1423. | 909. | 614. | 423. | 271. | 152. |

STATISTICS FOR HARTFORD, CONN.                    1959-68

RADIATION UNITS ARE WATTS/SQ.M OR MJOULES/SQ.M
SLOPE = 30.0 DEG    AZIMUTH = 180.0 DEG
DEGREE-DAY BASE TEMP = 18.3 DEGREES C

**JAN**

| SOLAR RAD GROUP | ALL | 0+ | 95+ | 189+ | 284+ | 378+ | 473+ |
|---|---|---|---|---|---|---|---|
| AV.THRSH.RAD.LEVL | 0. | 39. | 138. | 235. | 326. | 428. | 518. |
| AV.THRESHOLD TEMP | -4.8 | -2.6 | -1.1 | -3.0 | -1.3 | -1.5 | -2.3 |
| GL RAD OVER THRSH | 302.8 | 257.2 | 163.2 | 129.3 | 91.9 | 58.7 | 36.4 |
| TIME AVG.TEMP | -3.8 | -2.4 | -2.3 | -2.7 | -2.6 | -2.9 | -3.3 |
| TOTAL TIME MSEC | 2.6784 | 1.1639 | 0.7452 | 0.5605 | 0.4115 | 0.3244 | 0.2473 |
| GLOBAL RAD TOTAL | 302.9 | 302.8 | 286.4 | 260.8 | 225.9 | 197.5 | 164.5 |
| DIRECT RAD TOTAL | 202.3 | 202.3 | 199.4 | 193.1 | 175.8 | 161.8 | 149.4 |
| RAD AV TIME MSEC | 0.0864 | 0.0378 | 0.0299 | 0.0274 | 0.0234 | 0.0216 | 0.0198 |
| DEG-C DAYS | 686. | 279. | 178. | 136. | 99. | 79. | 62. |

**FEB**

| SOLAR RAD GROUP | ALL | 0+ | 95+ | 189+ | 284+ | 378+ | 473+ |
|---|---|---|---|---|---|---|---|
| AV.THRSH.RAD.LEVL | 0. | 39. | 138. | 239. | 330. | 425. | 526. |
| AV.THRESHOLD TEMP | -4.4 | -2.0 | -0.8 | -1.1 | -1.8 | -0.8 | -0.8 |
| GL RAD OVER THRSH | 351.3 | 306.4 | 229.2 | 168.2 | 124.4 | 88.2 | 57.0 |
| TIME AVG.TEMP | -3.1 | -1.6 | -1.3 | -1.5 | -1.6 | -1.6 | -1.7 |
| TOTAL TIME MSEC | 2.4451 | 1.1408 | 0.7816 | 0.6062 | 0.4824 | 0.3798 | 0.3082 |
| GLOBAL RAD TOTAL | 351.4 | 351.3 | 337.3 | 313.0 | 283.4 | 249.6 | 219.1 |
| DIRECT RAD TOTAL | 224.0 | 224.0 | 221.1 | 216.6 | 208.1 | 192.7 | 177.5 |
| RAD AV TIME MSEC | 0.0864 | 0.0407 | 0.0317 | 0.0292 | 0.0270 | 0.0245 | 0.0227 |
| DEG-C DAYS | 606. | 263. | 178. | 139. | 112. | 88. | 72. |

**MAR**

| SOLAR RAD GROUP | ALL | 0+ | 95+ | 189+ | 284+ | 378+ | 473+ |
|---|---|---|---|---|---|---|---|
| AV.THRSH.RAD.LEVL | 0. | 37. | 143. | 235. | 330. | 426. | 517. |
| AV.THRESHOLD TEMP | 0.4 | 1.9 | 3.1 | 3.8 | 4.5 | 4.5 | 5.2 |
| GL RAD OVER THRSH | 461.4 | 411.8 | 305.5 | 236.4 | 179.3 | 131.0 | 94.6 |
| TIME AVG.TEMP | 2.1 | 3.7 | 4.4 | 4.8 | 5.1 | 5.2 | 5.3 |
| TOTAL TIME MSEC | 2.6784 | 1.3507 | 0.9958 | 0.7595 | 0.5904 | 0.5503 | 0.4018 |
| GLOBAL RAD TOTAL | 461.4 | 461.4 | 448.4 | 414.3 | 376.4 | 345.5 | 302.2 |
| DIRECT RAD TOTAL | 271.2 | 271.2 | 269.0 | 264.1 | 254.5 | 245.7 | 224.5 |
| RAD AV TIME MSEC | 0.0864 | 0.0439 | 0.0353 | 0.0328 | 0.0295 | 0.0284 | 0.0256 |
| DEG-C DAYS | 505. | 229. | 161. | 119. | 92. | 77. | 61. |

**JUL**

| SOLAR RAD GROUP | ALL | 0+ | 95+ | 189+ | 284+ | 378+ | 473+ |
|---|---|---|---|---|---|---|---|
| AV.THRSH.RAD.LEVL | 0. | 42. | 147. | 237. | 336. | 424. | 520. |
| AV.THRESHOLD TEMP | 19.4 | 21.2 | 21.7 | 24.8 | 25.0 | 26.2 | 27.3 |
| GL RAD OVER THRSH | 590.2 | 524.7 | 393.0 | 298.1 | 213.5 | 152.9 | 100.9 |
| TIME AVG.TEMP | 22.6 | 24.9 | 25.8 | 26.6 | 27.0 | 27.5 | 27.8 |
| TOTAL TIME MSEC | 2.6784 | 1.5624 | 1.2589 | 1.0454 | 0.8600 | 0.6862 | 0.5418 |
| GLOBAL RAD TOTAL | 590.2 | 590.2 | 577.4 | 546.2 | 502.2 | 443.8 | 382.6 |
| DIRECT RAD TOTAL | 274.2 | 274.2 | 272.2 | 266.9 | 258.2 | 242.3 | 222.2 |
| RAD AV TIME MSEC | 0.0864 | 0.0504 | 0.0421 | 0.0374 | 0.0335 | 0.0295 | 0.0263 |
| DEG-C DAYS | 1. | 0. | 0. | 0. | 0. | 0. | 0. |

**AUG**

| SOLAR RAD GROUP | ALL | 0+ | 95+ | 189+ | 284+ | 378+ | 473+ |
|---|---|---|---|---|---|---|---|
| AV.THRSH.RAD.LEVL | 0. | 45. | 140. | 236. | 332. | 427. | 520. |
| AV.THRESHOLD TEMP | 18.4 | 20.3 | 21.2 | 23.9 | 23.8 | 25.8 | 25.5 |
| GL RAD OVER THRSH | 527.0 | 458.8 | 344.4 | 252.1 | 180.9 | 126.3 | 83.6 |
| TIME AVG.TEMP | 21.3 | 23.6 | 24.4 | 25.3 | 25.7 | 26.2 | 26.3 |
| TOTAL TIME MSEC | 2.6784 | 1.5127 | 1.2060 | 0.9569 | 0.7466 | 0.5760 | 0.4590 |
| GLOBAL RAD TOTAL | 527.0 | 527.0 | 513.2 | 478.3 | 428.6 | 372.0 | 322.1 |
| DIRECT RAD TOTAL | 250.7 | 250.7 | 248.8 | 242.0 | 232.4 | 214.9 | 198.6 |
| RAD AV TIME MSEC | 0.0864 | 0.0490 | 0.0410 | 0.0356 | 0.0313 | 0.0274 | 0.0248 |
| DEG-C DAYS | 8. | 2. | 1. | 1. | 0. | 0. | 0. |

**SEP**

| SOLAR RAD GROUP | ALL | 0+ | 95+ | 189+ | 284+ | 378+ | 473+ |
|---|---|---|---|---|---|---|---|
| AV.THRSH.RAD.LEVL | 0. | 44. | 140. | 235. | 326. | 427. | 526. |
| AV.THRESHOLD TEMP | 14.3 | 16.8 | 18.5 | 19.0 | 19.6 | 22.0 | 21.3 |
| GL RAD OVER THRSH | 505.0 | 445.8 | 344.2 | 263.6 | 200.3 | 143.8 | 97.9 |
| TIME AVG.TEMP | 17.2 | 20.6 | 20.8 | 21.4 | 21.9 | 22.5 | 22.6 |
| TOTAL TIME MSEC | 2.5920 | 1.3514 | 1.0519 | 0.8525 | 0.6930 | 0.5616 | 0.4619 |
| GLOBAL RAD TOTAL | 505.0 | 505.0 | 491.9 | 463.9 | 426.4 | 383.5 | 341.0 |
| DIRECT RAD TOTAL | 316.8 | 316.8 | 314.1 | 307.9 | 299.0 | 279.3 | 257.3 |
| RAD AV TIME MSEC | 0.0864 | 0.0454 | 0.0378 | 0.0333 | 0.0313 | 0.0277 | 0.0248 |
| DEG-C DAYS | 71. | 18. | 10. | 7. | 4. | 3. | 2. |

**APR**

| SOLAR RAD GROUP | ALL | 0+ | 95+ | 189+ | 284+ | 378+ | 473+ |
|---|---|---|---|---|---|---|---|
| AV.THRSH.RAD.LEVL | 0. | 47. | 141. | 233. | 334. | 424. | 520. |
| AV.THRESHOLD TEMP | 6.8 | 8.7 | 9.4 | 11.3 | 11.2 | 13.5 | 13.7 |
| GL RAD OVER THRSH | 512.4 | 445.2 | 340.0 | 260.6 | 193.6 | 145.7 | 104.7 |
| TIME AVG.TEMP | 9.4 | 11.6 | 12.4 | 13.3 | 13.9 | 14.6 | 14.9 |
| TOTAL TIME MSEC | 2.5920 | 1.4324 | 1.1131 | 0.8636 | 0.6653 | 0.5342 | 0.4252 |
| GLOBAL RAD TOTAL | 512.5 | 512.5 | 497.5 | 462.2 | 415.9 | 372.1 | 325.9 |
| DIRECT RAD TOTAL | 263.9 | 263.9 | 261.8 | 256.5 | 248.0 | 233.9 | 219.7 |
| RAD AV TIME MSEC | 0.0364 | 0.0479 | 0.0400 | 0.0353 | 0.0317 | 0.0284 | 0.0266 |
| DEG-C DAYS | 267. | 114. | 79. | 54. | 37. | 27. | 21. |

**MAY**

| SOLAR RAD GROUP | ALL | 0+ | 95+ | 189+ | 284+ | 378+ | 473+ |
|---|---|---|---|---|---|---|---|
| AV.THRSH.RAD.LEVL | 0. | 39. | 149. | 235. | 332. | 424. | 520. |
| AV.THRESHOLD TEMP | 11.7 | 13.4 | 13.8 | 16.6 | 17.7 | 18.6 | 20.6 |
| GL RAD OVER THRSH | 598.6 | 536.9 | 399.4 | 310.1 | 229.5 | 166.7 | 114.8 |
| TIME AVG.TEMP | 15.0 | 17.4 | 18.3 | 19.3 | 20.5 | 20.5 | 21.0 |
| TOTAL TIME MSEC | 2.6784 | 1.5606 | 1.2604 | 1.0336 | 0.8287 | 0.6822 | 0.5396 |
| GLOBAL RAD TOTAL | 598.6 | 598.6 | 586.7 | 553.0 | 504.9 | 456.2 | 395.7 |
| DIRECT RAD TOTAL | 271.2 | 271.2 | 269.9 | 264.1 | 255.8 | 246.0 | 224.9 |
| RAD AV TIME MSEC | 0.0364 | 0.0504 | 0.0421 | 0.0371 | 0.0331 | 0.0302 | 0.0259 |
| DEG-C DAYS | 120. | 43. | 28. | 17. | 11. | 7. | 4. |

**JUN**

| SOLAR RAD GROUP | ALL | 0+ | 95+ | 189+ | 284+ | 378+ | 473+ |
|---|---|---|---|---|---|---|---|
| AV.THRSH.RAD.LEVL | 0. | 43. | 146. | 234. | 333. | 422. | 522. |
| AV.THRESHOLD TEMP | 17.0 | 18.7 | 19.1 | 21.1 | 22.8 | 24.0 | 25.6 |
| GL RAD OVER THRSH | 574.0 | 509.5 | 382.7 | 294.7 | 216.8 | 158.9 | 106.5 |
| TIME AVG.TEMP | 20.2 | 22.5 | 23.4 | 24.3 | 25.2 | 25.7 | 26.2 |
| TOTAL TIME MSEC | 2.5920 | 1.5120 | 1.2254 | 1.0030 | 0.7862 | 0.6534 | 0.5220 |
| GLOBAL RAD TOTAL | 574.0 | 574.0 | 561.8 | 529.2 | 478.6 | 434.3 | 378.9 |
| DIRECT RAD TOTAL | 258.2 | 258.2 | 256.6 | 253.1 | 244.1 | 235.2 | 214.1 |
| RAD AV TIME MSEC | 0.0864 | 0.0504 | 0.0425 | 0.0382 | 0.0338 | 0.0313 | 0.0270 |
| DEG-C DAYS | 21. | 8. | 4. | 2. | 1. | 0. | 0. |

**OCT**

| SOLAR RAD GROUP | ALL | 0+ | 95+ | 189+ | 284+ | 378+ | 473+ |
|---|---|---|---|---|---|---|---|
| AV.THRSH.RAD.LEVL | 0. | 36. | 143. | 236. | 329. | 425. | 521. |
| AV.THRESHOLD TEMP | 8.9 | 11.1 | 13.1 | 13.1 | 14.8 | 15.5 | 14.6 |
| GL RAD OVER THRSH | 438.2 | 390.6 | 288.0 | 217.7 | 162.6 | 116.6 | 78.8 |
| TIME AVG.TEMP | 11.3 | 13.8 | 14.9 | 15.3 | 15.9 | 16.2 | 16.4 |
| TOTAL TIME MSEC | 2.6870 | 1.3219 | 0.9533 | 0.7614 | 0.5969 | 0.4784 | 0.3917 |
| GLOBAL RAD TOTAL | 438.2 | 438.2 | 424.9 | 397.5 | 353.7 | 319.8 | 282.9 |
| DIRECT RAD TOTAL | 291.3 | 291.3 | 287.8 | 283.2 | 267.2 | 249.2 | 226.0 |
| RAD AV TIME MSEC | 0.0364 | 0.0428 | 0.0338 | 0.0320 | 0.0284 | 0.0259 | 0.0234 |
| DEG-C DAYS | 219. | 75. | 46. | 34. | 24. | 19. | 15. |

**NOV**

| SOLAR RAD GROUP | ALL | 0+ | 95+ | 189+ | 284+ | 378+ | 473+ |
|---|---|---|---|---|---|---|---|
| AV.THRSH.RAD.LEVL | 0. | 45. | 137. | 235. | 332. | 428. | 518. |
| AV.THRESHOLD TEMP | 3.9 | 6.3 | 6.8 | 7.7 | 7.3 | 7.3 | 6.8 |
| GL RAD OVER THRSH | 276.3 | 226.0 | 159.9 | 111.8 | 76.7 | 50.2 | 32.0 |
| TIME AVG.TEMP | 5.3 | 7.2 | 7.4 | 7.7 | 7.7 | 7.8 | 7.9 |
| TOTAL TIME MSEC | 2.5920 | 1.1070 | 0.7189 | 0.4900 | 0.3636 | 0.2765 | 0.2027 |
| GLOBAL RAD TOTAL | 276.4 | 276.3 | 258.7 | 227.2 | 197.5 | 168.5 | 137.0 |
| DIRECT RAD TOTAL | 173.8 | 173.8 | 171.4 | 162.2 | 149.8 | 135.4 | 114.9 |
| RAD AV TIME MSEC | 0.0664 | 0.0371 | 0.0317 | 0.0270 | 0.0245 | 0.0223 | 0.0198 |
| DEG-C DAYS | 392. | 143. | 91. | 61. | 45. | 34. | 24. |

**DEC**

| SOLAR RAD GROUP | ALL | 0+ | 95+ | 189+ | 284+ | 378+ | 473+ |
|---|---|---|---|---|---|---|---|
| AV.THRSH.RAD.LEVL | 0. | 44. | 137. | 231. | 328. | 429. | 516. |
| AV.THRESHOLD TEMP | -3.0 | -0.4 | -0.4 | -0.8 | 0.1 | -0.1 | -1.3 |
| GL RAD OVER THRSH | 267.3 | 219.5 | 155.7 | 110.0 | 75.6 | 47.2 | 28.4 |
| TIME AVG.TEMP | -2.0 | -0.6 | -0.7 | -0.8 | -0.8 | -1.0 | -1.3 |
| TOTAL TIME MSEC | 2.6784 | 1.1142 | 0.6790 | 0.4882 | 0.3557 | 0.2804 | 0.2156 |
| GLOBAL RAD TOTAL | 268.0 | 263.0 | 249.0 | 222.8 | 192.2 | 167.5 | 139.7 |
| DIRECT RAD TOTAL | 183.5 | 183.5 | 180.2 | 172.1 | 155.9 | 142.2 | 122.7 |
| RAD AV TIME MSEC | 0.0664 | 0.0360 | 0.0299 | 0.0270 | 0.0234 | 0.0220 | 0.0202 |
| DEG-C DAYS | 631. | 244. | 149. | 103. | 79. | 63. | 49. |

**ALL MONTHS**

| SOLAR RAD GROUP | ALL | 0+ | 95+ | 189+ | 284+ | 378+ | 473+ |
|---|---|---|---|---|---|---|---|
| AV.THRSH.RAD.LEVL | 0. | 43. | 146. | 234. | 333. | 422. | 522. |
| AV.THRESHOLD TEMP | 9.7 | 12.7 | 14.2 | 15.1 | 15.8 | 16.2 | 16.4 |
| TOTAL TIME MSEC | 31.5705 | 16.1301 | 11.9894 | 9.4198 | 7.3868 | 5.9364 | 4.7167 |
| GLOBAL RAD TOTAL | 5405.6 | 5405.3 | 5233.0 | 4868.5 | 4390.5 | 3910.4 | 3391.7 |
| DIRECT RAD TOTAL | 2981.1 | 2981.1 | 2953.2 | 2882.3 | 2749.0 | 2578.5 | 2344.7 |
| RAD AV TIME MSEC | 0.0664 | 0.0457 | 0.0377 | 0.0339 | 0.0304 | 0.0276 | 0.0247 |
| DEG-C DAYS | 3527. | 1417. | 926. | 673. | 504. | 397. | 310. |

STATISTICS FOR HARTFORD, CONN.  1959-68

RADIATION UNITS ARE WATTS/SQ.M OR MJOULES/SQ.M
SLOPE = 40.0 DEG  AZIMUTH = 180.0 DEG
DEGREE-DAY BASE TEMP = 18.3 DEGREES C

### JAN

| | ALL | 0+ | 95+ | 189+ | 284+ | 378+ | 473+ |
|---|---|---|---|---|---|---|---|
| SOLAR RAD GROUP | | | | | | | |
| AV.THRSH.RAD.LEVL | 0. | 44. | 138. | 237. | 326. | 429. | 523. |
| AV.THRESHOLD TEMP | -4.8 | -2.5 | -1.1 | -3.2 | -1.7 | -0.7 | -2.7 |
| GL RAD OVER THRSH | 321.1 | 272.1 | 202.7 | 146.5 | 108.9 | 74.7 | 49.4 |
| TIME AVG.TEMP | -3.8 | -2.4 | -2.3 | -2.7 | -2.6 | -2.8 | -3.3 |
| TOTAL TIME MSEC | 2.6784 | 1.1142 | 0.7412 | 0.5638 | 0.4223 | 0.3344 | 0.2669 |
| GLOBAL RAD TOTAL | 321.2 | 321.1 | 304.7 | 280.3 | 246.7 | 218.1 | 190.0 |
| DIRECT RAD TOTAL | 223.9 | 223.9 | 221.0 | 214.8 | 197.4 | 181.9 | 164.4 |
| RAD AV TIME MSEC | 0.0864 | 0.0360 | 0.0299 | 0.0277 | 0.0241 | 0.0220 | 0.0209 |
| DEG-C DAYS | 686. | 267. | 177. | 137. | 102. | 82. | 67. |

### FEB

| | ALL | 0+ | 95+ | 189+ | 284+ | 378+ | 473+ |
|---|---|---|---|---|---|---|---|
| SOLAR RAD GROUP | | | | | | | |
| AV.THRSH.RAD.LEVL | 0. | 42. | 138. | 236. | 330. | 423. | 522. |
| AV.THRESHOLD TEMP | -4.2 | -2.2 | -1.0 | -0.6 | -1.9 | -1.2 | -0.1 |
| GL RAD OVER THRSH | 363.5 | 317.9 | 243.3 | 183.8 | 138.0 | 102.2 | 70.8 |
| TIME AVG.TEMP | -3.1 | -1.6 | -1.4 | -1.5 | -1.7 | -1.7 | -1.8 |
| TOTAL TIME MSEC | 2.4451 | 1.0872 | 0.7780 | 0.6030 | 0.4682 | 0.3874 | 0.3168 |
| GLOBAL RAD TOTAL | 363.6 | 363.5 | 350.5 | 326.4 | 299.3 | 266.0 | 236.1 |
| DIRECT RAD TOTAL | 240.6 | 240.6 | 238.7 | 233.6 | 226.2 | 210.1 | 194.7 |
| RAD AV TIME MSEC | 0.0864 | 0.0392 | 0.0320 | 0.0292 | 0.0277 | 0.0248 | 0.0230 |
| DEG-C DAYS | 606. | 251. | 178. | 138. | 113. | 89. | 74. |

### MAR

| | ALL | 0+ | 95+ | 189+ | 284+ | 378+ | 473+ |
|---|---|---|---|---|---|---|---|
| SOLAR RAD GROUP | | | | | | | |
| AV.THRSH.RAD.LEVL | 0. | 41. | 143. | 235. | 332. | 430. | 519. |
| AV.THRESHOLD TEMP | 0.5 | 1.7 | 3.2 | 3.7 | 4.8 | 4.9 | 4.8 |
| GL RAD OVER THRSH | 462.7 | 409.7 | 308.9 | 240.2 | 162.7 | 134.2 | 98.6 |
| TIME AVG.TEMP | 2.1 | 3.7 | 4.4 | 4.8 | 5.1 | 5.1 | 5.2 |
| TOTAL TIME MSEC | 2.6784 | 1.3064 | 0.9868 | 0.7466 | 0.5893 | 0.4946 | 0.4003 |
| GLOBAL RAD TOTAL | 462.8 | 462.7 | 449.8 | 415.5 | 378.6 | 347.1 | 306.5 |
| DIRECT RAD TOTAL | 278.6 | 278.6 | 276.6 | 272.3 | 261.9 | 252.6 | 232.8 |
| RAD AV TIME MSEC | 0.0864 | 0.0428 | 0.0353 | 0.0328 | 0.0295 | 0.0284 | 0.0259 |
| DEG-C DAYS | 505. | 221. | 159. | 117. | 91. | 76. | 61. |

### JUL

| | ALL | 0+ | 95+ | 189+ | 284+ | 378+ | 473+ |
|---|---|---|---|---|---|---|---|
| SOLAR RAD GROUP | | | | | | | |
| AV.THRSH.RAD.LEVL | 0. | 54. | 142. | 235. | 333. | 428. | 521. |
| AV.THRESHOLD TEMP | 19.4 | 22.1 | 22.2 | 24.9 | 24.8 | 27.3 | 26.8 |
| GL RAD OVER THRSH | 559.7 | 480.0 | 368.9 | 274.6 | 193.1 | 132.3 | 85.8 |
| TIME AVG.TEMP | 22.6 | 25.2 | 25.8 | 26.7 | 27.1 | 27.7 | 27.8 |
| TOTAL TIME MSEC | 2.6784 | 1.4636 | 1.2618 | 1.0098 | 0.8294 | 0.6433 | 0.5022 |
| GLOBAL RAD TOTAL | 559.9 | 559.8 | 547.9 | 512.1 | 469.7 | 407.6 | 347.3 |
| DIRECT RAD TOTAL | 252.8 | 252.8 | 252.4 | 246.6 | 239.7 | 221.5 | 202.7 |
| RAD AV TIME MSEC | 0.0864 | 0.0479 | 0.0421 | 0.0364 | 0.0328 | 0.0281 | 0.0252 |
| DEG-C DAYS | 1. | 0. | 0. | 0. | 0. | 0. | 0. |

### AUG

| | ALL | 0+ | 95+ | 189+ | 284+ | 378+ | 473+ |
|---|---|---|---|---|---|---|---|
| SOLAR RAD GROUP | | | | | | | |
| AV.THRSH.RAD.LEVL | 0. | 56. | 138. | 235. | 329. | 428. | 523. |
| AV.THRESHOLD TEMP | 18.3 | 21.0 | 21.7 | 24.0 | 24.1 | 25.6 | 25.4 |
| GL RAD OVER THRSH | 508.4 | 427.3 | 328.8 | 238.7 | 170.8 | 116.1 | 74.8 |
| TIME AVG.TEMP | 21.3 | 23.9 | 24.5 | 25.3 | 25.7 | 26.2 | 26.4 |
| TOTAL TIME MSEC | 2.6784 | 1.4508 | 1.1995 | 0.9274 | 0.7232 | 0.5544 | 0.4331 |
| GLOBAL RAD TOTAL | 500.5 | 508.5 | 494.4 | 456.8 | 408.8 | 353.3 | 301.4 |
| DIRECT RAD TOTAL | 240.7 | 240.7 | 239.1 | 232.0 | 223.6 | 206.4 | 188.4 |
| RAD AV TIME MSEC | 0.0864 | 0.0468 | 0.0410 | 0.0346 | 0.0310 | 0.0274 | 0.0241 |
| DEG-C DAYS | 8. | 2. | 1. | 1. | 0. | 0. | 0. |

### SEP

| | ALL | 0+ | 95+ | 189+ | 284+ | 378+ | 473+ |
|---|---|---|---|---|---|---|---|
| SOLAR RAD GROUP | | | | | | | |
| AV.THRSH.RAD.LEVL | 0. | 47. | 141. | 235. | 327. | 429. | 527. |
| AV.THRESHOLD TEMP | 14.4 | 16.5 | 18.6 | 19.2 | 19.6 | 22.0 | 21.8 |
| GL RAD OVER THRSH | 503.0 | 440.3 | 342.8 | 263.9 | 201.0 | 144.4 | 100.0 |
| TIME AVG.TEMP | 17.2 | 19.9 | 20.9 | 21.9 | 22.5 | 22.5 | 22.6 |
| TOTAL TIME MSEC | 2.5920 | 1.3295 | 1.0408 | 0.8352 | 0.6851 | 0.5537 | 0.4554 |
| GLOBAL RAD TOTAL | 503.0 | 503.0 | 489.5 | 460.5 | 425.2 | 382.3 | 340.0 |
| DIRECT RAD TOTAL | 320.1 | 320.1 | 317.4 | 310.7 | 302.7 | 282.3 | 259.5 |
| RAD AV TIME MSEC | 0.0864 | 0.0446 | 0.0378 | 0.0331 | 0.0313 | 0.0277 | 0.0248 |
| DEG-C DAYS | 71. | 17. | 10. | 7. | 4. | 3. | 2. |

**APR**

| SOLAR RAD GROUP | ALL | 0+ | 95+ | 189+ | 284+ | 378+ | 473+ |
|---|---|---|---|---|---|---|---|
| AV.THRSH.RAD.LEVL | 0. | 51. | 141. | 230. | 333. | 425. | 518. |
| AV.THRESHOLD TEMP | 6.8 | 8.9 | 9.6 | 11.2 | 11.4 | 14.3 | 13.5 |
| GL RAD OVER THRSH | 498.6 | 426.4 | 328.0 | 253.0 | 185.9 | 139.2 | 100.5 |
| TIME AVG.TEMP | 9.4 | 11.7 | 12.5 | 13.4 | 14.0 | 14.7 | 14.8 |
| TOTAL TIME MSEC | 2.5920 | 1.4040 | 1.0994 | 0.8388 | 0.6527 | 0.5112 | 0.4126 |
| GLOBAL RAD TOTAL | 498.6 | 498.6 | 482.9 | 446.2 | 403.3 | 356.2 | 314.3 |
| DIRECT RAD TOTAL | 257.7 | 257.7 | 255.8 | 250.1 | 243.2 | 227.7 | 214.4 |
| RAD AV TIME MSEC | 0.0864 | 0.0468 | 0.0400 | 0.0349 | 0.0317 | 0.0281 | 0.0263 |
| DEG-C DAYS | 267. | 111. | 77. | 52. | 37. | 26. | 20. |

**MAY**

| SOLAR RAD GROUP | ALL | 0+ | 95+ | 189+ | 284+ | 378+ | 473+ |
|---|---|---|---|---|---|---|---|
| AV.THRSH.RAD.LEVL | 0. | 51. | 144. | 232. | 335. | 429. | 521. |
| AV.THRESHOLD TEMP | 11.6 | 14.3 | 14.4 | 16.8 | 17.2 | 20.2 | 20.4 |
| GL RAD OVER THRSH | 570.6 | 495.3 | 377.9 | 289.8 | 207.3 | 147.3 | 101.0 |
| TIME AVG.TEMP | 15.0 | 17.8 | 18.4 | 19.4 | 20.1 | 20.8 | 21.0 |
| TOTAL TIME MSEC | 2.6784 | 1.4818 | 1.2618 | 0.9961 | 0.8075 | 0.6347 | 0.5065 |
| GLOBAL RAD TOTAL | 570.8 | 570.7 | 559.5 | 521.3 | 477.4 | 419.6 | 364.6 |
| DIRECT RAD TOTAL | 253.5 | 253.5 | 253.0 | 247.0 | 241.4 | 223.5 | 207.8 |
| RAD AV TIME MSEC | 0.0864 | 0.0479 | 0.0421 | 0.0360 | 0.0324 | 0.0281 | 0.0248 |
| DEG-C DAYS | 120. | 37. | 27. | 16. | 10. | 6. | 4. |

**JUN**

| SOLAR RAD GROUP | ALL | 0+ | 95+ | 189+ | 284+ | 378+ | 473+ |
|---|---|---|---|---|---|---|---|
| AV.THRSH.RAD.LEVL | 0. | 47. | 144. | 233. | 337. | 430. | 522. |
| AV.THRESHOLD TEMP | 17.0 | 18.7 | 19.6 | 21.7 | 22.4 | 25.6 | 25.2 |
| GL RAD OVER THRSH | 542.6 | 474.0 | 353.9 | 269.9 | 190.8 | 135.2 | 90.3 |
| TIME AVG.TEMP | 20.2 | 22.7 | 23.4 | 24.6 | 25.3 | 26.1 | 26.2 |
| TOTAL TIME MSEC | 2.5920 | 1.4684 | 1.2287 | 0.9526 | 0.7607 | 0.5983 | 0.4864 |
| GLOBAL RAD TOTAL | 542.7 | 542.6 | 531.4 | 491.5 | 446.9 | 392.2 | 344.1 |
| DIRECT RAD TOTAL | 236.1 | 236.1 | 235.8 | 230.5 | 225.3 | 208.9 | 193.4 |
| RAD AV TIME MSEC | 0.0864 | 0.0493 | 0.0428 | 0.0364 | 0.0335 | 0.0288 | 0.0259 |
| DEG-C DAYS | 21. | 7. | 4. | 2. | 1. | 0. | 0. |

**OCT**

| SOLAR RAD GROUP | ALL | 0+ | 95+ | 189+ | 284+ | 378+ | 473+ |
|---|---|---|---|---|---|---|---|
| AV.THRSH.RAD.LEVL | 0. | 40. | 142. | 237. | 325. | 425. | 525. |
| AV.THRESHOLD TEMP | 9.1 | 10.6 | 13.4 | 13.5 | 14.1 | 15.4 | 14.9 |
| GL RAD OVER THRSH | 452.1 | 400.8 | 304.0 | 232.5 | 178.9 | 130.3 | 90.0 |
| TIME AVG.TEMP | 11.3 | 13.8 | 14.9 | 14.9 | 15.3 | 16.2 | 16.3 |
| TOTAL TIME MSEC | 2.6870 | 1.2751 | 0.9475 | 0.7556 | 0.6066 | 0.4900 | 0.4032 |
| GLOBAL RAD TOTAL | 452.2 | 452.1 | 438.9 | 411.6 | 376.3 | 338.3 | 301.5 |
| DIRECT RAD TOTAL | 309.7 | 309.7 | 306.3 | 301.7 | 287.7 | 269.0 | 246.8 |
| RAD AV TIME MSEC | 0.0864 | 0.0418 | 0.0342 | 0.0324 | 0.0295 | 0.0266 | 0.0241 |
| DEG-C DAYS | 219. | 72. | 46. | 34. | 26. | 19. | 16. |

**NOV**

| SOLAR RAD GROUP | ALL | 0+ | 95+ | 189+ | 284+ | 378+ | 473+ |
|---|---|---|---|---|---|---|---|
| AV.THRSH.RAD.LEVL | 0. | 45. | 138. | 234. | 329. | 431. | 521. |
| AV.THRESHOLD TEMP | 3.9 | 6.7 | 6.7 | 7.6 | 7.6 | 6.7 | 7.0 |
| GL RAD OVER THRSH | 291.0 | 241.1 | 174.9 | 127.2 | 91.7 | 62.1 | 42.1 |
| TIME AVG.TEMP | 5.3 | 7.1 | 7.3 | 7.6 | 7.6 | 7.6 | 7.9 |
| TOTAL TIME MSEC | 2.5920 | 1.1059 | 0.7139 | 0.4964 | 0.3755 | 0.2898 | 0.2232 |
| GLOBAL RAD TOTAL | 291.1 | 291.1 | 273.4 | 243.4 | 215.1 | 186.9 | 158.2 |
| DIRECT RAD TOTAL | 191.9 | 191.9 | 189.6 | 180.4 | 167.7 | 153.5 | 134.6 |
| RAD AV TIME MSEC | 0.0864 | 0.0371 | 0.0320 | 0.0281 | 0.0252 | 0.0234 | 0.0212 |
| DEG-C DAYS | 392. | 143. | 91. | 62. | 47. | 36. | 27. |

**DEC**

| SOLAR RAD GROUP | ALL | 0+ | 95+ | 189+ | 284+ | 378+ | 473+ |
|---|---|---|---|---|---|---|---|
| AV.THRSH.RAD.LEVL | 0. | 43. | 138. | 233. | 329. | 427. | 526. |
| AV.THRESHOLD TEMP | -3.0 | -0.4 | -0.4 | -0.8 | 0.3 | 0.2 | -1.5 |
| GL RAD OVER THRSH | 285.0 | 237.1 | 173.2 | 126.9 | 91.7 | 63.3 | 39.7 |
| TIME AVG.TEMP | -2.0 | -0.6 | -0.7 | -0.8 | -0.8 | -1.1 | -1.3 |
| TOTAL TIME MSEC | 2.6784 | 1.1138 | 0.6718 | 0.4864 | 0.3661 | 0.2916 | 0.2390 |
| GLOBAL RAD TOTAL | 285.0 | 285.0 | 266.0 | 240.4 | 212.3 | 187.8 | 165.3 |
| DIRECT RAD TOTAL | 203.3 | 203.3 | 199.8 | 191.6 | 176.5 | 161.7 | 146.2 |
| RAD AV TIME MSEC | 0.0864 | 0.0360 | 0.0302 | 0.0274 | 0.0241 | 0.0223 | 0.0212 |
| DEG-C DAYS | 631. | 244. | 148. | 108. | 81. | 66. | 54. |

**ALL MONTHS**

| SOLAR RAD GROUP | ALL | 0+ | 95+ | 189+ | 284+ | 378+ | 473+ |
|---|---|---|---|---|---|---|---|
| TIME AVG.TEMP | 9.7 | 12.8 | 14.2 | 15.0 | 15.6 | 15.9 | 15.9 |
| TOTAL TIME MSEC | 31.5705 | 15.6208 | 11.9311 | 9.2117 | 7.3066 | 5.7834 | 4.6476 |
| GLOBAL RAD TOTAL | 5359.3 | 5358.8 | 5189.0 | 4806.0 | 4359.6 | 3655.2 | 3369.3 |
| DIRECT RAD TOTAL | 3008.8 | 3008.8 | 2965.5 | 2911.3 | 2793.5 | 2599.0 | 2385.9 |
| RAD AV TIME MSEC | 0.0864 | 0.0441 | 0.0377 | 0.0332 | 0.0302 | 0.0269 | 0.0244 |
| DEG-C DAYS | 3527. | 1372. | 918. | 673. | 511. | 403. | 326. |

STATISTICS FOR HARTFORD, CONN.          1959-68

RADIATION UNITS ARE WATTS/SQ.M OR MJOULES/SQ.M
SLOPE = 50.0 DEG   AZIMUTH = 180.0 DEG
DEGREE-DAY BASE TEMP = 18.3 DEGREES C

### JAN

| SOLAR RAD GROUP | ALL | 0+ | 95+ | 189+ | 284+ | 378+ | 473+ |
|---|---|---|---|---|---|---|---|
| AV.THRSH.RAD.LEVL | 0. | 36. | 138. | 239. | 325. | 427. | 524. |
| AV.THRESHOLD TEMP | -4.9 | -2.5 | -1.1 | -2.9 | -2.0 | -0.8 | -1.9 |
| GL RAD OVER THRSH | 332.1 | 288.7 | 213.8 | 157.7 | 120.7 | 86.0 | 59.0 |
| TIME AVG.TEMP | -3.8 | -2.4 | -2.3 | -2.7 | -2.7 | -2.8 | -3.3 |
| TOTAL TIME MSEC | 2.6784 | 1.2186 | 0.7308 | 0.5591 | 0.4280 | 0.3395 | 0.2790 |
| GLOBAL RAD TOTAL | 332.1 | 332.1 | 314.8 | 291.0 | 259.8 | 231.0 | 205.2 |
| DIRECT RAD TOTAL | 238.3 | 238.3 | 235.2 | 228.9 | 212.4 | 195.7 | 179.1 |
| RAD AV TIME MSEC | 0.0864 | 0.0396 | 0.0302 | 0.0281 | 0.0243 | 0.0223 | 0.0209 |
| DEG-C DAYS | 686. | 292. | 175. | 136. | 104. | 83. | 70. |

### FEB

| SOLAR RAD GROUP | ALL | 0+ | 95+ | 189+ | 284+ | 378+ | 473+ |
|---|---|---|---|---|---|---|---|
| AV.THRSH.RAD.LEVL | 0. | 40. | 139. | 237. | 331. | 423. | 522. |
| AV.THRESHOLD TEMP | -4.5 | -1.7 | -1.2 | -0.9 | -2.0 | -1.2 | -0.4 |
| GL RAD OVER THRSH | 377.4 | 329.9 | 250.2 | 191.1 | 145.0 | 109.3 | 77.9 |
| TIME AVG.TEMP | -3.1 | -1.6 | -1.5 | -1.6 | -1.8 | -1.7 | -1.8 |
| TOTAL TIME MSEC | 2.4451 | 1.1797 | 0.8046 | 0.6077 | 0.4907 | 0.3864 | 0.3172 |
| GLOBAL RAD TOTAL | 377.5 | 377.4 | 362.3 | 334.9 | 307.2 | 273.4 | 243.3 |
| DIRECT RAD TOTAL | 257.1 | 257.1 | 254.7 | 247.0 | 237.8 | 220.5 | 203.6 |
| RAD AV TIME MSEC | 0.0864 | 0.0425 | 0.0342 | 0.0302 | 0.0284 | 0.0256 | 0.0234 |
| DEG-C DAYS | 606. | 272. | 184. | 140. | 114. | 90. | 74. |

### MAR

| SOLAR RAD GROUP | ALL | 0+ | 95+ | 189+ | 284+ | 378+ | 473+ |
|---|---|---|---|---|---|---|---|
| AV.THRSH.RAD.LEVL | 0. | 37. | 142. | 233. | 331. | 429. | 519. |
| AV.THRESHOLD TEMP | 0. | 1.7 | 3.1 | 3.6 | 4.8 | 4.9 | 4.8 |
| GL RAD OVER THRSH | 458.5 | 407.8 | 304.3 | 236.4 | 179.8 | 132.0 | 97.1 |
| TIME AVG.TEMP | 2.1 | 3.6 | 4.3 | 4.7 | 5.1 | 5.1 | 5.2 |
| TOTAL TIME MSEC | 2.6784 | 1.3907 | 0.9839 | 0.7434 | 0.5782 | 0.4853 | 0.3920 |
| GLOBAL RAD TOTAL | 458.6 | 458.6 | 443.7 | 409.6 | 371.1 | 340.4 | 300.3 |
| DIRECT RAD TOTAL | 280.7 | 280.7 | 278.5 | 273.6 | 260.8 | 250.9 | 231.5 |
| RAD AV TIME MSEC | 0.0864 | 0.0450 | 0.0356 | 0.0331 | 0.0295 | 0.0284 | 0.0259 |
| DEG-C DAYS | 505. | 238. | 160. | 117. | 89. | 74. | 60. |

### JUL

| SOLAR RAD GROUP | ALL | 0+ | 95+ | 189+ | 284+ | 378+ | 473+ |
|---|---|---|---|---|---|---|---|
| AV.THRSH.RAD.LEVL | 0. | 51. | 139. | 242. | 329. | 428. | 520. |
| AV.THRESHOLD TEMP | 19.8 | 22.8 | 21.8 | 24.3 | 25.5 | 27.3 | 26.9 |
| GL RAD OVER THRSH | 496.8 | 427.8 | 332.5 | 235.0 | 166.8 | 108.0 | 66.6 |
| TIME AVG.TEMP | 22.6 | 25.4 | 26.1 | 26.7 | 27.2 | 27.8 | 27.9 |
| TOTAL TIME MSEC | 2.6784 | 1.3392 | 1.0915 | 0.9432 | 0.7844 | 0.5962 | 0.4496 |
| GLOBAL RAD TOTAL | 496.8 | 496.8 | 484.0 | 463.0 | 425.0 | 363.0 | 300.3 |
| DIRECT RAD TOTAL | 222.7 | 222.7 | 220.3 | 219.2 | 213.2 | 194.3 | 175.6 |
| RAD AV TIME MSEC | 0.0864 | 0.0432 | 0.0364 | 0.0342 | 0.0317 | 0.0270 | 0.0238 |
| DEG-C DAYS | 1. | 0. | 0. | 0. | 0. | 0. | 0. |

### AUG

| SOLAR RAD GROUP | ALL | 0+ | 95+ | 189+ | 284+ | 378+ | 473+ |
|---|---|---|---|---|---|---|---|
| AV.THRSH.RAD.LEVL | 0. | 44. | 146. | 239. | 326. | 426. | 518. |
| AV.THRESHOLD TEMP | 18.7 | 21.1 | 21.8 | 23.8 | 24.6 | 25.5 | 25.8 |
| GL RAD OVER THRSH | 460.8 | 401.2 | 295.3 | 213.9 | 153.9 | 100.9 | 64.5 |
| TIME AVG.TEMP | 21.3 | 24.0 | 24.8 | 25.4 | 25.8 | 26.2 | 26.4 |
| TOTAL TIME MSEC | 2.6784 | 1.3392 | 1.0436 | 0.8798 | 0.6387 | 0.5252 | 0.3985 |
| GLOBAL RAD TOTAL | 460.8 | 460.8 | 447.7 | 423.8 | 378.2 | 324.9 | 270.9 |
| DIRECT RAD TOTAL | 219.2 | 219.2 | 216.6 | 215.2 | 206.5 | 190.8 | 170.6 |
| RAD AV TIME MSEC | 0.0864 | 0.0432 | 0.0349 | 0.0331 | 0.0302 | 0.0266 | 0.0230 |
| DEG-C DAYS | 8. | 2. | 1. | 1. | 0. | 0. | 0. |

### SEP

| SOLAR RAD GROUP | ALL | 0+ | 95+ | 189+ | 284+ | 378+ | 473+ |
|---|---|---|---|---|---|---|---|
| AV.THRSH.RAD.LEVL | 0. | 42. | 142. | 236. | 326. | 429. | 527. |
| AV.THRESHOLD TEMP | 14.5 | 16.5 | 18.7 | 19.2 | 19.7 | 22.3 | 21.5 |
| GL RAD OVER THRSH | 484.9 | 429.9 | 330.6 | 254.2 | 193.6 | 138.1 | 94.9 |
| TIME AVG.TEMP | 17.2 | 19.9 | 21.0 | 21.5 | 22.0 | 22.6 | 22.6 |
| TOTAL TIME MSEC | 2.5920 | 1.3140 | 0.9904 | 0.8158 | 0.6696 | 0.5407 | 0.4424 |
| GLOBAL RAD TOTAL | 484.9 | 484.9 | 471.3 | 446.5 | 412.1 | 370.0 | 327.9 |
| DIRECT RAD TOTAL | 312.3 | 312.3 | 308.7 | 304.8 | 295.8 | 276.0 | 252.8 |
| RAD AV TIME MSEC | 0.0864 | 0.0439 | 0.0356 | 0.0331 | 0.0313 | 0.0277 | 0.0245 |
| DEG-C DAYS | 71. | 18. | 9. | 7. | 4. | 3. | 2. |

### APR

| SOLAR RAD GROUP | ALL | 0+ | 95+ | 189+ | 284+ | 378+ | 473+ |
|---|---|---|---|---|---|---|---|
| AV.THRSH.RAD.LEVL | 0. | 42. | 143. | 236. | 332. | 425. | 523. |
| AV.THRESHOLD TEMP | 7.1 | 8.8 | 9.7 | 10.9 | 11.9 | 14.6 | 13.0 |
| GL RAD OVER THRSH | 459.5 | 404.7 | 304.4 | 231.2 | 170.9 | 125.6 | 87.5 |
| TIME AVG.TEMP | 9.4 | 11.8 | 12.7 | 13.4 | 14.1 | 14.8 | 14.8 |
| TOTAL TIME MSEC | 2.5920 | 1.2931 | 0.9976 | 0.7906 | 0.6275 | 0.4842 | 0.3881 |
| GLOBAL RAD TOTAL | 459.5 | 459.5 | 446.9 | 417.4 | 379.0 | 331.1 | 290.6 |
| DIRECT RAD TOTAL | 239.8 | 239.8 | 237.4 | 235.5 | 230.1 | 214.3 | 199.6 |
| RAD AV TIME MSEC | 0.0864 | 0.0432 | 0.0356 | 0.0331 | 0.0310 | 0.0277 | 0.0256 |
| DEG-C DAYS | 267. | 101. | 69. | 48. | 34. | 24. | 19. |

### MAY

| SOLAR RAD GROUP | ALL | 0+ | 95+ | 189+ | 284+ | 378+ | 473+ |
|---|---|---|---|---|---|---|---|
| AV.THRSH.RAD.LEVL | 0. | 46. | 140. | 239. | 328. | 425. | 524. |
| AV.THRESHOLD TEMP | 12.1 | 15.1 | 13.3 | 16.2 | 17.9 | 20.3 | 20.2 |
| GL RAD OVER THRSH | 512.3 | 450.3 | 345.8 | 253.1 | 184.1 | 126.0 | 79.6 |
| TIME AVG.TEMP | 15.0 | 17.9 | 18.5 | 19.5 | 20.2 | 20.9 | 21.1 |
| TOTAL TIME MSEC | 2.6784 | 1.3392 | 1.1189 | 0.9374 | 0.7700 | 0.6005 | 0.4676 |
| GLOBAL RAD TOTAL | 512.3 | 512.3 | 502.1 | 476.7 | 436.8 | 381.2 | 324.7 |
| DIRECT RAD TOTAL | 226.5 | 226.5 | 224.9 | 223.0 | 218.4 | 201.5 | 185.9 |
| RAD AV TIME MSEC | 0.0864 | 0.0432 | 0.0371 | 0.0342 | 0.0317 | 0.0274 | 0.0241 |
| DEG-C DAYS | 120. | 33. | 24. | 14. | 9. | 6. | 4. |

### JUN

| SOLAR RAD GROUP | ALL | 0+ | 95+ | 189+ | 284+ | 378+ | 473+ |
|---|---|---|---|---|---|---|---|
| AV.THRSH.RAD.LEVL | 0. | 52. | 138. | 235. | 327. | 427. | 521. |
| AV.THRESHOLD TEMP | 17.4 | 20.3 | 18.5 | 21.0 | 23.1 | 25.7 | 24.6 |
| GL RAD OVER THRSH | 480.3 | 412.5 | 320.6 | 233.7 | 167.4 | 110.9 | 69.3 |
| TIME AVG.TEMP | 20.2 | 23.0 | 23.6 | 24.6 | 25.4 | 26.1 | 26.2 |
| TOTAL TIME MSEC | 2.5920 | 1.2960 | 1.0757 | 0.8896 | 0.7243 | 0.5623 | 0.4446 |
| GLOBAL RAD TOTAL | 480.3 | 480.3 | 468.7 | 443.1 | 404.2 | 351.2 | 300.9 |
| DIRECT RAD TOTAL | 206.2 | 206.2 | 204.3 | 203.2 | 199.3 | 183.3 | 168.2 |
| RAD AV TIME MSEC | 0.0864 | 0.0432 | 0.0371 | 0.0342 | 0.0324 | 0.0277 | 0.0245 |
| DEG-C DAYS | 21. | 6. | 4. | 1. | 1. | 0. | 0. |

### OCT

| SOLAR RAD GROUP | ALL | 0+ | 95+ | 189+ | 284+ | 378+ | 473+ |
|---|---|---|---|---|---|---|---|
| AV.THRSH.RAD.LEVL | 0. | 43. | 141. | 237. | 326. | 428. | 525. |
| AV.THRESHOLD TEMP | 8.8 | 11.2 | 13.4 | 13.4 | 14.2 | 15.6 | 15.6 |
| GL RAD OVER THRSH | 463.5 | 405.2 | 310.0 | 238.1 | 184.3 | 134.5 | 95.6 |
| TIME AVG.TEMP | 11.3 | 13.8 | 14.9 | 15.3 | 15.8 | 16.2 | 16.3 |
| TOTAL TIME MSEC | 2.6870 | 1.3579 | 0.9698 | 0.7474 | 0.6048 | 0.4878 | 0.4036 |
| GLOBAL RAD TOTAL | 463.5 | 463.5 | 446.8 | 415.5 | 381.6 | 343.4 | 307.4 |
| DIRECT RAD TOTAL | 324.5 | 324.5 | 319.3 | 311.0 | 297.3 | 276.9 | 254.3 |
| RAD AV TIME MSEC | 0.0864 | 0.0439 | 0.0356 | 0.0324 | 0.0299 | 0.0270 | 0.0245 |
| DEG-C DAYS | 219. | 77. | 47. | 34. | 26. | 19. | 16. |

### NOV

| SOLAR RAD GROUP | ALL | 0+ | 95+ | 189+ | 284+ | 378+ | 473+ |
|---|---|---|---|---|---|---|---|
| AV.THRSH.RAD.LEVL | 0. | 39. | 137. | 235. | 330. | 431. | 519. |
| AV.THRESHOLD TEMP | 3.8 | 6.7 | 7.0 | 7.4 | 7.8 | 6.9 | 6.7 |
| GL RAD OVER THRSH | 299.0 | 253.0 | 184.0 | 135.5 | 99.9 | 70.5 | 49.7 |
| TIME AVG.TEMP | 5.3 | 7.1 | 7.4 | 7.6 | 7.6 | 7.6 | 7.7 |
| TOTAL TIME MSEC | 2.5920 | 1.1740 | 0.7024 | 0.4972 | 0.3737 | 0.2927 | 0.2362 |
| GLOBAL RAD TOTAL | 299.1 | 299.0 | 280.6 | 252.3 | 223.3 | 196.6 | 172.2 |
| DIRECT RAD TOTAL | 203.8 | 203.8 | 201.3 | 192.4 | 178.3 | 164.6 | 148.1 |
| RAD AV TIME MSEC | 0.0864 | 0.0396 | 0.0320 | 0.0284 | 0.0252 | 0.0238 | 0.0220 |
| DEG-C DAYS | 392. | 152. | 89. | 62. | 47. | 37. | 29. |

### DEC

| SOLAR RAD GROUP | ALL | 0+ | 95+ | 189+ | 284+ | 378+ | 473+ |
|---|---|---|---|---|---|---|---|
| AV.THRSH.RAD.LEVL | 0. | 42. | 139. | 234. | 330. | 424. | 525. |
| AV.THRESHOLD TEMP | -3.0 | -0.5 | -0.2 | -0.7 | -0.4 | 0.6 | -0.7 |
| GL RAD OVER THRSH | 295.9 | 248.8 | 184.7 | 138.8 | 103.6 | 75.0 | 49.9 |
| TIME AVG.TEMP | -2.0 | -0.6 | -0.7 | -0.8 | -0.9 | -1.0 | -1.3 |
| TOTAL TIME MSEC | 2.6784 | 1.1261 | 0.6574 | 0.4838 | 0.3694 | 0.3010 | 0.2488 |
| GLOBAL RAD TOTAL | 296.0 | 295.9 | 276.3 | 252.1 | 225.3 | 202.8 | 180.6 |
| DIRECT RAD TOTAL | 217.5 | 217.5 | 213.9 | 205.6 | 191.0 | 176.4 | 161.0 |
| RAD AV TIME MSEC | 0.0864 | 0.0364 | 0.0302 | 0.0274 | 0.0245 | 0.0227 | 0.0220 |
| DEG-C DAYS | 631. | 247. | 145. | 107. | 82. | 67. | 57. |

### ALL MONTHS

| SOLAR RAD GROUP | ALL | 0+ | 95+ | 189+ | 284+ | 378+ | 473+ |
|---|---|---|---|---|---|---|---|
| TIME AVG.TEMP | 9.7 | 12.2 | 13.8 | 14.8 | 15.4 | 15.6 | 15.4 |
| TOTAL TIME MSEC | 31.5705 | 15.3677 | 11.1665 | 8.8949 | 7.1093 | 5.6038 | 4.4676 |
| GLOBAL RAD TOTAL | 5121.3 | 5121.0 | 4945.2 | 4626.3 | 4203.5 | 3709.3 | 3224.2 |
| DIRECT RAD TOTAL | 2948.5 | 2948.5 | 2915.1 | 2859.3 | 2741.1 | 2545.7 | 2330.3 |
| RAD AV TIME MSEC | 0.0864 | 0.0426 | 0.0350 | 0.0323 | 0.0298 | 0.0265 | 0.0239 |
| DEG-C DAYS | 3527. | 1437. | 908. | 668. | 510. | 403. | 331. |

STATISTICS FOR HARTFORD, CONN.    1959-68

RADIATION UNITS ARE WATTS/SQ.M OR MJOULES/SQ.M
SLOPE = 60.0 DEG   AZIMUTH = 180.0 DEG
DEGREE-DAY BASE TEMP = 18.3 DEGREES C

### JAN

| SOLAR RAD GROUP | ALL | 0+ | 95+ | 189+ | 284+ | 378+ | 473+ |
|---|---|---|---|---|---|---|---|
| AV.THRSH.RAD.LEVL | 0. | 36. | 138. | 238. | 325. | 425. | 519. |
| AV.THRESHOLD TEMP | -4.9 | -2.4 | -1.2 | -2.9 | -2.0 | -0.9 | -1.4 |
| GL RAD OVER THPSH | 334.9 | 291.4 | 218.6 | 162.6 | 125.9 | 91.8 | 65.4 |
| TIME AVG.TEMP | -3.8 | -2.4 | -2.4 | -2.7 | -2.8 | -2.8 | -3.2 |
| TOTAL TIME MSEC | 2.6784 | 1.2168 | 0.7153 | 0.5562 | 0.4230 | 0.3395 | 0.2826 |
| GLOBAL RAD TOTAL | 334.9 | 334.9 | 317.0 | 295.1 | 263.3 | 236.2 | 212.0 |
| DIRECT RAD TOTAL | 245.7 | 245.7 | 242.4 | 236.1 | 219.0 | 202.4 | 186.3 |
| RAD AV TIME MSEC | 0.0864 | 0.0396 | 0.0302 | 0.0281 | 0.0248 | 0.0227 | 0.0212 |
| DEG-C DAYS | 686. | 292. | 172. | 136. | 103. | 83. | 71. |

### FEB

| SOLAR RAD GROUP | ALL | 0+ | 95+ | 189+ | 284+ | 378+ | 473+ |
|---|---|---|---|---|---|---|---|
| AV.THRSH.RAD.LEVL | 0. | 40. | 139. | 238. | 333. | 426. | 522. |
| AV.THRESHOLD TEMP | -4.4 | -1.6 | -1.2 | -1.2 | -1.8 | -1.0 | -1.2 |
| GL RAD OVER THPSH | 374.2 | 326.9 | 248.2 | 189.9 | 144.2 | 108.6 | 79.0 |
| TIME AVG.TEMP | -3.1 | -1.6 | -1.6 | -1.7 | -1.8 | -1.8 | -1.9 |
| TOTAL TIME MSEC | 2.4451 | 1.1722 | 0.7949 | 0.5936 | 0.4806 | 0.3805 | 0.3100 |
| GLOBAL RAD TOTAL | 374.3 | 374.3 | 359.0 | 331.0 | 304.1 | 270.8 | 240.7 |
| DIRECT RAD TOTAL | 259.7 | 259.7 | 257.4 | 249.1 | 239.7 | 222.3 | 204.5 |
| RAD AV TIME MSEC | 0.0864 | 0.0421 | 0.0342 | 0.0302 | 0.0284 | 0.0256 | 0.0234 |
| DEG-C DAYS | 606. | 270. | 183. | 137. | 112. | 88. | 73. |

### MAR

| SOLAR RAD GROUP | ALL | 0+ | 95+ | 189+ | 284+ | 378+ | 473+ |
|---|---|---|---|---|---|---|---|
| AV.THRSH.RAD.LEVL | 0. | 36. | 141. | 233. | 334. | 430. | 513. |
| AV.THRESHOLD TEMP | 0.4 | 1.8 | 3.2 | 3.4 | 5.4 | 5.1 | 4.7 |
| GL RAD OVER THRSH | 441.2 | 390.5 | 289.4 | 222.3 | 166.5 | 121.7 | 90.9 |
| TIME AVG.TEMP | 2.1 | 3.6 | 4.3 | 4.7 | 5.1 | 5.1 | 5.1 |
| TOTAL TIME MSEC | 2.6784 | 1.3982 | 0.9655 | 0.7236 | 0.5580 | 0.4666 | 0.3726 |
| GLOBAL RAD TOTAL | 441.3 | 441.3 | 425.5 | 391.4 | 352.8 | 322.3 | 281.9 |
| DIRECT RAD TOTAL | 271.9 | 271.9 | 269.7 | 264.3 | 251.1 | 240.2 | 219.6 |
| RAD AV TIME MSEC | 0.0864 | 0.0450 | 0.0356 | 0.0331 | 0.0295 | 0.0281 | 0.0256 |
| DEG-C DAYS | 505. | 239. | 156. | 114. | 86. | 72. | 58. |

### JUL

| SOLAR RAD GROUP | ALL | 0+ | 95+ | 189+ | 284+ | 378+ | 473+ |
|---|---|---|---|---|---|---|---|
| AV.THRSH.RAD.LEVL | 0. | 56. | 142. | 239. | 329. | 426. | 523. |
| AV.THRESHOLD TEMP | 19.8 | 22.3 | 22.8 | 24.1 | 26.9 | 26.6 | 27.9 |
| GL RAD OVER THPSH | 453.2 | 378.6 | 286.5 | 196.5 | 132.8 | 79.9 | 43.5 |
| TIME AVG.TEMP | 22.6 | 25.4 | 26.2 | 26.8 | 27.6 | 27.8 | 28.3 |
| TOTAL TIME MSEC | 2.6784 | 1.3392 | 1.0721 | 0.9212 | 0.7142 | 0.5429 | 0.3766 |
| GLOBAL RAD TOTAL | 453.2 | 453.2 | 438.3 | 416.9 | 367.4 | 311.1 | 240.3 |
| DIRECT RAD TOTAL | 191.2 | 191.2 | 189.3 | 188.4 | 176.9 | 163.8 | 137.6 |
| RAD AV TIME MSEC | 0.0864 | 0.0432 | 0.0356 | 0.0338 | 0.0292 | 0.0256 | 0.0209 |
| DEG-C DAYS | 1. | 0. | 0. | 0. | 0. | 0. | 0. |

### AUG

| SOLAR RAD GROUP | ALL | 0+ | 95+ | 189+ | 284+ | 378+ | 473+ |
|---|---|---|---|---|---|---|---|
| AV.THRSH.RAD.LEVL | 0. | 46. | 146. | 239. | 331. | 429. | 521. |
| AV.THRESHOLD TEMP | 18.7 | 21.0 | 22.3 | 23.6 | 25.6 | 25.4 | 26.3 |
| GL RAD OVER THRSH | 427.9 | 365.9 | 262.6 | 184.1 | 125.8 | 79.1 | 47.4 |
| TIME AVG.TEMP | 21.3 | 24.0 | 24.9 | 25.4 | 26.1 | 26.3 | 26.6 |
| TOTAL TIME MSEC | 2.6784 | 1.3392 | 1.0314 | 0.8510 | 0.6282 | 0.4792 | 0.3434 |
| GLOBAL RAD TOTAL | 427.9 | 427.9 | 413.6 | 387.2 | 334.0 | 284.6 | 226.4 |
| DIRECT RAD TOTAL | 197.7 | 197.7 | 195.4 | 193.8 | 180.7 | 166.5 | 142.8 |
| RAD AV TIME MSEC | 0.0664 | 0.0432 | 0.0349 | 0.0328 | 0.0281 | 0.0248 | 0.0209 |
| DEG-C DAYS | 8. | 2. | 1. | 1. | 0. | 0. | 0. |

### SEP

| SOLAR RAD GROUP | ALL | 0+ | 95+ | 189+ | 284+ | 378+ | 473+ |
|---|---|---|---|---|---|---|---|
| AV.THRSH.RAD.LEVL | 0. | 41. | 144. | 239. | 325. | 426. | 522. |
| AV.THRESHOLD TEMP | 14.4 | 16.6 | 19.0 | 19.4 | 19.9 | 22.1 | 21.6 |
| GL RAD OVER THRSH | 462.6 | 408.5 | 307.9 | 232.9 | 177.7 | 124.8 | 84.6 |
| TIME AVG.TEMP | 17.2 | 19.9 | 21.1 | 22.1 | 22.6 | 22.6 | 22.7 |
| TOTAL TIME MSEC | 2.5920 | 1.3219 | 0.9774 | 0.7884 | 0.6426 | 0.5224 | 0.4205 |
| GLOBAL RAD TOTAL | 462.6 | 462.6 | 448.5 | 421.3 | 386.5 | 347.4 | 304.0 |
| DIRECT RAD TOTAL | 297.3 | 297.3 | 293.7 | 289.4 | 278.0 | 259.9 | 236.4 |
| RAD AV TIME MSEC | 0.0864 | 0.0443 | 0.0353 | 0.0328 | 0.0306 | 0.0270 | 0.0245 |
| DEG-C DAYS | 71. | 18. | 9. | 6. | 4. | 3. | 2. |

## OCT

| SOLAR RAD GROUP | ALL | 0+ | 95+ | 189+ | 284+ | 378+ | 473+ |
|---|---|---|---|---|---|---|---|
| AV.THRSH.RAD.LEVL | 0. | 44. | 141. | 237. | 324. | 428. | 525. |
| AV.THRESHOLD TEMP | 8.8 | 11.2 | 13.7 | 13.6 | 14.0 | 15.6 | 15.2 |
| GL RAD OVER THRSH | 457.1 | 398.2 | 304.9 | 234.4 | 182.8 | 132.1 | 93.9 |
| TIME AVG.TEMP | 11.3 | 13.8 | 14.9 | 15.3 | 15.7 | 16.1 | 16.2 |
| TOTAL TIME MSEC | 2.6870 | 1.3540 | 0.9565 | 0.7326 | 0.5972 | 0.4849 | 0.3953 |
| GLOBAL RAD TOTAL | 457.2 | 457.1 | 439.8 | 408.2 | 376.1 | 339.8 | 301.4 |
| DIRECT RAD TOTAL | 324.3 | 324.3 | 318.9 | 310.2 | 296.8 | 276.1 | 251.6 |
| RAD AV TIME MSEC | 0.0864 | 0.0436 | 0.0356 | 0.0324 | 0.0299 | 0.0270 | 0.0245 |
| DEG-C DAYS | 219. | 77. | 46. | 33. | 26. | 19. | 16. |

## NOV

| SOLAR RAD GROUP | ALL | 0+ | 95+ | 189+ | 284+ | 378+ | 473+ |
|---|---|---|---|---|---|---|---|
| AV.THRSH.RAD.LEVL | 0. | 39. | 138. | 236. | 330. | 430. | 517. |
| AV.THRESHOLD TEMP | 3.8 | 6.8 | 6.8 | 7.6 | 7.8 | 6.7 | 6.6 |
| GL RAD OVER THRSH | 299.7 | 253.9 | 186.3 | 138.0 | 103.5 | 74.3 | 53.5 |
| TIME AVG.TEMP | 5.3 | 7.1 | 7.3 | 7.6 | 7.6 | 7.5 | 7.7 |
| TOTAL TIME MSEC | 2.5920 | 1.1696 | 0.6869 | 0.4903 | 0.3679 | 0.2920 | 0.2401 |
| GLOBAL RAD TOTAL | 299.8 | 299.7 | 280.8 | 253.8 | 224.9 | 199.8 | 177.5 |
| DIRECT RAD TOTAL | 209.3 | 209.3 | 206.7 | 197.6 | 182.9 | 169.4 | 153.7 |
| RAD AV TIME MSEC | 0.0864 | 0.0396 | 0.0320 | 0.0284 | 0.0256 | 0.0238 | 0.0223 |
| DEG-C DAYS | 392. | 152. | 88. | 61. | 46. | 37. | 30. |

## DEC

| SOLAR RAD GROUP | ALL | 0+ | 95+ | 189+ | 284+ | 378+ | 473+ |
|---|---|---|---|---|---|---|---|
| AV.THRSH.RAD.LEVL | 0. | 41. | 139. | 233. | 330. | 424. | 523. |
| AV.THRESHOLD TEMP | -3.0 | -0.5 | -0.4 | -0.5 | -0.7 | 0.4 | -0.7 |
| GL RAD OVER THRSH | 300.2 | 254.1 | 191.0 | 145.8 | 109.6 | 81.4 | 56.2 |
| TIME AVG.TEMP | -2.0 | -0.6 | -0.7 | -0.8 | -0.9 | -1.0 | -1.3 |
| TOTAL TIME MSEC | 2.6784 | 1.1239 | 0.6466 | 0.4799 | 0.3704 | 0.3024 | 0.2527 |
| GLOBAL RAD TOTAL | 300.3 | 300.2 | 280.6 | 257.7 | 232.1 | 209.6 | 188.5 |
| DIRECT RAD TOTAL | 225.6 | 225.6 | 222.1 | 213.5 | 199.3 | 184.0 | 169.0 |
| RAD AV TIME MSEC | 0.0864 | 0.0364 | 0.0302 | 0.0274 | 0.0248 | 0.0227 | 0.0220 |
| DEG-C DAYS | 631. | 246. | 143. | 107. | 83. | 68. | 57. |

## APR

| SOLAR RAD GROUP | ALL | 0+ | 95+ | 189+ | 284+ | 378+ | 473+ |
|---|---|---|---|---|---|---|---|
| AV.THRSH.RAD.LEVL | 0. | 45. | 144. | 237. | 331. | 425. | 522. |
| AV.THRESHOLD TEMP | 7.1 | 8.7 | 9.9 | 11.0 | 12.8 | 13.7 | 13.8 |
| GL RAD OVER THRSH | 429.6 | 371.9 | 275.4 | 204.0 | 148.8 | 106.3 | 71.9 |
| TIME AVG.TEMP | 9.4 | 11.8 | 12.8 | 13.6 | 14.3 | 14.8 | 15.1 |
| TOTAL TIME MSEC | 2.5920 | 1.2956 | 0.9742 | 0.7668 | 0.5361 | 0.4522 | 0.3560 |
| GLOBAL RAD TOTAL | 429.6 | 429.6 | 415.3 | 385.5 | 342.7 | 298.4 | 257.6 |
| DIRECT RAD TOTAL | 220.3 | 220.3 | 217.8 | 216.1 | 206.5 | 193.5 | 177.6 |
| RAD AV TIME MSEC | 0.0864 | 0.0432 | 0.0349 | 0.0328 | 0.0295 | 0.0266 | 0.0241 |
| DEG-C DAYS | 267. | 101. | 67. | 46. | 31. | 22. | 17. |

## MAY

| SOLAR RAD GROUP | ALL | 0+ | 95+ | 189+ | 284+ | 378+ | 473+ |
|---|---|---|---|---|---|---|---|
| AV.THRSH.RAD.LEVL | 0. | 55. | 146. | 242. | 332. | 430. | 519. |
| AV.THRESHOLD TEMP | 12.1 | 14.3 | 14.8 | 16.1 | 19.4 | 20.1 | 20.9 |
| GL RAD OVER THRSH | 469.7 | 396.7 | 298.6 | 210.9 | 148.2 | 95.0 | 58.6 |
| TIME AVG.TEMP | 15.0 | 17.9 | 18.8 | 19.6 | 20.6 | 20.9 | 21.2 |
| TOTAL TIME MSEC | 2.6784 | 1.3392 | 1.0775 | 0.9122 | 0.6970 | 0.5440 | 0.4057 |
| GLOBAL RAD TOTAL | 469.7 | 469.7 | 455.4 | 431.4 | 379.4 | 328.6 | 269.2 |
| DIRECT RAD TOTAL | 197.5 | 197.5 | 195.6 | 194.7 | 183.7 | 171.2 | 152.9 |
| RAD AV TIME MSEC | 0.0864 | 0.0432 | 0.0356 | 0.0338 | 0.0292 | 0.0252 | 0.0223 |
| DEG-C DAYS | 120. | 33. | 22. | 14. | 7. | 5. | 3. |

## JUN

| SOLAR RAD GROUP | ALL | 0+ | 95+ | 189+ | 284+ | 378+ | 473+ |
|---|---|---|---|---|---|---|---|
| AV.THRSH.RAD.LEVL | 0. | 57. | 142. | 236. | 327. | 431. | 523. |
| AV.THRESHOLD TEMP | 17.4 | 19.3 | 19.9 | 21.5 | 24.5 | 24.6 | 26.3 |
| GL RAD OVER THRSH | 436.1 | 362.5 | 273.4 | 192.0 | 132.7 | 79.4 | 45.6 |
| TIME AVG.TEMP | 20.2 | 23.0 | 23.9 | 24.7 | 25.8 | 26.1 | 26.7 |
| TOTAL TIME MSEC | 2.5920 | 1.2960 | 1.0440 | 0.8640 | 0.6616 | 0.5155 | 0.3658 |
| GLOBAL RAD TOTAL | 436.1 | 436.1 | 421.8 | 396.2 | 346.0 | 301.5 | 237.0 |
| DIRECT RAD TOTAL | 174.6 | 174.6 | 173.1 | 172.5 | 162.5 | 152.4 | 128.7 |
| RAD AV TIME MSEC | 0.0864 | 0.0432 | 0.0360 | 0.0338 | 0.0295 | 0.0263 | 0.0212 |
| DEG-C DAYS | 21. | 6. | 4. | 2. | 1. | 0. | 0. |

## ALL MONTHS

| SOLAR RAD GROUP | ALL | 0+ | 95+ | 189+ | 284+ | 378+ | 473+ |
|---|---|---|---|---|---|---|---|
| TIME AVG.TEMP | 9.7 | 12.2 | 13.9 | 14.8 | 15.3 | 15.3 | 14.9 |
| TOTAL TIME MSEC | 31.5705 | 15.3659 | 10.9422 | 8.6300 | 6.7169 | 5.3219 | 4.1213 |
| GLOBAL RAD TOTAL | 4886.9 | 4886.6 | 4695.8 | 4375.6 | 3909.3 | 3450.1 | 2936.5 |
| DIRECT RAD TOTAL | 2815.3 | 2815.3 | 2782.2 | 2725.6 | 2577.2 | 2401.5 | 2160.7 |
| RAD AV TIME MSEC | 0.0864 | 0.0425 | 0.0345 | 0.0320 | 0.0286 | 0.0257 | 0.0229 |
| DEG-C DAYS | 3527. | 1436. | 888. | 656. | 498. | 397. | 326. |

STATISTICS FOR HARTFORD, CONN.    1959-68

RADIATION UNITS ARE WATTS/SQ.M OR MJOULES/SQ.M
SLOPE = 90.0 DEG   AZIMUTH = 180.0 DEG
DEGREE-DAY BASE TEMP = 18.3 DEGREES C

**JAN**

| | ALL | 0+ | 95+ | 189+ | 284+ | 378+ | 473+ |
|---|---|---|---|---|---|---|---|
| SOLAR RAD GROUP | ALL | 0+ | 95+ | 189+ | 284+ | 378+ | 473+ |
| AV.THRSH.RAD.LEVL | 0. | 36. | 141. | 238. | 327. | 424. | 528. |
| AV.THRESHOLD TEMP | -4.9 | -2.3 | -1.1 | -3.2 | -1.2 | -1.8 | -2.3 |
| GL RAD OVER THRSH | 300.1 | 257.4 | 188.3 | 138.8 | 104.5 | 74.5 | 47.9 |
| TIME AVG.TEMP | -3.8 | -2.5 | -2.9 | -2.8 | -3.2 | -3.2 | -3.4 |
| TOTAL TIME MSEC | 2.6784 | 1.2038 | 0.6552 | 0.5108 | 0.3856 | 0.3092 | 0.2556 |
| GLOBAL RAD TOTAL | 300.2 | 300.1 | 280.6 | 260.3 | 230.5 | 205.6 | 182.8 |
| DIRECT RAD TOTAL | 227.1 | 227.1 | 223.2 | 215.6 | 196.8 | 179.3 | 161.9 |
| RAD AV TIME MSEC | 0.0864 | 0.0392 | 0.0299 | 0.0261 | 0.0248 | 0.0223 | 0.0209 |
| DEG-C DAYS | 686. | 289. | 158. | 126. | 94. | 77. | 64. |

**FEB**

| | ALL | 0+ | 95+ | 189+ | 284+ | 378+ | 473+ |
|---|---|---|---|---|---|---|---|
| SOLAR RAD GROUP | ALL | 0+ | 95+ | 189+ | 284+ | 378+ | 473+ |
| AV.THRSH.RAD.LEVL | 0. | 42. | 138. | 238. | 332. | 425. | 523. |
| AV.THRESHOLD TEMP | -4.3 | -1.6 | -1.2 | -0.7 | -2.4 | -1.2 | -0.5 |
| GL RAD OVER THRSH | 317.3 | 269.6 | 198.7 | 145.2 | 105.3 | 75.4 | 49.8 |
| TIME AVG.TEMP | -3.1 | -1.6 | -1.6 | -1.8 | -2.1 | -1.9 | -2.1 |
| TOTAL TIME MSEC | 2.4451 | 1.1462 | 0.7330 | 0.5364 | 0.4237 | 0.3222 | 0.2621 |
| GLOBAL RAD TOTAL | 317.5 | 317.4 | 300.2 | 273.0 | 246.1 | 212.4 | 186.8 |
| DIRECT RAD TOTAL | 223.2 | 223.2 | 220.1 | 211.0 | 199.1 | 178.4 | 161.3 |
| RAD AV TIME MSEC | 0.0864 | 0.0410 | 0.0338 | 0.0302 | 0.0277 | 0.0245 | 0.0230 |
| DEG-C DAYS | 606. | 264. | 169. | 125. | 100. | 76. | 62. |

**MAR**

| | ALL | 0+ | 95+ | 189+ | 284+ | 378+ | 473+ |
|---|---|---|---|---|---|---|---|
| SOLAR RAD GROUP | ALL | 0+ | 95+ | 189+ | 284+ | 378+ | 473+ |
| AV.THRSH.RAD.LEVL | 0. | 38. | 141. | 236. | 335. | 419. | 523. |
| AV.THRESHOLD TEMP | 0.4 | 1.9 | 3.4 | 4.8 | 5.5 | 4.7 | 6.2 |
| GL RAD OVER THRSH | 337.8 | 285.1 | 194.4 | 137.1 | 90.4 | 61.5 | 34.7 |
| TIME AVG.TEMP | 2.1 | 3.6 | 4.5 | 5.0 | 5.1 | 4.9 | 4.9 |
| TOTAL TIME MSEC | 2.6784 | 1.3864 | 0.8791 | 0.6023 | 0.4720 | 0.3449 | 0.2581 |
| GLOBAL RAD TOTAL | 337.8 | 337.8 | 318.5 | 279.4 | 248.7 | 206.0 | 169.7 |
| DIRECT RAD TOTAL | 198.2 | 198.2 | 196.0 | 185.6 | 175.3 | 154.9 | 133.6 |
| RAD AV TIME MSEC | 0.0864 | 0.0450 | 0.0353 | 0.0302 | 0.0281 | 0.0248 | 0.0220 |
| DEG-C DAYS | 505. | 237. | 141. | 93. | 73. | 54. | 40. |

**JUL**

| | ALL | 0+ | 95+ | 189+ | 284+ | 378+ | 473+ |
|---|---|---|---|---|---|---|---|
| SOLAR RAD GROUP | ALL | 0+ | 95+ | 189+ | 284+ | 378+ | 473+ |
| AV.THRSH.RAD.LEVL | 0. | 52. | 144. | 235. | 332. | 426. | 498. |
| AV.THRESHOLD TEMP | 20.2 | 22.9 | 24.6 | 26.5 | 27.6 | 28.7 | 28.8 |
| GL RAD OVER THRSH | 262.9 | 205.1 | 122.6 | 61.3 | 21.1 | 2.7 | 0.0 |
| TIME AVG.TEMP | 22.6 | 26.0 | 26.8 | 27.5 | 28.1 | 28.7 | 28.8 |
| TOTAL TIME MSEC | 2.6784 | 1.1160 | 0.8939 | 0.6707 | 0.4172 | 0.1948 | 0.0382 |
| GLOBAL RAD TOTAL | 262.9 | 262.9 | 251.4 | 219.2 | 159.6 | 85.7 | 19.0 |
| DIRECT RAD TOTAL | 70.1 | 70.1 | 69.5 | 67.1 | 56.9 | 34.8 | 8.5 |
| RAD AV TIME MSEC | 0.0864 | 0.0360 | 0.0302 | 0.0263 | 0.0205 | 0.0137 | 0.0076 |
| DEG-C DAYS | 1. | 0. | 0. | 0. | 0. | 0. | 0. |

**AUG**

| | ALL | 0+ | 95+ | 189+ | 284+ | 378+ | 473+ |
|---|---|---|---|---|---|---|---|
| SOLAR RAD GROUP | ALL | 0+ | 95+ | 189+ | 284+ | 378+ | 473+ |
| AV.THRSH.RAD.LEVL | 0. | 47. | 141. | 238. | 333. | 423. | 516. |
| AV.THRESHOLD TEMP | 18.9 | 21.8 | 23.0 | 25.0 | 26.2 | 26.7 | 26.9 |
| GL RAD OVER THRSH | 277.2 | 220.0 | 134.6 | 72.8 | 34.1 | 12.3 | 2.5 |
| TIME AVG.TEMP | 21.3 | 24.3 | 25.1 | 26.0 | 26.6 | 26.8 | 26.9 |
| TOTAL TIME MSEC | 2.6784 | 1.2096 | 0.9094 | 0.6383 | 0.4079 | 0.2405 | 0.1062 |
| GLOBAL RAD TOTAL | 277.2 | 277.2 | 263.0 | 224.7 | 169.9 | 114.2 | 57.3 |
| DIRECT RAD TOTAL | 99.6 | 99.6 | 98.7 | 94.3 | 81.0 | 62.2 | 36.1 |
| RAD AV TIME MSEC | 0.0864 | 0.0392 | 0.0320 | 0.0270 | 0.0216 | 0.0173 | 0.0133 |
| DEG-C DAYS | 8. | 1. | 0. | 0. | 0. | 0. | 0. |

**SEP**

| | ALL | 0+ | 95+ | 189+ | 284+ | 378+ | 473+ |
|---|---|---|---|---|---|---|---|
| SOLAR RAD GROUP | ALL | 0+ | 95+ | 189+ | 284+ | 378+ | 473+ |
| AV.THRSH.RAD.LEVL | 0. | 40. | 142. | 233. | 333. | 425. | 519. |
| AV.THRESHOLD TEMP | 14.4 | 16.9 | 19.1 | 20.2 | 22.2 | 21.9 | 23.1 |
| GL RAD OVER THRSH | 341.0 | 288.0 | 194.6 | 133.1 | 81.9 | 47.9 | 24.1 |
| TIME AVG.TEMP | 17.2 | 19.9 | 21.2 | 21.9 | 22.5 | 22.6 | 22.9 |
| TOTAL TIME MSEC | 2.5920 | 1.3319 | 0.9122 | 0.6775 | 0.5137 | 0.3686 | 0.2527 |
| GLOBAL RAD TOTAL | 341.0 | 341.0 | 324.6 | 291.1 | 252.9 | 204.6 | 155.4 |
| DIRECT RAD TOTAL | 201.1 | 201.1 | 198.1 | 190.8 | 174.9 | 149.8 | 119.8 |
| RAD AV TIME MSEC | 0.0864 | 0.0443 | 0.0346 | 0.0310 | 0.0266 | 0.0227 | 0.0194 |
| DEG-C DAYS | 71. | 18. | 8. | 5. | 3. | 2. | 1. |

## APR

| SOLAR RAD GROUP | ALL | 0+ | 95+ | 189+ | 284+ | 378+ | 473+ |
|---|---|---|---|---|---|---|---|
| AV.THRSH.RAD.LEVL | 0. | 44. | 141. | 239. | 329. | 425. | 519. |
| AV.THRESHOLD TEMP | 7.2 | 8.9 | 10.5 | 13.0 | 13.8 | 15.6 | 15.7 |
| GL RAD OVER THRSH | 293.6 | 238.3 | 152.2 | 94.4 | 56.9 | 29.0 | 11.5 |
| TIME AVG.TEMP | 9.4 | 11.8 | 13.0 | 14.3 | 14.8 | 15.3 | 15.1 |
| TOTAL TIME MSEC | 2.5920 | 1.2481 | 0.6928 | 0.5915 | 0.4162 | 0.2876 | 0.1872 |
| GLOBAL RAD TOTAL | 293.6 | 293.6 | 277.8 | 235.4 | 193.6 | 151.4 | 108.7 |
| DIRECT RAD TOTAL | 124.4 | 124.4 | 123.4 | 117.5 | 107.8 | 91.7 | 72.6 |
| RAD AV TIME MSEC | 0.0864 | 0.0418 | 0.0338 | 0.0284 | 0.0252 | 0.0209 | 0.0173 |
| DEG-C DAYS | 267. | 97. | 59. | 32. | 21. | 13. | 9. |

## MAY

| SOLAR RAD GROUP | ALL | 0+ | 95+ | 189+ | 284+ | 378+ | 473+ |
|---|---|---|---|---|---|---|---|
| AV.THRSH.RAD.LEVL | 0. | 57. | 142. | 239. | 330. | 425. | 514. |
| AV.THRESHOLD TEMP | 12.5 | 15.3 | 16.1 | 18.8 | 20.9 | 21.9 | 21.2 |
| GL RAD OVER THRSH | 280.2 | 216.2 | 141.2 | 75.9 | 35.3 | 10.8 | 1.0 |
| TIME AVG.TEMP | 15.0 | 18.6 | 19.4 | 20.4 | 21.3 | 21.0 | 21.1 |
| TOTAL TIME MSEC | 2.6784 | 1.1160 | 0.8663 | 0.6714 | 0.4486 | 0.2578 | 0.1094 |
| GLOBAL RAD TOTAL | 280.2 | 280.2 | 267.0 | 236.5 | 183.2 | 120.3 | 57.3 |
| DIRECT RAD TOTAL | 80.8 | 80.8 | 79.9 | 78.3 | 67.6 | 51.0 | 28.7 |
| RAD AV TIME MSEC | 0.0864 | 0.0360 | 0.0299 | 0.0266 | 0.0216 | 0.0173 | 0.0115 |
| DEG-C DAYS | 120. | 24. | 16. | 8. | 4. | 2. | 1. |

## JUN

| SOLAR RAD GROUP | ALL | 0+ | 95+ | 189+ | 284+ | 378+ | 473+ |
|---|---|---|---|---|---|---|---|
| AV.THRSH.RAD.LEVL | 0. | 41. | 144. | 236. | 331. | 425. | 498. |
| AV.THRESHOLD TEMP | 17.8 | 19.3 | 21.7 | 24.5 | 25.6 | 27.1 | 27.9 |
| GL RAD OVER THRSH | 249.7 | 205.9 | 114.7 | 58.8 | 21.4 | 3.3 | 0.0 |
| TIME AVG.TEMP | 20.2 | 23.6 | 24.5 | 25.8 | 26.4 | 27.3 | 27.9 |
| TOTAL TIME MSEC | 2.5920 | 1.0800 | 0.8831 | 0.6044 | 0.3942 | 0.1933 | 0.0454 |
| GLOBAL RAD TOTAL | 249.7 | 249.7 | 241.7 | 201.6 | 152.0 | 85.4 | 22.6 |
| DIRECT RAD TOTAL | 58.0 | 58.0 | 57.9 | 55.2 | 48.5 | 29.9 | 6.2 |
| RAD AV TIME MSEC | 0.0864 | 0.0360 | 0.0313 | 0.0263 | 0.0209 | 0.0148 | 0.0086 |
| DEG-C DAYS | 21. | 4. | 3. | 1. | 0. | 0. | 0. |

## OCT

| SOLAR RAD GPOUP | ALL | 0+ | 95+ | 189+ | 284+ | 378+ | 473+ |
|---|---|---|---|---|---|---|---|
| AV.THRSH.RAD.LEVL | 0. | 44. | 138. | 237. | 333. | 426. | 519. |
| AV.THRESHOLD TEMP | 8.8 | 11.4 | 14.1 | 13.6 | 14.8 | 15.4 | 15.5 |
| GL RAD OVER THRSH | 378.7 | 319.6 | 236.0 | 170.0 | 120.2 | 81.8 | 51.8 |
| TIME AVG.TEMP | 11.3 | 13.8 | 15.1 | 15.4 | 15.9 | 16.2 | 16.4 |
| TOTAL TIME MSEC | 2.6870 | 1.3406 | 0.8928 | 0.6660 | 0.5177 | 0.4154 | 0.3197 |
| GLOBAL RAD TOTAL | 378.7 | 378.7 | 358.6 | 327.7 | 292.6 | 258.6 | 217.8 |
| DIRECT RAD TOTAL | 267.5 | 267.5 | 261.6 | 250.8 | 232.6 | 210.8 | 182.6 |
| RAD AV TIME MSEC | 0.0864 | 0.0432 | 0.0353 | 0.0317 | 0.0288 | 0.0256 | 0.0230 |
| DEG-C DAYS | 219. | 76. | 42. | 30. | 22. | 17. | 13. |

## NOV

| SOLAR RAD GROUP | ALL | 0+ | 95+ | 189+ | 284+ | 378+ | 473+ |
|---|---|---|---|---|---|---|---|
| AV.THRSH.RAD.LEVL | 0. | 39. | 139. | 234. | 325. | 426. | 521. |
| AV.THRESHOLD TEMP | 3.8 | 7.0 | 6.9 | 7.5 | 7.5 | 6.7 | 6.1 |
| GL RAD OVER THRSH | 263.8 | 219.5 | 156.7 | 114.7 | 84.7 | 57.5 | 37.4 |
| TIME AVG.TEMP | 5.3 | 7.2 | 7.2 | 7.3 | 7.5 | 7.5 | 7.7 |
| TOTAL TIME MSEC | 2.5920 | 1.1462 | 0.6232 | 0.4450 | 0.3290 | 0.2682 | 0.2113 |
| GLOBAL RAD TOTAL | 263.9 | 263.8 | 243.6 | 218.8 | 191.6 | 171.9 | 147.6 |
| DIRECT RAD TOTAL | 190.5 | 190.5 | 187.4 | 176.5 | 160.9 | 147.5 | 128.8 |
| RAD AV TIME MSEC | 0.0864 | 0.0389 | 0.0320 | 0.0284 | 0.0252 | 0.0238 | 0.0216 |
| DEG-C DAYS | 392. | 148. | 79. | 56. | 42. | 34. | 26. |

## DEC

| SOLAR RAD GROUP | ALL | 0+ | 95+ | 189+ | 284+ | 378+ | 473+ |
|---|---|---|---|---|---|---|---|
| AV.THRSH.RAD.LEVL | 0. | 38. | 140. | 230. | 330. | 420. | 527. |
| AV.THRESHOLD TEMP | -3.0 | -0.5 | -0.1 | -0.5 | -0.8 | 0.5 | -1.2 |
| GL RAD OVER THRSH | 273.7 | 231.1 | 170.2 | 130.0 | 95.2 | 70.1 | 44.8 |
| TIME AVG.TEMP | -2.0 | -0.6 | -0.7 | -0.9 | -1.1 | -1.1 | -1.4 |
| TOTAL TIME MSEC | 2.6784 | 1.1142 | 0.5972 | 0.4453 | 0.3481 | 0.2797 | 0.2376 |
| GLOBAL RAD TOTAL | 273.8 | 273.7 | 254.0 | 232.6 | 210.3 | 187.7 | 169.9 |
| DIRECT RAD TOTAL | 212.6 | 212.6 | 203.5 | 198.4 | 183.6 | 166.9 | 153.1 |
| RAD AV TIME MSEC | 0.0864 | 0.0360 | 0.0302 | 0.0277 | 0.0248 | 0.0227 | 0.0220 |
| DEG-C DAYS | 631. | 244. | 132. | 99. | 78. | 63. | 54. |

## ALL MONTHS

| SOLAR RAD GROUP | ALL | 0+ | 95+ | 189+ | 284+ | 378+ | 473+ |
|---|---|---|---|---|---|---|---|
| TIME AVG.TEMP | 9.7 | 11.9 | 13.9 | 14.4 | 14.0 | 12.7 | 9.9 |
| TOTAL TIME MSEC | 31.5705 | 14.4292 | 9.7582 | 7.0596 | 5.0738 | 3.4823 | 2.2835 |
| GLOBAL RAD TOTAL | 3576.5 | 3576.1 | 3381.4 | 3000.5 | 2530.9 | 2003.7 | 1494.9 |
| DIRECT RAD TOTAL | 1953.2 | 1953.2 | 1924.3 | 1841.2 | 1685.0 | 1457.4 | 1193.1 |
| RAD AV TIME MSEC | 0.0864 | 0.0401 | 0.0326 | 0.0287 | 0.0252 | 0.0220 | 0.0204 |
| DEG-C DAYS | 3527. | 1403. | 808. | 574. | 436. | 337. | 271. |

STATISTICS FOR  ATLANTA,GA          1971-1972          RSV020   FILE  4

RADIATION UNITS ARE WATTS/SQ.M OR MJOULES/SQ.M
SLOPE = 0.0 DEG   AZIMUTH = 0.0 DEG
DEGREE-DAY BASE TEMP = 18.3 DEGREES C

**JAN**

| SOLAR RAD GROUP | ALL | 0+ | 95+ | 189+ | 284+ | 378+ | 473+ |
|---|---|---|---|---|---|---|---|
| AV.THRSH.RAD.LEVL | 0. | 51. | 135. | 238. | 322. | 430. | 520. |
| AV.THRESHOLD TEMP | 6.0 | 9.3 | 10.0 | 6.0 | 8.1 | 4.3 | 6.4 |
| GL RAD OVER THRSH | 244.9 | 187.7 | 128.2 | 82.6 | 53.9 | 27.7 | 11.7 |
| TIME AVG.TEMP | 6.9 | 8.2 | 7.6 | 6.2 | 6.2 | 5.5 | 5.9 |
| TOTAL TIME MSEC | 2.6784 | 1.1160 | 0.7092 | 0.4428 | 0.3402 | 0.2448 | 0.1782 |
| GLOBAL RAD TOTAL | 244.9 | 244.9 | 224.1 | 188.1 | 163.6 | 132.9 | 104.3 |
| DIRECT RAD TOTAL | 131.0 | 131.0 | 128.3 | 123.8 | 115.1 | 101.7 | 82.0 |
| RAD AV TIME MSEC | 0.0864 | 0.0360 | 0.0288 | 0.0266 | 0.0230 | 0.0202 | 0.0166 |
| DEG-C DAYS | 353. | 131. | 88. | 62. | 48. | 36. | 26. |

**FEB**

| SOLAR RAD GROUP | ALL | 0+ | 95+ | 189+ | 284+ | 378+ | 473+ |
|---|---|---|---|---|---|---|---|
| AV.THRSH.RAD.LEVL | 0. | 50. | 143. | 234. | 337. | 422. | 526. |
| AV.THRESHOLD TEMP | 4.8 | 7.6 | 6.8 | 8.3 | 8.5 | 11.5 | 7.8 |
| GL RAD OVER THRSH | 312.4 | 258.7 | 180.5 | 129.4 | 85.1 | 56.5 | 29.3 |
| TIME AVG.TEMP | 6.1 | 7.8 | 7.9 | 8.4 | 8.5 | 8.5 | 7.7 |
| TOTAL TIME MSEC | 2.4624 | 1.0800 | 0.8370 | 0.5616 | 0.4320 | 0.3348 | 0.2628 |
| GLOBAL RAD TOTAL | 312.5 | 312.5 | 300.4 | 260.9 | 230.6 | 197.8 | 167.4 |
| DIRECT RAD TOTAL | 162.0 | 162.0 | 160.6 | 153.6 | 148.1 | 133.6 | 119.5 |
| RAD AV TIME MSEC | 0.0864 | 0.0378 | 0.0328 | 0.0274 | 0.0259 | 0.0227 | 0.0202 |
| DEG-C DAYS | 349. | 131. | 101. | 65. | 49. | 39. | 33. |

**MAR**

| SOLAR RAD GROUP | ALL | 0+ | 95+ | 189+ | 284+ | 378+ | 473+ |
|---|---|---|---|---|---|---|---|
| AV.THRSH.RAD.LEVL | 0. | 47. | 138. | 238. | 327. | 430. | 519. |
| AV.THRESHOLD TEMP | 8.4 | 9.7 | 10.2 | 10.6 | 12.1 | 13.0 | 12.1 |
| GL RAD OVER THRSH | 512.3 | 449.4 | 352.0 | 262.5 | 200.8 | 142.5 | 98.9 |
| TIME AVG.TEMP | 9.9 | 11.4 | 11.8 | 12.1 | 12.6 | 12.7 | 12.6 |
| TOTAL TIME MSEC | 2.6784 | 1.3320 | 1.0782 | 0.8910 | 0.6930 | 0.5652 | 0.4696 |
| GLOBAL RAD TOTAL | 512.3 | 512.3 | 500.5 | 474.5 | 427.4 | 385.6 | 353.1 |
| DIRECT RAD TOTAL | 300.1 | 300.1 | 296.6 | 291.3 | 261.4 | 263.2 | 250.5 |
| RAD AV TIME MSEC | 0.0864 | 0.0432 | 0.0367 | 0.0328 | 0.0302 | 0.0270 | 0.0252 |
| DEG-C DAYS | 262. | 108. | 83. | 66. | 48. | 39. | 33. |

**JUL**

| SOLAR RAD GROUP | ALL | 0+ | 95+ | 189+ | 284+ | 378+ | 473+ |
|---|---|---|---|---|---|---|---|
| AV.THRSH.RAD.LEVL | 0. | 48. | 141. | 238. | 335. | 425. | 519. |
| AV.THRESHOLD TEMP | 22.0 | 22.5 | 23.4 | 24.5 | 25.1 | 26.0 | 26.5 |
| GL RAD OVER THRSH | 596.4 | 521.4 | 401.5 | 295.7 | 207.2 | 142.4 | 88.9 |
| TIME AVG.TEMP | 23.9 | 25.3 | 25.9 | 26.4 | 26.8 | 27.2 | 27.6 |
| TOTAL TIME MSEC | 2.6784 | 1.5624 | 1.2888 | 1.0962 | 0.9072 | 0.7200 | 0.5670 |
| GLOBAL RAD TOTAL | 596.4 | 596.4 | 583.2 | 556.1 | 511.2 | 448.5 | 383.4 |
| DIRECT RAD TOTAL | 258.2 | 258.2 | 255.7 | 250.7 | 239.8 | 225.9 | 204.6 |
| RAD AV TIME MSEC | 0.0864 | 0.0504 | 0.0425 | 0.0374 | 0.0324 | 0.0288 | 0.0248 |
| DEG-C DAYS | 0. | 0. | 0. | 0. | 0. | 0. | 0. |

**AUG**

| SOLAR RAD GROUP | ALL | 0+ | 95+ | 189+ | 284+ | 378+ | 473+ |
|---|---|---|---|---|---|---|---|
| AV.THRSH.RAD.LEVL | 0. | 37. | 142. | 229. | 336. | 428. | 519. |
| AV.THRESHOLD TEMP | 22.4 | 23.8 | 24.0 | 25.1 | 25.8 | 26.4 | 27.7 |
| GL RAD OVER THRSH | 586.6 | 531.0 | 400.8 | 308.7 | 209.6 | 143.8 | 92.4 |
| TIME AVG.TEMP | 24.4 | 26.1 | 26.5 | 26.9 | 27.2 | 27.6 | 27.9 |
| TOTAL TIME MSEC | 2.6784 | 1.4868 | 1.2456 | 1.0584 | 0.9216 | 0.7218 | 0.5616 |
| GLOBAL RAD TOTAL | 586.6 | 586.6 | 577.6 | 551.0 | 519.7 | 452.5 | 383.9 |
| DIRECT RAD TOTAL | 310.0 | 310.0 | 308.4 | 300.7 | 289.7 | 266.9 | 236.6 |
| RAD AV TIME MSEC | 0.0864 | 0.0482 | 0.0414 | 0.0367 | 0.0328 | 0.0284 | 0.0248 |
| DEG-C DAYS | 0. | 0. | 0. | 0. | 0. | 0. | 0. |

**SEP**

| SOLAR RAD GROUP | ALL | 0+ | 95+ | 189+ | 284+ | 378+ | 473+ |
|---|---|---|---|---|---|---|---|
| AV.THRSH.RAD.LEVL | 0. | 55. | 141. | 242. | 329. | 433. | 516. |
| AV.THRESHOLD TEMP | 21.1 | 21.1 | 23.0 | 23.4 | 24.4 | 25.6 | 25.7 |
| GL RAD OVER THRSH | 487.3 | 416.6 | 322.0 | 231.7 | 168.0 | 105.2 | 67.9 |
| TIME AVG.TEMP | 22.8 | 24.5 | 25.2 | 25.6 | 26.1 | 26.5 | 26.8 |
| TOTAL TIME MSEC | 2.5920 | 1.2960 | 1.0854 | 0.9018 | 0.7362 | 0.5994 | 0.4536 |
| GLOBAL RAD TOTAL | 487.3 | 487.3 | 475.8 | 450.0 | 409.9 | 364.9 | 301.8 |
| DIRECT RAD TOTAL | 272.0 | 272.0 | 269.0 | 262.0 | 250.6 | 230.8 | 201.1 |
| RAD AV TIME MSEC | 0.0864 | 0.0432 | 0.0378 | 0.0328 | 0.0299 | 0.0252 | 0.0216 |
| DEG-C DAYS | 1. | 1. | 1. | 1. | 0. | 0. | 0. |

**APR**

| SOLAR RAD GROUP | ALL | 0+ | 95+ | 189+ | 284+ | 378+ | 473+ |
|---|---|---|---|---|---|---|---|
| AV.THRSH.RAD.LEVL | 0. | 39. | 146. | 236. | 334. | 428. | 525. |
| AV.THRESHOLD TEMP | 13.4 | 16.7 | 14.2 | 16.9 | 17.7 | 19.2 | 20.6 |
| GL RAD OVER THRSH | 662.2 | 609.4 | 478.5 | 383.9 | 297.1 | 225.8 | 163.8 |
| TIME AVG.TEMP | 16.1 | 18.6 | 19.7 | 20.1 | 19.7 | 20.1 | 20.2 |
| TOTAL TIME MSEC | 2.5920 | 1.3698 | 1.2132 | 1.0548 | 0.8874 | 0.7614 | 0.6354 |
| GLOBAL RAD TOTAL | 662.2 | 662.2 | 656.2 | 633.0 | 593.5 | 551.4 | 497.5 |
| DIRECT RAD TOTAL | 395.4 | 395.4 | 394.3 | 385.6 | 375.6 | 363.4 | 331.4 |
| RAD AV TIME MSEC | 0.0864 | 0.0457 | 0.0414 | 0.0357 | 0.0331 | 0.0310 | 0.0266 |
| DEG-C DAYS | 89. | 30. | 25. | 18. | 13. | 10. | 8. |

**MAY**

| SOLAR RAD GROUP | ALL | 0+ | 95+ | 189+ | 284+ | 378+ | 473+ |
|---|---|---|---|---|---|---|---|
| AV.THRSH.RAD.LEVL | 0. | 44. | 139. | 240. | 331. | 427. | 519. |
| AV.THRESHOLD TEMP | 16.6 | 17.6 | 18.4 | 18.8 | 21.1 | 20.8 | 22.1 |
| GL RAD OVER THRSH | 678.8 | 609.4 | 486.9 | 370.9 | 286.1 | 210.2 | 151.5 |
| TIME AVG.TEMP | 18.9 | 20.5 | 21.1 | 21.4 | 22.1 | 22.2 | 22.6 |
| TOTAL TIME MSEC | 2.6784 | 1.5624 | 1.2942 | 1.1502 | 0.9324 | 0.7884 | 0.6354 |
| GLOBAL RAD TOTAL | 678.8 | 678.8 | 666.9 | 646.9 | 594.6 | 547.0 | 481.6 |
| DIRECT RAD TOTAL | 345.4 | 345.4 | 341.7 | 338.3 | 321.3 | 312.2 | 291.4 |
| RAD AV TIME MSEC | 0.0864 | 0.0504 | 0.0428 | 0.0392 | 0.0342 | 0.0313 | 0.0284 |
| DEG-C DAYS | 24. | 7. | 4. | 4. | 2. | 2. | 1. |

**JUN**

| SOLAR RAD GROUP | ALL | 0+ | 95+ | 189+ | 284+ | 378+ | 473+ |
|---|---|---|---|---|---|---|---|
| AV.THRSH.RAD.LEVL | 0. | 61. | 143. | 244. | 330. | 431. | 519. |
| AV.THRESHOLD TEMP | 21.0 | 21.0 | 24.0 | 22.9 | 24.5 | 25.6 | 25.7 |
| GL RAD OVER THRSH | 710.1 | 618.0 | 511.1 | 396.4 | 311.3 | 226.3 | 162.4 |
| TIME AVG.TEMP | 23.3 | 25.0 | 25.7 | 25.9 | 26.3 | 26.7 | 26.8 |
| TOTAL TIME MSEC | 2.5920 | 1.5120 | 1.2942 | 1.1412 | 0.9936 | 0.8383 | 0.7272 |
| GLOBAL RAD TOTAL | 710.1 | 710.1 | 696.8 | 674.9 | 638.9 | 587.8 | 539.7 |
| DIRECT RAD TOTAL | 399.5 | 399.5 | 395.5 | 389.8 | 375.3 | 356.1 | 337.6 |
| RAD AV TIME MSEC | 0.0864 | 0.0504 | 0.0443 | 0.0407 | 0.0364 | 0.0320 | 0.0295 |
| DEG-C DAYS | 1. | 0. | 0. | 0. | 0. | 0. | 0. |

**OCT**

| SOLAR RAD GROUP | ALL | 0+ | 95+ | 189+ | 284+ | 378+ | 473+ |
|---|---|---|---|---|---|---|---|
| AV.THRSH.RAD.LEVL | 0. | 42. | 146. | 231. | 337. | 429. | 518. |
| AV.THRESHOLD TEMP | 15.3 | 17.2 | 18.5 | 19.1 | 19.6 | 20.5 | 21.1 |
| GL RAD OVER THRSH | 393.7 | 341.7 | 240.7 | 177.5 | 118.4 | 77.5 | 48.0 |
| TIME AVG.TEMP | 17.2 | 19.4 | 20.9 | 20.4 | 20.9 | 21.2 | 21.4 |
| TOTAL TIME MSEC | 2.6784 | 1.2438 | 0.9702 | 0.7452 | 0.5562 | 0.4428 | 0.3330 |
| GLOBAL RAD TOTAL | 393.7 | 393.7 | 382.3 | 349.5 | 305.8 | 267.6 | 220.5 |
| DIRECT RAD TOTAL | 222.6 | 222.2 | 220.3 | 214.7 | 201.5 | 186.8 | 156.5 |
| RAD AV TIME MSEC | 0.0864 | 0.0410 | 0.0342 | 0.0306 | 0.0266 | 0.0241 | 0.0205 |
| DEG-C DAYS | 57. | 13. | 8. | 6. | 4. | 3. | 2. |

**NOV**

| SOLAR RAD GROUP | ALL | 0+ | 95+ | 189+ | 284+ | 378+ | 473+ |
|---|---|---|---|---|---|---|---|
| AV.THRSH.RAD.LEVL | 0. | 54. | 138. | 239. | 327. | 432. | 524. |
| AV.THRESHOLD TEMP | 8.3 | 9.9 | 10.7 | 10.7 | 13.6 | 13.4 | 14.1 |
| GL RAD OVER THRSH | 312.7 | 253.9 | 183.0 | 121.6 | 79.3 | 41.4 | 17.4 |
| TIME AVG.TEMP | 9.8 | 11.0 | 12.5 | 13.2 | 13.9 | 14.0 | 14.2 |
| TOTAL TIME MSEC | 2.5920 | 1.0800 | 0.8496 | 0.6084 | 0.4806 | 0.3618 | 0.2592 |
| GLOBAL RAD TOTAL | 312.7 | 312.7 | 300.1 | 266.9 | 236.4 | 197.5 | 153.3 |
| DIRECT RAD TOTAL | 184.3 | 184.3 | 181.5 | 173.6 | 162.6 | 142.5 | 116.2 |
| RAD AV TIME MSEC | 0.0864 | 0.0360 | 0.0310 | 0.0263 | 0.0238 | 0.0202 | 0.0169 |
| DEG-C DAYS | 258. | 83. | 61. | 38. | 27. | 19. | 13. |

**DEC**

| SOLAR RAD GROUP | ALL | 0+ | 95+ | 189+ | 284+ | 378+ | 473+ |
|---|---|---|---|---|---|---|---|
| AV.THRSH.RAD.LEVL | 0. | 49. | 137. | 236. | 331. | 424. | 527. |
| AV.THRESHOLD TEMP | 3.9 | 11.8 | 11.9 | 11.4 | 13.0 | 10.4 | 10.1 |
| GL RAD OVER THRSH | 202.9 | 147.8 | 91.9 | 55.4 | 31.3 | 13.7 | 3.0 |
| TIME AVG.TEMP | 10.0 | 11.6 | 11.4 | 11.0 | 10.8 | 10.1 | 9.8 |
| TOTAL TIME MSEC | 2.6784 | 1.1160 | 0.6408 | 0.3654 | 0.2556 | 0.1890 | 0.1044 |
| GLOBAL RAD TOTAL | 202.9 | 202.9 | 179.5 | 141.8 | 115.8 | 93.8 | 57.9 |
| DIRECT RAD TOTAL | 85.8 | 85.8 | 82.6 | 80.4 | 69.6 | 62.2 | 42.8 |
| RAD AV TIME MSEC | 0.0864 | 0.0360 | 0.0263 | 0.0248 | 0.0198 | 0.0180 | 0.0126 |
| DEG-C DAYS | 259. | 90. | 53. | 32. | 22. | 18. | 11. |

**ALL MONTHS**

| SOLAR RAD GROUP | ALL | 0+ | 95+ | 189+ | 284+ | 378+ | 473+ |
|---|---|---|---|---|---|---|---|
| TIME AVG.TEMP | 15.8 | 18.3 | 19.1 | 20.0 | 20.7 | 21.1 | 21.4 |
| TOTAL TIME MSEC | 31.5791 | 15.7572 | 12.5064 | 10.0170 | 8.1360 | 6.5682 | 5.2074 |
| GLOBAL RAD TOTAL | 5700.4 | 5700.4 | 5543.1 | 5193.4 | 4747.3 | 4227.3 | 3644.5 |
| DIRECT RAD TOTAL | 3066.4 | 3066.4 | 3034.5 | 2964.6 | 2830.4 | 2645.2 | 2372.2 |
| RAD AV TIME MSEC | 0.0864 | 0.0451 | 0.0368 | 0.0348 | 0.0311 | 0.0278 | 0.0248 |
| DEG-C DAYS | 1653. | 594. | 425. | 292. | 213. | 167. | 127. |

STATISTICS FOR  ATLANTA,GA          1971-1972          RSV020    FILE 4

RADIATION UNITS ARE WATTS/SQ.M OR MJOULES/SQ.M
SLOPE = 30.0 DEG    AZIMUTH = 180.0 DEG
DEGREE-DAY BASE TEMP = 18.3 DEGREES C

**JAN**

| SOLAR RAD GROUP | ALL | 0+ | 95+ | 189+ | 284+ | 378+ | 473+ |
|---|---|---|---|---|---|---|---|
| AV.THRSH.RAD.LEVL | 0. | 46. | 136. | 240. | 332. | 426. | 533. |
| AV.THRESHOLD TEMP | 5.8 | 9.8 | 9.0 | 7.3 | 5.7 | 7.1 | 6.5 |
| GL RAD OVER THRSH | 335.1 | 279.2 | 210.4 | 160.8 | 123.3 | 92.7 | 63.4 |
| TIME AVG.TEMP | 6.9 | 8.3 | 7.3 | 6.3 | 6.2 | 6.3 | 6.1 |
| TOTAL TIME MSEC | 2.6784 | 1.2276 | 0.7632 | 0.4752 | 0.4068 | 0.3276 | 0.2736 |
| GLOBAL RAD TOTAL | 335.1 | 335.1 | 313.9 | 274.9 | 258.4 | 232.1 | 209.1 |
| DIRECT RAD TOTAL | 225.6 | 225.6 | 223.3 | 214.3 | 209.7 | 195.5 | 178.7 |
| RAD AV TIME MSEC | 0.0864 | 0.0396 | 0.0346 | 0.0299 | 0.0292 | 0.0263 | 0.0234 |
| DEG-C DAYS | 353. | 143. | 98. | 66. | 57. | 46. | 39. |

**FEB**

| SOLAR RAD GROUP | ALL | 0+ | 95+ | 189+ | 284+ | 378+ | 473+ |
|---|---|---|---|---|---|---|---|
| AV.THRSH.RAD.LEVL | 0. | 45. | 141. | 232. | 331. | 426. | 515. |
| AV.THRESHOLD TEMP | 4.5 | 7.2 | 6.7 | 8.8 | 11.9 | 8.3 | 10.6 |
| GL RAD OVER THRSH | 384.9 | 332.5 | 248.8 | 195.6 | 149.2 | 110.5 | 82.2 |
| TIME AVG.TEMP | 6.1 | 7.9 | 8.1 | 8.8 | 8.8 | 8.3 | 8.3 |
| TOTAL TIME MSEC | 2.4624 | 1.1700 | 0.8712 | 0.5850 | 0.4662 | 0.4086 | 0.3186 |
| GLOBAL RAD TOTAL | 384.9 | 384.9 | 371.5 | 331.2 | 303.7 | 284.6 | 246.3 |
| DIRECT RAD TOTAL | 240.5 | 240.5 | 238.6 | 230.3 | 223.8 | 217.1 | 194.7 |
| RAD AV TIME MSEC | 0.0864 | 0.0410 | 0.0356 | 0.0310 | 0.0288 | 0.0277 | 0.0238 |
| DEG-C DAYS | 349. | 142. | 103. | 65. | 53. | 48. | 38. |

**MAR**

| SOLAR RAD GROUP | ALL | 0+ | 95+ | 189+ | 284+ | 378+ | 473+ |
|---|---|---|---|---|---|---|---|
| AV.THRSH.RAD.LEVL | 0. | 31. | 143. | 231. | 328. | 426. | 519. |
| AV.THRESHOLD TEMP | 8.1 | 9.6 | 9.6 | 11.1 | 12.1 | 13.4 | 12.2 |
| GL RAD OVER THRSH | 573.3 | 529.6 | 406.0 | 324.7 | 258.2 | 199.0 | 148.9 |
| TIME AVG.TEMP | 9.9 | 11.4 | 12.0 | 12.4 | 12.9 | 13.0 | 12.9 |
| TOTAL TIME MSEC | 2.6784 | 1.4274 | 1.0962 | 0.9234 | 0.6894 | 0.6030 | 0.5364 |
| GLOBAL RAD TOTAL | 573.4 | 573.4 | 563.2 | 538.0 | 484.3 | 455.9 | 427.6 |
| DIRECT RAD TOTAL | 370.1 | 370.1 | 367.7 | 363.3 | 345.5 | 336.8 | 324.2 |
| RAD AV TIME MSEC | 0.0864 | 0.0464 | 0.0374 | 0.0353 | 0.0310 | 0.0292 | 0.0277 |
| DEG-C DAYS | 262. | 114. | 83. | 66. | 46. | 39. | 36. |

**JUL**

| SOLAR RAD GROUP | ALL | 0+ | 95+ | 189+ | 284+ | 378+ | 473+ |
|---|---|---|---|---|---|---|---|
| AV.THRSH.RAD.LEVL | 0. | 36. | 152. | 245. | 330. | 428. | 517. |
| AV.THRESHOLD TEMP | 21.9 | 23.0 | 23.3 | 24.3 | 25.7 | 26.6 | 26.9 |
| GL RAD OVER THRSH | 537.7 | 485.7 | 348.0 | 251.0 | 179.6 | 114.8 | 70.2 |
| TIME AVG.TEMP | 23.9 | 25.7 | 26.3 | 26.7 | 27.2 | 27.7 | 28.0 |
| TOTAL TIME MSEC | 2.6784 | 1.4508 | 1.1826 | 1.0458 | 0.8442 | 0.6570 | 0.5004 |
| GLOBAL RAD TOTAL | 537.7 | 537.7 | 528.1 | 507.3 | 457.9 | 396.2 | 329.1 |
| DIRECT RAD TOTAL | 226.6 | 226.6 | 225.5 | 223.6 | 212.5 | 199.8 | 177.3 |
| RAD AV TIME MSEC | 0.0864 | 0.0468 | 0.0385 | 0.0356 | 0.0306 | 0.0270 | 0.0234 |
| DEG-C DAYS | 0. | 0. | 0. | 0. | 0. | 0. | 0. |

**AUG**

| SOLAR RAD GROUP | ALL | 0+ | 95+ | 189+ | 284+ | 378+ | 473+ |
|---|---|---|---|---|---|---|---|
| AV.THRSH.RAD.LEVL | 0. | 31. | 147. | 234. | 338. | 429. | 515. |
| AV.THRESHOLD TEMP | 22.3 | 23.5 | 24.1 | 25.1 | 26.1 | 27.2 | 27.6 |
| GL RAD OVER THRSH | 563.9 | 518.8 | 383.9 | 292.0 | 203.8 | 140.1 | 94.8 |
| TIME AVG.TEMP | 24.4 | 26.2 | 26.9 | 27.2 | 27.7 | 28.0 | 28.3 |
| TOTAL TIME MSEC | 2.6784 | 1.4490 | 1.1628 | 1.0566 | 0.8514 | 0.7002 | 0.5274 |
| GLOBAL RAD TOTAL | 564.0 | 564.0 | 555.0 | 539.4 | 491.4 | 440.3 | 366.2 |
| DIRECT RAD TOTAL | 304.5 | 304.5 | 302.7 | 300.7 | 281.8 | 268.9 | 237.2 |
| RAD AV TIME MSEC | 0.0864 | 0.0468 | 0.0385 | 0.0367 | 0.0306 | 0.0284 | 0.0252 |
| DEG-C DAYS | 0. | 0. | 0. | 0. | 0. | 0. | 0. |

**SEP**

| SOLAR RAD GROUP | ALL | 0+ | 95+ | 189+ | 284+ | 378+ | 473+ |
|---|---|---|---|---|---|---|---|
| AV.THRSH.RAD.LEVL | 0. | 30. | 149. | 228. | 334. | 430. | 534. |
| AV.THRESHOLD TEMP | 20.8 | 21.5 | 22.8 | 23.5 | 24.7 | 25.6 | 25.8 |
| GL RAD OVER THRSH | 518.9 | 476.7 | 347.6 | 274.6 | 196.9 | 135.8 | 87.2 |
| TIME AVG.TEMP | 22.8 | 24.5 | 25.4 | 25.8 | 26.4 | 26.7 | 27.1 |
| TOTAL TIME MSEC | 2.5920 | 1.4004 | 1.0836 | 0.9270 | 0.7362 | 0.6336 | 0.4680 |
| GLOBAL RAD TOTAL | 518.9 | 518.9 | 509.3 | 486.0 | 442.5 | 408.2 | 337.0 |
| DIRECT RAD TOTAL | 313.6 | 313.6 | 311.7 | 305.0 | 289.4 | 277.3 | 239.5 |
| RAD AV TIME MSEC | 0.0864 | 0.0468 | 0.0378 | 0.0342 | 0.0299 | 0.0274 | 0.0227 |
| DEG-C DAYS | 1. | 1. | 1. | 1. | 1. | 1. | 0. |

### APR

| SOLAR RAD GROUP | ALL | 0+ | 95+ | 189+ | 284+ | 378+ | 473+ |
|---|---|---|---|---|---|---|---|
| AV.THRSH.RAD.LEVL | 0. | 28. | 150. | 238. | 322. | 426. | 519. |
| AV.THRESHOLD TEMP | 13.4 | 14.5 | 15.6 | 15.6 | 20.1 | 19.4 | 20.0 |
| GL RAD OVER THRSH | 662.5 | 622.6 | 484.1 | 391.0 | 318.5 | 242.4 | 181.3 |
| TIME AVG.TEMP | 16.1 | 18.3 | 19.2 | 19.5 | 20.4 | 20.4 | 20.6 |
| TOTAL TIME MSEC | 2.5920 | 1.4040 | 1.1394 | 1.0584 | 0.8622 | 0.7308 | 0.6588 |
| GLOBAL RAD TOTAL | 662.5 | 662.5 | 655.0 | 642.8 | 596.2 | 553.8 | 523.1 |
| DIRECT RAD TOTAL | 409.2 | 409.2 | 407.1 | 406.1 | 391.0 | 375.9 | 365.8 |
| RAD AV TIME MSEC | 0.0864 | 0.0468 | 0.0389 | 0.0371 | 0.0328 | 0.0299 | 0.0281 |
| DEG-C DAYS | 89. | 32. | 21. | 17. | 11. | 9. | 8. |

### MAY

| SOLAR RAD GROUP | ALL | 0+ | 95+ | 189+ | 284+ | 378+ | 473+ |
|---|---|---|---|---|---|---|---|
| AV.THRSH.RAD.LEVL | 0. | 33. | 152. | 244. | 329. | 428. | 519. |
| AV.THRESHOLD TEMP | 16.5 | 18.1 | 18.4 | 19.0 | 21.4 | 22.1 | 22.2 |
| GL RAD OVER THRSH | 618.3 | 571.4 | 429.3 | 329.3 | 254.7 | 184.8 | 131.5 |
| TIME AVG.TEMP | 18.9 | 20.9 | 21.6 | 21.9 | 22.6 | 22.8 | 23.0 |
| TOTAL TIME MSEC | 2.6784 | 1.4508 | 1.1934 | 1.0800 | 0.8784 | 0.7110 | 0.5850 |
| GLOBAL RAD TOTAL | 618.9 | 618.9 | 610.5 | 593.3 | 544.0 | 488.9 | 435.0 |
| DIRECT RAD TOTAL | 308.8 | 308.8 | 306.9 | 305.3 | 291.7 | 200.4 | 265.8 |
| RAD AV TIME MSEC | 0.0864 | 0.0468 | 0.0389 | 0.0371 | 0.0324 | 0.0288 | 0.0274 |
| DEG-C DAYS | 24. | 6. | 3. | 3. | 1. | 1. | 1. |

### JUN

| SOLAR RAD GROUP | ALL | 0+ | 95+ | 189+ | 284+ | 378+ | 473+ |
|---|---|---|---|---|---|---|---|
| AV.THRSH.RAD.LEVL | 0. | 38. | 150. | 248. | 329. | 434. | 515. |
| AV.THRESHOLD TEMP | 20.8 | 22.3 | 22.2 | 23.6 | 25.3 | 26.4 | 26.4 |
| GL RAD OVER THRSH | 624.8 | 572.1 | 441.3 | 336.1 | 263.0 | 182.2 | 129.9 |
| TIME AVG.TEMP | 23.3 | 25.4 | 26.1 | 26.4 | 26.9 | 27.2 | 27.4 |
| TOTAL TIME MSEC | 2.5920 | 1.4040 | 1.1592 | 1.0818 | 0.9000 | 0.7704 | 0.6444 |
| GLOBAL RAD TOTAL | 624.8 | 624.8 | 615.6 | 604.0 | 558.9 | 516.3 | 461.7 |
| DIRECT RAD TOTAL | 338.3 | 338.3 | 336.3 | 334.9 | 318.3 | 305.3 | 284.5 |
| RAD AV TIME MSEC | 0.0864 | 0.0468 | 0.0392 | 0.0382 | 0.0328 | 0.0295 | 0.0266 |
| DEG-C DAYS | 1. | 0. | 0. | 0. | 0. | 0. | 0. |

### OCT

| SOLAR RAD GROUP | 473+ | 378+ | 284+ | 189+ | 95+ | 0+ | ALL |
|---|---|---|---|---|---|---|---|
| AV.THRSH.RAD.LEVL | 517. | 427. | 338. | 235. | 144. | 37. | 0. |
| AV.THRESHOLD TEMP | 21.5 | 19.5 | 20.2 | 19.5 | 18.2 | 17.1 | 15.1 |
| GL RAD OVER THRSH | 97.6 | 134.3 | 179.4 | 241.3 | 309.8 | 418.1 | 467.7 |
| TIME AVG.TEMP | 21.7 | 21.3 | 21.1 | 20.8 | 20.1 | 19.4 | 17.2 |
| TOTAL TIME MSEC | 0.4068 | 0.5076 | 0.5994 | 0.7560 | 1.0152 | 1.3320 | 2.6784 |
| GLOBAL RAD TOTAL | 307.8 | 350.9 | 381.9 | 418.6 | 455.9 | 467.7 | 467.7 |
| DIRECT RAD TOTAL | 239.7 | 265.2 | 280.0 | 290.7 | 300.7 | 303.2 | 303.2 |
| RAD AV TIME MSEC | 0.0241 | 0.0274 | 0.0292 | 0.0320 | 0.0364 | 0.0439 | 0.0864 |
| DEG-C DAYS | 2. | 3. | 4. | 6. | 8. | 14. | 57. |

### NOV

| SOLAR RAD GROUP | 473+ | 378+ | 284+ | 189+ | 95+ | 0+ | ALL |
|---|---|---|---|---|---|---|---|
| AV.THRSH.RAD.LEVL | 524. | 429. | 339. | 228. | 136. | 46. | 0. |
| AV.THRESHOLD TEMP | 16.8 | 13.5 | 11.3 | 12.2 | 9.8 | 9.9 | 8.1 |
| GL RAD OVER THRSH | 87.5 | 123.6 | 166.1 | 229.4 | 291.2 | 374.7 | 429.7 |
| TIME AVG.TEMP | 14.3 | 14.2 | 13.7 | 13.4 | 12.4 | 11.9 | 9.8 |
| TOTAL TIME MSEC | 0.3798 | 0.4698 | 0.5706 | 0.6750 | 0.9270 | 1.1880 | 2.5920 |
| GLOBAL RAD TOTAL | 286.7 | 325.3 | 359.5 | 383.3 | 417.7 | 429.7 | 429.7 |
| DIRECT RAD TOTAL | 237.9 | 264.2 | 265.2 | 291.1 | 303.2 | 305.7 | 305.7 |
| RAD AV TIME MSEC | 0.0227 | 0.0263 | 0.0288 | 0.0299 | 0.0353 | 0.0396 | 0.0864 |
| DEG-C DAYS | 19. | 24. | 33. | 41. | 66. | 92. | 258. |

### DEC

| SOLAR RAD GROUP | 473+ | 378+ | 284+ | 189+ | 95+ | 0+ | ALL |
|---|---|---|---|---|---|---|---|
| AV.THRSH.RAD.LEVL | 521. | 423. | 332. | 243. | 134. | 45. | 0. |
| AV.THRESHOLD TEMP | 13.3 | 13.3 | 9.4 | 12.8 | 11.8 | 11.8 | 8.6 |
| GL RAD OVER THRSH | 38.8 | 57.6 | 81.4 | 109.7 | 153.1 | 214.4 | 269.4 |
| TIME AVG.TEMP | 10.5 | 11.2 | 10.9 | 11.3 | 11.5 | 11.6 | 10.0 |
| TOTAL TIME MSEC | 0.1926 | 0.2610 | 0.3186 | 0.3996 | 0.6840 | 1.2276 | 2.6784 |
| GLOBAL RAD TOTAL | 139.1 | 168.0 | 187.1 | 206.8 | 245.1 | 269.4 | 269.4 |
| DIRECT RAD TOTAL | 118.4 | 132.4 | 142.4 | 147.3 | 154.2 | 157.4 | 157.4 |
| RAD AV TIME MSEC | 0.0223 | 0.0245 | 0.0277 | 0.0288 | 0.0324 | 0.0396 | 0.0864 |
| DEG-C DAYS | 18. | 22. | 28. | 33. | 56. | 98. | 259. |

### ALL MONTHS

| SOLAR RAD GROUP | ALL | 0+ | 95+ | 189+ | 284+ | 378+ | 473+ |
|---|---|---|---|---|---|---|---|
| TIME AVG.TEMP | 15.8 | 18.1 | 19.0 | 20.1 | 20.6 | 20.7 | 20.7 |
| TOTAL TIME MSEC | 31.5791 | 16.1316 | 12.2778 | 10.0638 | 8.1234 | 6.7806 | 5.4918 |
| GLOBAL RAD TOTAL | 5987.0 | 5987.0 | 5840.8 | 5525.9 | 5065.7 | 4620.6 | 4068.7 |
| DIRECT RAD TOTAL | 3503.5 | 3503.5 | 3477.8 | 3412.5 | 3269.4 | 3119.3 | 2863.6 |
| RAD AV TIME MSEC | 0.0864 | 0.0449 | 0.0375 | 0.0347 | 0.0307 | 0.0281 | 0.0253 |
| DEG-C DAYS | 1653. | 642. | 438. | 297. | 233. | 192. | 161. |

STATISTICS FOR    ATLANTA,GA    1971-1972    RSV020    FILE 4

RADIATION UNITS ARE WATTS/SQ.M OR MJOULES/SQ.M
SLOPE = 40.0 DEG    AZIMUTH = 180.0 DEG
DEGREE-DAY BASE TEMP = 18.3 DEGREES C

### JAN

| SOLAR RAD GROUP | ALL | 0+ | 95+ | 189+ | 284+ | 378+ | 473+ |
|---|---|---|---|---|---|---|---|
| AV.THRSH.RAD.LEVL | 0. | 44. | 135. | 240. | 336. | 436. | 524. |
| AV.THRESHOLD TEMP | 5.8 | 9.9 | 8.8 | 8.2 | 5.7 | 5.8 | 7.7 |
| GL RAD OVER THRSH | 349.9 | 295.3 | 226.6 | 176.4 | 137.6 | 103.2 | 78.5 |
| TIME AVG.TEMP | 6.9 | 8.3 | 7.3 | 6.4 | 6.1 | 6.1 | 6.2 |
| TOTAL TIME MSEC | 2.6784 | 1.2276 | 0.7614 | 0.4788 | 0.4032 | 0.3420 | 0.2826 |
| GLOBAL RAD TOTAL | 349.9 | 349.9 | 329.2 | 291.1 | 273.0 | 252.4 | 226.5 |
| DIRECT RAD TOTAL | 243.7 | 243.7 | 241.5 | 231.7 | 226.3 | 215.5 | 195.6 |
| RAD AV TIME MSEC | 0.0864 | 0.0396 | 0.0346 | 0.0302 | 0.0292 | 0.0274 | 0.0238 |
| DEG-C DAYS | 353. | 143. | 97. | 66. | 57. | 48. | 39. |

### FEB

| SOLAR RAD GROUP | ALL | 0+ | 95+ | 189+ | 284+ | 378+ | 473+ |
|---|---|---|---|---|---|---|---|
| AV.THRSH.RAD.LEVL | 0. | 44. | 141. | 230. | 335. | 436. | 518. |
| AV.THRESHOLD TEMP | 4.5 | 7.1 | 6.8 | 9.3 | 11.8 | 7.5 | 10.5 |
| GL RAD OVER THRSH | 392.7 | 341.2 | 257.4 | 205.7 | 156.8 | 115.5 | 89.2 |
| TIME AVG.TEMP | 6.1 | 7.9 | 8.2 | 8.8 | 8.7 | 8.2 | 8.4 |
| TOTAL TIME MSEC | 2.4624 | 1.1682 | 0.8676 | 0.5814 | 0.4662 | 0.4014 | 0.3294 |
| GLOBAL RAD TOTAL | 392.7 | 392.7 | 379.4 | 339.2 | 312.7 | 291.1 | 257.6 |
| DIRECT RAD TOTAL | 252.7 | 252.7 | 251.1 | 241.5 | 234.9 | 227.6 | 209.6 |
| RAD AV TIME MSEC | 0.0864 | 0.0410 | 0.0360 | 0.0313 | 0.0288 | 0.0281 | 0.0252 |
| DEG-C DAYS | 349. | 141. | 102. | 64. | 53. | 48. | 39. |

### MAR

| SOLAR RAD GROUP | ALL | 0+ | 95+ | 189+ | 284+ | 378+ | 473+ |
|---|---|---|---|---|---|---|---|
| AV.THRSH.RAD.LEVL | 0. | 31. | 145. | 230. | 324. | 424. | 524. |
| AV.THRESHOLD TEMP | 8.1 | 9.6 | 9.9 | 11.0 | 12.1 | 14.3 | 11.7 |
| GL RAD OVER THRSH | 570.0 | 526.4 | 401.7 | 324.1 | 260.0 | 199.6 | 146.9 |
| TIME AVG.TEMP | 9.9 | 11.4 | 12.0 | 12.4 | 12.9 | 13.0 | 12.8 |
| TOTAL TIME MSEC | 2.6784 | 1.4274 | 1.0926 | 0.9054 | 0.6876 | 0.5994 | 0.5310 |
| GLOBAL RAD TOTAL | 570.0 | 570.0 | 559.8 | 532.7 | 482.5 | 454.0 | 424.9 |
| DIRECT RAD TOTAL | 372.2 | 372.2 | 369.7 | 365.3 | 348.3 | 338.4 | 327.4 |
| RAD AV TIME MSEC | 0.0864 | 0.0464 | 0.0374 | 0.0353 | 0.0313 | 0.0292 | 0.0281 |
| DEG-C DAYS | 262. | 114. | 82. | 64. | 46. | 39. | 36. |

### JUL

| SOLAR RAD GROUP | ALL | 0+ | 95+ | 189+ | 284+ | 378+ | 473+ |
|---|---|---|---|---|---|---|---|
| AV.THRSH.RAD.LEVL | 0. | 36. | 151. | 241. | 332. | 427. | 523. |
| AV.THRESHOLD TEMP | 21.9 | 23.0 | 23.2 | 24.9 | 25.6 | 26.7 | 27.4 |
| GL RAD OVER THRSH | 505.7 | 454.0 | 317.9 | 225.9 | 153.3 | 94.8 | 52.1 |
| TIME AVG.TEMP | 23.9 | 25.7 | 26.3 | 26.8 | 27.3 | 27.8 | 28.2 |
| TOTAL TIME MSEC | 2.6784 | 1.4508 | 1.1808 | 1.0170 | 0.7992 | 0.6192 | 0.4428 |
| GLOBAL RAD TOTAL | 505.7 | 505.7 | 496.1 | 471.3 | 418.8 | 359.0 | 283.7 |
| DIRECT RAD TOTAL | 204.5 | 204.5 | 203.7 | 201.0 | 190.2 | 178.1 | 150.4 |
| RAD AV TIME MSEC | 0.0864 | 0.0468 | 0.0385 | 0.0349 | 0.0292 | 0.0259 | 0.0209 |
| DEG-C DAYS | 0. | 0. | 0. | 0. | 0. | 0. | 0. |

### AUG

| SOLAR RAD GROUP | ALL | 0+ | 95+ | 189+ | 284+ | 378+ | 473+ |
|---|---|---|---|---|---|---|---|
| AV.THRSH.RAD.LEVL | 0. | 34. | 156. | 230. | 340. | 433. | 517. |
| AV.THRESHOLD TEMP | 22.4 | 23.4 | 24.4 | 25.3 | 26.5 | 26.9 | 28.5 |
| GL RAD OVER THRSH | 538.1 | 489.7 | 348.5 | 273.7 | 182.4 | 121.0 | 81.6 |
| TIME AVG.TEMP | 24.4 | 26.2 | 26.9 | 27.3 | 27.7 | 28.1 | 28.5 |
| TOTAL TIME MSEC | 2.6784 | 1.4454 | 1.1502 | 1.0188 | 0.8316 | 0.6552 | 0.4698 |
| GLOBAL RAD TOTAL | 538.2 | 538.2 | 528.3 | 507.7 | 464.7 | 404.8 | 324.5 |
| DIRECT RAD TOTAL | 286.1 | 286.1 | 284.6 | 280.8 | 264.9 | 247.6 | 209.9 |
| RAD AV TIME MSEC | 0.0864 | 0.0468 | 0.0385 | 0.0356 | 0.0302 | 0.0274 | 0.0227 |
| DEG-C DAYS | 0. | 0. | 0. | 0. | 0. | 0. | 0. |

### SEP

| SOLAR RAD GROUP | ALL | 0+ | 95+ | 189+ | 284+ | 378+ | 473+ |
|---|---|---|---|---|---|---|---|
| AV.THRSH.RAD.LEVL | 0. | 30. | 149. | 226. | 332. | 426. | 533. |
| AV.THRESHOLD TEMP | 20.8 | 21.5 | 22.4 | 24.0 | 24.5 | 25.7 | 26.7 |
| GL RAD OVER THRSH | 509.1 | 467.2 | 338.5 | 268.3 | 191.0 | 132.8 | 83.5 |
| TIME AVG.TEMP | 22.8 | 25.4 | 25.9 | 26.3 | 26.4 | 26.8 | 27.2 |
| TOTAL TIME MSEC | 2.5920 | 1.4004 | 1.0800 | 0.9126 | 0.7272 | 0.6228 | 0.4608 |
| GLOBAL RAD TOTAL | 509.1 | 509.1 | 499.5 | 474.6 | 432.6 | 397.9 | 329.0 |
| DIRECT RAD TOTAL | 309.6 | 309.6 | 307.7 | 300.8 | 285.0 | 272.6 | 234.7 |
| RAD AV TIME MSEC | 0.0864 | 0.0468 | 0.0378 | 0.0338 | 0.0295 | 0.0270 | 0.0223 |
| DEG-C DAYS | 0. | 1. | 1. | 1. | 1. | 0. | 0. |

**APR**

| SOLAR RAD GROUP | ALL | 0+ | 95+ | 189+ | 284+ | 378+ | 473+ |
|---|---|---|---|---|---|---|---|
| AV.THRSH.RAD.LEVL | 0. | 29. | 150. | 232. | 320. | 430. | 520. |
| AV.THRESHOLD TEMP | 13.4 | 14.5 | 15.8 | 16.5 | 18.9 | 20.1 | 20.6 |
| GL RAD OVER THRSH | 638.0 | 597.9 | 460.4 | 373.7 | 300.8 | 221.8 | 164.7 |
| TIME AVG.TEMP | 16.1 | 18.3 | 19.2 | 19.5 | 20.3 | 20.5 | 20.6 |
| TOTAL TIME MSEC | 2.5920 | 1.4022 | 1.1358 | 1.0494 | 0.8334 | 0.6840 | 0.6372 |
| GLOBAL RAD TOTAL | 638.0 | 638.0 | 630.4 | 617.5 | 567.3 | 529.9 | 495.8 |
| DIRECT RAD TOTAL | 391.2 | 391.2 | 389.2 | 388.0 | 371.0 | 359.0 | 347.6 |
| RAD AV TIME MSEC | 0.0364 | 0.0468 | 0.0385 | 0.0371 | 0.0320 | 0.0295 | 0.0277 |
| DEG-C DAYS | 89. | 32. | 20. | 17. | 11. | 8. | 7. |

**MAY**

| SOLAR RAD GROUP | ALL | 0+ | 95+ | 189+ | 284+ | 378+ | 473+ |
|---|---|---|---|---|---|---|---|
| AV.THRSH.RAD.LEVL | 0. | 33. | 152. | 236. | 329. | 429. | 517. |
| AV.THRESHOLD TEMP | 16.5 | 17.9 | 18.8 | 19.7 | 21.3 | 21.5 | 23.2 |
| GL RAD OVER THRSH | 582.4 | 533.8 | 393.2 | 303.3 | 225.8 | 157.5 | 111.1 |
| TIME AVG.TEMP | 18.9 | 20.9 | 21.6 | 21.9 | 22.6 | 22.8 | 23.2 |
| TOTAL TIME MSEC | 2.6784 | 1.4508 | 1.1880 | 1.0620 | 0.8334 | 0.6840 | 0.5292 |
| GLOBAL RAD TOTAL | 582.4 | 582.4 | 573.8 | 554.4 | 500.4 | 451.1 | 384.7 |
| DIRECT RAD TOTAL | 281.2 | 281.2 | 279.8 | 278.3 | 263.5 | 255.4 | 230.9 |
| RAD AV TIME MSEC | 0.0864 | 0.0468 | 0.0389 | 0.0364 | 0.0306 | 0.0284 | 0.0248 |
| DEG-C DAYS | 24. | 6. | 3. | 3. | 1. | 1. | 1. |

**JUN**

| SOLAR RAD GROUP | ALL | 0+ | 95+ | 189+ | 284+ | 378+ | 473+ |
|---|---|---|---|---|---|---|---|
| AV.THRSH.RAD.LEVL | 0. | 36. | 152. | 234. | 333. | 427. | 521. |
| AV.THRESHOLD TEMP | 20.8 | 22.3 | 22.4 | 24.2 | 25.0 | 26.0 | 27.9 |
| GL RAD OVER THRSH | 580.5 | 529.6 | 396.0 | 309.0 | 224.4 | 154.0 | 100.4 |
| TIME AVG.TEMP | 23.3 | 25.4 | 26.1 | 26.4 | 27.0 | 27.3 | 27.7 |
| TOTAL TIME MSEC | 2.5920 | 1.4040 | 1.1556 | 1.0602 | 0.8532 | 0.7470 | 0.5706 |
| GLOBAL RAD TOTAL | 580.5 | 580.5 | 571.5 | 557.0 | 508.6 | 473.2 | 397.8 |
| DIRECT RAD TOTAL | 301.6 | 301.6 | 300.3 | 298.0 | 281.9 | 272.5 | 239.1 |
| RAD AV TIME MSEC | 0.0864 | 0.0468 | 0.0392 | 0.0374 | 0.0313 | 0.0292 | 0.0241 |
| DEG-C DAYS | 1. | 0. | 0. | 0. | 0. | 0. | 0. |

**OCT**

| SOLAR RAD GROUP | ALL | 0+ | 95+ | 189+ | 284+ | 378+ | 473+ |
|---|---|---|---|---|---|---|---|
| AV.THRSH.RAD.LEVL | 0. | 38. | 144. | 235. | 339. | 428. | 519. |
| AV.THRESHOLD TEMP | 15.1 | 17.1 | 18.3 | 19.3 | 20.1 | 19.3 | 22.0 |
| GL RAD OVER THRSH | 472.2 | 422.2 | 314.6 | 246.4 | 184.5 | 139.0 | 101.5 |
| TIME AVG.TEMP | 17.2 | 19.4 | 20.1 | 20.7 | 21.1 | 21.3 | 21.8 |
| TOTAL TIME MSEC | 2.6784 | 1.3302 | 1.0080 | 0.7524 | 0.5940 | 0.5130 | 0.4104 |
| GLOBAL RAD TOTAL | 472.2 | 472.2 | 460.1 | 423.2 | 386.0 | 358.5 | 314.6 |
| DIRECT RAD TOTAL | 312.3 | 312.3 | 309.8 | 299.8 | 287.5 | 274.1 | 247.9 |
| RAD AV TIME MSEC | 0.0364 | 0.0439 | 0.0364 | 0.0320 | 0.0292 | 0.0277 | 0.0245 |
| DEG-C DAYS | 57. | 14. | 8. | 6. | 4. | 4. | 2. |

**NOV**

| SOLAR RAD GROUP | ALL | 0+ | 95+ | 189+ | 284+ | 378+ | 473+ |
|---|---|---|---|---|---|---|---|
| AV.THRSH.RAD.LEVL | 0. | 47. | 141. | 231. | 332. | 430. | 509. |
| AV.THRESHOLD TEMP | 8.1 | 9.8 | 9.9 | 12.2 | 12.2 | 13.4 | 16.3 |
| GL RAD OVER THRSH | 449.2 | 393.5 | 306.4 | 246.2 | 168.6 | 139.8 | 108.0 |
| TIME AVG.TEMP | 9.8 | 11.9 | 12.5 | 13.5 | 13.7 | 13.9 | 14.2 |
| TOTAL TIME MSEC | 2.5920 | 1.1880 | 0.9216 | 0.6678 | 0.5724 | 0.4986 | 0.4014 |
| GLOBAL RAD TOTAL | 449.2 | 449.2 | 436.7 | 400.8 | 378.8 | 354.3 | 312.5 |
| DIRECT RAD TOTAL | 328.5 | 328.5 | 325.7 | 312.2 | 304.8 | 292.6 | 263.2 |
| RAD AV TIME MSEC | 0.0664 | 0.0396 | 0.0353 | 0.0299 | 0.0292 | 0.0277 | 0.0241 |
| DEG-C DAYS | 258. | 92. | 66. | 40. | 32. | 27. | 21. |

**DEC**

| SOLAR RAD GROUP | ALL | 0+ | 95+ | 189+ | 284+ | 378+ | 473+ |
|---|---|---|---|---|---|---|---|
| AV.THRSH.RAD.LEVL | 0. | 44. | 133. | 237. | 336. | 430. | 525. |
| AV.THRESHOLD TEMP | 8.6 | 11.8 | 11.7 | 13.3 | 8.7 | 13.4 | 15.0 |
| GL RAD OVER THRSH | 279.3 | 225.1 | 165.0 | 123.1 | 91.2 | 66.4 | 47.7 |
| TIME AVG.TEMP | 10.0 | 11.6 | 11.4 | 11.3 | 10.8 | 11.2 | 10.5 |
| TOTAL TIME MSEC | 2.6784 | 1.2258 | 0.6768 | 0.4014 | 0.3222 | 0.2646 | 0.1980 |
| GLOBAL RAD TOTAL | 279.3 | 279.3 | 255.0 | 218.4 | 199.6 | 180.2 | 151.6 |
| DIRECT RAD TOTAL | 171.1 | 171.1 | 167.8 | 160.7 | 156.2 | 146.3 | 130.7 |
| RAD AV TIME MSEC | 0.0864 | 0.0396 | 0.0324 | 0.0292 | 0.0284 | 0.0256 | 0.0223 |
| DEG-C DAYS | 259. | 98. | 56. | 33. | 28. | 22. | 18. |

**ALL MONTHS**

| SOLAR RAD GROUP | ALL | 0+ | 95+ | 189+ | 284+ | 378+ | 473+ |
|---|---|---|---|---|---|---|---|
| TIME AVG.TEMP | 15.8 | 18.1 | 19.0 | 20.1 | 20.5 | 20.6 | 20.4 |
| TOTAL TIME MSEC | 31.5791 | 16.1208 | 12.2184 | 9.9072 | 7.9236 | 6.6636 | 5.2632 |
| GLOBAL RAD TOTAL | 5867.2 | 5867.1 | 5719.5 | 5387.9 | 4925.0 | 4506.5 | 3905.4 |
| DIRECT RAD TOTAL | 3454.7 | 3454.7 | 3430.9 | 3358.1 | 3214.4 | 3079.6 | 2766.9 |
| RAD AV TIME MSEC | 0.0864 | 0.0448 | 0.0374 | 0.0343 | 0.0302 | 0.0280 | 0.0246 |
| DEG-C DAYS | 1653. | 641. | 435. | 294. | 233. | 198. | 163. |

STATISTICS FOR ATLANTA,GA    1971-1972    RSV020    FILE  4

RADIATION UNITS ARE WATTS/SQ.M OR MJOULES/SQ.M
SLOPE = 50.0 DEG    AZIMUTH = 180.0 DEG
DEGREE-DAY BASE TEMP = 18.3 DEGREES C

### JAN

| SOLAR RAD GROUP | ALL | 0+ | 95+ | 189+ | 284+ | 378+ | 473+ |
|---|---|---|---|---|---|---|---|
| AV.THRSH.RAD.LEVL | 0. | 44. | 133. | 236. | 331. | 427. | 514. |
| AV.THRESHOLD TEMP | 5.8 | 9.9 | 8.7 | 7.9 | 8.6 | 3.8 | 8.2 |
| GL RAD OVER THRSH | 356.2 | 302.1 | 235.5 | 185.3 | 148.2 | 114.1 | 88.3 |
| TIME AVG.TEMP | 6.9 | 8.3 | 7.2 | 6.4 | 6.1 | 5.8 | 6.2 |
| TOTAL TIME MSEC | 2.6784 | 1.2276 | 0.7470 | 0.4806 | 0.3964 | 0.3564 | 0.2970 |
| GLOBAL RAD TOTAL | 356.2 | 356.2 | 335.0 | 299.5 | 279.4 | 266.3 | 240.9 |
| DIRECT RAD TOTAL | 253.9 | 253.9 | 251.6 | 242.2 | 235.4 | 229.1 | 209.7 |
| RAD AV TIME MSEC | 0.0864 | 0.0396 | 0.0349 | 0.0306 | 0.0288 | 0.0288 | 0.0248 |
| DEG-C DAYS | 353. | 143. | 96. | 67. | 56. | 52. | 42. |

### FEB

| SOLAR RAD GROUP | ALL | 0+ | 95+ | 189+ | 284+ | 378+ | 473+ |
|---|---|---|---|---|---|---|---|
| AV.THRSH.RAD.LEVL | 0. | 46. | 139. | 226. | 332. | 432. | 515. |
| AV.THRESHOLD TEMP | 4.6 | 7.0 | 6.8 | 8.7 | 12.3 | 8.5 | 9.2 |
| GL RAD OVER THRSH | 391.7 | 338.6 | 259.3 | 208.7 | 159.7 | 120.0 | 91.5 |
| TIME AVG.TEMP | 6.1 | 7.8 | 8.1 | 8.7 | 8.7 | 8.2 | 8.1 |
| TOTAL TIME MSEC | 2.4624 | 1.1556 | 0.8532 | 0.5814 | 0.4608 | 0.3996 | 0.3402 |
| GLOBAL RAD TOTAL | 391.8 | 391.7 | 377.8 | 340.1 | 312.8 | 292.5 | 266.8 |
| DIRECT RAD TOTAL | 257.2 | 257.2 | 255.6 | 247.3 | 238.6 | 231.1 | 217.1 |
| RAD AV TIME MSEC | 0.0864 | 0.0407 | 0.0360 | 0.0320 | 0.0288 | 0.0284 | 0.0263 |
| DEG-C DAYS | 349. | 141. | 102. | 65. | 52. | 48. | 41. |

### MAR

| SOLAR RAD GROUP | ALL | 0+ | 95+ | 189+ | 284+ | 378+ | 473+ |
|---|---|---|---|---|---|---|---|
| AV.THRSH.RAD.LEVL | 0. | 31. | 147. | 233. | 330. | 426. | 521. |
| AV.THRESHOLD TEMP | 8.1 | 9.7 | 9.9 | 11.4 | 12.4 | 12.5 | 12.5 |
| GL RAD OVER THRSH | 554.0 | 510.1 | 384.3 | 309.1 | 244.6 | 188.2 | 139.0 |
| TIME AVG.TEMP | 9.9 | 11.4 | 12.0 | 12.5 | 12.8 | 12.9 | 12.9 |
| TOTAL TIME MSEC | 2.6784 | 1.4238 | 1.0854 | 0.8743 | 0.6660 | 0.5868 | 0.5184 |
| GLOBAL RAD TOTAL | 554.1 | 554.0 | 543.6 | 512.7 | 464.1 | 438.0 | 408.9 |
| DIRECT RAD TOTAL | 363.0 | 363.0 | 360.5 | 355.2 | 337.5 | 329.0 | 315.5 |
| RAD AV TIME MSEC | 0.0864 | 0.0464 | 0.0374 | 0.0349 | 0.0310 | 0.0292 | 0.0277 |
| DEG-C DAYS | 262. | 114. | 82. | 61. | 44. | 39. | 34. |

### JUL

| SOLAR RAD GROUP | ALL | 0+ | 95+ | 189+ | 284+ | 378+ | 473+ |
|---|---|---|---|---|---|---|---|
| AV.THRSH.RAD.LEVL | 0. | 61. | 147. | 239. | 333. | 421. | 521. |
| AV.THRESHOLD TEMP | 21.8 | 24.0 | 24.1 | 24.9 | 25.9 | 26.6 | 27.8 |
| GL RAD OVER THRSH | 464.5 | 383.1 | 282.0 | 193.0 | 121.5 | 72.9 | 33.2 |
| TIME AVG.TEMP | 23.9 | 26.1 | 26.3 | 26.8 | 27.3 | 27.9 | 28.4 |
| TOTAL TIME MSEC | 2.6784 | 1.3392 | 1.1790 | 0.9612 | 0.7596 | 0.5526 | 0.3960 |
| GLOBAL RAD TOTAL | 464.5 | 464.5 | 454.7 | 422.0 | 374.6 | 305.7 | 239.7 |
| DIRECT RAD TOTAL | 176.2 | 176.2 | 175.8 | 171.2 | 163.5 | 146.1 | 124.1 |
| RAD AV TIME MSEC | 0.0864 | 0.0432 | 0.0385 | 0.0328 | 0.0281 | 0.0238 | 0.0202 |
| DEG-C DAYS | 0. | 0. | 0. | 0. | 0. | 0. | 0. |

### AUG

| SOLAR RAD GROUP | ALL | 0+ | 95+ | 189+ | 284+ | 378+ | 473+ |
|---|---|---|---|---|---|---|---|
| AV.THRSH.RAD.LEVL | 0. | 37. | 158. | 232. | 333. | 424. | 519. |
| AV.THRESHOLD TEMP | 22.3 | 23.7 | 24.5 | 25.6 | 26.7 | 27.3 | 28.2 |
| GL RAD OVER THRSH | 501.8 | 450.1 | 311.4 | 240.4 | 158.9 | 103.1 | 63.2 |
| TIME AVG.TEMP | 24.4 | 26.3 | 26.9 | 27.4 | 27.8 | 28.1 | 28.5 |
| TOTAL TIME MSEC | 2.6784 | 1.4148 | 1.1466 | 0.9522 | 0.8064 | 0.6138 | 0.4194 |
| GLOBAL RAD TOTAL | 501.9 | 501.8 | 492.0 | 461.4 | 427.6 | 363.4 | 280.9 |
| DIRECT RAD TOTAL | 259.0 | 259.0 | 257.8 | 249.3 | 240.3 | 220.9 | 183.7 |
| RAD AV TIME MSEC | 0.0864 | 0.0461 | 0.0385 | 0.0331 | 0.0299 | 0.0266 | 0.0223 |
| DEG-C DAYS | 0. | 0. | 0. | 0. | 0. | 0. | 0. |

### SEP

| SOLAR RAD GROUP | ALL | 0+ | 95+ | 189+ | 284+ | 378+ | 473+ |
|---|---|---|---|---|---|---|---|
| AV.THRSH.RAD.LEVL | 0. | 30. | 152. | 228. | 337. | 419. | 528. |
| AV.THRESHOLD TEMP | 20.8 | 21.4 | 23.0 | 23.9 | 25.1 | 25.3 | 26.8 |
| GL RAD OVER THRSH | 488.5 | 446.7 | 316.2 | 249.4 | 172.2 | 123.2 | 74.3 |
| TIME AVG.TEMP | 22.8 | 25.4 | 25.4 | 26.0 | 26.5 | 26.8 | 27.2 |
| TOTAL TIME MSEC | 2.5920 | 1.4004 | 1.0728 | 0.8766 | 0.7092 | 0.5922 | 0.4518 |
| GLOBAL RAD TOTAL | 483.5 | 483.5 | 478.7 | 449.0 | 410.9 | 371.5 | 312.6 |
| DIRECT RAD TOTAL | 296.2 | 296.2 | 294.4 | 285.5 | 271.8 | 255.2 | 223.0 |
| RAD AV TIME MSEC | 0.0864 | 0.0468 | 0.0378 | 0.0331 | 0.0292 | 0.0263 | 0.0220 |
| DEG-C DAYS | 1. | 1. | 1. | 1. | 0. | 0. | 0. |

## APR

| SOLAR RAD GROUP | ALL | 0+ | 95+ | 189+ | 284+ | 378+ | 473+ |
|---|---|---|---|---|---|---|---|
| AV.THRSH.RAD.LEVL | 0. | 29. | 158. | 232. | 330. | 432. | 521. |
| AV.THRESHOLD TEMP | 13.3 | 14.8 | 15.4 | 17.3 | 19.0 | 19.0 | 22.3 |
| GL RAD OVER THRSH | 600.1 | 560.2 | 413.9 | 339.3 | 262.7 | 191.2 | 140.1 |
| TIME AVG.TEMP | 16.1 | 18.4 | 19.2 | 19.7 | 20.4 | 20.6 | 20.9 |
| TOTAL TIME MSEC | 2.5920 | 1.3968 | 1.1322 | 1.0008 | 0.7848 | 0.7002 | 0.5778 |
| GLOBAL RAD TOTAL | 600.1 | 600.1 | 592.6 | 571.8 | 521.6 | 493.7 | 440.8 |
| DIRECT RAD TOTAL | 361.2 | 361.2 | 359.5 | 355.0 | 337.8 | 331.5 | 303.9 |
| RAD AV TIME MSEC | 0.0864 | 0.0468 | 0.0385 | 0.0356 | 0.0306 | 0.0292 | 0.0252 |
| DEG-C DAYS | 89. | 31. | 20. | 16. | 9. | 8. | 6. |

## MAY

| SOLAR RAD GROUP | ALL | 0+ | 95+ | 189+ | 284+ | 378+ | 473+ |
|---|---|---|---|---|---|---|---|
| AV.THRSH.RAD.LEVL | 0. | 52. | 153. | 234. | 332. | 420. | 525. |
| AV.THRESHOLD TEMP | 16.4 | 19.1 | 18.1 | 20.9 | 21.5 | 22.1 | 22.7 |
| GL RAD OVER THRSH | 534.8 | 464.2 | 344.7 | 264.5 | 186.6 | 130.3 | 80.6 |
| TIME AVG.TEMP | 18.9 | 21.3 | 21.6 | 22.3 | 22.6 | 22.9 | 23.2 |
| TOTAL TIME MSEC | 2.6784 | 1.3590 | 1.1808 | 0.9918 | 0.7974 | 0.6372 | 0.4752 |
| GLOBAL RAD TOTAL | 534.8 | 534.8 | 525.5 | 496.6 | 451.1 | 399.0 | 329.9 |
| DIRECT RAD TOTAL | 245.0 | 245.0 | 244.2 | 238.1 | 230.7 | 218.7 | 192.2 |
| RAD AV TIME MSEC | 0.0864 | 0.0439 | 0.0389 | 0.0342 | 0.0299 | 0.0274 | 0.0227 |
| DEG-C DAYS | 24. | 5. | 3. | 2. | 1. | 1. | 1. |

## JUN

| SOLAR RAD GROUP | ALL | 0+ | 95+ | 189+ | 284+ | 378+ | 473+ |
|---|---|---|---|---|---|---|---|
| AV.THRSH.RAD.LEVL | 0. | 60. | 145. | 233. | 342. | 425. | 523. |
| AV.THRESHOLD TEMP | 20.7 | 24.4 | 23.2 | 24.4 | 25.7 | 26.7 | 28.3 |
| GL RAD OVER THRSH | 534.5 | 446.2 | 347.1 | 262.6 | 172.7 | 118.3 | 68.1 |
| TIME AVG.TEMP | 23.3 | 25.9 | 26.1 | 26.7 | 27.1 | 27.5 | 27.7 |
| TOTAL TIME MSEC | 2.5920 | 1.2960 | 1.1664 | 0.9648 | 0.8280 | 0.6498 | 0.5112 |
| GLOBAL RAD TOTAL | 524.5 | 524.5 | 516.7 | 487.4 | 455.5 | 394.6 | 335.7 |
| DIRECT RAD TOTAL | 255.8 | 255.8 | 255.3 | 247.8 | 242.1 | 217.1 | 196.9 |
| RAD AV TIME MSEC | 0.0864 | 0.0432 | 0.0396 | 0.0342 | 0.0306 | 0.0256 | 0.0223 |
| DEG-C DAYS | 1. | 0. | 0. | 0. | 0. | 0. | 0. |

## OCT

| SOLAR RAD GROUP | ALL | 0+ | 95+ | 189+ | 284+ | 378+ | 473+ |
|---|---|---|---|---|---|---|---|
| AV.THRSH.RAD.LEVL | 0. | 40. | 145. | 232. | 342. | 431. | 523. |
| AV.THRESHOLD TEMP | 15.1 | 17.1 | 18.3 | 19.6 | 20.1 | 19.4 | 21.3 |
| GL RAD OVER THRSH | 466.0 | 413.3 | 309.9 | 244.4 | 179.9 | 134.9 | 97.7 |
| TIME AVG.TEMP | 17.2 | 19.4 | 20.2 | 20.8 | 21.1 | 21.3 | 21.7 |
| TOTAL TIME MSEC | 2.6784 | 1.3230 | 0.9882 | 0.7488 | 0.5886 | 0.5022 | 0.4050 |
| GLOBAL RAD TOTAL | 466.1 | 466.0 | 452.7 | 418.1 | 380.9 | 351.4 | 309.5 |
| DIRECT RAD TOTAL | 311.9 | 311.9 | 309.5 | 300.0 | 286.2 | 272.1 | 245.7 |
| RAD AV TIME MSEC | 0.0864 | 0.0439 | 0.0364 | 0.0324 | 0.0292 | 0.0277 | 0.0245 |
| DEG-C DAYS | 57. | 14. | 8. | 6. | 4. | 3. | 2. |

## NOV

| SOLAR RAD GROUP | ALL | 0+ | 95+ | 189+ | 284+ | 378+ | 473+ |
|---|---|---|---|---|---|---|---|
| AV.THRSH.RAD.LEVL | 0. | 46. | 141. | 228. | 333. | 429. | 509. |
| AV.THRESHOLD TEMP | 8.1 | 9.6 | 9.9 | 12.2 | 13.0 | 11.9 | 16.1 |
| GL RAD OVER THRSH | 457.7 | 403.2 | 315.3 | 257.7 | 198.0 | 149.7 | 116.0 |
| TIME AVG.TEMP | 9.8 | 11.9 | 12.6 | 13.6 | 13.8 | 13.9 | 14.3 |
| TOTAL TIME MSEC | 2.5920 | 1.1880 | 0.9198 | 0.6660 | 0.5706 | 0.5022 | 0.4212 |
| GLOBAL RAD TOTAL | 457.7 | 457.7 | 445.4 | 409.5 | 387.7 | 365.3 | 330.3 |
| DIRECT RAD TOTAL | 341.0 | 341.0 | 338.4 | 324.6 | 317.2 | 305.3 | 281.3 |
| RAD AV TIME MSEC | 0.0864 | 0.0396 | 0.0356 | 0.0306 | 0.0295 | 0.0281 | 0.0259 |
| DEG-C DAYS | 258. | 92. | 65. | 40. | 32. | 27. | 22. |

## DEC

| SOLAR RAD GROUP | ALL | 0+ | 95+ | 189+ | 284+ | 378+ | 473+ |
|---|---|---|---|---|---|---|---|
| AV.THRSH.RAD.LEVL | 0. | 44. | 133. | 237. | 341. | 432. | 521. |
| AV.THRESHOLD TEMP | 8.6 | 11.8 | 11.8 | 13.2 | 8.5 | 13.8 | 14.3 |
| GL RAD OVER THRSH | 282.6 | 228.3 | 170.2 | 129.3 | 96.1 | 72.1 | 54.3 |
| TIME AVG.TEMP | 10.0 | 11.6 | 11.4 | 11.2 | 10.8 | 11.3 | 10.5 |
| TOTAL TIME MSEC | 2.6784 | 1.2258 | 0.6534 | 0.3924 | 0.3204 | 0.2628 | 0.2016 |
| GLOBAL RAD TOTAL | 282.6 | 282.6 | 257.2 | 222.5 | 205.4 | 185.8 | 159.3 |
| DIRECT RAD TOTAL | 179.2 | 179.2 | 175.8 | 168.3 | 163.9 | 153.7 | 138.4 |
| RAD AV TIME MSEC | 0.0864 | 0.0396 | 0.0328 | 0.0292 | 0.0288 | 0.0256 | 0.0227 |
| DEG-C DAYS | 259. | 98. | 54. | 33. | 28. | 22. | 18. |

## ALL MONTHS

| SOLAR RAD GROUP | ALL | 0+ | 95+ | 189+ | 284+ | 378+ | 473+ |
|---|---|---|---|---|---|---|---|
| TIME AVG.TEMP | 15.8 | 18.0 | 19.1 | 20.1 | 20.4 | 20.3 | 20.1 |
| TOTAL TIME MSEC | 31.5791 | 15.7500 | 12.1248 | 9.4914 | 7.6878 | 6.3558 | 5.0148 |
| GLOBAL RAD TOTAL | 5622.7 | 5622.5 | 5472.0 | 5091.3 | 4671.7 | 4225.8 | 3655.4 |
| DIRECT RAD TOTAL | 3299.6 | 3278.1 | 3278.3 | 3184.5 | 3064.9 | 2909.8 | 2631.5 |
| RAD AV TIME MSEC | 0.0364 | 0.0437 | 0.0374 | 0.0331 | 0.0297 | 0.0273 | 0.0241 |
| DEG-C DAYS | 1653. | 639. | 431. | 290. | 229. | 199. | 166. |

STATISTICS FOR ATLANTA,GA    1971-1972    RSV020    FILE   4

RADIATION UNITS ARE WATTS/SQ.M OR MJOULES/SQ.M
SLOPE = 60.0 DEG   AZIMUTH = 180.0 DEG
DEGREE-DAY BASE TEMP = 18.3 DEGREES C

**JAN**

| SOLAR RAD GROUP | ALL | 0+ | 95+ | 189+ | 284+ | 378+ | 473+ |
|---|---|---|---|---|---|---|---|
| AV.THRSH.RAD.LEVL | 0. | 43. | 132. | 238. | 328. | 427. | 520. |
| AV.THRESHOLD TEMP | 5.8 | 9.9 | 8.7 | 7.9 | 7.8 | 4.7 | 7.1 |
| GL RAD OVER THRSH | 354.1 | 300.7 | 236.6 | 186.3 | 150.8 | 115.5 | 87.8 |
| TIME AVG.TEMP | 6.9 | 8.3 | 7.2 | 6.3 | 6.0 | 5.8 | 6.1 |
| TOTAL TIME MSEC | 2.6784 | 1.2276 | 0.7290 | 0.4734 | 0.3906 | 0.3582 | 0.2968 |
| GLOBAL RAD TOTAL | 354.1 | 354.1 | 332.4 | 298.8 | 279.1 | 268.5 | 243.1 |
| DIRECT RAD TOTAL | 256.5 | 256.5 | 254.1 | 245.0 | 237.3 | 231.7 | 212.2 |
| RAD AV TIME MSEC | 0.0864 | 0.0396 | 0.0349 | 0.0513 | 0.0295 | 0.0288 | 0.0256 |
| DEG-C DAYS | 353. | 143. | 94. | 66. | 56. | 52. | 43. |

**FEB**

| SOLAR RAD GROUP | ALL | 0+ | 95+ | 189+ | 284+ | 378+ | 473+ |
|---|---|---|---|---|---|---|---|
| AV.THRSH.RAD.LEVL | 0. | 47. | 140. | 226. | 331. | 433. | 514. |
| AV.THRESHOLD TEMP | 4.6 | 7.1 | 6.7 | 9.0 | 12.9 | 9.1 | 8.2 |
| GL RAD OVER THRSH | 382.4 | 328.4 | 251.0 | 202.4 | 154.8 | 115.1 | 87.6 |
| TIME AVG.TEMP | 6.1 | 7.8 | 8.1 | 8.8 | 8.7 | 8.0 | 7.8 |
| TOTAL TIME MSEC | 2.4624 | 1.1574 | 0.8334 | 0.5616 | 0.4554 | 0.3888 | 0.3384 |
| GLOBAL RAD TOTAL | 382.5 | 382.4 | 367.3 | 329.4 | 305.4 | 283.4 | 261.5 |
| DIRECT RAD TOTAL | 254.1 | 254.1 | 252.4 | 244.1 | 234.3 | 226.6 | 215.3 |
| RAD AV TIME MSEC | 0.0864 | 0.0407 | 0.0360 | 0.0324 | 0.0288 | 0.0284 | 0.0270 |
| DEG-C DAYS | 349. | 141. | 99. | 63. | 52. | 47. | 42. |

**MAR**

| SOLAR RAD GROUP | ALL | 0+ | 95+ | 189+ | 284+ | 378+ | 473+ |
|---|---|---|---|---|---|---|---|
| AV.THRSH.RAD.LEVL | 0. | 32. | 148. | 231. | 330. | 434. | 520. |
| AV.THRESHOLD TEMP | 8.1 | 9.6 | 10.0 | 11.8 | 12.8 | 11.9 | 12.7 |
| GL RAD OVER THRSH | 526.0 | 480.9 | 356.8 | 287.0 | 222.3 | 162.9 | 122.0 |
| TIME AVG.TEMP | 9.9 | 11.4 | 12.1 | 12.6 | 12.8 | 13.0 | 13.0 |
| TOTAL TIME MSEC | 2.6784 | 1.4238 | 1.0710 | 0.8406 | 0.6516 | 0.5688 | 0.4768 |
| GLOBAL RAD TOTAL | 526.0 | 526.0 | 514.8 | 480.8 | 437.2 | 409.9 | 370.8 |
| DIRECT RAD TOTAL | 342.7 | 342.7 | 340.1 | 334.5 | 317.4 | 307.7 | 287.1 |
| RAD AV TIME MSEC | 0.0864 | 0.0464 | 0.0374 | 0.0349 | 0.0306 | 0.0292 | 0.0263 |
| DEG-C DAYS | 262. | 114. | 80. | 58. | 44. | 38. | 32. |

**JUL**

| SOLAR RAD GROUP | ALL | 0+ | 95+ | 189+ | 284+ | 378+ | 473+ |
|---|---|---|---|---|---|---|---|
| AV.THRSH.RAD.LEVL | 0. | 43. | 136. | 242. | 328. | 430. | 516. |
| AV.THRESHOLD TEMP | 22.1 | 24.0 | 23.6 | 25.3 | 26.0 | 27.4 | 28.0 |
| GL RAD OVER THRSH | 392.9 | 340.0 | 244.9 | 149.7 | 90.2 | 42.4 | 16.6 |
| TIME AVG.TEMP | 23.9 | 26.1 | 26.6 | 26.9 | 27.4 | 28.1 | 28.5 |
| TOTAL TIME MSEC | 2.6784 | 1.2294 | 1.0170 | 0.9000 | 0.6930 | 0.4698 | 0.2983 |
| GLOBAL RAD TOTAL | 392.9 | 392.9 | 383.7 | 367.8 | 317.6 | 244.3 | 170.8 |
| DIRECT RAD TOTAL | 141.1 | 141.1 | 140.1 | 139.3 | 131.2 | 112.2 | 85.8 |
| RAD AV TIME MSEC | 0.0864 | 0.0396 | 0.0331 | 0.0306 | 0.0270 | 0.0209 | 0.0173 |
| DEG-C DAYS | 0. | 0. | 0. | 0. | 0. | 0. | 0. |

**AUG**

| SOLAR RAD GROUP | ALL | 0+ | 95+ | 189+ | 284+ | 378+ | 473+ |
|---|---|---|---|---|---|---|---|
| AV.THRSH.RAD.LEVL | 0. | 49. | 145. | 244. | 331. | 421. | 520. |
| AV.THRESHOLD TEMP | 22.3 | 24.8 | 24.6 | 26.2 | 26.8 | 27.7 | 26.4 |
| GL RAD OVER THRSH | 455.5 | 389.3 | 280.1 | 191.2 | 125.7 | 77.7 | 41.7 |
| TIME AVG.TEMP | 24.4 | 26.6 | 26.9 | 27.6 | 27.8 | 28.3 | 28.6 |
| TOTAL TIME MSEC | 2.6784 | 1.3446 | 1.1376 | 0.9036 | 0.7524 | 0.5328 | 0.3636 |
| GLOBAL RAD TOTAL | 455.6 | 455.5 | 445.3 | 411.4 | 374.5 | 301.9 | 230.7 |
| DIRECT RAD TOTAL | 224.0 | 224.0 | 223.2 | 214.3 | 205.9 | 178.2 | 151.1 |
| RAD AV TIME MSEC | 0.0864 | 0.0436 | 0.0382 | 0.0317 | 0.0288 | 0.0238 | 0.0216 |
| DEG-C DAYS | 0. | 0. | 0. | 0. | 0. | 0. | 0. |

**SEP**

| SOLAR RAD GROUP | ALL | 0+ | 95+ | 189+ | 284+ | 378+ | 473+ |
|---|---|---|---|---|---|---|---|
| AV.THRSH.RAD.LEVL | 0. | 30. | 150. | 231. | 337. | 413. | 521. |
| AV.THRESHOLD TEMP | 20.8 | 21.5 | 22.7 | 24.7 | 25.4 | 25.0 | 27.3 |
| GL RAD OVER THRSH | 457.8 | 416.5 | 287.9 | 220.5 | 147.9 | 105.8 | 59.7 |
| TIME AVG.TEMP | 22.8 | 24.5 | 25.4 | 26.2 | 26.6 | 26.8 | 27.3 |
| TOTAL TIME MSEC | 2.5920 | 1.3968 | 1.0656 | 0.8334 | 0.6858 | 0.5508 | 0.4302 |
| GLOBAL RAD TOTAL | 457.8 | 457.8 | 448.0 | 413.1 | 379.0 | 333.5 | 283.7 |
| DIRECT RAD TOTAL | 273.8 | 273.8 | 272.0 | 261.0 | 249.9 | 229.2 | 202.1 |
| RAD AV TIME MSEC | 0.0864 | 0.0468 | 0.0374 | 0.0320 | 0.0284 | 0.0252 | 0.0216 |
| DEG-C DAYS | 1. | 1. | 1. | 1. | 1. | 0. | 0. |

## OCT

| SOLAR RAD GROUP | ALL | 0+ | 95+ | 189+ | 284+ | 378+ | 473+ |
|---|---|---|---|---|---|---|---|
| AV.THRSH.RAD.LEVL | 0. | 39. | 143. | 230. | 336. | 430. | 521. |
| AV.THRESHOLD TEMP | 15.2 | 16.9 | 18.3 | 19.7 | 20.2 | 19.7 | 21.0 |
| GL RAD OVER THRSH | 449.7 | 398.7 | 296.3 | 232.9 | 170.7 | 125.0 | 89.0 |
| TIME AVG.TEMP | 17.2 | 19.3 | 20.2 | 20.8 | 21.1 | 21.3 | 21.7 |
| TOTAL TIME MSEC | 2.6784 | 1.3230 | 0.9828 | 0.7290 | 0.5832 | 0.4896 | 0.3924 |
| GLOBAL RAD TOTAL | 449.7 | 449.7 | 436.6 | 400.3 | 366.9 | 335.4 | 293.6 |
| DIRECT RAD TOTAL | 302.2 | 302.2 | 299.6 | 289.5 | 275.9 | 261.1 | 234.2 |
| RAD AV TIME MSEC | 0.0864 | 0.0439 | 0.0364 | 0.0324 | 0.0292 | 0.0274 | 0.0241 |
| DEG-C DAYS | 57. | 14. | 8. | 6. | 4. | 3. | 2. |

## NOV

| SOLAR RAD GROUP | ALL | 0+ | 95+ | 189+ | 284+ | 378+ | 473+ |
|---|---|---|---|---|---|---|---|
| AV.THRSH.RAD.LEVL | 0. | 46. | 141. | 227. | 328. | 430. | 522. |
| AV.THRESHOLD TEMP | 8.1 | 9.9 | 9.7 | 12.5 | 12.8 | 12.6 | 15.8 |
| GL RAD OVER THRSH | 454.9 | 400.4 | 313.9 | 257.8 | 200.3 | 148.9 | 111.0 |
| TIME AVG.TEMP | 9.8 | 11.9 | 12.5 | 13.6 | 13.7 | 13.8 | 14.1 |
| TOTAL TIME MSEC | 2.5920 | 1.1880 | 0.9072 | 0.6570 | 0.5670 | 0.5040 | 0.4140 |
| GLOBAL RAD TOTAL | 454.9 | 454.9 | 442.0 | 406.7 | 386.3 | 365.6 | 326.9 |
| DIRECT RAD TOTAL | 343.0 | 343.0 | 340.5 | 327.4 | 319.3 | 307.0 | 281.0 |
| RAD AV TIME MSEC | 0.0864 | 0.0396 | 0.0356 | 0.0310 | 0.0299 | 0.0284 | 0.0259 |
| DEG-C DAYS | 258. | 92. | 64. | 39. | 32. | 28. | 22. |

## DEC

| SOLAR RAD GROUP | ALL | 0+ | 95+ | 189+ | 284+ | 378+ | 473+ |
|---|---|---|---|---|---|---|---|
| AV.THRSH.RAD.LEVL | 0. | 44. | 131. | 232. | 341. | 423. | 514. |
| AV.THRESHOLD TEMP | 8.6 | 11.8 | 11.9 | 12.6 | 9.8 | 13.7 | 13.4 |
| GL RAD OVER THRSH | 280.1 | 226.3 | 171.6 | 132.2 | 97.7 | 76.4 | 57.3 |
| TIME AVG.TEMP | 10.0 | 11.6 | 11.4 | 11.2 | 10.8 | 11.1 | 10.4 |
| TOTAL TIME MSEC | 2.6784 | 1.2240 | 0.6300 | 0.3868 | 0.3168 | 0.2610 | 0.2088 |
| GLOBAL RAD TOTAL | 280.1 | 280.1 | 254.0 | 222.4 | 205.7 | 186.7 | 164.6 |
| DIRECT RAD TOTAL | 182.1 | 182.1 | 178.8 | 171.3 | 166.8 | 157.0 | 143.9 |
| RAD AV TIME MSEC | 0.0864 | 0.0396 | 0.0328 | 0.0295 | 0.0284 | 0.0259 | 0.0238 |
| DEG-C DAYS | 259. | 98. | 52. | 33. | 28. | 22. | 19. |

## APR

| SOLAR RAD GROUP | ALL | 0+ | 95+ | 189+ | 284+ | 378+ | 473+ |
|---|---|---|---|---|---|---|---|
| AV.THRSH.RAD.LEVL | 0. | 36. | 154. | 228. | 338. | 431. | 522. |
| AV.THRESHOLD TEMP | 13.3 | 15.0 | 15.1 | 18.9 | 18.8 | 20.9 | 21.1 |
| GL RAD OVER THRSH | 549.6 | 502.1 | 368.4 | 293.9 | 215.2 | 154.9 | 107.3 |
| TIME AVG.TEMP | 16.6 | 16.1 | 19.3 | 19.3 | 20.4 | 20.7 | 20.6 |
| TOTAL TIME MSEC | 2.5920 | 1.3374 | 1.1268 | 0.9396 | 0.7632 | 0.6498 | 0.5184 |
| GLOBAL RAD TOTAL | 549.7 | 549.6 | 542.1 | 513.3 | 473.0 | 434.7 | 378.1 |
| DIRECT RAD TOTAL | 320.2 | 320.2 | 318.9 | 310.7 | 300.0 | 287.8 | 256.7 |
| RAD AV TIME MSEC | 0.0864 | 0.0446 | 0.0349 | 0.0335 | 0.0302 | 0.0277 | 0.0234 |
| DEG-C DAYS | 89. | 29. | 20. | 13. | 9. | 7. | 6. |

## MAY

| SOLAR RAD GROUP | ALL | 0+ | 95+ | 189+ | 284+ | 378+ | 473+ |
|---|---|---|---|---|---|---|---|
| AV.THRSH.RAD.LEVL | 0. | 48. | 145. | 239. | 329. | 425. | 527. |
| AV.THRESHOLD TEMP | 16.6 | 19.9 | 18.0 | 20.5 | 21.6 | 23.0 | 23.1 |
| GL RAD OVER THRSH | 462.0 | 400.9 | 297.7 | 212.1 | 144.4 | 92.2 | 48.8 |
| TIME AVG.TEMP | 18.9 | 21.4 | 21.7 | 22.3 | 22.7 | 23.2 | 23.2 |
| TOTAL TIME MSEC | 2.6784 | 1.2618 | 1.0656 | 0.9144 | 0.7542 | 0.5418 | 0.4248 |
| GLOBAL RAD TOTAL | 462.0 | 462.0 | 452.5 | 430.5 | 392.2 | 322.4 | 272.7 |
| DIRECT RAD TOTAL | 199.7 | 199.7 | 198.1 | 195.3 | 191.1 | 169.6 | 153.8 |
| RAD AV TIME MSEC | 0.0864 | 0.0410 | 0.0349 | 0.0313 | 0.0292 | 0.0241 | 0.0216 |
| DEG-C DAYS | 24. | 4. | 3. | 2. | 1. | 1. | 1. |

## JUN

| SOLAR RAD GROUP | ALL | 0+ | 95+ | 189+ | 284+ | 378+ | 473+ |
|---|---|---|---|---|---|---|---|
| AV.THRSH.RAD.LEVL | 0. | 48. | 136. | 252. | 331. | 432. | 517. |
| AV.THRESHOLD TEMP | 21.1 | 23.7 | 22.3 | 25.1 | 25.6 | 28.2 | 27.5 |
| GL RAD OVER THRSH | 437.8 | 380.3 | 294.5 | 191.5 | 131.9 | 73.9 | 38.2 |
| TIME AVG.TEMP | 23.3 | 26.0 | 26.5 | 26.9 | 27.2 | 27.7 | 27.6 |
| TOTAL TIME MSEC | 2.5920 | 1.1880 | 0.9738 | 0.8910 | 0.7542 | 0.5724 | 0.4230 |
| GLOBAL RAD TOTAL | 437.8 | 437.8 | 427.4 | 416.1 | 381.6 | 321.4 | 256.8 |
| DIRECT RAD TOTAL | 201.1 | 201.1 | 199.0 | 198.5 | 183.3 | 169.1 | 143.8 |
| RAD AV TIME MSEC | 0.0864 | 0.0396 | 0.0331 | 0.0317 | 0.0284 | 0.0234 | 0.0202 |
| DEG-C DAYS | 0. | 0. | 0. | 0. | 0. | 0. | 0. |

## ALL MONTHS

| SOLAR RAD GROUP | ALL | 0+ | 95+ | 189+ | 284+ | 378+ | 473+ |
|---|---|---|---|---|---|---|---|
| TIME AVG.TEMP | 15.8 | 17.9 | 19.0 | 20.1 | 20.3 | 20.0 | 19.6 |
| TOTAL TIME MSEC | 31.5791 | 15.3018 | 11.5398 | 9.0324 | 7.3674 | 5.8878 | 4.5900 |
| GLOBAL RAD TOTAL | 5203.0 | 5202.9 | 5046.3 | 4690.7 | 4298.6 | 3807.8 | 3253.6 |
| DIRECT RAD TOTAL | 3040.4 | 3040.4 | 3016.7 | 2930.7 | 2818.8 | 2637.1 | 2367.1 |
| RAD AV TIME MSEC | 0.0864 | 0.0424 | 0.0359 | 0.0320 | 0.0291 | 0.0263 | 0.0235 |
| DEG-C DAYS | 1653. | 637. | 422. | 281. | 227. | 198. | 166. |

STATISTICS FOR ATLANTA,GA     1971-1972     RSV020     FILE 4

RADIATION UNITS ARE WATTS/SQ.M OR MJOULES/SQ.M
SLOPE = 90.0 DEG    AZIMUTH = 180.0 DEG
DEGREE-DAY BASE TEMP = 18.3 DEGREES C

**JAN**

| SOLAR RAD GROUP | ALL | 0+ | 95+ | 189+ | 284+ | 378+ | 473+ |
|---|---|---|---|---|---|---|---|
| AV.THRSH.RAD.LEVL | 0. | 43. | 136. | 222. | 351. | 430. | 528. |
| AV.THRESHOLD TEMP | 5.8 | 10.1 | 7.1 | 7.7 | 5.9 | 6.1 | 6.1 |
| GL RAD OVER THRSH | 301.8 | 248.9 | 192.0 | 153.8 | 106.2 | 80.5 | 54.0 |
| TIME AVG.TEMP | 6.9 | 8.3 | 6.4 | 6.2 | 5.9 | 5.8 | 5.8 |
| TOTAL TIME MSEC | 2.6784 | 1.2276 | 0.6156 | 0.4392 | 0.3708 | 0.3258 | 0.2682 |
| GLOBAL RAD TOTAL | 301.8 | 301.8 | 275.4 | 251.5 | 236.3 | 220.5 | 195.7 |
| DIRECT RAD TOTAL | 220.3 | 220.3 | 217.6 | 207.5 | 200.2 | 188.9 | 169.4 |
| RAD AV TIME MSEC | 0.0864 | 0.0396 | 0.0353 | 0.0310 | 0.0295 | 0.0274 | 0.0245 |
| DEG-C DAYS | 353. | 143. | 84. | 62. | 53. | 47. | 39. |

**FEB**

| SOLAR RAD GROUP | ALL | 0+ | 95+ | 189+ | 284+ | 378+ | 473+ |
|---|---|---|---|---|---|---|---|
| AV.THRSH.RAD.LEVL | 0. | 46. | 135. | 234. | 330. | 427. | 522. |
| AV.THRESHOLD TEMP | 4.5 | 7.3 | 7.5 | 11.0 | 10.8 | 8.0 | 9.0 |
| GL RAD OVER THRSH | 307.1 | 253.1 | 186.5 | 137.8 | 99.7 | 66.3 | 41.7 |
| TIME AVG.TEMP | 6.1 | 7.9 | 8.2 | 8.6 | 8.1 | 7.6 | 7.5 |
| TOTAL TIME MSEC | 2.4624 | 1.1736 | 0.7452 | 0.4932 | 0.3996 | 0.3438 | 0.2592 |
| GLOBAL RAD TOTAL | 307.1 | 307.1 | 287.4 | 253.3 | 231.4 | 213.0 | 176.9 |
| DIRECT RAD TOTAL | 200.6 | 200.6 | 197.8 | 188.2 | 173.6 | 170.2 | 143.8 |
| RAD AV TIME MSEC | 0.0864 | 0.0414 | 0.0356 | 0.0313 | 0.0292 | 0.0277 | 0.0230 |
| DEG-C DAYS | 349. | 142. | 88. | 57. | 48. | 43. | 33. |

**MAR**

| SOLAR RAD GROUP | ALL | 0+ | 95+ | 189+ | 284+ | 378+ | 473+ |
|---|---|---|---|---|---|---|---|
| AV.THRSH.RAD.LEVL | 0. | 39. | 142. | 236. | 334. | 428. | 527. |
| AV.THRESHOLD TEMP | 8.1 | 9.6 | 10.7 | 13.3 | 12.2 | 13.1 | 13.3 |
| GL RAD OVER THRSH | 381.3 | 325.8 | 223.7 | 158.5 | 103.9 | 63.2 | 30.4 |
| TIME AVG.TEMP | 9.9 | 11.4 | 12.2 | 12.9 | 12.8 | 12.9 | 12.9 |
| TOTAL TIME MSEC | 2.6784 | 1.4166 | 0.9936 | 0.6930 | 0.5580 | 0.4338 | 0.3294 |
| GLOBAL RAD TOTAL | 381.4 | 381.3 | 364.8 | 322.1 | 290.2 | 248.7 | 204.1 |
| DIRECT RAD TOTAL | 224.9 | 224.9 | 220.9 | 209.8 | 198.2 | 177.7 | 153.0 |
| RAD AV TIME MSEC | 0.0864 | 0.0461 | 0.0364 | 0.0317 | 0.0284 | 0.0248 | 0.0209 |
| DEG-C DAYS | 262. | 114. | 72. | 46. | 38. | 29. | 22. |

**JUL**

| SOLAR RAD GROUP | ALL | 0+ | 95+ | 189+ | 284+ | 378+ | 473+ |
|---|---|---|---|---|---|---|---|
| AV.THRSH.RAD.LEVL | 0. | 54. | 152. | 240. | 332. | 410. | 486. |
| AV.THRESHOLD TEMP | 22.4 | 24.9 | 25.7 | 26.9 | 28.3 | 29.0 | 30.7 |
| GL RAD OVER THRSH | 202.3 | 152.3 | 75.4 | 25.0 | 4.2 | 0.1 | 0.0 |
| TIME AVG.TEMP | 23.9 | 26.7 | 27.1 | 27.6 | 28.5 | 29.1 | 30.7 |
| TOTAL TIME MSEC | 2.6784 | 0.9270 | 0.7848 | 0.5742 | 0.2250 | 0.0522 | 0.0018 |
| GLOBAL RAD TOTAL | 202.3 | 202.3 | 194.6 | 162.7 | 78.9 | 21.5 | 0.9 |
| DIRECT RAD TOTAL | 32.1 | 32.1 | 31.9 | 29.8 | 14.8 | 3.1 | 0.1 |
| RAD AV TIME MSEC | 0.0864 | 0.0299 | 0.0263 | 0.0212 | 0.0144 | 0.0108 | 0.0036 |
| DEG-C DAYS | 0. | 0. | 0. | 0. | 0. | 0. | 0. |

**AUG**

| SOLAR RAD GROUP | ALL | 0+ | 95+ | 189+ | 284+ | 378+ | 473+ |
|---|---|---|---|---|---|---|---|
| AV.THRSH.RAD.LEVL | 0. | 44. | 148. | 238. | 327. | 413. | 509. |
| AV.THRESHOLD TEMP | 22.7 | 24.9 | 25.5 | 27.1 | 28.3 | 28.9 | 28.2 |
| GL RAD OVER THRSH | 259.1 | 209.6 | 116.1 | 53.8 | 19.6 | 3.9 | 0.0 |
| TIME AVG.TEMP | 24.4 | 26.8 | 27.3 | 27.9 | 28.5 | 28.7 | 28.2 |
| TOTAL TIME MSEC | 2.6784 | 1.1304 | 0.8982 | 0.6930 | 0.3852 | 0.1800 | 0.0414 |
| GLOBAL RAD TOTAL | 259.1 | 259.1 | 249.0 | 218.6 | 145.4 | 78.4 | 21.1 |
| DIRECT RAD TOTAL | 87.0 | 67.0 | 86.2 | 81.6 | 63.1 | 39.8 | 13.7 |
| RAD AV TIME MSEC | 0.0864 | 0.0367 | 0.0302 | 0.0256 | 0.0202 | 0.0148 | 0.0104 |
| DEG-C DAYS | 0. | 0. | 0. | 0. | 0. | 0. | 0. |

**SEP**

| SOLAR RAD GROUP | ALL | 0+ | 95+ | 189+ | 284+ | 378+ | 473+ |
|---|---|---|---|---|---|---|---|
| AV.THRSH.RAD.LEVL | 0. | 37. | 139. | 238. | 330. | 426. | 516. |
| AV.THRESHOLD TEMP | 20.8 | 22.1 | 23.4 | 25.1 | 26.4 | 27.3 | 27.2 |
| GL RAD OVER THRSH | 315.0 | 265.3 | 166.2 | 95.7 | 49.5 | 17.7 | 6.5 |
| TIME AVG.TEMP | 22.8 | 24.7 | 25.6 | 26.4 | 27.0 | 27.3 | 27.4 |
| TOTAL TIME MSEC | 2.5920 | 1.3284 | 0.9756 | 0.7128 | 0.5022 | 0.3294 | 0.1242 |
| GLOBAL RAD TOTAL | 315.0 | 315.0 | 301.8 | 265.3 | 215.2 | 158.2 | 70.7 |
| DIRECT RAD TOTAL | 161.5 | 161.5 | 159.3 | 149.1 | 128.8 | 100.6 | 52.2 |
| RAD AV TIME MSEC | 0.0864 | 0.0443 | 0.0353 | 0.0292 | 0.0234 | 0.0184 | 0.0151 |
| DEG-C DAYS | 1. | 1. | 1. | 1. | 0. | 0. | 0. |

## APR

| SOLAR RAD GROUP | ALL | 0+ | 95+ | 189+ | 284+ | 378+ | 473+ |
|---|---|---|---|---|---|---|---|
| AV.THRSH.RAD.LEVL | 0. | 41. | 149. | 234. | 333. | 422. | 519. |
| AV.THRESHOLD TEMP | 13.7 | 16.0 | 17.1 | 18.5 | 21.6 | 21.2 | 20.8 |
| GL RAD OVER THRSH | 334.4 | 286.0 | 182.0 | 116.9 | 62.6 | 27.6 | 5.1 |
| TIME AVG.TEMP | 16.1 | 18.9 | 19.6 | 20.2 | 20.9 | 20.2 | 20.2 |
| TOTAL TIME MSEC | 2.5920 | 1.1844 | 0.9648 | 0.7596 | 0.5508 | 0.3942 | 0.2304 |
| GLOBAL RAD TOTAL | 334.4 | 334.4 | 325.4 | 294.9 | 246.0 | 193.9 | 124.8 |
| DIRECT RAD TOTAL | 147.7 | 147.7 | 146.5 | 142.4 | 126.2 | 107.2 | 76.6 |
| RAD AV TIME MSEC | 0.0864 | 0.0396 | 0.0331 | 0.0295 | 0.0234 | 0.0198 | 0.0144 |
| DEG-C DAYS | 89. | 24. | 17. | 10. | 6. | 4. | 3. |

## MAY

| SOLAR RAD GROUP | ALL | 0+ | 95+ | 189+ | 284+ | 378+ | 473+ |
|---|---|---|---|---|---|---|---|
| AV.THRSH.RAD.LEVL | 0. | 47. | 148. | 237. | 331. | 421. | 507. |
| AV.THRESHOLD TEMP | 17.1 | 19.9 | 20.6 | 22.1 | 23.6 | 23.3 | 22.7 |
| GL RAD OVER THRSH | 239.2 | 193.1 | 109.2 | 53.9 | 18.5 | 3.7 | 0.0 |
| TIME AVG.TEMP | 18.9 | 22.3 | 22.3 | 22.9 | 23.4 | 23.2 | 22.7 |
| TOTAL TIME MSEC | 2.6784 | 0.9864 | 0.8280 | 0.6228 | 0.3780 | 0.1638 | 0.0432 |
| GLOBAL RAD TOTAL | 239.2 | 239.2 | 231.8 | 201.4 | 143.5 | 72.6 | 21.9 |
| DIRECT RAD TOTAL | 53.6 | 53.6 | 53.0 | 50.9 | 40.0 | 17.2 | 21.9 |
| RAD AV TIME MSEC | 0.0864 | 0.0320 | 0.0270 | 0.0230 | 0.0187 | 0.0144 | 0.0101 |
| DEG-C DAYS | 24. | 3. | 2. | 1. | 1. | 0. | 0. |

## JUN

| SOLAR RAD GROUP | ALL | 0+ | 95+ | 189+ | 284+ | 378+ | 473+ |
|---|---|---|---|---|---|---|---|
| AV.THRSH.RAD.LEVL | 0. | 35. | 141. | 242. | 330. | 413. | 500. |
| AV.THRESHOLD TEMP | 21.8 | 26.0 | 25.5 | 27.1 | 27.7 | 28.6 | 28.9 |
| GL RAD OVER THRSH | 180.2 | 152.7 | 83.0 | 32.4 | 8.0 | 1.2 | 0.0 |
| TIME AVG.TEMP | 23.3 | 26.9 | 27.1 | 27.6 | 27.9 | 28.6 | 28.9 |
| TOTAL TIME MSEC | 2.5920 | 0.7830 | 0.6606 | 0.5004 | 0.2754 | 0.0810 | 0.0144 |
| GLOBAL RAD TOTAL | 180.2 | 180.2 | 175.9 | 153.3 | 99.0 | 34.7 | 7.2 |
| DIRECT RAD TOTAL | 36.2 | 36.2 | 35.8 | 33.6 | 20.0 | 7.2 | 0.5 |
| RAD AV TIME MSEC | 0.0864 | 0.0263 | 0.0230 | 0.0184 | 0.0137 | 0.0112 | 0.0072 |
| DEG-C DAYS | 0. | 0. | 0. | 0. | 0. | 0. | 0. |

## OCT

| SOLAR RAD GROUP | ALL | 0+ | 95+ | 189+ | 284+ | 378+ | 473+ |
|---|---|---|---|---|---|---|---|
| AV.THRSH.RAD.LEVL | 0. | 43. | 137. | 237. | 332. | 424. | 523. |
| AV.THRESHOLD TEMP | 15.1 | 17.3 | 18.8 | 19.8 | 19.8 | 21.7 | 21.0 |
| GL RAD OVER THRSH | 347.9 | 289.8 | 203.7 | 140.7 | 92.5 | 57.0 | 29.6 |
| TIME AVG.TEMP | 17.2 | 20.3 | 20.9 | 20.9 | 21.2 | 21.7 | 21.7 |
| TOTAL TIME MSEC | 2.6784 | 1.3446 | 0.9180 | 0.6300 | 0.5058 | 0.3852 | 0.2790 |
| GLOBAL RAD TOTAL | 348.0 | 348.0 | 329.5 | 290.0 | 260.6 | 220.5 | 175.4 |
| DIRECT RAD TOTAL | 223.0 | 223.0 | 217.6 | 204.9 | 190.7 | 166.4 | 138.8 |
| RAD AV TIME MSEC | 0.0864 | 0.0443 | 0.0356 | 0.0310 | 0.0281 | 0.0241 | 0.0209 |
| DEG-C DAYS | 57. | 14. | 8. | 5. | 3. | 2. | 1. |

## NOV

| SOLAR RAD GROUP | ALL | 0+ | 95+ | 189+ | 284+ | 378+ | 473+ |
|---|---|---|---|---|---|---|---|
| AV.THRSH.RAD.LEVL | 0. | 47. | 143. | 238. | 342. | 429. | 508. |
| AV.THRESHOLD TEMP | 8.1 | 9.4 | 11.3 | 12.5 | 12.7 | 15.1 | 13.3 |
| GL RAD OVER THRSH | 385.0 | 329.4 | 249.2 | 192.1 | 138.2 | 99.1 | 71.5 |
| TIME AVG.TEMP | 9.8 | 11.9 | 12.9 | 13.6 | 13.7 | 13.9 | 13.6 |
| TOTAL TIME MSEC | 2.5920 | 1.1880 | 0.8298 | 0.6012 | 0.5220 | 0.4464 | 0.3510 |
| GLOBAL RAD TOTAL | 385.0 | 385.0 | 368.3 | 335.5 | 316.6 | 290.7 | 249.8 |
| DIRECT RAD TOTAL | 289.5 | 289.5 | 286.4 | 270.0 | 260.7 | 245.1 | 215.1 |
| RAD AV TIME MSEC | 0.0864 | 0.0396 | 0.0356 | 0.0302 | 0.0288 | 0.0281 | 0.0245 |
| DEG-C DAYS | 258. | 92. | 54. | 35. | 29. | 24. | 20. |

## DEC

| SOLAR RAD GROUP | ALL | 0+ | 95+ | 189+ | 284+ | 378+ | 473+ |
|---|---|---|---|---|---|---|---|
| AV.THRSH.RAD.LEVL | 0. | 42. | 137. | 236. | 331. | 431. | 521. |
| AV.THRESHOLD TEMP | 8.6 | 11.6 | 12.7 | 13.2 | 10.7 | 12.9 | 12.5 |
| GL RAD OVER THRSH | 239.5 | 187.8 | 133.9 | 104.8 | 77.5 | 54.5 | 38.9 |
| TIME AVG.TEMP | 10.0 | 11.6 | 11.6 | 11.1 | 10.7 | 10.7 | 9.9 |
| TOTAL TIME MSEC | 2.6784 | 1.2222 | 0.5184 | 0.3420 | 0.2898 | 0.2286 | 0.1728 |
| GLOBAL RAD TOTAL | 239.5 | 239.5 | 209.7 | 185.6 | 173.3 | 153.1 | 129.0 |
| DIRECT RAD TOTAL | 160.6 | 160.6 | 156.8 | 148.8 | 144.3 | 131.9 | 113.7 |
| RAD AV TIME MSEC | 0.0864 | 0.0396 | 0.0331 | 0.0295 | 0.0288 | 0.0256 | 0.0227 |
| DEG-C DAYS | 259. | 98. | 42. | 29. | 26. | 21. | 17. |

## ALL MONTHS

| SOLAR RAD GROUP | ALL | 0+ | 95+ | 189+ | 284+ | 378+ | 473+ |
|---|---|---|---|---|---|---|---|
| TIME AVG.TEMP | 15.8 | 17.5 | 18.9 | 19.6 | 18.6 | 17.0 | 14.6 |
| TOTAL TIME MSEC | 31.5791 | 13.9122 | 9.7326 | 7.0614 | 4.9626 | 3.3642 | 2.1150 |
| GLOBAL RAD TOTAL | 3493.0 | 3492.9 | 3313.6 | 2934.2 | 2436.2 | 1905.8 | 1377.4 |
| DIRECT RAD TOTAL | 1837.2 | 1837.2 | 1809.9 | 1716.2 | 1565.5 | 1351.2 | 1079.9 |
| RAD AV TIME MSEC | 0.0864 | 0.0394 | 0.0330 | 0.0284 | 0.0256 | 0.0237 | 0.0212 |
| DEG-C DAYS | 1653. | 632. | 368. | 246. | 204. | 171. | 136. |

STATISTICS FOR DENVER,COLO 1971-1972 IR INCLUDED RSV023 FILE 7

RADIATION UNITS ARE WATTS/SQ.M OR MJOULES/SQ.M
SLOPE = 0.0 DEG  AZIMUTH = 0.0 DEG
DEGREE-DAY BASE TEMP = 18.3 DEGREES C

**JAN**

| SOLAR RAD GROUP | ALL | 0+ | 95+ | 189+ | 284+ | 378+ | 473+ |
|---|---|---|---|---|---|---|---|
| AV.THRSH.RAD.LEVL | 0. | 59. | 142. | 237. | 324. | 428. | 516. |
| AV.THRESHOLD TEMP | -2.0 | -2.3 | 1.4 | 4.9 | 2.9 | 5.0 | 2.3 |
| GL RAD OVER THRSH | 240.9 | 182.0 | 115.6 | 61.0 | 30.2 | 9.6 | 1.5 |
| TIME AVG.TEMP | -0.4 | 2.2 | 3.3 | 4.1 | 3.5 | 3.9 | 2.7 |
| TOTAL TIME MSEC | 2.6784 | 1.0044 | 0.7956 | 0.5760 | 0.3546 | 0.1980 | 0.0918 |
| GLOBAL RAD TOTAL | 240.9 | 240.9 | 228.7 | 197.5 | 145.0 | 94.3 | 48.9 |
| DIRECT RAD TOTAL | 118.4 | 118.4 | 116.4 | 106.2 | 91.7 | 69.6 | 41.9 |
| RAD AV TIME MSEC | 0.0864 | 0.0324 | 0.0231 | 0.0220 | 0.0184 | 0.0155 | 0.0119 |
| DEG-C DAYS | 582. | 188. | 138. | 95. | 61. | 33. | 17. |

**FEB**

| SOLAR RAD GROUP | ALL | 0+ | 95+ | 189+ | 284+ | 378+ | 473+ |
|---|---|---|---|---|---|---|---|
| AV.THRSH.RAD.LEVL | 0. | 40. | 140. | 238. | 332. | 426. | 520. |
| AV.THRESHOLD TEMP | -1.5 | 0.8 | 1.5 | 3.1 | 5.1 | 6.2 | 5.9 |
| GL RAD OVER THRSH | 309.7 | 264.9 | 179.6 | 115.9 | 69.4 | 35.4 | 13.3 |
| TIME AVG.TEMP | 0.7 | 3.3 | 4.1 | 4.9 | 5.4 | 5.6 | 5.2 |
| TOTAL TIME MSEC | 2.4624 | 1.1124 | 0.8514 | 0.6534 | 0.4968 | 0.3618 | 0.2322 |
| GLOBAL RAD TOTAL | 309.7 | 309.7 | 299.2 | 271.4 | 234.1 | 189.3 | 134.2 |
| DIRECT RAD TOTAL | 160.6 | 160.6 | 157.7 | 152.1 | 137.0 | 122.9 | 93.0 |
| RAD AV TIME MSEC | 0.0864 | 0.0389 | 0.0313 | 0.0263 | 0.0220 | 0.0194 | 0.0148 |
| DEG-C DAYS | 504. | 193. | 140. | 102. | 74. | 53. | 35. |

**MAR**

| SOLAR RAD GROUP | ALL | 0+ | 95+ | 189+ | 284+ | 378+ | 473+ |
|---|---|---|---|---|---|---|---|
| AV.THRSH.RAD.LEVL | 0. | 48. | 146. | 240. | 340. | 427. | 519. |
| AV.THRESHOLD TEMP | 2.5 | 4.4 | 5.2 | 7.6 | 8.6 | 10.3 | 11.7 |
| GL RAD OVER THRSH | 457.9 | 396.3 | 290.0 | 206.1 | 133.3 | 83.9 | 47.8 |
| TIME AVG.TEMP | 5.5 | 8.7 | 9.6 | 10.5 | 11.2 | 11.9 | 12.6 |
| TOTAL TIME MSEC | 2.6784 | 1.2924 | 1.0854 | 0.8928 | 0.7272 | 0.5652 | 0.3906 |
| GLOBAL RAD TOTAL | 457.9 | 457.9 | 448.0 | 420.0 | 380.3 | 325.3 | 250.7 |
| DIRECT RAD TOTAL | 206.2 | 206.2 | 205.0 | 196.8 | 188.5 | 170.3 | 142.0 |
| RAD AV TIME MSEC | 0.0864 | 0.0418 | 0.0364 | 0.0313 | 0.0270 | 0.0234 | 0.0198 |
| DEG-C DAYS | 397. | 145. | 112. | 83. | 63. | 44. | 28. |

**JUL**

| SOLAR RAD GROUP | ALL | 0+ | 95+ | 189+ | 284+ | 378+ | 473+ |
|---|---|---|---|---|---|---|---|
| AV.THRSH.RAD.LEVL | 0. | 40. | 144. | 232. | 337. | 424. | 527. |
| AV.THRESHOLD TEMP | 17.6 | 18.4 | 19.5 | 21.8 | 23.7 | 24.4 | 26.7 |
| GL RAD OVER THRSH | 680.7 | 614.4 | 472.9 | 372.5 | 275.1 | 206.6 | 143.1 |
| TIME AVG.TEMP | 21.1 | 23.1 | 24.2 | 25.1 | 25.8 | 26.2 | 26.7 |
| TOTAL TIME MSEC | 2.6784 | 1.6722 | 1.3518 | 1.1484 | 0.9270 | 0.7866 | 0.6174 |
| GLOBAL RAD TOTAL | 680.7 | 680.7 | 668.0 | 638.7 | 587.3 | 540.1 | 468.0 |
| DIRECT RAD TOTAL | 386.7 | 386.7 | 383.7 | 378.7 | 365.3 | 355.8 | 323.0 |
| RAD AV TIME MSEC | 0.0864 | 0.0540 | 0.0446 | 0.0400 | 0.0346 | 0.0317 | 0.0259 |
| DEG-C DAYS | 19. | 9. | 6. | 2. | 1. | 0. | 0. |

**AUG**

| SOLAR RAD GROUP | ALL | 0+ | 95+ | 189+ | 284+ | 378+ | 473+ |
|---|---|---|---|---|---|---|---|
| AV.THRSH.RAD.LEVL | 0. | 56. | 140. | 244. | 336. | 426. | 526. |
| AV.THRESHOLD TEMP | 18.2 | 20.2 | 21.1 | 24.3 | 23.5 | 25.4 | 24.3 |
| GL RAD OVER THRSH | 666.2 | 583.8 | 474.2 | 358.6 | 270.0 | 198.8 | 134.1 |
| TIME AVG.TEMP | 21.5 | 24.2 | 24.7 | 25.3 | 25.5 | 25.9 | 26.1 |
| TOTAL TIME MSEC | 2.6784 | 1.4634 | 1.3032 | 1.1142 | 0.9648 | 0.7866 | 0.6498 |
| GLOBAL RAD TOTAL | 666.2 | 666.2 | 657.1 | 630.6 | 594.1 | 534.3 | 475.9 |
| DIRECT RAD TOTAL | 412.0 | 412.0 | 409.8 | 399.7 | 390.2 | 363.6 | 341.3 |
| RAD AV TIME MSEC | 0.0864 | 0.0472 | 0.0428 | 0.0371 | 0.0335 | 0.0284 | 0.0252 |
| DEG-C DAYS | 6. | 1. | 1. | 0. | 0. | 0. | 0. |

**SEP**

| SOLAR RAD GROUP | ALL | 0+ | 95+ | 189+ | 284+ | 378+ | 473+ |
|---|---|---|---|---|---|---|---|
| AV.THRSH.RAD.LEVL | 0. | 38. | 146. | 236. | 332. | 430. | 515. |
| AV.THRESHOLD TEMP | 12.3 | 14.2 | 13.6 | 16.3 | 17.5 | 20.5 | 19.8 |
| GL RAD OVER THRSH | 542.6 | 492.5 | 370.2 | 283.4 | 205.8 | 141.9 | 97.3 |
| TIME AVG.TEMP | 15.3 | 18.2 | 18.8 | 19.8 | 20.4 | 21.1 | 21.3 |
| TOTAL TIME MSEC | 2.5920 | 1.3248 | 1.1340 | 0.9612 | 0.8064 | 0.6570 | 0.5202 |
| GLOBAL RAD TOTAL | 542.6 | 542.6 | 535.4 | 510.2 | 473.7 | 424.1 | 365.3 |
| DIRECT RAD TOTAL | 358.6 | 358.6 | 356.3 | 347.9 | 334.3 | 315.2 | 287.2 |
| RAD AV TIME MSEC | 0.0864 | 0.0443 | 0.0385 | 0.0346 | 0.0306 | 0.0274 | 0.0238 |
| DEG-C DAYS | 108. | 38. | 29. | 18. | 12. | 7. | 6. |

**APR**

| SOLAR RAD GROUP | ALL | 0+ | 95+ | 189+ | 284+ | 378+ | 473+ |
|---|---|---|---|---|---|---|---|
| AV.THRSH.RAD.LEVL | 0. | 53. | 145. | 240. | 333. | 424. | 516. |
| AV.THRESHOLD TEMP | 6.2 | 7.6 | 8.7 | 9.5 | 13.1 | 15.3 | 14.0 |
| GL RAD OVER THRSH | 526.0 | 451.5 | 343.7 | 253.2 | 182.0 | 128.0 | 84.1 |
| TIME AVG.TEMP | 9.1 | 11.4 | 12.2 | 13.0 | 13.9 | 14.1 | 13.8 |
| TOTAL TIME MSEC | 2.5920 | 1.4040 | 1.1700 | 0.9576 | 0.7650 | 0.5886 | 0.4770 |
| GLOBAL RAD TOTAL | 526.0 | 526.0 | 513.6 | 482.3 | 436.6 | 377.6 | 330.5 |
| DIRECT RAD TOTAL | 255.9 | 255.9 | 252.9 | 248.3 | 238.2 | 222.6 | 207.5 |
| RAD AV TIME MSEC | 0.0864 | 0.0468 | 0.0403 | 0.0356 | 0.0313 | 0.0270 | 0.0238 |
| DEG-C DAYS | 278. | 114. | 86. | 62. | 42. | 31. | 27. |

**MAY**

| SOLAR RAD GROUP | ALL | 0+ | 95+ | 189+ | 284+ | 378+ | 473+ |
|---|---|---|---|---|---|---|---|
| AV.THRSH.RAD.LEVL | 0. | 40. | 143. | 230. | 333. | 421. | 522. |
| AV.THRESHOLD TEMP | 9.5 | 11.0 | 12.8 | 13.1 | 15.6 | 16.2 | 18.4 |
| GL RAD OVER THRSH | 638.7 | 574.3 | 437.7 | 342.4 | 248.8 | 182.8 | 125.6 |
| TIME AVG.TEMP | 13.1 | 15.4 | 16.3 | 17.1 | 17.9 | 18.4 | 19.1 |
| TOTAL TIME MSEC | 2.6784 | 1.5984 | 1.3302 | 1.0908 | 0.9144 | 0.7488 | 0.5634 |
| GLOBAL RAD TOTAL | 638.7 | 638.7 | 627.9 | 593.6 | 553.0 | 497.9 | 419.9 |
| DIRECT RAD TOTAL | 317.5 | 317.5 | 315.5 | 308.1 | 299.8 | 289.0 | 258.2 |
| RAD AV TIME MSEC | 0.0864 | 0.0518 | 0.0443 | 0.0385 | 0.0346 | 0.0313 | 0.0263 |
| DEG-C DAYS | 166. | 64. | 45. | 29. | 20. | 14. | 8. |

**JUN**

| SOLAR RAD GROUP | ALL | 0+ | 95+ | 189+ | 284+ | 378+ | 473+ |
|---|---|---|---|---|---|---|---|
| AV.THRSH.RAD.LEVL | 0. | 48. | 144. | 236. | 335. | 432. | 524. |
| AV.THRESHOLD TEMP | 16.4 | 18.2 | 21.1 | 20.8 | 23.2 | 22.9 | 24.6 |
| GL RAD OVER THRSH | 726.6 | 649.2 | 518.1 | 407.4 | 305.7 | 222.2 | 160.4 |
| TIME AVG.TEMP | 20.3 | 22.7 | 23.6 | 23.9 | 24.4 | 24.7 | 25.2 |
| TOTAL TIME MSEC | 2.5920 | 1.6200 | 1.3662 | 1.2042 | 1.0224 | 0.8658 | 0.6714 |
| GLOBAL RAD TOTAL | 726.6 | 726.6 | 714.5 | 691.2 | 648.4 | 595.9 | 512.0 |
| DIRECT RAD TOTAL | 425.5 | 425.5 | 421.0 | 416.9 | 400.0 | 389.2 | 351.2 |
| RAD AV TIME MSEC | 0.0864 | 0.0540 | 0.0457 | 0.0407 | 0.0353 | 0.0317 | 0.0259 |
| DEG-C DAYS | 11. | 1. | 0. | 0. | 0. | 0. | 0. |

**ALL MONTHS**

| SOLAR RAD GROUP | ALL | 0+ | 95+ | 189+ | 284+ | 378+ | 473+ |
|---|---|---|---|---|---|---|---|
| TIME AVG.TEMP | 9.7 | 13.7 | 14.8 | 16.1 | 17.2 | 18.3 | 19.4 |
| TOTAL TIME MSEC | 31.5791 | 15.7374 | 12.8790 | 10.4688 | 8.2908 | 6.4386 | 4.7376 |
| GLOBAL RAD TOTAL | 5645.9 | 5645.9 | 5509.8 | 5166.4 | 4650.1 | 4033.4 | 3309.8 |
| DIRECT RAD TOTAL | 3117.8 | 3117.8 | 3085.9 | 2998.4 | 2837.3 | 2624.6 | 2284.0 |
| RAD AV TIME MSEC | 0.0864 | 0.0459 | 0.0394 | 0.0345 | 0.0304 | 0.0273 | 0.0234 |
| DEG-C DAYS | 3467. | 1196. | 862. | 597. | 399. | 261. | 160. |

**OCT**

| SOLAR RAD GROUP | ALL | 0+ | 95+ | 189+ | 284+ | 378+ | 473+ |
|---|---|---|---|---|---|---|---|
| AV.THRSH.RAD.LEVL | 0. | 47. | 142. | 232. | 353. | 426. | 523. |
| AV.THRESHOLD TEMP | 7.3 | 8.5 | 9.7 | 13.4 | 16.1 | 16.3 | 16.1 |
| GL RAD OVER THRSH | 383.7 | 326.3 | 234.8 | 165.3 | 104.3 | 60.8 | 29.2 |
| TIME AVG.TEMP | 10.1 | 13.3 | 14.6 | 15.8 | 16.4 | 16.6 | 16.7 |
| TOTAL TIME MSEC | 2.6784 | 1.2276 | 0.9576 | 0.7740 | 0.6066 | 0.4644 | 0.3258 |
| GLOBAL RAD TOTAL | 383.7 | 383.7 | 371.1 | 344.9 | 306.1 | 258.8 | 199.7 |
| DIRECT RAD TOTAL | 216.1 | 216.1 | 213.2 | 207.9 | 195.0 | 177.4 | 150.1 |
| RAD AV TIME MSEC | 0.0864 | 0.0396 | 0.0335 | 0.0292 | 0.0252 | 0.0212 | 0.0187 |
| DEG-C DAYS | 257. | 79. | 49. | 31. | 21. | 16. | 10. |

**NOV**

| SOLAR RAD GROUP | ALL | 0+ | 95+ | 189+ | 284+ | 378+ | 473+ |
|---|---|---|---|---|---|---|---|
| AV.THRSH.RAD.LEVL | 0. | 53. | 144. | 240. | 329. | 425. | 523. |
| AV.THRESHOLD TEMP | -0.2 | 1.4 | 3.4 | 5.6 | 6.9 | 8.2 | 9.6 |
| GL RAD OVER THRSH | 240.6 | 187.2 | 117.5 | 67.2 | 35.4 | 14.8 | 3.1 |
| TIME AVG.TEMP | 1.8 | 4.9 | 6.1 | 7.4 | 8.2 | 9.1 | 9.8 |
| TOTAL TIME MSEC | 2.5920 | 1.0134 | 0.7650 | 0.5220 | 0.3564 | 0.2160 | 0.1188 |
| GLOBAL RAD TOTAL | 240.6 | 240.6 | 227.5 | 192.6 | 152.8 | 106.6 | 65.3 |
| DIRECT RAD TOTAL | 126.3 | 126.3 | 124.5 | 114.7 | 101.3 | 78.2 | 54.2 |
| RAD AV TIME MSEC | 0.0864 | 0.0346 | 0.0292 | 0.0234 | 0.0205 | 0.0169 | 0.0140 |
| DEG-C DAYS | 494. | 157. | 108. | 67. | 42. | 23. | 12. |

**DEC**

| SOLAR RAD GROUP | ALL | 0+ | 95+ | 189+ | 284+ | 378+ | 473+ |
|---|---|---|---|---|---|---|---|
| AV.THRSH.RAD.LEVL | 0. | 57. | 133. | 241. | 331. | 415. | 493. |
| AV.THRESHOLD TEMP | -4.4 | -1.9 | -0.7 | 1.7 | 4.2 | 1.8 | 0.2 |
| GL RAD OVER THRSH | 232.3 | 174.9 | 116.4 | 54.3 | 23.0 | 6.2 | 0.0 |
| TIME AVG.TEMP | -2.5 | 0.7 | 1.4 | 4.3 | 6.2 | 1.2 | 0.2 |
| TOTAL TIME MSEC | 2.6784 | 1.0044 | 0.7686 | 0.5742 | 0.3492 | 0.1998 | 0.0792 |
| GLOBAL RAD TOTAL | 232.3 | 232.3 | 218.8 | 192.9 | 138.6 | 89.2 | 39.1 |
| DIRECT RAD TOTAL | 134.1 | 134.1 | 130.0 | 121.1 | 96.0 | 70.8 | 34.4 |
| RAD AV TIME MSEC | 0.0364 | 0.0324 | 0.0270 | 0.0220 | 0.0158 | 0.0144 | 0.0090 |
| DEG-C DAYS | 645. | 206. | 150. | 108. | 64. | 39. | 0. |

STATISTICS FOR DENVER,COLO    1971-1972    IR INCLUDED    RSV023    FILE 7

RADIATION UNITS ARE WATTS/SQ.M OR MJOULES/SQ.M
SLOPE = 30.0 DEG   AZIMUTH = 180.0 DEG
DEGREE-DAY BASE TEMP = 18.3 DEGREES C

**JAN**

| SOLAR RAD GROUP | ALL | 0+ | 95+ | 189+ | 284+ | 378+ | 473+ |
|---|---|---|---|---|---|---|---|
| AV.THRSH.RAD.LEVL | 0. | 50. | 150. | 243. | 329. | 433. | 514. |
| AV.THRESHOLD TEMP | -2.4 | -2.3 | 2.1 | 6.1 | 4.0 | 2.9 | 3.1 |
| GL RAD OVER THRSH | 346.5 | 291.1 | 204.4 | 144.5 | 105.4 | 69.0 | 49.1 |
| TIME AVG.TEMP | -0.4 | 2.3 | 3.6 | 4.1 | 3.3 | 3.1 | 3.1 |
| TOTAL TIME MSEC | 2.6784 | 1.1160 | 0.8640 | 0.6462 | 0.4536 | 0.3492 | 0.2448 |
| GLOBAL RAD TOTAL | 346.5 | 346.5 | 334.0 | 301.3 | 254.6 | 220.2 | 175.0 |
| DIRECT RAD TOTAL | 228.9 | 228.9 | 226.3 | 212.0 | 199.1 | 182.4 | 153.2 |
| RAD AV TIME MSEC | 0.0864 | 0.0360 | 0.0317 | 0.0256 | 0.0245 | 0.0230 | 0.0202 |
| DEG-C DAYS | 582. | 208. | 147. | 106. | 79. | 62. | 43. |

**FEB**

| SOLAR RAD GROUP | ALL | 0+ | 95+ | 189+ | 284+ | 378+ | 473+ |
|---|---|---|---|---|---|---|---|
| AV.THRSH.RAD.LEVL | 0. | 42. | 139. | 237. | 323. | 427. | 526. |
| AV.THRESHOLD TEMP | -1.8 | 0.3 | 1.4 | 2.9 | 6.6 | 6.1 | 4.6 |
| GL RAD OVER THRSH | 402.8 | 354.0 | 267.8 | 197.3 | 147.7 | 100.2 | 65.9 |
| TIME AVG.TEMP | 0.7 | 3.4 | 4.4 | 5.1 | 5.6 | 5.3 | 5.1 |
| TOTAL TIME MSEC | 2.4624 | 1.1736 | 0.8838 | 0.7236 | 0.5760 | 0.4536 | 0.3474 |
| GLOBAL RAD TOTAL | 402.9 | 402.8 | 390.8 | 368.5 | 333.6 | 294.1 | 248.1 |
| DIRECT RAD TOTAL | 260.3 | 260.3 | 256.2 | 250.4 | 237.2 | 221.6 | 200.0 |
| RAD AV TIME MSEC | 0.0864 | 0.0414 | 0.0331 | 0.0299 | 0.0263 | 0.0238 | 0.0220 |
| DEG-C DAYS | 504. | 203. | 143. | 112. | 85. | 68. | 53. |

**MAR**

| SOLAR RAD GROUP | ALL | 0+ | 95+ | 189+ | 284+ | 378+ | 473+ |
|---|---|---|---|---|---|---|---|
| AV.THRSH.RAD.LEVL | 0. | 38. | 135. | 239. | 336. | 426. | 518. |
| AV.THRESHOLD TEMP | 2.1 | 3.9 | 5.5 | 6.8 | 9.4 | 10.7 | 10.7 |
| GL RAD OVER THRSH | 509.3 | 458.9 | 351.0 | 256.6 | 181.7 | 127.5 | 86.2 |
| TIME AVG.TEMP | 5.5 | 8.9 | 9.9 | 10.8 | 11.6 | 12.2 | 12.7 |
| TOTAL TIME MSEC | 2.6784 | 1.3320 | 1.1088 | 0.9108 | 0.7722 | 0.6030 | 0.4482 |
| GLOBAL RAD TOTAL | 509.3 | 509.3 | 500.9 | 474.1 | 441.0 | 384.2 | 318.3 |
| DIRECT RAD TOTAL | 270.4 | 270.4 | 268.4 | 261.9 | 255.3 | 234.3 | 211.3 |
| RAD AV TIME MSEC | 0.0864 | 0.0432 | 0.0371 | 0.0324 | 0.0295 | 0.0252 | 0.0223 |
| DEG-C DAYS | 397. | 148. | 111. | 82. | 64. | 46. | 32. |

**JUL**

| SOLAR RAD GROUP | ALL | 0+ | 95+ | 189+ | 284+ | 378+ | 473+ |
|---|---|---|---|---|---|---|---|
| AV.THRSH.RAD.LEVL | 0. | 35. | 146. | 229. | 337. | 432. | 523. |
| AV.THRESHOLD TEMP | 17.2 | 19.0 | 20.1 | 22.0 | 23.9 | 25.5 | 27.4 |
| GL RAD OVER THRSH | 638.0 | 583.2 | 440.8 | 349.6 | 258.0 | 187.8 | 136.6 |
| TIME AVG.TEMP | 21.1 | 23.8 | 24.8 | 25.6 | 26.7 | 27.1 | 27.6 |
| TOTAL TIME MSEC | 2.6784 | 1.5624 | 1.2816 | 1.0998 | 0.8532 | 0.7344 | 0.5634 |
| GLOBAL RAD TOTAL | 638.0 | 638.0 | 628.2 | 601.6 | 545.1 | 505.1 | 431.2 |
| DIRECT RAD TOTAL | 360.5 | 360.5 | 359.6 | 356.9 | 343.0 | 336.0 | 299.5 |
| RAD AV TIME MSEC | 0.0864 | 0.0504 | 0.0421 | 0.0385 | 0.0320 | 0.0295 | 0.0238 |
| DEG-C DAYS | 19. | 9. | 5. | 2. | 1. | 0. | 0. |

**AUG**

| SOLAR RAD GROUP | ALL | 0+ | 95+ | 189+ | 284+ | 378+ | 473+ |
|---|---|---|---|---|---|---|---|
| AV.THRSH.RAD.LEVL | 0. | 25. | 151. | 239. | 335. | 427. | 521. |
| AV.THRESHOLD TEMP | 17.7 | 19.9 | 21.9 | 23.5 | 25.2 | 24.0 | 26.9 |
| GL RAD OVER THRSH | 671.3 | 632.0 | 471.0 | 377.7 | 288.2 | 215.7 | 157.4 |
| TIME AVG.TEMP | 21.5 | 24.2 | 25.2 | 25.8 | 26.2 | 26.3 | 26.9 |
| TOTAL TIME MSEC | 2.6784 | 1.5606 | 1.2798 | 1.0620 | 0.9288 | 0.7884 | 0.6228 |
| GLOBAL RAD TOTAL | 671.3 | 671.3 | 664.2 | 631.3 | 599.5 | 552.5 | 481.7 |
| DIRECT RAD TOTAL | 427.5 | 427.5 | 426.8 | 414.3 | 406.1 | 390.6 | 353.8 |
| RAD AV TIME MSEC | 0.0864 | 0.0504 | 0.0418 | 0.0353 | 0.0320 | 0.0288 | 0.0241 |
| DEG-C DAYS | 6. | 1. | 1. | 0. | 0. | 0. | 0. |

**SEP**

| SOLAR RAD GROUP | ALL | 0+ | 95+ | 189+ | 284+ | 378+ | 473+ |
|---|---|---|---|---|---|---|---|
| AV.THRSH.RAD.LEVL | 0. | 38. | 137. | 232. | 333. | 418. | 526. |
| AV.THRESHOLD TEMP | 11.8 | 14.5 | 14.4 | 14.5 | 18.0 | 20.2 | 21.5 |
| GL RAD OVER THRSH | 620.8 | 569.7 | 457.3 | 366.2 | 280.4 | 221.1 | 160.4 |
| TIME AVG.TEMP | 15.3 | 18.5 | 19.2 | 20.1 | 20.8 | 21.5 | 21.8 |
| TOTAL TIME MSEC | 2.5920 | 1.3428 | 1.1376 | 0.9612 | 0.8514 | 0.6894 | 0.5670 |
| GLOBAL RAD TOTAL | 620.9 | 620.9 | 613.0 | 588.9 | 563.5 | 509.6 | 458.4 |
| DIRECT RAD TOTAL | 443.4 | 443.4 | 441.8 | 432.6 | 426.7 | 404.5 | 378.7 |
| RAD AV TIME MSEC | 0.0864 | 0.0450 | 0.0392 | 0.0346 | 0.0331 | 0.0295 | 0.0266 |
| DEG-C DAYS | 108. | 36. | 26. | 17. | 11. | 7. | 6. |

### APR

| SOLAR RAD GROUP | ALL | 0+ | 95+ | 189+ | 284+ | 378+ | 473+ |
|---|---|---|---|---|---|---|---|
| AV.THRSH.RAD.LEVL | 0. | 31. | 142. | 236. | 331. | 425. | 532. |
| AV.THRESHOLD TEMP | 5.7 | 8.0 | 8.6 | 9.7 | 14.3 | 14.7 | 16.1 |
| GL RAD OVER THRSH | 536.7 | 491.7 | 359.2 | 272.5 | 202.1 | 145.1 | 97.6 |
| TIME AVG.TEMP | 9.1 | 11.6 | 12.4 | 13.6 | 14.5 | 14.6 | 14.5 |
| TOTAL TIME MSEC | 2.5920 | 1.4706 | 1.1934 | 0.9162 | 0.7452 | 0.6030 | 0.4464 |
| GLOBAL RAD TOTAL | 536.8 | 536.7 | 528.2 | 489.0 | 448.6 | 401.5 | 335.0 |
| DIRECT RAD TOTAL | 279.4 | 279.4 | 278.8 | 269.6 | 264.2 | 252.7 | 226.5 |
| RAD AV TIME MSEC | 0.0864 | 0.0490 | 0.0410 | 0.0338 | 0.0313 | 0.0281 | 0.0238 |
| DEG-C DAYS | 278. | 117. | 84. | 54. | 37. | 30. | 22. |

### MAY

| SOLAR RAD GROUP | ALL | 0+ | 95+ | 189+ | 284+ | 378+ | 473+ |
|---|---|---|---|---|---|---|---|
| AV.THRSH.RAD.LEVL | 0. | 32. | 142. | 229. | 334. | 431. | 529. |
| AV.THRESHOLD TEMP | 9.2 | 11.4 | 12.8 | 13.7 | 16.0 | 16.8 | 19.4 |
| GL RAD OVER THRSH | 607.3 | 557.3 | 417.4 | 324.6 | 236.3 | 168.9 | 119.5 |
| TIME AVG.TEMP | 13.1 | 15.8 | 16.8 | 17.6 | 18.6 | 19.1 | 19.9 |
| TOTAL TIME MSEC | 2.6784 | 1.5624 | 1.2744 | 1.0638 | 0.8442 | 0.6912 | 0.5058 |
| GLOBAL RAD TOTAL | 607.3 | 607.3 | 598.1 | 568.2 | 518.0 | 466.9 | 387.0 |
| DIRECT RAD TOTAL | 303.5 | 303.5 | 302.8 | 299.5 | 288.5 | 279.2 | 246.9 |
| RAD AV TIME MSEC | 0.0864 | 0.0504 | 0.0421 | 0.0378 | 0.0324 | 0.0295 | 0.0245 |
| DEG-C DAYS | 166. | 61. | 40. | 26. | 16. | 11. | 6. |

### JUN

| SOLAR RAD GROUP | ALL | 0+ | 95+ | 189+ | 284+ | 378+ | 473+ |
|---|---|---|---|---|---|---|---|
| AV.THRSH.RAD.LEVL | 0. | 33. | 146. | 228. | 338. | 438. | 530. |
| AV.THRESHOLD TEMP | 16.1 | 18.7 | 21.8 | 21.4 | 23.8 | 24.0 | 24.8 |
| GL RAD OVER THRSH | 666.7 | 616.2 | 471.0 | 376.5 | 272.2 | 193.6 | 138.2 |
| TIME AVG.TEMP | 20.3 | 23.4 | 24.2 | 24.5 | 25.2 | 25.4 | 25.9 |
| TOTAL TIME MSEC | 2.5920 | 1.5120 | 1.2852 | 1.1520 | 0.9486 | 0.7900 | 0.6012 |
| GLOBAL RAD TOTAL | 666.7 | 666.7 | 659.2 | 639.7 | 593.2 | 539.6 | 456.9 |
| DIRECT RAD TOTAL | 383.4 | 383.4 | 362.5 | 380.3 | 365.0 | 354.6 | 315.3 |
| RAD AV TIME MSEC | 0.0864 | 0.0504 | 0.0428 | 0.0389 | 0.0328 | 0.0292 | 0.0234 |
| DEG-C DAYS | 11. | 1. | 0. | 0. | 0. | 0. | 0. |

### ALL MONTHS

| SOLAR RAD GROUP | ALL | 0+ | 95+ | 189+ | 284+ | 378+ | 473+ |
|---|---|---|---|---|---|---|---|
| TIME AVG.TEMP | 9.7 | 13.5 | 14.8 | 15.9 | 16.7 | 17.2 | 17.8 |
| TOTAL TIME MSEC | 31.5791 | 16.1568 | 13.0104 | 10.6272 | 8.6724 | 7.0200 | 5.3388 |
| GLOBAL RAD TOTAL | 6206.5 | 6206.6 | 6087.1 | 5745.7 | 5286.2 | 4737.1 | 4015.9 |
| DIRECT RAD TOTAL | 3799.7 | 3799.6 | 3776.2 | 3677.7 | 3552.1 | 3361.0 | 3007.3 |
| RAD AV TIME MSEC | 0.0864 | 0.0455 | 0.0385 | 0.0338 | 0.0303 | 0.0273 | 0.0235 |
| DEG-C DAYS | 3467. | 1268. | 897. | 633. | 482. | 368. | 261. |

### OCT

| SOLAR RAD GROUP | ALL | 0+ | 95+ | 189+ | 284+ | 378+ | 473+ |
|---|---|---|---|---|---|---|---|
| AV.THRSH.RAD.LEVL | 0. | 39. | 140. | 240. | 329. | 420. | 523. |
| AV.THRESHOLD TEMP | 7.0 | 13.3 | 9.3 | 14.5 | 15.2 | 18.8 | 15.7 |
| GL RAD OVER THRSH | 488.4 | 438.0 | 337.5 | 255.5 | 195.7 | 147.2 | 101.7 |
| TIME AVG.TEMP | 10.1 | 13.3 | 14.9 | 16.1 | 16.4 | 16.8 | 16.4 |
| TOTAL TIME MSEC | 2.6784 | 1.3014 | 0.9936 | 0.8154 | 0.6768 | 0.5310 | 0.4446 |
| GLOBAL RAD TOTAL | 488.5 | 488.5 | 476.5 | 451.6 | 418.3 | 370.3 | 334.0 |
| DIRECT RAD TOTAL | 328.3 | 328.3 | 324.4 | 320.4 | 309.8 | 287.7 | 278.2 |
| RAD AV TIME MSEC | 0.0864 | 0.0425 | 0.0346 | 0.0320 | 0.0292 | 0.0248 | 0.0248 |
| DEG-C DAYS | 257. | 83. | 49. | 31. | 24. | 17. | 15. |

### NOV

| SOLAR RAD GROUP | ALL | 0+ | 95+ | 189+ | 284+ | 378+ | 473+ |
|---|---|---|---|---|---|---|---|
| AV.THRSH.RAD.LEVL | 0. | 48. | 145. | 237. | 332. | 435. | 517. |
| AV.THRESHOLD TEMP | -0.4 | 1.2 | 3.2 | 6.2 | 6.4 | 6.4 | 7.1 |
| GL RAD OVER THRSH | 343.0 | 289.8 | 209.6 | 155.1 | 109.8 | 74.0 | 53.3 |
| TIME AVG.TEMP | 1.8 | 4.9 | 6.2 | 7.3 | 7.6 | 8.1 | 8.7 |
| TOTAL TIME MSEC | 2.5920 | 1.1070 | 0.8244 | 0.5940 | 0.4752 | 0.3510 | 0.2520 |
| GLOBAL RAD TOTAL | 343.1 | 343.0 | 329.4 | 296.0 | 267.8 | 226.5 | 183.5 |
| DIRECT RAD TOTAL | 233.2 | 233.2 | 230.9 | 218.7 | 208.2 | 189.4 | 161.4 |
| RAD AV TIME MSEC | 0.0864 | 0.0374 | 0.0328 | 0.0274 | 0.0252 | 0.0241 | 0.0223 |
| DEG-C DAYS | 494. | 172. | 116. | 76. | 59. | 42. | 28. |

### DEC

| SOLAR RAD GROUP | ALL | 0+ | 95+ | 189+ | 284+ | 378+ | 473+ |
|---|---|---|---|---|---|---|---|
| AV.THRSH.RAD.LEVL | 0. | 47. | 144. | 239. | 329. | 429. | 525. |
| AV.THRESHOLD TEMP | -4.6 | -2.0 | -2.4 | 4.2 | 2.4 | 0.0 | 7.4 |
| GL RAD OVER THRSH | 375.5 | 323.1 | 237.1 | 172.4 | 123.2 | 79.7 | 51.2 |
| TIME AVG.TEMP | -2.5 | 0.5 | 1.2 | 2.2 | 1.7 | 1.6 | 2.3 |
| TOTAL TIME MSEC | 2.6784 | 1.1160 | 0.8838 | 0.6822 | 0.5472 | 0.4356 | 0.2952 |
| GLOBAL RAD TOTAL | 375.5 | 375.5 | 364.6 | 335.5 | 303.3 | 266.5 | 206.3 |
| DIRECT RAD TOTAL | 280.8 | 280.8 | 277.8 | 261.0 | 248.9 | 228.1 | 182.5 |
| RAD AV TIME MSEC | 0.0864 | 0.0360 | 0.0324 | 0.0263 | 0.0245 | 0.0234 | 0.0184 |
| DEG-C DAYS | 645. | 230. | 176. | 127. | 106. | 84. | 55. |

STATISTICS FOR DENVER,COLO    1971-1972    IR INCLUDED    RSV023    FILE 7

RADIATION UNITS ARE WATTS/SQ.M OR MJOULES/SQ.M
SLOPE = 40.0 DEG    AZIMUTH = 180.0 DEG
DEGREE-DAY BASE TEMP = 18.3 DEGREES C

**JAN**

| SOLAR RAD GROUP | ALL | 0+ | 95+ | 189+ | 284+ | 378+ | 473+ |
|---|---|---|---|---|---|---|---|
| AV.THRSH.RAD.LEVL | 0. | 48. | 147. | 234. | 328. | 425. | 519. |
| AV.THRESHOLD TEMP | -2.4 | -2.2 | 2.4 | 5.6 | 2.9 | 5.4 | 0.0 |
| GL RAD OVER THRSH | 366.5 | 312.5 | 226.9 | 167.9 | 124.1 | 88.8 | 62.0 |
| TIME AVG.TEMP | -0.4 | 2.3 | 3.6 | 3.9 | 3.1 | 3.2 | 2.6 |
| TOTAL TIME MSEC | 2.6784 | 1.1160 | 0.8694 | 0.6768 | 0.4662 | 0.3636 | 0.2862 |
| GLOBAL RAD TOTAL | 366.5 | 366.5 | 354.6 | 326.3 | 277.0 | 243.4 | 210.5 |
| DIRECT RAD TOTAL | 252.7 | 252.7 | 250.2 | 239.9 | 222.2 | 205.1 | 186.1 |
| RAD AV TIME MSEC | 0.0864 | 0.0360 | 0.0320 | 0.0281 | 0.0248 | 0.0241 | 0.0230 |
| DEG-C DAYS | 582. | 208. | 149. | 113. | 82. | 64. | 52. |

**FEB**

| SOLAR RAD GROUP | ALL | 0+ | 95+ | 189+ | 284+ | 378+ | 473+ |
|---|---|---|---|---|---|---|---|
| AV.THRSH.RAD.LEVL | 0. | 50. | 135. | 235. | 326. | 428. | 525. |
| AV.THRESHOLD TEMP | -1.5 | -0.8 | 1.8 | 2.7 | 6.2 | 6.3 | 3.7 |
| GL RAD OVER THRSH | 416.8 | 360.7 | 284.8 | 212.0 | 159.1 | 113.3 | 78.4 |
| TIME AVG.TEMP | 0.7 | 3.3 | 4.3 | 4.9 | 5.4 | 5.2 | 4.9 |
| TOTAL TIME MSEC | 2.4624 | 1.1214 | 0.8892 | 0.7290 | 0.5832 | 0.4500 | 0.3600 |
| GLOBAL RAD TOTAL | 416.8 | 416.8 | 405.2 | 383.5 | 349.2 | 305.8 | 267.3 |
| DIRECT RAD TOTAL | 278.5 | 278.5 | 274.9 | 269.2 | 255.4 | 236.6 | 219.1 |
| RAD AV TIME MSEC | 0.0864 | 0.0392 | 0.0335 | 0.0302 | 0.0270 | 0.0238 | 0.0227 |
| DEG-C DAYS | 504. | 195. | 144. | 113. | 87. | 68. | 56. |

**MAR**

| SOLAR RAD GROUP | ALL | 0+ | 95+ | 189+ | 284+ | 378+ | 473+ |
|---|---|---|---|---|---|---|---|
| AV.THRSH.RAD.LEVL | 0. | 57. | 137. | 239. | 339. | 423. | 518. |
| AV.THRESHOLD,TEMP | 2.6 | 0.4 | 5.3 | 7.2 | 9.5 | 10.8 | 11.3 |
| GL RAD OVER THRSH | 505.5 | 436.1 | 348.1 | 255.8 | 179.7 | 130.3 | 88.3 |
| TIME AVG.TEMP | 5.5 | 8.9 | 9.9 | 10.9 | 11.6 | 12.2 | 12.6 |
| TOTAL TIME MSEC | 2.6784 | 1.2276 | 1.0980 | 0.9000 | 0.7614 | 0.5886 | 0.4446 |
| GLOBAL RAD TOTAL | 505.6 | 505.5 | 498.2 | 471.1 | 438.0 | 379.3 | 318.4 |
| DIRECT RAD TOTAL | 275.8 | 275.8 | 274.4 | 267.7 | 261.4 | 238.6 | 217.3 |
| RAD AV TIME MSEC | 0.0864 | 0.0396 | 0.0371 | 0.0320 | 0.0295 | 0.0256 | 0.0227 |
| DEG-C DAYS | 397. | 136. | 109. | 80. | 63. | 45. | 33. |

**JUL**

| SOLAR RAD GROUP | ALL | 0+ | 95+ | 189+ | 284+ | 378+ | 473+ |
|---|---|---|---|---|---|---|---|
| AV.THRSH.RAD.LEVL | 0. | 28. | 151. | 236. | 339. | 418. | 522. |
| AV.THRESHOLD TEMP | 17.8 | 22.3 | 18.4 | 22.3 | 24.7 | 25.4 | 27.1 |
| GL RAD OVER THRSH | 583.2 | 547.5 | 400.3 | 314.9 | 228.7 | 173.8 | 118.0 |
| TIME AVG.TEMP | 21.1 | 24.6 | 24.8 | 25.9 | 26.7 | 27.1 | 27.6 |
| TOTAL TIME MSEC | 2.6784 | 1.2798 | 1.1952 | 1.0098 | 0.8316 | 0.6966 | 0.5364 |
| GLOBAL RAD TOTAL | 583.3 | 583.2 | 580.8 | 552.9 | 510.9 | 465.1 | 398.1 |
| DIRECT RAD TOTAL | 328.9 | 328.9 | 328.8 | 322.4 | 318.6 | 307.7 | 277.2 |
| RAD AV TIME MSEC | 0.0864 | 0.0418 | 0.0392 | 0.0349 | 0.0313 | 0.0281 | 0.0234 |
| DEG-C DAYS | 19. | 7. | 5. | 2. | 1. | 0. | 0. |

**AUG**

| SOLAR RAD GROUP | ALL | 0+ | 95+ | 189+ | 284+ | 378+ | 473+ |
|---|---|---|---|---|---|---|---|
| AV.THRSH.RAD.LEVL | 0. | 37. | 148. | 245. | 335. | 418. | 526. |
| AV.THRESHOLD TEMP | 18.5 | 20.0 | 19.9 | 24.8 | 25.1 | 24.2 | 26.7 |
| GL RAD OVER THRSH | 631.4 | 585.2 | 452.0 | 351.6 | 269.1 | 206.5 | 140.9 |
| TIME AVG.TEMP | 21.5 | 25.0 | 25.1 | 26.0 | 26.2 | 26.5 | 26.9 |
| TOTAL TIME MSEC | 2.6784 | 1.2366 | 1.2096 | 1.0332 | 0.9108 | 0.7578 | 0.6066 |
| GLOBAL RAD TOTAL | 631.6 | 631.5 | 630.5 | 604.5 | 574.5 | 523.2 | 460.0 |
| DIRECT RAD TOTAL | 404.5 | 404.5 | 404.5 | 395.2 | 389.2 | 370.1 | 339.0 |
| RAD AV TIME MSEC | 0.0864 | 0.0400 | 0.0392 | 0.0346 | 0.0313 | 0.0277 | 0.0238 |
| DEG-C DAYS | 6. | 1. | 1. | 0. | 0. | 0. | 0. |

**SEP**

| SOLAR RAD GROUP | ALL | 0+ | 95+ | 189+ | 284+ | 378+ | 473+ |
|---|---|---|---|---|---|---|---|
| AV.THRSH.RAD.LEVL | 0. | 65. | 139. | 233. | 331. | 412. | 522. |
| AV.THRESHOLD,TEMP | 12.6 | 9.4 | 12.7 | 14.7 | 18.6 | 19.9 | 21.4 |
| GL RAD OVER THRSH | 613.9 | 536.2 | 455.7 | 365.4 | 283.5 | 227.5 | 166.0 |
| TIME AVG.TEMP | 15.3 | 18.4 | 19.2 | 20.2 | 20.9 | 21.4 | 21.7 |
| TOTAL TIME MSEC | 2.5920 | 1.2276 | 1.0998 | 0.9522 | 0.8406 | 0.6876 | 0.5634 |
| GLOBAL RAD TOTAL | 614.0 | 613.9 | 608.2 | 587.7 | 561.6 | 511.0 | 459.8 |
| DIRECT RAD TOTAL | 443.2 | 443.2 | 442.0 | 436.0 | 429.9 | 409.7 | 382.1 |
| RAD AV TIME MSEC | 0.0864 | 0.0396 | 0.0378 | 0.0346 | 0.0328 | 0.0299 | 0.0266 |
| DEG-C DAYS | 108. | 34. | 25. | 17. | 11. | 7. | 6. |

**OCT**

| SOLAR RAD GROUP | ALL | 0+ | 95+ | 189+ | 284+ | 378+ | 473+ |
|---|---|---|---|---|---|---|---|
| AV.THRSH.RAD.LEVL | 0. | 48. | 136. | 237. | 332. | 418. | 524. |
| AV.THRESHOLD TEMP | 7.3 | 6.1 | 9.3 | 15.3 | 15.6 | 17.9 | 16.5 |
| GL RAD OVER THRSH | 502.7 | 443.9 | 355.2 | 273.1 | 208.1 | 162.7 | 116.0 |
| TIME AVG.TEMP | 10.1 | 13.3 | 14.8 | 16.2 | 16.3 | 16.6 | 16.3 |
| TOTAL TIME MSEC | 2.6784 | 1.2258 | 1.0030 | 0.8118 | 0.6658 | 0.5292 | 0.4410 |
| GLOBAL RAD TOTAL | 502.7 | 502.0 | 492.2 | 465.6 | 435.7 | 383.7 | 346.9 |
| DIRECT RAD TOTAL | 347.0 | 347.0 | 344.6 | 336.6 | 330.8 | 305.1 | 293.4 |
| RAD AV TIME MSEC | 0.0864 | 0.0396 | 0.0356 | 0.0324 | 0.0302 | 0.0252 | 0.0248 |
| DEG-C DAYS | 257. | 79. | 49. | 30. | 24. | 18. | 16. |

**NOV**

| SOLAR RAD GROUP | ALL | 0+ | 95+ | 189+ | 284+ | 378+ | 473+ |
|---|---|---|---|---|---|---|---|
| AV.THRSH.RAD.LEVL | 0. | 48. | 143. | 233. | 335. | 424. | 522. |
| AV.THRESHOLD TEMP | -0.4 | 0.9 | 3.3 | 6.0 | 6.1 | 5.9 | 6.9 |
| GL RAD OVER THRSH | 362.4 | 309.0 | 230.8 | 175.3 | 126.2 | 94.1 | 65.0 |
| TIME AVG.TEMP | 1.8 | 4.8 | 6.2 | 7.1 | 7.4 | 7.9 | 8.3 |
| TOTAL TIME MSEC | 2.5920 | 1.1016 | 0.8244 | 0.6192 | 0.4788 | 0.3618 | 0.2970 |
| GLOBAL RAD TOTAL | 362.4 | 362.4 | 348.9 | 319.5 | 286.8 | 247.6 | 220.1 |
| DIRECT RAD TOTAL | 255.8 | 255.2 | 253.7 | 244.7 | 229.1 | 211.5 | 194.1 |
| RAD AV TIME MSEC | 0.0864 | 0.0371 | 0.0331 | 0.0295 | 0.0256 | 0.0248 | 0.0241 |
| DEG-C DAYS | 494. | 172. | 116. | 81. | 61. | 44. | 35. |

**DEC**

| SOLAR RAD GROUP | ALL | 0+ | 95+ | 189+ | 284+ | 378+ | 473+ |
|---|---|---|---|---|---|---|---|
| AV.THRSH.RAD.LEVL | 0. | 47. | 148. | 239. | 330. | 431. | 502. |
| AV.THRESHOLD TEMP | -4.6 | -2.0 | -2.5 | 4.5 | 2.5 | 1.6 | 1.9 |
| GL RAD OVER THRSH | 402.0 | 349.3 | 261.0 | 193.5 | 148.0 | 102.0 | 76.7 |
| TIME AVG.TEMP | -2.5 | 0.5 | 1.2 | 2.2 | 1.7 | 1.6 | 1.6 |
| TOTAL TIME MSEC | 2.6784 | 1.1160 | 0.8802 | 0.6822 | 0.5580 | 0.4554 | 0.3528 |
| GLOBAL RAD TOTAL | 402.0 | 402.0 | 390.9 | 361.7 | 332.0 | 298.1 | 254.0 |
| DIRECT RAD TOTAL | 309.9 | 309.9 | 306.7 | 289.0 | 277.0 | 258.4 | 226.2 |
| RAD AV TIME MSEC | 0.0664 | 0.0360 | 0.0324 | 0.0263 | 0.0248 | 0.0238 | 0.0216 |
| DEG-C DAYS | 645. | 230. | 174. | 127. | 107. | 88. | 68. |

**APR**

| SOLAR RAD GROUP | ALL | 0+ | 95+ | 189+ | 284+ | 378+ | 473+ |
|---|---|---|---|---|---|---|---|
| AV.THRSH.RAD.LEVL | 0. | 62. | 142. | 239. | 330. | 418. | 527. |
| AV.THRESHOLD TEMP | 6.6 | 3.8 | 7.2 | 10.5 | 14.7 | 14.4 | 16.5 |
| GL RAD OVER THRSH | 508.2 | 434.4 | 343.9 | 258.5 | 191.7 | 141.0 | 93.1 |
| TIME AVG.TEMP | 9.1 | 11.9 | 12.4 | 13.8 | 14.5 | 14.4 | 14.4 |
| TOTAL TIME MSEC | 2.5920 | 1.1880 | 1.1268 | 0.8892 | 0.7290 | 0.5796 | 0.4374 |
| GLOBAL RAD TOTAL | 508.4 | 508.3 | 504.5 | 470.6 | 432.4 | 383.1 | 323.7 |
| DIRECT RAD TOTAL | 267.7 | 267.7 | 267.5 | 261.7 | 257.2 | 245.3 | 221.2 |
| RAD AV TIME MSEC | 0.0864 | 0.0396 | 0.0385 | 0.0335 | 0.0310 | 0.0277 | 0.0238 |
| DEG-C DAYS | 278. | 91. | 81. | 51. | 36. | 29. | 22. |

**MAY**

| SOLAR RAD GROUP | ALL | 0+ | 95+ | 189+ | 284+ | 378+ | 473+ |
|---|---|---|---|---|---|---|---|
| AV.THRSH.RAD.LEVL | 0. | 37. | 148. | 241. | 335. | 418. | 527. |
| AV.THRESHOLD TEMP | 9.9 | 13.1 | 11.6 | 14.0 | 16.7 | 16.6 | 19.5 |
| GL RAD OVER THRSH | 559.9 | 512.9 | 380.8 | 289.5 | 212.3 | 158.4 | 105.3 |
| TIME AVG.TEMP | 13.1 | 16.6 | 16.8 | 17.9 | 18.7 | 19.2 | 20.0 |
| TOTAL TIME MSEC | 2.6784 | 1.2654 | 1.1880 | 0.9792 | 0.8172 | 0.6516 | 0.4896 |
| GLOBAL RAD TOTAL | 559.2 | 559.1 | 556.3 | 525.4 | 486.4 | 430.9 | 363.1 |
| DIRECT RAD TOTAL | 279.5 | 279.5 | 279.4 | 272.8 | 269.7 | 257.4 | 231.3 |
| RAD AV TIME MSEC | 0.0864 | 0.0410 | 0.0392 | 0.0346 | 0.0320 | 0.0284 | 0.0241 |
| DEG-C DAYS | 166. | 42. | 37. | 23. | 15. | 9. | 5. |

**JUN**

| SOLAR RAD GROUP | ALL | 0+ | 95+ | 189+ | 284+ | 378+ | 473+ |
|---|---|---|---|---|---|---|---|
| AV.THRSH.RAD.LEVL | 0. | 7. | 158. | 239. | 335. | 420. | 524. |
| AV.THRESHOLD TEMP | 16.7 | 23.3 | 20.1 | 21.6 | 24.1 | 23.9 | 24.7 |
| GL RAD OVER THRSH | 603.9 | 595.6 | 416.0 | 332.0 | 242.5 | 179.2 | 118.6 |
| TIME AVG.TEMP | 20.3 | 24.2 | 24.3 | 24.8 | 25.2 | 25.5 | 25.9 |
| TOTAL TIME MSEC | 2.5920 | 1.2528 | 1.1844 | 1.0440 | 0.9306 | 0.7470 | 0.5814 |
| GLOBAL RAD TOTAL | 604.0 | 604.0 | 603.5 | 581.3 | 554.2 | 492.7 | 423.2 |
| DIRECT RAD TOTAL | 346.5 | 346.5 | 346.5 | 338.9 | 335.5 | 320.6 | 289.0 |
| RAD AV TIME MSEC | 0.0864 | 0.0418 | 0.0396 | 0.0349 | 0.0324 | 0.0277 | 0.0230 |
| DEG-C DAYS | 11. | 0. | 0. | 0. | 0. | 0. | 0. |

**ALL MONTHS**

| SOLAR RAD GROUP | ALL | 0+ | 95+ | 189+ | 284+ | 378+ | 473+ |
|---|---|---|---|---|---|---|---|
| TIME AVG.TEMP | 9.7 | 13.2 | 14.5 | 15.8 | 16.5 | 17.0 | 17.2 |
| TOTAL TIME MSEC | 31.5791 | 14.3190 | 12.5730 | 10.3266 | 8.5932 | 6.8688 | 5.3964 |
| GLOBAL RAD TOTAL | 6056.5 | 6055.9 | 5973.8 | 5650.1 | 5238.7 | 4663.9 | 4045.0 |
| DIRECT RAD TOTAL | 3790.1 | 3790.0 | 3773.2 | 3676.8 | 3576.8 | 3366.1 | 3076.0 |
| RAD AV TIME MSEC | 0.0864 | 0.0396 | 0.0370 | 0.0327 | 0.0300 | 0.0268 | 0.0238 |
| DEG-C DAYS | 3467. | 1194. | 891. | 636. | 487. | 374. | 293. |

STATISTICS FOR DENVER,COLO    1971-1972    IR INCLUDED    RSV023  FILE 7

RADIATION UNITS ARE WATTS/SQ.M OR MJOULES/SQ.M
SLOPE = 50.0 DEG   AZIMUTH = 180.0 DEG
DEGREE-DAY BASE TEMP = 18.3 DEGREES C

**JAN**

| SOLAR RAD GROUP | ALL | 0+ | 95+ | 189+ | 284+ | 378+ | 473+ |
|---|---|---|---|---|---|---|---|
| AV.THRSH.RAD.LEVL | 0. | 47. | 146. | 234. | 330. | 426. | 525. |
| AV.THRESHOLD TEMP | -2.4 | -2.0 | 2.3 | 5.2 | 3.4 | 5.4 | -1.3 |
| GL RAD OVER THRSH | 377.0 | 324.4 | 238.5 | 179.3 | 135.1 | 99.4 | 70.9 |
| TIME AVG.TEMP | -0.4 | 2.3 | 3.5 | 3.8 | 3.2 | 3.2 | 2.6 |
| TOTAL TIME MSEC | 2.6784 | 1.1160 | 0.8676 | 0.6732 | 0.4626 | 0.3690 | 0.2898 |
| GLOBAL RAD TOTAL | 377.0 | 377.0 | 365.3 | 336.9 | 287.6 | 256.7 | 223.0 |
| DIRECT RAD TOTAL | 267.8 | 267.8 | 265.4 | 254.7 | 235.3 | 218.8 | 198.6 |
| RAD AV TIME MSEC | 0.0864 | 0.0360 | 0.0324 | 0.0284 | 0.0252 | 0.0241 | 0.0234 |
| DEG-C DAYS | 582. | 208. | 149. | 113. | 81. | 65. | 53. |

**FEB**

| SOLAR RAD GROUP | ALL | 0+ | 95+ | 189+ | 284+ | 378+ | 473+ |
|---|---|---|---|---|---|---|---|
| AV.THRSH.RAD.LEVL | 0. | 42. | 133. | 236. | 329. | 426. | 518. |
| AV.THRESHOLD TEMP | -2.0 | 0.7 | 1.7 | 3.2 | 6.3 | 6.4 | 4.3 |
| GL RAD OVER THRSH | 423.7 | 372.8 | 291.1 | 216.9 | 163.2 | 119.9 | 86.7 |
| TIME AVG.TEMP | 0.7 | 3.4 | 4.3 | 4.9 | 5.4 | 5.1 | 4.8 |
| TOTAL TIME MSEC | 2.4624 | 1.2078 | 0.8946 | 0.7254 | 0.5778 | 0.4428 | 0.3618 |
| GLOBAL RAD TOTAL | 423.7 | 423.7 | 410.5 | 387.9 | 353.1 | 308.7 | 274.2 |
| DIRECT RAD TOTAL | 290.1 | 290.1 | 285.5 | 278.4 | 265.0 | 244.6 | 226.6 |
| RAD AV TIME MSEC | 0.0864 | 0.0425 | 0.0342 | 0.0306 | 0.0274 | 0.0241 | 0.0227 |
| DEG-C DAYS | 504. | 209. | 145. | 113. | 87. | 68. | 57. |

**MAR**

| SOLAR RAD GROUP | ALL | 0+ | 95+ | 189+ | 284+ | 378+ | 473+ |
|---|---|---|---|---|---|---|---|
| AV.THRSH.RAD.LEVL | 0. | 47. | 136. | 240. | 336. | 422. | 517. |
| AV.THRESHOLD TEMP | 2.3 | 3.4 | 5.5 | 7.3 | 10.0 | 10.7 | 12.0 |
| GL RAD OVER THRSH | 497.4 | 435.4 | 337.4 | 245.6 | 173.6 | 124.4 | 84.5 |
| TIME AVG.TEMP | 5.5 | 8.9 | 9.9 | 11.0 | 12.7 | 12.7 | 12.7 |
| TOTAL TIME MSEC | 2.6784 | 1.3104 | 1.1070 | 0.8820 | 0.7470 | 0.5760 | 0.4176 |
| GLOBAL RAD TOTAL | 497.5 | 497.5 | 487.8 | 457.3 | 424.9 | 367.3 | 300.5 |
| DIRECT RAD TOTAL | 276.0 | 276.0 | 274.0 | 265.4 | 258.3 | 236.8 | 211.6 |
| RAD AV TIME MSEC | 0.0864 | 0.0425 | 0.0378 | 0.0317 | 0.0292 | 0.0256 | 0.0223 |
| DEG-C DAYS | 397. | 145. | 111. | 78. | 61. | 44. | 30. |

**JUL**

| SOLAR RAD GROUP | ALL | 0+ | 95+ | 189+ | 284+ | 378+ | 473+ |
|---|---|---|---|---|---|---|---|
| AV.THRSH.RAD.LEVL | 0. | 47. | 135. | 235. | 335. | 430. | 521. |
| AV.THRESHOLD TEMP | 17.6 | 21.2 | 19.0 | 23.1 | 24.5 | 27.6 | 26.3 |
| GL RAD OVER THRSH | 538.6 | 476.1 | 373.6 | 275.7 | 195.1 | 137.6 | 91.9 |
| TIME AVG.TEMP | 21.1 | 24.5 | 25.0 | 26.2 | 26.8 | 27.6 | 27.6 |
| TOTAL TIME MSEC | 2.6784 | 1.3392 | 1.1628 | 0.9738 | 0.8100 | 0.6066 | 0.5004 |
| GLOBAL RAD TOTAL | 538.6 | 538.6 | 530.3 | 504.9 | 466.3 | 398.2 | 352.6 |
| DIRECT RAD TOTAL | 292.9 | 292.9 | 291.4 | 288.2 | 284.7 | 258.4 | 245.0 |
| RAD AV TIME MSEC | 0.0864 | 0.0432 | 0.0382 | 0.0338 | 0.0310 | 0.0245 | 0.0223 |
| DEG-C DAYS | 19. | 7. | 4. | 2. | 1. | 0. | 0. |

**AUG**

| SOLAR RAD GROUP | ALL | 0+ | 95+ | 189+ | 284+ | 378+ | 473+ |
|---|---|---|---|---|---|---|---|
| AV.THRSH.RAD.LEVL | 0. | 29. | 131. | 244. | 340. | 427. | 524. |
| AV.THRESHOLD TEMP | 18.1 | 22.7 | 20.5 | 24.6 | 24.3 | 25.8 | 26.5 |
| GL RAD OVER THRSH | 594.6 | 555.4 | 434.4 | 318.8 | 234.8 | 175.3 | 119.3 |
| TIME AVG.TEMP | 21.5 | 24.9 | 25.2 | 26.0 | 26.2 | 26.8 | 26.9 |
| TOTAL TIME MSEC | 2.6784 | 1.3392 | 1.1916 | 1.0224 | 0.8784 | 0.6786 | 0.5778 |
| GLOBAL RAD TOTAL | 594.6 | 594.6 | 590.3 | 568.2 | 533.0 | 465.2 | 422.1 |
| DIRECT RAD TOTAL | 374.2 | 374.2 | 373.3 | 366.9 | 358.5 | 325.5 | 311.5 |
| RAD AV TIME MSEC | 0.0864 | 0.0432 | 0.0389 | 0.0342 | 0.0306 | 0.0248 | 0.0234 |
| DEG-C DAYS | 6. | 1. | 1. | 0. | 0. | 0. | 0. |

**SEP**

| SOLAR RAD GROUP | ALL | 0+ | 95+ | 189+ | 284+ | 378+ | 473+ |
|---|---|---|---|---|---|---|---|
| AV.THRSH.RAD.LEVL | 0. | 43. | 137. | 243. | 337. | 412. | 518. |
| AV.THRESHOLD TEMP | 12.1 | 14.1 | 13.2 | 16.4 | 19.0 | 19.1 | 21.5 |
| GL RAD OVER THRSH | 602.3 | 546.8 | 442.4 | 344.4 | 267.5 | 218.1 | 160.6 |
| TIME AVG.TEMP | 15.3 | 16.5 | 19.2 | 20.3 | 20.9 | 20.7 | 21.8 |
| TOTAL TIME MSEC | 2.5920 | 1.2798 | 1.1106 | 0.9233 | 0.8136 | 0.6570 | 0.5472 |
| GLOBAL RAD TOTAL | 602.3 | 602.3 | 594.9 | 570.0 | 542.0 | 489.1 | 443.9 |
| DIRECT RAD TOTAL | 435.1 | 435.1 | 433.1 | 425.9 | 418.2 | 393.0 | 370.9 |
| RAD AV TIME MSEC | 0.0864 | 0.0428 | 0.0385 | 0.0342 | 0.0324 | 0.0288 | 0.0266 |
| DEG-C DAYS | 108. | 35. | 26. | 15. | 10. | 7. | 5. |

**APR**

| SOLAR RAD GROUP | ALL | 0+ | 95+ | 189+ | 284+ | 378+ | 473+ |
|---|---|---|---|---|---|---|---|
| AV.THRSH.RAD.LEVL | 0. | 35. | 138. | 240. | 333. | 414. | 525. |
| AV.THRESHOLD TEMP | 6.1 | 8.8 | 7.7 | 11.3 | 14.5 | 14.5 | 17.3 |
| GL RAD OVER THRSH | 483.4 | 438.3 | 324.1 | 235.5 | 171.1 | 127.0 | 81.2 |
| TIME AVG.TEMP | 9.1 | 12.0 | 12.6 | 13.9 | 14.5 | 14.5 | 14.5 |
| TOTAL TIME MSEC | 2.5920 | 1.2942 | 1.1052 | 0.8658 | 0.6984 | 0.5400 | 0.4140 |
| GLOBAL RAD TOTAL | 483.4 | 483.4 | 476.8 | 443.7 | 403.5 | 350.8 | 298.6 |
| DIRECT RAD TOTAL | 252.4 | 252.4 | 251.7 | 246.8 | 241.3 | 224.8 | 206.1 |
| RAD AV TIME MSEC | 0.0864 | 0.0432 | 0.0382 | 0.0331 | 0.0302 | 0.0263 | 0.0230 |
| DEG-C DAYS | 278. | 98. | 77. | 49. | 34. | 27. | 21. |

**MAY**

| SOLAR RAD GROUP | ALL | 0+ | 95+ | 189+ | 284+ | 378+ | 473+ |
|---|---|---|---|---|---|---|---|
| AV.THRSH.RAD.LEVL | 0. | 44. | 133. | 238. | 336. | 417. | 521. |
| AV.THRESHOLD TEMP | 9.7 | 13.3 | 11.9 | 14.2 | 16.6 | 17.5 | 20.1 |
| GL RAD OVER THRSH | 520.3 | 461.2 | 359.0 | 257.7 | 180.2 | 134.0 | 85.4 |
| TIME AVG.TEMP | 13.1 | 16.4 | 16.9 | 17.9 | 18.8 | 19.6 | 20.1 |
| TOTAL TIME MSEC | 2.6784 | 1.3392 | 1.1556 | 0.9630 | 0.7866 | 0.5688 | 0.4698 |
| GLOBAL RAD TOTAL | 520.3 | 520.3 | 512.2 | 486.6 | 444.7 | 371.5 | 330.1 |
| DIRECT RAD TOTAL | 251.4 | 251.4 | 250.7 | 246.8 | 242.9 | 217.9 | 208.3 |
| RAD AV TIME MSEC | 0.0864 | 0.0432 | 0.0382 | 0.0342 | 0.0313 | 0.0259 | 0.0238 |
| DEG-C DAYS | 166. | 46. | 34. | 23. | 14. | 7. | 5. |

**JUN**

| SOLAR RAD GROUP | ALL | 0+ | 95+ | 189+ | 284+ | 378+ | 473+ |
|---|---|---|---|---|---|---|---|
| AV.THRSH.RAD.LEVL | 0. | 55. | 132. | 246. | 331. | 426. | 524. |
| AV.THRESHOLD TEMP | 16.4 | 21.9 | 20.5 | 23.6 | 23.3 | 25.2 | 24.6 |
| GL RAD OVER THRSH | 553.7 | 482.9 | 395.6 | 278.8 | 203.4 | 140.5 | 88.3 |
| TIME AVG.TEMP | 16.4 | 24.2 | 24.6 | 25.0 | 25.2 | 25.9 | 26.1 |
| TOTAL TIME MSEC | 2.5920 | 1.2960 | 1.1340 | 1.0224 | 0.8856 | 0.6588 | 0.5364 |
| GLOBAL RAD TOTAL | 553.7 | 553.7 | 544.8 | 530.1 | 496.5 | 421.5 | 369.3 |
| DIRECT RAD TOTAL | 304.0 | 304.0 | 301.4 | 299.2 | 294.0 | 267.4 | 249.8 |
| RAD AV TIME MSEC | 0.0864 | 0.0432 | 0.0378 | 0.0342 | 0.0310 | 0.0248 | 0.0220 |
| DEG-C DAYS | 11. | 0. | 0. | 0. | 0. | 0. | 0. |

**OCT**

| SOLAR RAD GROUP | ALL | 0+ | 95+ | 189+ | 284+ | 378+ | 473+ |
|---|---|---|---|---|---|---|---|
| AV.THRSH.RAD.LEVL | 0. | 44. | 135. | 239. | 337. | 418. | 519. |
| AV.THRESHOLD TEMP | 6.9 | 7.9 | 10.5 | 15.2 | 15.8 | 17.7 | 17.3 |
| GL RAD OVER THRSH | 510.2 | 451.9 | 358.7 | 276.2 | 209.8 | 168.0 | 124.1 |
| TIME AVG.TEMP | 10.1 | 13.3 | 14.9 | 16.2 | 16.3 | 16.5 | 16.3 |
| TOTAL TIME MSEC | 2.6784 | 1.3320 | 1.0170 | 0.7992 | 0.6768 | 0.5148 | 0.4338 |
| GLOBAL RAD TOTAL | 510.2 | 510.2 | 496.4 | 466.9 | 437.7 | 383.2 | 349.3 |
| DIRECT RAD TOTAL | 358.8 | 358.8 | 354.9 | 346.6 | 338.6 | 311.2 | 299.3 |
| RAD AV TIME MSEC | 0.0864 | 0.0432 | 0.0367 | 0.0324 | 0.0302 | 0.0256 | 0.0248 |
| DEG-C DAYS | 257. | 85. | 49. | 29. | 24. | 18. | 15. |

**NOV**

| SOLAR RAD GROUP | ALL | 0+ | 95+ | 189+ | 284+ | 378+ | 473+ |
|---|---|---|---|---|---|---|---|
| AV.THRSH.RAD.LEVL | 0. | 46. | 144. | 236. | 335. | 424. | 521. |
| AV.THRESHOLD TEMP | -0.5 | 1.4 | 3.7 | 5.7 | 6.3 | 5.9 | 6.2 |
| GL RAD OVER THRSH | 372.8 | 320.8 | 241.3 | 184.5 | 137.5 | 104.7 | 75.0 |
| TIME AVG.TEMP | 1.8 | 4.9 | 6.2 | 7.0 | 7.4 | 7.7 | 8.1 |
| TOTAL TIME MSEC | 2.5920 | 1.1196 | 0.8118 | 0.6192 | 0.4770 | 0.3654 | 0.3078 |
| GLOBAL RAD TOTAL | 372.3 | 372.3 | 358.5 | 330.7 | 297.1 | 259.8 | 235.3 |
| DIRECT RAD TOTAL | 270.3 | 270.3 | 267.9 | 259.3 | 242.3 | 225.4 | 208.3 |
| RAD AV TIME MSEC | 0.0864 | 0.0382 | 0.0331 | 0.0302 | 0.0259 | 0.0256 | 0.0241 |
| DEG-C DAYS | 494. | 174. | 114. | 81. | 61. | 45. | 37. |

**DEC**

| SOLAR RAD GROUP | ALL | 0+ | 95+ | 189+ | 284+ | 378+ | 473+ |
|---|---|---|---|---|---|---|---|
| AV.THRSH.RAD.LEVL | 0. | 48. | 150. | 237. | 327. | 428. | 510. |
| AV.THRESHOLD TEMP | -4.6 | -2.0 | -2.3 | 4.5 | 1.8 | 2.0 | 1.4 |
| GL RAD OVER THRSH | 418.0 | 364.8 | 276.0 | 217.1 | 166.3 | 119.4 | 88.3 |
| TIME AVG.TEMP | -2.5 | 0.5 | 1.2 | 2.2 | 1.8 | 1.8 | 1.7 |
| TOTAL TIME MSEC | 2.6784 | 1.1160 | 0.8694 | 0.6768 | 0.5652 | 0.4644 | 0.3780 |
| GLOBAL RAD TOTAL | 413.0 | 418.0 | 406.3 | 377.4 | 351.0 | 318.0 | 281.1 |
| DIRECT RAD TOTAL | 329.1 | 329.1 | 325.8 | 308.6 | 298.0 | 278.9 | 252.0 |
| RAD AV TIME MSEC | 0.0864 | 0.0360 | 0.0324 | 0.0263 | 0.0248 | 0.0241 | 0.0230 |
| DEG-C DAYS | 645. | 230. | 172. | 126. | 108. | 89. | 73. |

**ALL MONTHS**

| SOLAR RAD GROUP | ALL | 0+ | 95+ | 189+ | 284+ | 378+ | 473+ |
|---|---|---|---|---|---|---|---|
| TIME AVG.TEMP | 9.7 | 13.3 | 14.6 | 15.8 | 16.5 | 16.7 | 16.9 |
| TOTAL TIME MSEC | 31.5791 | 15.0894 | 12.4272 | 10.1520 | 8.3790 | 6.4442 | 5.2344 |
| GLOBAL RAD TOTAL | 5892.0 | 5891.9 | 5774.1 | 5460.5 | 5037.3 | 4390.6 | 3879.9 |
| DIRECT RAD TOTAL | 3702.1 | 3702.1 | 3675.3 | 3586.3 | 3477.1 | 3202.8 | 2988.1 |
| RAD AV TIME MSEC | 0.0864 | 0.0418 | 0.0367 | 0.0323 | 0.0295 | 0.0255 | 0.0236 |
| DEG-C DAYS | 3467. | 1238. | 882. | 628. | 481. | 370. | 295. |

STATISTICS FOR DENVER,COLO    1971-1972    IR INCLUDED    RSV023    FILE 7

RADIATION UNITS ARE WATTS/SQ.M OR MJOULES/SQ.M
SLOPE = 60.0 DEG   AZIMUTH = 180.0 DEG
DEGREE-DAY BASE TEMP = 18.3 DEGREES C

**JAN**

| SOLAR RAD GROUP | ALL | 0+ | 95+ | 189+ | 284+ | 378+ | 473+ |
|---|---|---|---|---|---|---|---|
| AV.THRSH.RAD.LEVL | 0. | 46. | 147. | 235. | 333. | 424. | 524. |
| AV.THRESHOLD TEMP | -2.4 | -2.0 | 3.0 | 5.2 | 2.8 | 5.6 | -1.9 |
| GL RAD OVER THRSH | 378.7 | 326.9 | 240.0 | 182.4 | 138.4 | 105.2 | 75.7 |
| TIME AVG.TEMP | -0.4 | 2.3 | 3.6 | 3.7 | 3.1 | 3.1 | 2.5 |
| TOTAL TIME MSEC | 2.6784 | 1.1160 | 0.8604 | 0.6552 | 0.4518 | 0.3654 | 0.2934 |
| GLOBAL RAD TOTAL | 378.7 | 378.7 | 366.8 | 336.6 | 288.7 | 260.0 | 229.5 |
| DIRECT RAD TOTAL | 274.7 | 274.7 | 272.2 | 261.1 | 240.1 | 224.3 | 205.1 |
| RAD AV TIME MSEC | 0.0864 | 0.0360 | 0.0324 | 0.0284 | 0.0252 | 0.0241 | 0.0238 |
| DEG-C DAYS | 582. | 208. | 147. | 111. | 80. | 64. | 54. |

**FEB**

| SOLAR RAD GROUP | ALL | 0+ | 95+ | 189+ | 284+ | 378+ | 473+ |
|---|---|---|---|---|---|---|---|
| AV.THRSH.RAD.LEVL | 0. | 44. | 137. | 235. | 327. | 424. | 520. |
| AV.THRESHOLD TEMP | -2.0 | 0.6 | 2.4 | 3.9 | 5.9 | 5.7 | 5.8 |
| GL RAD OVER THRSH | 418.6 | 365.6 | 284.4 | 213.9 | 162.1 | 119.7 | 85.4 |
| TIME AVG.TEMP | 0.7 | 3.4 | 4.4 | 4.9 | 5.2 | 4.9 | 4.8 |
| TOTAL TIME MSEC | 2.4624 | 1.2078 | 0.8748 | 0.7164 | 0.5652 | 0.4374 | 0.3546 |
| GLOBAL RAD TOTAL | 418.6 | 418.6 | 404.0 | 382.3 | 346.8 | 305.0 | 269.9 |
| DIRECT RAD TOTAL | 291.3 | 291.3 | 286.8 | 278.9 | 265.8 | 246.0 | 226.4 |
| RAD AV TIME MSEC | 0.0864 | 0.0425 | 0.0342 | 0.0306 | 0.0274 | 0.0248 | 0.0230 |
| DEG-C DAYS | 504. | 209. | 141. | 112. | 87. | 68. | 56. |

**MAR**

| SOLAR RAD GROUP | ALL | 0+ | 95+ | 189+ | 284+ | 378+ | 473+ |
|---|---|---|---|---|---|---|---|
| AV.THRSH.RAD.LEVL | 0. | 46. | 136. | 239. | 336. | 425. | 521. |
| AV.THRESHOLD TEMP | 2.2 | 4.0 | 5.5 | 7.5 | 10.3 | 11.1 | 13.1 |
| GL RAD OVER THRSH | 475.8 | 415.3 | 317.0 | 228.1 | 157.2 | 110.0 | 73.6 |
| TIME AVG.TEMP | 5.5 | 8.9 | 9.9 | 11.1 | 11.8 | 12.3 | 12.8 |
| TOTAL TIME MSEC | 2.6784 | 1.3266 | 1.0908 | 0.8640 | 0.7290 | 0.5292 | 0.3798 |
| GLOBAL RAD TOTAL | 475.8 | 475.8 | 465.0 | 434.2 | 402.0 | 334.9 | 271.4 |
| DIRECT RAD TOTAL | 265.5 | 265.5 | 263.3 | 254.7 | 248.0 | 220.2 | 197.0 |
| RAD AV TIME MSEC | 0.0864 | 0.0428 | 0.0374 | 0.0317 | 0.0288 | 0.0245 | 0.0223 |
| DEG-C DAYS | 397. | 146. | 108. | 75. | 58. | 40. | 27. |

**JUL**

| SOLAR RAD GROUP | ALL | 0+ | 95+ | 189+ | 284+ | 378+ | 473+ |
|---|---|---|---|---|---|---|---|
| AV.THRSH.RAD.LEVL | 0. | 56. | 148. | 241. | 323. | 428. | 513. |
| AV.THRESHOLD TEMP | 17.6 | 20.5 | 19.7 | 23.6 | 25.4 | 26.7 | 26.9 |
| GL RAD OVER THRSH | 484.0 | 408.9 | 309.4 | 221.2 | 161.5 | 101.7 | 64.8 |
| TIME AVG.TEMP | 21.1 | 24.5 | 25.4 | 26.3 | 27.1 | 27.6 | 27.8 |
| TOTAL TIME MSEC | 2.6784 | 1.3392 | 1.0854 | 0.9486 | 0.7218 | 0.5742 | 0.4320 |
| GLOBAL RAD TOTAL | 484.0 | 484.0 | 469.7 | 449.5 | 394.9 | 347.2 | 286.4 |
| DIRECT RAD TOTAL | 248.5 | 248.5 | 246.5 | 245.9 | 231.4 | 220.0 | 197.4 |
| RAD AV TIME MSEC | 0.0864 | 0.0432 | 0.0356 | 0.0335 | 0.0277 | 0.0238 | 0.0205 |
| DEG-C DAYS | 19. | 7. | 4. | 2. | 1. | 0. | 0. |

**AUG**

| SOLAR RAD GROUP | ALL | 0+ | 95+ | 189+ | 284+ | 378+ | 473+ |
|---|---|---|---|---|---|---|---|
| AV.THRSH.RAD.LEVL | 0. | 51. | 147. | 243. | 324. | 425. | 525. |
| AV.THRESHOLD TEMP | 18.1 | 21.3 | 22.8 | 25.0 | 23.7 | 26.9 | 25.4 |
| GL RAD OVER THRSH | 545.3 | 476.6 | 371.4 | 274.2 | 207.3 | 141.5 | 88.3 |
| TIME AVG.TEMP | 21.5 | 24.9 | 25.7 | 26.0 | 26.2 | 26.9 | 26.9 |
| TOTAL TIME MSEC | 2.6784 | 1.3392 | 1.1034 | 1.0080 | 0.8826 | 0.6516 | 0.5328 |
| GLOBAL RAD TOTAL | 545.3 | 545.3 | 533.2 | 519.2 | 474.1 | 418.7 | 368.1 |
| DIRECT RAD TOTAL | 332.8 | 332.8 | 328.8 | 327.1 | 314.5 | 289.6 | 268.7 |
| RAD AV TIME MSEC | 0.0864 | 0.0432 | 0.0360 | 0.0338 | 0.0292 | 0.0241 | 0.0220 |
| DEG-C DAYS | 6. | 1. | 1. | 0. | 0. | 0. | 0. |

**SEP**

| SOLAR RAD GROUP | ALL | 0+ | 95+ | 189+ | 284+ | 378+ | 473+ |
|---|---|---|---|---|---|---|---|
| AV.THRSH.RAD.LEVL | 0. | 45. | 135. | 245. | 342. | 436. | 521. |
| AV.THRESHOLD TEMP | 12.1 | 13.9 | 13.1 | 17.2 | 19.3 | 20.9 | 20.4 |
| GL RAD OVER THRSH | 573.7 | 515.6 | 417.4 | 316.7 | 241.0 | 184.8 | 141.2 |
| TIME AVG.TEMP | 15.3 | 18.5 | 19.3 | 20.5 | 21.1 | 21.6 | 21.7 |
| TOTAL TIME MSEC | 2.5920 | 1.2816 | 1.0962 | 0.9144 | 0.7812 | 0.5958 | 0.5130 |
| GLOBAL RAD TOTAL | 573.7 | 573.7 | 565.3 | 540.7 | 508.1 | 444.7 | 408.6 |
| DIRECT RAD TOTAL | 411.8 | 411.8 | 409.3 | 402.7 | 393.1 | 359.2 | 344.2 |
| RAD AV TIME MSEC | 0.0864 | 0.0428 | 0.0382 | 0.0342 | 0.0320 | 0.0270 | 0.0259 |
| DEG-C DAYS | 108. | 35. | 24. | 14. | 9. | 6. | 5. |

**APR**

| SOLAR RAD GROUP | ALL | 0+ | 95+ | 189+ | 284+ | 378+ | 473+ |
|---|---|---|---|---|---|---|---|
| AV.THRSH.RAD.LEVL | 0. | 47. | 138. | 237. | 336. | 433. | 517. |
| AV.THRESHOLD TEMP | 6.1 | 8.4 | 8.0 | 11.9 | 14.4 | 16.2 | 15.3 |
| GL RAD OVER THRSH | 449.2 | 388.2 | 292.5 | 208.1 | 143.1 | 98.2 | 66.0 |
| TIME AVG.TEMP | 9.1 | 12.0 | 12.8 | 13.9 | 14.6 | 14.6 | 14.3 |
| TOTAL TIME MSEC | 2.5920 | 1.2960 | 1.0512 | 0.8532 | 0.6534 | 0.4662 | 0.3834 |
| GLOBAL RAD TOTAL | 449.2 | 449.2 | 437.6 | 410.3 | 362.9 | 300.0 | 264.1 |
| DIRECT RAD TOTAL | 229.4 | 229.4 | 227.1 | 224.7 | 217.1 | 194.8 | 182.7 |
| RAD AV TIME MSEC | 0.0864 | 0.0432 | 0.0364 | 0.0328 | 0.0292 | 0.0241 | 0.0223 |
| DEG-C DAYS | 278. | 98. | 71. | 48. | 33. | 23. | 20. |

**MAY**

| SOLAR·RAD GROUP | ALL | 0+ | 95+ | 189+ | 284+ | 378+ | 473+ |
|---|---|---|---|---|---|---|---|
| AV.THRSH.RAD.LEVL | 0. | 56. | 147. | 243. | 325. | 432. | 520. |
| AV.THRESHOLD TEMP | 9.7 | 12.8 | 12.0 | 15.5 | 16.5 | 18.8 | 19.5 |
| GL RAD OVER THRSH | 472.3 | 396.9 | 299.0 | 210.4 | 151.3 | 96.1 | 59.9 |
| TIME AVG.TEMP | 13.1 | 16.4 | 17.3 | 18.2 | 18.9 | 19.9 | 20.2 |
| TOTAL TIME MSEC | 2.6784 | 1.3392 | 1.0746 | 0.9288 | 0.7164 | 0.5184 | 0.4122 |
| GLOBAL RAD TOTAL | 472.3 | 472.3 | 457.4 | 435.9 | 384.3 | 319.9 | 274.1 |
| DIRECT RAD TOTAL | 216.1 | 216.1 | 214.0 | 213.2 | 203.6 | 186.3 | 173.9 |
| RAD AV TIME MSEC | 0.0864 | 0.0432 | 0.0356 | 0.0338 | 0.0292 | 0.0241 | 0.0223 |
| DEG-C DAYS | 166. | 46. | 31. | 21. | 12. | 6. | 4. |

**JUN**

| SOLAR RAD GROUP | ALL | 0+ | 95+ | 189+ | 284+ | 378+ | 473+ |
|---|---|---|---|---|---|---|---|
| AV.THRSH.RAD.LEVL | 0. | 58. | 149. | 249. | 333. | 434. | 517. |
| AV.THRESHOLD TEMP | 16.4 | 20.6 | 24.2 | 23.1 | 24.3 | 24.9 | 25.0 |
| GL RAD OVER THRSH | 493.6 | 418.6 | 320.2 | 220.1 | 156.3 | 95.2 | 58.6 |
| TIME AVG.TEMP | 20.3 | 24.2 | 24.9 | 25.0 | 25.6 | 25.9 | 26.3 |
| TOTAL TIME MSEC | 2.5920 | 1.2960 | 1.0800 | 1.0026 | 0.7614 | 0.6030 | 0.4392 |
| GLOBAL RAD TOTAL | 493.6 | 493.6 | 481.1 | 469.5 | 409.5 | 356.4 | 285.8 |
| DIRECT RAD TOTAL | 253.2 | 253.2 | 251.6 | 250.9 | 232.0 | 221.8 | 191.3 |
| RAD AV TIME MSEC | 0.0864 | 0.0432 | 0.0360 | 0.0338 | 0.0266 | 0.0230 | 0.0191 |
| DEG-C DAYS | 11. | 0. | 0. | 0. | 0. | 0. | 0. |

**OCT**

| SOLAR RAD GROUP | ALL | 0+ | 95+ | 189+ | 284+ | 378+ | 473+ |
|---|---|---|---|---|---|---|---|
| AV.THRSH.RAD.LEVL | 0. | 43. | 136. | 238. | 337. | 425. | 524. |
| AV.THRESHOLD TEMP | 6.9 | 7.7 | 10.8 | 16.3 | 16.0 | 17.6 | 17.1 |
| GL RAD OVER THRSH | 501.1 | 443.4 | 350.1 | 270.2 | 204.4 | 161.2 | 120.0 |
| TIME AVG.TEMP | 10.1 | 13.3 | 15.1 | 16.3 | 16.3 | 16.4 | 16.2 |
| TOTAL TIME MSEC | 2.6784 | 1.3320 | 1.0080 | 0.7812 | 0.6678 | 0.4914 | 0.4158 |
| GLOBAL RAD TOTAL | 501.1 | 501.1 | 487.1 | 456.3 | 429.2 | 369.9 | 337.8 |
| DIRECT RAD TOTAL | 356.2 | 356.2 | 352.4 | 343.3 | 337.1 | 307.0 | 294.7 |
| RAD AV TIME MSEC | 0.0864 | 0.0432 | 0.0371 | 0.0324 | 0.0306 | 0.0259 | 0.0252 |
| DEG-C DAYS | 257. | 85. | 47. | 28. | 24. | 17. | 15. |

**NOV**

| SOLAR RAD GROUP | ALL | 0+ | 95+ | 189+ | 284+ | 378+ | 473+ |
|---|---|---|---|---|---|---|---|
| AV.THRSH.RAD.LEVL | 0. | 45. | 143. | 240. | 335. | 429. | 521. |
| AV.THRESHOLD TEMP | -0.5 | 1.4 | 3.8 | 5.9 | 5.9 | 6.0 | 6.0 |
| GL RAD OVER THRSH | 373.8 | 323.2 | 244.5 | 185.2 | 141.3 | 106.9 | 78.9 |
| TIME AVG.TEMP | 1.8 | 4.9 | 6.2 | 7.0 | 7.3 | 7.7 | 8.1 |
| TOTAL TIME MSEC | 2.5920 | 1.1178 | 0.8064 | 0.6102 | 0.4644 | 0.3636 | 0.3060 |
| GLOBAL RAD TOTAL | 373.8 | 373.8 | 359.7 | 331.7 | 296.7 | 262.9 | 238.2 |
| DIRECT RAD TOTAL | 276.0 | 276.0 | 273.6 | 264.4 | 246.5 | 229.4 | 211.6 |
| RAD AV TIME MSEC | 0.0864 | 0.0382 | 0.0331 | 0.0302 | 0.0259 | 0.0256 | 0.0241 |
| DEG-C DAYS | 494. | 174. | 113. | 81. | 59. | 44. | 37. |

**DEC**

| SOLAR RAD GROUP | ALL | 0+ | 95+ | 189+ | 284+ | 378+ | 473+ |
|---|---|---|---|---|---|---|---|
| AV.THRSH.RAD.LEVL | 0. | 47. | 150. | 235. | 327. | 429. | 517. |
| AV.THRESHOLD TEMP | -4.6 | -1.9 | -2.2 | 4.4 | 2.0 | 1.9 | 0.9 |
| GL RAD OVER THRSH | 424.5 | 371.8 | 283.2 | 226.2 | 174.6 | 127.1 | 93.2 |
| TIME AVG.TEMP | -2.5 | 0.5 | 1.2 | 2.2 | 1.7 | 1.7 | 1.6 |
| TOTAL TIME MSEC | 2.6784 | 1.1160 | 0.8604 | 0.6732 | 0.5598 | 0.4644 | 0.3852 |
| GLOBAL RAD TOTAL | 424.5 | 424.5 | 412.4 | 384.3 | 357.7 | 326.5 | 292.5 |
| DIRECT RAD TOTAL | 339.2 | 339.2 | 335.8 | 318.5 | 307.7 | 288.9 | 263.1 |
| RAD AV TIME MSEC | 0.0864 | 0.0360 | 0.0324 | 0.0263 | 0.0252 | 0.0245 | 0.0234 |
| DEG-C DAYS | 645. | 230. | 171. | 126. | 108. | 89. | 74. |

**ALL MONTHS**

| SOLAR RAD GROUP | ALL | 0+ | 95+ | 189+ | 284+ | 378+ | 473+ |
|---|---|---|---|---|---|---|---|
| TIME AVG.TEMP | 9.7 | 13.3 | 14.6 | 15.8 | 16.3 | 16.6 | 16.5 |
| TOTAL TIME MSEC | 31.5791 | 15.1074 | 11.9916 | 9.9558 | 7.8948 | 6.0606 | 4.8474 |
| GLOBAL RAD TOTAL | 5590.4 | 5590.3 | 5439.3 | 5150.6 | 4655.1 | 4046.5 | 3526.4 |
| DIRECT RAD TOTAL | 3494.8 | 3494.8 | 3461.3 | 3385.3 | 3236.8 | 2987.4 | 2756.1 |
| RAD AV TIME MSEC | 0.0864 | 0.0417 | 0.0356 | 0.0321 | 0.0283 | 0.0247 | 0.0229 |
| DEG-C DAYS | 3467. | 1239. | 857. | 616. | 471. | 358. | 292. |

STATISTICS FOR DENVER,COLO   1971-1972   IR INCLUDED   RSV023   FILE 7

RADIATION UNITS ARE WATTS/SQ.M OR MJOULES/SQ.M
SLOPE = 90.0 DEG   AZIMUTH = 180.0 DEG
DEGREE-DAY BASE TEMP = 18.3 DEGREES C

**JAN**

| SOLAR RAD GROUP | ALL | 0+ | 95+ | 189+ | 284+ | 378+ | 473+ |
|---|---|---|---|---|---|---|---|
| AV.THRSH.RAD.LEVL | 0. | 42. | 144. | 234. | 327. | 424. | 528. |
| AV.THRESHOLD TEMP | -2.4 | -1.8 | 5.1 | 2.7 | 5.2 | 2.8 | -0.1 |
| GL RAD OVER THRSH | 334.0 | 286.9 | 202.6 | 151.7 | 113.6 | 82.6 | 55.1 |
| TIME AVG.TEMP | -0.4 | 2.3 | 3.7 | 3.0 | 3.1 | 2.5 | 2.4 |
| TOTAL TIME MSEC | 2.6784 | 1.1160 | 0.8316 | 0.5616 | 0.4104 | 0.3186 | 0.2664 |
| GLOBAL RAD TOTAL | 334.0 | 334.0 | 322.0 | 283.3 | 247.8 | 217.8 | 195.7 |
| DIRECT RAD TOTAL | 248.6 | 248.6 | 245.7 | 232.4 | 210.5 | 192.0 | 174.8 |
| RAD AV TIME MSEC | 0.0864 | 0.0360 | 0.0320 | 0.0288 | 0.0252 | 0.0245 | 0.0234 |
| DEG-C DAYS | 582. | 208. | 141. | 99. | 72. | 58. | 49. |

**FEB**

| SOLAR RAD GROUP | ALL | 0+ | 95+ | 189+ | 284+ | 378+ | 473+ |
|---|---|---|---|---|---|---|---|
| AV.THRSH.RAD.LEVL | 0. | 43. | 143. | 235. | 332. | 430. | 523. |
| AV.THRESHOLD TEMP | -1.9 | 0.6 | 3.3 | 6.5 | 4.1 | 5.9 | 6.5 |
| GL RAD OVER THRSH | 349.1 | 297.0 | 215.1 | 156.2 | 109.4 | 74.7 | 48.3 |
| TIME AVG.TEMP | 0.7 | 3.4 | 4.7 | 5.1 | 4.6 | 4.7 | 4.4 |
| TOTAL TIME MSEC | 2.4624 | 1.2006 | 0.8262 | 0.6390 | 0.4788 | 0.3546 | 0.2862 |
| GLOBAL RAD TOTAL | 349.1 | 349.1 | 332.9 | 306.2 | 268.6 | 227.3 | 197.9 |
| DIRECT RAD TOTAL | 243.6 | 243.6 | 238.2 | 228.3 | 212.8 | 187.6 | 167.9 |
| RAD AV TIME MSEC | 0.0864 | 0.0421 | 0.0335 | 0.0295 | 0.0266 | 0.0234 | 0.0220 |
| DEG-C DAYS | 504. | 208. | 131. | 98. | 77. | 56. | 46. |

**MAR**

| SOLAR RAD GROUP | ALL | 0+ | 95+ | 189+ | 284+ | 378+ | 473+ |
|---|---|---|---|---|---|---|---|
| AV.THRSH.RAD.LEVL | 0. | 48. | 138. | 241. | 329. | 425. | 516. |
| AV.THRESHOLD TEMP | 2.2 | 4.4 | 6.6 | 9.8 | 11.4 | 11.6 | 12.9 |
| GL RAD OVER THRSH | 358.7 | 295.3 | 204.9 | 125.6 | 77.6 | 44.8 | 24.7 |
| TIME AVG.TEMP | 5.5 | 8.9 | 10.4 | 11.5 | 12.2 | 12.7 | 13.3 |
| TOTAL TIME MSEC | 2.6784 | 1.3302 | 0.9990 | 0.7722 | 0.5436 | 0.3420 | 0.2214 |
| GLOBAL RAD TOTAL | 358.7 | 358.7 | 342.9 | 311.6 | 256.5 | 190.2 | 138.9 |
| DIRECT RAD TOTAL | 168.2 | 168.2 | 184.3 | 177.1 | 158.9 | 134.8 | 106.2 |
| RAD AV TIME MSEC | 0.0864 | 0.0428 | 0.0353 | 0.0299 | 0.0252 | 0.0223 | 0.0202 |
| DEG-C DAYS | 397. | 147. | 94. | 64. | 42. | 24. | 15. |

**JUL**

| SOLAR RAD GROUP | ALL | 0+ | 95+ | 189+ | 284+ | 378+ | 473+ |
|---|---|---|---|---|---|---|---|
| AV.THRSH.RAD.LEVL | 0. | 45. | 148. | 234. | 328. | 422. | 500. |
| AV.THRESHOLD TEMP | 18.2 | 21.2 | 23.2 | 25.8 | 27.3 | 28.8 | 28.6 |
| GL RAD OVER THRSH | 257.4 | 209.1 | 119.3 | 62.3 | 21.9 | 2.4 | 0.0 |
| TIME AVG.TEMP | 21.1 | 25.3 | 26.3 | 27.2 | 28.0 | 28.8 | 28.6 |
| TOTAL TIME MSEC | 2.6784 | 1.0692 | 0.8694 | 0.6660 | 0.4302 | 0.2070 | 0.0306 |
| GLOBAL RAD TOTAL | 257.4 | 257.4 | 248.5 | 218.1 | 163.0 | 89.8 | 15.3 |
| DIRECT RAD TOTAL | 84.9 | 84.9 | 84.5 | 81.4 | 71.1 | 45.8 | 8.5 |
| RAD AV TIME MSEC | 0.0864 | 0.0349 | 0.0292 | 0.0245 | 0.0180 | 0.0119 | 0.0076 |
| DEG-C DAYS | 19. | 5. | 2. | 1. | 0. | 0. | 0. |

**AUG**

| SOLAR RAD GROUP | ALL | 0+ | 95+ | 189+ | 284+ | 378+ | 473+ |
|---|---|---|---|---|---|---|---|
| AV.THRSH.RAD.LEVL | 0. | 43. | 147. | 241. | 330. | 426. | 523. |
| AV.THRESHOLD TEMP | 18.5 | 21.9 | 23.4 | 25.6 | 26.3 | 27.0 | 27.8 |
| GL RAD OVER THRSH | 329.6 | 278.0 | 176.1 | 103.3 | 52.8 | 18.8 | 2.4 |
| TIME AVG.TEMP | 21.5 | 25.2 | 25.9 | 26.6 | 27.0 | 27.4 | 27.9 |
| TOTAL TIME MSEC | 2.6784 | 1.1952 | 0.9828 | 0.7776 | 0.5552 | 0.3528 | 0.1692 |
| GLOBAL RAD TOTAL | 329.6 | 329.6 | 320.4 | 290.3 | 239.2 | 169.1 | 90.9 |
| DIRECT RAD TOTAL | 156.6 | 156.6 | 155.4 | 148.9 | 134.7 | 105.4 | 63.0 |
| RAD AV TIME MSEC | 0.0864 | 0.0385 | 0.0324 | 0.0266 | 0.0212 | 0.0158 | 0.0115 |
| DEG-C DAYS | 6. | 1. | 0. | 0. | 0. | 0. | 0. |

**SEP**

| SOLAR RAD GROUP | ALL | 0+ | 95+ | 189+ | 284+ | 378+ | 473+ |
|---|---|---|---|---|---|---|---|
| AV.THRSH.RAD.LEVL | 0. | 46. | 142. | 235. | 330. | 422. | 526. |
| AV.THRESHOLD TEMP | 12.1 | 13.5 | 15.5 | 19.1 | 21.1 | 21.3 | 20.9 |
| GL RAD OVER THRSH | 415.1 | 355.3 | 256.9 | 181.8 | 124.5 | 80.0 | 42.1 |
| TIME AVG.TEMP | 15.3 | 18.5 | 19.9 | 20.8 | 21.5 | 21.6 | 21.7 |
| TOTAL TIME MSEC | 2.5920 | 1.2906 | 1.0080 | 0.8298 | 0.6030 | 0.4806 | 0.3654 |
| GLOBAL RAD TOTAL | 415.1 | 415.1 | 402.0 | 376.7 | 323.4 | 283.1 | 234.4 |
| DIRECT RAD TOTAL | 271.6 | 271.6 | 267.0 | 262.1 | 239.9 | 222.4 | 187.6 |
| RAD AV TIME MSEC | 0.0864 | 0.0432 | 0.0356 | 0.0328 | 0.0274 | 0.0256 | 0.0209 |
| DEG-C DAYS | 108. | 36. | 19. | 11. | 6. | 5. | 4. |

## APR

| SOLAR RAD GROUP | ALL | 0+ | 95+ | 189+ | 284+ | 378+ | 473+ |
|---|---|---|---|---|---|---|---|
| AV.THRSH.RAD.LEVL | 0. | 43. | 144. | 237. | 333. | 426. | 511. |
| AV.THRESHOLD TEMP | 6.2 | 8.5 | 10.5 | 14.6 | 15.9 | 13.4 | 13.6 |
| GL RAD OVER THRSH | 298.3 | 245.2 | 149.9 | 88.6 | 46.1 | 20.4 | 7.7 |
| TIME AVG.TEMP | 9.1 | 12.2 | 13.3 | 14.6 | 14.6 | 13.8 | 14.2 |
| TOTAL TIME MSEC | 2.5920 | 1.2312 | 0.9486 | 0.6552 | 0.4428 | 0.2754 | 0.1512 |
| GLOBAL RAD TOTAL | 298.3 | 298.3 | 286.1 | 244.0 | 193.6 | 137.8 | 84.9 |
| DIRECT RAD TOTAL | 122.6 | 122.6 | 121.6 | 115.9 | 105.7 | 85.7 | 57.5 |
| RAD AV TIME MSEC | 0.0864 | 0.0414 | 0.0335 | 0.0277 | 0.0230 | 0.0176 | 0.0148 |
| DEG-C DAYS | 278. | 90. | 59. | 33. | 22. | 16. | 3. |

## MAY

| SOLAR RAD GROUP | ALL | 0+ | 95+ | 189+ | 284+ | 378+ | 473+ |
|---|---|---|---|---|---|---|---|
| AV.THRSH.RAD.LEVL | 0. | 53. | 143. | 233. | 334. | 419. | 507. |
| AV.THRESHOLD TEMP | 10.2 | 13.4 | 14.6 | 17.1 | 19.2 | 21.5 | 20.5 |
| GL RAD OVER THRSH | 272.2 | 213.2 | 132.3 | 70.6 | 27.6 | 7.8 | 1.0 |
| TIME AVG.TEMP | 13.1 | 17.1 | 17.9 | 19.0 | 20.2 | 20.9 | 19.8 |
| TOTAL TIME MSEC | 2.6784 | 1.1160 | 0.8982 | 0.6822 | 0.4266 | 0.2340 | 0.0774 |
| GLOBAL RAD TOTAL | 272.2 | 272.2 | 260.7 | 229.8 | 170.2 | 105.8 | 40.2 |
| DIRECT RAD TOTAL | 81.7 | 81.7 | 80.8 | 79.0 | 68.6 | 51.6 | 24.8 |
| RAD AV TIME MSEC | 0.0360 | 0.0360 | 0.0302 | 0.0266 | 0.0198 | 0.0151 | 0.0112 |
| DEG-C DAYS | 166. | 35. | 23. | 12. | 5. | 2. | 1. |

## JUN

| SOLAR RAD GROUP | ALL | 0+ | 95+ | 189+ | 284+ | 378+ | 473+ |
|---|---|---|---|---|---|---|---|
| AV.THRSH.RAD.LEVL | 0. | 17. | 157. | 240. | 328. | 416. | 486. |
| AV.THRESHOLD TEMP | 17.7 | 25.0 | 23.2 | 24.8 | 26.1 | 27.1 | 19.3 |
| GL RAD OVER THRSH | 229.8 | 214.3 | 105.2 | 52.1 | 14.7 | 0.1 | 0.0 |
| TIME AVG.TEMP | 20.3 | 25.3 | 25.4 | 25.9 | 26.4 | 27.1 | 19.3 |
| TOTAL TIME MSEC | 2.5920 | 0.8946 | 0.7812 | 0.6372 | 0.4266 | 0.1656 | 0.0018 |
| GLOBAL RAD TOTAL | 229.8 | 229.8 | 227.4 | 205.2 | 154.6 | 69.0 | 0.9 |
| DIRECT RAD TOTAL | 74.1 | 74.1 | 74.0 | 70.0 | 60.4 | 32.6 | 0.2 |
| RAD AV TIME MSEC | 0.0864 | 0.0299 | 0.0263 | 0.0216 | 0.0166 | 0.0101 | 0.0036 |
| DEG-C DAYS | 11. | 0. | 0. | 0. | 0. | 0. | 0. |

## OCT

| SOLAR RAD GROUP | ALL | 0+ | 95+ | 189+ | 284+ | 378+ | 473+ |
|---|---|---|---|---|---|---|---|
| AV.THRSH.RAD.LEVL | 0. | 45. | 135. | 240. | 324. | 435. | 528. |
| AV.THRESHOLD TEMP | 6.8 | 8.5 | 12.9 | 16.8 | 15.8 | 16.5 | 15.5 |
| GL RAD OVER THRSH | 408.7 | 349.3 | 264.8 | 189.8 | 142.7 | 97.0 | 63.4 |
| TIME AVG.TEMP | 10.1 | 13.3 | 15.4 | 16.2 | 16.0 | 16.1 | 16.0 |
| TOTAL TIME MSEC | 2.6784 | 1.3302 | 0.9360 | 0.7146 | 0.5616 | 0.4104 | 0.3618 |
| GLOBAL RAD TOTAL | 408.7 | 408.7 | 391.1 | 361.2 | 324.5 | 275.6 | 254.4 |
| DIRECT RAD TOTAL | 286.5 | 286.5 | 281.7 | 272.7 | 261.7 | 233.7 | 219.2 |
| RAD AV TIME MSEC | 0.0864 | 0.0432 | 0.0360 | 0.0313 | 0.0299 | 0.0256 | 0.0245 |
| DEG-C DAYS | 257. | 84. | 41. | 27. | 21. | 15. | 14. |

## NOV

| SOLAR RAD GROUP | ALL | 0+ | 95+ | 189+ | 284+ | 378+ | 473+ |
|---|---|---|---|---|---|---|---|
| AV.THRSH.RAD.LEVL | 0. | 42. | 139. | 239. | 329. | 419. | 531. |
| AV.THRESHOLD TEMP | -0.5 | 1.6 | 5.0 | 5.7 | 5.8 | 5.5 | 7.1 |
| GL RAD OVER THRSH | 326.8 | 279.6 | 205.9 | 150.2 | 113.4 | 83.3 | 52.5 |
| TIME AVG.TEMP | 1.8 | 4.9 | 6.4 | 6.9 | 7.4 | 7.8 | 8.3 |
| TOTAL TIME MSEC | 2.5920 | 1.1178 | 0.7596 | 0.5616 | 0.4050 | 0.3366 | 0.2754 |
| GLOBAL RAD TOTAL | 326.8 | 326.8 | 311.7 | 284.1 | 246.8 | 224.2 | 198.6 |
| DIRECT RAD TOTAL | 245.6 | 245.6 | 242.8 | 232.5 | 210.9 | 195.5 | 176.4 |
| RAD AV TIME MSEC | 0.0864 | 0.0382 | 0.0331 | 0.0302 | 0.0259 | 0.0252 | 0.0248 |
| DEG-C DAYS | 494. | 174. | 105. | 74. | 51. | 41. | 32. |

## DEC

| SOLAR RAD GROUP | ALL | 0+ | 95+ | 189+ | 284+ | 378+ | 473+ |
|---|---|---|---|---|---|---|---|
| AV.THRSH.RAD.LEVL | 0. | 43. | 147. | 236. | 335. | 422. | 512. |
| AV.THRESHOLD TEMP | -4.6 | -1.5 | -1.3 | 4.1 | 1.5 | 2.2 | 1.2 |
| GL RAD OVER THRSH | 386.3 | 338.5 | 251.0 | 195.3 | 144.9 | 107.3 | 75.4 |
| TIME AVG.TEMP | -2.5 | 0.5 | 1.2 | 2.0 | 1.5 | 1.5 | 1.3 |
| TOTAL TIME MSEC | 2.6784 | 1.1142 | 0.8406 | 0.6282 | 0.5094 | 0.4320 | 0.3528 |
| GLOBAL RAD TOTAL | 386.3 | 386.3 | 374.6 | 343.4 | 315.3 | 289.4 | 256.0 |
| DIRECT RAD TOTAL | 314.0 | 314.0 | 310.2 | 291.7 | 276.1 | 257.5 | 230.9 |
| RAD AV TIME MSEC | 0.0864 | 0.0360 | 0.0324 | 0.0263 | 0.0252 | 0.0245 | 0.0230 |
| DEG-C DAYS | 645. | 229. | 167. | 119. | 99. | 84. | 69. |

## ALL MONTHS

| SOLAR RAD GROUP | ALL | 0+ | 95+ | 189+ | 284+ | 378+ | 473+ |
|---|---|---|---|---|---|---|---|
| TIME AVG.TEMP | 9.7 | 12.9 | 14.5 | 15.5 | 15.4 | 14.3 | 12.0 |
| TOTAL TIME MSEC | 31.5791 | 14.0058 | 10.6812 | 8.1252 | 5.8032 | 3.9096 | 2.5596 |
| GLOBAL RAD TOTAL | 3966.1 | 3966.0 | 3820.6 | 3453.9 | 2903.6 | 2279.2 | 1708.1 |
| DIRECT RAD TOTAL | 2318.0 | 2317.9 | 2286.3 | 2191.9 | 2011.2 | 1744.5 | 1417.1 |
| RAD AV TIME MSEC | 0.0864 | 0.0390 | 0.0328 | 0.0284 | 0.0245 | 0.0221 | 0.0214 |
| DEG-C DAYS | 3467. | 1218. | 783. | 537. | 395. | 302. | 238. |

STATISTICS FOR LOS ANGELES, CALIF 1971-1972   IR INCLUDED RSV020 FILE

RADIATION UNITS ARE WATTS/SQ.M OR MJOULES/SQ.M
SLOPE = 0.0 DEG   AZIMUTH = 0.0 DEG
DEGREE-DAY BASE TEMP = 18.3 DEGREES C

### JAN

| SOLAR RAD GROUP | ALL | 0+ | 95+ | 189+ | 284+ | 378+ | 473+ |
|---|---|---|---|---|---|---|---|
| AV.THRSH.RAD.LEVL | 0. | 40. | 148. | 237. | 336. | 430. | 514. |
| AV.THRESHOLD TEMP | 10.8 | 10.8 | 12.9 | 13.1 | 14.4 | 14.5 | 16.4 |
| GL RAD OVER THRSH | 326.6 | 281.9 | 187.3 | 127.8 | 73.6 | 35.4 | 11.4 |
| TIME AVG.TEMP | 12.1 | 13.9 | 14.7 | 15.3 | 15.8 | 16.2 | 16.9 |
| TOTAL TIME MSEC | 2.6784 | 1.1142 | 0.8784 | 0.6696 | 0.5436 | 0.4086 | 0.2862 |
| GLOBAL RAD TOTAL | 326.6 | 326.6 | 317.2 | 286.3 | 256.5 | 211.1 | 158.5 |
| DIRECT RAD TOTAL | 187.4 | 187.4 | 185.8 | 174.1 | 164.1 | 138.4 | 111.9 |
| RAD AV TIME MSEC | 0.0864 | 0.0360 | 0.0302 | 0.0248 | 0.0223 | 0.0176 | 0.0144 |
| DEG-C DAYS | 193. | 61. | 42. | 28. | 21. | 14. | 9. |

### FEB

| SOLAR RAD GROUP | ALL | 0+ | 95+ | 189+ | 284+ | 378+ | 473+ |
|---|---|---|---|---|---|---|---|
| AV.THRSH.RAD.LEVL | 0. | 47. | 144. | 236. | 331. | 425. | 521. |
| AV.THRESHOLD TEMP | 11.9 | 12.7 | 13.0 | 14.7 | 13.9 | 15.7 | 15.8 |
| GL RAD OVER THRSH | 387.3 | 333.9 | 248.4 | 181.0 | 124.0 | 78.2 | 41.3 |
| TIME AVG.TEMP | 13.2 | 14.7 | 15.2 | 15.7 | 15.9 | 16.3 | 16.5 |
| TOTAL TIME MSEC | 2.4624 | 1.1286 | 0.8820 | 0.7362 | 0.5976 | 0.4878 | 0.3834 |
| GLOBAL RAD TOTAL | 387.3 | 387.3 | 375.6 | 354.6 | 321.9 | 285.5 | 241.2 |
| DIRECT RAD TOTAL | 232.6 | 232.6 | 228.9 | 224.5 | 210.9 | 192.9 | 167.7 |
| RAD AV TIME MSEC | 0.0864 | 0.0396 | 0.0320 | 0.0292 | 0.0252 | 0.0223 | 0.0187 |
| DEG-C DAYS | 147. | 50. | 35. | 26. | 20. | 14. | 11. |

### MAR

| SOLAR RAD GROUP | ALL | 0+ | 95+ | 189+ | 284+ | 378+ | 473+ |
|---|---|---|---|---|---|---|---|
| AV.THRSH.RAD.LEVL | 0. | 44. | 144. | 230. | 329. | 419. | 521. |
| AV.THRESHOLD TEMP | 12.8 | 13.1 | 13.9 | 14.6 | 15.5 | 15.8 | 16.2 |
| GL RAD OVER THRSH | 500.0 | 443.1 | 335.4 | 261.5 | 192.9 | 139.3 | 88.7 |
| TIME AVG.TEMP | 14.1 | 15.2 | 15.8 | 16.7 | 16.9 | 17.2 | 17.2 |
| TOTAL TIME MSEC | 2.6784 | 1.2906 | 1.0818 | 0.8550 | 0.6966 | 0.5904 | 0.4968 |
| GLOBAL RAD TOTAL | 500.0 | 500.0 | 490.8 | 458.2 | 421.8 | 386.9 | 347.6 |
| DIRECT RAD TOTAL | 302.9 | 302.9 | 301.7 | 293.2 | 285.3 | 270.1 | 245.6 |
| RAD AV TIME MSEC | 0.0864 | 0.0418 | 0.0367 | 0.0313 | 0.0288 | 0.0263 | 0.0230 |
| DEG-C DAYS | 132. | 45. | 33. | 22. | 16. | 12. | 9. |

### JUL

| SOLAR RAD GROUP | ALL | 0+ | 95+ | 189+ | 284+ | 378+ | 473+ |
|---|---|---|---|---|---|---|---|
| AV.THRSH.RAD.LEVL | 0. | 35. | 141. | 229. | 339. | 418. | 539. |
| AV.THRESHOLD TEMP | 19.0 | 19.0 | 20.1 | 20.1 | 21.2 | 22.0 | 22.0 |
| GL RAD OVER THRSH | 816.2 | 762.5 | 618.9 | 518.3 | 401.8 | 329.3 | 232.0 |
| TIME AVG.TEMP | 20.6 | 21.7 | 22.1 | 22.4 | 22.7 | 22.9 | 23.0 |
| TOTAL TIME MSEC | 2.6784 | 1.5408 | 1.3482 | 1.1538 | 1.0548 | 0.9126 | 0.8064 |
| GLOBAL RAD TOTAL | 816.2 | 816.2 | 809.5 | 782.0 | 759.4 | 711.2 | 666.7 |
| DIRECT RAD TOTAL | 597.6 | 597.6 | 595.9 | 584.8 | 575.6 | 550.6 | 522.0 |
| RAD AV TIME MSEC | 0.0864 | 0.0497 | 0.0439 | 0.0382 | 0.0353 | 0.0313 | 0.0284 |
| DEG-C DAYS | 0. | 0. | 0. | 0. | 0. | 0. | 0. |

### AUG

| SOLAR RAD GROUP | ALL | 0+ | 95+ | 189+ | 284+ | 378+ | 473+ |
|---|---|---|---|---|---|---|---|
| AV.THRSH.RAD.LEVL | 0. | 58. | 138. | 243. | 324. | 422. | 519. |
| AV.THRESHOLD TEMP | 20.7 | 21.0 | 22.3 | 22.4 | 22.9 | 24.1 | 24.0 |
| GL RAD OVER THRSH | 745.8 | 661.6 | 560.5 | 443.5 | 362.3 | 279.9 | 204.7 |
| TIME AVG.TEMP | 22.3 | 23.6 | 24.0 | 24.2 | 24.4 | 24.7 | 24.8 |
| TOTAL TIME MSEC | 2.6784 | 1.4508 | 1.2618 | 1.1196 | 0.9972 | 0.8424 | 0.7776 |
| GLOBAL RAD TOTAL | 745.8 | 745.8 | 734.8 | 715.2 | 685.5 | 635.3 | 608.0 |
| DIRECT RAD TOTAL | 580.3 | 580.3 | 576.0 | 567.8 | 555.1 | 526.7 | 511.1 |
| RAD AV TIME MSEC | 0.0864 | 0.0468 | 0.0414 | 0.0371 | 0.0342 | 0.0299 | 0.0281 |
| DEG-C DAYS | 0. | 0. | 0. | 0. | 0. | 0. | 0. |

### SEP

| SOLAR RAD GROUP | ALL | 0+ | 95+ | 189+ | 284+ | 378+ | 473+ |
|---|---|---|---|---|---|---|---|
| AV.THRSH.RAD.LEVL | 0. | 38. | 152. | 238. | 338. | 429. | 522. |
| AV.THRESHOLD TEMP | 19.2 | 20.1 | 20.6 | 21.5 | 22.4 | 22.4 | 22.8 |
| GL RAD OVER THRSH | 559.6 | 509.1 | 383.4 | 302.4 | 226.2 | 167.0 | 116.6 |
| TIME AVG.TEMP | 20.7 | 21.2 | 22.4 | 22.8 | 23.1 | 23.2 | 23.3 |
| TOTAL TIME MSEC | 2.5920 | 1.3230 | 1.1070 | 0.9378 | 0.7668 | 0.6498 | 0.5382 |
| GLOBAL RAD TOTAL | 559.6 | 559.6 | 551.4 | 525.7 | 485.0 | 445.5 | 397.7 |
| DIRECT RAD TOTAL | 389.0 | 389.0 | 387.0 | 378.5 | 363.1 | 348.9 | 319.8 |
| RAD AV TIME MSEC | 0.0864 | 0.0443 | 0.0378 | 0.0331 | 0.0288 | 0.0266 | 0.0234 |
| DEG-C DAYS | 2. | 0. | 0. | 0. | 0. | 0. | 0. |

**APR**

| SOLAR RAD GROUP | ALL | 0+ | 95+ | 189+ | 284+ | 378+ | 473+ |
|---|---|---|---|---|---|---|---|
| AV.THRSH.RAD.LEVL | 0. | 52. | 135. | 242. | 330. | 433. | 523. |
| AV.THRESHOLD TEMP | 13.3 | 14.4 | 14.1 | 15.4 | 15.3 | 16.9 | 16.9 |
| GL RAD OVER THRSH | 710.3 | 637.2 | 537.5 | 420.8 | 336.9 | 252.9 | 188.6 |
| TIME AVG.TEMP | 15.2 | 16.8 | 17.2 | 17.5 | 17.8 | 18.2 | 18.4 |
| TOTAL TIME MSEC | 2.5920 | 1.4040 | 1.2042 | 1.0854 | 0.9576 | 0.8118 | 0.7200 |
| GLOBAL RAD TOTAL | 710.3 | 710.3 | 699.9 | 683.9 | 652.9 | 604.8 | 565.0 |
| DIRECT RAD TOTAL | 494.4 | 494.4 | 489.6 | 484.2 | 470.2 | 444.9 | 427.9 |
| RAD AV TIME MSEC | 0.0864 | 0.0468 | 0.0407 | 0.0374 | 0.0338 | 0.0295 | 0.0281 |
| DEG-C DAYS | 97. | 29. | 21. | 16. | 13. | 9. | 7. |

**MAY**

| SOLAR RAD GROUP | ALL | 0+ | 95+ | 189+ | 284+ | 378+ | 473+ |
|---|---|---|---|---|---|---|---|
| AV.THRSH.RAD.LEVL | 0. | 46. | 143. | 239. | 331. | 423. | 528. |
| AV.THRESHOLD TEMP | 14.8 | 14.4 | 16.1 | 15.9 | 17.5 | 17.9 | 17.8 |
| GL RAD OVER THRSH | 657.2 | 588.3 | 464.3 | 364.9 | 284.5 | 220.1 | 155.2 |
| TIME AVG.TEMP | 16.3 | 17.6 | 18.1 | 18.6 | 19.4 | 19.4 | 19.7 |
| TOTAL TIME MSEC | 2.6784 | 1.5012 | 1.2744 | 1.0404 | 0.8730 | 0.7020 | 0.6156 |
| GLOBAL RAD TOTAL | 657.3 | 657.3 | 646.8 | 613.3 | 573.4 | 516.8 | 480.3 |
| DIRECT RAD TOTAL | 419.1 | 419.1 | 417.4 | 407.3 | 398.7 | 377.2 | 356.4 |
| RAD AV TIME MSEC | 0.0864 | 0.0486 | 0.0425 | 0.0360 | 0.0331 | 0.0295 | 0.0266 |
| DEG-C DAYS | 71. | 23. | 15. | 9. | 5. | 3. | 2. |

**JUN**

| SOLAR RAD GROUP | ALL | 0+ | 95+ | 189+ | 284+ | 378+ | 473+ |
|---|---|---|---|---|---|---|---|
| AV.THRSH.RAD.LEVL | 0. | 34. | 141. | 240. | 331. | 418. | 522. |
| AV.THRESHOLD TEMP | 16.8 | 17.3 | 18.0 | 18.5 | 19.0 | 19.7 | 20.1 |
| GL RAD OVER THRSH | 654.5 | 601.2 | 463.8 | 360.5 | 280.5 | 218.1 | 158.2 |
| TIME AVG.TEMP | 18.3 | 19.3 | 19.8 | 20.2 | 20.5 | 20.8 | 21.1 |
| TOTAL TIME MSEC | 2.5920 | 1.5606 | 1.2852 | 1.0494 | 0.8730 | 0.7182 | 0.5778 |
| GLOBAL RAD TOTAL | 654.6 | 654.6 | 645.1 | 611.9 | 569.6 | 518.4 | 459.7 |
| DIRECT RAD TOTAL | 424.3 | 424.3 | 422.5 | 413.4 | 404.0 | 389.0 | 357.4 |
| RAD AV TIME MSEC | 0.0864 | 0.0522 | 0.0436 | 0.0371 | 0.0335 | 0.0302 | 0.0263 |
| DEG-C DAYS | 16. | 4. | 3. | 2. | 1. | 0. | 0. |

**ALL MONTHS**

| SOLAR RAD GROUP | ALL | 0+ | 95+ | 189+ | 284+ | 378+ | 473+ |
|---|---|---|---|---|---|---|---|
| TIME AVG.TEMP | 16.5 | 18.3 | 18.9 | 19.4 | 19.8 | 20.2 | 20.6 |
| TOTAL TIME MSEC | 31.5791 | 15.6852 | 13.0842 | 10.7910 | 9.0792 | 7.4142 | 6.1560 |
| GLOBAL RAD TOTAL | 6443.8 | 6443.7 | 6328.5 | 5999.6 | 5593.8 | 5040.5 | 4504.4 |
| DIRECT RAD TOTAL | 4343.6 | 4343.7 | 4202.3 | 4068.7 | 3797.2 | 3490.1 | 3102.1 |
| RAD AV TIME MSEC | 0.0864 | 0.0447 | 0.0386 | 0.0338 | 0.0308 | 0.0272 | 0.0247 |
| DEG-C DAYS | 990. | 299. | 208. | 142. | 103. | 72. | 51. |

**OCT**

| SOLAR RAD GROUP | ALL | 0+ | 95+ | 189+ | 284+ | 378+ | 473+ |
|---|---|---|---|---|---|---|---|
| AV.THRSH.RAD.LEVL | 0. | 54. | 138. | 241. | 336. | 428. | 521. |
| AV.THRESHOLD TEMP | 16.6 | 16.8 | 18.7 | 18.8 | 19.9 | 19.8 | 21.3 |
| GL RAD OVER THRSH | 444.7 | 378.9 | 291.9 | 205.3 | 142.5 | 93.8 | 55.6 |
| TIME AVG.TEMP | 18.2 | 20.1 | 20.7 | 21.1 | 21.7 | 22.2 | 22.9 |
| TOTAL TIME MSEC | 2.6784 | 1.2276 | 1.0332 | 0.8388 | 0.6624 | 0.5310 | 0.4086 |
| GLOBAL RAD TOTAL | 444.7 | 444.7 | 434.2 | 407.5 | 364.9 | 320.8 | 268.5 |
| DIRECT RAD TOTAL | 285.9 | 285.9 | 282.0 | 273.3 | 261.0 | 237.5 | 207.8 |
| RAD AV TIME MSEC | 0.0864 | 0.0396 | 0.0346 | 0.0302 | 0.0277 | 0.0238 | 0.0212 |
| DEG-C DAYS | 39. | 9. | 6. | 4. | 3. | 2. | 1. |

**NOV**

| SOLAR RAD GROUP | ALL | 0+ | 95+ | 189+ | 284+ | 378+ | 473+ |
|---|---|---|---|---|---|---|---|
| AV.THRSH.RAD.LEVL | 0. | 37. | 149. | 230. | 339. | 428. | 524. |
| AV.THRESHOLD TEMP | 13.5 | 13.6 | 15.8 | 16.1 | 17.2 | 17.4 | 18.5 |
| GL RAD OVER THRSH | 337.6 | 296.5 | 197.8 | 142.6 | 84.5 | 48.7 | 19.6 |
| TIME AVG.TEMP | 16.6 | 16.6 | 17.4 | 17.9 | 18.4 | 18.8 | 19.3 |
| TOTAL TIME MSEC | 2.5920 | 1.1250 | 0.8802 | 0.6786 | 0.5328 | 0.4050 | 0.3006 |
| GLOBAL RAD TOTAL | 337.6 | 337.6 | 328.3 | 298.7 | 265.2 | 221.8 | 177.2 |
| DIRECT RAD TOTAL | 230.8 | 230.8 | 228.3 | 218.0 | 203.5 | 178.4 | 149.8 |
| RAD AV TIME MSEC | 0.0864 | 0.0374 | 0.0306 | 0.0270 | 0.0234 | 0.0198 | 0.0166 |
| DEG-C DAYS | 105. | 26. | 14. | 9. | 5. | 3. | 2. |

**DEC**

| SOLAR RAD GROUP | ALL | 0+ | 95+ | 189+ | 284+ | 378+ | 473+ |
|---|---|---|---|---|---|---|---|
| AV.THRSH.RAD.LEVL | 0. | 55. | 145. | 239. | 327. | 440. | 521. |
| AV.THRESHOLD TEMP | 11.1 | 12.6 | 12.7 | 14.7 | 14.3 | 15.9 | 15.7 |
| GL RAD OVER THRSH | 303.7 | 248.1 | 171.6 | 112.7 | 66.5 | 26.5 | 6.7 |
| TIME AVG.TEMP | 12.3 | 14.3 | 14.6 | 15.3 | 15.4 | 15.9 | 15.9 |
| TOTAL TIME MSEC | 2.6784 | 1.0188 | 0.8478 | 0.6264 | 0.5238 | 0.3546 | 0.2448 |
| GLOBAL RAD TOTAL | 303.7 | 303.7 | 294.4 | 262.3 | 237.8 | 182.5 | 134.2 |
| DIRECT RAD TOTAL | 199.3 | 199.3 | 198.4 | 182.6 | 177.1 | 142.7 | 112.6 |
| RAD AV TIME MSEC | 0.0864 | 0.0328 | 0.0302 | 0.0241 | 0.0227 | 0.0169 | 0.0137 |
| DEG-C DAYS | 187. | 51. | 39. | 26. | 21. | 13. | 9. |

STATISTICS FOR LOS ANGELES, CALIF 1971-1972 IR INCLUDED RSV020 FILE

RADIATION UNITS ARE WATTS/SQ.M OR MJOULES/SQ.M
SLOPE = 20.0 DEG   AZIMUTH = 180.0 DEG
DEGREE-DAY BASE TEMP = 18.3 DEGREES C

**JAN**

| SOLAR RAD GROUP | ALL | 0+ | 95+ | 189+ | 284+ | 378+ | 473+ |
|---|---|---|---|---|---|---|---|
| AV.THRSH.RAD.LEVL | 0. | 40. | 140. | 232. | 340. | 428. | 522. |
| AV.THRESHOLD TEMP | 10.6 | 11.1 | 12.5 | 13.0 | 14.3 | 14.7 | 15.5 |
| GL RAD OVER THRSH | 424.0 | 375.5 | 280.6 | 212.1 | 145.0 | 98.9 | 61.7 |
| TIME AVG.TEMP | 12.1 | 13.9 | 14.7 | 15.3 | 15.8 | 16.1 | 16.5 |
| TOTAL TIME MSEC | 2.6784 | 1.2096 | 0.9486 | 0.7416 | 0.6246 | 0.5256 | 0.3942 |
| GLOBAL RAD TOTAL | 424.0 | 424.0 | 413.5 | 384.5 | 357.3 | 323.7 | 267.5 |
| DIRECT RAD TOTAL | 287.1 | 287.1 | 284.2 | 271.0 | 260.5 | 245.7 | 213.8 |
| RAD AV TIME MSEC | 0.0864 | 0.0392 | 0.0331 | 0.0277 | 0.0256 | 0.0238 | 0.0194 |
| DEG-C DAYS | 193. | 66. | 44. | 31. | 24. | 19. | 13. |

**FEB**

| SOLAR RAD GROUP | ALL | 0+ | 95+ | 189+ | 284+ | 378+ | 473+ |
|---|---|---|---|---|---|---|---|
| AV.THRSH.RAD.LEVL | 0. | 36. | 142. | 236. | 329. | 434. | 519. |
| AV.THRESHOLD TEMP | 11.7 | 12.3 | 13.5 | 14.1 | 14.3 | 15.7 | 15.6 |
| GL RAD OVER THRSH | 465.9 | 422.5 | 323.5 | 248.0 | 188.2 | 131.2 | 91.9 |
| TIME AVG.TEMP | 13.2 | 14.6 | 15.3 | 15.7 | 16.0 | 16.3 | 16.4 |
| TOTAL TIME MSEC | 2.4624 | 1.2240 | 0.9324 | 0.7866 | 0.6516 | 0.5454 | 0.4626 |
| GLOBAL RAD TOTAL | 466.0 | 465.9 | 455.6 | 434.9 | 402.8 | 367.9 | 331.9 |
| DIRECT RAD TOTAL | 314.4 | 314.4 | 310.6 | 305.7 | 290.8 | 272.9 | 254.0 |
| RAD AV TIME MSEC | 0.0864 | 0.0432 | 0.0346 | 0.0317 | 0.0281 | 0.0248 | 0.0227 |
| DEG-C DAYS | 147. | 55. | 36. | 27. | 21. | 16. | 13. |

**MAR**

| SOLAR RAD GROUP | ALL | 0+ | 95+ | 189+ | 284+ | 378+ | 473+ |
|---|---|---|---|---|---|---|---|
| AV.THRSH.RAD.LEVL | 0. | 31. | 140. | 232. | 329. | 432. | 528. |
| AV.THRESHOLD TEMP | 12.6 | 13.7 | 13.3 | 14.8 | 15.1 | 16.2 | 16.3 |
| GL RAD OVER THRSH | 551.5 | 509.9 | 392.6 | 311.4 | 240.3 | 178.4 | 127.0 |
| TIME AVG.TEMP | 14.1 | 15.4 | 15.9 | 16.4 | 16.8 | 17.2 | 17.3 |
| TOTAL TIME MSEC | 2.6784 | 1.3536 | 1.0746 | 0.8820 | 0.7362 | 0.5976 | 0.5364 |
| GLOBAL RAD TOTAL | 551.5 | 551.5 | 542.9 | 516.0 | 482.2 | 436.7 | 410.2 |
| DIRECT RAD TOTAL | 359.8 | 359.8 | 358.6 | 353.2 | 346.9 | 321.8 | 307.1 |
| RAD AV TIME MSEC | 0.0864 | 0.0439 | 0.0364 | 0.0324 | 0.0310 | 0.0263 | 0.0245 |
| DEG-C DAYS | 132. | 47. | 32. | 22. | 16. | 11. | 10. |

**JUL**

| SOLAR RAD GROUP | ALL | 0+ | 95+ | 189+ | 284+ | 378+ | 473+ |
|---|---|---|---|---|---|---|---|
| AV.THRSH.RAD.LEVL | 0. | 37. | 157. | 231. | 336. | 431. | 514. |
| AV.THRESHOLD TEMP | 18.8 | 19.3 | 20.5 | 20.4 | 22.9 | 21.6 | 22.6 |
| GL RAD OVER THRSH | 789.2 | 731.8 | 579.3 | 494.5 | 391.6 | 304.1 | 241.4 |
| TIME AVG.TEMP | 20.6 | 21.8 | 22.3 | 22.9 | 22.9 | 22.9 | 23.2 |
| TOTAL TIME MSEC | 2.6784 | 1.5624 | 1.2726 | 1.1340 | 0.9846 | 0.9216 | 0.7560 |
| GLOBAL RAD TOTAL | 789.2 | 789.2 | 778.6 | 756.9 | 722.3 | 701.1 | 629.8 |
| DIRECT RAD TOTAL | 574.8 | 574.8 | 572.1 | 564.1 | 548.2 | 539.0 | 494.2 |
| RAD AV TIME MSEC | 0.0864 | 0.0504 | 0.0414 | 0.0374 | 0.0328 | 0.0317 | 0.0266 |
| DEG-C DAYS | 0. | 0. | 0. | 0. | 0. | 0. | 0. |

**AUG**

| SOLAR RAD GROUP | ALL | 0+ | 95+ | 189+ | 284+ | 378+ | 473+ |
|---|---|---|---|---|---|---|---|
| AV.THRSH.RAD.LEVL | 0. | 26. | 151. | 233. | 338. | 424. | 517. |
| AV.THRESHOLD TEMP | 20.6 | 21.1 | 22.5 | 21.9 | 23.6 | 23.7 | 24.9 |
| GL RAD OVER THRSH | 758.1 | 717.4 | 559.3 | 470.4 | 368.4 | 294.1 | 227.2 |
| TIME AVG.TEMP | 22.3 | 23.5 | 24.1 | 24.3 | 24.6 | 24.7 | 24.9 |
| TOTAL TIME MSEC | 2.6784 | 1.5624 | 1.2672 | 1.0800 | 0.9684 | 0.8730 | 0.7128 |
| GLOBAL RAD TOTAL | 758.1 | 758.1 | 750.4 | 722.2 | 696.1 | 663.9 | 596.0 |
| DIRECT RAD TOTAL | 594.6 | 594.6 | 592.9 | 578.9 | 569.0 | 551.1 | 505.1 |
| RAD AV TIME MSEC | 0.0864 | 0.0504 | 0.0414 | 0.0356 | 0.0331 | 0.0310 | 0.0259 |
| DEG-C DAYS | 0. | 0. | 0. | 0. | 0. | 0. | 0. |

**SEP**

| SOLAR RAD GROUP | ALL | 0+ | 95+ | 189+ | 284+ | 378+ | 473+ |
|---|---|---|---|---|---|---|---|
| AV.THRSH.RAD.LEVL | 0. | 28. | 141. | 241. | 331. | 424. | 526. |
| AV.THRESHOLD TEMP | 19.2 | 20.0 | 20.4 | 22.2 | 21.5 | 22.9 | 22.5 |
| GL RAD OVER THRSH | 607.1 | 570.3 | 445.7 | 350.8 | 277.4 | 216.4 | 157.6 |
| TIME AVG.TEMP | 20.7 | 22.1 | 22.5 | 22.8 | 22.9 | 23.3 | 23.3 |
| TOTAL TIME MSEC | 2.5920 | 1.3374 | 1.0098 | 0.9504 | 0.8118 | 0.6588 | 0.5760 |
| GLOBAL RAD TOTAL | 607.2 | 607.1 | 600.6 | 579.6 | 546.2 | 495.5 | 460.5 |
| DIRECT RAD TOTAL | 441.1 | 441.1 | 439.5 | 433.1 | 425.4 | 397.1 | 380.5 |
| RAD AV TIME MSEC | 0.0864 | 0.0450 | 0.0374 | 0.0335 | 0.0310 | 0.0270 | 0.0248 |
| DEG-C DAYS | 2. | 0. | 0. | 0. | 0. | 0. | 0. |

## APR

| SOLAR RAD GROUP | ALL | 0+ | 95+ | 189+ | 284+ | 378+ | 473+ |
|---|---|---|---|---|---|---|---|
| AV.THRSH.RAD.LEVL | 0. | 22. | 147. | 232. | 337. | 426. | 523. |
| AV.THRESHOLD TEMP | 13.2 | 13.8 | 14.5 | 14.9 | 16.5 | 16.3 | 16.5 |
| GL RAD OVER THRSH | 730.8 | 698.6 | 546.3 | 457.3 | 356.8 | 283.0 | 215.8 |
| TIME AVG.TEMP | 15.2 | 16.7 | 17.3 | 17.4 | 18.0 | 18.2 | 18.6 |
| TOTAL TIME MSEC | 2.5920 | 1.4760 | 1.2150 | 1.0494 | 0.9540 | 0.8334 | 0.6912 |
| GLOBAL RAD TOTAL | 730.9 | 730.9 | 725.1 | 700.8 | 678.6 | 638.0 | 577.4 |
| DIRECT RAD TOTAL | 519.3 | 519.3 | 518.0 | 505.9 | 498.0 | 479.8 | 443.1 |
| RAD AV TIME MSEC | 0.0864 | 0.0493 | 0.0410 | 0.0360 | 0.0335 | 0.0310 | 0.0266 |
| DEG-C DAYS | 97. | 33. | 20. | 14. | 12. | 8. | 6. |

## MAY

| SOLAR RAD GROUP | ALL | 0+ | 95+ | 189+ | 284+ | 378+ | 473+ |
|---|---|---|---|---|---|---|---|
| AV.THRSH.RAD.LEVL | 0. | 35. | 151. | 230. | 332. | 428. | 525. |
| AV.THRESHOLD TEMP | 14.6 | 15.0 | 15.8 | 16.8 | 17.5 | 17.8 | 18.7 |
| GL RAD OVER THRSH | 640.6 | 586.2 | 443.0 | 361.7 | 277.0 | 208.8 | 154.2 |
| TIME AVG.TEMP | 16.3 | 17.6 | 18.2 | 18.7 | 19.2 | 19.4 | 19.9 |
| TOTAL TIME MSEC | 2.6784 | 1.5624 | 1.2348 | 1.0224 | 0.8334 | 0.7128 | 0.5598 |
| GLOBAL RAD TOTAL | 640.6 | 640.6 | 629.2 | 597.1 | 553.6 | 513.6 | 448.2 |
| DIRECT RAD TOTAL | 407.6 | 407.6 | 405.6 | 398.1 | 385.4 | 375.1 | 335.6 |
| RAD AV TIME MSEC | 0.0864 | 0.0504 | 0.0407 | 0.0356 | 0.0313 | 0.0299 | 0.0248 |
| DEG-C DAYS | 71. | 24. | 13. | 8. | 4. | 3. | 2. |

## JUN

| SOLAR RAD GROUP | ALL | 0+ | 95+ | 189+ | 284+ | 378+ | 473+ |
|---|---|---|---|---|---|---|---|
| AV.THRSH.RAD.LEVL | 0. | 37. | 146. | 233. | 335. | 431. | 510. |
| AV.THRESHOLD TEMP | 16.8 | 17.3 | 18.0 | 18.9 | 19.2 | 20.2 | 20.2 |
| GL RAD OVER THRSH | 623.2 | 567.0 | 433.7 | 343.6 | 259.5 | 194.7 | 152.3 |
| TIME AVG.TEMP | 18.3 | 19.4 | 19.9 | 20.3 | 20.7 | 21.0 | 21.2 |
| TOTAL TIME MSEC | 2.5920 | 1.5120 | 1.2294 | 1.0260 | 0.8280 | 0.6786 | 0.5328 |
| GLOBAL RAD TOTAL | 623.2 | 623.2 | 612.7 | 583.1 | 536.9 | 486.8 | 424.1 |
| DIRECT RAD TOTAL | 398.8 | 398.8 | 397.1 | 391.4 | 373.2 | 366.2 | 331.5 |
| RAD AV TIME MSEC | 0.0864 | 0.0504 | 0.0414 | 0.0367 | 0.0313 | 0.0295 | 0.0248 |
| DEG-C DAYS | 16. | 4. | 2. | 1. | 1. | 0. | 0. |

## OCT

| SOLAR RAD GROUP | ALL | 0+ | 95+ | 189+ | 284+ | 378+ | 473+ |
|---|---|---|---|---|---|---|---|
| AV.THRSH.RAD.LEVL | 0. | 37. | 138. | 241. | 324. | 428. | 527. |
| AV.THRESHOLD TEMP | 16.4 | 17.2 | 18.1 | 18.9 | 20.0 | 20.0 | 20.4 |
| GL RAD OVER THRSH | 527.9 | 478.4 | 371.2 | 278.5 | 219.0 | 156.4 | 106.5 |
| TIME AVG.TEMP | 18.2 | 19.9 | 20.7 | 21.1 | 21.7 | 22.0 | 22.4 |
| TOTAL TIME MSEC | 2.6784 | 1.3392 | 1.0602 | 0.9018 | 0.7200 | 0.6012 | 0.5022 |
| GLOBAL RAD TOTAL | 527.9 | 527.9 | 517.6 | 495.7 | 451.9 | 413.5 | 371.2 |
| DIRECT RAD TOTAL | 372.3 | 372.3 | 367.9 | 362.9 | 346.0 | 323.5 | 301.3 |
| RAD AV TIME MSEC | 0.0864 | 0.0432 | 0.0356 | 0.0331 | 0.0302 | 0.0263 | 0.0252 |
| DEG-C DAYS | 39. | 10. | 6. | 5. | 3. | 3. | 2. |

## NOV

| SOLAR RAD GROUP | ALL | 0+ | 95+ | 189+ | 284+ | 378+ | 473+ |
|---|---|---|---|---|---|---|---|
| AV.THRSH.RAD.LEVL | 0. | 36. | 144. | 238. | 334. | 430. | 524. |
| AV.THRESHOLD TEMP | 13.3 | 13.5 | 15.9 | 15.7 | 16.7 | 17.3 | 17.0 |
| GL RAD OVER THRSH | 448.5 | 404.9 | 302.2 | 231.7 | 174.0 | 122.8 | 84.4 |
| TIME AVG.TEMP | 14.8 | 16.6 | 17.4 | 17.9 | 18.4 | 18.6 | 19.0 |
| TOTAL TIME MSEC | 2.5920 | 1.2114 | 0.9540 | 0.7434 | 0.6048 | 0.5328 | 0.4086 |
| GLOBAL RAD TOTAL | 448.5 | 448.5 | 439.2 | 409.0 | 376.0 | 351.9 | 298.5 |
| DIRECT RAD TOTAL | 343.0 | 343.0 | 339.3 | 328.0 | 312.4 | 300.0 | 262.8 |
| RAD AV TIME MSEC | 0.0664 | 0.0407 | 0.0338 | 0.0299 | 0.0266 | 0.0256 | 0.0216 |
| DEG-C DAYS | 105. | 28. | 15. | 9. | 6. | 5. | 3. |

## DEC

| SOLAR RAD GROUP | ALL | 0+ | 95+ | 189+ | 284+ | 378+ | 473+ |
|---|---|---|---|---|---|---|---|
| AV.THRSH.RAD.LEVL | 0. | 46. | 136. | 243. | 340. | 421. | 529. |
| AV.THRESHOLD TEMP | 10.9 | 12.4 | 12.5 | 14.4 | 15.0 | 14.1 | 14.5 |
| GL RAD OVER THRSH | 418.3 | 366.0 | 282.0 | 206.5 | 148.3 | 107.7 | 66.5 |
| TIME AVG.TEMP | 12.3 | 13.5 | 14.7 | 15.3 | 15.5 | 15.6 | 16.1 |
| TOTAL TIME MSEC | 2.6784 | 1.1286 | 0.9324 | 0.7092 | 0.5994 | 0.4986 | 0.3834 |
| GLOBAL RAD TOTAL | 418.3 | 418.3 | 409.2 | 378.8 | 352.1 | 317.8 | 269.3 |
| DIRECT RAD TOTAL | 315.3 | 315.3 | 313.3 | 296.9 | 289.6 | 271.2 | 236.6 |
| RAD AV TIME MSEC | 0.0864 | 0.0364 | 0.0338 | 0.0274 | 0.0266 | 0.0248 | 0.0205 |
| DEG-C DAYS | 187. | 56. | 43. | 28. | 24. | 19. | 13. |

## ALL MONTHS

| SOLAR RAD GROUP | ALL | 0+ | 95+ | 189+ | 284+ | 378+ | 473+ |
|---|---|---|---|---|---|---|---|
| TIME AVG.TEMP | 16.5 | 18.2 | 18.9 | 19.4 | 19.7 | 20.0 | 20.3 |
| TOTAL TIME MSEC | 31.5791 | 16.4790 | 13.2210 | 11.0268 | 9.3168 | 7.9794 | 6.5160 |
| GLOBAL RAD TOTAL | 6985.3 | 6985.2 | 6874.7 | 6558.5 | 6156.1 | 5710.3 | 5084.4 |
| DIRECT RAD TOTAL | 4928.2 | 4928.1 | 4899.7 | 4789.2 | 4650.6 | 4443.4 | 4065.7 |
| RAD AV TIME MSEC | 0.0864 | 0.0461 | 0.0382 | 0.0338 | 0.0307 | 0.0283 | 0.0245 |
| DEG-C DAYS | 990. | 323. | 212. | 146. | 110. | 84. | 62. |

STATISTICS FOR   LOS ANGELES, CALIF 1971-1972   IR INCLUDED RSV020 FILE

RADIATION UNITS ARE WATTS/SQ.M OR MJOULES/SQ.M
SLOPE = 30.0 DEG   AZIMUTH = 180.0 DEG
DEGREE-DAY BASE TEMP = 18.3 DEGREES C

**JAN**

| | ALL | 0+ | 95+ | 189+ | 284+ | 378+ | 473+ |
|---|---|---|---|---|---|---|---|
| SOLAR RAD GROUP | ALL | 0+ | 95+ | 189+ | 284+ | 378+ | 473+ |
| AV.THRSH.RAD.LEVL | 0. | 41. | 143. | 230. | 339. | 434. | 522. |
| AV.THRESHOLD TEMP | 10.7 | 10.7 | 12.9 | 12.5 | 14.3 | 15.1 | 14.7 |
| GL RAD OVER THRSH | 458.2 | 408.9 | 311.2 | 244.9 | 176.0 | 123.7 | 85.6 |
| TIME AVG.TEMP | 12.1 | 13.9 | 14.7 | 15.2 | 15.7 | 15.9 | 16.2 |
| TOTAL TIME MSEC | 2.6784 | 1.2060 | 0.9540 | 0.7632 | 0.6336 | 0.5490 | 0.4338 |
| GLOBAL RAD TOTAL | 458.2 | 458.2 | 447.9 | 420.6 | 390.7 | 362.1 | 312.0 |
| DIRECT RAD TOTAL | 324.0 | 324.0 | 321.0 | 308.5 | 295.6 | 284.0 | 255.3 |
| RAD AV TIME MSEC | 0.0864 | 0.0389 | 0.0335 | 0.0288 | 0.0259 | 0.0245 | 0.0212 |
| DEG-C DAYS | 193. | 66. | 45. | 33. | 24. | 20. | 15. |

**FEB**

| | ALL | 0+ | 95+ | 189+ | 284+ | 378+ | 473+ |
|---|---|---|---|---|---|---|---|
| SOLAR RAD GROUP | ALL | 0+ | 95+ | 189+ | 284+ | 378+ | 473+ |
| AV.THRSH.RAD.LEVL | 0. | 34. | 139. | 242. | 333. | 428. | 519. |
| AV.THRESHOLD TEMP | 11.7 | 12.6 | 13.1 | 14.4 | 14.0 | 15.1 | 16.0 |
| GL RAD OVER THRSH | 490.4 | 449.0 | 348.9 | 268.8 | 207.3 | 154.9 | 110.6 |
| TIME AVG.TEMP | 13.2 | 14.7 | 15.2 | 15.2 | 16.3 | 16.3 | 16.4 |
| TOTAL TIME MSEC | 2.4624 | 1.2078 | 0.9522 | 0.7848 | 0.6696 | 0.5544 | 0.4878 |
| GLOBAL RAD TOTAL | 490.4 | 490.4 | 481.7 | 453.3 | 430.5 | 392.1 | 363.6 |
| DIRECT RAD TOTAL | 341.6 | 341.6 | 339.1 | 331.8 | 319.3 | 297.3 | 284.4 |
| RAD AV TIME MSEC | 0.0864 | 0.0428 | 0.0356 | 0.0317 | 0.0288 | 0.0252 | 0.0241 |
| DEG-C DAYS | 147. | 54. | 37. | 27. | 22. | 17. | 14. |

**MAR**

| | ALL | 0+ | 95+ | 189+ | 284+ | 378+ | 473+ |
|---|---|---|---|---|---|---|---|
| SOLAR RAD GROUP | ALL | 0+ | 95+ | 189+ | 284+ | 378+ | 473+ |
| AV.THRSH.RAD.LEVL | 0. | 29. | 140. | 233. | 330. | 426. | 526. |
| AV.THRESHOLD TEMP | 12.6 | 13.7 | 13.3 | 15.1 | 15.7 | 14.9 | 16.5 |
| GL RAD OVER THRSH | 561.9 | 522.3 | 403.5 | 322.3 | 250.7 | 192.6 | 138.5 |
| TIME AVG.TEMP | 14.1 | 15.4 | 15.9 | 16.5 | 16.8 | 17.0 | 17.3 |
| TOTAL TIME MSEC | 2.6784 | 1.3482 | 1.0782 | 0.8730 | 0.7326 | 0.6102 | 0.5382 |
| GLOBAL RAD TOTAL | 561.9 | 561.9 | 554.0 | 525.3 | 492.7 | 452.3 | 421.6 |
| DIRECT RAD TOTAL | 373.8 | 373.8 | 372.9 | 366.9 | 360.9 | 338.2 | 319.9 |
| RAD AV TIME MSEC | 0.0864 | 0.0439 | 0.0367 | 0.0324 | 0.0310 | 0.0274 | 0.0248 |
| DEG-C DAYS | 132. | 47. | 32. | 22. | 16. | 12. | 10. |

**JUL**

| | ALL | 0+ | 95+ | 189+ | 284+ | 378+ | 473+ |
|---|---|---|---|---|---|---|---|
| SOLAR RAD GROUP | ALL | 0+ | 95+ | 189+ | 284+ | 378+ | 473+ |
| AV.THRSH.RAD.LEVL | 0. | 28. | 155. | 228. | 342. | 418. | 520. |
| AV.THRESHOLD TEMP | 19.0 | 19.0 | 20.4 | 20.2 | 22.3 | 22.0 | 22.3 |
| GL RAD OVER THRSH | 740.7 | 699.8 | 545.8 | 467.1 | 355.9 | 288.0 | 214.8 |
| TIME AVG.TEMP | 20.6 | 21.8 | 22.4 | 22.6 | 22.9 | 22.9 | 23.2 |
| TOTAL TIME MSEC | 2.6784 | 1.4508 | 1.2114 | 1.0872 | 0.9738 | 0.8874 | 0.7218 |
| GLOBAL RAD TOTAL | 740.8 | 740.7 | 734.0 | 714.7 | 688.9 | 659.3 | 590.0 |
| DIRECT RAD TOTAL | 533.7 | 533.6 | 533.1 | 526.9 | 517.1 | 502.8 | 459.4 |
| RAD AV TIME MSEC | 0.0864 | 0.0472 | 0.0392 | 0.0360 | 0.0324 | 0.0306 | 0.0256 |
| DEG-C DAYS | 0. | 0. | 0. | 0. | 0. | 0. | 0. |

**AUG**

| | ALL | 0+ | 95+ | 189+ | 284+ | 378+ | 473+ |
|---|---|---|---|---|---|---|---|
| SOLAR RAD GROUP | ALL | 0+ | 95+ | 189+ | 284+ | 378+ | 473+ |
| AV.THRSH.RAD.LEVL | 0. | 17. | 160. | 238. | 344. | 421. | 527. |
| AV.THRESHOLD TEMP | 20.6 | 21.5 | 21.8 | 22.7 | 23.5 | 23.8 | 24.3 |
| GL RAD OVER THRSH | 725.5 | 701.4 | 533.1 | 451.7 | 350.1 | 285.7 | 211.2 |
| TIME AVG.TEMP | 22.3 | 23.7 | 24.2 | 24.6 | 24.8 | 24.8 | 24.9 |
| TOTAL TIME MSEC | 2.6784 | 1.4472 | 1.1736 | 1.0494 | 0.9594 | 0.8298 | 0.7038 |
| GLOBAL RAD TOTAL | 725.5 | 725.5 | 720.9 | 701.0 | 679.6 | 635.1 | 582.1 |
| DIRECT RAD TOTAL | 568.6 | 568.6 | 567.9 | 560.1 | 554.2 | 526.2 | 492.7 |
| RAD AV TIME MSEC | 0.0864 | 0.0468 | 0.0382 | 0.0346 | 0.0328 | 0.0295 | 0.0259 |
| DEG-C DAYS | 0. | 0. | 0. | 0. | 0. | 0. | 0. |

**SEP**

| | ALL | 0+ | 95+ | 189+ | 284+ | 378+ | 473+ |
|---|---|---|---|---|---|---|---|
| SOLAR RAD GROUP | ALL | 0+ | 95+ | 189+ | 284+ | 378+ | 473+ |
| AV.THRSH.RAD.LEVL | 0. | 28. | 143. | 238. | 330. | 421. | 522. |
| AV.THRESHOLD TEMP | 19.2 | 20.3 | 20.3 | 22.1 | 22.0 | 22.0 | 22.9 |
| GL RAD OVER THRSH | 612.6 | 576.2 | 450.9 | 360.6 | 286.5 | 226.1 | 168.1 |
| TIME AVG.TEMP | 20.7 | 22.1 | 22.5 | 22.8 | 22.9 | 23.2 | 23.3 |
| TOTAL TIME MSEC | 2.5920 | 1.3212 | 1.0890 | 0.9432 | 0.8136 | 0.6606 | 0.5760 |
| GLOBAL RAD TOTAL | 612.7 | 612.7 | 606.3 | 585.5 | 554.6 | 504.1 | 468.5 |
| DIRECT RAD TOTAL | 449.2 | 449.2 | 447.7 | 441.6 | 434.5 | 406.9 | 389.1 |
| RAD AV TIME MSEC | 0.0864 | 0.0443 | 0.0371 | 0.0331 | 0.0310 | 0.0274 | 0.0248 |
| DEG-C DAYS | 2. | 0. | 0. | 0. | 0. | 0. | 0. |

## APR

| SOLAR RAD GROUP | ALL | 0+ | 95+ | 189+ | 284+ | 378+ | 473+ |
|---|---|---|---|---|---|---|---|
| AV.THRSH.RAD.LEVL | 0. | 14. | 153. | 246. | 338. | 416. | 520. |
| AV.THRESHOLD TEMP | 13.2 | 14.3 | 13.8 | 15.9 | 16.0 | 16.3 | 17.2 |
| GL RAD OVER THRSH | 709.2 | 689.1 | 530.1 | 435.8 | 349.0 | 284.7 | 213.3 |
| TIME AVG.TEMP | 15.2 | 16.8 | 17.4 | 17.8 | 18.0 | 18.3 | 18.7 |
| TOTAL TIME MSEC | 2.5920 | 1.3950 | 1.1466 | 1.0188 | 0.9396 | 0.8226 | 0.6858 |
| GLOBAL RAD TOTAL | 709.2 | 709.2 | 705.6 | 686.1 | 666.6 | 627.1 | 570.1 |
| DIRECT RAD TOTAL | 505.1 | 505.1 | 504.4 | 495.6 | 491.0 | 471.5 | 438.6 |
| RAD AV TIME MSEC | 0.0864 | 0.0468 | 0.0385 | 0.0349 | 0.0331 | 0.0306 | 0.0266 |
| DEG-C DAYS | 97. | 29. | 18. | 13. | 11. | 8. | 6. |

## MAY

| SOLAR RAD GROUP | ALL | 0+ | 95+ | 189+ | 284+ | 378+ | 473+ |
|---|---|---|---|---|---|---|---|
| AV.THRSH.RAD.LEVL | 0. | 27. | 150. | 232. | 336. | 419. | 534. |
| AV.THRESHOLD TEMP | 14.8 | 15.1 | 15.8 | 16.2 | 17.6 | 18.3 | 18.3 |
| GL RAD OVER THRSH | 599.0 | 561.9 | 419.6 | 339.2 | 254.3 | 197.9 | 136.0 |
| TIME AVG.TEMP | 16.3 | 17.7 | 18.3 | 18.7 | 19.2 | 19.6 | 19.9 |
| TOTAL TIME MSEC | 2.6784 | 1.3968 | 1.1538 | 0.9792 | 0.8190 | 0.6786 | 0.5382 |
| GLOBAL RAD TOTAL | 599.0 | 599.0 | 592.6 | 566.4 | 529.2 | 482.1 | 423.3 |
| DIRECT RAD TOTAL | 378.3 | 378.2 | 377.8 | 373.9 | 366.6 | 348.9 | 315.6 |
| RAD AV TIME MSEC | 0.0864 | 0.0454 | 0.0378 | 0.0342 | 0.0313 | 0.0284 | 0.0241 |
| DEG-C DAYS | 71. | 20. | 12. | 7. | 4. | 3. | 2. |

## JUN

| SOLAR RAD GROUP | ALL | 0+ | 95+ | 189+ | 284+ | 378+ | 473+ |
|---|---|---|---|---|---|---|---|
| AV.THRSH.RAD.LEVL | 0. | 32. | 141. | 234. | 335. | 421. | 519. |
| AV.THRESHOLD TEMP | 16.8 | 17.3 | 18.1 | 19.1 | 19.1 | 20.3 | 20.6 |
| GL RAD OVER THRSH | 589.6 | 541.3 | 408.3 | 316.3 | 235.0 | 179.8 | 130.3 |
| TIME AVG.TEMP | 18.3 | 19.4 | 19.9 | 20.4 | 20.7 | 21.1 | 21.3 |
| TOTAL TIME MSEC | 2.5920 | 1.5120 | 1.2222 | 0.9828 | 0.8046 | 0.6444 | 0.5040 |
| GLOBAL RAD TOTAL | 589.6 | 589.6 | 580.4 | 546.6 | 504.9 | 451.2 | 392.0 |
| DIRECT RAD TOTAL | 369.1 | 369.1 | 368.8 | 360.7 | 352.1 | 336.6 | 303.8 |
| RAD AV TIME MSEC | 0.0864 | 0.0504 | 0.0414 | 0.0349 | 0.0310 | 0.0284 | 0.0238 |
| DEG-C DAYS | 16. | 4. | 2. | 1. | 1. | 0. | 0. |

## OCT

| SOLAR RAD GROUP | ALL | 0+ | 95+ | 189+ | 284+ | 378+ | 473+ |
|---|---|---|---|---|---|---|---|
| AV.THRSH.RAD.LEVL | 0. | 33. | 134. | 240. | 329. | 423. | 526. |
| AV.THRESHOLD TEMP | 16.4 | 17.5 | 17.8 | 18.7 | 19.8 | 20.5 | 19.4 |
| GL RAD OVER THRSH | 552.8 | 508.3 | 399.6 | 303.4 | 237.2 | 180.3 | 126.5 |
| TIME AVG.TEMP | 18.2 | 19.9 | 20.5 | 21.1 | 21.6 | 21.9 | 22.2 |
| TOTAL TIME MSEC | 2.6784 | 1.3320 | 1.0854 | 0.9018 | 0.7434 | 0.6066 | 0.5238 |
| GLOBAL RAD TOTAL | 552.9 | 552.8 | 544.6 | 520.1 | 482.0 | 436.9 | 401.9 |
| DIRECT RAD TOTAL | 399.8 | 399.8 | 397.1 | 389.4 | 376.2 | 348.6 | 330.8 |
| RAD AV TIME MSEC | 0.0864 | 0.0432 | 0.0371 | 0.0331 | 0.0313 | 0.0270 | 0.0256 |
| DEG-C DAYS | 39. | 10. | 7. | 5. | 4. | 3. | 2. |

## NOV

| SOLAR RAD GROUP | ALL | 0+ | 95+ | 189+ | 284+ | 378+ | 473+ |
|---|---|---|---|---|---|---|---|
| AV.THRSH.RAD.LEVL | 0. | 38. | 144. | 238. | 325. | 430. | 513. |
| AV.THRESHOLD TEMP | 13.2 | 13.6 | 15.8 | 16.1 | 16.5 | 17.2 | 17.0 |
| GL RAD OVER THRSH | 487.5 | 441.7 | 339.6 | 268.5 | 214.0 | 155.9 | 117.0 |
| TIME AVG.TEMP | 14.8 | 16.7 | 18.3 | 17.9 | 18.3 | 18.5 | 18.7 |
| TOTAL TIME MSEC | 2.5920 | 1.2024 | 0.9612 | 0.7596 | 0.6228 | 0.5544 | 0.4716 |
| GLOBAL RAD TOTAL | 487.6 | 487.6 | 478.4 | 449.3 | 416.7 | 394.5 | 358.8 |
| DIRECT RAD TOTAL | 383.5 | 383.5 | 379.6 | 368.7 | 352.8 | 341.7 | 318.2 |
| RAD AV TIME MSEC | 0.0864 | 0.0403 | 0.0346 | 0.0310 | 0.0277 | 0.0266 | 0.0245 |
| DEG-C DAYS | 105. | 27. | 16. | 9. | 6. | 6. | 4. |

## DEC

| SOLAR RAD GROUP | ALL | 0+ | 95+ | 189+ | 284+ | 378+ | 473+ |
|---|---|---|---|---|---|---|---|
| AV.THRSH.RAD.LEVL | 0. | 45. | 149. | 240. | 333. | 435. | 506. |
| AV.THRESHOLD TEMP | 10.9 | 12.4 | 12.7 | 14.3 | 14.7 | 15.1 | 13.1 |
| GL RAD OVER THRSH | 461.1 | 409.9 | 312.3 | 247.7 | 190.7 | 136.1 | 104.9 |
| TIME AVG.TEMP | 12.3 | 14.3 | 14.7 | 15.3 | 15.4 | 15.6 | 15.7 |
| TOTAL TIME MSEC | 2.6784 | 1.1286 | 0.9378 | 0.7128 | 0.6120 | 0.5364 | 0.4374 |
| GLOBAL RAD TOTAL | 461.1 | 461.1 | 452.4 | 418.8 | 394.6 | 369.4 | 326.3 |
| DIRECT RAD TOTAL | 359.7 | 359.7 | 358.1 | 338.1 | 331.5 | 321.1 | 289.8 |
| RAD AV TIME MSEC | 0.0864 | 0.0364 | 0.0342 | 0.0277 | 0.0270 | 0.0266 | 0.0234 |
| DEG-C DAYS | 187. | 56. | 43. | 29. | 24. | 21. | 17. |

## ALL MONTHS

| SOLAR RAD GROUP | ALL | 0+ | 95+ | 189+ | 284+ | 378+ | 473+ |
|---|---|---|---|---|---|---|---|
| TIME AVG.TEMP | 16.5 | 18.2 | 18.8 | 19.3 | 19.7 | 19.9 | 20.2 |
| TOTAL TIME MSEC | 31.5791 | 15.9480 | 12.9654 | 10.8558 | 9.3240 | 7.9344 | 6.6222 |
| GLOBAL RAD TOTAL | 6989.2 | 6898.8 | 6898.0 | 6592.6 | 6231.0 | 5766.1 | 5210.4 |
| DIRECT RAD TOTAL | 4986.2 | 4986.2 | 4967.6 | 4862.6 | 4751.8 | 4523.8 | 4197.3 |
| RAD AV TIME MSEC | 0.0864 | 0.0443 | 0.0373 | 0.0331 | 0.0307 | 0.0280 | 0.0248 |
| DEG-C DAYS | 990. | 312. | 212. | 147. | 113. | 89. | 69. |

STATISTICS FOR LOS ANGELES, CALIF 1971-1972  IR INCLUDED RSV020 FILE

RADIATION UNITS ARE WATTS/SQ.M OR MJOULES/SQ.M
SLOPE = 40.0 DEG   AZIMUTH = 180.0 DEG
DEGREE-DAY BASE TEMP = 18.3 DEGREES C

### JAN

| SOLAR RAD GROUP | ALL | 0+ | 95+ | 189+ | 284+ | 378+ | 473+ |
|---|---|---|---|---|---|---|---|
| AV.THRSH.RAD.LEVL | 0. | 40. | 147. | 231. | 333. | 429. | 518. |
| AV.THRESHOLD TEMP | 10.6 | 11.1 | 13.1 | 12.4 | 13.8 | 14.8 | 15.6 |
| GL RAD OVER THRSH | 481.4 | 432.8 | 331.0 | 266.5 | 200.7 | 147.1 | 105.6 |
| TIME AVG.TEMP | 12.1 | 13.9 | 14.7 | 15.1 | 15.7 | 15.9 | 16.2 |
| TOTAL TIME MSEC | 2.6784 | 1.2132 | 0.9522 | 0.7704 | 0.6408 | 0.5580 | 0.4680 |
| GLOBAL RAD TOTAL | 481.5 | 481.5 | 471.0 | 444.3 | 414.4 | 386.8 | 348.1 |
| DIRECT RAD TOTAL | 350.9 | 350.9 | 347.8 | 336.0 | 321.6 | 310.4 | 292.2 |
| RAD AV TIME MSEC | 0.0864 | 0.0392 | 0.0335 | 0.0299 | 0.0266 | 0.0252 | 0.0238 |
| DEG-C DAYS | 193. | 66. | 45. | 33. | 25. | 21. | 16. |

### FEB

| SOLAR RAD GROUP | ALL | 0+ | 95+ | 189+ | 284+ | 378+ | 473+ |
|---|---|---|---|---|---|---|---|
| AV.THRSH.RAD.LEVL | 0. | 37. | 139. | 238. | 332. | 426. | 525. |
| AV.THRESHOLD TEMP | 11.8 | 12.1 | 13.2 | 14.2 | 14.5 | 14.5 | 16.7 |
| GL RAD OVER THRSH | 503.5 | 459.3 | 361.7 | 284.2 | 220.2 | 167.1 | 119.0 |
| TIME AVG.TEMP | 13.2 | 14.6 | 15.2 | 15.7 | 15.9 | 16.2 | 16.4 |
| TOTAL TIME MSEC | 2.4624 | 1.1898 | 0.9540 | 0.7848 | 0.6822 | 0.5670 | 0.4842 |
| GLOBAL RAD TOTAL | 503.5 | 503.5 | 494.7 | 471.2 | 446.7 | 408.5 | 373.2 |
| DIRECT RAD TOTAL | 358.5 | 358.5 | 355.9 | 348.0 | 338.2 | 315.6 | 298.2 |
| RAD AV TIME MSEC | 0.0864 | 0.0418 | 0.0360 | 0.0320 | 0.0299 | 0.0259 | 0.0245 |
| DEG-C DAYS | 147. | 53. | 37. | 27. | 23. | 18. | 14. |

### MAR

| SOLAR RAD GROUP | ALL | 0+ | 95+ | 189+ | 284+ | 378+ | 473+ |
|---|---|---|---|---|---|---|---|
| AV.THRSH.RAD.LEVL | 0. | 31. | 139. | 231. | 330. | 424. | 523. |
| AV.THRESHOLD TEMP | 12.6 | 13.9 | 13.5 | 14.9 | 15.7 | 14.9 | 16.5 |
| GL RAD OVER THRSH | 559.6 | 518.7 | 401.9 | 323.7 | 251.2 | 193.6 | 140.5 |
| TIME AVG.TEMP | 14.1 | 15.5 | 15.9 | 16.5 | 16.8 | 17.0 | 17.3 |
| TOTAL TIME MSEC | 2.6784 | 1.3284 | 1.0746 | 0.8586 | 0.7308 | 0.6084 | 0.5364 |
| GLOBAL RAD TOTAL | 559.6 | 559.6 | 551.8 | 521.6 | 492.2 | 451.8 | 421.3 |
| DIRECT RAD TOTAL | 376.3 | 376.3 | 375.5 | 369.5 | 363.9 | 340.8 | 322.0 |
| RAD AV TIME MSEC | 0.0864 | 0.0432 | 0.0367 | 0.0324 | 0.0313 | 0.0274 | 0.0248 |
| DEG-C DAYS | 132. | 46. | 32. | 21. | 16. | 12. | 9. |

### JUL

| SOLAR RAD GROUP | ALL | 0+ | 95+ | 189+ | 284+ | 378+ | 473+ |
|---|---|---|---|---|---|---|---|
| AV.THRSH.RAD.LEVL | 0. | 44. | 152. | 242. | 335. | 415. | 540. |
| AV.THRESHOLD TEMP | 18.9 | 20.3 | 19.8 | 21.4 | 22.0 | 21.9 | 22.9 |
| GL RAD OVER THRSH | 684.6 | 625.6 | 499.5 | 406.1 | 316.9 | 252.7 | 164.5 |
| TIME AVG.TEMP | 20.6 | 22.2 | 22.4 | 22.8 | 22.9 | 23.1 | 23.2 |
| TOTAL TIME MSEC | 2.6784 | 1.3392 | 1.1664 | 1.0368 | 0.9594 | 0.8082 | 0.7056 |
| GLOBAL RAD TOTAL | 684.6 | 684.6 | 677.0 | 657.3 | 638.5 | 587.8 | 545.3 |
| DIRECT RAD TOTAL | 483.5 | 483.5 | 481.9 | 475.7 | 470.3 | 441.6 | 417.6 |
| RAD AV TIME MSEC | 0.0864 | 0.0432 | 0.0378 | 0.0338 | 0.0320 | 0.0281 | 0.0252 |
| DEG-C DAYS | 0. | 0. | 0. | 0. | 0. | 0. | 0. |

### AUG

| SOLAR RAD GROUP | ALL | 0+ | 95+ | 189+ | 284+ | 378+ | 473+ |
|---|---|---|---|---|---|---|---|
| AV.THRSH.RAD.LEVL | 0. | 35. | 150. | 245. | 336. | 416. | 526. |
| AV.THRESHOLD TEMP | 20.7 | 22.0 | 21.6 | 22.9 | 23.7 | 24.1 | 23.9 |
| GL RAD OVER THRSH | 692.9 | 645.5 | 511.1 | 413.6 | 328.3 | 264.2 | 188.7 |
| TIME AVG.TEMP | 22.3 | 23.9 | 24.2 | 24.5 | 24.7 | 24.8 | 24.9 |
| TOTAL TIME MSEC | 2.6784 | 1.3392 | 1.1700 | 1.0332 | 0.9378 | 0.7938 | 0.6912 |
| GLOBAL RAD TOTAL | 692.9 | 692.9 | 686.9 | 666.4 | 643.0 | 594.7 | 552.0 |
| DIRECT RAD TOTAL | 536.3 | 536.3 | 534.7 | 527.3 | 520.1 | 489.7 | 463.7 |
| RAD AV TIME MSEC | 0.0864 | 0.0432 | 0.0382 | 0.0342 | 0.0324 | 0.0281 | 0.0256 |
| DEG-C DAYS | 0. | 0. | 0. | 0. | 0. | 0. | 0. |

### SEP

| SOLAR RAD GROUP | ALL | 0+ | 95+ | 189+ | 284+ | 378+ | 473+ |
|---|---|---|---|---|---|---|---|
| AV.THRSH.RAD.LEVL | 0. | 31. | 141. | 238. | 328. | 416. | 516. |
| AV.THRESHOLD TEMP | 19.2 | 20.1 | 20.5 | 22.2 | 21.7 | 22.4 | 22.9 |
| GL RAD OVER THRSH | 605.1 | 565.3 | 445.5 | 354.6 | 282.6 | 225.2 | 167.5 |
| TIME AVG.TEMP | 20.7 | 22.1 | 22.5 | 22.8 | 22.9 | 23.2 | 23.3 |
| TOTAL TIME MSEC | 2.5920 | 1.2996 | 1.0872 | 0.9342 | 0.8010 | 0.6534 | 0.5742 |
| GLOBAL RAD TOTAL | 605.2 | 605.2 | 598.6 | 577.1 | 545.4 | 496.9 | 464.0 |
| DIRECT RAD TOTAL | 444.5 | 444.5 | 443.0 | 436.6 | 429.3 | 402.6 | 384.7 |
| RAD AV TIME MSEC | 0.0864 | 0.0432 | 0.0371 | 0.0331 | 0.0310 | 0.0274 | 0.0248 |
| DEG-C DAYS | 2. | 0. | 0. | 0. | 0. | 0. | 0. |

**APR**

| SOLAR RAD GROUP | ALL | 0+ | 95+ | 189+ | 284+ | 378+ | 473+ |
|---|---|---|---|---|---|---|---|
| AV.THRSH.RAD.LEVL | 0. | 32. | 144. | 243. | 339. | 421. | 526. |
| AV.THRESHOLD TEMP | 13.2 | 14.7 | 14.3 | 15.5 | 16.4 | 16.3 | 16.9 |
| GL RAD OVER THRSH | 684.0 | 642.4 | 514.7 | 414.7 | 325.4 | 261.9 | 192.0 |
| TIME AVG.TEMP | 15.2 | 17.1 | 17.4 | 17.8 | 18.1 | 18.4 | 18.7 |
| TOTAL TIME MSEC | 2.5920 | 1.2960 | 1.1394 | 1.0152 | 0.9252 | 0.7722 | 0.6660 |
| GLOBAL RAD TOTAL | 684.0 | 684.0 | 679.0 | 661.1 | 639.2 | 587.3 | 542.6 |
| DIRECT RAD TOTAL | 483.0 | 483.0 | 481.6 | 474.2 | 468.4 | 440.8 | 416.8 |
| RAD AV TIME MSEC | 0.0864 | 0.0432 | 0.0385 | 0.0349 | 0.0328 | 0.0288 | 0.0263 |
| DEG-C DAYS | 97. | 23. | 18. | 13. | 11. | 8. | 6. |

**MAY**

| SOLAR RAD GROUP | ALL | 0+ | 95+ | 189+ | 284+ | 378+ | 473+ |
|---|---|---|---|---|---|---|---|
| AV.THRSH.RAD.LEVL | 0. | 45. | 148. | 233. | 331. | 420. | 531. |
| AV.THRESHOLD TEMP | 14.7 | 15.8 | 15.8 | 16.7 | 17.8 | 18.3 | 18.5 |
| GL RAD OVER THRSH | 563.3 | 503.5 | 385.8 | 305.6 | 227.2 | 172.2 | 114.0 |
| TIME AVG.TEMP | 16.3 | 17.9 | 18.3 | 18.8 | 19.2 | 19.7 | 19.9 |
| TOTAL TIME MSEC | 2.6784 | 1.3392 | 1.1358 | 0.9468 | 0.8028 | 0.6120 | 0.5256 |
| GLOBAL RAD TOTAL | 563.3 | 563.3 | 554.2 | 526.2 | 492.7 | 429.6 | 393.2 |
| DIRECT RAD TOTAL | 347.3 | 347.3 | 345.8 | 341.2 | 335.9 | 308.0 | 288.8 |
| RAD AV TIME MSEC | 0.0864 | 0.0432 | 0.0371 | 0.0328 | 0.0310 | 0.0263 | 0.0238 |
| DEG-C DAYS | 71. | 17. | 12. | 7. | 4. | 2. | 2. |

**JUN**

| SOLAR RAD GROUP | ALL | 0+ | 95+ | 189+ | 284+ | 378+ | 473+ |
|---|---|---|---|---|---|---|---|
| AV.THRSH.RAD.LEVL | 0. | 50. | 148. | 241. | 333. | 425. | 541. |
| AV.THRESHOLD TEMP | 16.8 | 18.2 | 18.4 | 18.8 | 19.6 | 20.5 | 20.7 |
| GL RAD OVER THRSH | 533.9 | 469.0 | 360.7 | 273.8 | 202.4 | 149.3 | 95.4 |
| TIME AVG.TEMP | 18.3 | 19.8 | 20.1 | 20.4 | 20.8 | 21.2 | 21.3 |
| TOTAL TIME MSEC | 2.5920 | 1.2960 | 1.1106 | 0.9324 | 0.7722 | 0.5760 | 0.4662 |
| GLOBAL RAD TOTAL | 533.9 | 533.9 | 524.6 | 498.3 | 459.7 | 394.3 | 347.6 |
| DIRECT RAD TOTAL | 327.4 | 327.4 | 326.1 | 322.3 | 314.4 | 290.1 | 267.9 |
| RAD AV TIME MSEC | 0.0864 | 0.0432 | 0.0374 | 0.0331 | 0.0302 | 0.0256 | 0.0234 |
| DEG-C DAYS | 16. | 3. | 2. | 1. | 1. | 0. | 0. |

**ALL MONTHS**

| SOLAR RAD GROUP | ALL | 0+ | 95+ | 189+ | 284+ | 378+ | 473+ |
|---|---|---|---|---|---|---|---|
| TIME AVG.TEMP | 16.5 | 18.3 | 18.8 | 19.3 | 19.7 | 19.9 | 20.1 |
| TOTAL TIME MSEC | 31.5791 | 15.2676 | 12.7818 | 10.7262 | 9.2556 | 7.6680 | 6.6402 |
| GLOBAL RAD TOTAL | 6880.2 | 6880.0 | 6783.8 | 6486.4 | 6138.2 | 5610.6 | 5177.0 |
| DIRECT RAD TOTAL | 4928.3 | 4928.2 | 4905.2 | 4809.6 | 4699.1 | 4425.1 | 4187.5 |
| RAD AV TIME MSEC | 0.0864 | 0.0420 | 0.0367 | 0.0338 | 0.0306 | 0.0272 | 0.0251 |
| DEG-C DAYS | 990. | 301. | 211. | 149. | 114. | 91. | 72. |

**OCT**

| SOLAR RAD GROUP | ALL | 0+ | 95+ | 189+ | 284+ | 378+ | 473+ |
|---|---|---|---|---|---|---|---|
| AV.THRSH.RAD.LEVL | 0. | 39. | 135. | 238. | 332. | 418. | 529. |
| AV.THRESHOLD TEMP | 16.4 | 17.6 | 17.6 | 19.1 | 19.5 | 20.6 | 19.7 |
| GL RAD OVER THRSH | 564.3 | 513.8 | 410.2 | 317.8 | 246.5 | 193.8 | 135.6 |
| TIME AVG.TEMP | 18.2 | 20.1 | 20.5 | 20.5 | 21.5 | 21.9 | 22.2 |
| TOTAL TIME MSEC | 2.6784 | 1.2816 | 1.0872 | 0.8982 | 0.7524 | 0.6138 | 0.5274 |
| GLOBAL RAD TOTAL | 564.4 | 564.3 | 556.7 | 531.2 | 496.6 | 450.5 | 414.4 |
| DIRECT RAD TOTAL | 414.9 | 414.8 | 412.8 | 404.3 | 392.6 | 363.7 | 343.7 |
| RAD AV TIME MSEC | 0.0864 | 0.0418 | 0.0371 | 0.0335 | 0.0317 | 0.0274 | 0.0259 |
| DEG-C DAYS | 39. | 9. | 7. | 5. | 4. | 3. | 2. |

**NOV**

| SOLAR RAD GROUP | ALL | 0+ | 95+ | 189+ | 284+ | 378+ | 473+ |
|---|---|---|---|---|---|---|---|
| AV.THRSH.RAD.LEVL | 0. | 38. | 144. | 239. | 327. | 427. | 524. |
| AV.THRESHOLD TEMP | 13.3 | 13.6 | 15.6 | 16.1 | 16.1 | 16.4 | 17.8 |
| GL RAD OVER THRSH | 514.8 | 468.9 | 366.2 | 293.2 | 238.0 | 182.1 | 133.3 |
| TIME AVG.TEMP | 14.8 | 16.6 | 17.4 | 17.8 | 18.2 | 18.5 | 18.7 |
| TOTAL TIME MSEC | 2.5920 | 1.2168 | 0.9648 | 0.7686 | 0.6300 | 0.5580 | 0.5040 |
| GLOBAL RAD TOTAL | 514.8 | 514.8 | 505.3 | 477.0 | 443.9 | 420.3 | 397.3 |
| DIRECT RAD TOTAL | 412.6 | 412.6 | 408.5 | 398.3 | 380.9 | 367.9 | 354.0 |
| RAD AV TIME MSEC | 0.0864 | 0.0407 | 0.0349 | 0.0317 | 0.0284 | 0.0270 | 0.0259 |
| DEG-C DAYS | 105. | 28. | 16. | 10. | 7. | 6. | 4. |

**DEC**

| SOLAR RAD GROUP | ALL | 0+ | 95+ | 189+ | 284+ | 378+ | 473+ |
|---|---|---|---|---|---|---|---|
| AV.THRSH.RAD.LEVL | 0. | 44. | 151. | 229. | 325. | 431. | 521. |
| AV.THRESHOLD TEMP | 10.9 | 12.3 | 13.2 | 13.1 | 14.6 | 15.1 | 14.0 |
| GL RAD OVER THRSH | 492.3 | 442.2 | 341.9 | 284.1 | 224.5 | 166.4 | 121.9 |
| TIME AVG.TEMP | 12.3 | 14.3 | 14.7 | 15.1 | 15.4 | 15.6 | 15.6 |
| TOTAL TIME MSEC | 2.6784 | 1.1286 | 0.9396 | 0.7470 | 0.6210 | 0.5472 | 0.4914 |
| GLOBAL RAD TOTAL | 492.3 | 492.3 | 483.9 | 454.8 | 426.0 | 402.1 | 378.0 |
| DIRECT RAD TOTAL | 393.1 | 393.1 | 391.6 | 376.0 | 363.0 | 354.1 | 338.1 |
| RAD AV TIME MSEC | 0.0864 | 0.0364 | 0.0342 | 0.0299 | 0.0270 | 0.0274 | 0.0259 |
| DEG-C DAYS | 187. | 56. | 43. | 32. | 24. | 22. | 19. |

STATISTICS FOR   LOS ANGELES, CALIF 1971-1972   IR INCLUDED RSV020 FILE

RADIATION UNITS ARE WATTS/SQ.M OR MJOULES/SQ.M
SLOPE = 50.0 DEG   AZIMUTH = 180.0 DEG
DEGREE-DAY BASE TEMP = 18.3 DEGREES C

**JAN**

| SOLAR RAD GROUP | ALL | 0+ | 95+ | 189+ | 284+ | 378+ | 473+ |
|---|---|---|---|---|---|---|---|
| AV.THRSH.RAD.LEVL | 0. | 40. | 142. | 228. | 330. | 430. | 522. |
| AV.THRESHOLD TEMP | 10.6 | 11.2 | 12.5 | 13.0 | 14.2 | 15.0 | 15.4 |
| GL RAD OVER THRSH | 493.2 | 444.8 | 347.9 | 279.7 | 213.9 | 158.3 | 115.6 |
| TIME AVG.TEMP | 12.1 | 13.9 | 14.7 | 15.2 | 15.7 | 15.9 | 16.1 |
| TOTAL TIME MSEC | 2.6784 | 1.2132 | 0.9486 | 0.7920 | 0.6426 | 0.5562 | 0.4680 |
| GLOBAL RAD TOTAL | 493.2 | 493.2 | 482.6 | 460.0 | 426.3 | 397.7 | 359.8 |
| DIRECT RAD TOTAL | 367.2 | 367.2 | 363.9 | 356.0 | 335.9 | 324.6 | 306.7 |
| RAD AV TIME MSEC | 0.0864 | 0.0392 | 0.0335 | 0.0313 | 0.0266 | 0.0252 | 0.0245 |
| DEG-C DAYS | 193. | 66. | 44. | 34. | 25. | 21. | 17. |

**FEB**

| SOLAR RAD GROUP | ALL | 0+ | 95+ | 189+ | 284+ | 378+ | 473+ |
|---|---|---|---|---|---|---|---|
| AV.THRSH.RAD.LEVL | 0. | 42. | 139. | 242. | 334. | 419. | 519. |
| AV.THRESHOLD TEMP | 11.7 | 12.6 | 12.9 | 14.6 | 14.7 | 14.5 | 16.1 |
| GL RAD OVER THRSH | 508.1 | 456.5 | 363.2 | 283.1 | 221.0 | 172.4 | 124.2 |
| TIME AVG.TEMP | 13.2 | 14.6 | 15.2 | 15.7 | 15.9 | 16.1 | 16.4 |
| TOTAL TIME MSEC | 2.4624 | 1.2312 | 0.9612 | 0.7740 | 0.6768 | 0.5706 | 0.4660 |
| GLOBAL RAD TOTAL | 508.1 | 508.1 | 496.8 | 470.7 | 447.2 | 411.7 | 376.2 |
| DIRECT RAD TOTAL | 367.1 | 367.1 | 362.6 | 353.2 | 343.6 | 322.5 | 302.8 |
| RAD AV TIME MSEC | 0.0864 | 0.0432 | 0.0364 | 0.0320 | 0.0299 | 0.0266 | 0.0241 |
| DEG-C DAYS | 147. | 55. | 38. | 27. | 22. | 18. | 14. |

**MAR**

| SOLAR RAD GROUP | ALL | 0+ | 95+ | 189+ | 284+ | 378+ | 473+ |
|---|---|---|---|---|---|---|---|
| AV.THRSH.RAD.LEVL | 0. | 37. | 138. | 233. | 329. | 426. | 525. |
| AV.THRESHOLD TEMP | 12.7 | 13.4 | 13.9 | 14.9 | 15.7 | 15.2 | 17.0 |
| GL RAD OVER THRSH | 546.7 | 498.2 | 389.3 | 309.5 | 241.1 | 182.4 | 130.8 |
| TIME AVG.TEMP | 14.1 | 15.5 | 15.9 | 16.6 | 16.8 | 17.1 | 17.3 |
| TOTAL TIME MSEC | 2.6784 | 1.3140 | 1.0800 | 0.8352 | 0.7182 | 0.6012 | 0.5220 |
| GLOBAL RAD TOTAL | 546.8 | 546.7 | 538.1 | 504.4 | 477.1 | 436.6 | 404.9 |
| DIRECT RAD TOTAL | 369.0 | 368.9 | 367.8 | 360.7 | 354.8 | 330.9 | 311.9 |
| RAD AV TIME MSEC | 0.0864 | 0.0426 | 0.0371 | 0.0324 | 0.0310 | 0.0270 | 0.0245 |
| DEG-C DAYS | 132. | 44. | 32. | 20. | 16. | 12. | 9. |

**JUL**

| SOLAR RAD GROUP | ALL | 0+ | 95+ | 189+ | 284+ | 378+ | 473+ |
|---|---|---|---|---|---|---|---|
| AV.THRSH.RAD.LEVL | 0. | 55. | 143. | 248. | 328. | 429. | 520. |
| AV.THRESHOLD TEMP | 18.9 | 20.4 | 19.7 | 21.8 | 21.7 | 22.3 | 23.1 |
| GL RAD OVER THRSH | 619.2 | 545.8 | 446.2 | 338.2 | 267.8 | 192.4 | 131.3 |
| TIME AVG.TEMP | 20.6 | 22.2 | 22.5 | 22.9 | 22.9 | 23.2 | 23.3 |
| TOTAL TIME MSEC | 2.6784 | 1.3392 | 1.1268 | 1.0260 | 0.8820 | 0.7524 | 0.6678 |
| GLOBAL RAD TOTAL | 619.2 | 619.2 | 607.6 | 593.1 | 557.4 | 514.8 | 478.6 |
| DIRECT RAD TOTAL | 421.4 | 421.4 | 417.8 | 415.2 | 397.8 | 376.8 | 357.8 |
| RAD AV TIME MSEC | 0.0864 | 0.0432 | 0.0364 | 0.0338 | 0.0295 | 0.0259 | 0.0245 |
| DEG-C DAYS | 0. | 0. | 0. | 0. | 0. | 0. | 0. |

**AUG**

| SOLAR RAD GROUP | ALL | 0+ | 95+ | 189+ | 284+ | 378+ | 473+ |
|---|---|---|---|---|---|---|---|
| AV.THRSH.RAD.LEVL | 0. | 41. | 135. | 247. | 328. | 421. | 531. |
| AV.THRESHOLD TEMP | 20.7 | 22.1 | 21.7 | 22.9 | 23.6 | 24.9 | 24.4 |
| GL RAD OVER THRSH | 642.9 | 587.8 | 478.5 | 364.5 | 291.8 | 222.8 | 148.6 |
| TIME AVG.TEMP | 22.3 | 23.9 | 24.2 | 24.5 | 24.7 | 24.9 | 24.9 |
| TOTAL TIME MSEC | 2.6784 | 1.3392 | 1.1610 | 1.0224 | 0.8946 | 0.7434 | 0.6714 |
| GLOBAL RAD TOTAL | 642.9 | 642.9 | 635.6 | 616.8 | 585.3 | 535.7 | 505.4 |
| DIRECT RAD TOTAL | 487.3 | 487.3 | 485.0 | 479.1 | 465.3 | 434.6 | 418.3 |
| RAD AV TIME MSEC | 0.0864 | 0.0432 | 0.0378 | 0.0338 | 0.0310 | 0.0263 | 0.0252 |
| DEG-C DAYS | 0. | 0. | 0. | 0. | 0. | 0. | 0. |

**SEP**

| SOLAR RAD GROUP | ALL | 0+ | 95+ | 189+ | 284+ | 378+ | 473+ |
|---|---|---|---|---|---|---|---|
| AV.THRSH.RAD.LEVL | 0. | 34. | 139. | 236. | 327. | 422. | 527. |
| AV.THRESHOLD TEMP | 19.2 | 19.9 | 20.8 | 21.9 | 22.1 | 22.4 | 23.7 |
| GL RAD OVER THRSH | 583.4 | 539.2 | 426.3 | 336.9 | 265.7 | 205.9 | 147.4 |
| TIME AVG.TEMP | 20.7 | 22.1 | 22.6 | 22.8 | 23.0 | 23.2 | 23.3 |
| TOTAL TIME MSEC | 2.5920 | 1.2942 | 1.0746 | 0.9270 | 0.7812 | 0.6318 | 0.5562 |
| GLOBAL RAD TOTAL | 583.5 | 583.5 | 575.9 | 555.9 | 521.0 | 472.2 | 440.3 |
| DIRECT RAD TOTAL | 426.3 | 426.3 | 424.5 | 418.4 | 410.7 | 381.4 | 364.3 |
| RAD AV TIME MSEC | 0.0864 | 0.0432 | 0.0367 | 0.0328 | 0.0310 | 0.0266 | 0.0248 |
| DEG-C DAYS | 2. | 0. | 0. | 0. | 0. | 0. | 0. |

**APR**

| SOLAR RAD GROUP | ALL | 0+ | 95+ | 189+ | 284+ | 378+ | 473+ |
|---|---|---|---|---|---|---|---|
| AV.THRSH.RAD.LEVL | 0. | 40. | 133. | 239. | 331. | 421. | 532. |
| AV.THRESHOLD TEMP | 13.2 | 14.6 | 14.6 | 15.5 | 16.3 | 16.8 | 17.7 |
| GL RAD OVER THRSH | 642.8 | 590.9 | 485.9 | 378.8 | 296.5 | 230.8 | 158.9 |
| TIME AVG.TEMP | 15.2 | 17.1 | 17.5 | 17.8 | 18.1 | 18.6 | 18.8 |
| TOTAL TIME MSEC | 2.5920 | 1.2960 | 1.1250 | 1.0098 | 0.9036 | 0.7290 | 0.6462 |
| GLOBAL RAD TOTAL | 642.8 | 642.8 | 636.0 | 620.6 | 595.2 | 537.5 | 502.6 |
| DIRECT RAD TOTAL | 446.2 | 446.2 | 444.0 | 438.4 | 431.6 | 399.2 | 384.4 |
| RAD AV TIME MSEC | 0.0864 | 0.0432 | 0.0382 | 0.0346 | 0.0324 | 0.0277 | 0.0259 |
| DEG-C DAYS | 97. | 23. | 17. | 13. | 10. | 7. | 6. |

**MAY**

| SOLAR RAD GROUP | ALL | 0+ | 95+ | 189+ | 284+ | 378+ | 473+ |
|---|---|---|---|---|---|---|---|
| AV.THRSH.RAD.LEVL | 0. | 51. | 146. | 242. | 325. | 427. | 515. |
| AV.THRESHOLD TEMP | 14.7 | 15.9 | 15.8 | 17.1 | 17.9 | 18.1 | 19.4 |
| GL RAD OVER THRSH | 515.4 | 446.8 | 342.0 | 253.3 | 193.5 | 135.0 | 91.5 |
| TIME AVG.TEMP | 16.3 | 17.9 | 18.4 | 18.9 | 19.4 | 19.8 | 20.1 |
| TOTAL TIME MSEC | 2.6784 | 1.3392 | 1.1034 | 0.9234 | 0.7254 | 0.5724 | 0.4914 |
| GLOBAL RAD TOTAL | 515.4 | 515.4 | 503.3 | 477.0 | 429.0 | 379.4 | 344.8 |
| DIRECT RAD TOTAL | 305.8 | 305.8 | 303.1 | 299.9 | 286.4 | 265.7 | 247.3 |
| RAD AV TIME MSEC | 0.0864 | 0.0432 | 0.0356 | 0.0324 | 0.0284 | 0.0245 | 0.0230 |
| DEG-C DAYS | 71. | 17. | 11. | 6. | 3. | 2. | 1. |

**JUN**

| SOLAR RAD GROUP | ALL | 0+ | 95+ | 189+ | 284+ | 378+ | 473+ |
|---|---|---|---|---|---|---|---|
| AV.THRSH.RAD.LEVL | 0. | 55. | 150. | 240. | 325. | 415. | 507. |
| AV.THRESHOLD TEMP | 16.8 | 18.2 | 18.3 | 19.4 | 19.5 | 20.5 | 20.9 |
| GL RAD OVER THRSH | 482.1 | 410.6 | 309.0 | 227.0 | 167.9 | 119.0 | 78.8 |
| TIME AVG.TEMP | 18.3 | 19.8 | 20.2 | 20.5 | 20.8 | 21.2 | 21.4 |
| TOTAL TIME MSEC | 2.5920 | 1.2960 | 1.0728 | 0.9090 | 0.6966 | 0.5400 | 0.4410 |
| GLOBAL RAD TOTAL | 482.1 | 482.1 | 469.7 | 445.2 | 394.2 | 343.3 | 302.2 |
| DIRECT RAD TOTAL | 281.5 | 281.5 | 279.2 | 276.8 | 262.0 | 245.7 | 225.8 |
| RAD AV TIME MSEC | 0.0864 | 0.0432 | 0.0360 | 0.0324 | 0.0274 | 0.0245 | 0.0227 |
| DEG-C DAYS | 16. | 3. | 2. | 1. | 1. | 0. | 0. |

**OCT**

| SOLAR RAD GROUP | ALL | 0+ | 95+ | 189+ | 284+ | 378+ | 473+ |
|---|---|---|---|---|---|---|---|
| AV.THRSH.RAD.LEVL | 0. | 47. | 133. | 236. | 335. | 418. | 527. |
| AV.THRESHOLD TEMP | 16.4 | 17.3 | 17.9 | 19.1 | 20.0 | 19.9 | 20.1 |
| GL RAD OVER THRSH | 567.5 | 504.4 | 410.3 | 318.4 | 244.5 | 193.9 | 137.0 |
| TIME AVG.TEMP | 18.2 | 19.9 | 20.5 | 21.1 | 21.5 | 21.8 | 22.2 |
| TOTAL TIME MSEC | 2.6784 | 1.3320 | 1.0998 | 0.8892 | 0.7470 | 0.6120 | 0.5220 |
| GLOBAL RAD TOTAL | 567.5 | 567.5 | 556.5 | 528.5 | 494.9 | 449.6 | 412.0 |
| DIRECT RAD TOTAL | 421.4 | 421.4 | 417.0 | 406.3 | 394.9 | 366.3 | 343.3 |
| RAD AV TIME MSEC | 0.0864 | 0.0428 | 0.0382 | 0.0335 | 0.0320 | 0.0281 | 0.0259 |
| DEG-C DAYS | 39. | 10. | 7. | 5. | 4. | 3. | 2. |

**NOV**

| SOLAR RAD GROUP | ALL | 0+ | 95+ | 189+ | 284+ | 378+ | 473+ |
|---|---|---|---|---|---|---|---|
| AV.THRSH.RAD.LEVL | 0. | 38. | 138. | 237. | 334. | 432. | 529. |
| AV.THRESHOLD TEMP | 13.3 | 13.6 | 15.7 | 16.2 | 15.7 | 16.4 | 17.8 |
| GL RAD OVER THRSH | 528.3 | 482.6 | 386.0 | 308.2 | 247.3 | 193.0 | 143.4 |
| TIME AVG.TEMP | 14.8 | 16.6 | 17.4 | 17.8 | 18.2 | 18.5 | 18.7 |
| TOTAL TIME MSEC | 2.5920 | 1.2150 | 0.9630 | 0.7866 | 0.6282 | 0.5526 | 0.5112 |
| GLOBAL RAD TOTAL | 528.4 | 528.3 | 518.8 | 494.5 | 457.0 | 431.8 | 413.9 |
| DIRECT RAD TOTAL | 428.5 | 428.3 | 424.4 | 417.5 | 396.2 | 381.6 | 370.1 |
| RAD AV TIME MSEC | 0.0864 | 0.0407 | 0.0349 | 0.0331 | 0.0284 | 0.0270 | 0.0263 |
| DEG-C DAYS | 105. | 28. | 15. | 11. | 7. | 6. | 4. |

**DEC**

| SOLAR RAD GROUP | ALL | 0+ | 95+ | 189+ | 284+ | 378+ | 473+ |
|---|---|---|---|---|---|---|---|
| AV.THRSH.RAD.LEVL | 0. | 45. | 144. | 226. | 322. | 426. | 524. |
| AV.THRESHOLD TEMP | 10.9 | 12.4 | 12.8 | 13.6 | 14.4 | 14.9 | 15.0 |
| GL RAD OVER THRSH | 511.1 | 460.8 | 367.9 | 302.9 | 244.3 | 186.7 | 137.3 |
| TIME AVG.TEMP | 12.3 | 14.3 | 14.7 | 15.0 | 15.4 | 15.5 | 15.6 |
| TOTAL TIME MSEC | 2.6784 | 1.1286 | 0.9342 | 0.7902 | 0.6138 | 0.5526 | 0.5040 |
| GLOBAL RAD TOTAL | 511.2 | 511.2 | 502.5 | 481.7 | 441.8 | 422.1 | 401.4 |
| DIRECT RAD TOTAL | 414.6 | 414.6 | 413.0 | 404.7 | 381.6 | 374.0 | 361.4 |
| RAD AV TIME MSEC | 0.0864 | 0.0364 | 0.0342 | 0.0324 | 0.0274 | 0.0274 | 0.0266 |
| DEG-C DAYS | 187. | 56. | 43. | 34. | 24. | 22. | 20. |

**ALL MONTHS**

| SOLAR RAD GROUP | ALL | 0+ | 95+ | 189+ | 284+ | 378+ | 473+ |
|---|---|---|---|---|---|---|---|
| TIME AVG.TEMP | 16.5 | 18.3 | 18.8 | 19.3 | 19.7 | 19.9 | 20.1 |
| TOTAL TIME MSEC | 31.5791 | 15.3378 | 12.6504 | 10.6848 | 8.9100 | 7.4142 | 6.4872 |
| GLOBAL RAD TOTAL | 6641.0 | 6640.9 | 6523.5 | 6248.4 | 5826.4 | 5334.4 | 4942.1 |
| DIRECT RAD TOTAL | 4736.4 | 4702.4 | 4626.4 | 4460.8 | 4203.3 | 3994.2 | 3994.2 |
| RAD AV TIME MSEC | 0.0864 | 0.0421 | 0.0363 | 0.0330 | 0.0298 | 0.0265 | 0.0249 |
| DEG-C DAYS | 990. | 303. | 209. | 150. | 112. | 89. | 73. |

STATISTICS FOR LOS ANGELES, CALIF 1971-1972   IR INCLUDED RSV020 FILE

RADIATION UNITS ARE WATTS/SQ.M OR MJOULES/SQ.M
SLOPE = 90.0 DEG   AZIMUTH = 180.0 DEG
DEGREE-DAY BASE TEMP = 18.3 DEGREES C

**JAN**

| SOLAR RAD GROUP | ALL | 0+ | 95+ | 189+ | 284+ | 378+ | 473+ |
|---|---|---|---|---|---|---|---|
| AV.THRSH.RAD.LEVL | 0. | 41. | 143. | 235. | 335. | 425. | 529. |
| AV.THRESHOLD TEMP | 10.6 | 11.2 | 13.1 | 13.8 | 14.9 | 13.7 | 15.6 |
| GL RAD OVER THRSH | 424.0 | 374.1 | 283.4 | 214.9 | 158.3 | 115.5 | 73.7 |
| TIME AVG.TEMP | 12.1 | 13.9 | 14.9 | 15.3 | 15.7 | 15.9 | 16.3 |
| TOTAL TIME MSEC | 2.6784 | 1.2060 | 0.8964 | 0.7398 | 0.5688 | 0.4734 | 0.4014 |
| GLOBAL RAD TOTAL | 424.1 | 424.0 | 411.2 | 388.5 | 348.7 | 316.7 | 286.1 |
| DIRECT RAD TOTAL | 321.8 | 321.8 | 318.0 | 308.5 | 286.2 | 270.6 | 250.1 |
| RAD AV TIME MSEC | 0.0864 | 0.0392 | 0.0331 | 0.0310 | 0.0263 | 0.0252 | 0.0245 |
| DEG-C DAYS | 193. | 65. | 41. | 31. | 22. | 17. | 13. |

**FEB**

| SOLAR RAD GROUP | ALL | 0+ | 95+ | 189+ | 284+ | 378+ | 473+ |
|---|---|---|---|---|---|---|---|
| AV.THRSH.RAD.LEVL | 0. | 44. | 135. | 245. | 333. | 426. | 518. |
| AV.THRESHOLD TEMP | 11.7 | 12.6 | 13.6 | 14.7 | 14.6 | 16.7 | 15.4 |
| GL RAD OVER THRSH | 399.7 | 345.6 | 263.6 | 184.4 | 131.9 | 89.0 | 53.4 |
| TIME AVG.TEMP | 13.2 | 14.6 | 15.3 | 15.8 | 16.0 | 16.4 | 16.3 |
| TOTAL TIME MSEC | 2.4624 | 1.2258 | 0.9018 | 0.7218 | 0.5940 | 0.4626 | 0.3852 |
| GLOBAL RAD TOTAL | 399.7 | 399.7 | 385.4 | 361.1 | 329.8 | 286.0 | 253.1 |
| DIRECT RAD TOTAL | 283.6 | 283.6 | 278.4 | 268.4 | 251.4 | 225.9 | 205.6 |
| RAD AV TIME MSEC | 0.0864 | 0.0432 | 0.0360 | 0.0313 | 0.0281 | 0.0248 | 0.0234 |
| DEG-C DAYS | 147. | 55. | 34. | 24. | 19. | 13. | 11. |

**MAR**

| SOLAR RAD GROUP | ALL | 0+ | 95+ | 189+ | 284+ | 378+ | 473+ |
|---|---|---|---|---|---|---|---|
| AV.THRSH.RAD.LEVL | 0. | 46. | 135. | 230. | 332. | 428. | 519. |
| AV.THRESHOLD TEMP | 12.7 | 13.6 | 14.3 | 15.6 | 15.9 | 17.1 | 17.2 |
| GL RAD OVER THRSH | 378.4 | 316.8 | 231.1 | 162.1 | 102.5 | 59.3 | 29.9 |
| TIME AVG.TEMP | 14.1 | 15.4 | 16.2 | 16.8 | 17.1 | 17.4 | 17.5 |
| TOTAL TIME MSEC | 2.6784 | 1.3392 | 0.9612 | 0.7254 | 0.5850 | 0.4500 | 0.3240 |
| GLOBAL RAD TOTAL | 378.4 | 378.4 | 361.0 | 329.1 | 296.8 | 252.0 | 198.0 |
| DIRECT RAD TOTAL | 230.8 | 230.8 | 226.8 | 220.3 | 203.2 | 178.7 | 148.6 |
| RAD AV TIME MSEC | 0.0864 | 0.0432 | 0.0346 | 0.0310 | 0.0270 | 0.0227 | 0.0194 |
| DEG-C DAYS | 132. | 46. | 27. | 16. | 12. | 8. | 6. |

**JUL**

| SOLAR RAD GROUP | ALL | 0+ | 95+ | 189+ | 284+ | 378+ | 473+ |
|---|---|---|---|---|---|---|---|
| AV.THRSH.RAD.LEVL | 0. | 56. | 151. | 240. | 332. | 414. | 506. |
| AV.THRESHOLD TEMP | 19.3 | 22.4 | 21.8 | 22.8 | 23.4 | 23.9 | 23.2 |
| GL RAD OVER THRSH | 225.4 | 174.5 | 102.8 | 46.7 | 10.8 | 1.7 | 0.0 |
| TIME AVG.TEMP | 20.6 | 22.9 | 23.0 | 23.2 | 23.5 | 23.8 | 23.2 |
| TOTAL TIME MSEC | 2.6784 | 0.9126 | 0.7506 | 0.6354 | 0.3888 | 0.1116 | 0.0180 |
| GLOBAL RAD TOTAL | 225.4 | 225.4 | 216.4 | 198.9 | 139.9 | 47.9 | 9.1 |
| DIRECT RAD TOTAL | 83.7 | 83.7 | 82.4 | 80.2 | 59.7 | 15.0 | 0.6 |
| RAD AV TIME MSEC | 0.0864 | 0.0295 | 0.0245 | 0.0216 | 0.0151 | 0.0115 | 0.0108 |
| DEG-C DAYS | 0. | 0. | 0. | 0. | 0. | 0. | 0. |

**AUG**

| SOLAR RAD GROUP | ALL | 0+ | 95+ | 189+ | 284+ | 378+ | 473+ |
|---|---|---|---|---|---|---|---|
| AV.THRSH.RAD.LEVL | 0. | 47. | 154. | 237. | 333. | 427. | 509. |
| AV.THRESHOLD TEMP | 20.8 | 22.4 | 23.3 | 24.5 | 24.9 | 25.4 | 24.9 |
| GL RAD OVER THRSH | 305.9 | 252.0 | 153.1 | 91.8 | 39.9 | 8.5 | 0.1 |
| TIME AVG.TEMP | 22.3 | 24.2 | 24.6 | 24.9 | 25.1 | 25.2 | 25.2 |
| TOTAL TIME MSEC | 2.6784 | 1.1538 | 0.9234 | 0.7344 | 0.5400 | 0.3348 | 0.1026 |
| GLOBAL RAD TOTAL | 305.9 | 305.9 | 295.2 | 266.1 | 220.0 | 151.5 | 52.3 |
| DIRECT RAD TOTAL | 166.7 | 166.7 | 165.0 | 157.2 | 137.9 | 101.3 | 36.2 |
| RAD AV TIME MSEC | 0.0864 | 0.0374 | 0.0299 | 0.0252 | 0.0202 | 0.0148 | 0.0112 |
| DEG-C DAYS | 0. | 0. | 0. | 0. | 0. | 0. | 0. |

**SEP**

| SOLAR RAD GROUP | ALL | 0+ | 95+ | 189+ | 284+ | 378+ | 473+ |
|---|---|---|---|---|---|---|---|
| AV.THRSH.RAD.LEVL | 0. | 42. | 148. | 232. | 332. | 429. | 518. |
| AV.THRESHOLD TEMP | 19.2 | 20.2 | 21.9 | 22.0 | 23.1 | 23.0 | 23.3 |
| GL RAD OVER THRSH | 375.2 | 320.3 | 215.5 | 153.0 | 94.5 | 50.6 | 21.5 |
| TIME AVG.TEMP | 20.7 | 22.1 | 22.7 | 23.0 | 23.3 | 23.3 | 23.4 |
| TOTAL TIME MSEC | 2.5920 | 1.3032 | 0.9900 | 0.7470 | 0.5850 | 0.4500 | 0.3258 |
| GLOBAL RAD TOTAL | 375.3 | 375.3 | 362.0 | 326.1 | 288.5 | 243.8 | 190.5 |
| DIRECT RAD TOTAL | 236.1 | 236.1 | 232.4 | 222.4 | 205.6 | 179.2 | 146.9 |
| RAD AV TIME MSEC | 0.0864 | 0.0436 | 0.0342 | 0.0292 | 0.0252 | 0.0212 | 0.0169 |
| DEG-C DAYS | 2. | 0. | 0. | 0. | 0. | 0. | 0. |

**APR**

| SOLAR RAD GROUP | ALL | 0+ | 95+ | 189+ | 284+ | 378+ | 473+ |
|---|---|---|---|---|---|---|---|
| AV.THRSH.RAD.LEVL | 0. | 41. | 153. | 236. | 332. | 433. | 518. |
| AV.THRESHOLD TEMP | 13.3 | 15.1 | 16.0 | 16.9 | 17.8 | 18.6 | 19.2 |
| GL RAD OVER THRSH | 346.4 | 297.5 | 191.0 | 126.3 | 68.1 | 26.1 | 5.8 |
| TIME AVG.TEMP | 15.2 | 17.4 | 17.9 | 18.4 | 18.8 | 19.3 | 19.8 |
| TOTAL TIME MSEC | 2.5920 | 1.1862 | 0.9558 | 0.7812 | 0.6012 | 0.4176 | 0.2394 |
| GLOBAL RAD TOTAL | 346.4 | 346.4 | 336.9 | 310.3 | 267.9 | 206.9 | 129.8 |
| DIRECT RAD TOTAL | 180.1 | 180.1 | 178.3 | 170.7 | 155.1 | 126.5 | 84.9 |
| RAD AV TIME MSEC | 0.0864 | 0.0400 | 0.0328 | 0.0277 | 0.0238 | 0.0184 | 0.0140 |
| DEG-C DAYS | 97. | 19. | 12. | 8. | 4. | 2. | 1. |

**MAY**

| SOLAR RAD GROUP | ALL | 0+ | 95+ | 189+ | 284+ | 378+ | 473+ |
|---|---|---|---|---|---|---|---|
| AV.THRSH.RAD.LEVL | 0. | 48. | 147. | 233. | 333. | 417. | 494. |
| AV.THRESHOLD TEMP | 15.1 | 17.3 | 17.0 | 18.5 | 20.0 | 20.5 | 19.0 |
| GL RAD OVER THRSH | 217.7 | 169.9 | 91.5 | 43.9 | 11.6 | 0.8 | 0.0 |
| TIME AVG.TEMP | 16.3 | 16.3 | 18.7 | 19.5 | 20.0 | 20.4 | 19.0 |
| TOTAL TIME MSEC | 2.6784 | 0.9900 | 0.7938 | 0.5508 | 0.3258 | 0.1278 | 0.0108 |
| GLOBAL RAD TOTAL | 217.7 | 217.7 | 208.2 | 172.5 | 120.0 | 54.1 | 5.3 |
| DIRECT RAD TOTAL | 70.6 | 70.6 | 69.8 | 66.2 | 51.3 | 22.1 | 1.0 |
| RAD AV TIME MSEC | 0.0864 | 0.0320 | 0.0263 | 0.0212 | 0.0162 | 0.0119 | 0.0083 |
| DEG-C DAYS | 71. | 10. | 7. | 2. | 1. | 0. | 0. |

**JUN**

| SOLAR RAD GROUP | ALL | 0+ | 95+ | 189+ | 284+ | 378+ | 473+ |
|---|---|---|---|---|---|---|---|
| AV.THRSH.RAD.LEVL | 0. | 72. | 158. | 232. | 318. | 425. | 509. |
| AV.THRESHOLD TEMP | 17.3 | 19.7 | 19.7 | 20.7 | 21.4 | 22.0 | 22.1 |
| GL RAD OVER THRSH | 176.6 | 114.2 | 56.2 | 19.7 | 3.5 | 0.5 | 0.0 |
| TIME AVG.TEMP | 18.3 | 20.4 | 20.7 | 21.0 | 21.5 | 22.0 | 22.1 |
| TOTAL TIME MSEC | 2.5920 | 0.8640 | 0.6732 | 0.4968 | 0.1872 | 0.0288 | 0.0054 |
| GLOBAL RAD TOTAL | 176.6 | 176.6 | 162.8 | 134.9 | 63.1 | 12.7 | 2.7 |
| DIRECT RAD TOTAL | 44.7 | 44.7 | 43.9 | 41.9 | 22.9 | 1.7 | 0.2 |
| RAD AV TIME MSEC | 0.0864 | 0.0288 | 0.0230 | 0.0191 | 0.0119 | 0.0133 | 0.0108 |
| DEG-C DAYS | 16. | 1. | 1. | 1. | 0. | 0. | 0. |

**OCT**

| SOLAR RAD GROUP | ALL | 0+ | 95+ | 189+ | 284+ | 378+ | 473+ |
|---|---|---|---|---|---|---|---|
| AV.THRSH.RAD.LEVL | 0. | 51. | 133. | 243. | 325. | 427. | 514. |
| AV.THRESHOLD TEMP | 16.4 | 17.7 | 18.2 | 20.2 | 19.7 | 21.4 | 21.1 |
| GL RAD OVER THRSH | 434.8 | 366.4 | 282.3 | 194.8 | 142.5 | 90.9 | 57.3 |
| TIME AVG.TEMP | 18.2 | 19.9 | 20.6 | 21.3 | 21.3 | 22.1 | 22.3 |
| TOTAL TIME MSEC | 2.6784 | 1.3392 | 1.0296 | 0.7938 | 0.6372 | 0.5058 | 0.3888 |
| GLOBAL RAD TOTAL | 434.8 | 434.8 | 419.0 | 387.7 | 349.6 | 306.9 | 257.0 |
| DIRECT RAD TOTAL | 311.0 | 311.0 | 304.5 | 293.3 | 272.8 | 244.8 | 213.1 |
| RAD AV TIME MSEC | 0.0864 | 0.0432 | 0.0371 | 0.0331 | 0.0299 | 0.0263 | 0.0241 |
| DEG-C DAYS | 39. | 10. | 6. | 4. | 3. | 2. | 2. |

**NOV**

| SOLAR RAD GROUP | ALL | 0+ | 95+ | 189+ | 284+ | 378+ | 473+ |
|---|---|---|---|---|---|---|---|
| AV.THRSH.RAD.LEVL | 0. | 42. | 139. | 235. | 322. | 426. | 529. |
| AV.THRESHOLD TEMP | 13.2 | 14.1 | 16.2 | 16.3 | 16.2 | 16.4 | 17.7 |
| GL RAD OVER THRSH | 452.2 | 401.3 | 313.9 | 244.3 | 193.1 | 138.1 | 89.9 |
| TIME AVG.TEMP | 16.2 | 16.7 | 17.6 | 17.9 | 18.3 | 18.5 | 18.8 |
| TOTAL TIME MSEC | 2.5920 | 1.2114 | 0.9000 | 0.7272 | 0.5850 | 0.5292 | 0.4680 |
| GLOBAL RAD TOTAL | 452.3 | 452.2 | 439.1 | 415.1 | 381.7 | 363.7 | 337.6 |
| DIRECT RAD TOTAL | 365.1 | 365.1 | 360.4 | 352.0 | 330.2 | 316.9 | 297.2 |
| RAD AV TIME MSEC | 0.0864 | 0.0407 | 0.0346 | 0.0328 | 0.0284 | 0.0270 | 0.0263 |
| DEG-C DAYS | 105. | 28. | 13. | 9. | 6. | 5. | 4. |

**DEC**

| SOLAR RAD GROUP | ALL | 0+ | 95+ | 189+ | 284+ | 378+ | 473+ |
|---|---|---|---|---|---|---|---|
| AV.THRSH.RAD.LEVL | 0. | 43. | 144. | 224. | 320. | 433. | 530. |
| AV.THRESHOLD TEMP | 10.9 | 12.4 | 13.9 | 13.5 | 13.2 | 13.6 | 15.5 |
| GL RAD OVER THRSH | 456.2 | 408.1 | 317.6 | 257.5 | 203.4 | 143.6 | 97.7 |
| TIME AVG.TEMP | 12.3 | 14.3 | 14.8 | 14.9 | 15.4 | 15.5 | 15.7 |
| TOTAL TIME MSEC | 2.6784 | 1.1286 | 0.8928 | 0.7470 | 0.5670 | 0.5274 | 0.4734 |
| GLOBAL RAD TOTAL | 456.2 | 456.2 | 446.2 | 425.2 | 384.0 | 372.1 | 348.7 |
| DIRECT RAD TOTAL | 373.7 | 373.7 | 371.5 | 363.4 | 336.7 | 329.4 | 311.0 |
| RAD AV TIME MSEC | 0.0864 | 0.0364 | 0.0342 | 0.0328 | 0.0277 | 0.0274 | 0.0263 |
| DEG-C DAYS | 187. | 56. | 41. | 33. | 23. | 21. | 18. |

**ALL MONTHS**

| SOLAR RAD GROUP | ALL | 0+ | 95+ | 189+ | 284+ | 378+ | 473+ |
|---|---|---|---|---|---|---|---|
| TIME AVG.TEMP | 16.5 | 18.2 | 18.9 | 19.3 | 19.4 | 19.2 | 18.8 |
| TOTAL TIME MSEC | 31.5791 | 13.8600 | 10.6686 | 8.4006 | 6.1650 | 4.4190 | 3.1428 |
| GLOBAL RAD TOTAL | 4192.9 | 4192.8 | 4043.5 | 3715.8 | 3190.7 | 2614.3 | 2070.3 |
| DIRECT RAD TOTAL | 2667.7 | 2667.7 | 2631.3 | 2544.8 | 2313.2 | 2011.9 | 1695.5 |
| RAD AV TIME MSEC | 0.0864 | 0.0392 | 0.0328 | 0.0294 | 0.0255 | 0.0235 | 0.0226 |
| DEG-C DAYS | 990. | 290. | 181. | 129. | 89. | 69. | 55. |

STATISTICS FOR BRADLEY INTL., AIRPORT WX. STATION 1959-1968

RADIATION UNITS ARE WATTS/SQ.M OR MJOULES/SQ.M
SLOPE = 40.0 DEG   AZIMUTH = 150.0 DEG
DEGREE-DAY BASE TEMP = 18.3 DEGREES C

**JAN**

| SOLAR RAD GROUP | ALL | 0+ | 95+ | 189+ | 284+ | 378+ | 473+ |
|---|---|---|---|---|---|---|---|
| AV.THRSH.RAD.LEVL | 0. | 41. | 141. | 233. | 331. | 427. | 521. |
| AV.THRESHOLD TEMP | -4.8 | -1.4 | -2.0 | -2.7 | -1.3 | -2.1 | -2.9 |
| GL RAD OVER THRSH | 301.6 | 256.2 | 183.2 | 135.2 | 96.2 | 65.6 | 42.7 |
| TIME AVG.TEMP | -3.8 | -2.4 | -2.9 | -3.2 | -3.4 | -3.9 | -4.4 |
| TOTAL TIME MSEC | 2.6784 | 1.1164 | 0.7297 | 0.5224 | 0.3946 | 0.3197 | 0.2437 |
| GLOBAL RAD TOTAL | 301.7 | 301.7 | 285.9 | 256.7 | 227.0 | 202.2 | 169.7 |
| DIRECT RAD TOTAL | 203.3 | 203.3 | 201.2 | 191.8 | 179.2 | 167.3 | 146.0 |
| RAD AV TIME MSEC | 0.0864 | 0.0360 | 0.0295 | 0.0256 | 0.0223 | 0.0212 | 0.0184 |
| DEG-C DAYS | 686. | 268. | 179. | 130. | 99. | 82. | 64. |

**FEB**

| SOLAR RAD GROUP | ALL | 0+ | 95+ | 189+ | 284+ | 378+ | 473+ |
|---|---|---|---|---|---|---|---|
| AV.THRSH.RAD.LEVL | 0. | 41. | 139. | 236. | 331. | 425. | 519. |
| AV.THRESHOLD TEMP | -4.0 | -1.9 | -0.4 | -1.2 | -1.8 | -2.2 | -1.6 |
| GL RAD OVER THRSH | 343.6 | 297.9 | 224.4 | 167.5 | 124.9 | 91.0 | 63.8 |
| TIME AVG.TEMP | -3.1 | -1.9 | -1.9 | -2.3 | -2.7 | -2.9 | -3.1 |
| TOTAL TIME MSEC | 2.4451 | 1.1138 | 0.7538 | 0.5828 | 0.4507 | 0.3589 | 0.2912 |
| GLOBAL RAD TOTAL | 343.6 | 343.6 | 328.8 | 305.1 | 273.9 | 243.6 | 214.8 |
| DIRECT RAD TOTAL | 220.8 | 220.8 | 217.8 | 214.2 | 204.0 | 191.2 | 176.8 |
| RAD AV TIME MSEC | 0.0864 | 0.0396 | 0.0310 | 0.0288 | 0.0259 | 0.0238 | 0.0220 |
| DEG-C DAYS | 606. | 261. | 176. | 139. | 109. | 88. | 72. |

**MAR**

| SOLAR RAD GROUP | ALL | 0+ | 95+ | 189+ | 284+ | 378+ | 473+ |
|---|---|---|---|---|---|---|---|
| AV.THRSH.RAD.LEVL | 0. | 42. | 142. | 234. | 328. | 426. | 524. |
| AV.THRESHOLD TEMP | 0.9 | 2.2 | 2.5 | 3.7 | 4.7 | 4.8 | 5.2 |
| GL RAD OVER THRSH | 450.0 | 396.6 | 301.3 | 234.8 | 179.3 | 133.8 | 95.2 |
| TIME AVG.TEMP | 2.1 | 3.4 | 3.8 | 4.2 | 4.2 | 4.1 | 4.1 |
| TOTAL TIME MSEC | 2.6784 | 1.2712 | 0.9466 | 0.7240 | 0.5904 | 0.4637 | 0.3949 |
| GLOBAL RAD TOTAL | 450.1 | 450.1 | 436.5 | 404.5 | 373.2 | 331.6 | 302.3 |
| DIRECT RAD TOTAL | 269.8 | 269.8 | 267.6 | 263.8 | 257.5 | 240.3 | 229.0 |
| RAD AV TIME MSEC | 0.0864 | 0.0414 | 0.0338 | 0.0317 | 0.0299 | 0.0263 | 0.0248 |
| DEG-C DAYS | 505. | 220. | 160. | 119. | 96. | 76. | 65. |

**APR**

| SOLAR RAD GROUP | ALL | 0+ | 95+ | 189+ | 284+ | 378+ | 473+ |
|---|---|---|---|---|---|---|---|
| AV.THRSH.RAD.LEVL | 0. | 45. | 144. | 232. | 337. | 427. | 523. |
| AV.THRESHOLD TEMP | 7.7 | 8.8 | 9.2 | 10.5 | 12.5 | 11.5 | 13.5 |
| GL RAD OVER THRSH | 485.9 | 427.5 | 329.7 | 259.1 | 190.3 | 143.7 | 104.7 |
| TIME AVG.TEMP | 9.4 | 11.2 | 12.0 | 12.6 | 13.1 | 13.3 | 13.8 |
| TOTAL TIME MSEC | 2.5920 | 1.2974 | 0.9853 | 0.8096 | 0.6534 | 0.5191 | 0.4028 |
| GLOBAL RAD TOTAL | 485.9 | 485.9 | 471.9 | 446.5 | 410.4 | 365.1 | 315.5 |
| DIRECT RAD TOTAL | 258.6 | 258.6 | 255.8 | 250.2 | 236.3 | 236.1 | 215.5 |
| RAD AV TIME MSEC | 0.0864 | 0.0432 | 0.0353 | 0.0331 | 0.0320 | 0.0284 | 0.0252 |
| DEG-C DAYS | 267. | 108. | 74. | 56. | 42. | 33. | 23. |

**JUL**

| SOLAR RAD GROUP | ALL | 0+ | 95+ | 189+ | 284+ | 378+ | 473+ |
|---|---|---|---|---|---|---|---|
| AV.THRSH.RAD.LEVL | 0. | 39. | 143. | 236. | 333. | 431. | 520. |
| AV.THRESHOLD TEMP | 20.6 | 21.7 | 21.0 | 22.7 | 25.7 | 25.6 | 26.2 |
| GL RAD OVER THRSH | 548.4 | 492.6 | 375.5 | 286.2 | 206.6 | 142.8 | 98.5 |
| TIME AVG.TEMP | 22.6 | 24.4 | 25.1 | 25.8 | 26.3 | 26.5 | 26.8 |
| TOTAL TIME MSEC | 2.6784 | 1.4213 | 1.1243 | 0.9662 | 0.8179 | 0.6523 | 0.4954 |
| GLOBAL RAD TOTAL | 548.4 | 548.4 | 536.7 | 514.1 | 479.1 | 423.9 | 356.3 |
| DIRECT RAD TOTAL | 265.1 | 265.1 | 263.7 | 260.6 | 253.1 | 238.5 | 215.2 |
| RAD AV TIME MSEC | 0.0864 | 0.0461 | 0.0378 | 0.0353 | 0.0324 | 0.0288 | 0.0248 |
| DEG-C DAYS | 1. | 0. | 1. | 1. | 0. | 0. | 0. |

**AUG**

| SOLAR RAD GROUP | ALL | 0+ | 95+ | 189+ | 284+ | 378+ | 473+ |
|---|---|---|---|---|---|---|---|
| AV.THRSH.RAD.LEVL | 0. | 44. | 145. | 236. | 331. | 424. | 521. |
| AV.THRESHOLD TEMP | 19.3 | 20.8 | 20.5 | 23.5 | 24.7 | 24.1 | 25.6 |
| GL RAD OVER THRSH | 491.8 | 432.5 | 326.4 | 245.4 | 177.6 | 125.2 | 84.3 |
| TIME AVG.TEMP | 21.3 | 23.3 | 24.0 | 24.6 | 24.9 | 24.9 | 25.2 |
| TOTAL TIME MSEC | 2.6784 | 1.3597 | 1.0403 | 0.8921 | 0.7139 | 0.5620 | 0.4219 |
| GLOBAL RAD TOTAL | 491.8 | 491.8 | 478.2 | 455.6 | 413.6 | 363.4 | 304.0 |
| DIRECT RAD TOTAL | 242.9 | 242.9 | 240.7 | 237.9 | 231.2 | 217.4 | 194.6 |
| RAD AV TIME MSEC | 0.0864 | 0.0439 | 0.0356 | 0.0335 | 0.0306 | 0.0277 | 0.0234 |
| DEG-C DAYS | 8. | 2. | 1. | 1. | 1. | 1. | 0. |

**SEP**

| SOLAR RAD GROUP | ALL | 0+ | 95+ | 189+ | 284+ | 378+ | 473+ |
|---|---|---|---|---|---|---|---|
| AV.THRSH.RAD.LEVL | 0. | 41. | 143. | 235. | 330. | 422. | 525. |
| AV.THRESHOLD TEMP | 15.0 | 17.0 | 18.0 | 19.6 | 20.0 | 21.4 | 21.6 |
| GL RAD OVER THRSH | 479.2 | 428.0 | 329.5 | 255.3 | 193.1 | 145.2 | 99.5 |
| TIME AVG.TEMP | 17.2 | 19.6 | 20.4 | 20.8 | 21.1 | 21.4 | 21.4 |
| TOTAL TIME MSEC | 2.5920 | 1.2654 | 0.9580 | 0.8064 | 0.6559 | 0.5202 | 0.4453 |
| GLOBAL RAD TOTAL | 479.3 | 479.2 | 466.8 | 445.1 | 409.6 | 364.8 | 333.2 |
| DIRECT RAD TOTAL | 305.9 | 305.9 | 303.2 | 300.9 | 291.6 | 271.0 | 256.5 |
| RAD AV TIME MSEC | 0.0864 | 0.0425 | 0.0342 | 0.0324 | 0.0302 | 0.0263 | 0.0245 |
| DEG-C DAYS | 71. | 19. | 11. | 8. | 6. | 4. | 4. |

**OCT**

| SOLAR RAD GROUP | ALL | 0+ | 95+ | 189+ | 284+ | 378+ | 473+ |
|---|---|---|---|---|---|---|---|
| AV.THRSH.RAD.LEVL | 0. | 41. | 142. | 238. | 329. | 427. | 522. |
| AV.THRESHOLD TEMP | 9.2 | 12.0 | 12.9 | 14.9 | 14.7 | 14.7 | 15.1 |
| GL RAD OVER THRSH | 411.7 | 361.3 | 271.8 | 205.1 | 155.6 | 111.9 | 76.7 |
| TIME AVG.TEMP | 11.3 | 13.9 | 14.6 | 15.1 | 15.2 | 15.3 | 15.4 |
| TOTAL TIME MSEC | 2.6870 | 1.2298 | 0.8896 | 0.6919 | 0.5443 | 0.4460 | 0.3719 |
| GLOBAL RAD TOTAL | 411.7 | 411.7 | 397.8 | 369.8 | 334.7 | 302.3 | 270.7 |
| DIRECT RAD TOTAL | 275.0 | 275.0 | 271.5 | 266.3 | 254.0 | 239.3 | 270.0 |
| RAD AV TIME MSEC | 0.0864 | 0.0396 | 0.0320 | 0.0295 | 0.0266 | 0.0245 | 0.0227 |
| DEG-C DAYS | 219. | 69. | 45. | 32. | 26. | 21. | 17. |

**MAY**

| SOLAR RAD GROUP | ALL | 0+ | 95+ | 189+ | 284+ | 378+ | 473+ |
|---|---|---|---|---|---|---|---|
| AV.THRSH.RAD.LEVL | 0. | 40. | 144. | 236. | 330. | 428. | 516. |
| AV.THRESHOLD TEMP | 12.9 | 13.8 | 12.9 | 14.3 | 17.8 | 18.4 | 19.8 |
| GL RAD OVER THRSH | 561.8 | 504.8 | 388.0 | 299.1 | 223.7 | 159.8 | 113.9 |
| TIME AVG.TEMP | 15.0 | 16.9 | 17.7 | 18.4 | 19.3 | 19.7 | 20.0 |
| TOTAL TIME MSEC | 2.6784 | 1.4423 | 1.1279 | 0.9695 | 0.7978 | 0.6520 | 0.5202 |
| GLOBAL RAD TOTAL | 561.8 | 561.8 | 550.3 | 527.5 | 487.1 | 438.9 | 382.5 |
| DIRECT RAD TOTAL | 264.7 | 264.7 | 263.3 | 260.2 | 252.6 | 241.2 | 222.4 |
| RAD AV TIME MSEC | 0.0864 | 0.0461 | 0.0378 | 0.0353 | 0.0324 | 0.0292 | 0.0256 |
| DEG-C DAYS | 120. | 43. | 29. | 21. | 12. | 9. | 7. |

**JUN**

| SOLAR RAD GROUP | ALL | 0+ | 95+ | 189+ | 284+ | 378+ | 473+ |
|---|---|---|---|---|---|---|---|
| AV.THRSH.RAD.LEVL | 0. | 38. | 145. | 232. | 333. | 432. | 519. |
| AV.THRESHOLD TEMP | 18.2 | 19.1 | 18.0 | 19.3 | 23.3 | 23.7 | 24.5 |
| GL RAD OVER THRSH | 534.0 | 480.8 | 364.6 | 283.2 | 207.5 | 147.2 | 104.0 |
| TIME AVG.TEMP | 20.2 | 21.9 | 22.7 | 23.6 | 24.6 | 24.8 | 25.1 |
| TOTAL TIME MSEC | 2.5920 | 1.3889 | 1.0919 | 0.9274 | 0.7502 | 0.6127 | 0.4943 |
| GLOBAL RAD TOTAL | 534.0 | 534.0 | 522.6 | 498.8 | 457.6 | 411.8 | 360.6 |
| DIRECT RAD TOTAL | 249.4 | 249.4 | 248.3 | 246.0 | 239.1 | 228.0 | 209.5 |
| RAD AV TIME MSEC | 0.0864 | 0.0464 | 0.0382 | 0.0356 | 0.0328 | 0.0299 | 0.0259 |
| DEG-C DAYS | 21. | 8. | 4. | 2. | 1. | 1. | 0. |

**NOV**

| SOLAR RAD GROUP | ALL | 0+ | 95+ | 189+ | 284+ | 378+ | 473+ |
|---|---|---|---|---|---|---|---|
| AV.THRSH.RAD.LEVL | 0. | 40. | 138. | 231. | 330. | 424. | 524. |
| AV.THRESHOLD TEMP | 3.9 | 6.7 | 7.4 | 7.5 | 7.3 | 8.0 | 6.6 |
| GL RAD OVER THRSH | 260.1 | 216.1 | 151.3 | 108.7 | 76.0 | 51.9 | 32.3 |
| TIME AVG.TEMP | 5.3 | 7.1 | 7.4 | 7.4 | 7.3 | 7.3 | 7.1 |
| TOTAL TIME MSEC | 2.5920 | 1.1041 | 0.6606 | 0.4572 | 0.3308 | 0.2574 | 0.1948 |
| GLOBAL RAD TOTAL | 260.1 | 260.1 | 242.5 | 214.4 | 185.2 | 160.9 | 134.4 |
| DIRECT RAD TOTAL | 164.8 | 164.8 | 162.6 | 155.1 | 142.7 | 131.2 | 114.1 |
| RAD AV TIME MSEC | 0.0864 | 0.0371 | 0.0292 | 0.0256 | 0.0227 | 0.0205 | 0.0184 |
| DEG-C DAYS | 392. | 143. | 84. | 58. | 42. | 33. | 26. |

**DEC**

| SOLAR RAD GROUP | ALL | 0+ | 95+ | 189+ | 284+ | 378+ | 473+ |
|---|---|---|---|---|---|---|---|
| AV.THRSH.RAD.LEVL | 0. | 40. | 138. | 232. | 333. | 425. | 522. |
| AV.THRESHOLD TEMP | -3.0 | -0.1 | -1.1 | 0.3 | 0.7 | -0.9 | -2.6 |
| GL RAD OVER THRSH | 261.5 | 217.2 | 153.1 | 111.4 | 77.7 | 52.7 | 33.3 |
| TIME AVG.TEMP | -2.0 | -0.6 | -1.0 | -0.9 | -1.3 | -1.8 | -2.2 |
| TOTAL TIME MSEC | 2.6784 | 1.1084 | 0.6505 | 0.4439 | 0.3373 | 0.2711 | 0.1987 |
| GLOBAL RAD TOTAL | 261.6 | 261.6 | 243.2 | 214.6 | 189.9 | 167.8 | 137.1 |
| DIRECT RAD TOTAL | 179.9 | 179.9 | 177.2 | 166.7 | 155.6 | 143.3 | 121.2 |
| RAD AV TIME MSEC | 0.0864 | 0.0360 | 0.0292 | 0.0245 | 0.0223 | 0.0212 | 0.0184 |
| DEG-C DAYS | 631. | 243. | 146. | 99. | 77. | 63. | 47. |

**ALL MONTHS**

| SOLAR RAD GROUP | ALL | 0+ | 95+ | 189+ | 284+ | 378+ | 473+ |
|---|---|---|---|---|---|---|---|
| TIME AVG.TEMP | 9.7 | 12.2 | 13.4 | 14.4 | 15.0 | 15.1 | 15.2 |
| TOTAL TIME MSEC | 31.5705 | 15.0887 | 10.9685 | 8.7934 | 7.0373 | 5.6351 | 4.4752 |
| GLOBAL RAD TOTAL | 5130.1 | 5129.9 | 4961.3 | 4652.8 | 4241.2 | 3776.4 | 3281.2 |
| DIRECT RAD TOTAL | 2900.2 | 2900.2 | 2872.9 | 2817.1 | 2710.7 | 2545.2 | 2332.9 |
| RAD AV TIME MSEC | 0.0864 | 0.0424 | 0.0345 | 0.0320 | 0.0295 | 0.0266 | 0.0237 |
| DEG-C DAYS | 3527. | 1383. | 909. | 665. | 510. | 411. | 326. |

STATISTICS FOR BRADLEY INTL., AIRPORT WX. STATION 1959-1968

RADIATION UNITS ARE WATTS/SQ.M OR MJOULES/SQ.M
SLOPE = 40.0 DEG   AZIMUTH = 135.0 DEG
DEGREE-DAY BASE TEMP = 18.3 DEGREES C

**JAN**

|                   | ALL | 0+ | 95+ | 189+ | 284+ | 378+ | 473+ |
|-------------------|-----|-----|-----|------|------|------|------|
| SOLAR RAD GROUP   |     |     |     |      |      |      |      |
| AV.THRSH.RAD.LEVL | 0. | 40. | 138. | 238. | 325. | 426. | 522. |
| AV.THRESHOLD TEMP | -4.5 | -1.6 | -2.4 | -1.9 | -1.8 | -1.7 | -3.5 |
| GL RAD OVER THRSH | 269.6 | 229.5 | 163.4 | 114.8 | 82.4 | 53.8 | 32.4 |
| TIME AVG.TEMP     | -3.8 | -2.5 | -2.9 | -3.2 | -3.6 | -4.1 | -4.7 |
| TOTAL TIME MSEC   | 2.6784 | 1.0076 | 0.6714 | 0.4856 | 0.3737 | 0.2826 | 0.2246 |
| GLOBAL RAD TOTAL  | 269.6 | 269.6 | 256.2 | 230.5 | 203.9 | 174.3 | 149.6 |
| DIRECT RAD TOTAL  | 176.2 | 176.2 | 174.4 | 167.4 | 158.2 | 142.9 | 128.3 |
| RAD AV TIME MSEC  | 0.0864 | 0.0324 | 0.0266 | 0.0234 | 0.0216 | 0.0187 | 0.0176 |
| DEG-C DAYS        | 686. | 243. | 166. | 121. | 94. | 73. | 60. |

**FEB**

|                   | ALL | 0+ | 95+ | 189+ | 284+ | 378+ | 473+ |
|-------------------|-----|-----|-----|------|------|------|------|
| SOLAR RAD GROUP   |     |     |     |      |      |      |      |
| AV.THRSH.RAD.LEVL | 0. | 37. | 139. | 237. | 336. | 422. | 520. |
| AV.THRESHOLD TEMP | -3.8 | -2.1 | -0.5 | -1.1 | -1.7 | -1.7 | -2.4 |
| GL RAD OVER THRSH | 311.3 | 273.7 | 204.2 | 151.4 | 108.6 | 79.9 | 54.3 |
| TIME AVG.TEMP     | -3.1 | -2.1 | -2.1 | -2.4 | -2.8 | -3.1 | -3.5 |
| TOTAL TIME MSEC   | 2.4451 | 1.0105 | 0.6793 | 0.5407 | 0.4334 | 0.3337 | 0.2614 |
| GLOBAL RAD TOTAL  | 311.3 | 311.3 | 299.0 | 279.7 | 254.2 | 220.7 | 190.2 |
| DIRECT RAD TOTAL  | 197.2 | 197.2 | 194.6 | 192.4 | 186.2 | 172.0 | 155.4 |
| RAD AV TIME MSEC  | 0.0864 | 0.0360 | 0.0274 | 0.0263 | 0.0252 | 0.0223 | 0.0193 |
| DEG-C DAYS        | 606. | 238. | 161. | 130. | 106. | 83. | 66. |

**MAR**

|                   | ALL | 0+ | 95+ | 189+ | 284+ | 378+ | 473+ |
|-------------------|-----|-----|-----|------|------|------|------|
| SOLAR RAD GROUP   |     |     |     |      |      |      |      |
| AV.THRSH.RAD.LEVL | 0. | 37. | 143. | 233. | 332. | 431. | 514. |
| AV.THRESHOLD TEMP | 1.0 | 2.0 | 3.0 | 4.6 | 3.5 | 5.2 | 5.0 |
| GL RAD OVER THRSH | 426.6 | 382.4 | 285.3 | 222.8 | 168.1 | 122.6 | 92.3 |
| TIME AVG.TEMP     | 2.1 | 3.4 | 3.8 | 4.1 | 4.0 | 4.1 | 3.8 |
| TOTAL TIME MSEC   | 2.6784 | 1.2020 | 0.9155 | 0.6912 | 0.5562 | 0.4597 | 0.3647 |
| GLOBAL RAD TOTAL  | 426.6 | 426.6 | 416.1 | 384.0 | 352.5 | 320.6 | 279.6 |
| DIRECT RAD TOTAL  | 252.4 | 252.4 | 251.1 | 247.3 | 241.3 | 230.6 | 210.8 |
| RAD AV TIME MSEC  | 0.0864 | 0.0396 | 0.0331 | 0.0302 | 0.0281 | 0.0263 | 0.0230 |
| DEG-C DAYS        | 505. | 208. | 154. | 114. | 92. | 76. | 61. |

**APR**

|                   | ALL | 0+ | 95+ | 189+ | 284+ | 378+ | 473+ |
|-------------------|-----|-----|-----|------|------|------|------|
| SOLAR RAD GROUP   |     |     |     |      |      |      |      |
| AV.THRSH.RAD.LEVL | 0. | 52. | 143. | 232. | 334. | 425. | 518. |
| AV.THRESHOLD TEMP | 7.7 | 7.8 | 10.2 | 11.5 | 12.1 | 12.9 | 12.2 |
| GL RAD OVER THRSH | 488.1 | 420.5 | 326.8 | 254.1 | 190.1 | 142.9 | 103.7 |
| TIME AVG.TEMP     | 9.4 | 11.2 | 12.1 | 12.6 | 12.9 | 13.1 | 13.2 |
| TOTAL TIME MSEC   | 2.5920 | 1.3000 | 1.0332 | 0.8176 | 0.6246 | 0.5177 | 0.4201 |
| GLOBAL RAD TOTAL  | 488.1 | 488.1 | 474.3 | 443.5 | 398.8 | 363.1 | 321.6 |
| DIRECT RAD TOTAL  | 254.3 | 254.3 | 251.5 | 249.7 | 242.1 | 235.5 | 222.9 |
| RAD AV TIME MSEC  | 0.0864 | 0.0436 | 0.0371 | 0.0338 | 0.0302 | 0.0284 | 0.0266 |
| DEG-C DAYS        | 267. | 109. | 77. | 57. | 41. | 33. | 27. |

**JUL**

|                   | ALL | 0+ | 95+ | 189+ | 284+ | 378+ | 473+ |
|-------------------|-----|-----|-----|------|------|------|------|
| SOLAR RAD GROUP   |     |     |     |      |      |      |      |
| AV.THRSH.RAD.LEVL | 0. | 42. | 141. | 240. | 330. | 429. | 520. |
| AV.THRESHOLD TEMP | 20.6 | 20.3 | 22.7 | 24.2 | 24.4 | 26.6 | 25.8 |
| GL RAD OVER THRSH | 564.1 | 503.2 | 385.1 | 286.8 | 213.8 | 149.2 | 102.8 |
| TIME AVG.TEMP     | 22.6 | 24.3 | 25.2 | 25.7 | 26.0 | 26.4 | 26.3 |
| TOTAL TIME MSEC   | 2.6784 | 1.4411 | 1.1959 | 0.9940 | 0.8136 | 0.6509 | 0.5101 |
| GLOBAL RAD TOTAL  | 564.1 | 564.1 | 553.8 | 525.3 | 482.0 | 428.4 | 368.0 |
| DIRECT RAD TOTAL  | 272.6 | 272.6 | 270.7 | 268.5 | 258.4 | 244.6 | 226.8 |
| RAD AV TIME MSEC  | 0.0864 | 0.0464 | 0.0403 | 0.0364 | 0.0324 | 0.0288 | 0.0256 |
| DEG-C DAYS        | 1. | 0. | 0. | 1. | 0. | 0. | 0. |

**AUG**

|                   | ALL | 0+ | 95+ | 189+ | 284+ | 378+ | 473+ |
|-------------------|-----|-----|-----|------|------|------|------|
| SOLAR RAD GROUP   |     |     |     |      |      |      |      |
| AV.THRSH.RAD.LEVL | 0. | 50. | 143. | 234. | 332. | 426. | 518. |
| AV.THRESHOLD TEMP | 19.4 | 19.4 | 22.4 | 23.8 | 24.7 | 24.7 | 24.4 |
| GL RAD OVER THRSH | 498.5 | 429.8 | 327.0 | 245.0 | 177.1 | 125.0 | 85.6 |
| TIME AVG.TEMP     | 21.3 | 23.2 | 24.1 | 24.5 | 24.7 | 24.7 | 24.7 |
| TOTAL TIME MSEC   | 2.6784 | 1.3644 | 1.1077 | 0.9032 | 0.6926 | 0.5526 | 0.4284 |
| GLOBAL RAD TOTAL  | 498.5 | 498.5 | 485.6 | 456.3 | 407.0 | 360.5 | 307.6 |
| DIRECT RAD TOTAL  | 242.3 | 242.3 | 239.9 | 237.4 | 228.3 | 218.2 | 200.0 |
| RAD AV TIME MSEC  | 0.0864 | 0.0443 | 0.0378 | 0.0342 | 0.0299 | 0.0274 | 0.0241 |
| DEG-C DAYS        | 8. | 2. | 1. | 1. | 1. | 1. | 1. |

**SEP**

|                   | ALL | 0+ | 95+ | 189+ | 284+ | 378+ | 473+ |
|-------------------|-----|-----|-----|------|------|------|------|
| SOLAR RAD GROUP   |     |     |     |      |      |      |      |
| AV.THRSH.RAD.LEVL | 0. | 42. | 144. | 237. | 333. | 428. | 522. |
| AV.THRESHOLD TEMP | 15.1 | 16.6 | 19.2 | 19.7 | 19.7 | 21.6 | 21.6 |
| GL RAD OVER THRSH | 464.1 | 411.8 | 314.6 | 246.3 | 182.9 | 134.4 | 95.8 |
| TIME AVG.TEMP     | 17.2 | 19.6 | 20.4 | 20.7 | 20.9 | 21.2 | 21.1 |
| TOTAL TIME MSEC   | 2.5920 | 1.2316 | 0.9590 | 0.7754 | 0.6260 | 0.5108 | 0.4104 |
| GLOBAL RAD TOTAL  | 464.1 | 464.1 | 452.5 | 426.1 | 391.5 | 353.1 | 310.1 |
| DIRECT RAD TOTAL  | 290.6 | 290.6 | 288.7 | 285.5 | 277.2 | 262.1 | 238.6 |
| RAD AV TIME MSEC  | 0.0864 | 0.0418 | 0.0346 | 0.0313 | 0.0288 | 0.0259 | 0.0227 |
| DEG-C DAYS        | 71. | 18. | 11. | 8. | 6. | 4. | 4. |

**OCT**

|                   | ALL | 0+ | 95+ | 189+ | 284+ | 378+ | 473+ |
|-------------------|-----|-----|-----|------|------|------|------|
| SOLAR RAD GROUP   |     |     |     |      |      |      |      |
| AV.THRSH.RAD.LEVL | 0. | 34. | 142. | 237. | 336. | 422. | 523. |
| AV.THRESHOLD TEMP | 9.6 | 12.2 | 12.7 | 13.5 | 15.6 | 15.0 | 14.4 |
| GL RAD OVER THRSH | 373.4 | 336.0 | 248.4 | 168.0 | 136.1 | 100.0 | 66.1 |
| TIME AVG.TEMP     | 11.3 | 13.8 | 14.4 | 14.8 | 15.1 | 15.0 | 15.0 |
| TOTAL TIME MSEC   | 2.6870 | 1.1160 | 0.8057 | 0.6361 | 0.5238 | 0.4187 | 0.3370 |
| GLOBAL RAD TOTAL  | 373.4 | 373.4 | 363.0 | 338.9 | 312.2 | 276.9 | 242.4 |
| DIRECT RAD TOTAL  | 247.4 | 247.4 | 244.6 | 240.6 | 233.8 | 217.4 | 197.4 |
| RAD AV TIME MSEC  | 0.0864 | 0.0360 | 0.0288 | 0.0266 | 0.0256 | 0.0234 | 0.0205 |
| DEG-C DAYS        | 219. | 64. | 42. | 31. | 25. | 21. | 17. |

## MAY

| SOLAR RAD GROUP | ALL | 0+ | 95+ | 189+ | 284+ | 378+ | 473+ |
|---|---|---|---|---|---|---|---|
| AV.THRSH.RAD.LEVL | 0. | 41. | 142. | 238. | 328. | 428. | 520. |
| AV.THRESHOLD TEMP | 12.9 | 11.8 | 15.1 | 15.7 | 16.9 | 19.2 | 20.1 |
| GL RAD OVER THRSH | 575.0 | 516.2 | 394.6 | 299.4 | 227.6 | 163.0 | 114.9 |
| TIME AVG.TEMP | 15.0 | 16.8 | 17.8 | 18.3 | 19.0 | 19.5 | 19.6 |
| TOTAL TIME MSEC | 2.6784 | 1.4288 | 1.2024 | 0.9976 | 0.7963 | 0.6442 | 0.5267 |
| GLOBAL RAD TOTAL | 575.1 | 575.0 | 565.7 | 536.6 | 488.7 | 438.2 | 388.7 |
| DIRECT RAD TOTAL | 269.4 | 269.4 | 268.0 | 265.3 | 255.6 | 242.9 | 230.9 |
| RAD AV TIME MSEC | 0.0864 | 0.0464 | 0.0403 | 0.0364 | 0.0324 | 0.0288 | 0.0263 |
| DEG-C DAYS | 120. | 44. | 29. | 21. | 14. | 9. | 8. |

## JUN

| SOLAR RAD GROUP | ALL | 0+ | 95+ | 189+ | 284+ | 378+ | 473+ |
|---|---|---|---|---|---|---|---|
| AV.THRSH.RAD.LEVL | 0. | 45. | 141. | 236. | 330. | 430. | 520. |
| AV.THRESHOLD TEMP | 18.3 | 17.4 | 19.6 | 21.0 | 22.1 | 24.7 | 24.3 |
| GL RAD OVER THRSH | 550.4 | 488.1 | 376.4 | 285.9 | 215.0 | 154.2 | 108.2 |
| TIME AVG.TEMP | 20.2 | 21.9 | 22.8 | 23.5 | 24.2 | 24.7 | 24.7 |
| TOTAL TIME MSEC | 2.5920 | 1.3990 | 1.1538 | 0.9547 | 0.7549 | 0.6102 | 0.5083 |
| GLOBAL RAD TOTAL | 550.4 | 550.4 | 539.5 | 511.3 | 464.2 | 416.4 | 372.6 |
| DIRECT RAD TOTAL | 258.0 | 258.0 | 256.3 | 254.6 | 246.5 | 234.1 | 221.2 |
| RAD AV TIME MSEC | 0.0864 | 0.0468 | 0.0403 | 0.0371 | 0.0331 | 0.0295 | 0.0266 |
| DEG-C DAYS | 21. | 8. | 2. | 1. | 1. | 1. | 1. |

## ALL MONTHS

| SOLAR RAD GROUP | ALL | 0+ | 95+ | 189+ | 284+ | 378+ | 473+ |
|---|---|---|---|---|---|---|---|
| TIME AVG.TEMP | 9.7 | 12.5 | 13.9 | 14.7 | 15.0 | 15.3 | 15.3 |
| TOTAL TIME MSEC | 31.5705 | 14.4997 | 10.9296 | 8.6396 | 6.8188 | 5.4436 | 4.3470 |
| GLOBAL RAD TOTAL | 4969.2 | 4958.1 | 4840.8 | 4518.2 | 4088.9 | 3634.1 | 3165.6 |
| DIRECT RAD TOTAL | 2758.9 | 2758.9 | 2734.4 | 2690.7 | 2589.6 | 2434.2 | 2236.2 |
| RAD AV TIME MSEC | 0.0864 | 0.0415 | 0.0349 | 0.0319 | 0.0289 | 0.0261 | 0.0235 |
| DEG-C DAYS | 3527. | 1283. | 857. | 633. | 491. | 387. | 310. |

## NOV

| SOLAR RAD GROUP | ALL | 0+ | 95+ | 189+ | 284+ | 378+ | 473+ |
|---|---|---|---|---|---|---|---|
| AV.THRSH.RAD.LEVL | 0. | 39. | 138. | 235. | 326. | 426. | 521. |
| AV.THRESHOLD TEMP | 4.1 | 6.8 | 6.9 | 7.8 | 8.1 | 7.2 | 7.5 |
| GL RAD OVER THRSH | 234.3 | 195.6 | 135.2 | 94.2 | 65.3 | 42.5 | 25.5 |
| TIME AVG.TEMP | 5.3 | 7.1 | 7.3 | 7.4 | 7.3 | 7.1 | 7.0 |
| TOTAL TIME MSEC | 2.5920 | 0.9950 | 0.6095 | 0.4252 | 0.3143 | 0.2293 | 0.1778 |
| GLOBAL RAD TOTAL | 234.3 | 234.3 | 219.3 | 193.9 | 167.9 | 140.1 | 118.2 |
| DIRECT RAD TOTAL | 144.3 | 144.3 | 142.5 | 136.5 | 127.5 | 113.4 | 100.1 |
| RAD AV TIME MSEC | 0.0864 | 0.0335 | 0.0263 | 0.0238 | 0.0216 | 0.0187 | 0.0173 |
| DEG-C DAYS | 392. | 129. | 78. | 54. | 40. | 30. | 23. |

## DEC

| SOLAR RAD GROUP | ALL | 0+ | 95+ | 189+ | 284+ | 378+ | 473+ |
|---|---|---|---|---|---|---|---|
| AV.THRSH.RAD.LEVL | 0. | 41. | 134. | 240. | 324. | 431. | 520. |
| AV.THRESHOLD TEMP | -2.8 | -0.0 | -1.0 | 0.6 | 0.2 | -1.3 | -2.4 |
| GL RAD OVER THRSH | 232.5 | 191.7 | 136.1 | 91.9 | 65.7 | 40.6 | 24.9 |
| TIME AVG.TEMP | -2.0 | -0.6 | -1.0 | -1.0 | -1.6 | -2.1 | -2.4 |
| TOTAL TIME MSEC | 2.6784 | 1.0037 | 0.5962 | 0.4183 | 0.3092 | 0.2351 | 0.1775 |
| GLOBAL RAD TOTAL | 232.5 | 232.5 | 215.9 | 192.1 | 166.0 | 141.9 | 117.1 |
| DIRECT RAD TOTAL | 154.3 | 154.3 | 151.9 | 145.6 | 134.3 | 120.6 | 103.9 |
| RAD AV TIME MSEC | 0.0864 | 0.0324 | 0.0259 | 0.0230 | 0.0209 | 0.0187 | 0.0180 |
| DEG-C DAYS | 631. | 220. | 133. | 94. | 71. | 56. | 43. |

# Index